Springer Collected Works in Mathematics

More information about this series at http://www.springer.com/series/11104

Kenkichi Iwasawa

Collected Papers

Editor in Chief

Ichiro Satake

Editors

Genjiro Fujisaki • Kazuya Kato • Masato Kurihara • Shoichi Nakajima

Reprint of the 2001 Edition

Springer

Author
Kenkichi Iwasawa (1917-1998)
Princeton University
Princeton, NJ, USA

Editor in Chief
Ichiro Satake (emeritus)
University of California, Berkeley
Berkeley, CA, USA
and
Tohoku University
Sendai, Japan

Editors
Genjiro Fujisaki (emeritus)
The University of Tokyo
Tokyo, Japan

Kazuya Kato
University of Chicago
Chicago, IL, USA

Masato Kurihara
Keio University
Yokohama, Japan

Shoichi Nakajima
Gakushuin University
Tokyo, Japan

Contributor
John Coates
University of Cambridge
Cambridge, United Kingdom

ISSN 2194-9875
ISBN 978-4-431-55055-6
Springer Tokyo Heidelberg New York Dordrecht London

Library of Congress Control Number: 2014957321

Printed on acid-free paper

Springer is part of Springer Science+Business Media (www.springer.com)

K. Iwasawa
Princeton, 1986

円分体に関するいくつかの問題

岩澤健吉 (Kenkichi Iwasawa)

pを任意の奇素数, kを円のp分体とし, kに関するいくつかの問題を簡単に説明します。殆んど皆よく知られたことばかりですが, この研究集会のような機会にそれらの問題を一応整理しておくのもいくらか意義があるかと考える次第です。

§1. 上記kの有理数体上のガロア群をΔ, Δの指標群を$\hat{\Delta}$, kのイデヤル類群をCとする。周知のように$\hat{\Delta}$は$\bmod p$のDirichlet指標の成す群と同一視される。ΔはCの上に自然に作用するから, kのcomplex conjugationをJとする時 ($J \in \Delta$) Cの部分群$C^{\pm}$を

(1) $$C^{+} = \mathrm{Ker}(1-J : C \to C), \quad C^{-} = \mathrm{Ker}(1+J : C \to C)$$

により定義する。kの最大実部分体をk^{+}とし, k^{+}のイデヤル類群をC_{+}とすれば, 自然な同型: $C/C^{-} \cong C_{+}$が存在する[1)]。よ

First page of Iwasawa's manuscript on cyclotomic fields [62]

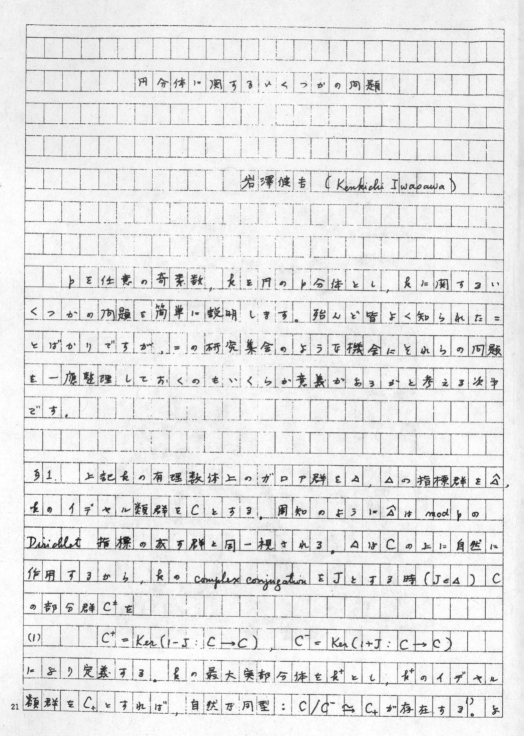

Contents

Preface

Kenkichi Iwasawa was born on September 11, 1917 in Shinshuku-mura (Kiryuu-shi) in Gunma prefecture. After elementary school he was educated in Tokyo, where he first attended Musashi High School, one of the few special high schools with seven grades (which combine four years of middle school education with three years of high school education). In 1937 he entered the University of Tokyo (then Tokyo Imperial University) to do undergraduate studies in mathematics. After graduation in 1940, he continued his graduate studies there, became Assistant and then Instructor, and obtained the degree of Doctor of Science in 1945 (this was during the World War II, 1941-1945). That year he became seriously ill with pleurisy and could return to his post only in April 1947. He was appointed Assistant Professor at the University of Tokyo in 1949.

In the summer of 1950, Iwasawa was invited to the International Congress of Mathematicians held in Cambridge, Massachusetts to give two talks ([23], [24]), and then spent two academic years as a member at the Institute for Advanced Study in Princeton. While preparing to return to Tokyo in the spring of 1952, he received an offer from the Massachusetts Institute of Technology, which he accepted; he was Assistant Professor there from 1952, Associate Professor from 1955, and Professor from 1958. In 1966 he moved to Princeton and was Professor at Princeton University from 1966 and H. B. Fine Professor of Mathematics from 1978 until his retirement in 1986. Meanwhile, he was also a member at the Institute for Advanced Study for 1957-58, 1966-67, Spring of 1971, Spring of 1979, and Spring of 1982.* Iwasawa and his wife returned to live in Tokyo in the fall of 1987. Though he no longer did university teaching, he regularly attended seminars at the University of Tokyo, which was a great inspiration to the younger generation in Japan. He died of acute pneumonia on October 26, 1998 at the age of 81.

Iwasawa was awarded the Asahi Cultural Prize in 1959, the Prize of the Japan Academy in 1962, the Frank Nelson Cole Prize in Algebra and in Number Theory of the American Mathematical Society in 1962, and the Fujiwara Prize in 1979.

Iwasawa made a number of fundamental contributions in group theory and in algebraic number theory; the main body of the latter, now known as Iwasawa theory, has many applications in a wider area of arithmetic algebraic geometry (including the solution of Fermat's Last Theorem). We are very fortunate to be able to include an expository article on Iwasawa theory by John Coates, to whom we are very grateful for his invaluable contribution to this volume.

Thanks are due to Springer-Verlag Tokyo for bringing out this publication, which, we hope, will enable future mathematicians to inherit the magnificent

---

* These data on the academic career of K. Iwasawa were kindly provided by the Faculty of Sciences, the University of Tokyo and the Department of Mathematics, Princeton University.

ideas and results of Kenkichi Iwasawa, one of the most original and influential mathematicians of our time.

April, 2001

Editorial Committee
G. Fujisaki, K. Kato, M. Kurihara,
S. Nakajima, and I. Satake (Chair)

Introduction

These two volumes contain all 66 published papers of Kenkichi Iwasawa ([1]-[66]), of which 47 are in English, 8 in German, and 11 in Japanese; the second volume contains also 5 unpublished papers in English ([U1]-[U5]). For each of these 11 papers in Japanese and 5 unpublished papers, either an (English) abstract or notes by the editors or other specialists are provided. Those papers in Japanese ([5], [10], [59], [61], [62]) which were originally photocopies of handwritten manuscripts were reproduced here from newly typed LaTeX files. For a few papers, whenever it seemed appropriate, the editors added some comments or corrections of misprints (indicated by asterisks). Most of the reprints and manuscripts were provided by the late Mrs. Aiko Iwasawa, to whom the editors are most grateful.

We note that there exist further some 19 articles by K. Iwasawa ([A1]-[A19]), including 10 papers in Japanese published in Zenkoku Shijo Sugaku Danwakai and Iso Sugaku (before 1945) and 6 book reviews, not included in these volumes.

The bibliography at the end of each volume gives the list of all publications of K. Iwasawa, including 5 books and lecture notes ([B1]-[B5]) as well as 19 articles mentioned above which are not reproduced here. We note that the numbering of the papers in this list from [1] to [62] coincides with that given in the following volume, published in commemoration of his seventieth birthday:

J. Coates, R. Greenberg, B. Mazur, and I. Satake (eds.), Algebraic Number Theory — in honor of Kenkichi Iwasawa, Adv. St. in Pure Math., Vol. **17**, Kinokuniya, Tokyo and Acad. Press, 1989.

Acknowledgements

The editors would like to thank the original publishers of Kenkichi Iwasawa's papers for granting permission to reprint them here.

The numbers following each publisher correspond to the numbering of the articles in the table of contents.

Academic Press Ltd. London, UK: [54]
The American Mathematical Society: [23], [24], [29], [35], [42], [50]
Centre National de la Recherche Scientifique: [45]
Department of Mathematics, Graduate School of Science, Osaka University: [5], [10]
Gauthier-Villars – Éditions Scientifiques et Médicales Elsevier: [31]
Graduate School of Mathematical Sciences, University of Tokyo: [3]
Graduate School of Mathematics, Nagoya University: [44]
Institute of Mathematics, Polish Academy of Sciences: [51]
The Japan Academy: [2], [8], [11], [12], [13], [14], [15], [63], [64]
Japan Society for the Promotion of Science: [56]
The Johns Hopkins University Press: [33], [60]
Kinokuniya Company Ltd.: [53]
The Mathematical Society of Japan: [1], [6], [7], [9], [16], [17], [18], [19], [20], [25], [28], [38], [40], [41], [43], [46], [66]
Mathematical Institute, Tohoku University: [58]
Princeton University Press: [21], [22], [26], [27], [34], [36], [37], [39], [48], [52]
Research Institute for Mathematical Sciences, Kyoto University: [59], [61], [62]
The Science Council of Japan: [30]
Springer-Verlag: [55], [65]
Tata Institute of Fundamental Research: [57]
Vandenhoeck & Ruprecht: [32]
Yeshiva University: [47]

Iwasawa's work in algebraic number theory

John Coates

Emmanuel College, Cambridge CB2 3AP, England

A brief glance at the articles in these volumes will quickly convince the reader that Iwasawa wrote mathematics, as indeed he also lectured it, with beauty, elegance, and precision. Nothing is a substitute for reading his original papers, and the aim of the present article is merely to provide, with the benefit of hindsight, a short and inevitably personal overview of the main themes of Iwasawa's work in algebraic number theory, together with a few comments on the much wider circle of problems in arithmetical algebraic geometry to which his ideas have been fruitfully applied. I have concentrated on his work on cyclotomic fields and $\mathbb{Z}_p$-extensions of number fields (these are essentially the papers from {26} onwards in the present volumes).† However, it should not be forgotten that Iwasawa also made important contributions to other areas of mathematics. Most notable amongst these are his celebrated paper {21} on locally compact groups and Hilbert's fifth problem, and his brief note {23} (see also {66}) in which, independently of J. Tate, he also discovered the adelic approach to Hecke's L-functions, which has so deeply influenced the modern theory of automorphic forms. Finally, we also urge the reader to consult the excellent and much more detailed article on Iwasawa's work by R. Greenberg [12].

Let F be a finite extension of the rational field $\mathbb{Q}$. Classical algebraic number theory seeks to describe the arithmetic of F (the ideal class group, the unit group, the set of rational primes which split completely in F, the Galois group of the maximal abelian extension of F which is unramified outside a given finite set of places, ...). Beginning with the work of Dirichlet, Dedekind, and Kummer, it has long been known that there is more than one connexion between these arithmetic problems and the complex zeta and L-functions attached to F. The new idea underlying Iwasawa's work was that one could obtain important arithmetic invariants of F, and establish deep new relations between the arithmetic of F and zeta values, by studying certain infinite towers of number fields over F, which he called *$\mathbb{Z}_p$-extensions*. Here p denotes an arbitrary prime number, and its choice gives a p-adic flavour to all of Iwasawa's work. By definition, a $\mathbb{Z}_p$-extension of F is an infinite tower of fields

$$(1) \qquad F = F_0 \subset F_1 \subset \cdots \subset F_n \subset \ldots,$$

where, for each $n \geqslant 0$, F_n is a cyclic extension of F of degree p^n. The reason for this terminology is that if we define F_∞ to be the union of all the $F_n (n \geqslant 0)$, then F_∞ is a Galois extension of F whose Galois group over F, which we shall always

† I use square brackets [1],... to denote the references at the end of this article, and curly brackets {26}, ... to denote the references to Iwasawa's papers as numbered in these volumes.

denote by Γ, is topologically isomorphic to the additive group of p-adic integers $\mathbb{Z}_p$. It is easy to see that every F has a unique $\mathbb{Z}_p$-extension contained in $F(\mu_{p^\infty})$, where μ_{p^∞} denotes the group of all p-power roots of unity; this $\mathbb{Z}_p$-extension is called the *cyclotomic* $\mathbb{Z}_p$-extension of F. However, if F is not totally real, class field theory shows that F admits infinitely many $\mathbb{Z}_p$-extensions distinct from the cyclotomic one. Some of these non-cyclotomic $\mathbb{Z}_p$-extensions also turn out to be very useful for the study of arithmetic questions. Of course, $\mathbb{Z}_p$-extensions are the simplest examples of Galois extensions of F whose Galois group is a p-adic Lie group of positive dimension, and it is a fascinating question to speculate on how much of Iwasawa's work might eventually be extended to more general p-adic Lie extensions of F.

There is one other important notion which underlies much of Iwasawa's work. Let G denote an arbitrary profinite group (we shall mainly be interested in the case $G = \Gamma$, but the remarks we shall make hold in complete generality). The Iwasawa algebra $\Lambda(G)$ of G is defined by

$$\Lambda(G) = \varprojlim_{U} \mathbb{Z}_p[G/U], \tag{2}$$

where U runs over all open normal subgroups of G, and $\mathbb{Z}_p[G/U]$ denotes the ordinary group ring of the finite group G/U with coefficients in $\mathbb{Z}_p$. The beauty of the Iwasawa algebra $\Lambda(G)$ is that it has both an algebraic and an analytic interpretation. The algebraic avatar of $\Lambda(G)$ is the fact that we can naturally extend the continuous action of G on any compact $\mathbb{Z}_p$-module X to an action of the whole Iwasawa algebra $\Lambda(G)$. This idea is explicit in Iwasawa's first papers {35}, {36} on $\mathbb{Z}_p$-extensions, but J-P. Serre (see [27], [28]) was the first to stress how one could exploit it more fully, noting in particular that, when $G = \mathbb{Z}_p{}^d$ for some integer $d \geqslant 1$, then $\Lambda(G)$ could be identified with the ring $\mathbb{Z}_p[[T_1, \ldots, T_d]]$ of formal power series in d variables, with coefficients in $\mathbb{Z}_p$. The analytic avatar of $\Lambda(G)$ is the fact that it can be identified with the algebra of measures on G with values in $\mathbb{Z}_p$, thus allowing us to define the integral

$$\int_G f d\mu$$

for all μ in $\Lambda(G)$ and all continuous functions f from G to $\mathbb{Q}_p$. Once again, this idea first appears implicitly in Iwasawa's work (see his paper {48} giving a new construction of the Kubota-Leopoldt p-adic L-functions). It was later systematically developed by B. Mazur (see [20]). Perhaps Iwasawa's most remarkable discovery is the fact that, at least in some important cases, there is a similar deep algebraic and analytic dichotomy in the arithmetic of $\mathbb{Z}_p$-extensions. A precise formulation of this dichotomy is often called "the main conjecture".

Algebraic Theory

Roughly speaking, the algebraic side of Iwasawa theory refers to the methods and results which do not involve making use of the special values of zeta and L-functions. Let F be a finite extension of $\mathbb{Q}$, and let F_∞ denote an arbitrary $\mathbb{Z}_p$-extension of F. As above, we write Γ for the Galois group of F_∞ over F, and $\Lambda(\Gamma)$ for the Iwasawa algebra of Γ. Compact $\Lambda(\Gamma)$-modules arise naturally in the theory of $\mathbb{Z}_p$-extensions as follows. Let N_∞ denote any abelian p-extension of F_∞, which is Galois over the base field F (typical examples are when N_∞ is the maximal unramified abelian p-extension of F_∞, or the maximal abelian p-extension of F_∞ which is unramified outside the primes of F_∞ lying above p). Write $G(N_\infty/F_\infty)$ for the Galois group of N_∞ which is clearly a compact $\mathbb{Z}_p$-module. The key point is that $G(N_\infty/F_\infty)$ has a natural continuous action of Γ given by

$$\sigma \cdot x = \tilde{\sigma} x \tilde{\sigma}^{-1} \tag{3}$$

for σ in Γ and x in $G(N_\infty/F_\infty)$, where $\tilde{\sigma}$ denotes any lifting of σ to the Galois group of N_∞ over F. But then, as remarked above, this Γ-action can be extended to a continuous action of the whole Iwasawa algebra $\Lambda(\Gamma)$ on $G(N_\infty/F_\infty)$. Much of the algebraic content of Iwasawa theory is concerned with the study of such Galois groups $G(N_\infty/F_\infty)$ as $\Lambda(\Gamma)$-modules. It is usually easy to show that $G(N_\infty/F_\infty)$ is finitely generated over $\Lambda(\Gamma)$, but much more delicate to determine its rank as a $\Lambda(\Gamma)$-module.

Iwasawa's first general algebraic result, which is valid for an arbitrary $\mathbb{Z}_p$-extension F_∞ over F, was the following asymptotic formula proven in {35}. Let A_n denote the p-primary subgroup of the ideal class group of F_n, and write p^{e_n} for the order of A_n. Then there exist integers λ, μ, and ν, depending only on the $\mathbb{Z}_p$-extension F_∞/F, such that, for all sufficiently large n, we have

$$e_n = \lambda n + \mu p^n + \nu. \tag{4}$$

Two basic facts lie behind the proof of (4). The first is that $G(L_\infty/F_\infty)$, where L_∞ denotes the maximal unramified abelian p-extension of F_∞, is a finitely generated torsion $\Lambda(\Gamma)$-module. The second is that there exists a structure theory for finitely generated torsion modules over $\Lambda(\Gamma)$, which is only marginally different from the structure theory for such modules over a principal ideal domain. In {35}, Iwasawa gave an *ad hoc* proof of this structure theory, but J-P. Serre [28] pointed out that it followed from known results in commutative algebra, since $\Lambda(\Gamma)$ is isomorphic to the ring $\mathbb{Z}_p[[T]]$ of formal power series in an indeterminate T with coefficients in $\mathbb{Z}_p$.

Iwasawa himself has said that the discovery of the asymptotic formula (4) suggested to him that there might be deep analogies between the p-primary subgroup of the ideal class group of a $\mathbb{Z}_p$-extension F_∞ of F and the p-primary subgroup of the group of points on the Jacobian variety of a curve over a finite field. This idea seems to have motivated much of his subsequent work. Of course, the existence of certain parallels between the arithmetic of number fields and curves over finite fields had been observed long before this, but Iwasawa's insights were to breathe new life into this old theme. Inspired by this analogy, he conjectured that the invariant μ appearing in (4) should always be zero when F_∞ is the cyclotomic $\mathbb{Z}_p$-extension of F. This conjecture was proved by B. Ferrero and L. Washington [8] (and a rather different proof was given later by W. Sinnott [30]) when F is an abelian extension of $\mathbb{Q}$, but it remains open in general. In the early 1970's, Iwasawa {53} discovered the first examples of non-cyclotomic $\mathbb{Z}_p$-extensions where $\mu > 0$. Today, there remain many interesting open questions about Iwasawa's invariants λ and μ, and their possible generalizations. We only mention three questions, all of which seem rather difficult. Firstly, if F is totally real and F_∞/F is the cyclotomic $\mathbb{Z}_p$-extension, R. Greenberg [10] has conjectured that we always have $\lambda = \mu = 0$ (this is equivalent to the assertion that when F is totally real, the maximal unramified abelian p-extension of F_∞ is always a finite extension of F_∞). Secondly, if we take F_∞ to be the cyclomotic $\mathbb{Z}_p$-extension of F and vary p over all primes, one can ask whether λ is bounded, by analogy with the Jacobian variety of a curve over a finite field (see [12])? Thirdly, one can ask if there is some generalization of (4) to arbitrary pro-p p–adic Lie extensions of F. Let K_∞ denote a Galois extension of F, whose Galois group G is a pro-p p-adic Lie group of positive dimension d. For each integer $n \geqslant 1$, let $P_n(G)$ denote the n-th subgroup of the lower central p-series of G, and define K_n to be the fixed field of $P_n(G)$. Again, we write p^{e_n} for the order of the p-primary subgroup of the ideal class group of K_n. Do there exist non-negative integers λ and μ such that

$$e_n = (\lambda n + \mu p^n)p^{(d-1)n} + O(p^{(d-1)n}) \tag{5}$$

as $n \to \infty$? When $G = \mathbb{Z}_p{}^d$, the formula (5) is true and was proven by A. Cuoco and P. Monsky [6].

Probably the most important of Iwasawa's subsequent papers on the algebraic side of his theory is {52}, which contains much material from his courses at Princeton. Let F_∞ be an arbitrary $\mathbb{Z}_p$-extension of F. The first six sections of {52} carry out a systematic study of $G(L_\infty/F_\infty)$ as a $\Lambda(\Gamma)$-module, where L_∞ again denotes the maximal unramified abelian p-extension of F_∞. One surprising new result proven there is that, if A_n denotes the p-primary subgroup of the ideal class group of F_n, then for all $m \geqslant n \geqslant 0$, the order of the kernel of the natural map from A_n to A_m is bounded independent of n and m. While this kernel is known to be non-zero in general (see [10]), it is still unknown if this can happen when $F = \mathbb{Q}(\mu_p)$ and $F_\infty = \mathbb{Q}(\mu_{p^\infty})$ for some odd prime p; here μ_p denotes the group of p-th roots of unity, and μ_{p^∞} the group of all p-power roots of unity. The next two sections of {52} specialize to the case of the cyclotomic $\mathbb{Z}_p$-extension F_∞/F. Let M_∞ denote the maximal abelian p-extension of F_∞ which is unramified outside p. Iwasawa then determines the exact $\Lambda(\Gamma)$-rank of $G(M_\infty/F_\infty)$, a result which turns out to be equivalent to establishing a weak form of H. Leopoldt's conjecture on the p-adic independence of the global units of F_n as $n \to \infty$. He also proves that $G(M_\infty/F_\infty)$ has no non-zero finite $\Lambda(\Gamma)$-submodule. The final part of {52} gives a detailed construction in the number field case of a skew-symmetric bilinear form, which is an analogue of the Weil pairing on the Tate module of the Jacobian of a curve over a finite field.

Analytic Theory

I believe that the most remarkable of all Iwasawa's discoveries on $\mathbb{Z}_p$-extensions was his work on what became known later as the "main conjecture". This work provided the key to our complete understanding of the connexion, partially discovered by Kummer, between the arithmetic of the field $\mathbb{Q}(\mu_{p^\infty})$ and the values of the zeta function at the odd negative integers. Today we have become so used to almost routinely formulating (but, alas, very rarely proving) "main conjectures" for elliptic curves, modular forms, motives, ..., that it is easy to forget how much difficulty Iwasawa had in uncovering even the statement of his "main conjecture".

The interested reader can see the evolution of his ideas at first hand in his papers {39}, {41}, {42}, {45}, {46}, {47}, {48}, especially {41} and {48}.

We now discuss Iwasawa's work on the "main conjecture" in some detail, and with the benefit of hindsight. In connexion with his work on Fermat's Last Theorem, Kummer defined an odd prime p to be *irregular* if p divides the class number of the field $\mathbb{Q}(\mu_p)$. Kummer himself proved a mysterious and unexpected connexion between the irregularity of p and the special values of the Riemann zeta function $\zeta(s)$. For $k = 2, 4, 6, \ldots$, we have $\zeta(1-k) = -B_k/k$, where the B_k are the Bernoulli numbers defined by the expression

$$\frac{t}{e^t - 1} = \sum_{n=0}^{\infty} B_n \frac{t^n}{n!}.$$

Kummer proved that p is irregular if and only if p divides the numerator of at least one of the rational numbers $\zeta(1-k)$, with $k = 2, 4, \ldots, p-3$.

Let K_∞ be the maximal real subfield of $\mathbb{Q}(\mu_{p^\infty})$, that is

$$K_\infty = \mathbb{Q}(\mu_{p^\infty}) \cap \mathbb{R},$$

and let G be the Galois group of K_∞ over $\mathbb{Q}$ (the structure of G is given by $G = \Gamma \times \Delta$, where Γ is isomorphic to $\mathbb{Z}_p$, and Δ is cyclic of order $(p-1)/2$). Let M_∞ be the maximal abelian p-extension of K_∞ which is unramified outside p, and put

$$X_\infty = G(M_\infty/K_\infty). \tag{6}$$

Now the exact analogue of (3) defines an action of G on X_∞, which we can then extend to an action of the whole Iwasawa algebra $\Lambda(G)$. Similar algebraic arguments to those mentioned earlier then show that X_∞ is finitely generated over $\Lambda(G)$, and has $\Lambda(G)$-rank equal to 0 (in the sense that X_∞ is annihilated by an element of $\Lambda(G)$ which is not a divisor of zero). But there is a simple structure theory for finitely generated $\Lambda(G)$-modules, coming essentially from the structure theory of $\Lambda(\Gamma)$-modules, and the fact that $G = \Delta \times \Gamma$ mentioned above. It follows that there exist an integer $\tau_p \geqslant 1$, and elements $f_1, \ldots, f_{\tau_p}$ of $\Lambda(G)$, which are not divisors of zero such that we have an exact sequence of $\Lambda(G)$-modules

$$0 \to X_\infty \to \bigoplus_{i=1}^{\tau_p} \Lambda(G)/f_i\Lambda(G) \to D \to 0, \tag{7}$$

where D is a $\Lambda(G)$-module of finite cardinality. It is not obvious that there is a connexion between the module X_∞ and the irregularity of a prime number p, but in fact it can be shown by algebraic arguments that p is irregular if and only if $X_\infty \neq 0$. Thus, in view of Kummer's criterion for the irregularity of p, it follows that $X_\infty \neq 0$ if and only if p divides at least one of the special values $\zeta(1-k)$, with $k = 2, 4, \ldots, p-3$. What is the connexion between this mysterious statement and the exact sequence (7) of $\Lambda(G)$-modules?

To formulate the answer, Iwasawa realized that one had to use the analytic avatar of $\Lambda(G)$. Earlier, T. Kubota and H. Leopoldt [17], using congruences going back to Kummer, had proven the existence of a p-adic analogue of $\zeta(s)$, but their construction gave no hint of any connexion with $\Lambda(G)$. Motivated by arithmetic arguments emerging from explicit reciprocity laws in {41}, Iwasawa had been led to study the classical Stickelberger elements in the tower of number fields $\mathbb{Q}(\mu_{p^\infty})$. In a beautifully simple paper {48}, Iwasawa brought these two lines of thought together, and proved the following result, modulo a slight change of language. We say that an element φ of the ring of fractions of $\Lambda(G)$ is a pseudo-measure if $(\sigma - 1)\varphi$ belongs to $\Lambda(G)$ for all σ in G (intuitively, one should think of such a φ as possibly having a simple pole at the trivial character of G). Let $G_\mathbb{Q}$ be the Galois group of some fixed algebraic closure of $\mathbb{Q}$, and write $\psi : G_\mathbb{Q} \to \mathbb{Z}_p^\times$ for the homomorphism giving the action of $G_\mathbb{Q}$ on μ_{p^∞}, that is $\sigma(\zeta) = \zeta^{\psi(\sigma)}$ for all σ in $G_\mathbb{Q}$ and all p-power roots of unity ζ. By Galois theory, G is a quotient of $G_\mathbb{Q}$, and the characters ψ^k, where k is any even integer, factor through G. Then Iwasawa proved that there is a unique pseudo-measure ρ_p on G such that

$$\int_G \psi^k d\rho_p = (1 - p^{k-1})\zeta(1-k) \qquad (k = 2, 4, 6, \dots); \tag{8}$$

this makes sense, since one can integrate any non-trivial p-adic homomorphism of G against any pseudo-measure.

Let $\Lambda(G)_0 = \ker(\Lambda(G) \to \mathbb{Z}_p)$ be the augmentation ideal of $\Lambda(G)$. Because ρ_p is a pseudo-measure, $\rho_p\Lambda(G)_0$ is an ideal in $\Lambda(G)$, and Iwasawa's "main conjecture" is the assertion that

$$\rho_p\Lambda(G)_0 = f_1 \dots f_{r_p}\Lambda(G). \tag{9}$$

Familiarity should not allow us to forget how revolutionary (9) is. It gives a bridge between the p-adic analytic world, represented by the ideal on the left of (9), with the p-adic arithmetic world represented by the ideal on the right of (9). Outside the p-adic world, no one has even come close at present to formulating an analogue of (9) for the Riemann zeta function $\zeta(s)$, nor of defining a good analogue of the right hand side of (9) for the maximal unramified abelian extension of the field generated over $\mathbb{Q}$ by all roots of unity.

The honour of giving the first unconditional proof of (9) fell in 1984 to B. Mazur and A. Wiles [21], using methods from the theory of modular curves. But there is still great interest, especially for applications of similar ideas to elliptic curves, in studying the methods which led Iwasawa to (9), especially those in his wonderful paper {41}. In this paper, Iwasawa used the classical cyclotomic units to construct a compact $\Lambda(G)$-module Y_∞, together with a natural $\Lambda(G)$-homomorphism

$$\varphi : Y_\infty \to X_\infty. \tag{10}$$

He then proved, by an ingenious use of an explicit reciprocity law going back to E. Artin and H. Hasse, that we have an isomorphism of $\Lambda(G)$-modules

$$Y_\infty \simeq \Lambda(G)/\rho_p\Lambda(G)_0; \tag{11}$$

here we are implicitly using the construction of ρ_p given in the later paper {43}. The precise definition of the module Y_∞ is as follows. Let $F_n = \mathbb{Q}(\zeta_n + \zeta_n^{-1})$, where ζ_n denotes a primitive p^{n+1}-th root of unity $(n = 0, 1, \dots)$. Let V_n be the subgroup of $F_n^\times$ generated by all conjugates of $(1 - \zeta_n)(1 - \zeta_n^{-1})$. The group of cyclotomic units C_n of F_n is defined to be the intersection of V_n with the unit group of the ring of integers of F_n. There is a unique prime v_n of F_n above p, and we write U_n for the group of local units in the completion of F_n at v_n, which are congruent to 1 modulo v_n. Let $\mathcal{C}_n$ be the completion of $C_n \cap U_n$ in the v_n-adic topology. Then Iwasawa's module Y_∞ is the projective limit of $U_n/\mathcal{C}_n$, taken with respect to the norm maps.

The unconditional proof of (11) was a great achievement. The beauty of (11) is that, via (8), it makes the values of the Riemann zeta function at the odd negative integers appear naturally in the arithmetic of cyclotomic fields. But (11) also leads inexorably to the formulation of the "main conjecture" (9). Indeed, assuming that the class number of the maximal real subfield of $\mathbb{Q}(\mu_p)$ is prime to p, Iwasawa showed that the map φ appearing in (10) is an isomorphism, and hence (11) implies (9) in this case. It is plausible, but unknown at present, that φ is injective and has finite cokernel for all odd primes p (this turns out to be equivalent to Greenberg's conjecture that $\lambda = 0$ for the cyclotomic $\mathbb{Z}_p$-extension F_∞ of the maximal real subfield of $\mathbb{Q}(\mu_p)$). Happily, the "main conjecture" is weaker, and boils down to proving that the kernel and cokernel of φ have roughly the same size (to be precise, have the same characteristic ideals) as $\Lambda(G)$-modules. In 1990, K. Rubin [24], using ideas on Euler systems inspired by the the work of F. Thaine [32] and V. Kolyvagin [16], succeeded in proving just enough about the kernel and cokernel of φ to be able to deduce the "main conjecture" from (11).

Another theme of Iwasawa's research, growing out of his work on (11), was his proof in {46} of a remarkable explicit formula for the Hilbert norm residue symbol in the field generated over $\mathbb{Q}_p$ by the p^n-th roots of unity for some integer $n \geqslant 1$. His paper has inspired a large body of work on explicit reciprocity laws in general over the last twenty years, beginning with the generalization of it to the fields generated by points of finite order on Lubin-Tate formal groups by A. Wiles [33].

Influence of Iwasawa's work

The heartland of arithmetic geometry is the study of the arithmetic of algebraic varieties defined over finite extensions of $\mathbb{Q}$, and the motives derived from them. Many of these arithmetic problems, at least conjecturally, are related to the special values of the L-functions attached to the cohomology of these varieties. The methods introduced by Iwasawa form the only general bridge known today between Galois cohomological problems and the special values of L-functions. As such, they have already deeply influenced the course of arithmetic geometry over the last forty years, and it is highly likely that they will continue to do so in the years ahead, especially in the search for a proof of exact formulae, typified by the conjecture of Birch and Swinnerton-Dyer on the arithmetic of elliptic curves. For lack of space, we only briefly mention in what follows several areas in which Iwasawa's ideas have already been fruitfully applied.

We begin with the arithmetic of elliptic curves. The first step in applying to these problems ideas from Iwasawa theory was made by B. Mazur [19]. Again, let F be a finite extension of $\mathbb{Q}$, and let E be an elliptic curve defined over F. Let K_∞ denote a Galois extension of F whose Galois group G is a p-adic Lie group of positive dimension (for example, K_∞ could be the field obtained by adjoining to F all p-power roots of unity, or the coordinates of all p-power division points on

E). As was first emphasized by Mazur [19], the natural Iwasawa module to study in this situation is the analogue for K_∞ of the classical Selmer group of E, which we denote by $S(E/K_\infty)$. By definition, $S(E/K_\infty)$ consists of all elements of $H^1(G(\overline{\mathbb{Q}}/K_\infty), E_{p^\infty})$ which, for each finite prime v of K_∞, localize to zero in the H^1 of the absolute Galois group of $K_{\infty,v}$ acting on $E(\overline{K}_{\infty,v})$; here E_{p^∞} denotes the Galois module of all p-power division points on E, and, as usual for infinite extensions, $K_{\infty,v}$ is the union of the completions at v of all finite extensions of F contained in K_∞. The natural action of G on $H^1(G(\overline{\mathbb{Q}}/K_\infty), E_{p^\infty})$ induces an action of G on $S(E/K_\infty)$. As $S(E/K_\infty)$ is discrete and p-primary, it is more convenient to work with its compact dual

$$X(E/K_\infty) = \mathrm{Hom}\,(S(E/K_\infty), \mathbb{Q}_p/\mathbb{Z}_p),$$

endowed with its action of the Iwasawa algebra $\Lambda(G)$. It is always true that $X(E/K_\infty)$ is a finitely generated $\Lambda(G)$-module, but it is usually a difficult question to determine its rank as a $\Lambda(G)$-module. Mazur [19] raised the following two general questions. If p is any prime, and $K_\infty = F(\mu_{p^\infty})$, prove that the group $E(K_\infty)$ of K_∞-rational points of E is always finitely generated as an abelian group? Secondly, if p is a prime such that E has good ordinary reduction at all primes v of F dividing p, and K_∞ is the cyclotomic $\mathbb{Z}_p$-extension of F, prove that $X(E/K_\infty)$ has rank zero as a module over the Iwasawa algebra $\Lambda(G)$? We now know that both of these results are true when $F = \mathbb{Q}$ thanks to the deep work of K. Kato [15] and the fact that all elliptic curves over $\mathbb{Q}$ are now known to be modular, but little has been proven about them over an arbitrary base field F. Kato's work is based on analytic methods in Iwasawa theory, and the first step towards creating such an analytic Iwasawa theory for elliptic curves over $\mathbb{Q}$ was made by B. Mazur and H. Swinnerton-Dyer [20]. When E is an elliptic curve defined over $\mathbb{Q}$, and p is a prime where E has good ordinary reduction, they constructed in $\Lambda(G)$, where K_∞ is the cyclotomic $\mathbb{Z}_p$-extension of $\mathbb{Q}$, a p-adic analogue of the complex L-function of E, and posed an analogue of the "main conjecture" (9) for $X(E/K_\infty)$.

It is remarkable that up until now, all major progress towards proving "main conjectures" for the Iwasawa theory of elliptic curves uses the same broad framework of ideas as those introduced by Iwasawa in his work on cyclotomic fields, complemented by additional arguments using the notion of Euler systems introduced by V. Kolyvagin [16]. Suppose first that F is an imaginary quadratic field with class number one, and E is an elliptic curve defined over F with complex multiplication by the ring O of integers of F (e.g. $F = \mathbb{Q}(\sqrt{-1})$, and $E : y^2 = x^3 - Dx$ for $D \neq 0$ in F). Let p be a prime $\geqslant 5$, which splits in K, say $p = \pi\bar{\pi}$, and assume that E has good reduction at both π and $\bar{\pi}$. Let E_{π^∞} be the group of all π-power division points on E, and take $K_\infty = K(E_{\pi^\infty})$. Now $S(E/K_\infty)$ is a p-primary O-module, and so it breaks up as the direct sum of its π-primary and $\bar{\pi}$-primary subgroups. In fact, the π-primary subgroup of $S(E/K_\infty)$ is equal to $\mathrm{Hom}\,(G(M_\infty/K_\infty), E_{\pi^\infty})$, where M_∞ is the maximal abelian p-extension of K_∞ unramified outside π. In [2], [3], Andrew Wiles and I showed that one could carry through a modified form of the arguments of {41} for the $\Lambda(G)$-module $G(M_\infty/K_\infty)$, provided one replaced the cyclotomic units by the elliptic units in the definition of Y_∞. We proved the desired analogue of (11), and later K. Rubin [25] used Euler system arguments to deduce the analogue of the "main conjecture" (9) in this case. In his thesis, R. Yager [36] extended Wiles and my method to the larger field $K_\infty = F(E_{p^\infty})$, proving a version of (11), and again K. Rubin [26] showed that the evident analogue of the "main conjecture" also holds in this case by rather more elaborate arguments with Euler systems. The motivation for proving these "main conjectures" for elliptic curves with complex multiplication is that even weak forms of them imply special cases of the conjecture of Birch and Swinnerton-Dyer (see [2], [3]). It was always known that it was a far more difficult problem to apply the ideas of {41} to elliptic curves E over $\mathbb{Q}$ without complex multiplication, largely because there was no longer a natural Euler system attached to E, which could also be proven to be related to the values at $s = 1$ of the twists of the complex L-functions of E by Dirichlet characters. However, in a tour de force comparable in both ideas and technical brilliance to Iwasawa's original work, K. Kato [15] has recently proven the existence of such an Euler system for all elliptic curves E over $\mathbb{Q}$. He has used this Euler system to prove that $E(\mathbb{Q}(\mu_{p^\infty}))$ is a finitely generated abelian group for all primes p (the proof uses also a result of D. Rohrlich [22]). Suppose that p is a prime $\geqslant 5$, where E has good ordinary reduction, and let K_∞ be the cyclotomic $\mathbb{Z}_p$-extension of $\mathbb{Q}$. Kato also proves that, in this case, $X(E/K_\infty)$ is $\Lambda(G)$-torsion, and that a weak form of the Mazur-Swinnerton-Dyer analogue of the "main conjecture" (9) holds, in the sense that the ideal on the left of (9) is contained in the ideal on the right. We have only touched here on few of the many results which have now been proven about the Iwasawa theory of elliptic curves (see R. Greenberg [11] for more details). As an example of a different type of result from those mentioned above, we mention the remarkable paper by R. Greenberg and G. Stevens [13], which determines the first derivatives at the trivial character of the p-adic L-function of an elliptic curve E over $\mathbb{Q}$, which has split multiplicative reduction at p, in terms of the Tate p-adic period q_E attached to E over $\mathbb{Q}_p$.

A quite different direction in which Iwasawa's work has been extended is the proof by A. Wiles [34] of the analogue of the "main conjecture" (9), when $\mathbb{Q}$ is replaced by an arbitrary totally real number field as base field. Here the proof of the existence of a good generalization of the pseudo-measure ρ_p for totally real fields F was made by P. Cassou-Noguès [1], and P. Deligne and K. Ribet [7], after earlier slightly weaker results by J-P. Serre [29]. See also the remarkable paper by P. Colmez [5], which proves, for any totally real F, that this pseudo-measure has the residue one would expect at the trivial character, provided Leopoldt's p-adic regulator for F is non-zero. It seems highly unlikely that there exist natural Euler systems attached to totally real number fields, and in fact Wiles [34] has followed a quite different path from that of Iwasawa to prove the "main conjecture" in this case. His ingenious and deep method seems to have been inspired by the earlier work of K. Ribet [23], and exploits ideas of H. Hida [14] on p-adic families of modular forms.

We end by indicating two other areas in which Iwasawa's ideas have played an important role. Firstly, nearly all that is known about the exact formulae for the orders of the higher even K-groups of the ring of integers of number fields in terms of zeta values, which have been conjectured by B. Birch and J. Tate [31], and S. Lichtenbaum [18], ultimately depends on some version of Iwasawa's "main conjecture". Secondly, in the work of M. Flach [9] and A. Wiles [35] on the exact order of the Selmer group of the symmetric square motive attached to a modular form, which plays such a key role in the proof that all elliptic curves over $\mathbb{Q}$ are modular, ideas from Iwasawa theory are a fertile ingredient.

References

[1] P. Cassou-Nogués, "Valeurs aux entiers négatifs des fonctions zêta p-adiques", *Inv. Math.* **51** (1979), 29-59.

[2] J. Coates, "Elliptic curves with complex multiplication and Iwasawa theory", *Bull. London Math. Soc.* **23** (1991), 321-350.

[3] J. Coates, A. Wiles, "On the conjecture of Birch and Swinnerton-Dyer", *Inv. Math.* **39** (1977), 223-251.

[4] J. Coates, A. Wiles, "On the p-adic L-functions and elliptic units", *J. Australian Math. Soc.* **26** (1978), 1-25.

[5] P. Colmez, "Résidu en $s = 1$ des fonctions zêta p-adiques", *Inv. Math.* **91** (1988), 371-389.

[6] A. Cuoco, P. Monsky, "Class numbers in $\mathbb{Z}_p{}^d$-extensions", *Math. Ann.* **225** (1981), 235-258.

[7] P. Deligne, K. Ribet, "Values at abelian L-functions at negative integers over totally real fields", *Inv. Math.* **59** (1980), 227-286.

[8] B. Ferrero, L. Washington, "The Iwasawa invariant μ_p vanishes for abelian number fields", *Ann. Math.* **109** (1979), 377-395.

[9] M. Flach, "A finiteness theorem for the symmetric square of an elliptic curve", *Inv. Math.* **109** (1992), 307-327.

[10] R. Greenberg, "On the Iwasawa invariants of totally real number fields", *Amer. J. Math.* **98** (1976), 263-284.

[11] R. Greenberg, "Iwasawa theory for elliptic curves", in Springer Lecture Notes 1716, 51-105.

[12] R. Greenberg, "Iwasawa theory-past and present", in Adv. Study in Pure Math., Vol. 30, Class Field Theory - its centenary and prospect, Math. Soc. Japan (2001), 335-385.

[13] R. Greenberg, G. Stevens, "p-adic L-functions and p-adic periods of modular forms", *Inv. Math.* (1993), 407-447.

[14] H. Hida, "Iwasawa modules attached to congruences of cusp forms", *Ann. Sci. Ec. Norm. Sup.* **19** (1986), 231-273.

[15] K. Kato, "p-adic Hodge theory and values of zeta functions of modular forms", to appear.

[16] V. Kolyvagin, "Euler systems", in *Progress in Mathematics* **87**, Birkhäuser (1990), 435-483.

[17] T. Kubota, H. Leopoldt, "Eine p-adische theorie der zetawerte", Crelle **214** (1964), 328-339.

[18] S. Lichtenbaum, "Values of zeta functions, étale cohomology, and algebraic K-theory", in Springer Lecture Notes **342**, 489-501.

[19] B. Mazur, "Rational points of abelian varieties with values in a tower of number fields", *Inv. Math.* **18** (1972), 173-266.

[20] B. Mazur, H. Swinnerton-Dyer, "Arithmetic of Weil curves", *Inv. Math.* **25** (1974), 1-61.

[21] B. Mazur, A. Wiles, "Class fields of abelian extensions of $\mathbb{Q}$", *Inv. Math.* **76** (1984), 179-330.

[22] Rohrlich, D., "On L-functions of elliptic curves and cyclotomic towers", *Inv. Math.* **75** (1984), 409-423.

[23] Ribet, K., "A modular construction of unramified p-extensions of $\mathbb{Q}(\mu_p)$", *Inv. Math.* **34** (1976), 151-162.

[24] K. Rubin, "The main conjecture", Appendix to "Cyclotomic Fields I-II" by S. Lang, *Graduate Texts Math.* **121**, Springer (1990).

[25] K. Rubin, "The one variable main conjecture for elliptic curves with complex multiplication", in London Math. Soc. Lecture Notes **153**, 353-371.

[26] "K. Rubin, "The "main conjecture" of Iwasawa theory for imaginary quadratic fields", *Inv. Math.* **103** (1991), 25-68.

[27] J-P. Serre, unpublished letters to K. Iwasawa dated 12th and 19th August, 1958.

[28] J-P. Serre, "Classes des corps cyclotomiques" (d'après K. Iwasawa), *Sém. Bourbaki* **174** (1959).

[29] J-P. Serre, "Formes modulaires et fonctions zêta p-adiques", Springer Lecture Notes **350**, 191-268.

[30] W. Sinnott, "On the μ-invariant of the Γ-transform of a rational function", *Inv. Math.* **75** (1984), 273-282.

[31] J. Tate, "Symbols in arithmetic", *Actes, Congrès Intern. Math.*, 1970, tome 1, 201-211.

[32] F. Thaine, "On the ideal class groups of real abelian number fields", *Ann. Math.* **128** (1988), 1-18.

[33] A. Wiles, "Higher explicit reciprocity laws", *Ann. Math.* **107** (1978), 235-254.

[34] A. Wiles, "The Iwasawa conjecture for totally real fields", *Ann. Math.* **131** (1990), 493-540.

[35] A. Wiles, "Elliptic curves and Fermat's Last Theorem", *Ann. Math.* **141** (1995), 443-551.

[36] R. Yager, "On two variable p-adic L-functions", *Ann. Math.* **115** (1982), 411-449.

[1] Ueber die Struktur der endlichen Gruppen, deren echte Untergruppen sämtlich nilpotent sind

Proc. Phys.-Math. Soc. Japan, 23-1 (1941), 1-4.

(Gelesen am 21. Dezember, 1940.)

Eine Gruppe $\mathfrak{G}$ heisst nilpotent, wenn die aufsteigende Zentralreihe von $\mathfrak{G}$ die ganze Gruppe als letztes Glied enthält, wenn also

$$1 = \mathfrak{z}_0 \subset \mathfrak{z}_1 \subset \ldots \subset \mathfrak{z}_c = \mathfrak{G}$$

ist, wobei $\mathfrak{z}_{i+1}/\mathfrak{z}_i$ immer das Zentrum von $\mathfrak{G}/\mathfrak{z}_i$ ist[1]. c bedeutet natürlich eine ganze rationale Zahl. Eine endliche nilpotente Gruppe ist bekanntlich das direkte Produkt ihrer Sylowgruppen und umgekehrt[2]. Im folgenden soll die Struktur derjenigen endlichen Gruppen untersucht werden, deren echte Untergruppen alle nilpotent sind.

1. $\mathfrak{G}$ sei also eine endliche Gruppe, deren echte Untergruppen alle nilpotent sind, und p sei ein Primteiler der Ordnung g von $\mathfrak{G}$. Wir nehmen zunächt an, dass $\mathfrak{G}$ nicht p-normal ist, d.h. dass das Zentrum $\mathfrak{z}_p$ einer p-Sylowgruppe $\mathfrak{P}$ von $\mathfrak{G}$ als Nicht-Normalteiler in einer anderen p-Sylowgruppe $\overline{\mathfrak{P}}$ enthalten ist[3]. Alsdann gibt es nach einem Satz von Burnside[4] eine solche p-Untergruppe $\mathfrak{H}$ von $\mathfrak{G}$, deren Normalisator $N(\mathfrak{H})$ in $\mathfrak{G}$ diejenigen Elemente enthält, die in $\mathfrak{H}$ Automorphismen von zu p primen Ordnungen induzieren. Wenn nun $N(\mathfrak{H})$ eine echte Untergruppe von $\mathfrak{G}$ wäre, dann würde es nach Annahme nilpotent sein, und kein Element aus $N(\mathfrak{H})$ sollte in $\mathfrak{H}$ einen Automorphismus von einer zu p primen Ordnung induzieren. In unserem Fall muss daher $N(\mathfrak{H}) = \mathfrak{G}$ sein. $\mathfrak{G}$ enthält also einen Normalteiler $\mathfrak{H}$ mit einer p-Potenz Ordnung.

Nunmehr sei $\mathfrak{G}$ p-normal, und $\mathfrak{z}_p$ nicht ein Normalteiler von $\mathfrak{G}$. Der Normalisator $N(\mathfrak{z}_p)$ von $\mathfrak{z}_p$ ist dann als echte Untergruppe von $\mathfrak{G}$ nilpotent, und seine p-Faktorkommutatorgruppe[5] ist nicht die Einheitsgruppe. Nun gilt nach Grün[6] die folgende Tatsache: Wenn die endliche Gruppe $\mathfrak{G}$ p-normal ist, dann ist die p-Faktorkommutatorgruppe von $\mathfrak{G}$ isomorph zu der p-Faktorkommutatorgruppe des Normalisators eines p-Zentrums $\mathfrak{z}_p$. Nach diesem Satz enthält $\mathfrak{G}$ also

(1) H. Zassenhaus, Lehrbuch der Gruppentheorie I, (1937), S. 105. Zitiert mit H.Z..
(2) H.Z., S. 107.
(3) O. Grün, J. reine u. angew. Math. 174, S. 1-14, od. H.Z., S. 135.
(4) W. Burnside, Theory of Groups of finite order, 2nd ed. (1911), S. 156.
(5) H.Z., S. 122.
(6) Vgl. (3).

einen Normalteiler von einem p-Potenz Index. In jedem Fall enthält $\mathfrak{G}$ daher einen Normalteiler entweder von einer p-Potenz Ordnung oder von einem p-Potenz Index. Da auch jede Faktorgruppe von $\mathfrak{G}$ die Eigenschaft besitzt, dass ihre sämtlichen echten Untergruppen nilpotent sind, so ergibt sich hieraus, dass $\mathfrak{G}$ auflösbar ist. Also gilt

Satz 1. Jede endliche Gruppe $\mathfrak{G}$, deren echte Untergruppen sämtlich nilpotent sind, ist auflösbar.

Nun sei $g=p_1^{e_1}p_2^{e_2}\ldots p_r^{e_r}$ mit $r\geqq 3$, wobei p_i verschiedene Primteiler von g sind. In der endlichen auflösbaren Gruppe $\mathfrak{G}$ gibt es nach P. Hall[7] für jedes Paar i, j, $(i\neq j,\ 1\leqq i, j\leqq r)$ eine Untergruppe $\mathfrak{G}_{i,j}$ von der Ordnung $p_i^{e_i}p_j^{e_j}$. Da $\mathfrak{G}_{i,j}$ eine echte Untergruppe von $\mathfrak{G}$ ist, so ist es nach unsrer Voraussetzung eine nilpotente Gruppe, also das direkte Produkt der p_i- und p_j-Sylowgruppen von $\mathfrak{G}$. p_i-Sylowgruppen $(i=1, 2, \ldots, r)$ sind deshalb untereinander elementweise vertauschbar; $\mathfrak{G}$ ist folglich in diesem Fall das direkte Produkt ihrer Sylowgruppen.

Satz 2. Wenn die echte Untergruppen von einer endlichen Gruppe $\mathfrak{G}$ alle nilpotent sind, und die Ordnung von $\mathfrak{G}$ durch mindestens drei verschiedene Primzahlen teilbar ist, dann ist $\mathfrak{G}$ selbst nilpotent.

Eine abelsche oder hamiltonsche Gruppe ist offenbar nilpotent. Daher folgt aus Satz 2 ohne weiters der folgende

Satz 3[8]. Wenn die echten Untergruppen von einer endlichen Gruppe $\mathfrak{G}$ alle abelsch bzw. hamiltonsch sind, und die Ordnung von $\mathfrak{G}$ durch mindestens drei verschiedene Primzahlen teilbar ist, dann ist $\mathfrak{G}$ selbst abelsch bzw. hamiltonsch.

Im Gegensatz zu Satz 3 gibt es nicht-abelsche Gruppen von den bloss zwei Primteiler p, q enthaltenden Ordnungen, deren echte Untergruppen alle abelsch sind. Eine solche Gruppe $\mathfrak{G}$ hat nach G. A. Miller folgende Struktur[8]:

1) Eine Sylowgruppe, z.B. eine p-Sylowgruppe $\mathfrak{P}$, ist zyklisch; $\mathfrak{P}=\{A\}$, und eine zum anderen Primfaktor gehörige Sylowgruppe, eine q-Sylowgruppe $\mathfrak{Q}$, ist ein abelscher Normalteiler von $\mathfrak{G}$, deren Elemente ausser Eins alle die Ordnungen q haben.

2) A induziert in $\mathfrak{Q}$ einen Automorphismus von der Ordnung p, der kein Element von $\mathfrak{Q}$ ausser Eins festlässt.

3) $\mathfrak{Q}$ wird durch ein beliebiges Element $B\neq 1$ und seine Konjugierten erzeugt, d.h. $\mathfrak{Q}$ ist ein minimaler Normalteiler von $\mathfrak{G}$.

Mit Hilfe dieser Eigenschaften soll die Struktur derjenigen Gruppen von der Ordnung $p^\alpha q^\beta$ bestimmt werden, deren echte Untergruppen sämtlich nilpotent sind, ohne jedoch selbst nicht nilpotent zu

(7) P. Hall, J. London Math. Soc. 3, S. 98-105.

(8) G.A. Miller u. H.C. Moreno, Trans. Amer. Math. Soc. 4, S. 398–404.

sein.

2. $\mathfrak{G}$ sei also eine endliche Gruppe von der Ordnung $p^\alpha q^\beta$, deren echte Untergruppen alle nilpotent sind, ohne selbst nicht nilpotent zu sein. Da $\mathfrak{G}$ nach Satz 1 auflösbar ist, gibt es einen Normalteiler vom Index p oder q. Nun sei $\mathfrak{G}_1$ ein solcher Normalteiler vom Index p. Nach Voraussetzung ist $\mathfrak{G}_1$ das direkte Produkt der Sylowgruppen, und eine q-Sylowgruppe $\mathfrak{Q}$ von $\mathfrak{G}_1$, die natürlich zugleich eine q-Sylowgruppe von $\mathfrak{G}$ ist, ist in $\mathfrak{G}_1$ als charakteristische Untergruppe enthalten. $\mathfrak{Q}$ ist daher ein Normalteiler von $\mathfrak{G}$. Enthält $\mathfrak{G}$ also einen Normalteiler vom Index p, so enthält es keinen vom Index q; denn sonst wäre auch eine p-Sylowgruppe ein Normalteiler von $\mathfrak{G}$, und $\mathfrak{G}$ selbst nilpotent.

Enthielte nun eine p-Sylowgruppe $\mathfrak{P}$ von $\mathfrak{G}$ zwei verschiedene Untergruppen $\mathfrak{P}_1$, $\mathfrak{P}_2$ vom Index p, dann würde $\mathfrak{Q}$ mit $\mathfrak{P}_1$, $\mathfrak{P}_2$, daher auch mit $\mathfrak{P}$, elementweise vertauschbar sein, weil die echten Untergruppen $\mathfrak{P}_1\mathfrak{Q}$, $\mathfrak{P}_2\mathfrak{Q}$ von $\mathfrak{G}$ nilpotent sind. Dann wäre also $\mathfrak{G}$ wieder selbst nilpotent gegen unsere Voraussetzung. Daher muss $\mathfrak{P}$ nur eine einzige Untergruppe vom Index p enthalten, folglich zyklisch sein: $\mathfrak{P}=\{A\}$. Bezeichnet man mit $\mathfrak{Q}'$ die Kommutatorgruppe von $\mathfrak{Q}$, so ergibt sich hieraus, dass jede echte Untergruppe von $\mathfrak{G}/\mathfrak{Q}'$ abelsch ist, da sie das direkte Produkt der abelschen Sylowgruppen ist. $\mathfrak{G}/\mathfrak{Q}'$ selbst ist aber nicht abelsch, denn sonst gäbe es in $\mathfrak{G}$ einen Normalteiler vom Index q, was, wie oben bemerkt, unmöglich ist. Ferner ist es klar, dass $\mathfrak{P}\mathfrak{Q}'$ und $\mathfrak{P}_1\mathfrak{Q}$ echte Untergruppen von $\mathfrak{G}$ sind, wenn $\mathfrak{P}_1=\{A^p\}$ gesetzt wird. Sie sind daher nach Voraussetzung beide nilpotent.

Satz 4. Eine Gruppe $\mathfrak{G}$ von der Ordnung $p^\alpha q^\beta$ hat dann und nur dann lauter nilpotente echte Untergruppen und selbst nicht nilpotent, wenn sie die folgenden Eigenschaften besitzt:

a) Eine Sylowgruppe, z.B. q-Sylowgruppe $\mathfrak{Q}$, ist ein Normalteiler von $\mathfrak{G}$, und jede zum anderen Primfaktor gehörige Sylowgruppe, also im oben angegebenen Fall p-Sylowgruppe $\mathfrak{P}$, ist zyklisch: $\mathfrak{P}=\{A\}$.

b) Bezeichnet man die Kommutatorgruppe von $\mathfrak{Q}$ mit $\mathfrak{Q}'$, so ist jede echte Untergruppe von $\mathfrak{G}/\mathfrak{Q}'$ abelsch. $\mathfrak{G}/\mathfrak{Q}'$ selbst ist es aber nicht.

c) Setzt man ferner $\mathfrak{P}_1=\{A^p\}$, dann sind die Gruppen $\mathfrak{P}\mathfrak{Q}'$ und $\mathfrak{P}_1\mathfrak{Q}$ beide nilpotent.

Beweis. Nun sei $\mathfrak{G}$ eine Gruppe mit der Eigenschaft a), b), c). $\mathfrak{H}$ sei eine maximale Untergruppe von $\mathfrak{G}$. Nach a), b) ist zunächst klar, dass $\mathfrak{G}$ nicht nilpotent ist. Es soll gezeigt werden, dass $\mathfrak{H}$ nilpotent ist. Wäre nun $\mathfrak{Q}'$ nicht in $\mathfrak{H}$ enthalten, so wäre $\mathfrak{G}=\mathfrak{Q}'\mathfrak{H}$, und hiernach

$\mathfrak{Q}=\mathfrak{Q}_1\mathfrak{Q}'$ mit einer q-Sylowgruppe $\mathfrak{Q}_1$ von $\mathfrak{H}$. Die q-Sylowgruppe $\overline{\mathfrak{Q}}=\mathfrak{Q}/\mathfrak{Q}'$ von $\overline{\mathfrak{G}}=\mathfrak{G}/\mathfrak{Q}'$ ist nach 1) eine *elementare* abelsche Gruppe. Nach dem Basissatz von Burnside[9] würde daher aus $\mathfrak{Q}=\mathfrak{Q}_1\mathfrak{Q}'$ folgen, dass $\mathfrak{Q}=\mathfrak{Q}_1$ wäre, was wegen $\mathfrak{H}\not\supset\mathfrak{Q}'$ ein Widerspruch ist. $\mathfrak{Q}'$ ist also in $\mathfrak{H}$ enthalten, und $\overline{\mathfrak{H}}=\mathfrak{H}/\mathfrak{Q}'$ ist eine maximale Untergruppe von $\overline{\mathfrak{G}}$. Um nun die maximale Untergruppe $\overline{\mathfrak{H}}$ von $\overline{\mathfrak{G}}$ näher zu bestimmen, nehmen wir zunächst an, dass $\overline{\mathfrak{H}}$ eine zyklische p-Sylowgruppe $\overline{\mathfrak{P}}=\{\overline{A}\}$ von $\overline{\mathfrak{G}}$ enthält. $\overline{\mathfrak{H}}$ fällt dann mit $\overline{\mathfrak{P}}$ zusammen, denn wenn $\overline{\mathfrak{H}}$ nur noch ein Element $\overline{B}$ von $\overline{\mathfrak{Q}}$ enthielte, dann wäre nach 3) die ganze Gruppe $\overline{\mathfrak{Q}}$ in $\overline{\mathfrak{H}}$ enthalten, folglich $\overline{\mathfrak{H}}=\overline{\mathfrak{G}}$. Wenn dagegen $\overline{\mathfrak{H}}$ nicht $\overline{\mathfrak{P}}$ enthält, so sieht man leicht ein, dass $\overline{\mathfrak{H}}$ mit $\overline{\mathfrak{P}}_1\cdot\overline{\mathfrak{Q}}$ zusammenfällt, wobei $\overline{\mathfrak{P}}_1=\{\overline{A}^p\}$ gesetzt wird. Die maximale Untergruppe $\mathfrak{H}$ von $\mathfrak{G}$ ist daher entweder $\mathfrak{P}\mathfrak{Q}'$ oder $\mathfrak{P}_1\mathfrak{Q}$. Da diese beiden Gruppen nach c) nilpotent sind, so ist $\mathfrak{H}$ nilpotent, w.z.b.w.

Eine nicht nilpotente endliche Gruppe muss immer eine solche Untergruppe enthalten, die im Satz 4 angegebene Struktur besitzt. Daher gilt

Satz 5. Eine endliche Gruppe ist dann und nur dann abelsch, hamiltonsch bzw. nilpotent, wenn ihre Untergruppen, deren Ordnungen durch höchstens zwei verschiedene Primzahlen teilbar sind, sämtlich abelsch, hamiltonsch bzw. nilpotent sind.

Kaiserliche Universität zu Tokyo.

(Eingegangen am 22. Dezember, 1940.)

---

(9) W. Burnside, Proc. London Math. Soc. 2, S. 13, od. H.Z., S. 45.

[2] Über die Einfachheit der speziellen projektiven Gruppen

Proc. Imp. Acad. Japan, 17-3 (1941), 57-59.

Mathematical Institute, Tokyo Imperial University.

(Comm. by T. TAKAGI, M.I.A., March 12, 1941.)

Die Gesamtheit der nicht singulären linearen Transformationen des n-dimensionalen Vektorraumes in sich über einem Körper K bildet eine Gruppe, welche die allgemeine lineare Gruppe $GL(n, K)$ heisst[1]. Wir schreiben nun die Transformationen aus $GL(n, K)$ in der Form

$$\begin{pmatrix} \xi_1' \\ \xi_2' \\ \vdots \\ \xi_n' \end{pmatrix} = \begin{pmatrix} a_{11} & a_{12} \cdots a_{1n} \\ a_{21} & \vdots \\ \vdots & \vdots \\ a_{n1} \cdots\cdots & a_{nn} \end{pmatrix} \begin{pmatrix} \xi_1 \\ \xi_2 \\ \vdots \\ \xi_n \end{pmatrix}$$

In $GL(n, K)$ bilden die Transformationen mit Determinante Eins einen Normalteiler, welcher die spezielle lineare Gruppe $SL(n, K)$ heisst. Diese Gruppe wird erzeugt durch die Transformationen

$$B_{r,s,\lambda}: \begin{cases} \xi_r' = \xi_r + \lambda \xi_s & (r \neq s) \\ \xi_\nu' = \xi_\nu & \nu \neq r, \end{cases}$$

wobei λ alle Elemente aus K durchläuft[2]. Das Zentrum von $SL(n, K)$ besteht aus allen λE mit $\lambda^n = 1$, wenn mit E die identische Transformation bezeichnet wird. Die Faktorgruppe von $SL(n, K)$ nach dem Zentrum heisst die spezielle projektive Gruppe $PSL(n, K)$.

Für $n > 1$ ist nun $PSL(n, K)$ eine *einfache* Gruppe, ausser in zwei Fällen $n=2$, $K=GF(2)$ und $n=2$, $K=GF(3)$. Dieser Satz wurde zuerst von L. E. Dickson für endliche Grundkörper K bewiesen[3]. Sein Beweis aber wird, wie van der Waerden bemerkt[4], durch eine kleine Abänderung auch für den Fall unendlicher Grundkörper K gültig, wenn K von der Charakteristik $\neq 2$ oder vollkommen ist.

In der vorliegenden Note soll dieser Satz ohne oben erwähnte Einschränkung über K neu bewiesen werden.

Hilfssatz 1. $SL(n, K)$ fällt mit seiner Kommutatorgruppe zusammen, ausser wenn $n=2$, $K=GF(2)$ oder $n=2$, $K=GF(3)$.

Beweis. Zunächst sei $n \geqq 3$ und i, j $(1 \leqq i, j \leqq n)$ beliebig gegeben. Für $k \neq i, j$ gilt dann

$$B_{i,k,\lambda} B_{k,j,1} B_{i,k,\lambda}^{-1} B_{k,j,1}^{-1} = B_{i,j,\lambda}.$$

Damit ist unsere Behauptung erledigt.

---

1) Bzgl. der Bezeichnungen $GL(n, K)$, $SL(n, K)$ u.s.w. vgl. B. L. van der Waerden: Gruppen von linearen Transformationen, 1935.

2) Vgl. L. E. Dickson: Linear groups, with an exposition of Galois Field theory, 1901, S. 78.

3) Vgl. loc. cit. 2), S. 83.

4) Vgl. loc. cit. 1), S. 7.

Nun sei $n=2$. Alsdann ist

$$\begin{pmatrix}\lambda & 0\\ 0 & \lambda^{-1}\end{pmatrix}\begin{pmatrix}1 & \mu\\ 0 & 1\end{pmatrix}\begin{pmatrix}\lambda^{-1} & 0\\ 0 & \lambda\end{pmatrix}\begin{pmatrix}1 & -\mu\\ 0 & 1\end{pmatrix}=\begin{pmatrix}1 & \mu(\lambda^2-1)\\ 0 & 1\end{pmatrix}.$$

Wenn K nicht $GF(2)$ oder $GF(3)$ ist, so gibt es in K mindestens ein Element $\lambda \neq 0$ mit $\lambda^2 \neq 1$. Da μ ein beliebiges Element aus K ist, so enthält die Kommutatorgruppe von $SL(2, K)$ alle $B_{1,2,\lambda}$ und ebenso alle $B_{2,1,\lambda}$: sie fällt also mit $SL(2, K)$ selbst zusammen.

Die Gruppe $PSL(n, K)$ besteht bekanntlich aus projektiven Transformationen des $(n-1)$-dimensionalen projektiven Raumes über K. Wir können also diese Gruppe als Permutationsgruppe von Punkten in $P_{n-1}(K)$ auffassen. Es gilt dann

Hilfssatz 2. $PSL(n. K)$ ist $(n>1)$, als Permutationsgruppe, zweifach transitiv.

Beweis. Es seien a, b zwei beliebige Punkte von $P_{n-1}(K)$:

$$a=(\alpha_1, \alpha_2, \dots, \alpha_n), \quad b=(\beta_1, \beta_2, \dots, \beta_n).$$

Es genügt ersichtlich nur zu zeigen, dass es eine solche Transformation σ in $PSL(n, K)$ gibt, die Punkte $e_1=(1, 0, 0, \dots, 0)$, $e_2=(0, 1, 0, \dots, 0)$ in a bzw. b überführt. Wegen $a \neq b$ gibt es dann eine geeignete Matrix

$$A_0=\begin{pmatrix}\alpha_1 & \beta_1 & \cdots\cdots\\ \alpha_2 & \beta_2 & \cdots\cdots\\ \vdots & \vdots & \\ \alpha_n & \beta_n & \cdots\cdots\end{pmatrix}$$

mit $|A_0| \neq 0$. σ wird dann wie ersichtlich durch die Matrix $A_0/|A_0|$ angegeben.

Nunmehr sei $\mathfrak{H}$ diejenige Untergruppe von $\mathfrak{G}=PSL(n, K)$ $(n>1)$, die aus allen Transformationen aus $\mathfrak{G}$ besteht, die e_1 festlassen. Eine Transformation aus $\mathfrak{H}$ ist daher von der Form

$$A=\begin{pmatrix}\alpha_1 & \alpha_2 \cdots\cdots \alpha_n\\ 0 & \\ \vdots & A_1\\ 0 & \end{pmatrix}$$

wobei A_1 eine $(n-1)$-reihige Matrix mit $\alpha_1|A_1|=1$ ist. In $\mathfrak{H}$ bilden alle Matrizen[1)] der Form

$$B=\begin{pmatrix}1 & \beta_2 \cdots\cdots \beta_n\\ 0 & \\ \vdots & E_{n-1}\\ 0 & \end{pmatrix}$$

einen Normalteiler $\mathfrak{N}_0$, denn genau diese Matrizen werden beim Homomorphismus $A \rightarrow A_1$ der $(n-1)$-reihigen Einheitsmatrix E_{n-1} zugeordnet. $B_{1,2,\lambda}$ sind sämtlich in $\mathfrak{N}_0$ enthalten.

1) Ausführlicher: die durch diese Matrizen vertretenen Restklassen von $SL(n, K)/\mathfrak{Z} \simeq PSL(n, K)$.

Nun sei $\mathfrak{N}$ ein Normalteiler von $\mathfrak{G}$, der keine Einheitsgruppe ist. Da $\mathfrak{G}$ nach Hilfssatz 2 zweifach transitiv ist, so ist $\mathfrak{N}$ als Permutationsgruppe von Punkten aus $P_{n-1}(K)$ sicherlich transitiv. Es ist also $\mathfrak{G}=\mathfrak{N}\mathfrak{H}$, und $\mathfrak{N}^*=\mathfrak{N}\mathfrak{N}_0$ ist natürlich ein Normalteiler von $\mathfrak{G}$. Da ein beliebiges $B_{i,j,\lambda}$ mit $B_{1,2,\pm\lambda}$ in $SL(n, K)$ konjugiert ist, so enthält $\mathfrak{N}^*$ mit $B_{1,2,\lambda}$ zugleich alle $B_{i,j,\lambda}$; $\mathfrak{N}^*$ fällt also mit $\mathfrak{G}$ zusammen. Aus $\mathfrak{G}=\mathfrak{N}\mathfrak{N}_0$ folgt $\mathfrak{G}/\mathfrak{N}\simeq\mathfrak{N}_0/\mathfrak{N}\cap\mathfrak{N}_0$ und $\mathfrak{N}$ ist mithin ein Normalteiler mit abelscher Faktorgruppe. Da aber $\mathfrak{G}$ nach Hilfssatz 1 ausser in zwei genannten Fällen mit seiner Kommutatorgruppe zusammenfällt, so ergibt sich hieraus $\mathfrak{G}=\mathfrak{N}$.

Es gilt also der

Hauptsatz. Die Gruppe $PSL(n, K)$ ist einfach $(n>1)$, ausser wenn $n=2$, $K=GF(2)$ oder $n=2$, $K=GF(3)$.

Für $n=2$, $K=GF(2)$ bzw. $n=2$, $K=GF(3)$ ist $PSL(n, K)$ bekanntlich isomorph mit der symmetrischen Gruppe von 3 Symbolen bzw. mit der Tetraedergruppe.

In ähnlicher Weise kann man auch die Einfachheit der projektiven Komplexgruppen $PC(2m, K)$[1] nachweisen.

---

1) $PC(2m, K)$ ist als Permutationsgruppe von Punkten aus $P_{2m-1}(K)$ nicht zweifach transitiv. Es lässt sich aber leicht beweisen, dass jeder Normalteiler von $PC(2m, K)$ transitiv ist, was für den Beweis der Einfachheit wesentlich ist.

[3] Über die endlichen Gruppen und die Verbände ihrer Untergruppen

J. Fac. Sci. Tokyo Imp. Univ., Sect I. 4-3 (1941), 171-199.

Die Untergruppen einer gegebenen Gruppe $\mathfrak{G}$ bilden bekanntlich einen Verband; bezeichnet man nämlich für zwei beliebige Untergruppen $\mathfrak{A}, \mathfrak{B}$ von $\mathfrak{G}$ die von $\mathfrak{A}$ und $\mathfrak{B}$ erzeugte Untergruppe bzw. den Durchschnitt von $\mathfrak{A}, \mathfrak{B}$ mit $\mathfrak{A} \cup \mathfrak{B}$ bzw. $\mathfrak{A} \cap \mathfrak{B}$ bezeichnet, so sieht man leicht ein, dass alle Verbandsaxiome erfüllt sind. Wir bezeichnen diesen Verband mit $V(\mathfrak{G})$ und nennen ihn den zu $\mathfrak{G}$ gehörigen Verband.

Nun wollen wir folgende Definition einführen:

Definition. Eine Gruppe $\mathfrak{G}$ heisst modular, distributiv, komplementär bzw. Boolesch, wenn der zugehörige Verband $V(\mathfrak{G})$ modular, distributiv, komplementär bzw. Boolesch ist. Wir bezeichnen oft solche Gruppen bzw. auch kurz mit *M*-, *D*-, *K* bzw. *B*-Gruppe. Wenn ferner $V(\mathfrak{G})$ modular komplementär und irreduzibel ist, also die „projektive Geometrie“ bildet, dann sagen wir, $\mathfrak{G}$ sei eine *P*-Gruppe.

Im folgenden soll die Relation zwischen $\mathfrak{G}$ und $V(\mathfrak{G})$ untersucht werden; insbesondere soll die Struktur der endlichen M-Gruppen (§§ 1 u. 4), D-, B- bzw. P-Gruppen (§ 2) vollständig bestimmt werden.

Daran anschliessend wird auch die Struktur der endlichen „quasi-Hamiltonschen“ Gruppen bestimmt (§ 3), die eine Verallgemeinerung der Hamiltonschen Gruppen bilden.

§ 1.

1. Ein Verband A heisst ausgeglichen, wenn alle Hauptkette, die zwei Elemente von A verbinden, gleiche Länge haben.[1] Da ein modularer Verband stets ausgeglichen ist, so untersuchen wir zuerst die Struktur derjenigen endlichen Gruppe, deren zugehöriger Verband ausgeglichen ist.

Hilfssatz 1.[2] Wenn der zugehörige Verband $V(\mathfrak{G})$ einer endlichen Gruppe $\mathfrak{G}$ ausgeglichen ist, so hat $\mathfrak{G}$ eine Hauptreihe

(1) $$1 = \mathfrak{G}_0 \subset \mathfrak{G}_1 \subset \cdots \subset \mathfrak{G}_r = \mathfrak{G} ,$$

1) G. Birkhoff, Proc. Camb. Phil. Soc. 29, 441-464, 1933.
2) Vgl. O. Ore, Duke Math. Jour., 5, 431-460, 1939.

wobei $[\mathfrak{G}_i:\mathfrak{G}_{i-1}]=p_i$ $(i=1,2,\ldots,r)$ Primzahlen mit $p_1 \geqq p_2 \geqq \cdots \geqq p_r$ sind.

Beweis. Für eine Gruppe von einer Primzahlordnung ist die Behauptung klar. Wir beweisen also den Hilfssatz durch vollständige Induktion nach der Ordnung von $\mathfrak{G}$. g sei also die Ordnung von $\mathfrak{G}$ und p_r der kleinste Primteiler von g. Wir nehmen zunächst an, dass $\mathfrak{G}$ p_r-normal[3] ist. Nach Grün[4] ist dann die p_r-Faktorkommutatorgruppe $\mathfrak{G}/\mathfrak{G}'(p_r)$ von $\mathfrak{G}$ zur p_r-Faktorkommutatorgruppe $N_z/N_z'(p_r)$ von N_z isomorph, wobei $\mathfrak{Z}$ das Zentrum einer p_r-Sylowgruppe $\mathfrak{P}$ von $\mathfrak{G}$, und N_z der Normalisator von $\mathfrak{Z}$ bedeutet. Ist $\mathfrak{Z}$ nicht ein Normalteiler von $\mathfrak{G}$, so ist N_z eine echte Untergruppe von $\mathfrak{G}$, und nach Induktionsvoraussetzung ist $\mathfrak{G}/\mathfrak{G}'(p_r) \simeq N_z/N_z(p_r)$ offenbar keine Einheitsgruppe. Sei nun $\mathfrak{G}$ nicht p_r-normal. Nach einem Satz von Burnside[5] gibt es dann eine p_r-Untergruppe $\mathfrak{H}$ von $\mathfrak{G}$, so dass der Normalisator $N(\mathfrak{H})$ von $\mathfrak{H}$ ein Element A mit einer zu p_r primen Ordnung enthält, das in $\mathfrak{H}$ einen nicht identischen Automorphismus σ;

$$H^\sigma = AHA^{-1}, \qquad H \in \mathfrak{H}, \tag{2}$$

induziert. Wäre nun $N(\mathfrak{H})$ eine echte Untergruppe von $\mathfrak{G}$, so folgte aus der Induktionsvoraussetzung, dass die Untergruppe $\mathfrak{H}_1=\{A,\mathfrak{H}\}$ direktes Produkt von $\mathfrak{H}$ und $\{A\}$ ist. A wäre dann also mit $\mathfrak{H}$ elementweise vertauschbar, was mit (2) in Widerspruch steht. $\mathfrak{H}$ ist daher ein Normalteiler von $\mathfrak{G}$. In jedem Fall enthält $\mathfrak{G}$ also einen Normalteiler $\mathfrak{N}$ von einer Primzahlpotenzordnung oder von einem Primzahlpotenzindex. Da offenbar mit $V(\mathfrak{G})$ auch $V(\mathfrak{N})$ und $V(\mathfrak{G}/\mathfrak{N})$ ausgeglichen sind, so sieht man nach Induktionsvoraussetzung sofort, dass $\mathfrak{G}$ auflösbar ist. Da $\mathfrak{G}$ auflösbar ist, so ist eine Kompositionsreihe von $\mathfrak{G}$ offenbar zugleich eine Hauptkette von $V(\mathfrak{G})$, die $\mathfrak{G}$ mit 1 verbindet, und die Länge r dieser Hauptkette ist gleich der Anzahl der Primfaktoren von $\mathfrak{G}$. Da jede Hauptkette, die $\mathfrak{G}$ mit 1 verbindet, stets gleiche Länge r hat, so ergibt sich hieraus, dass für zwei Untergruppen $\mathfrak{A}$, $\mathfrak{B}$ der Index $[\mathfrak{A}:\mathfrak{B}]$ eine Primzahl ist, wenn sie in einer Hauptkette von $V(\mathfrak{G})$ benachbart sind, d. h. wenn $\mathfrak{B}$ eine maximale Untergruppe von $\mathfrak{A}$ ist. Es sei nun p_1 der grösste Primteiler von g und $\mathfrak{P}_1$ sei eine p_1-Sylowgruppe von $\mathfrak{G}$. Wir verfeinern die Kette $1 \subset \mathfrak{P}_1 \subset \mathfrak{G}$ zu einer Hauptkette

$$1=\mathfrak{B}_0 \subset \mathfrak{B}_1 \subset \cdots \subset \mathfrak{B}_k = \mathfrak{P}_1 \subset \mathfrak{B}_{k+1} \subset \cdots \subset \mathfrak{B}_r = \mathfrak{G}.$$

$[\mathfrak{B}_{k+1}:\mathfrak{P}_1]=q$ ist dabei, wie schon bemerkt, eine Primzahl. Die

3) O. Grün, Jour. f. reine u. angew. Math. 174, 1-14, 1935.

4) Vgl. Grün, a. a. O. 3).

5) W. Burnside, Theory of groups of finite Order (1911), S. 156.

Anzahl s der p_1-Sylowgruppen von $\mathfrak{B}_{k+1}$ ist also entweder 1 oder q. Da aber nach Annahme $p_1 > q$, und nach dem Sylowsatz $s \equiv 1$ mod. p ist, so muss $s=1$ sein. $\mathfrak{P}_1$ ist also als Normalteiler in $\mathfrak{B}_{k+1}$ enthalten.

$[\mathfrak{B}_{k+2} : \mathfrak{B}_{k+1}] = q'$ ist nun auch eine Primzahl mit $p_1 > q'$, also ergibt sich in genau derselben Weise wie oben, dass $\mathfrak{P}_1$ als Normalteiler in $\mathfrak{B}_{k+2}$ enthalten ist. So fortfahrend beweist man schliesslich, dass $\mathfrak{P}_1$ ein Normalteiler von $\mathfrak{G}$ ist. Sei nun $\mathfrak{Z}_1$ das Zentrum von $\mathfrak{P}_1$, und $\mathfrak{H}_1$ eine Untergruppe von $\mathfrak{G}$ von der Ordnung $[\mathfrak{G} : \mathfrak{P}_1]$. Dass es wirklich ein solches $\mathfrak{H}_1$ gibt, folgt daraus dass $\mathfrak{G}$ auflösbar ist.[6] Verfeinert man wieder die Kette $1 \subset \mathfrak{H}_1 \subset \mathfrak{Z}_1\mathfrak{H}_1 \subset \mathfrak{G}$ zu einer Hauptkette

$$1 = \mathfrak{B}_0 \subset \mathfrak{B}_1 \subset \cdots \subset \mathfrak{B}_h = \mathfrak{H}_1 \subset \mathfrak{B}_{h+1} \subset \cdots \subset \mathfrak{B}_l = \mathfrak{Z}_1\mathfrak{H}_1 \subset \cdots \subset \mathfrak{B}_r = \mathfrak{G},$$

so ist der Index $[\mathfrak{B}_{h+1} : \mathfrak{H}_1]$ offenbar gleich p. $\mathfrak{N}_1 = \mathfrak{B}_{h+1} \cap \mathfrak{Z}_1$ ist also ein Normalteiler von $\mathfrak{B}_{h+1}$ mit der Ordnung p_1, und da es in $\mathfrak{Z}_1$ enthalten ist, so ist es auch ein Normalteiler von $\mathfrak{P}_1$, folglich von $\mathfrak{G}$. Wendet man nunmehr die Induktionsvoraussetzung auf $\mathfrak{G}/\mathfrak{N}_1$ an, so sieht man leicht ein, dass die Behauptung für $\mathfrak{G}$ richtig ist.

Hilfssatz 2.[7] Dann und nur dann ist der Verband $V(\mathfrak{G})$ einer endlichen Gruppe $\mathfrak{G}$ ausgeglichen, wenn für zwei Untergruppen $\mathfrak{A}$, $\mathfrak{B}$ von $\mathfrak{G}$ der Index $[\mathfrak{A} : \mathfrak{B}]$ stets eine Primzahl ist, sobald $\mathfrak{B}$ als maximale Untergruppe in $\mathfrak{A}$ enthalten ist.

Beweis. Dass die Bedingung notwendig ist, haben wir schon im Beweis von Hilfssatz 1 gesehen. Ist umgekehrt die Bedingung erfüllt, so ergibt sich sofort, dass die Länge einer Hauptkette, die zwei Untergruppen $\mathfrak{A}$, $\mathfrak{B}$ verbindet, immer der Anzahl der Primfaktoren vom Index $[\mathfrak{A} : \mathfrak{B}]$ gleich ist. $V(\mathfrak{G})$ ist also ausgeglichen.

Aus Hilfssätzen 1, 2 folgt nun

Satz 1. Es sei $\mathfrak{G}$ eine endliche Gruppe und $V(\mathfrak{G})$ der zugehörige Verband. $V(\mathfrak{G})$ ist dann und nur dann ausgeglichen, wenn $\mathfrak{G}$ eine primzahlstufige Hauptreihe besitzt, wenn es also eine solche Hauptreihe

$$1 = \mathfrak{N}_0 \subset \mathfrak{N}_1 \subset \cdots \subset \mathfrak{N}_{r-1} \subset \mathfrak{N}_r = \mathfrak{G} \tag{3}$$

gibt, wobei die Indizen $[\mathfrak{N}_i : \mathfrak{N}_{i-1}]$ $(i = 1, 2, \ldots, r)$ sämtlich Primzahlen sind.

Beweis. Dass $\mathfrak{G}$ eine solche Hauptreihe besitzt, falls $V(\mathfrak{G})$ ausgeglichen ist, folgt aus Hilfssatz 1. Nach Hilfssatz 2 haben wir also nur noch zu beweisen, dass für zwei Untergruppen $\mathfrak{A}$, $\mathfrak{B}$ einer Gruppe $\mathfrak{G}$ mit der Hauptreihe (3), der Index $[\mathfrak{A} : \mathfrak{B}]$ stets eine Primzahl ist,

6) P. Hall, J. of London Math. Soc. 12, 198–200, 1937.

7) Vgl. O. Ore, a. a. O. 2).

wenn $\mathfrak{B}$ als maximale Untergruppe in $\mathfrak{A}$ enthalten ist. Dazu wenden wir Induktionsschluss nach der Ordnung von $\mathfrak{G}$ an. Da mit $\mathfrak{G}$ auch jede Untergruppe von $\mathfrak{G}$ eine primzahlstufige Hauptreihe besitzt, so genügt es nach Induktionsvoraussetzung nur zu zeigen, dass eine maximale Untergruppe $\mathfrak{M}$ von $\mathfrak{G}$ einen Primzahlindex hat. Wenn $\mathfrak{M}$ $\mathfrak{N}_1$ enthält, so wenden wir Induktionsvoraussetzung auf $\mathfrak{G}/\mathfrak{N}_1$ an. Wenn aber $\mathfrak{M}$ $\mathfrak{N}_1$ nicht enthält, so ist $\mathfrak{M}\mathfrak{N}_1=\mathfrak{G}$, $\mathfrak{M}\frown\mathfrak{N}_1=1$. Der Index $[\mathfrak{G}:\mathfrak{M}]$ ist dann gleich der Ordnung von $\mathfrak{N}_1$, also eine Primzahl, w. z. b. w..

Aus diesem Satz folgt nun, dass eine Gruppe $\mathfrak{G}$ mit einer primzahlstufigen Hauptreihe sogar eine solche Hauptreihe besitzt, wie sie im Hilfssatz 1 angegeben ist. Das ist aber natürlich auch direkt leicht zu beweisen.

2. Wir untersuchen nun die Struktur der endlichen M-Gruppen und zeigen, dass eine beliebige endliche M-Gruppe ins direkte Produkt ebensolcher Gruppen zerlegbar ist, deren Ordnungen durch höchstens zwei verschiedene Primzahlen teilbar sind.

In dieser Nr. bedeutet $\mathfrak{G}$ immer eine endliche M-Gruppe. Die Primfaktoren der Ordnung von $\mathfrak{G}$ bezeichnen wir mit $p, q, \cdots$

Hilfssatz 3. Eine endliche M-Gruppe ist auflösbar.

Beweis. Folgt sofort aus Satz 1.

Hilfssatz 4. Wenn die Ordnungen, a, b von zwei Untergruppen $\mathfrak{A}, \mathfrak{B}$ von $\mathfrak{G}$ teilerfremd sind, so ist die Ordnung von $\mathfrak{A}\smile\mathfrak{B}$ gleich ab.

Beweis. Die Ordnung von $\mathfrak{A}\smile\mathfrak{B}$ ist offenbar durch ab teilbar. Da für einen modularen Verband $V(\mathfrak{G})$ die Verbandsisomorphie $\mathfrak{A}\smile\mathfrak{B}/\mathfrak{A}\simeq\mathfrak{B}/\mathfrak{A}\frown\mathfrak{B}$ gilt,[8] so ist die Länge der Hauptkette, die $\mathfrak{A}\smile\mathfrak{B}$ mit $\mathfrak{A}$ verbindet, gleich der Länge der Hauptkette, die $\mathfrak{B}$ mit $\mathfrak{A}\frown\mathfrak{B}=1$ verbindet. Nach Hilfssatz 2 ist sodann die Anzahl der Primfaktoren der Ordnung von $\mathfrak{A}\smile\mathfrak{B}$ gleich der Summe von der Anzahl der Primfaktoren der Ordnungen von $\mathfrak{A}$ und der von $\mathfrak{B}$. Die Ordnung von $\mathfrak{A}\smile\mathfrak{B}$ ist also gleich ab.

Hilfssatz 5. Es sei $\mathfrak{P}$ eine p-Untergruppe, und $\mathfrak{Q}$ eine q-Untergruppe von $\mathfrak{G}$. Wenn p grösser als q ist, dann ist $\mathfrak{Q}$ im Normalisator $N(\mathfrak{P})$ von $\mathfrak{P}$ enthalten.

Beweis. Wenn die Ordnung von $\mathfrak{P}$ bzw. $\mathfrak{Q}$ p^α bzw. q^β ist, so ist die Ordnung von $\mathfrak{P}\smile\mathfrak{Q}$ nach Hilfssatz 4 gleich $p^\alpha q^\beta$. Nach Hilfssatz 1 ist dann p-Sylowgruppe $\mathfrak{P}$ ein Normalteiler von $\mathfrak{P}\smile\mathfrak{Q}$, also ist $\mathfrak{Q}$ in $N(\mathfrak{P})$ enthalten.

Für modularen p-Gruppen sollen nun einige einfache Tatsachen

8) Vgl. G. Birkhoff, Lattice theory (1940), S. 37.

hergeleitet werden, die wir bald nötig haben. Eine ausführliche Untersuchung darüber soll in § 4 vorgenommen werden.

Hilfssatz 6. Für zwei beliebige Untergruppen $\mathfrak{A}$, $\mathfrak{B}$ einer modularen p-Gruppe $\mathfrak{G}$ gilt

$$(\mathfrak{A})(\mathfrak{B})=(\mathfrak{A}\cap\mathfrak{B})(\mathfrak{A}\cup\mathfrak{B}).$$

$(\mathfrak{A})$ bedeutet dabei die Ordnung von $\mathfrak{A}$. Gleiches gilt für $\mathfrak{B}$, u. s. w..

Beweis. Nach der Verbandsisomorphie $\mathfrak{A}\cup\mathfrak{B}/\mathfrak{A}\simeq\mathfrak{B}/\mathfrak{A}\cap\mathfrak{B}$ enthalten die Indizen $[\mathfrak{A}\cup\mathfrak{B}:\mathfrak{A}]$ und $[\mathfrak{B}:\mathfrak{A}\cap\mathfrak{B}]$ gleiche Anzahl von Primfaktoren. Da sie aber beide p-Potenzen sind, so sind sie einander gleich.

Hilfssatz 7. Zwei beliebige Elemente A, B von der Ordnung p aus einer modularen p-Gruppe $\mathfrak{G}$ sind stets vertauschbar.

Beweis. Die Gruppe $\{A, B\}$ ist stets abelsch, da ihre Ordnung nach Hilfssatz 6 p oder p^2 ist.

Nun soll die Struktur der nicht nilpotenten M-Gruppe untersucht werden, deren Ordnung durch genau zwei verschiedene Primzahlen teilbar ist. p und q seien vorläufig immer zwei verschiedene Primzahlen.

Hilfssatz 8. $\mathfrak{G}$ sei von der Ordnung $p^a q$. Falls $\mathfrak{G}$ nicht nilpotent ist und $q>p$ gilt, dann ist jede p-Sylowgruppe von $\mathfrak{G}$ zyklisch.

Beweis. Im Fall $a=1$ ist die Behauptung klar. Sei $a=2$. Nach Hilfssatz 1 folgt aus $q>p$, dass eine q-Sylowgruppe $\mathfrak{Q}$ ein Normalteiler von $\mathfrak{G}$ ist. Nach Voraussetzung ist dann eine p-Sylowgruppe kein Normalteiler von $\mathfrak{G}$. $\mathfrak{D}$ sei ein maximaler Durchschnitt zweier verschiedener p-Sylowgruppen: $\mathfrak{D}=\mathfrak{P}_1\cap\mathfrak{P}_2$. Nach der Verbandsisomorphie $\mathfrak{P}_1\cup\mathfrak{P}_2/\mathfrak{P}_1\simeq\mathfrak{P}_2/\mathfrak{D}$ ist $\mathfrak{D}$ eine maximale Untergruppe von $\mathfrak{P}_2$, d. h. $[\mathfrak{P}_2:\mathfrak{D}]=p$, da $\mathfrak{P}_1$ eine maximale Untergruppe von $\mathfrak{G}=\mathfrak{P}_1\cup\mathfrak{P}_2$ ist. $\mathfrak{D}$ ist als Durchschnitt sämtlicher p-Sylowgruppen von $\mathfrak{G}$ ein Normalteiler,[9)] folglich mit $\mathfrak{Q}$ elementweise vertauschbar. Wir nehmen nun an, dass eine p-Sylowgruppe $\mathfrak{P}$ nicht zyklisch ist. Setzt man

$$\mathfrak{Q}=\{Q\}\,,\quad \mathfrak{D}=\{D\}\,,\quad \mathfrak{P}=\{D, P\}\,,\quad \mathfrak{H}=\{P, Q\}\,,$$

so ist $\mathfrak{G}=\mathfrak{H}\times\mathfrak{D}$. Nach Voraussetzung ist $\mathfrak{H}$ nicht abelsch, also ist die Ordnung von PQ gleich p. Wenn $\mathfrak{A}=\{PQD\}$, $\mathfrak{B}=\{P\}$ gesetzt wird, so ergibt sich aus der Verbandsisomorphie $\mathfrak{A}\cup\mathfrak{B}/\mathfrak{A}\simeq\mathfrak{B}$, dass die Ordnung von $\mathfrak{A}\cup\mathfrak{B}$ gleich p^2 oder pq ist, was aber wegen

$$\mathfrak{A}\cup\mathfrak{B}=\{PQD, P\}=\{QD, P\}=\{Q, D, P\}=\mathfrak{G}$$

9) Vgl. z. B. H. Zassenhaus, Lehrbuch der Gruppentheorie (1937), S. 102. (zitiert mit H. Z.).

ein Widerspruch ist. $\mathfrak{P}$ muss daher zyklisch sein. Sei nun $\alpha \geqq 3$. Wie im Fall $\alpha=2$ ist der Durchschnitt $\mathfrak{D}$ aller p-Sylowgruppen ein Normalteiler von $\mathfrak{G}$. Es gilt wie vorher $[\mathfrak{P}:\mathfrak{D}]=p$, und $\mathfrak{D}$ ist mit $\mathfrak{Q}$ elementweise vertauschbar. Wir nehmen wieder an, dass $\mathfrak{P}$ nicht zyklisch ist, und wählen eine von $\mathfrak{D}$ verschiedene Untergruppe $\mathfrak{P}_1$ mit $[\mathfrak{P}:\mathfrak{P}_1]=p$. Setzt man nun $\mathfrak{D}_1=\mathfrak{D}\frown\mathfrak{P}_1$, so ist $\mathfrak{D}_1$ ein Normalteiler von $\mathfrak{G}$, und $\mathfrak{G}/\mathfrak{D}_1$ ist eine M-Gruppe von der Ordnung p^2q. Da eine p-Sylowgruppe $\mathfrak{P}/\mathfrak{D}_1$ von $\mathfrak{G}/\mathfrak{D}_1$ nicht zyklisch ist, so muss $\mathfrak{G}/\mathfrak{D}_1$, abelsch sein. $\mathfrak{P}$ ist folglich ein Normalteiler von $\mathfrak{G}$, was aber unserer Voraussetzung widerspricht. $\mathfrak{P}$ muss also wieder zyklisch sein.

Nach diesem Hilfssatz ergeben sich nun die folgenden wichtigen Hilfssätze 9, 10.

Hilfssatz 9. Wenn eine p-Sylowgruppe $\mathfrak{P}$ kein Normalteiler von $\mathfrak{G}$ ist, so ist $\mathfrak{P}$ zyklisch.

Beweis. $\mathfrak{D}$ sei der Durchschnitt zweier verschiedenen p-Sylowgruppen $\mathfrak{P}_1$, $\mathfrak{P}_2$ von $\mathfrak{G}$, und $\overline{\mathfrak{P}}_2$ sei eine $\mathfrak{D}$ enthaltende Untergruppe von $\mathfrak{P}_2$ mit dem Index $[\overline{\mathfrak{P}}_2:\mathfrak{D}]=p$. Setzt man nun $\mathfrak{H}=\mathfrak{P}_1\smile\overline{\mathfrak{P}}_2$, so folgt aus $\mathfrak{D}=\overline{\mathfrak{P}}_2\frown\mathfrak{P}_1$ und $\mathfrak{H}/\mathfrak{P}_1\simeq\overline{\mathfrak{P}}_2/\mathfrak{D}$, dass der Index $[\mathfrak{H}:\mathfrak{P}_1]=q$ eine Primzahl ist. (Vgl. Hilfssätze 1, 2). $\mathfrak{H}$ ist also eine Untergruppe von der Ordnung $p^\alpha q$, und $\mathfrak{P}_1$ ist kein Normalteiler von $\mathfrak{H}$. Nach Hilfssatz 8 ist $\mathfrak{P}_1$ zyklisch.

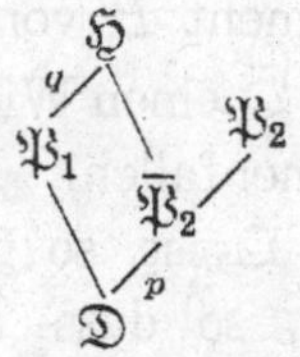

Hilfssatz 10. $\mathfrak{P}_1$ und $\mathfrak{P}_2$ seien zwei beliebige p-Sylowgruppen von $\mathfrak{G}$. Wenn $\mathfrak{P}_1$ nicht mit $\mathfrak{P}_2$ zusammenfällt, so ist der Index von $\mathfrak{D}=\mathfrak{P}_1\frown\mathfrak{P}_2$ in $\mathfrak{P}_1$ bzw. in $\mathfrak{P}_2$ gleich p. Es gilt dann $Q\mathfrak{P}_1Q^{-1}=\mathfrak{P}_2$ mit einem geeigneten Element Q, dessen Ordnung eine Primzahl q $(\neq p)$ ist.

Beweis. $\overline{\mathfrak{P}}_2$ sei eine $\mathfrak{D}$ enthaltende Untergruppe von $\mathfrak{P}_2$ mit dem Index $[\overline{\mathfrak{P}}_2:\mathfrak{D}]=p$. Wenn $\mathfrak{H}=\mathfrak{P}_1\smile\overline{\mathfrak{P}}_2$ gesetzt wird, dann zeigt sich genau wie im Beweis von Hilfssatz 9, dass $[\mathfrak{H}:\mathfrak{P}_1]=q$ eine Primzahl ist. Der Durchschnitt $\overline{\mathfrak{D}}$ aller p-Sylowgruppen von $\mathfrak{H}$ ist, wie beim Beweis von Hilfssatz 8 gezeigt wird, ein Normalteiler mit $[\mathfrak{P}:\overline{\mathfrak{D}}]=p$. Wenn $\overline{\mathfrak{P}}_2$ nicht mit $\mathfrak{P}_2$ zusammenfällt, so ist $\overline{\mathfrak{P}}_2$ in $\overline{\mathfrak{D}}$ enthalten, weil $\mathfrak{P}_2$ nach Hilfssatz 9 zyklisch ist. $\overline{\mathfrak{P}}_2$ muss dann auch in $\mathfrak{P}_1$ enthalten sein, was mit $\bar{\mathfrak{D}}=\bar{\mathfrak{P}}_2\frown\mathfrak{P}_1$ in Widerspruch steht. Es folgt also $\bar{\mathfrak{P}}_2=\mathfrak{P}_2$ und $[\mathfrak{P}_1:\mathfrak{D}]=[\mathfrak{P}_2:\mathfrak{D}]=p$. Da $\mathfrak{P}_1, \mathfrak{P}_2$ zwei p-Sylowgruppen von $\mathfrak{H}$ sind, so gibt es in $\mathfrak{H}$ ein geeignetes Element Q von der Ordnung q, so dass $Q\mathfrak{P}_1Q^{-1}=\mathfrak{P}_2$ ist.

Nun soll die Struktur der nicht nilpotenten M-Gruppen von der Ordnung $p^\alpha q^\beta$ bestimmt werden. Wenn p grösser als q ist, so ist eine p-Sylowgruppe $\mathfrak{P}$ ein Normalteiler von $\mathfrak{G}$. (Vgl. Hilfssatz 1). Nach Voraussetzung ist dann eine q-Sylowgruppe $\mathfrak{Q}$ kein Normalteiler, also nach Hilfssatz 9 zyklisch: $\mathfrak{Q}=\{Q\}$. Der Normalisator von $\mathfrak{Q}$ in $\mathfrak{G}$ sei mit $N(\mathfrak{Q})$ bezeichnet. Wir nehmen zunächst an, dass $N(\mathfrak{Q})\neq\mathfrak{Q}$ ist. Die Untergruppe $\mathfrak{D}=N(\mathfrak{Q})\frown\mathfrak{P}$ ist dann keine Einheitsgruppe und ist offenbar mit $\mathfrak{Q}$ elementweise vertauschbar. A sei nun ein Element aus $\mathfrak{D}$ und B ein beliebiges Element aus $\mathfrak{P}$, das nicht in $\{A\}$ enthalten ist. Wenn die Ordnungen von A und B beide gleich p sind, so ist auch die Ordnung von AB gleich p. (Vgl. Hilfssatz 7). Nach Hilfssatz 5 ist $\mathfrak{Q}$ im Normalisator von $\{B\}$ bzw. $\{AB\}$ enthalten. Ist nun also

$$QBQ^{-1}=B^\gamma, \qquad Q(AB)Q^{-1}=(AB)^\delta=A^\delta B^\delta,$$

so folgt aus $QAQ^{-1}=A$, dass $\gamma\equiv\delta\equiv1$ mod. p ist, d. h. dass B mit Q vertauschbar ist. Sei $\bar{\mathfrak{Q}}$ eine von $\mathfrak{Q}$ verschiedene q-Sylowgruppe von $\mathfrak{G}$, so ist $\bar{\mathfrak{Q}}=P\mathfrak{Q}P^{-1}$ mit einem Element P von der Ordnung p (Hilfssatz 10). Da andererseits jedes Element B von der Ordnung p mit Q vertauschbar ist, so ergibt $P\mathfrak{Q}P^{-1}=\bar{\mathfrak{Q}}$ einen Widerspruch. Der Normalisator $N(\mathfrak{Q})$ muss also mit $\mathfrak{Q}$ zusammenfallen. $P_1\neq1$ sei nun ein beliebiges Element von $\mathfrak{P}$. Da $P_1\mathfrak{Q}P_1^{-1}\neq\mathfrak{Q}$ ist, so gibt es nach Hilfssatz 10 ein Element P von der Ordnung p, so dass $P_1\mathfrak{Q}P_1^{-1}=P\mathfrak{Q}P^{-1}$ gilt. $P_1^{-1}P$ liegt daher in $N(\mathfrak{Q})\frown\mathfrak{P}=1$, also gilt $P_1=P$. Jedes Element von $\mathfrak{P}$ ausser Eins muss mithin die Ordnung p haben. Nach Hilfssatz 7 ist $\mathfrak{P}$ daher eine abelsche Gruppe vom Typus $(p, p, \ldots, p)$. Für zwei beliebige Elemente A_1, B_1 von $\mathfrak{P}$ beweist man genau in derselben Weise wie oben, dass aus $QA_1Q^{-1}=A_1^\mu$, $QB_1Q^{-1}=B_1^\nu$ folgt $\mu\equiv\nu$ mod. p. Es gilt also für ein beliebiges Element P von $\mathfrak{P}$ die Relation

$$QPQ^{-1}=P^r,$$

wobei r eine bestimmte nur von $\mathfrak{G}$ abhängige natürliche Zahl ist. Da $\mathfrak{P}$ nicht mit $\mathfrak{Q}$, wohl aber mit $\{Q^q\}$ elementweise vertauschbar ist, so muss dabei $r\not\equiv1$ mod. p, $r^q\equiv1$ mod. p sein.

Dass umgekehrt eine Gruppe $\mathfrak{G}$ von der Ordnung $p^\alpha q^\beta$ eine M-Gruppe ist, falls sie die oben genannte Struktur besitzt, kann man folgendermassen beweisen. $\mathfrak{A}, \mathfrak{B}, \mathfrak{C}$ seien beliebige Untergruppen von $\mathfrak{G}$ und $\mathfrak{A}$ sei in $\mathfrak{C}$ enthalten. Es soll dann gezeigt werden, dass $(\mathfrak{A}\cup\mathfrak{B})\cap\mathfrak{C}=\mathfrak{A}\cup(\mathfrak{B}\cap\mathfrak{C})$ gilt. Dazu genügt es offenbar nur zu beweisen, dass $(\mathfrak{A}\cup\mathfrak{B})\cap\mathfrak{C}$ in $\mathfrak{A}\cup(\mathfrak{B}\cap\mathfrak{C})$ enthalten ist, und kann man dabei noch annehmen, dass $\mathfrak{A}$ bzw. $\mathfrak{B}$ kein Normalteiler von $\mathfrak{G}$ ist, in unserem Fall also eine q-Sylowgruppe von $\mathfrak{G}$ enthält. Man setze

$$\mathfrak{A}=\mathfrak{A}_p\mathfrak{Q}\,, \quad \mathfrak{B}=\mathfrak{B}_p\mathfrak{Q}_1\,, \quad \mathfrak{C}=\mathfrak{C}_p\mathfrak{Q}\,,$$

wobei $\mathfrak{A}_p$, $\mathfrak{B}_p$ bzw. $\mathfrak{C}_p$ eine p-Sylowgruppe von $\mathfrak{A}, \mathfrak{B}$ bzw. $\mathfrak{C}$ mit $\mathfrak{A}_p\subseteqq\mathfrak{C}_p$ ist und $\mathfrak{Q}$ und $\mathfrak{Q}_1$ q-Sylowgruppen von $\mathfrak{G}$ sind. Ferner sei $\mathfrak{Q}_1=P\mathfrak{Q}P^{-1}$ mit einem Element P aus $\mathfrak{P}$. Jedes Element T aus $(\mathfrak{A}\cup\mathfrak{B})\cap\mathfrak{C}$ lässt sich dann in der Form

$$T=CQ^\alpha=ABP^\beta Q^\alpha$$

schrieben, wobei A, B bzw. C geeignetes Element von $\mathfrak{A}_p$, $\mathfrak{B}_p$ bzw. $\mathfrak{C}_p$ ist. Sei zunächst $\mathfrak{Q}_1$ in $\mathfrak{C}$ enthalten und P ein Element aus $\mathfrak{C}_p$. Dann ist $\mathfrak{B}\cap\mathfrak{C}=(\mathfrak{B}_p\cap\mathfrak{C}_p)\mathfrak{Q}_1$, also $\mathfrak{A}\cup(\mathfrak{B}\cap\mathfrak{C})=\{\mathfrak{A}_p, (\mathfrak{B}_p\cap\mathfrak{C}_p), P\}\cdot\mathfrak{Q}$, und T ist offenbar in $\mathfrak{A}\cup(\mathfrak{B}\cap\mathfrak{C})$ enthalten. Wenn aber $\mathfrak{Q}_1$ nicht in $\mathfrak{C}$ enthalten ist, so ist $\mathfrak{B}\cap\mathfrak{C}=(\mathfrak{B}_p\cap\mathfrak{C}_p)\cdot\{Q^q\}$, folglich $\mathfrak{A}\cup(\mathfrak{B}\cap\mathfrak{C})=\{\mathfrak{A}_p\cup(\mathfrak{B}_p\cap\mathfrak{E}_p)\}\mathfrak{Q}$. In diesem Fall liegt P nicht in $\mathfrak{C}_p\cup\mathfrak{B}_p$ (sonst kann man $\mathfrak{Q}_1$ in $\mathfrak{C}$ wählen), also $B^{-1}A^{-1}C=P^\beta=1$, und $B=A^{-1}C\in\mathfrak{C}_p\cap\mathfrak{B}_p$. $T=ABQ^\alpha$ ist wieder in $\mathfrak{A}\cup(\mathfrak{B}\cap\mathfrak{C})$ enthalten. Somit gilt der

Satz 2. Eine nicht nilpotente Gruppe $\mathfrak{G}$ von der Ordnung $p^\alpha q^\beta$ $(p>q)$ ist dann und nur dann modular, wenn sie folgende Struktur besitzt:

(1) Eine p-Sylowgruppe $\mathfrak{P}$ ist ein Normalteiler von $\mathfrak{G}$. Sie ist eine abelsche Gruppe vom Typus $(p, p, \ldots, p)$.

(2) Eine q-Sylowgruppe $\mathfrak{Q}$ ist zyklisch; $\mathfrak{Q}=\{Q\}$.

(3) Für ein beliebiges Element P von $\mathfrak{P}$ gilt

$$QPQ^{-1}=P^r\,,$$

wobei r eine bestimmte nur von $\mathfrak{G}$ abhängige natürliche Zahl ist, die den folgenden Bedingungen genügen:

$$r\not\equiv 1 \text{ mod. } p\,, \qquad r^q\equiv 1 \text{ mod. } p\,.$$

3. In dieser Nr. soll die Struktur derjenigen M-Gruppe untersucht werden, deren Ordnung durch genau drei verschiedene Primzahlen teilbar ist. Im folgenden seien p_1, p_2, p_3 immer verschiedene Primzahlen mit $p_1>p_2>p_3$.

Hilfssatz 11. $\mathfrak{G}$ sei eine M-Gruppe von der Ordnung $p_1^{a_1}p_2^{a_2}p_3^{a_3}$, und $\mathfrak{P}_i$ sei eine p_i-Sylowgruppe von $\mathfrak{G}$ $(i=1, 2, 3)$. Wenn $\mathfrak{P}_1 \cup \mathfrak{P}_2$ nilpotent ist, dann ist es auch entweder $\mathfrak{P}_1 \cup \mathfrak{P}_3$ oder $\mathfrak{P}_2 \cup \mathfrak{P}_3$.

Beweis. Zunächst bemerken wir, dass $\mathfrak{P}_i \cup \mathfrak{P}_j$ $(i \neq j)$ eine Untergruppe von der Ordnung $p_i^{a_i}p_j^{a_j}$ ist. (Vgl. Hilfssatz 4). Wenn $\mathfrak{P}_1 \cup \mathfrak{P}_3$ und $\mathfrak{P}_2 \cup \mathfrak{P}_3$ beide nicht nilpotent sind, so haben sie die im Satz 2 angegebenen Struktur. Daher fällt der Normalisator $N(\mathfrak{P}_3)$ von $\mathfrak{P}_3$ in $\mathfrak{G}$ mit $\mathfrak{P}_3$ zusammen: $N(\mathfrak{P}_3)=\mathfrak{P}_3$. P_1 bzw. P_2 sei von 1 verschiedenes Element von $\mathfrak{P}_1$ bzw. $\mathfrak{P}_2$. Da $P_1\mathfrak{P}_3P^{-1} \neq P_2\mathfrak{P}_3P_2^{-1}$ ist, so gibt es nach Hilfssatz 10, ein in $\mathfrak{P}_1 \cup \mathfrak{P}_2$ enthaltenes Element Q von einer Primzahlordnung, $(\neq p_3)$, so dass $QP_1\mathfrak{P}_3P_1^{-1}Q^{-1}=P_2\mathfrak{P}_3P_2^{-1}$ ist. Aus $N(\mathfrak{P}_3)=\mathfrak{P}_3$ folgt dann $QP_1=P_2$, was aber wegen $\mathfrak{P}_1 \cup \mathfrak{P}_2=\mathfrak{P}_1 \times \mathfrak{P}_2$ ein Widerspruch ist. Es muss also entweder $\mathfrak{P}_1 \cup \mathfrak{P}_3$ oder $\mathfrak{P}_2 \cup \mathfrak{P}_3$ nilpotent sein.

Hilfssatz 12. $\mathfrak{G}$ sei eine M-Gruppe von der Ordnung $p_1^{a_1}p_2^{a_2}p_3^{a_3}$ und $\mathfrak{P}_i$ sei eine p_i-Sylowgruppe von $\mathfrak{G}$ $(i=1, 2, 3)$. Wenn $\mathfrak{P}_1 \cup \mathfrak{P}_2$ nicht nilpotent ist, dann sind $\mathfrak{P}_1 \cup \mathfrak{P}_3$ und $\mathfrak{P}_2 \cup \mathfrak{P}_3$ beide nilpotent, d. h. es gilt $\mathfrak{G}=(\mathfrak{P}_1 \cup \mathfrak{P}_2) \times \mathfrak{P}_3$.

Beweis. Sei zunächst $\mathfrak{P}_1 \cup \mathfrak{P}_3$ nicht nilpotent. Nach Satz 2 ist dann $\mathfrak{P}_3$ zyklisch: $\mathfrak{P}_3=\{P_3\}$, und für jedes Element P_1 von $\mathfrak{P}_1$ gilt $P_3P_1P_3^{-1}=P_1^{\alpha}$, $\alpha \not\equiv 1$ mod. p_1, $\alpha^{p_3}\equiv 1$ mod. p_1.

Da nach Hilfssatz 5 der Normalisator $N(\mathfrak{P}_2)$ von $\mathfrak{P}_2$ jede p_3-Sylowgruppe von $\mathfrak{G}$ enthält, so liegt $(P_1^{-1}P_3P_1)P_3^{-1}=P_1^{\alpha-1}$ in $N(\mathfrak{P}_2)$. $\mathfrak{P}_1$ ist also in $N(\mathfrak{P}_2)$ enthalten. Da $\mathfrak{P}_1$ ein Normalteiler von $\mathfrak{P}_1 \cup \mathfrak{P}_2$ ist (vgl. Hilfssatz 5), so folgt hieraus $\mathfrak{P}_1 \cup \mathfrak{P}_2=\mathfrak{P}_1 \times \mathfrak{P}_2$, was unserer Voraussetzung widerspricht. Daher muss $\mathfrak{P}_1 \cup \mathfrak{P}_3$ nilpotent sein. Nun sei $\mathfrak{P}_2 \cup \mathfrak{P}_3$ wieder nicht nilpotent. Nach Satz 2 ist dann $\mathfrak{P}_2$ eine zyklische Gruppe von der Ordnung p_2: $\mathfrak{P}_2=\{P_2\}$. Wenn $\mathfrak{P}_3=\{P_3\}$ gesetzt wird, so gilt für jedes Element P_1 von $\mathfrak{P}_1$

(4) $\quad P_2P_1P_2^{-1}=P_1^{\alpha}, \quad \alpha \not\equiv 1 \text{ mod. } p_1, \quad \alpha^{p_2}\equiv 1 \text{ mod. } p_1,$

(5) $\quad P_3P_2P_3^{-1}=P_2^{\beta}, \quad \beta \not\equiv 1 \text{ mod. } p_2, \quad \beta^{p_3}\equiv 1 \text{ mod. } p_2,$

(6) $\quad P_1P_3P_1^{-1}=P_3.$

Daraus folgt

$$P_1^{\alpha^{\beta}}=P_1^{\alpha}, \quad \text{also} \quad \alpha^{\beta-1}\equiv 1 \text{ mod. } p_1, \quad \beta\equiv 1 \text{ mod. } p_2,$$

was aber mit zweiter Bedingung von (5) in Widerspruch steht. Daher muss auch $\mathfrak{P}_2 \cup \mathfrak{P}_3$ nilpotent sein, w. z. b. w..

Hilfssatz 13. $\mathfrak{G}$ sei eine M-Gruppe von der Ordnung $p_1^{a_1}p_2^{a_2}p_3^{a_3}$, und

$\mathfrak{P}_i$ sei eine p_i-Sylowgruppe von $\mathfrak{G}$ $(i=1,2,3)$. Dann besteht immer eine direkte Zerlegung

$$\mathfrak{G}=(\mathfrak{P}_{i_1}\cup\mathfrak{P}_{i_2})\times\mathfrak{P}_{i_3},$$

wobei (i_1, i_2, i_3) eine geeignete Permutation von $(1,2,3)$ ist.

Beweis. Wenn $\mathfrak{P}_1\cup\mathfrak{P}_2$ nilpotent ist, so gilt nach Hilfssatz 12 entweder $\mathfrak{G}=(\mathfrak{P}_1\cup\mathfrak{P}_3)\times\mathfrak{P}_2$, oder $\mathfrak{G}=(\mathfrak{P}_2\cup\mathfrak{P}_3)\times\mathfrak{P}_1$. Wenn aber $\mathfrak{P}_1\cup\mathfrak{P}_2$ nicht nilpotent ist, dann besteht nach Hilfssatz 12 auch eine direkte Zeregung $\mathfrak{G}=(\mathfrak{P}_1\cup\mathfrak{P}_2)\times\mathfrak{P}_3$.

Um die Struktur der allgemeinen endlichen M-Gruppe zu bestimmen, benutzen wir noch einen Hilfssatz:

Hilfssatz 14. $\mathfrak{G}$ sei das direkte Produkt von endlichen Gruppen $\mathfrak{G}_1, \mathfrak{G}_2, \ldots, \mathfrak{G}_r$, deren Ordnungen zueinander teilerfremd sind. Der zu $\mathfrak{G}$ gehörige Verband $V(\mathfrak{G})$ ist dann die direkte Summe der zu $\mathfrak{G}$ gehörigen Verbande $V(\mathfrak{G}_i)$ $(i=1,2,\ldots,r)$. Wenn also $\mathfrak{G}_i$ sämtlich M-Gruppen sind, dann ist auch $\mathfrak{G}$ eine M-Gruppe.

Beweis. Es ist leicht einzusehen, dass eine beliebige Untergruppe $\mathfrak{U}$ von $\mathfrak{G}$ das direkte Produkt von $\mathfrak{U}_i=\mathfrak{U}\frown\mathfrak{G}_i$ $(i=1,2,\ldots,r)$ ist. Nun seinen $\mathfrak{A}$ und $\mathfrak{B}$ zwei beliebige Untergruppen von $\mathfrak{G}$. Aus

$$\mathfrak{A}=\mathfrak{A}_1\times\mathfrak{A}_2\times\cdots\times\mathfrak{A}_r, \qquad \mathfrak{B}=\mathfrak{B}_1\times\mathfrak{B}_2\times\cdots\times\mathfrak{B}_r$$

folgt dann

$$\mathfrak{A}\cup\mathfrak{B}=(\mathfrak{A}_1\cup\mathfrak{B}_1)\times(\mathfrak{A}_2\cup\mathfrak{B}_2)\times\cdots\times(\mathfrak{A}_r\cup\mathfrak{B}_r) \quad \text{und}$$

$$\mathfrak{A}\frown\mathfrak{B}=(\mathfrak{A}_1\frown\mathfrak{B}_1)\times(\mathfrak{A}_2\frown\mathfrak{B}_2)\times\cdots\times(\mathfrak{A}_r\frown\mathfrak{B}_r).$$

Daraus folgt unmittelbar die Behauptung.

Das direkte Produkt zweier M-Gruppen ist aber nicht immer eine M-Gruppe, wenn ihre Ordnungen nicht teilerfremd sind. Z. B. ist das folgende $\mathfrak{G}$ keine M-Gruppe, obwohl $\mathfrak{A}$ und $\mathfrak{B}$ beide modular sind:

$$\mathfrak{G}=\mathfrak{A}\times\mathfrak{B}, \quad \mathfrak{A}=\{A, B\}, \quad A^{p^2}=B^p=1, \quad BAB^{-1}=A^{1+p},$$

$$\mathfrak{B}=\{C\}, \quad C^{p^2}=1, \quad (p>2).$$

Mit Hilfe der Hilfssatz 13, 14 kann man nun leicht die Struktur von endlichen M-Gruppen bestimmen.

$\mathfrak{G}$ sei eine M-Gruppe von der Ordnung $g=p_1^{a_1}p_2^{a_2}\ldots p_r^{a_r}$. Eine p_i-Sylowgruppe von $\mathfrak{G}$ sei mit $\mathfrak{P}_i$ bezeichnet $(i=1,2,\ldots,r)$. Es ist dann offenbar $\mathfrak{G}=\mathfrak{P}_1\cup\mathfrak{P}_2\cup\cdots\cup\mathfrak{P}_r$. Wenn $\mathfrak{P}_1\cup\mathfrak{P}_i$ $(i=2,\ldots,r)$ sämtlich nilpotent sind, so gilt $\mathfrak{G}=\mathfrak{P}_1\times(\mathfrak{P}_2\cup\mathfrak{P}_3\cup\cdots\cup\mathfrak{P}_r)$. Wenn aber z. B. $\mathfrak{P}_1\cup\mathfrak{P}_2$ nicht nilpotent ist, dann ist nach Hilfssatz 13 $\mathfrak{P}_1\cup\mathfrak{P}_2$ mit jedem

$\mathfrak{P}_j$ $(j=3, 4, \ldots, r)$ elementweise vertauschbar, also besteht eine direkte Zerlegung $\mathfrak{G}=(\mathfrak{P}_1 \cup \mathfrak{P}_2) \times (\mathfrak{P}_3 \cup \mathfrak{P}_4 \cup \cdots \cup \mathfrak{P}_r)$. Da $(\mathfrak{P}_2 \cup \mathfrak{P}_3 \cup \cdots \cup \mathfrak{P}_r)$ bzw. $(\mathfrak{P}_3 \cup \mathfrak{P}_4 \cup \cdots \cup \mathfrak{P}_r)$ wieder eine M-Gruppe ist, so kann man dieses Verfahren fortsetzen, bis endlich $\mathfrak{G}$ ins direkte Produkt einiger $\mathfrak{P}_i$ und einiger $\mathfrak{P}_j \cup \mathfrak{P}_k$ zerlegt wird. Dieses $\mathfrak{P}_i$ bzw. $\mathfrak{P}_j \cup \mathfrak{P}_k$ ist offenbar modular, und da ausserdem $\mathfrak{P}_j \cup \mathfrak{P}_k$ nicht nilpotent ist, so besitzt dieses die im Satz 2 angegebene Struktur. Wenn umgekehrt eine Gruppe $\mathfrak{G}$ das direkte Produkt von denjenigen M-Gruppen ist, deren Ordnungen durch höchstens zwei verschiedene Primzahlen teilbar, und zueinander teilerfremd sind, dann ist nach Hilfssatz 14 $\mathfrak{G}$ selbst modular.

Also haben wir

Satz 3. Eine endliche Gruppe $\mathfrak{G}$ ist dann und nur dann modular, wenn sie ins direkte Produkt von einigen modularen Sylowgruppen und einigen Gruppen mit der im Satz 2 angegebenen Struktur zerlegbar ist. Die direkten Faktoren sollen dabei zueinander teilerfremden Ordnungen besitzen.

§ 2.

Wir untersuchen nun die Struktur von endlichen D-Gruppen.[10] Da eine D-Gruppe offenbar zugleich eine M-Gruppe ist, so können wir alle Ergebnisse von § 1 brauchen.

Sei $\mathfrak{G}$ zunächst eine D-Gruppe von der Ordnung p^2, wobei p wie immer eine Primzahl bedeutet. Wenn $\mathfrak{G}$ nicht zyklisch wäre, und $\mathfrak{G}=\{A, B\}$, $\mathfrak{A}=\{A\}$, $\mathfrak{B}=\{B\}$, $\mathfrak{C}=\{AB\}$ gesetzt wird, so wäre

$$\mathfrak{A}=\mathfrak{A} \cup (\mathfrak{B} \cap \mathfrak{C}) \neq (\mathfrak{A} \cup \mathfrak{B}) \cap (\mathfrak{A} \cup \mathfrak{C}) = \mathfrak{G},$$

was unsrer Annahme widerspricht. $\mathfrak{G}$ ist also zyklisch. Hieraus folgt nun ohne Weiteres, dass eine distributive p-Gruppe entweder zyklisch oder eine verallgemeinerte Quaternionengruppe ist, da ihre Untergruppen von den Ordnungen p^2 sämtlich zyklisch sind.[11] Eine verallgemeinerte Quaternionengruppe ist aber wie leicht ersichtlich nicht distributiv. Jede distributive p-Gruppe muss daher zyklisch sein. $\mathfrak{G}$ sei nun eine D-Gruppe von der Ordnung $p^\alpha q^\beta$. Wenn $\mathfrak{G}$ nicht nilpotent wäre, so müsste es die im Satz 2 angegebene Struktur besitzen;

$$\mathfrak{G}=\{P, Q\}, \quad P^p=1, \quad Q^{q^n}=1 \quad QPQ^{-1}=P^r,$$
$$r \not\equiv 1 \text{ mod. } p, \quad r^q \equiv 1 \text{ mod. } p.$$

Setzt man $\mathfrak{A}=\{P\}$, $\mathfrak{B}=\{Q\}$, $\mathfrak{C}=\{PQP^{-1}\}$, dann gilt

---

10) Vgl. O. Ore, Duke Math. Jour., 4, 247–269, 1938.

11) Vgl. H. Z., a. a. O. S. 113.

$$\{P, Q^q\}=\mathfrak{A}\cup(\mathfrak{B}\cap\mathfrak{C})\neq(\mathfrak{A}\cup\mathfrak{B})\cap(\mathfrak{A}\cup\mathfrak{C})=\mathfrak{G},$$

was wieder nicht geht. $\mathfrak{G}$ muss also nilpotent sein. Hieraus folgt nach Satz 3, dass eine endliche D-Gruppe das direktes Produkt ihrer zyklischen Sylowgruppen, also selbst zyklisch ist.

Sei nun $\mathfrak{G}$ umgekehrt eine endliche zyklische Gruppe, und $\mathfrak{A}, \mathfrak{B}, \mathfrak{C}$ ihre beliebige Untergruppen. Wenn die Ordnung von $\mathfrak{A}, \mathfrak{B}, \mathfrak{C}$ bzw. mit a, b, c bezeichnet wird, so bedeutet die Bedingung

$$\mathfrak{A}\cup(\mathfrak{B}\cap\mathfrak{C})=(\mathfrak{A}\cup\mathfrak{B})\cap(\mathfrak{A}\cup\mathfrak{C})$$

nicht anders als eine elementar-zahlentheoretische Relation

$$\{a, (b, c)\}=(\{a, b\}, \{a, c\}).$$

$\mathfrak{G}$ ist also éine D-Gruppe. Damit ist der folgende Satz bewiesen.

Satz 4. Eine endliche Gruppe $\mathfrak{G}$ ist dann und nur dann distributiv, wenn es eine zyklische Gruppe ist.

Bevor wir die Struktur der endlichen B-Gruppen untersuchen, seien einige Bemerkungen über K-Gruppen vorausgeschickt. Die Struktur der allgemeinen endlichen K-Gruppen scheint sehr kompliziert zu sein. Z. B. ist eine K-Gruppe nicht immer auflösbar, und die Untergruppe einer endlichen K-Gruppe ist nicht immer eine K-Gruppe. (Z. B. $\mathfrak{A}_5$ bzw. $\mathfrak{S}_4$). Eine K-Gruppe von einer Primzahlpotenzordnung ist aber leicht zu bestimmen. $\mathfrak{G}$ sei also eine komplementäre p-Gruppe und $\Phi(\mathfrak{G})$ sei die Φ-Untergruppe von $\mathfrak{G}$[12]. Das Komplement $\mathfrak{G}_1$ von $\Phi(\mathfrak{G})$ erfüllt die Relationen

$$\Phi(\mathfrak{G})\cup\mathfrak{G}_1=\Phi(\mathfrak{G})\mathfrak{G}_1=\mathfrak{G}, \qquad \Phi(\mathfrak{G})\cap\mathfrak{G}_1=1.$$

Nach dem Basissatz von Burnside fällt dann $\mathfrak{G}_1$ mit $\mathfrak{G}$ zusammen, und aus $\mathfrak{G}_1\simeq\mathfrak{G}/\Phi(\mathfrak{G})$ folgt weiter, dass $\mathfrak{G}$ eine abelsche Gruppe vom Typus $(p, p, \dots, p)$ sein muss. Die Umkehrung ist klar.

P. Hall hat die Struktur einer endlichen K-Gruppe $\mathfrak{G}$, deren Verband $V(\mathfrak{G})$ zugleich ausgeglichen ist, vollständig bestimmt.[13] Dann ist nämlich $\mathfrak{G}$ eine Untergruppe eines direkten Produktes der Gruppen von quadratfreien Ordnungen, und umgekehrt.

Ferner sieht man leicht, dass die Untergruppen einer K-Gruppe immer dann komplementär sind, wenn $\mathfrak{G}$ zugleich eine M-Gruppe ist. Die Sylowgruppen einer B-Gruppe müssen daher elementare abelsche Gruppen sein. Nach Satz 4 folgt also

12) H. Z., a. a. O. S. 44.

13) P. Hall, J. of London Math. Soc. 12, 201–204, 1937.

Satz 5. Eine endliche Gruppe $\mathfrak{G}$ ist dann und nur dann Boolsch, wenn es eine zyklische Gruppe von quadratfreier Ordnung ist.

Sei nun $\mathfrak{G}$ eine endliche Gruppe, deren zugehöriger Verband $V(\mathfrak{G})$ modular komplementär und irreduzibel ist. Nach Satz 3 ist dann $\mathfrak{G}$ das direkte Produkt einiger einfächeren M-Gruppen. Nun muss aber $\mathfrak{G}$ nur einen einzigen direkten Faktor besitzen, da wir $V(\mathfrak{G})$ als irreduzibel vorausgesetzt haben. $\mathfrak{G}$ ist also entweder eine p-Gruppe, mithin nach dem soeben Gesagten eine abelsche Gruppe vom Typus $(p, p, \ldots, p)$, oder aber nach Satz 2 durch folgende Relationen gegeben:

(7) (i) $\mathfrak{G}=\mathfrak{P}\cdot\mathfrak{Q}$.

(ii) $\mathfrak{P}$ ist eine abelsche Gruppe vom Typus $(p, p, \ldots, p)$, $\mathfrak{Q}=\{Q\}$, $Q^q=1$.

(iii) Für ein beliebiges Element P aus $\mathfrak{P}$ gilt $QPQ^{-1}=P^r$, $r\not\equiv 1$ mod. p, $r^q\equiv 1$ mod. p.

Dass umgekehrt der zugehörige Verband $V(\mathfrak{G})$ solcher Gruppe $\mathfrak{G}$ modular, komplementär und irreduzibel ist, d. h. dem Verband der Unterräumen eines projektiven Raumes isomorph ist, lässt sich leicht beweisen. Für eine abelsche Gruppe vom Typus $(p, p, \ldots, p)$, ist das klar. Einer durch oben genannte Relationen (7) gegebenen Gruppe $\mathfrak{G}$ von der Ordnung $p^n q$ kann man z. B. folgendermassen einen projektiven Raum S_n entsprechen lassen. S_n sei ein n-dimensionaler projektiver Raum über dem Körper $GF(p)$. Jede Untergruppe von der Ordnung p von $\mathfrak{P}$ sei einer unendlichen Punkt von S_n, und jede Untergruppe $P\mathfrak{Q}P^{-1}$ $(P\in\mathfrak{P})$ sei einer endlichen Punkt von S_n zugeordnet. Durch diese Zuordnung wird dann, wie leicht ersichtlich, $V(\mathfrak{G})$ isomorph auf dem Verband der Unterräumen von S_n abgebildet. Also gilt

Satz 6. Eine endliche Gruppe $\mathfrak{G}$ ist dann und nur dann eine P-Gruppe, wenn sie entweder eine abelsche Gruppe vom Typus $(p, p, \ldots, p)$ ist, oder die in (7) angegebene Struktur besitzt.

Bemerkung[14]: Um die Struktur der endlichen D-Gruppen zu bestimmen, haben wir Satze 2, 3 in § 1 gebraucht. Das gleiche Ziel lässt sich aber leicht ohne diese Satze erreicht werden. Man kann zwar sämtliche D-Gruppen bestimmen: sie sind nämlich;

(i) die additive Gruppe aller rationalen Zahlen: $\mathfrak{R}$,

(ii) die Faktorgruppe von $\mathfrak{R}$ nach der Gruppe aller ganzen rationalen Zahlen: $\mathfrak{R}_1$,

(iii) die Untergruppen von $\mathfrak{R}$ oder $\mathfrak{R}_1$.

§ 3.

$\mathfrak{A}$ und $\mathfrak{B}$ seien nun zwei Untergruppen einer endlichen Gruppe $\mathfrak{G}$.

---

14) Vgl. O. Ore, a. a. O. 10).

Wenn der Komplex $\mathfrak{A}\mathfrak{B}$ mit $\mathfrak{B}\mathfrak{A}$ zusammenfällt, so sagt man, dass $\mathfrak{A}$ mit $\mathfrak{B}$ vertauschbar ist. Der Begriff der vertauschbaren Untergruppen hängt eng mit dem der M-Gruppen zusammen. In diesem Paragraphen soll dieser Zusammenhang untersucht werden.

Wenn die Ordnung von $\mathfrak{A}, \mathfrak{B}, \mathfrak{A} \frown \mathfrak{B}$ bzw. a, b, d ist, so ist die Anzahl der Elemente von $\mathfrak{A}\mathfrak{B}$ gleich ab/d. Da $\mathfrak{A}$ mit $\mathfrak{B}$ dann und nur dann vertauschbar ist, wenn $\mathfrak{A}\mathfrak{B} = \mathfrak{A} \smile \mathfrak{B}$ gilt, so folgt hieraus

Hilfssatz 15. Die Untergruppen $\mathfrak{A}, \mathfrak{B}$ einer endlichen Gruppe ist dann und nur dann vertauschbar, wenn $ab = dm$ gilt, wobei a, b, d, m bzw. die Ordnung von $\mathfrak{A}, \mathfrak{B}, \mathfrak{A} \frown \mathfrak{B}, \mathfrak{A} \smile \mathfrak{B}$ bedeutet.

Nach O. Ore[15)] nennen wir eine Untergruppe $\mathfrak{A}$ von $\mathfrak{G}$ „quasi-normal", wenn $\mathfrak{A}$ mit jeder Untergruppe $\mathfrak{B}$ von $\mathfrak{G}$ vertauschbar ist. Wir nennen dann eine Gruppe $\mathfrak{G}$ „quasi-Hamiltonsch", wenn jede Untergruppe von $\mathfrak{G}$ quasi-normal ist. Es ist klar, dass ein Normalteiler quasi-normal ist, und folglich eine abelsche Gruppe bzw. eine Hamiltonsche Gruppe immer quasi-Hamiltonsch ist.

Wenn nun $\mathfrak{A}$ mit $\mathfrak{B}$ vertauschbar ist, so gilt, wie bekannt ist und übrigens leicht zu sehen ist, für jedes $\mathfrak{C} \geqq \mathfrak{A}$ die Relation

$$\mathfrak{A} \smile (\mathfrak{B} \frown \mathfrak{C}) = (\mathfrak{A} \smile \mathfrak{B}) \frown \mathfrak{C}.$$

Eine quasi-Hamiltonsche Gruppe ist daher stets modular. Nach Hilfssatz 6 und Hilfssatz 15 ist umgekehrt eine modulare p-Gruppe auch quasi-Hamiltonsch. Also gilt

Hilfssatz 16. Eine p-Gruppe $\mathfrak{G}$ ist dann und nur dann quasi-Hamiltonsch, wenn es modular ist.

Für allgemeine endliche Gruppe $\mathfrak{G}$ gilt nun

Satz 7. Eine endliche Gruppe $\mathfrak{G}$ ist dann und nur dann quasi-Hamiltonsch, wenn sie das direkte Produkt ihrer modularen Sylowgruppen ist.

Beweis. Es sei $\mathfrak{G}$ eine quasi-Hamiltonsche Gruppe und $\mathfrak{P}_1, \mathfrak{P}_2$ seien ihre p-Sylowgruppen. Die Ordnung von $\mathfrak{P}_1 \smile \mathfrak{P}_2$ ist dann nach Hilfssatz 15 wieder eine p-Potenz, also $\mathfrak{P}_1 = \mathfrak{P}_2 = \mathfrak{P}_1 \smile \mathfrak{P}_2$. $\mathfrak{G}$ ist also nilpotent und aus Hilfssatz 16 folgt die Behauptung. Die Umkehrung ist auch nach Hilfssatz 14, 16 klar.

Nach diesem Satz ist auch die Untersuchung von endlichen quasi-Hamiltonschen Gruppen auf die von modularen p-Gruppen zurückgeführt werden.

Nach Satz 3, 7 folgt nun ohne Weiteres

---

15) O. Ore, Duke Math. J. 3, 149–174, 1936.

Satz 8. Eine endliche M-Gruppe $\mathfrak{G}$ ist dann und nur dann quasi-Hamiltonsch, wenn sie nilpotent ist.

Es gibt aber zufolge Satz 2 nicht nilpotente M-Gruppen; eine M-Gruppe ist also nicht immer quasi-Hamiltonsch. Zwei Untergruppen einer endlichen M-Gruppe sind aber nach Hilfssatz 4 stets miteinander vertauschbar, wenn ihre Ordnungen zueinander teilerfremd sind.

Allgemeiner gilt nun

Satz 9. $\mathfrak{G}$ sei eine endliche M-Gruppe und $\mathfrak{A}, \mathfrak{B}$ seien beliebige Untergruppen von $\mathfrak{G}$. Es gilt dann stets eine konjugierte Untergruppe von $\mathfrak{A}$, die mit $\mathfrak{B}$ vertauschbar ist.

Beweis. Nach Satz 3 und Hilfssatz 16 genügt es den Satz nur für die Gruppe mit der im Satz 2 angegebenen Struktur zu beweisen. Man wähle alsdann eine konjugierte Untergruppe $X\mathfrak{A}X^{-1}$ von $\mathfrak{A}$, so dass (in derselben Bezeichnungsweise wie im Satz 2) eine q-Sylowgruppe von $X\mathfrak{A}X^{-1}$ und eine q-Sylowgruppe von $\mathfrak{B}$ in einer und derselben q-Sylowgruppe von $\mathfrak{G}$ enthalten werden. Dann ist offenbar $X\mathfrak{A}X^{-1}$ mit $\mathfrak{B}$ vertauschbar.

Aus diesem Beweis ist leicht einzusehen, dass die Konjugierte von $\mathfrak{A}$ sogar in $\mathfrak{A} \cup \mathfrak{B}$ gewählt werden kann. Diese Bedingung ist aber nicht hinreichend dafür, dass eine endliche Gruppe modular ist. Z. B. ist die folgende Gruppe $\mathfrak{G}$ keine M-Gruppe, obgleich es die genannte Eigenschaft besitzt, (vgl. den Beweis von Hilfssatz 8):

$$\mathfrak{G} = \{A, B\} \times \{C\}, \quad A^p = 1, \quad B^q = C^q = 1, \quad BAB^{-1} = A^r,$$
$$r \not\equiv 1 \text{ mod. p}, \qquad r^q \equiv 1 \text{ mod. } p.$$

Für eine p-Gruppe ist jedoch auch diese Eigenschaft hinreichend. Vgl. Satz 12 in § 4.

Im Anschluss an Obigem soll ein Problem von O. Ore[16] betrachtet werden. Wenn zwei Untergruppen $\mathfrak{A}, \mathfrak{B}$ einer Gruppe G mit einander vertauschbar sind, so gilt für beliebige Untergruppen mit $\mathfrak{C} \geqq \mathfrak{A}$ bzw. $\mathfrak{D} \geqq \mathfrak{B}$ von $\mathfrak{G}$ die Relationen

(8) $(\mathfrak{A} \cup \mathfrak{B}) \frown \mathfrak{C} = \mathfrak{A} \cup (\mathfrak{B} \frown \mathfrak{C}) \qquad (\mathfrak{C} \geqq \mathfrak{A})$,

(9) $(\mathfrak{B} \cup \mathfrak{A}) \frown \mathfrak{D} = \mathfrak{B} \cup (\mathfrak{A} \frown \mathfrak{D}) \qquad (\mathfrak{D} \geqq \mathfrak{B})$.

O. Ore hat die Frage aufgeworfen, ob umgekehrt die Relationen (8), (9) dafür hinreichend sind, dass $\mathfrak{A}$ mit $\mathfrak{B}$ vertauschbar ist. Dass (8) und (9) dafür nicht immer hinreichend sind, ergibt sich sofort daraus, dass es eine nicht quasi-Hamiltonsche M-Gruppe gibt. Nun gilt

16) Vgl. O. Ore, a. a. O. 11).

Satz 10.[17] Dafür, dass zwei beliebige Untergruppen $\mathfrak{A}, \mathfrak{B}$ einer endlichen Gruppe G immer miteinander vertauschbar sind, sobald sie die Bedingungen (8) und (9) erfüllen, ist notwendig und hinreichend, dass $\mathfrak{G}$ nilpotent ist.

Beweis. Wir nehmen zunächst an, dass zwei beliebige Untergruppen $\mathfrak{A}, \mathfrak{B}$ von $\mathfrak{G}$ immer vertauschbar sind, wenn sie (8) und (9) erfüllen. g sei die Ordnung von $\mathfrak{G}$, und p ein Primteiler von g. Vorausgesetzt, dass eine p-Sylowgruppe $\mathfrak{P}$ kein Normalteiler von $\mathfrak{G}$ ist, bezeichnen wir eine den Normalisator $N(\mathfrak{P})$ von $\mathfrak{P}$ enthaltende maximale Untergruppe von $\mathfrak{G}$ mit $\mathfrak{M}$. (Evtl. $\mathfrak{M}=N(\mathfrak{P})$). Die Ordnung von $\mathfrak{P}$, $N(\mathfrak{P})$ sei bzw. p^a, n und der Index $[\mathfrak{G}:\mathfrak{M}]$, $[\mathfrak{M}:N(\mathfrak{P})]$ sei bzw. l, m. Da nach Sylowsatz der Normalisator $N(\mathfrak{M})$ von $\mathfrak{M}$ mit $\mathfrak{M}$ selbst zusammenfällt, so enthält $\mathfrak{G}$ genau l konjugierte Untergruppen von $\mathfrak{M}$, und von lm p-Sylowgruppen von $\mathfrak{G}$ sind je m in diesen Konjugierten enthalten. Für $\mathfrak{M}$ und eine Konjugierte $\mathfrak{M}_1$ von $\mathfrak{M}$ gilt, wie leicht ersichtlich, die Relationen (8), (9). Nach Voraussetzung ist dann $\mathfrak{M}$ mit $\mathfrak{M}_1$ vertauschbar, also gilt zufolge Hilfssatz 15 $gd=(mn)^2$, oder $ld=mn$, wobei d die Ordnung von $\mathfrak{M}\cap\mathfrak{M}_1$ bedeutet. Aus $p^a \mid n$ und $p^a \nmid l$ folgt nun $p^a \mid d$. $\mathfrak{M}$ und $\mathfrak{M}_1$ müssen also eine p-Sylowgruppe von $\mathfrak{G}$ gemeinsam enthalten, was aber oben erwähnter Tatsache widerspricht. $\mathfrak{P}$ ist also ein Normalteiler von $\mathfrak{G}$. Da p ein beliebiger Primteiler von g war, so folgt hieraus dass $\mathfrak{G}$ nilpotent ist.

Sei nun $\mathfrak{G}$ umgekehrt eine endliche nilpotente Gruppe und $\mathfrak{A}, \mathfrak{B}$ zwei Untergruppen von $\mathfrak{G}$, die (8) und (9) erfüllen. Wenn

$\mathfrak{G}=\mathfrak{P}_1\times\mathfrak{P}_2\times\cdots\times\mathfrak{P}_r$ und entsprechend $\mathfrak{A}=\mathfrak{A}_1\times\mathfrak{A}_2\times\cdots\times\mathfrak{A}_r$,

$\mathfrak{B}=\mathfrak{B}_1\times\mathfrak{B}_2\times\cdots\times\mathfrak{B}_r$ gesetzt wird, dann ist

$\mathfrak{A}\asymp\mathfrak{B}=(\mathfrak{A}_1\asymp\mathfrak{B}_1)\times(\mathfrak{A}_1\asymp\mathfrak{B}_1)\times\cdots\times(\mathfrak{A}_r\asymp\mathfrak{B}_r)$. (vgl. Hilfssatz 14).

Die Ordnung von $\mathfrak{A}, \mathfrak{B}, \mathfrak{A}\cap\mathfrak{B}, \mathfrak{A}\cup\mathfrak{B}$ sei bzw. mit a, b, d, m und die von $\mathfrak{A}_i, \mathfrak{B}_i, \mathfrak{A}_i\cap\mathfrak{B}_i, \mathfrak{A}_i\cup\mathfrak{B}_i$ bzw. mit a_i, b_i, d_i, m_i bezeichnet. ($i=1, 2, \ldots, r$). Aus $a=\prod_{i=1}^{r} a_i$, $b=\prod_{i=1}^{r} b_i$, $d=\prod_{i=1}^{r} d_i$, $m=\prod_{i=1}^{r} m_i$ und $a_ib_i \mid m_id_i$ folgt, dass $ab \mid md$ ist. Nach (8) gibt es nun ersichtlich eine eineindeutige Abbildung vom Quotientverband $\mathfrak{A}\cup\mathfrak{B}/\mathfrak{A}$ in $\mathfrak{B}/\mathfrak{A}\cap\mathfrak{B}$. Die Länge der Hauptkette, die $\mathfrak{A}\cup\mathfrak{B}$ mit $\mathfrak{A}$ verbindet, ist daher nicht grösser als diejenige der Hauptkette, die $\mathfrak{B}$ mit $\mathfrak{A}\cap\mathfrak{B}$ verbindet. Da $\mathfrak{G}$ nach Satz 1 ausgeglichen ist, so ist hiernach die Anzahl der Primfaktoren von $[\mathfrak{A}\cup\mathfrak{B}:\mathfrak{A}]=m/a$ höchstens gleich der Anzahl der Primfaktoren von

17) Vgl. O. Ore, a. a. O. 2).

$[\mathfrak{B}:\mathfrak{A}\frown\mathfrak{B}]=b/d$, und mit $ab|md$ zusammen folgt daraus, dass $ab=md$ ist. Nach Hilfssatz 15 ist dann $\mathfrak{A}$ mit $\mathfrak{B}$ vertauschbar, w. z. b. w..

Durch Satz 10 wird eine Charakterisierung von endlichen nilpotenten Gruppen gegeben. Ob es auch für unendliche Gruppen gültig ist, ist eine offene Frage.

§ 4.

In diesem Paragraphen soll die Struktur der modularen p-Gruppen bestimmt werden.

Eine Gruppe von der Ordnung p oder p^2 ist abelsch, also natürlich modular. Es gibt aber zwei nicht abelsche Typen von Gruppen mit der Ordnung p^3[18]; für $p>2$ sind sie nämlich durch die Relationen gegeben:

(10) $A^{p^2}=B^p=1, \quad BAB^{-1}=A^{1+p}$,

(11) $A^p=B^p=(A,B)^p=1, \quad A(A,B)=(A,B)A, \quad B(A,B)=(A,B)B$,
und für $p=2$ durch

(12) die Quaternionengruppe: $A^4=1, \quad A^2=B^2, \quad BAB^{-1}=A^{-1}$,

(13) die Diedergruppe: $A^4=1, \quad B^2=1, \quad BAB^{-1}=A^{-1}$.

Es ist nun leicht nachweisbar, dass unter diesen Gruppen die Typen (10) und (12) modular, die Typen (11) und (13) aber nicht modular sind. Es gilt also

Satz 11. Eine nichtabelsche Gruppe $\mathfrak{G}$ mit der Ordnung p^3 ist dann und nur dann modular, wenn es den Typus (10) bzw. (12) hat.

Folgender Hilfssatz ist nun oft nützlich, wenn man eine p-Gruppe als modular nachweisen will.

Hilfssatz 17. $\mathfrak{G}$ sei eine p-Gruppe, deren echte Untergruppen sämtlich modular sind. Wenn $\mathfrak{G}$ selbst doch keine M-Gruppe ist, so enthält $\mathfrak{G}$ einen Normalteiler $\mathfrak{N}$, so dass $\mathfrak{G}/\mathfrak{N}$ eine Gruppe vom Typus (11) oder (13) ist.

Beweis. Da $V(\mathfrak{G})$ nach Voraussetzung kein modularer Verband ist, so gibt es nach einem bekannten Satz der Verbandstheorie[19] solche Untergruppen $\mathfrak{A}, \mathfrak{B}$ von $\mathfrak{G}$, dass

$$\mathfrak{A}\smile\mathfrak{B}=\mathfrak{G}, \quad [\mathfrak{G}:\mathfrak{A}]\geqq p^2, \quad [\mathfrak{G}:\mathfrak{B}]\geqq p^2, \quad [\mathfrak{A}:\mathfrak{A}\frown\mathfrak{B}]=[\mathfrak{B}:\mathfrak{A}\frown\mathfrak{B}]=p$$

gilt. Ferner sei $\mathfrak{A}_1$ bzw. $\mathfrak{B}_1$ eine $\mathfrak{A}$ bzw. $\mathfrak{B}$ enthaltende Untergruppe von $\mathfrak{G}$ mit dem Index $[\mathfrak{G}:\mathfrak{A}_1]=p$ bzw. $[\mathfrak{G}:\mathfrak{B}_1]=p$. $\mathfrak{N}=\mathfrak{A}\frown\mathfrak{B}$ ist nun offenbar ein Normalteiler von $\mathfrak{G}$. Wir setzen also

$$\bar{\mathfrak{G}}=\mathfrak{G}/\mathfrak{N}, \quad \bar{\mathfrak{A}}=\mathfrak{A}/\mathfrak{N}, \quad \bar{\mathfrak{B}}=\mathfrak{B}/\mathfrak{N}, \quad \bar{\mathfrak{A}}_1=\mathfrak{A}_1/\mathfrak{N}, \quad \bar{\mathfrak{B}}_1=\mathfrak{B}_1/\mathfrak{N}.$$

18) Vgl. z. B. H. Z., a. a. O. S. 114.

19) Vgl. G. Birkhoff, a. a. O. 8).

Wenn ferner

$$\bar{\mathfrak{A}}=\{\bar{A}\}\,,\quad \bar{\mathfrak{B}}=\{\bar{B}\}\,,\quad \bar{\mathfrak{A}}_2=\{\bar{A}, \bar{B}\bar{A}\bar{B}^{-1}, \bar{B}^2\bar{A}\bar{B}^{-2}, \ldots\}$$

gesetzt wird, so ist $\bar{\mathfrak{A}}_2$ ein Normalteiler von $\bar{\mathfrak{G}}$ mit $\bar{\mathfrak{A}}\subset\bar{\mathfrak{A}}_2\subset\bar{\mathfrak{A}}_1$, $\bar{\mathfrak{A}}_2\cup\bar{\mathfrak{B}}=\bar{\mathfrak{G}}$, $\bar{\mathfrak{A}}_2\cap\bar{\mathfrak{B}}=1$.

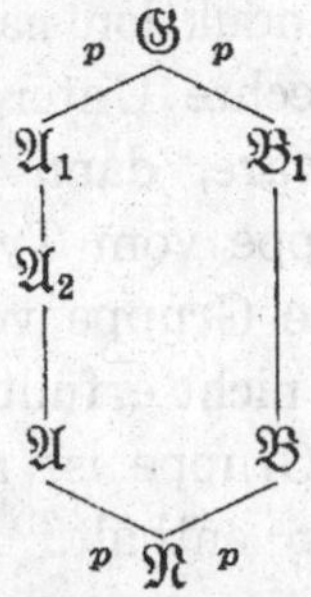

Daraus folgt $[\bar{\mathfrak{G}}:\bar{\mathfrak{A}}_2]=p$, also $\bar{\mathfrak{A}}_2=\bar{\mathfrak{A}}_1=\{\bar{A}, \bar{B}\bar{A}\bar{B}^{-1}, \bar{B}^2\bar{A}\bar{B}^{-2}, \ldots\}$. Genau so erhält man $\bar{\mathfrak{B}}_1=\{\bar{B}, \bar{A}\bar{B}\bar{A}^{-1}, \bar{A}^2\bar{B}\bar{A}^{-2}, \ldots\}$. Da nach Voraussetzung $\mathfrak{A}_1$, $\mathfrak{B}_1$ modular sind, so sind $\bar{\mathfrak{A}}_1$, $\bar{\mathfrak{B}}_1$ zufolge Hilfssatz 7 beide elementare abelsche Gruppen.[20] Wenn die Ordnung von $\bar{\mathfrak{A}}_1$ p^f ist, und wenn

$$\bar{X}^\sigma=\bar{B}\bar{X}\bar{B}^{-1}\,,\qquad \text{für}\quad \bar{X}\in\bar{\mathfrak{A}}_1\,,$$

gesetzt wird, so können wir $\bar{\mathfrak{A}}_1$ als f-dimensionalen Vektorraum über $GF(p)$, und σ als lineare Transformation in diesem Raum auffassen, und aus $\bar{\mathfrak{A}}_1=\{\bar{A}, \bar{A}^\sigma, \bar{A}^{\sigma^2}, \ldots\}$ und $\sigma^p=1$ folgt dann sofort, dass das Minimalpolynom von σ $(x-1)^f$ ist. Durch geeignete Wahl der Basis wird also σ durch die Matrix

$$\begin{pmatrix} 1 & 1 & & 0 \\ & 1 & 1 & \\ & & \ddots & \ddots \\ & & & \cdot\, 1 \\ 0 & & & 1 \end{pmatrix}$$

dargestellt, und daraus folgt leicht, dass die Gruppe $\bar{\mathfrak{A}}_1^*$, die aus allen Elementen $\bar{X}$ von $\bar{\mathfrak{A}}_1$ mit $\bar{X}^\sigma=\bar{X}$ besteht, die Ordnung p hat. Da die Elemente aus $\bar{\mathfrak{A}}_1\cap\bar{\mathfrak{B}}_1$ offenbar mit $\bar{\mathfrak{B}}$ elementweise vertauschbar, also in $\bar{\mathfrak{A}}_1^*$ enthalten sind, so ist die Ordnung von $\bar{\mathfrak{A}}_1\cap\bar{\mathfrak{B}}_1$ höchstens p. $\bar{\mathfrak{G}}$ ist daher eine nicht modulare Gruppe von der Ordnung p^3, also vom Typus (11) oder (13).

---

20) Eine elementare abelsche p-Gruppe heisst bekanntlich eine abelsche Gruppe vom Typus $(p, p, \ldots, p)$.

Mit diesem Hilfssatz ist z. B. der folgende im vorigen Paragraphen erwähnte Satz leicht zu beweisen.

Satz 12. $\mathfrak{G}$ sei eine p-Gruppe und $\mathfrak{A}, \mathfrak{B}$ beliebige Untergruppen von $\mathfrak{G}$. Wenn es immer ein Element X aus $\mathfrak{A} \cup \mathfrak{B}$ gibt, so dass $\mathfrak{A}$ mit $X\mathfrak{B}X^{-1}$ vertauschbar wird, so ist $\mathfrak{G}$ modular.

Beweis. Wir wenden die Induktion nach der Ordnung von $\mathfrak{G}$ an, und setzen voraus, dass jede echte Untergruppen von $\mathfrak{G}$ modular ist. Wenn $\mathfrak{G}$ selbst nicht modular wäre, dann enthielte es einen Normalteiler $\mathfrak{N}$, so dass $\mathfrak{G}/\mathfrak{N}$ eine Gruppe vom Typus (11) oder (13) ist. Das ist aber ein Widerspruch, da die Gruppe vom Typus (11) oder (13) die im Satz angegebene Bedingung nicht erfüllt. $\mathfrak{G}$ muss also modular sein.

Die Struktur derjenigen p-Gruppe ist nun bekannt, die einen zyklischen Normalteiler vom Index p enthält.[21] Danach kann man leicht den folgenden Satz beweisen.

Satz 13. Die modulare Gruppe von der Ordnung p^n, in denen ein Element mit der Ordnung p^{n-1} enthalten ist, hat einen der folgenden Typen:

(i) abelsche Gruppe,

(ii) die Quaternionengruppe,

(iii) $A^{p^{n-1}}=1,\ B^p=1,\ BAB^{-1}=A^{1+p^{n-2}}.$

Wir nennen nun eine Gruppe $\mathfrak{G}$ „halbzyklisch", wenn es einen solchen zyklischen Normalteiler $\mathfrak{N}$ enthält, dass die Faktorgruppe $\mathfrak{G}/\mathfrak{N}$ eine zyklische Gruppe ist. Mit $\mathfrak{G}$ ist dann auch jede Untergruppe und jede Faktorgruppe von $\mathfrak{G}$ halbzyklisch. Es gilt

Satz 14. Eine halbzyklische p-Gruppe $\mathfrak{G}$ ist modular, wenn $p>2$ ist. Es wird gegeben durch folgende Relationen:

$$(14)\quad A^{p^m}=1,\quad B^{p^n}=A^{p^t},\quad BAB^{-1}=A^{1+p^s},\quad t+s\geqq m,\quad n+s\geqq m.$$

Eine halbzyklische 2-Gruppe $\mathfrak{G}$ ist dann und nur dann modular, wenn sie entweder eine Quaternionengruppe ist oder durch (14) und noch eine zusätzliche Bedingung $s\geqq 2$ gegeben wird.

Beweis. Wir beweisen zunächst, dass eine halbzyklische p-Gruppe $\mathfrak{G}$ $(p>2)$ modular ist. Wir wenden dazu die Induktion nach der Ordnung von $\mathfrak{G}$ an, und setzen voraus, dass jede echte Untergruppe von $\mathfrak{G}$ modular ist. Wenn $\mathfrak{G}$ selbst nicht modular wäre, so enthielte sie nach Hilfssatz 17 einen Normalteiler $\mathfrak{N}$ so dass $\mathfrak{G}/\mathfrak{N}$ eine Gruppe von Typus (11) ist. Das ist aber unmöglich, da die Gruppe vom Typus (11) nicht halbzyklisch ist; $\mathfrak{G}$ muss also modular sein. Dass eine halbzyklische

21) Vgl. z. B. H. Z., a. a. O. S. 114.

p-Gruppe durch die Relationen (14) gegeben wird, ist eine bekannte Tatsache.[22] $\mathfrak{G}$ sei nun eine modulare halbzyklische 2-Gruppe, und $\mathfrak{N}$ der zyklische Normalteiler von $\mathfrak{G}$. Wir setzen

$$\mathfrak{N}=\{A\}, \quad \mathfrak{G}/\mathfrak{N}=\{\bar{B}\}, \quad A^{2^m}=1, \quad B^{2^n}=A^{2^t}, \quad BAB^{-1}=A^{\pm 1+2^s h},$$
$$s \geqq 2, \quad h \not\equiv 0 \text{ mod. } 2.$$

Falls $BAB^{-1}=A^{1+2^s h}$ ist, so sieht man leicht, dass $\mathfrak{G}$ von Typus (14) ist, wenn man evtl. B durch eine Potenz von B ersetzt. Sei nun $BAB^{-1}=A^{-1+2^s h}$. Wenn $t \geqq 2$ wäre, so wäre $\bar{\mathfrak{G}}=\mathfrak{G}/\{A^4, B^2\}$ eine Gruppe vom Typus (13), also nicht modular. Es ist daher $t \leqq 1$. $\mathfrak{G}$ hat dann einen zyklischen Normalteiler vom Index 2 und ist nach Satz 13 entweder eine Quaternionengruppe oder eine Gruppe mit der Relationen (14) mit $s \geqq 2$. Dass umgekehrt jede solche 2-Gruppe modular ist, kann man genau so wie im Fall $p > 2$ beweisen.

Mit Hilfe von Satz 13, 14 lässt sich nun folgenden Satz nach einigen Rechnungen ohne Schwierigkeit beweisen.

Satz 15. Die nichtabelsche M-Gruppen von der Ordnung p^4 haben folgende Typen:

(1) für $p > 2$,

(15) $A^{p^3}=B^p=1, \quad BAB^{-1}=A^{1+p^2}$,

(16) $A^{p^2}=B^p=C^p=1, \quad BAB^{-1}=A^{1+p}, \quad AC=CA, \quad BC=CB$,

(17) $A^{p^2}=B^{p^2}=1, \quad BAB^{-1}=A^{1+p}$,

(2) für $p=2$,

(18) $A^8=B^2=1, \quad BAB^{-1}=A^5$,

(19) $A^4=C^2=1, \quad A^2=B^2, \quad BAB^{-1}=A^{-1}, \quad AC=CA, \quad BC=CB$.

Es sei nun bemerkt, dass für zwei beliebige Elemente C, D in einer Gruppe $\mathfrak{G}$ vom Typus (16) oder (17), die Gleichung

$$(CD)^p=C^p D^p$$

gilt, was später benutzt werden wird.

Hilfssatz 18. Es sei $\mathfrak{G}$ eine modulare p-Gruppe und p^μ die maximale Ordnung eines Elementes aus $\mathfrak{G}$. Diejenige Elemente aus $\mathfrak{G}$, deren p^a-te Potenzen gleich 1 sind, bilden eine charakteristische Untergruppe Ω_a von $\mathfrak{G}$.[23] Es gilt dann

$$1=\Omega_0 \subset \Omega_1 \subset \cdots \subset \Omega_{\mu-1} \subset \Omega_\mu = G,$$

22) Vgl. z. B. W. Burnside, a. a. O. 5), S. 126.

23) Wir folgen die Bezeichnungen von P. Hall, Proc. London Math. Soc. (2) 36, 29–95, 1933.

und Ω_i/Ω_{i-1} $(i=1, 2, \ldots, \mu)$ sind sämtlich elementare abelsche Gruppen.

Beweis. Alle Elemente X aus $\mathfrak{G}$ mit $X^p=1$ bilden nach Hilfssatz 7 eine charakteristische Untergruppe Ω_1 von $\mathfrak{G}$. In $\bar{\mathfrak{G}}=\mathfrak{G}/\Omega_1$ bilden die Elemente $\bar{X}$ mit $\bar{X}^p=1$ wieder eine charakteristische Untergruppe $\bar{\Omega}_1$. Die in $\mathfrak{G}$ $\bar{\Omega}_1$ entsprechende Gruppe Ω_2 besteht offenbar genau aus allen Elementen Y von $\mathfrak{G}$ mit $Y^{p^2}=1$, und $\bar{\Omega}_1=\Omega_2/\Omega_1$ ist eine elementare abelsche Gruppe. In derselben Weise kann man der Reihe nach die Gruppen $\Omega_3, \Omega_4, \ldots$ bilden und beweisen, dass Ω_i/Ω_{i-1} alle elementar abelsch sind.

Im folgenden sollen $\mu=\mu(\mathfrak{G})$ und Ω_a immer den in diesem Hilfssatz angegebenen Sinn haben.

$\mathfrak{G}$ sei nun eine modulare 2-Gruppe mit $\mu(\mathfrak{G})=2$, und enthalte eine Quaternionengruppe $\mathfrak{Q}$:

$$\mathfrak{Q}=\{A, B\}, \quad A^4=1, \quad A^2=B^2, \quad BAB^{-1}=A^{-1}.$$

Wir nehmen zunächst an, dass $[\mathfrak{G}:\Omega_1]\geqq 2^3$ ist, und bezeichen mit C ein nicht in $\mathfrak{Q}\Omega_1$ enthaltenes Element von $\mathfrak{G}$. Wenn $C^2=A^2=B^2$ ist, dann ist $\{A, B, C\}$ eine Gruppe von der Ordnung 2^4, andererseits ist sie aber nicht vom Typus (19), was unmöglich ist. Es ist also $\{C\}\frown\{A, B\}=1$, und nach Satz 15 sind $\{A, C\}$, $\{B, C\}$ beide abelsch. Da $\mathfrak{G}$ nach Hilfssatz 16 quasi-Hamiltonsch ist, so gibt es ganze Zahlen x, y, so dass

$$B(AC)=(AC)^x B^y$$

gilt. Nun ist

$$B(AC)=A^{-1}BC=(AC)^x B^y=A^x C^x B^y=A^x B^y C^x,$$

also

$$A^{-1}B=A^x B^y, \qquad C=C^x,$$

folglich $x\equiv 1$, $y\equiv -1$ mod. 4,

$$(AC)^{-1}B(AC)=B^{-1}.$$

Da B^2 nicht mit $(AC)^2=A^2C^2$ zusammenfällt, und im Zentrum von $\{(AC), B\}$ enthalten ist, so ist $\{AC), B\}/\{B^2\}$ eine Diedergruppe, also nicht modular. In unserem Fall ist das nicht möglich ; es muss $\mathfrak{G}=\mathfrak{Q}\Omega_1$, namlich $[\mathfrak{G}:\Omega_1]=2$ sein. Wenn nun $A^2, C_1, C_2, \ldots, C_r$ eine Basis von Ω_1 ist, so sind $\{C_i, A\}$, $\{C_i, B\}$ nach Satz 11 beide abelsch. $\mathfrak{G}$ ist mithin das direkte Produkt von einer Quaternionengruppe und einer abelschen Gruppe vom Typus $(2, 2, \ldots, 2)$. Dass umgekehrt eine solche Gruppe wirklich eine M-Gruppe ist, kann man leicht beweisen. Nun sei $\mathfrak{G}$ eine modulare 2-Gruppe mit $\mu(\mathfrak{G})=2$, die keine Quaternionengruppe enthält, und C, D beliebige Elemente aus $\mathfrak{G}$. Die Ordnung von $\{C, D\}$ ist dann offenbar hochstens 2^4. Da es aber keine Quaternionengruppe enthält, so folgt aus Satzen 11, 15 dass $\{C, D\}$ abelsch ist. $\mathfrak{G}$ selbst muss daher abelsch sein.

Es gilt also

Hilfssatz 19. Eine modulare 2-Gruppe mit $\mu(\mathfrak{G})=2$ ist entweder abelsch oder direktes Produkt von einer Quaternionengruppe und einer abelschen Gruppe vom Typus $(2, 2, \ldots, 2)$.

Die letztere Gruppe wollen wir die spezielle 2-Gruppe nennen.[24)]

Nunmehr sei $\mathfrak{G}$ eine allgemeine modulare 2-Gruppe. $\bar{\mathfrak{G}}=\Omega_i/\Omega_{i-2}$ $(i=2, 3, \ldots, \mu)$ ist dabei natürlich eine modulare 2-Gruppe mit $\mu(\bar{\mathfrak{G}})=2$. Wir nehmen nun an, dass für einen Index i Ω_i/Ω_{i-2} nichtabelsch ist, dass also Ω_i/Ω_{i-2} eine spezielle 2-Gruppe ist.

Sei zunächst $i<\mu$. Falls die in $\bar{\mathfrak{G}}=\Omega_i/\Omega_{i-2}$ enthaltene Quaternionengruppe mit $\bar{\mathfrak{Q}}=\{\bar{A}, \bar{B}\}$ bezeichnet wird, so gilt für ein Element $\bar{X}$ aus $\bar{\mathfrak{G}}$ mit der Ordnung 2^3 etwa $\bar{X}\equiv\bar{A}$ (Ω_{i-1}), und hieraus folgt, dass $\bar{X}^4=\bar{A}^2=\bar{B}^2$ ist. $\bar{\mathfrak{A}}=\{\bar{X}, \bar{B}\}$ ist daher eine nichtabelsche Gruppe mit der Ordnung 2^4 und enthält ein Element $\bar{X}$ von der Ordnung 2^4; es muss also nach Satz 15 eine Gruppe vom Typus (18) sein. Das ist aber unmöglich, denn, während jede echte Untergruppe der Gruppe vom Typus (18) abelsch ist, $\bar{\mathfrak{A}}$ enthält doch eine nichtabelsche Untergruppe $\{\bar{X}^2, \bar{B}\}$. Es muss also $i=\mu$ sein.

Sei nun $i>2$. Wenn $\mathfrak{G}^*=\Omega_i/\Omega_{i-3}$ gesetzt wird, und wenn die $\bar{A}, \bar{B}$ entsprechende Elemente von $\mathfrak{G}^*$ mit A, B bezeichnet werden, so gilt $A^2\equiv B^2$ (Ω_{i-2}), d. h. $A^2=B^2C$ mit einem Element C aus Ω_{i-2}, und daraus folgt $A^4=B^4$. Die Ordnung von $\mathfrak{A}^*=\{A, B\}$ ist mithin höchstens 2^5. $\mathfrak{A}^*/\{A^4\}$ ist also eine nichtabelsche Gruppe von der Ordnung 2^3 oder 2^4. Nach Satz 11, 15 gilt nun aber $A^2\equiv B^2(A^4)$. Die Ordnung von $\mathfrak{A}^*$ ist also 2^4 und da $\mathfrak{A}^*$ ein Element A mit der Ordnung 2^3 enthält, so ist es eine Gruppe vom Typus (18), folglich ist $\mathfrak{A}^*/\{A^4\}$ abelsch. Das ist aber unmöglich, da $\mathfrak{A}^*/\{A^4\}$ nach Konstruktion eine Quaternionengruppe ist. Es muss also $i=2$ sein.

Die Faktorgruppen Ω_i/Ω_{i-2} $(i=2, 3, \ldots, \mu)$ von einer modularen 2-Gruppe $\mathfrak{G}$ sind also, sämtlich abelsch, wenn $\mathfrak{G}$ selbst keine spezielle 2-Gruppe ist.

Nun sei $\mathfrak{G}$ eine modulare p-Gruppe $(p>2)$ mit $\mu(\mathfrak{G})=2$, und C, D beliebige Elemente von $\mathfrak{G}$. Die Gruppe $\{C,D\}$ ist dann entweder

---

24) Eine spezielle 2-Gruppe ist also nicht anders als eine Hamiltonsche 2-Gruppe.

abelsch oder vom Typus (10), (16), bzw. (17). Es gilt also, wie schon bemerkt,

$$(CD)^p = C^p D^p. \tag{20}$$

(20) ist auch für eine modulare 2-Gruppe $\mathfrak{G}$ mit $\mu(\mathfrak{G})=2$ gültig, wenn es keine spezielle 2-Gruppe ist. Es gilt nun

Hilfssatz 20. Es sei $\mathfrak{G}$ eine modulare p-Gruppe ($p \geqq 2$) mit $\mu(\mathfrak{G})=\mu$, die keine spezielle 2-Gruppe ist. Für beliebige Elemente C, D aus $\mathfrak{G}$ gilt dann die gleichung

$$(CD)^{p^{\mu-1}} = C^{p^{\mu-1}} D^{p^{\mu-1}}.$$

Beweis. $\bar{\mathfrak{G}} = \Omega_\mu/\Omega_{\mu-2}$ ist eine modulare p-Gruppe mit $\mu(\bar{\mathfrak{G}})=2$. Es gilt also nach (20)

$$(CD)^p \equiv C^p D^p \quad (\Omega_{\mu-2}),$$

d. h.

$$(CD)^p = C^p D^p Z, \qquad Z \in \Omega_{\mu-2}.$$

Alle Elemente in dieser Gleichung sind in $\Omega_{\mu-1}$ enthalten. Wenn man also in $\Omega_{\mu-1}/\Omega_{\mu-3}$ rechnet, so folgt daraus wieder nach (20)

$$(CD)^{p^2} \equiv C^{p^2} D^{p^2} Z^p \equiv C^{p^2} D^{p^2} \quad (\Omega_{\mu-3}),$$

d. h.

$$(CD)^{p^2} = C^{p^2} D^{p^2} Z', \qquad Z' \in \Omega_{\mu-3}.$$

So fortfahrend erhält man schliesslich

$$(CD)^{p^{\mu-1}} = C^{p^{\mu-1}} D^{p^{\mu-1}},$$

w. z. b. w..

Sei $\mathfrak{G}$ wieder eine modulare Gruppe mit der Ordnung p^n, die keine spezielle 2-Gruppe ist. Nun setzen wir $[\Omega_\alpha : \Omega_{\alpha-1}] = p^{\omega_\alpha}$ ($\alpha = 1, 2, \dots, \mu$). Wir wählen Elemente $A_1, A_2, \dots, A_{\omega_\mu}$ die in $\Omega_\mu/\Omega_{\mu-1}$ eine Basis bilden. Indem man Hilfssatz 20 auf $\Omega_\mu/\Omega_{\mu-2}$ anwendet, sieht man leicht ein, dass $A_1^p, A_2^p, \dots, A_{\omega_\mu}^p$ zu einer Basis $A_1^p, A_2^p, \dots, A_{\omega_\mu}^p, A_{\omega_\mu+1}, \dots, A_{\omega_{\mu-1}}$ von $\Omega_{\mu-1}/\Omega_{\mu-2}$ ergänzt werden kann. Dieses Elementensystem lässt sich dann wieder zu einer Basis von $\Omega_{\mu-2}/\Omega_{\mu-3}$ ergänzt werden, und so fortfahrend kann man schliesslich ein Erzeugendensystem $A_1, \dots, A_{\omega_\mu}, \dots, A_{\omega_{\mu-1}}, \dots, A_{\omega_1}$ von $\mathfrak{G}$ erhalten. Ist die Ordnung von A_i gleich p^{μ_i}, so ist jedes Element aus $\mathfrak{G}$ in der Form

$$X = A_1^{x_1} A_2^{x_2} \dots A_{\omega_1}^{x_{\omega_1}}, \qquad 0 \leqq x_i < p^{\mu_i} \tag{21}$$

darstellbar, da $\mathfrak{G}$ nach Hilfssatz 16 zugleich quasi-Hamiltonsch ist. Die Anzahl der A_i, deren Ordnung gleich p^α ist, ist $\omega_\alpha - \omega_{\alpha+1}$ ($\omega_{\mu+1}=0$). Aus der Gleichung

$$\mu\omega_\mu+(\mu-1)(\omega_{\mu-1}-\omega_\mu)+\cdots+(\omega_1-\omega_2)=\omega_1+\omega_2+\cdots+\omega_\mu=n$$

folgt sofort, dass die rechte Seite von (21) durch X eindeutig bestimmt wird, d. h. dass die A_i eine Basis von $\mathfrak{G}$ bilden.[25]

Nun sei $B_1, B_2, \dots, B_\omega$ eine beliebige Basis von $\mathfrak{G}$. Jedes Element X aus $\mathfrak{G}$ wird dann eindeutig in der Form

$$X=B_1^{x_1}B_2^{x_2}\dots B_\omega^{x_\omega}, \qquad 0\leqq x_i<p^{\nu_i}$$

dargestellt, wenn die Ordnung von B_i p^{ν_i} ist. $(i=1, 2, \dots, \omega)$. Sei nun $\nu_1\geqq\nu_2\geqq\cdots\geqq\nu_\omega$, und insbesondere $\nu_1=\nu_2=\cdots=\nu_r=\mu$. Wenn das Bild von $\mathfrak{G}$ beim Homomorphismus $C\to C^{p^{\mu-1}}$ (C: beliebiges Element von $\mathfrak{G}$) mit $\mho_{\mu-1}$ bezeichnet wird, so ist offenbar

$$\mho_{\mu-1}=\{B_1^{p^{\mu-1}}, B_2^{p^{\mu-1}}, \dots, B_r^{p^{\mu-1}}\}=\{A_1^{p^{\mu-1}}, A_2^{p^{\mu-1}}, \dots, A_{\omega_\mu}^{p^{\mu-1}}\},$$

und mithin $r=\omega_\mu$,

$$\Omega_{\mu-1}=\{B_1^p, \dots, B_r^p, B_{r+1}, \dots, B_\omega\}=\{A_1^p, \dots, A_{\omega_\mu}^p, \dots, A_{\omega_1}\}.$$

Wenn $\nu_{r+1}=\nu_{r+2}=\cdots=\nu_{r'}=\mu-1$ ist, und auf $\Omega_{\mu-1}$ wieder obiger Schluss angewandt wird, so ergibt sich $r'=\omega_{\mu-1}$, u. s. w.. Man erkennt also, dass das zahlensystem $(\nu_1, \nu_2, \dots, \nu_\omega)$ eben mit $(\mu_1, \mu_2, \dots, \mu_{\omega_1})$ zusammenfällt.

Es gilt also

Satz 16. Eine modulare p-Gruppe $\mathfrak{G}$ besitzt eine Basis, wenn sie keine spezielle 2-Gruppe ist. Durch geeignete Wahl von Elementensystem $\{A_1, A_2, \dots, A_\omega\}$ ist nämlich jedes Element X von $\mathfrak{G}$ eindeutig in der Form

$$X=A_1^{x_1}A_1^{x_2}\dots A_\omega^{x_\omega}, \qquad 0\leqq x_i<p^{\mu_i}$$

zu schreiben, wobei p^{μ_i} die Ordnung von A_i bedeutet.

Wenn ferner $\mu_1\geqq\mu_2\geqq\cdots\geqq\mu_\omega$ vorausgesetzt wird, so ist das Zahlensystem $(\mu_1, \mu_2, \dots, \mu_\omega)$ für $\mathfrak{G}$ charakteristisch, und die Anzahl ω_a derjenigen μ_i, die $\mu_i\geqq a$ erfüllen, wird eben durch $[\Omega_a:\Omega_{a-1}]=p^{\omega_a}$ angegeben. Insbesondere gilt $\omega=\omega_1$.

Wir bestimmen nun die Struktur der modularen p-Gruppe mit $\omega_1\leqq 2$.

Hilfssatz 21. Für eine modulare p-Gruppe $\mathfrak{G}$, die keine spezielle 2-Gruppe ist, ist dann und nur dann $\omega_1\leqq 2$, wenn $\mathfrak{G}$ durch zwei Elemente erzeugt werden kann.

---

25) Vgl. P. Hall, a. a. O. 18).

Beweis. Dass eine Gruppe $\mathfrak{G}$ mit $\omega_1 \leqq 2$ durch zwei Elemente erzeugt wird, ist klar. Sei umgekehrt $\mathfrak{G}=\{A, B\}$. Wenn die Ordnung von A nicht kleiner als die von B ist, so kann man A, wie aus dem Beweis von Satz 16 ersichtlich ist, als ein Basiselement wählen. Es sei also $A=A_1, A_2, \dots, A_\omega$ eine Basis von $\mathfrak{G}$, und ferner sei $B=A_1^{e_1}A_2^{e_2}\dots A_\omega^{e_\omega} = A_1^{e_1}\bar{B}$. Es gilt dann offenbar $\mathfrak{G}=\{A, \bar{B}\}$, $\{A\} \frown \{\bar{B}\}=1$. Wenn also $\bar{B} \neq 1$ ist, bilden $A, \bar{B}$ eine Basis von $\mathfrak{G}$, woraus die Behauptung folgt.

Satz 17. Für eine modulare p-Gruppe $\mathfrak{G}$, die keine spezielle 2-Gruppe ist, gilt dann und nur dann $\omega_1 \leqq 2$, wenn $\mathfrak{G}$ halbzyklisch ist.

Beweis. Nach Hilfssatz 21 ist zunächst klar, dass für eine halbzyklische modulare p-Gruppe $\omega_1 \leqq 2$ gilt. $\mathfrak{G}$ sei nun umgekehrt eine modulare p-Gruppe mit $\omega_1=2$.

a) Wir nehmen zunächst an, dass $\mathfrak{G}'$ (die Kommutatorgruppe von $\mathfrak{G}$) zyklisch ist; $\mathfrak{G}'=\{C\}$. Da $\mathfrak{G}$ durch zwei Elemente erzeugt werden kann, so ist $\mathfrak{G}/\mathfrak{G}'$ eine abelsche Gruppe mit höchstens zwei Basiselemente. Falls $\mathfrak{G}/\mathfrak{G}'$ zyklisch ist, dann ist $\mathfrak{G}$ schon zyklisch. Wir nehmen also an, dass $\mathfrak{G}/\mathfrak{G}'$ eine abelsche Gruppe vom Typus (p^m, p^n) mit der Basiselemente A, B ist:

$$A^{p^m}=C^{hp^{m'}}, \quad B^{p^n}=C^{kp^{n'}} \quad hk \not\equiv 0 \text{ mod. } p, \quad m' \geqq n'.$$

Sei $n'>0$. $\{C^{p^{n'}}\}=\mathfrak{G}_1$ ist ein Normalteiler von $\mathfrak{G}$, und $\bar{\mathfrak{G}}=\mathfrak{G}/\mathfrak{G}_1$ ist wieder eine modulare p-Gruppe mit $\omega_1 \leqq 2$. $A^{p^{m-1}}, B^{p^{n-1}}, C^{p^{n'-1}}$ haben alle die Ordnung p in $\bar{\mathfrak{G}}$, also ist $A^{p^{m-1}}$ in $\{B^{p^{n-1}}, C^{p^{n'-1}}\}$ enthalten, was damit in Widerspruch steht, dass A, B eine Basis von $\mathfrak{G}/\mathfrak{G}'$ bilden. Es gilt daher $n=0$, und $\{B\}$ ist ein Normalteiler von $\mathfrak{G}$; $\mathfrak{G}$ ist also halbzyklisch.

b) Wir nehmen nun an, dass $\mathfrak{G}'$ mit Ω_1 zusammenfällt, also eine abelsche Gruppe vom Typus (p, p) ist. Ein in Ω_1 enthaltenen Normalteiler mit der Ordnung p sei mit $\mathfrak{Z}=\{Z)$ bezeichnet. Da die Kommutatorgruppe $\mathfrak{G}'/\mathfrak{Z}$ von $\mathfrak{G}/\mathfrak{Z}$ eine Gruppe von der Ordnung p ist, so ist $\mathfrak{G}/\mathfrak{Z}$ nach a) halbzyklisch, und es gilt nach Satz 14

$$\mathfrak{G}/\mathfrak{Z}=\{\bar{A}, \bar{B}\}, \quad \bar{A}^{p^m}=1, \quad \bar{B}^{p^n}=\bar{A}^{p^t}, \quad \bar{B}\bar{A}\bar{B}^{-1}=\bar{A}^{1+p^{m-1}}, \quad m \geqq 2.$$
$$(\text{für } p=2,\ m \geqq 3).$$

Wählt man A, B in der Restklasse $\bar{A}$ bzw. $\bar{B}$, so gilt, wie leicht ersichtlich, $\mathfrak{G}=\{A, B\}$. $\{A, Z\}$ ist dann ein Normalteiler von $\mathfrak{G}$ und es gilt

$$BAB^{-1}=A^{1+p^{m-1}}Z^k.$$

Da $A^{p^{m-1}}Z^k$ im Zentrum von $\mathfrak{G}$ enthalten ist, so folgt hieraus $\mathfrak{G}'=\{A^{p^{m-1}}Z^k\}$, was mit $G'=\Omega_1$ in Widerspruch steht. Für modulare p-Gruppe mit $\omega_1=2$ ist also immer $\mathfrak{G}'\neq\Omega_1$.

c) Um nun $\mathfrak{G}$ als halbzyklisch zu erkennen, genügt es nach a) nur zu beweisen, dass $\mathfrak{G}'$ zyklisch ist. Dazu wenden wir die Induktion nach der Ordnung von $\mathfrak{G}$ an. $\mathfrak{Z}$ sei ein in $\mathfrak{G}'$ enthaltener Normalteiler von $\mathfrak{G}$ mit der Ordnung p. Die Kommutatorgruppe $\mathfrak{G}'/\mathfrak{Z}$ von $\mathfrak{G}/\mathfrak{Z}$ ist nach Voraussetzung zyklisch. Da $\mathfrak{Z}$ im Zentrum von $\mathfrak{G}$ enthalten ist, so ist $\mathfrak{G}'$ entweder zyklisch oder eine abelsche Gruppe etwa vom Typus (p^l, p). Im letzten Fall bezeichnen wir mit $\mathfrak{G}_1$ den aus allen p-ten Potenzen der Elemente von $\mathfrak{G}'$ erzeugten Normalteiler von $\mathfrak{G}$. Dann wird aber die Kommutatorgruppe $\mathfrak{G}'/\mathfrak{G}_1$ von $\mathfrak{G}/\mathfrak{G}_1$ eine abelsche Gruppe vom Typus (p, p), was wegen b) nicht möglich ist. $\mathfrak{G}'$ muss also zyklisch sein, w. z. b. w..

Um die Struktur der allgemeinen modularen p-Gruppen zu bestimmen, schicken wir noch einen Hilfssatz voraus.

Hilfssatz 22. Die modulare p-Gruppe $\mathfrak{G}$, die keine spezielle 2-Gruppe ist, und deren Kommutatorgruppe $\mathfrak{G}'$ die Ordnung p hat, besitzt folgende Struktur:

(i) $\mathfrak{G}=\mathfrak{A}_1\times\mathfrak{A}_2$

(ii) $\mathfrak{A}_1=\{A_1, A_2\}$, $A_1^{p^m}=A_2^{p^n}=1$, $A_2A_1A_2^{-1}=A_1^{1+p^{m-1}}$,

(iii) $\mathfrak{A}_2$ ist abelsch und die Ordnung eines beliebigen Elementes aus $\mathfrak{A}_2$ ist höchstens p^{m-1}.

Beweis. Nach Satz 17 besitzt $\mathfrak{G}$ eine Basis $A_1, A_2, \dots, A_r$. Man kann dabei, wie leicht ersichtlich, ohne Verlust der Allgemeinheit annehmen, dass $\mathfrak{G}'$ in $\{A_1\}$ enthalten ist.

Aus $\{A_i, A_j\}\frown\mathfrak{G}'=1$ $(i, j\geqq 2)$ folgt dann $A_iA_j=A_jA_i$, d. h. dass $\{A_2, A_3, \dots, A_r\}$ abelsch ist. Unter denjenigen A_i $(i\geqq 2)$, die mit A_1 nicht vertauschbar sind, sei A_2 von kleinster Ordnung. Man erkennt alsdann, dass nachdem evtl. A_i $(i\geqq 2)$ durch $A_iA_2^g$ ersetzt wird, $\mathfrak{G}=\{A_1, A_2\}\times\{A_3, \dots, A_r\}$, $A_2A_1A_2^{-1}=A_1^{1+p^{m-1}}$, $A_iA_j=A_jA_i$ $(i, j\geqq 3)$ gilt. Aus $A_2(A_1A_i)A_2^{-1}=A_1^{1+p^{m-1}}A_i=(A_1A_i)^x$ $(i\geqq 3)$ folgt dann sofort, dass die Ordnung von A_i $(i\geqq 3)$ höchstens p^{m-1} ist.

Nun gilt

Satz 18. Eine modulare p-Gruppe $\mathfrak{G}$, die keine spezielle 2-Gruppe ist, hat folgende Struktur und umgekehrt:

(i) $\mathfrak{G}$ hat einen abelschen Normalteiler $\mathfrak{A}$,

(ii) $\mathfrak{G}/\mathfrak{A}$ ist zyklisch,

(iii) Für einen Vertreter T einer geeigneten Erzeugende von $\mathfrak{G}/\mathfrak{A}$ gilt $TAT^{-1}=A^{1+p^s}$,

wobei A beliebiges Element aus $\mathfrak{A}$, und s eine von A unabhängige natürliche Zahl ($s\geqq 2$ für $p=2$) ist.

Beweis. Um zu beweisen, dass eine Gruppe $\mathfrak{G}$ von oben angegebenen Struktur eine modulare p-Gruppe ist, wenden wir die Induktion nach der Ordnung von $\mathfrak{G}$ an. Da mit $\mathfrak{G}$ auch jede Untergruppe von $\mathfrak{G}$ offenbar die Eigenschaften (i), (ii), (iii) besitzt, so können wir annehmen, dass jede echte Untergruppe von $\mathfrak{G}$ modular ist. Wenn $\mathfrak{G}$ selbst nicht modular wäre, dann gäbe es nach Hilfssatz 17 einen Normalteiler $\mathfrak{N}$, so dass $\mathfrak{G}/\mathfrak{N}$ eine Gruppe vom Typus (11) oder (13) wäre. Das ist aber nicht möglich, da eine Gruppe vom Typus (11) oder (13) nicht die oben angegebene Struktur besitzt. $\mathfrak{G}$ muss also modular sein. Dass $\mathfrak{G}$ keine spezielle 2-Gruppe ist, ist auch klar.

Wir beweisen nun umgekehrt, dass eine modulare p-Gruppe $\mathfrak{G}$, die keine spezielle 2-Gruppe ist, die Eigenschaften (i), (ii), (iii) besitzt. Wir wenden wieder die Induktion nach der Ordnung von $\mathfrak{G}$ an. Die Bezeichnungen $\mu=\mu(\mathfrak{G})$, Ω_a, bzw. $\mho_{\mu-1}$ sollen dieselben Bedeutungen haben wie beim Beweis von Satz 16. Wenn ferner mit $\mathfrak{Z}=\{Z\}$ ein in $\mho_{\mu-1}$ enthaltener Normalteiler von $\mathfrak{G}$ mit der Ordnung p bezeichnet wird, dann ist $\mathfrak{G}/\mathfrak{Z}$ nach Induktionsvoraussetzung eine Gruppe mit den Eigenschaften (i), (ii), (iii). $\mathfrak{A}_1/\mathfrak{Z}$ sei also der in (i) angegebene abelsche Normalteiler von $\mathfrak{G}/\mathfrak{Z}$ und $\bar{T}$ sei die in (iii) angegebene Vertreter der Erzeugende von $\mathfrak{G}/\mathfrak{A}_1$. T sei ferner ein Vertreter der Restklasse von $\bar{T}$ mod. $\mathfrak{Z}$. Es ist dann offenbar entweder $\mathfrak{A}_1'=1$ oder $\mathfrak{A}_1'=\mathfrak{Z}$[26] und für ein beliebiges Element A von $\mathfrak{A}_1$ gilt die Gleichung

$$TAT^{-1}=A^{1+p^s}Z^k. \quad (k: \text{eine natürliche Zahl}).$$

Wir unterscheiden nun zwei Fälle.

(I) $\mathfrak{A}_1'=1$, d.h. $\mathfrak{A}_1$ abelsch.

(a) Sei $\mathfrak{Z}\frown\{T\}=1$. Wir wählen eine geeignete Basis $A_1, A_2, \ldots, A_r$ von $\mathfrak{A}_1$, so dass $\mathfrak{Z}$ in $\{A_1\}$ und $\mathfrak{A}_1\frown\{T\}$ in $\{A_2\}$ enthalten ist. Für $i\geqq 2$ folgt aus $\{T, A_i\}\frown\mathfrak{Z}=1$, dass $TA_iT^{-1}=A_i^{1+p^s}$ ist. Wenn die Ordnung von A_1 p^m ist, so ist $\mathfrak{Z}=\{A_1^{p^{m-1}}\}$, folglichen $TA_1T^{-1}=A_1^{1+p^s+hp^{m-1}}$, wobei h eine geeignete natürliche Zahl bedeutet. Da Z nach Definition $p^{\mu-1}$-te Potenz eines Elementes aus $\mathfrak{G}$ ist, so folgt hier-

26) Mit $\mathfrak{A}_1'$ bezeichnen wir die Kommutatorgruppe von $\mathfrak{A}_1$.

aus ohne Weiteres $m=\mu$. Wenn die Ordnung eines A_i ($i\geqq 3$), z. B. die von A_3, p^μ ist, so folgt aus

$$T(A_1A_3)T^{-1}=A_1^{1+p^s+hp^{m-1}}A_3^{1+p^s}=(A_1A_3)^x=A_1^xA_3^x\,,$$

dass $x\equiv 1+p^s$ mod. p^μ, folglich $h\equiv 0$ mod. p ist. $\mathfrak{G}$ hat also die im Satz angegebene Struktur. Wenn nun die Ordnung von A_2 gleich p^μ ist, so kann man statt T ein geeignetes Element $T^*=TA_2^q$ nehmen, so dass $\{A_2\}\frown\{T^*\}=1$ gilt, und dann folgt wieder aus $T^*(A_1A_2)T^*=(A_1A_2)^x$, dass $T^*A_1T^{*-1}=A_1^{1+p^s}$ ist. Wir nehmen also an, dass die Ordnung von A_i ($i\geqq 2$) sämtlich kleiner als p^μ ist. Falls $\mu-1\leqq s$ ist, so ist $\mathfrak{G}'=1$ oder $\mathfrak{G}'=\mathfrak{Z}$. Nach Hilfssatz 22 sind wir fertig. Wenn aber $\mu-1>s$ ist, so ersetzen wir T durch $T^*=T^{1-hp^{\mu-1-s}}$. Es gilt dann wieder $T^*A_iT^{*-1}=A_i^{1+p^s}$ $(i=1, 2, \ldots, r)$.

b) Sei nun $\mathfrak{Z}\in\{T\}$. Wir wählen eine Basis $A_1, \ldots, A_r$, so dass $\{T\}\frown\mathfrak{A}_1$ in $\{A_1\}$ enthalten ist. Die Ordnung von T sei p^n. Wenn die Ordnung eines A_i nicht kleiner als p^n ist, so kann man T durch $T^*=TA_i^q$ ersetzen und den Beweis auf dem Fall a) zurückführen. Wir nehmen daher an, dass $n=\mu$ ist, und die Ordnung von A_i $(i=1, 2, \ldots, r)$ sämtlich höchstens $p^{\mu-1}$ ist. Es ist nun leicht ersichtlich, dass man für $i\geqq 3$ $TA_iT^{-1}=A_i^{1+p^s}$ und für $i=2$ $TA_2T^{-1}=A_2^{1+p^s}$ oder $TA_2T^{-1}=A_2^{1+p^s}T^{p^{\mu-1}}$ ohne Beschränkung der Allgemeinheit annehmen kann.

b₁) Es sei $TA_1T^{-1}=A_1^{1+p^s+hp^{n-1}}$, $TA_iT^{-1}=A_i^{1+p^s}$ ($i\geqq 2$). Wenn man ein geeignetes a wählt und A_1 durch $A_1^*=A_1T^a$ ersetzt, dann wird $A_1^*, A_2, \ldots, A_r$ offenbar wieder abelsch und es gilt $TA_1^*T^{-1}=A_1^{*1+p^s}$, $TA_iT^{-1}=A_i^{1+p^s}$ ($i\geqq 2$).

b₂) Sei nun $TA_1T^{-1}=A_1^{1+p^s+hp^{m-1}}$, $TA_2T^{-1}=A_2^{1+p^s}T^{p^{\mu-1}}$, $TA_iT^{-1}=A_i^{1+p^s}$ ($i\geqq 3$). Wenn $s<\mu-1$ ist, so ersetzte man A_2 durch $A_2^*=A_2T^{p^{\mu-1-s}}$, und reduziere den Beweis auf dem Fall b_1). Wenn aber $s\geqq\mu-1$ ist, dann ist entweder $\mathfrak{G}'=1$ oder $\mathfrak{G}'=\mathfrak{Z}$. Nach Hilfssatz 22 sind wir auch dann fertig.

(II) $\mathfrak{A}_1'=\mathfrak{Z}$. Nach Hilfssatz 22 gilt

$$\mathfrak{A}_1=\{A_1, A_2\}\times\{A_3, A_4, \ldots, A_r\}\,,\quad A_1^{p^m}=1\,,\quad A_2A_1A_2^{-1}=A_1^{1+p^{m-1}}\,,$$

$$A_iA_j=A_jA_i\ \ (i, j\geqq 3).$$

Ausserdem kann man noch annehmen, dass $TA_1T^{-1}=A_1^{1+p^s}$, $TA_2T^{-1}=A^{1+p^s}$ ist, indem man evtl. T durch $T^*=TA_1^aA_2^\beta$ ersetzt. Z ist nun nach Definition die $p^{\mu-1}$-te Potenz eines Elementes von $\mathfrak{G}$. Sei also $Z=Z_0^{p^{\mu-1}}$, $Z_0=A_1^{e_1}A_2^{e_2}\dots A_r^{e_r}T^f$. Nach Hilfssatz 20 gilt dann

$$A_1^{p^{m-1}}=Z=A_1^{e_1p^{\mu-1}}A_2^{e_2p^{\mu-1}}T^{fp^{\mu-1}}.$$

Wenn also $A_1^{p^{m-1}}=1$, so muss $T^{p^{\mu-1}}\neq 1$ sein.

a) Sei zunächst $A_1^{p^{m-1}}\neq 1$. Es gilt dann offenbar $\mu=m$. Wenn $s\geqq\mu$ ist, so ist $\mathfrak{G}'=\mathfrak{Z}$ und sind wir fertig. Sei $s=\mu-1$. Wenn $A_2^{p^{\mu-1}}=1$ ist, dann ist wieder $\mathfrak{G}'=\mathfrak{Z}$. Es sei also $A^{p^{\mu-1}}\neq 1$. Nach einiger Rechnung folgt dann $TA_iT^{-1}=A_i$ $(i\geqq 3)$ und man kann den Beweis auf dem Fall (I) reduzieren, indem man A_2 durch $A_2^*=A_2T^{-1}$ ersetzt. Sei endlich $s<\mu-1$. Der Beweis wird wieder auf dem Fall (I) zurückgeführt, indem man A_2 durch $A_2^*=A_2T^{-p^{\mu-1-s}}$ ersetzt.

b) Es sei nun $A_1^{p^{\mu-1}}=1$, $(m<\mu)$, $T^{p^{\mu-1}}\neq 1$. Wenn der Index von $\{A_1, A_2\}$ in $\{T, A_1, A_3\}$ p^l ist, und wenn $T^{p^l}=A_2^{-hp^f}A^{-p^k}$ gesetzt wird, so ist nach Annahme $l>k$. Wir nehmen zunächst an, dass $k\geqq 1$ ist. Mann kann alsdann in $X=T^{p^{l-1}}A^{xp^{k-1}}A^{yp^{f-1}}$ bzw. $X^*=T^{p^{l-k}}A_1^z$ x, y bzw. $z\not\equiv 0 \pmod{p}$ so wählen, dass $X^p=1$, bzw. $X^{*p^{k-1}}=T^{p^{l-1}}A_1^{xp^{k-1}}$ gilt. Wenn die Ordnung von A_2 p^t ist, so bilden $A_2^{p^{t-1}}$, X eine Basis der Ω_1 entsprechende Untergruppe der modularen p-Gruppe $\mathfrak{H}=\{X^*, A_2\}$. Nun gilt $A_2X^*A_2^{-1}=A_2^uX^*A_1^{zp^{m-1}}$, folglich $A_1^{p^{m-1}}\in\{A_2^{p^{t-1}}, X_2\}$, was offenbar ein Widerspruch ist. Es ist also $k=0$ und $\mathfrak{G}=\{T, A_2, \dots, A_r\}$. Der Beweis wird nun wieder auf dem Fall (I) zurückgeführt, und damit sind alle mögliche Fälle erschöpft worden.

Durch diesen Satz, zusammen mit Satz 3 bzw. Satz 7, ist die Struktur der endlichen M-Gruppen bzw. quasi-Hamiltonschen Gruppen vollständig bestimmt worden.[27] Danach gilt z. B. der

Satz 19. Die endliche M-Gruppe ist meta-abelsch.

Mathematisches Institut,
Kaiserliche Universität zu Tokyo.

---

27) Auch für unendlichen M-Gruppen kann man den zum Satz 18 analogen Satz aufstellen. Insbesondere kann die Struktur der unendlichen quasi-Hamiltonschen Gruppen völlig bestimmt werden.

[4] On almost periodic functions

Isô Sûgaku, 4-2 (10/1942), 56-60.

Abstract

It is shown that some (discrete) groups appearing in topology are "maximally almost periodic" (m.a.p. for short).

1. Given a group G, the sequence defined by

$$(1) \qquad Z_n = [G, Z_{n-1}], \quad Z_1 = G \supset Z_2 \supset \ldots \supset Z_n \supset \ldots$$

is called the *descending central series* of G; in particular, when $Z_n = \{1\}$ for some n, G is called *nilpotent*. It is known (Magnus) that if G is a finitely generated free group, then the intersection of all Z_n's is equal to 1. The following theorem is proved:

Theorem 1. Let G be a finitely generated nilpotent group. Then the intersection of all normal subgroups N of G with finite index is equal to 1.

This implies that any finitely generated nilpotent group is m.a.p. (Theorem 2). It is remarked that without the assumption of finite generation Theorem 1 does not hold in general (e.g., for the additive group **Q**), but it is not known if every nilpotent group is m.a.p. or not.

2. Combining Theorem 1 with Magnus's theorem, one obtains

Theorem 3. For any free group F the intersection of all normal subgroups with finite index is equal to 1.

As corollaries to this theorem, one obtains the following

Theorem 4. Given a finite number of words $W_1(S_1,...,S_n)$, ..., $W_k(S_1,...,S_n)$ in S_1, ..., S_n, there exists a finite group G generated by S_1, ..., S_n such that $W_1 \neq 1$, ..., $W_k \neq 1$.
(According to Prof. Tan'naka, this result was already given by E. Artin.)

Theorem 5. Any free group is maximally almost periodic.

3. For an arbitrary group G one can find a free group F and a normal subgroup N such that one has $G \cong F/N$. In this notation one has

Theorem 6. Let $\bar{N}$ be the closure of N in the weak topology (w.r.t. almost periodic functions) of F and let

$$G^* = G/N^* \cong F/\bar{N}.$$

Then one has
(i) The totality of almost periodic functions on G coincides with that of G^*.
(ii) G^* is m.a.p.
(iii) G is m.a.p. if and only if $G = G^*$, i.e., if and only if N is closed in F.
(iv) G is minimally almost periodic if and only if $G^* = 1$, i.e., if and only if N is dense in F.

This follows from Theorem 5 and the lemmas (Lem. 2, 3) saying that, if H is m.a.p., A is a normal subgroup of H, and $\bar{A}$ is the closure of A in the weak topology of H, then the totality of almost periodic functions on H/A coincides with that of $H/\bar{A}$, and H/A is m.a.p. if and only if A is closed.

4. Now we consider a group G generated by $S_1, S_2, ..., S_n$ with the relations

$$(4) \qquad S_1^{e_1} = S_2^{e_2} = ... = S_n^{e_n} = 1, \quad e_i > 0, \quad i = 1, 2, ..., n.$$

If one denotes the commutator subgroup of G by G', then G/G' is an abelian group of order $e = e_1 e_2 ... e_n$. On the other hand, it can easily be proved by a method of Schreier that G' is a free group. It follows that the intersection of all subgroups with finite index in G is equal to 1. But, in general, this property of G is equivalent to saying that the intersection of all normal subgroups with finite index in G is equal to 1 (Lemma 4). Thus one has

Theorem 7. Let G be a group generated by $S_1, S_2, ..., S_n$ with the relation (4). Then the intersection of all normal subgroups of G with finite index is equal to 1. In particular, G is maximally almost periodic.
(The special case of $S_1^2 = 1$, $S_2^3 = 1$, for which the group G is the classical modular group, was given by J. von Neumann as an example of a m.a.p. group.)

Next we consider a group with the relations

$$(5) \qquad S_1^{e_1} = S_2^{e_2} = ... = S_n^{e_n}, \quad e_i > 0, \quad i = 1, 2, ..., n$$

with $n \geq 2$.
By Theorem 6, if N_1 and N_2 are normal subgroups of G with $N_1 \cap N_2 = 1$ and if both G/N_1 and G/N_2 are m.a.p., then so is also G (Lemma 5). Now let G be a group with the relation (5) and set $Z = S_1^{e_1} = S_2^{e_2} = ... = S_n^{e_n}$. Let G' be the commutator subgroup of G and put $G'\{Z\} = G_1$, $G' \cap \{Z\} = G_2$. Then, since $n \geq 2$, it can easily be shown that $[G_1 : G'] = [\{Z\} : G_2]$ is infinite, and so $G_2 = 1$. Since by Theorem 7 both $G/\{Z\}$ and G/G' are m.a.p., so is also G. This proves

Theorem 8. A group G generated by $S_1, S_2, ..., S_n$ with the relation (5) is maximally almost periodic.

5. Modifying V. L. Nisnewitsch's method (Rec. Math. **51** (1940)), one can show that a free product of any number of m.a.p. groups is also m.a.p. (Theorem 9). It follows that in Theorems 7, 8 the conditions $e_i > 0$ may be replaced by $e_i \geq 0$ and also the number of generators n can be infinite.
Combining the above results, one can show that there are many other groups which are m.a.p. However, according to Fuchs-Rabinowitsch (Comptes Rendus

(Doklady), **27**, **29** (1940)), the group with the relations

$$S_1S_2 = S_2S_1, \quad S_2^2 = 1, \quad S_3S_1 = S_1S_2S_3, \quad S_4S_1 = S_1^2S_4$$

is not m.a.p. It seems an interesting problem to consider under what conditions the groups of similar types are m.a.p.

概週期函數に就いて

東京帝大 岩 澤 健 吉

discrete な群の上に於ける概週期函數について，　特に位相幾何學によく現はれる二三の群が，所謂

"maximally almost periodic" であることを述べて見ようと思ひます．

1. 群 $\mathfrak{G}$ が與へられた時 $\mathfrak{Z}_n=[\mathfrak{G},\mathfrak{Z}_{n-1}]$(1) で定義される系列

$$\mathfrak{Z}_1=\mathfrak{G}\supset\mathfrak{Z}_2\supset\mathfrak{Z}_3\supset\cdots\supset\mathfrak{Z}_n\supset\cdots\cdots\quad(1)$$

を $\mathfrak{G}$ の降核心群列 (absteigende Zentralreihe) と呼び特に (1) が有限回で切れて單位群 1 に到達する時 $\mathfrak{G}$ を零巾群 (nilpotente Gruppe) と呼びます．定義から直ちに分る様に $\mathfrak{Z}_{i-1}/\mathfrak{Z}_i$ は $\mathfrak{G}/\mathfrak{Z}_i$ の核心に含まれ，從つて abel 群ですが，特に $\mathfrak{G}$ が有限個の生成元を持つならば $\mathfrak{Z}_{i-1}/\mathfrak{Z}_i$ も同樣になります．又 $\mathfrak{G}$ が有限個の生成元を持つ自由群であれば，すべての $\mathfrak{Z}_n$ の共通部分群 (Durchschnitt) は 1 であることが證明されてゐます (Magnus の定理)(2)．さて次の定理が成立します．

定理 1. 有限個の生成元を有する零巾群 $\mathfrak{G}$ に於ては

指數 $[\mathfrak{G}:\mathfrak{N}]$ が有限なる如き，凡ての不變部分群 $\mathfrak{N}$ の Durchschnitt は 1 である．

$$\underset{[\mathfrak{G}:\mathfrak{N}]=\text{fin}}{D}\mathfrak{N}=1,$$

この定理を證明する爲に先づ次の補助定理を證明しておきます．

補助定理 1. $\mathfrak{G}$ は任意の群，$\mathfrak{N}$ は有限個の生成元を有する $\mathfrak{G}$ の不變部分群で $\mathfrak{Z}=[\mathfrak{G}:\mathfrak{N}]$ は $\mathfrak{G}$ の核心に含まれる有限群とする．然らば $\mathfrak{N}$ は $[\mathfrak{N}:\mathfrak{M}]$ が有限となる如き $\mathfrak{G}$ の核心に含まれる部分群 $\mathfrak{M}$ を含む．且つ $\mathfrak{M}$ は有限個の生成元をもつとしてよい．

證明. $\mathfrak{Z}$ の位數を l とすれば，假定により $\mathfrak{Z}$ は $\mathfrak{G}$ の核心に含まれる故，任意の $g\in\mathfrak{G}$, $n\in\mathfrak{N}$ に對し

$$1=(g,n)^l=(g,n^l)\ (3)$$

卽ち n^l は $\mathfrak{G}$ の核心に屬する．よつて $\mathfrak{M}=\{n^l\}$ とおけばよい．$[\mathfrak{N}:\mathfrak{M}]$ が有限となることは $\mathfrak{N}/\mathfrak{Z}$ が有限個の生成元を有する abel 群なることから容易に分る．

定理 1 の證明. $\mathfrak{G}=\mathfrak{Z}_1\supset\mathfrak{Z}_2\supset\cdots\supset\mathfrak{Z}_n=1$

$$\mathscr{F}_i=\mathfrak{Z}_{i-1}/\mathfrak{Z}_i$$

とし F_i の内無限なるものを

$$\mathscr{F}_{i_1},\ \mathscr{F}_{i_2},\ \ldots,\ \mathscr{F}_{i_k},\qquad i_1<i_2<\cdots<i_k$$

とする．$k=0$ ならば $\mathfrak{G}$ は有限群で定理は明かですから，k に關して歸納法を用ひることにします．先づ $i_k\neq n$ とする．$\mathfrak{G}/\mathfrak{Z}_{i_k+1}=\bar{\mathfrak{G}}$,

$$\mathfrak{Z}_{i_k-1}/\mathfrak{Z}_{i_k+1}=\bar{\mathfrak{N}}\quad \bar{\mathfrak{Z}}=[\bar{\mathfrak{G}},\bar{\mathfrak{N}}]=\mathfrak{Z}_{i_k}/\mathfrak{Z}_{i_k+1}$$

に補助定理を適用すれば $[\mathfrak{Z}_{i_k-1}:\mathfrak{M}]$ が有限で且つ $\bar{\mathfrak{M}}=\mathfrak{M}/\mathfrak{Z}_{i_k+1}$ が $\bar{\mathfrak{G}}$ の核心に屬する樣な群 $\mathfrak{M}$ が存在する．次に $\mathfrak{Y}/\mathfrak{Z}_{i_k+2}=\bar{\bar{\mathfrak{Y}}}$, $\mathfrak{M}/\mathfrak{Z}_{i_k+2}=\bar{\bar{\mathfrak{M}}}$ とすれば $\bar{\mathfrak{M}}$ が $\bar{\mathfrak{G}}$ の核心に屬することから $[\bar{\bar{\mathfrak{G}}},\bar{\bar{\mathfrak{M}}}]=\bar{\bar{\mathfrak{Z}}}$ は $\mathfrak{Z}_{i_k+1}/\mathfrak{Z}_{i_k+2}$ に含まれる．再び補助定理 1 により $[\mathfrak{M}:\mathfrak{M}^*]$ が有限で，且つ $\bar{\bar{\mathfrak{M}}}^*=\mathfrak{M}^*/\mathfrak{Z}_{i_k+2}$ が $\bar{\bar{\mathfrak{G}}}$ の核心に屬する樣な部分の $\mathfrak{M}^*$ が存在する．よつて同樣のことを繰返せば，結局 $[\mathfrak{Z}_{i_k-1}:\mathfrak{K}]$ が有限で且つ $\mathfrak{G}$ の核心に含まれる樣な部分群 $\mathfrak{K}$ が存在する．$i_k=n$ の時は始めから $\mathfrak{K}=\mathfrak{Z}_{i_k-1}$ とすればよい．

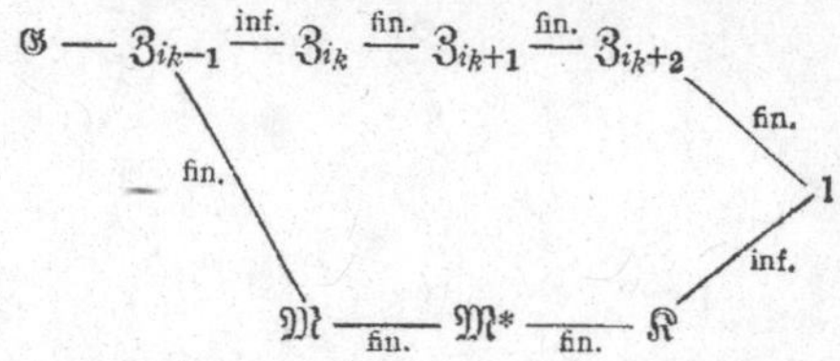

さて $[\mathfrak{K}:\mathfrak{K}^*]$ が有限なる如き任意の部分群 $\mathfrak{K}^*$ をとれば $\mathfrak{G}/\mathfrak{K}^*$ に對しては歸納法の假定により定理が成立してゐるから

$$\underset{\substack{\mathfrak{N}\geqq\mathfrak{K}^*\\ [\mathfrak{G}:\mathfrak{N}]=\text{fin.}}}{D}\mathfrak{N}=\mathfrak{K}^*\qquad(2)$$

又一方 $\mathfrak{K}$ は有限個の生成元を有する abel 群であるから Basis をとつて考へれば容易に知られる樣に

$$D\,\mathfrak{K}^*=1\qquad(3)$$

(2) (3) から

$$\underset{[\mathfrak{G}:\mathfrak{N}]=\text{fin.}}{D}\mathfrak{N}=1$$

を得て定理は證明された．

定理 1 から直ちに次の定理が得られます．

定理 2. 有限個の生成元を有する零巾群は maximally almost periodic である．

定理 1 に於ては $\mathfrak{G}$ が有限個の生成元を有することが essential なのであります．例へば有理數全體の加法群をとれば，それは abel 群，從つて零巾ですが指數有限の部分群は trivial なものを除けば一つも存在しません．然し **定理 2** ではどうでせうか．**有限個の生成元を持つ**といふ條件がなくても，零巾群は maximally a.p. なのではないでせうか．これに關して問題は澤山ある樣ですが未だよく分りません．御教示を得たいと思ひます．

(1) $[\mathfrak{G},\mathfrak{Z}_{n-1}]$ は $\mathfrak{G}$ と $\mathfrak{Z}_{n-1}$ との交換子群．

(2) 之等の事については H. Zassenhaus の教科書又は E. Witt; Treue Darstellung Liescher Ringe, Crelle J. **177** (1937), S. 153–160 を參照されたい．

(3) $(a,b)=aba^{-1}b^{-1}$ とする．上記計算は (g,n) が $\mathfrak{G}$ の核心に屬することから出る．

次に自由群を考察することにします．$\mathfrak{G}$ を有限個の生成元をもつ自由群とし $\mathfrak{Z}_n$ は始めに定義した通りとします．Magnus の定理により $\underset{n}{D}\,\mathfrak{Z}_n=1$, 且つ $\mathfrak{G}/\mathfrak{Z}_n$ に對しては**定理 1** が成立してゐるのですからこれから $\mathfrak{G}$ に對しても，**定理 1** と同樣なことが成立することが分ります．

次に $\mathcal{F}$ を任意個の生成元 $\{S_i\}$ を有する自由群とし

$$A=S_{i_1}^{\pm1}\cdots\cdots S_{i_k}^{\pm1}\neq 1$$

を $\mathcal{F}$ の任意の要素，又 $S_{i_1},\cdots,S_{i_k}$ から生成される群を $\mathfrak{G}$ とかくことにします．さて

$$S_i\to 1\qquad i\neq i_j,\qquad j=1,2,\cdots k$$

なる準同型寫像により，單位元に移る F の不變部分群を $\mathfrak{N}$ とすれば

$$\mathfrak{N}\cdot\mathfrak{G}=\mathcal{F},\qquad \mathfrak{N}\cap\mathfrak{G}=1,\qquad \mathcal{F}/\mathfrak{N}\cong\mathfrak{G}.$$

$\mathfrak{G}$ については**定理 1** が成立してゐるのですから，適當に $\mathfrak{N}$ を含む有限指數の $\mathcal{F}$ の不變部分群 $\mathfrak{N}^*$ をとれば，$\mathfrak{N}^*\notin A$ となります．$A\neq1$ は $\mathcal{F}$ の任意の要素ですから，これから

$$\underset{[\mathcal{F}:\mathfrak{N}^*]=\text{fin.}}{D}\mathfrak{N}^*=1$$

即ち

定理 3. $\mathcal{F}$ を任意の自由群とすれば，$\mathcal{F}$ に於て有限指數を有する凡ての不變部分群の Durchschnitt は 1 である．又この系として

定理 4(4). $S_1,\cdots\cdots,S_n$ に關する有限個のWort $W_1(S_1,\cdots,S_n),\cdots,W_k(S_1,\cdots,S_n)$ が與へられた時 $S_1,\cdots,S_n$ から生成され，且つ $W_1\neq1,\cdots,W_k\neq1$ となる如き有限群 $\mathfrak{G}$ が常に存在する．

定理 5. 任意の自由群は maximally almost periodic である．

3. 任意の群 $\mathfrak{G}$ は適當に自由群 $\mathcal{F}$ 及びその不變部分群 $\mathfrak{N}$ をとれば

$$\mathfrak{G}\simeq\mathcal{F}/\mathfrak{N}$$

となる(5)．よつて上の結果を用ひて $\mathfrak{G}$ 上の概週期函數に對して多少考察を試みて見ようと思ひます．

補助定理 2. $\mathfrak{H}$ を maximally a.p. なる任意の群とし $\mathfrak{A}$ を $\mathfrak{H}$ の不變部分群とすれば，$\mathfrak{H}/\mathfrak{A}$ が maximally a.p. なる爲に必要且つ十分なる條件は $\mathfrak{A}$ が $\mathfrak{H}$ に於ける weak topology で閉ぢてゐる事である．

證明(6). 必要であること： $\mathfrak{H}/\mathfrak{A}$ の a.p.f. は又 $\mathfrak{H}$ の a.p.f. と考へられる．假定により $\mathfrak{H}/\mathfrak{A}$ の凡ての a.p.f. $f(x)$ に對し $f(a)=f(1)$ となる a の全體が丁度 $\mathfrak{A}$ であるから $\mathfrak{A}$ は閉ぢてゐる．逆に $\mathfrak{A}$ が $\mathfrak{H}$ の閉ぢた部分群とせよ．$x_0\notin\mathfrak{A}$ とすれば $\mathfrak{H}$ は completely regular で，$x_0\mathfrak{A}$ で 1 となり 1 に於て 0, その他で $0\leqq\varphi(x)\leqq1$ となる樣な $\mathfrak{H}$ 上の uniformly continnous f. $\varphi(x)$ が存在する．$\varphi(x)$ は $\mathfrak{H}$ 上の a.p.f. であるが

$$\varphi_0(x)=\underset{a\in\mathfrak{A}}{M}\varphi(xa)$$

とすれば $\varphi_0(x)$ も亦 a.p.f. である．(但し $\varphi(xa)$ は $\mathfrak{H}\times\mathfrak{A}$ 上の a.p.f. で上記平均は a に關し $\mathfrak{A}$ 上でとるのである)．然も $\varphi_0(x)$ は $\mathfrak{H}/\mathfrak{A}$ の各 Restklasse に對し，一定で，そのつくり方から分る樣に $\varphi_0(x_0)=1,\ \varphi_0(1)\neq1$.

即ち $\qquad \varphi_0(x_0\mathfrak{A})\neq\varphi_0(\mathfrak{A})$.

これで $\mathfrak{H}/\mathfrak{A}$ が maximally a.p. なる事が分つた(7)．

補助定理 3. $\mathfrak{H}$ を maximally a.p. なる群，$\mathfrak{A}$ を $\mathfrak{H}$ の任意の部分群，$\bar{\mathfrak{A}}$ を $\mathfrak{H}$ に於ける weak topology に對する $\mathfrak{A}$ の closure とする．然らば $\mathfrak{H}/\mathfrak{A}$ 上の a.p.f. の全體は $\mathfrak{H}/\bar{\mathfrak{A}}$ 上の a.p.f. の全體と一致する．

證明. $\mathfrak{H}/\bar{\mathfrak{A}}$ 上の a.p.f. が又 $\mathfrak{H}/\mathfrak{A}$ 上の a.p.f. であることは明か．逆に $\mathfrak{H}/\mathfrak{A}$ 上の a.p.f. $f(x)$ をとる．x_0 を一つ定めておけば $f(x_0a)=f(x_0)$ となる如き a の全體は閉集合を作るが，それは $\mathfrak{A}$ を含む故，又 $\bar{\mathfrak{A}}$ を含む．即ち $f(x_0\bar{\mathfrak{A}})$ は一定で $f(x)$ は $\mathfrak{H}/\bar{\mathfrak{A}}$ 上の a.p.f. である．

さて任意の群 $\mathfrak{G}$ に對して $\mathfrak{G}\simeq\mathcal{F}/\mathfrak{N}$ とすれば，**定理 5** 及び上記**補助定理**から直ちに次の定理が得られる．

定理 6. $\mathfrak{G}\simeq\mathcal{F}/\mathfrak{N}$ とし $\mathcal{F}$ の a.p.f. による weak topology に對する $\mathfrak{N}$ の closure を $\bar{\mathfrak{N}}$ とする．又 $\mathfrak{G}^*=\mathfrak{G}/\mathfrak{N}\simeq\mathcal{F}/\bar{\mathfrak{N}}$ とすれば

(i) $\mathfrak{G}$ 上の a.p.f. の全體は $\mathfrak{G}^*=\mathfrak{G}/\mathfrak{N}^*$ 上の a.p.f. の全體と一致する．

(ii) $\mathfrak{G}^*$ は maximally almost periodic である．

(iii) 從つて $\mathfrak{G}$ が maximally almost periodic なる爲に必要且つ十分なる條件は $\mathfrak{G}=\mathfrak{G}^*$ なる事即ち $\mathfrak{N}$ が $\mathcal{F}$ に於て閉ぢてゐることである．

(iv) 又 $\mathfrak{G}$ が minimally almost periodic なる爲に必要且つ十分なる條件は $\mathfrak{G}^*=1$ なること，即ち $\mathfrak{N}$ が $\mathcal{F}$ に於て稠密なることである．

4. 次に $S_1,S_2,\cdots,S_n$ から生成され

(4) 淡中氏の御注意によれば，これは既に E. Artin により與へられた結果であるさうです．

(5) この $\mathfrak{Y}$, $\mathfrak{N}$ は勿論前§の $\mathfrak{Y}$, $\mathfrak{N}$ とは意味が異る．

(6) この樣な事けよく知られてゐる事なのかも知れませんが一應證明しておきます．

(7) この定理は $\mathfrak{U}$ が $\mathfrak{H}$ の不變部分群でなくとも成立する．

$S_1^{e_1}=1,\ S_2^{e_2}=1, \cdots, S_n^{e_n}=1,$

$$e_i>0,\ i=1,2,\cdots,n, \cdots (4)$$

なる關係式によつて與へられる群 $\mathfrak{G}$ を考へます．$\mathfrak{G}$ の交換子群を $\mathfrak{G}'$ とすれば $\mathfrak{G}/\mathfrak{G}'$ は次數 $e=e_1e_2\cdots e_n$ なる abel 群となりますが，一方 $\mathfrak{G}'$ は自由群であることが Schreier の方法により容易に計算されます．從つて**定理3**によれば，$[\mathfrak{G}':\mathfrak{N}]=$有限なる如き $\mathfrak{G}'$ の不變部分群 $\mathfrak{N}$ の Durchschnitt は1となります．$[\mathfrak{G}:\mathfrak{N}]=e[\mathfrak{G}':\mathfrak{N}]$ なる故，$[\mathfrak{G}:\mathfrak{N}]=$有限なる如き $\mathfrak{G}$ の**部分群**の全體をとれば，その Durchschnitt も亦1となりますが，一般に次のことが成立します．

補助定理 4. 群 $\mathfrak{G}$ に於て有限なる指數を有する凡ての部分群の Durchschnitt が1ならば，有限なる指數を有する凡ての**不變部分群**の Durchschnitt も亦1である．

證明. $\mathfrak{U}$ を指數有限なる $\mathfrak{G}$ の任意の部分群とする時，$\mathfrak{U}$ に含まれて矢張り指數が有限である樣な $\mathfrak{G}$ の不變部分群 $\mathfrak{N}$ が存在する事を云へばよいわけですが $[\mathfrak{G}:\mathfrak{U}]=$有限 なる故 $\mathfrak{U}$ の共軛部分群は有限個で，その Durchschnitt を $\mathfrak{N}$ とおけば明かに條件を滿足します．

この**補助定理**により直ちに次の定理が得られます．

定理 7. $S_1, S_2, \cdots, S_n$ から生成され，關係式

$S_1^{e_1}=1,\ S_2^{e_2}=1, \cdots, S_n^{e_n}=1,$

$$e_i>0,\ i=1,2,\cdots,n$$

によつて定義される群を $\mathfrak{G}$ とすれば $\mathfrak{G}$ に於て有限指數を有する凡ての不變部分群の Durchschnitt は1である．從つて $\mathfrak{G}$ は勿論 maximally almost periodic である．

$S_1^2=1,\ S_2^3=1$ なる場合には $\mathfrak{G}$ は普通の modular group となり，それは J. v. Neumann によつて與へられた例の一つであります[(8)]．今度は $n\leqq 2$ とし

$S_1^{e_1}=S_2^{e_2}=\cdots=S_n^{e_n},\ e_i>0,\ i=1,2,\cdots,n \cdots (5)$

なる關係式を持つ群を考察することにしますが，矢張り初めに一般的な定理を一つ述べておきます．

補助定理 5. $\mathfrak{N}_1, \mathfrak{N}_2$ を任意の群 $\mathfrak{G}$ の不變部分群とし，且つ $\mathfrak{N}_1\cap\mathfrak{N}_2=1$ とする．この時もしも $\mathfrak{G}/\mathfrak{N}_1$, $\mathfrak{G}/\mathfrak{N}_2$ が共に maximally almost periodic ならば $\mathfrak{G}$ 自身も亦同樣である．

證明. $\mathcal{F}$ を自由群

$$\mathfrak{G}=\mathcal{F}/\mathfrak{M},\quad \mathfrak{N}_1\simeq\mathfrak{M}_1/\mathfrak{M},\quad \mathfrak{N}_2\simeq\mathfrak{M}_2/\mathfrak{M}$$

とおけば假定により $\mathfrak{M}_1\cap\mathfrak{M}_2=\mathfrak{M}$ で且つ**定理6**によれば，$\mathfrak{M}_1, \mathfrak{M}_2$ は $\mathcal{F}$ の weak topology で閉ぢてゐる．よつて $\mathfrak{M}_1\cap\mathfrak{M}_2=\mathfrak{M}$ も亦閉集合で，再び**定理6**により $\mathcal{F}/\mathfrak{M}$ は maximally almost periodic である．

さて (5) なる關係式を持つ群を $\mathfrak{G}$ とすれば

$$Z=S_1^{e_1}=\cdots\cdots=S_n^{e_n}$$

は $\mathfrak{G}$ の核心に屬し $\mathfrak{G}/\{Z\}$ は關係式 (4) によつて與へられる群となります．$\mathfrak{G}$ の交換子群を $\mathfrak{G}'$, $\mathfrak{G}'\cup\{Z\}=\mathfrak{G}_1$, $\mathfrak{G}'\cap\{Z\}=\mathfrak{G}_2$ とすれば，$n\geqq 2$ なる故，容易に分る樣に指數 $[\mathfrak{G}_1:\mathfrak{G}']=[\{z\}:\mathfrak{G}_2]$ は無限，從つて $\mathfrak{G}_2=1$, 即ち $\mathfrak{G}'\cap\{Z\}=1$ でなければなりません．**定理7**により $\mathfrak{G}/\{Z\}$ は maximally almost priodic, 又 $\mathfrak{G}/\mathfrak{G}'$ は abel 群ですから勿論 maximally almost periodic で，**補助定理5**を用ひれば，これから $\mathfrak{G}$ 自身 maximally almost periodic なることが分ります．

定理 8. $S_1, S_2, \cdots, S_n$ から生成され

$S_1^{e_1}=S_2^{e_1}=\cdots=S_n^{e_n},\ e_i>0\quad i=1,2,\cdots,n$

によつて定義された群 $\mathfrak{G}$ は maximally almost periodic である．

上の**定理 7, 8** に於る條件 $e_i>0$ を條件 $e_i\geqq 0$ でおきかへてもよいことは，次の§の結果を用ひれば，直ちに證明されます．

5. 今迄主として群論的考察によつて**定理 5, 7**等を證明して來ましたが單にそれらの群が maximally almost periodic である事を云ふのでしたら，直接に次の樣にも證明されます．實際 V. L. Nisnewitsch[(9)] は任意個の群 $\mathfrak{G}_\nu$ が同一の可換體 $\mathfrak{K}$ の上の同じ次數の matrix で isomorph に表現されてゐれば，それらの free product も亦適當な可換體の上の matrix により isomorph に表現出來ることを證明しましたが，その證明を多少變形すれば，次の定理が證明されます[(10)]．

定理 9. Maximally almost periodic な群の任意個の free product は又 maximally almost periodic である．

これによれば，又前§の最後に述べた注意は明かであり，一方**定理 7, 8** に於て生成元 S_i の數が必ずしも有限でなくともよいといふ事も分ります．

以上述べました諸定理を組合はせて用ひれば，倘

(8) J. v. Neumann and E. P. Wigner, Minimally almost periodic groups, Ann. Math. **41** (1940).

(9) V. L. Nisnewitsch, Über Gruppen, die durch Matrizen über einem kommutativen Feld isomorph darstellbar sind, Rec. Math. **51** (1940).

(10) 證明は稍々長くなりますから省略します．

多くの群が maximally almost periodic であること が證明されます．然し比較的簡單な關係式によつて與へられる群であつて，しかも maximally almost periodic でないものも存在します．例へば Fuchs-Rabinowitsch(11) によれば

$S_1 S_2 = S_2 S_1$, $S_2^2 = 1$, $S_3 S_1 = S_1 S_2 S_3$,

$S_4 S_1 = S_1^2 S_4$

なる關係式で與へられる群は maximally almost periodic になりません．

一般にどの様な條件があれば，上の如き種類の群が maximally almost periodic になるかと云ふことは興味ある問題であると思ひます．（終）

(11) Fuchs-Rabinowitsch, Über eine Gruppe mit endlichvielen Erzeugenden und Relationen, die keine isomorphe Darstellung durch Matrizen von endlicher Ordnung zulässt, Comptes Rendus (Doklady) **27** (1940), Beispiel einer diskreten Gruppe mit endichvielen Erzeugenden und Relationen, die keine vollständiges System der isomorphen Darstellungen zulässt, Comptes Rendus (Doklady), **29** (1940).

---

*[4] 正誤表 (編者)

P.57, 左 ℓ.24: $[\mathfrak{G}:\mathfrak{N}] \to [\mathfrak{G},\mathfrak{N}]$; ℓ.1 ff.: $F_i \to \mathfrak{F}_i$

右 ℓ.8: $\mathfrak{Y}/\mathfrak{Z}_{i_k+2} = \overline{\overline{\mathfrak{Y}}} \to \mathfrak{G}/\mathfrak{Z}_{i_k+2} = \overline{\overline{\mathfrak{G}}}$

脚註 (2): 数科書 → 教科書

P.58, 右 ℓ.21: (二つ目の) $\mathfrak{H}/\bar{\mathfrak{A}} \to \mathfrak{H}/\mathfrak{A}$; ℓ.12 ff.: $\mathfrak{N} \to \overline{\mathfrak{N}}$

右 ℓ.11 ff.: $\mathfrak{G}/\mathfrak{N} \cong \mathfrak{F}/\mathfrak{N} \to \mathfrak{G}/\mathfrak{N}^* \cong \mathfrak{F}/\overline{\mathfrak{N}}$

脚註 (5): $\mathfrak{Y} \to \mathfrak{G}$, $\mathfrak{Y} \to \mathfrak{G}$; 脚註 (7): $\mathfrak{U} \to \mathfrak{A}$

P.59, 左 ℓ.10 ff.: $n \leq 2 \to n \geq 2$; 右 ℓ.11: $\{z\} \to \{Z\}$

[5] On normed rings and a theorem of Segal, I

Zenkoku Shijo Sugaku Danwakai, 246 (12/1942), 1522-1555

Abstract (by Y. Kawahigashi)

We first note that a summary of [5] and [10] is also given in [11] (1944).

I. E. Segal [S] has defined the group ring $R(G)(= R^{(1,1)}(G)$ in our notation) for a locally compact separable group G and studied its various interesting properties, particularly when the group G is compact or abelian. Replacing $R(G)$ by $L(G) = L^{(1,p)}(G)$ to be defined below, we study relationships between representations of G and those of $L(G)$ (and $R(G)$), and generalize Segal's theorem ([S], Th.3). Though the details of Segal's proof are not available to us, our methods are probably similar to his.

We use the notion of a normed ring in the sense of Gelfand [G] which has been used for studies of abelian groups by Gelfand-Raikov. In the first chapter of [5] we study a normed ring R satisfying the following assumption:

(A) For any left [resp. right] maximal ideal I_l [resp. I_r] of R, the quotient R/I_l [resp. R/I_r] is always finite dimensional.

We obtain some results (Theorem 1-8) on the ideal structure of such a normed ring R, which are useful in the second chapter. For instance, we show (Theorem 8) that, if R satisfies the condition (A), then the "radical" of R (i.e., the intersection of all maximal two-sided ideals of R) is the largest left (or right, hence two-sided) ideal contained in the set of all "generalized nilpotent elements" in R (i.e., those elements x in R with $\lim \sqrt[n]{\|x^n\|} = 0$).

Now, let G be a locally compact separable group and μ a right invariant Haar measure of G. We study $L^{(1,p)}(G) = L^1(G) \cap L^p(G)$ for $p \geq 1$. With the convolution product it has a ring structure. We denote by $R^{(1,p)}(G)$ the ring obtained from $L^{(1,p)}(G)$ by adjoining a multiplicative unit. For representations by finite dimensional matrices, we obtain the following main theorem (Theorem 9).

Theorem. We have a bijective correspondence between continuous representations of $L^{(1,p)}(G)$ and bounded continuous representations of G. That is, for a given continuous representation $x(g) \mapsto T(x)$ of $L^{(1,p)}(G)$, we have a unique bounded continuous representation $a \mapsto D_T(a)$ of G such that we have

$$T(x) = \int_G x(g) D_T(g) dg$$

for any $x(g) \in L^{(1,p)}(G)$. Moreover, for a given bounded continuous representation $a \mapsto D(a)$ of G, the above integral formula gives a continuous representation $x(g) \mapsto T_D(x)$ of $L^{(1,p)}(G)$. These constructions are the inverse to each other and preserve the equivalence relation of representations.

As corollaries (of the proof), we obtain the following two results (Theorems 10, 11).

Corollary. Any bounded measurable representation of G is continuous.

Corollary. Any continuous representation of $L^{(1,p)}(G)$ is completely reducible, and conversely, any completely reducible representation of $L^{(1,p)}(G)$ is continuous.

We have similar results for $R^{(1,p)}(G)$ (Theorems 15, 16), which imply a generalization of Segal's theorem (Theorem 18). As an application of the above main result, we also have the following theorem (Theorem 19).

Theorem. If G is not compact, then we have no non-zero representation of G belonging to $L^p(G)$, $p \geq 1$. If G is compact, then any representation belonging to $L^p(G)$, $p \geq 1$, is bounded and continuous.

References

[G] I. M. Gelfand, Normierte Ringe, Rec. Math. **51** (1941), 3-24.
[S] I. E. Segal, The group ring of a locally compact group, I, Proc. Nat. Acad. Sci., U.S.A., **27** (1941), 348-352.

1088. ノルム環トSegalノ定理ニツイテ I

I. E. Segal ハサキニ任意の locally compact, separable ナ群 G ニ對シ group ring $R(G)$ ヲ定義シ，特ニ G ガ abel 群又ハ compact 群ナル場合ニ種々ノ興味アル結果ヲ得マシタ. [1] ソノ compact 群ニ関スル主ナ結果ハ深宮氏ニヨッテ簡単ナ証明ガ與ヘラレマシタ. [2]

ココデハ G ガ一般ニ locally compact ナル場合ヲ考へ，特ニ G ノ表現ト $R(G)$ ノ表現トノ関係ヲ述ベテ見ヨウト思ヒマス．ソレハ Segal ノ論文ニ於ケル定理 3 ノ拡張ニ當ルワケデスガ我々ハ $R(G)$ ヨリモムシロ $L(G)$ ヲ目標ニシテ進ミマシタ（後述）．然シ Segal ノ証明ハ知ル由モアリマセンガ恐ラク以下述ベルモノト大同小異ナモノデアリマセウカラ，以下ノ結果モ亦既知ノモノト思ハレマス．

I. ノルム環ニ関スル注意

§ 1．我々ハ Gelfand ノノルム環 [3] ヲ用ヒテ Gelfand，Raikov [4] ガ abel 群ノ場合ニ用ヒタ方法ニ従フコトニシマスノデ，始メニ一般ノ非可換ナノルム環ニツイテ若干ノ簡単ナ注意ヲ述ベルコトニシマス．勿論非可換ノルム環ノ一般論ト云フヤウナモノデハアリマセン．

先ヅ一般ノノルム環 R ヲ Gelfand, N. R. § 1 ニヨリ定義シマス．但シコノ際乗法ノ可換性ハ假定シマセン．可換性ハナクトモ N. R. ニ於ケル代数的ナ定理ノ多クハ成立スルコトガ容易ニ確カメラレマス．唯逆元 x^{-1} ト云フトキ左逆元ト右逆元トヲ区別セネバナラヌコト，及ビ Ideal ニツイテモ左-Ideal，右-Ideal，両側-Ideal ノ別ヲ考ヘネバナラヌコト等ヲ注意スレバヨイワケデ，ソノヤウナ点ヲ考慮ニ入レサヘスレバ N. R. ノ§1－§5ノ定理ハ凡テソノママ成立シマス．

特ニ I_ℓ ヲ任意ノ閉ヂタ左-Ideal トスレバ R/I_ℓ ハ一般ニ環デハアリマセンガ兎モ角 Banach 空間デ R ノ各元 x ハソノ（左カラノ） Operator ト考ヘラレ $x \in R$, $Y \in R/I_\ell$ ニ對シ $\|xY\| \leq \|x\|\|Y\|$ トナリマス．(N. R., Hilfssatz 4) 然シ可換ナ場合ト本質的ニ異ナル点ハ單純ノルム環ガ必ズシモ複素数体（或ハソノ上ノ有限次ノ Algebra）トナラヌコトデ，コノタメ R ヲ maximal Ideal デ割ッタ R/I ノ様子ガワカラヌコトデス．

§2．ヨッテ我々ハ先ヅ一番簡單ナ有限次元ノノルム環ヲ調ベテ見ルコトニシマス．
複素数体 K 上ノ n 次ノ全行列環ヲ K_n トカクコトニシ

$$x \in K_n, \quad x = \sum_{i,j} \alpha_{i,j} e_{i,j}, \quad \alpha_{i,j} \in K, \quad e_{i,j}\text{ハ行列ノ單位}$$

ニ對シテ

$$\|x\|^* = \left(\sum_{i,j} |\alpha_{i,j}|^2\right)^{\frac{1}{2}} \tag{1}$$

トオケバコノ norm ニヨリ K_n ガノルム環トナルコトハ明カデス．(但シ $\|e\|^* = 1$ ニハナッテヰナイ.) ヨッテ K_n ノ部分環ハ矢張リスベテ上ノ norm ニヨリノルム環トナリマスガ実ハ有限次元ノノルム環ハコレ以外ニハ存在シナイコトガ証明サレマス．ソノ爲ニ先ヅ

補助定理 1.[5] M ヲ一ツノ有限 K-加群トシソノ Basis ヲ $u_1, u_2, \ldots, u_n$ トスル．B ハ Banach 空間デ K-加群トシテ M ニ含マレテヰルモノトスレバ，B ニ於ケル norm $\|x\|$ ハ M ニ於テ定義サレル norm

$$\|x\|^* = \|\alpha_1 u_1 + \cdots + \alpha_n u_n\|^* = (|\alpha_1|^2 + \cdots + |\alpha_n|^2)^{\frac{1}{2}}, \quad \alpha_i \in K$$

ト等值デアル．即チ適當ニ $C_1, C_2 > 0$ ヲトレバ B ニ属スル任意ノ x ニ対シテ

$$C_1\|x\|^* \leq \|x\| \leq C_2\|x\|^*$$

トナル．

証明. $B = M$ ノトキハ

$$\|x\| \leq |\alpha_1|\,\|u_1\| + \cdots + |\alpha_n|\,\|u_n\|$$

$$\leq \left(\sum_{i=1}^n |\alpha_i|^2\right)^{\frac{1}{2}} \left(\sum_{i=1}^n \|u_i\|^2\right)^{\frac{1}{2}} = C_2\|x\|^*$$

ナル故 $\|x\|^*$ カラ $\|x\|$ ヘノ寫像ハ連続トナリコレハ明白デス．

一般ニハ B ノ K-Basis $v_1, v_2, \ldots, v_m$ ヲトリコレヲ補充シテ M ノ Basis $v_1, v_2, \ldots, v_m, w_1, \ldots, w_{n-m}$ ヲツクリマス．

$$x \in M,\ x = \alpha_1 u_1 + \cdots + \alpha_n u_n = \beta_1 v_1 + \cdots + \beta_m v_m + \beta_{m+1} w_1 + \cdots + \beta_n w_{n-m}$$

ナルトキ

$$\|x\|^{**} = (|\beta_1|^2 + \cdots + |\beta_m|^2 + |\beta_{m+1}|^2 + \cdots + |\beta_n|^2)^{\frac{1}{2}}$$

トオケバ $B = M$ ナルトキノ結果ニヨリ $\|x\|^*$ ト $\|x\|^{**}$ トハ等値．特ニ x ガ B ニ含マレテヰレバ $\beta_{m+1} = \cdots = \beta_n = 0$ デ $\|x\|^{**}$ ハ B ニ於ケル一ツノ norm ヲ与ヘマスガ，再ビ $B = M$ ナルトキノ結果ヲ用ヒレバコレハ B ニ於ケル $\|x\|$ ト等値．ヨッテ結局 norm $\|x\|$ ハ B 内ニ induce サレタ norm $\|x\|^*$ ト等値ニナリ定理ハ証明サレマシタ．

サテ任意ノ有限次元ノノルム環ハ環トシテハ勿論行列環ノ部分環トナリマスカラ上ノ補助定理ニヨリ直チニ次ノ定理ガ得ラレマス．

定理 1. 有限次元ノノルム環ハ (1) ナル norm ニ関スル全行列環 K_n ノアル部分環ト一致スル．

コノ定理ニヨリ有限次元ノノルム環ノ様子ハワカッタワケデスガ，尚後デ用ヒル定理ヲココデ証明シテオキマス．

定理 2. $R = K_n$ ニ於テ $\sqrt[n]{\|x^n\|} \to 0$ $(n \to \infty)$ ナル為メニ必要且ツ十分ナル條件ハ零巾元ナルコトデアル．

証明. 十分ノ方ハ明白デス．ヨッテ x ガ零巾デナイトスレバ適当ニ y ヲトリ標準型

$$z = yxy^{-1} = \begin{pmatrix} \rho_1 & & * \\ & \ddots & \\ 0 & & \rho_n \end{pmatrix}$$

ヲツクッタトキ少クトモ一ツノ ρ_i ハ 0 デハアリマセン．コレヲ $\rho_k \neq 0$ トスレバ $\|z^n\|^* \geq |\rho_k|^n$．ヨッテ $\underline{\lim}\sqrt[n]{\|z^n\|^*} \geq |\rho_k| > 0$. 即チ $\lim \sqrt[n]{\|z^n\|^*} \neq 0$．任意ノ norm ハ $\|x\|^*$ ト等値ナル故 $\lim \sqrt[n]{\|z^n\|} \neq 0$．ヨッテ $\lim \sqrt[n]{\|x^n\|} \neq 0$. (norm ガ規準化サレテアレバ $\|z^n\| = \|yx^ny^{-1}\| \leq \|y\|\,\|x^n\|\,\|y^{-1}\|$ ナル故.)

§ 3. 前節ノ結果ヲ用ヒテノルム環ノ表現ニツイテ注意ヲ述ベテオキマス．但シココデ表現トハ普通ノヤウニノルム環 R カラ有限次ノ行列ヘノ K 上ノ環トシテノ準同型寫像ヲ云フモノトシマス．一般ニ表現ヲ

(2) $$x \to T(x) = \{t_{ij}(x)\} \quad i, j = 1, 2, \ldots, n$$

トシ，ソノ核即チ $T(x) = 0$ ナル如キ x ノツクル両側 Ideal ヲ M_T ト書クコトニシマス．

$\{T(x)\}$ 全体ハ行列環トシテ又一ツノノルム環ト考ヘラレマス．コノトキ (2) ナル寫像ガ連続ナラバ M_T ハ R デ閉ヂタ Ideal トナリマスガ，コノ逆モ成立シマス．何トナレバ M_T ガ 閉ヂテヰレバ N. R. Hilfssatz 4 ノ方法デ $R/M_T \simeq \{T(x)\}$ ニ norm $\|X\|$ ヲ入レレバコノ norm ニ對シテハ (2) ハ明カニ連続寫像トナリマス．(*編者註．X は x の属する剰余類をあらわす記号である．) 然ルニ定理 1 ニヨレバ $\{T(x)\}$ ノ norm ハスベテ等値デスカラ $\|x\|$ カラ $\|T(x)\|^*$ ヘノ寫像モ連続トナリマス．即チ適當ニ常数 $C > 0$ ヲトレバ

(3) $$\|x\| \geq \|X\| \geq \frac{1}{C}\|T(x)\|^*.$$

特ニ

$$C\|x\| \geq |t_{ij}(x)|. \tag{4}$$

マトメテ云ヘバ

定理 3. ノルム環 R ノ表現 (2) ガ連続ナルタメニ必要且ツ十分ナル條件ハ核 M_T ガ R デ閉ヂテヰルコトデアル．コノトキ R ノ任意ノ x ニ対シ適當ニ常数 $C > 0$ ヲトレバ (4) ガ成立スル．

§ 4．一般のノルム環デハトテモ手ガ出ナイノデ以下我々ハ常ニ次ノ如キ假定ヲオイテ進ムコトニシマス．

假定 A. I_ℓ, I_r ヲ R ノ任意ノ maximal 左 乃至右-Ideal トスレバ $R/I_\ell, R/I_r$ ハ常ニ有限次元デアル．

サウスレバ先ヅ

定理 4. 假定Aガ満足サレテヰレバ R ノ任意ノ maximal 両側-Ideal I ニ對シ R/I モ亦常に有限次元，従ッテソレハ K 上ノ全行列環ニ同型デアル．

$$R/I \simeq K_n. \tag{5}$$

証明. I ヲ含ム maximal 左-Ideal ヲ I_ℓ トセヨ．假定ニヨリ R/I_ℓ ハ有限次元デソレハ R ノ元ヲ左-Operator トシテ有スルカラ R ノ表現加群ト考ヘラレマス．コノ表現デ 0 ニ対應スル両側 Ideal ヲ I' トスレバ明カニ $I' \supset I$ デスガ I ハ maximal ナル故 $I = I'$. $R/I = R/I'$ ガ有限次元ナルコトハコレカラワカリマスガソレハ K 上ノ單純環デスカラ $R/I \simeq K_n$.

コノ証明カラ又直チニ次ノコトガワカリマス．

定理 5. 假定Aノモトニ任意ノ maximal 左-Ideal I_ℓ ハアル maximal 両側-Ideal I ヲ含ム．maximal 右-Ideal I_r ニ関シテモ同様．

注意. R ニ於テ定理 5 ガ成立シテヰテモ必ズシモ逆ニAガ満足サレテハヰマセン．(サウ云フ実例ヲツクルコトガ出来マス) 又定理 5 ヲ云フニハ假定Aノウチ I_ℓ カ I_r カドチラカダケガ成立シテヰレバ十分デスガ，ソレラノ條件ガ果シテ独立ナモノカドウカハヨクワカリマセン．

定理 6. 假定Aノモトニ R ノ元 x ノ左逆元ハ又 x ノ右逆元デアリ逆モ成立スル．ヨッテコノ唯一ノ逆元ヲ x^{-1} トカクコトガ出来ル．

証明. x ガ左 (右)-逆元ヲ有スルタメニ必要且ツ十分ナル條件ハ x ガ任意ノ maximal 左 (右)-Ideal ニ含マレヌコトデアリマス. (N. R. Satz 6) 定理 5 ニヨレバソレハ又任意ノ maximal 両側 Ideal I ヲトルトキ x ガ R/I ニ於テ maximal 左 (右)-Ideal ニ含マレヌコトト同ジデスガ $R/I \simeq K_n$ デハソレハ又 maximal 右 (左)-Ideal ニ含マレヌコトト等値 (即チ K_n ニ於テ行列式ガ 0 デナク逆元ヲ有スルコト) デスカラコレ

サテ可換ノ場合ト同様ニ $\sqrt[n]{\|x^n\|} \to 0\ (n \to \infty)$ ナル R ノ元 x ノコトヲ一般零巾元ト呼ブコトニスレバ次ノ定理ガ成立シマス.[6)]

定理 7. 假定Aノモトニ x ガ一般零巾元ナルタメニ必要且ツ十分ナル條件ハ R ノ任意ノ maximal 両側 Ideal I ヲトッタトキ x ガ R/I ニ於テ普通ノ意味デ零巾トナルコトデアル.

証明. maximal 両側 Ideal I ヲトッタトキ (5) ニヨル R/I ノ既約表現ノ一ツヲ $T_I(x)$ トカクコトニシマス. maximal Ideal ハ閉ヂテヰマスカラ定理3ニヨリ $T_I(x)$ ハ連続表現デス.

サテ任意ノ $x \in R$ ニ對シ

$$\|x^n\| \geq \|T_I(x^n)\| = \|T_I(x)^n\|$$

ナル故 $T_I(x)$ ガ零巾ナルコト, 即チ x ガ R/I デ零巾ナルコトハ必要條件デアリマス. 逆ニ $T_I(x)$ ガ零巾デアレバ任意ノ $\lambda \in K$ ニ對シ $T_I(\lambda x) = \lambda T_I(x)$ モ亦零巾ナル故 m ヲ十分大キクトレバ $(\lambda x)^m \in I$. コレカラ $e - \lambda x$ ハ決シテ maximal 左-Ideal, I_ℓ ニ含マレヌコトガ証明サレマス.

何トナレバ $e - \lambda x \in I_\ell$ トスレバ, 定理2ニヨリ $I_\ell \supset I$ ナル maximal 両側-Ideal I ヲトリ $(\lambda x)^m \in I$ トシマス. サウスレバ $(\lambda x)^{m-1}(e - \lambda x) = (\lambda x)^{m-1} - (\lambda x)^m \in I_\ell$ カラ $(\lambda x)^{m-1} \in I_\ell$ ヲ得, 同様ニシテ $(\lambda x)^{m-2}, \cdots, (\lambda x), e \in I_\ell$ トナリ, コレハ不合理. 故ニ $e - \lambda x \notin I_\ell$. 從ッテ $e - \lambda x$ ハ左逆元ヲ有シマスガ定理6ニヨリ

$$(e - \lambda x)^{-1}$$

ハ凡テノ $\lambda \in K$ ニ対シ存在スルコトニナリマス. コレカラハ可換環ノ場合ト全ク同様デ $(e - \lambda x)^{-1}$ ガ λ ノ整函数ナルコトカラ展開

$$(e - \lambda x)^{-1} = e + \lambda x + \lambda^2 x^2 + \cdots$$

ニ於テ $\|\lambda^n x^n\| \to 0$. ヨッテ $\overline{\lim} \sqrt[n]{\|x^n\|} \leq \frac{1}{|\lambda|}$, λ ハ任意ダカラ

$$\lim \sqrt[n]{\|x^n\|} = 0$$

トナリマス.

サテ再ビ可換環ノ場合ト同様ニ R ニ於ケル凡テノ maximal 両側-Ideal ノ共通 Ideal ヲ R ノ根基 (Radikal) ト呼ブコトニスレバ上ノ一般零巾元ノ全体ガ必ズシモ根基ト一致スルワケデハアリマセンガ (例ヘバ $R = K_n$ ノ場合ヲ見レバヨイ) 次ノ定理ガ成立シマス.

定理 8. 假定Aノモトニ R ニ於ケル一般零巾元ノ全体ヲ N トスレバ根基 N_0 ハ N ニ含マレル最大ノ左 (右)-Ideal. 從ッテ最大ノ両側 Ideal デアル.

証明. N_0 ガ N ニ含マレルコトハ前定理ニヨリ明カデスカラ N ニ含マレル任意ノ左-Ideal I_ℓ ガ N_0 ニ含マレルコトヲ云ヘバ十分デス. サテ任意ノ maximal 両側 Ideal I ヲトルトキ $I_\ell \cdot I/I$ ハ R/I ノ左-Ideal デスガ $I_\ell \subset N$ ナル假定ニヨリソレハ零巾元バカリカラ成ルモノデ, 從ッテ $I_\ell \cdot I = I$,[7)] $I_\ell \subset I$. I ハ任意デスカラ $I_\ell \subset N_0$.

定理8ニヨリ特ニ假定Aノ満足サレテヰルノルム環デ一般零巾元ガ 0 以外ニ存在シナイ場合ニハ $N_0 = 0$. 即チ R ハ準單純環デアルコトガ云ヘルワケデス.

§5． 以上ノ如ク假定Aヲ承認スレバ，ノルム環ノ理論ハ代数的ナ部分ニ関スル限リアル程度マデ可換ナ場合ト同様ニ進ムコトガ出来マスガ，位相的部分，即チ maximal Ideal ニ Topologie ヲ入レテ論ズルトイウヤウナ部分ハソノママデハ巧ク行キマセン．コレハ特ニ一般ノルム環ノ表現ノ問題ト関聯シテ今後ニ残サレタ問題デアルト思ヒマス．

次ニ假定Aニ関シ少シバカリ注意ヲ述ベテ見マス．假定Aヲ満足シナイノルム環ガ存在スルコトハ Hilbert 空間ニ於ケル有界作用素カラ容易ニソノ例ヲツクルコトガ出来マス．又假定Aヲ満足スル場合トシテハ次ノ例ガ知ラレテキマス．

1．R ガ可換ナル場合

2．R/I_0 ガ有限次元ナル如キ一ツノ両側 Ideal I_0 ガ存在シテ I_0 ニ含マレル x ニ対シテハ $A_x \cdot y = xy, A'_x \cdot y = yx$ ナル Operator A_x, A'_x ガ完全連続ナル場合．

1ハ Gelfand ノ可換ノルム環ノ場合デアッテ，2ハ深宮氏ガ compact 群ノ群環ニ於テ述ベラレタ場合デアリマス．[8)]

一般ニドノヤウナ條件ガアレバ假定Aガ満足サレルカト云フコトハ難カシイ問題デアラウト思ヒマス．

II. 群環 $R(G)$

§1． 群 G ハ locally compact, separable トシ，μ ヲ G 上ノ右-invariant ナ Haar ノ測度トシマス．[9)] G 上ノ可測函数ニツイテ L^p $(p \geq 1)$ ノ定義及ビソコニ於ケル norm

(6) $$\|x(g)\|_p = \left\{ \int_G |x(g)|^p dg \right\}^{\frac{1}{p}}$$

ハ周知ノ通リ．

$x(g) \in L^p$, $y(g) \in L^1$ ニ對シ

(7) $$x \times y(g) = z(g) = \int_G x(gh^{-1}) y(h) dh$$

トオケバ簡單ナ計算ニヨリ [10)]

(8). $$\|z(g)\|_p \leq \|x(g)\|_p \cdot \|y(g)\|_1$$

即チ $z(g)$ ハ實際存在シテ L^p ニ属スルコトガ知ラレマス．

サテ L^1 ト L^p $(p \geq 1)$ トノ共通部分ヲ $L^{(1,p)}$ トシ $x, y \in L^{(1,p)}$ ニ対シ (7) ニヨリ積ヲ定義スレバソレガ環ニナルコトハ容易ニワカリマス。ソコデ更ニ

(9) $$\|x\| = \mathrm{Max}(\|x(g)\|_p, \|x(g)\|_1)$$

トオケバ (8) ニヨリ

(10) $$\|x \times y\| \leq \|x\| \|y\|$$

デ $L^{(1,p)}$ ハ單位を有スルト云フ條件ヲ除ケバノルム環ノ他ノ條件ヲ凡ベテ満シマス．ヨッテ $L^{(1,p)}$ ニ單位 e ヲ附加シテ

$$\lambda e + x(g), \quad \lambda \in K, \quad x(g) \in L^{(1,p)}$$

ナル全体ヲツクレバコレハ普通ノ様ニシテノルム環トナリマス．コレヲ $R = R^{(1,p)}(G)$ トカクコトニシマス．G ガ discrete ナ場合ニハ既ニ $L^{(1,p)}$ ニ単位ガ存在シ $L^{(1,p)}$ 自身ノルム環トナルワケデスガ我々ハコレラノ場合ヲ區別シナイコトニシマス．

Segal ハ $R^{(1,1)}(G)$ ヲ G ノ群環ト呼ンデヰルノデスガ $R^{(1,p)}(G)$ ハムシロ補助的ナ意味ヲモツモノデ $L^{(1,p)}(G)$ ノ方ガ（有限群ノ場合ノ拡張ト考ヘテモ）群環ノ名ニ相應シイノデハナイデセウカ．以下ノ我々ノ考察モ目標ハ $L^{(1,p)}(G)$ ノ方ニ置イテ進ミタイト思ヒマス．

$R^{(1,p)}(G)$ ノ中デ考ヘレバ $L^{(1,p)}(G)$ ハ一ツノ maximal 両側 Ideal M_0 ヲツクリマス．又 $x(g) \in L^{(1,p)}, a \in G$ ニ対シ

$$ax(g) = x(ga), \quad xa(g) = x(ag) \tag{11}$$

ト定義スレバ ax, xa モ亦 $L^{(1,p)}$ ニ属シ特ニ $\|ax\| = \|x\|$.

§2．以下ニ於テハ $L^{(1,p)}(G)$ ノ表現ト G ノ表現トノ関係ヲ考ヘルコトニシマス．ココニ表現トハ普通ノ意味ノ有限次行列ニヨル表現ヲ云フワケデスガ特ニ G ノ表現トシテ 0-表現ヲモ許スコトニシマス．即チ G ノ任意ノ元 a, b ニ対シ $D(a)D(b) = D(ab)$ ヲ満足スル行列系 $\{D(a)\}$ ハスベテコレヲ G ノ表現ト呼ブコトニシ G ノ單位ガ單位行列ニ対應スルト云フコトハ要求シナイコトニシマス．

サテ我々ノ目標ハ次ノ定理デアリマス．

定理 9. $L^{(1,p)}(G)$ ノ連續表現 $x(g) \to T(x)$ ト G ノ有界連續表現 $a \to D(a)$ トハ次ノ如キ意味デ一對一ニ対應スル．

i) $L^{(1,p)}(G)$ ノ連續表現 $x(g) \to T(x)$ ガ與ヘラレタトキ適當ニ G ノ有界連續表現 $a \to D(a)$ ヲトレバ任意ノ $x(g) \in L^{(1,p)}$ ニ対シ

$$T(x) = \int_G x(g)D(g)dg \tag{12}$$

トナル．コノ様ナ $D(a)$ ハ $T(x)$ ニヨリ一意的ニ定マリ，コレヲ $D_T(a)$ トカクコトニスル．

ii) G ノ有界連續表現 $a \to D(a)$ ガ與ヘラレタトキ (12) ニヨリ $T(x)$ ヲツクレバ $x(g) \to T(x)$ ハ $L^{(1,p)}(G)$ ノ連續表現ヲ與ヘル．$D(a)$ ニ對應スルコノ表現ヲ $T_D(x)$ トカクコトニスル．

iii) i), ii) ノ對應ハ互ニ他ノ逆ニナッテヰル．即チ任意ノ $T(x), D(a)$ ニ對シ

$$T_{D_T} = T, \quad D_{T_D} = D.$$

iv) 上ノ對應デ等値ナ表現ハ等値ナ表現ニ對應スル．即チ $AT_1(x)A^{-1} = T_2(x)$ ナラバ $AD_{T_1}(a)A^{-1} = D_{T_2}(a)$, 又 $BD_1(a)B^{-1} = D_2(a)$ ナラバ $BT_{D_1}(x)B^{-1} = T_{D_2}(x)$.

証明. 各段ニ分ケテ証明シマスガ面倒ナノハ i) ダケデアトハコレカラ直チニ出マス．

i) $L^{(1,p)}(G)$ ノ表現ヲ $x(g) \to T(x) = \{t_{ij}(x(g))\}$ トオイテ $t_{ij}(x(g))$ ヲ考ヘテ見マス．

$T(x)$ ハ連續表現ナル故定理3ニヨリ $|t_{ij}(x(g))| \leq C\|x(g)\|$ ナル常数 C ガ存在シマス．特ニ有限ナ測度ヲ有スル場合 E ノ特性函数 $x_E(g)$ ヲトレバ

$$|t_{ij}(x_E(g))| \leq C\|x_E(g)\| = C\max(\mu(E), \mu(E)^{\frac{1}{p}}). \tag{13}$$

ヨッテ適當ニ可測函数 $u_{ij}(g)$ ヲトレバ [11] 上ノ如キ任意ノ E ニ對シ

$$t_{ij}(x_E(g)) = \int_E u_{ij}(g)dg = \int_G x_E(g)u_{ij}(g)dg. \tag{14}$$

$t_{ij}(x(g))$ ガ連續ナルコトヲ用ヒレバコレカラ一般ニ

$$t_{ij}(x(g)) = \int_G x(g)u_{ij}(g)dg.$$

即チ $U(g) = \{u_{ij}(g)\}$ トオケバ

$$T(x) = \int_G x(g)U(g)dg. \tag{15}$$

$T(x)$ ハ表現デスカラ $L^{(1,p)}(G)$ ニ含マレル任意ノ $x(g)$, $y(g)$ ニ對シテ $T(x)T(y) = T(x \times y)$, 即チ

$$\int x(g)U(g)dg \int y(h)U(h)dh = \int x(gh^{-1})y(h)U(g)dgdh.$$

ヨッテ

$$\int x(g)y(h)U(g)U(h)dgdh = \int x(g)y(h)U(gh)dgdh.$$

有限ノ測度ヲ有スル集合ノ特性函数ハ凡テ $L^{(1,p)}(a)$ ニ含マレマスカラ上式カラ (g,h)-測度 0 ヲ除キ

$$U(g)U(h) = U(gh) \tag{16}$$

ヲ得マス．ヨッテ適當ニ測度 0 ナル集合 X_0 ヲトレバ $h \notin X_0$ ナラバ測度 0 ナル集合 X_h ガ定ッテ $g \notin X_h$ ニ對シ (16) ガ成立シマス．故ニ $a \notin X_0$ ノトキハ

$$\begin{aligned} T(a^{-1}x) &= \int x(ga^{-1})U(g)dg = \int x(g)U(ga)dg = \int x(g)U(g)U(a)dg \\ &= T(x)U(a). \end{aligned}$$

即チ

$$T(a^{-1}x) = T(x)U(a), \quad a \notin X_0. \tag{17}$$

次ニ適當ニ $x_0(g) \in L^{(1,p)}(G)$ ヲトレバ任意ノ $a \notin X_0$ ニ對シ $T(a^{-1}x_0) = U(a)$ トナルコトヲ証明シマス． $\{T(x)\}$ ハ K 上ノ行列環デスカラ適當ニ變形スレバ

$$\begin{pmatrix} T_1 & & & & \\ & T_2 & & 0 & \\ & & \ddots & & \\ & 0 & & T_r & \\ & & & & 0 \end{pmatrix} \tag{18}$$

ナル形ヲシタ準單純ナ部分ト

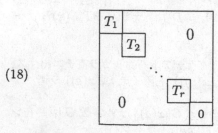

ナル形ヲシタ準單純ナ部分ト

$$
(19)\qquad
\left(\begin{array}{c|c|c|c|c}
0 & & & & \\ \hline
 & 0 & & \ast & \\ \hline
 & & \ddots & & \\ \hline
 & & & 0 & \\ \hline
 & 0 & & & \begin{matrix} 0 & & \ast \\ & 0 & \\ 0 & & \ddots\ 0 \end{matrix}
\end{array}\right)
$$

ナル形ヲシタ根基ノ部分トノ直和 (K-加群トシテノ) トナリマス. 但シココニ $T_1, \dots, T_r$ ハ 0 ナラザル既約部分, 最後ノ $\begin{pmatrix} 0 & & \ast \\ & 0 & \\ 0 & & \ddots\ 0 \end{pmatrix}$ ハ 0 ナル既約部分ヲマトメタモノデス.

ヨッテ特ニ

$$
(20)\qquad T(x_0) =
\left(\begin{array}{c|c|c|c|c}
E_{n_1} & & & & \\ \hline
 & E_{n_2} & & 0 & \\ \hline
 & & \ddots & & \\ \hline
 & & & E_{n_r} & \\ \hline
 & 0 & & & 0
\end{array}\right)
\quad ; \ E_{n_k}\ \text{ハ}\ n_k\ \text{次ノ單位行列}
$$

ナル $x_0(g)$ ガ存在シマス. サテ $\{T(x)\}$ ノ行列ハ一般ニ (18), (19) ノ和デ

$$
(21)\qquad T(x) =
\left(\begin{array}{c|c|c|c|c}
\ast & & & & \\ \hline
 & \ast & & \ast & \\ \hline
 & & \ddots & & \\ \hline
 & & & \ast & \\ \hline
 & 0 & & & \begin{matrix} 0 & & \ast \\ & 0 & \\ 0 & & \ddots\ 0 \end{matrix}
\end{array}\right)
$$

ナル形ヲシテヰマスカラ $U(g)$ モ亦同ジヤウナ形ヲ有スルモノト考ヘテ差支ヘアリマセン. $(t_{ij}(x(g))$ ガ恒等的ニ 0 ナラバ $U_{ij}(g) = 0$ トオイテヨイカラ) $a \notin X_0$ トスレバ (17) ニヨリ

$$
(22)\qquad T(a^{-m}x) = T(x)U(a)^m.
$$

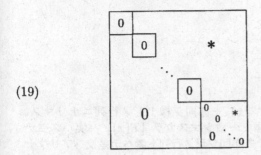

(23)

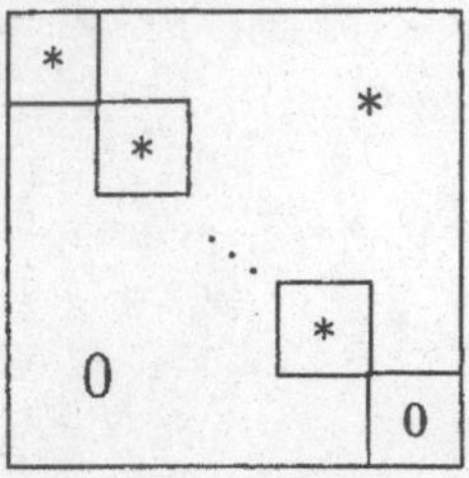

ナル形トナリマスガ (22) ニヨッテ $T(a^{-m}x)$ モ亦同ジ様ナ形ノ行列ニナリマス．$\{T(a^{-m}x)\}$ ノ全体ハ勿論 $\{T(x)\}$ ノ全体ト一致シマスカラ $\{T(x)\}$ ハ実ハスベテ (23) ナル形ヲシテキタワケデ從ッテ $U(g)$ モ亦同ジ形ノ行列ト考ヘテ差支ヘアリマセン．故ニ (20) ノ x_0 ヲ用ヒレバ

$$T(a^{-1}x_0) = T(x_0)U(a) = U(a), \quad a \notin X_0. \tag{24}$$

サテ上ノ如キ x_0 ニ對シ一般ニ

$$T(a^{-1}x_0) = D(a) \tag{25}$$

トオケバ (24) カラ特ニ $a \notin X_0$ ナラバ $D(a) = U(a)$ トナリマスカラ $\mu(X_0) = 0$ ニ注意スレバ

$$T(x) = \int_G x(g)D(g)dg. \tag{26}$$

又 $a, b \notin X_0$ ナラバ

$$D(ab) = T((ab)^{-1}x_0) = T(b^{-1}a^{-1}x_0) = T(a^{-1}x_0)U(b)$$
$$= T(x_0)U(a)U(b) = U(a)U(b) = D(a)D(b).$$

即チ

$$D(ab) = D(a)D(b). \tag{27}$$

$T(x)$ ハ連続表現デ $a^{-1}x_0$ ハ x_0 ヲ定メレバ a ニツイテ norm $\|x\|$ ニ関シ連続デスカラ $D(a)$ ハ a ニ関シ連續デス．又 $\mu(X_0) = 0$ ナル故 $G - X_0$ ハ G デ稠密．コレラノコトカラ (27) ハ任意ノ $a, b \in G$ ニツイテ成立スルコトガ知ラレマス．即チ $D(a)$ ハ連続表現デスガ (3) ニヨリ $\|D(a)\|^* = \|T(a^{-1}x_0)\|^* \leq C\|a^{-1}x_0\| = C\|x_0\|$ ナル故ソレハ又有界デアリマス．

ヨッテ i) ニ於ケル有界連続表現ノ存在ハワカリマシタ．次ニソノ一意性デスガ $D(a), D'(a)$ ガ (12) ヲ満足スル二ツノ可測表現デアルトスレバ,[12] 任意ノ $x(g) \in L^{(1,p)}(G)$ ニ對シ

$$\int_G x(g)D(g)dg = \int_G x(g)D'(g)dg.$$

コレカラ測度 0 ヲ除イテ

$$D(g) = D'(g), \quad g \notin X_1, \ \mu(X_1) = 0.$$

G ノ任意ノ元 a ハ $a = a_1a_2$, $a_1, a_2 \notin X_1$ とカケマスカラコレカラ一般ニ

$$D(a) = D'(a)$$

ヲ得マス．

ii) G ノ任意ノ有界連續表現 $a \to D(a)$ ニ對シ (12) ニヨリ $T(x)$ ヲツクレバコレガ linear ナ對應デアルコトハ明ラカデスガ, 又

$$T(x \times y) = \int_{G\times G} x(gh^{-1})y(h)D(g)dgdh = \int x(g)y(h)D(gh)dgdh$$

$$= \int x(g)D(g)dg \cdot \int y(h)D(h)dh = T(x)T(y)$$

ナル故 $x(g) \to T(x)$ ハ確カニ一ツノ表現ヲ與ヘマス. 又 $\|D(g)\|^* \leq C$ トスレバ

$$\|T(x)\|^* \leq \|D(g)\|^* \|x(g)\|_1 \leq C\|x\|$$

ナル故ソレハ連續ナ表現デアリマス.

iii) $T_{D_T} = T$ ノ方ハソノ意味カラ明白. $D_{T_D} = D$ モ i) ニ於ケル一意性カラ明カデス.

iv) コレモ云フ迄モアリマセン.

カクシテ定理ハ証明サレマシタ. コノ定理カラ種々ノ結果ガ得ラレマス. 先ヅ任意ノ有界可測表現 $D_1(a)$ ガ與ヘラレタ場合, コノ $D_1(a)$ カラ (12) ニヨッテ $T(x)$ ヲツクレバ前定理ノ ii) ノ証明ト全ク同様ニシテソレガ $L^{(1,p)}(G)$ ノ連続表現トナルコトガワカリマス. ヨッテ前定理 i) ニヨリ $T(x) = \int_G x(g)D(g)dg$ ナル有界連続表現 $D(a)$ ガ存在シマス. 即チ

$$\int_G x(g)D_1(g)dg = \int_G x(g)D(g)dg.$$

前定理 i) ノ一意性ノ証明ニヨリ $D_1(g) = D(g)$, 即チ $D_1(g)$ ハ又連續デアリマス. ヨッテ

定理 10. G ノ任意ノ有界可測表現ハ連續デアル.

次ニ上ノ對應ノサセ方カラワカルヤウニ表現 $T(x)$ ト表現 $D(a)$ トハ全ク同ジヤウニ分解サレマス. ヨッテ有界表現 $D(a)$ ガ完全可約デアルコトヲ用ヒレバ,[13] $T(x)$ モ亦完全可約ナルコトガワカリマス. 又 $T(x)$ ヲ 0 ナラザル $L^{(1,p)}(G)$ ノ既約表現トシ $T(x) = 0$ ナル如キ x 全体ノツクル両側 Ideal ヲ M_T トスレバ $T(x)$ ガ既約ナルコトカラ M_T ハ maximal トナリ, 従ッテ $L^{(1,p)}(G)$ デ閉ヂテヰマス. (後節参照) ヨッテ定理 3 ニヨリ $T(x)$ ハ連続デアリマス. 故ニ一般ニ完全可約ナ $L^{(1,p)}(G)$ ノ表現ハ常ニ連続トナリ次ノ定理ヲ得マス.

定理 11. $L^{(1,p)}(G)$ ノ連続表現ハ常ニ完全可約デアリ, 逆ニ完全可約ナ $L^{(1,p)}(G)$ ノ表現ハ常ニ連続デアル.[14]

コノ定理ヲ Ideal ヲツカッテ云ヒ直セバ

定理 12. M ハ $L^{(1,p)}(G)$ ノ両側 Ideal デ $L^{(1,p)}/M$ ハ有限次元ナルモノトスル. コノトキ M ガ $L^{(1,p)}(G)$ デ閉ヂテヰレバ $L^{(1,p)}/M$ ハ準單純環デアリ, 逆ニ $L^{(1,p)}/M$ ガ準單純ナラバ M ハ $L^{(1,p)}(G)$ デ閉ヂテヰル.

コレラハイヅレモ有限群ノ群環ニ関スル定理ノ拡張ト考ヘラレマス.

サテ $L^{(1,p)}(G)$ ノ既約表現ノ各類 $\{T(x)\}$ ニハソレゾレ一定ノ maximal 両側 Ideal M_T ガ對應シマスガ $\{T(x)\}$ ハ又定理 9 ニヨリ G ノ有界既約表現類 $\{D(a)\}$ ト一對一對應シマスカラ次ノ定理ヲ得マス.

定理 13. $\{D\}$ ヲ G ノ有界可測既約表現ノ一ツノ類, $D(a)$ ヲソレニ属スル一ツノ既約表現トスルトキ

$$\int_G x(g)D(g)dg = 0$$

ナル如キ $x(g)$ 全体ハ $L^{(1,p)}(G)$ ニ於テ $\{D\}$ ニヨッテ一定スル maximal 両側 Ideal $M\{D\}$ ヲツクル. $\{D\} \to M_{\{D\}}$ ナルコノ對應ニヨリ G ノスベテノ有界可測既約表現ノ類 $\{D\}$ ト $L^{(1,p)}/M$ ガ有限次元トナル如キ $L^{(1,p)}(G)$ ノスベテノ maximal 両側 Ideal トハ一對一對應スル.

一般ニ G ノ二ツノ有界可測表現 $D_1(a), D_2(a)$ ガ與ヘラレタ場合, ソレラニ含マレル 0 ナラザル既約部分が全体トシテ一致スルトキ (即チ $D_1(a)$ ニ含マレル既約成分ハ凡テ $D_2(a)$ ニ含マレ, 逆モ亦成立スルトキ) $D_1(a)$ ト $D_2(a)$ トヲ同ジ類ニ入レルコトニスレバ, コノ様ニシテ G ノ凡テノ有界可測表現ヲ類別スルコトガ出来マスガ, コノヤウナ表現類トソレニ對應スル $L^{(1,p)}(G)$ ノ表現ノ核トナル $L^{(1,p)}(G)$ ノ両側 Ideal トガ一對一ニ對應スルコトハ明カデス. 實際 $D(a)$ ニ含マレル既約表現(類) ヲ $D_1(a), \ldots, D_k(a)$ トスレバ $a \to D(a)$ ニ對應スル $L^{(1,p)}(G)$ ノ表現 $x \to T_D(x)$ ノ核ハ $M_{\{D_1\}}, \ldots, M_{\{D_k\}}$ ノ共通 Ideal トナリマス.

$L^{(1,p)}$, $p = 1, 2, \ldots$ ニ於イテ p ヲ変ヘテモ G ノ表現類ハ一定デスカラソノ對應ノサセ方カラ次ノ定理ヲ得マス.

定理 14. $L^{(1,1)}(G) = L^1(G)$ ニ於テ L^1/M ガ有限次元トナル如キ任意ノ閉ヂタ両側 Ideal M ヲトリ

$$M \cap L^{(1,p)}(G) = M_p, \quad p = 1, 2, \ldots$$

トオケバ M_p ハ $L^{(1,p)}(G)$ ニ於テ閉ヂタ両側 Ideal デ, $L^{(1,p)}/M_p$ ハ有限次元デアル. カクノ如クニシテ $L^{(1,p)}/M_p$ ガ有限次元トナル如キ $L^{(1,p)}(G)$ ノ凡テノ閉両側 Ideal ハ $p = 1, 2, \ldots$ ニ對シ互ニ一對一ニ對應スル. コレラハ又 G ノスベテノ有限可測表現の類 $\{D\}$ ト一對一ニ對應スル.

ヨッテ $L^{(1,p)}(G)$ 或ハ G ノ表現ヲ問題ニシテヰル限リ $L^{(1,p)}$ $(p = 1, 2, \ldots)$ ノドレヲトッテモ同ジコトデ, ソノ場合ニ應ジテ便利ナモノヲ選ベバヨイワケデス.

§ 3. 次ニ始メニ述ベマシタノルム環 $R^{(1,p)}(G)$ 即チ

$$\lambda e + x(g), \quad x(g) \in L^{(1,p)}(G)$$

ノ様子ヲ調ベテ見ルコトニシマス. ソレニハ前節ノ結果ヲ用ヒルコトニスレバ次ノ補助定理ダケデ十分デス.

補助定理 2. M ハ $L^{(1,p)}(G)$ ノ両側 Ideal, $L^{(1,p)}/M$ ハ單位元ヲ有シ且有限次元トスル. 然ラバ適當ニ $R^{(1,p)}(G)$ ノ両側 Ideal M' ヲトレバ

$$M = M' \cap L^{(1,p)}(G) \tag{28}$$

$$R^{(1,p)}(G) = M' + L^{(1,p)}(G) \tag{29}$$

トナル．而モコノ様ナ M' ハ唯一ツ定マル．

証明. M ハ $L^{(1,p)}(G)$ ノ両側 Ideal デスガ，掛ヶ算ノ定義ノ仕方カラソレハ又 $R^{(1,p)}(G)$ ノ両側 Ideal デモアリマス．仮定ニヨリ $L^{(1,p)}/M$ ハ單位元ヲ有シマスカラソレニ對應スル $L^{(1,p)}(G)$ ノ函数ヲ $x_0(g)$ トシ，$\boldsymbol{y}_0 = e - x_0(g)$ トオケバ容易ニワカルヤウニ $R^{(1,p)}/M$ ハ $L^{(1,p)}/M$ ト $\{\boldsymbol{y}_0, M\}/M$ トノ直和トナリマス．ヨッテ $\{\boldsymbol{y}_0, M\} = M'$ トオケバ M' ハ $R^{(1,p)}(G)$ ノ両側 Ideal デ (28)，(29) ヲ満足スルコトハ明カデス．M' ノ一意性ハ $R^{(1,p)}/M$ ノ分解ノ一意性ニヨリ明カデス．

サテ上ノ M' ヲトレバ (28)，(29) カラ $R^{(1,p)}/M' \simeq L^{(1,p)}/M$ トナリマスカラ特ニ M ガ $L^{(1,p)}$ デ maximal デアレバ M' ハ $R^{(1,p)}$ デ maximal トナリマス．$R^{(1,p)}(G)$ ハノルム環デスカラソコデノ maximal Ideal M' ハ閉ヂテヰマス．ヨッテ (28) ニヨリ M モ $R^{(1,p)}(G)$, 從ッテ $L^{(1,p)}(G)$ デ閉ヂタ Ideal トナリマス．(前節參照)

又コノ補助定理ニヨリ $L^{(1,p)}(G)$ ノ閉ヂタ両側 Ideal M[15] ト $R^{(1,p)}(G)$ ノ $M_0 = L^{(1,p)}(G)$ ニ含マレナイ閉ヂタ両側 Ideal トガ (28)，(29) ノ關係ニヨリ互ニ一對一ニ對應スルコトモワカリマス．コレラノコトカラ前節ノ結果ヲ用ヒレバ容易ニ次ノ諸定理ヲ導クコトガ出来マス．

定理 15. $R^{(1,p)}(G)$ ノ任意ノ連続表現 $\boldsymbol{y} \to T'(\boldsymbol{y})$ ニ對シテハ G ノ連続有限表現 $a \to D(a)$ ガ一意的ニ定マッテ $\boldsymbol{y} = \lambda e + x(g)$ ナルトキ

$$T'(\boldsymbol{y}) = T'(\lambda e + x(g)) = \lambda D(e_0) + \int_G x(g) D(g) dg^{16)} \tag{30}$$

トナル．逆ニ G ノ任意ノ連続有界表現 $a \to D(a)$ ニ對シ (30) ニヨリ $T'(\boldsymbol{y})$ ヲツクレバ $\boldsymbol{y} \to T'(\boldsymbol{y})$ ハ $R^{(1,p)}(G)$ ノ連続表現ヲ與ヘル．カクシテ表現 $T'(\boldsymbol{y})$ ト $D(a)$ トハ一對一ニ對應スル．

定理 16. $R^{(1,p)}(G)$ ノ連続表現ハ完全可約デアリ逆ニ完全可約ナ $R^{(1,p)}(G)$ ノ表現は凡テ連続デアル．

定理 17. M_1 ハ $R^{(1,p)}(G)$ ノ両側 Ideal デ $R^{(1,p)}/M_1$ ハ有限次元ナモノトスル．コノトキ M_1 ガ $R^{(1,p)}(G)$ デ閉ヂテヰレバ $R^{(1,p)}/M_1$ ハ準單純環デアリ，逆モ亦成立スル．

定理 18. $\{D\}$ ヲ G ノ有界可測既約表現ノ一ツノ類，$D(a)$ ヲソレニ属スル任意ノ既約表現トスルトキ

$$(\boldsymbol{y}, D) = \lambda E_n + \int_G x(g) D(g) dg = 0 \quad (n \text{ ハ } D(a) \text{ ノ次数})$$

ナル如キ $\boldsymbol{y} = \lambda e + x(g)$ ノ全体ハ $R^{(1,p)}(G)$ ニ於テ $\{D\}$ ニヨッテ一定スル maximal 両側 Ideal $M\{D\}$ ヲツクル．コノ對應ニヨリ G ノ凡テノ有界可測既約表現ノ類 $\{D\}$ ト $R^{(1,p)}/M_1$ ガ有限次元デ $M_1 \neq M_0$ ナル如キ $R^{(1,p)}(G)$ ノ凡テノ maximal 両側 Ideal トハ一對一ニ對應スル．

コレガ Segal ノ論文ニ於ケル定理 3 デス．[17] 上記ノ定理ハ Segal ト同ジヤウニ云ヒ表ハシタノデスガ，モシモ前ノ様ニ 0-表現 ヲモ既約表現ノ仲間ニ入レレバ定理ノ後半ニ於ケル $M_1 \neq M_0$ ナル條件ハ取除クコトガ出来マス．

同様ニシテ定理 14 ニ相當スル定理モ得ラレマス．

以上ノ如ク我々ハ前節ノ定理 9 − 13 (14) ヲ用ヒテ定理 15 − 18 ヲ証明シタノデスガ，コレラノ諸定理ハ勿論直接ニモ証明サレマス．証明ノ方法ハ前節ト全ク同様デ，シカモコノ際 $R^{(1,p)}(G)$ ニハ単位元ガ存在シマスカラ $L^{(1,p)}(G)$ ノ場合ヨリムシロ簡単ニ出来マス．

然シ逆ニ定理 15 − 18 カラ前節ノ結果ヲ導カウトスルト結局前節ニ於テシタト同様ナ考察ヲ繰返スコトニナル様ニ思ハレマス．ソシテ定理ノ本質的ナ部分ハ $R^{(1,p)}(G)$ ヨリ $L^{(1,p)}(G)$ ニアルト思ハレマスノデ $L^{(1,p)}(G)$ ニツイテ稍々面倒ナ証明ヲシタワケデス．

§4．定理 9 ノ應用トシテ $L^p(G)$ ニ属スル表現ヲ考察シテ見マス．ココニ表現 $D(a) = \{d_{ij}(a)\}$ ガ $L^p(G)$ ニ属ストハ各函数 $d_{ij}(a)$ ガ $L^p(G)$ ニ属スコトヲ略稱シタノデス．

先ヅ $D(a)$ ガ $L^1(G)$ ニ属スル表現デアレバ

$$\int_{G\times G} |d_{ij}(gh)|dgdh = \int_{G\times G} |\sum_k d_{ik}(g)d_{kj}(h)|dgdh$$

$$\leq \sum_k \int |d_{ik}(g)d_{kj}(h)|dgdh = \sum_k \int |d_{ik}(g)|dg \cdot \int |d_{kj}(h)|dh < \infty.$$

一方

$$\int_{G\times G} |d_{ij}(gh)|dgdh = \int_{G\times G} |d_{ij}(g)|dgdh$$

$$= \mu(G) \cdot \int_G |d_{ij}(g)|dg.$$

故ニ $\mu(G) = \infty$ ト假定スレバ $\int_G |d_{ij}(g)|dg = 0$. 即チ測度 0 ヲ除イテ $d_{ij}(g) = 0$. i, j ハ任意デスカラ適當ニ $g = g_0$ ヲトレバ凡テノ i, j ニ関シ，$d_{ij}(g_0) = 0$, 即チ $D(g_0) = 0$. ヨッテ任意ノ $g \in G$ ニ對シ $D(g) = D(gg_0^{-1})D(g_0) = 0$. 即チ $D(a)$ ハ 0 表現デナケレバナリマセン．

$D(a)$ ガ L^p $(p \geq 2)$ ニ属スル場合モ Hölder ノ不等式

$$|d_{ij}(gh)|^p = |\sum_k d_{ik}(g)d_{kj}(h)|^p \leq \left\{\sum_k |d_{jk}(g)d_{kj}(h)|^p\right\}^{\frac{1}{p}} n^{\frac{p}{q}} \quad (\frac{1}{p} + \frac{1}{q} = 1)$$

ヲ用ヒレバ同様ナ結果ヲ得マス．

即チ 0-表現デナイ $D(a)$ ハ $\mu(G) < \infty$, 即チ G ガ compact ナ場合ダケ存在スルワケデ，以下ソノ場合ヲ考ヘテ見マス．

先ヅ $D(a)$ ヲ $L^1(G)$ ニ属スル任意ノ表現トスルトキ凡テノ $x(g) \in L^1(G)$ ニ對シ $\int x(g)D(g)dg$ ガ存在シ且 $x(g)$ ニ関シテ連続デアルコトヲ証明シマス．

$0 < c_1 < \frac{\mu(G)}{2n^2}$ ナル條件ヲ満足スル常数 c_1 ヲトリ，又 $\mathrm{Max}_{i,j}\{\|d_{ij}(g)\|_1\} = c_2$ トオケバ

$$\int |d_{ij}(gh)x(h)|dgdh = \|d_{ij}(g)\|_1 \|x(g)\|_1$$

ナル故

$$\mu E\left(g; \int |d_{ij}(gh)x(h)|dh \geq \frac{1}{c_1}\|d_{ij}(g)\|_1 \|x(g)\|_1\right) \leq c_1.$$

又

$$\int |d_{ij}(g^{-1})|dg = \int |d_{ij}(g)|dg = \|d_{ij}(g)\|,^{18)}$$

ナル故 $\mu E(g; |d_{ij}(g^{-1})| \geq \frac{1}{c_1}\|d_{ij}(g)\|) \leq c_1$. $2n^2c_1 < \mu(G)$ ナル故適當ニ $g = g_0$ ヲトレバ上式カラ

$$\int |d_{ij}(g_0h)x(h)|dh \leq \frac{1}{c_1}\|d_{ij}(g)\|_1\|x(g)\|_1 \leq \frac{c_2}{c_1}\|x(g)\|_1$$

$$|d_{ij}(g_0^{-1})| \leq \frac{1}{c_1}\|d_{ij}(g_0)\|_1 \leq \frac{c_2}{c_1}, \quad i, j = 1, 2, \ldots, n$$

ガ同時ニ成立シマス．即チ

$$\|\int d(g_0h)x(h)dh\|^* \leq \frac{nc_2}{c_1}\|x(g)\|_1$$

$$\|D(g_0^{-1})\|^* \leq \frac{nc_2}{c_1}.$$

ヨッテ

$$\|\int D(g)x(g)dg\|^* = \|D(g_0^{-1})\int D(g_0h)x(h)dh\|^*$$

$$\leq \|D(g_0^{-1})\|^*\|\int D(g_0h)x(h)dh\|^* \leq \left(\frac{nc_2}{c_1}\right)^2\|x(g)\|_1.$$

コレハ $\int D(g)x(g)dg$ ガ存在シテ且ツ $x(g)$ ニ関シ連続ナルコトヲ示シテヰマス．

故ニ定理９ノ (12) ニヨリ $D(a)$ カラ $T(x)$ ヲツクレバコレハ $L^{(1,p)}$ ノ連続ナ表現ヲ與ヘマスカラ定理１０ヲ導イタト同様ナ方法デ $D(a)$ ガ実は有界連続ナルコトガ示サレマス．

次ニ $D(a)$ ガ $L^p(G)$ $(p \geq 2)$ ニ属シテヰルトキハ $L^{(1,q)}(G)$ $\left(\frac{1}{p} + \frac{1}{q} = 1\right)$ ヲトッテ考ヘレバ $x(g) \in L^{(1,q)}(G)$ ナルトキ $\int D(g)x(g)dg$ ガ存在シテ $x(g)$ ニ関シ連続ナルコトハ Hölder ノ不等式カラ明白デスカラ上ト同様ニシテ $D(a)$ ハ有界連続トナリマス．ヨッテ

定理 19. G ガ compact デナケレバ 0-表現以外 $L^p(G)$ $(p \geq 1)$ ニ属スル表現ハ存在シナイ．又 G ガ compact ナラバ $L^p(G)$ $(p \geq 1)$ ニ属スル表現ハスベテ有界連続デアル．

G ガ compact ナラバ逆ニ勿論任意ノ有界連続表現ハ凡テノ $L^p(G)$ ニ含マレマス．

§５． 再ビ $R^{(1,p)}(G)$ ヲ考察スルコトニシ先ヅソレガ **I** ノ假定Ａヲ満足スル場合ヲ考ヘテ見マス．$R^{(1,p)}(G)$ 根基ヲ N_0 トスレバ G ノ測度ガ両側不変デ $N_0 \neq 0$ ナラバソレハ有界連続且エルミート型函数 $z(g) \neq 0$, $z(g) = \overline{z(g^{-1})}$ ヲ含ムコトガ可換群ノ場合ト同様ニ容易ニ示サレマス．[19] 然ルニ矢張リ可換ノ時ト全ク同様ニソノヤウナ $z(g)$ ハ一般零巾元トハナリマセン．コレハ我々ノ定理８ト矛盾シマス．ヨッテ G ノ測度ガ両側不変ノトキニハ次ノ定理ヲ得マス．

定理 20. $R^{(1,p)}(G)$ ガ假定Ａヲ満足スレバソレハ準單純デアル．[20]

abel 群 compact 群等ノ場合 $R^{(1,p)}(G)$ ガ準單純ニナルトイフコトハソレガ假定Aヲ滿足スルコトノ結果デアッタワケデス．

次ノ定理モ abel 群ノ場合ト全ク同樣ニシテ證明サレマス．[21)]

定理 21． $R^{(1,p)}/M_1$ ガ有限次元ナル如キ凡テノ maximal 両側 Ideal M_1 ノ共通 Ideal ガ 0 ナラバ G ノ任意ノ相異ナル元 a, b ニ對シ $D(a) \neq D(b)$ ナル如キ有界連続既約表現ガ存在スル．ヨッテ特ニ $R^{(1,p)}(G)$ ガ假定Aヲ滿足スレバ G ハ maximally almost periodic デアル．

サテコレラノ定理ニ述ベタ條件ガ滿足サレルノハドウイフ場合カト云フトコレハ非常ニ難カシイ問題デハナイカト思ヒマス．

唯 $L^{(1,2)}(G)$ ヲ用ヒマストソレガ Hilbert 空間ノ有界作用素ノ環トシテ表現サレルノデ Hilbert 空間ノ作用素論トモ関聯シテ来ルワケデスガ，コレラノコトハ又後ニ述ベテ見タイト思ヒマス．

脚註

1) I. E. Segal; The group ring of a locally compact group, I, Proc. Nat. Acad. Sci., U.S.A. **27** (1941)
2) 深宮政範; Bicompact ナ群ノ群環ニツイテ， 全国紙上談話会，第 240 号
3) I. Gelfand; Normierte Ringe, Rec. Math., **51** (1941). コノ論文ヲ以下 N.R. トスル．
4) I. Gelfand and D. Raikov, C.R. Acad. Sci. URSS, **28** (1940)
5) コンナコトハ周知ノコトデアリマセウガ一應証明シテオキマス．
6) N.R. Satz 8.
7) $R/I \simeq K_n$ ナル故 R/I ノ Ideal ハ 0 デナケレバ idempotent ナ元ヲ含ム．
8) 註 2) 參照．ソコデノ証明ハ両側 Ideal ヲ考ヘラレタノデスガ片側 Ideal デモ同様ニ成立シマス．
9) 勿論 G ガ適當ナ性質ヲ有スル測度ヲ持ツ群デアレバ locally compact, separable デナクトモヨイノデスガ，簡單ノタメニ上ノ如クシテオキマス．
10) Segal, ℓ.c.
11) $p = 1$ ナラバ $u_{ij}(g)$ ハ有界，$p \geq 2$ ナラバ差當リ適當ニ $\mu(E) \leq 2$ ナル E ガ存在シテ E ノ外デハ $u_{ij}(g)$ ハ有界．E 上デハ L_E^q $(\frac{1}{p} + \frac{1}{q} = 1)$ ニ属スルコトガワカル．実ハコノトキモ $u_{ij}(g)$ ガ有界ナルコトハ以下ノ証明ニ示ス通リデス．
12) 後の應用ノタメニ我々ハ $D(a), D'(a)$ ガ可測デアルコトダケヲ假定シテオキマス．
13) ココデコノコトヲ用ヒルノハマヅイヤリ方デアッテ我々ノ立場トシテハ逆ニ $T(x)$ ガ完全可約ナルコトカラ $D(a)$ ノ完全可約性ヲ出シタイ様ニ思ヒマス．ソシテコレハ特ニ G ガ compact ナル場合ニハ比較的簡單ニ証明サレマス．コレラノコトニ関シテハ又別ニ述べテ見タイト思ヒマス．
14) 一般ニ連續デナイ表現ガ存在スルモノナノカドウカヨクワカリマセン．
15) M ガ閉ヂテキレバ定理 12 ニヨリ $L^{(1,p)}/M$ ハ準單純，從ッテ單位ガ存在スルコトニ注意．
16) 群 G ノ單位ヲ $R^{(1,p)}(G)$ ノ單位ト區別シテ e_0 トカキマシタ．
17) Segal ハソコデ G ガ compact 或ハ abelian ナル場合ダケヲ云ッテヰルノデスガ，ソレハ恐ラク $R^{(1,1)}(G)$ ノ凡テノmaximal ideal ヲ問題ニシテヰルタメデアリマセウ．
18) G ハ compact ナル故 $\mu(E) = \mu(E^{-1})$ トナル．
19) 註 4) ノ論文ニ於ケル定理 5 參照．
20) Segal ノ定理 1．又深宮氏ノ論文參照．ソコデ直接簡單ナ証明ヲサレテヰマス．
21) 註 4) ノ論文ニ於ケル定理 6．

[6] On the structure of infinite M-groups

Japanese J. Math., 18 (1943), 709-728.

(Received February 21st, 1942.)

§1. Introduction.

In a previous publication[1], I have studied the structure of finite M-groups. The purpose of the present paper is to extend the results of that paper to the case of infinite groups.

A group $\mathfrak{G}$ is called M-group when the lattice $L(\mathfrak{G})$ formed by all subgroups of $\mathfrak{G}$ is a modular lattice[2]: it holds namely for arbitrary subgroups $\mathfrak{A}$, $\mathfrak{B}$, $\mathfrak{C}$ with $\mathfrak{A} \subseteqq \mathfrak{C}$, of $\mathfrak{G}$ the modularity equation

$$\mathfrak{A} \smile (\mathfrak{B} \frown \mathfrak{C}) = (\mathfrak{A} \smile \mathfrak{B}) \frown \mathfrak{C}. \tag{1}$$

In this case, if we make each subgroup $\mathfrak{X}$ lying between $\mathfrak{A} \smile \mathfrak{B}$ and $\mathfrak{B}$ correspond to the subgroup $\mathfrak{X} \frown \mathfrak{A}$, and each subgroup $\mathfrak{Y}$ lying between $\mathfrak{A}$ and $\mathfrak{A} \frown \mathfrak{B}$ to the subgroup $\mathfrak{Y} \smile \mathfrak{B}$, we have a lattice isomorphism between quotient lattices $\mathfrak{A} \smile \mathfrak{B}/\mathfrak{B}$ and $\mathfrak{A}/\mathfrak{A} \frown \mathfrak{B}$:

$$\mathfrak{A} \smile \mathfrak{B}/\mathfrak{B} \simeq \mathfrak{A}/\mathfrak{A} \frown \mathfrak{B}. \tag{2}$$

It follows then, that the length of any principal chain which connects two subgroups of $\mathfrak{G}$ is always equal to each other[3]. A group is called "group of finite length", when all its principal chains which connect the whole group with the identity, have always the same and finite length. For example any finite M-group is a group of finite length. But it is not yet decided whether a group of finite length is finite or not. This is a special case of the unproved assumption, that a group is finite when its subgroups satisfy the both chain conditions[4].

The structure of a finite M-group is given by the following theorems[5]:

---

(1) K. Iwasawa. Über die endlichen Gruppen und die Verbände ihrer Untergruppen, Journal of the faculty of science, Tokyo Imperial University, I, vol. IV, part 3. (1941)—referred to as G.V.

(2) See G. V. p. 171.

(3) For the general lattice theory see G. Birkhoff, Lattice theory (1940).

(4) This assumption is indeed valid for those groups which can be isomorphically represented with matrices on some commutative field. Cf. I. Schur, Sitzungsber. Preuss. Acad. Wiss. 1911, p. 619—627.

(5) See G. V. Satz 2, 3, 18.

1. A p-group ($p>2$) $\mathfrak{P}$ is an M-group if and only if it has the following structure:

(i) $\mathfrak{P}$ contains an abelian normal subgroup $\mathfrak{A}$,

(ii) $\mathfrak{P}/\mathfrak{A}$ is cyclical,

(iii) for a suitable generator T of $\mathfrak{P}/\mathfrak{A}$ we have for any element A of $\mathfrak{A}$

$$TAT^{-1} = A^{1+p^s}, \tag{3}$$

where s means an integer which is uniquely determined by $\mathfrak{P}$ and $\mathfrak{A}$, and is independent of A.

2. A 2-group is an M-group, if and only if it has either the same structure as in 1 with an additional condition $s \geqq 2$, or it is a Hamiltonean 2-group, i.e. it is the direct product of a quaternion-group and an abelian group of type (2, 2 ..., 2).

3. Let p, q be two different prime numbers, $p>q$. A group $\mathfrak{G}$ of order $p^\alpha q^\beta$, which is not the direct product of their Sylowgroups, is an M-group if and only if it has the following structure:

(i) A p-Sylowgroup $\mathfrak{P}$ is normal in $\mathfrak{G}$ and is an abelian group of type (p, p..., p).

(ii) A q-Sylowgroup $\mathfrak{Q}$ is cyclic; $\mathfrak{Q} = \{Q\}$.

(iii) For an arbitrary element P of $\mathfrak{P}$ it holds

$$QPQ^{-1} = P^r \tag{4}$$

where r is a definite integer which depends only on the group $\mathfrak{G}$ and satisfies following conditions

$$r \not\equiv 1 \text{ mod. } p\,, \qquad r^q \equiv 1 \text{ mod. } p\,. \tag{5}$$

4. A finite group $\mathfrak{G}$ is an M-group if and only if it is the direct product of groups of mutually prime orders which have structures mentioned above in 1-3.

Now let $\mathfrak{G}$ be an arbitrary M-group and A and B its elements of finite orders. We prove first that AB is also of finite order. Let $\mathfrak{A} = \{A\}$, $\mathfrak{B} = \{B\}$. From (2) it follows easily that $\mathfrak{A} \smile \mathfrak{B}$ is of finite length; consequently it contains no element of infinite order, because an element C of infinite order would give rise to an infinite sequence of subgroups $\{C\}$, $\{C^2\}$, $\{C^4\}$, ... As an element of $\mathfrak{A} \smile \mathfrak{B}$ AB is of finite order. Thus we obtain the following

Theorem 1. In an M-group $\mathfrak{G}$, elements of finite orders form together a characteristic subgroup $\mathfrak{E}$ of $\mathfrak{G}$. $\mathfrak{G}/\mathfrak{E}$ is then an M-group whose all elements except the identity are of infinite orders.

In the following we treat two cases separately, $\mathfrak{G}=\mathfrak{C}$, and $\mathfrak{G}\neq\mathfrak{C}$. In §2 we study the case $\mathfrak{G}=\mathfrak{C}$. On that occasion we must employ the assumption mentioned above, i.e. that a group of finite length is finite. Therefore the theorems which are to be obtained in §2 are available only for those groups whose subgroups of finite length have finite orders, for example locally finite groups, quasi-Hamiltonean groups etc. But if that assumption could have been proved anyhow, then our results would give the structure of general M-groups whose all elements are of finite orders.

In §3, we determine the structure of an M-group which has at least one element of infinite order. It is to be noted that in this case $\mathfrak{G}\neq\mathfrak{C}$, we can carry out our investigation without any assumption. This may deserve some interest. In fact we prove a special case of the above assumption and by help of this lemma we can avoid the relating difficulties after some delicate considerations.

§2. M-groups without element of infinite order.

1. In this paragraph we always consider such an M-group $\mathfrak{G}$ that its elements are all of finite orders and its subgroups of finite length are finite. We have then at once.

Lemma 1. If $\mathfrak{A}$ and $\mathfrak{B}$ are finite subgroups of $\mathfrak{G}$, then $\mathfrak{A}\cup\mathfrak{B}$ is also finite. Finite subgroups of $\mathfrak{G}$ thus form a sublattice of $L(\mathfrak{G})$.

Proof. From (2) it follows easily that $\mathfrak{A}\cup\mathfrak{B}$ is of finite length, and so is finite by the assumption(6).

Now let p be any prime number. We assume first that any element of p-power order is commutative with all elements which have orders prime to p. Let A, B be any element of order p^α, p^β respectively. According to Lemma 1 $\{A, B\}$ is a finite group and by the assumption its p-Sylowgroup is normal. While A, B are contained in this p-Sylowgroup, $\{A, B\}$ itself is a p-group and AB has also a p-power order. All elements of p-power orders form therefore a characteristic subgroup of $\mathfrak{G}$: the p-component $\mathfrak{P}$ of $\mathfrak{G}$.

We assume next that an element P of order p^m is not commutative with an element Q of different prime power order q^n. As before it is easy to see that $\{P, Q\}$ is a finite group of type 3 in §1. We may suppose for instance, exchanging P and Q if necessary, that a p-Sylowgroup $\mathfrak{P}_0$ is an abelian normal subgroup of type $(p, p, \dots, p)$, and a q-Sylowgroup is cyclic. Now let P' be another element of p-power

---

(6) The converse is also true; that is to say, this lemma is equivalent with the assumption mentioned above.

order. $\{P, P', Q\}$ is also a group of type 3 in §1. It follows therefore

$$PP' = P'P, \quad P^p = P'^p = 1,$$

that is, all elements of $\mathfrak{G}$ of p-power orders form an abelian characteristic subgroup $\mathfrak{P}$, any of whose elements satisfy the relation

$$X^p = 1. \tag{6}$$

Then it is easy to see that for an arbitrary element P_0 of $\mathfrak{P}$ it holds

$$QP_0Q^{-1} = P_0^r, \quad r \not\equiv 1, \text{ mod. } p, \quad r^q \equiv 1, \text{ mod. } p. \tag{7}$$

If we take another Q' of q-power order then $\{P, Q, Q'\}$ is also of type 3 in §1. We denote the normal p-Sylowgroup of this group by $\mathfrak{P}_1$ and a cyclic q-Sylowgroup by $\mathfrak{Q}_1 = \{Q^*\}$. It follows then

$$Q^*P_1Q^{*-1} = P_1^{r_1}, \quad r_1 \not\equiv 1, \text{ mod. } p, \quad r_1^q \equiv 1, \text{ mod. } p, \tag{8}$$

for all P_1 in $\mathfrak{P}_1$.
If we have

$$Q^{*u} \equiv Q \qquad \text{mod. } \mathfrak{P}_1,$$

it follows that

$$P^{r_1^u} = Q^{*u}PQ^{*-u} = QPQ^{-1} = P^r,$$

consequently (cf. (7), (8))

$$r_1^u \equiv r, \text{ mod. } p, \qquad u \not\equiv 0, \text{ mod, } q.$$

Q generates therefore also a q-Sylowgroup and it is

$$\{P, Q, Q'\} = \{\mathfrak{P}_1, Q\}.$$

Any element of q-power order is thus contained in the subgroup $\{\mathfrak{P}, Q\}$. $\{\mathfrak{P}, Q\}$ is therefore a characteristic subgroup of $\mathfrak{G}$. From above considerations we obtain the following theorem which corresponds to 4 in §1.

Theorem 2. An M-group $\mathfrak{G}$ which satisfies the assumption stated at the beginning of this paragraph is the direct product of following groups:

(α) The M-group in which any element has an order which is a power of a fixed prime number.

(β) The group $\{\mathfrak{P}, Q\}$, where $\mathfrak{P}$ is an abelian group whose elements satisfy the relation

$$X^p = 1,$$

and Q is an element of q-power order. p, q means then two different prime numbers. Between Q and an arbitrary element P of $\mathfrak{P}$ it holds

$$QPQ^{-1} = P^r, \quad r \not\equiv 1, \text{ mod. } p, \quad r^q \equiv 1, \text{ mod. } p,$$

where r is an integer depending only upon $\{\mathfrak{P}, Q\}$.
The orders of elements contained in different direct factors must be prime to each other.
Conversely any group, which has the structure mentioned above, is an M-group which satisfies our assumption.

The last part of the theorem is easily seen from 3, 4 in §1.

2. According to Theorem 2 it suffices to investigate only such M-groups any of whose elements has an order which is a power of a fixed prime number.

We first consider a 2-group $\mathfrak{G}$ which contains a quaternion-group $\mathfrak{Q}$. Let A, B be any two elements of the group and put $\mathfrak{G}_1 = \{\mathfrak{Q}, A, B\}$. According to 2 in §1, $\mathfrak{G}_1$ is a Hamiltonean 2-group and it holds

$$BAB^{-1} = A^{\alpha}, \qquad ABA^{-1} = B^{\beta}.$$

The whole group $\mathfrak{G}$ is therefore also Hamiltonean and is the direct product of $\mathfrak{Q}$ and an abelian group of exponent 2. Conversely any such group is obviously an M-group.

Now let $\mathfrak{G}$ be a modular p-group which is neither abelian nor Hamiltonean. By the assumption $\mathfrak{G}$ contains a finite non-abelian subgroup $\mathfrak{G}_1$. We consider any finite subgroup $\mathfrak{G}_2$ of $\mathfrak{G}$ containing $\mathfrak{G}_1$

The structure of $\mathfrak{G}_2$ is given by (i), (ii), (iii) in 1 of §1. We will write

$$\mathfrak{G}_2 = \{\mathfrak{A}_2, T, s\}, \tag{9}$$

if $\mathfrak{G}_2$ has an abelian normal subgroup $\mathfrak{A}_2$ and

$$\mathfrak{G}_2/\mathfrak{A}_2 = \{T\}, \; TAT^{-1} = A^{1+p^s} \quad \text{for all } A \text{ in } \mathfrak{A}_2.$$

The expression (9) may not be unique for $\mathfrak{G}_2$, that is to say, it may be also written in the form

$$\mathfrak{G}_2 = \{\mathfrak{A}_2^*, T^*, s^*\}$$

with some other $\mathfrak{A}_2^*$, T^*, s^*. But, as $\mathfrak{G}_2$ is not abelian the maximum of these $s, s^*, \ldots$ is surely determined by $\mathfrak{G}_2$ and we denote it by $s(\mathfrak{G}_2)$.

Now let s_0 be the minimum of all $s(\mathfrak{G}_2)$ when $\mathfrak{G}_2$ goes round over all finite subgroups of $\mathfrak{G}$ containing $\mathfrak{G}_1$. We choose such $\mathfrak{G}_3$, that

$$\mathfrak{G}_3 = \{\mathfrak{A}_3, T_3, s_3\}, \qquad s_3 = s(\mathfrak{G}_3) = s_0.$$

Take any finite subgroup $\mathfrak{G}_4$ containing $\mathfrak{G}_3$, and let

$$\mathfrak{G}_4 = \{\mathfrak{A}_4, T_4, s_4\}, \qquad s_4 = s(\mathfrak{G}_4).$$

If we put

$$\mathfrak{G}_3 \frown \mathfrak{A}_4 = \mathfrak{A}_3^*, \quad \mathfrak{G}_3 \smile \mathfrak{A}_4 = \mathfrak{G}_4^*, \quad [\mathfrak{G}_4 : \mathfrak{G}_4^*] = p^u,$$

we have then

$$\mathfrak{G}_4^*/\mathfrak{A}_4 \cong \mathfrak{G}_3/\mathfrak{A}_3^*, \qquad \mathfrak{G}_3 = \{\mathfrak{A}_3^*, T_3^*, s_4+u\}.$$

From

$$s_4+u \leqq s_3, \qquad s_3 = s_0 \leqq s_4$$

it follows

$$s_3 = s_4 = s_0, \qquad u = 0,$$

that is to say,

$$s(\mathfrak{G}_4) = s_0, \quad \mathfrak{G}_3 \smile \mathfrak{A}_4 = \mathfrak{G}_4, \quad \mathfrak{G}_4/\mathfrak{A}_4 \cong \mathfrak{G}_3/\mathfrak{A}_3^*,$$

$$\mathfrak{G}_4 = \{\mathfrak{A}_4, T_3^*, s_0\}.$$

If the order of T_3^* is p^l, we have for any element A_4 of $\mathfrak{A}_4$

$$A_4 = T_3^{*p^l} A_4 T_3^{*-p^l} = A_4^{(1+p^{s_0})p^l}$$

The order of A_4 is therefore at most p^{s_0+l}, and the orders of elements of $\mathfrak{G}_4$ also do not exceed p^{s_0+l} (7) But the number p^{s_0+l} depends only on $\mathfrak{G}_3$, and $\mathfrak{G}_4$ could be arbitrarily chosen. The orders of elements of $\mathfrak{G}$ are therefore bounded.

Now let $\mathfrak{G}_4$ be, as before, any finite subgroup containing $\mathfrak{G}_3$. We suppose that $\mathfrak{G}_4$ can be expressed in two different ways:

$$\mathfrak{G}_4 = \{\mathfrak{A}_4, T_4, s_0\}$$

$$= \{\mathfrak{A}_4^*, T_4^*, s_0\}.$$

It is then easy to see, that the commutator group $\mathfrak{G}_4'$ of $\mathfrak{G}_4$ is given by

$$\mathfrak{G}_4' = \mathfrak{A}_4^{p^{s_0}} = \mathfrak{A}_4^{*p^{s_0}},$$

where $\mathfrak{A}_4^{p^{s_0}}$ or $\mathfrak{A}_4^{*p^{s_0}}$ is the group generated by all p^{s_0}-power of elements of $\mathfrak{A}_4$ or $\mathfrak{A}_4^*$. The maximum of orders of elements of $\mathfrak{A}_4$ and $\mathfrak{A}_4^*$ is therefore equal to each other and is uniquely determined by $\mathfrak{G}_4$. We denote this by $o(\mathfrak{G}_4)$. Let p^k be the maximum of $o(\mathfrak{G}_4)$, for all $\mathfrak{G}_4$ containing $\mathfrak{G}_3$. This is surely determined, as the orders of elements of $\mathfrak{G}$ are bounded. For some fixed $\mathfrak{G}_5 \geqq \mathfrak{G}_3$, we have then $o(\mathfrak{G}_5) = p^k$.

(7) See G. V. Hilfssatz 18.

Now let

$$(10)\qquad \mathfrak{A}_5^{(1)},\quad \mathfrak{A}_5^{(2)},\quad \ldots,\quad \mathfrak{A}_5^{(r)}$$

be all the abelian normal subgroups of $\mathfrak{G}_5$, such that

$$\mathfrak{G}_5 = \{\mathfrak{A}_5^{(i)}, T_5^{(i)}, s_0\}, \quad i = 1, 2, \ldots, r.$$

We prove, that if we choose suitable $\mathfrak{A}_5^{(i_0)}$, we can find for any $\mathfrak{G}_6 \geqq \mathfrak{G}_5$ an abelian normal subgroup $\mathfrak{A}_6$, such that

$$\mathfrak{A}_6 \geqq \mathfrak{A}_5^{(i_0)}, \quad \mathfrak{G}_6 = \{\mathfrak{A}_6, T_6, s_0\}.$$

Suppose by the contrary that there exists no such $\mathfrak{A}_5^{(i_0)}$ in (10). Then we can find for any $\mathfrak{A}_5^{(i)}$ a finite subgroup $\mathfrak{G}_6^{(i)}$, which has no abelian normal subgroup $\mathfrak{A}_6^{(i)}$ with conditions

$$\mathfrak{G}_6^{(i)} = \{\mathfrak{A}_6^{(i)}, T_6^{(i)}, s_0\}, \quad \mathfrak{A}_6^{(i)} \geqq \mathfrak{A}_5^{(i)}.$$

Put

$$\mathfrak{G}_6^* = \mathfrak{G}_6^{(1)} \smile \mathfrak{G}_6^{(2)} \smile \cdots \smile \mathfrak{G}_6^{(r)}, \quad \mathfrak{G}_6^* = \{\mathfrak{A}_6^*, T_6^*, s_0\}, \quad \mathfrak{A}_5^* = \mathfrak{A}_6^* \frown \mathfrak{G}_5.$$

It follows then

$$\mathfrak{G}_5 = \{\mathfrak{A}_5^*, T_5^*, s_0\},$$

and $\mathfrak{A}_5^*$ is therefore one of $\mathfrak{A}_5^{(i)}$ in (10): $\mathfrak{A}_5^* = \mathfrak{A}_5^{(i_0)}$. Now $\mathfrak{G}_6^{(i_0)}$ can be written in the form

$$\mathfrak{G}_6^{(i_0)} = \{\mathfrak{A}_6^{(i_0)}, T_6^{(i_0)}, s_0\},$$

with

$$\mathfrak{A}_6^{(i_0)} = \mathfrak{A}_6^* \frown \mathfrak{G}_6^{(i_0)} \geqq \mathfrak{A}_6^* \frown \mathfrak{G}_5 = \mathfrak{A}_5^{(i_0)},$$

which contradicts our assumption.

We have thus found a finite subgroup $\mathfrak{G}_5 = \mathfrak{G}^*$ with following properties:

(α) $\mathfrak{G}^*$ is expressed in the form

$$\mathfrak{G}^* = \{\mathfrak{A}^*, T, s_0\}.$$

(β) Any finite subgroup $\mathfrak{G}_6$, which contains $\mathfrak{G}^*$, can be expressed in the form

$$\mathfrak{G}_6 = \{\mathfrak{A}_6, T_6, s_0\}, \quad \mathfrak{A}_6 \geqq \mathfrak{A}^*, \quad \mathfrak{G}_6 = \mathfrak{A}_6 \smile \mathfrak{G}^*$$

(γ) $o(\mathfrak{G}^*) = p^k = \text{Max.}\, o(\mathfrak{G}_4)$.

As $\mathfrak{G}_6 = \mathfrak{A}_6 \smile \mathfrak{G}^* = \{\mathfrak{A}_6, T\}$ we can put moreover in (β) $T_6 = T$.

Now let $\mathfrak{G}_7$, $\mathfrak{G}_8$ be any finite subgroups such that

$$\mathfrak{G}_8 \geqq \mathfrak{G}_7 \geqq \mathfrak{G}^*,$$

$$\mathfrak{G}_7 = \{\mathfrak{A}_7, T, s_0\}, \quad \mathfrak{A}_7 \geqq \mathfrak{A}^*$$

$$\mathfrak{G}_8 = \{\mathfrak{A}_8, T, s_0\}, \quad \mathfrak{A}_8 \geqq \mathfrak{A}^*.$$

Put

$$\mathfrak{A}_7 \smile \mathfrak{A}_8 = \{\mathfrak{A}_8, T^{p^u}\}, \quad T^{p^u} = A_7 A_8, \quad A_7 \in \mathfrak{A}_7, \quad A_8 \in \mathfrak{A}_8.$$

Take an element A_0 of order p^k from $\mathfrak{A}^*$. As $\mathfrak{A}^*$ is contained in $\mathfrak{A}_7 \frown \mathfrak{A}_8$, we have

$$A_0^{(1+p^{s_0})p^u} = T^{p^u} A_0 T^{-p^u} = A_7 A_8 A_0 A_8^{-1} A_7^{-1} = A_0,$$

and consequently

(11) $$s_0 + u \geqq k.$$

As the order of any element $\mathfrak{A}_8$ is at most p^k by the assumption, it is easy to see from (11) that T^{p^u} is commutative with any element contained in $\mathfrak{A}_8$. $\mathfrak{A}_7 \smile \mathfrak{A}_8$ is therefore an abelian group and we can write

$$\mathfrak{G}_8 = \{\mathfrak{A}_7 \smile \mathfrak{A}_8, T, s_0\}.$$

If we take $\mathfrak{G}_7 = \mathfrak{G}_8$, above consideration shows that there is unique maximal abelian normal subgroup $\mathfrak{A}_8$ such that

$$\mathfrak{G}_8 = \{\mathfrak{A}_8^*, T, s_0\}, \quad \mathfrak{A}_8^* \geqq \mathfrak{A}^*.$$

In the following if any $\mathfrak{G}_9 \geqq \mathfrak{G}^*$ is expressed in the form

$$\mathfrak{G}_9 = \{\mathfrak{A}_9, T, s_0\},$$

we take then alawys as $\mathfrak{A}_9$ the maximal abelian normal subgroup containing $\mathfrak{A}^*$.

For any finite subgroups $\mathfrak{G}_9$, $\mathfrak{G}_{10}$ with

$$\mathfrak{G}_{10} \geqq \mathfrak{G}_9 \geqq \mathfrak{G}^*,$$

$$\mathfrak{G}_9 = \{\mathfrak{A}_9, T, s_0\}, \quad \mathfrak{G}_{10} = \{\mathfrak{A}_{10}, T, s_0\},$$

it is then easy to see from above consideration, that

(12) $$\mathfrak{A}_9 = \mathfrak{G}_9 \frown \mathfrak{A}_{10}.$$

Now for arbitrary elements X, Y in $\mathfrak{G}$, put

$$\mathfrak{G}_X = \{\mathfrak{G}^*, X\} = \{\mathfrak{A}_X, T, s_0\}$$

$$\mathfrak{G}_Y = \{\mathfrak{G}^*, Y\} = \{\mathfrak{A}_Y, T, s_0\}$$

$$\mathfrak{G}_{X,Y} = \{\mathfrak{G}^*, X, Y\} = \{\mathfrak{A}_{X,Y}, T, s_0\}.$$

It follows from (12) that

$$\mathfrak{A}_X = \mathfrak{A}_{X,Y} \frown \mathfrak{G}_X, \qquad \mathfrak{A}_Y = \mathfrak{A}_{X,Y} \frown \mathfrak{G}_Y.$$

As $\mathfrak{A}_{X,Y}$ is abelian, elements of $\mathfrak{A}_X$ and $\mathfrak{A}_Y$ are commutative with one another. The group

$$\mathfrak{A} = \underset{X \in \mathfrak{G}}{U} \mathfrak{A}_X$$

is accordingly an abelian group, and for any element A in $\mathfrak{A}$ we have

$$TAT^{-1} = A^{1+p^{s_0}}.$$

It is also clear that

$$\mathfrak{G} = \{\mathfrak{A}, T\}.$$

We have thus proved the first half of the following theorem, which is an extension of results 1, 2 of §1 in the case of infinite groups.

Theorem 3. A non-abelian p-group[8] $\mathfrak{G}$, whose subgroups of finite length are all finite, is an M-group, if and only if it has the following structure.

(a) $p > 2$.

(i) $\mathfrak{G}$ contains an abelian normal subgroup $\mathfrak{A}$. The order of elements of $\mathfrak{A}$ is bounded.

(ii) $\mathfrak{G}/\mathfrak{A}$ is a cyclical group of order p^m.

(iii) For a suitable generator T of $\mathfrak{G}/\mathfrak{A}$ we have for any element A in $\mathfrak{A}$

$$TAT^{-1} = A^{1+p^s},$$

where s means an integer which is uniquely determined by $\mathfrak{G}$ and $\mathfrak{A}$ and satisfies the inequality

$$s + m \geqq n,$$

if the maximum of the order of elements in $\mathfrak{A}$ is p^n.

(iv) $T^m = T_0$ is an element of $\mathfrak{A}$ and it holds

$$T_0^{p^s} = 1.$$

---

(8) A group is called p-group generally when the orders of its elements are all p-powers.

(β) $p=2$.

$\mathfrak{G}$ is either a Hamiltonean group, or a group which has the same structure as mentioned in (α) with an additional condition $s \geqq 2$.

Corollary. A modular p-group, whose subgroups of finite length are finite, is either abelian or meta-abelian.

Proof. We have only to prove the second half. It is easy to see that in any group $\mathfrak{G}$, which has the structure mentioned above, subgroups of finite length are all finite. Let A, B be any element of $\mathfrak{G}$. The finite group $\{A, B\}$ is then of type 1 (or 2) in §1. There exists consequently such integers x, y that

$$BA = A^x B^y .$$

$\mathfrak{G}$ is therefore quasi-Hamiltonean(9), accordingly modular a fortiori.

In a quasi-Hamiltonean group any subgroup of finite length is, as readily to be seen, finite. If we remark therefore that a finite quasi-Hamiltonean group is the direct product of their Sylowgroups and the groups given in Theorem 3 are all quasi-Hamiltonean, we have immediately the following

Theorem 4. A group, whose elements have all finite orders, is quasi-Hamiltonean, if and only if it is the direct product of an abelian group and such groups, which have the structure stated in Theorem 3, where the orders of elements of different direct factors are prime to each other. A quasi-Hamiltonean group is therefore always either abelian or meta-abelian.

§ 3. M-groups which contain elements of infinite orders.

1. We now determine the structure of M-groups, which contain elements of infinite orders. This time we can carry out our investigation without any assumption.

Lemma 2. Let A, B be any two elements of an M-group $\mathfrak{G}$, of which A has an infinite order. If $\{A\} \frown \{B\} = 1$, then

(13) $\{A, B\}, \{A^2, B\}, \{A^3, B\}, \dots, \{A^n, B\}, \dots, \{B\}$

are different from one another and exhaust all subgroups of $\mathfrak{G}$ which lie between $\{A, B\}$ and $\{B\}$.

(9) A group $\mathfrak{G}$ is called "quasi-Hamiltonean", if for any subgroups $\mathfrak{A}$, $\mathfrak{B}$ we have

$$\mathfrak{A}\mathfrak{B} = \mathfrak{B}\mathfrak{A} = \mathfrak{A} \smile \mathfrak{B} .$$

For the relation between an M-group and a quasi-Hamiltonean group see G. V. § 3.

Proof. Put $\mathfrak{A} = \{A\}$, $\mathfrak{B} = \{B\}$ and apply (2). As $\{A\} \frown \{B\} = 1$ and the subgroups of $\{A\}$ are given by

$$\{A\}, \{A^2\}, \{A^3\}, \ldots, \{A^n\}, \ldots, \{1\},$$

we can immediately obtain the lemma.

By help of this we prove the following

Lemma 3. Let A be any element of infinite order and B any element of finite order of an M-group $\mathfrak{G}$. It holds then in taking suitably an integer r

$$(14) \qquad ABA^{-1} = B^r, \quad (r, m) = 1,$$

where m is the order of B.

Proof. The group $\{B, ABA^{-1}\}$ lies between $\{A, B\}$ and $\{B\}$ and is therefore one of the subgroups (13). But as it is generated by elements of finite orders, it contains no element of infinite order (cf. Theorem 1). It must be then $\{B, ABA^{-1}\} = \{B\}$ and we can readily obtain (14).

Lemma 4. Let A, B be any two elements of an M-group $\mathfrak{G}$, of which A has an infinite order. If it holds for suitable integers α, β

$$(15) \qquad BA^{\alpha}B^{-1} = A^{\beta},$$

then it must be $\alpha = \beta$ and A^{α} commutes with B.

Proof. If $\{A\} \frown \{B\} = \{A^{\gamma}\} \neq 1$, then it follows from (15)

$$A^{\alpha\gamma} = BA^{\alpha\gamma}B^{-1} = A^{\beta\gamma}$$

whence $\alpha\gamma = \beta\gamma$, $\alpha = \beta$. Let $\{A\} \frown \{B\} = 1$. It is

$$\{A^{\alpha}, B\} = \{A^{\beta}, B\}$$

and from Lemma 2 we conclude $\alpha = \pm\beta$. Assume $\alpha = -\beta$. From Lemma 3 it follows easily that B has an infinite order. The group $\{A^{\alpha}, B\}/\{A^{4\alpha}, B^2\}$ is then a dihedral group, which is not modular. This contradicts to our assumption that $\mathfrak{G}$ is an M-group. It must be therefore $\alpha = \beta$ and $A^{\alpha}B = BA^{\alpha}$.

We now study the structure of M-groups which contain elements of infinite orders.

Theorem 5. Let $\mathfrak{G}$ be an M-group which contains elements of infinite orders. The group $\mathfrak{E}$, which consists of all elements of finite orders of $\mathfrak{G}$, is then an abelian group.

Proof. Let Z be an element of infinite order and A, B be any two elements of $\mathfrak{E}$. AZ is then also of infinite order: otherwise the order of Z would be finite (cf. Theorem 1). If we denote the order of B by m, it is then from Lemma 3

$$ZBZ^{-1} = B^r, \quad (AZ)B(AZ)^{-1} = B^{r'}, \quad (rr', m) = 1.$$

Consequently

$$AB^rA^{-1} = B^{r'}. \tag{16}$$

This shows that $\mathfrak{E}$ is either abelian or Hamiltonean. If $\mathfrak{E}$ is Hamiltonean it contains a quaternion group $\mathfrak{Q}$. By Lemma 3 Z gives an automorphism of finite order in $\mathfrak{Q}$:

$$Z\mathfrak{Q}Z^{-1} = \mathfrak{Q}.$$

A suitable power Z_0 of Z is therefore commutative with every element of $\mathfrak{Q}$. $\{Z_0, \mathfrak{Q}\}/\{Z_0^4\}$ is then a 2-group which is not modular (cf. 2 in §1). We thus come to a contradiction and $\mathfrak{E}$ must be abelian.

2. We consider next $\mathfrak{G}/\mathfrak{E}$, that is to say, such M-groups, which contain no element of finite order except the identity. We first prove a lemma, which is a special case of the assumption stated in §§1, 2.

Lemma 5. An M-group which is generated by two elements of order 2 is a finite group.

Proof. Let $\mathfrak{G} = \{A, B\}$, $A^2 = B^2 = 1$. As any element of $\mathfrak{G}$ has a finite order according to Theorem 1, put

$$C = AB, \qquad C^m = 1.$$

It is then $\mathfrak{G} = \{A, C\}$ and

$$A^{-1}CA = C^{-1}.$$

The order of $\mathfrak{G}$ is therefore at most $2m$.

The structure of $\mathfrak{G}/\mathfrak{E}$ is then given by the following theorem.

Theorem 6. An M-group which contains no element of finite order except the identity is abelian.

The proof is rather complicated. We devide it into two parts, the latter of which is available for any M-group.

Lemma 6. Let A and B be elements of an M-group which contains no element of finite order except the identity. If $\{A\} \frown \{B\} \neq 1$, then the subgroup $\{A, B\}$ is cyclical.

Lemma 7. Let A and B be elements of infinite orders of an M-group. If $\{A\}\frown\{B\} = 1$, then A and B are commutative with one another;

$$AB = BA\,.$$

Proof of Lemma 6. Clearly it is sufficient to prove that $\{A, B\}$ is abelian.
Let

$$\{A\}\frown\{B\} = \{Z\}\,, \quad A^{\alpha} = Z$$

and suppose $A \neq BAB^{-1}$. As

$$A^{\alpha} = (BAB^{-1})^{\alpha},\ \{A\}\frown\{BAB^{-1}\} \neq 1\,,$$

we put

$$\{A\}\frown\{BAB^{-1}\} = \{C\}\,, \quad C = A^{\alpha} = BA^{\alpha'}B^{-1}\,.$$

From Lemma 4 we have $\alpha = \alpha'$. Therefore if we denote $A_1 = A$, $A_2 = BAB^{-1}$, it holds

$$A_1^{\alpha} = A_2^{\alpha} = C\,, \quad \{A_1\}\frown\{A_2\} = \{C\}\,.$$

As the M-group $\{A_1, A_2\}/\{C\}$ is generated by two elements of finite orders, it contains no element of infinite order. There is consequently such $\gamma(\neq 0)$, that

$$(A_1A_2^{-1})^{\beta} = C^{\gamma}\,.$$

Let

$$\mathfrak{A} = \{(A_1C^{\gamma})\}\,, \quad \mathfrak{B} = \{(A_2C^{\gamma})\}\,, \quad \mathfrak{C} = \{A_1\}\,.$$

It is then $\mathfrak{A} \subsetneqq \mathfrak{C}$ and from $\mathfrak{B}\frown\mathfrak{C} \subseteqq \{A_1\}\frown\{A_2\} = \{C\}$ we can readily see that $\mathfrak{B}\frown\mathfrak{C} = \{(A_2C^{\gamma})^{\alpha}\} = \{C^{1+\alpha\gamma}\}$. As $(A_1C^{\gamma})^{\alpha} = C^{1+\alpha\gamma}$ it is also $\mathfrak{B}\frown\mathfrak{A} = \{C^{1+\alpha\gamma}\}$. Now $\mathfrak{A}\smile\mathfrak{B}$ contains $(A_1C^{\gamma})(A_2C^{\gamma})^{-1} = A_1A_2^{-1}$, therefore $\mathfrak{A}\smile\mathfrak{B} = \{A_1C^{\gamma}, A_2C^{\gamma}, C^{\gamma}\} = \{A_1, A_2\} = \mathfrak{B}\smile\mathfrak{C}$. We have thus

$$\mathfrak{A} \subseteqq \mathfrak{C}\,, \quad \mathfrak{A}\frown\mathfrak{B} = \mathfrak{C}\frown\mathfrak{B}\,, \quad \mathfrak{A}\smile\mathfrak{B} = \mathfrak{C}\smile\mathfrak{B}\,.$$

From (1) we conclude $\mathfrak{A} = \mathfrak{C}$, but this is clearly a contradiction.

Remark that γ is not zero as $A_1A_2^{-1}$ is of infinite order. We used here essentially the assumption that the group has no element of finite order except the identity.

Proof of Lemma 7. We put $\mathfrak{G} = \{A, B\}$ and prove that $\mathfrak{G}$ is abelian. For that purpose we must first prove the following

Lemma 8. Let $\mathfrak{G} = \{A, B\}$ be an M-group as above defined. If $\mathfrak{G}$ has an abelian normal subgroup $\mathfrak{A}$ with a finite index, then $\mathfrak{G}$ itself is abelian.

Proof. We remark first that $\mathfrak{G}$ contains no element of finite order except the identity: for if $Z \neq 1$ is any of such elements we have

$$\{A, Z\} = \{A, B^m\},$$

and there exist infinitely many subgroups between $\{A, Z\}$ and $\{A\}$ (cf. Lemma 2). But this contradicts to the lattice isomorphism

$$\{A, Z\}/\{A\} \cong \{Z\}/\{A\} \frown \{Z\}.$$

Now according to the assumption there is suitable integers α, β such that

$$A^\alpha, B^\beta \in \mathfrak{A}, \quad A^\alpha B^\beta = B^\beta A^\alpha.$$

Hence we have $\{A\} \frown \{B^\beta A B^{-\beta}\} \neq 1$. If $A \neq B^\beta A B^{-\beta}$, then we come to a contradiction as in the proof of Lemma 6. We must have therefore

$$A = B^\beta A B^{-\beta} \quad \text{or} \quad B^\beta = A^{-1} B^\beta A.$$

From $\{B\} \frown \{A^{-1}BA\} \neq 1$ we have again similarly

$$B = A^{-1}BA \quad \text{or} \quad AB = BA.$$

We now prove Lemma 7. Denote

$$\mathfrak{G} = \{A, B\}, \ \mathfrak{G}_1 = \{ABA^{-1}, B\}, \ \mathfrak{G}_2 = \{A, B^2\}, \ \mathfrak{G}_3 = \{AB^2A^{-1}, B^2\}.$$

By Lemma 2 we have

$$\mathfrak{G}_1 = \{A^\alpha, B\} \quad \text{or} \quad \mathfrak{G}_1 = \{B\}. \tag{17}$$

As ABA^{-1} is contained in $\mathfrak{G}_1$, A is a normaliser of $\mathfrak{G}_1$ and $\mathfrak{G}_1$ is accordingly normal in $\mathfrak{G}$. In the same way we can see that $\mathfrak{G}_3$ is normal in $\mathfrak{G}_2$. From $\mathfrak{G}_3 = \{(AB)B^2(AB)^{-1}, B^2\}$ it is also normal in $\{AB, B^2\}$, consequently in $\mathfrak{G} = \{A, AB, B^2\}$. As $\mathfrak{G}_1/\mathfrak{G}_3$ is generated by two elements of order 2 it is a finite group according to Lemma 5.

We first assume that $\mathfrak{G}/\mathfrak{G}_3$ is infinite. $\mathfrak{G}/\mathfrak{G}_1$ is then also infinite and from (17) we have $\mathfrak{G}_1 = \{B\}$: otherwise $\mathfrak{G}/\mathfrak{G}_1$ would be of order α. It follows $ABA^{-1} = B^\beta$ and by Lemma 4 $AB = BA$, that is to say, $\mathfrak{G}$ is abelian. Next let $\mathfrak{G}/\mathfrak{G}_3$ be finite. As $\mathfrak{G}_3 \subseteqq \mathfrak{G}_2$ and $\mathfrak{G} \neq \mathfrak{G}_2$ (cf. Lemma 2), $\mathfrak{G}/\mathfrak{G}_3$ is not the identical group. As any finite M-group is meta-abelian (cf. §1) we see from this that the commutator group $\mathfrak{G}'$ of $\mathfrak{G}$ does not coincide with $\mathfrak{G}: \mathfrak{G} \neq \mathfrak{G}'$. If $\mathfrak{G}' \subseteqq \{A\}$ and $\mathfrak{G}' \subseteqq \{B\}$ it follows then from $\{A\} \frown \{B\} = 1$ that $\mathfrak{G}' = 1$ and that $\mathfrak{G}$ is abelian. We assume therefore for instance $\mathfrak{G}' \nsubseteqq \{A\}$. From $\{A\} \neq \{A\} \smile \mathfrak{G}'$ we have then by Lemma 2

$$\{A\} \smile \mathfrak{G}' = \{A, B^v\}, \quad v \neq 0.$$

As $\{A\}\smile\mathfrak{G}'/\mathfrak{G}'$ is a cyclic group generated by $\bar{A}$, there exists suitable u such that

$$A^u \equiv B^v \quad \text{mod. } \mathfrak{G}'.$$

$B_1 = A^u B^{-v}$ is then contained in $\mathfrak{G}'$ and $\{A\}\smile\mathfrak{G}' = \{A, B^v\} = \{A, B_1\}$. If $\{A\}\frown\{B_1\} \neq 1$, then it follows from Lemma 6 that $\{A\}\smile\mathfrak{G}'$ is abelian (cf. the above remark), and as $\mathfrak{G}/\{A\}\smile\mathfrak{G}'$ is finite, $\mathfrak{G}$ is also abelian according to Lemma 8
Let therefore be

$$\{A\}\frown\{B_1\} = 1\,.$$

As $\mathfrak{G}'$ is a subgroup of $\{A, B_1\}$ and contains $\{B_1\}$ it is

$$\mathfrak{G}' = \{A^n, B_1\} \qquad \text{or} \qquad \mathfrak{G}' = \{B_1\}$$

according to Lemma 2.
Assume first $\mathfrak{G}' = \{B_1\}$. By Lemma 4 it is easy to see that B_1 is contained in the center of $\mathfrak{G}$ and that $\{A\}\smile\mathfrak{G}'$ is abelian. $\mathfrak{G}$ itself is then an abelian group according to Lemma 8. We assume therefore

$$\mathfrak{G}' = \{A^n, B\}\,.$$

$\mathfrak{G}/\mathfrak{G}'$ is then a finite group. In this case. $\mathfrak{G}'$ is not abelian: otherwise $\mathfrak{G}$ would be abelian and $\mathfrak{G}' = 1$ according to Lemma 8.
Now $\mathfrak{G}'$ have the same structure as $\mathfrak{G}$. If we denote the commutator group of $\mathfrak{G}'$ by $\mathfrak{G}''$, $\mathfrak{G}'/\mathfrak{G}''$ is also finite, but not the identical group. $\mathfrak{G}''/\mathfrak{G}'''$ is again finite, where $\mathfrak{G}'''$ is the commutator group of $\mathfrak{G}''$. $\mathfrak{G}/\mathfrak{G}'''$ is then also a finite M-group. But according to §1, 1−4 any finite M-group is meta-abelian: we have thus met a contradiction. $\mathfrak{G}$ must be consequently abelian in any case.

Theorem 6 is now clear from these two lemmas.

Let $\mathfrak{G}$ be any M-group, which contains elements of infinite orders, and $\mathfrak{E}$ be the normal subgroup of $\mathfrak{G}$ which consists of all elements of finite order in $\mathfrak{G}$. By Theorem 5, 6 $\mathfrak{E}$ and $\mathfrak{G}/\mathfrak{E}$ are both abelian groups. It remains therefore only to investigate how $\mathfrak{E}$ is extended by an abelian group which contains no element of finite order except the identity.

Suppose first that the rank of $\mathfrak{G}/\mathfrak{E}$ is at least 2. Then there exists in $\mathfrak{G}$ two elements A, B of infinite orders such that

$$\{A\}\frown\{B\} = 1, \qquad AB = BA. \tag{18}$$

Let C be any element of infinite order in $\mathfrak{G}$. It is then by (18)

$$\{A\}\frown\{C\} = 1 \qquad \text{or} \qquad \{B\}\frown\{C\} = 1\,.$$

Suppose for instance

$$\{A\}\frown\{C\}=1\,,\qquad AC=CA\,. \tag{19}$$

Let D be another element of infinite order. If $\{C\}\frown\{D\}=1$, we have

$$CD=DC\,.$$

If $\{C\}\frown\{D\}\neq 1$, it follows then from (19) $\{AD\}\frown\{C\}=1$ and $(AD)C=C(AD)$. We have again by (19)

$$CD=DC\,. \tag{20}$$

Elements of infinite orbers are therefore commutative with one another. Now let E be any element of finite order. Then AE is of infinite order (cf. Theorem 1) and from (20) we obtain

$$(AE)C=C(AE)$$

that is,

$$EC=CE\,. \tag{21}$$

By (20), (21) and Theorem 5 we see that $\mathfrak{G}$ is abelian.

Theorem 7. Let $\mathfrak{G}$ be any M-group and $\mathfrak{E}$ be the normal subgroup of $\mathfrak{G}$ which consists of all elements of finite orders in $\mathfrak{G}$. If the abelian group $\mathfrak{G}/\mathfrak{E}$ has a rank $\geqq 2$, then $\mathfrak{G}$ itself is abelian.

It is sufficient therefore to observe the case that the rank of $\mathfrak{G}/\mathfrak{E}$ is 1. Suppose first that $\mathfrak{G}/\mathfrak{E}$ is an infinite cyclic group: $\mathfrak{G}/\mathfrak{E}=\{\bar{Z}\}$.

Let $\mathfrak{P}$ be the p-component of $\mathfrak{E}$, that is to say, the subgroup of $\mathfrak{E}$ which consists of all elements of p-power order in $\mathfrak{E}$. We denote further by $\mathfrak{P}_k(k=0,1,2,\ldots)$ the subgroup of $\mathfrak{P}$ which consists of elements satisfying the relation

$$X^{p^k}=1\,.$$

We prove that for any element in $\mathfrak{P}_n$ we have

$$ZAZ^{-1}=A^{r_n}, \tag{22}$$

where r_n is uniquely determined mod. p^n by $\mathfrak{P}_n$. This is clear for $n=0$. We prove therefore by induction, assuming (22) for $n=k$. Take any element B of order p^{k+1}. From Lemma 3

$$ZBZ^{-1}=B^r\,.$$

According to

$$B^p\in\mathfrak{P}_k\,,\quad ZB^pZ^{-1}=(B^p)^r=(B^p)^{r_k}\,,$$

it is then

$$(23)\qquad r \equiv r_k \quad \text{mod.} \quad p^k.$$

Hence for an arbitrary element U of $\{B, \mathfrak{P}_k\}$ we have

$$ZUZ^{-1} = U^r.$$

Let C be another element of order p^{k+1}. If $\{B\} \frown \{C\} \neq 1$, then C is contained in $\{B, \mathfrak{P}_k\}$ and we have

$$ZCZ^{-1} = C^r.$$

Let

$$\{B\} \frown \{C\} = 1, \quad ZCZ^{-1} = C^{r'}, \quad Z(BC)Z^{-1} = (BC)^{r''},$$

then

$$B^{r''}C^{r''} = B^r C^{r'}, \qquad r'' \equiv r' \equiv r \quad \text{mod.} \quad p^{k+1},$$

that is to say,

$$ZCZ^{-1} = C^r.$$

For any element V of $\mathfrak{P}_{k+1}$ we have thus

$$ZVZ^{-1} = V^{r_{k+1}}$$

with $r_{k+1} = r$.

Now for an element A or order p it holds especially

$$ZAZ^{-1} = A^{r_1}.$$

If $r_1 \neq 1$ mod. p then by taking a suitable power Z_1 of Z we have

$$Z_1 A Z_1^{-1} = A^t, \quad t \neq 1 \quad \text{mod. } p, \quad t^q \equiv 1 \quad \text{mod. } p,$$

where q is a prime number different from p.
But in this case we can easily find by simple calculation that the group $\{Z_1, A\}/Z_1^{pq}$ of order p^2q is not an M-group (cf. 3 in §1). This is a contradiction and it must be

$$r_1 \equiv 1 \qquad \text{mod. } p.$$

In a similar way we find also that

$$r_1 \equiv 1 \qquad \text{mod. } 4$$

for $p = 2$.

Now if it holds $\mathfrak{P} = \mathfrak{P}_k$ for a suitable k we put $\alpha = \alpha(p) = r_k$. But if $\mathfrak{P} \neq \mathfrak{P}_k$ for any k, put

$$\alpha = \alpha(p) = \lim_{k\to\infty} r_k$$

as the sequence $\{r_k\}$ is always convergent in p-adic sence according to $r_k \equiv r_{k+1}$ mod. p^k (cf. (23)). In either case it is then

$$ZXZ^{-1} = X^{\alpha(p)} \tag{24}$$

for an element X in $\mathfrak{P}$. (Even in the case, where $\alpha(p)$ is a p-adic number the meaning of above equation will be clear.) We come to the conclusion: there exists for any component, say p-component, $\mathfrak{P}$ of $\mathfrak{E}$ a p-adic number $\alpha(p)$ such that

$$ZXZ^{-1} = X^{\alpha(p)},\ \alpha(p) \equiv 1 \quad \text{mod. } p \quad (\alpha(2) \equiv 1 \quad \text{mod. } 4 \text{ for } p = 2)$$

for any element X of $\mathfrak{P}$.

We prove next that any group $\mathfrak{G} = \{\mathfrak{E}, Z\}$ given by above relation is always an M-group. For that purpose it is sufficient to prove that for arbitrary A, B of $\mathfrak{G}$ there are suitable integers x, y such that

$$AB = B^x A^y . \tag{25}$$

If A or B is contained in $\mathfrak{E}$ this is almost clear. We suppose therefore

$$A = A_1 Z^u, \quad B = B_1 Z^v, \quad A_1,\ B_1 \in \mathfrak{E}, \quad u,\ v \neq 0 .$$

The group $\mathfrak{H} = \{Z, A_1, B_1\}$ is an infinite cyclic extention of the finite abelian group $\mathfrak{E}_1 = \{A_1, B_1\}$. If we denote by $\mathfrak{P}_i$ a p_i-Sylowgroup of $\mathfrak{E}_1$ $(i = 1, 2, \ldots, s)$ we have

$$ZA_iZ^{-1} = A_i^{r_i},\ r_i \equiv 1 \text{ mod. } p_i \ (r_i \equiv 1 \text{ mod. } 4 \text{ for } p_i = 2), \tag{26}$$

for any A_i of $\mathfrak{P}_i$ according to (24).
The order of automorphism given by Z in $\mathfrak{P}_i$ is consequently a power of p_i, say $p_i^{n_i}$. $Z_0 = Z^{p_1^{n_1} \cdots p_s^{n_s}}$ is then commutative with all elements of $\mathfrak{E}_1$ and $\{\mathfrak{E}_1, Z_0\}$ is the centraliser of $\mathfrak{E}_1$ in $\mathfrak{H}$. Suppose $\{A\} \frown \{Z\} = \{Z^\mu\}$, $\{B\} \frown \{Z\} = \{Z^\nu\}$. $Z^{\mu\nu}$ is then commutative with A, B, accordingly with A_1, B_1.
It follows

$$Z^{\mu\nu} \in \{\mathfrak{E}_1,\ Z_0\}, \qquad Z^{\mu\nu} \in \{Z_0\},$$

$$\mu\nu = p_1^{m_1} \ldots p_s^{m_s} k,\ (k,\ p_i) = 1,\ m_i \geqq n_i \quad (i = 1, 2, \ldots, s).$$

If we put $\bar{\mathfrak{H}} = \mathfrak{H}/\{Z^{\mu\nu}\}$, it is easy to see that $\bar{\mathfrak{H}}$ is the direct product of its Sylowgroups:

$$\bar{\mathfrak{H}} = \{\bar{\mathfrak{P}}_1,\ \bar{Z}^{p_2^{m_2} \cdots p_s^{m_s} k}\} \times \{\bar{\mathfrak{P}}_2,\ \bar{Z}^{p_1^{m_1} p_3^{m_3} \cdots p_s^{m_s} k}\} \times \cdots ,$$

and these direct factors have the structure stated in §1, 1 (or 2) according to (26). As the nilpotent M-group $\bar{\mathfrak{H}}$ is quasi-Hamiltonean[10] there are suitable integers x_1, y_1 such that

$$\bar{A}\bar{B} = \bar{B}^{x_1}\bar{A}^{y_1},$$

that is,

$$AB \equiv B^{x_1}A^{y_1} \quad \text{mod. } \{Z^{\mu\nu}\},$$

$$AB = B^{x_1}A^{y_1}Z^{h\mu\nu}.$$

If $Z^{\mu\nu} = A^l$, then

$$AB = B^{x_1}A^{y+hl} = B^x A^y, \quad x = x_1, \quad y = y_1 + hl.$$

We have thus proved the following theorem.

Theorem 8. Let $\mathfrak{G}$ be an M-group and $\mathfrak{E}$ be the normal subgroup of $\mathfrak{G}$ which consists of all elements of finite order in $\mathfrak{G}$. If $\mathfrak{G}/\mathfrak{E}$ is an infinite cyclic group then $\mathfrak{G}$ has the following structure:
Let Z be a representative of the co-set which generates $\mathfrak{G}/\mathfrak{E}$ and $\mathfrak{P}$ be any component, say p-component, of $\mathfrak{E}$. It holds then for any A of $\mathfrak{P}$

(27) $ZAZ^{-1} = A^{\alpha(p)}$, $\alpha(p) \equiv 1$ mod. p ($\alpha(2) \equiv 1$ mod. 4 for $p = 2$)

where $\alpha(p)$ is a p-adic number which is uniquely determined mod. p^n by $\mathfrak{P}$ and Z if the maximum of order of elements in $\mathfrak{P}$ is p^n (evtl. $n = \infty$). Conversely if we extend an arbitrary abelian group $\mathfrak{E}$ which contains only elements of finite orders by an infinite cyclic group $\{Z\}$ with the relation (27) then we obtain a quasi-Hamiltonean group which is then of course an M-group.

We assume next that $\mathfrak{G}/\mathfrak{E}$ is a general abelian group of rank 1. In this case we can find such normal series of $\mathfrak{G}$

$$\mathfrak{E} < \mathfrak{G}_1 < \mathfrak{G}_2 < \cdots < \mathfrak{G}, \tag{28}$$

that $\mathfrak{G}_1/\mathfrak{E}$ is an infinite cyclic group and $\mathfrak{G}_{i+1}/\mathfrak{G}_i$ ($i = 1, 2, \ldots$) is a cyclic group of a prime order; $[\mathfrak{G}_{i+1} : \mathfrak{G}_i] = p_i$ ($i = 1, 2, \ldots$). Now any $\mathfrak{G}_i$ has the structure given in Theorem 8. We put therefore

$$\mathfrak{G}_i/\mathfrak{E} = \{\bar{Z}_i\}, \quad Z_i A Z_i^{-1} = A^{\alpha_i(p)} \tag{29}$$

$$\alpha_i(p) \equiv 1 \text{ mod. } p \ (\alpha_i(p) \equiv 1 \text{ mod. } 4 \text{ for } p = 2),$$

for any A of the p-component $\mathfrak{P}$ of $\mathfrak{E}$. $\alpha_i(p)$ is uniquely determined mod. p^n if the maximum of order of elements in $\mathfrak{P}$ is p^n (evtl. $n = \infty$). Between two successive $\mathfrak{G}_i$, $\mathfrak{G}_{i+1}$ it holds

---

(10) See G. V. Satz 7.

(30) $$a_{i+1}(p)^{p_i} \equiv a_i(p) \mod. p^n.$$

(31) $$Z_{i+1}^{p_i} = Z_i E_i$$

(32) $$Z_{i+1} Z_i Z_{i+1}^{-1} = Z_i E_i^{1-p_{i+1}} \quad (Z_{i+1} E_i Z_{i+1}^{-1} = E_i^{p_{i+1}}),$$

where E_i are elements of $\mathfrak{E}$.

Conversely if we extend an abelian group $\mathfrak{E}$ according to (28)—(32) successively, then the whole group $\mathfrak{G}$ thus obtained is quasi-Hamiltonean, because any two elements A, B of $\mathfrak{G}$ are contained in some $\mathfrak{G}_i$ and they satisfy consequently the relation (25). We have therefore

Theorem 9. Let $\mathfrak{G}$ be an M-group and $\mathfrak{E}$ be the normal subgroup of $\mathfrak{G}$ which consists of all elements of finite orders in $\mathfrak{G}$. If $\mathfrak{G}/\mathfrak{E}$ is an abelian group of rank 1 then $\mathfrak{G}$ is generated with $\mathfrak{E}$ and with elements $Z_1, Z_2, \ldots$ of infinite orders by the relations (28)—(32).
Conversely any group of such structure is quasi-Hamiltonean, and therefore modular.

We have further

Theorem 10. For groups which contain elements of infinite orders the notions "modular" and "quasi-Hamiltonean" are equivalent.

The structure of an M-group which contains elements of infinite orders is thus thoroughly determined.

Mathematical Institute,
Tokyo Imperial University.

[7] On the structure of conditionally complete lattice-groups

Japanese J. Math., 18 (1943), 777-789.

(Received September 14, 1942.)

A multiplicative group is called a lattice group, when it is at the same time a lattice such that the order is preserved under left and right multiplication: $a \leqq b$ implies $ac \leqq bc$ and $ca \leqq cb$, or, what is the same thing,

$$ac \asymp bc = (a \asymp b)c\,, \qquad ca \asymp cb = c(a \asymp b)\text{(1)}\,.$$

A lattice-group is called conditionally complete when every bounded set in it has a supremum and an infimum.

The object of this paper is to prove a conjecture of G. Birkhoff(2), that any conditionally complete lattice-group will be abelian. We shall prove that such a group is the direct product of two subgroups, one of which is a vector lattice and the other is a restricted direct product of cyclic groups. To determine the structure of the former part, we shall follow the method of H. Nakano in the spectral-theory for vector lattices(3). We shall also prove that this direct-product-decomposition is unique for a given conditionally complete lattice-group.

§1. In the following we mean by $\mathfrak{G}$ always a conditionally complete lattice-group. We prove first some lemmas which may be also valid for lattice-groups with some weaker conditions.

Lemma 1(4). Let

$$a_1 \leqq a_2 \leqq \cdots\cdots\cdots \leqq a_\omega \leqq \cdots\cdots\cdots \leqq a\,,$$

$$b_1 \leqq b_2 \leqq \cdots\cdots\cdots \leqq b_\omega \leqq \cdots\cdots\cdots \leqq b\,,$$

be two bounded transfinite sequences of the same order-type in $\mathfrak{G}$. We have then

$$\bigvee_\xi a_\xi \bigvee_\xi b_\xi = \bigvee_\xi a_\xi b_\xi\,.$$

---

(1) Cf. T. Nakayama, Note on Lattice-ordered Groups, Proc. Imp. Acad. Japan, 18, (1942).

(2) See Nakayama, l.c.

(3) H. Nakano, Teilweise geordnete Algebra, Jap. Jour. Math. 17 (1941). This paper is referred to as N. T. in the following.

(4) For Lemma 1, 2 see H. Freudental, Teilweise geordnete Moduln. Proc. Amsterdam 39 (1936).

Proof. We have

$$\bigvee a_\xi b_\xi \geqq a_\eta b_\eta \geqq a_\xi b_\eta \qquad (\xi \leqq \eta),$$

therefore

$$\bigvee a_\xi b_\xi \geqq a_\xi \bigvee b_\eta, \qquad \bigvee a_\xi b_\xi \geqq \bigvee a_\xi \bigvee b_\eta = \bigvee a_\xi \bigvee b_\xi .$$

On the other hand,

$$\bigvee a_\xi \bigvee b_\xi \geqq a_\xi b_\xi, \quad \text{whence} \quad \bigvee a_\xi \bigvee b_\xi \geqq \bigvee a_\xi b_\xi .$$

Lemma 2. For any sequence

$$a_1 \leqq a_2 \leqq \dots \leqq a_\omega \leqq \dots \leqq a$$

and arbitrary b of $\mathfrak{G}$, we have

$$(\bigvee a_\xi) \frown b = \bigvee (a_\xi \frown b).$$

Proof. As it is sufficient to prove $\bigvee(a_\xi b^{-1}) \frown 1 = \bigvee(a_\xi b^{-1} \frown 1)$, We may assume $b=1$. It is then by Lemma 1

$$\bigvee a_\xi = ((\bigvee a_\xi) \smile 1)((\bigvee a_\xi) \frown 1)(^5) = (\bigvee (a_\xi \smile 1))((\bigvee a_\xi) \frown 1)$$

and

$$\bigvee a_\xi = \bigvee ((a_\xi \smile 1)(a_\xi \frown 1)) = \bigvee (a_\xi \smile 1) \bigvee (a_\xi \frown 1),$$

therefore

$$(\bigvee a_\xi) \frown 1 = \bigvee (a_\xi \frown 1).$$

Lemma 3. Let $\mathfrak{M}$ be any set, which is bounded from above. It holds then

$$(\bigvee_{a \in \mathfrak{M}} a) \frown b = \bigvee_{a \in \mathfrak{M}} (a \frown b).$$

Proof. Let

$$a_1, a_2, \dots, a_\omega, \dots, a_\eta, \dots, \qquad \eta < \xi$$

be elements of $\mathfrak{M}$, well-ordered in a transfinite sequence. We prove

$$(\bigvee_{\eta<\xi} a_\eta) \frown b = \bigvee_{\eta<\xi} (a_\eta \frown b).$$

Assuming

$$(\bigvee_{\eta<\xi'} a_\eta) \frown b = \bigvee_{\eta<\xi'} (a_\eta \frown b), \qquad \xi' < \xi,$$

---

(5) As in the abelian case, one can prove easily

$$a = (a \smile 1)(a \frown 1)$$

for any a in $\mathfrak{G}$.

we use the transfinite induction. If $\xi = \xi_0+1$ we have from the distributivity of lattice-groups([6])

$$(\bigvee_{\eta<\xi} a_\eta)\frown b = ((\bigvee_{\eta<\xi_0} a_\eta)\smile a_{\xi_0})\frown b = ((\bigvee_{\eta<\xi_0} a_\eta)\frown b)\smile(a_{\xi_0}\frown b) = \bigvee_{\eta<\xi}(a_\eta\frown b).$$

If ξ is a limit number, put

$$m_\eta = \bigvee_{\eta'<\eta} a_{\eta'},$$

then we have

$$m_1 \leqq m_2 \leqq \ldots \leqq m_\omega \leqq \ldots, \qquad \bigvee_{\eta<\xi} a_\eta = \bigvee_{\eta<\xi} m_\eta,$$

and in virtue of Lemma 2

$$(\bigvee_{\eta<\xi} a_\eta)\frown b = (\bigvee_{\eta<\xi} m_\eta)\frown b = \bigvee_{\eta<\xi}(m_\eta\frown b) = \bigvee_{\eta<\xi}(\bigvee_{\eta'<\eta}(a_{\eta'}\frown b)) = \bigvee_{\eta'<\xi}(a_{\eta'}\frown b).$$

We define, as usual, the positive and negative part and the absolute of a by

$$a_+ = (a\smile 1), \qquad a_- = (a\frown 1)^{-1}, \qquad |a| = a_+a_-.$$

It holds then as in the abelian case

$$a_+\frown a_- = 1, \qquad a_+(a_-)^{-1} = a.$$

Lemma 4. $|a|\frown|b| = 1$ implies $ab = ba$.
Proof. If $a\geqq 1$, $b\geqq 1$, $a\frown b = 1$, we have

$$a^{-1}b^{-1} = a^{-1}(a\frown b)b^{-1} = b^{-1}\frown a^{-1} = a^{-1}\frown b^{-1} = b^{-1}(a\frown b)a^{-1} = b^{-1}a^{-1},$$

therefore

$$ab = ba.$$

In general from

$$a_+,\ a_- \leqq |a|, \qquad b_+,\ b_- \leqq |b|, \qquad |a|\frown|b| = 1,$$

it follows

$$a_+\frown b_+ = a_+\frown b_- = a_-\frown b_+ = a_-\frown b_- = 1$$

and from above

$$ab = ba.$$

Lemma 5. $|a|\frown|p| = 1$, $|b|\frown|p| = 1$ implies $|ab|\frown|p| = 1$.
Proof. Assume first $a, b\geqq 1$. Then

$$ab\frown|p| = a(b\frown a^{-1}|p|) \leqq a(b\frown|p|) = a,$$

([6]) See Nakayama, l.c.

therefore

$$ab \frown |p| = ab \frown |p| \frown a = 1 .$$

In general, we have from the assumption

$$a_+ \frown |p| = 1, \quad a_- \frown |p| = 1, \quad b_+ \frown |p| = 1, \quad b_- \frown |p| = 1 .$$

Now

$$(ab)_+ = (ab \smile 1) \leqq (a \smile 1)(b \smile 1) = a_+ b_+ ,$$

therefore from

$$a_+ \frown |p| = 1, \quad b_+ \frown |p| = 1, \quad a_+ b_+ \frown |p| = 1 ,$$

it follows

$$(ab)_+ \frown |p| = 1 .$$

In the same way we have from $(ab)_- \leqq b_- a_-$

$$(ab)_- \frown |p| = 1 ,$$

whence

$$|ab| \frown |p| = 1 .$$

Now, let an element p of $\mathfrak{G}$ be given. Then all elements a of $\mathfrak{G}$, which is orthogonal to p, i.e. such that $|a| \frown |p| = 1$, form a subgroup $\mathfrak{K}_p$ in $\mathfrak{G}$ by the last lemma. The elements, which is orthogonal to any element in $\mathfrak{K}_p$ also form a subgroup, which we denote by $\mathfrak{H}_p$. Making use of Lemma 4 we can then prove in the same way as in N. T.(7) that $\mathfrak{G}$ is the direct product of $\mathfrak{K}_p$ and $\mathfrak{H}_p$:

$$\mathfrak{G} = \mathfrak{K}_p \times \mathfrak{H}_p .$$

Following N. T. we define the projector $[p]$ by

$$[p]a = a'', \quad \text{if} \quad a = a'a'', \quad a' \in \mathfrak{K}_p , \quad a'' \in \mathfrak{H}_p .$$

For $a \geqq 1$ it holds then(8)

$$[p]a = \bigvee_{n=1}^{\infty} (|p|^n \frown a) ,$$

whence

$$a' \geqq 1, \qquad a'' = [p]a \geqq 1 .$$

The lattice $\mathfrak{G}$ is therefore the direct sum of the lattices $\mathfrak{K}_p$ and $\mathfrak{H}_p$.

Theorem 1. For any p in $\mathfrak{G}$ we have

---

(7) Cf. N.T. Satz 3.2, Satz 3.3.
(8) Cf. N.T. Satz 3.8.

$$\mathfrak{G} = \mathfrak{K}_p \times \mathfrak{H}_p ,$$

and the lattice $\mathfrak{G}$ is the direct sum of the lattices $\mathfrak{K}_p$ and $\mathfrak{H}_p$. $\mathfrak{K}_p$ and $\mathfrak{H}_p$ are consequently conditionally complete lattice-groups.

We now extend this theorem to the following form([9]).

Theorem 2. Let $\mathfrak{M}$ be any subset of $\mathfrak{G}$. Then all elements of $\mathfrak{G}$, which are orthogonal to any element m in $\mathfrak{M}$, form a subgroup $\mathfrak{K}_{\mathfrak{M}}$, and all elements of $\mathfrak{G}$, which are orthogonal to any element of $\mathfrak{K}_{\mathfrak{M}}$ form a subgroup $\mathfrak{H}_{\mathfrak{M}}$. It is then

$$\mathfrak{G} = \mathfrak{K}_{\mathfrak{M}} \times \mathfrak{H}_{\mathfrak{M}}$$

and the lattice $\mathfrak{G}$ is the direct sum of the lattices $\mathfrak{K}_{\mathfrak{M}}$ and $\mathfrak{H}_{\mathfrak{M}}$. $\mathfrak{K}_{\mathfrak{M}}$ and $\mathfrak{H}_{\mathfrak{M}}$ are consequently conditionally complete lattice-groups.

Proof. Obviously, it is sufficient to prove that any element $x \geqq 1$ can be decomposed into a product of $k \geqq 1$ and $h \geqq 1$ of $\mathfrak{K}_{\mathfrak{M}}$ and $\mathfrak{H}_{\mathfrak{M}}$ respectively. Put

$$h = \bigvee_{m \in \mathfrak{M}} ([m]x) .$$

The join in the right-hand side always exists as $x \geqq [m]x$, and for any $k' \in \mathfrak{K}_{\mathfrak{M}}$ we have by Lemma 3

$$h \frown | k' | = (\bigvee([m]x)) \frown | k' | = \bigvee([m]x \frown | k' |) = 1 .$$

Therefore h is contained in $\mathfrak{H}_{\mathfrak{M}}$ and $h \geqq 1$. If we put further $xh^{-1} = k$, we have $k \geqq 1$ and

$$| m | \frown k = | m | \frown xh^{-1} \leqq | m | \frown x([m]x)^{-1} = 1 ,$$

whence

$$| m | \frown k = 1, \qquad k \in \mathfrak{K}_{\mathfrak{M}} ,$$

which completes the proof.

With respect to the projector $[p]$ it holds almost every theorems in N. T.; for instance

$$[p]ab = [p]a[p]b , \qquad [p](a \smile b) = [p]a \smile [p]b .$$

More generally we have

$$[p] \bigvee a = \bigvee [p]a ,$$

if the join $\bigvee a$ exists. Theoremes 3.5, 3.6, 3.7, 3.8, 3.14, 3.15, 3.16, 3.17,

---

([9]) Cf. S. Bochner and S. R. Phillips, Additive set functions and vector lattices, Ann. of Math. 42 (1941).

3.18, 3.19, 3.20, 3.21, 3.22, 3.23, 3.24, 3.25 in N.T. are also valid, but we omit the proofs, as those given in N.T. can be easily transfered to our non-abelian case with minor modifications.

§2. We now prove some more lemmas which are to be used afterwards.

Lemma 6. Let $a \geqq 1$. Then any x, such that $1 \leqq x \leqq a^n$ (n; any natural number), can be described as product of elements y_i with $1 \leqq y_i \leqq a$.

Proof. Induction with respect to n. Put $y = x(x \frown a^{n-1})^{-1}$, then

$$y = x(x^{-1} \smile a^{-(n-1)}) = 1 \smile xa^{-(n-1)} \leqq 1 \smile a^n \cdot a^{-(n-1)} = a .$$

Therefore

$$1 \leqq y \leqq a, \qquad x = y(x \frown a^{n-1}) .$$

By the assumption of the induction $(x \frown a^{n-1})$ is products of such y', that $1 \leqq y' \leqq a$, which completes the proof.

Lemma 7. No element of $\mathfrak{G}$ except the identity has a finite order.

Proof. Let $a^m = 1$ and put $b = (1 \frown a)$. It is then for any natural number n

$$b^n = (1 \frown a \frown \ \ldots \ \frown a^n) .$$

Therefore there are only finite different elements among the powers of b. b is consequently also of a finite order, $b^l = 1$. b is obviously $\leqq 1$, but if $b < 1$, then $b^{l-1} < 1$, which contradicts to $b^{l-1} = b^{-1} > 1$. It must be therefore $b = 1$. In the same way we have $b' = (1 \smile a) = 1$, whence $a = 1$.

Lemma 8. In $\mathfrak{G}$ $aba^{-1} \frown b = 1$ implies $b = 1$.

Proof. As $\mathfrak{H}_b$ is a normal subgroup of $\mathfrak{G}$, b and aba^{-1} are both contained in $\mathfrak{H}_b$. But by the assumption $aba^{-1} \in \mathfrak{K}_b$, therefore $aba^{-1} \in \mathfrak{H}_b \frown \mathfrak{K}_b = 1$, $b = 1$.

Lemma 9. In $\mathfrak{G}$ $aba^{-1} = b^{-1}$ implies $b = 1$.

Proof. We have $a(1 \smile b)a^{-1} = (1 \smile b^{-1}) = b_-$, that is $ab_+a^{-1} = b_-$. As $b_+ \frown b_- = 1$, it follows from Lemma 8 that $b_+ = b_- = 1$, $b = 1$.

Lemma 10. In $\mathfrak{G}$ $a^2 = b^2$ implies $a = b$.

Proof. Put

$$a^2 = b^2 = z, \quad b = ac, \quad \{a, b\} = \mathfrak{A}, \quad \{a, b\}/\{z\} = \bar{\mathfrak{A}} .$$

In $\bar{\mathfrak{A}}$ it holds

$$\bar{a}^2 = \bar{b}^2 = 1, \quad \bar{b} = \bar{a}\bar{c}, \quad \bar{a}\bar{c}\bar{a}^{-1} = \bar{c}^{-1} ,$$

therefore

$$aca^{-1} = c^{-1}z^{\alpha}$$

with some α. Putting $c^2 z^{-\alpha} = c'$, we have

$$ac'a^{-1} = c'^{-1},$$

and by Lemma 9

$$c' = 1, \quad c^2 = z^\alpha.$$

From above and Lemma 7, it is easy to see that the abelian group $\{c, z\}$ is cyclic: $\{c, z\} = \{d\}$, and $\mathfrak{A} = \{a, d\}$. As $\{d\}$ is a normal subgroup of $\mathfrak{A}$, it holds either $ada^{-1} = d$ or $ada^{-1} = d^{-1}$. In the latter case we have by Lemma 9 $d = 1$, $\mathfrak{A} = \{a\}$, $a = b$. But if $ada^{-1} = d$, $\mathfrak{A}$ is abelian and from $a^2 = b^2$ we have $a = b$ by Lemma 7.

Lemma 11. In $\mathfrak{G}$ $ab^2 = b^2 a$ implies $ab = ba$.

Proof. Putting $a' = b^2 a b^{-2}$, we have $a'^2 = b^2 a^2 b^{-2} = a^2$, whence $a = a'$ by Lemma 10.

§3. Now let $p \geqq 1$. We call an element e a part(10) of p if it holds

$$e \frown pe^{-1} = 1.$$

From $e \frown pe^{-1} = 1$ we see by Lemma 4 $e(pe^{-1}) = (pe^{-1})e$, i.e. $pe = ep$. With e, pe^{-1} is also a part of p. If e_1, e_2 are parts of p then $(e_1 \frown e_2)$ is also a part of p:

$$(e_1 \frown e_2) \frown p(e_1 \frown e_2)^{-1} = (e_1 \frown e_2) \frown (pe_1^{-1} \smile pe_1^{-1})$$

$$= (e_1 \frown e_2 \frown pe_1^{-1}) \smile (e_1 \frown e_2 \frown pe_2^{-1}) = 1.$$

In the same way we see that $e_1 \smile e_2$ is a part of p. Now if $e_1 \leqq e_2$, then we have $e_1 e_2 = e_2 e_1$, as $1 \leqq e_1 \frown e_2 e_1^{-1} \leqq e_1 \frown pe_1^{-1} = 1$. On the other hand from

$$e_2 \smile pe_1^{-1} \geqq e_2 \smile pe_2^{-1} = p$$

it follows $e_2 \smile pe_1^{-1} = p$, and $e_2(pe_1^{-1}) = (e_2 \smile pe_1^{-1})(e_2 \frown pe_1^{-1})$(11) $= p(e_2 \frown pe_1^{-1})$, therefore $e_2 e_1^{-1} = (e_2 \frown pe_1^{-1})$. $e_2 e_1^{-1}$ is consequently also a part of p.

Lemma 12. If e_1, e_2 are parts of p, then they are commutative with each other: $e_1 e_2 = e_2 e_1$.

Proof. Put $e_1 \frown e_2 = e_3$. As e_2 is also a part of p and $e_1 \geqq e_3$, $e_2 \geqq e_3$, we have from above consideration $e_1 e_3 = e_3 e_1$, $e_2 e_3 = e_3 e_2$. From $e_1 e_3^{-1} \frown e_2 e_3^{-1} = 1$, it follows also that $e_1 e_3^{-1}$ is commutative with $e_2 e_3^{-1}$. It is therefore $e_1 e_2 = e_2 e_1$.

---

(10) "Teil" in N.T. Definition 5.1.

(11) If a, b are commutative we have

$$ab = (a \frown b)(a \smile b).$$

Cf. (5).

We call an element $p \geqq 1$ "singular", if all a, such that $p \geqq a \geqq 1$, are parts of p. It is easy to see that if p is a singular element and $p \geqq p' \geqq 1$, then p' is also singular and they are commutative by Lemma 12.

Lemma 13. Let p_1, p_2 be singular elements. Then any x, y, such that $1 \leqq x \leqq p_1^m$, $1 \leqq y \leqq p_2^n$ (m, n are natural numbers), are commutative with each other.

Proof. By Lemma 6 it is sufficient to prove that any x, y satisfying $1 \leqq x \leqq p_1$, $1 \leqq y \leqq p_2$ are commutative. We can prove it in the same way as in the proof of Lemma 12.

Now let $a \geqq 1$, $b \geqq 1$. From

$$[p_1]a = \bigvee_{m=1}^{\infty} (p_1^m \frown a), \qquad [p_2]b = \bigvee_{n=1}^{\infty} (p_2^n \frown b)$$

we see in virtue of Lemma 13 that $[p_2]a$ and $[p_2]b$ are commutative.

If we denote by $\mathfrak{S}$ the set of all singular elements in $\mathfrak{G}$, then we have by Theorem 2

$$\mathfrak{G} = \mathfrak{K}_{\mathfrak{S}} \times \mathfrak{H}_{\mathfrak{S}} .$$

But as $\mathfrak{H}_{\mathfrak{S}}$ is generated by $\bigvee_{p \in \mathfrak{S}} [p]x$, $(x \geqq 1)$, it follows from what is mentioned just above, that $\mathfrak{H}_{\mathfrak{S}}$ is an abelian group. It is also evident that $\mathfrak{K}_{\mathfrak{S}}$ is a conditionally complete lattice-group, which has no singular element except the identity.

4. To prove that $\mathfrak{G}$ is abelian it is therefore sufficient to prove that $\mathfrak{K}_{\mathfrak{S}}$ is an abelian group. We assume accordingly for a time that $\mathfrak{G} = \mathfrak{K}_{\mathfrak{S}}$, that is to say, that $\mathfrak{G}$ contains no singular element except that identity.

Lemma 14. Let $a > 1$, $a > x^2$, $x \geqq 1$. Then there exists an element x', such that $x' > x$, $a \geqq x'^2$.

Proof. We prove first that if $a > 1$, there is an element x', satisfying $a \geqq x'^2$, $x' > 1$. As a is not singular by the assumption, there is an element a', such that $a > a' > 1$, $x' = a' \frown a a'^{-1} \neq 1$. It follows then $x'^2 \leqq a a'^{-1} a' = a$. Now let $a > x^2$ and put $a' = x^{-2} a > 1$. From what is just mentioned above there is such z, that $a' \geqq z^2$, $z > 1$. Put $xzx^{-1} \frown z = y$, $x' = xy$. Then we have $y > 1$ by Lemma 8 and it holds

$$x' > x, \quad x'^2 = xyxy = x^2 x^{-1} y x y \leqq x^2 z^2 \leqq x^2 a' = a ,$$

which completes the proof.

Lemma 15. If $a > 1$, there exists an element x such that $a = x^2$.

Proof. We construct a sequence,

$$x_1 = 1 < x_2 < x_3 < \ldots, \quad x_\eta^2 < a,$$

as follows. Let x_η, $\eta < \xi$, be already determined and let $\xi = \xi_0 + 1$. As $x_{\xi_0}^2 > a$, we can choose an element x by Lemma 14, such that $x_{\xi_0} < x$, $x^2 \leqq a$. If ξ is a limit number, we put $x = \bigvee_{\eta < \xi} x_\eta$ and see easily that $x_\eta < x$, $x^2 \leqq a$. In any case put $x_\xi = x$, if $x^2 < a$. But this process can not be continued ad infinitum; we must have therefore $x^2 = a$ for some x.

Now let a be an arbitrary element of $\mathfrak{G}$. As $a = a_+(a_-)^{-1}$, $a_+ \geqq 1$, $a_- \geqq 1$, we have $a_+ = x^2$, $a_- = y^2$ for some x, y. But these x, y are commutative by Lemma 11 and putting $z = xy^{-1}$, we have $a = z^2$. Moreover according to Lemma 10 such z is uniquely determined by a. If we denote $z = a^{\frac{1}{2}}$ and define $a^{\frac{1}{4}} = (a^{\frac{1}{2}})^{\frac{1}{2}}$, $a^{\frac{1}{8}} = (a^{\frac{1}{4}})^{\frac{1}{2}}$, $\cdots$, $a^{\frac{k}{2^n}} = (a^{\frac{1}{2^n}})^k$, we have a^λ for any diadic rational number $\lambda = \frac{k}{2^n}$, $n = 0, 1, 2, \cdots$, $k = 0, \pm 1, \pm 2, \cdots$. These "powers" are commutative with each other and satisfy ordinary rules of power:

$$a^0 = 1, \quad a^\lambda a^{\lambda'} = a^{\lambda+\lambda'}, \quad (a^\lambda)^{\lambda'} = {}^{\lambda\lambda'}.$$

If a and b are commutative, then $a^{\frac{1}{2}}$ and $b^{\frac{1}{2}}$ are also commutative by Lemma 11. Accordingly a^λ and $b^{\lambda'}$ are commutative in general and it holds

$$(ab)^\lambda = a^\lambda b^\lambda.$$

Lemma 16. For any p, a in $\mathfrak{G}$, and a diadic rational number λ, we have

$$[p](a^\lambda) = ([p]a)^\lambda.$$

Proof. It is sufficient to prove it for $\lambda = \frac{1}{2}$, but this follows at once from $([p]a^{\frac{1}{2}})^2 = [p](a^{\frac{1}{2}})^2 = [p]a$.

Lemma 17. If $a \geqq 1$ and $\lambda > \lambda'$ (λ, λ': diadic rational number), we have $a^\lambda \geqq a^{\lambda'}$. If moreover there is an element b such that $a^\lambda \leqq b$ for all $\lambda > 0$, one must have $a = 1$.

Proof. The first part is obvious from $a \geqq a^{\frac{1}{2}} \geqq 1$. Let $a^\lambda \leqq b$ for all $\lambda > 0$. Then the join $\bigvee_\lambda a^\lambda = a_0$ exists and from $a_0 a = \bigvee a^{\lambda+1} = \bigvee a^\lambda = a_0$, we have $a = 1$.

After these preparations we can now easily obtain the spectral representation of an element in $\mathfrak{G}$ $(= \mathfrak{R}_{\mathfrak{S}})$. The proof is quite analogous to that in vector lattices and we give here therefore only the outline of it([12]). Let $p>1$ and take any $a\geqq 1$ in $\mathfrak{H}_p$. We define the resolution of p by

$$e_\lambda=[(p^\lambda a^{-1})_+]p\,,$$

for every diadic rational number $\lambda\geqq 0$. It holds then:

(1) e_λ are all parts of p and are commutative with one another by Lemma 12,

(2) $\lambda>\mu$ implies $e_\lambda\geqq e_\mu$; $e_\lambda e_\mu^{-1}$ is therefore also a part of p,

(3) $[e_\lambda]\,a\leqq e_\lambda^\lambda$, $[pe_\lambda^{-1}]\,a\geqq(pe_\lambda^{-1})^\lambda$ for all $\lambda\geqq 0$,

(4) from (3) we have $(pe_\lambda^{-1})^\lambda\leqq a$ and then in virtue of Lemma 17 $\bigwedge_{\lambda>0}(pe_\lambda^{-1})=1$, that is to say,

$$\bigvee_{\lambda>0}e_\lambda=p\,,$$

(5) from (3) it follows by simple calculations $(e_\lambda e_\mu^{-1})^\mu\leqq[e_\lambda e_\mu^{-1}]\,a$ $=[e_\lambda]\,a\,([e_\mu]a)^{-1}\leqq(e_\lambda e_\mu^{-1})^\lambda$ for $\lambda>\mu$.

Now given any natural numbers $m, n,$ we devide equally the interval $[0, m]$ into $\kappa=2^n m$ subintervals of length $\dfrac{1}{2^n}$;

$$0=\lambda_0<\lambda_1<\ldots\ldots<\lambda_\kappa=m\,,\quad \lambda_\nu=\frac{\nu}{2^n}\,.$$

From (5) we have then

$$s_{m,n}=\prod_{\nu=1}^{\kappa}(e_{\lambda_\nu}e_{\lambda_{\nu-1}}^{-1})^{\lambda_{\nu-1}}\leqq[e_m]a([e_0]a)^{-1}$$
$$=[e_m]a\leqq\prod_{\nu=1}^{\kappa}(e_{\lambda_\nu}e_{\lambda_{\nu-1}}^{-1})^{\lambda_\nu}=s'_{m,n}\,,$$

and whence

$$1\leqq s_{m,n}^{-1}[e_m]a\leqq s_{m,n}^{-1}\cdot s'_{m,n}=\prod_{\nu=1}^{\kappa}(e_{\lambda_\nu}e_{\lambda_{\nu-1}}^{-1})^{\lambda_\nu-\lambda_{\nu-1}}=e_m^{\frac{1}{2^n}}\,.$$

It follows

$$1\leqq\bigwedge_{n=1}^{\infty}s_{m,n}^{-1}[e_m]a\leqq\bigwedge_{n=1}^{\infty}e_m^{\frac{1}{2^n}}=1\,,\quad \bigvee_{n=1}^{\infty}s_{m,n}=[e_m]a\,,$$

and as $\bigvee_{m=1}^{\infty}e_m=p$ by (4), we have finally([13])

$$\bigvee_{m,n=1}^{\infty}s_{m,n}=\bigvee_{m=1}^{\infty}[e_m]a=[p]a=a\,.$$

---

(12) Cf. the proof of Satz 5.15 in N.T.

(13) Cf. N.T. Satz 3.23.

If we take another $b \geqq 1$ in $\mathfrak{H}_p$, we have similarly

$$b = \bigvee_{m, n=1}^{\infty} \bar{s}_{m, n},$$

where $\bar{s}_{m, n}$ is like $s_{m, n}$ a product of powers of some parts of p. $s_{m, n}$ and $\bar{s}_{m', n'}$ are therefore commutative with one another and so are a and b consequently. As $\mathfrak{H}_p$ is generated by such elements a, and $b, \cdots$, we have thus proved that $\mathfrak{H}_p$ is an abelian group. Now $\mathfrak{G}/\mathfrak{K}_p$ is isomorphic to $\mathfrak{H}_p$ according to Theorem 1; the commutator group $\mathfrak{G}'$ of $\mathfrak{G}$ is therefore contained in $\mathfrak{K}_p$. But it is easy to see that the meet of all $\mathfrak{K}_p$, $p \geqq 1$, is the identical group. We have thus $\mathfrak{G}' = 1$, that is to say, that $\mathfrak{G}$ is abelian.

Theorem 3. Any conditionally complete lattice-group is abelian.

§5. We now investigate further the structure of conditionally complete lattice-groups which are just proved to be abelian. We return to the decomposition

$$\mathfrak{G} = \mathfrak{K}_{\mathfrak{S}} \times \mathfrak{H}_{\mathfrak{S}},$$

where $\mathfrak{S}$ is the set of all singular elements in $\mathfrak{G}$. It is almost evident from the consideration in §4, that $\mathfrak{K}_{\mathfrak{S}}$ is a complete vector lattice, the structure of which has been studied by many authors. We have therefore only to consider $\mathfrak{H}_{\mathfrak{S}}$. Now, by Lorenzen, Clifford and Nakayama(14), $\mathfrak{H}_{\mathfrak{S}}$ can be imbedded lattice-group-isomorphically in a direct product of linear groups $\mathfrak{L}_\tau$: We have the representation for $a \in \mathfrak{G}$

$$a \longleftrightarrow (\ldots\ldots, a_\tau, \ldots\ldots), \quad a_\tau \in \mathfrak{L}_\tau.$$

We may assume of course that $\mathfrak{L}_\tau$ consists of such a_τ, which corresponds to some a in $\mathfrak{G}$. We denote by $\mathfrak{T}_2$ the set of all coordinates τ such that $s_\tau = 1$ for every s in $\mathfrak{S}$, and by $\mathfrak{T}_1$ the set of the resting coordinates. If we denote further by $\mathfrak{N}_i$ $(i = 1, 2)$ the direct product of $\mathfrak{L}_\tau$, $\tau \in \mathfrak{T}_i$ $(i = 1, 2)$, $\mathfrak{H}_{\mathfrak{S}}$ is imbedded isomorphically in $\mathfrak{N}_1 \times \mathfrak{N}_2$ and we have

$$a \longleftrightarrow (a_1, a_2), \quad a_1 \in \mathfrak{N}_1, \quad a_2 \in \mathfrak{N}_2.$$

We prove then that the lattice-homomorphic mapping $a \longrightarrow a_1$ is an isomorphism. Suppose

$$a_0 \longleftrightarrow (1, n_2^*), \quad n_2^* \in \mathfrak{N}_2,$$

---

(14) P. Lorenzen, Abstrakte Begründung der multiplikativen Idealtheorie, Math. Zeitschr. 45 (1939); A. H. Clifford, Partially ordered abelian groups, Ann. Math. 41 (1940.) Cf. also Nakayama, l. c.

then

$$|a_0| \rightarrow (1, |n_2^*|).$$

As

$$s \longleftrightarrow (s_1, 1)$$

for every $s \in \mathfrak{S}$, we have $|a_0| \frown s = 1$, $[s]|a_0| = 1$, whence

$$|a_0| = \bigvee_{s \in \mathfrak{S}} [s]|a_0| = 1, \quad a_0 = 1.$$

We may therefore assume from the beginning that $\mathfrak{T}_2$ is an empty set and that for any coordinate τ there is some s in $\mathfrak{S}$ such that $s_\tau > 1$. For such s_τ, assume that there is some a_τ satisfying

$$s_\tau > a_\tau > 1.$$

Putting $a' = (a \smile 1) \frown s$, we have

$$1 \leqq a' \leqq s, \quad a'_\tau = a_\tau.$$

But as s is a singular element it holds $a' \frown s a'^{-1} = 1$, and whence $a'_\tau \frown s_\tau a'^{-1}_\tau = a_\tau \frown s_\tau a_\tau{}^{-1} = 1$, which contradicts to the linearity of $\mathfrak{L}_\tau$. Thus each $\mathfrak{L}_\tau$, having a minimal element s_τ, is isomorphic to the lattice-group of all rational integers.

Now we call two coordinates τ_1, τ_2 are associated when there is no element a in $\mathfrak{G}$ such that $a_{\tau_1} > 1$, $a_{\tau_2} = 1$. It is then easy to see that there is also no element a in $\mathfrak{G}$ such that $a_{\tau_1} = 1$, $a_{\tau_2} > 1$ and that all coordinates can be thus divided into classes of mutually associated ones. We denote these classes by $\mathfrak{C}_\nu$ and select from each $\mathfrak{C}_\nu$ a representative τ_ν. It is then clear that the correspondence

$$a \rightarrow (\ldots\ldots, a_{\tau_\nu}, \ldots\ldots)$$

gives an isomorphic mapping of $\mathfrak{G}$ into the direct product of $\mathfrak{L}_{\tau_\nu}$. If we now take two different coordinates τ_τ, τ_μ, then there exists by definition an element a in $\mathfrak{G}$ such that $a_{\tau\nu} > 1$, $a_{\mu_\tau} = 1$. Substituting a by $a \smile 1$ if necessary, we may suppose that $a > 1$. Form then the meet $a^* = \wedge a$, where a is selected as above corresponding to each $\mathfrak{C}_{\tau_\mu}$ with $\tau_\nu \neq \tau_\mu$. The meet exists always according to $a > 1$, and as $a_{\tau_\nu} > 1$ and $\mathfrak{L}_\tau$ is cyclic, it follows $a^*_{\tau_\nu} > 1$ and $a^*_{\tau_\mu} = 1$ for $\tau_\nu \neq \tau_\mu$. We have thus proved that $\mathfrak{H}_{\mathfrak{S}}$ is a restricted direct product of groups which are isomorphic to the group of all rational integers ordered as usual and that any conditionally complete lattice-group $\mathfrak{G}$ is the direct product of such a group with a vector lattice. It is also clear that such a decomposition of $\mathfrak{G}$ into the direct product is unique, for in a vector lattice there is no singular element except the identity, whereas any restricted direct product of cyclic groups is generated by its singular elements. We have thus proved the following final theorem:

Theorem 4. Any conditionally complete lattice-group is the direct product of two conditionally complete lattice-groups; one is a vector lattice and the other is a restricted direct product of groups of all rational integers.

Mathematical Institute,
Faculty of Science.
Tokyo Imperial University.

[8] Einige Sätze über freie Gruppen

Proc. Imp. Acad. Japan, 19-6 (1943), 272-274.

Mathematisches Institut, Kaiserliche Universität zu Tokyo.
(Comm. by T. TAKAGI, M.I.A., June 12, 1943.)

1. Es sei $\mathfrak{G}$ eine beliebige Gruppe. Die durch $\mathfrak{Z}_1=\mathfrak{G}$, $\mathfrak{Z}_n=[\mathfrak{G}, \mathfrak{Z}_{n-1}]$[1] ($n=2, 3, \ldots$) definierte Kette von Untergruppen von $\mathfrak{G}$

$$\mathfrak{Z}_1 > \mathfrak{Z}_2 > \mathfrak{Z}_3 > \cdots > \mathfrak{Z}_n > \cdots \tag{1}$$

heisst die absteigende Zentralreihe von $\mathfrak{G}$[2]. $\mathfrak{Z}_{i-1}/\mathfrak{Z}_i$ ist dann im Zentrum von $\mathfrak{G}/\mathfrak{Z}_i$ enthalten, also abelsch und hat endlich viele Erzeugende, wenn $\mathfrak{G}$ selbst endlich viele Erzeugende besitzt. Wenn $\mathfrak{Z}_n$ für irgendein n mit der Einheitsgruppe zusammenfällt, so nennt man $\mathfrak{G}$ nilpotent[3].

Es gilt dann

Satz 1. *$\mathfrak{G}$ sei eine nilpotente Gruppe mit endlich vielen Erzeugenden. Der Durchschnitt aller Normalteiler mit endlichen Indizen ist dann die Einheitsgruppe.*

Zum Beweis schicken wir einen Hilfssatz voraus.

Hilfssatz. Es sei $\mathfrak{G}$ eine beliebige Gruppe und $\mathfrak{N}$ ein Normalteiler von $\mathfrak{G}$ mit endlich vielen Erzeugenden. Ist $[\mathfrak{G}, \mathfrak{N}]$ eine endliche Gruppe und im Zentrum $\mathfrak{z}$ von $\mathfrak{G}$ enthalten, so enthält $\mathfrak{N}$ eine Untergruppe $\mathfrak{M}$ mit einem endlichen Index $[\mathfrak{N}:\mathfrak{M}]$, die in $\mathfrak{z}$ enthalten ist und durch endlich viele Elemente erzeugt wird.

Beweis. l sei die Ordnung von $[\mathfrak{G}, \mathfrak{N}]$. Da $[\mathfrak{G}, \mathfrak{N}]$ in $\mathfrak{z}$ enthalten ist, so gilt für beliebige $A \in \mathfrak{G}$, $N \in \mathfrak{N}$

$$(A, N)^l = (A, N^l) = 1\text{[4]}.$$

Daher ist N^l ein Element aus $\mathfrak{z}$ und $\mathfrak{M}=\{N^l;\ N \in \mathfrak{N}\}$ hat alle behaupteten Eigenschaften.

Beweis von Satz 1. Wir setzen

$$\mathfrak{G}=\mathfrak{Z}_1 > \mathfrak{Z}_2 > \cdots > \mathfrak{Z}_n = 1$$

$$\mathfrak{F}_i = \mathfrak{Z}_{i-1}/\mathfrak{Z}_i, \qquad i=2, \ldots, n.$$

$\mathfrak{F}_{i_1}, \mathfrak{F}_{i_2}, \ldots, \mathfrak{F}_{i_k}$ ($i_1 < i_2 < \cdots < i_k$) seien sämtliche unendliche Gruppen unter $\mathfrak{F}_i$. Der Satz ist offenbar richtig falls $k=0$. Wir wenden also die Induktion nach k an.

Sei zunächst $i_k \neq n$. Nach dem obigen Hilfssatz, angewandt auf

$$\bar{\mathfrak{G}}=\mathfrak{G}/\mathfrak{Z}_{i_k+1}, \quad \bar{\mathfrak{N}}=\mathfrak{Z}_{i_k-1}/\mathfrak{Z}_{i_k+1}, \quad [\bar{\mathfrak{G}}, \bar{\mathfrak{N}}]=\mathfrak{Z}_{i_k}/\mathfrak{Z}_{i_k+1},$$

gibt es eine solche Gruppe $\mathfrak{M}$ in $\mathfrak{Z}_{i_k-1}$, so dass $[\mathfrak{Z}_{i_k-1}:\mathfrak{M}]$ endlich ist und $\bar{\mathfrak{M}}=\mathfrak{M}/\mathfrak{Z}_{i_k+1}$ im Zentrum von $\bar{\mathfrak{G}}$ enthalten ist. Setzt man alsdann

1) Die Klammer soll die Kommutatorbildung bedeuten.
2) Vgl. H. Zassenhaus, Lehrbuch der Gruppentheorie (1937), S. 118.
3) Vgl. H. Zassenhaus, l. c., S. 105, 119.
4) (A, N) bedeutet den Kommutator von $A, N: (A, N)=ANA^{-1}N^{-1}$. Für diese Gleichheit vgl. H. Zassenhaus, l. c., S. 57.

$$\overline{\overline{\mathfrak{G}}} = \mathfrak{G}/\mathfrak{Z}_{i_k+2}, \qquad \overline{\overline{\mathfrak{M}}} = \mathfrak{M}/\mathfrak{Z}_{i_k+2},$$

so folgt aus der letzten Eigenschaft von $\mathfrak{M}$ dass $[\overline{\overline{\mathfrak{G}}}, \overline{\overline{\mathfrak{M}}}]$ in $\mathfrak{Z}_{i_k+1}/\mathfrak{Z}_{i_k+2}$ enthalten ist. Nach nochmaliger Anwendung des Hilfssatzes ergibt sich daraus, dass $\mathfrak{M}$ eine Untergruppe $\mathfrak{M}^*$ mit endlichem $[\mathfrak{M}:\mathfrak{M}^*]$ enthält, die im Zentrum von $\overline{\overline{\mathfrak{G}}}$ enthalten ist. So fortfahrend bekommt man schliesslich eine Gruppe $\mathfrak{K}$ in $\mathfrak{Z}_{i_k-1}$ mit einem endlichen Index $[\mathfrak{Z}_{i_k-1}:\mathfrak{K}]$, die im Zentrum von $\mathfrak{G}$ liegt und endlich viele Erzeugende hat.

$$
\begin{array}{ccccccc}
 & & \text{unend.} & \text{end.} & \text{end.} & \text{end.} & \\
\mathfrak{G}\!-\!\!-\!\mathfrak{Z}_{i_k-1} & \!-\!\!-\! & \mathfrak{Z}_{i_k} & \!-\!\!-\!\mathfrak{Z}_{i_k+1} & \!-\!\!-\!\mathfrak{Z}_{i_k+2} & \!-\!\cdots\!-\! & \mathfrak{Z}_n \\
\text{end.}\;\diagdown & & & & & & \diagdown\;\text{end.} \\
\mathfrak{M}\!-\!\!-\!\!-\! & \mathfrak{M}^* & \!-\!\!-\!\!-\!\!-\! & \cdots\cdots & \!-\!\!-\! & \mathfrak{K} & \!-\!\!-\!1 \\
 & \text{end.} & & \text{end.} & & \text{unend.} &
\end{array}
$$

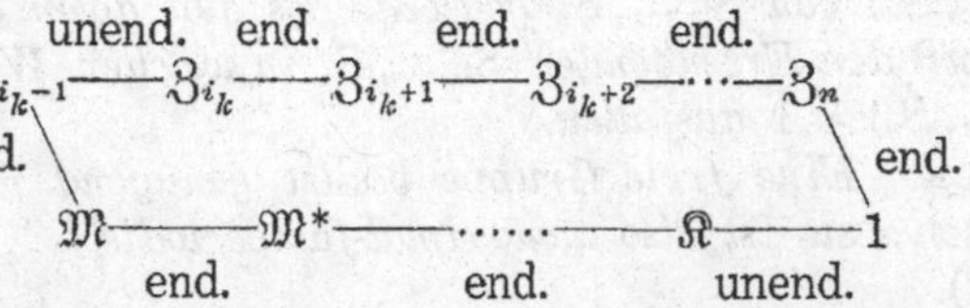

Auch im Fall $i_k = n$ bleibt die letzte Aussage gültig, wenn man als $\mathfrak{K}$ $\mathfrak{Z}_{i_k-1}$ selbst nimmt.

Nun sei $\mathfrak{K}^*$ eine beliebige Untergruppe von $\mathfrak{K}$ mit einem endlichen Index $[\mathfrak{K}:\mathfrak{K}^*]$. Der Satz ist dann gültig für $\mathfrak{G}/\mathfrak{K}^*$ nach der Induktionsvoraussetzung: Der Durchschnitt aller Normalteiler $\mathfrak{N}$ mit endlichen $[\mathfrak{G}:\mathfrak{N}]$, welche $\mathfrak{K}^*$ enthalten, ist also gerade $\mathfrak{K}^*$. Andereseits ist aber $\mathfrak{K}$ eine abelsche Gruppe mit endlich vielen Erzeugenden, also fällt der Durchschnitt aller solchen $\mathfrak{K}^*$ mit der Einheitsgruppe zusammen. Der Satz ist somit für $\mathfrak{G}$ selbst bewiesen.

In diesem Satz ist diejenige Bedingung wesentlich, dass $\mathfrak{G}$ durch endlich viele Elemente erzeugbar ist: z. B. gilt der Schluss des Satzes nicht für die additive Gruppe aller rationalen Zahlen.

2. Nun betrachten wir freie Gruppen. Es sei $\mathfrak{F}$ eine freie Gruppe mit beliebig vielen Erzeugenden S_i, und

$$A = S_{i_1}^{\pm 1} \cdots S_{i_k}^{\pm 1} \neq 1$$

sei ein beliebiges Wort in $\mathfrak{F}$. Bezeichnet man ferner mit $\mathfrak{G}$ bzw. $\mathfrak{N}$ die durch $S_{i_1}, \ldots, S_{i_k}$ erzeugte Untergruppe von $\mathfrak{F}$ bzw. den durch sämtliche übrige S_i erzeugten Normalteiler von $\mathfrak{F}$, so gilt, wie leicht ersichtlich,

$$(2) \qquad \mathfrak{F} = \mathfrak{G} \cdot \mathfrak{N}, \qquad \mathfrak{G} \cap \mathfrak{N} = 1, \qquad \mathfrak{F}/\mathfrak{N} \cong \mathfrak{G}.$$

Die absteigende Zentralreihe von $\mathfrak{G}$ sei wieder durch (1) gegeben. $\mathfrak{G}/\mathfrak{Z}_n$ ist dann offenbar eine nilpotente Gruppe mit endlich vielen Erzeugenden $S_{i_1}, \ldots, S_{i_k}$. Nach Satz 1 fällt also der Durchschnitt aller $\mathfrak{Z}_n$ enthaltenden Normalteiler in $\mathfrak{G}$ mit endlichen Indizen mit $\mathfrak{Z}_n$ zusammen. Da $\mathfrak{G}$ eine freie Gruppe ist, so ist der Durchschnitt aller $\mathfrak{Z}_n$ nach einem Satz von Magnus[1] die Einheitsgruppe. Satz 1 ist somit auch für $\mathfrak{G}$ richtig. Es gibt also in $\mathfrak{G}$ einen Normalteiler mit

---

1) W. Magnus, Über Beziehungen zwischen höheren Kommutatoren, Crelles Journ. 177 (1937).

einem endlichen Index, welcher das Element A nicht enthält. Nach (2) gibt es dann auch in $\mathfrak{F}$ einen Normalteiler mit einem endlichen Index, in welchem A nicht liegt. Da aber $A \neq 1$ beliebig in $\mathfrak{F}$ ausgewäht worden ist, so gilt Satz 1 auch für $\mathfrak{F}$ statt $\mathfrak{G}$. Somit is also bewiesen der

Satz 2. *$\mathfrak{F}$ sei eine freie Gruppe mit beliebig vielen Erzeugenden. Der Durchschnitt aller Normalteiler in $\mathfrak{F}$ mit endlichen Indizen fällt dann mit der Einheitsgruppe zusammen.*

Aus Satz 2 ergeben sich ohne Weiteres folgende Sätze.

Satz 3. *Es seien nicht leere, reduzierte Worte $W_1(S_1, \ldots, S_n), \ldots, W_k(S_1, \ldots, S_n)$ von $S_1, \ldots, S_n$ gegeben. Es gibt dann eine solche endliche Gruppe mit den Erzeugenden $S_1, \ldots, S_n$ in welcher $W_1(S_1, \ldots, S_n) \neq 1, \ldots, W_k(S_1, \ldots, S_n) \neq 1$ ausfallen.*

Satz 4. *Eine freie Gruppe besitzt genügend viele fastperiodische Funktionen; sie ist also „maximal-fastperiodisch" (maximally almost periodic*[1]).

Mit Hilfe der fastperiodischen Funktionen lässt sich die „schwache" Topologie in eine beliebige Gruppe einführen. Nach Satz 4 genügt diese Topologie im Fall der freien Gruppen dem Trennungsaxiom. Man kann daraus sofort den folgenden Satz herleiten[2].

Satz 5. *Es sei $\mathfrak{F}$ eine freie Gruppe und $\mathfrak{N}$ ein Normalteiler in $\mathfrak{F}$. Die abgeschlossene Hülle von $\mathfrak{N}$ in bezug auf der oben genannten Topologie sei mit $\mathfrak{N}^*$ bezeichnet. Setzt man ferner*

$$\mathfrak{G} = \mathfrak{F}/\mathfrak{N}\,; \qquad \mathfrak{G}^* = \mathfrak{F}/\mathfrak{N}^*,$$

so gilt:

(i) die Gesamtheit der fastperiodischen Funktionen von $\mathfrak{G}$ fällt mit der der fastperiodischen Funktionen von $\mathfrak{G}^*$ zusammen;

(ii) $\mathfrak{G}^*$ ist maximal-fastperiodisch;

(iii) $\mathfrak{G}$ ist folglich genau dann maximal-fastperiodisch, wenn $\mathfrak{N}$ in $\mathfrak{F}$ abgeschlossen ist, und genau dann minimal-fastperiodisch, wenn $\mathfrak{N}$ in $\mathfrak{F}$ überall dicht liegt.

---

1) Vgl. J. v. Neumann, Almost periodic functions in a group. I, Trans. Amer. Math. Soc. 36 (1934).

2) Vgl. eine Note des Verfassers in „Isō-Sūgaku" 4, 2 (1942) (japanisch).

[9] On one-parameter families of probability laws

J. Phys.-Math. Soc. Japan, 17-6 (1943), 217-220

Abstract (by S. Kusuoka)

Let $\mathfrak{F}$ be the set of all probability distribution functions on the real line $\mathbf{R}$. Then $\mathfrak{F}$ becomes a topological semigroup with the Lévy metric and with the convolution product. Let E_x, $x \in \mathbf{R}$, denote the probability measure on $\mathbf{R}$ concentrated at x. Then E_0 is the unit element in $\mathfrak{F}$. We say that $\{F_t;\ t \in I\}$ is a *one-parameter family* if I is an interval and if the map from I into $\mathfrak{F}$ given by $t \mapsto F_t$ is bijective and bicontinuous. We also say that $\{F_t;\ t \in I\}$ is a *one-parameter semigroup* if $\{F_t;\ t \in I\}$ is a one-parameter family, contains E_0, and is a subsemigroup of $\mathfrak{F}$.

We define a partial order $\geq$ in $\mathfrak{F}$ as follows: one has $F \geq G$ if there exists an $H \in \mathfrak{F}$ satisfying $F * H = G$. We say that a one-parameter family $\{F_t;\ t \in I\}$ is *linear* if $F_t \geq F_s$ for all $t, s \in I$ with $t \leq s$, or if $F_t \leq F_s$ for all $t, s \in I$ with $t \leq s$.

Theorem 1. F is an infinitely divisible distribution (cf. [L], Ch.7) if and only if there is a linear one-parameter family connecting E_0 and F.

Let us define an equivalence relation $\sim$ on $\mathfrak{F}$ as follows: one has $F \sim G$ if $F \geq G$ and $F \leq G$. Then we see that $F \sim G$ if and only if $F = G * E_x$ for some $x \in \mathbf{R}$. K. Itô [I] proved that the topology induced by the Lévy metric and the order topology coincide on a linearly ordered subset of $\mathfrak{F}/\sim$. So Theorem 1 seems of some interest because it gives a characterization of infinitely divisible distributions only by the order.

By Lévy's representation theorem, we see that for any infinitely divisible probability measure $F \in \mathfrak{F}$ and $s > 0$, there is a $F^{(s)} \in \mathfrak{F}$ such that

$$\varphi(z; F^{(s)}) = \varphi(z, F)^s, \ z \in \mathbf{R},$$

$\varphi(z, F)$ denoting the characteristic function of F.

The second result is the following

Theorem 2. Any one-parameter semigroup $\mathfrak{G}$ in $\mathfrak{F}$ coincides, after a suitable topological transformation of the parameter, with one of the following standard one-parameter semigroups.

(1) $\mathfrak{G} = \{E_0\}$.
(2) $\mathfrak{G} = \{F^{(s)}; 0 \leq s < \infty\}$, where F is an infinitely divisible distribution.
(3) $\mathfrak{G} = \{E_s; -\infty < s < \infty\}$.

References

[I] K. Itô, On stochastic processes (I), Jap. J. Math. **18** (1942), 261-301.
[L] P. Lévy, Théorie de l'addition des veriables aléatoires, Gauthier-Villars, 1937.

日本數學物理學會誌

第十七卷　第六號

原　著

確率法則の一徑數集合について

（昭和 18 年 3 月 13 日講演）

岩　澤　健　吉

§1. 實數軸上の確率法則（分布函數）の全體を $\mathfrak{F}$ とせよ．F_1, F_2 を任意の分布函數とすれば F_1 と F_2 との convolution $F_1 * F_2$ も又一つの分布函數であるから $\mathfrak{F}$ は convolution に關して可換な半群をつくる[1]．又 $y=F_1(x)$, $y=F_2(x)$ なる二つの曲線を $x+y=a$ $(-\infty<a<\infty)$ なる直線で截つた時の二交點間の距離の a に關する下限を $\rho(F_1, F_2)$ とすればこの ρ により $\mathfrak{F}$ は距離空間となり，しかも上の $F_1 * F_2$ なる演算はこの ρ に關して連續である[2]．かくして得られる位相的半群 $\mathfrak{F}$ に關しては P. Lévy, A. Khintchine 等により種々の研究が成されてゐる．

さて平均値 0, 標準偏差 $\sqrt{t}\,(t\geqq 0)$ なる Gauss 分布を G_t, 平均値 $t\,(t\geqq 0)$ なる Poisson 分布を P_t とすれば明かに

$$G_t * G_{t'} = G_{t+t'},$$
$$P_t * P_{t'} = P_{t+t'}$$

なる關係が成立する．即ち G_t 乃至 P_t の全體は丁度 Lie 群に於ける一徑數部分群の如き性質を有する．よつてこれを一般化して次の如く定義する．

定義 1. 實數軸上の任意の區間 I（開區間でも，閉區間でも又半開區間でもよい）から $\mathfrak{F}$ の中への一對一兩側連續な寫像

(1)　　$t \rightarrow F_t$

による I の像 $\mathfrak{S}=\{F_t;\ t\in I\}$ を $\mathfrak{F}$ の一徑數集合 (one parameter set) と呼ぶ．即ちそれは區間 I の位相的寫像である．特に $\mathfrak{S}$ が E を含む $\mathfrak{F}$ の部分半群であればこれを一徑數部分半群と呼ぶ．但しこゝに E は半群 $\mathfrak{F}$ の單位元で $x=0$ に確率 1 を與へる法則である．

§2. $\mathfrak{F}$ の元の間には又“順序”を與へることが出來る：即ち F, G を分布函數とするとき $F*H=G$ なる如き H が存在するならば

(2)　　$F\geqq G$

とするのである[3]．かくすれば $F\geqq F$ なること及び $F\geqq G$, $G\geqq H$ ならば $F\geqq H$ となることは明白であるが

(3)　　$F\geqq F'$　且　$F'\geqq F$

であつても必ずしも $F=F'$ とはならない．よつて一般に (3) が成立するとき F と F' とを同等,

(4)　　$F\sim F'$

と云ふことにして $\mathfrak{F}$ を類別し，この類全體の集合を $\mathfrak{E}$ とする．さて $x=m$ に確率 1 を與へる様な法則を E_m とすれば（從つて特に $E_0=E$）容易に證明される様に上の類別に於て E を含む類 e は E_m $(-\infty<m<\infty)$

1) §1, §2 に關しては特に K. Ito: On stochastic processes (I), Jap. Jour. Math., Vol. 18 (1942) 第二章參照．こゝに用ひた記號等は大部分上記論文のそれに從つた．尙本稿を草するに當り伊藤淸氏より種々有益なる御注意を賜つた．こゝに記して深謝の意を表する．

2) 今後 $\mathfrak{F}$ の“位相”について考へるときは何時もこの意味の距離による位相をとるものとする．

3) このことは勿論一般の半群に於て可能である．以下も同樣．

の全體から成り又 $\mathfrak{L}$ の各類は $\mathfrak{F}$ の部分半群 e による剩餘類と一致する．即ち F と F' とが同一の類 f に含まれる爲には $F'=F*E_m$ なる如き適當な實數 m が存在することが必要且十分で確率論的に云へばそれは F, F' に從ふ確率變數が常數値 m だけの差を有すると云ふことである．

商半群 $\mathfrak{F}/e$ として $\mathfrak{L}$ のなかにも演算 $*$ 及び順序 $\geqq$ を導入することが出來るが今度は $f\geqq f'$ 且 $f'\geqq f$ ならば $f=f'$ となるから，それは所謂“半順序を有する集合”である．

f に屬する任意の F と g に屬する任意の G との距離 $\rho(F, G)$ の下限を $\rho(f, g)$ とすればこれにより $\mathfrak{L}$ も又距離空間となり $f*g$ はこの距離に關し連續である．$\mathfrak{L}$ は半順序を有する集合であるからこの外に尙上に述べた順序によつて位相を導入することが出來る[4]．この順序による位相は一般に上の距離 ρ による位相よりも“强い”のであるが，特に線型に順序附けられたの部分集合の上では兩者は一致する[5]．

さて一徑數集合のうちでも次の如きものが特に重要である．

定義 2. 一徑數集合 $\mathfrak{S}$ に於て $t\leqq s$ なるとき常に $F_t\geqq F_s$ 又は $F_t\leqq F_s$ が成立するならば $\mathfrak{S}$ を線型一徑數集合と呼ぶ．$\mathfrak{S}$ が部分半群である場合も同樣である．

定義 1, 2 に於ける一徑數集合，一徑數半群等の概念は勿論 $\mathfrak{L}$ に於ても定義されるがそれは $\mathfrak{F}$ に於けると全く同樣であるから特に述べる必要もあるまい．

§3. 先づ次の定理を證明する．

定理 1. F が無限に分解可能な法則[6]である爲に必要十分な條件は E と F とを結ぶ線型一徑數集合が存在することである．類 f についても同樣．

證明. F に對して證明する．f でも同樣である．

必要なこと： Lévy の定理[7]により無限に分解可能な F の特性函數 $\varphi_F(z)$ は

$$(5)\quad \varphi_F(z)=\exp\Big(imz-\frac{\sigma^2}{2}z^2+\oint_{-\infty}^{\infty}\Big(e^{izu}-1-\frac{izu}{1+u^2}\Big)dn(u)\Big)$$

なる形に書ける．よつて

$$(6)\quad \varphi_t(z)=\exp\Big(imtz-\frac{(\sigma\sqrt{t})^2}{2}z^2+\oint_{-\infty}^{\infty}\Big(e^{izu}-1-\frac{izu}{1+u^2}\Big)d(tn(u))\Big)$$

を特性函數とする法則を F_t とすれば $F_0=E,\ F_1=F$ で $\mathfrak{S}=\{F_t;\ 0\leqq t\leqq 1\}$ が求むる線型一徑數集合であることは明白である．

十分なこと：$\mathfrak{S}=\{F_t;\ a\leqq t\leqq b\}$ を $F_a=E,\ F_b=F$ なる如き線型一徑數集合とせよ．先づ任意に與へられた $\varepsilon>0$ に對し次の如き $\delta=\delta(\varepsilon)>0$ が存在することを證明する．即ち $\delta>t'-t\geqq 0$ なる如き $[a, b]$ 中の任意の二數 t, t' をとれば假定により $F_t\geqq F_{t'}$ であるから $F_{t'}=F_t*G$ を滿足する G が存在するがこの樣な任意の G に對し $\rho(E, G)<\varepsilon$ が成立する．歸謬法を用ひることにして上の如き δ が存在しないと假定すれば $n=1, 2, \cdots$ に對し

$$(7)\quad \frac{1}{n}>t'_n-t_n\geqq 0,\ F_{t'_n}=F_{t_n}*G_n,\ \rho(E,G_n)\geqq\varepsilon$$

なる如き $t_n,\ t'_n\ G_n$ が存在する．$t_n,\ t_n'$ は $[a, b]$ に含まれてゐるから適當に部分列に移ることにより

$$(8)\quad \lim_{n\to\infty} t_n=\lim_{n\to\infty} t'_n=t_0,\quad a\leqq t_0\leqq b$$

なる t_0 が存在するものと考へてよい．さて假定により (1) なる寫像は連續であるから (8) により

$$(9)\quad \lim_{n\to\infty} F_{t_n}=\lim_{n\to\infty} F_{t'_n}=F_{t_0}.$$

よつて[8]

$$(10)\quad \lim_{n\to\infty} G_n=E.$$

これは (7) の $\rho(E, G_n)\geqq\varepsilon$ に反する．故に $\delta=\delta(\varepsilon)$ の存在が證明された．この $\delta(\varepsilon)$ を用ひて區間 $[a, b]$ を幅が $\delta(\varepsilon)$ より小なる如き小區間に分割しその分點を $a=t_0<t_1<\cdots<t_k=b$ とせよ．

$$(11)\quad F_{t_i}=F_{t_{i-1}}*G_i,\quad i=1, 2, \cdots, k$$

とおけば $\delta(\varepsilon)$ の性質から

$$(12)\quad \rho(E, G_i)<\varepsilon,\quad i=1, 2, \cdots k.$$

$F=G_1*G_2*\cdots*G_k$ で $\varepsilon>0$ は任意であるから (12) により F が無限に分解可能なることが知られる．證明終．

定理 1 は無限に分解可能な法則の意味から考へれば極めて當然の結果と云へるが次の如き意味に於て若干

4) G. Birkhoff: *Lattice theory* (1940), p. 28 參照．

5) 伊藤氏の論文，定理 2. 2 參照．

6) P. Lévy: *Théorie de l'addition des variables aréatoires* (1937), 第七章參照．

7) 註 6), P. Lévy, p. 180 參照．

8) 註 1), 伊藤氏の論文の補題 3. 1 參照．

の興味が存する．即ち §2 に於て述べた如く $\mathfrak{L}$ の線型部分集合に於ては距離 ρ による位相を順序による位相によつて置換へることが出來るから線型一徑數集合なる概念は ρ を用ひずに順序だけによつて定義することが出來る．よつて上の定理1によれば無限に分解可能な法則類は順序だけにより特徴付けられる．類の概念も又順序 $\geqq$ により定義されたのであるからこれによつて無限に分解可能な法則自身順序だけによつて特徴付け得ることがわかるのである．

§4. $\mathfrak{S}=\{F_t;\ t\in I\}$ を任意の一徑數半群とせよ．$\mathfrak{S}$ は E を含むから適當に座標をづらせることにより一般性を失ふことなく

(13) $$F_0=F$$

と假定しても差支へない．

t, t' を I に含まれる任意の實數とし

(14) $$F_t*F_{t'}=F_{t''}$$

とすれば t'' は t, t' により一意的に定まるが t を一定しておけばそれは t' の函數であるからこれを

(15) $$t''=f_t(t')$$

としよう．假定によりこれは t' の連續函數である．又定義から直ちに

(16) $$f_t(t')=f_{t'}(t).$$

さて I は正數を含むものと假定してその一つを $t>0$ とせよ．然らば任意の $t'\in I$ に對し常に

(17) $$f_t(t')>t'$$

となる．何となれば $f_t(t')\leqq t'$ なる如き t' が存在したとすれば $f_t(0)=t>0$ で且 f_t は連續であるから $f_t(t'')=t''$ を滿足する t'' が存在する．即ち

(18) $$F_t*F_{t''}=F_{t''}$$

これから容易に $F_t=E$ なることが知られる[9]．即ち $t>0$ で $F_t=F_0$ であるからこれは $\mathfrak{S}$ の定義に矛盾する．よつて常に (17) が成立する．換言すれば $F_t*F_{t'}=F_{t''}$ 且 $t>0$ であれば $t''>t'$ なることがわかつた．

區間 I の非負なる部分を I' とし $\mathfrak{S}'=\{F_t;\ t\in I'\}$ とすれば上の考察により特に $\mathfrak{S}'$ も半群であることが知られるがそこでは convolution により徑數が單調に增大する．一方 $\mathfrak{S}'$ に於ては又 Archimedes の公理も滿足されてゐる：即ち $t>0$ とし F_t の n 個の積を $F_t*F_t*\cdots*F_t=F_{t_n}$ とするとき n を十分大きくとれば，任意の t' に對し $t_n>t'$ となる．何となれば

[9] 註 6) の P. Lévy, p. 91, 定理 29, 1 參照.

若し凡ての $n=1, 2, \cdots$ に對し $t_n\leqq t'$ とすれば t_n は單調增大であるから

$$\lim_{n\to\infty} t_n=t''$$

が存在するがこれから

$$\lim_{n\to\infty} F_{t_n}=F_{t''}.$$

$F_t\neq E$ であるからこれは不合理である[10]．

故に $\mathfrak{S}'$ は徑數 t の大小によつて線型に順序附けられた可換な Archimedes 的半群でしかも順序の意味で連結されてゐるからそれは正の實數全體のつくる半群と同型である．よつて I' 上で定義され $(0, \infty)$ なる區間の凡ての値をとる單調增大函數

(19) $$s=g(t)$$

が存在してこれにより徑數を變換すれば

(20) $$\mathfrak{S}'=\{F_s;\ 0\leqq s<\infty\}$$
$$F_s*F_{s'}=F_{s+s'}$$

となることが知られる．よつて特に $s\leqq s'$ ならば $F_s\geqq F_{s'}$ である．

さて任意の $s>0$ に對し

$$\mathfrak{M}_s=\{F_{s'};\ 0\leqq s'\leqq s\}$$

とおけば $\mathfrak{M}_s$ は明かに E と F_s とを結ぶ線型一徑數集合であるから定理1により F_s は無限に分解可能である．よつて F_s の特性函數を $\varphi_s(z)$ とすれば (5) により

(21) $$\varphi_s(z)=\exp\{\psi_s(z)\},\quad \psi_s(0)=0$$

とおくことが出來る．(20) により

(22) $$\varphi_s(z)\varphi_{s'}(z)=\varphi_{s+s'}(z)$$

であるから (21) を代入して

(23) $$\psi_s(z)+\psi_{s'}(z)=\psi_{s+s'}(z)$$

$\psi_s(z)$ は s について連續であるから (23) から直ちに

(24) $$\psi_s(z)=s\cdot\psi_1(z).$$

を得る．即ち $\varphi_1(z)=\varphi(z)$ とおけば

(25) $$\varphi_s(z)=\varphi(z)^s,\quad 0\leqq s<\infty$$

となる．特性函數の間の (25) の如き關係を簡單に

(26) $$F_s=F^{(s)}$$

と記すこととする．

§5. 今迄我々は始めの區間 I が正數を含むものと假定して考察を進めて來たがこゝで更めて場合を分けて考へよう．

a) I が唯一點 0 から成る場合．

[10] 例へば Lévy-Doeblin の補助定理 (P. Lévy, p. 155, 定理 48) を用ひればよい．

このときは明かに $\mathfrak{S}=\{E\}$ で問題はない.

b) I が正數を含み負數を含まぬ場合.

§4 に述べた考察により適當に徑數を變換すれば

(27) $$\mathfrak{S}=\{F_s;\ 0\leqq s<\infty\}$$
$$F_s=F^{(s)}$$

となる. こゝに F はある無限に分解可能な法則である.

c) I が負數を含み正數を含まぬ場合.

このときは $t\to-t$ なる變換により b) の場合に歸着する.

d) I が正數も負數をも含む場合,

$t_0>0$ を一つとり (15) の函數 $f_{t_0}(t')$ を考察する. $f_{t_0}(0)=t_0>0$ であり且 $f_{t_0}(t')$ は t' に關し連續であるから t' が負で十分小であれば $f_{t_0}(t')>0$ となる. この樣な t' の一つを t_1 とし $f_{t_1}(t)$ を考へる.

$$f_{t_1}(0)=t_1<0,\quad f_{t_1}(t_0)=f_{t_0}(t_1)>0$$

であるから連續性により

(28) $$f_{t_1}(t_2)=0,\quad 0<t_2<t$$

なる t_2 が存在する. 即ち

(29) $$F_{t_1}*F_{t_2}=F_0=E,\quad t_1<0,\quad t_2>0.$$

(29) から適當な實數 m をとれば

(30) $$F_{t_1}=E_{-m},\quad F_{t_2}=E_m$$

となることが知られる[11]. 然るに §4 の結果によれば I の正, 負の部分はそれぞれその中の一つの法則の冪の全體と一致するから (30) により

(31) $$\mathfrak{S}=\{E_s;\ -\infty<s<\infty\}$$

なることがわかる.

かくして凡ての場合について $\mathfrak{S}$ の構造が完全に決定されたが逆に (27) 乃至 (31) で與へられる $\mathfrak{S}$ が一徑數半群であることは明白であるから次の定理が成立する.

定理 2. $\mathfrak{F}$ の任意の一徑數部分半群 $\mathfrak{S}$ は徑數の適當な位相的變換により次の如き標準型のいづれかと一致する.

I. $\mathfrak{S}=\{E\}$,

II. $\mathfrak{S}=\{F^{(s)};\ 0\leqq s<\infty\}$, こゝに F は任意の無限に分解可能な法則とする.

III. $\mathfrak{S}=\{E_s;\ -\infty<s<\infty\}$.

同樣な定理が $\mathfrak{L}$ に於ける一徑數部分半群に對しても成立する. この場合には $\mathfrak{S}=\{e\}$ であるか又はある無限に分解可能な f に對し $\mathfrak{S}=\{f^{(s)};\ 0\leqq s<\infty\}$ となる.

さて定理2から直ちに一徑數半群に含まれる法則は常に無限に分解可能であると云ふことがわかるが逆に又任意の無限に分解可能な F をとればそれを含む一徑數半群が存在することも II により知られる. しかも特に $F\neq E_m$ とすれば徑數の變換を除きその樣な F を含む一徑數半群は唯一つしか存在しない. 即ち $\mathfrak{F}$ に於て $E_m(-\infty<m<\infty)$ 以外の無限に分解可能な法則の全體の集合は一徑數半群により隙間なく一重に埋められてゐるのである.

東京帝國大學理學部數學教室

(昭和十八年三月十三日受理)

11) 註 9) 參照.

[10] On normed rings and a theorem of Segal, II

Zenkoku Shijo Sugaku Danwakai, 251 (3/1943), 167-186

Abstract (by Y. Kawahigashi)

We generalize the results of [5] to representations on Hilbert spaces. Let $L^{(1,p)}(G)$ be as in [5], $\mathfrak{H}$ a Hilbert space, and B the ring of all bounded linear operators on $\mathfrak{H}$.

We say that a representation $g \mapsto D(g) \in B$ of G into B is *proper* if $D(e) = I$, where e is the unit element of the group G and I is the identity operator on $\mathfrak{H}$. We say that a representation $x(g) \mapsto A(x)$ of $L^{(1,p)}(G)$ into B is *proper* when $A(x)f = 0$, $f \in \mathfrak{H}$, for all $x(g) \in L^{(1,p)}(G)$ implies $f = 0$. We say that such a representation of $L^{(1,p)}(G)$ is *continuous* when it is a continuous map from $L^{(1,p)}(G)$ with L^1-norm into B with uniform topology.

Now as a generalization of the result in [5], we have the following theorem (Theorem 1).

Theorem. We have a bijective correspondence between continuous proper representations of $L^{(1,p)}(G)$ into B and bounded measurable proper representations of G into B. That is, for a given continuous proper representation $x \mapsto A(x)$ of $L^{(1,p)}(G)$, we have a unique bounded measurable proper representation $g \mapsto D(g)$ of G such that we have

$$A(x) = \int_G x(g)D(g)dg$$

for any $x(g) \in L^{(1,p)}(G)$. Moreover, for a given such representation $g \mapsto D(g)$ of G, the above integral formula gives a continuous proper representation $x \mapsto A(x)$ of $L^{(1,p)}(G)$. These constructions are the inverse to each other and preserve the equivalence relation of representations.

Next assume that G is unimodular, i.e., G has a bi-invariant Haar measure. Then the correspondence $x(g)(\in L^{(1,p)}(G)) \mapsto x^*(g) = \overline{x(g^{-1})}$ gives an involution of $L^{(1,p)}(G)$. We assume that a representation $x(g) \mapsto A(x)$ satisfies $A(x)^* = A(x^*)$ for $x \in L^{(1,p)}(G)$, but do not assume that the representation is proper. Then we have a bijective correspondence between such continuous representation A of $L^{(1,p)}(G)$ into B and bounded measurable representations D of G into B satisfying $D(g)^* = D(g^{-1})$ in the above way. If $A(x)$ is proper, then the corresponding $D(g)$ is a unitary representation, and the converse also holds (Theorem 2).

Returning to the general case, we now consider for convenience a left invariant measure on G and define $L^{(1,p)}(G)$ as above. Let $\mathfrak{M}$ be the set of all $x(g) \in L^{(1,p)}(G)$ satisfying $x^*(g) \in L^{(1,p)}(G)$. Though this $\mathfrak{M}$ is not equal to $L^{(1,p)}(G)$ in general, we know that $\mathfrak{M}$ is a dense subring of $L^{(1,p)}(G)$ with respect to the L^1-norm. Let $D(g)$ be a unitary representation of G and $A(x)$ the corresponding representation of $L^{(1,p)}(G)$:

$$A(x) = \int_G x(g)D(g)dg.$$

We set $\boldsymbol{A} = \{A(x);\ x \in L^{(1,p)}(G)\}$ and $\boldsymbol{M} = \{A(x);\ x \in \mathfrak{M}\}$. Then $\boldsymbol{M}$ is dense in $\boldsymbol{A}$ in the uniform topology, hence also in the strong topology. We then know that the operator ring $R(\boldsymbol{M})$ generated by $\boldsymbol{M}$ in the sense of von Neumann [N] is equal to $\overline{\boldsymbol{M}}$, the closure of $\boldsymbol{M}$ in the strong topology. We also have $\overline{\boldsymbol{M}} = \overline{\boldsymbol{A}} = R(\boldsymbol{M}) = R(\boldsymbol{A})$. Furthermore, we have $R(A(x)) = R(D(g))$ (Theorem 3). We also prove (Theorem 4) that any measurable unitary representation $D(g)$ of G is continuous with respect to the strong topology, which was first proved by K. Kodaira [K].

References

[K] K. Kodaira, Über die Gruppe der messbaren Abbildungen, Proc. Imp. Acad. Tokyo **17** (1941), 18-23.

[N] J. von Neumann, Zur Algebra der Funktionaloperatoren und Theorie der normalen Operatoren, Math. Ann. **102** (1930), 370-427.

1109. ノルム環トSegalノ定理ニツイテ　II

§1. 談話1088[1]ノツヅキトシテ，ソコニ述ベタ結果ヲ Hilbert 空間ニ於ケル operator ニヨル表現ノ場合ニ拡張シマス．

群 G ノ群環 $L(G)$ 等ハ (I) ニ於ケルト同じモノトシ又 $\mathfrak{H}$ ヲ Hilbert 空間，$\boldsymbol{B}$ ヲ $\mathfrak{H}$ ニ於ケル凡テノ bounded operaor ガツクル環トシマス．

始メニ言葉の説明ヲシテオキマス．

1） $\boldsymbol{B}$ ニ於ケル G ノ表現 $g \to D(g)$ $(D(g) \in \boldsymbol{B})$ ガ固有デアルトハ $D(e) = 1$ ナルコト．

ココニ e ハ G ノ単位元，1 ハ $\mathfrak{H}$ ニ於ケル unit operator デス．

又 $D(g)$ ガ 一様ニ有界ナトキ即チ

$$|||D(g)||| \leq C, \quad g \in G$$

ナル常数 C ガ存在スルトキ，表現ハ有界デアルト云フコトニシマス．

2） $\boldsymbol{B}$ ニ於ケル $L(G)$ $(= L^{(1,p)}(G))$ ノ表現 $x(g) \to A(x)$ $(A(x) \in \boldsymbol{B})$ ガ固有デアルトハ凡テノ $x(g) \in L(G)$ ニ対シ $A(x)f = 0$ トナル如キ $\mathfrak{H}$ ノ元 f ハ $f = 0$ ニ限ルコト．

又，ソレガ連続デアルト云フノハ

$$|||A(x)||| \leq C\|x\|_1 \quad x(g) \in L(G)$$

ナル如キ常数 C ガ存在スルコト，即チ $L(G)$ ニ於ケル norm $\|x\|_1$ カラ $\boldsymbol{B}$ ノ uniform topology ヘ寫像ガ連続デアルコトヲ意味シマス．

サテ (I) ノ定理 9 ニ對應シテ次ノ定理ガ成立シマス.

定理 1. $L^{(1,p)}(G)$ ノ B ニ於ケル連続固有表現 $x \to A(x)$ ト G ノ B ニ於ケル有界可測 [2)] 固有表現 $g \to D(g)$ トハ次ノ意味デ一対一ニ對應スル：

i) $L^{(1,p)}(G)$ ノ表現 $x \to A(x)$ ニ對シテ適當ナ G ノ表現 $g \to D(g)$ ガ存在シテ

$$A(x) = \int_G x(g)D(g)dg \tag{1}$$

トナル. 且コノ様ナ $D(g)$ ハ唯一ツ定マル.

ii) $g \to D(g)$ ガ G ノ表現デアレバ (1) ニヨリ $A(x)$ ヲ定義スレバ $x \to A(x)$ ハ $L^{(1,p)}(G)$ ノ表現トナル.

iii) i), ii) ノ對應ハ互ニ他ノ逆デアル.

iv) 共軛ナ表現ニハ共軛ナ表現ガ對應スル.

注意. (1) ノ意味ハ任意ノ $f, f' \in \mathfrak{H}$ ニ対シ

$$(A(x)f, f') = \int_G x(g)(D(g)f, f')dg \tag{2}$$

ガ成立スルコト.

以下同様ナ記法ヲ用ヒルコトニシマス.

証明. iii), iv) ハ i), ii) 殊ニ表現ノ一意性カラ容易ニワカリマスカラ i), ii) ヲ証明シマス.

ii) ノ証明：

假定ニヨリ　$|||D(g)||| < C$ トスレバ

$$\left|\int_G x(g)(D(g)f, f')dg\right| \leq \int_G |x(g)|\,|(D(g)f, f')|dg \tag{3}$$

$$\leq \int_G |x(g)| \cdot C\|f\|\,\|f'\|dg = C\|x\|_1\|f\|\,\|f'\|.$$

ヨッテ任意ノ $f, f' \in \mathfrak{H}$ ニ對シ $\int_G x(g)(D(g)f, f')dg$ ハ常ニ存在シ, 明ニソレハ f ニ対シ linear, 又 f' ニ對シ conjugate linear デスカラ (3) カラ F. Riess ノ定理ニヨリ (2) ヲ満足スル operator $A(x)$ ガ存在シテ

$$|||A(x)||| \leq C\|x\| \tag{4}$$

トナリマス. $x \to A(x)$ ガ $L(G)$ ノ表現ニナッテヰルコトハ (I) ニ於ケルト同様ニ計算ニヨリ確カメラレマス. (4) ニヨリソレハ又連続デス. サテ凡ベテノ $x(g) \in L(G)$ 対シ $A(x)f_0 = 0$ ナル如キ $f_0 \in \mathfrak{H}$ ガ存在シタトスレバ任意ノ $f' \in \mathfrak{H}$ ニ対シ $(A(x)f_0, f') = 0$, 即チ (2) カラ

$$\int_G x(g)(D(g)f_0, f')dg = 0.$$

ココデ $x(g) \in L(G)$ ハ任意デスカラ測度 0 ノ g-集合ヲ除イテ

$$(D(g)f_0, f') = 0.$$

除外スベキ測度 0 ノ集合ハ勿論一般ニ f' ニ関係シマスカラコレヲ $E_{f'}$ トカクコト

ニシマス.

$\mathfrak{H}$ ノ一ツノ完全正規直交系ヲ $\varphi_1, \varphi_2, \ldots$ トシ $E_0 = \sum_{i=1}^{\infty} E_{\varphi_i}$ トオケバ E_0 ノ測度も亦 0 デスカラ $g_0 \notin E_0$ ナル g_0 ガ存在シマス. ヨッテ

$$(D(g_0)f_0, \varphi_i) = 0, \quad i = 1, 2, \ldots .$$

コレカラ $D(g_0)f_0 = 0$. ヨッテ $D(g_0^{-1})D(g_0)f_0 = D(e)f_0 = 1 \cdot f_0 = 0$, 即チ $f_0 = 0$ デ $x \to A(x)$ ハ固有表現デス.

i) ノ証明: $x(g) \to A(x)$ ヲ與ヘラレタ連続固有表現トシ $|||A(x)||| \leq C\|x\|$, トシマス. 任意ノ $f, f' \in \mathfrak{H}$ ニ對シ

$$|(A(x)f, f')| \leq |||A(x)||| \, \|f\| \, \|f'\| \leq C\|x\|_1 \|f\| \, \|f'\| \tag{5}$$

ナル故 (I) ニ於ケルト同様ニシテ

$$(A(x)f, f') = \int_G x(g) u_{f,f'}(g) dg \tag{6}$$

ヲ満足スル可測函数 $u_{f,f'}(g)$ ガ存在シ (5) カラ

$$|u_{f,f'}(g)| \leq C\|f\| \cdot \|f'\| \tag{7}$$

トナリマス. f, f' ヲ與ヘレバ (6) ニヨリ函数 $u_{f,f'}(g)$ ハ測度 0 ヲ除イテ定マリマスカラ任意ノ $f_1, f_2, f'_1, f'_2 \in \mathfrak{H}$, $\alpha, \beta \in K$ (複素数体) ニ対シ

$$\begin{aligned} u_{f_1+f_2,f'}(g) &\sim u_{f_1,f'}(g) + u_{f_2,f'}(g) \\ u_{f,f'_1+f'_2}(g) &\sim u_{f,f'_1}(g) + u_{f,f'_2}(g) \\ u_{\alpha f,f'}(g) &\sim \alpha u_{f,f'}(g) \\ u_{f,\beta f'}(g) &\sim \bar{\beta} u_{f,f'}(g). \end{aligned} \tag{8}$$

但シ $\sim$ ハ両辺ガ測度 0 ヲ除イテ一致スルコトヲ示シマス.

サテ $\mathfrak{H}$ ノ一ツノ完全正規直交系ヲ $\varphi_1, \varphi_2, \ldots$ トシ, 又有理複素数ノ全体ヲ $\rho_1, \rho_2, \cdots$ トシマス. ρ_j ヲ係数トスル有限個ノ φ_i ノ一次結合の全体ヲ $\mathfrak{L}$ トスレバ $\mathfrak{L}$ ハ可附番集合デス.

ヨッテ (8) ニ於テ $f, f', f_1, f_2, f'_1, f'_2$ ガ $\mathfrak{L}$ ノ凡テノ元ヲ動キ α, β ガ凡テノ有理複素数ヲ動イタトシテモ, カクシテ得ラレル (8) ノ如キ関係式ハ可附番個デスカラ適當ニ測度 0 ナル集合 $E_0(\mu(E_0) = 0)$ ヲトレバ $g \notin E_0$ ニナルトキ $\mathfrak{L}$ ニ属スル $f, f', \ldots$ 及ビ有理複素数 α, β ニ対シテハ (8) ハ等式トナリマス.

ヨッテ $g \notin E_0$ ナル g ヲ一ツ定メレバ $u_{f,f'}(g)$ ハ $\mathfrak{L}$ ニ属スル f ニ對シ linear, 又 f' ニ對シテハ conjugate linear トナリ且ツ (7) ガ成立シマス. $\mathfrak{L}$ ハ $\mathfrak{H}$ ニ於テ dense デスカラ, コレカラ $u_{f,f'}(g)$ ヲ凡テノ $f, f' \in \mathfrak{H}$ ニ対シ定義サレ, f ニ対シ linear, f' ニ對シ, conjugate linear ナ functional $u^*_{f,f'}(g)$ ニ拡張スルコトガ出来マス. (7) ニヨリ

$$|u^*_{f,f'}(g)| \leq C\|f\| \, \|f'\|.$$

ヨッテ $g \notin E_0$ ナル假定ノモトニ

$$(D_1(g)f, f') = u^*_{f,f'}(g) \tag{9}$$

$$|||D_1(g)||| \leq C \tag{10}$$

ナル如キ $\boldsymbol{B}$ ノ operator $D_1(g)$ ガ存在スルコトガワカリマス. $g \in E_0$ ノトキニハ

$D_1(g)=0$ ト定義シテオキマス.

サテ $f, f' \in \mathfrak{L}$ トスレバ, $g \notin E_0$ ナルトキ

$$(D_1(g)f, f') = u^*_{f,f'}(g) = u_{f,f'}(g)$$

ナル故 $\mu(E_0)=0$ ニ注意スレバ

$$(A(x)f, f') = \int_G x(g)(D_1(g)f, f')dg. \tag{11}$$

上ノ等式ノ両辺ハ f, f' ニ関シ連続デスカラ $\mathfrak{L}$ ガ $\mathfrak{H}$ デ dense ナルコトヲ用ヒレバ上記等式ハ任意ノ $f, f' \in \mathfrak{H}$ ニ対シ成立スルコトガ知レマス.

$A(x \times y) = A(x) \cdot A(y)$ ナル関係ヲ (11) 式ニ代入シテ計算スレバ

$$\int_{G\times G} x(g)y(h)(D_1(g)D_1(h)f, f')dgdh$$

$$= \int_{G\times G} x(g)y(h)(D_1(gh)f, f')dgdh.$$

ヨッテ $G \times G$ ニ於テ測度 0 ナル集合 $E_{f,f'}$ ヲ除ケバ

$$(D_1(g)D_1(h)f, f') = (D_1(gh)f, f'). \tag{12}$$

f, f' ガ $\mathfrak{L}$ ノ凡テノ元ヲ動クトキ, $E_{f,f'}$ ノ和ヲ E_1 トスレバ E_1 ノ測度モ 0 デ且ツ $(g, h) \notin E_1$ ナラバ $\mathfrak{L}$ ニ属スル任意ノ f, f' ニ対シ (12) ガ成立シマス. 然ルニ (12) ノ両辺ハ f, f' ニ関シ連続デスカラ, ソレハ又 $\mathfrak{H}$ ノスベテノ f, f' ニ対シ成立シマス. 即チ $D_1(g)D_1(h) = D_1(gh)$. ヨッテ G ニ於テ測度 0 ナル適當ナ集合 E_2 ヲトレバ $h \notin E_2$ ナルトキ $\mu(E_h)=0$ ナル集合 E_h ガ定マッテ $g \notin E_h$ ナルトキ

$$D_1(g)D_1(h) = D_1(gh) \tag{13}$$

トナリマス. (Fubini ノ定理)

故ニ $a \notin E_2$ ナラバ

$$(A(a^{-1}x)f, f') = \int_G x(ga^{-1})(D_1(g)f, f')dg = \int_G x(g)(D_1(ga)f, f')dg$$

$$= \int_G x(g)(D_1(g)D_1(a)f, f')dg = (A(x)D_1(a)f, f').$$

即チ

$$A(a^{-1}x) = A(x)D_1(a), \quad a \notin E_2. \tag{14}$$

サテ $\mu(E_2)=0$ ナル故 $\mu(E_2^{-1})=0$. ヨッテ $E_3 = E_2 \cup E_2^{-1}$ トオケバ $E_3^{-1} = E_3$. 且ツ $\mu(E_3)=0$.

又 G ノ任意ノ元 g ハ $g = g_1g_2, g_1, g_2 \notin E_3$ ナル形ニカクコトガ出来マス. ソコデ

$$D(g) = D_1(g_1)D_1(g_2), \quad g = g_1g_2, \quad g_1, g_2 \in E_3 \tag{15}$$

ト定義シマス.

先ヅ (15) ノ定義ニ於テ右辺ガ g ニノミ関係シテ定マルコトヲ云ハネバナリマセンガ,

モット一般ニ $g=g_1g_2\cdots g_r=g_1'g_2'\cdots g_s', g_i, g_j'\notin E_3$ ナラバ $D_1(g_1)D_1(g_2)\cdots D_1(g_r)=D_1(g_1')\cdots D_1(g_s')$ ナルコトヲ証明シマス．$x(g)\in L(G)$ ヲ任意ニトレバ (14) ニヨリ

$$A(g^{-1}x)=A(g_r^{-1}g_{r-1}^{-1}\cdots g_2^{-1}g_1^{-1}x)=A(g_r^{-1}(g_{r-1}^{-1}\cdots g_1^{-1}x))$$
$$=A(g_{r-1}^{-1}\cdots g_1^{-1}x)D_1(g_r)=\cdots=A(x)D_1(g_1)D_1(g_2)\cdots D_1(g_r).$$

同様ニ

$$A(g^{-1}x)=A(x)D_1(g_1')D_1(g_2')\cdots D_1(g_s').$$

ヨッテ $f\in\mathfrak{H}$ ヲ任意ニトルトキ

$$\begin{aligned}0 &= A(g^{-1}x)f-A(g^{-1}x)f\\ &= A(x)D_1(g_1)\cdots D_1(g_r)f-A(x)D_1(g_1')\cdots D_1(g_s')f\\ &= A(x)\{D_1(g_1)\cdots D_1(g_r)-D_1(g_1')\cdots D_1(g_s')\}f.\end{aligned}$$

$x(g)\to A(x)$ ハ假定ニヨリ固有表現デスカラ，コレカラ

$$\{D_1(g_1)\cdots D_1(g_r)-D_1(g_1')\cdots D_1(g_s')\}f=0,$$

$f\in\mathfrak{H}$ ハ任意ナル故

$$D_1(g_1)\cdots D_1(g_r)=D_1(g_1')\cdots D_1(g_s').$$

ヨッテ (15) ニヨリ実際一意的ニ $D(g)$ ナル operator ガ凡テノ $g\in G$ ニ對シ定義シマス．

(10) ニヨリ

$$|||D(g)|||\le|||D_1(g_1)|||\cdot|||D_1(g_2)|||\le c^2.$$

又 $g=g_1g_2, g'=g_1'g_2', gg'=g_1''g_2'', g_1,g_2,g_1',g_2',g_1'',g_2''\notin E_3$ トスレバ $g_1g_2g_1'g_2'=g_1''g_2''$ ナル故

$$D(gg')=D_1(g_1'')D_1(g_2'')=D_1(g_1)D_1(g_2)D_1(g_1')D_1(g_2')=D(g)D(g')$$

ヨッテ $g\to D(g)$ ハ有界表現デス．

又 $e=gg^{-1}, g\notin E_3$ トスレバ定義ニヨリ

$$D(e)=D_1(g)D_1(g^{-1})$$

デスガ

$$A(ex)=A(x)=A(g^{-1}gx)=A(x)D_1(g)D_1(g^{-1})$$

カラ前ト同様ニシテ $D_1(g)D_1(g^{-1})=1$，即チ $D(e)=1$ デソレガ固有表現デアルコトガ知ラレマス．

サテ $h_0\notin E_3$，從ッテ勿論 $h_0\notin E_2$ ナル如キ h_0 ヲトリ $(E_3\cup E_{h_0})h_0=E_4$ トシマス．

$$\mu(E_4)=\mu((E_3\cup E_{h_0})h_0)=\mu(E_3\cup E_{h_0})\le\mu(E_3)+\mu(E_{h_0})=0.$$

即チ $\mu(E_4)=0$.

$g\notin E_4$ ヲ任意ニトリ

$$g=g'h_0$$

トオケバ $g'\notin E_3\cup E_{h_0}$，即チ $g', h_0\notin E_3$ デスカラ定義ニヨリ

$$D(g)=D_1(g')D(h_0).$$

一方 $g' \notin E_{h_0}$ ナル故 (13) カラ

$$D_1(g) = D_1(g'h_0) = D_1(g')D_1(h_0).$$

ヨッテ

$$D(g) = D_1(g), \quad g \notin E_4 \quad (\mu(E_4) = 0). \tag{16}$$

ツクリ方カラワカルヤウニ $(D_1(g)f, f')$ ハ可測デスカラ (16) ニヨリ $D(g)$ ハ可測表現デアルコトガワカリ又 (11) カラ

$$A(x) = \int_G x(g)D(g)dg$$

トナリマス．ヨッテ $D(g)$ ノ存在ハ証明サレマシタ．

次ニソノ一意性ヲ云ヒマス．今有界可測表現 $D'(g)$ ニヨリ

$$A(x) = \int_G x(g)D'(g)dg$$

トナッタトスレバ任意ノ $f, f' \in \mathfrak{H}$ ニ對シ

$$\int_G x(g)(D(g)f, f')dg = \int_G x(g)(D'(g)f, f')dg.$$

ヨッテ $\mu(E_{f,f'}) = 0$ ナル集合 $E_{f,f'}$ ヲ除キ

$$(D(g)f, f') = (D'(g)f, f'). \tag{17}$$

f, f' ガ $\mathfrak{H}$ ノ完全正規直交系 $\varphi_1, \varphi_2, \ldots,$ ノ全体ヲ動クトキ $E_{f,f'}$ ノ和ヲ E_5 トスレバ矢張リ $\mu(E_5) = 0$ デ $g \notin E_5$ ナラバ $f, f' = \varphi_1, \varphi_2, \ldots$ ニ對シ (17) ガ成立シマス．然ルニ (17) ノ両辺ハ f, f' ニ関シ linear 且ツ連続デスカラ (17) ハ又 $g \notin E_5$ ニ対シテハ凡テノ $f, f' \in \mathfrak{H}$ ニ対シ成立シマス．即チ

$$D(g) = D'(g), \quad g \notin E_5, \quad (\mu(E_5) = 0).$$

コレカラ凡テノ $g \in G$ ニ対シ $D(g) = D'(g)$ トナルコトハ (I) ニ於ケルト同様デス．

§2．コノ§デハ G ハ左右-不変ノ測度ヲ有スルモノト仮定シマス．サウスレバ $x(g)$ ガ $L^{(1,p)}(G)$ ニ含マレレバ $\overline{x(g^{-1})} = x^*(g)$ ナル函数モ亦 $L^{(1,p)}(G)$ ニ含マレマス．$x \to x^*$ ハ $L^{(1,p)}(G)$ ノ逆同型ヲ與ヘマスカラ $L^{(1,p)}(G)$ ヲ B 内デ表現スル場合 $A(x)$ ト $A(x^*)$ トガ互ニ adjoint ニナッテヰル場合ヲ考ヘルコトハ自然デアリマセウ．ヨッテ以下 $x(g) \to A(x)$ ハ

$$A(x)^* = A(x^*) \tag{18}$$

ナル條件ヲ満足スル連続表現トシ，但シ固有デアルト云フコトハ假定シナイコトニシマス．サテ $\mathfrak{H}$ ニ於テ凡テノ $x(g) \in L(G)$ ニ対シ

$$A(x)f = 0$$

トナルヤウナ f ノ全体ヲ $\mathfrak{N}$ トスレバ $\mathfrak{N}$ ハ明カニ閉線状部分空間トナリマス．ソノ complement ヲ $\mathfrak{M}$ トシマス．$\mathfrak{M} = \mathfrak{H} - \mathfrak{N}$. $\{A(x)\}$ ナル集合ハ各 operator ト共ニソノ adjoint ヲ含ム故 $\mathfrak{M}, \mathfrak{N}$ ハ $\{A(x)\}$ ヲ "reduce" シマス．ソノ $\mathfrak{N}$ ニ於ケル部分

ノ表現ハ 0-表現デスガ $\mathfrak{M}$ ニ於ケル部分ハ固有表現 $A_1(x)$ ヲ與ヘマスカラ $\mathfrak{M}$ ニ対シテ定理 1 ヲ用ヒレバ

$$A_1(x) = \int_G x(g)D_1(g)dg$$

ナル G ノ表現 $D_1(g)$ ガ存在シマス．サテ，

$$A_1(x^*) = \int_G \overline{x(g^{-1})}D_1(g)dg$$

$$(A_1(x^*)f, f') = \int_G \overline{x(g^{-1})}(D_1(g)f, f')dg = \overline{\int_G x(g^{-1})(D_1^*(g)f', f)dg}$$

$$= \overline{\int_G x(g)(D_1^*(g^{-1})f', f)dg}.$$

一方

$$(A_1(x^*)f, f') = (A_1(x)^*f, f') = \overline{(A_1(x)f', f)} = \overline{\int_G x(g)(D_1(g)f', f)dg}.$$

ヨッテ

$$\int_G x(g)(D_1^*(g^{-1})f', f)dg = \int_G x(g)(D_1(g)f', f)dg.$$

表現ノ一意性ヲ用ヒレバ，コレカラ

$$D_1^*(g^{-1}) = D_1^*(g)^{-1} = D_1(g).$$

即チ $D_1(g)$ ハスベテ unitary operator デアルコトガワカリマス．

逆ニ $D_1(g)$ ガ unitary ナラバ $A_1(x^*) = A_1(x)^*$ トナルコトハ上ノ計算ヲ逆ニタドッテ見レバワカリマス．

サテ $\mathfrak{N}$ ニ於テ $D_2(g) = 0$ トシ

$$D(g) = \begin{pmatrix} D_1(g) & \\ & D_2(g) \end{pmatrix} \quad \begin{matrix} \mathfrak{M} & \mathfrak{N} \end{matrix}$$

トオケバ明カニ

$$A(x) = \int_G x(g)D(g)dg,$$

且ツ

$$D(g)^* = D(g^{-1})$$

トナリマス．ヨッテ

定理 2. G ノ測度ガ左右-不変トスレバ $A(x^*) = A(x)^*$ ノ満足スル $L^{(1,p)}(G)$ ノ B ニ於ケル連続表現 $x \to A(x)$ ト $D(g)^* = D(g^{-1})$ ヲ満足スル G ノ B ニ於ケル有界可測表現トハ定理 1 ノ意味デ一對一ニ對應スル．特ニ $A(x)$ ガ固有表現デアレバソレニ對應スル $D(g)$ ハ unitary 表現デアリ逆モ亦成立スル．

§ 3．今マデ G ノ測度ハ右-不変（又ハ特ニ左右-不変）ト考ヘテ来タノデスガ，コノ

§デハ左-不変ノ測度ヲトルコトニシマス.[0)] サウシテモ今マデノ議論ハスベテ成立シマス. 唯

$$x \times y(g) = \int_G x(h)y(h^{-1}g)dh$$

トナリ (I) ノ (8) ノ代リニ

(19) $$\|x \times y\|_p \leq \|x\|_1 \cdot \|y\|_p; \quad x(g) \in L^1(G), \quad y(g) \in L^p(G)$$

ガ成立シマス.

サテ $L^{(1,2)}(G)$ ハ Hilbert 空間 $L^2(G)$ 内デ dense ナ線状部分空間トナッテヰマスガ, ココデ

(20) $$A_x \cdot y = x \times y, \quad x, y \in L^{(1,2)}(G)$$

トオケバ (19) ニヨリ

(21) $$\|A_x \cdot y\|_2 \leq \|x\|_1 \cdot \|y\|_2.$$

ヨッテ A_x ハ bounded ナ operator トナリマス. 故ニ A_x ヲ $L^{(1,2)}(G)$ カラ $L^2(G)$ 全体ニマデ拡張スルコトガ出来マスガ, ソノ結果ハ明カニ

$$A_x \cdot y = x \times y = \int x(h)y(h^{-1}g)dh, \quad x \in L^{(1,2)}(G), \ g \in L^2(G)$$

ニヨリ與ヘラレマス. (21) ハソノママ成立シマスカラ

(22) $$|||A_x||| \leq \|x\|_1.$$

又

$$A_{x \times y} \cdot z = (x \times y) \times z = x \times (y \times z) = A_x \cdot A_y z.$$

即チ

(23) $$A_{x \times y} = A_x \cdot A_y.$$

(22), (23) ニヨリ $x(g) \to A(x)$ ハ $L^{(1,2)}(G)$ ノ連続表現デアルコトガ分リマス.

ソレハ有限次ノ Algebra ニ於ケル左カラノ正規表現ノ拡張ト考ヘラレマス. (同様ニシテモシ右-不変ナ測度ヲ用ヒテ $A'_x y = y \times x$ トオケバ右カラノ正規表現ヲ得マスガ, ソレハ有限次元ノ場合と同様ニ逆同型ヲ與ヘルコトニナリマスノデ, ソレヲ避ケテ左-不変ナ測度ヲトリマシタ.)

サテ, コノ $A(x)$ ニ§1ノ定理1ノ意味デ對應シテヰル G ノ表現ハ何カト云フトソレハ

$$(A_x y, z) = \int_G A_x \cdot y(g) \cdot \overline{z(g)} dg = \int_G (\int_G x(h)y(h^{-1}g)dh) \overline{z(g)} dg$$

$$= \iint_{G \times G} x(h)y(h^{-1}g)\overline{z(g)} dg dh = \int_G x(h) \left\{ \int_G y(h^{-1}g)\overline{z(g)} dg \right\} dh$$

$$= \int_G x(h)(U(h)y, z)dh.$$

コノ計算カラワカル様ニ $y(g) \to y(h^{-1}g)$ ナル unitary operator $U(h)$ デアリマス. コノ意味デ $U(h)$ ヲ G ノ正規表現ト云ッテモヨイカト思ヒマス.

§ 4．ココデモ左-不変ノ測度ヲトルコトニシマス．サテ $L^{(1,p)}(G)$ ニ含マレル函数 $x(g)$ ニ對シ函数 $x^*(g)=\overline{x(g^{-1})}$ ヲ考ヘレバ G ノ測度ガ左右不変デナイトキハ $x^*(g)$ ハ必ズシモ $L^{(1,p)}(G)$ ニ属シマセンガ $x(g)$, $x^*(g)$ ガ共ニ $L^{(1,p)}(G)$ ニ含マレルヤウナ $x(g)$ ノ全体 $\mathfrak{M}$ ハ norm $\|x\|_1$ ニ関シテ dense ナ $L^{(1,p)}(G)$ ノ部分環ヲツクリマス．

ソレガ部分環ヲツクルコトハ明ラカデスガ dense デアルコトハ例ヘバ G ノ bi-compact (compact) ナ部分集合ノ特性函数ガスベテ $\mathfrak{M}$ ニ含マレルコトカラ知ラレマス．

G ノ一ツノ unitary 表現 $D(g)$ ガ與ヘラレタモノトシソレニ対應スル $L^{(1,p)}(G)$ ノ表現ヲ $A(x)$ トシマス．

$$A(x)=\int_G x(g)D(g)dg.$$

又 $\boldsymbol{B}$ ニ於テ

$$\boldsymbol{A}=\{A(x);\ x\in L^{(1,p)}(G)\},\quad \boldsymbol{M}=\{A(x);\ x\in\mathfrak{M}\}$$

トオケバ $\mathfrak{M}$ ハ $L^{(1,p)}(G)$ デ dense デ $x\to A(x)$ ハ連續デスカラ $\boldsymbol{M}$ ハ $\boldsymbol{A}$ ニ於テ uniform topology ニ関シ dense, 従ッテ勿論 strong topology ニ関シ dense トナリマス．

§ 2 ノ計算カラ明カナヤウニ，一般ニ，A, B ナル operator ガ $\boldsymbol{M}$ ニ含マレレバ $A+B$, αA, AB, A^* モ亦 $\boldsymbol{M}$ ニ含マレマスカラ strong topology ニ関スル $\boldsymbol{M}$ ノ closure ヲ $\overline{\boldsymbol{M}}$ トスレバ

$$R(\boldsymbol{M})=\overline{\boldsymbol{M}}. \tag{24}$$

但シココニ $R(\boldsymbol{M})$ ハ $\boldsymbol{M}$ カラ生成サレタ Neumann ノ意味ノ operator ring デアリマス．[4)]

$\boldsymbol{M}$ ハ $\boldsymbol{A}$ 内デ dense ナル故 $\overline{\boldsymbol{M}}\supset\boldsymbol{A}$, 即チ $R(\boldsymbol{M})\supset\boldsymbol{A}$, ヨッテ $R(\boldsymbol{M})\supset R(\boldsymbol{A})$. 一方 $R(\boldsymbol{M})\subset R(\boldsymbol{A})$ ハ明カデスカラ

$$\overline{\boldsymbol{M}}=\overline{\boldsymbol{A}}=R(\boldsymbol{M})=R(\boldsymbol{A}). \tag{25}$$

サテ任意ノ $x(g)\in L^{(1,p)}(G)$ ヲトルトキ (14) ニ於ケルト同様ノ計算デ

$$A(xa^{-1})=D(a)A(x).$$

但シ $xa^{-1}(g)=x(a^{-1}g)$ トシマス．[5)]　又

$$|||D(a)A(x)-A(x)|||=|||A(xa^{-1})-A(x)|||=|||A(xa^{-1}-x)|||\le C\|xa^{-1}-x\|_1.$$

xa^{-1} ハ x ヲ定メレバ a ニ関シ norm $\|x\|_1$ デ連続デスカラ $D(a)A(x)$ モ亦 x ヲ定メレバ a ニ関シ uniform topology デ連続デス．

$R(\boldsymbol{A})$ ニ含マレル任意ノ operator A 及ビ任意ノ $f\in\mathfrak{H}$, $\epsilon>0$ ガ與ヘラレタトシマス．(25) ニヨリ $R(\boldsymbol{A})$ ハ $\boldsymbol{A}$ ノ strong topology ニ関スル closure デスカラ

$$\|(A-A(x))f\|<\frac{\epsilon}{3}$$

ナル $A(x)$ ガ存在シマス．又 G ニ於ケル単位 e ノ近傍 $\mathfrak{U}(e)$ ヲ適當ニトレバ $a\in\mathfrak{U}(e)$ ナルトキ

$$\|(D(a)A(x)-A(x))f\|<\frac{\epsilon}{3}.$$

ヨッテ $D(a)$ ガ unitary ナルコトカラ

$$\|(D(a)A-A)f\| \leq \|(D(a)A-D(a)A(x))f\| + \|(D(a)A(x)-A(x))f\|$$
$$+\|(A(x)-A)f\| \leq \frac{\epsilon}{3}+\frac{\epsilon}{3}+\frac{\epsilon}{3}=\epsilon.$$

即チ $f, \epsilon > 0$ ニ對シ $\mathfrak{U}(e)$ ヲ適當ニトレバ $a \in \mathfrak{U}(e)$ ナルトキ

$$\|(D(a)A-A)f\| \leq \epsilon,$$

ヨッテ $D(a)A$ ハ A ヲ定メタ場合 strong topology ニ関シ連続デアルコトガワカリマス.

サテ $R(\boldsymbol{A})$ ニ含マレル最大ノ projection ヲ E_0 トスレバ任意ノ $A \in R(\boldsymbol{A})$ ニ對シ $E_0A = AE_0 = A.$ [6] ヨッテ特ニ

$$A(x) = A(x)E_0 = \int_G x(g)D(g)E_0 dg. \tag{26}$$

$D(g)E_0 = \bar{D}(g)$ トオケバ $\bar{D}(g)$ モ明カニ有界可測, 且ツソレハ $R(\boldsymbol{A})$ ニ含マレマス. (何トナレバ任意ノ $A(x)$ ニ對シ $D(a)A(x) = A(xa^{-1}) \in \boldsymbol{A}.$ ヨッテソノ closure ヲトッテ見レバ任意ノ $A \in R(\boldsymbol{A})$ ト共ニ $D(a)A$ モ亦 $R\boldsymbol{A})$ ニ含マレマス.) ヨッテ $E_0\bar{D}(g) = \bar{D}(g)$, 故ニ

$$\bar{D}(g)\bar{D}(h) = D(g)E_0\bar{D}(h) = D(g)\bar{D}(h) = D(g)D(h)E_0 = D(gh)E_0 = \bar{D}(gh).$$

即チ $g \to \bar{D}(g)$ ハ G ノ表現デ且 (26) カラ

$$A(x) = \int_G x(g)\bar{D}(g)dg$$

トナリマスカラ一意性ニヨリ $\bar{D}(g) = D(g)$, 即チ

$$D(g)E_0 = D(g). \tag{27}$$

故ニ $D(g)$ ハ $R(\boldsymbol{A})$ ニ含マレマス. 一方

$$(A(x)f, f') = \int_G x(g)(D(g)f, f')dg$$

カラ明ラカナヤウニ $A(x)$ ハ $D(g)$ カラ生成サレル Neumann ノ意味ノ operator ring $R(D(g))$ (ソレハ weakly closed ナル故) ニ含マレマスカラ次ノ定理ヲ得マス:

定理 3.

$$R(A(x)) = R(D(g)).$$

サテ $D(g)E_0$ ハ E_0 ヲ定メレバ g ニ関シ連続 (strongly) デアッタノデスカラ (27) ニヨリ

定理 4. G ノ可測ナ unitary 表現 $D(g)$ ハスベテ strong topology ニ関シ連続デアル.

コトガ証明サレマシタ. [7]

脚註

1) コレヲ以下 (I) トシテ引用シマス.
2) 任意ノ $f, f' \in \mathfrak{H}$ ニ対シ $(D(g)f, f')$ ガ可測ナルコト.
3) ソノ理由ハ以下参照. 始メカラ全部左-不変ノ測度デ議論シテオケバヨカッタノデス.
4) J. v. Neumann: Zur Algebra der Funktionaloperatoren und Theorie der normalen Operatoren, Math. Ann. **102** (1930)
5) ココデハ左-不変ノ測度ヲ用キテキルコトニ注意.
6) 脚註 4) ノ論文参照.
7) コノ定理ハ measure ト topology ノ関係ヲ用ヒテ小平サンガ始メテ証明サレタモノデス: K. Kodaira: Über die Gruppe der messbaren Abbildungen, Proc. Imp. Acad. Tokyo **17** (1941), 18-23.

[11] On group rings of topological groups

Proc. Imp. Acad. Japan, 20-2 (1944), 67-70.

Mathematical Institute, Tokyo Imperial University.

(Comm. by T. TAKAGI, M.I.A., Feb. 12, 1944.)

§ 1. Let G be a locally compact topological group, satisfying the second axiom of countability and μ a left invariant Haar measure on G. We denote as usual by $L^p(G)$ $(p \geqq 1)$ the set of all μ-measurable functions $x(g)$ of G with finite

$$\| x(g) \|_p = \left\{ \int_G | x(g) |^p \mu(dg) \right\}^{\frac{1}{p}} .$$

For arbitrary $x(g) \in L^1(G)$, $y(g) \in L^p(G)$ and

(1) $$z(g) = x \times y(g) = \int_G x(h) y(h^{-1}g) \mu(dh) ,$$

we have

(2) $$\| z \|_p = \| x \times y \|_p \leqq \| x \|_1 \| y \|_p .$$

Defining the multiplication by (1) and putting

(3) $$\| x \| = \text{Max.} (\| x \|_1, \| x \|_p) ,$$

the intersection $L^{(1,p)}(G)$ of $L^1(G)$ and $L^p(G)$ thus becomes a non-commutative normed ring[1]. But, generally speaking, $L^{(1,p)}(G)$ has not a unit element. Adjoining therefore formally the unit e, I. E. Segal considered the set of all

$$\mathfrak{z} = \lambda e + x(g) ; \quad \lambda = \text{complex number}, \quad x(g) \in L^{(1,p)}(G) ,$$

and called it the group ring $R^{(1,p)}(G)$ of G[2]. But we would rather prefer to call $L^{(1,p)}(G)$ itself the group ring of G. We shall give in this paper certain close relations between G and $L^{(1,p)}(G)$, some of which are generalizations of the results of I. E. Segal.

§ 2. We consider representations of G and $L^{(1,p)}(G)$, i. e. homomorphic mappings of G and $L^{(1,p)}(G)$ into matrices, whose components are complex numbers[3].

Our main theorem is then:

Theorem 1. There is a one-to-one correspondencc between continuous[4] representations of $L^{(1,p)}(G)$ and bounded continuous representations of G in the following sense:

i) For a given continuous representation $x(g) \rightarrow T(x)$ of $L^{(1,p)}(G)$, there corresponds uniquely a bounded continuous representation $a \rightarrow D(a)$ of G, so that it holds

---

1) For normed rings cf. I. Gelfand: Normierte Ringe, Rec. Math., **51** (1941), 37–58.

2) I. E. Segal: The group ring of a locally compact group, I, Proc. Nat. Acad. Sci., U. S. A. **27** (1941).

3) For the representation of G, we do not require that the unit of G corresponds to the unit matrix.

4) The topology in $L^{(1,p)}(G)$ is of course given by the norm $\| x \|$ in (3).

(4) $$T(x)=\int_G x(g)D(g)\mu(dg)^{5)}$$

for all $x(g)$ in $L^{(1,p)}(G)$. We denote this representation of G by $D_T(a)$.

ii) Conversely, if $a\to D(a)$ is an arbitrary bounded continuous representation of G and if we define $T(x)$ by (4), the mapping $x(g)\to T(x)$ gives a continuous representation for $x(g)$ in $L^{(1,p)}(G)$. We denote this representation by $T_D(x)$.

iii) i) and ii) give mutually inverse correspondences: if $D_1=D_{T_1}$, then $T_1=T_{D_1}$ and if $T_2=T_{D_2}$, then $D_2=D_{T_2}$.

iv) Equivalent representations correspond to each other: if $AT_1(x)A^{-1}=T_2(x)$, then $AD_{T_1}(a)A^{-1}=D_{T_2}(a)$ and if $BD_1(a)B^{-1}=D_2(a)$, then $BT_{D_1}(x)B^{-1}=T_{D_2}(x)$.

From Theorem 1 follow immediately some corollaries: Let $a\to D(a)$ be a bounded measurable representation of G. If we put

$$T(x)=\int_G x(g)D(g)\mu(dg)\,,$$

$x(g)\to T(x)$ gives, as before, a continuous representation of $L^{(1,p)}(G)$. By Theorem 1 we have thus $T(x)=T_{D_1}(x)$ with some bounded continuous representation $D_1(a)$ of G and hence $D(a)=D_1(a)$. Thus

Theorem 2. Any bounded measurable representation of G is continuous.

In a similar way we obtain by a simple calculation the following

Theorem 3. If G is locally compact, but not compact, then there is no representation of G belonging to $L^p(G)$ $(p\geqq 1)$ except the zero representation, which maps every element of G to the zero matrix. On the other hand if G is compact, any representation belonging to $L^p(G)$ $(p\geqq 1)$ is bounded and continuous.

Now, as a bounded representation of G is always completely reducible, it follows from Theorem 1, iv) that a continuous representation of $L^{(1,p)}(G)$ is completely reducible. But the converse is also true. It holds namely

Theorem 4. A representation $T(x)$ of $L^{(1,p)}(G)$ is continuous if and only if it is completely reducible. Especially a irreducible representation of $L^{(1,p)}(G)$ is always continuous.

This theorem is equivalent to the following

Theorem 5. Let M be a two-sided ideal of $L^{(1,p)}(G)$, such that the rest class ring $L^{(1,p)}/M$ is of finite dimension. $L^{(1,p)}/M$ is then semi-simple if and only if M is closed in $L^{(1,p)}(G)$. Especially a maximal ideal M is always closed in $L^{(1,p)}(G)$.

These theorems can be regarded as a generalization of the complete reducibility of the group ring of a finite group.

The above mentioned relation between ideals and representations of $L^{(1,p)}(G)$ is explicitly given by

Theorem 6. Let $\{D\}$ be a class of equivalent bounded continuous

5) The right-hand side means a matrix with (i,j)-component $\int_G x(g)d_{ij}(g)\mu(dg)$, where $D(a)=\{d_{ij}(a)\}$.

irreducible representations of G and $D(a)$ be a representant of it. Then all functions $x(g)$ of $L^{(1,p)}(G)$ satisfying

$$\int_G x(g)D(g)\mu(dg)=0,$$

constitute a maximal two-sided ideal $M_{\{D\}}$ in $L^{(1,p)}(G)$. $\{D\}\to M_{\{D\}}$ thus gives a one-to-one correspondence between all classes of equivalent bounded continuous irreducible representations of G and all maximal two-sided ideals M of $L^{(1,p)}(G)$, for which $L^{(1,p)}/M$ is of finite dimension.

If we define the classes of (not necessarily irreducible) representations of G suitably, then the result of Theorem 6 can be extended to those classes of representations of G and all closed two-sided ideals M of $L^{(1,p)}(G)$, for which $L^{(1,p)}/M$ is of finite dimension. It follows then also, that for $L^{(1,p)}(G)$, $p=1,2,\dots$ the ideals of that kind correspond one-to-one to each other.

§ 3. In order to establish corresponding theorems for Segal's group ring $R^{(1,p)}(G)$, we have only to prove the following

Lemma. Let M be a two-sided ideal in $L^{(1,p)}(G)$ such that the rest class ring $L^{(1,p)}/M$ is of finite dimension and has a unit element. Then a two-sided ideal M' of $R^{(1,p)}(G)$ can be uniquely determined, so that it holds

$$M=M'\frown L^{(1,p)}(G), \qquad R^{(1,p)}(G)=M'\smile L^{(1,p)}(G).$$

From this lemma it follows that closed ideals in $R^{(1,p)}(G)$ which are not contained in $L^{(1,p)}(G)$ and closed ideals in $L^{(1,p)}(G)$ correspond one-to-one to each other. Making use of this fact and Theorem 1 we obtain

Theorem 7. For a continuous representation[6)] $\mathfrak{z}\to T(\mathfrak{z})$ of $R^{(1,p)}(G)$ there is a continuous bounded representation $a\to D(a)$ of G, so that for any $\mathfrak{z}=\lambda e+x(g)$ in $R^{(1,p)}(G)$ it holds

$$T(\mathfrak{z})=T\big(\lambda e+x(g)\big)=\lambda D(1)+\int_G x(g)D(g)\mu(dg)^{7)}. \tag{5}$$

Conversely, for any continuous bounded representation $a\to D(a)$ of G, $T(\mathfrak{z})$ in (5) gives a continuous representation $\mathfrak{z}\to T(\mathfrak{z})$ of $R^{(1,p)}(G)$ and thus continuous representations of $R^{(1,p)}(G)$ and continuous bounded representations of G correspond one-to-one to each other.

We can also prove similar theorems to Theorems 4, 5, 6. Especially a one-to-one correspondence is to be established between all classes of irreducible bounded continuous representations of G and all maximal ideals $M'\neq L^{(1,p)}(G)$ of $R^{(1,p)}(G)$, for which $R^{(1,p)}/M'$ is of finite dimensions[8)].

§ 4. We now extend our Theorem 1 to representations of G and $L^{(1,p)}(G)$ by bounded operators in a Hilbert space $\mathfrak{H}$. Let $\boldsymbol{B}$ be the

6) We define the norm in $R^{(1,p)}(G)$ by $\|\mathfrak{z}\|=\|\lambda e+x(g)\|=|\lambda|+\|x\|$ and say "continuous" in the sense of this norm.

7) $D(1)$ is the matrix corresponding to the unit of G.

8) Cf. Segal, l. c. 2).

ring of all bounded operators in $\mathfrak{H}$. A representation $a \to D(a)\big(D(a) \in \boldsymbol{B}\big)$ of G in $\boldsymbol{B}$ is called a "proper" representation, when it holds $D(1)=E$, where 1 means the unit in G and E is the unit operator in $\boldsymbol{B}$. It is called bounden, if there exists a constant C so that

$$|||D(a)||| \leqq C, \qquad \text{for all } a \in G^{9)}.$$

On the other hand we call a representation $x(g) \to T(x)\big(T(x) \in \boldsymbol{B}\big)$ of $L^{(1,p)}(G)$ in $\boldsymbol{B}$ "proper", if $T(x)f=0 (f \in \mathfrak{H})$ for all $x(g) \in L^{(1,p)}(G)$ implies $f=0$, and we call it "continuous", if there is a constant C' so that

$$|||T(x)||| \leqq C' \|x\|, \qquad \text{for all } x(g) \in L^{(1,p)}(G)^{10)}.$$

We can now prove the following

Theorem 8. There is a one-to-one correspondence between continuous proper representations of $L^{(1,p)}(G)$ in $\boldsymbol{B}$ and bounded measurable[11] proper representations of G in $\boldsymbol{B}$ in the following sense:

i) For such a representation $x(g) \to T(x)$ of $L^{(1,p)}(G)$ in $\boldsymbol{B}$, there is a bounded measurable proper representation $a \to D(a)$ of G, so that

$$T(x) = \int_G x(g) D(g) \mu(dg)^{12)} \tag{6}$$

for all $x(g)$ in $L^{(1,p)}(G)$. Such $D(a)$ is uniquely determined by $T(x)$.

ii) Conversely, if $a \to D(a)$ is such a representation of G, $T(x)$ in (6) gives a continuous proper representation of $L^{(1,p)}(G)$ in $\boldsymbol{B}$.

iii) Above correspondences are mutually inverse.

iv) Equivalent representations correspond to each other[13].

If the measure μ is not only left invariant, but also right invariant, then we can obtain some more precise results. We can thus prove for example the following theorem.

Theorem 9. If G has an invariant Haar measure, then measurable unitary representations of G in $\boldsymbol{B}$ are all strongly continuous[14].

Detailed proofs of above theorems will appear elsewhere. They need some considerations on non-commutative normed rings[15] and rings of operators in a Hilbert space, as will be also discussed there precisely[16].

---

9) $|||A|||$ means the bound of the operator A.

10) Thus we consider in $\boldsymbol{B}$ the uniform topology.

11) That is to say, that $(D(g)f, f')$ is μ-measurable for any f, f' in $\mathfrak{H}$.

12) (6) means $(T(x)f, f') = \int_G x(g) (D(g)f, f') \mu(dg)$ for any f, f' in $\mathfrak{H}$.

13) Cf. Theorem 1, iii), iv).

14) Cf. K. Kodaira: Über die Gruppen der messbaren Abbildungen, Proc. **17** (1941), 18-23.

15) Some of the theorems, obtained by I. Gelfand, concerning commutative normed rings can be transferred to our non-commutative case.

16) Cf. also author's note in Zenkoku Sijo Sugaku Danwakai, **246** (1942), 1522-1555, **251** (1943), 167-186.

[12] Über nilpotente topologische Gruppen I

Proc. Japan Acad., 21-3 (1945), 124-137.

Mathematisches Institut, Kaiserliche Universität zu Tokyo.
(Comm. by T. TAKAGI, M.I.A., March 12, 1945.)

In der abstrakten Gruppentheorie spielen bekanntlich die höheren Kommutatorgruppen, die absteigenden bzw. die aufsteigenden Zentralreihen wichtige Rollen. Sie charakterisieren besondere Klassen von Gruppen, nämlich auflösbare bzw. nilpotente Gruppen, deren Struktur, vor allem bei endlichen Gruppen, eingehend untersucht worden ist[1]. Dementsprechend sollen im folgenden die Kommutatorgruppen und die Zentralgruppen auch für topologische Gruppen definiert werden und dann die Struktur der auflösbaren bzw. der nilpotenten topologischen, insbesondere kompakten Gruppen untersucht werden.

Es sei $\mathfrak{G}$ eine separable topologische Gruppe und $\mathfrak{H}$, $\mathfrak{A}$,…… (nicht notwendig abgeschlossene) Untergruppen von $\mathfrak{G}$. Wir bezeichnen mit $\overline{\mathfrak{H}}$, $\overline{\mathfrak{A}}$,……bzw. $[\mathfrak{H}, \mathfrak{A}]$,……die abgeschlossenen Hüllen bzw. die abstrakten Kommutatorgruppen (im Sinne der abstrakten Gruppentheorie)[2] solcher Untergruppen. Es gilt dann wie leicht ersichtlich

$$(1) \qquad \overline{[\mathfrak{H}, \mathfrak{A}]} = \overline{[\overline{\mathfrak{H}}, \overline{\mathfrak{A}}]}.$$

Wir nennen $\overline{[\mathfrak{H}, \mathfrak{A}]}$ die *topologische Kommutatorgruppe* von $\mathfrak{H}$ und $\mathfrak{A}$.

Nun ersetzten wir die gewöhnlichen Kommutatorgruppen durch die topologischen und definieren die topologischen höheren Kommutatorgruppen $\mathfrak{D}_\xi^*(\mathfrak{G})$ bzw. die topologischen absteigenden Zentralgruppen $\mathfrak{Z}_\xi^*(\mathfrak{G})$ folgendermassen. Es sei

$$\mathfrak{D}_0^*(\mathfrak{G}) = \mathfrak{G}, \qquad \mathfrak{Z}_0^*(\mathfrak{G}) = \mathfrak{G}$$

und für jede Ordnungszahl η mit $\eta < \xi$ sei $\mathfrak{D}_\eta^*(\mathfrak{G})$ bzw. $\mathfrak{Z}_\eta^*(\mathfrak{G})$ schon definiert. Ist $\xi = \eta + 1$, so setze man $\mathfrak{D}_\xi^*(\mathfrak{G})$ bzw. $\mathfrak{Z}_\xi^*(\mathfrak{G})$ gleich der topologischen Kommutatorgruppe von $\mathfrak{D}_\eta^*(\mathfrak{G})$ und $\mathfrak{D}_\eta^*(\mathfrak{G})$ bzw. von $\mathfrak{G}$ und $\mathfrak{Z}_\eta^*(\mathfrak{G})$, und wenn ξ eine Limeszahl ist, so sei $\mathfrak{D}_\xi^*(\mathfrak{G})$ bzw. $\mathfrak{Z}_\xi^*(\mathfrak{G})$ der Durchschnitt aller $\mathfrak{D}_\eta^*(\mathfrak{G})$ bzw. $\mathfrak{Z}_\eta^*(\mathfrak{G})$ mit $\eta < \xi$.

Nach (1) kann man durch Induktion leicht beweisen, dass $\mathfrak{D}_n^*(\mathfrak{G})$ bzw. $\mathfrak{Z}_n^*(\mathfrak{G})$ für eine natürliche Zahl n der Abschliessung der gewöhnlichen höheren

---

1) Vgl. z. B. P. Hall, A contribution to the theory of groups of primepower orders, Proc. Lond. Math. Soc. II **36** (1933); R. Raer, Nilpotent groups and their generalizations, Trans. Amer. Math. Soc. Vol. **47** (1940).

2) $[\mathfrak{H}, \mathfrak{A}]$ ist die von $(h, u) = huh^{-1}u^{-1}$, $h \varepsilon \mathfrak{H}$, $u \varepsilon \mathfrak{A}$ erzeugte Gruppe.

Kommutatorgruppe $\mathfrak{D}_n(\mathfrak{G})$ bzw. der gewöhnlichen absteigenden Zentralgruppe $\mathfrak{Z}_n(\mathfrak{G})$ gleich ist:

$$(2)\qquad \mathfrak{D}^*_n(\mathfrak{G})=\overline{\mathfrak{D}_n(\mathfrak{G})}\,,\qquad \mathfrak{Z}^*_n(\mathfrak{G})=\overline{\mathfrak{Z}_n(\mathfrak{G})}\,,\qquad n=1,2,\ldots\ldots$$

Nun definieren wir die topologischen aufsteigenden Zentralgruppen $\mathfrak{z}^*_\xi(\mathfrak{G})$ von $\mathfrak{G}$. Es sei $\mathfrak{z}^*_0(\mathfrak{G})$ die Einheitsgruppe und $\mathfrak{z}^*_\eta(\mathfrak{G})$ mit $\eta<\xi$ seien schon definiert. Wenn $\xi=\eta+1$ ist, so sei $\mathfrak{z}^*_\xi(\mathfrak{G})/\mathfrak{z}^*_\eta(\mathfrak{G})$ das Zentrum von $\mathfrak{G}/\mathfrak{z}^*_\eta(\mathfrak{G})$ und wenn ξ eine Limeszahl ist, so sei $\mathfrak{z}^*_\xi(\mathfrak{G})$ die Abschliessung der Vereinigung aller $\mathfrak{z}^*_\eta(\mathfrak{D})$ mit $\eta<\xi$.

Zunächst beweisen wir einige Hilfssätze für diese $\mathfrak{G}^*_\xi(\mathfrak{G})$, $\mathfrak{Z}^*_\xi(\mathfrak{G})$, $\mathfrak{z}^*_\xi(\mathfrak{G})$.

Hilfssatz 1. Für eine abgeschlossene Untergruppe $\mathfrak{H}$ von $\mathfrak{G}$ gilt

$$(3)\qquad \mathfrak{D}^*_\xi(\mathfrak{G})\geqq\mathfrak{D}^*_\xi(\mathfrak{H})\,,\qquad \mathfrak{Z}^*_\xi(\mathfrak{G})\geqq\mathfrak{Z}^*_\xi(\mathfrak{H})\,.$$

Wenn $\mathfrak{N}$ ein abgeschlossener Normalteiler von $\mathfrak{G}$ ist, so ist

$$(4)\qquad \mathfrak{D}^*_\xi(\mathfrak{G}/\mathfrak{N})\geqq\mathfrak{D}^*_\xi(\mathfrak{G})\mathfrak{N}/\mathfrak{N}\,,\quad \mathfrak{Z}^*_\xi(\mathfrak{G}/\mathfrak{N})\geqq\mathfrak{Z}^*_\xi(\mathfrak{G})\mathfrak{N}/\mathfrak{N}\,,$$

$$(4')\qquad \mathfrak{D}^*_n(\mathfrak{G}/\mathfrak{N})=\mathfrak{D}^*_n(\mathfrak{G})\mathfrak{N}/\mathfrak{N}\,,\quad \mathfrak{Z}^*_n(\mathfrak{G}/\mathfrak{N})=\mathfrak{Z}^*_n(\mathfrak{G})\mathfrak{N}/\mathfrak{N}\,,\ \text{für}\ n=1,2,\ldots$$

$$(5)\qquad \mathfrak{z}^*_\xi(\mathfrak{G}/\mathfrak{N})\geqq\mathfrak{z}^*_\xi(\mathfrak{G})\mathfrak{N}/\mathfrak{N}\,.$$

Beweis. (3) ist nach Definition ohne weiters klar. (4), (4′), (5) lassen sich durch Induktion nach ξ (n) leicht beweisen. Um z. B. (4) zu beweisen setzen wir voraus dass $\mathfrak{D}^*_\eta(\mathfrak{G}/\mathfrak{N})\geqq\mathfrak{D}^*_\eta(\mathfrak{G})\mathfrak{N}/\mathfrak{N}$ schon für $\eta<\xi$ als gültig bewiesen ist. Wenn $\xi=\eta+1$ ist, so ist

$$\mathfrak{D}^*_\xi(\mathfrak{G}/\mathfrak{N})=\overline{[\mathfrak{D}^*_\eta(\mathfrak{G}/\mathfrak{N}),\ \mathfrak{D}^*_\eta(\mathfrak{G}/\mathfrak{N})]}\geqq\overline{[\mathfrak{D}^*_\eta(\mathfrak{G})\mathfrak{N}/\mathfrak{N}\,,\ \mathfrak{D}^*_\eta(\mathfrak{G})\mathfrak{N}/\mathfrak{N}]}$$
$$\geqq\overline{[\mathfrak{D}^*_\eta(\mathfrak{G}),\ \mathfrak{D}^*_\eta(\mathfrak{G})]}\mathfrak{N}/\mathfrak{N}=\mathfrak{D}^*_\xi(\mathfrak{G})\mathfrak{N}/\mathfrak{N}\,.$$

Wenn ξ eine Limeszahl ist, so gilt

$$\mathfrak{D}^*_\xi(\mathfrak{G}/\mathfrak{N})=\bigwedge_{\eta<\xi}\mathfrak{D}^*_\eta(\mathfrak{G}/\mathfrak{N})\geqq\bigwedge_{\eta<\xi}(\mathfrak{D}^*_\eta(\mathfrak{G})\mathfrak{N}/\mathfrak{N})\geqq(\bigwedge_{\eta<\xi}\mathfrak{D}^*_\eta(\mathfrak{G}))\mathfrak{N}/\mathfrak{N}$$
$$=\mathfrak{D}^*_\xi(\mathfrak{G})\mathfrak{N}/\mathfrak{N}\,.$$

Hilfssatz 2. Wenn $\mathfrak{G}$ eine kompakte Gruppe ist, so ist

$$\mathfrak{D}^*_\omega(\mathfrak{G})=\mathfrak{D}^*_{\omega+1}(\mathfrak{G})=\ldots\ldots,\quad \mathfrak{Z}^*_\omega(\mathfrak{G})=\mathfrak{Z}^*_{\omega+1}(\mathfrak{G})=\ldots\ldots$$

Für einen abgeschlossenen Normalteiler $\mathfrak{N}$ gilt dann ausserdem

$$\mathfrak{D}^*_\omega(\mathfrak{G}/\mathfrak{N})=\mathfrak{D}^*_\omega(\mathfrak{G})\mathfrak{N}/\mathfrak{N}\,,\quad \mathfrak{Z}^*_\omega(\mathfrak{G}/\mathfrak{N})=\mathfrak{Z}^*_\omega(\mathfrak{G})\mathfrak{N}/\mathfrak{N}\,.$$

Beweis. Wir beweisen $\mathfrak{D}^*_\omega(\mathfrak{G})=\mathfrak{D}^*_{\omega+1}(\mathfrak{G})=\ldots\ldots$ $\mathfrak{Z}^*_\omega(\mathfrak{G})=\mathfrak{Z}^*_{\omega+1}(\mathfrak{G})=\cdots\cdot$ lässt sich in analoger Weise beweisen. Zum Beweis der ersteren schicken wir einen Hilfssatz voraus, welcher auch an sich von Interesse sein mag.

Hilfssatz 3. In einer kompakten Lieschen Gruppe $\mathfrak{L}$ gilt der minimale Kettensatz für abgeschlossene Untergruppen. Es folgt nämlich aus

$$\mathfrak{L}\geqq\mathfrak{H}_1\geqq\mathfrak{H}_2\geqq\ldots\ldots,$$

dass es von einem gewissen n an

$$\mathfrak{H}_n=\mathfrak{H}_{n+1}=\mathfrak{H}_{n+2}=\ldots\ldots$$

gilt.

Bemerkung. Der maximale Kettensatz ist dagegen im allgemein nicht gültig. Das sieht man sofort an dem Beispiel der mod. 1 reduzierten additiven Gruppe $\mathfrak{K}$ aller reellen Zahlen[1].

Beweis. Für Dimensionen der Lieschen Gruppen $\mathfrak{H}_i$ gilt offenbar

$$\dim.\ \mathfrak{H}_1 \geqq \dim.\ \mathfrak{H}_2 \geqq \cdots\cdots$$

Von einem gewissen m an ist also

$$\dim.\ \mathfrak{H}_m = \dim.\ \mathfrak{H}_{m+1} = \cdots\cdots$$

und die 1-Komponente[2] von $\mathfrak{H}_m, \mathfrak{H}_{m+1}, \cdots\cdots$ fallen zusammen. Wenn man diese Gruppe mit $\mathfrak{N}_0$ bezeichnet, so sind $\mathfrak{H}_m/\mathfrak{N}_0, \mathfrak{H}_{m+1}/\mathfrak{N}_0, \cdots\cdots$ sämtlich kompakt und diskret, also endliche Gruppen und es gilt

$$\mathfrak{H}_m/\mathfrak{N}_0 \geqq \mathfrak{H}_{m+1}/\mathfrak{N}_0 \geqq \cdots\cdots$$

Daher gibt es ein n $(\geqq m)$ mit

$$\mathfrak{H}_n/\mathfrak{N}_0 = \mathfrak{H}_{n+1}/\mathfrak{N}_0 = \cdots\cdots, \text{ also } \mathfrak{H}_n = \mathfrak{H}_{n+1} = \cdots\cdots,$$

w. z. b. w.

Nun zeigen wir $\mathfrak{D}^*_\omega(\mathfrak{G}) = \mathfrak{D}^*_{\omega+1}(\mathfrak{G}) = \cdots\cdots$ Es genügt offenbar nur $\mathfrak{D}^*_\omega(\mathfrak{G}) = \mathfrak{D}^*_{\omega+1}(\mathfrak{G})$ nachzuweisen. Wir setzen voraus $\mathfrak{D}^*_\omega(\mathfrak{G}) \neq \mathfrak{D}^*_{\omega+1}(\mathfrak{G})$ und wählen ein Element g, mit $g \in \mathfrak{D}^*_\omega(\mathfrak{G})$, $g \notin \mathfrak{D}^*_{\omega+1}(\mathfrak{G})$. Da $\mathfrak{G}/\mathfrak{D}^*_{\omega+1}(\mathfrak{G})$ eine kompakte Gruppe ist, gibt es eine stetige Darstellung D von $\mathfrak{G}/\mathfrak{D}^*_{\omega+1}(\mathfrak{G})$, welche g nicht auf die Einheitsmatrix E abbildet. Es ist also

$$(6) \qquad D(\mathfrak{D}^*_\omega(\mathfrak{G})) \neq \{E\}, \qquad D(\mathfrak{D}^*_{\omega+1}(\mathfrak{G})) = \{E\}.$$

Aus der Kompaktheit von $\mathfrak{G}$ und aus $\mathfrak{D}^*_\omega(\mathfrak{G}) = \bigwedge_n \mathfrak{D}^*_n(\mathfrak{G})$ sieht man

$$D(\mathfrak{D}^*_\omega(\mathfrak{G})) = \bigwedge_n D(\mathfrak{D}^*_n(\mathfrak{G})).$$

Da es aber

$$D(\mathfrak{D}^*_0(\mathfrak{G})) \geqq D(\mathfrak{D}^*_1(\mathfrak{G})) \geqq D(\mathfrak{D}^*_2(\mathfrak{G})) \geqq \cdots\cdot$$

ist und $D(\mathfrak{D}^*_0(\mathfrak{G})) = D(\mathfrak{G})$ eine kompakte Liesche Gruppe ist, so folgt nach Hilfssatz 3

$$(7) \qquad D(\mathfrak{D}^*_n(\mathfrak{G})) = D(\mathfrak{D}^*_{n+1}(\mathfrak{G})) = \cdots\cdots \quad = D(\mathfrak{D}^*_\omega(\mathfrak{G})).$$

Andererseits ist aber $\mathfrak{D}^*_\omega(\mathfrak{G})/D^*_{\omega+1}(\mathfrak{G})$ eine abelsche Gruppe. Es ist also nach Hilfssatz 1, (4′) und (7)

$$\begin{aligned} D(\mathfrak{D}^*_\omega(\mathfrak{G})) &= D(\mathfrak{D}^*_{n+1}(\mathfrak{G})) = \mathfrak{D}^*_{n+1}(D(\mathfrak{G})) = \overline{[\mathfrak{D}^*_n(D(\mathfrak{G})),\ \mathfrak{D}^*_n(D(\mathfrak{G}))]} \\ &= \overline{[D(\mathfrak{D}^*_n(\mathfrak{G})),\ D(\mathfrak{D}^*_n(\mathfrak{G}))]} = \overline{[D(\mathfrak{D}^*_\omega(\mathfrak{G})),\ D(\mathfrak{D}^*_\omega(\mathfrak{G}))]} \\ &\leqq D(\mathfrak{D}^*_{\omega+1}(\mathfrak{G})) = \{E\}, \end{aligned}$$

was mit (6) in Widerspruch steht. Es muss also $\mathfrak{D}^*_\omega(\mathfrak{G}) = \mathfrak{D}^*_{\omega+1}(\mathfrak{G})$ sein.

---

1) Im folgenden soll $\mathfrak{K}$ immer solche Gruppe bedeuten.

2) Die zusammenhändgende Komponente, welche das Einselement enthält. Sie bildet bekanntlich einen Normalteiler.

Man beweist nun $\mathfrak{D}^*_\omega(\mathfrak{G}/\mathfrak{N})=\mathfrak{D}^*_\omega(\mathfrak{G})\mathfrak{N}/\mathfrak{N}$ folgendermassen:

$$\begin{aligned}\mathfrak{D}^*_\omega(\mathfrak{G}/\mathfrak{N}) &= \bigwedge_n \mathfrak{D}^*_n(\mathfrak{G}/\mathfrak{N}) = \bigwedge_n \mathfrak{D}^*_n(\mathfrak{G})\mathfrak{N}/\mathfrak{N} \quad \text{(nach (4'))} \\ &= (\bigwedge_n \mathfrak{D}^*_n(\mathfrak{G}))\mathfrak{N}/\mathfrak{N} \quad \text{(nach der Kompaktheit von } \mathfrak{G}\text{)} \\ &= \mathfrak{D}^*_\omega(\mathfrak{G})\mathfrak{N}/\mathfrak{N}\,.\end{aligned}$$

Genan so zeigt man $\mathfrak{Z}^*_\omega(\mathfrak{G}/\mathfrak{N})=\mathfrak{Z}^*_\omega(\mathfrak{G})\mathfrak{N}/\mathfrak{N}$ und damit ist der Hilfssatz 2 in allen Teilen bewiesen.

Nun definieren wir auflösbare bzw. nilpotente topologische Gruppen. $\mathfrak{G}$ heisst *topologisch auflösbar* wenn für irgendeine Ordnungszahl ξ $\mathfrak{D}^*_\xi(\mathfrak{G})=e$[1] ist, und *topologisch nach unten bzw. nach oben nilpotent* wenn für irgendeine Ordnungszahl ξ $\mathfrak{Z}^*_\xi(\mathfrak{G})=e$ bzw. $\mathfrak{z}^*_\xi(\mathfrak{G})=\mathfrak{G}$ gilt. Gilt zwar $\mathfrak{D}^*_\xi(\mathfrak{G})=e$ ($\mathfrak{Z}^*_\xi(\mathfrak{G})=e$ oder $\mathfrak{z}^*_\xi(\mathfrak{G})=\mathfrak{G}$), $\mathfrak{D}^*_\eta\neq e$ ($\mathfrak{Z}^*_\eta(\mathfrak{G})\neq e$ oder $\mathfrak{z}^*_\eta(\mathfrak{G})\neq\mathfrak{G}$) für $\eta<\xi$, so soll ξ die Klasse der auflösbaren (nach unten oder nach oben nilpotente) Gruppe $\mathfrak{G}$ heissen. Falls es nur um endliche Gruppen handelt, so fallen bekanntlich die beiden Begriffen der nach unten und der nach oben nilpotenten Gruppen zusammen und nilpotente Gruppe besitzt nach unten dieselbe Klasse wie nach oben. Das gilt aber in unserem Falle nicht. Wir geben in der Tat nachher Beispiele von Gruppen, die nach unten aber nicht nach oben nilpotent sind und umgekehrt.

Da wir im folgenden immer mit topologischen Gruppen zu tun haben, wollen wir zukünftig das Wort "topologisch" in den Ausdrücken wie "topologische auflösbare oder nilpotente Gruppen" und den Stern in den Bezeichnungen wie $\mathfrak{D}^*_\xi$ bzw. $\mathfrak{Z}^*_\xi$, $\mathfrak{z}^*_\xi$ fallen lassen, wenn es kein Furcht von Misverständnis steht.

Aus Hilfssätzen 1, 2 schliesst man sofort folgende Sätze.

Satz 1. Eine abgeschlossene Untergruppe einer auflösbaren bzw. nach unten nilpotenten topologischen Gruppe ist wieder auflösbar bzw. nach unten nilpotent. Eine Faktorgruppen nach einem abgeschlossenen Normalteiler einer nach oben nilpotenten topologischen Gruppen ist wieder nach oben nilpotent.

Satz 2. Eine kompakte Gruppe $\mathfrak{G}$ ist genau dann auflösbar bzw. nach unten nilpotent, wenn $\mathfrak{D}_\omega(\mathfrak{G})=e$ bzw. $\mathfrak{Z}_\omega(\mathfrak{G})=e$ gilt. Faktorgruppen nach abgeschlossenen Normalteiler einer kompakten auflösbaren bzw. nach unten nilpotenten Gruppe sind wieder auflösbar bzw. nach unten nilpotent.

Aus der Definition folgt ferner leicht der

Satz 3. Das direkte Prokukt endlichvieler oder abzählbarer kompakten auflösbaren Gruppen ist wieder auflösbar. Dasselbe gilt auch für nach unter oder nach oben nilpotente Gruppen.

Wir untersuchen nun die Struktur der kompakten nach unten nilpotenten Gruppen. Dazu beweisen wir zunächst den

---

1) e bedeutet das Einselement und zugleich auch die Einheitsgruppe.

Satz 4. Die 1-Komponente $\mathfrak{G}_0$ einer kompakten auflösbaren Gruppe $\mathfrak{G}$ ist abelsch.

Beweis. Nach Satz 2 ist $\mathfrak{G}_0$ auch auflösbar. Wir können also $\mathfrak{G}=\mathfrak{G}_0$ annehmen, und haben nur zu beweisen, dass eine zusammenhängende kompakte auflösbare Gruppe abelsch ist. Da es dann aber für ein beliebiges Element $g \neq e$ eine stetige Darstellung D von $\mathfrak{G}$ mit $D(g) \neq E$ gibt, so genügt es zu zeigen, dass jede stetige Darstellung von $\mathfrak{G}$ abelsch ist. Wir können also wieder von vornherein annehmen, dass $\mathfrak{G}=D(\mathfrak{G})$ ist, d.h. dass $\mathfrak{G}$ eine kompakte Liesche Gruppe ist. Nun gilt nach der Voraussetzung

$$\mathfrak{G}=\mathfrak{D}_0(\mathfrak{G}) \geqq \mathfrak{D}_1(\mathfrak{G}) \geqq \mathfrak{D}_2(\mathfrak{G}) \geqq \ldots\ldots \geqq \mathfrak{D}_\omega(\mathfrak{G})=e,$$

$$\bigwedge_n \mathfrak{D}_n(\mathfrak{G})=\mathfrak{D}_\omega(\mathfrak{G})=e.$$

Nach Hilfssatz 3 gilt also von einem n an

$$\mathfrak{D}_n(\mathfrak{G})=\mathfrak{D}_{n+1}(\mathfrak{G})=\ldots = \mathfrak{D}_\omega(\mathfrak{G})=e.$$

$\mathfrak{G}$ ist mithin eine zusammenhängende, kompakte, und im Sinne der Lieschen Theorie auflösbare Gruppe, ist also bekanntlich abelsch.

Für eine nach unten nilpotenten Gruppe können wir noch mehr sagen. Es gilt nämlich

Satz 5. Die 1-Komponent $\mathfrak{G}_0$ einer kompakten nach unten nilpotenten Gruppe $\mathfrak{G}$ ist im Zentrum von $\mathfrak{G}$ enthalten.

Beweis. Genau wie im vorigen Beweis dürfen wir $\mathfrak{G}$ als Liesche Gruppe annehmen. Da $\mathfrak{G}$ offenbar auflösbar ist, so ist $\mathfrak{G}_0$ nach Satz 4 eine zusammenhängende kompakte abelsche Liesche Gruppe, also direktes Produkt endlichvieler Gruppen $\mathfrak{K}_i$, welche der mod. 1 reduzierten additiven Gruppe $\mathfrak{K}$ aller reellen Zählen isomorph sind[1].

$$\mathfrak{G}_0=\mathfrak{K}_1 \times \mathfrak{K}_2 \times \mathfrak{K}_3 \times \ldots \times \mathfrak{K}_m, \quad \mathfrak{K}_i \cong \mathfrak{K} \quad (i=1, 2, \ldots, \mathrm{m}).$$

Die Elemente von $\mathfrak{G}_0$ können also durch "Koordinaten" dargestellt werden:

$$h \leftrightarrow (h_1, h_2, \ldots, h_m) \quad (h_i, \text{ mod. } 1). \tag{8}$$

Die Transformation xhx^{-1} durch ein Element x aus $\mathfrak{G}$ induziert dann in $\mathfrak{G}_0$ einen Automorphismus, welcher durch eine ganzzahlige Matrix $A(x)$ dargestellt wird:

$$xhx^{-1} \leftrightarrow (\mathrm{h}_1\, h_2, \ldots\ldots, h_m)A(x). \tag{9}$$

Aus (8), (9) ergibt sich

$$(x, h)=xhx^{-1}h^{-1} \leftrightarrow (h_1, h_2, \ldots\ldots, h_m)\,(A(x)-E) \tag{10}$$

und daraus durch n-malige Kommutatorbildung

$$(x, (x, \ldots (x, h)) \ldots) \leftrightarrow (h_1, h_2, \ldots, h_m)\,(A(x)-E)^n.$$

Man sieht aber genau so wie im Beweis des Satzes 4, dass für genügend grosses n

1) Vgl. L. Pontrjagin, Topological groups (1939), S. 170.

$\mathfrak{Z}_n(\mathfrak{G})=e$ stattfindet. Für beliebige h, x ist dann der n-te Kommutator $(x, (x,\ldots(x, h)\ldots)$ immer gleich dem Einselement und daraus folgt

$$(A(x)-E)^n=0\,.$$

Die Eigenwerten von $A(x)$ sind mithin sämtlich gleich 1. Da aber $x\to A(x)$ offenbar eine Darstellung der endlichen Gruppe $\mathfrak{G}/\mathfrak{G}_0$ gibt, so schliesst man sofort

$$A(x)=E, \quad \text{also} \quad xhx^{-1}=h\,,$$

w.z.b.w.

Wir untersuchen nun die Gruppe $\mathfrak{G}/\mathfrak{G}_0$, d.h. 0-dimensionale kompakte nach unten nilpotente Gruppe.

Eine 0-dimensionale kompakte Gruppe heisst eine p-Gruppe, wenn es der Limes einer $\mathfrak{G}_\nu$-adischen Reihe von endlichen p-Gruppen ist. Eine Untergruppe $\mathfrak{P}$ einer 0-dimensionalen kompakten Gruppe $\mathfrak{G}$ heisst eine p-Sylowgruppe von $\mathfrak{G}$, wenn $\mathfrak{P}$ eine maximale p-Gruppe in $\mathfrak{G}$ ist. Unter diesen Begriffsbildungen hat v. Dantzig die Sylowsätze in der Theorie der endlichen Gruppen auf die topologischen Gruppen übertragen.[1] Er hat z. B. bewiesen, dass alle p-Sylowgruppen von $\mathfrak{G}$ einarder konjugiert sind. Für die nach unten nilpotenten Gruppen beweisen wir nun folgende Sätze über Sylowgruppen.

Hilfssatz 4. Wenn $\mathfrak{G}$ nach unten nilpotent ist, so ist $\mathfrak{G}$ das direkte Produkt seiner Sylowgruppen.

Beweis. Es sei $\mathfrak{P}$ eine p-Sylowgruppe von $\mathfrak{G}$ und

$$(11) \qquad \mathfrak{N}_1>\mathfrak{N}_2>\mathfrak{N}_3>\cdots, \qquad \bigwedge_n \mathfrak{N}_n=e$$

eine Reihe offener (und zugleich abgeschlossener) Normalteiler von $\mathfrak{G}$, welche ein Umgebungssystem in $\mathfrak{G}$ bestimmt. Es lässt sich dann leicht zeigen, dass $\mathfrak{P}\mathfrak{N}_i/\mathfrak{N}_i$ eine p-Sylowgruppe der endlichen Gruppe $\mathfrak{G}/\mathfrak{N}_i$ ist und dass man umgekehrt jede p-Sylowgruppe von $\mathfrak{G}$ in solcher Weise als $\mathfrak{G}_\nu$-adischen Limes der p-Sylowgruppen von $\mathfrak{G}/\mathfrak{N}_i$ erhält. Setzt man nun $\mathfrak{G}$ als nach unten nilpotent voraus, so sind $\mathfrak{G}/\mathfrak{N}_i$ sämtlich endliche nilpotente Gruppen, also direkte Produkte ihrer Sylowgruppen. Daraus und aus dem oben Gesagten folgt sofort, dass $\mathfrak{G}$ selbst direktes Produkt der Sylowgruppen ist.

Hilfssatz 5. Jede p-Gruppe ist nach unten nilpotent.

Beweis. Es sei $\mathfrak{P}$ eine p-Gruppe und $\mathfrak{N}_i$ seien wie oben in (11) ein Umgebungssystem bildende, offene Normalteiler von $\mathfrak{P}$. Nach Definition sind $\mathfrak{P}/\mathfrak{N}_i$ immer endliche p-Gruppen. Daraus folgt.

$$\mathfrak{Z}_\omega(\mathfrak{P})\leqq\mathfrak{N}_i \qquad (i=1, 2,\ldots)\,, \text{ also } \mathfrak{Z}_\omega(\mathfrak{P})=e\,.$$

Satz 2 und Hilfssätze 4, 5 können wir in folgendem Satz zusammenfassen.

---

1) D. van Dantzig, Zur topologischen Algebra III. Comp. Math. Vol. 3 (1936).

Satz 6. Dafür, dass eine 0-dimensionale kompakte Gruppe nach unten nilpotent sei, ist notwendig und hinreichend, dass sie direktes Produkt ihrer Sylowgruppen ist.

Nach diesem Satz reduziert sich die Untersuchung der Strukten der 0-dimensionalen kompakten nach unten nilpotenten Gruppe in einem gewissen Sinne auf die der Struktur der endlichen p-Gruppen.

Bei dieser Gelegenheit sei noch darauf aufmerksam gemacht, dass sich verschiedene Definitionen und Sätze in der Theorie der endlichen Gruppen auf dem Falle der 0-dimensionalen kompakten Gruppen übertragen lassen. So sind z. B. die Definition der Φ-Untergruppe und daran anschliessendes Kriterium für das Nilpotentsein einer Gruppe[1] auf unserem Fall anwendbar.

Nun gehen wir zu nach oben nilpotenten Gruppen über.

Satz 7. Eine 0-dimensionale kompakte nach oben nilpotente Gruppe ist auch nach unten nilpotent.

Beweis. Es sei $\mathfrak{G}$ eine solche Gruppe und eine Reihe offener Normalteiler $\mathfrak{N}_i$ sei wie oben im Beweis von Hilfssatz 4 gewält. Nach Satz 1 ist jede Faktorgruppe $\mathfrak{G}/\mathfrak{N}_i$ nach oben nilpoient. Da es aber eine endliche Gruppe ist, so ist es auch nach unten nilpotent. Es ist also $\mathfrak{Z}_\omega(\mathfrak{G})\leqq\mathfrak{N}_i$ und daraus folgt $\mathfrak{Z}_\omega(\mathfrak{G})=e$.

Es gibt aber 0-dimensionale kompakte Gruppen, welche nach unten nilpotent, aber nicht nach oben nilpotent sind. Wir zeigen tatsächlich nachher, dass es eine p-Gruppe gibt, welche kein Zentrum (ausser dem Einselement) besitzt. Um die Untersuchung der Struktur der kompakten nach oben nilpotenten Gruppen weiter zu führen, studieren wir zunächst Liesche Gruppen.

Hilfssatz 6. Eine zusammenhängende kompakte nach oben nilpotente Liesche Gruppe ist abelsch.

Beweis. $\mathfrak{G}$ sei eine solche Gruppe. Wir zeigen zunächst, dass dim. $\mathfrak{z}_1(\mathfrak{G})\geqq 1$ gilt, falls $\mathfrak{G}$ nicht endlich ist. Man setze voraus dim. $\mathfrak{z}_1(\mathfrak{G})=0$. $\mathfrak{z}_1(\mathfrak{G})$ ist dann diskret und kompakt, also eine endliche Gruppe. Es ist dann sicher $\mathfrak{z}_1(\mathfrak{G})\neq\mathfrak{z}_2(\mathfrak{G})$, denn sonst wäre $\mathfrak{z}_1(\mathfrak{G})=\mathfrak{z}_2(\mathfrak{G})=\dots\ =\mathfrak{G}$ endlich entgegen der Voraussetzung. Es gibt also ein Element g_0 mit $g_0\notin\mathfrak{z}_1(\mathfrak{G})$, $g_0\in\mathfrak{z}_2(\mathfrak{G})$, und $g\to(g_0, g)$[2] für beliebiges g aus $\mathfrak{G}$ liefert dann eine stetige Abbildung von $\mathfrak{G}$ in $\mathfrak{z}_1(\mathfrak{G})$. Das Bild von $\mathfrak{G}$ ist dabei einerseits zusammenhängend, andererseits aber als Teilmenge von $\mathfrak{z}_1(\mathfrak{G})$ diskret und enthält sicher das Einselement. Es muss also mit diesem zusammenfallen und $(g_0, g)=e$ sein für alle g in $\mathfrak{G}$. Das bedeutet aber dass g_0 $\mathfrak{z}_1(\mathfrak{G})$ angehört entgegen der Wahl von g_0.

---

1) Vgl. z. B. H. Zassenhaus, Lehrbuch der Gruppentheorie (1937), S. 44, 107.

2) Klammer bedeutet Kommutatorbildung.

Wenn wir statt $\mathfrak{G}$ $\mathfrak{G}/\mathfrak{Z}_1(\mathfrak{G})$ betrachten, so folgt nach Obigem, dass entweder $\mathfrak{G}/\mathfrak{Z}_1(\mathfrak{G})$ endlich ist oder dim $\mathfrak{Z}_2(\mathfrak{G})/\mathfrak{Z}_1(\mathfrak{G})\geqq 1$ gilt. So fortfahrend für $\mathfrak{G}/\mathfrak{Z}_1(\mathfrak{G})$, $\mathfrak{G}/\mathfrak{Z}_2(\mathfrak{G}),\dots$ erkennt man schliesslich, dass es eine natürliche Zahl n mit $\mathfrak{Z}_n(\mathfrak{G})=\mathfrak{G}$ gibt. Eine nach oben nilpotente Gruppe mit einer endlichen Klasse ist aber auch nach unten nilpotent und umgekehrt.[1] $\mathfrak{G}$ ist also auch nach unten nilpotent, folglich nach Satz 5 abelsch.

Hilfssatz 7 Die 1-Komponente $\mathfrak{G}_0$ einer kompakten nach oben nilpotenten Lieschen Gruppe $\mathfrak{G}$ ist abelsch.

Beweis. Nach Hilfssatz 6 genügt es zu zeigen, dess auch $\mathfrak{G}_0$ nach oben nilpotent ist. Wir konstruieren dazu eine Reihe von Normalteiler $\mathfrak{H}_\xi$ in $\mathfrak{G}_0$ folgendermassen. Es sei $\mathfrak{H}_0=e$ und $\mathfrak{H}_\eta(\eta<\xi)$ sei schon definiert. Wenn $\xi=\eta+1$ ist, so soll $\mathfrak{H}_\xi/\mathfrak{H}_\eta$ aus denjenigen Elementen von $\mathfrak{G}_0/\mathfrak{H}_\eta$ bestehen, welche mit $\mathfrak{G}/\mathfrak{H}_\eta$ elementweise vertauschbar sind. Ist dagegen ξ eine Limeszahl, so setzen wir $\mathfrak{H}_\xi=\overline{\bigvee_{\eta<\xi}\mathfrak{H}_\eta}$. Es gilt dann wie leicht ersichtlich

$$\mathfrak{Z}_\xi(\mathfrak{G}_0)\geqq\mathfrak{H}_\xi\,.$$

Es genügt daher zu beweisen, dass für irgendein ξ $\mathfrak{H}_\xi=\mathfrak{G}_0$ wird. Wir setzen also voraus dass dies nicht der Fall ist und leiten daraus einen Widerspruch her. Nach der Annahme gibt es eine Ordnungszahl η mit

$$\mathfrak{H}_\eta=\mathfrak{H}_{\eta+1}=\cdots,\qquad \mathfrak{H}_\eta\neq\mathfrak{G}_0\,. \tag{12}$$

Setzt man $\mathfrak{G}/\mathfrak{H}_\eta=\mathfrak{G}^*$, $\mathfrak{G}_0/\mathfrak{H}_\eta=\mathfrak{G}_0^*$, so folgt aus (12)

$$\mathfrak{Z}_1(\mathfrak{G}^*)\frown\mathfrak{G}_0^*=e.$$

Nun sei $\mathfrak{Z}_{n-1}(\mathfrak{G}^*)\frown\mathfrak{G}_0^*=e$ schon bewiesen. Aus $h\in\mathfrak{Z}_n(\mathfrak{G}^*)\frown\mathfrak{G}_0^*$ ergibt sich dann für beliebiges $g\in\mathfrak{G}^*$

$$(h,g)\in\mathfrak{Z}_{n-1}(\mathfrak{G}^*)\frown\mathfrak{G}_0^*=e,\qquad (h,g)=e,$$

also

$$h\in\mathfrak{Z}_1(\mathfrak{G}^*)\frown\mathfrak{G}_0^*=e,\qquad h=e.$$

Es gilt somit

$$\mathfrak{Z}_n(\mathfrak{G}^*)\frown\mathfrak{G}_0^*=e,\qquad n=1,2,\dots$$

Aus

$$\mathfrak{Z}_n(\mathfrak{G}^*)\mathfrak{G}_0^*/\mathfrak{G}_0^*\cong\mathfrak{Z}_n(\mathfrak{G}^*)/\mathfrak{Z}_n(\mathfrak{G}^*)\frown\mathfrak{G}_0^*=\mathfrak{Z}_n(\mathfrak{G}^*)$$

folgt dann

$$\text{Ordnung }\mathfrak{G}^*/\mathfrak{G}_0^*\geqq\text{Ordnung }\mathfrak{Z}_n(\mathfrak{G}^*)\,.$$

Man beachte dabei dass $\mathfrak{G}^*/\mathfrak{G}_0^*\cong\mathfrak{G}/\mathfrak{G}_0$ eine endliche Gruppe ist. Da aber andererseits $\mathfrak{G}^*$ nach oben nilpotent ist und

$$\mathfrak{Z}_1(\mathfrak{G}^*)<\mathfrak{Z}_2(\mathfrak{G}^*)<\cdots,\quad \text{Ordnung }\mathfrak{Z}_n(\mathfrak{G}^*)\to\infty$$

gilt, so ergibt sich ein Widerspruch.

1) Vgl. H. Zassenhaus, loc, cit. S. 119.

Aus diesem Hilfssatz folgert man sofort, dass eine kompakte nach oben nilpotente Liesche Gruppe immer auflösbar (sogar mit einer endlichen Klasse) ist.

Satz 8. Eine kompakte nach oben nilpotente Gruppe ist auflösbar. Ihre 1-Komponente ist also abelsch.

Beweis. Jede stetige Darstellung solcher Gruppe ist nach Hilfssatz 7 auflösbar und hieraus folgt der Satz unmittelbar.

Wir kehren wieder zu kompakten Lieschen Gruppen zurück.

Hilfssatz 8. Es sei $\mathfrak{G}$ eine kompakte nach oben nilpotente Liesche Gruppe und $\mathfrak{G}_0$ ihre 1-Komponente. Es ist dann

$$\mathfrak{Z}_\omega(\mathfrak{G}) \geq \mathfrak{G}_0$$

und für eine gewisse natürliche Zahl n gilt

$$\mathfrak{Z}_{\omega+n}(\mathfrak{G}) = \mathfrak{G}\,.$$

Beweis. Wir zeigen zunächst durch Induktion

$$\text{(13)} \qquad \mathfrak{Z}_{\omega+n}(\mathfrak{G}) \frown \mathfrak{G}_0 = \mathfrak{Z}_\omega(\mathfrak{G}) \frown \mathfrak{G}_0\,, \quad n=1, 2, \dots\,.$$

Dazu setzen wie voraus $\mathfrak{Z}_{\omega+1}(\mathfrak{G}) \frown \mathfrak{G}_0 \neq \mathfrak{Z}_\omega(\mathfrak{G}) \frown \mathfrak{G}_0$ und leiten daraus einen Widerspruch her. Da $\mathfrak{Z}_{\omega+1}(\mathfrak{G}) \frown \mathfrak{G}_0/\mathfrak{Z}_\omega(\mathfrak{G}) \frown \mathfrak{G}_0$ eine kompakte abelsche Liesche Gruppe ist, so gibt es ein Element a, mit

$$\text{(14)} \qquad a \in \mathfrak{Z}_{\omega+1}(\mathfrak{G}) \frown \mathfrak{G}_0\,, \quad a^p \in \mathfrak{Z}_\omega(\mathfrak{G}) \frown \mathfrak{G}_0\,, \quad a \notin \mathfrak{Z}_\omega(\mathfrak{G}) \frown \mathfrak{G}_0\,,$$

wobei p eine gewisse Primzahl ist. Wenn man die 1-Komponente von $\mathfrak{Z}_\omega(\mathfrak{G})$ mit $\mathfrak{Z}_{\omega,0}(\mathfrak{G})$ bezeichnet, so ist bekanntlich $\mathfrak{Z}_\omega(\mathfrak{G})/\mathfrak{Z}_{\omega,0}(\mathfrak{G})$ eine endliche Gruppe und aus $\mathfrak{Z}_\omega(\mathfrak{G}) = \overline{\bigvee_l \mathfrak{Z}_l(\mathfrak{G})}$ folgt dann

$$\text{(15)} \qquad \mathfrak{Z}_\omega(\mathfrak{G}) = \mathfrak{Z}_{\omega,0}(\mathfrak{G}) \cdot \mathfrak{Z}_l(\mathfrak{G})$$

für gewisse natürliche Zahl l. $\mathfrak{Z}_{\omega,0}(\mathfrak{G})$ ist aber eine zusammenhängende kompakte abelsche Liesche Gruppe, also direktes Prokukt von mit $\mathfrak{K}$ isomorphen Gruppen. Indem wir also a durch geeignetes $a' = ah$, $h \in \mathfrak{Z}_{\omega,0}(\mathfrak{G})$ ersetzen, können wir sogar annehmen dass a^p in $\mathfrak{Z}_l(\mathfrak{G}) \frown \mathfrak{G}_0$ enthalten ist. Wir bezeichnen die Elemente von $\mathfrak{F} = \mathfrak{G}/\mathfrak{G}_0$ mit $\sigma, \tau, \dots$ Die Transformation durch ein Element aus der Restklasse σ gibt in $\mathfrak{G}_0$ einen Automorphismus, den wir wieder mit σ bezeichnen. $\mathfrak{F}$ ist als endliche nilpotente Gruppe direktes Prokukt von einer p-Sylowgruppe $\mathfrak{F}_p$ und einer Gruppe $\mathfrak{F}'$ deren Ordnung f' zu p prim ist:

$$\mathfrak{F} = \mathfrak{F}_p \times \mathfrak{F}'\,.$$

Nach (14) gilt nun für beliebiges σ

$$a^\sigma \equiv a \qquad \text{mod. } \mathfrak{Z}_\omega(\mathfrak{G}) \frown \mathfrak{G}_0\,,$$

und daraus folgt

$$\prod_{\sigma \in \mathfrak{F}'} a^\sigma \equiv a^{f'} \not\equiv e \qquad \text{mod. } \mathfrak{Z}_\omega(\mathfrak{G}) \frown \mathfrak{G}_0\,.$$

$b = \prod_{\sigma \in \mathfrak{F}'} a^\sigma$ erfüllt also

(16) $\quad b \notin \mathfrak{z}_\omega(\mathfrak{G}), \quad b^p \in \mathfrak{z}_l(\mathfrak{G}) \frown \mathfrak{G}_0, \quad b^\sigma = b$ für beliebiges σ aus $\mathfrak{F}'$.

Nun bilden diejenige Elemente h aus $\mathfrak{G}_0$, welche $h^p \in \mathfrak{z}_l(\mathfrak{G}) \frown \mathfrak{G}_0$, $\; h^\sigma \equiv h$ mod. $\mathfrak{z}_l(\mathfrak{G}) \frown \mathfrak{G}_0$ für beliebiges σ aus $\mathfrak{F}'$ erfüllen, einen $\mathfrak{z}_l(\mathfrak{G}) \frown \mathfrak{G}_0$ enthaltenden Normalteiler $\mathfrak{H}$. Da aber $\mathfrak{G}_0/\mathfrak{z}_l(\mathfrak{G}) \frown \mathfrak{G}_0$ als kompakte abelsche Liesche Gruppe nur endlichviele Elemente von der Ordnung p enthält, ist $\mathfrak{H}/\mathfrak{z}_l(\mathfrak{G}) \frown \mathfrak{G}_0$ eine endliche p-Gruppe, deren einzelene Elemente gegen den Automorphismen von $\mathfrak{F}'$ invariant bleiben. Transformationen durch beliebige Elemente aus $\mathfrak{G}$ geben also für $\mathfrak{H}/\mathfrak{z}_l(\mathfrak{G}) \frown \mathfrak{G}_0$ nur Automorphismen von p-Potenzordnungen. Daraus ergibt sich sofort, dass es eine natürliche Zahl m gibt, so dass $\mathfrak{H}$ in $\mathfrak{z}_{l+m}(\mathfrak{G})$ enthalten ist. b muss dann als Element von $\mathfrak{H}$ in $\mathfrak{z}_{l+m}(\mathfrak{G})$, also natürlich in $\mathfrak{z}_\omega(\mathfrak{G})$ enthalten sein, was mit (16) in Widerspruch steht. Wir haben also

(17) $$\mathfrak{z}_{\omega+1}(\mathfrak{G}) \frown \mathfrak{G}_0 = \mathfrak{z}_\omega(\mathfrak{G}) \frown \mathfrak{G}_0$$

bewiesen. Es sei nun schon

$$\mathfrak{z}_{\omega+n-1}(\mathfrak{G}) \frown \mathfrak{G}_0 = \mathfrak{z}_\omega(\mathfrak{G}) \frown \mathfrak{G}_0$$

bewiesen. Wir wählen beliebig h aus $\mathfrak{z}_{\omega+n}(\mathfrak{G}) \frown \mathfrak{G}_0$. Es ist dann für beliebiges g

$$(h, g) \in \mathfrak{z}_{\omega+n-1}(\mathfrak{G}) \frown \mathfrak{G}_0 = \mathfrak{z}_\omega(\mathfrak{G}) \frown \mathfrak{G}_0 .$$

Dies zeigt aber dass h in $\mathfrak{z}_{\omega+1}(\mathfrak{G}) \frown \mathfrak{G}_0 = \mathfrak{z}_\omega(\mathfrak{G}) \frown \mathfrak{G}_0$ enthalten ist, also

$$\mathfrak{z}_{\omega+n}(\mathfrak{G}) \frown \mathfrak{G}_0 = \mathfrak{z}_\omega(\mathfrak{G}) \frown \mathfrak{G}_0 .$$

Damit ist (13) bewiesen. Nach (13) gilt nun

(18) $$\mathfrak{z}_{\omega+n}(\mathfrak{G})\mathfrak{G}_0/\mathfrak{G}_0 \cong \mathfrak{z}_{\omega+n}(\mathfrak{G})/\mathfrak{z}_{\omega+n}(\mathfrak{G}) \frown \mathfrak{G}_0 = \mathfrak{z}_{\omega+n}(\mathfrak{G})/\mathfrak{z}_\omega(\mathfrak{G}) \frown \mathfrak{G}_0 .$$

Da aber $\mathfrak{G}/\mathfrak{G}_0$ endlich ist, muss es einmal

$$\mathfrak{z}_{\omega+n}(\mathfrak{G})\mathfrak{G}_0 = \mathfrak{z}_{\omega+n+1}(\mathfrak{G})\mathfrak{G}_0 = \dots .$$

eintreten. Mit (18) zusammen ergibt sich dann

$$\mathfrak{z}_{\omega+n}(\mathfrak{G}) = \mathfrak{z}_{\omega+n+1}(\mathfrak{G}) = \dots\dots = \mathfrak{G}$$

und nach (13) haben wir

$$\mathfrak{z}_\omega(\mathfrak{G}) \frown \mathfrak{G}_0 = \mathfrak{z}_{\omega+n}(\mathfrak{G}) \frown \mathfrak{G}_0 = \mathfrak{G} \frown \mathfrak{G}_0 = \mathfrak{G}_0 ,$$

also

$$\mathfrak{z}_\omega(\mathfrak{G}) \geqq \mathfrak{G}_0 ,$$ w. z. b. w.

Hilfssatz 10. Keine kompakte nach oben nilpotente Liesche Gruppe besitzt die Klasse ω.

Bemerkung. Es gibt eine 0-dimensionale kompakte nach oben nilpotente Gruppe von der Klasse ω; z. B. das direkte Produkt aller endlichen p-Gruppen.

Beweis. Es sei $\mathfrak{z}_\omega(\mathfrak{G}) = \mathfrak{G}$. Aus $\mathfrak{G} \geqq \mathfrak{z}_n(\mathfrak{G})\mathfrak{G}_0 \geqq \mathfrak{G}_0$ und aus der Endlichkeit von $\mathfrak{G}/\mathfrak{G}_0$ folgt

$$\mathfrak{z}_n(\mathfrak{G})\mathfrak{G}_0 = \mathfrak{z}_{n+1}(\mathfrak{G})\mathfrak{G}_0 = \dots$$

für eine gewisse natürliche Zahl n. $\mathfrak{z}_\omega(\mathfrak{G})=\bigvee_i \mathfrak{z}_i(\mathfrak{G})$ ergibt dann

$$\mathfrak{z}_n(\mathfrak{G})\mathfrak{G}_0=\mathfrak{G}.$$

Nun ist aber $\mathfrak{G}/\mathfrak{z}_n(\mathfrak{G}) \cong \mathfrak{G}_0/\mathfrak{z}_n(\mathfrak{G}) \frown \mathfrak{G}_0$ nach Satz 8 abelsh, somit $\mathfrak{z}_{n+1}(\mathfrak{G})=\mathfrak{G}$.

Wir zeigen nun, dass es immer eine kompakte nach oben nilpotente Liesche Gruppe gibt, welche die Klasse $\xi=\omega+n$, $n=1, 2,\dots$ besitzt.

Es sei $\mathfrak{K}$ wie vorher die Gruppe der reellen Zahlen mod. 1, und σ derjenige Automorphismus von $\mathfrak{K}$, welcher jedes x von $\mathfrak{K}$ mit seinem Inverse $-x$ vertauscht. Wir setzen

$$\mathfrak{G}=\{\mathfrak{K}, a\}^{1)},\ a^2=e,\ axa^{-1}=x^\sigma=(-x),\ \text{für } x\in\mathfrak{K}.$$

Es zeigt sich leicht

$$\mathfrak{z}_\omega(\mathfrak{G})=\mathfrak{K}=\mathfrak{G}_0, \qquad \mathfrak{z}_{\omega+1}(\mathfrak{G})=\mathfrak{G},$$

d. h. dass $\mathfrak{G}$ die Klasse $\omega+1$ besitzt. Dieses Beispiel zeigt zugleich dass eine kompakte nach oben nilpotente Gruppe nicht immer nach unten nilpotent ist, denn man beweist leicht

$$\mathfrak{Z}_1(\mathfrak{G})=\mathfrak{Z}_2(\mathfrak{G})=\dots\quad=\mathfrak{K}.$$

Dagegen ist eine kompakte nach unten nilpotente Liesche Gruppe immer auch nach oben nilpotent, da solche Gruppe eine endliche Klasse besitzt, wie es im Beweis von Satz 5 bemerkt ist.

Wir konstruieren nun eine Gruppe mit der Klasse $\omega+n$. Es sei $\mathfrak{F}$ eine endliche 2-Gruppe mit der Klasse n. Das Zentsum $\mathfrak{z}$ von $\mathfrak{F}$ bestehe aus zwei Elemente e, σ. Durch

$$e\to 1, \qquad \sigma\to -1$$

wird eine Darstellung von $\mathfrak{z}$ gegeben. Die von dieser Darstellung induzierte Darstellung von $\mathfrak{F}$ sei mit $D(\mathfrak{F})$ bezeichnet. Der Grad von $D(\mathfrak{F})$ sei m. Die Matrizen von $D(\mathfrak{F})$ haben lauter ganzrationale Komponenten. Wir setzen also

$$\mathfrak{G}_0=\mathfrak{K}_1\times\mathfrak{K}_2\times\ \dots\ \times\mathfrak{K}_m,\quad \mathfrak{K}_i\cong\mathfrak{K}\ \ (i=1, 2,\dots, m)$$

und definieren für beliebiges σ aus $\mathfrak{F}$ einen Automorphismus von $\mathfrak{G}_0$ durch

$$(x_1, x_2,\dots, x_m)^\sigma=(x_1, x_2,\dots, x_m)D(\sigma)\quad (x_i \text{ mod. } 1).$$

Indem man dann $\mathfrak{G}_0$ durch $\mathfrak{F}$ mit einem zerfallenden Faktorensystem erweitert, erhält man eine Gruppe $\mathfrak{G}$, welche die verlangte Eigenschaft besitzt. Für $\mathfrak{G}_1=\{\sigma, \mathfrak{G}_0\}$ beweist man nämlich durch Induktion

$$\mathfrak{z}_l(\mathfrak{G}_1)\geqq\mathfrak{z}_l(\mathfrak{G}), \qquad l=1, 2, \dots$$

Es ist aber wie leicht ersichtlich $\mathfrak{z}_\omega(\mathfrak{G}_1)=\mathfrak{G}_0$, also

$$\mathfrak{G}_0\geqq\mathfrak{z}_\omega(\mathfrak{G}).$$

Andererseits folgt aus Hilfssatz 9

$$\mathfrak{z}_\omega(\mathfrak{G})\geqq\mathfrak{G}_0,$$

---

1) $\{\mathfrak{K}, a\}$ bedeutet die durch a erweiterte Obergruppe von $\mathfrak{K}$.

folglich

$$\mathfrak{z}_\omega(\mathfrak{G})=\mathfrak{G}_0 .$$

Die Klasse von $\mathfrak{G}$ muss also gerade $\omega+n$ sein.

Wir haben also bewiesen den

Satz 9. Die Klasse ξ einer kompakten nach oben nilpotenten Lieschen Gruppe ist gleich einer der Zahlen

$$1, 2, 3, \dots, \qquad \omega+1, \omega+2, \dots$$

Ist umgekehrt ξ einer solcher Zahlen gleich, so gibt es eine kompakte nach oben nilpotente Liesche Gruppe mit der Klasse ξ.

Indem wir direktes Produkt von Gruppen mit der Klassen $\omega+n$ $(n=1, 2, \dots)$ bilden, erhalten wir eine kompakte nach oben nilpotente Gruppe mit der Klasse $\omega 2$. Wenn man in oben erwähnten Beispielen die Gruppe $\mathfrak{K}$ durch geeignete kompakte abelsche 2-Gruppe ersetzt, kann man auch 0-dimensionale kompakte nach oben nilpotente Gruppen mit der Klasse $\omega+n (n=1, 2, \dots)$ bzw. $\omega 2$ konstruieren.

Schliesslich geben wir ein Beispiel der kompakten p-Gruppe, welche kein Zentrum besitzt. Dass es bei abstrakten Gruppen solche p-Gruppen gibt, ist schon von R. Baer bemerkt worden[1]. Die von ihm gegebene p-Gruppe ist aber nicht nach unten nilpotent und man kann in ihn keine kompakte Topologie einführen. Wir konstruieren eine Reihe von endlichen p-Gruppen $\mathfrak{G}_\nu$ $(\nu=1, 2, \dots)$ mit Homomorphismen

$$\mathfrak{G}_1 \leftarrow \mathfrak{G}_2 \leftarrow \mathfrak{G}_3 \leftarrow \cdots$$

und als deren Limes eine kompakte p-Gruppe $\mathfrak{G}$, welche kein Zentrum besitzt.

Wenn es erlaubt ist, dass $\mathfrak{G}$ Elemente mit der Ordnung p^∞ enthalten darf, so kann man sehr leicht solche Gruppe bilden[2]. Im folgenden soll aber sogar ein solches $\mathfrak{G}$ konstruiert werden, dass ihre Elemente wirklich endliche p-Potenzorddungen besitzen, und zwar deren Ordnungen p^2 nicht übersteigen.

Nun sei $\mathfrak{V}_1$ die p-dimensionale Vektorgruppe über dem Galoisfeld $GF(p)$, nämlich die additive Gruppe aller $x=(x_1, x_2, \dots, x_p)$, $x_i \in GF(p)$ und A_1 eine Matrix vom Grad p über $GF(p)$ der Form

---

1) Vgl. R. Baer, loc. cit.

2) Man nehme als $\mathfrak{G}_\nu$ die Gruppe aller Matrizen vom Grad ν über $GF(p)$ mit der Form

$$\begin{Bmatrix} 1 & a_{12} & a_{13} & \cdots & a_{1\nu} \\ 0 & 1 & a_{23} & & \vdots \\ 0 & 0 & 1 & & \\ \vdots & \vdots & & \ddots & a_{\nu-1\,\nu} \\ 0 & 0 & 0 & \cdots\; 0 & 1 \end{Bmatrix}$$

wobei $a_{ij} (i<j)$ beliebiges Element von $GF(p)$ ist. Ordnet man jeder Matrix A von $\mathfrak{G}_{\nu+1}$ die von ersten ν Zeilen und ersten ν Spalten von A gebildeten Matrix A' von $\mathfrak{G}_\nu$ zu, so erhält man eine homomorphe Abbildung $\mathfrak{G}_\nu \leftarrow \mathfrak{G}_{\nu+1}$.

$$A_1=\left.\begin{pmatrix} 1 & 1 & 0 & \cdots & 0 & 0 \\ 0 & 1 & 1 & & & \vdots \\ 0 & 0 & 1 & 1 & & \vdots \\ \vdots & \vdots & & \ddots & 1 & 0 \\ \vdots & \vdots & & & 1 & 1 \\ 0 & 0 & \cdots & & 0 & 1 \end{pmatrix}\right\} p$$

Die Einheitsmatrix vom Grad p über $GF(p)$ sei mit E_1 bezeichnet. Indem wir für beliebiges $x=(x_1, x_2 \ldots, x_p)$ aus $\mathfrak{V}_1$

$$x^{\sigma_1}=(x_1, x_2, \ldots, x_p)A_1$$

setzen, erhalten wir einen Automorphismus σ_1 von $\mathfrak{V}_1$. Wir erweitern dann $\mathfrak{V}_1$ durch A_1 mit einem zerfallenden Faktorensystem und bezeichnen die so gewonnene Gruppe mit $\mathfrak{G}_1$:

$$\mathfrak{G}_1=\{A_1, \mathfrak{V}_1\}\,.$$

Aus $A_1^p=E_1$ folgt, dass $\mathfrak{G}_1$ eine endliche p-Gruppe ist und dass die Ordnung eines beliebigen Elementes von $\mathfrak{G}_1$ höchstens p^2 ist.

Nun seien $A_2^{(1)}$, $A_2^{(2)}$ Matrizen vom Grad p^2 über $GF(p)$, welche die Formen

$$A_2^{(1)}=\left.\begin{pmatrix} A_1 & & & \\ & A_1 & & \\ & & \ddots & \\ & & & A_1 \end{pmatrix}\right\} p\text{-mal}, \qquad A_2^{(2)}=\left.\begin{pmatrix} E_1 & E_1 & & & \\ & E_1 & E_1 & & \\ & & \ddots & \ddots & \\ & & & E_1 & E_1 \\ & & & & E_1 \end{pmatrix}\right\} p\text{-mal}$$

besitzen, d. h.

$$A_2^{(1)}=A_1\times E_1, \qquad A_2^{(2)}=E_1\times A_1 \qquad (\text{Kroneckersches Produkt !})$$

$A_2^{(1)}$ und $A_2^{(2)}$ sind einander vertauschbar und beide haben die Ordnung p. Sie erzeugen also eine abelsche Gruppe $\{A_2^{(1)}, A_2^{(2)}\}$ mit der Ordnung p^2. Wir erweitern nun genau so wie oben die p^2-dimensionale Vektorgruppe $\mathfrak{V}_2$ über $GF(p)$ durch $\{A_2^{(1)}, A_2^{(2)}\}$ und erhalten eine Gruppe

$$\mathfrak{G}_2=\{A_2^{(1)}, A_2^{(2)}, \mathfrak{V}_2\}\,.$$

$\mathfrak{G}_2$ ist offenbar eine endliche p-Gruppe und die Ordnung eines beliebigen Elementes von $\mathfrak{G}_2$ ist wieder hochstens p^2. Ausserdem kann man einen Homomorphismus von $\mathfrak{G}_2$ auf $\mathfrak{G}_1$ herstellen, indem man jeder Matrix von $\{A_2^{(1)}, A_2^{(2)}\}$ die von ihren ersten p-Zeilen und p-Spalten gebildeten Matrix zuordnet und jedem $x=(x_1, x_2, \ldots, x_{p^2})$ von $\mathfrak{V}_2$ den Vektor $x'=(x_1, x_2, \ldots, x_p)$ von $\mathfrak{V}_1$ zuordnet. Es gibt also einen Homomorphismus von $\mathfrak{G}_2$ auf $\mathfrak{G}_1$:

$$\mathfrak{G}_1 \leftarrow \mathfrak{G}_2\,.$$

Aus

$$A_3^{(1)}=A_1\times E_1\times E_1,\ A_3^{(2)}=E_1\times A_1\times E_1,\ A_3^{(3)}=E_1\times E_1\times A_1$$

und aus der p^3-dimensionalen Vektorgruppe $\mathfrak{V}_3$ über $GF(p)$ erhalten wir die Gruppe

$$\mathfrak{G}_3 = \{A_3^{(1)}, A_3^{(2)}, A_3^{(3)}, \mathfrak{B}_3\}$$

und dann gegau so wie oben einen Homomorphismus

$$\mathfrak{G}_2 \leftarrow \mathfrak{G}_3 .$$

Die so definierte Reihe von endlichen p-Gruppe

$$\mathfrak{G}_1 \leftarrow \mathfrak{G}_2 \leftarrow \mathfrak{G}_3 \leftarrow \cdots$$

bestimmt bekanntlich eine kompakte Gruppe $\mathfrak{G}$. Da aber die Ordnung eines Elementes von $\mathfrak{G}_\nu$ immer hochstens p^2 ist, so gilt dasselbe auch für $\mathfrak{G}$.

Nun zeigen wir, dass das Zentrum von $\mathfrak{G}$ nur aus dem Einselement besteht. Nach Konstruktion gibt es offene Normalteiler $\mathfrak{N}_\nu (\nu = 1, 2, \ldots)$, so dass

$$(19) \qquad \mathfrak{G}/\mathfrak{N}_\nu \cong \mathfrak{G}_\nu , \qquad \wedge\, \mathfrak{N}_\nu = e .$$

Wir betrachten z. B. $\mathfrak{G}_2$ und zeigen dass das Zentrum von $\mathfrak{G}_2$ der von

$$z_2 = (0, 0, \ldots, 0, 1)$$

erzeugte Untergruppe von $\mathfrak{B}_2$ gleich ist. Zunächst muss ein Zentrumselement von $\mathfrak{G}_2$ in $\mathfrak{B}_2$ enthalten sein, weil $\mathfrak{G}_2/\mathfrak{B}_2$ einer Automorphismengruppe von $\mathfrak{B}_2$ isomorph ist. Aus

$$xA_2^{(1)} = x , \qquad xA_2^{(2)} = x$$

folgt dann sofort dass der Vektor x ein Vielfaches von z_2 ist. Genau so beweist man allgemein dass das Zentrum von $\mathfrak{G}_\nu$ die von

$$z_\nu = (0, 0, \ldots, 0, 1)$$

erzeugte Untergruppe von $\mathfrak{B}_\nu$ ist.

Nun sei z ein Element aus dem Zentrum von $\mathfrak{G}$. z muss natürlich in $\mathfrak{G}/\mathfrak{N}_\nu \cong \mathfrak{G}_\nu$ in der Restklasse von $\{z_\nu\}$ enthalten sein. Da aber diese Restklasse beim Homomorphismus $\mathfrak{G}_{\nu-1} \leftarrow \mathfrak{G}_\nu$ ins Einselement von $\mathfrak{G}_{\nu-1}$ abgebildet wird, so ist z in $\mathfrak{N}_{\nu-1}$ enthalten. Aus (19) folgt dann

$$z = e , \qquad \text{w. z. b. w.}$$

[13] Der Bezoutsche Satz in zweifach projektiven Räumen

Proc. Japan Acad., 21-4 (1945), 213-222.

Mathematisches Institut, Kaiserliche Universität zu Tokio.
(Comm. by T. TAKAGI, M.I.A., April 26, 1945.)

Der Bezoutsche Satz über Schnittpunkte einer algebraischen Kurve mit Hyperflächen in einem projektiven Raum lässt sich sofort auf den Fall einer Kurve in einem mehrfach projektiven Raum verallgemeinern, wenn die Charakteristik des Grundkörpers Null ist[1]. Ist aber die Charakteristik nicht gleich Null, dann treten inseparable Körpererweiterungen auf und der Sachverhalt wird etwas umständlicher. Im folgenden soll gezeigt werden, dass der Bezoutsche Satz auch auf diesen Fall erweitert werden kann. Wir beschränken uns dabei der Einfachheit halber auf den Fall der zweifach projektiven Räume. Dieser Fall besitzt anderseits wegen Anwendung auf die Korrespondenztheorie der algebraischen Kurven eine besondere Wichtigkeit[2]. Daran anschliessend werden wir einige Bemerkungen über Schnittkurven von einer 2-dimensionalen Mannigfaltigkeit mit Hyperflächen angeben.

Es sei P_l bzw. P_m ein l- bzw. m-dimensionaler projektiver Raum über einem Grundkörper k mit der Charakteristik p (=Null oder eine Primzahl). Wir nehmen k als algebraisch abgeschlossen an. Die Punkte von P_l bzw. P_m sind bis auf beliebige Faktoren λ bestimmte, nicht sämtlich verschwindende $(l+1)$- bzw. $(m+1)$-tupel $\xi=(\xi_0, \xi_1, \ldots\ldots, \xi_l)$ bzw. $\eta=(\eta_0, \eta_1, \ldots\ldots, \eta_m)$, wobei ξ_i bzw. η_j Elemente in einem gewissen Erweiterungskörper von k sind.[3] Wir bezeichnen dann mit $P_{l,m}$ den zweifach projektiven Raum, der das Produkt von P_l und P_m ist. Die Punkte von $P_{l,m}$ sind also die Gesamtheit der Paare (ξ, η).

Nun sei $K=k(w_1, w_2, \ldots\ldots, w_\alpha)$ ein rationaler Funktionenkörper der Unbestimmten $w_1, w_2, \ldots\ldots, w_\alpha$ über k und ξ ein in bezug auf diesem Körper algebraischer Punkt in P_l, d.h. die Verhältinisse der Koordinaten ξ_i $(i=0, 1,$

---

1) Vgl. B. L. van der Waerden, Zur algebraische Geometrie I, Math. Ann. 108 (1933), XIII, Math. Ann. 115 (1939).

2) Algebraische Korrespondenzen zwischen algebraischen Kurven können bekanntlich durch Kurven in geeigneten zweifach projektiven Räumen dargestellt werden. Vgl. die nachfolgende Arbeit des Verfassers in diesen Proceedings.

3) Nach v. d. Waerden nehmen wir solchen Erweiterungskörper nicht als fest an, sondern vielmehr als ein wachsender Körper im Laufe einer geometrischen Betrachtung. Vgl. B. L. van der Waerden, Einführung in die algebraische Geometrie, Berlin (1939), S. 105 oder l.c.1).

......, l) von ξ seien über K algebraisch. Den durch die Adjunktion dieser Verhältnisse entstehenden Erweiterungskörper von K bezeichnen wir mit

$$K_1 = K(\xi) = k(w_1, w_2, \ldots\ldots, w_a; \xi).$$

Der reduzierte Grad bzw. der Exponent der Erweiterung K_1/K sei n_1 bzw. p^{e_1}[1]. Wir bilden nun mit Unbestimmten $u_0, u_1, \ldots\ldots, u_n$ den Ausdruck

$$(1) \qquad \prod_{\nu=1}^{n_1}(u, \xi^{(\nu)})^{p^{e_1}} = \prod_{\nu=1}^{n_1}(u_0\xi_0^{(\nu)} + u_1\xi_1^{(\nu)} + \ldots\ldots + u_n\xi_n^{(\nu)})^{p^{e_1}},$$

wobei $\xi^{(\nu)}$ alle verschiedene konjugierte Punkte von ξ in bezug auf K[2] durchläuft. Nach W.-L. Chow[3] kann man durch geeignete Normierung der Koordinaten von ξ erreichen, dass der Ausdruck (1) eine irreduzible Form von u_i und w_j über k wird. Wir heissen in diesem Fall, dass die Koordinaten von ξ in bezug auf $k(w_1, w_2, \ldots\ldots, w_a)$ normiert sind.

Es seien ξ, η wie oben über $k(w_1, w_2, \ldots\ldots, w_a)$ algebraische Punkte in P_l bzw. P_m und

$$(2) \qquad f(x, y, a) = a_0x_0^sy_0^t + a_1x_0^{s-1}x_1y_0^t + \ldots\ldots, \quad st > 0$$

eine allgemeine Form von $x = (x_0, x_1, \ldots\ldots, x_l)$ und $y = (y_0, y_1, \ldots\ldots, y_m)$ mit dem Grad (s, t). Die Koeffizienten a_λ sind also Unbestimmten. Mit normierten Koordinaten von ξ, η bilde man

$$F(w, a) = \prod_{\nu=1}^{n} f(\xi^{(\nu)}, \eta^{(\nu)}, a)^{p^e},$$

wo n bzw. p^e den reduzierten Grad bzw. den Exponent von $k(w_1, w_2, \ldots\ldots, w_a, \xi, \eta)/k(w_1, w_2, \ldots\ldots, w_a)$ bedeutet und ν alle verschiedenen Konjugierten dieser Erweiterung durchläuft. Man kann dann sofort beweisen, dass $F(w, a)$ eine irreduzible Form von w, a über k ist[4].

Nun sei C eine irreduzible Kurve in $P_{l,m}$ und $f(x, y, a)$ die in (2) angegebene allgemeine Form vom Grad (s, t). Ferner sei $g(x, y, b)$ eine andere allgemeine Form vom Grad (s', t'), $s't' > 0$ mit unbestimmten Koeffizienten b_μ. Die durch $f(x, y, a) = 0$ bzw. $g(x, y, b) = 0$ bestimmte Hyperfläche H_a bzw. H_b in $P_{m,n}$ schneidet die Kurve C in ihren allgemeinen Punkten $(\xi^{(\lambda)}, \eta^{(\lambda)})$ bzw. $(\xi'^{(\mu)}, \eta'^{(\mu)})$, welche über $k(a)$ bzw. $k(b)$[5] algebraisch und einander konjugiert sind. Für normierte Koordinaten von ξ, η sind dann die Ausdrücke

---

1) Wir wollen nicht e, sondern p^e als Exponent bezeichnen. Es soll $p^e = 1$ bedeuten, wenn $p = 0$.

2) Das bedeutet, dass die Verhältnisse der Koordinaten in bezug auf K einander konjugiert sind.

3) W.-L. Chow, Die geometrische Theorie der algebraischen Funktionen für beliebige vollkommene Korper, Math. Ann. 114 (1937).

4) Vgl. Chow, l.c.

5) Den Körper $k\left(\frac{a_1}{a_0}, \frac{a_2}{a_0}, \ldots\right)$ bezeichnen wir kurz mit $k(a)$. Ebenso ist mit $k(b)$.

$$F(a,b)=\prod_{\mu=1}^{n'} f(\xi'^{(\mu)}, \eta'^{(\mu)}, a)^{p^{e'}}$$

$$G(a,b)=\prod_{\lambda=1}^{n} g(\xi^{(\lambda)}, \eta^{(\lambda)}, b)^{p^e}$$

nach dem oben Bemerkten beide irreduzible Formen in $k[a,b]$. n, p^e und $n', p^{e'}$ sind dabei reduzierte Grade und Exponenten der Erweiterungen $k(a,\xi,\eta)/k(a)$ und $k(b,\xi',\eta')/k(b)$. Genau wie in einem gewöhnlichen (einfachen) projectiven Raum[1] schliesst man dann sogleich, dass $F(a,b)$ und $G(a,b)$ (bis auf einem Faktor in k) zusammenfallen und dass die Exponenten $p^e, p^{e'}$ gleich 1 sind. $k(a,\xi,\eta)/k(a)$ und $k(b,\xi,\eta)/k(b)$ sind also separable Erweiterungen. Hier tritt kein besonderer Umstand im Vergleich mit dem Fall der gewöhnlichen projektiven Räume auf.

Wir untersuchen nun die Schnittpunkte von C mit der Hyperflache H_{a_1}, welche durch eine allgemeine Form $f_1(x,a_1)$ in $P_{l,m}$ bestimmt wird, die in x den Grad $s(>0)$ besitzt, aber y nicht enthält $(t=0)$. Ausgenommen der ausgeartete Fall, dass C mit einem Punkt ζ in P_l und einer Kurve C_2 in P_m in der Form

$$(\zeta)\times C_2$$

dargestellt werden kann, besitzt C immer Schnittpunkte $(\bar\xi^{(\nu)}, \bar\eta^{(\nu)})$ mit H_{a_1}, welche über $k(a_1)$ algebraisch und einander konjugiert sind[2]. Für normierte Koordinaten sind dann die beiden Ausdrücke

$$F_1(a_1,b)=\prod_{\mu=1}^{n''} f_1(\xi'^{(\mu)}, a_1)^{p^{e''}},$$

$$G(a_1,b)=\prod_{\nu=1}^{n_1} g(\bar\xi^{(\nu)}, \bar\eta^{(\nu)}, b)^{p^{e_1}}$$

irreduzible Formen in $k[a_1,b]$, wo μ bzw. ν alle verschiedene Konjugierten von ξ' bzw. $(\bar\xi,\bar\eta)$ in bezug auf $k(b)$ bzw. $k(a_1)$ durchläuft und $p^{e''}, p^{e_1}$ die Exponenten von $k(b,\xi')/k(b)$ und $k(a_1,\bar\xi,\bar\eta)/k(a_1)$ sind. Man beweist dann genau wie oben, dass $F_1(a_1,b)$ mit $G(a_1,b)$ bis auf einen konstanten Faktor zusammenfällt. $k(b,\xi',\eta')$ ist aber, wie schon bemerkt, über $k(b)$ separabel. Der Exponent $p^{e''}$ von $k(b,\xi')/k(b)$ ist folglich gleich 1. Ferner fällt $k(b,\xi')$ in diesem Fall mit $k(b,\xi',\eta')$ zusammen. Zum Beweis setze man

$$n^*=[k(b,\xi',\eta') : k(b,\xi')], \text{ also } n'=n^*n''.^{3)}$$

Es gibt dann n^* Schnittpunkte von C mit H_b, die erste Koordinate ξ' haben. Eine durch (ξ',η') hindurchgehende allgemeine Hyperfläche vom Grad (s',t')

1) Vgl. Chow, l. c.

2) Vgl. v. d. Waerden, l. c. 3), S. 146.

3) n'' war der reduzierte Grad von $k(b,\xi')/k(b)$. Da aber $k(b,\xi')/k(b)$ eine separable Erweiterung ist, so ist es gleich $[k(b,\xi') : k(b)]$.

schneidet aber die Kurve C ausser in (ξ', η') in keinem Punkt mit der ersten Koordinate ξ' mehr, denn die Hyperebene $\xi'_1 x_0 - \xi'_0 x_1 = 0$ in $P_{l,m}$ schneidet C nur in endlichvielen Punkten, falls C nicht von der Form $(\zeta) \times C_2$ ist. Es muss also $n^* = 1$ und folglich $k(b, \xi') = k(b, \xi', \eta')$ sein und es gilt

$$(3) \qquad F_1(a_1, b) = \prod_{\mu=1}^{n'} f_1(\xi'^{(\mu)}, a_1), \quad n' = [k(b, \xi', \eta') : k(b)].$$

Wir betrachten nun den Exponent p^{e_1}. Nach der Theorie der algebraischen Korrespondenzen haben $\bar{\xi}, \bar{\eta}, a_1$ isomorphe algebraische Relationen über k wie ξ^*, η^*, a_1^*, die man erhält, wenn man durch einen allgemeinen Punkt (ξ^*, η^*) von C eine allgemeinste Hyperfläche $f_1(x, a_1^*)$ vom Grad s legt[1]. Setzt man

$$f_1(x, a_1) = a_{10} x_0^s + a_{11} x_0^{s-1} x_1 + a_{12} x_0^{s-2} x_1^2 + \ldots\ldots$$
$$= a_{10} x_0^s + f_1'(x, a_{11}, a_{12}, \ldots\ldots),$$

so lassen sich $a_{1\nu}^*$ z.B. so bestimmen, dass für unbestimmte $a_{11}^*, a_{12}^*, \ldots\ldots$

$$a_{10}^* = -\frac{f_1'(\xi^*, a_{11}^*, a_{12}^*, \ldots\ldots)}{\xi_0^{*s}}$$

gesetzt wird[2]. Es gilt dann

$$(4) \qquad k(a_1, \bar{\xi}, \bar{\eta}) \cong k(a_1^*, \xi^*, \eta^*) = k(\xi^*, \eta^*, a_{11}^*/a_{10}^*, a_{12}^*/a_{10}^* \ldots\ldots)$$
$$= k(\xi^*, \eta^*, a_{12}^*/a_{11}^*, a_{13}^*/a_{11}^*, \ldots\ldots)$$

und ähnlich

$$(5) \qquad k(a_1, \bar{\xi}) \cong k(a_1^*, \xi^*) = k(\xi^*, a_{12}^*/a_{11}^*, a_{13}^*/a_{11}^*, \ldots\ldots).$$

Da ξ^*, η^* und $a_{12}^*/a_{11}^*, a_{13}^*/a_{11}^*, \ldots\ldots$ von einander algebraisch unabhängig sind, so folgt aus (4), (5) dass der Grad und Exponent von $k(a, \bar{\xi}, \bar{\eta})/k(a_1, \xi)$ denjenigen von $k(\xi^*, \eta^*)/k(\xi^*)$ gleich sind.

Nun ist $\bar{\xi}$ als Schnittpunkt der auf P_l projizierenden Bildkurve von C mit der allgemeinen Hyperflache $f_1(x, a_1) = 0$ in P_l sicher separabel über $k(a_1)$. p^{e_1} ist daher gleich dem Exponent von $k(a_1, \bar{\xi}, \bar{\eta})/k(a_1, \bar{\xi})$, also nach oben denjenigen von $k(\xi^*, \eta^*)/k(\xi^*)$. Setzt man alsdann

$$r_1 = [k(\xi^*, \eta^*) : k(\xi^*)] = [k(a_1, \bar{\xi}, \bar{\eta}) : k(a_1, \bar{\xi})] = \bar{r}_1 p^{e_1},$$

so fällt $\bar{r}_1$ mit dem reduzierten Grad von $k(\xi^*, \eta^*)/k(\xi^*)$ zusammen, da $k(\xi^*)$ ein algebraischer Funktionenkörper von einer Veränderlichen ist[3]. Daraus folgt auch

$$(6) \qquad n_1 = r_1 [k(a_1, \bar{\xi}) : k(a_1)].$$

Nun definieren wir die Multiplizität eines Schnittpunktes einer irreduziblen Kurve C mit einer allgemeiner Hyperfläche $f(x, y, a) = 0$ vom Grad (s, t), $st \geqq 0$ folgendermassen: es ist gleich

1) Vgl. z.B. v.d. Waerden, l. c. 3), Kap. V.

2) Man kann natürlich $\xi_0 \neq 0$ annehmen.

3) Vgl. z.B. M. Deuring, Arithmetische Theorie der Korrespondenzen algebraischer Funktionenkörper, I, Crelle's Jour. 177 (1937).

i) 1, für $st>0$,

ii) dem Expenent p^{e_1} von $k(\xi^*, \eta^*)/k(\xi^*)$, für $t=0$,

iii) dem Exponent p^{e_2} von $k(\xi^*, \eta^*)/k(\eta^*)$, für $s=0$,

wobei (ξ^*, η^*) ein allgemeinen Punkt von C über k bedeutet.

Es sei bemerkt, dass die so definierte Multiplizität allenfalls nur von C abhängig ist.

Nach dieser Definition kann man den oben Bewiesenen in folgender einfacher Form zusammenfassen.

Satz 1. Es seien $f(x, y, a)$, $g(x, y, b)$ allgemeine Formen von Graden (s, t), (s', t') und C eine irreduzible Kurve in $P_{l,m}$. Schnittpunkte von C mit den Hyperflächen $f(x, y, a)=0$, $g(x, y, b)=0$ seien, mit Multiplizitäten versehen, $(\xi^{(\lambda)}, \eta^{(\lambda)})$, $\lambda=1, \ldots\ldots, \alpha$, $(\xi'^{(\mu)}, \eta'^{(\mu)})$, $\mu=1, \ldots\ldots, \beta$. Ist dann $st>0$ oder $s't'>0$, so gilt nach geeigneter Normierung der Koordinaten

$$\prod_{\mu=1}^{\beta} f(\xi'^{(\mu)}, \eta'^{(\mu)}, a) = \prod_{\lambda=1}^{\alpha} g(\xi^{(\lambda)}, \eta^{(\lambda)}, b) . \tag{7}$$

Satz 2. Die Multiplizität eines Schnittpunktes (ξ, η) von C mit einer allgemeinen Hyperfläche $f(x, y, a)=0$ ist gleich dem Exponent von $k(a, \xi, \eta)/k(a)$.

Satz 3. Es sei (ξ, η) ein allgemeiner Punkt von C und r_1, p^{e_1} der Grad und der Exponent von $k(\xi, \eta)/k(\xi)$. Ferner sei C_1 die auf P_l projizierende Bildkurve von C, welche wir nicht nulldimensional annehmen[1]. Die Schnittpunkte von C_1 mit einer allgemeinen Hyperfläche $f_1(x, a_1)=0$ in P_l seien $\xi^{(1)}, \ldots\ldots, \xi^{(g)}$, $g=[k(a_1, \xi^{(1)}):k(a_1)]$. Es gibt dann für jedes $\xi^{(\lambda)}$ genau r_1 Schnittpunkte von C mit $f_1(x, a_1)=0$ in $P_{l,m}$, deren erste Koordinaten gleich $\xi^{(\lambda)}$ sind und von diesen Schnittpunkten fallen je p^{e_1} einander zusammen.

Im folgenden bezeichnen wir die Gesamtheit der mit Vielfachheiten versehenen Schnittpunkte von C mit $f(x, y, a)=0$ immer mit

$$[C, f(x, y, a)] .$$

Nun seien $f(x, y, a)$, $f_1(x, y, a_1)$, $f_2(x, y, a_2)$ allgemeine Formen von Graden (s, t), (s_1, t_1), (s_2, t_2) und es gelte

$$s=s_1+s_2, \quad t=t_1+t_2, \quad st \geqq 0, \quad s_1t_1 \geqq 0, \quad s_2t_2 \geqq 0 .$$

Es gilt dann der

Satz 4. Setzt man

$$f(x, y, a^*)=f_1(x, y, a_1)f_2(x, y, a_2) ,$$

so geht $[C, f(x, y, a)]$ bei der relationstreuen Spezialisierung $a \to a^*$ in die Summe von $[C, f_1(x, y, a_1)]$ und $[C, f_2(x, y, a_2)]$.

---

1) Genau dann ist ja $k(\xi, \eta)/k(\xi)$ eine endliche Erweiterung.

Beweis. Wir nehmen zunächst an, dass C wirklich $f_1(x, y, a_1)=0$ und $f_2(x, y, a_2)=0$ schneidet und setzen

$$[C, f(x, y, a)] = \{(\xi^{(\rho)}, \eta^{(\rho)}), \rho=1, \dots\dots, \alpha\},$$
$$[C, f_1(x, y, a_1)] = \{(\xi'^{(\sigma)}, \eta'^{(\sigma)}), \sigma=1, \dots\dots, \beta\},$$
$$[C, f_2(x, y, a_2)] = \{(\xi''^{(\tau)}, \eta''^{(\tau)}), \tau=1, \dots\dots, \gamma\}.$$

Setzt man ferner

$$g(x, y, b) = \sum_{i=0}^{l}\sum_{j=0}^{m} b_{ij}x_i y_j, \ [C, g(x, y, b)] = \{(\xi^{*(\omega)}, \eta^{*(\omega)}), \omega=1, \dots\dots, \delta\},$$

so gilt nach Satz 1, (7)

$$F(a, b) = \prod_\rho g(\xi^{(\rho)}, \eta^{(\rho)}, b) = \prod_\omega f(\xi^{*(\omega)}, \eta^{*(\omega)}, a),$$
$$F(a^*, b) = \prod_\omega f(\xi^{*(\omega)}, \eta^{*(\omega)}, a^*) = \prod_\omega f_1(\xi^{*(\omega)}, \eta^{*(\omega)}, a_1)\prod_\omega f_2(\xi^{*(\omega)}, \eta^{*(\omega)}, a_2)$$
$$= \prod_\sigma g(\xi'^{(\sigma)}, \eta'^{(\sigma)}, b)\prod_\tau g(\xi''^{(\tau)}, \eta''^{(\tau)}, b).$$

Geht $[C, f(x, y, a)]$ bei der Spezialsierung relationstreu in $\{(\bar\xi^{(\rho)}, \bar\eta^{(\rho)}), \rho=1, \dots\dots, \alpha\}$ über, so gilt[1]

$$F(a, b) = \prod_\rho g(\xi^{(\rho)}, \eta^{(\rho)}, b) \rightarrow F(a^*, b) = \prod_\rho g(\bar\xi^{(\rho)}, \bar\eta^{(\rho)}, b),$$

folglich

$$(8) \qquad \prod_\rho g(\bar\xi^{(\rho)}, \bar\eta^{(\rho)}, b) = \prod_\sigma g(\xi', \eta', b)\prod_\tau g(\xi'', \eta'', b).$$

und daraus folgt sofort die Bahauptung.

Wir betrachten nun den Fall, dass C und $f_1(x, y, a_1)=0$ keinen Schnittpunkt haben. Dies geschieht genau dann, wenn $t_1=0$ und C von der Form $(\zeta)\times C_2$ ist. Ist ausserdem $t_2=0$, so ist auch $t=0$ und $[C, f(x, y, a)]$, $[C, f_1(x, y, a_1)]$, $[C, f_2(x, y, a_2)]$ sind sämtlich Nullmengen. Es bleibt also nichts zu beweisen.

Wir nehmen also an, dass $t>0$, $t_2>0$ gilt und dass C mit $f(x, y, a)=0$, $f_2(x, y, a_2)=0$ Schnittpunkte besitzt. Es gilt dann wie (8)

$$(9) \qquad \prod_\rho g(\bar\xi^{(\rho)}, \bar\eta^{(\rho)}, b) = \prod_\omega f_1(\xi^{*(\omega)}, a_1)\prod_\tau g(\xi''^{(\tau)}, \eta''^{(\tau)}, b).$$

$F_1(a_1, b) = \prod_\omega f_1(\xi^{*(\omega)}, a_1)$ ist dabei eine Form in $k[a_1, b]$. Wir wollen nun zeigen, dass $F_1(a_1, b)$ nicht von b abhängt, indem wir aus der gegenteilige Voraussetzung einen Widerspruch ableiten. Man nehme also an, dass $F_1(a_1, b)$ von b abhänge. Man bestimmte alsdann $\bar b$ und die entsprechenden Spezialisierungen $\bar\xi^{*(\omega)}$ von $\xi^{*(\omega)}$ so, dass es für unbestimmte a_1

$$F_1(a_1, \bar b) = \prod_\omega f_1(\bar\xi^{*(\omega)}, a_1) = 0$$

gelte. So würde einer von $\bar\xi^{*(\omega)}$ einerseits auf $f_1(x, a_1)=0$, andererseits aber als

1) Vgl. v. d. Waerden, l.c. 3), S. 165 oder Chow, l.c. 6).

Spezialisierung von $\xi^{*(\omega)}$ auch auf C liegen, entgegen der Voraussetzung, dass C und $f_1(x, a_1)=0$ keinen Punkt gemein haben. $\prod_\omega f_1(\xi^{*(\omega)}, a_1)$ ist also von b unabhängig. Vergleicht man die beiden Seiten von (9) als Formen von b, so sieht man sofort, dass $[C, f(x, y, a)]$ bei $a \to a^*$ zu $[C, f_2(x, y, a_2)]$ übergeht. Genau so erledigt man den Fall, dass C mit $f_2(x, y, a_2)=0$ keinen Schnittpunkt besitzt. Der Satz ist also in allen Fällen bewiesen.

Nun definieren wir die Multiplizität eines Schnittpunktes (ξ^0, η^0) von C mit einer beliebigen Hyperfläche $f(x, y, a')=0$ wie üblich folgendermassen. Wir nehmen eine allgemeine Form $f(x, y, a)$ mit demselben Grad wie $f(x, y, a')$ und heissen dann die Zahl, die angibt, wie viele Punkte von $[C, f(x, y, a)]$ sich bei $a \to a'$ relationstreu zu (ξ^0, η^0) spezialisieren, die Multiplizität oder Vielfachheit von (ξ^0, η^0). Es ist bekannt, dass diese Multiplizität unabhängig von dem Spezialisierungsprozess eindeutig durch $a \to a'$ bestimmt wird[1]. Die Punktgruppe je mit Vielfachheiten versehener Schnittpunkte von C mit $f(x, y, a')=0$ bezeichnen wir wieder mit

$$[C, f(x, y, a')]\,.$$

Nach Definition ist dann bei $a \to a'$

$$[C, f(x, y, a)] \to [C, f(x, y, a')] \tag{10}$$

und es gilt das Prinzip der Erhaltung der Anzahl[2].

Es sei noch bemerkt, dass bei einer ausgearteten Kurve C, Schnittpunkte mit der Multiplizität Null auftreten können. Z.B. kann $(\zeta)\times C_2$ wohl eine Hyperfläche $f_1(x, a_1')=0$ schneiden (d.h. $f_1(\zeta, a_1')=0$), obgleich $[(\zeta)\times C_2, f_1(x, a_1')]$ eine Nullmenge ist. Mag dieser Umstand einmal unangenehm scheinen, so wird es sich später doch als zweckmässig erweisen.

Satz 5. Es seien $f_1(x, y, a_1'), f_2(x, y, a_2')$ beliebige Formen und

$$f(x, y, a')=f_1(x, y, a_1')f_2(x, y, a_2')\,.$$

Es gilt dann für eine irreduzible Kurve C

$$[C, f(x, y, a')]=[C, f_1(x, y, a_1')]+[C, f_2(x, y, a_2')]\,.$$

Beweis. Die $f(x, y, a'), f_1(x, y, a_1'), f_2(x, y, a_2')$ entsprechende allgemeine Formen bezeichnen wir mit $f(x, y, a)$, $f_1(x, y, a_1)$, $f_2(x, y, a_2)$. Es sei ferner

$$f(x, y, a^*)=f_1(x, y, a_1)f_2(x, y, a_2)\,.$$

Aus (10) folgt

$$\begin{aligned} &[C, f(x, y, a)] \to [C, f(x, y, a')], && \text{bei } a \to a'\,, \\ &[C, f_1(x, y, a_1)] \to [C, f_1(x, y, a_1')], && \text{bei } a_1 \to a_1'\,, \\ &[C, f_2(x, y, a_2)] \to [C, f_2(x, y, a_2')], && \text{bei } a_2 \to a_2'\,. \end{aligned} \tag{11}$$

1) Vgl. v. d. Waerden, l.c. 3), S. 165.

2) Vgl. v. d. Waerden, l. c. 3), S. 164.

Nach Satz 4 gilt andererseits

(12) $[C, f(x, y, a)] \rightarrow [C, f_1(x, y, a_1)] + [C, f_2(x, y, a_2)]$, bei $a \rightarrow a^*$.

Denkt man sich also die Spezialisierung $a \rightarrow a'$ so geschehen, dass man erst die Spezialisierung $a \rightarrow a^*$, und dann $a^* \rightarrow a'$ $(a_1 \rightarrow a_1', a_2 \rightarrow a_2')$ vorgenommen hat, so erhält man aus (11), (12)

$$[C, f(x, y, a')] = [C, f_1(x, y, a_1')] + [C, f_2(x, y, a_2')],$$

w.z.b.w.

Wir bezeichnen nun mit $A(s, t)$ die Anzahl der Schnittpunkte von C mit einer Hyperfläche $f(x, y, a') = 0$ vom Grad (s, t). Nach dem Prinzip der Erhaltung der Anzahl hängt $A(s, t)$ nur von C und (s, t) ab und es gilt nach Satz 5

$$A(s_1 + s_2, t_1 + t_2) = A(s_1, t_1) + A(s_2, t_2).$$

Daraus folgt

$$A(s, t) = sA(1, 0) + tA(0, 1).$$

Nun sei (ξ, η) ein allgemeiner Punkt von C und C_1 bzw. C_2 die auf P_l bzw. P_m projizierende Bildkurve von C. ξ bzw. η ist also ein allgemeiner Punkt von C_1 bzw. C_2. Der Grad von C_i $(i = 1, 2)$ bezeichnen wir mit γ_i $(i = 1, 2)$[1]. Setzt man ferner

$$r_1 = [k(\xi, \eta) : k(\xi)], \qquad r_2 = [k(\xi, \eta) : k(\eta)]^{2)},$$

so folgt aus Satz 3

$$A(1, 0) = \gamma_1 r_1, \qquad A(0, 1) = \gamma_2 r_2.$$

Wir haben also die folgende Erweiterung des Satzes von Bezout:

Satz 6. Die Anzahl der Schnittpunkte einer irreduziblen Kurve C mit einer Hyperfläche $f(x, y, a') = 0$ vom Grad (s, t) wird gegeben durch

$$A(s, t) = s\gamma_1 r_1 + t\gamma_2 r_2,$$

wo $\gamma_1, \gamma_2, r_1, r_2$ wie oben angegeben durch C eindeutig bestimmte Zahlen sind.

Nun sei Γ_1 bzw. Γ_2 irreduzible algebraische Kurve in P_l bzw. P_m uud

$$\Gamma = \Gamma_1 \times \Gamma_2$$

das Produkt von Γ_1, Γ_2 in $P_{l,m}$. Es ist eine 2-dimensionale irreduzible Mannigfaltigkeit in $P_{l,m}$. Die Schnittkurven von Γ mit allgemeinen Hyperflächen $f(x, y, a) = 0$, $g(x, y, b) = 0$ bezeichnen wir mit C_a bzw. C_b. Diese Kurven sind über $k(a)$ bzw. $k(b)$ irreduzibel und ihr Schnitt ist eine über $k(a, b)$ irreduzible Punktgruppe[3], welche wir mit

(13) $[\Gamma, f(x, y, a), g(x, y, b)]$

---

1) Z. B. ist γ_1 die Schnittpunktsanzahl von C_1 mit einer allgemeinen Hyperebene in P_l. Wenn C_i nulldimensional ist, so setze man $\gamma_i = 0$. Vgl. v. d. Waerden, l.c. 3), S. 145.

2) r_1, r_2 haben eine wichtige Bedeutung in der Theorie der algebraischen Korrespondenzen. Vgl. die nachfolgende Arbeit des Verfassers.

3) Vgl. l.c. 3), S. 144.

bezeichnen. Jeder Schnittpunkt in (13) soll einfach gezählt werden. Das ist gerechtfertigt, da, wie gezeigt werden wird, jeder Punkt von (13) über $k(a, b)$ separabel ist. Zum Beweis unterscheiden wir verschiedene Fälle entsprechend den Graden (s, t), (s', t') von $f(x, y, a)$ und $g(x, y, b)$.

1) Fall $st>0$, $s't'>0$. (ξ, η) ist als Schnittpunkt von C_a mit $g(x, y, b)=0$ separabel über $\overline{k(a)}(b)$, wo $\overline{k(a)}$ eine algebraisch abgeschlossene Hülle von $k(a)$ bedeutet[1]. Ebenso ist es separabel über $\overline{k(b)}(a)$, wo $\overline{k(b)}$ eine algebraisch abgeschlossene Hülle von $k(b)$ ist. Daraus schliesst man sofort, dass (ξ, η) über $k(a, b)$ separabel ist.

2) Fall $st>0$, $s'>0$, $t'=0$. C_b ist dann die Vereinigung von $(\xi^{(\nu)})\times\Gamma_2$ $(\nu=1, \dots, a)$, wo $\xi^{(\nu)}$ Schnittpunkte von Γ_1 mit $g(x, b)=0$ bedeuten. Es sei also $\xi=\xi^{(1)}$. ξ ist bekanntlich über $k(b)$ separabel. η ist dann ein Schnittpunkt von $(\xi)\times\Gamma_2$ mit $f(\xi, y, a)=0$. Die Koeffizienten von y in $f(\xi, y, a)$ sind ersichtlich von einander algebraisch unabhängig über $k(\xi)$. $f(\xi, y, a)$ ist daher eine allgemeine Form von y. η ist also über $k(\xi, a)$ separabel und (ξ, η) ist folglich separabel über $k(a, b)$. Genau so erledigt man die Fälle $st>0$, $s'=0$, $t'>0$; $s>0$, $t=0$, $s't'>0$; $s=0$, $t>0$, $s't'>0$; $s=0$, $t>0$, $s'>0$, $t'=0$; $s>0$, $t=0$, $s'=0$, $t'>0$.

3) Fall $s=0$, $t>0$, $s'=0$, $t'>0$ oder $s>0$, $t=0$, $s'>0$, $t'=0$. In diesem Fall gibt es keinen Schnittpunkt.

Damit haben wir in allen Fällen bewiesen, dass (ξ, η) über $k(a, b)$ separabel ist.

Daraus und aus der Definition der Schnittpunktgruppe einer Kurve mit Hyperflächen, erhält man

$$(14)\quad [\Gamma, f(x, y, a), g(x, y, b)]=[C_a, g(x, y, b)]=[C_b, f(x, y, a)].$$

Nach Satz 4 folgt dann, dass $[\Gamma, f(x, y, a), g(x, y, b)]$ in die Summe von $[\Gamma, f(x, y, a), g_1(x, y, b_1)]$ und $[\Gamma, f(x, y, a), g_2(x, y, b_2)]$ spezialisiert wird, wenn man g ins Produkt g_1g_2 von allgemeinen Formen g_1 und g_2 spezialisiert.

Wir wollen nun die Vielfachheiten von irreduziblen Bestandteilen C_ν $(\nu=1, \dots, a)$ der Schnittkurve einer beliebigen Hyperfläche $f(x, y, a')$ mit Γ definieren. Um z.B. die Vielfachheit von C_1 zu bestimmen, nehme man eine $f(x, y, a')$ entsprechende allgemeine Form $f(x, y, a)$ und eine beliebige allgemeine Form $g(x, y, b)$, welche sicher mit C_1 Schnittpunkte besitzt. Man beweist dann leicht, dass $[\Gamma, f(x, y, a), g(x, y, b)]$ bei der Spezialisierung $a\to a'$ in eine Punktgruppe der Form

$$\sum_{\nu=1}^{a} \lambda_\nu[C_\nu, g(x, y, b)] \qquad (\lambda_\nu\geqq 0)$$

1) Vgl. die am Anfang gegebene Bemerkung.

übergeht, worin λ_1 sicher von Null verschieden ist. Ausserdem ist λ_1 von den Wahl von $g(x, y, b)$ unabhängig. Es seien nämlich $g_1(x, y, b_1)$, $g_2(x, y, b_2)$ beliebige C_1 schneidende allgemeine Formen und $g_3(x, y, b_3)$ eine allgemeine Form mit demselben Grad wie $g_1 g_2$. Ferner sei bei der Spezialisierung

$$[\Gamma, f(x, y, a), g_i(x, y, b_i)] \rightarrow \sum_{\nu=1}^{a} \lambda_{i\nu}[C_\nu, g_i(x, y, b_i)], \nu=1, 2, 3.$$

Es gilt aber bei $g_3 \rightarrow g_1 g_2$

$$[\Gamma, f(x, y, a), g_3(x, y, b_3)] \rightarrow [\Gamma, f(x, y, a), g_1(x, y, b_1)] + [\Gamma, f(x, y, a), g_2(x, y, b_2)],$$

$$[C_\nu, g_3(x, y, b_3)] \rightarrow [C_\nu, g_1(x, y, b_1)] + [C_\nu, g_2(x, y, b_2)], \nu=1, \dots, a.$$

Man erhält also wegen der Eindeutigkeit der Spezialisierung aus $[\Gamma, f(x, y, a), g_3(x, y, b_3)]$ durch $a \rightarrow a'$, $g_3 \rightarrow g_1 g_2$ dieselbe Punktgruppe

$$(15) \quad \sum_{\nu=1}^{a} \lambda_{3\nu}([C_\nu, g_1(x, y, b_1)] + [C_\nu, g_2(x, y, b_2)])$$
$$= \sum_{\nu=1}^{a} \lambda_{1\nu}[C_\nu, g_1(x, y, b_1)] + \sum_{\nu=1}^{a} \lambda_{2\nu}[C_\nu, g_2(x, y, b_2)].$$

Da $[C_\nu, g_i(x, y, b_i)]$ $(i=1, 2; \nu=1, \dots, a)$ wie leicht ersichtlich, einander frémde Punktgruppen sind, so bekommt man aus (15)

$$\lambda_{11} = \lambda_{21} = \lambda_{31},$$

wie behauptet wurde.

Die so für jede C_ν definierte Zahl λ_ν nennen wir die Vielfachheit von C_ν[1] und bezeichnen das mit diesen Vielfachheiten versehne Kurvensystem mit $[\Gamma, f(x, y, a')]$:

$$[\Gamma, f(x, y, a')] = \sum_{\nu=1}^{a} \lambda_\nu C_\nu.$$

Nun sei $g(x, y, b')$ eine beliebige Spezialisierung von $g(x, y, b)$. Es gilt nach der Definition der Schnittpunktgruppen

$$[C_a, g(x, y, b)] = [\Gamma, f(x, y, a), g(x, y, b)] \rightarrow [C_a, g(x, y, b')],$$
$$[C_\nu, g(x, y, b)] \rightarrow [C_\nu, g(x, y, b')]$$

bei $b \rightarrow b'$. Andererseits gilt auch nach der Definition

$$[\Gamma, f(x, y, a), g(x, y, b)] \rightarrow \sum_{\nu=1}^{a} \lambda_\nu [C_\nu, g(x, y, b)], \text{ bei } a \rightarrow a'.$$

Daraus folgt dann sofort der folgende

Satz 7. Für beliebige Formen $f(x, y, a')$, $g(x, y, b')$ gilt

$$[\Gamma, f(x, y, a'), g(x, y, b')] = [[\Gamma, f(x, y, a')], g(x, y, b')]$$
$$= [[\Gamma, g(x, y, b')], f(x, y, a')].$$

1) Vgl. v. d. Waerden, Zur algebraische Geometrie, VI, Math. Ann. 110 (1935).

[14] Zur Theorie der algebraischen Korrespondenzen I, Schnittpunktgruppen von Korrespondenzen

Proc. Japan Acad., 21-4 (1945), 204-212.

Mathematisches Institut, Kaiserliche Universität zu Tokio.
(Comm. by T. TAKAGI, M.I.A., April 26, 1945.)

Die Schnittpunktgruppen von algebraischen Korrespondenzen der algebraischen Kurven sind im klassischen Fall erst von F. Severi zweckmässig definiert worden[1]. In dieser Note wollen wir dieselben auch für Kurven über einem beliebigen algebraisch abgeschlossenen Grundkörper definieren und einige Eigenschaften davon ableiten. Wir folgen dabei durchaus der algebraisch-geometrischen Methode, welche von v.d. Waerden streng begründet worden ist[2].

Es sei also Γ_1 eine irreduzible singularitätenfreie algebraische Kurve in einem l-dimensionalen projektiven Raum P_l über einem beliebigen algebraisch abgeschlossenen Grundkörper k und Γ_2 eine ebensolche Kurve in einem m-dimensionalen Raum P_m über k. Es sei ferner

$$P_{l,m}=P_l\times P_m\,,\qquad \Gamma_{12}=\Gamma_1\times\Gamma_2\,.$$

$P_{l,m}$ ist ein zweifach projektiver Raum über k und Γ_{12} eine irreduzible 2-dimensionale Mannigfaltigkeit in $P_{l,m}$[3]. Algebraische Korrespondenzen zwischen Γ_1 und Γ_2 werden alsdann durch Systeme von endlichvielen, mit beliebigen Vielfachheiten versehenen irreduziblen Kurven über Γ_{12} gegeben. Wir wollen zunächst die Schnittpunktgruppe von zwei verschiedenen irreduziblen Korrespondenzen, d.h. die von zwei irreduziblen Kurven auf Γ_{12} definieren[4].

Es seien C, D verschiedene irreduzible Kurven auf Γ_{12}. Wir nehmen einen allgemeine $(l-2)$-bzw. $(m-2)$-dimensionalen linearen Raum $L^{(1)}_{l-2}$ bzw. $L^{(2)}_{m-2}$ in P_l bzw. P_m und einen beliebigen Punkt (a, b) in C. Verbindet man a mit einem Punkt a' in $L^{(1)}_{l-2}$ und b mit einem Punkt b' in $L^{(2)}_{m-2}$ und bezeichnet man diese Linien mit $L^{(1)}_1$, $L^{(2)}_1$, so erzeugt, wie ersichtlich, das Produkt

$$L_2=L^{(1)}_1\times L^{(2)}_1$$

---

1) F. Severi, Trattato di geometria algebrica, vol. 1, Bologna (1926). Siehe auch F. Severi, Über die Grundlagen der algebraischen Geometrie, Abh. Hamb., 9 (1933).

2) Vgl. eine Serie von Abhandlungen "Zur algebraischen Geometrie" in Math. Ann. Man vergleiche auch v.d. Waerden "Einführung in die algebraische Geometrie," Berlin (1939).

3) Vgl. K. Iwasawa, Der Bezoutsche Satz in zweifach projektiven Räumen, in diesen "Proceedings," zitiert mit "B". Wir benutzen im folgenden oft dieselbe Bezeichnungen oder Schreibweisen wie in "B".

4) Vgl. v. d. Waerden, Zur algebraischen Geometrie XIV, Math. Ann. 115 (1938).

eine irreduzible Hyperfläche H in $P_{l,m}$, wenn (a, b) in C, a' in $L^{(1)}_{l-2}$ und b' in $L^{(2)}_{m-2}$ von einander unabhängig durchlaufen. Als Multiplizität (Vielfachheit) eines Schnittpunktes von C, D nehmen wir die Multiplizität dieses Punktes in $[D, H]$[1] und bezeichnen die mit diesen Vielfachheiten versehene Schnittpunktgruppe von C, D mit

$$[C, D].$$

Für allgemeine

$$C=\sum_{i=1}^{\alpha}\lambda_i C_i, \quad D=\sum_{j=1}^{\beta}\mu_j D_j \quad (C_i, D_j\text{: irreduzibel und } C_i \neq D_j),$$

definieren wir

$$[C, D]=\sum_{i=1}^{\alpha}\sum_{j=1}^{\beta}\lambda_i\mu_j[C_i, D_j].$$

Wir nehmen C, D wieder als irreduzibel an und setzen mit obigen H

$$(1) \qquad [\Gamma_{12}, H]=\sum_{i=1}^{\alpha}\lambda_i C_i, \qquad C_1=C^{2)}.$$

Man beweist dann wie im gewöhnlichen Fall[3], dass kein von $L^{(1)}_{l-2}$, $L^{(2)}_{m-2}$ unabhängiger Punkt von $[\Gamma_{12}, H]$ in C_i $(i \geqq 2)$ liegt. Es gilt ferner

$$\lambda_1 = 1.$$

Zum Beweis nehmen wir zunächst an, dass C ausgeartet ist, d.h. dass $C=(a)\times\Gamma_2$ oder $C=\Gamma_1\times(b)$ ist. Im Fall $C=(a)\times\Gamma_2$ wird H ersichtlich durch $l(x, u')=0$ gegeben, wo $l(x, u')$ eine durch a hindurchgehende allgemeinste Form in P_l bedeutet. λ_1 ist dann nach Definition[4] die Multiplizität eines in C_1 enthaltenen Schnittpunktes von Γ_{12}, $l(x, u')=0$ mit einer allgemeinen Hyperfläche $\sum_{i=0}^{l}\sum_{j=0}^{m} c_{ij}x_iy_j=0$ in $P_{l,m}$. Nun gehen die Schnittpunkte von Γ_1 mit einer allgemeinen linearen Form $l(x, u)$ bei der Spezialisierung $u \rightarrow u'$ wegen der Singularitätenfreiheit von Γ_1 alle in verschiedene Schnittpunkte von Γ_1 mit $l(x, u')$ über. λ_1 ist also gleich der Multiplizität eines Schnittpunktes von Γ_2 mit $\sum_{j=0}^{m}(\sum_{i=0}^{l} c_{ij}a_i)y_j=0$, wenn $a=(a_0, a_1, \ldots\ldots, a_l)$ gesetzt wird[5]. Da aber $\sum_{j=0}^{m}(\sum_{i=0}^{l} c_{ij}a_i)y_j=0$ auch eine allgemeine Hyperebene in I_m ist, so folgt daraus $\lambda_1=1$ in bekannter Weise. Man erledigt ebenso den Fall $C=\Gamma_1\times(b)$.

Nun sei C nicht ausgeartet. Man nehme $l-1$ $L^{(1)}_{l-2}$ bestimmende allgemeine Punkte $a^{(1)}, \ldots\ldots, a^{(l-1)}$ und einen anderen allgemeinen Punkt q in P_l. Die von $L^{(1)}_{l-2}$ und q erzeugte Hyperebene in P_l wird dann durch

---

1) $[D, H]$ bedeutet die Schnittpunktgruppe von D mit H, vgl. "B".

2) $[\Gamma_{12}, H]$ bedeutet das Schnittkurvensystem von Γ_{12} mit H, vgl. "B".

3) Vgl. v. d. Waerden, l.c. 4).

4) Vgl. "B".

5) Vgl. "B". Man vergleiche auch den folgenden Schluss für den nicht ausgearteten Fall.

$$(2)\quad \begin{vmatrix} x_0, x_1, \ldots\ldots\ldots, x_l \\ q_0, q_1, \ldots\ldots\ldots, q_l \\ a_0^{(1)}\, a_1^{(1)}, \ldots\ldots, a_l^{(1)} \\ \vdots \\ a_0^{(l-1)}, a_1^{(l-1)}, \ldots, a_l^{(l-1)} \end{vmatrix} \equiv u_0x_0 + u_1x_1 + \ldots\ldots + u_lx_l = 0\,,$$

$$(3)\quad u_i = \pi_{i0}q_0 + \ldots + \pi_{i,i-1}q_{i-1} + \pi_{i,i+1}q_{i+1} + \ldots + \pi_{i,l}q_l \quad (i=0,1,\ldots,l)$$

gegeben, wobei $\pi_{i,j}$ die Plückerschen Koordinaten von $L_{l-2}^{(1)}$ sind. Ebenso wird die von $L_{m-2}^{(2)}$ und einem allgemeinen Punkt q' von P_m erzeugte Hyperebene von P_m durch

$$(4)\quad v_0y_0 + v_1y_1 + \ldots\ldots \quad + v_my_m = 0$$

$$(5)\quad v_i = \pi_{i0}'q_0' + \ldots\ldots + \pi_{i,i-1}'q_{i-1}' + \pi_{i,i+1}'q_{i+1}' + \ldots\ldots + \pi_{i,m}'q_m'$$

gegeben, wobei π'_{ij} wieder die Plückerschen Koordinaten von $L_{m-2}^{(2)}$ bedeuten. Da (2) eine allgemeine Hyperebene in $P_{l,m}$ gibt, so sind die Schnittpunkte von (2) mit C sämtlich allgemeine Punkte von C über k und einander konjugiert über $k(u)$. Wir nehmen einen solchen Punkt (ξ', η'). ξ' ist dann als Schnittpunkt von (2) mit der auf P_l projizierenden Bildkurve von C separabel über $k(u)$, also umsomehr über $k(\pi, q)$. Der Exponent p^{e}[1] von η' über $k(\pi, q)$ ist folglich gleich denjenigen von η' über $k(\xi', \pi, q)$, welcher andererseits dem Exponent von η' über $k(\xi')$ gleich ist[2]. Man bilde mit über $k(\pi, q)$ normierten Koordinaten[3]

$$(6)\quad F(q,q',\pi,\pi') = \prod_{\nu=1}^{n'} (v, \eta^{(\nu)})^{p^e} = \prod_{\nu=1}^{n'} (v_0q_0^{(\nu)} + v_1\eta_1^{(\nu)} + \ldots\ldots + v_m\eta_m^{(\nu)})^{p^e},$$

wobei $\eta^{(\nu)}$ alle verschiedenen konjugierten Punkte von η' über $k(\pi, q)$ durchläuft. Man beweist dann leicht, dass $F(q, q', \pi, \pi')$ eine irreduzible Form in $k[q, q', \pi, \pi']$ ist und dass jeder Punkt von $F(x, y, \pi, \pi')=0$ in H liegt. Da aber H und $F(x, y, \pi, \pi')=0$ beide irreduzible Hyperflächen in $P_{l,m}$ sind, so fallen sie zusammen, d.h. H wird durch $F(x, y, \pi, \pi')=0$ gegeben.

Um λ_1 zu bestimmen, schneiden wir nun $[\Gamma_{12}, F(x, y, \pi, \pi')]$ mit einer allgemeinen Hyperebene $l(x, c) \equiv c_0x_0 + c_1x_1 + \ldots\ldots + c_lx_l = 0$. Es gilt

$$(7)\quad [\Gamma_{12}, F(x, y, \pi, \pi'), l(x, c)] = \sum_{i=1}^{a} \lambda_i [C_i, l(x, c)]\,.$$

Wir nehmen einen Punkt (ξ, η) aus $[C_1, l(x, c)]$. Es ist ein allgemeiner Punkt von C_1 über k und seine Vielfachheit in $[C_1, l(x, c)]$ ist gleich dem Exponent von η über $k(\xi)$[4]. Dieser ist aber gerade p^e, da (ξ, η) und (ξ', η') als allgemeiner Punkt von C_1 dieselbe algebraische Eigenschaften über k besitzen. Bezeichnet man mit $F(x, y, w)$ die $F(x, y, \pi, \pi')$ entsprechende allgemeine Form, so

1) Wir bezeichnen die Charakteristik von k mit p, wie in "B"
2) Vgl. "B".
3) Vgl. "B".
4) Vgl. "B", Satz 2.

besagt (7), dass genau $\lambda_1 p^e$ Punkte von $[\Gamma, F(x, y, w), l(x, c)]$ bei der Spezialisierung $F(x, y, w) \rightarrow F(x, y, \pi, \pi')$ in (ξ, η) übergehen. Die Punkte von $[\Gamma, F(x, y, w), l(x, c)]$ erhält man folgendermassen: zunächst suche man alle Schnittpunkte $\xi^{(1)}, \xi^{(2)}, \ldots\ldots, \xi^{(\sigma)}$ von Γ_1 mit $l(x, c)=0$ und dann für jeden $\xi^{(\nu)}$ die Schnittpunkte von Γ_2 mit $F(\xi^{(\mu)}, y, w)=0$. Es muss also z.B. $\xi=\xi^{(1)}$ sein und $\lambda_1 p^e$ ist die Vielfachheit von η als Schnittpunkt von Γ_2 mit $F(\xi, y, \pi, \pi')=0$. Gehen nun die Punkte $\eta^{(\nu)}$ in (6) bei der Spezialisierung $q \rightarrow \xi$, $\xi' \rightarrow \xi$ relationstreu zu $\eta'^{(\nu)}$ über, so gilt

$$(8) \qquad F(\xi, y, \pi, \pi')=\prod_{\nu=1}^{n'}(v', \eta'^{(\nu)})^{pe}=\prod_{\nu=1}^{n'}(v'_0\eta'^{(\nu)}_0+v'_1\eta'^{(\nu)}_1+\ldots+v'_m\eta'^{(\nu)}_m)^{pe}$$

mit

$$(9) \qquad v'_i=\pi'_{i0}y_0+\ldots\ldots+\pi'_{i,i-1}y_{i-1}+\pi'_{i,i+1}y_{i+1}+\ldots\ldots+\pi'_{i,l}y_l\,.$$

Da die durch q und $L^{(1)}_{l-2}$ hindurchgehende Hyperebene (2), folglich auch die Schnittpunkte $\eta^{(\nu)}$ von (2) und C_1 bei der Spezialisierung $q \rightarrow \xi'$, $\xi' \rightarrow \xi'$ fest bleiben und da ξ dieselbe algebraische Eigenschaften wie ξ' hat, so sind $\eta'^{(\nu)}$ alle von einander verschieden. η fällt ersichtlich mit einem von $\eta'^{(\nu)}$ zusammen, z.B. $\eta=\eta'^{(1)}$. In (8) ist $(v', \eta'^{(\nu)})=0$ die von $\eta'^{(\nu)}$ und $L^{(2)}_{m-2}$ erzeugte Hyperebene in P_m, d.h. eine durch $\eta'^{(\nu)}$ hindurchgehende allgemeinste Hyperebene in P_m. Die von $\eta'^{(\nu)}$ verschiedene Schnittpunkte von $(v', \eta'^{(\nu)})=0$ mit Γ_2 sind also sämtlich von $L^{(2)}_{m-2}$ abhängig und η fällt folglich mit keinem von ihnen zusammen. Ferner ist $\eta=\eta'^{(1)}$ als Schnittpunkt von $(v', \eta'^{(1)})=0$ mit Γ_2 einfach gezählt, da Γ_2 eine singularitätenfreie Kurve ist. Hieraus und aus (8) sieht man sofort, dass die Vielfachheit von η als Schnittpunkt von Γ_2 mit $F(\xi, y, \pi, \pi')=0$ gleich p^e ist. Nach dem oben Bemerkten ist es andererseits gleich $\lambda_1 p^e$. Es ist also $\lambda_1=1$, wie behauptet wurde.

Es sei noch bemerkt, dass in diesem nicht ausgearteten Fall alle $C_i (i \geqq 2)$ in (1) auch nicht ausgeartet sind. Wäre nämlich z.B. $C_2=(a)\times\Gamma_2$ und $\eta^{(\nu)} \rightarrow \bar{\eta}^{(\nu)}$ bei $q \rightarrow a$, so würde die Hyperfläche

$$F(a, y, \pi, \pi')=\prod_{\nu=1}^{n'}(v', \bar{\eta}^{(\nu)})^{pe}=0$$

Γ_2 enthalten, was aber offenbar ein Widerspruch ist, da $(v', \bar{\eta}^{(\nu)})=0$ eine durch $\bar{\eta}^{(\nu)}$ hindurchgehende allgemeinste Hyperebene in P_m ist und sicher nicht Γ_2 enthält. Andererseits ist es auch klar, dass $C_i (i \geqq 2)$ sämtlich ausgeartet sind, wenn C_1 eine ausgeartete Kurve ist.

Nun gilt der

Satz 1. Es sei $g(x, y, b')=0$ eine beliebige Hyperfläche in $P_{l,m}$ und C eine beliebige Korrespondenz auf Γ_{12}. Es gilt dann

$$[C, [\Gamma_{12}, g(x, y, b')]] = [C, g(x, y, b')]^{1)}.$$

Beweis. Man kann offenbar C als irreduzibel annehmen. Die C mit $L_{l-2}^{(1)} \times L_{m-2}^{(2)}$ verbindende Hyperfläche in $P_{l,m}$ wird wie oben mit $H\ (l(x, u')=0$ oder $F(x, y, \pi, \pi')=0)$ bezeichnet. Aus $\lambda_1=1$ folgt

$$[\Gamma_{12}, H] = C_1 + \sum_{i=2}^{\alpha} \lambda_i C_i, \qquad (C_1 = C)$$

und es gilt[2]

$$[[\Gamma_{12}, g(x, y, b')], H] = [[\Gamma_{12}, H], g(x, y, b')]$$
$$= [C_1, g(x, y, b')] + \sum_{i=2}^{\alpha} \lambda_i [C_i, g(x, y, b')].$$

Da jeder Punkt von C_i $(i \geqq 2)$ von $L_{l-2}^{(1)} \times L_{m-2}^{(2)}$ abhängt, so liegen die (von $L_{l-2}^{(1)} \times L_{m-2}^{(2)}$ unabhängige) Schnittpunkte von C und $[\Gamma_{12}, g(x, y, b')]$ nur in $[C_1, g(x, y, b')]$. Nach Definition von $[C, [\Gamma_{12}, g(x, y, b')]]$ folgt daraus die Behauptung.

Satz 2. Für beliebige Korrespondenzen C, D gilt

$$[C, D] = [D, C].$$

Beweis. Wir können natürlich C, D als irreduzibel annehmen. Bezeichnet man die D mit einem allgemeinen linearen Raum $L'^{(1)}_{l-2} \times L'^{(2)}_{m-2}$ verbindende Hyperfläche in $P_{l,m}$ mit H' und setzt

$$[\Gamma_{12}, H'] = D_1 + \sum_{j=2}^{\beta} \mu_j D_j, \qquad (D_1 = D),$$

so folgt aus Satz 1

$$(10) \qquad [C, H'] = [C, [\Gamma_{12}, H']] = [C, D_1] + \sum_{j=2}^{\beta} \mu_j [C, D_j].$$

Nach Definition besitzt ein Schnittpunkt in $[D, C]$ dieselbe Vielfächheit wie in $[C, H']$. Da aber jeder Punkt von $\sum_{j=2}^{\beta} \mu_j [C, D_j]$ von $L'^{(1)}_{l-2} \times L'^{(2)}_{m-2}$ abhängt, so folgt aus (10) $[C, D] = [D, C]$.

Nun untersuchen wir $[C, D]$ in dem Fall, wo $C = (a) \times \Gamma_2$ und D irreduzibel und nicht von der Form $(a') \times \Gamma_2$ ist. H ist dann eine durch a hindurchgehende allgemeinste Hyperebene $l(x, u')=0$. Für eine $l(x, u')$ entsprechende allgemeine Form $l(x, u)$ gilt dann[3]:

$$(11) \qquad [D, l(x, u)] \to [D, l(x, u')], \text{ bei } u \to u'$$

und man erhält $[D, l(x, u)]$ folgendermassen: zunächst nehme man alle Schnittpunkte $\xi^{(1)}, \xi^{(2)}, \ldots\ldots, \xi^{(\rho)}$ von der auf P_l projizierende Bildkurve von D mit $l(x, u)=0$ in P_l und für jeden $\xi^{(\mu)}$ je einen Punkt $\eta^{(\mu)}$ in P_m, so dass $(\xi^{(\mu)}, \eta^{(\mu)})$

1) Für allgemeine Korrespondenz $C = \Sigma \lambda_i C_i$ soll $[C, g(x, y, b')] = \Sigma \lambda_i [C_i, g(x, y, b')]$ sein.

2) Vgl. "B", Satz 7.

3) Vgl. (10) in "B".

ein allgemeiner Punkt von D wird. $[D, l(x, u)]$ wird dann gegeben durch

$$\{(\xi^{(\mu)}, \eta^{(\mu,\nu)}), \quad \mu=1, 2,\ldots\ldots, g, \quad \nu=1, 2,\ldots\ldots, \gamma\}$$

wobei $\gamma=[k(\xi^{(\mu)}, \eta^{(\mu)}) : k(\xi^{(\mu)})]$ $(\mu=1, 2,\ldots\ldots, g)$ gesetzt wird und $\eta^{(\mu,\nu)}$ alle nicht notwendig verschiedenen Konjugierten von $\eta^{(\mu)}$ in bezug auf $k(\xi^{(\mu)})$ durchläuft. Wegen der Singularitätenfreiheit von Γ_1 geht bei (11) nur ein Punkt von $\xi^{(\mu)}$ zu a über: z.B. $\xi^{(1)}\to a$. Gehen dabei zugleich $\eta^{(1,\nu)}$ $(\nu=1,2,\ldots,\gamma)$ zu $b^{(\nu)}$ über, so folgt nach Definition

$$[C, D]=\{(a, b^{(\nu)}); \quad \nu=1,2,\ldots\ldots, \gamma\}. \tag{12}$$

Damit ist bewiesen der

Satz 3. Es sei D eine irreduzible Korrespondenz, welche nicht die Form $(a')\times\Gamma_2$ hat und (ξ, η) ein allgemeiner Punkt von D. Die sämtlichen (nicht notwendig von einander verschiedenen) Konjugierten von η in bezug auf $k(\xi)$ seien $\eta^{(1)},\ldots\ldots, \eta^{(\gamma)}$ $(\gamma=[k(\xi, \eta) : k(\xi)])$[1]. Gehen $\eta^{(\mu)}$ $(\mu=1,\ldots\ldots, \gamma)$ bei der Spezialisierung $\xi\to a$ relationstreu zu $b^{(\mu)}$ $(\mu=1,\ldots\ldots, \gamma)$ über, so wird die Schnittpunktgruppe von D mit $C=(a)\times\Gamma_2$ durch

$$[C, D]=\{(a, b^{(\mu)}); \mu=1,\ldots\ldots, \gamma\}$$

gegeben. In analoger Weise erhält man die Schnittpunktgruppe von D mit $\Gamma_1\times(b)$, falls D nicht von der Form $\Gamma_1\times(b')$ ist.

Satz 4. Setzt man für eine beliebige Form $g(x, y, b')$ und einen beliebigen Punkt a in P_l

$$C=(a)\times\Gamma_2, D=[\Gamma_{12}, g(x, y, b')], \quad [\Gamma_2, g(a, y, b')]=\{b^{(\mu)}; \mu=1,\ldots, \delta\},$$

so gilt

$$[C, D]=\{(a, b^{(\mu)}); \quad \mu=1,\ldots\ldots, \delta\}.$$

Beweis. Nach Satz 1 gilt

$$[C, [\Gamma_{12}, g(x, y, b')]]=[C, g(x, y, b')]. \tag{13}$$

Bezeichnet man die $g(x, y, b')$ entsprechende allgemeine Form mit $g(x, y, b)$, so sieht man leicht

$$[C, g(x, y, b)]=\{(a, b^{*(\mu)}); \quad b^{*(\mu)}\in[\Gamma_2, g(x, y, b)]\}.$$

Durch die Spezialisierung $b\to b'$ erhält man (13).

Nun beweisen wir eine wichtige Eigenschaft von der oben definierten Schnittpunktgruppe, nämlich deren birationale Invarianz. Es seien also $\varphi^{(1)}, \varphi^{(2)}$, $\varphi=(\varphi^{(1)}, \varphi^{(2)})$ birationale Abbildungen, welche Γ_1, Γ_2 bzw. $\Gamma_{12}=\Gamma_1\times\Gamma_2$ je auf andere ebenso singularitätenfreie Γ'_1, Γ'_2 bzw. $\Gamma'_{12}=\Gamma'_1\times\Gamma'_2$ abbilden. Eine irreduzible Korrespondenz C auf Γ_{12} wird dabei auf eine wieder irreduzible Korrespondenz $C'=\varphi(C)$ abgebildet. Für eine beliebige Korrespondenz $\Sigma\lambda_i C_i$ setzen wir $\varphi(\Sigma\lambda_i C_i)=\Sigma\lambda_i\varphi(C_i)$. Es gilt dann

1) Die Endlichkeit von γ folgt daraus, dass D nicht die Form $(a')\times\Gamma_2$ hat.

Satz 5. Für beliebige Korrespondenzen C, D auf Γ_{12} gilt

$$[\varphi(C), \varphi(D)] = \varphi([C, D]).$$

Beweis. Die Abbildung $\varphi = (\varphi^{(1)}, \varphi^{(2)})$ und ihre Umkehrung $\psi = (\psi^{(1)}, \psi^{(2)})$ seien durch

$$(14) \quad x'_0 : x'_1 : \ldots\ldots : x'_l = \varphi^1_0(x) : \varphi^1_1(x) : \ldots\ldots : \varphi^1_l(x);$$
$$y'_0 : y'_1 : \ldots\ldots : y'_m = \varphi^2_0(y) : \varphi^2_1(y) : \ldots\ldots \varphi^2_m(y),$$

$$(15) \quad x_0 : x_1 : \ldots\ldots : x_l = \psi^1_0(x') : \psi^1_1(x') : \ldots\ldots \psi^1_l(x'),$$
$$y_0 : y_1 : \ldots\ldots : y_m = \psi^2_0(y') : \psi^2_1(y') : \ldots\ldots : \psi^2_m(y').$$

gegeben. Man hat offenbar den Satz nur für irreduzible C, D zu beweisen. Wenn C oder D ausgeartet ist, so folgt der Satz unmittelbar aus Satz 3, da die Abbildung φ auf Γ_{12} eineindeutig ist. Wir nehmen also an, dass C, D beide irreduzibel und nicht ausgeartet sind und schicken zum Beweis einige Hilfssätze voraus.

Hilfssatz 1. Es sei $F(x, y, w)$ eine allgemeine Form vom Grade (s, t), $st > 0$ und (a, b) ein beliebiger Punkt in $[\Gamma_{12}, F(x, y, w)]$. Dann hängt entweder a oder b von w ab. Dasselbe gilt auch für einen beliebigen Punkt in $\varphi([\Gamma_{12}, F(x, y, w)])$.

Beweis. Die erste Hälfte ist ohne weiters klar. Die zweite folgt aus $k(a) = k(\varphi^{(1)}(a))$, $k(b) = k(\varphi^{(2)}(b))$.

Hilfssatz 2. $F(x, y, w)$ sei dieselbe Form wie in Hilfssatz 1. Setzt man

$$F^*(x', y', w) = F(\psi^{(1)}(x'), \psi^{(2)}(y'), w),$$

so fällt $\varphi([D, F(x, y, w)])$ für eine (von w unabhängige) nicht ausgeartete irreduzible Korrespondenz D auf Γ_{12} genau mit dem von w abhängigen Teil von $[\varphi(D), F^*(x', y', w)]$ zusammen.

Beweis. Es seien $(\xi'^{(1)}, \eta'^{(1)}), \ldots\ldots, (\xi'^{(\alpha)}, \eta'^{(\alpha)})$ die von w abhängige Punkte von $[\varphi(D), F^*(x', y', w)]$. Diese Punkte sind nach W.-L.Chow[1] einander konjugiert über $k(w)$ und fallen je p^e zusammen, falls der Exponent von $k(\xi'^{(\nu)}, \eta'^{(\nu)}, w)/k(w)$ p^e ist. Ferner liegt weder $\xi'^{(\mu)}$ noch $\eta'^{(\mu)}$ in k, da $(\xi'^{(\mu)}, \eta'^{(\mu)})$ ein allgemeiner Punkt der nicht ausgearteten Kurve $\varphi(D)$ ist. $\{\psi^{(1)}_i(\xi'^{(\mu)}); i = 0, 1, \ldots\ldots, l\}$ und $\{\psi^{(2)}_j(\eta'^{(\mu)}); j = 0, 1, \ldots\ldots, m\}$ geben also die Koordinaten von $\psi^{(1)}(\xi'^{(\mu)}), \psi^{(2)}(\eta'^{(\mu)})$ und hieraus folgt sogleich, dass $(\psi^{(1)}(\xi'^{(\mu)}), \psi^{(2)}(\eta'^{(\mu)}))$ in $[D, F(x, y, w)]$ liegt, d.h. dass $(\xi'^{(\mu)}, \eta'^{(\mu)})$ in $\varphi([D, F(x, y, w)])$ liegt. Umgekehrt zeigt man in analoger Weise, dass ein beliebigen Punkt $\varphi((\xi, \eta))$ von $\varphi([D, F(x, y, w)])$ in $[\varphi(D), F^*(x', y', w)]$ liegt. Dieser ist aber nach Hilfssatz 1 von w abhängig, also fällt mit einem von $(\xi'^{(\mu)}, \eta'^{(\mu)})$

1) W.-L. Chow, Die geometrische Theorie der algebraischen Funktionen für beliebige voliebige vollkommene Körper, Math. Ann. 114 (1937).

zusammen: z.B. $\varphi((\xi,\eta))=(\xi'^{(1)},\eta'^{(1)})$. Es bleibt also nur zu zeigen $p^e=1$, da jeder Punkt in $[D, F(x,y,w)]$ einfach gezählt ist[1]. Aus $\varphi((\xi,\eta))=(\xi'^{(1)},\eta'^{(1)})$ folgt aber $k(\xi'^{(1)},\eta'^{(1)},w)=k(\xi,\eta,w)$ und es muss $p^e=1$ sein, da $k(\xi'^{(1)}, \eta'^{(1)},w)/k(w)=k(\xi,\eta,w)/k(w)$ separabel ist[2].

Hilfssatz 3. $F(x,y,w)$, $F^*(x',y',w)$ seien dieselbe Formen wie in Hilfssatz 2. Es gilt dann

$$(16)\quad [\Gamma'_{12}, F^*(x',y',w)]=\varphi([\Gamma, F(x,y,w)])+\Sigma\lambda^i((a'^{(i)})\times\Gamma'_2) + \Sigma\mu_j(\Gamma'_1\times(b'^{(j)}))\,,$$

wobei $a'^{(i)}$ bzw. $b'^{(j)}$ je

$$\psi^1_0(a'^{(i)})=\psi^1_1(a'^{(i)})=\ldots\ldots=\psi^1_l(a'^{(i)})=0 \text{ bzw. } \psi^2_0(b'^{(j)})=\psi^2_1(b'^{(j)})=\ldots\ldots=\psi^2_m(b'^{(j)})=0$$

genügende Punkte in $P_{l'}$ bzw. $P_{m'}$ sind.

Beweis. Um $[\Gamma'_{12}, F^*(x',y',w)]$ zu bestimmen, schneiden wir Γ'_{12} mit einer allgemeinen Hyperfläche $\Sigma v_{ij}x'_iy'_j=0$ und wenden Hilfssatz 2 auf $D=\psi([\Gamma'_{12},\Sigma v_{ij}x'_iy'_j])$ an. Der von w abhängige Teil von $[\psi(D), F^*(x',y',w)] =[[\Gamma'_{12},\Sigma v_{ij}x'_iy'_j], F^*(x',y',w)]=[[\Gamma'_{12}, F^*(x',y',w)], \Sigma v_{ij}x'_iy'_j]$ liegt auf $\varphi([\Gamma_{12}, F(x,y,w)])$ und jeder Punkt in ihm ist einfach gezählt. Nimmt man nun ein von w unabhängiger Punkt (a',b') von $[[\Gamma'_{12},\Sigma v_{ij}x'_iy'_j], F^*(x',y',w)]$, so liegt es nach Hilfssatz 1 nicht in $\varphi([\Gamma_{12}, F(x,y,w)])$. Ausserdem gilt entweder $\psi^1_1(a')=\psi^1_1(a')=\ldots=\psi^1_l(a')=0$ oder $\psi^2_0(b')=\psi^2_1(b')=\ldots=\psi^2_m(b')=0$ denn sonst würde $\{\psi^1_i(a'),\psi^2_j(b')\}$ die Koordinaten von $\psi((a',b'))$ geben und $\psi((a',b'))$ würde auf $[\Gamma_{12}, F(x,y,w)]$, folglich (a',b') auf $\varphi([\Gamma_{12}, F(x,y,w)])$ liegen, entgegen der obigen Bemerkung. Nach Definition[3] von $[\Gamma'_{12}, F^*(x',y',w)]$ erhält man dann sofort (16).

Hilfssatz 4. $F(x,y,w)$, $F^*(x',y',w)$ und D mögen dieselbe Bedeutungen haben wie in Hilfssatz 3. Es gilt dann

$$(17)\quad \varphi([D, F(x,y,w)])=[\varphi(D), \varphi([\Gamma_{12}, F(x,y,w)])]\,.$$

Beweis. Aus (16) folgt.

$$\begin{aligned}(18)\quad [\varphi(D), F^*(x',y',w)]&=[\varphi(D), [\Gamma'_{12}, F^*(x',y',w)]]\\ &=[\varphi(D), \varphi([\Gamma_{12}, F(x,y,w)])]\\ &\quad+\Sigma\lambda_i[\varphi(D), (a'^{(i)})\times\Gamma'_2]\\ &\quad+\Sigma\mu_j[\varphi(D), \Gamma'_1\times(b'^{(j)})]\end{aligned}$$

Da aber Punkte in $[\varphi(D), \varphi([\Gamma_{12}, F(x,y,w)])]$ nach Hilfssatz 1 von w abhängig, dagegen Punkte in $[\varphi(D), (a'^{(1)})\times\Gamma'_2]$ oder in $[\varphi(D), \Gamma'_1\times(b'^{(j)})$ von w unabhängig sind, so folgt aus Hilfssatz 2 und (18) die Behauptung.

1) Vgl. "B". Man bemerke, dass $F(x,y,w)$ einen Grad (s,t), $st>0$ hat.
2) Vgl. "B".
3) Vgl. "B".

Nun wollen wir zum Beweis von Satz 5 zurückkehren. Es seien also C, D nicht ausgeartete irreduzible Kurven auf Γ_{12}. Die C mit einem allgemeinen $L_{l-2}^{(1)} \times L_{m-2}^{(2)}$ verbindende Hyperfläche bezeichnen wir wie vorher mit $F(x, y, \pi, \pi')=0$ und die $F(x, y, \pi, \pi')$ entsprechende allgemeine Form mit $F(x, y, w)$. Die Multiplizitat ϵ eines Punktes (a, b) in $[C, D]$ ist mit derjenigen von (a, b) in $[D, F(x, y, \pi, \pi')]$ identisch. Bei der Spezialisierung $F(x, y, w) \to F(x, y, \pi, \pi')$ gehen also genau ϵ Punkte von $[D, F(x, y, w)]$ in (a, b) über, folglich nach (17) genau ϵ Punkte von $[\varphi(D), \varphi([\Gamma_{12}, F(x, y, w)])]$ in $\varphi((a, b))$ über[1]. Wenn $[\Gamma'_{12}, F^*(x', y', w)]$ bei derselben Spezialisierung in $\Sigma v_\rho C'_\rho + \Sigma \lambda_i((a'^{(i)}) \times \Gamma'_2) + \Sigma \mu_j(\Gamma'_1 \times (b'^{(j)}))$ also $\varphi([\Gamma_{12}, F(x, y, w)])$ in $\Sigma v_\rho C'_\rho$ übergeht, so ist die Anzahl von $\varphi((a, b))$ in $\Sigma \nu_\rho(\varphi(D), C'_\rho)$ gleich ϵ. Da aber $\varphi(C)$ sicher in C'_ρ auftritt, so muss die Vielfachheit von $\varphi((a, b))$ in $[\varphi(C), \varphi(D)] = [\varphi(D), \varphi(C)]$ nicht grösser als ϵ sein. Es ist also bewiesen, dass $[\varphi(C), \varphi(D)]$ in $\varphi([C, D])$ enthalten ist. Genau so beweist man, dass $[\psi(\varphi(C)), \psi(\varphi(D))] = [C, D]$ in $\psi([\varphi(C), \varphi(D)])$, folglich $\varphi([C, D])$ in $\varphi(\psi([\varphi(C), \varphi(D)])) = [\varphi(C), \varphi(D)]$ enthalten ist. Es gilt also

$$\varphi([C, D]) = [\varphi(C), \varphi(D)] .$$

Satz 6. $\Gamma_{12}, \Gamma'_{12}, \varphi, \psi$ mögen dieselbe Bedeutungen haben wie oben. Es sei ferner $F(x, y, w)$ eine allgemeine Form von Grade $(s, t), st > 0$ und $F(x, y, w')$ eine beliebige Spezialisierung von $F(x, y, w)$. Setzt man

$$F^*(x', y', w') = F(\psi^{(1)}(x'), \psi^{(2)}(y'), w') ,$$

so gilt

$$(19) \qquad [\Gamma'_{12}, F^*(x', y', w')] = \varphi([\Gamma, F(x, y, w')]) + \sum_{i=1}^{\tau} \lambda_i C'_i ,$$

mit geeigneten ausgearteten Korrespondenzen C'_i.

Beweis. Nach Satz 5 und (18) gilt

$$\begin{aligned} [\varphi(D), [\Gamma'_{12}, F^*(x', y', w)]] &= [\varphi(D), \varphi([\Gamma_{12}, F(x, y, w)])] + \Sigma \lambda_i [\varphi(D), C'_i] \\ &= \varphi([D, [\Gamma_{12}, F(x, y, w)]]) + \Sigma \lambda_i [\varphi(D), C'_j] . \end{aligned}$$

Spezialisiert man $w \to w'$, so ist

$$\begin{aligned} [\varphi(D), [\Gamma'_{12}, F^*(x', y', w')]] &= \varphi([D, [\Gamma_{12}, F(x, y, w')]]) + \Sigma \lambda_i [\varphi(D), C'_i] \\ &= [\varphi(D), \varphi([\Gamma_{12}, F(x, y, w')])] + \Sigma \lambda_i [\varphi(D), C'_i] \\ &= [\varphi(D), \varphi([\Gamma_{12}, F(x, y, w')]) + \Sigma \lambda_i C'_i] \end{aligned}$$

Daraus erhalt man (19), da $\varphi(D)$ eine beliebige irreduzible Kurve auf Γ'_{12} sein kann.

---

1) Man bemerke dabei, dass $\varphi(\xi)$ bei der Spezialisierung $\xi \to a$ immer zu $\varphi(a)$ übergeht.

[15] Zur Theorie der algebraischen Korrespondenzen II, Multiplikation der Korrespondenzen

Proc. Japan Acad., 21-9 (1945), 411-418.

Mathematisehes Institut, Kaiserliche Universität zu Tokio.

(Comm. by T. TAKAGI, M.I.A., Nov. 12, 1945.)

In dieser Note wollen wir die Multiplikation der algebraischen Korrespondenzen auf algebraischen Kurven definieren und einige Sätze darüber herleiten. Der Multiplikatorenmodul und der Multiplikatorenring der algebraischen Kurve sollen auch eingeführt werden. Die Bezeichnungen wie k, P_l, P_m, $P_{l,m}$, Γ_1, Γ_2, Γ_{12} u.s.w. mögen dieselbe Bedeutungen haben wie im Teil I[1], wenn nicht das Gegenteilige angemerkt wird.

Zunächst führen wir noch einige Bezeichnungen ein. Für eine irreduzible Korrespondenz C auf Γ_{12} und einen beliebigen Punkt a in Γ_1 definieren wir eine Punktgruppe $C^{(2)}(a)$ in Γ_2 mit

$$C^{(2)}(a)=\{b^{(\mu)};\ (a, b^{\mu})\in[(a)\times\Gamma_2, C]\},^{2)} \quad \text{falls } C\neq(a)\times\Gamma_2,$$

$$C^{(2)}(a)=\text{die Nullgruppe, falls } C=(a)\times\Gamma_2.$$

Allgemein setze man für eine beliebige Korrespondenz $C=\Sigma\lambda_i C_i$ (C_i: irreduzibel) auf Γ_{12} und eine beliebige Punktgruppe $G_1=\Sigma\mu_j a^{(j)}$ in Γ_1

$$C^{(2)}(G_1)=\Sigma\lambda_i\mu_j C_i^{(2)}(a^{(j)}).$$

Analog definieren wir $C^{(1)}(b)$ bzw. $C^{(1)}(G_2)$ für einen Punkt b bzw. eine Punktgruppe G_2 in Γ_2. Es gilt dann offenbar

$$C^{(2)}(G_1)\to C^{(2)}(G_1'),\ C^{(1)}(G_2)\to C^{(1)}(G_2'),$$

wenn man $G_i (i=1, 2)$ in G_i' spezialisiert.

Nun beweisen wir den

Satz 1. Es gilt für jeden Punkt a in Γ_1

$$C^{(2)}(a)=D^{(2)}(a)$$

genau dann, wenn die Korrespondenzen C, D bis auf Korrespondenzen der Form $(a^{(j)})\times\Gamma_2$ mit einander zusammenfallen. Analoges gilt für $C^{(1)}$, $D^{(1)}$.

Beweis. Es sei

$$C=\Sigma\lambda_i C_i,\ D=\Sigma\mu_j D_j$$

und es gelte für einen allgemeinen Punkt ξ von Γ_1

$$\Sigma\lambda_i C_i^{(2)}(\xi)=\Sigma\mu_j D_j^{(2)}(\xi).$$

---

1) Zur Theorie der algebraischen Korrespondenzen I, Proc. Imp. Acad. Jap. Vol. 21 (1945), zitiert im folgenden mit I.

2) [C, D] bedeutet die Schnittpunktgruppe von C, D, vgl. I. (α; E) bedeutet die Menge von α, welches die Eigenschaft E besitzt.

Wenn C_i nicht die Form $(a^{(i)})\times\Gamma_2$ hat, so wird C_i nach Satz 3 in I durch $C_i^{(2)}(\xi)=[(\xi)\times\Gamma_2, C_i]$ eindeutig bestimmt. Dasselbe gilt natürlich auch für D_j und daraus folgt offenbar die Behauptung.

Nun sei Γ_3 eine singularitätenfreie irreduzible Kurve in einem n-dimensionalen projektiven Raum P_n über k. Wir setzen

$$P_{l,n}=P_l\times P_n,\ P_{m,n}=P_m\times P_n,$$
$$\Gamma_{1,3}=\Gamma_1\times\Gamma_3,\ \Gamma_{2,3}=\Gamma_2\times\Gamma_3.$$

Für eine beliebige Korrespondenz $C=C_{12}$ auf Γ_{12} und ein $D=D_{23}$ auf Γ_{23} wollen wir eine solche Korrespondenz $E=E_{13}$ auf Γ_{13} suchen, so dass es für beliebige Punkte a, c in Γ_1 bzw. Γ_3 immer

$$(1)\qquad E_{13}^{(3)}(a)=D_{23}^{(3)}(C_{12}^{(2)}(a)),\ E_{13}^{(1)}(c)=C_{12}^{(1)}(D_{23}^{(2)}(c))$$

gilt. Es gibt, wie wir zeigen werden, eine einzige solche Korrespondenz E_{13} und wir heissen diese Korrespondenz das Produkt von C und D:

$$E_{13}=C_{12}\times D_{23}$$

Die Eindeutigkeit folgt ohne Weiters aus Satz 1. Beim Beweis der Existenz kann man C, D als irreduzibel annehmen, denn für allgemeine $C=\Sigma\lambda_i C_i$, $D=\Sigma\mu_j D_j$ hat man nur zu setzen

$$E=C\times D=\Sigma\lambda_i\mu_j(C_i\times D_j),$$

wenn $C_j\times D_j$ schon bestimmt ist. Wenn eines von C, D ausgeartet ist, so können wir ohne Weiters ein solches E angeben. Z.B. setze man für $C=\Gamma_1\times(b)$

$$E=C\times D=\Sigma(\Gamma_1\times(c^{(j)})),$$

wobei $c^{(j)}$ alle Punkte in $D^{(3)}(b)$ durchläuft. Man bemerke dabei, dass E in diesem Fall immer nur aus ausgearteten irreduziblen Korrespondenzen besteht und dass in gewissen Fallen die Nullkorrespondenz θ_{13} auftritt; es gilt z.B.

$$((a)\times\Gamma_2)\times(\Gamma_2\times(c))=\theta_{13}.$$

Wir nehmen also an, dass C, D beide irreduzible nicht ausgeartete Korrespondenzen sind und beweisen die Existenz von $E=C\times D$. Wir bezeichnen mit $f(x, y, u)=0$ diejenige Hyperfläche in $P_{l,m}$, die C mit einem allgemeinen linearen Raum $L_{l-2}^{(1)}\times L_{m-2}^{(2)}$ verbindet und mit $g(u, z, v')$[3] $=0$ die Hyperfläche in $P_{m,n}$, die D mit einem ebensolchen Raum $L_{m-2}'^{(2)}\times L_{n-2}'^{(3)}$ verbindet. Es gilt dann

$$[\Gamma_{12}, f(x, y, u')]=C+\sum_{i=2}^{\alpha}\lambda_i C_i,\ [\Gamma_{23}, g(x, y, v')]=D+\sum_{j=2}^{\beta}\mu_j D_j,$$

wobei $C_i(i\geqq 2)$, $D_j(j\geqq 2)$ je von $L_{l-2}^{(1)}\times L_{m-2}^{(2)}$ bzw. $L_{m-2}'^{(1)}\times L_{n-2}'^{(3)}$ abhängige irreduzible Korrespondenzen sind.[4] Wir schneiden nun in P_m die Hyperflächen

3) Vgl. I. Wir bezeichnen die Unbestimmten in P_n mit $z=(z_0, z_1, \ldots, z_n)$.

4) Vgl. I.

$f(x, y, u')=0$, $g(y, z, v')=0$, wobei f, g als Form von y betrachtet werden sollen, mit der Kurve Γ_2. Es sei

$$(2)\qquad [\Gamma_2, f(x, y, u')] = \{\eta^{(\lambda)};\ \lambda=1,\dots,\gamma\},$$
$$[\Gamma_2, g(y, z, v')] = \{\eta'^{(\mu)};\ \mu=1,\dots,\delta\}.$$

Mit normierten Koordinaten gilt dann[5]

$$\prod_{\mu=1}^{\delta} f(x, \eta'^{(\mu)}, u') = \prod_{\lambda=1}^{\gamma} g(\eta^{(\lambda)}, z, v')$$

und dies ist eine Form von x, z, welche wir mit $h(x, z, w')$ bezeichnen wollen. Es sei E' der von $L_{l-2}^{(1)}\times L_{m-2}^{(2)}$, $L_{m-2}^{(2)}\times L_{n-2}'^{(3)}$ unabhängige Teil von $[\Gamma_{13}, h(x, z, w')]$ und E'' der Rest:

$$[\Gamma_{13}, h(x, z, w')] = E' + E''.$$

Wir wollen nun aus der Annahme, dass $[\Gamma_{13}, h(x, z, w')]$ eine ausgeartete irreduzible Korrespondenz enthält, einen Widerspruch ableiten. Es sei also z.B. $(a')\times\Gamma_3$ in $[\Gamma_{13}, h(x, z, w')]$ enthalten und es gelte $\eta^{(\lambda)}\to\bar{\eta}^{(\lambda)}$ bei der Spezialisierung $x\to a'$. Es gilt dann

$$h(a', z, w') = \Pi g(\bar{\eta}^{(\lambda)}, z, w')$$

und eines von $g(\bar{\eta}^{(\lambda)}, z, v')=0$ muss Γ_3 enthalten, d.h. eines von $(\bar{\eta}^{(\lambda)}\times\Gamma_3)$ muss in $[\Gamma_{23}, g(y, z, v')]$ liegen, was aber unmöglich ist, wie wir in I bemerkt haben. $[\Gamma_{13}, h(x, z, w')] = E' + E''$ enthält also keine ausgeartete Korrespondenz und folglich wird nach Satz 1 durch das Bild $E'^{(3)}(\xi)+E''^{(3)}(\xi)$ von einem allgemeinen Punkt ξ von Γ_1 völlig bestimmt. $E'^{(3)}(\xi)$ ist dabei offenbar von $L_{l-1}^{(1)}\times L_{m-2}^{(2)}$, $L_{m-2}'^{(2)}\times L_{n-2}'^{(3)}$ unabhängig, während jeder Punkt von $E''^{(3)}(\xi)$ wirklich von einem solcher Räume abhängt. Wenn nun $\eta^{(\lambda)}$ bei der Spezialisierung $x\to\xi$ relationstreu in $\eta^{*(\lambda)}$ übergeht, so ist

$$[\Gamma_2, f(\xi, y, u')] = \{\eta^{*(\lambda)};\ \lambda=1,\dots\gamma\},\quad h(\xi, z, w') = \Pi g(\eta^{*(\lambda)}, z, v').$$

Es gilt andererseits nach Satz 4 in I

$$E'^{(3)}(\xi)+E''^{(3)}(\xi) = [\Gamma_{13}, h(x, z, w')]^{(3)}(\xi) = \{\zeta^{(\omega)};\ (\xi, \zeta^{(\omega)})\epsilon[(\xi)\times\Gamma_3,$$
$$h(x, z, w')]\} = [\Gamma_3, h(\xi, z, w')],$$

ebenso

$$[\Gamma_{23}, g(y, z, v')]^{(3)}(\eta^{*(\lambda)}) = [\Gamma_3, g(\eta^{*(\lambda)}, z, v')],$$
$$[\Gamma_{12}, f(x, y, u')]^{(2)}(\xi) = [\Gamma_2, f(\xi, y, u')].$$

Es folgt also

$$E'^{(3)}(\xi)+E''^{(3)}(\xi) = [\Gamma_3, h(\xi, z, w')] = [\Gamma_3, \prod_{\lambda=1}^{\gamma} g(\eta^{*(\lambda)}, z, v')]$$

---

5) Vgl. W.-L. Chow, Die geometrische Theorie der algebraischen Funktionen für beliebige vollkommene Körper, Math. Ann. 114 (1937).

$$
\begin{aligned}
&=\sum_{\lambda=1}^{\tau}[\Gamma_3, g(\eta^{*(\lambda)}, z, v')]=\sum_{\lambda=1}^{\tau}[\Gamma_{23}, g(y, z, v')]^{(3)}(\eta^{*(\lambda)}) \\
&=[\Gamma_{23}, g(y, z, v')]^{(3)}(\sum_{\lambda=1}^{\tau}\eta^{*(\lambda)}) \\
&=[\Gamma_{23}, g(y, z, v')]^{(3)}([\Gamma_2, f(\xi, y, u')]) \\
&=[\Gamma_{23}, g(y, z, v')]^{(3)}([\Gamma_{12}, f(x, y, u')]^{(2)}(\xi)) \\
&=(D+\sum_{j=2}^{\beta}\eta_j D_j)^{(3)}((C+\sum_{i=2}^{\alpha}\lambda_i C_i)^{(2)}(\xi)).
\end{aligned}
$$

Mithin haben wir

$$
\begin{aligned}
(3)\quad E'^{(3)}(\xi)+E''^{(3)}(\xi)&=D^{(3)}(C^{(2)}(\xi))+\sum_{i=2}^{\alpha}\lambda_i D^{(3)}(C_i^{(2)}(\xi))+\sum_{j=2}^{\beta}\mu_j D_j^{(3)}(C^{(2)}(\xi)) \\
&+\sum_{i=2}^{\alpha}\sum_{j=2}^{\beta}\lambda_i\mu_j D_j^{(3)}(C_i^{(2)}(\xi)).
\end{aligned}
$$

Offenbar hängen die Punkte in $D^{(3)}(C^{(2)}(\xi))$ nicht von $L_{l-2}^{(1)}\times L_{m-2}^{(2)}$, $L_{m-2}'^{(2)}\times L_{n-2}'^{(3)}$ ab. Wir zeigen nun, dass jeder Punkt von $D^{(3)}(C_i^{(2)}(\xi))$, $D_j^{(3)}(C^{(2)}(\xi))$, $D_j^{(3)}(C_i^{(2)}(\xi))$ von einem solcher linearen Räume wirklich abhängt. Z.B. nehme man einem Punkt $\zeta^{(\omega)}$ in $D^{(3)}(C_i^{(2)}(\xi))$. Nach Konstruktion gibt es einem Punkt $\eta^{*(\lambda)}$, so dass $(\xi, \eta^{*(\lambda)})$ ein allgemeiner Punkt von C_i und $(\eta^{*(\lambda)}, \zeta^{(\omega)})$ ein allgemeiner Punkt von D ist. $(\xi, \eta^{*(\lambda)})$ und $\eta^{*(\lambda)}$ hängen also nicht von $L_{m-2}'^{(2)}\times L_{n-2}'^{(3)}$ ab, während $(\eta^{*(\lambda)}, \zeta^{(\omega)})$ wirklich von diesem Raum abhängt. $\zeta^{(\omega)}$ muss folglich von $L_{m-2}'^{(2)}\times L_{n-2}'^{(3)}$ abhängen. Aus (3) erhält man also

$$E'^{(3)}(\xi)=D^{(3)}(C^{(2)}(\xi)).$$

Genau so beweist man

$$E'^{(2)}(\zeta)=C^{(1)}(D^{(2)}(\zeta))$$

für einen allgemeinen Punkt ζ in Γ_3. $E'=E$ genügt also der Bedingung (1). Zusammenfassend haben wir

Satz 2. Für beliebige Korrespondenzen C, D auf Γ_{12} bzw. Γ_{23} gibt es eine und nur eine Korrespondenz E auf Γ_{13}, so dass es für beliebige Punkte a, c in Γ_1 bzw. Γ_3

$$E^{(3)}(a)=D^{(3)}(C^{(2)}(a)),\quad E^{(1)}(c)=C^{(1)}(D^{(2)}(c))$$

gilt. Wir nennen E das Produkt von C und D:

$$E=C\times D.$$

Wenn eines von C, D nur aus ausgearteten irreduziblen Korrespondenzen besteht, so gilt dasselbe auch für $E=C\times D$ und wenn C und D keine ausgeartete Korrespondenz enthalten, so enthält auch $E=C\times D$ keine solche Korrespondenz.

Nun bildet bekanntlich die Gesamtheit aller Korrespondenzen auf $\Gamma_{12}=\Gamma_1\times\Gamma_2$ einen Modul, den Korrespondenzenmodul $\mathfrak{C}(\Gamma_1, \Gamma_2)$ von Γ_1, Γ_2. In $\mathfrak{C}(\Gamma_1, \Gamma_2)$ erzeugen die Korrespondenzen der Form $(a)\times\Gamma_2$ bzw. $\Gamma_1\times(b)$ je einen Teilmodul $\mathfrak{A}_1(\Gamma_1, \Gamma_2)$ bzw. $\mathfrak{A}_2(\Gamma_1, \Gamma_2)$. Die Summe von $\mathfrak{A}_1(\Gamma_1, \Gamma_2)$, $\mathfrak{A}_2(\Gamma_1, \Gamma_2)$ bezeichnen wir mit $\mathfrak{A}(\Gamma_1, \Gamma_2)$. Ebenso erzeugen alle nicht ausgeartete irreduzible Korrespondenzen einen Teilmodul $\mathfrak{C}(\Gamma_1, \Gamma_2)$. Es gilt offenbar die Zerlegungen in direkte Summen:

$$\mathfrak{C}(\Gamma_1, \Gamma_2) = \mathfrak{C}(\Gamma_1, \Gamma_2) + \mathfrak{A}(\Gamma_1, \Gamma_2),$$

$$\mathfrak{A}(\Gamma_1, \Gamma_2) = \mathfrak{A}_1(\Gamma_1 \Gamma_2) + \mathfrak{A}_2(\Gamma_1, \Gamma_2).$$

Diese Moduln haben ersichtlich birational invariante Bedeutungen, d.h. das Bild von einem solchen Moduln, z.B. das von $\mathfrak{C}(\Gamma_1, \Gamma_2)$ bei einer birationalen Transformation, welche $\Gamma_{12} = \Gamma_1 \times \Gamma_2$ auf $\Gamma'_{12} = \Gamma'_1 \times \Gamma'_2$ abbildet, fällt mit $\mathfrak{C}(\Gamma'_1, \Gamma'_2)$ zusammen. Wir bezeichnen ferner den von $\mathfrak{A}$ und von der Korrespondenzen der Form $[\Gamma_{12}, f(x, y, u')]$ ($f(x, y, u') =$ beliebige Form über k) erzeugten Teilmodul mit $\mathfrak{N}(\Gamma_1, \Gamma_2)$ und den Restklassenmodul $\mathfrak{C}(\Gamma_1, \Gamma_2)/\mathfrak{N}(\Gamma_1, \Gamma_2)$ mit $\mathfrak{M}(\Gamma_1, \Gamma_2)$. $\mathfrak{M}(\Gamma_1, \Gamma_2)$ nennt man den Multiplikatorenmodul von Γ_1, Γ_2. Dass $\mathfrak{N}(\Gamma_1, \Gamma_2)$ und somit auch $\mathfrak{M}(\Gamma_1, \Gamma_2)$ birational invariante Moduln sind, sieht man unmittelbar aus Satz 6 in I ein.

Späterer Anwendung wegen schalten wir hier einige Hilfssätze ein.

Hilfssatz 1. In jeder Restklasse von $\mathfrak{M}(\Gamma_1, \Gamma_2)$ gibt es eine solche Korrespoodenz

$$C = \sum_{i=1}^{\alpha} C_i,$$

welche folgende Bedingungen erfüllt:

i) C_i sind von einander verschieden,

ii) der allgemeine Punkt $(\xi^{(i)}, \eta^{(i)})$ von C_i ist so beschaffen, dass $k(\xi^{(i)}, \eta^{(i)})/k(\xi^{(i)})$ separabel ist,

iii) C_i $(i=1, \dots, \alpha)$ fallen mit keiner von den von vornherein angegebenen endlichvielen irreduziblen Korrespondenzen D_j $(j=1, \dots, \beta)$ zusammen: $C_i \neq D_j$ $(i=1, \dots, \alpha, j=1, \dots, \beta)$.

Beweis. Es sei C^* eine beliebige Korrespondenz in einer Restklasse von $\mathfrak{M}(\Gamma_1, \Gamma_2)$ und ξ ein allgemeiner Punkt von Γ_1. Wir wählen eine Punktgruppe $G = \{b^{(\rho)}; \rho = 1, \dots, \gamma\}$ in Γ_2 mit genügend grosser Ordnung γ. Wir nehmen an dass jedes $b^{(\rho)}$ in k liegt und dass die Ordnung δ von $G + C^{*(2)}(\xi)$ zu p prim ist. Wir bezeichnen mit A_λ die allgemeine Punktgruppe der Vollschar der mit $G + C^{*(2)}(\xi)$ äquivalenten effektiven Divisoren und setzen

$$A_\lambda + B = [\Gamma_2, \lambda_0 g_0(y) + \lambda_1 g_1(y) + \dots + \lambda_r g_r(y)],$$

wobei B einen geeigneten festen Divisor und $g_i(y)$ Formen in $k(\xi)[y]$ bedeuten. Wenn γ genügend gross gewählt ist, so folgt nach dem Riemann-Rochschen Satz, dass A_λ genau der von λ abhängige Teil von $[\Gamma_2, \Sigma\lambda_i g_i(y)]$ ist. Nach dem verallgemeinerten Satz von Bertini[6] sind also die Punkte $\eta^{(\mu)}$ $(\mu=1, \ldots, \delta)$ von A_λ einander konjugiert über $k(\xi, \lambda)$ und fallen je p^e zusammen, falls der Exponent von $k(\xi, \lambda, \eta^{(1)}),/k(\xi, \lambda)$ p^e ist. Da aber δ zu p prim ist, so folgt $p^e=1$, d.h. dass $\eta^{(\mu)}$ alle voneinander verschieden sind. Durch geeignete Spezialisierung $\lambda\to\lambda'$ $(\lambda'\epsilon k(\xi))$ kann man also eine Punktgruppe $A_{\lambda'}$ erhalten, welche aus voneinander verschiedenen, über $k(\xi)$ algebraischen Punkten $\eta^{(\mu)}(\mu=1, \ldots, \delta)$ besteht. Da

$$G+C^{*(2)}(\xi)\equiv A_{\lambda'}{}^{7)}$$

ist, gibt es andereseits Formen $\varphi_1(y)$, $\varphi_2(y)$ von demselben Grad in $k(\xi)[y]$, so dass

$$[\Gamma_2, \varphi_1(y)]+G+C^{*(2)}(\xi)=[\Gamma_2, \varphi_2(y)]+A_{\lambda'}$$

gilt. Folglich gibt es zwei Formen $\Phi_1(x, y)$, $\Phi_2(x, y)$ in $k[x, y]$, welche dieselben Grade in x und y besitzen und

$$\frac{\Phi_1(\xi, y)}{\Phi_2(\xi, y)}=\frac{\varphi_1(y)}{\varphi_2(y)}, \quad [\Gamma_2, \Phi_1(\xi, y)]+G+C^{*(2)}(\xi)=[\Gamma_2, \Phi_2(\xi, y)]+A_{\lambda'}$$

erfüllen. Setzt man also

$$C=C^*+\sum_{\rho=1}^{\tau}\Gamma_1\times(b^{(\rho)})+[\Gamma_{12}, \Phi_1(x, y)]-[\Gamma_{12}, \Phi_2(x, y)],$$

so liegt C in derselben Restklasse in $\mathfrak{M}(\Gamma_1, \Gamma_2)$ wie C^* und es gilt

$$C^{(2)}(\xi)=A_{\lambda'}=\{\eta^{(\mu)}; \mu=1,\ldots, \delta\}. \tag{4}$$

Lässt man aus C irreduzible Korrespondenzen in $\mathfrak{A}_1(\Gamma_1, \Gamma_2)$ weg und bezeichnet den Rest wieder mit C, so bleibt (4) noch gültig. Da aber $\eta^{(\mu)}$ alle voneinander verschieden sind, so folgt hieraus, dass die Bedingungen i), ii) erfüllt sind. Man bemerke dabei, dass ein $(\xi, \eta^{(\mu)})$ einen allgemeinen Punkt einer in C enthaltenen irreduziblen Korrespondenz darstellt. Spezialisiert man $\eta^{(\mu)}$ ausserdem so, dass $\eta^{(\mu)}$ mit keinem Punkt von $D_\xi^{(2)}(\xi)$ zusammenfällt, so wird auch die Bedingung iii) erfüllt.[8]

Hilfssatz 2. Es sei C bzw. D eine Korrespondenz auf Γ_{12} bzw. Γ_{23}. Wenn C in $\mathfrak{N}(\Gamma_1, \Gamma_2)$ oder D in $\mathfrak{N}(\Gamma_2, \Gamma_3)$ liegt, so liegt $C\times D$ in $\mathfrak{N}(\Gamma_1, \Gamma_3)$.

Beweis. Man hat offenbar den Satz nur in dem Fall zu beweisen, wo C eine nicht ausgeartete irreduzible Korrespondenz auf Γ_{12} ist und $D=[\Gamma_{23}, g(y, z, v')]$ mit einer geeigneten Form $g(y, z, v')$ gilt. Es sei (ξ, η) ein allgemeiner Punkt

---

6) Vgl. l.c. 5).

7) $G\equiv F$ bedeutet, dass die Punktgruppen G, F äquivalent sind.

8) Dieser Hilfssatz lässt sich auch arithmetisch bewiesen, vgl. M. Deuring, Arithmetische Theorie der Korrespondenzen algebraischer Funktionenkörper II, Crelle's Jour. 183 (1940). Siehe auch eine spätere Abhandlung des Verfassers.

von C und

$$C^{(2)}(\xi)=\{\eta^{(\rho)}\},\ \eta^{(1)}=\eta.$$

$F(\xi, z)=\prod_{\rho} g(\eta^{(\rho)}, z, v')$ ist dann offenbar eine Form in $k(\xi)[z]$. Durch geeignete Normierung von $\eta^{(\rho)}$ kann man sogar erreichen, dass es in $k[\xi, z]$ liegt. Wie oben oft gezeigt worden ist, gilt dann für $E=[\Gamma_{13}, F(x, z)]$

$$E^{(3)}(\xi)=D^{(3)}(C^{(2)}(\xi))=(C\times D)^{(3)}(\xi).$$

Nach Satz 1 ist also

$$E\equiv C\times D \quad \text{mod. } \mathfrak{A}_1(\Gamma_1, \Gamma_3).$$

Aus $E\in\mathfrak{R}(\Gamma_1, \Gamma_3)$ folgt dann $C\times D\in\mathfrak{R}(\Gamma_1, \Gamma_3)$, wie behauptet wurde.

Wir beweisen nun den folgenden wichtigen

Satz 3[9]. Eine Korrespondenz C auf Γ_{12} liegt dann und nur dann in $\mathfrak{R}(\Gamma_1, \Gamma_2)$, wenn für beliebige Punkte a, a' in Γ_1 immer

(5) $$C^{(2)}(a)\equiv C^{(2)}(a')$$

gilt. Man sagt in diesem Fall, dass C die Wertigkeit Null besitzt[10]. Es gilt dann für beliebige Punkte b, b' in Γ_2 immer

(6) $$C^{(1)}(b)\equiv C^{(1)}(b').$$

Beweis. Wenn C in $\mathfrak{A}(\Gamma_1, \Gamma_2)$ liegt, so gilt offenbar für beliebige Punkte a, a' in Γ_1

$$C^{(2)}(a)=C^{(2)}(a')$$

und wenn $C=[\Gamma_{12}, f(x, y, u')]$ ist, so ist

$$C^{(2)}(a)=[\Gamma_2, f(a, y, u')\equiv[\Gamma_2, f(a', y, u')]=C^{(2)}(a')$$

Daraus folgt, dass (5) für eine beliebige Korrespondenz in $\mathfrak{R}(\Gamma_1, \Gamma_2)$ gilt. Nun sei umgekehrt C eine Korrespondenz in $\mathfrak{C}(\Gamma_1, \Gamma_2)$, für welche (5) gilt. Nach dem eben Bewiesenen kann man C durch eine andere Korrespondenz ersetzen, welche in der Restklasse von C in $\mathfrak{M}(\Gamma_1, \Gamma_2)$ liegt, ohne die Gültigkeit von (5) zu verletzen. Nach Hilfssatz 1 können wir also annehmen, dass C die Bedingungen i), ii) in demselben Hilfssatz erfüllt. Nun gibt es nach (5) eine Schar $G(\lambda,y)=\lambda_0 g_0(y)+\lambda_1 g_1(y)+\ldots+\lambda_r g_r(y)$ über k, so dass für einen beliebigen in k liegenden Punkt a auf Γ_1 immer

(7) $$C^{(2)}(a)+B=[\Gamma_2, G(\lambda(a), y)],\ B=\{b^{(\mu)};\ \mu=1,\ldots, \beta\}$$

---

9) Dieser Satz entspricht Satz 8, $\bar{8}$ in M. Deuring, Arithmetische Theorie der Korrespondenzen algebraischer Funktionenkörper I, Crelle's Jour. 177 (1937). Deuring hat diesen Satz aus seiner Sätzen 7, $\bar{7}$ abgeleitet, welche eine Verallgemeinerung des klassischen Additionstheorems (nebst Umkehrung) bilden. Man siehr aber leicht, dass auch umgekehrt das Additionstheorem nebst Umkehrung (Satz 7, $\bar{7}$) sich aus unserem Satz 3 unmittelbar ableiten lässt. Der Zusammenhang der Deuringschen arithmetischen Theorie mit unserer geometrischen Theorie soll in einer späteren Abhandlung ausführlich dargestellt werden.

10) Vgl. z.B. Severi-Löffler, Vorlesungen über algebraische Geometrie, Berlin (1921), S. 162. Allgemeine Wertigkeitskorrespondenzen werden wir nachher untersuchen.

gilt, wobei $\lambda(a)$ eine geeignete Spezialisierung von λ und B eine feste Punktgruppe in k ist. Da aber die Dimension einer Vollscher bei der Körpererweiterung $k(\xi)/k$ des Grundkörpers k invariant bleibt, so gilt (7) auch für einen allgemeinen Punkt ξ von Γ_1. Setzt man also

$$C^{(2)}(\xi)=\{\eta^{(\rho)};\ \rho=1,\dots,\alpha\},$$

so besitzt das lineare Gleichungssystem

$$(8)\qquad G(\lambda(\xi),\eta^{(\rho)})=\lambda_0(\xi)g_0(\eta^{(\rho)})+\lambda_1(\xi)g_1(\eta^{(\rho)})+\dots+\lambda_r(\xi)g_r(\eta^{(\rho)}),\quad \rho=1,\dots,\alpha$$

mindestens eine nicht triviale Lösung $\lambda(\xi)=(\lambda_0(\xi),\lambda_1(\xi),\dots,\lambda_r(\xi))$ in $K=k(\xi,\eta^{(1)},\dots,\eta^{(\alpha)})$. Nach Voraussetzung ist K eine separable galoissche Erweiterung von $k(\xi)$ und das Gleichungssystem (8) ist ersichtlich invariant gegenüber den Galoisautomorphismen von $K/k(\xi)$. Durch Spurbildung kann man also eine Lösung in $k(\xi)$ erhalten. Es gibt also Formen $\lambda(x)=(\lambda_0(x),\lambda_1(x),\dots,\lambda_r(x))$ in $k[x]$, so dass $G(\lambda(x),y)$ für $x=\xi$, $y=\eta^{(\rho)}$ verschwindet und

$$C^{(2)}(\xi)+D^{(2)}(\xi)=[\Gamma_2, G(\lambda(\xi),y)]\ \text{mit}\ D=\Sigma\Gamma_1\times(b^{(\mu)})$$

gilt. Daraus sieht man sofort, dass C mit $[\Gamma_{12}, G(\lambda(x),y)]$ bis auf Korrespondenzen in $\mathfrak{A}(\Gamma_1,\Gamma_2)$ übereinstimmt, folglich in $\mathfrak{R}(\Gamma_1,\Gamma_2)$ liegt, w.z.b.w.

Nun seien Γ_1, Γ_2 birational äquivalent. Wir können also annehmen,

$$P_l=P_m,\ \Gamma_1=\Gamma_2=\Gamma.$$

Wendet man nun Satz 2 auf $\Gamma_1=\Gamma_2=\Gamma_3=\Gamma$ an, so sieht man sofort, dass $\mathfrak{C}(\Gamma,\Gamma)$ in diesem Fall einen Ring bildet: den " Korrespondenzenring " $\mathfrak{C}(\Gamma)$ von Γ. Es ist auch leicht zu sehen, dass $\mathfrak{E}(\Gamma)=\mathfrak{E}(\Gamma,\Gamma)$ einen Teilring und $\mathfrak{A}(\Gamma)=\mathfrak{A}(\Gamma,\Gamma)$, $\mathfrak{A}_1(\Gamma)=\mathfrak{A}_1(\Gamma,\Gamma)$, $\mathfrak{A}_2(\Gamma)=\mathfrak{A}_2(\Gamma,\Gamma)$ sogar Ideale in $\mathfrak{C}(\Gamma)$ bilden. Es gilt dann z.B.

$$\mathfrak{C}(\Gamma)/\mathfrak{A}(\Gamma)\cong\mathfrak{E}(\Gamma).$$

Aus Hilfssatz 2 folgt ferner, dass $\mathfrak{R}(\Gamma)=\mathfrak{R}(\Gamma,\Gamma)$ auch ein Ideal von $\mathfrak{C}(\Gamma)$ ist. Der Restklassenring $\mathfrak{M}(\Gamma)=\mathfrak{M}(\Gamma,\Gamma)$ nach diesem Ideal nennt man den " Multiplikatorenring " von Γ. $\mathfrak{M}(\Gamma)$ oder allgemeiner $\mathfrak{M}(\Gamma_1,\Gamma_2)$ bildet eine Verallgemeinerung des Rings oder Moduls der klassischen komplexen Multiplikationen von Γ oder von Γ_1, Γ_2[11]. Die Struktur von $\mathfrak{M}(\Gamma)$ soll in einer späteren Arbeit ausführlich untersucht werden.

11) Vgl. M. Deuring, l.c. 9).

[16] On the representation of Lie algebras

Japanese J. Math., 19 (1948), 405-426.

(Received January 21, 1947)

§ 1. Introduction.

Let L be a Lie algebra over a commutative field F and A an associative algebra over F. L and A may have finite or infinite ranks over F. A mapping

$$a \longrightarrow a'$$

from L into A is called a representation of L in A, if

$$\lambda a+\mu b \longrightarrow \lambda a'+\mu b',$$

$$[a, b] \longrightarrow [a', b'] = a'b'-b'a',$$

$$a, b \in L, \quad a', b' \in A, \quad \lambda, \mu \in F.$$

The case is particularly important, where the mapping is one-to-one. As to such "faithful" representation, the following two remarkable results are known:

I. If F is the field of all real or complex numbers and if L has a finite basis over F, we can represent L faithfully by finite matrices over F. (A is here the algebra of all square matrices of some fixed degree over F).

II. Any Lie algebra over F can be represented faithfully in some associative algebra A over F. F is here an arbitrary field.

The theorem I, which is particularly important according to its various applications in the theory of Lie groups(¹), was first proved by Ado [2]. However the proof was considerably complicated and Cartan [5] gave a more simple and elegant proof by making use of the analytical theory of Lie groups.

The theorem II was proved independently by Birkhoff [3] and Witt [10] in similar ways and Birkhoff has thence algebraically deduced the proof of I in the case, when L is a nilpotent Lie algebra. But according to the rational character of the proof, the field F can be quite arbitrary

(¹) See Birkhoff [3], Cartan [5].

in Birkhoff's result. This suggests to us that the theorem I would be proved in a purely algebraic manner and thus it would be generalized to the case of an arbitrary field F.

On the other hand Lie algebras over a field of characteristic $p(\neq o)$ were treated in detail in connection with the study of p-groups by Zassenhaus [12], [13], who gave a more precise form to the result of Birkhoff in this modular case. Jacobson [9] also studied such Lie algebras and proved the representability of some special types of Lie algebras by matrices.

However it seems not yet proved in a purely algebraic way, that any Lie algebra with a finite basis over an arbitrary field F can be represented faithfully by finite matrices over F. In this paper we shall give such a proof by making use of a theorem for splitting Lie algebras, which corresponds to E. Artin's theorem of splitting groups in the abstract group theory(2). It was already pointed out by Zassenhaus [12], that O. Schreier's extension theory for abstract groups, including the existence of splitting groups, can be transferred to the case of Lie algebras. We shall prove in fact in §2, that for any Lie algebra L over F, containing an abelian ideal N, there exists a splitting Lie algebra L_0 with respect to the extension L/N. The Lie algebra L_0 thus constructed corresponds in some sense to the Birkhoff-Witt's universal representing associative algebra to L and has in general an infinite rank over F, even if L has a finite basis over F. So if we want to obtain a splitting Lie algebra with a finite basis for such L, we must seek for a proper ideal in L_0 and take the residue class algebra with respect to it. The essential part of the paper is devoted to this procedure.

We treat the cases separately, according as the characteristic of L is o or a prime number p. In the former case we use a special type of basis of L over F, which is similar to that, given in the proof of Cartan [5]. In the modular case we follow the idea of Jacobson [9]. Here the representability of L by matrices may be proved directly by a lemma(3), without making use of splitting Lie algebras. But we do not follow this way, for the structure of Lie algebras over a field of characteristic p seems not yet so clarified as in the case of characteristic o and the theorem for splitting algebras may be of interest in this case, also considered for itself(4).

(2) See Iyanaga [7] or Zassenhaus [11].

(3) Lemma 2.

(4) It is noted, that the analogy of E. Levi's theorem is not valid in the modular case. By Theorem 2 in §6. we can also show that there are representations of semi-simple Lie algebras with characteristic p, which are not completely reducible.

In any case, when the existence of splitting Lie algebras is once proved, the representability of L follows immediately. From the existence theorem in general, we have the result II of Birkhoff-Witt and from that for splitting Lie algebras with a finite basis over F, we can obtain the theorem for the representability by finite matrices over F, including the result I of Ado-Cartan as a special case. As was said above, the field F may be quite arbitrary in our result.

§ 2. The existence of splitting Lie algebras in general.

Following the Schreier's principle in the group theory([5]), we define first the factor sets for the extension of Lie algebras. But in view of later applications, we shall confine ourselves in the case of the extension of abelian ideals, whereas the general theory goes as well([6]).

Now let L be a Lie algebra over a field F and N its abelian ideal and let us suppose that $\bar{x}_1, \bar{x}_2, \ldots$ constitute a basis of $\bar{L} = L/N$ over F, well-ordered in some fixed way. Further put

$$(2.1) \qquad [\bar{x}_i, \bar{x}_j] = \gamma_{ij}^k \bar{x}_k, \qquad \gamma_{ij}^k \in F,$$

where γ_{ij}^k are the structural constants of $\bar{L}$. If we then choose an element x_i of L from each residue class $\bar{x}_i$, we have from (2.1)

$$(2.2) \qquad [x_i, x_j] = \gamma_{ij}^k x_k + a_{ij}, \qquad a_{ij} \in N.$$

If we put further

$$(2.3) \qquad [x_i, a] = X_i a$$

for any a in N, X_i gives obviously a linear transformation in N and between a_{ij} and X_i hold following relations:

$$(2.4) \qquad a_{ij} + a_{ji} = o, \qquad a_{ii} = o,$$

$$(2.5) \qquad X_i X_j - X_j X_i = \gamma_{ij}^k X_k,$$

$$(2.6) \qquad \gamma_{ij}^l a_{kl} + \gamma_{jk}^l a_{il} + \gamma_{ki}^l a_{jl} + X_i a_{jk} + X_j a_{ki} + X_k a_{ij} = o.$$

These follow immediately from (2.2), (2.3) and the well-known identities of Lie-Jacobi in L, namely

$$[u, v] + [v, u] = o, \; [u, u] = o, \; [[u, v], w] + [[v, w], u] + [[w, u], v] = o.$$

([5]) See for example Zassenhaus [11].

([6]) It was first precisely given by Abe [1], where interesting applications are also given, particularly to a proof of E. Levi's theorem.

Now conversely let $\bar{L}$ be a Lie algebra over F with a basis $\bar{x}_1, \bar{x}_2, \ldots$ and γ_{ij}^k its structural constants and suppose that there is a linear space N over F, elements a_{ij} in N and linear transformations X_i of N, which satisfy (2.4), (2.5), (2.6). If we then denote by L the linear space spanned by $x_1, x_2, \ldots$ and N and define in L a distributive multiplication by

$$[x_i, x_j] = \gamma_{ij}^k x_k + a_{ij},$$

$$[x_i, a] = -[a, x_i] = X_i a,$$

$$[a, a'] = 0, \qquad a, a' \in N,$$

L becomes a Lie algebra over F, N an abelian ideal in L and L/N is isomorphic to $\bar{L}$.

Thus an extension L of an abelian ideal N is completely determined, when the combination (a_{ij}, X_i) is known, the so-called *factor set* for N. Two such factor sets are called associated, when they give isomorphic extensions over N. We can give explicitely the relations which must hold between associated factor sets, but here we do not enter in this problem(7).

In the following the case is particularly important, when *L splits* over N. We call namely that an extension L splits over N, if there is a subalgebra L_1 in L, so that

$$L = L_1 + N. \qquad L_1 \wedge N = 0.$$

It is easy to see that a necessary and sufficient condition for the splitting is the existence of elements a_i in N, satisfying

$$a_{ij} = \gamma_{ij}^k a_k - X_i a_j + X_j a_i. \tag{2.7}$$

Of course an extension L over N does not always split. But, as it will be showh soon afterwards, we can construct a closely related Lie algebra L^*, called *a splitting Lie algebra* for L/N, defined as follows.

Definition. A Lie algebra L^* is called a splitting Lie algebra for the extension L/N, if

i) L^* contains L and an abelian ideal N^*, so that

ii) $L^* = L + N^*$, $L \wedge N^* = N$ (consequently $L^*/N^* \cong L/N$) and

iii) L^* splits over N^*

We prove then

(7) See Abe [1].

Theorem 1. *For any extension L over N, there is always a splitting Lie algebra L_0.*

Proof. Let x_i, γ_{ij}^k, a_{ij}, X_i be as before. Take a linear space M over F with a basis

$$z_{i_1 i_2 \cdots\cdots i_\alpha},$$

where α is an arbitrary integer and $i_1, i_2, \cdots, i_\alpha$ run over the set of all indexes i of x_i, preserving the order

$$i_1 \leqq i_2 \leqq \cdots\cdots \leqq i_\alpha.$$

We put then

$$N_0 = N + M \qquad \text{(direct sum)}$$

and extend the linear transformation X_i into the whole space N_0 by the following definition. Let

$$(2.8) \qquad X_i z_j = z_{ij}, \qquad \text{for} \quad i \leqq j$$
$$= \gamma_{ij}^k z_k - a_{ij} + z_{ji}, \qquad \text{for} \quad i > j$$

and for $\alpha \geqq 2$ by induction with respect to α and i

$$(2.9) \qquad X_i z_{i_1 i_2 \cdots\cdots i_\alpha} = z_{i i_1 i_2 \cdots\cdots i_\alpha}, \qquad \text{for} \quad i \leqq i_1$$
$$= X_{i_1} X_i z_{i_2 \cdots\cdots i_\alpha} + \gamma_{i i_1}^k X_k z_{i_2 \cdots\cdots i_\alpha}, \qquad \text{for} \quad i > i_1.$$

By (2.9) $X_1 z_{i_1 \cdots\cdots i_\alpha}$ is always well defined. If we consider therefore that all $X_k z_{i_1 \cdots\cdots i_\beta}$ $(\beta < \alpha)$ and $X_j z_{i_1 \cdots\cdots i_\alpha}$ $(j < i)$ are already defined, the right side of the second equality in (2.9) contains only known terms and the definition is thus complete. Here we remark that in general $X_i z_{i_1 \cdots\cdots i_\alpha}$ is linear combination of elements of N and some z with at most $\alpha+1$ suffixes.

Now we can prove easily by induction with respect to α and i, j

$$(2.10) \qquad X_i X_j z_{i_1 \cdots\cdots i_\alpha} - X_j X_i z_{i_1 \cdots\cdots i_\alpha} = \gamma_{ij}^k X_k z_{i_1 \cdots\cdots i_\alpha}.$$

For the verification (2.4) and (2.5) are essentially used. The extended X_i therefore also fulfil the relation (2.5) as linear transformations in N_0 and we have thus a facfor set (a_{ij}, X_i) for N_0. We can construct accordingly a Lie algebra L_0 over N_0 by the inverse process mentioned before. It is then obvious that i), ii) are fulfilled for L_0, N_0. But from (2.8) (and $\gamma_{ij}^k + \gamma_{ji}^k = o$, $\gamma_{ii}^k = o$ and (2.4)) the splitting condition (2.7) is also satisfied with $a_i = z_i$ and L_0 is really a splitting Lie algebra for L/N, q. e. d.

The splitting Lie algebra L_0 in the above proof is, as the construction shows, in some sense a "*universal splitting algebra.*" Namely, any splitting Lie algebra L^* for L/N, in which N^* is generated by N and a_i satisfying (2.7), is isomorphic to some L_0/M_0, where M_0 is a proper ideal in L_0, so that

$$(2.11) \qquad M_0 \leq N_0, \qquad M_0 \wedge N = O.$$

Conversely any such M_0 offers a splitting algebra L_0/M_0 for L/N. For in virtue of (2.11) residue classes in L_0/M_0, which are represented by elements of L, form a subalgebra isomorphic to L and thus can be identified with L.

Now we suppose that L has a finite basis over F. Even in this case above L_0 has, except for the trivial case $L = N$, an infinite rank over F. If we want therefore to obtain a splitting Lie algebra L^*, which has also a finite basis over F, we must seek for a linear subspace M_0, which satisfies (2.11) and

$$(2.12) \qquad [N_0 : M_0] < \infty \text{ (8)}$$

and which is invariant under all X_i.

We prove the existence of such M_0 in following sections.

§ 3. Nilpotent algebras.

From now on until the end of § 5, we assume that F is algebraically closed and has the charrcteristic o, although we do not use these properties in the present section. We prove in this section the existence of a space M_0 in the case, where L is a nilpotent Lie algebra with a finite basis over F. To do this, we adjoin a base z_0 to N_0 and put

$$(3.1) \qquad N_1 = \{z_0, N_0\}, \qquad X_i z_0 = z_i.$$

We know already by (2.10) that linear transformations X_i in N_0 give a representation $\bar{x}_i \rightarrow X_i$ of $\bar{L} = L/N$. Now by the definitions (2.8), (3.1), we have

$$(3.2) \qquad X_i X_j z_0 - X_j X_i z_0 = \gamma_{ij}^k X_k z_0 - a_{ij}, \qquad a_{ij} \in N,$$

so that X_i, when considered as transformations in N_1/N, also give a representation of $\bar{L}$.

(8) $[T : S]$ means the rank of the residue space T/S over F

Let us now suppose that the "*class*" of the nilpotent Lie algebra L is c, that is to say that any product of more than $c+1$ factors in L always vanishes([9]). It holds then

$$(3.3)\quad X_{i_1}\ldots X_{i_{k-1}}X_{i_k}\ldots X_{i_\alpha}z_0-X_{i_1}\ldots X_{i_k}X_{i_{k-1}}\ldots X_{i_\alpha}z_0=\gamma^l_{i_{k-1}i_k}X_{i_1}\ldots X_l\ldots X_{i_\alpha}z_0$$

for any $\alpha\geqq c+2$ and $i_1, i_2, \ldots i_\alpha$. To prove this, suppose first $k<\alpha$. Then $X_{i_{k+1}}\ldots X_{i_\alpha}z_0$ is contained in N_0 and (3.3) is a consequence of the fact that X_i in N_0 give a represenation of L. Next let $k=\alpha$. (3.3) follows then from (3,2) and

$$(3.4)\qquad X_{i_1}\ldots X_{i_{\alpha-2}}a=[x_{i_1}[\ldots[x_{i_{\alpha-2}}, a]]\ldots]=o$$

for all a in N, according to $\alpha-2\geqq c$.

Now as $\overline{L}$ is also nilpotent, we can choose a basis $\bar{x}_1, \ldots, \bar{x}_n$ of $\overline{L}$ in such a manner, that

$$(3.5)\qquad \gamma^k_{ij}=o,\qquad \text{if } k\geqq i \text{ or } k\geqq j.$$

Under such conditions, when we *straighten*([10]) any term

$$(3.6)\qquad X_{i_1}\ldots\ldots X_{i_\alpha}z_0$$

by making use of (3.3), it becomes a linear combination of terms of canonical form

$$(3.7)\qquad z_{j_1,\ldots\ldots,j_\beta}=X_{j_1}\ldots X_{j_\beta}z_0,\qquad j_1\leqq j_2\leqq\ldots\leqq j_\beta,$$

where β obeys the condition

$$(3.8)\qquad \beta\geqq\alpha/2^{n-1}.$$

We can prove it just as in Birkhoff [3], in ascribing a numerical *weight*

$$\Omega=2^{-i_1}+\cdots\cdots+2^{-i_\alpha}$$

to every term of type (3.6) and observing that, when (3.5) holds, the weight of any term in the right side of (3.3) is not less then that of a term in the left side. The applicability of (3.3) at each step of such straightening is indeed assured by (3.8), when α is sufficiently large, namely when

$$\alpha\geqq\alpha_0=2^{n-1}(c+2).$$

---

(9) See Zassenhaus [12].

(10) See Birkhoff [3].

If we take therefore as M_0 the linear set over F, generated by elements of the form (3.6) with $a \geqq a_0$, we have

$$M_0 \leqq M, \qquad M_0 \wedge N = O.$$

Moreover it is obvious that such M_0 is invariant under all X_i and that $N_0 \geqq M_0$. On the other hand, as any $z_{i_1 \ldots\ldots i_a}$, $a \geqq a_0$ is contained in M_0, we have

$$[M : M_0] \leqq n + n^2 + \ldots + n^{a_0 - 1} < \infty$$

and consequently

$$[N_0 : M_0] < \infty .$$

We have thus proved the existence of desired M_0 in the case of nilpotent L.

§ 4. Soluble algebras.

We now consider the case, where L is a soluble Lie algebra with a finite basis over F. Notations in § 2 are preserved in the same sense, if not particularly mentioned. Here the problem is again to find a suitable subspace M_0 in N_0. For that purpose we must choose a special basis $x_1, \ldots, x_n$ for L/N.

Take the maximal nilpotent ideal K in L(11). It is well known that the derived algebra $L' = [L, L]$ is contained in K:

(4.1) $$L' \leqq K.$$

We show first that there is a suitable nilpotent subalgebra H in L, so that

(4.2) $$L = H + K.$$

For the purpose take a regular element h_0 in L, that is to say an element which has as many different characteristic roots as possible in the regular representation of L and make the Cartan-decomposition of L by h_0(12):

(4.3) $$L = H + L_\alpha + L_\beta + \ldots .$$

As h_0 induces non-singular linear transformations in L_α, L_β, ..., we have

(11) See Zassenhaus [12]. In this section we can carry out the proof with the help of L' instead of K. But we take K in relation to the next section.

(12) See Cartan [4] or Zassenhaus [12].

$$L_\alpha = [h_0, L_\alpha], \quad L_\beta = [h_0, L_\beta], \ldots\ldots$$

and this implies

(4.4) $$L_\alpha + L_\beta + \ldots\ldots \leqq L'.$$

(4.1) and (4.4) then give immediately (4.2)

Now take $x_1, \ldots, x_n$ as follows([13]). By (4.2) we first take $x_{\omega+1}, \ldots, x_n$ from H, so that they represent a basis L/K. Then put

$$H_0 = \{x_{\omega+1}, \ldots, x_n, H \frown K\}.$$

As $H \frown K$ is invariant by x_i and $[x_i, x_j] \in H \frown K$ for $i, j \geqq \omega+1$, H_0 is a nilpotent subalgebra of H. We then adjoin $x_1, \ldots, x_\omega$ as follows:

a) $\{x_1, \ldots, x_i, N\}$ is an ideal of L for any $i = 1, \ldots, \omega$. The same condition is already satisfied for $i = \omega+1, \ldots, n$, as L/K is abelian:

b) we have particularly

$$K = \{x_1, \ldots, x_\omega, N\}$$

and $x_1, \ldots, x_\omega$ are all contained in K:

c) for suitable $t_1, \ldots, t_\kappa$ $(\leqq \omega)$

$$x_{t_1}, \ldots, x_{t_\kappa}, x_{\omega+1}, \ldots, x_n$$

are contained in H_0 and form a basis of $H_0 + N/N$.

We can choose such $x_1, \ldots x_\omega$ by a procedure of Chevalley [6], applied to $H \frown K$ and a series of ideals

$$K_\omega = K > K_{\omega-1} > \ldots > K_1 > K_0 = N,$$

where

$$[K_i : K_{i-1}] = 1, \qquad i = 1, 2, \ldots, \omega.$$

For such $x_1, \ldots, x_n$, γ_{ij}^k and a_{ij} have following properties:

a′) $\gamma_{ij}^k = o$, if $k > i$ or $k > j$ (from a),

b′) $\gamma_{ij}^k = o$, if $i, j \leqq \omega$ and $k \geqq i$ or $k \geqq j$ (form the nilpotency of K and a)),

c′) $\gamma_{ij}^k = o$, if $x_i, x_j \in H_0$ and $k \geqq i$ or $k \geqq j$ (from the nilpotency of H_0 and a)),

d′) $a_{ij} \in K$ in general and $a_{ij} \in H_0 \frown K$ for $x_i, x_j \in H_0$ (from (4.1) and b), c)).

([13]) For the choice of such a basis see Cartan [5].

Now let us consider that we have constructed L_0, N_0 as in §2 for such x_i and denote in particular $X_{t_1}, \dots, X_{t_\kappa}, X_{\omega+1}, \dots, X_n$ by

$$Y_{\omega-\gamma+1}, \dots, Y_\omega, Y_{\omega+1}, \dots, Y_n$$

respectively. We take as M_0 the linear space generated by elements

$$(4.5) \qquad X_{i_1} \dots X_{i_\alpha} Y_{j_1} \dots Y_{j_\beta} z_0, \qquad i_1, \dots, i_\alpha \leqq \omega, \ \alpha \geqq n_1$$

and

$$(4.6) \qquad X_{i_1} \dots X_{i_\alpha} Y_{j_1} \dots Y_{j_\beta} z_0, \qquad i_1, \dots, i_\alpha \leqq \omega, \ \beta \geqq n_2,$$

where n_1 and n_2 are sufficiently large integers, which we shall fix precisely later on. We prove first that M_0 is invariant under all X_i. For $i \leqq \omega$. this is evident. If $i > \omega$, then $X_i = Y_i$ and we have

$$\begin{aligned} Y_i X_{i_1} \dots X_{i_\alpha} Y_{j_1} \dots Y_{j_\beta} z_0 & \\ = [Y_i, X_{i_1}] X_{i_2} \dots X_{i_\alpha} Y_{j_1} \dots Y_{j_\beta} z_0 + X_{i_1} [Y_i, X_{i_2}] X_{i_3} \dots X_{i_\alpha} Y_{j_1} \dots Y_{j_\beta} z_0 + \dots \\ + X_{i_1} \dots X_{i_{\alpha-1}} [Y_i, X_{i_\alpha}] Y_{j_1} \dots Y_{j_\beta} z_0 + X_{i_1} \dots X_{i_\alpha} Y_i Y_{j_1} \dots Y_{j_\beta} z_0 . \end{aligned}$$

In the right side all terms except $X_{i_1} \dots X_{i_{\alpha-1}} [Y_i, X_{i_\alpha}] Y_{j_1} \dots Y_{j_\beta} z_0$ is clearly contained in M_0, for $\bar{x}_i \to X_i$ is a homomorphic mapping, when we consider X_i as linear transformations in N_0, and we can substituse $[Y_i, X_{i_1}], \dots, [Y_i, X_{i_{\alpha-1}}]$ for linear combinations of X_i $(i \leqq \omega)$. If $\beta > o$ $X_{i_1} \dots X_{i_{\alpha-1}} [Y_i, X_{i_\alpha}] Y_{j_1} \dots Y_{j_\beta} z_0$ is also contained in M_0 by the same reason. But if $\beta = o$, we have according to (3.2)

$$X_{i_1} \dots X_{i_{\alpha-1}} [Y_i, Y_{i_\alpha}] z_0 = \gamma^k_{ii_\alpha} X_{i_1} \dots X_{i_{\alpha-1}} X_k z_0 - X_{i_1} \dots X_{i_{\alpha-1}} a_{i, i_\alpha} .$$

Here $\gamma^k_{ii_\alpha} X_{i_1} \dots X_{i_{\alpha-1}} X_k z_0 \in M_0$ and just as in (3.4) we have

$$(4.7) \qquad X_{i_1} \dots X_{i_{\alpha-1}} a_{i, i_\alpha} = [x_{i_1} [\dots [x_{i_{\alpha-1}}, a_{i, i_\alpha}]] \dots] = o,$$

if

$$(4.8) \qquad n_1 \geqq c_1 + 1,$$

where c_1 is the class of nilpotent K. Note that $x_{i_1}, \dots, x_{i_{\alpha-1}}, a_{i, i_\alpha}$ are all contained in K. (by b) and d')). Thus the invariance of M_0 is proved.

Next we want to prove that M_0 is contained in M and for that purpose straighten the term $Y_{j_1} \dots Y_{j_\beta} z_0$ in (4.6). As Y_j correspond to elements in H_0, we can change the order of Y_i just as in § 3, (3.3) and thus $Y_{j_1}\dots Y_{j_\beta}z_0$ becomes a linear combination of

$$(4.9)\qquad Y_{k_1}\dots Y_{k_\gamma}z_0, \quad k_1 \leqq k_2 \leqq \dots \leqq k_\gamma, \quad \gamma \geqq \gamma_0 = n_2/2^{n-\omega+\varkappa-1}$$

as far as n_2 satisfies

$$(4.10)\qquad n_2 \geqq 2^{n-\omega+\varkappa-1}(c_2+2),$$

where we denote by c_2 the class of nilpotent H_0. We can therefore express (4.6) linearly by

$$(4.11)\qquad X_{l_1}\dots\dots X_{l_{\delta-1}}X_{l_\delta}z_0, \quad l_1, \dots\dots, l_{\delta-1} \leqq l_\delta$$

or

$$(4.12)\qquad X_{l_1}\dots\dots X_{l_\delta}z_0, \quad l_1, \dots\dots, l_\delta \leqq \omega, \ \delta \geqq \gamma_0,$$

according as $k_\gamma > \omega$ or $k_\gamma \leqq \omega$ respectively.

Now consider (4.5). $Y_{j_1}\dots Y_{j_\beta}z_0$ is here again composed linearly of terms of the form (4.9) with arbitrary γ and eventually some of a_{ij}. But as we have

$$X_{i_1}\dots\dots X_{i_\alpha}a_{ij} = o$$

just as in (4.7), we can see immediately that (4.5) is also a linear combination of (4.11) or (4.12), though the condition $\delta \geqq \gamma_0$ in (4.12) must be replaced this time by $\delta \geqq n_1$. In any case it is therefore sufficient to prove that (4.11) or (4.12) (with $\delta \geqq \gamma_0$ or $\delta \geqq n_1$) is contained in M. As to (4.11) this follows from the fact that in straightening $X_{l_1}\dots X_{l_{\delta-1}}$, we have only such X_i with $i \leqq l_\delta$, according to a′). As to (4.12) it can be proved, as $X_{l_1}, \dots, X_{l_\delta}$ all correspond to x_i in K and $a_{ij} \in K$ by d′), by the same consideration as in § 3, as far as γ_0 and n_1 are sufficiently large, namely

$$(4.13)\qquad \gamma_0, n_1 \geqq 2^{\omega+1}(c_1+2).$$

We have thus proved that M_0 is contained in M and consequently that

$$M_0 \frown N = O$$

Conditions for n_1, n_2 are (4.8), (4.10), (4.13) and are of course satisfied for sufficiently large n_1, n_2. For example we may put

$$n_1 = n_2 = 2^{2g}(g+2)\,,$$

if g is the rank of L over F.

Now take any

(4.14) $$z_{m_1 m_2 \ldots\ldots m_\varepsilon} \qquad (m_1 \leqq m_2 \leqq \ldots \leqq m_\varepsilon).$$

If $\varepsilon \geqq n_1 + n_2$, then (4.14) can be written in the form (4.5) or (4.6). Therefore we have

$$[M : M_0] \leqq n + n^2 + \ldots\ldots + n^{n_1+n_2-1}, \quad [N_0 : M_0] < \infty$$

and we see that M_0 satisfies all conditions mentioned at the end of § 2

The existence of a splitting Lie algebra with a finite basis is thus proved for any soluble L.

§ 5. General case with the characteristic 0.

In this section we consider a general Lie algebra L with a finite basis over F. This time we must modify the general procedure of constructing L_0 explained in § 2 and change the sense of the notations accordingly.

By a theorem of E. Levi, L can be written in the form

(5.1) $$L = S + R.$$

where S is a semi-simple subalgebra of L and R the radical of L. An abelian ideal N of L is of course contained in R. We suppose that $s_1, s_2, \ldots\ldots, s_m$ constitute a basis of S over F and $x_1, \ldots, x_n$ represent that of R/N over F. Thus we consider the soluble ideal R just as L in the preceeding section. We assume that $\bar{x}_i$, x_i, a_{ij}, X_i satisfy (2.1)—(2.6) in § 2. Further let

(5.2) $$[s_\sigma, s_\tau] = \alpha^\rho_{\sigma\tau} s_\rho\,, \qquad \alpha^\rho_{\sigma\tau} \in F,$$

(5.3) $$[s_\sigma, x_i] = -[x_i, s_\sigma] = \beta^k_{\sigma i} x_k + b_{\sigma i}\,, \quad \beta^k_{\sigma i} \in F, \quad b_{\sigma i} \in N,$$

(5.4) $$[s_\sigma, a] = S_\sigma a, \qquad a \in N.$$

Then between $\alpha^\rho_{\sigma\tau}$, $\beta^k_{\sigma i}$, γ^k_{ij}, a_{ij}, $b_{\sigma i}$, X_i, S_σ hold as (2.4)—(2.6)

(5.5) $$b_{\sigma i} + b_{i\sigma} = o\,,$$

(5.6) $$S_\sigma S_\tau - S_\tau S_\sigma = \alpha^\rho_{\sigma\tau} S_\rho\,, \qquad S_\sigma X_i - X_i S_\sigma = \beta^k_{\sigma i} X_k\,,$$

(5.7) $$\alpha^\rho_{\sigma\tau} b_{i\rho} + \beta^l_{\tau i} b_{\sigma l} + \beta^l_{i\sigma} b_{\tau l} + S_\sigma b_{\tau i} + S_\tau b_{i\sigma} = o,$$

$$\gamma^l_{ij} b_{\sigma l} + \beta^l_{j\sigma} a_{il} + \beta^l_{\sigma i} a_{jl} + X_i b_{j\sigma} + X_j b_{\sigma i} + S_\sigma a_{ij} = o.$$

Now take $N_0 = N+M$ as before, namely the linear space generated by N and

$$z_{i_1 i_2 \ldots\ldots i_\alpha}, \qquad i_1 \leqq i_2 \leqq \ldots\ldots \leqq i_\alpha$$

and define X_i for M by (2.8), (2.9). In a similar way we define here the linear transformation S_σ for any z in M as follows: put

(5.8) $$S_\sigma z_i = \beta^k_{\sigma i} z_k - b_{\sigma i}$$

and in general by induction with respect to α

(5.9) $$S_\sigma z_{i_1 i_2 \ldots\ldots i_\alpha} = X_{i_1} S_\sigma z_{i_2 \ldots\ldots i_\alpha} + \beta^k_{\sigma i_1} X_k z_{i_2 \ldots\ldots i_\alpha}.$$

We can then verify easily by making use of (5.5), (5.6), (5.7), that (5.6) holds as well, even when we condsider X_i and S_σ as linear transformations in N_0. Therefore $(a_{ij}, b_{\sigma i}, X_i, S_\sigma)$ forms a factor set for N_0 and we obtain an extension L_0 over N_0 corresponding to this factor set, which turns out again to be a splitting Lie algebra for the extension L/N. The problem is then, when L has a finite basis over F, to find a linear space M_0, which satisfies (2.11), (2.12) and is invariand under all X_i and S_σ. In order to do it, we must choose a suitable basis $x_1, \ldots, x_n$ for R/N, as we have used for L/N in § 4 and for that purpose we first prove a lemma([14]).

Lemma 1. *If K is the maximal nilpotent ideal in R. then we have*

(5.10) $$[s, r] \in K$$

for any $s \in S$ and $r \in R$.

Corollary. *The radical of the derived algebra L' of any Lie algebra L with a finite basis over F is always nilpotent.*

Proof. Let $y_1, \ldots, y_\nu$ form a basis for R/K and

$$r = \xi^i y_i + r_1, \qquad \xi^i \in F,\ r_1 \in K.$$

Then the characteristic polynomial $f_R(x\,;r)$ of r in the regular representation of R can be written in the form

(5.11) $$f_R(x\,;r) = \prod (x - \varphi_i)^{e_i},$$

where

(5.12) $$\varphi_i = \eta_{ij} \xi^j, \qquad \eta_{ij} \in F$$

([14]) See Cartan [5].

are linear forms of ξ^i.

Now let

(5.13) $$dr = [s, r] = \xi'^i y_i + r'_1 \qquad \xi'^i \in F,\ r'_1 \in K.$$

As $f_R(x\,;r)$ is invariant when r changes by the increment dr, we have

$$df_R = \frac{\partial f_R}{\partial \xi^i}\,\xi'^i = o,$$

consequently

(5.14) $$d\varphi_i = \eta_{ij}\xi'^j = o.$$

But according to the maximal nilpotency of K, there are just ν independent forms among φ_i and it follows from (5.14)

$$\xi'^i = o,$$

i. e. by (5.13)

$$[s, r] \in K, \qquad \text{q. e. d.}$$

By Lemma 1 we see that K is an ideal in L and that R/K, considered as a representation module of S, gives a o-representation of S. But as any representation of semi-simple S is completely reducible there is always such r in any given residue class $\bar{r}$ of R/K, for which we have

$$[s, r] = o, \qquad \text{for all } s \in S.$$

Now we choose a particular basis $x_1, \ldots, x_n$ for K/N, just as we have done in § 4 with respect to soluble L/N. This time we suppose, as it is allowed from the above, that the regular element h_0 in R satisfies

$$[s, h_0] = o, \qquad \text{for all } s \in S.$$

It then follows immediately, that the nilpotent algebras H and H_0 are invariant by all s in S:

(5.15) $$[s, H] \leqq H, \qquad [s, H_0] \leqq H, \qquad s \in S.$$

For such choice of $x_1, \ldots, x_n$ following relations hold for $\beta^k_{\sigma i}$ and $b_{\sigma i}$:

e') $\beta^k_{\sigma i} = o$ for $k \geqq \omega + 1$ (σ and i arbitrary) or for $x_i \in H_0$, $x_k \notin H_0$,

f') $b_{\sigma i} \in K$ in general and $b_{\sigma i} \in H_0 \frown K$ for $x_i \in H_0$.

These follow from Lemma 1 and (5.15).

Under such circumstances, we prove that the space M_0, generated as before by the elements (4.5) and (4.6), is invariant by all S_σ. As it was already proved in § 4, that M_0 is invariant by X_i and that

$$M_0 \frown N = o, \qquad [N_0 : M_0] < \infty,$$

the proof of the existence of desired M_0 will be then completed.

To prove the invariance by S_σ, take N_1 as in (3.15) and define

(5.16) $$S_\sigma z_0 = o, \qquad \sigma = 1, \ldots, m.$$

It follows then from (3.8)

(5.17) $$S_\sigma X_i z_0 - X_i S_\sigma z_0 = \beta^k_{\sigma i} X_k z_0 - b_{\sigma i}.$$

For the element (4.5) or (4.6) we have

$$S_\sigma X_{i_1} \ldots X_{i_\alpha} Y_{j_1} \ldots Y_{j_\beta} z_0 = [S_\sigma, X_{i_1}] X_{i_2} \ldots Y_{j_\beta} z_0 + X_{i_1}[S_\sigma, X_{i_2}] \ldots Y_{i_\beta} z_0 + \ldots$$
$$+ X_{i_1} \ldots X_{i_\alpha} Y_{j_1} \ldots [S_\sigma, Y_{j_\beta}] z_0 + X_{i_1} \ldots X_{i_\alpha} Y_{j_1} \ldots Y_{j_\beta} S_\sigma z_0.$$

The last term is o by (5.16) and if we remark that $[S_\sigma, X_{i_k}]$ and $[S_\sigma, Y_{j_k}]$ can be expressed linearly by the same kind of X_i or Y_j according to e′), we see that all terms in the right side are contained in M_0, except for

(5.18) $$X_{i_1} \ldots X_{i_\alpha} Y_{j_1} \ldots Y_{j_{\beta-1}} [S_\sigma, Y_{j_\beta}] z_0,$$

or in case $\beta = o$

(5.19) $$X_{i_1} \ldots X_{i_{\alpha-1}} [S_\sigma, X_{i_\alpha}] z_0.$$

In (5.18) or (5.19) can appear from (5.17) a term

(5.20) $$X_{i_1} \ldots X_{i_\alpha} Y_{j_1} \ldots Y_{j_{\beta-1}} b_{\sigma, j} \quad (Y_{j_\beta} = X_j)$$

or

(5.21) $$X_{i_1} \ldots X_{i_{\alpha-1}} b_{\sigma, i_\alpha}$$

respectively. But from $b_{\sigma, j} \in H_0$ or $b_{\sigma, i_\alpha} \in K$ (see f′)) and

$$\alpha \geqq n_1 \geqq c_1 + 1, \qquad \beta \geqq n_2 \geqq c_2 + 1,$$

(5.20) and (5.21) vanish just as we have shown in (3.4) or (4.7). (5.18) or (5.19) is therefore also contained in M_0 and we have thus proved the invariance of M_0 by S_σ.

By the considerations in §§ 3–5, we have now reached the conclusion, that, with respect to an abelian ideal N, any Lie algebra L with a finite basis over an algebraically closed field F of characteristic o has always a splitting Lie algebra L, which has also a finite basis over F.

Now let us suppose the case that L is a Lie algebra with a finite basis over a field F, which is of characteristic o, but not necessarily algebraically closed. Denote by $\tilde{F}$ an algebraic closure over F. Then the Lie algebra

$$\tilde{L} = L \times \tilde{F}$$

over F has with respect to its abelian ideal

$$\tilde{N} = N \times \tilde{F}$$

a splitting Lie algebra $\tilde{L}^*$ with a finite basis over $\tilde{F}$. It is then easy to see, that we can find a subfield F_1 of the coefficient field $\tilde{F}$, which is a finite extension over F, so that the Lie algebra $L \times F_1$ over F_1 has also with respect to $N \times F_1$ a splitting algebra L_1^* with a finite basis over F_1. From L_1^* we then obtain immediately a splitting Lie algebra L^* for the extension L/N, which has a finite basis over F.

The existence of a splitting algebra with a finite basis is thus proved for any field F of characteristic o.

§6. The modular case.

In this section let us suppose that the characteristic of the field F is a prime number p. We shall prove the existence of a splitting Lie algebra with a finite basis over F.

We prove first a lemma.

Lemma 2. *Let L be a Lie algebra with a finite basis over F. Then there exists a Lie algebra W over F with following properties:*

i) $W \geqq L$ and W has a finite basis over F,

ii) any ideal in L is also invariant in W,

iii) if we denote by $\underline{x}$ the inner derivation of W, defined by an element x in W, then $\underline{x}^p$ is also an inner derivation of W——in other words, for given x, there is always an element $x^{[p]}$ in W, so that we have

$$(6.1) \qquad \overbrace{[x[x \ldots [x}^{p}, y]] \ldots] = [x^{[p]}, y]$$

for all y in W(15).

Proof. By II in §1, which we shall prove in the next section by making use of Theorem 1, we may suppose that L is imbedded in an associative algebra A over F, which is generated by elements of L. Let Z be the centre of L and Z_1 that of A and let

(15) W is thus p-invariant in the sense of Zassenhaus [13]. Obviously $x^{[p]}$ is determined only modulo the centre of W.

$$Z_1 = Z + Z_0$$

its direct decomposition. It is then easy to see that linear space

$$W^* = \left\{\sum_i \lambda_i w_i^{p^{e_i}};\ w_i \in L+Z_1,\ \ \lambda_i \in F,\ \ e_i = 0, 1, 2, \dots\right\}$$

forms a Lie algebra containing L, with respect to alternants

$$[x, y] = xy - yx.$$

For the proof it is sufficient to note that $w^{p^e} = x^{p^e} + z^{p^e}$ for $x \in L$, $z \in Z_1$ and that

$$(6.2)\qquad [x_1^{p^{e_1}}, x_2^{p^{e_2}}] = [\cdots[[\overbrace{x_1[x_1[\cdots[x_1}^{p^{e_1}}, x_2]]\cdots]\overbrace{x_2]x_2]\cdots x_2]}^{p^{e_2}-1}$$

is contained in L. Moreover W^* contains with an element t the p-th power t^p of it; we have namely by a fundamental identity in a non-commutative ring of characteristic p

$$(6.3)\qquad t^p = (\sum_i \lambda_i w_i^{p^{e_i}})^p = \sum_i \lambda_i^{p^{e_i+1}} w_i^{p^{e_i+1}} + \Lambda(\dots, \lambda_i w_i^{p^{e_i}}, \dots),$$

where Λ is some bracket product of $\lambda_i w_i^{p^{e_i}}$ and is thus contained in L([16]). W^* has accodingly, as readily seen, all properties i), ii), iii), except that it has not yet a finite basis over F.

Now as any x in W^* gives a derivation $\underline{x}$ in L, we see immediately that W^*/Z_1 is isomorphic to a subalgebra of the derivation algebra of L and it has accordingly a finite rank over F. If we put therefore

$$W = W^*/Z_0,$$

W satisfies all our conditions i), ii), iii).

Let L be again a Lie algebra with a finite basis over F and N its abelian ideal. In the Lie algebra W of Lemma 2, N is also an ideal and a splitting algebra for W/N gives immediately that for L/N. In seeking for a splitting Lie algebra, we may therefore suppose, if necessary by replacing L by W, that the p-th power of any inner derivation of L gives also an inner derivation for L, i.e. that for any x in L, there exists a suitabie element $x^{[p]}$, so that (6.1) holds for all y in L.

Now we return to § 2 and use the same notations as there. According to above assumption there exist suitable $x_1^{[p]}, \dots, x_n^{[p]}$ for $x_1, \dots, x_n$, so that

([16]) For the formulas (6.2), (6.3) see Jacobson [8] or Zassenhaus [12].

$$(6.4)\qquad \overbrace{[x_i[x_i[\ldots[x_i}^{p}, y]]\cdots] = [x_i^{[p]}, y]$$

for all y in L. Let

$$(6.5)\qquad x_i^{[p]} = \xi_i^j\ x_j + d_i\,, \qquad \xi_i^j \in F, \quad d_i \in N$$

and define linear transformations $X_i^{[p]}$ by

$$(6.6)\qquad X_i^{[p]} = \xi_i^j X_j\,.$$

Then from (6.4) we see that the transformation

$$(6.7)\qquad U_i = X_i^p - X_i^{[p]},$$

when considered as a linear transformation in N_0 or in N_1/N, commutes with all X_j:

$$(6.8)\qquad U_i X_j = X_j U_i$$

and in particular we have

$$(6.9)\qquad U_i a = o$$

for all a in N.

Then, in virtue of (6.7), (6.8), we can see that N_0 is generated by N and elements of the form

$$(6.10)\qquad f(U_i) X_1^{e_1} \ldots X_n^{e_n} z_0\,, \qquad (o \leqq e_i < p)$$

where $f(U_i)$ is a polynomial of U_i.

If we therefore take as M_0 the linear space generated by all elements of the form (6.10), where $f(U_i)$ contains however only terms of degree $\geqq 3$ in U_i, then

$$[N_0 : M_0] < \infty\,.$$

Next, we want to prove the invariance of M_0 by X_i and consider the term

$$(6.11)\qquad X_i(\prod_{k=1}^{\alpha} U_{i_k}) X_1^{e_1} \ldots X_n^{e_n} z_0\,, \qquad \alpha \geqq 3, \quad o \leqq e_i < p.$$

If $e_1 + \ldots + e_n > o$ and $X_1^{e_1} \ldots X_n^{e_n} z_0$ is in N_0, (6.11) is equal to

$$(6.12)\qquad (\prod_{k=1}^{\alpha} U_{i_k}) X_i X_1^{e_1} \ldots X_n^{e_n} z_0$$

by (6.8) If $e_1 + \ldots + e_n = o$ we have by (6.8), (6.9)

$$X_i(\prod_{k=1}^{a} U_{i_k})z_0 = X_i(\prod_{k=1}^{a-1} U_{i_k})U_{ia}z_0 = (\prod_{k=1}^{a-1} U_{i_k})X_iU_{ia}z_0$$
$$= (\prod_{k=1}^{a-1} U_{i_k})U_{ia}X_iz_0 + (\prod_{k=1}^{a-1} U_{i_k})a = (\prod_{k=1}^{a} U_{i_k})X_iz_0\,,$$

where a is a suitable element in N. Therefore (6.11) is equal to (6.12) in any case. But as

$$X_iX_1^{e_1}\dots X_n^{e_n}z_0 = \sum g_{e'_1,\dots\dots,e'_n}(U_i)X_1^{e_1'}\dots X_n^{e_n'}z_0 + a',$$

with some $o \leqq e'_i < p$, $a' \in N$ and as $(\prod_{k=1}^{a} U_{i_k})a' = o$ by (6.9), (6.12) is equal to

$$\sum_{e_i'}(\prod_{k=1}^{a} U_{i_k})g_{e'_1,\dots\dots,e'_n}(U_i)X_1^{e_1'}\dots X_n^{e_n'}z_0\,,$$

which proves that M_0 is invariant under all X_i(17).

In a similar way we can prove, that in

$$(\prod_{k=1}^{a} U_{i_k})X_1^{e_1}\dots X_n^{e_n}z_0\,,\quad \alpha \geqq 3,\quad O \leqq e_i < p,$$

we may remove U_{i_k} to any other place, so that we can consider that M_0 is genarated by the elements

(6.13) $$U_1^{f_1}X_1^{e_1}\dots\dots U_n^{f_n}X_n^{e_n}z_0\,,$$

where

$$o \leqq e_i < p,\quad f_i = 0,1,2,\dots,\quad f_1 + \dots + f_n \geqq 3.$$

We can then conclude just as in Jacobson [9], that different terms of the form (6.13) are linearly independent over F. This follows from the fact, that, when we express (6.13) linearly by the basis $z_{i_1\dots\dots i_\alpha}$ of N_0, it contains the leading term

$$z_{i_1\dots\dots i_\alpha}\,,$$

where each i appears just $pf_i + e_i$ times in $i_1,\dots,i_\alpha$, while all other z have less number of indexes, and that two such leading terms for (6.13) and for

$$U_1^{f_1'}X_1^{e_1'}\dots U_n^{f_n'}X_n^{e_n'}z_0$$

(17) Here we have only used $\alpha \geqq 2$ in (6.11). The condition $\alpha \geqq 3$ is used in the next proof.

coincide if and only if

$$e_i = e'_i, \quad f_i = f'_i, \qquad i = 1, \ldots, n.$$

If follows then

$$M_0 \frown N = O$$

and we see that the conditions required for M_0 are all fulfilled. The existence of a splitting Lie algebra for L/N is thus proved also for a field of characteristic p.

We can now resume the results of §§ 3–6 in the following main theorem.

Theorem 2. *Let L be a Lie algebra with a finite basis over a field F and N an abelian ideal in L. Then there is a splitting Lie algebra L^* for the extension L/N, which has also a finite basis over F. Here the field F may be quite arbitrary.*

§ 7. The representation of Lie algebras.

From Theorem 1 and 2 we can prove immediately the isomorphic representability of a Lie algebra in an associative algebra.

Suppose first that N^* is a linear space with a finite or infinite rank over a field F and T the set of all linear transformations in N^*. Form the direct sum

$$A = T + N^*.$$

It is also a linear space over F. We then define a multiplication between elements of A as follows: take

$$u_1 = (\sigma_1, v_1), \qquad u_2 = (\sigma_2, v_2), \qquad \sigma_i \in T, \quad v_i \in N^*$$

in A and put

$$u_1 u_2 = (\sigma_1 \sigma_2, \sigma_1(v_2))$$

where $\sigma_1(v_2)$ is the image of v_2 by the linear transformation σ_1. It can be readily seen, that A becomes an associative algebra over F by this multiplication. If particularly N^* has a finite rank, say l, over F and if we suppose that σ and v are represented by matrices

$$\begin{pmatrix} a_{11} \cdots\cdots a_{1l} \\ \vdots \qquad \vdots \\ a_{l1} \cdots\cdots a_{ll} \end{pmatrix}, \qquad \begin{pmatrix} v_1 \\ \vdots \\ v_l \end{pmatrix}, \quad a_{ij}, \; v_i \in F,$$

respectively, then A is nothing else as the algebra of all matrices[18]

$$\begin{pmatrix} a_{11} \cdots\cdots a_{1l} & v_1 \\ \vdots \qquad\quad \vdots & \vdots \\ a_{l1} \cdots\cdots a_{ll} & v_l \\ 0 \cdots\cdots 0 & 0 \end{pmatrix}$$

Now let L be an arbitrary Lie algebra over F and Z the centre of L. We denote by $\tau(x)$ the regular representation of L:

$$\tau(x)y = [x, y], \qquad \text{for} \quad y \in L$$

and by B the associative algebra generated by the linear transformations $\tau(x)$. Then construct by Theorem 1 a splitting Lie algebra L^* over N^* for the extension L/Z:

$$L^* = L + N^* = L_1 + N^*, \quad L_1 \frown N^* = 0, \quad L \frown N^* = N, \quad L_1 \cong L/Z.$$

As $x \to \tau(x)$ gives a faithful representation of L/Z, L_1 is also represented faithfully in B.

On the other hand we define the linear transformation $\sigma(w)$ in N^* corresponding to any w in L_1 by

$$\sigma(w)v = [w, v], \qquad v \in N^*.$$

$\sigma(w)$ gives obviously a representation of L_1. We then consider the mapping

$$u = w + v \to (\sigma(w), v), \qquad w \in L_1, \quad v \in N^*, \tag{7.1}$$

from L^* into the associative algebra A defined above. As N^* is abelian, we have

$$\begin{aligned}[u_1, u_2] = [w_1+v_1, w_2+v_2] &= [w_1, w_2]+[w_1, v_2]+[v_1, w_2]+[v_1, v_2] \\ &= [w_1, w_2]+\sigma(w_1)v_2-\sigma(w_2)v_1\end{aligned}$$

and this corresponds to

$$\begin{aligned}&(\sigma([w_1, w_2]), \sigma(w_1)v_2-\sigma(w_2)v_1) \\ &\qquad = (\sigma(w_1)\sigma(w_2)-\sigma(w_2)\sigma(w_1), \sigma(w_1)v_2-\sigma(w_2)v_1) \\ &\qquad = (\sigma(w_1)\sigma(w_2), \sigma(w_1)v_2)-(\sigma(w_2)\sigma(w_1), \sigma(w_2)v_1) \\ &\qquad = (\sigma(w_1), v_1)(\sigma(w_2), v_2)-(\sigma(w_2), v_2)(\sigma(w_1), v_1).\end{aligned}$$

(18) The same is of course also true for the general case, if we consider infinite matrices.

The mapping (7.1) gives therefore a repesesentation of the Lie algebra L^* in A, in which N^* is represented faithfully. Accordingly, combining it with $\tau(x)$, we can obtain a faithful representation of L^* in the direct sum of A and B. As L is a subalgebra of L^*, L is also represented faithfully in $A+B$.

Thus we have proved the following theorem of Birkhoff-Witt (II in § 1).

Theorem 3. *A Lie algebra L over a field F can be represented faithfully in a suitable associative algebra over F.*

Now if L has a finite basis over F, then we can suppose that L^* has also a finite rank over F by Theorem 2. Associative algebras A and B have accordingly finite ranks over F and we have finally

Theorem 4. *A Lie algebra L with a finite basis over a field F can be represented faithfully in a suitable associative algebra, which has also a finite rank over F. Particularly it admits a faithful representation by finite matrices over F. Here the field F may be qutie arbitrary.*

A precise calculation would give us an upper bound of the minimal degree of faithful representations of L over F as a function of the rank of L over F.

Mathematical Institute,
Tokyo University.

Bibliography

[1] Abe, M. On factor sets of Lie algebras, Zenkoku sijō Sūgaku Danwakai, **226** (1941) (in japanese).

[2] Ado, I. On the representation of finite continuous groups by means of linear transformations, Izvestia Kazan, **7** (1934–35) (in russian).

[3] Birkhoff, G. D. Representability of Lie algebras and Lie groups by matrices, Ann. of Math., **38** (1947).

[4] Cartan, E. Sur la structure des groupes de transformations finis et continus, (Thèse), Paris (1894).

[5] Cartan, E. Les representations linéaires des groupes de Lie, Jour. de Math. pures et appliq., **17** (1938).

[6] Chevalley, C. On the topological structure of soluvable groups, Ann. of Math., **42** (1941).

[7] Iyanaga, S. Zum Beweis des Hauptidealsatzes, Hamb. Abh., **10** (1934).

[8] Jacobson, N. Abstract derivation and Lie algebras, Trans. Amer. Math. Soc., **42** (1937)

[9] Jacobson, N. Restricted Lie algebras of characteristic p, Trans. Amer. Math. Soc., **50** (1941).

[10] Witt, E. Treue Darstellung Liescher Ringe, Crelle's Jour., **117** (193?).

[11] Zassenhaus, H. Lehrbuch der Gruppentheorie, I, Leipzig u. Berlin (1937).

[12] Zassenhaus, H. Über Liesche Ringe mit Primzahlcharakteristik, Hamb. Abh., **13** (1939)

[13] Zassenhaus, H. Endliche p-Gruppe und Lie-Ring mit der Charakteristik p, Hamb. Abh., **13** (1939).

[17] On linearly ordered groups

J. Math. Soc. Japan, 1-9 (1948), 1-9.

(Received Sept. 1, 1947.)

A group G is called a *linearly ordered group* (=l. o. group), when in G is defined a linear order $a>b$, preserved under the group multiplication:

$$a>b \text{ implies } ac>bc \text{ and } ca>cb \text{ for all } c \text{ in } G.$$

A typical example is the additive group R of all real numbers with respect to the usual order. Subgroups of R are also linearly ordered and they are, as is well known, characterized among other l. o. groups by the condition that their linear orders are archimedean, that is to say, that for any positive elements[1)] a, b there is a positive integer n, so that it holds

$$a^n>b, \ b^n>a.$$

Everett and Ulam have proved that we can define a linear order in a free group with two generators, so that it becomes a l. o. group[2)]. In the following we shall generalize this theorem in the form that any l. o. group can be obtained by an order homomorphism from a proper l. o. free group, and then study the general character of group- and order-structure of these groups. Finally we shall add some examples which will illustrate our theorems.

We prove frst some lemmas.

Lemma 1. Let G be a l. o. group and P the set of all positive elements in G. P has then following properties:

i) *$e \notin P$, and if $x \neq e$ either $x \in P$ or $x^{-1} \in P$.*

ii) *$x \in P$ and $y \in P$ implies $xy \in P$.*

iii) *if $x \in P$, then $axa^{-1} \in P$ for all a in G.*

Conversely, if a group G contains a subset P, having the properties i), ii), iii), we can then introduce a linear order in G by defining,

$$a>b, \text{ if } ab^{-1} \in P. \tag{1}$$

Proof. The former part is almost obviovs. We have only to note that iii) follows from $axa^{-1}>aea^{-1}=e$ for $x>e$. We prove the latter part Ac-cording to i) and $ba^{-1}=(ab^{-1})^{-1}$ it can be seen that one and only one of the relations

$$a=b,\ a>b,\ b>a$$

holds for any a, b in G. The transitivity follows then from ii). Moreover $a>b$ implies $(ac)(bc)^{-1}=ab^{-1}\epsilon P$, i e $ac>bc$ and also from iii) $(ca)(cb)^{-1}=c(ab^{-1})c^{-1}\epsilon P$, namely $ca>cb$, which completes the proof.

Lemma 2. Let $\{G_\alpha\}$ be a set of l. o. groups and G the (restricted or complete) direct product of G_α. Then one can introduce a linear order in G so that it becomes a l. o. group.

Proof. We may consider that the groups G_α are well-ordered, where α runs over all transfinite numbers $\alpha<\Omega$. Elements of G are then given by their components:

$$x=\{x_\alpha\},\ x_\alpha\epsilon G_\alpha.$$

Now let P be the set of all such $x=\{x_\alpha\}$, that

$$x_\alpha=e_\alpha{}^{3)} \text{ for all } \alpha<\beta \text{ and } x_\beta>e_\beta \text{ in } G_\beta$$

for some transfinite number $\beta(<\Omega)$ We see readily that P satisfies i), ii), iii) of Lemma 1 and thus we can introduce a linear order in G[4].

We now consider abelian groups and prove

Lemma 3. A linear order can be defined in an abelian group A, if and only if A contains no element of finite order except the unit[5].

Proof. The condition is necessary; if $x>e$ and $x^n=e$ with some integer $n>1$, then it would be $x^{-1}=x^{n-1}>e$, what is a contradiction. Now let A be an abelian group with no element of finite order. We can then imbed A in a direct product G of groups G_α, which are isomorphic to the additive group of all rational numbers. But as the latter is obviously a l. o. group, so is G according to Lemma 2. The subgroup A of G can be therefore also linearly ordered, q e.d.

We now return to the general case and study the order homomorphism. We mean here by an *order homomorphism* (*isomorphism*) a homomorphic (isomorphic) mapping $a\rightarrow a'$ between two l. o. groups G, G', so that $a\geqq b$ in G implies $a'\geqq b'$ in G'. We can then easily prove the following two lemmas.

Lemma 4. Let $a\rightarrow a'$ be an order homomorphism between G, G' and let $G'\cong G/N$, where N is a normal subgroup of G. Then

$$a\geqq b\geqq e \text{ and } a\epsilon N \text{ implies } b\epsilon N \qquad (2)$$

and the set P of positive elements in G consists of positive elements in N and the elements whose homomorphic maps in G' are positive. Conversely if a normal subgroup N of a l. o. group G has the above property (2), then elements of any coset aN(≠N) of G/N are all positive or all negative. Defining positive or negative accordingly, G/N becomes a l. o. group and the natural mapping a→aN then gives an order homomorphism between G and G/N.

Lemma 5. Let G be a group and N a normal subgroup of G. We suppose that N and G/N are both linearly ordered, and the order in N is invariant under the inner transformations of G, namely

$$a\in N,\ a\geqq e \text{ implies } bab^{-1}\geqq e \text{ for all } b\in G. \qquad (3)$$

Then, if we take as positive elements of G positive elements in N and the elements belonging positive cosets in G/N, we can define a linear order in G. The natural mapping G→G/N gives then an order homomorphism.

It is to be noted here particularly that the condition (3) is necessary in order that G becomes a l. o. group, when N and G/N are so. But if N is contained in the centre of G, (3) is trivially fulfiled, so that by a central extension we always obtain a l. o. group G. This remark will be useful soon afterwards.

Now we prove the following

Theorem 1. For a given linearly ordered group G there is a linearly ordered free group F, so that G is the image of an order homomorphic mapping from F to G.

Proof. We take a free group F and a normal subgroup N of F, so that $G\simeq F/N$. By transferring the linear order in G, we can make F/N a l. o. group. Now let $F_1=F, F_2, F_3, \ldots\ldots$ be the descending central chain of F[6] and put $N_i=F_i\cap N$. These groups are then all normal in F and we have

$$N_1=N\geqq N_2\geqq N_3\geqq \ldots\ldots, \quad \cap_{i=1}^{\infty}N_i=e,$$

as $\cap_{i=1}^{\infty}F_i=e$[7]. From

$$[F,\ N_{i-1}]^{8)}=[F,\ F_{i-1}\cap N]\leqq[F,\ F_{i-1}]\cap[F,\ N]\leqq F_i\cap N=N_i,$$

$$N_{i-1}/N_i=(F_{i-1}\cap N)/(F_i\cap N)=(F_{i-1}\cap N)/(F_i\cap(F_{i-1}\cap N))\simeq(F_i(F_{i-1}\cap N))/F_i,$$

we see that N_{i-1}/N_i is contained in the centre of F/N_i and has no element of finite order, because $F_i(F_{i-1} \cap N)/F_i$ is a subgroup of F_{i-1}/F_i, which is known to be free abelian[9]. We can therefore define a linear order in N_{i-1}/N_i according to Lemma 3. By making use of the remark stated just before the theorem, a linear order can be defined, starting from $F/N_1 = F/N$, step by step to all factor groups F/N_i, so that

$$F/N_i \rightarrow F/N_{i-1} \tag{4}$$

gives always an order homomorphism. Now take an element $x \neq e$ in F. According to $\cap_{i=1}^{\infty} N_i = e$ there is an N_i, which does not contain x. We then call x positive (or negative), if the coset of x in F/N_i is positive (or negative) in the sense of above defined linear order in F/N_i. By the successive order homomorphism (4), such a definition is uniquely determined independently of the choice of N_i, which does not contain x. It is then easy to see that F thus becomes a l. o. group and $F \rightarrow F/N$ gives an order homomorphism (Lemma 4).

The above theorem shows at the same time that a free group F admits various linear orders. If we take as G the unit group in the above proof, we shall have a particular linear order in F, perhaps the simplest one. Positive elements in F are then defined as follows. We first define an arbitrary linear order in each central factor group F_{i-1}/F_i. For an element $x \neq e$ in F there is an index i such that x is contained in F_{i-1} but not in F_i. We call then x positive, if x is in a positive coset in F_{i-1}/F_i. Now, when F is thus ordered, we can define a topology in F taking the subsets $\{x; a > x > b\}$ as neighbourhoods (order topology), and F becomes then a topological group (every l. o. group can be thus considered as a topological group). It is note-worthy that this topology in F just coincides with the one, which was defined by G. Birkhoff by making use of congruences with respect to F_i[10].

Now let G be an arbitrary l. o. group. We define the absolute $|x|$ of x in G by

$$|x| = x, \text{ if } x \geqq e$$
$$= x^{-1}, \text{ if } x < e.$$

It is then easily seen that

$$|x| = |x^{-1}|, \qquad |x||y| \geqq |xy|;$$

We denote by $G(x)$ the set of all y in G, for which we have $|x|^n \geqq |y|$ with some positive integer n. From (5) it follows immediately that $G(x)$ forms a subgroup of G and if $y \in G(x)$ and $|z| \leqq |y|$, z is also contained in $G(x)$ On the other hand we call x and y equivalent: $x \sim y$, when it holds $|x|^m \geqq |y|$, $|y|^n \geqq |x|$ with some integers m, n and define a linear order in the set of these equivalent classes $\lambda, \mu, \ldots\ldots$ by putting

$$\lambda > \mu, \quad \text{if} \quad x > y \quad \text{for all} \quad x \in \lambda,\ y \in \mu.$$

From the definition it follows immediately that the unit of G forms a class with only one element. We denote by L the set of all equivalent classes, which are different from this unit class.

It is clear that $G(x) = G(y)$ if and only if $x \sim y$ and we may therefore write

$$G(x) = G_\lambda, \text{ if } x \in \lambda.$$

Now let G_λ^* be the composite of all groups G_μ with $\mu < \lambda$ (if λ is the first ele : ent in L, we take the unit group as G_λ^*). We can prove easily that if $G_\lambda = G(x)$, G_λ^* is the set of all elements in G_λ, which are not equivalent to x and consequently that G_λ^* is normal in G_λ. Moreover as G_λ and G_λ^* satisfy (3) in Lemma 5, we can induce the linear order of G into G_λ/G_λ^*. But this order in G_λ/G_λ^* is archimedean, for all elements in G_λ, not contained in G_λ^*, are mutually equivalent. There is accordingly an order isomorphism I_λ between G_λ/G_λ^* and a subgroup R_λ of the group of all real numbers R.

Next take an element x and consider the inner automorphi m I_x of G by x. If $G_\lambda = G(y)$, then $xG_\lambda x^{-1} = G(xyx^{-1}) = G_{\lambda'}$, so that $\lambda \to \lambda'$ gives a one-to-one order preserving correspondence in L. As G_λ^* is transformed into $G_{\lambda'}^*$ by the same automorphism, I_x induces an order isomorphism between G_λ/G_λ^* and $G_{\lambda'}/G_{\lambda'}^*$, which we denote by $I_{x,\lambda}$. $I_{\lambda'} I_{x,\lambda} I_\lambda^{-1}$ then gives an order isomorphism between R_λ and $R_{\lambda'}$, so that

$$r' = I_{\lambda'} I_{x,\lambda} I_\lambda^{-1}(r) = r_{x,\lambda} r, \text{ for } r \in R_\lambda,\ r' \in R_{\lambda'},$$

with some positive real number $r_{x,\lambda} > 0$.

We have thus obtained the following

Theorem 2. For any linearly ordered gronp G, there is a linearly ordered set $L = \{\lambda, \mu, \ldots\ldots\}$ and a corresponding sequence of subgroups $\{G_\lambda; \lambda \in L\}$ of

G, which have following properties:

1) $G_\lambda > G_\mu$, *if* $\lambda > \mu$ ($G_\lambda \neq e$, *if* λ *is the first element in* L);

2) for any x *in* G, *there is an element* λ *in* L, *such that* x *is contained in* G_λ, *but not in* G_μ *for all* $\mu < \lambda$;

3) let $G_\lambda{}^*$ *be the composite of all* G_μ *with* $\mu < \lambda$ ($G_\lambda{}^* = e$, *if* λ *is the first element in* L). $G_\lambda{}^*$ *is then normal in* G_λ *and there is an isomorphism* I_λ *between* $G_\lambda/G_\lambda{}^*$ *and a subgroup* R_λ *of the group of all real numbers* R;

4) if we denote by I_x *the inner automorphism of* G *by the element* x, *each* G_λ *goes to some other* $G_{\lambda'}$ *under* I_x. $\lambda \to \lambda'$ *then gives an order preserving one-to-one correspondence in* L. $G_\lambda{}^*$ *is transformed consequently to* $G_{\lambda'}{}^*$ *and* I_x *induces an isomorphism* $I_{x,\lambda}$ *between* $G_\lambda/G_\lambda{}^*$ *and* $G_{\lambda'}/G_{\lambda'}{}^*$. *Further the isomorphism* $I_{\lambda'} I_{x,\lambda} I_\lambda^{-1}$ *between* R_λ *and* $R_{\lambda'}$ *is given by*

$$r' = I_{\lambda'} I_{x,\lambda} I_\lambda^{-1}(r) = r_{x,\lambda} r, \text{ fo } r \in R_\lambda,\ r' \in R_{\lambda'},$$

where $r_{x,\lambda}$ *is a certain positive number;*

5) let $x \in G_\lambda$, $x \notin G_\mu$ *for* $\mu < \lambda$ (*cf.* 2). x *is then positive if and only if the coset of* x *in* $G_\lambda/G_\lambda{}^*$ *is mapped by* I_λ *into a positive number in* R_λ.

For a given linearly ordered group G, *such* L *and* G_λ *are uniquely determined by the relations 1)—4) (up to an isomorphism)*

Conversely if G *is a group and there exists a sequence of subgroups* G_λ *of* G, *corresponding to a linearly ordered set* L, *and satisfying above conditions 1)—4) and if we define positive elements by 5), then* G *becomes a linearly ordered group.*

As L and the corresponding R_λ are uniquely determined for a given l. o. group, we may classify the set of all l. o. groups by grouping together those groups into a class, which have the same L and R_λ in the sense of above theorem. We note here some remarks on this classification. First there is no restriction upon L and R_λ. Namely, if a linearly ordered set L and corresponding subgroups R_λ of R are arbitrarily given, there is always a l. o group G, to which L and R belong in the above sense: to obtain such G, we have only to construct the restricted direct product of all R_λ ($\lambda \in L$) and take as G_λ the restricted direct product of all R_μ $\mu \leqq \lambda$. We have thus proved in the same time that every such class contains an abelian group.

Next let us suppose that L is such a linearly ordered set that it admits no order preserving one-to-one correspondence except the identity. This is particularly the case, when L is well-ordered or has an order inverse to a well-ordered set. Then we have from 4) in Therem 2 that each G_λ must be normal in G and that $r \to r' = r_{\lambda,x} \cdot r$ $(r_{\lambda,x} > 0)$ gives an automorphism of R_λ G is thus a solvable group in a generalized sense[11]. If further R_λ admits no automorphism of that type except the identity, G_λ/G_λ^* is contained in the centre of G/G_λ^*. For example if L is well-ordered (or has an order inverse to it) and all R_λ are the group of rational integers (namely free cyclic groups), every group of type $\{L, R_\lambda\}$ is nilpotent in a generalized sense[12]. It may be possible to determine in this manner the group-theoretical structure of l. o. groups of various types more explicitely. Here we do not enter in this problem[13].

We give finally some examples of l. o. groups of various types.

Example 1. Let $G^{(1)}$ be the restricted direct product of denumerable number of free cyclic groups $C_1, C_2 \ldots\ldots$ and let G_n be the direct product of $C_1, C_2, \ldots\ldots, C_n$. By theorem 2 we can define a linear order in $G^{(1)}$. L is then of type $\omega = 1, 2, 3, \ldots\ldots$ and R_λ are the group of all rational integers I.

Example 2. Again, let $G^{(2)}$ be the (complete or restricted) direct product of $C_1, C_2, \ldots\ldots$ Now take as G_n the direct product of $C_n, C_{n+1}, \ldots\ldots$ L is here of type $\omega^* = \ldots\ldots, 3, 2, 1$ and R_λ are again I. The linear order considered here coincides with the one stated in the proof of Lemma 2.

Example 3. Let $G^{(3)}$ be generated by three elements a, b, c with relations

$$ab = ba, \quad cac^{-1} = a, \quad cbc^{-1} = ab.$$

An element in G can be written uniquely in the normal form $a^x b^y c^z (x, y, z = 0, \pm 1, \pm 2, \ldots\ldots)$. If we put $G_1 = \{a\}$, $G_2 = \{a, b\}$, $G_3 = \{a, b, c\}$, $G^{(3)}$ becomes a l. o. group. Here $L = \{1, 2, 3\}$, $R_1 = R_2 = R_3 = I$. $G^{(3)}$ may be perhaps one of the simplest non-abelian l. o. group.

Example 4. As a generalization of $G^{(3)}$ let us consider $G^{(4)} = \{z, a_1, a_2, \ldots\ldots\}$ with relations

$$za_i = a_i z, \quad a_i a_j = a_j a_i, \text{ for } |j-i| \neq 1, \quad a_i a_{i+1} = a_{i+1} a_i z.$$

Putting $G_n = \{z, a_1, \ldots\ldots, a_{n-1}\}$ we have a non-abelian l. o. group $G^{(4)}$ of the same type with $G^{(1)}$.

Example 5. A free group F with a finite number of generators has the same type with $G^{(2)}$, if we define the order by making use of F_i.

Example 6. Let $G^{(6)}$ be the set of all linear functions $f(x)=ax+b$, where $a>0$, $-\infty<b<\infty$. Defining the group composition $h=f\times g$ by $h(x)=f(g(x))$ and the order $f>g$ by the value of these functions for sufficiently large x, $G^{(6)}$ becomes a l. o. group. Here $G_1=\{f(x)=x+b\,;\,-\infty<b<\infty\}$, $G_2=G^{(6)}$, $L=\{1, 2\}$ and $R_1=R_2=R$. Moreover we have

$$r_{f,1}=a, \text{ for } f(x)=ax+b,$$

whereas $r_{f,2}=1$ for all f.

Example 7. Take the additive group $G^{(7)}$ of all such real functions $f(x)$ which are defined on the interval $[0, 1]$ and are represented in the x-y plane by a broken line through the origin $(0, 0)$. We define $f(x)>g(x)$, if there is such a $(0\leqq a<1)$, so that $f(x)=g(x)$ for $0\leqq x\leqq a$ and $f(a+\varepsilon)>g(a+\varepsilon)$ for sufficiently small $\varepsilon>0$. Putting

$$G_\lambda=\{f(x)\,;\,f(x)=0 \text{ for } 0\leqq x\leqq 1-\lambda\},\ 0<\lambda\leqq 1,$$

we see that L coincides with the interval $(0, 1]$ and $R_\lambda=R$ for all λ.

Example 8. Let $G^{(8)}$ be the subset of all those functions considered in Example 7, which are moreover monotone increasing (in the strict sense) and go through the point $(1, 1)$. We define the group composition $h=f\times g$ by $h(x)=f(g(x))$. By the same order as in Example 7, $G^{(8)}$ becomes a non-abelian l. o. group. Here G_λ is given by

$$G_\lambda=\{f(x)\,;\,f(x)=x \text{ for } 0\leqq x\leqq 1-\lambda\},\ 0<\lambda\leqq 1,$$

and again $L=(0, 1]$. $R=R_\lambda$ for all λ. But in this case no G_λ except $G_1=G^{(8)}$ is normal in $G^{(8)}$. In fact we have

$$fG_\lambda f^{-1}=G_{f(\lambda)}.$$

On the other hand $r_{f,\lambda}$ are here all 1. If we want to obtain an example in which $r_{f,\lambda}$ is not constant, we have only to extend[14] the group $G^{(7)}$ by $G^{(8)}$, putting

$$gfg^{-1}=f(g^{-1}(x)), \text{ for } f\in G^{(7)},\ g\in G^{(8)}.$$

Mathematical Institute,
Tokyo University.

Revised February 17, 1948.

References.

1) An element a is called positive in a l. o. group, if $a > e$, where e is the unit of the group.

2) Everett, C. J. and Ulam, S. On ordered groups, Trans. Amer. Math. Soc. 57 (1945).

3) e_α is the unit of G_α.

4) In the case of the restricted direct product we can define a linear order in another way. Cf. the first example at the end of the paper.

5) We can prove more generally (by making use of subsequent lemmas) that a linear order can be defined in such a group, which has no element of finite order and has a finite or transfinite ascending central chain.

6) Cf. H. Zassenhaus, Lehrbuch der Gruppentheorie (1937), p. 119 or E. Witt, Treue Darstellung Liescher Ringe, Crelle Jour., 177 (1937).

7) E. Witt, l. c. 6).

8) The bracket means the commutator group.

9) E. Witt, l. c. 6).

10) G. Birkhoff, Moore-Smith convergence in general topology, Ann. Math., 38 (1937).

11) If L is particularly a finite set, G is surely solvable in the usual sense. Cf. Example 3, 6 below.

12) For the generalization of nilpotent groups cf. R. Baer, Nilpotent groups and their generalizations, Trans. Math. Amer. Math. Soc., 47 (1940).

13) In this connection hold particularly remarkable relations between l. o. groups and nilpotent groups, cf. 5).

14) Cf. Lemma 5.

[18] Finite groups and compact groups

Sûgaku, 1-2 (3/1948), 94-95

Abstract

Some theorems concerning finite groups can be extended to the case of compact topological groups with a minor modification. The following theorem is an example of such phenomena.

Theorem. Let G be a compact topological semigroup satisfying the condition:

(1) If one has $ax = ay$ or $xa = ya$ with $a, x, y \in G$, then one has $x = y$.

Then G is a compact topological group.

Here semigroups are assumed to be associative. This theorem follows easily from the following two lemmas.

Lemma 1. Any compact topological semigroup (not necessarily satisfying the condition (1)) contains a subgroup.

Lemma 2. A semigroup G satisfying the condition (1) contains at most one idempotent element. If an idempotent exists, then it is the unit element of G.

To prove Lemma 1, it suffices to observe that a minimal compact subsemigroup H (which exists by Zorn's lemma) is a subgroup, (because one has $hH = Hh = H$ for any $h \in H$ by the minimality of H). To prove Lemma 2, let e be an idempotent and a an arbitrary element in G. Then from $e(ea) = e^2a = ea$ one can deduce by (1) that $ea = a$; similarly one has $ae = a$. Hence e is the unit element of G, which is uniquely determined.

Finally, to prove the Theorem, let G be a compact topological semigroup satisfying the condition (1). Then, by Lemmas 1 and 2, G contains the unit element e. Since, for any $a \in G$, aG is a compact subsemigroup of G, one has (again by Lemmas 1, 2) $e \in aG$ and similarly $e \in Ga$. Hence there exist elements $x, y \in G$ such that $ax = e$ and $ya = e$, which proves that G is a group. The continuity of the map $a \mapsto a^{-1}$ can easily be verified.

寄　　書

有限群と compact 群

岩澤健吉（東大理）

（昭和21年7月10日受理）

有限群に關する定理には些少の變更によつてこれを compact な位相群に擴張し得る場合がある．以下述べるのはその一例である．Gを有限半群一即ち associative な乘法の定義されてゐる有限集合とせよ，このとき (1) $ax=ay$ 又は $xa=ya$ なるとき常に $x=y$ となるならば，G は有限群となる．之はよく知られて居る定理であるが，その證明には G が有限集合であるといふことが essential に用ひられてゐる．實際無限半群では上の定理は必ずしも成立しない．例へば——正の實數の全體のつくる加法半群を考へてみればよい．所が G が同時に compact space で乘法がこの topology に關し連續であるといふ條件を入れると定理は無限集合の場合にも成立するのである．即ち

定理　上の意味で compact な半群 G が (1) を滿足すればそれは compact 位相群である．

これを證明する爲にはGが抽象的な意味で群になること（單位元及び逆元の存在）及び逆元 a^{-1} が a に關して連續なること云はねばならぬが，後者は容易であるから前者の證明のみを述べることにする．そのため先づ二つの lemma を證明する．

Lemma 1. 任意の compact な半群は部分群を含む（こゝでは (1) は必要でない）．

證明　G の凡ての compact 部分半群の集合 S を考へる．S の任意の線型系列

$$H_1 \geqq H_2 \geqq H_3 \geqq \cdots \geqq H_n \geqq \cdots$$

は G の compact 性から空集合でない共通部分を有し，それは明かに compact な部分半群である．よつて Zorn の lemma を用ひれば S には minimal な compact 部分半群 H が存在する．この H が部分群になる．何となれば，H の任意の h に對し $h \cdot H$ は H に含まれる compact 部分半群となるから H の minimal 性から $hH=H$，同樣に $Hh=H$，よつて任意の $h' \in H$ に對し $hx=h'$ 或は $yh=h'$ を滿足する x, y がそれぞれ H 中に存在し H は群となる．（尙 topology に關する證明が要るわけだが，これは省略する）．

Lemma 2. (1) を滿足する半群 G には $e^2=e$ なる元　は高々一つしか存在しない. ──もし存在すればそれは G の單位元である.

證明　$e^2=e$ とすれば任意の $a\epsilon G$ に對し

$$e(ea)=e^2a=ea$$

(1) を用ひれば $ea=a$ 同様に $ae=a$ 卽ち e は G の單位元で, 從つて unique に定る.

さて上の定理はこれらの lemma から容易に出る. G を compact な半群で且 (1) を滿足するものとせよ. H を G の任意の compact 部分半群とすれば lemma 1 により H は更にその中に compact 部分群を含むから, この部分群の單位元である $e^2=e$ なる元 e を含む. 然るに lemma 2 に依りこの樣な e は G の單位元に外ならない. 卽ち G の任意の a に對し aG は明かに G の compact 部分半群であるから, $e\epsilon aG$, 同様に $e\epsilon Ga$, よつて

$$ax=e \qquad ya=e$$

なる x, y が G 中に存在し, G は群となることがわかる. 證明終.

序に有限群と compact 群とで大分樣子の異る群論的性質の一例を揭げておかう. G を次數 $r\geqq 2$ なる凡ての unitary matrix のつくる compact 群, H を diagonal matrix 全體から成るその部分群とする. 然らば任意の unitary matrix は同じく適當な unitary matrix により diagonal form に變形できるから, H とその共軛部分群の全體により G の凡ての元が盡される. この樣なことは勿論有限群では起らない.

[19] Hilbert's fifth problem

Sûgaku, 1-3 (11/1948), 161-171

Abstract

This note with the subtitle "On the structure of solvable topological groups" is a preliminary report of the results in [21] (1949) given at the annual meeting of the Mathematical Society of Japan, October 1947. In the introduction (§1), after explaining the meaning of Hilbert's fifth problem and its development up to 1940 (due to J. von Neumann and L. Pontrjagin), the author writes,

"... According to recent information from Mathematical Reviews, etc., it seems that there has been important progress in this area during 1941-46, namely, the results of C. Chevalley and A. Malcev on solvable topological groups, yielding an affirmative solution of the Hilbert problem for this case, which gives a complete characterization of a solvable Lie group as a topological group. It is our purpose here to explain these results along with some generalizations, in particular, to give some results on the relation between general topological groups and Lie groups."

The main results of [21] (§4,5) on (L)-groups and general locally compact groups are given without proofs in §5 of this note; in particular, the solution of the Hilbert problem for (L)-groups ([21] Th. 12), the local and global structure theorems for a connected (L)-group ([21] Th. 11, 13), and the existence of the radical for a connected locally compact group ([21] Th. 15) are given here as Theorems 10, 11, 12, and 13, respectively. The conjecture (C_1) and its equivalence with the modified Chevalley's conjecture (C_2) are also stated. However, the extension theorem for Lie groups ([21] Th. 7) is given in this note only for the case of solvable Lie groups (Theorem 5).

The definitions and basic results on solvable topological groups are given (with proofs) in §2 similar to those in §§ 1,2 of [21]; in particular, Theorems 2′, 3, and 4 here are the same as Theorems 1, 4, and 5 in [21]. § 3 concerning with solvable Lie groups contains the extension theorem (Theorem 5) mentioned above and a structure theorem for solvable Lie groups (Theorem 6 = [21] Lem. 3.6). In §4 the following important theorem is proved:

Theorem 7. Let G be a connected locally compact solvable group. Then there exists a compact abelian normal subgroup N such that G/N is a solvable Lie group.

Here the subgroup N becomes automatically central (Theorem 3) and can be chosen arbitarily small (Theorem 7′). As in the case of compact groups (cf. Pontrjagin [10], §§ 4,5) it follows that every locally Euclidean solvable group is a Lie group (Theorem 9). Theorem 7 also implies that every connected locally compact solvable group is an (L)-group ([21] Th. 10).

論　　説

Hilbertの第五の問題[1)]

可解位相群の構造について

東京大學　岩澤健吉

（昭和22年12月5日受理）

§1　問題の説明

1900年Parisの萬國數學者會議における'Mathematische Probleme'と題する講演のなかでD. Hilbertは，あたかも世紀の變り目に當り所謂'回顧と展望'を行つて，來るべき新世紀において數學者の研究の目標となるべき問題を數學の各分野に亘つて23箇提出した[2)]．これらの問題のうちには，例えば高木先生の類體論のように，當初Hilbertが豫想していた以上に一般に且美麗な形で解決されたものもあるのであるが，然し今日においても尚未解決のまま殘されているものも少くない．次に述べようとするのはHilbertが揭げた23問中の丁度5番目に當るので普通Hilbertの第五の問題と呼ばれているものであるが，これもその未解決なものの一つである．

先ず問題を説明しよう．それは簡單に次の如く述べられる．即ち

「凡てのlocally Euclidean groupはLie groupではないか？」

こゝにlocally Euclidean groupと言うのはあるEuclid空間内の開集合にhomeomorphicな單位近傍を有する位相群のことである．今そのEuclid空間の次元をnとすればこのような群Gの單位元の近傍には$(x_1, \cdots, x_n)$なる實數の座標を導入することができる．一般に座標$(x_1 \cdots, x_n)$を有するGの元を

$$g(x_1, \cdots, x_n)$$

と書くこととし，$g(x_1, \cdots, x_n)$と$g(y_1, \cdots, y_n)$との積を$g(z_1, \cdots, z_n)$とすればz_kは$x_1, \cdots \cdots, x_n ; y_1, \cdots, y_n$の連續函數となるであろう．即ち

$$(1)\quad g(z_1, \cdots, z_n) = g(x_1, \cdots, x_n)g(y_1, \cdots, y_n),$$
$$z_k = \varphi_k(x, \cdots, x_n ; y_1, \cdots, y_n),$$
$$k=1, \cdots, n.$$

さてこのようなGの單位近傍における座標系のとり方は幾通りもあるわけであるが，ここでHilbertは「適當な座標系を選べば(1)における函數φ_kを$x_1, \cdots, x_n ; y_1, \cdots, y_n$の解析函數とすることができるのではないか？」と豫想するのである．これが上に述べた問題の意味である．

さて今世紀に入つてから，O. Schreierによつて確立された一般位相群の概念がその後益々重要性を加えつつある一方，E. Cartan等の研究により古典的Lie群論にも又大きな進歩がなされた．かくしてこの二つの理論，即ち一般の位相群論，特にlocally compact groupの理論とそのspecial caseである所のLie groupの理論とを如何に結び付けるかというのが極めて重要な問題となつて來た．そうして上に述べたHilbert[3)]の問題の解決こそその關聯を解明すべき第一の鍵なのである．實際若しもHilbertの問題が完全に解決されればこれによりLie groupはlocally Euclidean groupとして位相群論的に特徴付けられ，一般位相群としてのLie groupの位置も明かになるわけであるが，これと同時に一方解析を縱横に用いて得られるLie群論の深い結果を一般位相群の研究に應用する途も又開かれるわけである．

先に述べたようにこの問題は今のところ未だ完全には解決されていないのであるが位相群論の中心的課題である爲多くの數學者の研究を俟つて今日次第に解決に近づきつつあるのである．次にこれまでに知られている主な結果について極く簡單に述べてみよう．先ず1933年J. v. Neumannは位相群の連續表現の理論を用いて群Gがcompact groupである場合にHilbertの問題が解決されることを示した．次いで1934年L. Pontrjagin は今日彼の名をもつて呼ばれている locally compact abelian group の理論の中で同じくこの問題を G が abelian group の場合に證明した．後に H. Freudenthal はこれら compact group 乃至 abelian group に對する Hilbert の問題の解決の根據がこれらの群が所謂 maximally almost periodic group であることに存することを明かにした．この他にも尙 H. Cartan の analytic transformation group に關する研究等優れた結果も多いがそれ等は省略する．兎も角 compact group と abelian group，或はそれらを一まとめにして言えば maximally almost periodic group に對する問題の解決が我々がこの戰爭前迄に知つていた Hilbert の問題に關する最も重要な結果であつたと言つてもよいであろう[4]．ところが最近 Mathematical Reviews 等に依つて知り得た所によるとこの 1941—1946 年間にこの方面で又重要な進歩がなされた模樣である．卽ち solvable topological group（可解位相群）に關する C. Chevalley, A. Malcev の研究がそれで，これによつて Hilbert の問題は solvable group に對しても解決され，從つて solvable Lie group は位相群としての性質だけで完全に特徵付けられたわけである．そこで次にこれらの結果を紹介し，その擴張，特に一般の locally compact group と Lie group との關係について二三述べて見ようと思う[5]．

§ 2 Solvable topological group

先づ solvable topological group の定義を與えよう．その爲次のようなやや一般的な考察から初める．G を任意の topological group とし H_1, H_2 を G の subgroup とせよ．任意の $h_1 \in H_1$, $h_2 \in H_2$ の commutator $[h_1, h_2]=h_1h_2h_1^{-1}h_2^{-1}$ から generate される G の abstract subgroup を $[H_1, H_2]$ とする．$[H_1, H_2]$ は必ずしも G で閉じていないからその closure $\overline{[H_1, H_2]}$ をとつて H_1, H_2 の G における topological commutator group と名付け $C(H_1, H_2)$ と記すこととする．この $C(H_1, H_2)$ に關し次の lemma が後に役立つ．

Lemma 1. H_1 或は H_2 が connected group ならば $C(H_1, H_2)$ も又 connected である．

證明 $C(H_1, H_2)$ の connected component を M とせよ．H_1 が connected ならば H_2 の定まつた元 h_2 に對し

$$h_1 \to [h_1, h_2], \qquad h_1 \in H_1$$

なる H_1 から $C(H_1, H_2)$ 内への寫像を考える．この寫像が連續であることは明かだからこれによる H_1 の像は $C(H_1, H_2)$ の connected subset で $e=[e, h_2]$ を含む．從つてそれは M に含まれる．このことは H_2 の任意の h_2 について言えるから結局任意の $h_1 \in H_1$, $h_2 \in H_2$ に對し $[h_1, h_2]$ が M に含まれる．M は closed subgroup であるからこれから $\overline{[H_1, H_2]} \subseteq M$，卽ち $C(H_1, H_2)=M$．これは $C(H_1, H_2)$ が connected であるということに外ならない．證終．

さて G を任意の topological group とし，ξ を任意の超限數とするとき，G の subgroup $C_\xi(G)$ を次の如く定義する．卽ち先ず

$$C_0(G) = G$$

とし，次に $\eta<\xi$ なる任意の η に對し $C_\eta(G)$ が旣に定義されたものとすれば

$\xi=\xi'+1$ ならば $C_\xi(G)=C(C_{\xi'}(G), C_{\xi'}(G))$,

ξ が極限數ならば $C_\xi(G)=\bigcap_{\eta<\xi} C_\eta(G)$

とおく．かくすれば超限歸納法により $C_\xi(G)$ が一意的に定義される．この定義から直ちに各 $C_\xi(G)$ は凡て G の normal subgroup なること，$\eta \leqq \xi$ ならば $C_\eta(G) \supseteq C_\xi(G)$ なること，又 $C_\xi(G)=C_{\xi+1}(G)$ ならば凡ての $\zeta \geqq \xi$ に對し $C_\zeta(G)=C_\xi(G)$ となることなどは容易に證明される．よつて特に $\eta<\xi_0$ なるとき $C_\eta(G) \neq C_{\xi_0}(G)$，又 $\eta \geqq \xi_0$ ならば $C_\eta(G)=C_{\xi_0}(G)$ となるような超限數 ξ_0 が唯一つ存在することがわかる．

さて上の如くにして定義された $G=C_0(G)$ から始まる G の normal subgroup の系列

$$C_0(G) \supseteq C_1(G) \supseteq \cdots \supseteq C_\xi(G) \supseteq \cdots$$

を G の commutator series と呼び，ξ_0 をこの series の長さ，或は簡單に G の長さと略稱することにする．

定義 $\{C_\xi(G)\}$ を topological group G の commutator series, ξ_0 をその長さとするとき

$$C_{\xi_0}(G)=e \qquad (e=\text{單位元})$$

ならば G を solvable topological group と呼

ぶ.

これで solvable topological group の意義は確定したが次に後章で用いる solvable group の性質を二三證明しておく.

定理 1. Connected compact group G の長さは高々1である. よつて特に G が solvable ならばそれは abelian group である.

この定理を證明するためには先ず一般に compact group K の automorphism の群 $A(K)$ を考察せねばならぬ. 我々は次の様にして $A(K)$ に topology を導入する. 即ち V が K の一つの單位近傍系 $\{V\}$ を動くとき

$$U(V)=\{\sigma\,;\,\sigma\in A(K),\ K^{\sigma-1}\subseteq V\}$$

によつて定義される subset の集合 $\{U(V)\}$ をもつて $A(K)$ の單位近傍系を定めるのである. 之によつて $A(K)$ が實際 topological group になることは容易に證明される. さてこの時次の定理が成立する.

定理 2. K が compact abelian group ならば $A(K)$ は totally disconnected (0-dimensional) group である.

證明 K の charactor group を X とし, X の automorphism の群を $A(X)$ とせよ. 然らば

$$\chi(s^{\sigma})=\chi^{\sigma'}(s),\ s\in K,\ \chi\in X,\ \sigma\in A(K),\ \sigma'\in A(X)$$

なる關係により $A(K)$ の元 σ と $A(X)$ の元 σ' とは1對1に對應しこの對應によつて $A(K)$ と $A(X)$ とは代數的に同型となることがわかる. 然るに $\chi_1,\cdots,\chi_n$ を任意に X からとる時

$$U(\chi_1,\cdots,\chi_n)=\{\sigma'\,;\,\sigma'\in A(X),\ \chi_i^{\sigma'}=\chi_i;\ i=1,\cdots,n\}$$

によつて與えられる $A(X)$ の subset の集合 $\{U(\chi_1,\cdots,\chi_n)\}$ をもつて $A(X)$ の單位近傍系と定めるならばこれにより $A(X)$ も又 topological group となつて上に與えた同型對應 $\sigma\leftrightarrow\sigma'$ が實は $A(K)$ と $A(X)$ との間の topological group としての同型を與えることが證明される[6]. 所が $U(\chi_1,\cdots,\chi_n)$ は容易に分るように $A(X)$ において同時に open 且 closed な集合であるから $A(X)$ は totally disconnected, 從つて $A(K)$ も亦 totally disconnected である. 證終.

尙この定理より一般に次の定理が成立するが證明は省略する.

定理 2′. K を任意の compact group, $I(K)$ を K の inner automorphism の群をすれば $I(K)$ は $A(K)$ の compact normal subgroup で $A(K)/I(K)$ は totally disconnected である.

さて定理2から容易に次の定理が導かれる.

定理 3. G を任意の connected topoloigcal group, K を G の compact abelian normal subgroup とすれば K は G の centre に含まれる.

證明 G の任意の元 a に對し

$$s^{\sigma(a)}=a^{-1}sa,\ s\in K$$

とおけば $\sigma(a)$ は K の automorphism を與えるがここで

$$a\rightarrow\sigma(a)$$

は容易にわかるように G から $A(K)$ 内への連續寫像となる. G は connected なる故集合 $\{\sigma(a)\,;\,a\in G\}$ は $A(K)$ の connected subset でなければならぬが一方定理2により $A(K)$ は totally disconnected なる故 $\{\sigma(a)\}$ は凡て K の identical automorphism と一致することがわかる. 即ち

$$s^{\sigma(a)}=s,\quad as=sa,\quad s\in K.$$

a は G の任意の元であるから K は G の centre に含まれねばならぬ. 證終.

さてこの定理3を用いて定理1が次のように證明される.

$$G=C_0(G)\supseteq C_1(G)\supseteq C_2(G)\supseteq\cdots$$

を G の commutator series とし

$$G'=C_0(G)/C_2(G),\ N'=C_1(G)/C_2(G)$$

とせよ. a' を G' の任意の元とするとき, a' と N' とから generate される G' の subgroup を M' とすれば定理3により N' は G' の centre に含まれるから a' と N' の各元とは commutative で從つて M' は compact abelian group である. 又 M' は N' を含み, G'/N' は abelian であるから M' は G' の normal subgroup となる. よつて又定理3により M', 特に a' は G' の centre に含まれる. a' は任意であつたから從つて G' は abelian group で $N'=e$ でなければならぬ. 即ち

$$C_1(G)=C_2(G).$$

よつて定義により G の長さは高々1である．特に G が solvable ならば $C_1(G)=e$, 即ち G は abelian group でなければならぬ．

これと同様な方法で次の定理が證明できる．

定理 1′. G を任意の connected topological group, K を G の任意の必ずしも connected でない compact normal subgroup とすれば K の長さは高々1である．

さて定理1を locally compact group に擴張するために先ず次の lemma を證明する．

Lemma 2. G を locally compact group, G_i $(i=0, 1, 2, \cdots)$ を

$$G=G_0\supseteqq G_1\supseteqq G_2\supseteqq\cdots$$

なる如き G の connected subgroup の系列で $\bigcap_{i=0}^{\infty} G_i=e$ なるものとすれば適當に m をとれば $i\geqq m$ なるとき G_i は凡て compact group となる．

證明 U を G の單位近傍で closure $\bar{U}$ が compact なるものとせよ．$R=\bar{U}-U$ を U の境界とすれば $\bigcap G_i=e$ なる故十分大なる $i(i\geqq m)$ に對しては R と G_i とは共通點をもたない．G_i は connected なる故このような G_i は全く U に含まれる．$\bar{U}$ は compact だから從つて G_i も又 compact でなければならぬ．證終．

この lemma により定理1が次のように擴張される．

定理 4. Connected locally compact group G の長さは有限である．特に G が solvable ならば適當な自然數 n をとるとき

$$C_n(G)=e$$

となる．

證明 $G=C_0(G)\supseteqq C_1(G)\supseteqq\cdots\supseteqq C_\omega(G)\supseteqq\cdots$ を G の commutator series とせよ．$C_\omega(G)=N$, $G/N=G'$ とおけば

$$C_i(G')=C_i(G)/N,\ \bigcap C_i(G')=e,\ i=1\ 2\cdots$$

なることは容易にわかる． Lemma 1 により各 $C_i(G')$ は connected だから Lemma 2 を用いれば $i\geqq m$ なる時 $C_i(G')$ は凡て compact となる．$C_i(C_m(G'))=C_{i+m}(G')$ であるから特に $C_m(G')$ は connected compact solvable group であつて從つて定理1によりその長さは高々1である．よつて

$$C_{m+1}(G')=C_{m+2}(G'),$$

$$C_{m+1}(G)=C_{m+2}(G).$$

即ち G の長さは高々 $n=m+1$ である．證終．

定理 1′ と同様にこの定理 4 を擴張することも勿論可能である．又定理 4 により connected locally compact group に對しては初め極めて一般に與えた solvable topological group の定義が實際には例えば有限群の場合の定義の極く natural な擴張になつていることがわかる[7]．

3. Solvable Lie group

この章では次章の證明に必要な solvable Lie group の性質を證明しておく．

L を connected Lie group, R を L に對應する Lie algebra とせよ．L を solvable と假定し L の commutator series を

(1) $L=L_0\supset L_1\supset\cdots\supset L_n=e,\ L_i=C_i(L)$

として，L_i に對應する Lie algebra を R_i とすれば

$$R=R_0\supset R_1\supset\cdots\supset R_n=0$$

となる．ここで R_i は R の ideal で R_{i-1}/R_i は abelian なる故 R は Lie algebra として solvable となる．逆に R が solvable ならばそれに對應する Lie group L が §2 の意味で solvable となり又その長さが R の derivation algebra の series の長さと一致することも容易に證明される．即ち connected Lie group に對しては我々が先に與えた solvable group の定義が Lie algebra を用いて與えられる普通の定義と全く一致することがわかる．

次にここでついでに後になつで用いられる lemma を一つ揭げておこう．

Lemma 3. L を connected solvable Lie group とすれば L は G_{i-1}/G_i が vector group 或は toroidal group となるような characteristic subgroup の series

$$L=G_0\supset G_1\supset\cdots\supset G_n=e$$

を有する．

證明 以下特に斷らぬ限り V は vector group を，又 T は toroidal group を表わすものと約束する。特に次元を明示したい時には V_n, T_n 等と

書くことにする．さて L の commutator series (2) は明かに L の charactieristic subgroup の series を與えるが，その factor group L_{i-1}/L_i はいずれも Lemma 1 により connected abelian Lie group であるから

$$L_{i-1}/L_i=V'\times T'$$

と vector group V' と toroidal group T' との直積に分解される．しかもここに T' は L_{i-1}/L の characteristic subgroup となるから $T'=M/L_i$ となるような subgroup M は又 L の characteristic subgroup である．各 i についてこのような M を L_{i-1} と L_i との間に挿入すれば求むる series を得ることは明かであろう．證終．

さてこの章の主な目的は次の定理の證明にある．

定理 5. G を connected locally compact group, N を G の normal subgroup とするとき N 及び G/N が共に solvable Lie group ならば G 自身 solvable Lie group である．

このために先ず次の lemma を證明する．

Lemma 4. G を locally compact group, N を G の normal subgroup をせよ．この時 N が vector group 乃至 toroidal group で又 G/N が1次元の vector group V_1 乃至 toroidal group T_1 に同型ならば G のなかに

$$G=HN,\quad H\frown N=e,\quad H\cong G/N$$

なる如き subgroup H が存在する．

證明　先ず $N=T$ とせよ．然らば定理3により N は G の centre に含まれる．これと $G/N\cong V_1$ 或は $G/N\cong T_1$ なることから容易に G が abelian group なることが知られる．そうすれば connected locally compact abelian group の構造に關する Pontrjagin の定理[8]から直ちに lemma の成立することが知られる．次に $N=V$ とせよ．この場合にも若しも N が G の centre に含まれているならば上と同様にして證明される．一般の場合には $N=V_n$ として n に關する induction を用いるとよい．$n=0$, $N=e$ ならば lemma は明白だから N の次元が n より小なる場合には既に證明がすんでいるものと假定する．$N=V_n$ とすれば N の元 u は

$$u=u(\xi_1,\cdots,\xi_n),\ -\infty<\xi_i<\infty,\ i=1,\cdots,n$$

と real parameter ξ_i を以て表わすことができる．ここに勿論

$$u(\xi_1,\cdots,\xi_n)\,u(\eta_1,\cdots,\eta_n)=u(\xi_1+\eta_1,\cdots,\xi_n+\eta_n)$$

とする．今 G の任意の元 a により

$$a^{-1}u(\xi_1,\cdots,\xi_n)\,a=u(\xi_1',\cdots,\xi_n')$$

とおけば $a^{-1}ua$ は vector group N に automorphism をひき起すから

$$(\xi_1',\cdots,\xi_n')=(\xi_1,\cdots,\xi_n)A$$

となる．ここに $A=A(a)$ は a に依存する n 行 n 列の實係數の matrix で

(3) $\qquad a\to A(a)$

なる對應は明かに G の連續表現を與える．a が N に含まれれば $A(a)=E$[9] であるから對應 (3) は又 G/N の表現とも考えられる．從つて集合 $\{A(a);\,a\in G\}$ に含まれる matrix は互に commutative である．よつて今かりに G の凡ての元 a に對し行列式

$$|A(a)-E|=0$$

となつたとすれば，卽ち凡ての matrix $A(a)$ が1を固有値として有するならば少くとも1次元の subspace V' があつてこの上では凡ての transformation $A(a)$ が恒等變換となる．V' を $N=V$ の subgroup と考えればこれは V' か G の centre に含まれていることを意味する．よつても しも $V'=V$ ならば初めの注意により lemma は證明されたことになる．又 $V'\neq V$ とすれば V', V/V' の次元は共に n より小となるから先ず G/V' と N/V' とに induction の假定を適用すれば

$$H'\supset V',\ G=H'N,\ H'\frown N=V',\ H'/V\cong G/N$$

なる G の subgroup H' が存在することがわかる．よつて再び H, V' に induction の假定を用いれば

$$H'=HV',\quad H\frown V'=e,\quad H\cong H'/V'$$

なる如き H の存在が言われる．この H が lemma の條件を滿たすことは明白であろう．

故に結局 G の適當なる元 a_0 に對し

(4) $\qquad |A(a_0)-E|\neq 0$

なるものとして lemma を證明すればよい．この場合には u の座標を $\xi_1\ \cdots,\xi_n$ とし又 $u^{-1}a_0^{-1}ua_0$ の座標を $\xi_1',\cdots,\xi_n'$ とすれば

$$(\xi_1',\cdots,\xi_n')=(\xi_1,\cdots,\xi_n)\,(A(a_0)-E)$$

であるから u が N の凡ての元を動けば $u^{-1}a_0^{-1}ua_0$ も又 N の凡ての元を動くことがわかる．さて b を G の任意の元とする時 G/N は abelian なる故 $b^{-1}a_0^{-1}ba_0$ は N に含まれる．よつて上の注意により

$$b^{-1}a_0^{-1}ba_0=u^{-1}a_0^{-1}ua_0$$

なる如き N の元 u を定めることができる．即ち

(5) $\quad (bu^{-1})\,a_0=a_0(bu^{-1})$

從つて今 a_0 と commutative な G の元の全體を H とすれば H は明かに G の subgroup であるが (5) により

$$G=H\,N$$

となる．次に $H\frown N$ に含まれる任意の元 v をとれ．v の座標を $\eta_1,\cdots,\eta_n$ とすれば $a_0v=va_0$ より

$$(\eta_1,\cdots,\eta\;)(A(a_0)-E)=0$$

よつて (4) から

$$(\eta_1\;\cdots,\eta_n)=(0\cdots,0),\ \text{即ち}\ v=e,$$

$$H\frown N=e.$$

これで l mma が完全に證明された．

Lemma 5. G は locally compact group, N は G の normal subgroup とせよ．N が Lie group で且 $G=HN$, $H\frown N=e$, $H\cong V_1$ 或は $H\cong T_1$ となるような subgroup H が存在すれば G も又 Lie group である．

證明 N の第一種の cononical parameter を $\xi_1,\cdots,\xi_n$ とし又

$$H=\{h(\lambda)\},\ h(\lambda)h(\mu)=h(\lambda+\mu),$$
$$-\infty<\lambda<\infty,\ (\lambda \bmod. 1)$$

とせよ[10]．N の十分 e に近い元 u をとり

$$u=u(\xi_1,\cdots,\xi_n)$$
$$g=g(\lambda;\xi_1,\cdots,\xi_n)=h(\lambda)\,u(\xi_1,\cdots,\xi_n)$$

とおけば $\lambda;\xi_1,\cdots,\xi_n$ が G の單位近傍に座標を與えることは明かである．次にこの座標が G の乘法に對し analytical に變換されることを示そう．實際

$$g(\lambda;\xi_1,\cdots,\xi_n)g(\mu;\eta_1,\cdots,\eta_n)$$
$$=h(\lambda)\,u(\xi_1,\cdots,\xi_n)h(\mu)u(\eta_1,\cdots,\eta_n)$$
$$=h(\lambda)\;h(\mu)\;(h(\mu)^{-1}u(\xi_1,\cdots,\xi_n)h(\mu))\;u(\eta_1,\cdots,\eta_n)$$
$$=h(\lambda+\mu)\,\mu(\xi_1',\cdots,\xi_1')\,u(\eta_1,\cdots,\eta_1)$$
$$=h(\lambda+\mu)\,u(\zeta_1,\cdots,\zeta_n)$$
$$=g(\lambda+\mu;\zeta_1,\cdots,\zeta_n)$$

但しここに

(6) $\quad u(\xi_1',\cdots,\xi_n')=h(\mu)^{-1}u(\xi_1,\cdots,\xi_n)\,h(\mu)$,

(7) $\quad u(\zeta,\cdots,\zeta_n)=u(\xi_1',\cdots,\xi_n')\,u(\eta_1,\cdots,\eta_n)$

この第二式 (7) において $\zeta_1,\cdots,\zeta_n$ が $\xi_1',\cdots,\xi_n'$; $\eta_1,\cdots,\eta_n$ に analytical に依存することは ξ_i が canonical parameter であることから當然である．よつて (6) において $\xi_1',\cdots,\xi_n'$ が μ と $\xi_1,\cdots,\xi_n$ との解析函數であることさえ證明できれば lemma は完全に證明される．さて第一種の canonical parameter の性質により前 lemma の證明における と同様に

(8) $\quad (\xi_1',\cdots,\xi_n')=(\xi_1,\cdots,\xi_n)A$

と書ける．ここに $A=A(\mu)$ は $h(\mu)$ によつて定まる n 行 n 列の實係數の matrix でしかも

$$h(\mu)\to A(\mu)$$

は H の連續表現を與える．(8) により $\xi_1',\cdots,\xi_n'$ は $\xi_1,\cdots,\xi_n$ の一次函數だから勿論それ等について analytical である．又任意の one-parameter group の連續表現が analytical になることを用いれば

$$A(\mu)=(a_{ij}(\mu))$$

とするとき各 $a_{ij}(\mu)$ は μ の解析函數となることが知られる．よつて ξ_k' は $\mu;\xi_1,\cdots,\xi_n$ について analytical である．證終．

これら二つの lemma を用いて定理5を證明することができる．次にそれを述べよう．

G, N は定理5における條件を滿足する群とし，N' を N の connnected component とせよ．然らば N' は又 G の normal subgroup であるが N は Lie group であるから N/N' は discrete，從つて G/N' は G/N に locally isomorphic でこれも又 solvable Lie group でなければならぬ．よつて N' を改めて N と書くことにすれば結局 N が connected であると假定して定理を證明すれば十分であることがわかる．そのような假定のもとに G/N の次元を n とせよ．$n=0$ ならば定理は明白だから n に關する induction によつて證明する．Lemma 3 を G/N に適用すれば G は

$N_1 \supseteq N$, $G/N_1 \cong V_1$ 或は $G/N_1 \cong T_1$

となるような connected normal subgroup N_1 を含む．N_1/N の次元は G/N のそれより 1 だけ小であるから N_1, N に對し induction の假定を用いれば N_1 が solvable Lie group であることが知られる．よつて Lemma 3 により

$$N_1 \supset N_2 \supset \cdots \supset N_r = e$$

なる N_1 の characteristic subgroup の series で N_{i-1}/N_i が凡て vector group か troidal group となるようなものが存存する．N_i は N_1 のなかで characteristic であるから又 G の normal subgroup となる．さて G/N_2, N_1/N_2 に Lemma 4 を適用すれば

$$G = H_1N_1,\ H_1 \frown N_1 = N_2,\ H_1/N_2 \cong G/N_1$$

なる G の subgroup H_1 が存存する．次に H_1/N_3, N_2/N_3 に同じ lemma を用いて

$$H_1 = H_2N_2,\ H_2 \frown N_2 = N_3;\ \text{従つて}$$

$$G = H_2N_1,\ H_2 \frown N_1 = N_3$$

となる subgroup H_2 の存存がわかる．以下同樣にして結局適當に $H = H_{r-1}$ をとれば

$$G = HN_1,\ H \frown N_1 = e$$

となることが知られる．N_1 は Lie group であるから Lemma 5 を用いれば G も又 Lie group となることがわかる．それが solvable であることは殆ど自明であろう．

これで目標とした定理は完全に證明されたわけであるが上の證明によつて同時に次のことがわかる．即ち G を任意の connected solvable Lie group とすれば G は $G/N_1 \cong V_1$ 或は $G/N_1 \cong T_1$ となるような connected normal subgroup N_1 を含み，しかも同時に

$$G = H_1N_1,\ H_1 \frown N_1 = e,\ H_1 \cong V_1 \text{ 或は } H_1 \cong T_1$$

となる subgroup H_1 が存在する．次に N_1 について同樣な分解を行えば

$$N_1 = H_2N_2,\ H_2 \frown N_2 = e,\ H \cong V_1 \text{ 或は } H_2 \cong T_1$$

なる H_2, N_2 が存存する．これを繰返せば結局次の定理が得られるであろう．

定理 6. G を n 次元の connected solvable Lie group とすれば G は次の如き subgroup $H_1, \cdots, H_n$ を含む．即ち

i) H_i は凡て V_1 或は T_1 に同型である：$H_1 = \{h_i(\lambda_i)\}$, $h_i(\lambda_i)\ h_i(\mu_i) = h_i(\lambda_i + \mu_i)$ とする，

ii) $G_i = H_{i+1} \cdots H_n$ とおけば G_i は $n-i$ 次元の G の subgroup でしかも G_{i-1} の normal subgroup である，

iii) 特に $G = G_0 = H_1 \cdots H_n$ と分解され

$$g = g(\lambda_1, \cdots, \lambda_n) = h_1(\lambda_1) \cdots h_n(\lambda_n)$$

とすれば $\lambda_1, \cdots, \lambda_n$ は G の analytical coordinate を與える．

この定理から直ちに G は topological space としては $H_1, \cdots, H_n$ の直積 (cartesian product) となることが知られる．H_i のうち V_1 に同型なものの數を r, T_1 に同型なものの數を $n-r$ とすれば G は $V_r \times T_{n-r}$ に homeomorphic であつて特に G が simply connected ならばそれは V_n に homeomorphic である[11]．

§ 4. Locally compact solvable group の構造

本章では connected な locally compact solvable group の構造を調べて Hilbert の問題をこれ等の群に對して解決することとする．

Lemma 6. G は connected locally compact group, N, V は G の normal subgroup で $N \supseteq V$ 且 N/V は compact abelian group をする．この時 V の各元と commutative な N の元の全體を M とすれば M は G の normal subgroup で，N/M は compact abelian Lie group, 又 M は V と G の normal subgroup たる compact abelian group K との直積になる．

證明 Lemma 4 の證明におけるように $V = V_n$ の元 u を座標であらわし

$$u = u(\xi_1, \cdots \xi_n),\ -\infty < \xi_i < \infty$$

とせよ．又同じように N の任意の元 a に對し

$$a^{-1}u(\xi_1, \cdots, \xi_n)a = u(\xi', \cdots, \xi'_n)$$

$$(\xi_1', \cdots, \xi_n') = (\xi_1, \cdots, \xi_n)A(a)$$

とおけば

$$a \to A(a)$$

によつて N の連續表現が與えられる．この hommorphism の kernel を M とすれば M は丁度 lemma に述べたように V の各元と commutative な N の元の全體となる．又明かに $M \supseteq V$ で N/V

は compact abelian だから N/M も compact abelian，よつてそれは matrix の群 $\{A(a)\,;\,a\in N\}$ に同型で，従つて abelian Lie group でなければならない．M が G の normal subgroup なることは N, V が共に G の normal subgroup なることから明かである．次に M が abelian group なることを證明しよう．それができれば M が V と compact abelian group K との直積になること及びこの K が G の normal subgroup になることは Pontrjagin の定理[12]から容易に知られる．さて a, b, c を M の任意の元とするとき M/V は abelian かから $[a, b]=aba^{-1}b^{-1}$, $[a, c]=aca^{-1}c^{-1}$ は共に V に含まれるが M の定義から V は M の centre に入つているから簡單な計算で

$$[a, bc]=a(bc)a^{-1}(bc)^{-1}=[a, b]\,[a, c]$$

となる．よつて a を定めておいて

$$b\to[a, b],\qquad b\in M$$

なる對應を考へればこれは M から V の中への連續な homomorphism を與える．この寫像の kernel を M' とすれば $M'\supseteq V$ で M/M' は compact であるから $\{[a, b]\,;\,b\in M\}$ は M/M' に同型，從つて compact group である．然るに一方それは vector group V の subgroup であるから結局 $\{[a, b]\,;\,b\in M\}=e$ でなければならぬ．卽ち $[a, b]=e$. a も b も M の任意の元でよかつたからこれで M が abelian なることが證明された．證終．

この Lemma 6 と定理5を用いて次の重要な定理が證明される．

定理 7. G を任意の connected locally compact solvble group とすれば G は G/N が solvable Lie group となるような compact abelian normal subgroup N を含む．

證明　$G=G_0\supset G_1\supset\cdots\supset G_n=e$, $G_i=C_i(G)$ を G の commutator series とせよ．series がこのように有限で切れることは定理4によるのである．さて $n=0$, 卽ち $G=e$ の時には問題はないから G の長さ n に關する induction で定理を證明することにする．$G'=G/G_{n-1}$ とすれば

$$G_i/G_{n-1}=C_i(G'),\quad i=0, 1, \cdots, n-1$$

となることは容易にわかるから G は長さ $n-1$ の solvable group である．よつて induction の假定により適當に G_{n-1} を含む G の normal subgroup N_1 をとれば G/N_1 は solvable Lie group で N_1/G_{n-1} は compact abelian group となる．さて G_{n-1} は Lemma 1 により connected locally compact abelian group なる故

$$G_{n-1}=V\times K$$

と vector group V と compact abelian group K との直積となる[13]．ここで K は又 G の normal subgroup となるから $G''=G/K$ とおいて，$N''=N_1/K$, $V''=G_{n-1}/K\cong V$ に Lemma 6 を適用すれば G'' 中に $M''=M/K$ なる normal subgroup が存存し $N_1''/M''\cong N_1/M$ は compact abelian Lie group となり，又 M'' は V'' と compact abelian な $N''=N/K$ との直積となる．さて

$$G\supseteq N_1\supseteq M\supseteq N\supseteq e$$

なる series において G/N_1 は假定により solvable Lie group, N_1/M は compact abelian Lie group, 又 M/N は V'' と同型なる故矢張 abelian Lie group である．よつて定理5により G/N は solvable Lie group となる．又 N/K, K は共に compact abelian なる故 N 自身 compact で且 solvable となる．よつて定理 1' を用いればそれは abelian でなければならぬ．これで定理は完全に證明された．

さて上記定理における N は定理3を用いれば G の centre に含まれていることがわかる．一方それは compact abelian group であるから G の單位近傍 U を任意に與えた時に U に含まれる N の subgroup N^* で N/N^* が abelian Lie group となるようなものが存存する[14]．定理5を用いれば G/N^* も又 solvable Lie group であるから定理7よりも一層精密に次のことが言えるわけである．卽ち

定理 7'. 定理7における N は更に與えられた任意の G の單位近傍 U の中に含まれていると假定しても差支えない．

定理 7, 7' が言えればこれから後は全く compact group の場合と同樣にして次のことが證明される．

定理 8. Connected locally compact solvable group G の單位近傍 U で $\overline{U}$ が compact 且

finite dimensional となるようなものが存存すればGはlocallyにはlocal Lie groupとtotally disconnected compact abelian groupとの直積となる.

證明は例えばPontrjagin[10], §45を参照されたい. 但しそこではGに可附番公理を假定しているがこれがなくとも差支えないことは容易に知られるであろう.

定理8から直ちに我々の目標であつた次の定理, 即ちsolvable groupに對するHilbertの問題の解答を得ることができる[15].

定理 9. Locally compact solvable group Gがlocally EuclideanならばそれはLie groupである[16].

Locally Euclideanの代りにlocally connected且finite dimensionalと言うような條件でもよいことは勿論である.

§5 (L)-group

先ず初めに定義を述べておく方が便利である.

定義 Locally compact group Gが次の二つの條件を滿足するnormal subgroupのsystem $\{N_\alpha\}$を含むときGを(L)-groupと呼ぶ. 即ち

i) G/N_αは凡てLie groupである,

ii) 凡てのN_αの共通部分はeである.

さて連續表現の理論を用いれば任意のcompact groupが上の意味で(L)-groupとなることは明かであろう[17]. 又前章の定理7′により任意のconnected locally compact solvable groupも又(L)-groupなることがわかる. そうしてこれから見られるようにcompact groupやsolvable groupに對してHilbertの問題が解決されたと言うことは結局これらの群が(L)-groupであると言ふことに基いているのである. よつてこのような(L)-groupの性質を研究すればHilbertの問題の一般的解決に何等かの寄與をなし得るであろうと言うことは容易に豫想される所である. 次にそれについて少しく述べて見よう.

先ず當然豫想されるように次の定理が容易に證明される.

定理 10. (L)-group Gがlocally EuclideanならばそれはLie groupである.

然しながらこれよりもつと精密に次の定理が成立する.

定理 11. Gを任意のconnected (L)-group, UをGの任意の單位近傍とするとき適當なloc l Lie group L及びUに含まれるGのcompact normal subgroup NをとるときGはlocallyにLとNとの直積となる. 逆にlocallyにlocal Lie groupとcompact groupとの直積となるようなconn cted locally compact groupは(L)-groupである.

この定理はGがsolvableである場合にはMalcevによつて與えられたものであるが兎も角これにより(L)-groupのlocal structureは十分に解明されたと言つてもよいであろう.

次に(L)-groupのglobalな性質を與えるものとして次の定理がある.

定理 12. Gをconnected (L)-groupとすればGの任意のcompact subgroupは必らずGのあるmaximal compact subgroupに含まれる. そうしてこのようなGのmaximal compact subgroupは凡てconnectedで且互にconjugateである. 今その一つをKとせよ. 然らばGはV_1に同型な有限個のsubgroup $H_1, \cdots, H_n$を含み

$$G=KH_1\cdots H_n$$

となる. 即ちGの任意の元gは

$$g=sh_1\cdots h_n,\quad s\in K,\ h_i\in H_i$$

なる形に一意的に且連續的に分解される. よつて特にGのtopological space (G)はKのspace (K)と直線(H_i) $(i=1,\cdots,n)$の直積となる. 即ち(G)はcompact space (K)とn次元Euclid空間E_nとの直積にhomeomorphicである.

特にGがsolvableな場合にはKは定理1によりabelianとなるから結局connected locally compact solvable groupのtopological structureはcompact abeli n groupのそれに歸着することが知られる. 尙又上の定理は定理6或はCartan, Malcev等のLie groupのtopological structureに關する定理[18]の擴張になつていることを注意しておく. この定理11, 12及びこれから述べるこの章の定理の證明には多くの準備を要

するので凡てここではそれを省略させて頂くことにする。

さて上の二つの定理，特に solvable group に對するそれ等の結果を用いて次のことが證明される。

定理 13. G を任意の connected locally compact group とすれば G は maximal solvale normal subgroup N を含む．ここに maximal とは凡ての G の solvable normal subgroup が N に含まれると言う意味である．N の connected component を N_0 とすれば N_0 は同様な意味で G の maximal connected solvable normal subgroup となる．

定義 我々は上の定理における N_0 を Lie group の場合に倣つて G の radical と呼ぶこととする．

この radical と言う概念を用いれば次のことが言える．

定理 14. G を connected locally compact group，N_0 を G の radical とせよ．このとき若しも G/N_0 が compact ならば G は (L)-group である．よつてこのような G を particular (L)-group と呼ぶこととする．

定理 10, 14 から

定理 15. Particlar (L)-group G が locally Euclidean ならばそれは Lie group である．

Compact group，solvable group 等は定義から直ちにわかるようにいずれも particular (L)-group であるから上の定理はこれらの群に對する Hilbert の問題の解答を凡て含んでいるわけである．今 L を一つの Lie group，R を L に屬する Lie algebra，R_1 を R の radical とせよ．然らば R/R_1 は勿論 semi-simple Lie algebra であるが，Weyl の定理[19]によれば任意の semi-simple Lie algebra は compact Lie group に對應するものと然らざるものとに區別されるから，この意味で上の R/R_1 が compact Lie group に屬する場合我々は初めの Lie group L を particular Lie group と呼ぶことにしよう．そうすれば定理 15 の意味は換言すればこのような particular Lie group が位相群論的に完全に特徴付けられると言うことに外ならない．

さて最後に最も重要な問題，即ち (L)-group，或は particular (L)-group が一般の locally compact group に對してどう言ふ關係にあるかと言うことについて述べて見たい．それに關して先ず次の定理が成立する．

定理 16. G を任意の connected locally compact group とする時定理 13 におけると同樣な意味で G の中に maximal normal (L)-group N_1 ((L)-group で G の normal subgroup なるもの）が存存する．又 factor group が (L)-group となるような minimal normal subgroup N_2 が存存する．particular (L)-group についても同樣なことが言える．

この定理によつて定まる群 $N_1, G/N_2$ は我々が G に對し Lie 群論を應用し得る最大の限界を與えるものと言うことができよう．そうして (L)-group が一般の locally compact group とどれ程異るかと言うことは結局 G/N_1 或は N_2 かどう言う構造の群であるかと言うことに懸つているわけである．さてこれから先は自分一人の豫想なのであるが實際には G/N_1, N_2 は共に單位群なのではなかろうか？換言すれば

(C_1) 「凡ての connected locally compact group は (L)-group であろう」

と豫想されるのである．次に (C_1) が正しいと思われる根據を一二擧げて見よう．先ずその一は我々に現存よく知られている群，例えば Lie group，compact group，それから solvable group 等はいずれもこの (L)-group の範圍を出ないと言うことである．これは初めの定義の仕方からある意味では當然のこととも言われようが兎も角若しも (C_1) に反例があかるとしてもそれは餘程變つた example でなければならぬことを示している．それからもう一つの理由として次のような事情がある．1933 年 C. Chevalley は Hilbert 空間の unitary operator を用いて locally compact group の構造を研究し次の如き豫想を與えた[20]．

(C_2') 「Connected locally compact group G が任意に小なる subgroup を含まないならば——精しく言えば G の適當な單位近傍をとるときこの中に含まれる subgroup が（e 自身を除いては）最早存存しないならば——G は Lie group

であろう」

この豫想はその後種々の場合に正しいことが確かめられているのであるが我々はこれよりも少し強い次の命題をとる．卽ち

(C_2) 「Connected locally compact group G が任意に小なる normal subgroup を含まないならば G は Lie group であろう」

さてこの (C_2) は (C_2') と殆ど同じ程確からしく思われる豫想であるが實はこの (C_2) と先の (C_1) とが同値であることが證明されるのである．從つて (C_2) が若し證明できれば (C_1) も成立し，これにより Hilbert の問題が一般の場合に完全に解決されるばかりでなく定理 11, 12 により一般の connected locally compact group の構造及びそれと Lie group との關係も大分明かになるわけである．

いずれにせよ，たとえ (C_1) 或は (C_2) が成立しないとしても，上に述べた (L)-group と言う群の class が一般の locally compact group の集合のなかで極めて重要な地位を占めていると言うことだけは先ず確實に言われるであろう．（昭和22年10月秋季例會特別講演）

文　獻

[1] Abe, M. Über Automorphismen der lokal-kompakten abelschen Gruppen, Proc. Imp. Acad. Tokyo, vol. 16 (19.0)

[2] Cartan, E. Groupes simples clos et ouverts et géometrie riemanniene, Jour. Math. pures et appl., t. 8 (1929)

[3] Cartan, H. Sur les groupes de transformations analytiques, Actualités sci. et indus., 198 (1935)

[4] Chevalley, C. Génération d'un groupe topologique par des transformations infinitésimales, C. R. Acad. Sci. Paris, t. 196 (1933)

[5] Chevalley, C. On the topopological structure of solvable groups, Ann. of Math., vol. 42 (1941)

[6] Chevalley, C. Two theorems on solvable topological groups, Lectures in topology, University of Michigan Press, 1941

[7] Hilbert, D. Gesammelte Werke, III

[8] Iwasawa, K. Über nilpotente topologische Grupp n, Proc. Imp. Acad. Tokyo, vol. 21 (1945) (未刊)

[9] Malcev, A. On the theory of Lie groups in the large, Rec. Math., N. S. 16 (1945)

[10] Pontrjagin, L. Topological groups, Princeton, 1939

[11] Yosida, K. 完全に閉じた群に關する Hilbert の問題，日本數學物理學會誌，11巻（昭和12年）

註

1) 本稿は昭和22年秋の日本數學例會における同じ表題の特別講演の內容を多少敷衍したものである．

2) この講演の精しい內容については例えば Hilbert [7] 參照．（括孤內の數字は本文の終にある文獻表の番號を示す．

3) 以下 Hilbert の第五の問題を單に Hilbert の問題と略稱する．他の Hilbert の問題について觸れることはないから．

4) これらの結果の證明及び精しい文獻については吉田 [11] 或は Pontrjagin [10] 參照．

5) Chevalley, Malcev 等の結果については今日我々はこれを唯 abstract によつてのみ知つているに過ぎないのであるから紹介と言つても次に述べる證明が例えば Chevalley のそれと全く同じであるかどうかはわからない．然し恐らくは大同小異のものであろうと考えている次第である．

6) Abe [1] 參照．

7) Connected でない群に對しては定理 1, 4 は成立しない．例えば凡ての有限群の直積である compact group をとればわかる．然し compact group の長さは一般の場合でも高々 ω 以下となることが證明される．これらについては Iwasawa [8] 參照．

8) Pontrjagin [10], § 45, Theorem 41.

9) E は單位行列．

10) この證明に用いられる canonical parameter の性質等については Pontrjagin [10], Chapter VI 參照．

11) Chevalley [5]

12) l. c. 8)

13) l. c. 8)

14) Pontrjagin [10], Chapter VII 參照．

15) 證明については同じく Pontrjagin [10], Chapter VII 參照．

16) Chevalley [7]

17) l. c. 14)

18) Cartan [2], Malcev [9].

19) Pontrjagin [10], § 58, Theorem 80.

20) Cartan [3], Chevalley [4].

[20] (with T. Tamagawa)
Automorphisms of a function field

Sûgaku, 1-4 (1/1949), 315-316.

Abstract

This is a preliminary report of the joint work with T. Tamagawa on the group of automorphisms of an algebraic function field. A detailed account with full proofs was published later in [25] (1951).

代数函数体の自己同型置換

岩沢健吉（東大理）
玉河恒夫（東大理）

函数体の自己同型置換の成す群について，次の H. L. Schmid の定理[1]の別証を與える．

定理 k を標数 $p\neq 0$ の代数的閉体とする時，k を係数体とする genus $g\geqq 2$ なる一変数代数函数体 K の自己同型置換で，k の元を個別に動かさないもの全体は，有限群を成す．

以下次の記号を使用する．

A, B 等；K の divisor, $L(A)$；A^{-1} の multiples 全体の成す modul, $l(A)$；$L(A)$ の k 上の次元，

$f_1, f_2, \cdots, f_g$；K の第一種微分の底．

以下定理1迄は，$g\geqq 1$ であればよい．以下 $g\neq 0$．

Lemma 1.[2] K の prime divisor P を定め，K の同型置換で，$P^\sigma=P$ なる σ の次数は有限である．その次数を n とする時，$n=n'p^a$, $(n', p)=1$ と分解する時，

$$n'\leqq 6(2g-1),\quad n\leqq \max(6(2g-1)(g+1), p(g+1))$$

である．

Lemma 2. 上の如き σ 全体の成す群を $G(P)$ とす．$G(P)$ は次の如き構造をもつ．

1. $G(P)$ の元で次数が p の巾なるものは正規部分群 $N(P)$ を成し，$G(P)/N(P)$ は cyclic，その次数は p と素で，$6(2g-1)$ を越えない．

2. $\nu=\min(n\,;\, l(P^n)\geqq 2)$, $(x)=A/P^\nu$ なる ν, x をとる時，$x^\sigma=x$ なる σ 全体は $G(P)$ の正規部分群 $N_0(P)$ である．$(N(P)/N_0(P))$ は Abel 群で，その元の次数はすべて p．$N_0(P)$ 次数が ν を越えない有限群．明らかに $\nu\leqq g+1$．

Lemma 3. $G(P)$ の部分群 H が次の條件を充たすとす．

1. $N(P)\supseteqq H$

2. H の任意の次数 p の部分群 $\{\sigma\}$, $\sigma^p=1$ で，$z^\sigma=z$ なる z 全体の成す K の部分体．（これをこの部分群に対應する部分体をいう．）の genus は常に 0.

その時 H は高々 p^2 次の巡回群か又は $p=2$ で quotanion group である．

以上三つの Lemma は容易に証明される．証明の鍵になるのは特に Lemma 3 である．

定理 1. $G(P)$ は有限群である．

$N(P)$ が有限なることをいえばよい．$N(P)$ の有限部分群でそれに対應する体の genus g_0 が 0 ならざる最小の値をとるようなものを $N^{(0)}$, $N^{(0)}$ に対應する体を K_0 とす．$N(P)$ における $N^{(0)}$ の normalizer N^* をとる．$N^*/N^{(0)}$ は K_0 の自己同型の群で Lemma 3 の條件を満す．故に有限．$[N(P):N^*]$ は $N(P)$ における $N^{(0)}$ の共軛の数である．$N^{(0)}N_0(P)$ は $N(P)$ の不変部分群なることより，$N^{(0)}$ の共軛はすべてこれに含まれる．故に $[N(P):N^*]$ は有限，故に $G(P)$ の有限性が結論される．簡單な計算により

$$[G(P):e]\leqq 12(2g-1)^2 p\cdot\nu^d$$

$\nu \leqq g+1$, 又は d は $N^{(0)}$ の生成元の数で $[(\log(2g-1))/\log p] \geqq d$.

次に最初に述べた定理の証明を与える. $f_1, f_2, \ldots, f_g$ は K の第一種微分の底とす. $\sigma \in G$ (G は K の自己同型全体の群) をとる時,

$$\begin{pmatrix} f_1^{\sigma} \\ \vdots \\ f_g^{\sigma} \end{pmatrix} = A(\sigma) \begin{pmatrix} f_1 \\ \vdots \\ f_g \end{pmatrix}$$

により G の表現 $A(\sigma)$ を得. $g \geqq 2$ の時, この表現は $p=2$, K: hyperelliptic の場合を除き同型表現であり, 例外の時も, $A(\sigma)=E_g$ なる σ は 2 次の群をなす. $A(\sigma)$ が有限なることをいう. $A(\sigma)$ を既約に分解する:

$$A(\sigma) = \begin{pmatrix} A_1(\sigma) & & 0 \\ & \ddots & \\ * & & A_s(\sigma) \end{pmatrix}$$

σ を任意にとる時, $f^{\sigma}=f$ なる differential f が存在するから, $n(f)=2g-2$ より σ の次数は, 有界である. 故に既約な一次変換群に関する定理により $A_1(\sigma), \ldots, A_s(\sigma)$ は有限な表現である. 故に $A_1(\sigma)=E_{n_1}$ なる σ 全体は G の正規部分群 N_1 をなし, G/N_1 は有限群となる. $N_1 \ni \sigma$ とする時, $f_1^{\sigma}=f_1$ であるから, $P|(f_1)$ なる P をとる時, $N_1 \ni \sigma$ で, $P^{\sigma}=P$ なるもの全体の群 $N_1(P)$ は, $[N_1 : N_1(P)] \leqq 2g-2$ を満足する. $N_1(P)$ は $G(P)$ の部分群であるから有限群. 故に G は有限群である.

以上の証明では F. K. Schmidt の Weierstrass point の理論を使用して居ない. 以上 $p \neq 0$ を仮定したが, $p=0$ なる時は定理 1 の証明は非常に簡単になり, 又 $A(\sigma)$ を既約に分解する必要がない. この時, $[G:e] \leqq 84(g-1)$[3]

(昭和 23 年 5 月年会講演)

文献

1) H. L. Schmid. Über die Automorphismen eines algebraischen Funktionenkörpers von Primzahlcharakteristik. Crelle **179**.

2) A. Hurwitz. Über diejenigen algebraischen Gebilde, welche eindeutige Transformation in sich zulassen. Werke Band I. S. 241.

3) A. Hurwitz. Über algebraische Gebilde mit eindeutige Transformationen in sich. Werke Band I. S. 391.

[21] On some types of topological groups

Ann. of Math., (2) 50 (1949), 507-558.

(Received March 9, 1948)

Introduction

As is well-known the so-called fifth problem of Hilbert on continuous groups was solved by J. v. Neumann [14][2] for compact groups and by L. Pontrjagin [15] for abelian groups. More recently, it is reported, C. Chevalley [6] solved it for solvable groups.[3]

Now it seems, as H. Freudenthal [7] clarified for maximally almost periodic groups, that the essential source of the proof of Hilbert's problem for these groups lies in the fact that such groups can be *approximated by Lie groups*. Here we say that a locally compact group G can be approximated by Lie groups, if G contains a system of normal subgroups $\{N_\alpha\}$ such that G/N_α are Lie groups and that the intersection of all N_α coincides with the identity e. For the brevity we call such a group a *group of type* (L) or an (L)*-group*. In the present paper we shall study the structure of such (L)-groups, and apply the result to solve the Hilbert's problem for a certain class of groups, which contains both compact and solvable groups as special cases. We shall be able to characterize a Lie group G, for which the factor group G/N of G modulo its radical N is compact, completely by its structure as a topological group.

The outline of the paper is as follows. In §1 we study the topological structure of the group of automorphisms of a compact group and prove theorems concerning compact normal subgroups of a connected topological group, which are to be used repeatedly in succeeding sections. In §2 come some preliminary considerations on solvable groups, whereas finer structural theorems on these groups are, as special cases of (L)-groups, given later. In §3 we prove some theorems on Lie groups. The theorems here stated are not all new, but we give them here for the sake of completeness, and thereby refine and modify these theorems so as to be applied appropriately in succeeding sections.[4] After these preparations we study in §4 the structure of (L)-groups. In particular, it is shown that the study of the local structure and the global topological structure

---

[1] In preparing the present paper the author is greatly indebted to Prof. S. Iyanaga for his kind encouragements and advices. The author expresses here his best thanks to Prof. Iyanaga for his kindness.

[2] The numbers in brackets refer to the bibliography at the end of the paper.

[3] Some of the books and papers as Chevalley [6], Malcev [12], which were published in foreign lands during the last war, are known to the author only by the abstracts in Mathematical Reviews. Recently he was also informed that A. Malcev had proved in a paper of 1946, theorems on solvable groups, which are similar to that for (L)-groups in §4 of the present paper. However the details of these books and papers were not attainable to the author.

[4] This is also due to the circumstances as stated in the footnote 3.

of these groups can be reduced to that of compact and Lie groups, groups, which we know, it may be said much about. The solution of Hilbert's problem for (L)-groups then follows immediately. We also prove as a corollary of a general theorem that any connected locally compact solvable group is an (L)-group, so that our result covers Chevalley's theorem referred to above. In the last section, §5, we study the general locally compact groups in making use of the results obtained in preceding sections. Particularly it is proved that any connected locally compact group has a *radical*, i.e. a uniquely determined maximal connected solvable normal subgroup. By this notion of the radical we define a special type of topological groups. We say namely that a connected locally compact group G is a *(C)-group*, when the factor group G/N modulo its radical N is compact. It is then proved that any (C)-group is an (L)-group, so that Hilbert's problem is also solved for (C)-groups. This contains both results of v. Neumann and of Chevalley. After these considerations there remains a question: what is the situation of the class of (L)-groups in the set of all locally compact groups? It seems likely to us that any connected locally compact group is an (L)-group. In connection with this conjecture we also show that it is equivalent to a slightly strengthened *hypothesis of C. Chevalley*.[5] If this conjecture turns out to be valid, results in §4 would give us the structure of general connected locally compact groups and the Hilbert's problem would be then completely solved.

Although our investigations are not complete in various view-points, we hope that the present paper makes a contribution to the foundation of the theory of locally compact groups.

1. On the group of automorphisms of a compact group

Let K be a compact topological group and $A(K)$ the set of all continuous automorphisms of K. $A(K)$ forms an abstract group; we shall introduce in the following a topology in $A(K)$, so that it becomes a topological group. For that purpose take any V in a system $\{V\}$ of neighborhoods of the identity in K and put

$$U(V) = \{\sigma;\ \sigma \in A(K),\ s^{\sigma-1} \in V, \text{ for all } s \text{ in } K\}. \tag{1.1}$$

If we take the family $\{U(V)\}$ as a system of neighborhoods of the identity in $A(K)$, it is easy to see that $A(K)$ becomes a topological group. In the following we always consider $A(K)$ with such a topology.

The same topology can be introduced also as follows. Let $C(K)$ be the set of all continuous (complex-valued) functions on K and define for any $f(s)$ in $C(K)$ and σ in $A(K)$ the function $f^\sigma(s)$ by

$$f^\sigma(s) = f(s^\sigma).$$

Obviously f^σ is also contained in $C(K)$, and we have

$$\|f^\sigma\| = \|f\|,$$

[5] Cf. H. Cartan [4].

where the norm $\|f\|$ is defined by

$$\|f\| = \text{Max}\,|f(s)|, \qquad s \text{ in } K.$$

Now take any $f_1, \cdots, f_n$ in $C(K)$ and $\epsilon > 0$, and put

$$U(f_1, \cdots, f_n; \epsilon) = \{\sigma; \sigma \in A(K), \|f_i^\sigma - f_i\| < \epsilon, i = 1, \cdots, n\}.$$

The family $\{U(f_1, \cdots, f_n; \epsilon)\}$ also defines a system of neighborhoods of the identity in $A(K)$ and we can see immediately that this topology is equivalent to the one defined above by $\{U(V)\}$.

It is here also to be noted that, if K is connected (1.1) can be replaced by the family

$$U'(V) = \{\sigma; \sigma \in A(K), s^{\sigma-1} \in V, \text{ for all } s \text{ in } V_0\}, \tag{1.2}$$

where V_0 is an arbitrary fixed neighborhood of the identity in K.

Now let G be a topological group, which contains K as a normal subgroup. Any g in G then defines an automorphism $\sigma(g)$ in K by

$$s^{\sigma(g)} = g^{-1}sg, \qquad s \in K.$$

It is then easy to see that the homomorphic mapping

$$g \rightarrow \sigma(g) \tag{1.3}$$

from G into $A(K)$ is continuous in the sense of above defined topology in $A(K)$. If we take in particular $G = K$, we see that the image $I(K)$ of the mapping (1.3), i.e. the group of inner automorphisms of K is a compact normal subgroup of $A(K)$. We can therefore consider $A(K)/I(K)$ also as a topological group.

Now the main purpose of the present section is to prove the following theorem.

THEOREM 1. *$A(K)/I(K)$ is a totally disconnected* (0-*dimensional*) *group.*

We first prove a special case of the theorem, namely

LEMMA 1.1. *If K is a compact Lie group, $A(K)/I(K)$ is discrete.*

PROOF. We shall reduce the proof step by step to the case where K has a more simple structure.

i) Let K_0 be the connected component of the identity in K. K/K_0 is then a finite group of order, say, n. We prove first that if the lemma is valid for K_0, it is also valid for K. Denote by A_1 the set of all automorphisms of K which leave invariant every coset of K/K_0, and by $I^*(K_0)$ the set of inner automorphisms of K induced by elements in K_0. A_1 is a closed normal subgroup of $A(K)$ with a finite index, for $A(K)/A_1$ is ismorphic to a subgroup of the group of automorphisms of finite K/K_0. A_1 is consequently open in $A(K)$. $I^*(K_0)$ is also a normal subgroup of $A(K)$ and is contained in $I(K)$. Therefore $A(K)/I(K)$ will be surely discrete if we can prove that $A_1/I^*(K_0)$ is discrete.

As K_0 is a characteristic subgroup of K, any σ in A_1 induces an automorphism σ' in K_0 and the homomorphic mapping

$$\sigma \rightarrow \sigma' = \varphi(\sigma)$$

from A_1 into $A(K_0)$ is continuous. Let N be the kernel of this homomorphism.

Now from the definition of A_1, we have for σ in A_1

$$s^\sigma = su, \qquad s \in K, \qquad u \in K_0,$$

where $u = u(s)$ is a continuous function of s. If in particular σ is contained in N, $u(s)$ is an element of the center Z of K_0 and it depends only upon the coset of K/K_0 containing s.[6] From $(ss')^\sigma = s^\sigma s'^\sigma$ we have then

$$u(ss') = (s'^{-1}u(s)s')u(s') = u(s)^{s'}u(s'). \tag{1.4}$$

If we put

$$\prod u(s) = v, \qquad s \text{ mod. } K_0,$$

it follows from (1.4) that

$$v^{1-s'} = u(s')^n, \qquad \text{for all } s' \text{ in } K. \tag{1.5}$$

As Z is a compact abelian (not necessarily connected) Lie group, we can find a finite characteristic subgroup Z_1 of Z, so that Z/Z_1 is connected, i.e. a toroidal group of some finite dimension. Therefore there is an element w in Z such that

$$w^n \equiv v^{-1} \qquad \text{mod. } Z_1.$$

If we denote by σ_1 the inner automorphism induced by the element w in K and if we put

$$\sigma_2 = \sigma_1\sigma, \qquad s^{\sigma_2} = su'(s) \qquad s \in K, \qquad u'(s) \in Z,$$

a simple calculation shows that we have

$$(u'(s))^n \in Z_1$$

in virtue of (1.5). Let Z_2 be the set of all elements in Z, whose n^{th} power fall in Z_1. Z_2 is then also a finite group and $u'(s)$ can be considered as a function from the finite group K/K_0 into finite Z_2. As there exists only a finite number of such functions, it follows that the index

$$[N:N \frown I^*(K_0)]$$

is finite.

On the other hand we have obviously

$$\varphi(I^*(K_0)) = I(K_0)$$

and it follows that the index

$$[\varphi^{-1}(I(K_0)):I^*(K_0)]$$

is finite. We see that $I^*(K_0)$ is at the same time closed and open in $\varphi^{-1}(I(K_0))$.

Now suppose that $A(K_0)/I(K_0)$ is discrete. $\varphi(A_1)/I(K_0)$ is then also discrete and as the mapping φ is continuous $A_1/\varphi^{-1}(I(K_0))$ is discrete. Combining with what we have proved above, it follows that $A_1/I^*(K_0)$, consequently $A(K)/I(K)$ is discrete. The reduction from K to K_0 is thus completed.

[6] Cf. for example Zassenhaus [19], p. 74.

ii) By i) we may suppose $K = K_0$, i.e. that K is a connected Lie group. Let Z be again the center of K and Z_0 the connected component of Z. Further we denote by S the maximal connected semi-simple normal subgroup in K. It is well-known that

$$K = Z_0 S. \tag{1.6}$$

As Z_0 and S are characteristic subgroups of K, any automorphism σ in $A(K)$ induces automorphisms σ_1 and σ_2 in Z_0 and S respectively. The homomorphic mapping

$$\sigma \to (\sigma_1, \sigma_2) = \varphi(\sigma)$$

from $A(K)$ into the direct product $A(Z_0) \times A(S)$ is continuous and is one-to-one according to (1.6). Obviously we have

$$\varphi(I(K)) = I(Z_0) \times I(S) = e \times I(S).^7$$

Therefore $\varphi(A(K))/\varphi(I(K))$ is isomorphic to a group of the form

$$(A_1/I(Z_0)) \times (A_2/I(S)), \qquad A_1 \subset A(Z_0), \qquad A_2 \subset A(S). \tag{1.7}$$

If we assume that $A(Z_0)/I(Z_0)$ and $A(S)/I(S)$ are both discrete, then the group (1.7) is also discrete and by taking $\overset{-1}{\varphi}$ we see that $A(K)/I(K)$ is discrete. We have thus reduced the general case to the case, where K is either abelian or semi-simple.

iii) Consider first $K = Z_0$, i.e. the case where K is a connected compact abelian Lie group, namely a toroidal group of some finite dimension, say m. Elements of K are determined uniquely by their coordinates:

$$u = u(\xi_1, \cdots, \xi_m),$$

where each ξ_i takes real values mod. 1. An automorphism σ of K is then given by

$$(u(\xi_1, \cdots, \xi_m))^\sigma = u(\xi_1^\sigma, \cdots, \xi_m^\sigma),$$

$$(\xi_1^\sigma, \cdots, \xi_m^\sigma) = (\xi_1, \cdots, \xi_m)D, D = (d_{ij}),$$

where $D = D(\sigma)$ is a matrix of degree m with integral components d_{ij}. It is obvious that the correspondence

$$\sigma \leftrightarrow D(\sigma)$$

gives an algebraic isomorphism between $A(K)$ and the group of all integral matrices with determinants ± 1. The isomorphism holds also, when we consider them as topological groups, the latter group being topologized in the usual way as a matrix group. That follows for example from the fact that we can take (1.2) for $A(K)$ as neighborhoods of the identity. As the matrix group is obviously discrete, so is $A(K) = A(K)/I(K)$ and the lemma is proved for such case.

iv) Finally we consider the case $K = S$, i.e. the case, where K is a connected

---

[7] In the following we denote always by e the identities of various groups and do not remark it particularly, if there is no risk of misunderstanding.

semi-simple Lie group. Let R be the Lie algebra which corresponds to K and let $A(R)$ be the group of automorphisms of R. It is well-known that $A(R)$ can be isomorphically represented by real matrices and thus forms a Lie group, whose connected component A_0 just coincides with the so-called adjoint group of K, namely the group of automorphisms of R, which are induced by the inner automorphisms of K.

Now let σ be any automorphism in $A(K)$ and let σ' be the automorphism of R, induced by σ on R. The mapping

$$\sigma \rightarrow \sigma' = \varphi(\sigma)$$

from $A(K)$ into $A(R)$ is then, as is readily seen, continuous and we have

$$I(K) = \varphi^{-1}(A_0).$$

As $A(R)$ is a Lie group and $A(R)/A_0$ is discrete, it follows immediately, just as in the case i), that $A(K)/I(K)$ is also discrete and the lemma is thus completely proved.

Proof of Theorem 1. We now proceed to prove Theorem 1. Let

$$s \rightarrow D(s), \qquad s \in K \tag{1.8}$$

be a continuous representation and let $\chi(s)$ be the character of D. As $\chi(s)$ is obviously a continuous function on K, we see by the second definition of the topology in $A(K)$ that the group

$$A\chi = \{\sigma;\, \sigma \in A(K),\, \chi^{\sigma} = \chi\}$$

is a closed subgroup of $A(K)$. We shall prove next that $A\chi$ is at the same time open in $A(K)$. For that purpose, let

$$\chi = n_1\chi_1 + \cdots + n_r\chi_r \qquad (n_i > 0)$$

be the decomposition of χ into irreducible characters $\chi_i (i = 1, \cdots, r)$ and let

$$\chi^{\tau} = n'_1\chi_1 + \cdots + n'_r\chi_r + \sum \chi' \qquad (n'_i \geqq 0)$$

be that of χ^{τ}, where τ is an automorphism in $A(K)$, n'_i is either 0 or equal to some n_j and χ' are characters different from χ_i. With respect to the Haar measure μ on K, with the total measure 1, we now integrate these functions and obtain

$$(\chi, \chi) = \int_K \chi(s)\bar{\chi}(s)\, d\mu(s) = n_1^2 + \cdots + n_r^2,$$

$$(\chi, \chi^{\tau}) = \int_K \chi(s)\bar{\chi}^{\tau}(s)\, d\mu(s) = n_1 n'_1 + \cdots + n_r n'_r.$$

It follows then immediately that

$$(\chi, \chi) \geqq (\chi, \chi^{\tau}),$$

where the equality holds only when $n_i = n'_i (i = 1, \cdots, r)$, i.e. when $\chi = \chi^{\tau}$. As (χ, χ^{τ}) takes only integral values, we come to the conclusion that

(1.9) $\quad (\chi, \chi^{\tau}) > (\chi, \chi) - 1$ implies $\tau \in A\chi$.

Now put

$$\alpha = \int_K | \chi(s) | \, d\mu(s)$$

and let τ be an automorphism in the neighborhood

(1.10) $$U\left(\chi, \frac{1}{\alpha}\right) = \left\{\sigma; \sigma \in A(K), \| \chi - \chi^{\sigma} \| < \frac{1}{\alpha}\right\}.$$

It follows then that

$$| (\chi, \chi) - (\chi, \chi^{\tau}) | \leqq \| \chi - \chi^{\tau} \| \int_K | \chi(s) | \, d\mu(s) < 1$$

and in virtue of (1.9) we have

$$\tau \in A_{\chi}.$$

$U(\chi, 1/\alpha)$ is therefore contained in A_{χ} and A_{χ} is consequently open in $A(K)$.

Now take any σ in A_{χ}. From $\chi = \chi^{\sigma}$ we see in particular that σ leaves invariant the kernel $N = N_{\chi}$ of the homomorphism (1.8):

$$N^{\sigma} = N.$$

σ induces therefore an automorphism σ' in K/N and the mapping

$$\sigma \to \sigma' = \varphi(\sigma)$$

is a continuous homomorphism from A_{χ} into $A(K/N)$. As the image $\varphi(A_{\chi})$ of this mapping obviously contains the group $I(K/N)$ of inner automorphisms of K/N, we put

$$A'_{\chi} = \varphi^{-1}(I(K/N)).$$

Here K/N is isomorphic to the group of matrices $D(K)$, consequently it is a compact Lie group (not necessarily connected). It follows then by Lemma 1.1, that $A(K/N)/I(K/N)$ is discrete and we see that A'_{χ} is at the same time closed and open in A_{χ}, consequently also in $A(K)$.

We now prove that if χ varies over all continuous characters of K, the intersection of corresponding A'_{χ} coincides with the group $I(K)$ of the inner automorphisms of K. It is clear that the latter group is contained in the former. We have only to show therefore that if σ is contained in $\bigcap_{\chi} A'_{\chi}$, it is an inner automorphism of K. Now we denote by $M(\chi)$ the set of all t in K, such that

(1.11) $\quad s^{\sigma} \equiv t^{-1}st \quad \text{mod. } N_{\chi}, \quad$ for all s in K.

By the definition of A'_{χ}, $M(\chi)$ is then a non-empty closed subset of K for any character χ. If $\chi_1, \cdots, \chi_r$ are (here not necessarily irreducible) characters of K and if we put

$$\chi = \chi_1 + \cdots + \chi_r.$$

$M(\chi)$ is obviously contained in the intersection of all $M(\chi_i)$ $(i = 1, \cdots, r)$ and it follows that the intersection of any finite number of $M(\chi)$ is a non-empty set. As K is compact, we see then immediately that there exists an element t_0 which is contained in all $M(\chi)$. Such t_0 satisfies

$$s^\sigma \equiv t_0^{-1} s t_0 \quad \text{mod. } N_\chi \tag{1.12}$$

for all characters χ. However as the intersection of all N_χ contains only e, it follows from (1.12) that

$$s^\sigma = t_0^{-1} s t_0 ,$$

which shows that σ is an inner automorphism induced by t_0. We have thus proved

$$\bigcap_\chi A'_\chi = I(K). \tag{1.13}$$

Now it has already been proved that A'_χ is at the same time closed and open in $A(K)$. By (1.13) we can see then immediately that $A(K)/I(K)$ is totally disconnected and the proof of Theorem 1 is thus completed.

Now consider the group $I^*(K_0)$, which is the group of all inner automorphisms induced by elements in the connected component K_0 of K. As the mapping (1.3) is continuous $I^*(K_0)$ is connected, while $I(K)/I^*(K_0)$ is totally disconnected. By making use of Theorem 1 we can obtain therefore the following more precise result.

THEOREM 1′. *The connected component of the group of automorphisms $A(K)$ of a compact group K is the group $I^*(K_0)$, consisting of all inner automorphisms induced by elements in the connected component K_0 of K.*

COROLLARY. *If K is a compact abelian group, $A(K)$ is totally disconnected.*

Although this corollary is an immediate consequence of Theorem 1, we shall give here another simple proof for this case. Let D be the discrete group, which is dual to K in the sense of Pontrjagin,[8] i.e. the character group of K. We denote by $A(D)$ the group of all automorphisms of D and introduce a topology in $A(D)$ in the following way. Take any finite number of elements $a_1, \cdots, a_n$ in D and put

$$U(a_1, \cdots, a_n) = \{\sigma;\ \sigma \in A(D),\ a_i^\sigma = a_i,\ i = 1, \cdots, n\}. \tag{1.14}$$

As is readily seen, $A(D)$ thus becomes a topological group with $\{U(a_1, \cdots, a_n)\}$ as a system of the neighborhoods of e. Now K and D form a pair so that the product sa is defined for $s \in K$ and $a \in D$. We can see then that

$$s^\sigma a = s a^{\sigma'}$$

gives a one-to-one correspondence $\sigma \leftrightarrow \sigma'$ between $A(K)$ and $A(D)$. It is then proved that this correspondence is an isomorphism of $A(K)$ and $A(D)$, even

[8] Cf. Pontrjagin [15] or [16]. Throughout this paper we have often used without any explicit reference those results on topological groups, in particular on Lie groups, which are stated in Pontrjagin's book [16].

when we consider them as topological groups.[9] It is sufficient therefore to prove that $A(D)$ is totally disconnected. For that purpose take an automorphism τ which is not contained in (1.14). Some a_i, for example a_1 is not fixed by τ:

$$a_1^\tau = b \neq a_1 .$$

Consequently any τ' of the form

$$\tau' = \tau\tau'', \qquad \tau'' \in U(b)$$

is also not contained in (1.14). The neighborhood $U(a_1, \cdots, a_n)$ is therefore closed and $A(D)$ is consequently totally disconnected, q.e.d.

Now let G be a connected topological group which contains K as a compact normal subgroup. As stated above there is a continuous mapping (1.3) from G into $A(K)$. The image $\varphi(G)$ of G by this mapping is, as we have assumed G to be connected, a connected subgroup of $A(K)$. By Theorem 1 it is therefore contained in $I(K)$ and this means that for any given g in G there is some x in K, so that

$$g^{-1}sg = x^{-1}sx, \qquad \text{for all } s \text{ in } K. \tag{1.15}$$

We have thus proved the following theorem.

THEOREM 2. *Let G be a connected topological group and K a compact normal subgroup of G. If we denote by H the normal subgroup of G, consisting of all elements in G which are commutative with every element in K, we have*

$$G = HK. \tag{1.16}$$

An immediate consequence of Theorem 2 is

THEOREM 3. *Let G and K be groups as in Theorem 2. If K' is a normal subgroup of K, it is also normal in G.*

From the definition of H, it is obvious that the intersection

$$Z = H \frown K \tag{1.17}$$

is the center of K and it follows from (1.18) that G/Z is the direct product of H/Z and K/Z. This shows that the situation of a compact normal subgroup in a connected topological group is considerably simple. For example, if K has no center (except e) K is always a direct factor in a connected group, in which it is contained as a normal subgroup.

Next suppose that K is abelian. We have then $Z = K$ and consequently $H = G$. This proves

THEOREM 4. *A compact abelian normal subgroup of a connected topological group G is contained in the center of G.*

For example the group Z in (1.17) is contained in the center of G.

---

[9] Cf. Abe [1].

2. Solvable groups

Let G be an arbitrary topological group and let H_1 and H_2 be arbitrary subgroups of G. We denote by $[H_1, H_2]$ the subgroup of G generated by elements of the form

$$[h_1, h_2] = h_1 h_2 h_1^{-1} h_2^{-1}, \qquad h_1 \in H_1, \qquad h_2 \in H_2.$$

Obviously $[H_1, H_2]$ is not always closed in G (see the example in §4). We denote therefore the closure of $[H_1, H_2]$ by $C(H_1, H_2)$ and call it the *topological commutator group* of H_1 and H_2. If H_1 is normal in G, $C(H_1, H_2)$ is contained in H_1, and if moreover H_2 is also normal in G, $C(H_1, H_2)$ is a normal subgroup of G contained in $H_1 \frown H_2$.

LEMMA 2.1. *If one of H_i, for example H_1 is connected, $C(H_1, H_2)$ is also a connected subgroup of G.*

PROOF. Let M be the connected component of $C(H_1, H_2)$. M is a closed subgroup of G. Take an element h_2 in H_2 and consider the continous mapping

$$h_1 \longrightarrow [h_1, h_2], \qquad h_1 \in H_1$$

from H_1 into G. As the image $[H_1, h_2]$ of this mapping is a connected subset containing $e = [e, h_2]$, it follows

$$[H_1, h_2] \subset M.$$

Since h_2 is an arbitrary element in H_2, we have then $[H_1, H_2] \subset M$, whence $C(H_1, H_2) \subset M$, $C(H_1, H_2) = M$, q.e.d.

Now we define by induction the *series of commutator subgroups* $\{D_\xi(G)\}$ of G as follows. First we put

$$D_0(G) = G$$

and suppose that $D_\eta(G)$ are already defined for all ordinal numbers $\eta < \xi$. We put then

$$D_\xi(G) = C(D_{\xi'}(G), D_{\xi'}(G)),$$

if $\xi = \xi' + 1$ and

$$D_\xi(G) = \textstyle\bigcap_{\eta<\xi} D_\eta(G),$$

if ξ is a limit number.

As $\{D_\xi(G)\}$ is a descending sequence of normal subgroups in G, there exists a number ξ_0 such as

$$\text{(2.1)} \qquad D_{\xi_0}(G) = D_{\xi_0+1}(G) = \cdots, \qquad D_\eta(G) \neq D_{\xi_0}(G) \quad \text{for} \quad \eta < \xi_0.$$

We say such ξ_0 is the length of the series of commutator subgroups of G, or short the *length of* G.

If in particular the group $D_{\xi_0}(G)$ in (2.1) is equal to e we say that G is *solvable*, or more precisely *topologically solvable*. (In a similar way we can define the *series of central subgroups* of a topological group, and *nilpotent topological groups*. However we do not discuss these subjects[10] here.)

[10] Cf. Iwasawa [10].

We now prove some lemmas.

LEMMA 2.2. *The length of the series of commutator subgroups of a connected compact group G is at most* 1. *In particular a connected compact solvable group is abelian.*

PROOF. We have to prove $D_1(G) = D_2(G)$. Replacing G by $G/D_2(G)$ if necessary, we may suppose $D_2(G) = e$. $D_1(G)$ is then a compact abelian normal subgroup of G and it is contained in the center of G by Theorem 4. Now take an arbitrary element u in G and denote by M the closure of the group generated by u and $D_1(G)$. As $D_1(G)$ is a central subgroup and as $G/D_1(G)$ is abelian, M is also an abelian normal subgroup of G. By Theorem 4 it is therefore contained in the center of G, and as u may be arbitrary chosen, this shows that G is abelian, i.e. $D_1(G) = D_2(G) = e$, q.e.d.

LEMMA 2.3. *Let G be a locally compact group and let* $\{H_\alpha\}$ *be a system of subgroups of G with following properties:*

i) H_α *are all connected,*

ii) *the intersection H of all* H_α *is a discrete subgroup of G,*

iii) *for any finite number of* H_α, *say* $H_{\alpha_1}, \cdots, H_{\alpha_k}$, *there exists a subgroup* H_β *in the system which is contained in the intersection of* H_{α_i} $(i = 1, \cdots, k)$.

Then the system contains a subgroup which is compact.

PROOF.[11] Take a neighborhood U of e in G such that the closure $\bar{U}$ of U is compact and $\bar{U} \frown H = e$. Let $R = \bar{U} - U$. R is then compact and has no element common with H. By ii), iii) there exists consequently a subgroup H_β which does not meet with R. As H_β is connected by i) we see that it must be contained in U. The group H_β is then obviously compact.

We now prove the following

THEOREM 5. *The length of the series of commutator subgroups of a connected locally compact group G is finite. In particular if G is solvable, we have*

$$D_n(G) = e,$$

with some integer n.

PROOF. We prove $D_n(G) = D_{n+1}(G)$ for some integer n. Replacing G by $G/D_\omega(G)$ if necessary, we may suppose $D_\omega(G) = e$, i.e. that the intersection of all $D_m(G)$ $(m = 0, 1, 2, \cdots)$ is e. By Lemma 2.1 $D_m(G)$ are all connected groups and we can apply Lemma 2.3 to the system $\{D_m(G)\}$. It then follows that some $D_m(G)$, say $D_{n-1}(G)$ is compact and from Lemma 2.2

$$D_1(D_{n-1}(G)) = D_2(D_{n-1}(G)), \qquad \text{i.e.} \quad D_n(G) = D_{n+1}(G),$$

q.e.d.

REMARK. In a similar way we can prove that the length of the series of commutator subgroups of a not necessarily connected group G is finite, if G is contained as a normal subgroup in a connected locally compact group G'. Lemma 2.2 can be also extended for a not necessarily connected compact normal

[11] Cf. Freudenthal [7].

subgroup of a connected group. It is to be noted however that without any condition on the connectivity of the group, the conclusion of these theorems does not hold. For example the length of the direct product of all finite groups (considered as a compact group) is just ω. On the other hand we can prove that the length of any compact group is at most ω.[12]

Here we add one more lemma which will be used afterwards.

LEMMA 2.4. *Let G be a connected topological group and N a compact normal subgroup of G. Put further $N_1 = D_1(N)$ and let Z be the center of N. Then we have $N = N_1Z$, and $N_1 \frown Z$ is a totally disconnected group.*

PROOF. Let H be the group consisting of all elements in G, which commute with every element in N_1. By Theorem 2 we have

$$G = HN_1.$$

If we put $H_1 = H \frown N$, we have also

$$N = H_1N_1, \quad H_1 \supset Z.$$

$Z_0 = H_1 \frown N_1$ is then the center of N_1. As $H_1/Z_0 \cong N/N_1$ is abelian, we see that H_1 is solvable, consequently abelian by the above remark. It follows then immediately $H_1 = Z$. We shall show next that Z_0 is a totally disconnected group. Take a normal subgroup M of N_1, so that N_1/M is a Lie group. By the above remark we have $D_1(N_1) = D_2(N) = N_1$ and we see that N_1/M is a semi-simple compact Lie group. Z_0M/M is consequently a finite group and the same is true for $Z_0/(Z_0 \frown M)$. As the intersection of all such normal subgroups M is e, we have

$$\bigcap(Z_0 \frown M) = e,$$

which proves that Z_0 is totally disconnected.

REMARK. $G = N$ gives us in particular the structure of a connected compact group.[13]

3. On Lie groups

In this section we shall prove theorems concerning Lie groups, which will be used in succeeding sections.

Let G be a Lie group and L its corresponding Lie algebra. It is well-known that local subgroups of G and subalgebras of L correspond in one-to-one way each other, so that we may use in the following the same letter for the corresponding local subgroup and subalgebra.

Now let R be the radical of L. We denote by N the normal subgroup of G generated by the local Lie group R and by R_1 the group germ of N, which is also a local Lie group. It is then easy to see that N is a topological solvable group in the sense of the preceding section and R_1 is consequently a solvable ideal in L. As R_1 obviously contains R, it follows from the definition of the

12 Cf. Iwasawa [10].

13 Cf. Kampen [11].

radical $R_1 = R$. From these considerations we can prove immediately the following lemmas.

LEMMA 3.1. *Let G be a Lie group and L its corresponding Lie algebra. G contains then a connected solvable normal subgroup N, which locally coincides with the radical of L. Any connected solvable normal subgroup of G is contained in N and we call N the radical of G.*

LEMMA 3.2. *A connected Lie group G is solvable if and only if its corresponding Lie algebra L is solvable. The length of G is then the same as the length of the series of derived algebras of L.*

LEMMA 3.3. *A connected solvable Lie group G has a series of characteristic subgroups*

$$G = M_0 \supset M_1 \supset \cdots \supset M_r = e,$$

such as M_{i-1}/M_i is either a vector group or a toroidal group.

PROOF. Let

$$G = G_0 \supset G_1 \supset \cdots \supset G_n = e, \qquad G_i = D_i(G) \tag{3.1}$$

be the series of commutator subgroups of G. As G_{i-1}/G_i is a connected abelian Lie group, it is the direct product of a vector group and a toroidal group, of which the latter is a characteristic subgroup of G/G_i. By putting the corresponding characteristic subgroup between G_{i-1} and G_i we can refine (3.1) to a series as stated in the lemma, q.e.d.

In the following the letters V and T are reserved for a vector group or a toroidal group respectively. If the dimension of these groups is to be given precisely, we denote it with lower index, as V_n or T_n.

LEMMA 3.4. *Let G be a locally compact group and N a normal subgroup of G. If $G/N \cong V_1$ or $\cong T_1$ and if N is a connected solvable Lie group, G contains a subgroup H such as*

$$G = HN, \qquad H \frown N = e.$$

In other words the extension G of N by V_1 or T_1 always splits.

PROOF. We use the induction with respect to the dimension of N and assume that the lemma is already proved for a normal subgroup of less dimension than that of N. Suppose that N contains a proper connected normal subgroup N_1 or G. We then apply the induction assumption to G/N_1 and see that there exists a subgroup H_1 such that

$$G = H_1N, \qquad H_1 \frown N = N_1, \qquad H_1/N_1 \cong G/N.$$

Again the induction assumption for H_1 and N_1 gives us a subgroup H such as

$$H_1 = HN, \qquad H \frown N_1 = e$$

and this H obviously satisfies the assertion of the lemma.

It is therefore sufficient to prove the lemma for the case where N has no proper connected normal subgroup of G. By Lemma 3.3 N is then isomorphic either to a vector group or to a toroidal group.

We note first that if G is abelian the lemma follows immediately from a theorem of Pontrjagin on the structure of connected locally compact abelian groups[14] and we prove then that this is surely the case when N is contained in the center of G. For that purpose take an arbitrary a_0 in G which is not contained in N. As $G/N \cong V_1$ or $\cong T_1$ there exists an element a_1 such as

$$a_1^2 \equiv a_0 \qquad \text{mod. } N.$$

In the same way we take $a_2, a_3, \cdots$, so that

$$a_2^2 \equiv a_1, \qquad a_3^2 \equiv a_2, \cdots \qquad \text{mod. } N.$$

It is then

$$a_{i+j}^{2^i} = a_j u, \qquad u \in N$$

and we see that $a_0, a_1, a_2, \cdots$ commute with each other. As the group generated by $a_0, a_1, \cdots$ and N is dense in G, it follows immediately that G is abelian.

Now if N is a toroidal group, it is contained in the center of G by Theorem 4. It remains therefore only to prove the case where N is a vector group V_n and is not contained in the center of G. In such a case elements of N are given by their coordinates:

$$u = u(\xi_1, \cdots, \xi_n),$$

where ξ_i varies over all real numbers. The transformation by an element a in G induces an automorphism in N, given by

$$a^{-1}ua = u' = u(\xi_1', \cdots, \xi_n')$$
$$(\xi_1', \cdots, \xi_n') = (\xi_1, \cdots, \xi_n)A,$$

where $A = A(a)$ is a real matrix of degree n. The correspondence

$$a \rightarrow A(a)$$

gives obviously a representation of G/N.

We prove next that there exists a suitable a_0 in G such that the determinant $|A(a_0) - E| \neq 0$. As N is not contained in the center of G, we can find an element a_0 in G, which does not commute with some element in N. Consider then the eigen-space V' for the eigen-value 1 of the matrix $A(a_0)$. As the set of matrices $\{A(a); a \in G\}$ is commutative, V' is invariant under all $A(a)$. This means that the subgroup N' of N, which corresponds to V' in V_n, is normal in G. By the assumption we must have either $N' = N$ or $N' = e$. The former case does not hold, for a_0 is not commutative with some element in N. Therefore $N' = e$ and this means that V' is 0-dimensional, i.e. $|A(a_0) - E| \neq 0$. It follows then that $a_0ua_0^{-1}u^{-1}$ runs over all elements of N, if u runs through N.

[14] Cf. Pontrjagin [16].

Now take an arbitrary b in G and put

$$a_0 b a_0^{-1} b^{-1} = v.$$

As G/N is abelian v is contained in N. Take then an element u in N such that $a_0 u a_0^{-1} u^{-1} = v$ and put $c = u^{-1} b$. It follows that

$$a_0 c = c a_0 .$$

This shows that in any coset mod. N there exists an element which is commutative with a_0. Therefore if we denote by H the set of all elements in G commuting with a_0, we have

$$G = HN.$$

Next take an arbitrary w in $H \frown N$. We have then $a_0 w a_0^{-1} w^{-1} = e$ and $| A(a_0) - E | \neq 0$ implies $w = e$, i.e. $H \frown N = e$. The lemma is thus completely proved.

A minor modification of this proof gives the following

LEMMA 3.5. *Let G be a locally compact group and N a normal subgroup of G. If N is a connected solvable Lie group and if L' is a local one-parameter subgroup of the group $G' = G/N$, G contains a local one-parameter subgroup L such as*

$$LN/N = L', \qquad L \frown N = e.$$

Now by Lemma 3.4 we can obtain immediately the following

LEMMA 3.6. *Let G be a connected solvable Lie group of dimension n. G contains then n subgroups $H_1, \cdots, H_n$ with the following properties*:

i) *each H_i is isomorphic either to V_1 or to T_1*,

ii) *the product $G_i = H_{i+1} \cdots H_n$ ($i = 0, 1, \cdots, n-1$) forms a $(n-i)$-dimensional subgroup of G and G_i is normal in G_{i-1}*,

iii) *we have in particular $G = G_0$ and any g in G can be decomposed uniquely and continuously in the form*

$$g = h_1 h_2 \cdots h_n , \qquad h_i \in H_i .$$

From this lemma it follows particularly that the space of G is the cartesian product of a Euclidean space and a torus[15] and that a system of canonical coordinates can be introduced, which are analytical in the whole space.

LEMMA 3.7. *Let G be a locally compact group and N a normal subgroup of G. If N is a vector group and G/N a compact Lie group, G contains a compact subgroup K such as $KN = G$, $K \frown N = e$. Moreover, if K' is any other compact subgroup with $K'N = G$, $K' \frown N = e$, K and K' are conjugate in G.*

PROOF. We denote by φ the natural homomorphism from G onto $G' = G/N$. By Lemma 3.5 we can find a subset U in G with the following properties:

i) $U' = \varphi(U)$ is an open neighborhood of e in G',

ii) φ is a homeomorphism on U,

iii) the closure $\bar{U}$ of U is compact.

[15] Cf. Chevalley [5].

Take then a finite number of elements $\sigma_1, \cdots, \sigma_r$ in G', so that $\sigma_i U' (i = 1, \cdots, r)$ cover the whole group G' and put

$$W'_i = \sum_{j=1}^{i} \sigma_j U' - \sum_{j=1}^{i-1} \sigma_j U'$$
$$= \sigma_i U'_i, \qquad U'_i \subset U', i = 1, \cdots, r.$$

We take r elements $s_1, \cdots, s_r$ in G such as $\sigma_i = \varphi(s_i)$ and define a mapping ψ from G' into G by

$$\psi(\sigma) = \psi(\sigma_i\tau) = s_i\varphi^{-1}(\tau), \qquad \tau \in U'_i, \varphi^{-1}(\tau) \in U,$$

where $\sigma = \sigma_i\tau$ is assumed to be contained in W'_i. Since τ is contained in U' and φ is a homeomorphism on U, $\varphi^{-1}(\tau)$ is uniquely determined as an element in U. It is obvious that ψ is an inverse mapping of φ and that the set $M = \psi(G')$ forms a complete system of representatives for $G' = G/N$ in G.

Now as before, elements of $N = V_n$ are given by

$$u = u(\xi_1, \cdots, \xi_n), \qquad -\infty < \xi_i < \infty.$$

If we put therefore

$$\psi(\sigma)\psi(\tau) = u(\sigma, \tau)\psi(\sigma\tau)$$
$$u(\sigma, \tau) = u(\xi_1(\sigma, \tau), \cdots, \xi_n(\sigma, \tau)), \qquad \sigma, \tau \in G',$$

we obtain n real functions $\xi_i(\sigma, \tau)$ on $G' \times G'$. From the construction of ψ it is then easy to see that these functions are bounded and measurable with respect to the Haar measure on $G' \times G'$.

On the other hand we see easily that

$$u(\sigma, \tau)u(\sigma\tau, \rho) = (su(\tau, \rho)s^{-1})u(\sigma, \tau\rho), \qquad s = \psi(\sigma).\text{[16]} \tag{3.2}$$

Putting

$$su(\xi_1, \cdots, \xi_n)s^{-1} = u(\xi'_1, \cdots, \xi'_n),$$
$$(\xi'_i) = A(\xi_i), \qquad A = A(s) = A(\sigma) = (\alpha_{ij}(\sigma)),$$

where (ξ_i) and (ξ'_i) are columns with components ξ_i or ξ'_i, it follows from (3.2) that

$$\xi_i(\sigma, \tau) + \xi_i(\sigma\tau, \rho) = \sum_{j=1}^{n} \alpha_{ij}(\sigma)\xi_j(\tau, \rho) + \xi_i(\sigma, \tau\rho). \tag{3.3}$$

If we put further

$$\eta_i(\sigma) = \int_{G'} \xi_i(\sigma, \tau)\, d\mu(\tau), \tag{3.4}$$

where μ is the Haar measure on G' with the total measure 1, the integration of (3.3) gives us

$$\xi_i(\sigma, \tau) + \eta_i(\sigma\tau) = \sum_{j=1}^{n} \alpha_{ij}(\sigma)\eta_j(\tau) + \eta_i(\sigma).$$

[16] Cf. Zassenhaus [19].

The elements

$$v(\sigma) = u(\eta_1(\sigma), \cdots, \eta_n(\sigma))$$
$$\psi'(\sigma) = v(\sigma)^{-1}\psi(\sigma), \qquad \sigma \in G'$$

satisfy then

$$u(\sigma, \tau)v(\sigma\tau) = (sv(\tau)s^{-1})v(\sigma),$$
$$\psi'(\sigma)\psi'(\tau) = \psi'(\sigma\tau).$$

The set $K = \{\psi'(\sigma); \sigma \in G'\}$ forms therefore an abstract group such as $KN = G$, $K \frown N = e$. As the closure of $M = \{\psi(\sigma); \sigma \in G'\}$ is compact by the construction of $\psi(\sigma)$ and as $\eta_i(\sigma)$ is bounded according to (3.4), the closure $\bar{K}$ of K is a compact group, whence follows immediately $\bar{K} \frown N = e$, $\bar{K} = K$. The existence of a compact subgroup K is thus proved.

In order to prove the second part, it is sufficient to show that any compact subgroup K^* of G is contained in some conjugate subgroup of K. Now any element g in G can be written uniquely in the form

$$g = uv, \qquad u \in N, \qquad v \in K.$$

In particular we put

$$s = u(s)v(s), \qquad u(s) \in N, \qquad v(s) \in K,$$

for any s in K^*. It holds then

$$u(s_1s_2) = u(s_1)(s_1u(s_2)s_1^{-1})$$

and as before, by putting $\xi_i^*(s) = \xi_i(u(s))$, we have

$$\xi_i^*(s_1s_2) = \xi_i^*(s) + \textstyle\sum_{j=1}^n \alpha_{ij}(s_1)\xi_j^*(s_2).$$

As the functions are here all continuous on K^*, the integration with respect to the Haar measure μ^* on K^* gives

(3.5) $$\eta_i = \xi_i^*(s) + \textstyle\sum_{j=1}^n \alpha_{ij}(s)\eta_j,$$

where

$$\eta_i = \int_{K^*} \xi_i^*(s)\, d\mu^*(s).$$

If we put

$$u_0 = u(\eta_1, \cdots, \eta_n),$$

(3.5) means

$$u_0 = u(s)(su_0s^{-1}), \qquad \text{or} \qquad s = u_0v(s)u_0^{-1}.$$

K^* is therefore contained in $u_0Ku_0^{-1}$ and the lemma is completely proved.

LEMMA 3.8. *The preceding lemma holds also, when G/N is an arbitrary compact group, N being again a vector group.*

PROOF. In the above proof we have used the fact that G/N is a Lie group only to prove the existence of the set U in G. Therefore, if we could prove the existence of such U for arbitrary compact G/N, the proof would be immediately extended to the general case. However here we follow another way. Let M be the subgroup of G, consisting of all elements in G, which are commutative with every element in N. M is obviously a normal subgroup containing N. Corresponding to any a in G there is the automorphism

$$u \to aua^{-1}, \qquad u \in N$$

of N, and it can be seen that G/M is isomorphic to a compact group of linear transformations in a vector space $V = N$. G/M is consequently a compact Lie group. Now by the definition of M, N is contained in the center of M. As N contains two discrete subgroups D_1, D_2, such that N/D_1, N/D_2 are compact and $D_1 \frown D_2 = e$, we see that M is a maximally almost periodic group. Moreover the factor group M/M_0 of M modulo its connected component M_0 is compact, for M_0 obviously contains N. M is consequently the direct product of N and a characteristic compact subgroup K_1.[17] Consider then G/K_1 and apply Lemma 3.7 with respect to its normal subgroup $M/K_1 \cong N$. It follows then immediately that there exists a compact subgroup K such that

$$G = KN, \qquad K \frown N = e.$$

The second part of the lemma, i.e. the conjugacy of such K holds also in the general case, for the corresponding part of the proof of Lemma 3.7 has not used the fact that G/N is a Lie group.

LEMMA 3.9. *Let G be a connected topological group and N a normal subgroup of G. If N is a connected semi-simple Lie group, G contains a subgroup H, such as $HN = G$ and $H \frown N$ is a discrete group. If moreover G/N is a simply connected Lie group, we may take such H as $H \frown N = e$.*

PROOF. If we take a system of canonical coordinates of the first kind $\xi_1, \cdots, \xi_n$ in N, any automorphism of N induces a linear transformation in ξ_i. In particular the automorphisms

$$u \to aua^{-1}, \qquad u \in N,$$

induced by elements a in G, can be represented as a connected linear group on ξ_i. Now it is well-known that the connected component of the group of automorphisms of a semi-simple Lie group just coincides with the group of inner automorphisms. If we denote therefore by H the set of elements in G, which commute with every element in N, we have, as in the proof of Theorem 2, $G = HN$. The group $H \frown N$ is discrete, for it is the center of semi-simple N. Suppose next that G/N is a simply connected Lie group and let H_0 be the connected component of H. It is then easy to see that we have $G = H_0N$. It follows then

$$G/N \cong H_0/H_0 \frown N$$

[17] Cf. Freudenthal [7].

and this shows that H_0 is locally isomorphic to simply connected G/N. As H_0 is connected, it follows then $H_0 \frown N = e$, as asserted.

LEMMA 3.10. *Let G be a locally compact group and N a normal subgroup of G. If N is a connected Lie group and if G/N is isomorphic to V_1, G contains a subgroup H such that $G = HN$, $H \frown N = e$.*

PROOF. Let N_1 be the radical of N. Applying the previous lemma to G/N_1 and N/N_1, we see that there exists a subgroup H_1 in G such that $G = H_1N$, $H_1 \frown N = N_1$, $H_1/N_1 \cong V_1$. By Lemma 3.4 there is then a subgroup H so that $H_1 = HN_1$, $H \frown N_1 = e$. It is then easy to see that $G = HN$, $H \frown N = e$.

LEMMA 3.11. *Let G be the adjoint group of a real semi-simple Lie algebra L. Then there exists a connected, simply connected solvable subgroup H and a maximal compact subgroup K of G, so that*

$$G = HK, \qquad H \frown K = e,$$

i.e. that any element g in G can be written uniquely in the form

$$g = hk, \qquad h \in H,\ k \in K$$

and h and k depend continuously on g. In particular the space of G is the cartesian product of the spaces of H and K, of which the former is homeomorphic to a Euclidean space.

PROOF.[18] Let $\tilde{L}$ be the complex form of the semi-simple Lie algebra L of degree s and L_0 the compact form of $\tilde{L}$, i.e. the Lie algebra which corresponds to a compact semi-simple Lie group and whose complex form is $\tilde{L}$. By a theorem of E. Cartan[19] the algebra L, or more rigorously speaking, an algebra which is isomorphic to L can be obtained from L_0 in the following way. Take a suitable involutive automorphism σ of L_0 and choose a real basis of L_0

$$u_1, \cdots, u_s,$$

such that

$$\sigma(u_i) = u_i, \qquad \text{for } i = 1, \cdots, r,$$
$$= -u_i, \qquad \text{for } i = r + 1, \cdots, s,$$

for some integer r. A real basis of L can be then obtained by putting

$$v_1 = u_1, \cdots, v_r = u_r, \qquad v_{r+1} = \sqrt{-1}u_{r+1}, \cdots, v_s = \sqrt{-1}u_s.$$

In other words, if we put

$$\begin{aligned} u &= u' + u'', \\ u' &= \tfrac{1}{2}(u + \sigma(u)), \qquad u'' = \tfrac{1}{2}(u - \sigma(u)), \\ \sigma(u') &= u', \qquad \sigma(u'') = -u'', \end{aligned} \tag{3.6}$$

for any u in L_0, then

[18] The author's original proof was only valid for complex semi-simple Lie algebras. The present proof is due to Prof. C. Chevalley, to whom the author wishes to express his hearty thanks.

[19] Cf. E. Cartan [3] or Gantmacher [8], [9].

$$L = \{\bar{u}; \bar{u} = u' + \sqrt{-1}u''\} \tag{3.7}$$

and consequently

$$L_0' = L \frown L_0 = \{u'; u' \in L_0, \sigma(u') = u'\}. \tag{3.8}$$

In the following we denote also by σ the automorphism of $\bar{L}$, which is induced by σ on the complex form $\bar{L}$.

Now we take the maximum number of mutually commutative linearly independent elements $h_1', \cdots, h_m'$ in L_0 such that

$$\sigma(h_i') = -h_i', \qquad i = 1, \cdots, m.$$

Let H_0 be a maximal abelian subalgebra of L_0 containing these h_i'. If h is an arbitrary element in H_0, h, $\sigma(h)$ and consequently $h' = h - \sigma(h)$ are all commutative with h_i' $(i = 1, \cdots, m)$ and we have moreover $\sigma(h') = -h'$. According to the choice of $h_1', \cdots, h_m'$, h' must be contained in the linear set $\{h_1', \cdots, h_m'\}$ and this shows that $\sigma(h) \in H_0$ and H_0 is invariant by the automorphism σ.

After Weyl[20] we can then choose a complex basis

$$h_1, \cdots, h_n, e_\alpha, e_{-\alpha}, \cdots, e_\rho, e_{-\rho}$$

of $\bar{L}$ as follows:

$$(3.9)\quad \begin{aligned} &[h_i, h_j] = 0, \\ &[\textstyle\sum_{i=1}^n \lambda_i h_i, e_\alpha] = (\alpha\lambda)e_\alpha, \qquad (\alpha\lambda) = \sum_{i=1}^n \alpha_i\lambda_i, \\ &[e_\alpha, e_{-\alpha}] = -\textstyle\sum_{i=1}^n \alpha_i h_i, \\ &[e_\alpha, e_\beta] = N_{\alpha,\beta}e_{\alpha+\beta}, \qquad (\alpha + \beta \neq 0). \end{aligned}$$

Here $\alpha = (\alpha_1, \cdots, \alpha_n)$ are real vectors, called root vectors of $\bar{L}$ and $N_{\alpha,\beta} \neq 0$ if and only if $\alpha + \beta$ is again a root vector. Moreover L_0 can be identified with the set of elements

$$x = \textstyle\sum_{i=1}^n \sqrt{-1}\lambda_i h_i + \sum_\alpha \tau_\alpha e_\alpha, \qquad \lambda_i = \text{real}, \quad \tau_{-\alpha} = \bar{\tau}_\alpha$$

and in particular

$$H_0 = \{h; h = \textstyle\sum_{i=1}^n \sqrt{-1}\lambda_i h_i, \lambda_i = \text{real}\}.$$

We may also suppose that $\{h_1', \cdots, h_m'\} = \{\sqrt{-1}h_1, \cdots, \sqrt{-1}h_m\}$ and

$$(3.10)\quad \begin{aligned} \sigma(\sqrt{-1}h_i) &= -\sqrt{-1}h_i, \qquad i = 1, \cdots, m, \\ &= \sqrt{-1}h_i, \qquad i = m+1, \cdots, n. \end{aligned}$$

If we therefore consider $(\lambda_1, \cdots, \lambda_n)$ as coordinates of $h = \sum_{i=1}^n \sqrt{-1}\lambda_i h_i$, σ induces a linear transformation

[20] Cf. Weyl [18].

(3.11) $$\sigma((\lambda_1, \cdots, \lambda_n)) = (-\lambda_1, \cdots, -\lambda_m, \lambda_{m+1}, \cdots, \lambda_n)$$

in the λ-space. In particular for any root vector α of $\tilde{L}$ we have from (3.9)

$$[\sigma(\textstyle\sum_{i=1}^{n} \sqrt{-1}\lambda_i h_i), \sigma(e_\alpha)] = (\alpha\lambda)\sigma(e_\alpha),$$

$$[-\textstyle\sum_{i=1}^{m} \sqrt{-1}\lambda_i h_i + \sum_{i=m+1}^{n} \sqrt{-1}\lambda_i h_i, \sigma(e_\alpha)] = (\sigma(\alpha)\sigma(\lambda))\sigma(e_\alpha),$$

which shows that $\alpha' = \sigma(\alpha)$ is again a root vector of $\tilde{L}$ and that $\sigma(e_\alpha)$ is an eigen-element belonging to α'.

In order to see the relation between α and $\alpha' = \sigma(\alpha)$, we define a linear order in the above λ-space, so that it becomes a linearly ordered abelian group. We put namely

$$(\lambda_1, \cdots, \lambda_n) > (\lambda_1', \cdots, \lambda_n'),$$

if and only if $\lambda_1 = \lambda_1', \cdots, \lambda_{i-1} = \lambda_{i-1}', \lambda_i > \lambda_i'$ for some i. It then follows immediately that for any root vector $\alpha > 0$, we have either $\sigma(\alpha) < 0$ or $\sigma(\alpha) = \alpha$. We denote by E the set of all $\alpha > 0$ with $\sigma(\alpha) < 0$ and by E' that of $\alpha > 0$ with $\sigma(\alpha) = \alpha$. For any $\alpha \in E'$ it holds then

(3.12) $$\sigma(e_\alpha) = e_\alpha, \quad \sigma(e_{-\alpha}) = e_{-\alpha}.$$

For $\sigma(\alpha) = \alpha$ means by (3.11) that the first m coordinates of α vanish and this implies that e_α and $e_{-\alpha}$ are both commutative with every element in $\{h_1', \cdots, h_m'\} = \{\sqrt{-1}h_1, \cdots, \sqrt{-1}h_m\}$. Therefore the elements $u_1 = (e_\alpha + e_{-\alpha}) - \sigma(e_\alpha + e_{-\alpha})$ and $u_2 = \sqrt{-1}(e_\alpha - e_{-\alpha}) - \sigma(\sqrt{-1}(e_\alpha - e_{-\alpha}))$ of L_0 are also commutative with h_i' $(i = 1, \cdots, m)$ and satisfy $\sigma(u_1) = -u_1$, $\sigma(u_2) = -u_2$. By the choice of h_i' it follows $u_1 = u_2 = 0$, $\sigma(e_\alpha) = e_\alpha$, $\sigma(e_{-\alpha}) = e_{-\alpha}$, as stated above.

Now let R be the set of all elements

(3.13) $$y = \textstyle\sum_{i=1}^{m} \kappa_i h_i + \sum_{\alpha\in E} \tau_\alpha e_\alpha,$$

where κ_i and τ_α run through arbitrary complex numbers. If we note that $\alpha, \beta \in E$ implies $\alpha + \beta \in E$ and $\alpha + \beta > \alpha, \beta$, we can see easily that R is a solvable subalgebra of $\tilde{L}$. Putting $R' = L \frown R$, we shall prove then

(3.14) $$L = R' + L_0',$$

where L_0' was given by (3.8). It is of course sufficient to prove $L \subset R + L_0'$ or by (3.6), (3.7), (3.8) to show that $\sqrt{-1}(u - \sigma(u))$ is contained in $R + L_0'$ for any u in L_0. As

$$\sqrt{-1}h_i, \qquad i = 1, \cdots, n,$$

$$e_\alpha + e_{-\alpha}, \quad \sqrt{-1}(e_\alpha - e_{-\alpha}), \qquad \alpha > 0$$

form a real basis of L_0, we have only to prove that the elements

$$\sqrt{-1}(\sqrt{-1}h_i - \sigma(\sqrt{-1}h_i)), \qquad i = 1, \cdots, n,$$

$$\sqrt{-1}((e_\alpha + e_{-\alpha}) - \sigma(e_\alpha + e_{-\alpha})),$$

$$\sqrt{-1}(\sqrt{-1}(e_\alpha - e_{-\alpha}) - \sigma(\sqrt{-1}(e_\alpha - e_{-\alpha}))), \qquad \alpha > 0$$

are contained in $R + L_0'$. Again by (3.10), (3.12) it suffices to show that

$$h_i, \qquad i = 1, \cdots, m,$$

$$\sqrt{-1}((e_\alpha + e_{-\alpha}) - \sigma(e_\alpha + e_{-\alpha})), \qquad (e_\alpha - e_{-\alpha}) - \sigma(e_\alpha - e_{-\alpha}), \ \alpha \in E$$

are elements of $R + L_0'$. As to h_i $(i = 1, \cdots, m)$ this is obvious. For the other elements we have

$$\sqrt{-1}((e_\alpha + e_{-\alpha}) - \sigma(e_\alpha + e_{-\alpha})) \equiv \sqrt{-1}(e_{-\alpha} - \sigma(e_\alpha))$$
$$\equiv -(\sqrt{-1}(e_\alpha - e_{-\alpha}) + \sigma(\sqrt{-1}(e_\alpha - e_{-\alpha}))) \qquad \text{mod. } R$$

and the last side is obviously an element of L_0'. $\sqrt{-1}((e_\alpha + e_{-\alpha}) - \sigma(e_\alpha + e_{-\alpha}))$ is therefore contained in $R + L_0'$ and so is $(e_\alpha - e_{-\alpha}) - \sigma(e_\alpha - e_{-\alpha})$ by a similar reason. We have thus proved (3.14).

Now let $\tilde{G}$ be the adjoint group of $\tilde{L}$. The adjoint group G of L is then a closed subgroup of $\tilde{G}$. Let further H and K be connected closed subgroups of G, which are generated by local subgroups of G, corresponding to the Lie algebras R' and L_0' respectively. As K is clearly contained in the compact adjoint group of L_0, it is also a compact subgroup of G. On the other hand the coefficients κ_i in (3.13) of any element y in R' are, as one readily sees, all real and the matrix for y in the regular representation of $\tilde{L}$ with respect to the basis

$$h_1, \cdots, h_n, e_\alpha, e_\beta, \cdots, e_\rho; \qquad \alpha < \beta < \cdots < \rho$$

has the form

$$\begin{pmatrix} 0 & & & & & & 0 \\ & \ddots & & & & & \\ & & 0 & & & & \\ & & & (\alpha\kappa) & & & \\ & & & & \ddots & & \\ * & & & & & & (\rho\kappa) \end{pmatrix},$$

By integrating these matrices we can see then immediately that H is a group of matrices of the form

$$\begin{pmatrix} \xi_1 & & & & 0 \\ & \xi_2 & & & \\ & & \ddots & & \\ & & & \ddots & \\ * & & & & \xi_s \end{pmatrix}, \qquad \xi_i > 0.$$

H is consequently a simply connected solvable Lie group and we have $H \frown K = e$, for H contains no proper compact subgroup.

Now according to (3.14) there exist neighborhoods U_1 and U_2 of e in H and K respectively, such that for an arbitrary $g = hk$, $h \in U_1$, $k \in U_2$, we have

$$hk = k'h', \qquad h' \in H, \qquad k' \in K.$$

Here h' and k' are uniquely determined and continuously depend on h, k. Take then a neighborhood U_3 of e in K so that $\bar{U}_3 \subset U_2$. As K is compact we can determine an integer p such that any element in K can be written as the product of p elements in U_3. On the other hand $h' = h'(h, k)$ is near to e for arbitrary k in $\bar{U}_3$, if h is sufficiently near to e, for $h'(h, k)$ is continuous and $h'(e, k) = e$. It follows that there exists a neighborhood U_4 of e in H such that any element $g = hk$, $h \in U_4$, $k \in K$ can be written in the form

$$hk = k'h', \qquad h' \in H, \qquad k' \in K. \tag{3.15}$$

As H is connected and is generated by elements in U_4, (3.15) is also true for arbitrary h in H. This means that the set

$$HK = KH$$

forms a group and this group coincides just with G, for some neighborhood of e in G is contained in HK. We have thus

$$G = HK, \qquad H \frown K = e \tag{3.16}$$

and any g in G can be written uniquely in the form

$$g = hk, \qquad h \in H, \qquad k \in K.$$

We can prove then easily that the mapping

$$(h, k) \rightarrow hk$$

from $H \times K$ on G is open, so that G is, as a topological space, homeomorphic to the cartesian product of H and K, of which the former is homeomorphic to a Euclidean space by Lemma 3.6.

Finally K is a maximal compact subgroup of G, as follows from (3.16) and the fact that H contains no proper compact subgroup.

Lemma 3.12. *Let G be a connected semi-simple Lie group. Maximal compact subgroups of G are then connected and conjugate to each other. Let K be one of them. Then there exists subgroups H and M such that*

i) $G = HM$, $H \frown M = e$,

ii) *H is a connected simply connected solvable Lie group,*

iii) *either $M = K$ or $M = K \times V$ with a vector group V, and M contains the discrete center of G.*

The space of G is the cartesian product of the compact space of K and that of H or HV, which is in either case homeomorphic to a Euclidean space.

Proof. Let L be the Lie algebra of G and let Z be the center of G. The factor group G/Z is then isomorphic to the adjoint group of L. Let further H', K' be the subgroups of G containing Z, such that H'/Z and K'/Z have the same property for G/Z as H and K in the preceding lemma. For the connected component H'_0 of H' we see immediately from the simply-connectedness of H'/Z, that

$$H'_0 Z = H', \qquad H'_0 \frown Z = e, \qquad H'_0 \cong H'/Z.$$

Consequently we have

$$G = H'_0 K', \qquad H'_0 \frown K' = e.$$

Here K' is a connected group, which is locally isomorphic to the compact group K'/Z. It is therefore either itself compact or the direct product of a compact group K and a vector group V. Moreover it is easy to see that K' or K is a maximal compact subgroup of G. Next let K^* be an compact subgroup of G. K^*Z/Z is then also a compact subgroup of G/Z and by a theorem of E. Cartan[21] some conjugate of K^*Z/Z is contained in K'/Z. A conjugate subgroup of K^* must be then contained in K' and even in K, if K' is not compact. Thus it can be seen that maximal compact subgroups of G are conjugate to each other. By putting $H'_0 = H$, $K' = M$ i), ii), iii) are proved and the assertion on the topological structure follows also immediately.

THEOREM 6.[22] *Let G be an arbitrary connected Lie group. Then maximal compact subgroups of G are connected and conjugate to each other. Let K be one of them. There exist then subgroups $H_1, \cdots, H_r$ of G, which are all isomorphic to V_1 and are such that any element g in G can be decomposed uniquely and continuously in the form*

$$(3.17) \qquad g = h_1 \cdots h_r k, \qquad h_i \in H_i, \qquad k \in K,$$

or briefly expressed:

$$G = H_1 \cdots H_r K.$$

In particular the space of G is the cartesian product of the compact space of K and that of $H_1 \cdots H_r$, which is homeomorphic to the r-dimensional Euclidean space E_r.

PROOF. If G is a simple abelian group, namely if G is isomorphic to V_1 or T_1, the theorem is evident. If G is simple and semi-simple, the theorem follows immediately from Lemma 3.6 and Lemma 3.12. Therefore if we apply the induction with respect to the dimension of G, the proof for the general case will be completed by the following lemma, which we shall prove in a slightly more generalized form than is required for the present proof, for the sake of later applications.

LEMMA 3.13. *Let G be a locally compact group and N a normal subgroup of G. If N and G/N are both connected Lie groups and if Theorem 6 is already proved to be true for these groups, then the same propositions are also valid for G, even when*

[21] Cf. E. Cartan [3].

[22] Cf. Malcev [12].

we do not assume that G is a Lie group. Moreover if K_1 is a subgroup of G containing N such that K_1/N is a maximal compact subgroup of G/N and if K_2 is a maximal compact subgroup of N, we can find a maximal compact subgroup K of G, which satisfies

$$K_1 = KN, \qquad K_2 = K \frown N. \tag{3.18}$$

PROOF. Suppose first that N is semi-simple. By Lemma 3.12 N has a subgroup $M = K_2 \times V$, in which the center of N is contained. Now by Lemma 3.9 the set of all elements in G, which commute with every element in N, form a normal subgroup H, for which we have $G = HN$ and $Z = H \frown N$ is the discrete center of N. Z is therefore a subgroup of M and we put $Z_1 = Z \frown V$. By a suitable choice of V we may suppose that Z/Z_1 is a finite group.

Now as is readily seen, H has a subgroup K_3 containing Z, such that

$$K_1 = K_3N, \qquad Z = K_3 \frown N, \qquad K_3/Z \cong K_1/N.$$

As K_3/Z is compact and Z/Z_1 is finite, K_3/Z_1 is also a compact group. Combining this with $K_3 \frown V = Z_1$ we see that K_3V is a closed subgroup, V is normal in K_3V and that K_3V/V is compact. By Lemma 3.8 there exists consequently a compact group K_4 such that

$$K_3V = K_4V, \qquad K_4 \frown V = e.$$

Now the elements of K_2 are commutative with elements of both K_3 and V, consequently with that of K_4. $K = K_2K_4$ is therefore a compact subgroup of G. As it contains K_2 we see immediately that $K_2 = K \frown N$. On the other hand KN contains K_3, consequently K_1. As K_1/N is a maximal compact subgroup of G/N, it follows $K_1 = KN$. The existence of a compact group K as (3.18) is thus proved for the semi-simple group N.

Now under the same assumption for N, let K^* be any compact subgroup of G. As K^*N/N is a compact subgroup of G/N, some conjugate of K^*N/N must be contained in K_1/N by the assumption of G/N. We can find accordingly an element in G, by which K^* is transformed in K_1. We may suppose therefore from the beginning that K^* is contained in K_1. Now consider K^*Z/Z in $G/Z = H/Z \times N/Z$. It is then easy to see that K^*Z/Z lies in the direct product of K_3/Z and a maximal compact subgroup of N/Z. As N/Z is the adjoint group of the semi-simple Lie group N and as one of maximal compact subgroups of N/Z is given by M/Z,[23] we may suppose, if necessary after a suitable transformation, that the latter direct factor is just M/Z. K^* is consequently contained in K_3M. On the other hand this group is equal to $K_3VK_2 = VK_4K_2 = VK = V \times K$. The compact group K^* must be therefore contained in K. This proves that K is a maximal compact subgroup of G and that all such groups are conjugate to each other.

To prove the like propositions on K for the general case, we use the induction

---

[23] See the proof of Lemma 3.12.

with respect to the dimension of N. By the above considerations we may suppose that N is not semi-simple. By Lemma 3.3 N contains then a normal subgroup N_1 of G, which is isomorphic either to a vector group or to a toroidal group. Let us consider first the former case. By making use of Lemma 3.7, we can see readily that K_2N_1/N_1 is then a maximal compact subgroup of N/N_1. By the induction assumption applied to G/N_1 and N/N_1, there exists then a subgroup K_3 of G containing N_1, such that K_3/N_1 is a maximal compact subgroup of G/N_1 and that

$$K_1 = K_3N, \qquad K_2N_1 = K_3 \frown N.$$

By Lemma 3.8 K_3 contains a compact subgroup K, such as

$$K_3 = KN_2, \qquad K \frown N_1 = e.$$

As K_2 is a compact subgroup of K_3, we may suppose moreover, if necessary by taking a conjugate group, that K contains K_2. We can see then immediately that such K satisfies the relations (3.18). Now let K^* be an arbitrary compact subgroup of G. As K^*N_1/N_1 is a compact subgroup of G/N_1, some conjugate of it must be contained in K_3/N_1. Consequently some conjugate of K^* lies in K_3 and even in K after a suitable transformation, as one readily sees by Lemma 3.8. This completes the proof for the first case.

Next we consider the case, where N_1 is a toroidal group. In this case N_1 is contained in K_2, for this is a maximal compact subgroup of N. By the definition N_1 is also a subgroup of K_1. Therefore all considerations can be easily reduced to that of the factor group G/N_1, for which the propositions are proved to be valid by the induction assumption. Our assertions on the maximal conjugate subgroups of G are thus completely proved.

Now let $H_1', \cdots, H_s'$ be subgroups of G containing N, such that G/N is the product of H_i'/N and K_1/N as stated in Theorem 6, and let $H_{s+1}, \cdots, H_r$ be subgroups of N, such that $N = H_{s+1} \cdots H_rK_2$. By Lemma 3.10 there exist subgroups $H_1, \cdots, H_s$ in G, so that

$$H_i' = H_iN, \qquad H_i \frown N = e, \qquad i = 1, \cdots, s.$$

It is then easy to see that any g in G can be decomposed uniquely in the form

$$g = h_1 \cdots h_sk_1, \qquad h_i \in H_i, \quad k_1 \in K_1 \tag{3.19}$$

and that h_i and k_1 depend continuously on g. Now we have already proved that $K_1 = KN$, $K_2 = K \frown N$, whence follows $K_1 = NK = H_{s+1} \cdots H_rK_2K = H_{s+1} \cdots H_rK$. Any k_1 in K_1 can be written consequently in the form

$$k_1 = h_{s+1} \cdots h_rk, \qquad h_i \in H_i, \quad k \in K. \tag{3.20}$$

The uniqueness of this decomposition is obvious. If a sequence $k_1^{(i)}$ in K_1 converges to k_1, we can choose a subsequence so that its K-components $k^{(i)}$ converge to some element k in K. Then the remaining H-components also converge to some h. However as the set $H_{s+1} \cdots H_r$ is closed by the assumption, h must belong to $H_{s+1} \cdots H_r$. Therefore $k_1 = hk$ is the decomposition of k_1 into h in $H_{s+1} \cdots H_r$ and k in K. This shows the continuity of k; that of each h_i follows from the assumption on N. Combining (3.20) with (3.19), Lemma 3.13 is thus completely proved.

Now we shall give some consequences of Theorem 6 and Lemma 3.13.

LEMMA 3.14. *A connected Lie group G is simply connected, if and only if a maximal compact subgroup K of G is simply connected.*

PROOF. It follows at once from Theorem 6.

DEFINITION. *We call the integer r in Theorem 6, namely the dimension of the coset space of a connected Lie group G with respect to a maximal compact subgroup K, the characteristic index of G.*

LEMMA 3.15. *Let G be a connected Lie group and N a connected normal subgroup of G. If K is a maximal compact subgroup of G, $K \frown N$ and KN/N are maximal compact subgroups of N and G/N respectively and the characteristic index of G is the sum of the characteristic indices of N and G/N. Moreover G is simply connected, if and only if N and G/N are both simply connected.*

PROOF. By Lemma 3.13 there exists a suitable maximal compact subgroup K of G, so that $K \frown N$ and KN/N are maximal compact subgroups of N and G/N respectively. However the same is also true for any other maximal compact subgroup of G, for these groups are conjugate to each other. The assertion on the characteristic index follows also immediately from the considerations in the proof of Lemma 3.13. On the other hand it is easy to see by a structural theorem on compact Lie groups,[24] that the compact group K is simply connected if and only if $K \frown N$ and $K/K \frown N$ are both simply connected. As $KN/N \cong K/K \frown N$ the last assertion then follows from Lemma 3.14.

REMARK. Even when we do not assume that G is a Lie group, but only assume that N and G/N are both connected Lie groups, all above propositions hold as well. Note that we can prove in such a case that K is a compact Lie group, for it can be readily seen that K does not contain arbitrary small subgroups.

LEMMA 3.16. *Let G be a locally compact group and N a normal subgroup of G. Let further N and G/N be both connected Lie groups. Then the simply connected covering group $\tilde{G}$ of G contains a connected normal subgroup $\tilde{N}$, such that $\tilde{N}$ and $\tilde{G}/\tilde{N}$ are both simply connected and that they are locally isomorphic to N and G/N respectively.*

PROOF. By the above considerations it can be seen that G has a neighborhood of e, which is homeomorphic to a Euclidean space. Therefore we can construct really the simply connected covering group $\tilde{G}$ over G, which is locally isomorphic to G. Let φ be the homomorphic mapping from $\tilde{G}$ onto G and let us denote by $\tilde{N}$ the connected component of $\varphi^{-1}(N)$. Then N, $\varphi^{-1}(N)$ and $\tilde{N}$ are all locally isomorphic and the same is also true for G/N and $\tilde{G}/\tilde{N}$. $\tilde{N}$ and $\tilde{G}/\tilde{N}$ are

[24] See for example Pontrjagin [16], §55.

consequently both connected Lie groups. That they are moreover simply connected follows from Lemma 3.15 and the remark after it.

Now we shall give some lemmas, which will be used in the proof of the next theorem.

LEMMA 3.17. *If a connected locally compact group G contains a normal subgroup N and a subgroup H, which are both connected Lie groups and if moreover $G = HN$, then G itself is a Lie group.*

PROOF. Let $M = H \frown N$ and take a system of one-parameter groups

$$L_i = \{x_i(\lambda_i); |\lambda_i| < \epsilon\}, \qquad i = 1, \cdots, t,$$

such that

$$H \sim L_1 \cdots L_s, \quad N \sim L_{r+1} \cdots L_t, \quad M \sim L_{r+1} \cdots L_s \ (r \leqq s \leqq t),$$

where $H \sim L_1 \cdots L_s$ means for example that any h in H, which is sufficiently near to e, can be expressed uniquely in the product form

$$h = x_1(\lambda_1) \cdots x_s(\lambda_s), \qquad |\lambda_i| < \epsilon.$$

It is then easy to see that any element g in some neighborhood of e in G can be uniquely decomposed into the product

$$(3.21) \quad g = x_1(\lambda_1) \cdots x_r(\lambda_r)x_{r+1}(\lambda_{r+1}) \cdots x_s(\lambda_s) \cdots x_t(\lambda_t), \qquad |\lambda_i| < \epsilon.$$

We may take therefore $(\lambda_1, \cdots, \lambda_t)$ as a system of coordinates for such g. Now we prove that if g and g' are sufficiently near to e, so that g, g' and gg' are all of the form (3.21), the coordinates of gg' are analytic functions of that of g and g'. For that purpose let

$$g = g_1g_2, \qquad g' = g_1'g_2', \qquad g_1, g_1' \in L_1 \cdots L_r, \qquad g_2, g_2' \in L_{r+1} \cdots L_t.$$

We have then

$$gg' = (g_1g_1')(g_2''g_2'), \qquad g_1, g_1' \in H, \qquad g_2', g_2'' = g_1'^{-1}g_2g_1' \in N.$$

As $\lambda_1, \cdots, \lambda_s$ are canonical coordinates of the second kind in H, first s coordinates of g_1g_1' depend analytically on those of g_1 and g_1'. Similarly last $t - r$ coordinates of $g_2''g_2'$ also depend analytically on those of g_2'' and g_2'. By forming the product of g_1g_1' and $g_2''g_2'$ only $s - r$ intermediate coordinates are changed, the change being again analytical by the reason that $\lambda_{r+1}, \cdots, \lambda_s$ are canonical coordinates of the second kind in M. Therefore it is sufficient to prove that the coordinates of $g_2'' = g_1'^{-1}g_2g_1'$ are analytic functions of that of g_1' and g_2. To see it, let $\xi_1, \cdots, \xi_{t-r}$ be a system of canonical coordinates of the first kind in N and let $L = \{x(\lambda); |\lambda| < \epsilon\}$ be a one-parameter group in G. We put

$$u(\xi_1', \cdots, \xi_{t-r}') = x(\lambda)u(\xi_1, \cdots, \xi_{t-r})x(\lambda)^{-1}$$

for $u(\xi_1, \cdots, \xi_{t-r})$ in N. If we can then prove that $\xi_1', \cdots, \xi_{t-r}'$ are analytic functions of λ and ξ_i, the above assertion will be completely proved, for

$\lambda_{r+1}, \cdots, \lambda_t$ and $\xi_1, \cdots, \xi_{t-r}$ are mutually connected by analytic relations and we may take for $x(\lambda)$ any one of $x_i(\lambda_i)$, $i = 1, \cdots, r$.

Now as ξ_i are canonical coordinates of the first kind we have

$$(\xi_1', \cdots, \xi_{t-r}') = (\xi_1, \cdots, \xi_{t-r})A(\lambda)$$

with a real matrix $A(\lambda)$. ξ_i' therefore depend analytically, indeed linearly on ξ_i. However as $A(\lambda)$ gives a continuous representation of L and as any continuous representation of a one-parameter group is even analytical, ξ_i' depend also analytically on λ. The lemma is thus proved.

LEMMA 3.18. *Let G be a locally compact group and N a normal subgroup in G. If N is a vector group and G/N is a connected Lie group, G itself is a Lie group.*

PROOF.[25] Let $G' = G/N$ and let

$$M = \{\psi(\sigma); \sigma \in G'\}$$

be a complete system of representatives for G/N in G, such as $\psi(\sigma)$ depends continuously on σ. The existence of such system follows immediately from the considerations in the proof of Lemma 3.13. As in the proof of Lemma 3.7 we put

$$N = V_n, \qquad u = u(\xi_1, \cdots, \xi_n), \qquad -\infty < \xi_i < \infty, \qquad u \in N,$$

$$\psi(\sigma)\psi(\tau) = u(\sigma, \tau)\psi(\sigma\tau),$$

$$u(\sigma, \tau) = u(\xi_1(\sigma, \tau), \cdots, \xi_n(\sigma, \tau)), \qquad \sigma, \tau \in G'.$$

$\xi_i(\sigma, \tau)$ are then real continuous functions on $G' \times G'$ and satisfy the relation (3.3).

Now let $\lambda(\sigma)$ be a real function on G' with the following properties:

i) $\lambda(\sigma)$ vanishes outside a certain neighborhood U' of e in G', whose closure $\bar{U}'$ is compact,

ii) $\lambda(\sigma)$ is continuously differentiable with respect to a system of analytical coordinates $(\zeta_1(\sigma), \cdots, \zeta_m(\sigma))$ of σ in G',

iii) we have

$$\int_{G'} \lambda(\sigma)\, d\mu(\sigma) = 1, \tag{3.22}$$

where μ is a left-invariant Haar measure on G'. We put then, as in the proof of Lemma 3.7,

$$\begin{aligned} \eta_i(\sigma) &= \int_{G'} \xi_i(\sigma, \tau)\lambda(\tau)\, d\mu(\tau), \\ v(\sigma) &= u(\eta_1(\sigma), \cdots, \eta_n(\sigma)), \\ \psi'(\sigma) &= v(\sigma)^{-1}\psi(\sigma). \end{aligned} \tag{3.23}$$

[25] The author first proved the lemma for the special case, where G/N is a Lie group of type (C) (see §5). The present proof for the general case is due to Mr. M. Kuranishi. Cf. also his forthcoming paper.

By the property ii) the integral (3.23) surely exists and gives a continuous function of σ. Therefore the set $M' = \{\psi'(\sigma); \sigma \in G'\}$ is again a continuous complete system of representatives for G/N in G. The functions $\xi_i'(\sigma, \tau)$ for such $\psi'(\sigma)$, defined by

$$\psi'(\sigma)\psi'(\tau) = u'(\sigma, \tau)\psi'(\sigma\tau),$$
$$u'(\sigma, \tau) = u(\xi_1'(\sigma, \tau), \cdots, \xi_n'(\sigma, \tau)),$$

are related to the previous $\xi_i(\sigma, \tau)$ by

$$\xi_i'(\sigma, \tau) = \xi_i(\sigma, \tau) - \eta_i(\sigma) - \textstyle\sum_{j=1}^{n} \alpha_{ij}(\sigma)\eta_j(\tau) + \eta_i(\sigma\tau), \tag{3.24}$$

where $A(\sigma) = (\alpha_{ij}(\sigma))$ is, as before, the matrix, which represents the automorphism induced by $\psi(\sigma)$ on N. (3.14) follows immediately by calculating the equation

$$v(\sigma)^{-1}\psi(\sigma)v(\tau)^{-1}\psi(\tau) = u'(\sigma, \tau)v(\sigma\tau)^{-1}\psi(\sigma\tau).$$

By substituting (3.22), (3.23) in (3.24) we see that the right-hand side of (3.24) is equal to

$$\int_{G'} \{\xi_i(\sigma, \tau) - \xi_i(\sigma, \rho) - \textstyle\sum_{j=1}^{n} \alpha_{ij}(\sigma)\xi_j(\tau, \rho) + \xi_i(\sigma\tau, \rho)\} \lambda(\rho)\, d\mu(\rho).$$

In virtue of (3.3) we have therefore

$$\begin{aligned} \xi_i'(\sigma, \tau) &= \int_{G'} (\xi_i(\sigma, \tau\rho) - \xi_i(\sigma, \rho))\lambda(\rho)\, d\mu(\rho) \\ &= \int_{G'} \xi_i(\sigma, \tau)\lambda(\tau^{-1}\rho)\, d\mu(\rho) - \int_{G'} \xi_i(\sigma, \rho)\lambda(\rho)\, d\mu(\rho). \end{aligned}$$

As $\tau^{-1}\rho$ depends analytically on τ with respect to the coordinates $(\zeta_1, \cdots, \zeta_m)$ on G', the integrand of the first integral is a continuously differentiable function of $\zeta_i(\tau)$. Moreover if τ is contained in a compact subset W' of G', the same integrand vanishes outside the compact set $W'\bar{U}'$. It follows then immediately that $\xi_i'(\sigma, \tau)$ is a continuously differentiable function of $\zeta_i(\tau)$.

Now any element g in G can be written uniquely in the form

$$g = u\psi'(\sigma) = u(\xi_1, \cdots, \xi_n)\psi'(\zeta_1(\sigma), \cdots, \zeta_m(\sigma)), \qquad u \in N, \qquad \sigma \in G'.$$

We can therefore take $(\xi_1, \cdots, \xi_m; \zeta_1(\sigma), \cdots, \zeta_m(\sigma))$ as a system of coordinates in G and G becomes thus an analytical manifold. It is easy to see that the topology of G as a manifold coincides with the given one on G as a topological group. On the other hand we have for $g_1 = u_1\psi'(\sigma_1)$, $g_2 = u_2\psi'(\sigma_2)$,

$$g_1 g_2 = u_1 u_2^{\sigma_1} u'(\sigma_1, \sigma_2)\psi'(\sigma_1\sigma_2), \qquad u_2^{\sigma_1} = \psi'(\sigma_1)u_2\psi'(\sigma_1)^{-1}.$$

By what we have proved above, it follows that the coordinates of the product

$g_1 g_2$ are continuously differentiable functions of that of g_2. By making use of a theorem of I. E. Segal[26] we see finally that G is a Lie group.

We now prove the following

THEOREM 7. *Let G be a locally compact group and N a normal subgroup of G. If N and G/N are both Lie groups, then G itself is a Lie group.*

PROOF. It can be readily seen that we may suppose N and G/N both to be connected. If we make use of Lemma 3.16, we may assume moreover that N and G/N are both simply connected. Let n be the dimension of N. As the theorem is obviously true for the case $n = 0$, $N = e$, we prove it by the induction with respect to n. Suppose first that N is semi-simple. By Lemma 3.9 G contains then a subgroup H such as $HN = G$, $H \frown N = e$. As $H \cong G/N$, H is also a Lie group and the theorem follows immediately from Lemma 3.17 for such a case. We consider therefore the case, where N is not semi-simple. By Lemma 3.3 N contains then a normal subgroup N_1 of G, which is, as N is simply connected, isomorphic to a vector group. By the induction assumption applied to G/N_1 and N/N_1, we see that G/N_1 is a Lie group and the theorem then follows from Lemma 3.18.

Finally we shall prove some more lemmas on Lie groups, which will be used in §4.

Let G be a connected Lie group and N the radical of G. Let further Z be a subgroup of N, contained in the center of G. We denote by φ the homomorphic mapping from G onto $G' = G/Z$ and by φ' that from G' onto $G_1 = G/N$. We denote further by $\tilde{G}_1$ the simply connected covering group of G_1. Then, let L' be a semi-simple local Lie subgroup of G' such that $L' \frown \varphi(N) = e$ and $\varphi'(L')$ is open in G_1. We suppose that L' is composed of n one-parameter groups $L'_i = \{x'_i(\lambda_i); |\lambda_i| < \epsilon\}$, namely that any g' in L' is given uniquely as the product

$$g' = x'_1(\lambda_1) \cdots x'_n(\lambda_n), \qquad |\lambda_i| < \epsilon$$

and that, as it is always possible for sufficiently small ϵ, $\varphi'(L')$ is mapped isomorphically on some neighborhood U of e in $\tilde{G}_1$. We denote this mapping by φ_1. It is then obvious that L' is mapped isomorphically by $\varphi_1\varphi'$ onto U.

We then prove the following

LEMMA 3.19. *Under the above assumptions G contains n one-parameter groups $L_i = \{x_i(\lambda_i); |\lambda_i| < \epsilon\}$, such that the set L of all elements $g = x_1(\lambda_1) \cdots x_n(\lambda_n)$, $|\lambda_i| < \epsilon$, is mapped by φ homeomorphically on L' in the way*

$$\varphi(x_1(\lambda_1) \cdots x_n(\lambda_n)) = x'_1(\lambda_1) \cdots x'_n(\lambda_n), \qquad |\lambda_i| < \epsilon$$

and that for any g_1, g_2 in L, the product $g_1 g_2$ is again contained in L if and only if $\varphi(g_1)\varphi(g_2)$ is in L'. Moreover such a system of one-parameter groups is uniquely determined by the above conditions.

PROOF. We first prove the uniqueness. Let $L^*_i = \{x^*_i(\lambda_i); |\lambda_i| < \epsilon\}$ be another system of one-parameter groups in G with the same properties as L_i.

[26] Cf. Segal [17].

If we put

$$x_i(\lambda_i)x_i^*(\lambda_i)^{-1} = z_i(\lambda_i), \qquad z_i(\lambda_i) \in Z,$$

the mapping

$$x_1'(\lambda_1) \cdots x_n'(\lambda_n) \to z_1(\lambda_1) \cdots z_n(\lambda_n)$$

gives a local homomorphism from L' into Z. As L' is semi-simple and Z is abelian, $z_1(\lambda_1) \cdots z_n(\lambda_n)$ must be equal to e for sufficiently small λ_i. In particular $z_i(\lambda_i) = e$, if λ_i is small enough. However this relation must hold for all λ_i, $|\lambda_i| < \epsilon$, for $z_i(\lambda_i)$ obviously forms a one-parameter group. This shows that $L_i = L_i^*$, $i = 1, \cdots, n$.

Next we prove the existence of L_i. It is easy to see that $\varphi^{-1}(L')$ is a local Lie subgroup of G and Z is its radical in the local sense. Therefore by a theorem of E. Levi, we can find in G a local Lie group L_1 such that $L_1 \frown Z = e$ and that $\varphi(L_1)$ locally coincides with L'. We may suppose that $\varphi(L_1)$ is contained in L' and is open in it. Consequently $\varphi'\varphi(L_1)$ is also an open subset of $\varphi'(L')$. Therefore we can find an open subset U_1 of U containing e, such that L_1 is isomorphically mapped by $\psi^{-1} = \varphi_1\varphi'\varphi$ on U_1. Now as $\tilde{G}_1$ is simply connected this local isomorphism ψ from U_1 onto L_1 can be extended to a homomorphic mapping from $\tilde{G}_1$ into G, which we denote again by ψ. Consider the mapping $\psi_1 = \psi\varphi_1\varphi'$ defined on L' and put $L = \psi_1(L')$. It is then obvious that $\psi_1\varphi$ coincides with the identical mapping on L_1 and that if g_1', g_2' and $g_1'g_2'$ are contained in L', $\psi_1(g_1')\psi_1(g_2')$ is contained in L. If we put therefore

$$L_i = \{x_i(\lambda_i); |\lambda_i| < \epsilon\}, \qquad x_i(\lambda_i) = \psi_1(x_i'(\lambda_i)),$$

we see immediately that L_i are one-parameter groups and that L is the set of all elements $x_1(\lambda_1) \cdots x_n(\lambda_n)$, $|\lambda_i| < \epsilon$. Moreover if λ_i is sufficiently small, $x_i(\lambda_i)$ is contained in L_1, and as $\psi_1^{-1} = \varphi$ on L_1 we have $x_i'(\lambda_i) = \varphi(x_i(\lambda_i))$ for such λ_i. However this relation is also true for all λ_i, $|\lambda_i| < \epsilon$, for $x_i(\lambda_i)$ and $x_i'(\lambda_i)$ are both one-parameter groups and φ is a homomorphic mapping from G on G'. It follows then immediately that $\psi_1^{-1} = \varphi$ holds on whole L, not only on L_1. This proves that φ is a homeomorphism on L and that for any g_1, g_2 in L g_1g_2 is contained in L, if and only if $\varphi(g_1)\varphi(g_2)$ is in L'. The lemma is thus proved.

LEMMA 3.20. *Let $L = \{x(\lambda); |\lambda| < \epsilon\}$ be a one-parameter group and let*

$$x(\lambda) \to A_0(\lambda)$$

be a continuous representation of L, where $A_0(\lambda)$ are real matrices of some fixed degree, say n. If then

$$x(\lambda) \to A(\lambda) = \begin{pmatrix} & & a_1(\lambda) \\ & & \vdots \\ & A_0(\lambda) & \vdots \\ & & \vdots \\ & & a_n(\lambda) \\ 0 \cdots & 0 & 1 \end{pmatrix}$$

is also a continuous representation of L of degree $n+1$, the functions $a_1(\lambda), \cdots, a_n(\lambda)$ of λ can be expressed linearly by some fixed analytical functions $f_1(\lambda), \cdots, f_m(\lambda)$, which depend only upon the representation $A_0(\lambda)$, namely

$$a_i(\lambda) = \sum_{j=1}^{m} \alpha_{ij} f_j(\lambda), \qquad i = 1, \cdots, n,$$

where α_{ij} are real constants.

PROOF. We treat the various cases separately.

i) If $A_0(\lambda)$ is of the form

$$\begin{pmatrix} 1 & * & & & & & * \\ 0 & 1 & & & & & \\ & & \cdot & & & & \\ & & & \cdot & & & \\ & & & & \cdot & & \\ & & & & & 1 & * \\ 0 & & & & & 0 & 1 \end{pmatrix}$$

the same is true for $A(\lambda)$ and we have

$$A(\lambda) = \exp(\lambda B) = E + (\lambda/1!)B + (\lambda^2/2!)B^2 + \cdots,$$

where B is a matrix of the form

$$\begin{pmatrix} 0 & * & & & & & * \\ 0 & 0 & & & & & \\ & & \cdot & & & & \\ & & & \cdot & & & \\ & & & & \cdot & & \\ & & & & & 0 & * \\ 0 & & & & & 0 & 0 \end{pmatrix}$$

Hence $B^{n+1} = 0$ and $a_i(\lambda)$ is a polynomial of λ with a degree $\leqq n$. We may therefore take for $f_i(\lambda)$ the powers of λ: $1, \lambda, \lambda^2, \cdots, \lambda^n$.

ii) If $A_0(\lambda)$ does not contain 1 as a common eigen-value $A(\lambda)$ is equivalent to

$$A'(\lambda) = \begin{pmatrix} & & 0 \\ & A_0(\lambda) & \vdots \\ & & 0 \\ 0 & \cdots\cdots 0 & 1 \end{pmatrix}$$

Calculating

$$UA(\lambda)U^{-1} = A'(\lambda),$$

where U is a suitable constant matrix, we see immediately that $a_i(\lambda)$ are linear combinations of components of $(A_0(\lambda) - E)$.

iii) In the general case, transforming by a constant matrix if necessary, we may suppose that $A_0(\lambda)$ has the form

$$A_0(\lambda) = \begin{pmatrix} A_1(\lambda) & \\ & A_2(\lambda) \end{pmatrix},$$

where $A_1(\lambda)$ and $A_2(\lambda)$ are matrices as considered in i) and ii) respectively. If the degree of $A_1(\lambda)$ is s, we apply the results of i), ii) to the representations

$$\begin{pmatrix} & a_1(\lambda) \\ A_1(\lambda) & \vdots \\ & a_s(\lambda) \\ 0\cdots\cdots 0 & 1 \end{pmatrix}, \quad \begin{pmatrix} & a_{s+1}(\lambda) \\ A_2(\lambda) & \vdots \\ & a_n(\lambda) \\ 0\cdots\cdots 0 & 1 \end{pmatrix}$$

respectively and see that the lemma is also true for such a case.

LEMMA 3.21. *Let L be a semi-simple local Lie group and let*

$$u \to A_0(u), \qquad u \in L$$

be a continuous representation of L of degree n. If then

$$u \to A(u) = \begin{pmatrix} & a_1(u) \\ A_0(u) & \vdots \\ & a_n(u) \\ 0\cdots\cdots 0 & 1 \end{pmatrix}$$

is also a continuous representation of L, $a_1(u), \cdots, a_n(u)$ *are, when considered as functions of analytical coordinates* $\xi_1, \cdots, \xi_r$ *in* L, *linear combinations of some fixed analytical functions* $f_1(u), \cdots, f_m(u)$, *depending only upon* $A_0(u)$.

Proof runs in a similar way as in ii) in the previous proof.

REMARK. Combining these two lemmas we can easily extend Lemma 3.21 to an arbitrary local Lie group. However as we do not use later such a general case, we omit here the proof.

4. On (L)-groups

We now study the structure of (L)-groups. To make the conception clear we give here again the definition of (L)-groups.

Definition. *A locally compact group G is called an (L)-group or a group of type (L), when G contains a system of normal subgroups $\{N_\alpha\}$, such that*

i) *each G/N_α is a Lie group and*

ii) *the intersection of all N_α is equal to e.*

We say any such system of normal subgroups $\{N_\alpha\}$ a canonical system of normal subgroups in G.

Lemma 4.1. *Let G be a connected locally compact group. Then G is an (L)-group, if and only if G contains a compact normal subgroup N, such as G/N is a Lie group. Here we may assume moreover that N is contained in a given arbitrary small neighborhood of e in G.*

Proof. Suppose first that G is an (L)-group with a canonical system of normal subgroups $\{N_\alpha\}$. We may assume that the intersection of any finite number of N_α is again contained in that system. If we denote by N_α^0 the connected component of N_α containing e and apply Lemma 2.3 to the system of subgroups $\{N_\alpha^0\}$, it follows that some N_α^0 must be contained in a given arbitrary small neighborhood U of e in G, whose closure $\bar{U}$ may be assumed to be compact. As N_α/N_α^0 is a totally disconnected normal subgroup of G/N_α^0, it is contained in the center of G/N_α^0. We can find therefore a normal subgroup N in U such that $N_\alpha^0 \subset N \subset N_\alpha$ and N_α/N is discrete. N is then a compact normal subgroup of G and G/N is a Lie group, as it is locally isomorphic to G/N_α.

Next suppose that G contains a compact normal subgroup N, such that G/N is a Lie group. As N is compact, there exists a system of normal subgroups $\{N_\alpha\}$, such that each N/N_α is a Lie group and that the intersection of all N_α is equal to e. By Theorem 3 N_α are then also normal in G and by Theorem 7 G/N_α are Lie groups. Therefore the system $\{N_\alpha\}$ serves as a canonical system for G and G is thus an (L)-group.

As an corollary of Lemma 4.1 we can prove the following

Lemma 4.2. *A connected (L)-group G contains the maximal compact normal subgroup N, in which any other compact normal subgroup of G is contained, and the factor group G/N is a Lie group.*

Proof. We note first that the lemma is valid for any connected Lie group G, for in such a case any compact normal subgroup of G is contained in every maximal compact subgroup of G by Theorem 6 and we may therefore take as N the intersection of all maximal compact subgroups of G. Now let us consider the general case and let N_1 be a compact normal subgroup of G, such as G/N_1 is a Lie group. If we take as N the subgroup of G containing N_1, such that N/N_1 is the maximal compact normal subgroup of G/N_1, it is easy to see that N is the maximal compact normal subgroup of G and that G/N is a Lie group.

Theorem 8. *A subgroup of an (L)-group G is again an (L)-group. If G is moreover connected, any factor group of G is also an (L)-group.*

Proof. Let $\{N_\alpha\}$ be a canonical system of normal subgroups in G. If H

is any subgroup of G, $H/H \frown N_\alpha$ is a Lie group, for a compact neighborhood of e in $H/H \frown N_\alpha$ can be isomorphically mapped in the Lie group G/N_α. Therefore $\{H \frown N_\alpha\}$ forms a canonical system for H and H is thus an (L)-group. Next let N be a normal subgroup of connected G and N_1 a compact normal subgroup of G, such as G/N_1 is a Lie group. It is then easy to see that N_1N/N is a compact normal subgroup of G/N and that G/N_1N is a Lie group. By Lemma 4.1 G/N is therefore an (L)-group.

REMARK. It seems likely that any factor group of an (L)-group is, even when this is not connected, again an (L)-group, but we have not succeeded in proving it.

LEMMA 4.3. *Let G be a connected locally compact group and N a normal subgroup of G. If the connected component N_0 of N is a Lie group, there exists a compact totally disconnected subgroup N_1 in N, which is normal in G and is such that N/N_1 is a Lie group.*

PROOF. Let n be the dimension of N_0. We prove the theorem by the induction with respect to n. If $n = 0$, $N_0 = e$, N is a totally disconnected normal subgroup of G and is contained in the center of G. We can find therefore a compact subgroup N_1 in N, so that N/N_1 is discrete, and the theorem is thus true for such a case. Now we assume that the theorem is already proved for the case, where N_0 has a dimension less than n and shall prove it for the case of dimension n. Suppose first that N_0 is semi-simple and let Z be the discrete center of N_0. As the proof of Lemma 3.9 shows, N_0/Z is then a direct factor in G/Z and there exists a totally disconnected normal subgroup N' in G containing Z, such that $N = N_0N'$ and that N/N' is a Lie group. Let N_1 be a compact subgroup of N' such as N'/N_1 is discrete. As N' is contained in the center of G, N_1 is also normal in G and N/N_1 is a Lie group, for it is locally isomorphic to N/N'. The theorem is thus proved for such case. In the following we may therefore assume that N_0 is not semi-simple. By Lemma 3.3 N_0 contains then a characteristic subgroup M, which is isomorphic either to a vector group or to a toroidal group. If we apply the induction assumption to G/M and N/M, we see that there exists a normal subgroup M_1 containing M, which is contained in N and is such that N/M_1 is a Lie group and that M_1/M is a compact totally disconnected group. Suppose first that M is a vector group. If we then denote by M_1' the subgroup of M_1 consisting of all elements in M_1, which commute with every element in M, the proof of Lemma 3.8 shows that M_1/M_1' is a compact totally disconnected Lie group, i.e. a finite group and that M_1' is the direct product of M and a compact totally disconnected characteristic subgroup N_1 of M_1'. As M_1 and M_1' are both normal in G, N_1 is also a normal subgroup of G. Moreover if we make use of Theorem 7, we can see that N/N_1 is a Lie group, for N/M_1' and M_1'/N_1 are both Lie groups. Next suppose that M is a toroidal group. As M_1 is then a finite dimensional compact normal subgroup of G, we can find a totally disconnected normal subgroup N_1 in M_1, such that M_1/N_1 is a Lie group. By Theorem 3 N_1 is then also normal in G and as in the former case, N/N_1 is a Lie group by Theorem 7. The lemma is thus completely proved.

LEMMA 4.4. *Let G be a connected locally compact group and N a normal subgroup in G. Then N is an (L)-group, if and only if its connected component N_0*

is an (L)-group. Similarly G/N is an (L)-group, if and only if G/N_0 is an (L)-group.

Proof. If N is an (L)-group, so is N_0 by Theorem 8. Suppose conversely that N_0 is an (L)-group and let M be the maximal compact normal subgroup of N_0. M is then also normal in G and N_0/M is a Lie group by Lemma 4.2. If we therefore apply the previous lemma to G/M and N/M, we can find normal subgroup M_1 in N containing M, such that N/M_1 is a Lie group and that M_1/M is compact. M_1 is then itself a compact normal subgroup of G and if we note that any normal subgroup of M_1 is also normal in G by Theorem 3, the considerations in the proof of Lemma 4.1 shows immediately that N is an (L)-group.

Next we prove the second part. If G/N_0 is an (L)-group, so is G/N by Theorem 8. Suppose conversely that G/N is an (L)-group. Then there exists a system of normal subgroups $\{N_\alpha\}$ in G, such that each G/N_α is a Lie group and that the intersection of all N_α is equal to N. We may suppose moreover that the intersection of any finite number of N_α is also contained in that system. Now as N/N_0 is contained in the center of G/N_0, we can find a normal subgroup N_1 in N, which contains N_0 and is such that N_1/N_0 is compact and N/N_1 is discrete. If we consider G/N_1 and the system of normal subgroups $\{N_\alpha/N_1\}$ and make use of Lemma 2.3, it can be shown as in the proof of Lemma 4.1 that there exists a compact normal subgroup M/N_1 of G/N_1 such as G/M is a Lie group. As M/N_1 and N_1/N_0 are both compact, M/N_0 is also a compact normal subgroup of G/N_0 and it then follows from Lemma 4.1 that G/N_0 is an (L)-group.

We now prove one more lemma, which will be used in the proof of the next theorem.

Lemma 4.5. *Let G be a connected locally compact group and N a normal subgroup of G. If N is a connected Lie group and G/N is compact, then G is an (L)-group.*

Proof. Let n be the dimension of N. As the lemma is obvious for the case $n = 0$, $N = e$, we use the induction with respect to n. Suppose first that N is semi-simple. By Lemma 3.9 there exists a discrete normal subgroup Z in N, so that G/Z is the direct product of N/Z and a compact group. G/Z is therefore an (L)-group and by the previous lemma G itself is then an (L)-group. We may therefore assume that N is not semi-simple. By Lemma 3.3 N contains then a characteristic subgroup M, which is isomorphic either to a vector group or to a toroidal group. If we then apply the induction assumption to G/M and N/M, we see that G/M is an (L)-group and that consequently G has a normal subgroup M_1 which contains M and is such that G/M_1 is a Lie group and M_1/M is compact. Therefore if M is a toroidal group, M_1 itself is compact and the lemma follows immediately from Lemma 4.1. We may suppose therefore that M is a vector group and denote by M_1' the subgroup of M_1, consisting of all elements in M_1, commuting with every element in M. The proof of Lemma 3.8 shows that M_1/M_1' is a compact Lie group and that M_1' is the direct product of M and a compact characteristic subgroup N_1. As G/M_1, M_1/M_1', M_1'/N_1 are all Lie groups, G/N_1 is also a Lie group by Theorem 7 and the lemma follows again from Lemma 4.1.

We now prove the following

THEOREM 9. *Let G be a connected locally compact group and N a normal subgroup of G. Then, if N and G/N are both (L)-groups G itself is an (L)-group.*

For the sake of later applications we prove this theorem in the following slightly generalized form:

THEOREM 9′. *Let G be a connected locally compact group and M and N its normal subgroups. Then, if M contains N and N and M/N are both (L)-groups, M itself is an (L)-group.*

PROOF. Let M_0 and N_0 be the connected components of M and N respectively. Obviously M_0 contains N_0 and we can see immediately by Lemma 4.4 that N_0 and M_0/N_0 are both (L)-groups. It follows also from the same lemma that we have only to prove that M_0 is an (L)-group. Let K be the maximal compact normal subgroup of N_0, for which N_0/K is a Lie group by Lemma 4.2. In order to prove that M_0 is an (L)-group, it is then sufficient to show that M_0/K is such a group, for if this can be proved, there exists a normal subgroup K_1 in M_0 containing K, such that M_0/K_1 is a Lie group and K_1/K is compact. K_1 itself is then a compact group and M_0 is consequently an (L)-group by Lemma 4.1. In the following we may therefore only consider the group M_0/K, or in other words, we may assume that N_0 is a Lie group. Now let P be a normal subgroup of M_0 containing N_0, such as P/N_0 is compact and M_0/P is a Lie group. As M_0/N_0 is an (L)-group such P surely exists by Lemma 4.1. The connected component P_0 of P then contains N_0 and is an (L)-group by the previous lemma. Let Q be the maximal compact normal subgroup of P_0. P_0/Q is then the connected component of P/Q and is a Lie group by Lemma 4.2. Lemma 4.3 then shows that there exists a normal subgroup Q_1 of M_0 in P which contains Q and is such that P/Q_1 is a Lie group and Q_1/Q is compact. As Q is compact, Q_1 itself is a compact normal subgroup of M_0 and as M_0/P and P/Q_1 are both Lie groups, M_0/Q_1 is also a Lie group by Theorem 7. It then follows from Lemma 4.1 that M_0 is an (L)-group and the theorem is thus proved.

Since any locally compact abelian group is obviously an (L)-group, there follows from Theorem 5 and Theorem 9 the following

THEOREM 10. *A connected locally compact solvable group is an (L)-group.*

To study the local structure of an (L)-group, we shall give now some preliminary lemmas.

LEMMA 4.6. *Let G be a connected locally compact group and N a totally disconnected normal subgroup in G. If L' is a local Lie subgroup of $G' = G/N$, G contains then a local Lie subgroup L such that $L \frown N = e$ and that LN/N coincides locally with L'.*

PROOF. See Pontrjagin [16], §45. As we do not assume the validity of any of the countability axioms, minor modifications of the proof are necessary; we must namely make use of the generalized limiting process on directed sets instead of the ordinary limiting process. However we may omit the detailed account, as it brings no essential difficulty.

LEMMA 4.7. *Let G be a connected locally compact group and N a compact normal subgroup of G. If H_1 is a subgroup of G containing N and if $H_1/N \cong V_1$, G contains then a subgroup H such that $H_1 = HN$, $H \frown N = e$, $H \cong V_1$.*

PROOF. Let Z be the center of N. By the remark after Theorem 5, it is then easy to see that N/Z is a compact normal subgroup of G/Z, which has no center except e. By Theorem 2 it then follows that N/Z is a direct factor in G/Z and that there exists a subgroup H_1' containing Z, such as $H_1 = H_1'N$, $H_1' \frown N = Z$, $H_1'/Z \cong V_1$. Now Z is contained in the center of G by Theorem 4. Consequently we can prove just as in the proof of Lemma 3.4 that H_1' is an abelian group and we see then immediately that there exists a subgroup H such as $H_1' = HZ$, $H \frown Z = e$, $H \cong V_1$. Such H obviously satisfies all conditions in the lemma.

LEMMA 4.8. *Let G be a connected (L)-group and Z a compact abelian normal subgroup of G, which is, by Theorem 4, contained in the center of G. Let further G/Z be a Lie group and N a subgroup containing Z, such that N/Z is the radical of G/Z. G contains then a semi-simple local Lie group L, such as LN/N is open in G/N.*

PROOF. As N/Z is the radical of G/Z, there exists by a theorem of E. Levi, a semi-simple local Lie subgroup L' of $G' = G/Z$, which represents some neighborhood of e in G/N. We may assume that L' fulfils all conditions stated before Lemma 3.18 and that in particular L' is composed of n one-parameter groups $L_i' : L' = L_1' \cdots L_n'$. Let $\{N_\alpha\}$ be a system of subgroups of Z, such that each Z/N_α is a Lie group and that the intersection of all N_α is equal to e. By Theorem 7 G/N_α is then also a Lie group and we see that the system $\{N_\alpha\}$ forms a canonical system of normal subgroups in G. We apply then Lemma 3.18 to the factor group G/N_α and obtain for each α n one-parameter groups $L_i^{(\alpha)}$ in G/N_α, which are mapped on L_i' by the homomorphism φ_α from G/N_α onto G/Z. Let N_β be any subgroup of the canonical system, which is contained in N_α and let $\varphi_{\alpha,\beta}$ be the homomorphism from G/N_β on G/N_α. From the uniqueness of $L_i^{(\alpha)}$ and from $\varphi_\alpha(L_i^{(\alpha)}) = \varphi_\beta(L_i^{(\beta)}) = \varphi_\alpha\varphi_{\alpha,\beta}(L_i^{(\beta)})$ we have $L_i^{(\alpha)} = \varphi_{\alpha,\beta}(L_i^{(\beta)})$. If we therefore denote by ψ_α the homomorphism from G onto G/N_α, the intersection L_i of all $\psi_\alpha^{-1}(L_i^{(\alpha)})$ forms a one-parameter group in G for each i. Put $L = L_1 \cdots L_n$ and denote by ψ the homomorphism from G on G/Z. ψ maps then the set L homeomorphically on L' and for any g_1, g_2 in L the product g_1g_2 is again contained in L, if and only if $\psi(g_1)\psi(g_2)$ is in L'. Consequently L is surely a local Lie group which satisfies all our conditions.

DEFINITION. *We denote by T_Ω a compact group which is the direct product of groups X_α, $\alpha \in \Omega$, all isomorphic to T_1; here Ω may be a finite or infinite set of suffixes.*

LEMMA 4.9. *Let G be a connected locally compact group and M a normal subgroup of G, for which the factor group G/M is a simple or semi-simple Lie group. Let further N be a subgroup of M, which is contained in the center of G and is isomorphic to T_Ω. Suppose that there exists a local Lie group L_1 in G such that*

$L_1 \frown M = e$, L_1M/M is open in G/M, and that M is locally the direct product of a local Lie group L_2 and N. G itself is then locally the direct product of a local Lie group L and a subgroup $N' = T_{\Omega'}$ of N, where Ω' is a suitable subset of Ω.

PROOF. Let $\xi_1, \cdots, \xi_n$ be a system of canonical coordinates of the first kind in L_2 and put

$$N = T_\Omega = \Pi_\alpha X_\alpha, \qquad X_\alpha = \{x_\alpha(\lambda_\alpha); \lambda_\alpha \text{ mod. } 1\}, \qquad \alpha \in \Omega.$$

Any element a in $L_2 \times N$ can be then written in the form

$$a = v(\xi_1, \cdots, \xi_n)\Pi_\alpha x_\alpha(\lambda_\alpha).$$

We transform this element by an element u in L_1 and obtain

$$u^{-1}au = v(\xi_1', \cdots, \xi_n')\Pi_\alpha x_\alpha(\lambda_\alpha'). \tag{4.1}$$

Take then an arbitrary suffix τ from Ω and put

$$N_\tau = \Pi_{\alpha \neq \tau} X_\alpha .$$

As $\xi_1, \cdots, \xi_n, \lambda_\tau$ can be considered as canonical coordinates in M/N_τ, we have

$$(\xi_1', \cdots, \xi_n', \lambda_\tau') = (\xi_1, \cdots, \xi_n, \lambda_\tau)A_\tau(u), \tag{4.2}$$

where $A_\tau(u)$ gives a continuous representation of L_1. $A_\tau(u)$ has the form

$$A_\tau(u) = \begin{pmatrix} & a_{1\tau}(u) \\ & \vdots \\ A(u) & \vdots \\ & \vdots \\ & a_{n\tau}(u) \\ 0 \cdots\cdots 0 & 1 \end{pmatrix}, \tag{4.3}$$

with some fixed representation $A(u)$ independent of τ. By Lemma 3.20, 3.21 there exist therefore functions $f_1(u), \cdots, f_m(u)$, such as

$$a_{i\tau}(u) = \alpha_{i\tau}^1 f_1(u) + \cdots + \alpha_{i\tau}^m f_m(u).$$

Now consider the matrix

$$(\alpha_{i\tau}^j),$$

where (i, j) is the row index and τ is the column index. Let p be the rank of this matrix and suppose that

$$| \alpha_{i_r \tau_s}^{j_r} | \neq 0, \qquad r, s = 1, \cdots, p. \tag{4.4}$$

If we put then

$$q_{ij}(\lambda) = \Pi_\tau x_\tau(\alpha_{i\tau}^j \lambda), \qquad i = 1, \cdots, n; j = 1, \cdots, m,$$

we have

$$q_{ij}(\lambda) = \prod_{r=1}^p q_{i_r j_r}(\beta_{ijr}\lambda)$$

with some real β_{ijr}. Take a positive number ϵ and put

$$Q_r = \{q_{i_r j_r}(\lambda);\, |\lambda| < \epsilon\}, \qquad r = 1, \cdots, p.$$

For a suitable small $\delta > 0$ it follows then

$$\prod_{i=1}^{n} \prod_{j=1}^{m} q_{ij}(\mu_{ij}) \in Q_1 Q_2 \cdots Q_p, \qquad \text{for } |\mu_{ij}| < \delta. \tag{4.5}$$

On the other hand we have from (4.1), (4.2), (4.3)

$$u^{-1} v(\xi_1, \cdots, \xi_n) u = v(\xi_1', \cdots, \xi_n') \Pi_r x_r(\lambda_r'),$$

$$\lambda_r' = \sum_{i=1}^{n} a_{ir}(u) \xi_i .$$

It holds therefore that

$$x_r(\lambda_r') = x_r\left(\sum_{i=1}^{n} a_{ir}(u)\xi_i\right) = x_r\left(\sum_{i=1}^{n} \sum_{j=1}^{m} \alpha_{ir}^{j} f_j(u)\xi_i\right),$$

$$\Pi_r x_r(\lambda_r') = \prod_{i=1}^{n} \prod_{j=1}^{m} q_{ij}(f_j(u)\xi_i).$$

Since $f_j(u)$ is continuous there exists a positive number δ_1 and a neighborhood U of e in L_1, such that

$$|f_j(u)\xi_i| < \delta, \qquad \text{for } u \in U,\ |\xi_i| < \delta_1 .$$

From (4.5) we have finally

$$\Pi_r x_r(\lambda_r') \in Q_1 Q_2 \cdots Q_p, \qquad \text{for } u \in U,\ |\xi_i| < \delta_1 .$$

Therefore if L denotes the set of all

$$uv(\xi_1, \cdots, \xi_n) \prod_{r=1}^{p} q_{i_r j_r}(\lambda_r), \qquad u \in U, \quad |\xi_i| < \delta_1, \quad |\lambda_r| < \epsilon,$$

L is a local Lie group. We take then as Ω' the set of all τ in Ω different from $\tau_1, \cdots, \tau_p$ in (4.4) and see immediately that G is locally the direct product of L and $T_{\Omega'}$.

We now prove the following theorem, which characterizes the local structure of connected (L)-groups.

THEOREM 11. *Let G be a connected (L)-group and U an arbitrary neighborhood of e in G. G contains then a local Lie group L and a compact normal subgroup K contained in U, such that G is locally the direct product of L and K. Conversely a connected topological group G is an (L)-group, whenever it is locally the direct product of a local Lie group L and a compact group K.*

PROOF. The latter half is almost obvious, for any canonical system of normal subgroups for K serves also as such for G. We prove therefore the former part. We shall reduce the proof step by step to the case, where G has a more simple structure.

By Lemma 4.1 U contains a compact normal subgroup N of G such as G/N is a Lie group. Let $N_1 = D_1(N)$ be the commutator group of N and let Z be the center of N. By Lemma 2.4 $Z_0 = N_1 \frown Z$ is then a compact totally disconnected normal subgroup of G. Suppose first that the theorem is already proved for G/Z_0 and for the neighborhood U/Z_0. There exists then a local Lie group L' and a compact subgroup K' of G/Z_0 contained in U/Z_0, such that G/Z_0 is locally the direct product of L' and K'. Let $K' = K/Z_0$ and let L be a local Lie group in G such that $L \frown Z_0 = e$ and LZ_0/Z_0 is contained as an open set in L'; the existence of such L being assured by Lemma 4.6. It is then obvious that K is a compact group contained in U, LK contains some neighborhood of e in G and that $L \frown K = e$. Now we can assume, without loss of generality, that L is connected, for example homeomorphic to a Euclidean space. Then the set

$$[L, K] = \{usu^{-1}s^{-1}; u \in L, s \in K\}$$

is also connected and contains e. On the other hand, this set is contained in Z_0, for L' and K' are direct factors. As Z_0 is totally disconnected it follows then that $[L, K]$ consists of e alone and elements of L and K commute with each other. G is therefore locally the direct product of L and K.

It is therefore sufficient to prove the theorem for G/Z_0, or in other words, we may suppose $Z_0 = e$. In such a case, N_1 is a compact normal subgroup of G without center. By Theorem 2 it is therefore a direct factor:

$$G = G_1 \times N_1 .$$

Consequently we have only to prove the theorem for G_1, or on rewriting $G = G_1$, we may assume that U contains a compact abelian normal subgroup N such as G/N is a Lie group. By Theorem 4 N is then contained in the center of G. Let N_0 be a totally disconnected subgroup of N, such that N/N_0 is isomorphic to some T_Ω. A similar consideration as above shows that it is sufficient to consider G/N_0 instead of G, or in other words, we may suppose $N = T_\Omega$. We shall do it in the following.

Now let M', H_1', $\cdots$, H_n' be subgroups of G containing N such that M'/N is the radical of G/N and that H_i'/N generate the solvable group M'/N as stated in Lemma 3.6. We have namely $M' = M_0' = H_1' \cdots H_n'$ and that each $M_i' = H_{i+1}' \cdots H_n'$ is normal in $M_{i-1}' = H_i' \cdots H_n'$. If H_i'/N is isomorphic to V_1, we can find by Lemma 4.7 a subgroup H_i in G such as

$$H_i' = H_iN, \qquad H_i \frown N = e, \qquad H_i \cong V_1 .$$

Since $N = T_\Omega$, it is easy to see that we can find a similar subgroup H_i in G also in the case, where H_i'/N is isomorphic to T_1. Consequently we have

$$M_i' = H_{i+1} \cdots H_nN, \qquad M_{i-1}' = H_iM_i', \qquad H_i \frown M_i' = e.$$

On the other hand by Lemma 4.8 G contains a local Lie group L_1 such that $L_1 \frown M_0' = e$ and L_1M_0'/M_0' is open in G/M_0'. If we therefore apply Lemma 4.9 to M_{n-1}', $\cdots$, M_1', M_0', G successively, we see finally that there exists a local Lie group L and a subgroup $K = T_{\Omega'}$ of N such that G is locally the direct product of L and K. The theorem is thus completely proved.

From Theorem 11 there follows immediately the following

THEOREM 12. *If an (L)-group is locally connected and has a finite dimensional neighborhood of e, if in particular it is locally Euclidean, it is a Lie group.*

REMARK. To prove Theorem 12 only, the long considerations for the proof of Theorem 11 are not necessary. We can prove it, as one readily sees, by the help of Lemma 4.1 in a similar way as in the case of compact groups.[27]

We now return to general (L)-groups. An (L)-group is, as Theorem 11 shows, locally the direct product of a local Lie group and a compact group, but this is not true in the global sense. We shall show it by an example at the end of this section. However an (L)-group has, also when globaly considered, some remarkable properties. In particular its topological structure is considerably simple. It holds namely the following generalization of Theorem 6 for (L)-groups.

THEOREM 13. *Let G be a connected (L)-group. Any compact subgroup of G is then contained in some maximal compact subgroup of G, whereas such maximal compact subgroups of G are all connected and conjugate to each other. Let K be one of them. G contains then subgroups $H_1, \cdots, H_r$, which are all isomorphic to V_1 and are such that any element g in G can be decomposed uniquely and continuously in the form*

$$g = h_1 \cdots h_r k, \qquad h_i \in H_i,\ k \in K.$$

In particular the space of G is the cartesian product of the compact space of K and that of $H_1 \cdots H_r$, which is homeomorphic to the r-dimensional Euclidean space E_r. We call the integer r, as in the case of Lie groups, the characteristic index of G.

PROOF. Let N be a compact normal subgroup of G, for which G/N is a Lie group and let K' be an arbitrary compact subgroup of G. $K'N$ is then also a compact subgroup of G containing N. We can therefore find a compact subgroup K of G containing $K'N$ such that K/N is a maximal compact subgroup of G/N. It is then easy to see that K is a maximal compact subgroup of G and that conversely any such group can be obtained in this way from a maximal compact subgroup in G/N. It then follows immediately from Theorem 6 that maximal compact subgroups of G are conjugate to each other. Now let $H'_1, \cdots, H'_r$ be subgroups of G containing N, such that G/N is decomposed into the product of H'_i/N and K/N as stated in Theorem 6. By Lemma 4.7 we have then r subgroups $H_1, \cdots, H_r$ satisfying

$$H'_i = H_i N, \qquad H_i \frown N = e, \qquad H_i \cong V_1 .$$

We can see then immediately that such K and $H_1, \cdots, H_r$ fulfil the theorem, in a similar way as in the proof of Lemma 3.13.

REMARK. As the proof shows the characteristic index of G is equal to that of G/N, where N is a compact normal subgroup of G, for which G/N is a Lie group.

Now we prove two lemmas on characteristic indices, which will be used in the next section.

LEMMA 4.10. *Let G be a connected (L)-group and N a connected normal subgroup of G. Then the characteristic index of G is equal to the sum of that of N and*

---

[27] Cf. Pontrjagin [16], §45.

G/N, Moreover if K is a maximal compact subgroup of G, $K \frown N$ and KN/N are maximal compact subgroups for N and G/N respectively.

PROOF. We consider first the case where N is compact. Take then a compact normal subgroup N' of G containing N, such as G/N' is a Lie group. By the remark after Theorem 13 the characteristic indices of G and G/N are both equal to that of G/N'. As the characteristic index of compact N is obviously 0 and as any maximal compact subgroup of G contains N, the lemma is proved for such special cases. We note that in the above proof we have not used the fact that N is a connected group.

Now let us consider the general case and let N_1 be the maximal compact normal subgroup of N. By Lemma 4.2 N/N_1 is a Lie group and as N_1 is a characteristic subgroup of N, it is also normal in G. The above consideration implies that the characteristic indices of G and N are equal to that of G/N_1 and N/N_1 respectively. Therefore, replacing G by G/N_1 if necessary, we may suppose that N is a Lie group. Then all considerations in the proof of Lemma 3.13 can be transferred term by term to the present case and thus follows our lemma.

LEMMA 4.11. *Let G be a connected (L)-group and N the maximal compact normal subgroup in G. Then the dimension n of G/N does not exceed $\frac{1}{2}r(r+3)$, where r is the characteristic index of G.*

PROOF. As the characteristic index of G is equal to that of G/N, we may assume, replacing G by G/N if necessary, that G is a Lie group, which contains no compact normal subgroup except e and we have then only to prove that the dimension n of G is $\leqq \frac{1}{2}r(r+3)$. Let K be a maximal compact subgroup of G and n' the dimension of K. By Theorem 6 we have $n = n' + r$. For any given a in G the mapping

$$xK \rightarrow axK, \qquad x \in G$$

gives a homeomorphism $\sigma(a)$ on the right coset space of G with respect to K and G is thus represented as a transformation group $\{\sigma(a)\}$ on G/K. By such a representation K is represented faithfully, for otherwise G would contain a compact normal subgroup other than e. As the space G/K is homeomorphic to the r-dimensional Euclidean space E_r, we have

$$n' \leqq \tfrac{1}{2}r(r+1),\text{[28]}$$

consequently

$$n = n' + r \leqq \tfrac{1}{2}r(r+3), \qquad \text{q.e.d.}$$

Finally we shall give, as promised before, an example of an (L)-group, which is not the direct product of a Lie group and a compact group.

EXAMPLE. Let

$$H_1 = \{h_1(\lambda_1);\ -\infty < \lambda_1 < \infty\}, \qquad H_2 = \{h_2(\lambda_2);\ -\infty < \lambda_2 < \infty\}$$

be groups, which are isomorphic to V_1 and let $N = T_\Omega$ be the direct product of

[28] Cf. Montgomery-Zippin [13].

$$X_\alpha = \{x_\alpha(\mu_\alpha); \mu_\alpha \text{ mod. } 1\}, \qquad \alpha \in \Omega,$$

which are all isomorphic to T_1. We define G as the set of elements

$$g = h_1(\lambda_1)h_2(\lambda_2) \prod_\alpha x_\alpha(\mu_\alpha).$$

The topology in G is defined as the cartesian product of H_1, H_2 and N and the law of multiplication in G is given as follows: elements of N shall be commutative with every element in H_1 or in H_2 and

$$h_1(\lambda_1)h_2(\lambda_2)h_1(\lambda_1)^{-1}h_2(\lambda_2)^{-1} = \prod_\alpha x_\alpha(\xi_\alpha\lambda_1\lambda_2),$$

where ξ_α are real numbers, which are linearly independent over rational numbers.

It is then easy to see that in such way a connected locally compact solvable (even nilpotent) group is uniquely defined and that the commutator group $D_1(G)$ of G coincides with N. If G is the direct product of a Lie group L and a compact group K, the latter must be abelian by Lemma 2.2 and we have

$$D_1(G) = D_1(L) = N.$$

Therefore such a decomposition is surely impossible, if Ω is an infinite set and N has consequently an infinite dimension.

We note here also that above G gives, even for finite Ω, an example of groups, for which the abstract commutator group is not closed in G. G has also the remarkable property that in any continuous representation of G, elements of N are mapped to the unit matrix.[29]

5. On locally compact groups in general

In this last section we study the structure of locally compact groups in general and consider in particular the situation of (L)-groups in the set of all locally compact groups.

THEOREM 14. *In a locally compact group G there exists a connected compact normal subgroup N, which contains all other connected compact normal subgroups in G, namely the maximal connected compact normal subgroup of G.*

PROOF. If N_1, N_2 are connected compact normal subgroups (c. c. n. subgroups) of G, N_1N_2 is also such a group. It is sufficient therefore to show that if

$$N_1 \subset N_2 \subset \cdots \subset N_\xi \subset \cdots, \qquad \xi < \xi_0$$

is an ascending chain of c. c. n. subgroups of G, there exists a c. c. n. subgroup K, which contains all N_ξ. Let K be the closure of the composite of all N_ξ. As is readily seen, K is then a connected normal subgroup of G. We shall prove that K is compact. For that purpose, let H_ξ be the group of all elements in K, which commute with every element in N_ξ. By Theorem 2 we have

$$H_\xi N_\xi = K.$$

[29] Cf. Birkhoff [2].

It is also obvious that $H_\eta \subset H_\xi$, if $\xi < \eta$ and that the intersection $Z = \bigcap_\xi H_\xi$ is the center of K. Let H_ξ^0 be the connected component of H_ξ. As a direct factor of $K/H_\xi \frown N_\xi$, the group $H_\xi/H_\xi \frown N_\xi$ is connected and whence follows $H_\xi^0(H_\xi \frown N_\xi) = H_\xi$. Consequently we have

$$H_\xi^0 N_\xi = K. \tag{5.1}$$

Put

$$Z_1 = \bigcap_\xi H_\xi^0, \qquad Z_1 \subset Z$$

and let Z_0 be the connected component of Z_1. Take then a group W such as $Z_0 \subset W \subset Z_1$, W/Z_0 is compact and Z_1/W is discrete. If we apply Lemma 2.3 to $\{H_\xi^0/W\}$ we see that some H_ξ^0/W, say H_α^0/W, is compact. H_α^0/Z_0 will be then also a compact group. Now as Z_0 is a connected locally compact abelian group, put

$$Z_0 = K_1 \times V,$$

where K_1 is compact and V is a vector group. By Lemma 3.8 applied to the extension H_α^0/V, we have

$$H_\alpha^0 = K_2 \times V$$

with some compact K_2. From (5.1) we see then

$$K = (K_2 N_\alpha) \times V,$$

where $K_2 N_\alpha$ is also a compact group. As any compact subgroup, in particular N_ξ, is contained in $K_2 N_\alpha$, it follows finally

$$K = K_2 N_\alpha, \qquad V = e,$$

which proves K to be compact, q.e.d.

REMARK. It seems likely that a connected locally compact group contains the maximal (not neccessarily connected) compact normal subgroup. As we have already seen, this is surely the case for any connected (L)-groups.

In order to prove a similar theorem for solvable normal subgroups we first prove

LEMMA 5.1. *Let G be a connected locally compact group and N_1, N_2 solvable normal subgroups of G. Then the closure of N_1N_2 is also a solvable normal subgroup of G.*

PROOF. By the remark after Theorem 5 we see that the length of N_i is finite. It then follows immediately that N_1N_2 is a solvable group with a finite length in the sense of the abstract group theory and whence follows that the closure of N_1N_2 is also solvable as a topological group.[30]

THEOREM 15. *In a connected locally compact group G there exists a solvable normal subgroup N, in which any other solvable normal subgroup of G is contained.*

[30] Cf. Iwasawa [10].

N is the maximal solvable normal subgroup of G. The connected component N_0 of N is then the maximal connected solvable normal subgroup of G in a similar sense.

PROOF. Assume first that G contains connected solvable normal subgroups (c. s. n. subgroups) with arbitrary large characteristic indices. By Lemma 5.1 and Lemma 4.10 we have then a series of c. s. n. subgroups

$$H_1 \subset H_2 \subset \cdots \subset H_n \subset \cdots,$$

whose characteristic indices r_n diverge to infinity. As we may of course suppose $r_{n-1} < r_n$, there exists for each n a subgroup W_n, such as $W_n \cong V_1$, $W_n \subset H_n$, $W_n \frown H_{n-1} = e$ and $W_n H_{n-1}$ is closed. Now take a neighborhood U of e with the compact closure $\bar{U}$ and put $U_n = H_n \frown U$. Since $U_n H_{n-1}$ is open and $\bar{U}_n H_{n-1}$ is closed in H_n and since W_n is connected and $W_n H_{n-1}/H_{n-1} \cong V_1$, there exists an element a_n, which is contained in $W_n \frown \bar{U}_n H_{n-1}$, but not in $W_n \frown U_n H_{n-1}$. However as $\{a_n\}$ is a subset of compact $\bar{U}$, we can find integers $s, t (s < t)$, such as

$$a_t a_s^{-1} \in U.$$

$a_t a_s^{-1}$ is therefore an element in U_t and a_t is consequently contained in $U_t H_{t-1}$, which contradicts to the choice of a_t.

We have thus proved that the characteristic indices of c. s. n. subgroups of G have a maximum, say r. We take then a subgroup H with such a characteristic index r. If H' is any c. s. n. subgroup of G containing H, the characteristic index of H'/H is 0 by Lemma 4.10, which means that H'/H is a compact group. By Lemma 2.2 and Theorem 4 H'/H is then abelian and is contained in the center of G/H. If we take therefore a normal subgroup N_0 containing H, such as N_0/H is the connected component of the center of G/H, H' is contained in N_0 and we see that N_0 is the maximal c. s. n. subgroup of G. The maximal solvable normal subgroup N can be then, as one readily sees, determined by the condition that N/N_0 is the center of G/N_0.

DEFINITION. *We call the maximal connected solvable normal subgroup N_0 of a connected locally compact group G, the radical of G.*

By the notion of the radical we can define, as stated in the introduction, a special type of topological groups as follows.

DEFINITION. *A connected locally compact group G is called a connected (C)-group or a connected group of type (C), when the factor group G/N_0 modulo its radical N_0 is compact.*

As is readily seen from this definition, connected compact or solvable groups are connected (C) groups. As to a Lie group G, it is of type (C), if and only if the semi-simple factor group of G modulo its radical (in the usual sense) is the so called compact form of the corresponding complex semi simple Lie algebra. In particular, by a theorem of Weyl,[31] any group which is locally isomorphic to a Lie group of type (C) is again a (C)-group.

---

[31] Cf. Weyl [18].

Now we have immediately from Theorem 9 the following

THEOREM 16. *A connected (C)-group is an (L)-group.*

We can prove however the following more precise result.

THEOREM 17. *A connected locally compact group G is a (C)-group, if and only if there exists a system of normal subgroups $\{N_\alpha\}$ in G, such that*

i) each G/N_α is a Lie group of type (C) and

ii) the intersection of all N_α is equal to e.

PROOF. Let us suppose that G contains a system of normal subgroup $\{N_\alpha\}$ as stated above and show that the factor group G/N_0 by the radical N_0 of G is compact. In a similar way as in the proof of Lemma 4.1 we can prove then the existence of a compact normal subgroup N_1 in G such as G/N_1 is a Lie group of type (C). Put $M = N_0N_1$. M/N_0 is then compact and G/M is again a Lie group of type (C). Let M_1 be a subgroup containing N_0 such that M_1/N_0 is the center of M/N_0. M/M_1 is then a compact normal subgroup of G/M_1 without center and is consequently a direct factor in G/M_1 by Theorem 2. Therefore if M' is a subgroup containing M such that M'/M is the radical of G/M, there exists a normal subgroup M_2 containing M_1, which satisfies $M' = M_2M$, $M_1 = M_2 \frown M$, $M'/M \cong M_2/M_1$. Consequently M_2/M_1 and M_2/N_0 are both solvable and if $M' \neq M$, $M_2 \neq M_1$, G would contain a connected solvable normal subgroup greater than N_0. As N_0 is the maximal connected solvable normal subgroup of G, we must have $M' = M$ and we see that G/M is a semi-simple Lie group. As a Lie group of type (C), G/M is then compact and since M/N_0 was compact, G/N_0 must be also a compact group. This proves the first part of the theorem.

Now let us suppose conversely that G is a connected (C)-group. As it is an (L)-group by Theorem 15, there exists a compact normal subgroup N_1 in G such as G/N_1 is a Lie group. If N_0 denotes the radical of G, N_0N_1/N_1 is solvable and G/N_0N_1 is compact, for G/N_0 is a compact group by the definition. Therefore G/N_1 is a Lie group of type (C) and if we take a canonical system of normal subgroups $\{N_\alpha\}$ in N_1, we see immediately, as in the proof of Lemma 4.1, that $\{N_\alpha\}$ serves also as a canonical system for G and moreover that each G/N_α is a Lie group of type (C), for G/N_1 is of type (C) and N_1/N_α is compact. This proves the converse part of the theorem.

Now it seems natural to extend for not necessarily connected groups the concept of (C)-groups as follows.

DEFINITION. *A locally compact group G is called a (C)-group or a group of type (C), when G contains a system of normal subgroups $\{N_\alpha\}$ such that*

i) *each G/N_α is a Lie group of type (C)*[32] *and*

ii) *the intersection of all N_α is equal to e.*

By making use of the results of the previous sections or directly, starting from the above definition, in a similar way as in §4, we can prove analogous theorems

---

[32] For Lie groups we define that G is a (C)-group, when the connected component of G is a Lie group of type (C) in the former sense.

for (C)-groups as proved before for (L)-groups. Without proof we mention some of them here.

THEOREM 18. *Let G be a connected locally compact group and N a normal subgroup of G. If N and G/N are both (C)-groups then G itself is a (C)-group.*

From this theorem and Lemma 4.10 we have immediately the following characterization of connected (C)-groups.

THEOREM 19. *A connected locally compact group G is a (C)-group, if and only if G contains a series of subgroups*

$$G = G_0 \supset G_1 \supset \cdots \supset G_n = e,$$

such that each G_i is normal in G_{i-1} and that G_{i-1}/G_i is a connected maximally almost periodic group. If

$$G_{i-1}/G_i = K_i \times V_{r_i}$$

is the decomposition of G_{i-1}/G_i into the direct product of a compact group K_i and a vector group V_{r_i} of dimension r_i, the characteristic index of G is given by

$$r = r_1 + r_2 + \cdots + r_n.$$

Corresponding to Theorem 11 in §4 we have the following theorem, which gives the local structure of a connected (C)-group.

THEOREM 20. *A connected locally compact group G is a (C)-group, if and only if it is locally the direct product of a compact group and a local Lie group of type (C).*

In particular we have

THEOREM 21. *If a connected (C)-group is locally Euclidean, it is a Lie group of type (C).*

The first definition of connected (C)-groups being given by the only help of notions of topological and group-theoretical character, we see that, by Theorem 21, a Lie group of type (C) can be thoroughly characterized by its structure as a topological group. It may also be considered as a generalization of the results of v. Neumann and Chevalley on the Hilbert's problem.

Now we return to the general locally compact groups and prove a lemma, which will be used in the proof of the next theorem.

LEMMA 5.2. *A connected locally compact group G contains the maximal normal subgroup P of type (C) in G and G/P contains no normal subgroup of type (C) except e. If G is moreover an (L)-group, G/P is the direct product of the adjoint groups of simple, semi-simple, non-compact Lie groups.*

PROOF. Let N_0 be the radical of G and let P_0 be the subgroup containing N_0, such that P_0/N_0 is the maximal connected compact normal subgroup of G/N_0. The existence of such P_0 is assured by Theorem 14. Now let P_0' be an arbitrary connected normal subgroup of type (C) in G and N_0' the radical of P_0'. As N_0' is a connected solvable normal subgroup of G, it is contained in N_0. Moreover since P_0'/N_0' is compact we can see easily that N_0P_0' is a closed subgroup of G and that N_0P_0'/N_0 is a connected compact normal subgroup of G/N_0. By the definition of P_0 we have $P_0 \supset N_0P_0'$ and in particular $P_0 \supset P_0'$. We have thus proved that P_0 is the maximal connected normal subgroup of type (C) in G.

Now let P be the subgroup of G containing P_0 such as P/P_0 is the center of G/P_0 and let P' be an arbitrary normal subgroup of type (C) in G. By an analogous proposition for (C)-groups, which corresponds to Theorem 9′ in §4, we see that P is a (C)-group. On the other hand, since the connected component P_0' of P' is contained in P_0 and P'/P_0' is contained in the center of G/P_0', it can be seen immediately that $P'P_0/P_0$ is contained in the center of G/P_0, i.e. in P/P_0. P' is therefore a subgroup of P and P is proved to be the maximal normal subgroup of type (C) in G. That G/P contains no normal subgroup of type (C) except e follows from the fact that P is the maximal normal subgroup of type (C) in G and from the analogue of Theorem 9′.

Now suppose that G is an (L)-group. By Theorem 8 G/P is then also an (L)-group and as it contains no compact subgroup except e, G/P must be a Lie group by Lemma 4.1. That G/P has the structure as stated in the lemma follows immediately from the fact that it contains no abelian normal subgroup except e.

We now prove the following

THEOREM 22. *A connected locally compact group G contains the uniquely determined maximal normal subgroup Q of type (L) in G and G/Q contains no normal subgroup of type (L) except e.*

PROOF. Let P be the maximal normal subgroup of type (C) in G and let Q_0' be an arbitrary connected normal subgroup of type (L) in G, P' the maximal normal subgroup of type (C) in Q_0'. As P' is also normal in G, P' is contained in P and we can see easily that $P' = P \frown Q_0'$. On the other hand since Q_0'/P' is a semi-simple Lie group without center, it is a direct factor in G/P' by Lemma 3.9. It then follows immediately that PQ_0' is a closed subgroup of G and that PQ_0'/P is isomorphic to Q_0'/P'. By Theorem 9′ we see that PQ_0' is an (L)-group.

Now let $\{M\}$ be the class of all normal subgroups of type (L) in G containing P, such that M/P are connected. For example above PQ_0' belongs to such a class $\{M\}$. P is then obviously the maximal normal subgroup of type (C) in each M and M/P is consequently a semi-simple non-compact Lie group without center and is a direct factor in G/P. It then follows that if M_1 and M_2 belong to $\{M\}$, M_1M_2 is closed and is again contained in that system. We can prove therefore in a similar way as in the proof of Theorem 15, that the system contains the maximal group Q, in which all other M are contained. In particular any connected normal subgroup Q_0' of type (L) in G is a subgroup of Q.

Now let N be the subgroup of G containing Q, such as N/Q is the center of G/Q. Since Q/P is a direct factor in G/P, we can find a normal subgroup N_1 containing P, such that $N = N_1Q$, $P = N_1 \frown Q$ and that N_1/P is in the center of G/P. However as G/P has no center by Lemma 5.2 we must have $N_1 = P$, consequently $N = Q$. The group G/Q has therefore no center except e. Let Q' be any normal subgroup of type (L) in G and Q_0' the connected component of Q', which is contained in Q as proved above. Since Q'/Q_0' is contained in the center of G/Q_0', QQ'/Q is then also in the center of G/Q. Therefore, as we have shown that G/Q is without center, $QQ' = Q$, which proves that Q is the maximal normal subgroup of type (L) in G. The second part of the theorem follows immediately from Theorem 9′.

REMARK. The connected component of Q is the maximal connected normal subgroup of type (L) in G.

THEOREM 23. *A connected locally compact group G contains the uniquely determined minimal normal subgroup R, for which G/R is an (L)-group. R contains then no normal subgroup R' except R itself, such that R/R' is an (L)-group.*

PROOF. The existence of R is obvious, for we can take as R the intersection of all normal subgroups N_α such that G/N_α are Lie groups. Now let R'' be the minimal normal subgroup of R, for which R/R'' is an (L)-group. As R'' is a characteristic subgroup of R, it is also normal in G. By Theorem 9 it follows then that G/R'' is an (L)-group and consequently we have $R = R''$, which proves the latter half of the theorem.

Theorem 22 and Theorem 23 give us, in some sense, the limit of the effectiveness of our theory of (L)-groups in the theory of connected locally compact groups. Our theory can say nothing about G/Q or R. The problem remains unsolved: what is the structure of these groups? Now we conjecture that both G/Q and R reduce to the unit group, or in other words that

(C_1) *Any connected locally compact group is an (L)-group.*

This may be a bold conjecture. But one of the reasons, which support our conjecture, is the fact that connected locally compact groups, which are familiar to us, such as compact groups, solvable groups and Lie groups, are all (L)-groups and that, the system of (L)-groups being closed under the group-theoretical operations, such as taking subgroups, factor groups or group extensions, all groups which can be obtained by these operations from the familiar groups fall into the same system.

On the other hand, we have also another ground, namely we can prove that (C_1) is equivalent to the following conjecture, which is slightly stronger than a hypothesis uttered by C. Chevalley in 1934:[33]

(C_2) *A connected locally compact group, which contains no arbitrary small normal subgroup except e, is a Lie group.*

If we make use of Lemma 4.1 (C_2) follows immediately from (C_1). Conversely if (C_2) is valid and if Q denotes the maximal normal subgroup of type (L) in a connected locally compact group G, G/Q is, as it has no compact normal subgroup except e, a Lie group. By Theorem 9 we see that G is an (L)-group and consequently $G = Q$, which proves that (C_1) is true.

It can be also easily proved that (C_1) and (C_2) are equivalent to the following

(C_3) *A connected locally compact simple group is a simple Lie group.*

In any case even when our conjectures are not true, it seems to be sure that the class of all (L)-groups plays an important rôle in the theory of locally compact groups.

MATHEMATICAL INSTITUTE,
TOKYO UNIVERSITY.

[33] H. Cartan [4].

Bibliography

[1] ABE, M.: *Über Automorphismen der lokal-kompakten abelschen Gruppen.* Proc. Imp. Acad. Tokyo 16: 59–62 (1940).

[2] BIRKHOFF, G.: *Lie groups simply isomorphic with no linear group.* Bull. Amer. Math. Soc. 42: 883–888 (1936).

[3] CARTAN, E.: *Groupes simples clos et ouverts et géométrie riemannienne.* Jour. Math. pures et appl. 8: 1–33 (1929).

[4] CARTAN, H.: Sur les groupes de transformations analytiques, Actualités sci. et indus., 198 (1935).

[5] CHEVALLEY, C.: *On the topological structure of solvable groups.* Ann. of Math. 42: 668–675 (1941).

[6] CHEVALLEY, C.: *Two theorems on solvable topological groups,* Lectures in Topology, University of Michigan Press, 1941.

[7] FREUDENTHAL, H.: *Topologische Gruppen mit genügend vielen fastperiodischen Funktionen.* Ann. of Math. 37: 57–77 (1936).

[8] GANTMACHER, F.: *Canonical representation of automorphisms of a complex semi-simple Lie group.* Rec. Math. N. S. 5: 101–144 (1939).

[9] GANTMACHER, F.: *On the classification of real simple Lie groups.* Rec. Math. N. S. 5: 217–249 (1939).

[10] IWASAWA, K.: *Über nilpotente topologische Gruppen,* Proc. Imp. Acad. Tokyo, vol. 21 (1945) (not yet published).

[11] KAMPEN, E. VAN.: *The structure of a compact connected group.* Amer. Jour. Math. 57: 301–308 (1935).

[12] MALCEV, A.: *On the theory of Lie groups in the large.* Rec. Math. N. S. 16: 163–189 (1945).

[13] MONTGOMERY, D. AND ZIPPIN, L.: *Topological transformation groups.* Ann. of Math. 41: 778–791 (1940).

[14] NEUMANN, J. VON.: *Die Einführung analytischer Parameter in topologischen Gruppen.* Ann. of Math. 34: 170–190 (1933).

[15] PONTRJAGIN, L.: *The theory of topological commutative groups.* Ann. of Math. 35: 361–388 (1934).

[16] PONTRJAGIN, L.: Topological groups, Princeton University Press, 1939.

[17] SEGAL, I. E.: *Topological groups in which multiplication of one side is differentiable,* Bull. Amer. Math. Soc. 52: 481–487 (1946).

[18] WEYL, H.: *Theorie der Darstellung kontinuierlicher halb-einfachen Gruppen durch lineare Transformationen, I, II, III.* Math. Zeitschr. 23: 271–309 (1925); 24: 328–395 (1926).

[19] ZASSENHAUS, H.: Lehrbuch der Gruppentheorie, Leipzig und Berlin, Teubner, 1937.

[22] Topological groups with invariant compact neighborhoods of the identity

Ann. of Math., (2) 54 (1951), 345-348.

(Received January 20, 1951)

1. Let G be a topological group, and consider the following two conditions on G;

PI. *G contains a compact neighborhood of the identity e, which is invariant under all inner automorphisms of G,*

PII. *G contains arbitrary small compact neighborhoods of e, which are invariant under all inner automorphisms of G.*

Obviously PII implies PI. However, Mostow showed recently by an example, that PI does not always imply PII. He analyzed precisely the structure of connected Lie groups, which satisfy PI.[1]

In the following we shall study, in general, the structure of topological groups satisfying PI and obtain some results including a generalization of Mostow's theorem.

2. Let G be a topological group satisfying PI and let U be an invariant[2] open neighborhood of e, whose closure $\overline{U}$ is compact. We associate with U a function $\varphi = \varphi_U$ on G, defined as follows;

$$\varphi(\sigma) = m(U \cap \sigma U), \qquad \sigma \in G.$$

Here we denote by m a Haar measure on G, which is two-sided invariant because of the existence of the invariant subset U. $\varphi(\sigma)$ is a continuous function on G,[3] which takes the maximum value at $\sigma = e$ and satisfies

$$\varphi(\tau\sigma\tau^{-1}) = \varphi(\sigma), \qquad \sigma, \tau \in G.$$

The latter follows immediately from the invariance of U and m. Therefore, if we take a real number ξ, such that $0 < \xi < \varphi(e)$, and put

$$U_\xi = \{\sigma; \sigma \in G, \xi < \varphi(\sigma) \leqq \varphi(e)\},$$

U_ξ is again an invariant neighborhood of e.

Now, we prove the following lemma.

LEMMA. *The intersection N of all closed invariant neighborhoods of e in G is a compact normal subgroup of G.*

PROOF. Since, by the assumption, there exists at least one compact invariant neighborhood of e in G, N is an invariant compact subset of G. Moreover it

---

[1] Cf. G. D. Mostow, *On an assertion of Weil*, in these Annals, vol. 54, p. 339. As the title shows, his investigation is derived from a statement in A. Weil's book, *L'integration dans les groupes topologiques et ses applications*, p. 129.

[2] In the following "invariant" means always the invariance under the group of inner automorphisms of G.

[3] Cf. Weil, l.c. 1, p. 41.

345

satisfies $N^{-1} = N$, for the inverse of an invariant neighborhood of e is again such a neighborhood. Therefore, it suffices to show $NN = N$. Let σ be any element in N. Then, σ must be contained in every $\overline{U}_\xi$, $0 < \xi < \varphi(e)$, where φ and U_ξ are defined as above. It follows that $\varphi(\sigma) = \varphi(e)$, and this means $\sigma U = U$ or $\sigma\overline{U} = \overline{U}$ by the definition of φ. As N is obviously also the intersection of all compact closures of invariant open neighborhoods of e, we have $\sigma N = N$. This proves $NN = N$.

Now it is clear from the definition of N, that the factor group G/N satisfies PII. We have thus obtained the following

THEOREM 1. *A topological group G, which satisfies* PI, *contains a compact normal subgroup N, such that G/N satisfies* PII.[4]

Let us assume moreover that G is a connected group. G/N is then the direct product of a compact group and a vector group.[5] Therefore, if we denote by K the inverse image in G of that compact factor of G/N, K is a compact normal subgroup of G, such that G/K is a vector group. Consequently, the topological commutator group[6] $C(G)$ of G is contained in K and is compact.

On the other hand, if the topological commutator group $C(G)$ of a locally compact group G is compact, the inverse image in G of any compact neighborhood of the identity of $G/C(G)$ is clearly an invariant compact neighborhood of e in G. Therefore:

THEOREM 2. *A connected locally compact group G satisfies* PI, *if and only if the topological commutator group of G is compact.*

3. We shall study in more detail the structure of a connected locally compact group G satisfying PI. Let K be the compact normal subgroup as given above. Since G is connected and G/K is a vector group, K must be also connected and is the unique maximal compact subgroup of G, which contains any other compact subgroup of G. K contains, as a connected compact group, a uniquely determined maximal connected semi-simple[7] normal subgroup K_0, such that K/K_0 is abelian. K_0 is then also normal in G and the factor group G/K_0 is solvable, for G/K and K/K_0 are both abelian. Let H be the set of all elements in G, which commute with every element in K_0. H is a normal subgroup of G and we have

$$G = HK_0 = H_0K_0,$$

where H_0 is the connected component of e in H.[8] It is obvious from the definition of H, that $Z_0 = H_0 \cap K_0$ is contained in the center of K_0. Hence it is a totally disconnected abelian normal subgroup of G, for K_0 is semisimple and contains no connected abelian normal subgroup except e. From $G = H_0K_0$ we have

$$H_0/Z_0 \cong G/K_0, \qquad K_0/Z_0 \cong G/H_0.$$

[4] This theorem was also proved independently by H. Yamabe in a similar way.

[5] Cf. Weil, l.c. 1, p. 129.

[6] Cf. K. Iwasawa, *On some types of topological groups*, Ann. of Math. 50 (1949), p. 516.

[7] A connected compact group G is called semi-simple, when it is a projective limit of compact semi-simple Lie groups. This is equivalent to the radical of G being e.

[8] Cf. Iwasawa, l.c. 6, p. 515.

Since G/K_0 is solvable and Z_0 is abelian, we see from the first isomorphism that H_0 is a solvable group. Moreover, from the second one, it follows that H_0 is the maximal connected solvable normal subgroup of G, i.e. the radical of G, for K_0/Z_0 is a semi-simple group.

Now consider the compact normal subgroup N of our previous lemma. From the definition of N, it follows immediately that any normal subgroup of G, the factor group by which satisfies PII, contains N. In particular, since G/H_0 is compact, it satisfies PII so that N is contained in H_0, and H_0/N is a connected solvable group satisfying PII. Consequently, H_0/N must be abelian[9] and the topological commutator group $H_0' = C(H_0)$ of H_0 is contained in N. On the other hand, G/H_0' is the product of the abelian group H_0/H_0' and the compact group K_0H_0'/H_0', which commute elementwise with each other. Consequently G/H_0' satisfies PII and N is contained in H_0'. We have thus proved that $N = H_0'$. It follows that N is connected and is contained in the center of G, for it is the commutator group of connected H_0 and a compact solvable normal subgroup of G.[10]

We summarize our results as follows.[11]

THEOREM 3. *Let G be a connected group satisfying* PI *and let R be the radical of G. Then, there exists a connected compact semi-simple normal subgroup K_0, such that $G = RK_0$ and that R and K_0 are elementwise commutative to each other. The topological commutator group of R is a connected compact group contained in the center of G and coincides with the intersection of all closed invariant neighborhoods of e in G.*

It is obvious that any group G with above mentioned structure satisfies PI, for the topological commutator group of such G is obviously compact.

4. Let G be again a group satisfying PI, which is not necessarily connected, and let N be the intersection of all closed invariant neighborhoods of e in G. Take a small neighborhood V of e and consider the union U of all $\sigma V\sigma^{-1}$ for σ in G. Since U is clearly an invariant neighborhood of e, N is contained in the closure of U. This shows that, if ρ is any element in N, there exists suitable τ and σ in G, such that τ and $\sigma\tau\sigma^{-1}$ are arbitrarily near to e and ρ respectively.

Suppose, now, that G is a group of matrices over the field of complex numbers. As the eigen-values of the matrix τ are the same as those of $\sigma\tau\sigma^{-1}$, the eigen-values of ρ are arbitrarily near to those of the unit matrix e. Therefore the eigen-values of ρ must be all equal to 1. However, since ρ is contained in the compact group N, this implies that ρ is the unit matrix, i.e. $\rho = e$. Thus we have proved $N = e$ and obtained the following

THEOREM 4. *For a group of matrices over the field of complex numbers, two properties* PI *and* PII *are equivalent.*

5. Finally, we shall give an application of our results to two-ended groups.[12]

---

[9] Cf. l.c. 5.

[10] Cf. Iwasawa, l.c. 6, p. 516–517.

[11] Cf. Mostow, l.c. 1.

[12] This application was suggested to the author by Professor D. Montgomery.

Let G be a connected locally compact two-ended group and C the set of all elements in G, which are contained in some compact subgroups of G. It is then proved by Zippin[13] that C is compact and G is the product of C and a subgroup T, which is isomorphic to the additive group of real numbers:

$$G = CT.$$

We can see easily that any compact subset of G is contained in a set of the form CI, where I is a suitable compact subset of T. Consider, then, the union of all $\sigma(CI)\sigma^{-1}$ for σ in G. Since C is an invariant subset from the definition, this union is equal to the product of C and the union of all $\sigma I\sigma^{-1}$. However, the latter set is contained in the compact invariant set CIC^{-1}, for any σ in G can be written in the form $\gamma\tau$, where $\gamma \in C$, $\tau \in T$, and τ commutes with every element in I. This shows that any compact subset of G is contained in some compact invariant subset of G. In particular, G has an invariant compact neighborhood of e, i.e. G satisfies PI. Therefore G contains a unique maximal compact subgroup K which is normal, and it follows immediately that C coincides with K and $K \cap T = e$ so that $G/K = T$. Let H_1 be the group consisting of all elements in G, which commute with every element in K. We have then $G = H_1K$[14] so that

$$H_1/H_1 \cap K \cong G/K \cong T.$$

Since $H_1 \cap K$ is contained in the center of H_1 from the definition of H_1, it then follows immediately that H_1 is an abelian group[15] and contains a subgroup T_1, which satisfies

$$G = KT_1, \qquad K \cap T_1 = e, \qquad T_1 \cong T.$$

This shows that G is the direct product of K and T_1 and proves

THEOREM 5. *A connected locally compact two-ended group G is the direct product of a compact group and the additive group of real numbers.*

Obviously the converse of the theorem is also true. Thus the structure of two-ended groups is completely determined.

THE INSTITUTE FOR ADVANCED STUDY

[13] Cf. L. Zippin, *Two-ended topological groups*, Proc. Amer. Math. Soc. 1 (1950), p. 309-315.

[14] Cf. Iwasawa, l.c. 8.

[15] Cf. Iwasawa, l.c. 6, p. 520.

[23] A note on L-functions

"Proc. Int. Congress of Math." (Cambridge, Mass., 1950), Vol.1, Amer. Math. Soc., 1952, p. 322.

Let K be a finite algebraic number-field, J the idèle group of K and P the group of principal idèles of K. If we introduce a *natural topology* in J, it becomes a locally compact group and P a discrete subgroup of J, and the zeta-function of K with its Γ- and other factors can then be written as an integral over J:

$$\xi(s) = \int_J \varphi(\mathfrak{a}) V(\mathfrak{a})^s \, d\mu(\mathfrak{a}),$$

where $\varphi(\mathfrak{a})$ is a suitable continuous function on J, $V(\mathfrak{a})$ the volume of $\mathfrak{a}$ in the sense of Artin-Whaples, and $\mu(\mathfrak{a})$ a Haar measure on J. The same integral can also be transformed into an integral over $\bar{J} = J/P$:

$$\xi(s) = \int_{\bar{J}} \bar{\varphi}(\bar{\mathfrak{a}}) V(\bar{\mathfrak{a}})^s d\mu(\bar{\mathfrak{a}}), \qquad \bar{\varphi}(\bar{\mathfrak{a}}) = \textstyle\sum_{\alpha \epsilon P} \varphi(\mathfrak{a}\alpha),$$

and the functional equation $\xi(s) = \xi(1 - s)$ follows then immediately from the theta-formula for $\Theta(\bar{\mathfrak{a}}) = 1 + \bar{\varphi}(\bar{\mathfrak{a}})$ and from the invariance of $\mu(\bar{\mathfrak{a}})$. In carrying out this calculation, we obtain Dirichlet's unit theorem for K from the compactness of a subgroup of $\bar{J}$. The above method is also applicable for Hecke's L-functions "mit Grössencharakteren", for these characters form essentially a particular class of continuous characters of $\bar{J}$. (According to a letter of Professor A. Weil, he and Professor E. Artin also obtained these results some time ago.)

More generally, we can define L-functions for a division algebra A over the field of rational numbers by the help of the idèle group of A, and then prove in a similar way the functional equations for them.

Tokyo University,
Tokyo, Japan.

[24] Some properties of (L)-groups

Proc. Int. Congress of Math. (Cambridge, Mass., 1950), Vol.2, Amer. Math. Soc., 1952, pp. 447-450.

One of the main problems in the theory of topological groups at the present time is the study of the structure of locally compact groups in both algebraical and topological aspects. In some particular cases, for example, in the case of Lie groups, we already know much about them. There are indeed still many difficult problems left unsolved concerning Lie groups, but we are in possession of powerful methods from algebra and analysis by which we can research deeply in these groups. The situation is similar for compact groups or locally compact abelian groups, for they are projective limits of Lie groups and we can study their structure by approximating them as precisely as we want by Lie groups. Thus the fifth problem of Hilbert is solved for these groups.

However if we leave these particular cases and consider locally compact groups in general, our knowledge about them is, it may be said, as yet insufficient. Indeed we have no general method by which we can analyze the structure of these groups as we do in the case of compact or abelian groups. Therefore it seems advisable to me to consider first the class of those locally compact groups which can be studied by known methods just as in the special cases mentioned above, and then ask the relation between these groups and general locally compact groups. Such are the class of (L)-groups, namely the class of those locally compact groups which are projective limits of Lie groups.[2]

As one expects from this definition, (L)-groups, in particular connected (L)-groups, have properties similar to those of Lie groups in various aspects. First, it is easy to see that subgroups and factor groups of (L)-groups are also (L)-groups.[3] Moreover we can prove the following theorem.

THEOREM 1.[4] *Let G be a connected locally compact group and N a closed normal subgroup of G. If N and G/N are both* (L)*-groups, then G itself is an* (L)*-group.*

The proof is essentially based upon the following lemma.

[1] This address was listed on the printed program under the title *Locally compact groups*.

[2] In my paper, *On some types of topological groups*, Ann. of Math. vol. 50 (1949), I defined the (L)-group as a locally compact group G containing a system of normal subgroups N_α such that G/N_α are Lie groups and such that the intersection of all N_α is e. If G is a projective limit of Lie groups, G surely contains such a system of normal subgroups, and the converse can also be proved to be true when we assume that G is connected. But we do not know whether these two definitions are completely equivalent.

[3] If we take the former definition of (L)-groups, we cannot yet decide whether a factor group of an (L)-group is always an (L)-group or not.

[4] For the proof of following theorems see the paper of K. Iwasawa, *On some types of topological groups*, Ann. of Math. vol. 50 (1949). Cf. also A. M. Gleason, *On the structure of locally compact groups*, Proc. Nat. Acad. Sci. U. S. A. vol. 35 (1949).

LEMMA. *Let G and N be as above. If N and G/N are both Lie groups, G itself is also a Lie group.*

Theorem 1 and the foregoing remark show us that the class of connected (L)-groups is closed under the group-theoretical operations, such as taking subgroups, forming factor groups or group extensions. Now Lie groups are of course (L)-groups, and compact groups and locally compact abelian groups are also known to be (L)-groups. Therefore those connected locally compact groups, which can be obtained from these "known groups" by successive group-theoretical operations as stated above, are always (L)-groups. In particular, connected locally compact solvable groups are (L)-groups.

A connected (L)-group has a characteristic local structure. We have namely the following theorem.

THEOREM 2. *A connected locally compact group G is an* (L)*-group if and only if G is locally the direct product of a local Lie group L and a compact normal subgroup K.*[5]

This theorem implies in particular that a locally Euclidean (L)-group is a Lie group. Therefore, as every connected locally compact solvable group is an (L)-group, every locally Euclidean solvable group is a solvable Lie group.[6]

As to the global structure of (L)-groups we have the following theorem.

THEOREM 3. *Let G be a connected* (L)*-group. Any compact subgroup of G is then contained in some maximal compact subgroup of G, whereas such maximal compact subgroups of G are all connected and conjugate to each other. Let K be one of them. G contains then subgroups* $H_1, \cdots, H_r$, *which are all isomorphic to the additive group of real numbers and are such that any element g in G can be decomposed uniquely and continuously in the form*

$$g = h_1 \cdots h_r k, \qquad h_i \in H_i, k \in K.$$

In particular, the space of G is the cartesian product of the compact space of K and that of $H_1 \cdots H_r$, *which is homeomorphic to the r-dimensional Euclidean space.*

This theorem is obviously a generalization of Cartan-Malcev's theorem on the topological structure of Lie groups.

By Theorem 3 we see that the topological structure of a connected (L)-group G is completely determined by that of its maximal compact subgroup K. For example, the study of the topological structure of a connected locally compact solvable group G is reduced to that of a compact abelian group, for we can prove in this case that the maximal compact subgroup K is an abelian group.

[5] For a simple proof cf. A. Borel, *Limites projectives de groupes de Lie*, C. R. Acad. Sci. Paris vol. 230 (1950).

[6] Chevalley's theorem.

Now, by making use of Theorem 3, we can also prove the following theorem.

THEOREM 4. *A connected locally compact group G contains a uniquely determined maximal connected solvable normal subgroup R, in which any other connected solvable normal subgroups of G are contained.*

We call such an R, just as in the case of Lie groups, the *radical* of G. We also say that G is semi-simple if the radical R of G is equal to e. In general a connected locally compact group G is an extension of the solvable radical R by the semi-simple group $\bar{G} = G/R$.

For (L)-groups we have moreover the following generalization of E. E. Levi's theorem.

THEOREM 5. *Let G be a connected* (L)*-group and R the radical of G. Then G contains a subgroup S, such that*

(i) *S is the homomorphic image of a connected semi-simple* (L)*-group S′ in G,*

(ii) $RS = G$ *and* $R \cap S$ *is a totally disconnected normal subgroup of G.*[7]

Now it can be proved that a connected locally compact group G contains not only the radical R, but also a uniquely determined maximal connected compact normal subgroup N. By making use of these facts we have the following theorem.

THEOREM 6. *A connected locally compact group G contains a uniquely determined maximal normal subgroup Q of type (L), and G/Q contains no normal subgroup of type (L) other than e.*

On the other hand, it follows immediately from the definition of (L)-groups that a connected locally compact group G contains a uniquely determined minimal normal subgroup Q' of type (L). These facts show us some relations between the class of all connected (L)-groups and general connected locally compact groups. But we have as yet no essential result concerning the situation of connected (L)-groups in the set of all connected locally compact groups. However we have the following conjecture:[8]

(C_1) Any connected locally compact group is an (L)-group.

By theorem 6 this conjecture is easily seen to be equivalent to the following:

(C_2) A connected locally compact group which contains no arbitrary small normal subgroup is a Lie group.

[7] Cf. A. Borel, *Limites projectives de groupes de Lie*, C. R. Acad. Sci. Paris vol. 230 (1950) and Y. Matsushima, *On the decomposition of an* (L)*-group*, Journal of the Mathematical Society of Japan vol. 1 (1950).

[8] Cf. K. Iwasawa, *On some types of topological groups*, Ann. of Math. vol. 50 (1949) and A. M. Gleason, *On the structure of locally compact groups*, Proc. Nat. Acad. Sci. U. S. A vol. 35 (1949).

We can also prove that (C_1), (C_2) are equivalent to the following:

(C_3) A connected locally compact simple group is a simple Lie group.

If these conjectures could be proved to be true, we would have not only a complete solution of the fifth problem of Hilbert, but also decisive progress in the theory of locally compact groups.

INSTITUTE FOR ADVANCED STUDY,
PRINCETON, N. J., U. S. A.

[25] (with T. Tamagawa) On the group of automorphisms of a function field

J. Math. Soc. Japan, 3 (1951), 137-147. Corrections, ibid., 4 (1952), 100-101, 203-204.

§1. Let K be an algebraic function field over an algebraically closed constant field k. It is well-known that the group of automorphisms of K over k is a finite group, when the genus of K is greater than 1. In the classical case, where k is the field of complex numbers, this theorem was proved by Klein and Poincaré[1] by making use of the analytic theory of Riemann surfaces. On the other hand, Weierstrass and Hurwitz gave more algebraical proofs in the same case[2], which essentially depend upon the existence of so-called Weierstrass points of K. Because of its algebraic nature, the latter method is immediately applicable to the case of an arbitrary constant field of characteristic zero. In the case of characteristic $p \neq 0$, H. L. Schmid proved the theorem along similar lines[3]; the proof being based upon F. K. Schmidt's generalization of the classical theory of Weierstrass points in such a case[4].

Now it has been remarked, since Hurwitz, that the representation of the group G of automorphisms of K by the linear transformations, induced by G in the set of all differentials of the first kind of K, is very important for the study of the structure of G. The purpose of the present paper is to show that we can indeed prove the finiteness of G by the help of such a representation instead of the theorem on Weierstrass points. In the next paragraph we analyze the structure of the subgroup $G(p)$ of G, consisting of those automorphisms of K, which leave a given prime divisor P of K fixed, where K may be any function field of genus greater than zero. The finiteness of $G(p)$ is also proved by H. L. Schmid; but his proof depends essentially upon formal calculations of polynomials, whereas our method is more group-theoretical. In the last paragraph we then prove our theorem by considering the above mentioned representation of G and by using a theorem of Burnside on irreducible groups of linear transformations.

1) Cf. Poincaré [3]
2) Cf. Weierstrass [6] and Hurwitz [2]
3) Cf. H. L. Schmid [4]
4) Cf. F. K. Schmidt [5]

§ 2. Let K be an algebraic function field over an algebraically closed constant field K, whose characteristic p may be either zero or a prime number. In this paragraph we always assume that the genus g of K is different from zero.

Lemma 1. *Let σ be an automorphism of K*[5]*, which maps a rational subfield $K'=k(x)$ onto itself. If the degree $n=[K:K']$ is prime to p*[6]*, then σ has a finite order, which does not exceed $n(2n+2g-2)(2n+2g-3)(2n+2g-4)$.*

Proof. Let $P^{(1)}, \ldots\ldots, P^{(s)}$ be all the prime divisors of K, which divide the different of K/K', and let $Q^{(i)}(i=1,\ldots\ldots, s)$ be the projection of $P^{(i)}$ in K'. Choose any $Q=Q^{(i)}$, and consider the decomposition

$$Q=P_1^{e_1}\ldots\ldots\ldots P_t^{e_t}$$

of Q in K. As n is prime to p, the contribution of each P_i to the different of K/K' is given by

$$P_1^{e_1-1}\ldots\ldots\ldots P_t^{e_t-1}$$

whose degree is equal to

$$\sum_{i=1}^{t}(e_i-1)=\sum_{i=1}^{t}e_i-t=n-t\leqq n-1$$

On the other hand, the degree d of the different of K/K' is given by

$$d=2n+2(g-1), \tag{1}$$

which is greater than $2(n-1)$, since we have assumed $g>0$. Therefore there exist at least three, but at most d different prime divisors among $Q^{(i)}$.

Now σ obviously leaves the different of K/K' fixed, and it permutes $P^{(i)}$ and $Q^{(i)}$ among themselves. Therefore some of σ, say σ^l, where

$$l\leqq d(d-1)(d-2), \tag{2}$$

leaves three different $Q^{(i)}$'s invariant. However, an automorphism of a rational function field $K'=K(x)$, which leaves three different prime divisors

5) In the following we always consider only those automorphisms of K, which leave every element in k fixed.

6) If p is zero, n may be an arbitrary integer.

fixed, is the identity. Consequently σ^l leaves all elements of K' fixed. As there exist at most n relative automorphisms of K with respect to K', some power of σ^l, say σ^{lm} is the identity automorphism of K, where m is not greater than n. From (1), (2) we have

$$lm \leqq n(2n+2g-2)(2n+2g-3)(2n+2g-4),$$

which proves our lemma.

Now we study the structure of the group $G(P)$, consisting of all automorphisms of K, which leave a prime divisor P of K fixed. For that purpose, let us consider the set $L(P^n)$ of all elements in K whose denominators divide P^n. $L(P^n)$ is a finite dimensional linear space over k, and we denote its dimension by $l(P^n)$. We have then, obviously,

$$k=L(P^0)\subseteqq L(P^1)\subseteqq L(P^2)\subseteqq \ldots\ldots,$$

$$1=l(P^0)\leqq l(P^1)\leqq l(P^2)\leqq \ldots\ldots$$

However the Riemann-Roch theorem tells us that either $l(P^{n+1})=l(P^n)$ or $l(P^{n+1})=l(P^n)+1$ and that the latter case surely occurs if $n>2g-2$. It follows that we can choose a basis

$$x_1,\ x_2,\ \ldots\ldots,\ x_r \qquad (r=l(P^{2g+1}))$$

of $L(P^{2g+1})$ in such a way, that $x_i,\ x_{i+1},\ \ldots\ldots,\ x_r$ form a basis of some $L(P^{n_i})$ $(n_i\leqq 2g+1)$ for every $i\leqq r$. The denominators of x_1 and x_2 are then just P^{2g+1} and P^{2g} respectively.

Now any automorphism σ of $G(P)$ obviously induces a linear transformation in every $L(P^n)$. In particular we have, for $L(P^{2g+1})$,

$$\sigma(x_j)=\sum_{i=1}^{r} a_{ij}x_i, \qquad a_{ij}\in k, \qquad j=1,\ \ldots\ldots,\ r,$$

or simply in a matrix equation

$$(\sigma(x_1),\ \ldots\ldots,\ \sigma(x_r))=(x_1,\ \ldots\ldots,\ x)A_\sigma, \quad A_\sigma=(a_{ij}).$$

As a result of the particular choice of our basis, A_σ has the following triangular form

$$A_\sigma=\begin{pmatrix} a_1 & & & 0 \\ & a_2 & & \\ & & \ddots & \\ * & & & a_r \end{pmatrix} \qquad (a_i=a_{ii})$$

and $\sigma \rightarrow A_\sigma$ gives a representation of $G(P)$. Moreover this representation is an isomorphic one. In fact, if A_σ is the unit matrix, σ leaves x_1 and x_2 and, consequently, every element in $k(x_1, x_2)$ fixed. But this field $k(x_1, x_2)$ coincides with K, as one readily sees from the fact that the degree $[K: k(x_1, x_2)]$ divides both degrees $[K: k(x_1)]=2g+1$ and $[K: k(x_2)]=2g$. It follows that such σ is the identity automorphism of K.

By the help of this isomorphic representation we can prove the following

Lemma 2. *The order of any element σ in $G(P)$ is finite and has a bound which depends only upon g and p.*

Proof. Consider the eigen values $u_1, u_2, \ldots\ldots, u_r$ of A_σ and suppose first that all u_i are different from each other. By changing the basis suitably, we may then assume that A_σ is a diagonal matrix, or, in other words,

$$\sigma(x_i)=u_i x_i, \qquad i=1, 2, \ldots\ldots, r.$$

The subfields $k(x_1)$, $k(x_2)$ are, consequently, mapped onto itself by σ. As one of the degrees $[K: k(x_1)]=2g+1$ and $[K: k(x_2)]=2g$ is prime to p, it follows, from Lemma 1, that σ has a finite order, which is bounded by a number depending only upon g and $n=2g+1$ or $2g$.

Now assumme that some u_i and u_j coincide $(i\neq j)$. We can then find linearly independent elements x and y in $L(P^{2g+1})$, such that

$$\sigma(x)=u_i x, \qquad \sigma(y)=u_i(x+y)$$

For $z=\frac{y}{x}$ we have then

$$\sigma(z)=z+1,$$

and the field $k(z)$ is mapped onto itself by σ. Moreover, the degree $n=[K: k(z)]$ is not greater than $2(2g+1)$, for the degrees of the denominators of x and y are most $2g+1$ and that of z is, consequently, at most $2(2g+1)$. Therefore, if the characteristic p of k is zero, it follows again from Lemma 1 that the order of σ is finite and has a bound depending only upen g. On the other hand, if p is not zero, we have $\sigma^p(z)=z$, and σ^p is a relative automorphism of K with respect to $k(z)$. It follows that the order of σ^p dose not exceed n and that the order of σ is at most

$2p(2g+1)$.

Now take a prime element u for P, i. e. such an element u in K, which is divisible by P, but not by P^2. For any σ in $G(P)$, $\sigma(u)$ is again a prime element for P, and we have

$$(3) \qquad \sigma(u) \equiv \gamma u \mod \mathfrak{P}^2$$

where γ is a suitable constant and $\mathfrak{P}$ is the prime ideal in the valuation ring of P. As γ is uniquely determined by the above congruence, we may denote it by γ_σ. $\sigma \rightarrow \gamma_\sigma$ is then a representation of $G(P)$ in k, and, if we denote by N the kernel of this representation, $G(P)/N$ is isomorphic to the multiplicative group Γ of γ_σ. However, we know by Lemma 2 that the orders of elements in $G(P)$ are bounded. Therefore, the orders of elements in $G(P)/N$ or in Γ are also bounded. It follows that Γ is the group of all m-th roots of unity in k, where m is a suitable integer prime to p. Therefore $G(P)/N$ is also a cyclic group of order m and $G(P)$ contains an element of order m. As m is prime to p, we can then prove, by a standard argument[7], that

$$(4) \qquad m \leqq 6(2g-1).$$

We consider, now, the structure of the normal subgroup N. From (3) it follows immediately that the eigen values α_i of A_σ are powers of γ, and, in particular,

$$\alpha_1 = \gamma^{-(2g+1)}, \qquad \alpha_2 = \gamma^{-2g}$$

This shows that N consists of all those σ in $G(P)$, for which the matrix A_σ has the form

$$(5) \qquad \begin{pmatrix} 1 & & & 0 \\ & 1 & & \\ & & \ddots & \\ & * & & 1 \end{pmatrix}$$

However, if the characteristic of k is zero, such a matrix can not have a finite order unless it is the unit matrix. Therefore, we see, by Lemma 2,

7) Cf. H. L. Schmid [4]. Note that P ramifies completely in the extension of degree m and that the degree of the different of that extension is at least $m-1$. Cf. the proof of Lemma 4 below.

that N is the unit group if $p=0$. On the other hand, if p is not zero, the group N, which is isomorphic to a group of matrices of the form (5), is a nilpotent group and the order of any element in N is a power of p. In order to show that N is actually a finite p-group in such a case, we first prove some lemmas.

Lemma 3. *Let H be a group of automorphisms of a function field K of genus $g>0$, such that*

1) *H is abelian and the order of any element in H is a power of p,*
2) *every element in H leaves a prime divisor P fixed,*
3) *the fixed field*[8] *of any non.trivial finite subgroup of H is a rational function field.*

Then H is a cyclic group of order either 1, p *or* p^2.

Proof. Suppose that H is not the unit group, and take a subgroup $U=\{\sigma\}$ of order p. By assumption, the fixed field of U is a rational function field $k(x)$. We can take x in such a way that the denominator of x is P^p. As H is abelian, any τ in H then maps $k(x)$ onto itself, and as the denominator of x is invariant under τ and since the order of τ is a power of p, we have

$$\tau(x)=x+a \qquad a\epsilon k. \tag{6}$$

It follows that $\tau^p(x)=x$, $\tau^p \in U$, $\tau^{p^2}=e$, so that the order of any τ in H is at most p^2.

To prove the lemma, it is therefore sufficient to show that H contains no subgroup of order p other than U. Suppose, for a moment, that there exists such a subgroup $V=\{\tau\}$ of order p. We shall deduce a contradiction from this assumption. As τ is not in U, a is not zero in (6). Therefore, replacing x by $\frac{x}{a}$, we may assume

$$\tau(x)=x+1, \qquad \sigma(x)=x \tag{7}$$

In a similar way, we can find an element y such that the denominator of y is P^{p^2} and

---

8) The fixed field K' of a finite group G of automorphisms of K is the set of all elements of G. K/K' is then a Galois extension with the Galoisgroup G. In particular we have $[K:K']=[G:e]$

(8) $$\sigma(y)=y+1, \qquad \tau(y)=y.$$

As y is not contained in $k(x)$, we have $K=k(x, y)$. On the other hand x^p-x and y^p-y are both contained in the fixed field K' of the subgroup $UV=\{\sigma, \tau\}$ of order p^2. However, as these elements have the same denominators P^{p^2} and K' is a rational function field with $[K:K']=p^2$, we must have

$$y^p-y=\beta(x^p-x)+\gamma \qquad \beta, \gamma\epsilon k.$$

If we then put

(9) $$z=y-\beta^{\frac{1}{p}}x,$$

we have

(10) $$z^p-z-\gamma=(\beta^{\frac{1}{p}}-\beta)x.$$

Therefore, if $\beta^{\frac{1}{p}}-\beta=0$, z is constant in k, and (9) gives us $k(x)=k(y)$, which is obviously a contradiction. On the other hand, if $\beta^{\frac{1}{p}}-\beta\neq 0$ (9) and (10) show that x and y are both contained in $k(z)$. We have then $K=k(x, y)=k(z)$, which also contradicts the assumption that the genus of K is not zero. The lemma is thus proved[9].

Lemma 4. *Let K be a function field of genus $g>0$ and H group of automorphisms of K, which satisfies the conditions* 1), 2) *of the previous Lemma. H is then a finite group, and its order does not exceed $p^2(2g-1)$.*

Proof. Let U be an arbitrary finite subgroup of order n in H and K' its fixed field. The genus g' of K' is given by the following formula:

(11) $$2(g-1)=d+2n(g'-1),$$

where d is the degree of the different of K/K'. However, in the present case, d is always at least $n-1$, for the prime divisor P ramifies completely in the extension K/K'. Therefore if $2(g-1)<n-1$, namely if $2g\leqq n$, g' must be zero. It follows that there exists a maximal subgroup V of order less than $2g$, such that its fixed field K'' has a genus different from zero.

9) A slightly finer consideration shows us that the condition 2) is not necessary in the present lemma.

The factor group H/V, considered as a group of automorphisms of K'', obviously satisfies all conditions of the previous lemma. The order of H/V is, consequently, at most p^2, and the order of H itself does not exceed $p^2(2g-1)$.

Finally we prove a purely group-theoretical lemma.

Lemma 5. *Let G be a finite or infinite group of order $\geqq n$, containing a central subgroup Z of order p, such that the factor group G/Z is an elementary abelian p-group*[10]. *Then G contains an abelian subgroup of order at least $\sqrt{pn}$.*

Proof. We may assume that G is a finite group, for otherwise, we may replace G by a suitable finite subgroup of order $\geqq n$. Let U be a maximal abelian normal subgroup of G. Z is then contained in U, and U/Z is an elementary abelian p-group. We select $\sigma_1, \ldots\ldots, \sigma_s$ in U, such that the cosets of σ_i modulo Z form a basis of U/Z. For an arbitrary σ in G, we then put

$$\sigma\sigma_i\sigma^{-1}\sigma_i^{-1}=\zeta_i \qquad i=1, \ldots\ldots, s.$$

As G/Z is abelian, $\zeta_i=\zeta_i(\sigma)$ is contained in Z, and we see easily that the mapping

$$\sigma\rightarrow(\zeta_1(\sigma), \ldots\ldots, \zeta_s(\sigma))$$

is a homomorphism from G into the direct product of s copies of Z. Moreover the kernel of this homomorphism coincides with U, for U is a maximal abelian normal subgroup of G. It follows that the order of G/U is at most p^s. On the other hand, the order of U is equal to p^{s+1}. We have, consequently,

$$n\leqq[G:e]=[G:U][U:e]\leqq p^s\cdot p^{s+1},$$
$$\sqrt{pn}\leqq p^{s+1}=[U:e],$$

which proves our lemma.

We now return to the group $G(P)$ and show that the nilpotent normal subgroup N of $G(p)$ is a finite group. Let $x=x_{r-1}$ be the next to last element in the above chosen basis $x_1, \ldots\ldots, x_r$ or $L(P^{2g+1})$. Because of

10) A group is called an elementary abelian p-group, when it is abelian and the p-th power of any element of the group is the unit element.

the choice of our basis, x is an element in K, such that it has a denominator of the least possible positive power of P, say P^n, among all elements in K. From (5) we have

$$\sigma(x)=x+a_\sigma, \qquad a_\sigma \in k,$$

for any σ in N, and $\sigma \to a_\sigma$ gives a homomorphism from N into the additive group of k. Therefore, if we denote the kernel of this homomorphism by N_1, N/N_1 is an elementary abelian p-group. Moreover, as any σ in N_1 is a relative automophism of $K/k(x)$, the order of N_1 is at most $m=[K:k(x)]$. As N is nilpotent, we can find a subgroup N_2 of index p in N_1, such that it is normal in N and N_1/N_2 is contained in the center of N/N_2. Let K' be the fixed field of N_2. From the relation

$$p[K:K']=[N_1:N_2][N_2:e]=[N_1:e]\leqq[K:k(x)],$$

we see that the genus g' of K' is not zero, for othewise, K would contain a non-constant element whose denominator is a proper divisor of P^m. Since N/N_2 can be considered as a group of automorphisms of K', we see, from Lemma 4, that the order of any abelian subgroup of N/N_2 is at most $p^2(2g'-1)$. On the other hand, if we put $Z=N_1/N_2$, the group N/N_2 has the structure mentioned in Lemma 5. Therefore, if the order of N/N_2 is not less than n', it contains an abelian subgroup of order $\geqq\sqrt{pn'}$. It then follows that

$$\sqrt{pn'}\leqq p^2(2g'-1).$$

Consequently the order of N/N_2 is at most $p^3(2g'-1)^2$, and the order of N is not greater than $p^3(2g'-1)^2 \cdot mp^{-1}=p^2m(2g'-1)^2$. However, we know from (11) that

$$2(g-1)\geqq(m-1)+2m(g'-1),$$

or

$$2g-1\geqq m(2g'-1), \quad (2g-1)^2\geqq m(2g'-1)^2.$$

We have thus proved that the order of N is at most $p^2(2g-1)^2$ and obtained the following theorem[11].

---

11) An example in H. L. Schmid [4] shows that $p^2(2g-1)^2$ seems to be near to the best value of the bounds of the order of such N.

Theorem 1. *Let K be a function field of genus $g>0$ over an algebraically closed constant field k, aud let P be an arbitrary prime divisor of K. Then the group $G(P)$ of all automorphisms of K which leave P fixed has the following structure:*

1) *if the characteristic of k is zero, $G(P)$ is a cyclic group of order $\leqq 6(2g-1)$.*

2) *if the characteriatic of k is a prime number p, a p-Sylowgroup N of $G(P)$ is a normal subroup of order $\leqq p^2(2g-1)^2$ and the factor group $G(P)/N$ is a cyclic group of order $\leqq 6(2g-1)$.*

In any case the order of $G(P)$ has a bound depending only upon g and p.

§ 3. Let us now assume that the genus g of K is greater than 1 and denote the set of all differentials of the first kind of K by D. As is well-known D is a g-dimensional linear space over k and any automorphism of K induces a linear transformation in D. Thus the group G of all automorphisms of K can be represented by such linear transformations in D.

Now take an aibitrary automorphism σ in G. We can then find a differential $\omega \neq 0$ in D, such that

$$\sigma(\omega)=\alpha\omega, \qquad \alpha\in k.$$

If follows that σ permutes the $2g-2$ zeros of ω among themselves, and some power of α, say α^l, where

$$l\leqq 2g-2,$$

leaves one of these zeros of ω, say P, fixed. σ^l is therefore contained in $G(P)$ and Theorem 1 then shows us that the order of σ has a bound, depending only upon g and p.

Let M be an irreducible invariant subspace of D with respects to the above representation of G, We denote by G_0 the kernel of the irreducible representation of G in M, so that G/G_0 is isomorphic to the irreducible group of linear transformations. However, we know that the orders of elements in G, a fortiori the orders of elements in G/G_0, are bounded. If follows then from a theorem of Burnside[12] that G/G_0 is a finite group.

Now take a differential $\omega\neq 0$ in M. Since $\sigma(\omega)=\omega$ for any σ in G_0, each such σ permutes the $2g-2$ zeros of ω among themselves. These zeros are not necessarily different from each other, but there exists at

12) Cf. Burnside [1]

least one such zero of ω by the assumption $g>1$. Therefore there exists a subgroup G_1 of G_0, such that the index $[G_0 : G_1]$ is at most $(2g-2)$ and such that each σ in G_1 leaves a prime divisor P fixed. G_1 is thus contained in the finite group $G(P)$, and we see, finally, that the group G itself is a finite group.

We have thus proved the following

Theorem 2. *The group G of all automorphisms of a function field of genus $g>1$ over an algebraically closed field k, is always a finite group.*

From the above proof, we can also find a bound for the order of G, which depends only upon g and p, though it is much greater than the best value of such bounds in the case characteristic zero.

Bibliography

(1) Burnside, W. Theory of groups of finite order, 2nd. ed. Note J. p. 491—494.

(2) Hurwitz, A. Analytische Gebilde mit eindeutigen Transformationen in sich, Math. Ann, Bd. 41 (1893), p. 403—422 (Werke Be. I, p. 391—430).

(3) Poincaré, H. Sur un théorème de M. Fuchs, Acta Math, Bd. 7 (1885), p. 1—32.

(4) Schmid, H. L. Über die Automorphismen eines algebraischen Funktionenkörpers von Primzahlcharakteristik, Crelle's Jour. Bd. 179 (1938),p. 5—15.

(5) Schmidt, F. K. Zur arithmetische Theorie der algebraischen Funktionen. II. Allgemeine der Weierstrasspunkte, Math. Zeitschr., Bd. 45 (1939), p. 73—96.

(6) Weierstrass, K. Aus einem noch nicht veröffentlichten Briefe an Herrn Professor Schwarz, Werke Bd. II, p.235—244.

Journal of the Mathematical Society of Japan Vol. 4, No. 1, July, 1952.

Correction.

On the group of automorphisms of a function field.

This Journal Vol. 3, No. 1.

By Kenkiti IWASAWA and Tsuneo TAMAGAWA

A correction should be made for the proof of Lemma 1, as the one given in our paper is not valid, when the ramification order of some prime divisor of K over K' is divisible by the characteristic p of K. We shall show, however, that the conclusion of this lemma is also true in this exceptional case, although the estimation of the order of automorphisms given in our paper fails then to hold. First we shall prove the following local theorem.

THEOREM. Let k be an algebraically closed field with characteristic $p\neq 0$, $\Omega=k((t))$ a field of formal power series of one variable t, and Ω' a finite separable extension of Ω whose degree n is divisible by p. If a continuous automorphism σ of Ω' over k leaves invariant Ω, then the first coefficient α_1 of the expansion

$$t^{\sigma}=\alpha_1 t+\cdots$$

is a root of unity.

PROOF. We first show that we have only to prove our assertion in this special case: Ω' is a cyclic extension of Ω of degree p. Let Ω^* be a finite Galois extension of Ω containing Ω', and L the intermediate field between Ω^* and Ω corresponding to the p-Sylow group of the Galois group of Ω^*/Ω. As the degree e of L over Ω is prime to p, there is no other extension of Ω of degree e in an algebraic closure $\bar{\Omega}$ of Ω^*. Hence, if $\bar{\sigma}$ is any extension of σ to $\bar{\Omega}$, we have $(\Omega'L)^{\bar{\sigma}}=\Omega'L$. Let H be an intermediate field between $\Omega'L$ and L with degree p over L, then H is cyclic over L, and some power $\bar{\sigma}^h$ of σ leaves invariant H. Thus we have reduced our theorem to the special case. So we shall assume in the following that Ω'/Ω is cyclic of degree p. Then we can choose an Artin-Schreier's generater π of Ω' over Ω satisfying the following equation:

$$\wp(\pi)=\pi^{p}-\pi=\frac{\beta}{t^{\lambda}}+\cdots \qquad (\lambda, p)=1.$$

As π^{σ} is then also a generator of the same nature, we have

$$\wp(\pi^{\sigma})-\wp(\pi)\in\wp(\Omega), \quad \text{and} \quad \alpha_1^{\lambda}=1.$$

THE PROOF OF LEMMA 1 IN THE EXCEPTIONAL CASE.

Let P be a prime divisor of K, whose ramification order is divisible by p, and Q the prime divisor of K' such that $P\mid Q$. We may assume without loss of generality that $\nu_Q(x)>0$. Then some power σ^h of σ leaves invariant P, and we have

$$x^{\sigma^h}=\frac{\alpha x}{x+\beta} \qquad (\alpha \neq 0).$$

From the above theorem, we have $\alpha^{\lambda}=1$. Hence the order of σ is $\leq h\cdot\lambda[K:K']$.

Journal of the Mathematical Society of Japan Vol. 4, No. 2, October, 1952.

Correction: On the paper "On the group of automorphisms of a function field".

(This journal, vol. 3 (1951), pp. 137–147)

By Kenkichi IWASAWA and Tsuneo TAMAGAWA

A correction was made to our paper mentioned in the title, as the proof of Lemma 1 was not complete. We give here another correction, which will give more explicitly the bound of the order of the automorphism in question. We use the same notations as in Lemma 1 of the original paper, without mentioning explicitly their meanings. Here we are concerned with the modular case, i.e. the case where the characteristic

$$p \text{ of } k \text{ is } \neq 0 .$$

Obviously we can assume either $\sigma(x)=x+\alpha$ or $\sigma(x)=\alpha x$ with some α in k. In the first case, the order of σ does not exceed pn. Hence we assume $\sigma(x)=\alpha x$. If the divisor of x is of the form P^nQ^{-n}, where P, Q are completely ramified primes of K over K', the contributions of P and Q to the different of K/K' are P^{n-1} and Q^{n-1} respectively. The original proof of Lemma 1 can be applied to this case and we see that the order of σ is at most $n(2n+2g-2)(2n+2g-3)(2n+2g-4)$. Suppose, now, that either the numerator or the denominator of x, say the latter, contains two different primes P_1, P_2 of K. A suitable power $\tau=\sigma^l$, $l\leqq n$, leaves P_1 fixed, and we can find an element y in K such that $\tau(y)=\beta y+\gamma$, $\beta,\gamma\in k$, and that the denominator of y is P_1^r, $r\leqq g+1$ (cf. the proof in p. 139 of the original paper). If $\beta=1$, the order of τ is at most $p(g+1)$ and, consequently, the order of σ is at most $pn(g+1)$. We may therefore assume that $\beta\neq 1$ and $\tau(y)=\beta y$. Let $F(X, Y)=\sum \alpha_{ij}X^iY^j$ be an irreducible polynomial over k such that $F(x,y)=0$. Since $\tau(x)=\alpha^l x$, $\tau(y)=\beta y$, we have $F(\alpha^l X,\beta Y)=\xi F(X,Y)$, $\xi\in k$. Therefore, if $\alpha_{ij}\neq 0$, $\alpha_{st}\neq 0$, $(i,j)\neq(s,t)$, $z=x^{i-s}y^{j-t}$

is invariant under σ and is not contained in k, for the denominator of x contains P_1, P_2, while the denominator of y is P_1^r. Since the degree of F in X is at most $r=[K: k(y)]\leqq g+1$ and the degree of F in Y is at most $n=[K: k(x)]$, we have $|i-s|\leqq g+1$, $|j-t|\leqq n$ and, consequently $[K: k(z)]\leqq 2n(g+1)$. It then follows from $\tau=\sigma^l$, $\tau(z)=z$ that the order of σ is at most $2n^2(g+1)$, which is not greater than $n(2n+2g-2)(2n+2g-3)(2n+2g-4)$.

Thus Lemma 1 is completely proved, a bound of the order of σ being the maximum of $n(2n+2g-2)(2n+2g-3)(2n+2g-4)$ and $pn(g+1)$.

[26] On the rings of valuation vectors

Ann. of Math., (2) 57 (1953), 331-356.

(Received December 13, 1951)

Introduction

Let $\{R_\alpha\}$ be a system of topological rings and O_α open subrings of R_α. We consider the set R of all vectors $a = (a_\alpha)$, where a_α are elements in R_α and which belong to O_α, except for a finite number of α. By the usual definition of component-wise addition and multiplication, R forms a ring containing the direct sum O of all O_α. We can then define, in a unique manner, a topology in R so that R becomes a topological ring, O becomes an open subring of R and is the cartesian product of O_α as a topological space. We call R the local direct sum of R_α relative to O_α.[1]

Now, let k be a finite algebraic number field or an algebraic function field of one variable over a constant field F. In the following, we shall call such a field, simply, a number field or a function field respectively. We consider the set $\{K_P\}$ of all completions of K with respect to prime divisors P of K, which are trivial on F in the case of a function field, and denote by O_P the valuation ring of P in K_P, if P is non-archimedean, and the field K_P itself, if P is archimedean. Then, with respect to the usual topology of K_P induced by a valuation of P, each O_P is an open subring of K_P, and we can form the local direct sum R of K_P relative to O_P. We call this R the ring of valuation vectors of K. If we identify each element ξ of K with the vector $a = (a_\alpha)$, whose components a_α are all equal to ξ, K is isomorphically imbedded in R and becomes a discrete subfield of R, as we shall show later.

According to recent results in algebraic number theory, it has become clearer and clearer that the topological properties of R and of its related structures, in particular that of the multiplicative group of R, have essential relations to the arithmetic of the field K.[2] Hence, it seems to be of some interest to know what are the characteristic properties of R as a topological ring, for this would show us the sources of arithmetic theorems which can be deduced from the topological structure of R, and might possibly give us some suggestions for further developments in algebraic number theory.

The main purpose of the present paper is to give such a characterization of the rings of valuation vectors of number fields and function fields. After some preparations in §1, we shall do this in §2 and §3, and then show in §4, by some examples, how arithmetic properties of K are related to the topological structure of the ring R. For the proofs, we shall use Haar measures of locally compact groups, some fundamental properties of locally compact rings and, in particular,

[1] Cf. Kaplansky [9]. The number in the bracket refers to the bibliography at the end of the paper.

[2] Cf. Chevalley [6], Weil [12].

of locally compact fields, as well as elementary results in valuation theory. We shall also use in §4 the duality theorem of locally compact abelian groups and its analogue for locally linearly compact spaces.

§1. The norm of an automorphism of a locally compact group

Let G be a locally compact group and $\mu(E)$ a left-invariant Haar measure of G. For any automorphism σ of G, $\mu'(E) = \mu(\sigma(E))$ gives also a Haar measure of the same kind as $\mu(E)$, and we have

$$\mu'(E) = \kappa\mu(E),$$

with some positive constant κ. As is readily seen from the definition, κ does not depend upon the choice of the Haar measure μ and is uniquely determined by σ. Hence we call κ the norm of the automorphism σ and denote it by $N(\sigma, G)$.[3] For any integrable function $f(x)$ on G, we see that

$$\int_G f(x)\, d(\sigma(x)) = \int_G f(\sigma^{-1}(x))\, dx = N(\sigma, G) \int_G f(x)\, dx,$$

or, symbolically, that

$$d(\sigma(x)) = N(\sigma, G)\, dx, \tag{1.1}$$

which may be also regarded as the definition of $N(\sigma, G)$. We also note that, if G has open, compact subgroups U and V, such that V is contained in both U and $\sigma(U)$, then $N(\sigma, G)$ is given by the ratio of the indices:

$$N(\sigma, G) = [\sigma(U):V]\,[U:V]^{-1}. \tag{1.2}$$

Now, let H be a closed normal subgroup of G which is mapped onto itself by the automorphism σ. σ then induces automorphisms on H and G/H, which, for simplicity, we again denote by σ. Let x, ξ and P denote generic points of G, H and G/H respectively. Since we can take Haar measures on G, H and G/H in such a manner that the symbolic relation

$$dx = dP \cdot d\xi \tag{1.3}$$

holds for the corresponding integrals over G, H and G/H,[4] we see immediately from (1.1) that

$$N(\sigma, G) = N(\sigma, G/H)N(\sigma, H).^{5} \tag{1.4}$$

In particular, if G is the direct product of H_1 and H_2, which are both invariant under σ, we have

$$N(\sigma, G) = N(\sigma, H_1)N(\sigma, H_2). \tag{1.5}$$

---

[3] This is the same as "le module" of σ in Braconnier [5].

[4] Cf. Weil [11], p. 45.

[5] For a precise proof, cf. Braconnier [5].

From the definition of $N(\sigma, G)$, it is also clear that

$$N(\sigma_1\sigma_2, G) = N(\sigma_1, G)N(\sigma_2, G), \tag{1.6}$$

where σ_1, σ_2 are arbitrary automorphisms of G.

We now consider a locally compact ring R containing the unity element 1. For any regular element ξ of R, the mapping

$$\sigma: a \longrightarrow \xi a, \qquad a \in R$$

gives an automorphism of the additive group of R, which leaves every left ideal A of R invariant. Hence, we can define the norms $N(\sigma, R)$, $N(\sigma, A)$ or $N(\sigma, R/A)$ as above. In the following, we shall denote these norms by $N(\xi, R)$, $N(\xi, A)$ or $N(\xi, R/A)$ respectively.

That being said, we prove the following:

LEMMA 1. *Suppose that R contains a set of closed maximal two-sided ideals $\{M\}$ such that the intersection of all M is 0. Then, the norm $N(\xi, R/M)$ of a regular element ξ in R is equal to 1 except for a finite number of M, and we have*

$$N(\xi, R) = \prod N(\xi, R/M), \tag{1.7}$$

where the product runs over all M in $\{M\}$.

PROOF. Let R' be the connected component of 0 in R. As is well-known, R' is a closed two-sided ideal of R and contains a closed, nilpotent, two-sided ideal R'' of R, such that R'/R'' is a semi-simple algebra of finite rank over the real number field.[6] Since R/M is a simple ring with a unity element, we have $R'' + M = M$, i.e. $R'' \subset M$ for all M in $\{M\}$. From the assumption on $\{M\}$, it follows that $R'' = 0$ and that R' is a semi-simple algebra of finite rank over the real number field. Hencr R' contains a unity element and R is, consequently, the direct sum of a totally disconnected locally compact ring and a finite number of simple algebras over the real number field. Using (1.5), this reduces the proof of our lemma to the case, where R is totally disconnected. Now, in that case, there exist open, compact (additive) subgroups U and V in R, such that V is contained in both U and $\sigma(U)$, and that $N(\xi, R) = N(\sigma, R)$ is given by (1.2). Let V' be the union of the complements of V in U and in $\sigma(U)$. Since V' is compact and the intersection of all M is 0, there exist a finite number of closed maximal ideals $M_1, \cdots, M_r$ in $\{M\}$ such that the intersection of V' and $A = M_1 \cap \cdots \cap M_r$ is empty. It follows that

$$U \cap A = V \cap A, \qquad \sigma(U) \cap A = V \cap A,$$

and, consequently, that

$$[U:V] = [U + A:V + A], \qquad [\sigma(U):V] = [\sigma(U) + A:V + A].$$

Hence, we have by (1.2)

$$N(\xi, R) = [\sigma(U) + A:V + A][U + A:V + A]^{-1}.$$

---

[6] See, for instance, Kaplansky [8].

But, if we apply (1.2) to R/A, $U + A/A$ and $V + A/A$, we see that the right hand side of the above equality is nothing but $N(\xi, R/A)$, and we get

$$N(\xi, R) = N(\xi, R/A).$$

Since R/A is obviously isomorphic to the direct sum of R/M_i $(i = 1, \cdots, r)$, it follows from (1.5) that

$$N(\xi, R) = N(\xi, R/A) = \prod_{i=1}^{r} N(\xi, R/M_i).$$

Now, let M' be any other closed maximal ideal in $\{M\}$ different from $M_1, \cdots, M_r$. As the intersection of V' and $A' = M' \cap A$ is again empty, we see, in the same way as above, that

$$N(\xi, R) = N(\xi, R/A') = N(\xi, R/M')\prod_{i=1}^{r} N(\xi, R/M_i).$$

We have, therefore, $N(\xi, R/M') = 1$ and, consequently, that

$$N(\xi, \mathrm{R}) = \prod N(\xi, R/M),$$

as stated in the lemma.

LEMMA 2. *Under the same assumptions as in the previous lemma, suppose, furthermore, that $N(\xi, R/M_0) \neq 1$ for some fixed ideal M_0 in $\{M\}$. R is, then, the direct sum of M_0 and a closed two-sided ideal L:*

$$R = M_0 + L.$$

PROOF. Let L be the intersection of all ideals in $\{M\}$ different from M_0. If L were 0, we would have, by Lemma 1,

$$N(\xi, R) = \prod' N(\xi, R/M),$$

where the product $\prod'$ runs over all $M \neq M_0$ in $\{M\}$. But this contradicts (1.7) and $N(\xi, R/M_0) \neq 1$. Hence L must be different from 0. Since $M_0 \cap L = 0$ and M_0 is maximal, we have, immediately, the direct sum decomposition $R = M_0 + L$.

In the next section, we shall apply Lemmas 1 and 2 to a commutative ring R. In that case, the residue class ring R/M will be a locally compact field[7] and it will then become necessary to consider the norm $N(\xi, K)$ of an element $\xi \neq 0$ in such a locally compact field K. But, by well-known theorems on the structure of locally compact fields, we can see easily that $N(\xi, K)$ is given by

$$N(\xi, K) = \| \xi \|,$$

where $\| \xi \|$ denotes the so-called normed valuation of K.[8] Note that the topology of K is defined by the metric induced by the valuation $\| \xi \|$, and also that, if K

---

[7] Note that, for locally compact fields, the continuity of the inverse a^{-1} follows from other axioms of topological fields. Cf. also the remark at the end of this section.

[8] See Artin-Whaples [2].

is isomorphic to the complex number field, $\| \xi \|$ is not a valuation, but the square of a valuation in the usual sense.

Now, we consider a locally linearly compact linear space S over a field F and define the norm of an automorphism σ of S as follows.[9] We first consider a real-valued function $\nu(U, V)$, which is defined for all open, linearly compact subspaces U, V of S and satisfies the following conditions:

(1) $\nu(U, V) + \nu(V, W) = \nu(U, W)$,

(2) If V is contained in U, then $\nu(U, V)$ is equal to the dimension $[U:V]_F$ of the linear space U/V over F.

If such a function $\nu(U, V)$ is given and if W is any open, linearly compact subspace contained in both U and V, it follows, from (1), (2), that

$$(U, V) = [U:W]_F - [V:W]_F . \tag{1.8}$$

On the other hand, it is easy to see that the right-hand side of the above equality gives the same value for any open, linearly compact W contained in both U and V, and that the function $\nu(U, V)$ defined by (1.8) satisfies the above conditions, (1), (2). We see, therefore, that there exists a unique function $\nu(U, V)$ satisfying (1), (2). It follows, in particular, that, if σ is an automorphism of the linear space S, we have

$$\nu(\sigma(U), \sigma(V)) = \nu(U, V),$$

and, consequently, that

$$\nu(U, \sigma(U)) = \nu(U, V) + \nu(V, \sigma(V)) + \nu(\sigma(V), \sigma(U))$$

$$= \nu(V, \sigma(V)).$$

This shows that $\nu(U, \sigma(U))$ is independent of the choice of U and we, therefore, define the norm of σ by

$$N(\sigma, S) = \exp(-\gamma\nu(U, \sigma(U))), \tag{1.9}$$

where γ is an arbitrary, but fixed, positive real number. Note that, if G is a totally disconnected locally compact group, $N(\sigma, G)$ can be defined in a similar way, using (1.2) and without referring to Haar measures on G.

We now prove the analogue of (1.4) for the above defined $N(\sigma, S)$. Let T be a closed subspace of S, which is invariant under the automorphism σ, and let ν_1, ν_2 be the ν-functions defined in the same manner as above for the spaces S/T and T respectively. For any open, linearly compact subspaces U, V of S, we put

$$\nu'(U, V) = \nu_1(\bar{U}, \bar{V}) + \nu_2(U \cap T, V \cap T),$$

---

[9] For locally linearly compact spaces, see Lefschetz [10], p. 72. The definition of the norm of an automorphism of a locally linearly compact space, as given here, was suggested to the author by Prof. E. Artin, to whom the author wishes to express his gratitude.

where $\bar{U}$, $\bar{V}$ denote the canonical images of U, V in S/T, i.e. the spaces $U + T/T$ and $V + T/T$ respectively. It is clear that ν' satisfies the condition (1) of the ν-function for S. Moreover, if V is contained in U, $\bar{V}$ and $V \cap T$ are obviously contained in $\bar{U}$ and $U \cap T$ respectively, and we have

$$\nu'(U, V) = [\bar{U}:\bar{V}]_F + [U \cap T:V \cap T]_F$$
$$= [U + T:V + T]_F + [U \cap T:V \cap T]_F .$$

But, it is easy to see that the last sum is equal to $[U:V]_F$. Therefore, ν' also satisfies the condition (2), and, it follows from the uniqueness of the ν-function, that $\nu = \nu'$, namely that

(1.10) $$\nu(U, V) = \nu_1(\bar{U}, \bar{V}) + \nu_2(U \cap T, V \cap T).$$

Putting $V = \sigma(U)$ in the above formula, we get, immediately,

(1.11) $$N(\sigma, S) = N(\sigma, S/T)N(\sigma, T).$$

Here $N(\sigma, S/T)$ and $N(\sigma, T)$ are defined as in (1.9), using the same constant γ. In particular, if S is the direct sum of T_1 and T_2, which are both invariant under σ, we have

$$N(\sigma, S) = N(\sigma, T_1)N(\sigma, T_2).$$

On the other hand, it is clear that the analogue of (1.6) also holds, namely that

$$N(\sigma_1\sigma_2 , S) = N(\sigma_1 , S)N(\sigma_2 , S),$$

for any automorphisms σ_1, σ_2 of S.

Now, these formulae enable us to prove analogous results to Lemmas 1 and 2 for locally linearly compact rings. Namely, let R be a topological ring which contains a subfield F in its center, so that R is linearly topologized as a linear space over F and is locally linearly compact with respect to that linear topology. For any regular element ξ in R, $a \rightarrow \xi a$ again gives an automorphism σ of the additive group of R, and the norm $N(\xi, R) = N(\sigma, R)$ is defined accordingly. Then the same statement as in Lemma 1 holds also for such a locally linearly compact ring R. Since we have proved the analogous formulae to (1.4), (1.5) for the norm $N(\sigma, S)$, the main part of the proof of Lemma 1 is still valid for the present case without any change. The only point which needs a proof is the assertion that we can find $A = M_1 \cap \cdots \cap M_r$ such that $U \cap A = \sigma(U) \cap A = V \cap A$. For the proof, suppose, for instance, that $U \cap A \neq V \cap A$ for any finite intersection $A = M_1 \cap \cdots \cap M_r$. Since U/V is a finite dimensional linear space over F, this implies that there exists an element a of U outside V, such that $a + V$ intersects all $U \cap A$. Then the family of closed linear varieties $\{(a + V) \cap U \cap A\}$ forms a filter in the linearly compact space U, and the intersection of all these varieties is not empty. But this implies a contradiction, for the intersection of all A is 0 and 0 is not contained in $a + V$.

Since the proof of Lemma 2 is based in its entirety upon Lemma 1, it is clear that Lemma 2 is valid also for a locally linearly compact ring R.

We now consider a locally linearly compact field K and show that the norm

$N(\xi, K)$ of an element $\xi \neq 0$ in K is again given by a valuation of K which defines the topology of K. To do that, we first prove the following:

LEMMA 3. *Let R be a topological ring containing a subfield F in its center such that R is a linear space over F, is linearly topologized, and is locally linearly compact with respect to that linear topology. For any pair of open, linearly compact subspaces U, V of R, we can find an open, linearly compact subspace W, such that VW is contained in U.*

PROOF. By the continuity of multiplication in R, there exists an open, linearly compact subspace V_0, such that

$$V_0V_0 \subset U, \qquad V_0 \subset V.$$

Since V/V_0 is linearly compact and discrete, it is a finite dimensional linear space over F. We take $x_1, \cdots, x_n$ in V so that $V = V_0 + Fx_1 + \cdots + Fx_n$ and denote by $f_i(x)$ the residue class of x_ix in R/U for any x in R. $f_i(x)$ is obviously a linear, continuous function of x and, since R/U is discrete, $V_i = \{x; x \in R, f_i(x) = 0\}$ $(i = 1, \cdots, n)$ are open subspaces of R. It is then clear that the intersection W of all V_i $(i = 0, 1, \cdots, n)$ is an open, linearly compact subspace of R and that VW is contained in U.

It follows, immediately, from this lemma, that such a locally linearly compact ring R always contains an open, linarly compact subring S. For, if U and V are open, linearly compact subspaces of R such that $UV \subset U$, $V \subset U$, then the ring S generated by the elements of V is contained in U and, hence, is open and linearly compact.[10]

Now, let K be a field, which is a locally linearly compact ring over a subfield F in the sense described precisely in Lemma 3. We assume that K is not discrete. Let S be an open, linearly compact subring of K. Replacing S by $S + F$, if necessary, we may assume that S contains F. Since $S\,\xi \subset S$ for any ξ in S, we have

$$N(\xi, K) \leqq 1, \qquad \xi \neq 0, \xi \in S.$$

But, if ξ is sufficiently near to 0, $\xi\, S$ is also sufficiently small, by Lemma 3. Since K is not discrete, we see that there exists an element $x \neq 0$ in S, such that $N(x,K) < 1$. For such x, if η is contained in x^nS and $\eta = x^n\xi$, $\xi \in S$, we have

$$N(\eta, K) = N(x, K)^nN(\xi, K) \leqq N(x, K)^n.$$

It follows immediately that the interaction of all $x^nS (n = 1, 2, \cdots)$ is 0 and that x^nS tends to 0 as n goes to infinity.[11] This implies in particular that x is transcendental over F, for, otherwise, x^n would be contained in some finite dimensional, discrete subspace of K, and $x^n = 0$ for some integer n. Hence S contains the polynomial ring $F[x]$ of x over F, and we can see easily from $x^nS \to 0$ that the closure K_0 of the rational function field $F(x)$ in K is the power-series field $F((x))$ of x over F with its usual topology, i.e., K_0 is the completion of $F(x)$

---

[10] This is obviously the analogue of a key lemma in Kaplansky [8].

[11] Cf. the previous proof of the existence of $A = M_1 \cap \cdots \cap M_r$, such that $U \cap A = V \cap A$.

with respect to a valuation $| u |$ of $F(x)$, which is defined by the prime ideal (x) in $F[x]$. Let $\omega_1, \cdots, \omega_m$ be linearly independent elements of K over K_0. Since K_0 is a complete field, ω_i span, in K, a closed m-dimensional linear space T over K_0, which is invariant under the automorphism induced by multiplication with x. It then follows from (1.11) that

$$N(x, K) = N(x, K/T)N(x, T).$$

But, since the canonical image of S in K/T is mapped into itself by multiplication with x, we have $N(x, K/T) \leqq 1$, and, since T is an m-dimensional linear space over K_0, we have $N(x, T) = N(x, K_0)^m$, where $N(x, K_0)$ is obviously less than 1. It follows, therefore, that $N(x, K) \leqq N(x, K_0)^m$ and, consequently, that m is bounded. This shows, however, that K is a finite extension of K_0. By a well-known result in valuation theory, the valuation $| u |$ can be uniquely extended to a non-archimedean discrete valuation $| \xi |$ of K, and the topology of K induced by this valuation $| \xi |$ coincides with the given linear topology of K, for K is a finite dimensional linear space over K_0 in both topologies.[12]

Now, let O be the valuation ring of $| \xi |$ in K, i.e. the set of all ξ in K such that $| \xi | \leqq 1$, and let P be the prime ideal of O consisting of all ξ with $| \xi | < 1$. Since K is locally linearly compact, it is easy to see that O and P are both open, linearly compact subsets of K and that the residue class field $L = O/P$ is a finite extension of F. If we put $s = [L:F]$, we see immediately that

$$\nu(O, \xi O) = s\nu(\xi), \qquad \xi \neq 0,$$

where $\nu(\xi)$ denotes the normalized exponential valuation, which belongs to the discrete non-archimedean valuation $| \xi |$. It then follows from the definition of the norm $N(\xi, K)$, that

$$N(\xi, K) = \exp(-\gamma s\nu(\xi)).$$

The right-hand side of the above equality also gives a valuation of K, which is equivalent to $| \xi |$. We call it the normed valuation of K and denote it by $\| \xi \|$.

The converse of the above result is also true. Namely, if K is a complete field with respect to a non-trivial, discrete, non-archimedean valuation $| \xi |$, such that $| \xi |$ is trivial on a subfield F and such that the residue class field of $| \xi |$ is a finite extension of F, then, it is easy to see that K is a locally linearly compact field over F with respect to the linear topology defined by that valuation, and the norm $N(\xi, K)$ is given as above. We can thus summarize our results as follows.

LEMMA 4. *Let K be a field which is also a locally linearly compact ring over a subfield F, as described in Lemma 3. If K is not discrete, then there exists a discrete, non-archimedean valuation $\| \xi \|$ of K, such that it induces the given linear topology on K, it is trivial on F, the residue class field is a finite extension of F, and the norm $N(\xi, K)$ of an element $\xi \neq 0$ in K is equal to $\| \xi \|$:*

$$N(\xi, K) = \| \xi \|.$$

$\| \xi \|$ is called the normed valuation of K.

[12] See, for instance, Artin [1], p. 20.

Conversely, if K is a complete field with respect to a non-trivial, discrete, non-archimedean valuation $|\xi|$ and if the valuation is trivial on a subfield F and the residue class field of $|\xi|$ is a finite extension of F, then K is a locally linearly compact, non-discrete field over the subfield F with respect to the linear topology induced by the valuation $|\xi|$.

Note that if K is a discrete field, we have

$$N(\xi, K) = 1 = \|\xi\|, \qquad \xi \neq 0,$$

where $\|\xi\|$ denotes the trivial valuation of K.

Finally, we make some remarks concerning the proof of Lemma 4. We did not use the commutativity of the field K. Hence, we actually proved that a non-discrete, locally linearly compact division ring R over a subfield F contains a power-series field $K_0 = F((x))$ and is a finite dimensional linear space over K_0. Moreover, if the center of R is not discrete, x can be chosen from that center, and R becomes an algebra over K_0. If, in particular, R is a field K and K' is a non-discrete closed subfield of K containing F, then K is a finite extension of K'. We note also that we can prove in a similar way the well-known result on the structure of a non-discrete, totally disconnected, locally compact field K. Namely, if the characteristic of K is $p \neq 0$, K can be considered as a locally linearly compact field over the prime field of p elements and the structure of K is, therefore, given by Lemma 4. On the other hand, if K is a field of characteristic 0, we have only to note that K has an open, compact subring containing all rational integers and that we can find a prime number p such that $N(p, K) < 1$. K is then, obviously, a finite extension of the p-adic number field.[13]

§2. The structure of V-rings and V'-rings

We consider a topological ring R which satisfies following conditions:

(1) *R is a semi-simple commutative ring with the unity element* 1,

(2) *R is locally compact, but is neither compact nor discrete,*

(3) *R has a subfield K containing* 1, *such that K is discrete in R and the residue class space R/K is compact. K is called a base field of R.*

As we shall show in the following, such a ring R is nothing but the ring of valuation vectors of a number field or a function field over a finite constant field. But, for the moment, we call such R a *V-ring.*

Similarly, we define a V'-ring as follows. A topological ring R is called a *V'-ring*, when it satisfies the following conditions:

(1′) *R is a semi-simple commutative ring with the unity element* 1,

(2′) *R has a subfield F containing* 1 *and is linearly topologized as a linear space over F. It is locally linearly compact, but is neither linearly compact nor discrete,*

(3′) *R has a subfield K containing F, such that K is discrete in R and the residue class space R/K is linearly compact. K is again called a base field of R.*

We shall also prove that such a V'-ring R is the ring of valuation vectors of a function field over the constant field F. We note that, by a result of Kaplansky,

[13] Cf. Braconnier [4] and Kaplansky [7].

the intersection of all closed maximal ideals of a V-ring or a V′-ring R is 0[14] and that, in the following, we shall always use the semi-simplicity of R in this form. Therefore, this property of R might as well be used to define a V-ring or a V'-ring instead of the semi-simplicity of R, though the latter is a purely algebraic condition and seems to be simpler than the former.

We now begin the study of the structure of V-rings and V'-rings.

Lemma 5. *A V-ring (or a V'-ring) R is not a field.*

Proof. Suppose that R is a field. Since it is locally (linearly) compact, but not discrete, the topology of R is given by a non-trivial valuation $|\xi|$ of R. If $|\xi|$ is an archimedan valuation, then R is isomorphic to either the real number field or the complex number field, and no subfield of R can be discrete in R. Hence $|\xi|$ must be non-archimedean. Let O be the valuation ring of $|\xi|$ and P the prime ideal in O. As the base field K of R is discrete, the intersection of K and P is 0, and K can be considered as a subfield of the residue class field O/P. But, since O/P is (linearly) compact and is a finite field (a finite extension of F), the same must be true for K. From the (linear) compactness of R/K, it then follows that R itself is (linearly) compact, which contradicts the condition (2) of a V-ring (or (2′) of a V'-ring.)

Lemma 6. *A V-ring (or a V'-ring) R contains no (linearly) compact or open ideals other than* 0 *and* R.

Proof. Suppose that N is a (linearly) compact ideal of R. For any closed maximal ideal M of R, $M + N$ coincides with either R or M. But $M + N = R$ implies that R/M is (linearly) compact. Since the base field K of R is mapped faithfully into R/M by the canonical homomorphism from R onto R/M, K is also (linearly) compact, and this leads to a contradiction, as in the proof of Lemma 5. We see, therefore, that $M + N = M$, i.e. $N \subset M$. However, as the intersection of all M is 0, we have, immediately, $N = 0$.

Suppose, next, that N' is an open ideal of R. $N' + K$ is then a closed subgroup of R, and it follows from the isomorphism $N' + K/K = N'/N' \cap K$[15] that $N'/N' \cap K$ is (linearly) compact. Therefore, if $N' \cap K = 0$, N' is (linearly) compact and is 0 as proved above. But this contradicts the fact that R is not discrete. Hence, we have $N' \cap K \neq 0$ and, consequently, that $N' = R$.

Now, let M be a closed maximal ideal of a V-ring (or a V'-ring) R. Since M is not open, R/M is a non-discrete, locally (linearly) compact field. Let K' be the closure of the image of the base field K in R/M by the canonical homomorphism from R onto R/M. K' is a closed subfield of R/M and is not discrete, for otherwise, $M + K$ would be closed in R, and we would have, as in the proof of Lemma 6, either $M = 0$ or $M = R$. But $M = R$ is obviously excluded and $M = 0$ contradicts Lemma 5. Thus K' is a non-discrete subfield of R/K, and, consequently,

[14] Cf. Kaplansky [8], [9]. We can easily transfer Kaplansky's proof to the case of locally linearly compact rings. Note that, so far as we consider commutative rings, the proof is much easier than in the general case.

[15] Note that K is a discrete subgroup of R. In the following, we shall often use this kind of isomorphism.

the latter is a finite extension of the former.[16] Now the inverse image S of K' by the homomorphism from R onto R/M is a closed subgroup of R containing K. Hence R/S is (linearly) compact and so is the residue class space of R/M modulo K'. But, since R/M is a finite dimensional linear space over K' and since K' is not (linearly) compact, this can only happen when $R/M = K'$. We have thus proved that the canonical image of K in R/M is elverywhere dense in R/M. It then follows, in particular, that K contains an element $\xi \neq 0$ such that $N(\xi, R/M) \neq 1$ and, then, by Lemma 2, that R is the direct sum of M and a closed ideal K_M:

$$R = M + K_M .$$

K_M is obviously a locally (linearly) compact field isomorphic to R/M, and, if we denote by ϕ_M the projection from R onto K_M, the image $\phi_M(K)$ of K is an everywhere dense subfield of K_M. As we have remarked in §1, the norm, $N(a', K_M)$, of an element $a' \neq 0$ in K_M is given by the normed valuation $\| a' \|_M$ of K_M:

$$N(a', K_M) = \| a' \|_M , \qquad a' \neq 0, a' \in K_M .$$

For any element a in R, we, then, put

$$\| a \|_M = \| \phi_M(a) \|_M .$$

If a is regular in R, we see that

$$\| a \|_M = N(a, R/M),$$

and, in particular, that

$$\| \xi \|_M = N(\xi, R/M), \qquad \xi \neq 0, \xi \in K. \tag{2.1}$$

Since $\phi_M(K)$ is everywhere dense in K_M, the latter is the completion of $\phi_M(K)$ with respect to the valuation $\| a' \|_M$, and $\| \xi \|_M$ gives a valuation of the field K.

Now, let M and M' be different closed maximal ideals of R. We shall prove that the corresponding valuations $\| \xi \|_M$ and $\| \xi \|_{M'}$ of K are inequivalent. Suppose, for a moment, that $\| \xi \|_M$ and $\| \xi \|_{M'}$ are equivalent. Then, the one-to-one mapping

$$\phi_M(\xi) \to \phi_{M'}(\xi), \qquad \xi \in K,$$

from $\phi_M(K)$ onto $\phi_{M'}(K)$ is bicontinuous and can be uniquely extended to an isomorphosm ϕ between K_M and $K_{M'}$. On the other hand, since $M \neq M'$, R is the direct sum of K_M, $K_{M'}$, and $N = M \cap M'$:

$$R = K_M + K_{M'} + N.$$

Therefore, if we denote by L the set of all elements $a + \phi(a)$, $a \in K_M$, it is easy to see that L is a closed subfield of R and R is, as an additive group, the direct sum of K_M, L, and N:

$$R = K_M + L + N.$$

---

[16] Cf. the remark at the end of § 1.

But K is obviously contained in $L + N$. It then follows, from the (linear) compactness of R/K, that $K_M(\cong R/L + N)$ is (linearly) compact, and this contradicts the fact that $K_M(\cong R/M)$ is not (linearly) compact.

We now consider the norm $N(\xi, R)$ of an element $\xi \neq 0$ in K relative to the ring R. Since K is invariant under the automorphism induced by multiplication with ξ, it follows from (1.4), (1.11), that

$$N(\xi, R) = N(\xi, R/K)N(\xi, K).$$

But, as K is discrete and R/K is (linearly) compact, we can see easily, from the definition of the norm, that $N(\xi, K)$ and $N(\xi, R/K)$ are both equal to 1. It then follows from Lemma 1 and (2.1), that

$$\prod \| \xi \|_M = 1, \qquad \xi \neq 0, \xi \in K.$$

We have thus proved the following:

LEMMA 7. *Every closed maximal ideal M of a V-ring (or a V'-ring) R gives a non-trivial valuation $\| \xi \|_M$ of the base field K, and these valuations of K are mutually inequivalent. For any element $\xi \neq 0$ in K, $\| \xi \|_M = 1$ except for a finite number of M, and the product of $\| \xi \|_M$ for all closed maximal ideals of R is equal to 1:*

$$\prod \| \xi \|_M = 1, \qquad \xi \neq 0, \xi \in K \tag{2.2}$$

By Artin-Whaples [2], it then follows immediately that K is either a number field or a function field and that $\| \xi \|_M$ consists of all inequivalent valuations of K (which are trivial on F). But, since we make many more assumptions on K, we shall prove these results in another way, namely, by analyzing the structure of the containing topological ring R.

We first consider a V-ring R, whose base field K has the characteristic 0. We call a closed maximal ideal M finite or infinite according to whether the corresponding valuation $\| \xi \|_M$ of K is non-archimedean or archimedean, respectively. If M is infinite, $\| n \|_M > 1$ for any integer $n \neq 0, 1$, and if M is finite, $\| n \|_M \leqq 1$ for any integer n; and, in particular, $\| p \|_M < 1$ for some prime number p.[17] It then follows immediately from Lemma 7, that there exist both finite and infinite closed maximal ideals in R and, indeed, a finite number of infinite ones. If we denote by $M_1, \cdots, M_r$ all these infinite closed maximal ideals and put $K_i = K_{M_i}$, then R is the direct sum of $R_0 = M_1 \cap \cdots \cap M_r$ and these K_i $(i = 1, \cdots, r)$:

$$R = R_0 + R_\infty, \qquad R_\infty = K_1 + \cdots + K_r. \tag{2.3}$$

We denote by ϕ_0, ϕ_∞ the projections from R onto R_0 and R_∞, respectively.

For any infinite $M = M_i$, the corresponding K_i is isomorphic to either the real number field or the complex number field and we assume that r_1 of M_i are isomorphic to the real number field and r_2 of them to the complex number field. On the other hand, if M is finite, K_M is a locally compact field whose topology is

[17] Note that the valuation $\| \xi \|_M$ is not trivial on the rational number field contained in K.

given by the discrete, non-archimedean normed valuation $\| a' \|_M$ of K_M. We denote by O_M the valuation ring of $\| a' \|_M$ in K_M and prove that the intersection O of $\phi_M^{-1}(O_M)$ for all finite M is an open subring of R. Since K_M is clearly contained in R_0, for any finite M, and, since O is the direct sum of $O_0 = \phi_0(O)$ and R_∞, it suffices to show that O_0 is open in R_0. Now, R_0 is a totally disconnected, locally compact ring and contains an open, compact subring C. For any finite M, the projection $\phi_M(C)$ of C in K_M is also an open, compact subring of K_M and, hence, is contained in O_M. C is therefore a subring of O_0, and this proves that O_0 is open in R_0.

Putting $O_K = O \cap K$, we next prove that $\phi_\infty(O_K)$ is discrete in R_∞. Let ω be any element in O_K different from 0. We have then $\| \omega \|_M \leqq 1$ for any finite M, and it follows, from (2.2), that

$$\prod_{i=1}^{r} \| \omega \|_{M_i} \geqq 1.$$

But, since the topology of $R_\infty = K_1 + \cdots + K_r$ is induced by the valuations $\| a' \|_{M_i}$, the above inequality shows that $\phi_\infty(\omega)$ can not be arbitrarily near to 0 in R_∞. Therefore the subring $\phi_\infty(O_K)$ is discrete in R_∞.

Now, since $O + K$ is open in R, $O + K/K$ is a closed subgroup of R/K and, hence, is compact. It then follows from $O/O_K \cong O + K/K$ that O/O_K is also compact. On the other hand, since $\phi_\infty(O_K)$ is discrete in R_∞, $O_0 + O_K = O_0 + \phi_\infty(O_K)$ is a closed subgroup of O and is the direct sum of O_0 and O_K. Therefore, O_0 is isomorphic to $O_0 + O_K/O_K$ which is a closed subgroup of the compact group O/O_K, and this proves that O_0 is compact. We then make correspond to each a in O_0 the element $(\phi_M(a))$ in the direct sum $\sum' O_M$ of the compact rings O_M, where M runs over all finite closed maximal ideals in R. This mapping is obviously one-to-one and continuous, and since each K_M is a direct summand of R, the image of O_0 in $\sum' O_M$ is everythere dense in $\sum' O_M$. Therefore, using that O_0 is compact, it follows immediately that O_0 is mapped isomorphically onto $\sum' O_M$ and can be identified with $\sum' O_M$. We then take an arbitrary element a in R_0 and consider the vector $(\phi_M(a))$, where M again runs over all finite closed maximal ideals. By the continuity of multiplication in R_0, we can find a neighborhood U of 0 in R_0, such that

$$aU \subset O_0, \; U \subset O_0.$$

But, since $O_0 = \sum' O_M$, $\phi_M(U)$ coincides with O_M except for a finite number of M, and $\phi_M(a)$ is consequently contained in O_M except for a finite number of M. It then follows easily that, under the mapping $a \to (\phi_M(a))$, R_0 is isomorphically mapped onto the local direct sum of K_M relative to O_M, where M runs over all finite closed maximal ideals. Putting $O_M = K_M$ if M is infinite, we see, finally, that R is the local direct sum of all K_M relative to O_M.

Now, as we have proved before, $\phi_\infty(O_K)$ is a discrete subgroup of R_∞ and, hence, is a free abelian group of rank, say, n which does not exceed the dimension $n' = r_1 + 2r_2$ of R_∞ as a vector group over the real number field. Since O_K is mapped isomorphically onto $\phi_\infty(O_K)$ by ϕ_∞, O_K has a basis $\omega_1, \cdots, \omega_n$ such that any ω in

O_K can be written uniquely in the form $\omega = u_1\omega_1 + \cdots + u_n\omega_n$, where u_i are rational integers. Furthermore, if any ξ in K is given, we can find an integer u such that $u\xi$ is contained in O_K, for $\phi_M(\xi)$ are in O_M except for a finite number of M. It follows that the degree of K over the rational number field Q is equal to n:

$$[K:Q] = n \leqq n' = r_1 + 2r_2 .$$

On the other hand, as $\| \xi \|_{M_1}, \cdots, \| \xi \|_{M_r}$ are inequivalent archimedean valuations of K, $r_1 + 2r_2$ does not exceed the degree, n, of K over Q. We have therefore proved that $n = r_1 + 2r_2$ and that $\| \xi \|_{M_i}$ $(i = 1, \cdots, r)$ consists of all inequivalent archimedean valuations of K.

For any prime number p, let $M_1', \cdots, M_g'$ be all the finite closed maximal ideals such that $\| p \|_M < 1$, and let n_i be the degree of $K_{M_i'}$ over the p-adic number field Q_p. The absolute value $\| p \|_{M_i'}$ is then equal to p^{-n_i} and it follows from

$$\prod_{i=1}^{g} \| p \|_{M_i'} = (\prod_{i=1}^{r} \| p \|_{M_i})^{-1} = p^{-n},$$

that

$$n = \sum_{i=1}^{g} n_i.$$

By a result on the extension of non-archimedean valuations,[18] we then see that $\| \xi \|_{M_i'}$ gives all valuations of K, such that $| p | < 1$ and this shows that the set of valuations $\| \xi \|_M$ for all finite and infinite M consists of all inequivalent valuations of K. Since R is the local direct sum of all K_M relative to O_M and since K_M is obviously isomorphic to the completion of K with respect to $\| \xi \|_M$, we have thus proved that, if R is a V-ring with a base field K of characteristic 0, then K is a number field, and R is nothing but the ring of valuation vectors of K.

We now consider a V-ring R, in which a base field K has the characteristic $p \neq 0$. In this case, it is easy to see that R also satisfies the conditions (1′), (2′), (3′) of a V'-ring considered as a linear space over the finite prime field F of K. Hence, it is sufficient to study the structure of a V'-ring in general.

Suppose, therefore, that R be a V'-ring, K a base field of R, F a subfield of K which is used to define a linear topology in R. Let $x \neq 0$ be an element in K, such that $\| x \|_M > 1$ for some M, and let $M_1, \cdots, M_r$ be all the closed maximal ideals of R, such that $\| x \|_M > 1$. It is then clear that R is the direct sum of $R_0 = M_1 \cap \cdots \cap M_r$ and $R_\infty = K_1 + \cdots + K_r$, where $K_i = K_{M_i}$ is the subfield corresponding to M_i:

$$R = R_0 + R_\infty, \qquad R_\infty = K_1 + \cdots + K_r .$$

Again we denote by ϕ_0, ϕ_∞ the projections from R onto R_0 and R_∞ respectively. We say that $M_1, \cdots, M_r$ are infinite and all other closed maximal ideals are finite. Note that, for a V'-ring R, the valuation $\| a' \|_M$ of K_M is always non-archimedean. Defining O, O_K, etc. as before, we can prove, exactly in the same way as above, that O is open in R, that $\phi_\infty(O_K)$ is discrete in R_∞, that O_0 is linearly compact, and, finally, that R is the local direct sum of all K_M relative to O_M.

[18] See, for instance, Artin [1], p. 232.

Now, obviously, x is not algebraic over F, and the subfield $F(x)$ of K is the rational function field of x over the constant field F. Every valuation $\| \xi \|_{M_i}$ $(i = 1, \cdots, r)$ then induces the same valuation on $F(x)$ such that $|x| > 1$, and the completion L of $F(x)$ with respect to that valuation is the power-series field $F((x^{-1}))$ of x^{-1} over F with the usual topology of such a field. Since K_i is complete with respect to $\| a' \|_{M_i}$, the closure L_i of $\phi_{M_i}(K)$ in K_i is ismorphic to L, and K_i is a finite extension of L_i. We put

$$n_i = [K_i : L_i], \qquad i = 1, \cdots, r,$$

$$n' = n_1 + \cdots + n_r .$$

We can then consider R_∞ as an n'-dimensional linear space over L. Now, as we have noted above, $\phi_\infty(O_K)$ is a discrete subgroup of R_∞, and, if ω' is in $\phi_\infty(O_K)$, the product $f(x)\omega'$ of ω' and an arbitrary polynomial $f(x)$ is also contained in $\phi_\infty(O_K)$. It then follows easily that there exist n elements $\omega_1', \cdots, \omega_n'$ in $\phi_\infty(O_K)$, such that $n \leqq n'$ and that any ω' in $\phi_\infty(O_K)$ can be written uniquely in the form

$$\omega' = f_1(x)\omega_1' + \cdots + f_n(x)\omega_n' ,$$

where $f_i(x)$ are polynomials in $F[x]$.[19] Since ϕ_∞ is an isomorphism on O_K, we can then find n elements $\omega_1, \cdots, \omega_n$, such that any ω in O_K can be written uniquely in the form

$$\omega = f_1(x)\omega_1 + \cdots + f_n(x)\omega_n , \qquad f_i(x) \in F[x].$$

On the other hand, for any ξ in K, there exists a polynomial $f(x)$ such that $f(x)\xi$ is in O_K, and it follows that the degree of K over $F(x)$ is equal to n. But, since $\| \xi \|_{M_i}$ $(i = 1, \cdots, r)$ are extensions in K of the same valuation in $F(x)$, it follows from a result on the extension of non-archimedean valuations that $n' = n_1 + \cdots + n_r$ can not exceed the degree n of K over $F(x)$.[20] We see therefore that $n = n'$ and that $\| \xi \|_{M_i}$ $(i = 1, \cdots, r)$ gives all the inequivalent valuations of K such that $|x| > 1$. Since we can prove easily that $\| x \|_{M_i} = \exp(n_i \gamma)$, we have

$$\prod_{i=1}^{r} \| x \|_{M_i} = \exp(n\gamma),$$

where γ is the constant which is used to define the norm $N(\xi, R)$. We then take an arbitrary irreducible polynomial $p(x)$ in $F(x)$ and apply (2.2) to $\xi = p(x)$. It follows, exactly as before, that the set of valuations $\| \xi \|_M$ for all finite and infinite M gives all the inequivalent valuations of K, which are trivial on F. This proves that K is a function field over the constant field F and that R is the ring of valuation vectors of K. In particular, a V-ring R, whose base field K has the characteristic $p \neq 0$, is the ring of valuation vectors of a function field over a finite constant field of characteristic p.

Finally, we add some remarks concerning a V'-ring R. We change our previous notations somewhat and denote by O_M the valuation ring of K_M, even for an

[19] Cf. Bourbaki [3], p 62.
[20] Cf. footnote 18.

infinite ideal M. This is allowed, for the valuation $\| a' \|_M$ is always non-archimedean. We then denote by O the intersection of $\phi_M^{-1}(O_M)$ for all finite and infinite M. O is, then, the direct sum of all O_M and is a linearly compact subring of R. $F' = O \cap K$ is therefore a finite extension field of F, for it is a discrete and linearly compact subring of K over F. For any $\xi \neq 0$ in F', we have then $\| \xi \|_M \leqq 1$ for all M, and, hence, by the product formula (2.2), $\| \xi \|_M = 1$ for all M. Thus, an element $\xi \neq 0$ in K is contained in F' if and only if $\| \xi \|_M = 1$ for all M. It then follows immediately that F' is the algebraic closure of F in K, and K is also a function field over F'. Moreover, R also satisfies the conditions (1′), (2′), (3′) of a V'-ring when it is considered as a linear space over F'. Hence, if a V'-ring R is given, we may assume, if necessary, that F is algebraically closed in K.

§3. The existence of V-rings and V'-rings

In the preceding section, we have studied the structure of a V-ring or a V'-ring R and proved that a base field K of R is either a number field or a function field and that R is the ring of valuation vectors of K as defined in the introduction. In the present section, we shall prove that, for any number field or function field K, there exists a V-ring or a V'-ring R, which contains K as a base field. This will prove, in particular, that the ring of valuation vectors R of an arbitrary number field or function field is a V-ring or a V'-ring respectively, namely, that R satisfies the three conditions given at the beginning of §2.

We first consider the ring of valuation vectors R of the rational number field $K = Q$. R is, by definition, the local direct sum of all completions K_P of K relative to O_P, where P denotes an arbitrary prime divisor of K, O_P is the valuation ring of P if P is non-archimedean, and $O_P = K_P$ if P is archimedean. Since the valuation rings O_P are all compact, it is easy to see that R is a locally compact, commutative ring with the unity element 1, that R is neither compact nor discrete, and that the intersection of all closed maximal ideals of R is 0. Therefore, to show that R is a V-ring, it suffices to prove that K is a discrete subfield of R and that R/K is compact. Let U be the set of all vectors $u = (u_P)$ in R, such that $u_P \in O_P$ for all non-archimedean P and such that $| u_P | \leqq \frac{1}{2}$ for the archimedean $P = P_\infty$, where $| u |$ denotes the usual absolute value of u in the real number field K_{P_∞}. As is readily seen, U is a compact neighborhood of 0 in R, and the intersection of U and K is 0. K is therefore discrete in R. To show the compactness of R/K, take an arbitrary $a = (a_P)$ in R. For any non-archimedean $P = P_p$, which belongs to a prime number p, let

$$a_P = \sum_{m=n}^{\infty} \alpha_m^{(P)} p^m$$

be the p-adic expansion of $a_P \in K_P$, where $\alpha_m^{(P)}$ are taken from $0, 1, \cdots, p-1$. We define the principal part of a_P by

$$a'_P = \sum_{m=n}^{-1} \alpha_m^{(P)} p^m, \qquad \text{if } n < 0,$$

$$a'_P = 0, \qquad \text{if } n \geqq 0.$$

a'_P are, then, rational numbers and are 0 except for a finite number of P_p. Hence the sum, $\alpha' = \sum a'_P$, of a'_P for all non-archimedean P is a well-defined rational number. Putting

$$\alpha = \alpha' + [a_{P_\infty} - \alpha' + \tfrac{1}{2}],$$

where $[t]$ denotes the largest integer not exceeding t, we see immediately that α is in K and $a - \alpha$ is in U. This shows that $R = K + U$ and, consequently, that R/K is compact. We have thus proved that the ring of valuation vectors of the rational number field Q is a V-ring with the base field Q.

Now, let F be an arbitrary field and $K = F(x)$ the rational function field of x over the constant field F. We consider the ring of valuation vectors R of K. R is the local direct sum of all completions K_P relative to O_P, where P denotes an arbitrary prime divisor of K, which is trivial on F, and O_P is the valuation ring of P in K_P. It is easy to see that R is a locally linearly compact, commutative ring with the unity element 1, that R is neither linearly compact nor discrete, and that the intersection of all closed maximal ideals of R is 0. Then, we denote by U the set of all vectors $u = (u_P)$ in R such that u_P is contained in O_P for all P and such that, in particular, $| u_P |_P < 1$ for the prime divisor $P = P_\infty$ which is obtained from the ideal (x^{-1}) of the polynomial ring $F[x^{-1}]$. As is readily seen, U is a linearly compact neighborhood of 0 in R, and the intersection of U and K is 0. Hence K is discrete in R. Using polynomials of $F[x]$ instead of rational integers, we can see also, as above, that $R = K + U$ and that R/K is linearly compact. The ring of valuation vectors R of $K = F(x)$ is therefore a V'-ring with the base field $F(x)$. We note that if F is a finite field, then R is also a V-ring.

Now, let R be a V-ring (or a V'-ring) with a base field K (and a subfield F of K) as defined in §2, and let K' be an arbitrary finite extension field of K. We form the Kronecker product $R' = R \times K'$ over K. If $\omega_1, \cdots, \omega_n$ is a basis of K' over K, any element a' in R' can be written uniquely in the form

$$a' = a_1 \omega_1 + \cdots + a_n\omega_n, \qquad a_i \in R.$$

Therefore, the mapping which makes correspond to a' the element $(a_1, \cdots, a_n)$ of the cartesian product $R^{(n)}$ of n-copies of R, is one-to-one, and we can transfer, by this mapping, the product topology of $R^{(n)} = R \times \cdots \times R$ into R'. It is then easy to see that the topology of R', defined in that way, is independent of the choice of the basis ω_i and that R' is a topological ring with respect to that topology.

That being said, we prove the following:

LEMMA 8. *The topological ring $R' = R \times K'$ as defined above is a V-ring (or a V'-ring) with the base field K' (and with the same field F).*

PROOF. It is clear that R' is a commutative ring with the unity element and that R' is locally (linearly) compact, but is neither (linearly) compact nor discrete. Since we have

$$R' = R\omega_1 + \cdots + R\omega_n, \qquad K' = K\omega_1 + \cdots + K\omega_n$$

for a basis $\omega_1, \cdots, \omega_n$ of K' over K, and since, by definition, K is discrete in R and R/K is (linearly) compact, we see immediately that K' is discrete in R' and R'/K' is (linearly) compact. Therefore, we have only to prove that R' is semi-simple, or more explicitly, that the intersection of all closed maximal ideals of R' is 0. Let M be a closed maximal ideal of R. Using the same notation as in §2, R is the direct sum of M and K_M:

$$R = M + K_M.$$

It is then easy to see that R' is the direct sum of MK' and K_MK' and that the intersection of MK' for all closed maximal ideals of R is 0. Therefore, it suffices to show that the intersection of all closed maximal ideals of K_MK' is 0. But, since K_MK' is the Kronecker product of the fields K_M and $\phi_M(1)K'$ over $\phi_M(K)$ and, hence, is an algebra of rank n over K_M, the above assertion is equivalent to the fact that the radical A of the algebra K_MK' over K_M is 0. If K' is a separable extension of K, this can be proved easily in the usual manner without using the properties of R and R'. However, in general, we can proceed as follows. We note first that A is a closed ideal of R' and that $A^n = 0$, for A is the radical of an algebra of rank n. Take an arbitrary $\xi \neq 0$ in K' and a in A. Then, $\xi + a = \xi(1 + \xi^{-1}a)$ has an inverse in R', for $\xi^{-1}a$ is contained in A and $1 + \xi^{-1}a$ has the inverse $1 + (-\xi^{-1}a) + \cdots + (-\xi^{-1}a)^{n-1}$. Thus the norm $N(\xi + a, R')$ of $\xi + a$ in R' is defined, and, by (1.4), (1.11), we have

$$N(\xi + a, R') = N(\xi + a, R'/A)N(\xi + a, A/A^2) \cdots N(\xi + a, A^{n-1}/A^n).$$

But $\xi + a$ and ξ induce the same automorphism on R'/A. Hence we have $N(\xi + a, R'/A) = N(\xi, R'/A)$, and, similarly, $N(\xi + a, A^i/A^{i+1}) = N(\xi, A^i/A^{i+1})$ for $i = 1, \cdots, n - 1$. It then follows immediately that

$$\begin{aligned} N(\xi + a, R') &= N(\xi, R'/A)N(\xi, A/A^2) \cdots N(\xi, A^{n-1}/A^n) \\ &= N(\xi, R') \\ &= N(\xi, R'/K')N(\xi, K') \\ &= 1. \end{aligned}$$

Now, for any open, (linearly) compact neighborhood U of 0 in R', we can find a small neighborhood V of 0, so that UV is smaller U. It then follows from the definition of the norm that, if a regular element v of R' is contained in such a V, the norm $N(v, R')$ of v is less than 1. Combining this with the above result $N(\xi + a, R') = 1$, we see that

$$(K' + A) \cap V = A \cap V,$$

and this shows, in particular, that the group $K' + A$ is closed in R'. But, since R'/K' is (linearly) compact and A is isomorphic to $K' + A/K'$, it follows that A is (linearly) compact. As A is an algebra of finite rank over K_M and K_M is not (linearly) compact, this can only happen when $A = 0$. Thus the lemma is completely proved.

Note that, if R is a V'-ring over a finite field F and, hence, is also a V-ring, then $R' = R \times K'$ is again a V-ring.

Now, let K be an arbitrary number field or function field over a constant field F. K is, then, a finite extension of the rational number field Q or a rational function field $F(x)$ over F; and, by the previous lemma, the Kronecker product R of K and the ring of valuation vectors of Q or $F(x)$ is a V-ring or a V'-ring respectively, whose base field is K. But we have proved, in §2, that such an R is nothing but the ring of valuation vectors of K. Therefore, we may summarize our results of §2, §3 as follows:

A topological ring R is the ring of valuation vectors of a number field or a function field over a finite constant field, if and only if it satisfies the conditions (1), (2), (3) *of* §2, *and a linearly topologized topological ring R over a subfield F is the ring of valuation vectors of a function field over the constant field F, if and only if it satisfies the conditions* (1′), (2′), (3′) *of* §2.

From Lemma 8, it also follows that, if K is a number field or a function field and K' is a finite extension of K, then the ring of valuation vectors of K' is the Kronecker product of K' and the ring of valuation vectors of K over K.

§4. Some properties of the rings of valuation vectors

In this section, we shall study some properties of the ring of valuation vectors R of a number field or a function field K and show, by some examples, how the topologico-algebraic structure of R is related to arithmetic properties of the field K.

We first consider the case where K is a number field or a function field over a finite constant field, i.e., the case where R is a V-ring. Let χ be a non-trivial continuous character of the additive group of R, which vanishes identically on K. We make correspond to any element r in R the character $\chi_r(a) = \chi(ra)$ of R. It is easy to see that this mapping defines a representation ψ of the additive group of R into the character group $X(R)$ of R.[21] The kernel of ψ is given by

$$N = \{u;\, u \in R,\, \chi(uR) = 0\},$$

and it forms a closed ideal of R different from R. If we denote by S the closure of $N + K$ in R, S is a closed subring of R, and it follows from $\chi(K) = 0$ and $\chi(N) = 0$ that $\chi(S) = 0$. Now, those characters of R which vanish on S form a closed subgroup of $X(R)$, and we may identify that group with the character group $X(R/S)$ of R/S. It then follows, from $\chi(S) = 0$, that the image $\psi(S)$ of S by ψ is contained in $X(R/S)$. But, since S contains K and R/S is compact, $X(R/S)$ is a discrete group. Hence, the kernel N of ψ is open in S. It then follows that $N + K$ is open in S and, consequently, that $S = N + K$. N is then isomorphic to the compact group S/K and hence, is 0, by Lemma 6. This proves that the mapping ψ is one-to-one.

Now, by a reasoning similar to that above, we can see that $\psi(K)$ is contained in $X(R/K)$ and is a discrete subgroup of $X(R)$. Therefore, if we take a compact

[21] Following the terminology of Weil [11], we distinguish a representation and a homomorphism of a topological group G into another group H.

set C in R such that $R = K + C$, $\psi(C)$ is also compact, and $\psi(R) = \psi(K) + \psi(C)$ is closed in $X(R)$. On the other hand, if a is an element in R such that $\chi_r(a) = 0$ for all r in R, a is contained in N, and, consequently, $a = 0$. This means that $\psi(R)$ is everywhere dense in $X(R)$, and, since $\psi(R)$ is closed in $X(R)$, we have $\psi(R) = X(R)$. It then follows easily from $X(R) = \psi(K) + \psi(C)$, that any compact subset of $X(R)$ is contained in a set of the form $\psi(E) + \psi(C)$, where E is a suitable finite subset of K. In particular, a compact neighborhood of 0 in $X(R)$ is contained in the image of a suitable compact subset $E + C$ of R. But, as ψ is a homeomorphism on the compact set $E + C$, it follows that ψ^{-1} is also continuous. We have thus proved that the mapping ψ as defined above is an isomorphism of the additive group of R with its character group $X(R)$. We can state this self-duality of R more clearly as follows. Using the same character χ, we put

$$(a, b) = \chi(ab),$$

for any a, b in R. (a, b) then gives a dual pairing[22] of the additive group of R with itself; and, if we denote by H' the annihilator of a closed subgroup H of R by the above product (a, b), H' is (isomorphic to) the character group of R/H, H is (isomorphic to) the character group of R/H', and $(H')' = H$.

We shall now prove that the annihilator K' of K coincides with K:

$$K' = K.$$

It follows from $\chi(K) = 0$ that $K \subset K'$ and $KK' \subset K'$. As the character group of the compact group R/K, K' is discrete, and as a subgroup of R/K, K'/K is compact. Hence K'/K is a finite group. But, since K' is a K-module, we have $K' = K$. It follows, in particular, that the compact group R/K is the character group of K. For instance, if K is a number field of absolute degree n, R/K is the direct sum of n solenoids. The self-duality of R, together with $K' = K$, is essential for the validity of theta-formulae for the field K.[23]

Now, let A be an arbitrary closed ideal of R and a an element in $A \cap A'$. Since A' is obviously a closed ideal of R, aR is contained in both A and A', and we have

$$(a^2, R) = (aR, aR) = 0.$$

It then follows from $R' = 0$ that $a^2 = 0$ and, consequently, that $a = 0$, for the intersection of all closed maximal maximal ideals of R is 0. We have, therefore, $A \cap A' = 0$, and this implies that $R = 0' = (A \cap A')' = A' + A$. It is thus proved that, if A is any closed ideal of R, R is the direct sum of A and A', and A and A' are both self-dual. Using the same notation as in §2, we see, in particular, that

$$M' = K_M, \; K'_M = M; \qquad R'_0 = R_\infty, \; R'_\infty, = R_0.$$

[22] Cf. Lefschetz [10], p. 66 or Weil [11], p. 94.

[23] Cf. a forthcoming paper of J. Tate.

We note also that, unless $A = 0$, the character χ cannot vanish identically on a closed ideal A. For $\chi(A) = 0$ implies that $(A, R) = 0$, $A = R' = 0$. It then follows that, if $A \neq 0$, $K + A$ is always everywhere dense in R, for otherwise, we could take the character χ as vanishing identically on K and A. In particular, we have $\chi(K_M) \neq 0$ for any closed maximal ideal M, and $K + K_M$ is everywhere dense in R. This shows us the existence of some ξ in K, which satisfies suitable congruence and magnitude relations with respect to all but one prime divisor of K.

We now prove another proposition which also asserts the existence of some element ξ in K. Let U_1 be the compact neighborhood of 0 in R, defined by

$$U_1 = \{u;\, u \in R,\ \| u \|_M \leqq 1, \text{ for all } M\}, \tag{4.1}$$

where M and $\| u \|_M$ are as given in §2. We take a compact neighborhood V of 0 such that $V - V \subset U_1$ and assume that $aU_1 \cap K = 0$ for a regular element a in R. Then, the intersection of aV and $b + K$ contains at most one element for any b in R. We take Haar measures μ, μ' and μ'' of R, R/K and K, respectively, so that each point of K has the measure 1 by μ'' and so that the relation (1.3) holds for the corresponding integrals over R, R/K and K. If we then denote by $f(x)$ the characteristic function of aV in R and by $\bar{f}(P)$ the characteristic function of the canonical image of aV in R/K, we have

$$\mu(aV) = \int_R f(x)\, dx = \int_{R/K} dP \int_K f(b + \xi)\, d\xi = \int_{R/K} \bar{f}(P)\, dP$$

$$\leqq \int_{R/K} dP = \mu'(R/K).$$

From the definition of the norm, it follows immediately that

$$N(a, R)\, \mu(V) \leqq \mu'(R/K),$$

or

$$N(a, R) \leqq \delta, \tag{4.2}$$

where $\delta = \mu'(R/K)\, \mu(V)^{-1}$ is a positive constant independent of a. In other words, if the norm $N(a, R)$ of a regular element a in R is greater than a suitable constant $\delta > 0$, there exists an element $\xi \neq 0$ in K, such that

$$\| \xi \|_M \leqq \| a \|_M$$

for any closed maximal ideal M in R. This is obviously an analogue of Riemann's theorem in the theory of algebraic functions.

Now, let $J = J(R)$ be the set of all regular elements in R. J obviously forms a multiplicative group, and, since the mapping $r \rightarrow ar$ $(r \in R)$ defines an automorphism of the additive group of R for any a in J, J can be considered as a group of automorphisms of the additive group of R. Hence we can define a topology in J as such a group of automorphisms.[24] Namely, a system of neighbor-

[24] Cf. Braconnier [5].

hoods of 1 in J is given by those subsets $U(C, U)$ consisting of all elements a in J, such that

$$(1 - a)\, C \subset U, \qquad (1 - a^{-1})\, C \subset U.$$

Here C and U denote, respectively, arbitrary compact subsets of R and arbitrary neighborhoods of 0 in R. It is easy to see that a subset $\{a_\alpha\}$ of J converges to an element a of J in the above topology, if and only if $a_\alpha \to a$ and $a_\alpha^{-1} \to a^{-1}$ by the additive topology of R.

Let A be a closed ideal of R and let $R = A + B$ $(B = A')$ be the direct sum decomposition of R as proved above. We put

$$1 = e' + e'', \qquad e' \in A,\ e'' \in B$$

and also

$$A^* = J(A) + e'',$$

where $J(A)$ denotes the set of all regular elements of the ring A. A^* is then a closed subgroup of J and is isomorphic to $J(A)$, which is also topologized as a group of automorphisms of the additive group of A. We can define similarly for B a closed subgroup B^* of J and see that J is the direct product of A^* and B^*:

$$J = A^* \times B^*.$$

In particular, we have

$$J = M^* \times K_M^*$$

for any closed maximal ideal M in R. But K_M^* is isomorphic to the multiplicative group $J(K_M)$ of the locally compact field K_M, and the topology of $J(K_M)$ coincides with the one which is induced on $J(K_M)$ by the additive topology of K_M. Hence K_M^* is a locally compact group. Now, from the additive decomposition (2.3), we have also the multiplicative decomposition

$$J = J_0 \times J_\infty,$$

where we put

$$J_0 = R_0^*, \qquad J_\infty = R_\infty^* = K_1^* \times \cdots \times K_r^*.$$

By the above remark, J_∞ is a locally compact group. To see that J_0 is also locally compact, we consider the group $J(R_0)$ of regular elements of R_0 which is isomorphic to J_0. Using the same notation as in §2, R_0 contains an open, compact subring, O_0, and we can form a neighborhood $U(O_0, O_0)$ of the identity e in $J(R_0)$. By definition, it consists of those elements a in $J(R_0)$ for which

$$(e - a)O_0 \subset O_0, \qquad (e - a^{-1})O_0 \subset O_0.$$

But this merely means that a and a^{-1} are both contained in O_0. Hence, $U(O_0, O_0)$ is nothing but the group $J(O_0)$ of regular elements of O_0, and the topologies of $U(O_0, O_0)$ and $J(O_0)$ also coincide. As O_0 is a compact ring, the set of regular elements $J(O_0)$ is closed in O_0 by the additive topology of O_0; and the topology of $J(O_0)$ as a group of automorphisms of O_0 is the same as the induced topology on $J(O_0)$ by the additive topology of O_0. Hence $J(O_0)$ is an open, compact subgroup of $J(R_0)$, and this proves that $J(R_0)$ and J_0 are locally compact. We see, therefore, that J is also a locally compact group.

We now consider the mapping

$$a \to N(a, R)$$

from J into the multiplicative group of positive real numbers. As is readily seen, this mapping is a homomorphism of J onto the multiplicative group of positive real numbers or onto a suitable discrete subgroup of it, according to whether K is a number field or a function field over a finite constant field, respectively. Hence, if we denote the kernel of this homomorphism by

$$J_1 = \{a; a \in J, N(a, R) = 1\},$$

J/J_1 is isomorphic to either the additive group of all real numbers or that of all rational integers.

By the product formula (2.2), the multiplicative group P of all elements $\xi \neq 0$ in K is contained in J_1 and is a discrete subgroup of J_1, for the topology of P as a subgroup of J is stronger than the induced topology on P by the additive topology of R. We now prove that J_1/P is compact. Let a_0 be an element in J such that $N(a_0, R) > \delta$, where δ is a positive constant, as given in (4.2). For any a in J_1, we have $N(a^{-1}a_0, R) > \delta$. Hence, there exists an element $\xi \neq 0$ in $a^{-1} a_0 U_1 \cap K$, where U_1 is given by (4.1). Namely, we can find ξ such that

$$a\xi \in a_0 U_1, \qquad \xi \in P.$$

Since $(a\xi)^{-1}$ is contained in J_1, there also exists a ξ' such that

$$(a\xi)^{-1}\xi' \in a_0 U_1, \qquad \xi' \in P$$

It follows that $\xi' \in (a\xi)a_0 U_1$ and, consequently, that $\xi' \in a_0^2 U_1^2$. But, since $a_0^2 U_1^2$ is compact and K is discrete in R, there exists only a finite number of elements $\xi_1, \cdots, \xi_s$ in $a_0^2 U_1^2 \cap P$. Hence $(a\xi)^{-1}$ is contained in some $\xi_i^{-1} a_0 U_1$ $(1 \leqq i \leqq s)$. Therefore, if we denote by C the union of $a_0 U_1$ and all $\xi_i^{-1} a_0 U_1$ $(1 \leqq i \leqq s)$, and by C' the set of all b in J_1 such that b and b^{-1} are both contained in C, we have

$$J_1 = PC'.$$

Since C is a compact subset of R, it is easy to see that C' is a compact subset of J_1, and this implies immediately that J_1/P is compact.

Now, we put

$$E = \{a; a \in J, \| a \|_M = 1, \text{ for all } M\}.$$

It is then clear that E is a compact subgroup of J_1 and that

$$E = E_0 \times E_\infty, \qquad E_0 = E \cap J_0, \qquad E_\infty = E \cap J_\infty.$$

Here E_0 is an open, compact subgroup of J_0 corresponding to $J(O_0)$ by the natural isomorphism from J_0 onto $J(R_0)$.

We put, further,

$$J_{\infty,1} = J_\infty \cap J_1, \qquad H = (E_0 \times J_\infty) \cap J_1 = E_0 \times J_{\infty,1},$$

and consider the sequence of groups

$$J_1 \supset HP \supset EP \supset P.$$

Since E is a compact group, EP is closed in J_1, and, since $E_0 \times J_\infty$ is open in J, H and HP are open subgroups of J_1. It then follows from the compactness of J_1/P that J_1/HP and HP/EP are both compact groups. But, as HP is open in J_1 and J_1/HP is discrete, J_1/HP must be finite. Consequently, the group $J/P(E_0 \times J_\infty)$, which is easily seen to be isomorphic to J_1/HP, is a finite group, and this proves the finiteness of the ideal classes of the field K.

Now we assume that K is a number field. In that case, J_∞ is isomorphic to the direct product of r_1 copies of the multiplicative group of real numbers and r_2 copies of the multiplicative group of complex numbers. From

$$H/E \cong J_{\infty,1}/E_\infty,$$

it follows that H/E is (isomorphic to) an $(r-1)$-dimensional vector group over the real number field ($r = r_1 + r_2$). On the other hand, we see from the isomorphisms

$$HP/EP \cong H/E(H \cap P), \qquad E(H \cap P)/E = H \cap P/E \cap P,$$

that $H/E(H \cap P)$ is compact and that $E(H \cap P)/E$ is discrete. Thus, by the sequence

$$H \supset E(H \cap P) \supset E,$$

$H/E(H \cap P)$ is compact, $E(H \cap P)/E$ is discrete, and H/E is an $(r-1)$-dimensional vector group over the real number field. It follows that $E(H \cap P)/E$ is an $(r-1)$-dimensional lattice in H/E and, in particular, that it is a free abelian group with $r-1$ generators. Therefore $H \cap P/E \cap P$ also has the same structure. However, $H \cap P$ is nothing but the unit group of the number field K and $E \cap P$ is the group of all roots of 1 contained in K, for E is compact, P is discrete, and, consequently, $E \cap P$ is a finite group. We have thus proved the classical Dirichlet's unit theorem for K, namely that the unit group of K modulo the group of roots of 1 is a free abelian group with $r-1$ generators.[25] A similar reasoning holds also for the case of a function field.

We now consider the ring of valuation vectors R of a function field K over a constant field F. We assume that F is algebraically closed in K. In this case, instead of the usual characters, we consider continuous linear characters of the

[25] Using the compactness of J_1/P instead of Dirichlet's unit theorem, J. Tate gave an elegant proof for the first inequality in class field theory.

additive group of R which take values in F.[26] Let χ be such a non-trivial character of R which vanishes identically on K. We can then prove, almost in the same way as before, that $(a, b) = \chi(ab)$ gives a dual pairing of R with itself, that K coincides with its annihilator K', etc. We can also prove easily that, if the norm $N(a, R)$ of a regular element a is sufficiently large, then aO contains some field element $\xi \neq 0$. Here we denote by O, following the modified definition given at the end of §2, the compact subring of R which is the direct sum of all valuation rings O_M in K_M. But, as is well-known, this kind of result can be given more precisely by the theorem of Riemann-Roch, and we shall prove that theorem from our point of view.[27]

We call a subset A of R a divisor of K, when A is an open, linearly compact O-module in R. It is then easy to see that a divisor A can be written in the form aO, where a is a regular element in R, and, conversely, that any such aO is a divisor of K. If $A = aO$ and $B = bO$ are divisors of K, so is the product $AB = abO$. Under this multiplication, divisors of K form a group, the divisor group of the function field K. O is the unity element of that group, and the inverse A^{-1} of $A = aO$ is given by $a^{-1}O$. Using the ν-function of R as given in §2, we define the degree $n(A)$ of A by

$$n(A) = \nu(O, A) = -\nu(A, O).$$

From

$$\nu(O, abO) = \nu(O, aO) + \nu(aO, abO) = \nu(O, aO) + \nu(O, bO),$$

it follows that

$$n(AB) = n(A) + n(B), \qquad n(A^{-1}) = -n(A).$$

Now, the annihilator A' of A as defined above is also a divisor of K, for A' is linearly compact as the character group of discrete R/A, is open since R/A' is discrete as the character group of linearly compact A, and $OA' \subset A'$ follows from $OA \subset A$. The inverse of the annihilator O' of O is called a canonical divisor of K and will be denoted by D.[28] From $A = aO$, we see easily that $A' = A^{-1}O' = A^{-1}D^{-1}$.

Now, since A is linearly compact and K is discrete, $A \cap K$ is a finite dimensional linear space over F, and its annihilator is given by $(A \cap K)' = A' + K' = A' + K$. Hence $A \cap K$ and $R/A' + K$ are dual to each other and have the same finite dimension over F:

$$m(A) = [A \cap K:0]_F = [R:A' + K]_F.$$

[26] For linear characters and the duality theorem for these characters, cf. Lefschetz [10], p. 72.

[27] That the theorem of Riemann-Roch can be proved by the duality theorem for locally linearly compact spaces, was also remarked by T. Tamagawa.

[28] Note that D depends upon the character χ, which defines the product in R. If χ varies over all characters of R which vanish identically on K, the corresponding divisors D form the canonical divisor class of K.

On the other hand, if we apply (1. 10) to the groups R, K and R/K, we have

$$\begin{aligned} n(A) &= \nu(O, A) = \nu_1(O + K, A + K) + \nu_2(O \cap K, A \cap K) \\ &= \nu_1(R, A + K) - \nu_1(R, O + K) + \nu_2(O \cap K, 0) - \nu_2(A \cap K, 0) \\ &= [R:A + K]_F - [R:O + K]_F + [O \cap K:0]_F - [A \cap K:0]_F \\ &= m(A') - m(O') + m(O) - m(A). \end{aligned}$$

Since F is algebraically closed in K, we have $F = O \cap K$ and $m(O) = 1$. Therefore, if we put $m(O') = g \geqq 0$, we have

$$m(A) = -n(A) + m(A^{-1} D^{-1}) - g + 1.$$

Defining the dimension of A by

$$l(A) = m(A^{-1}),$$

we obtain the theorem of Riemann-Roch

$$l(A) = l(A^{-1} D) + n(A) - g + 1.$$

Note that $l(A^{-1} D) = m(AD^{-1}) = m((A^{-1})') = [R:A^{-1} + K]_F$ is the dimension of all linear characters of R which vanish identically on K and A^{-1}.

THE INSTITUTE FOR ADVANCED STUDY

BIBLIOGRAPHY

[1] ARTIN, E. Algebraic numbers and algebraic functions I. Lecture notes at Princeton University (1950–1951).

[2] ARTIN, E. and WHAPLES, G. *Axiomatic characterization of fields by the product formula for valuations*. Bull. Amer. Math. Soc., vol. 51 (1945), pp. 469–493.

[3] BOURBAKI, N. Éléments de Math., Topologie générale, Chap. V–VIII. Actualités sci. et indus., 1029 (1947).

[4] BRACONNIER, J. *Sur les espaces vectoriels localement compacts*. C. R. Acad. Sci. Paris, vol. 222 (1946), pp. 777–778.

[5] BRACONNIER, J. *Sur les groupes topologiques localement compacts*. J. Math. Pures Appl., vol. 27 (1948), pp. 1–85.

[6] CHEVALLEY, C. *La théorie du corps de classes*, Ann. of Math., vol. 41 (1940), pp. 394–418.

[7] KAPLANSKY, I. *Topological methods in valuation theory*. Duke Math. J., vol. 14 (1947), pp. 527–541.

[8] KAPLANSKY, I. *Locally compact rings*. Amer. J. Math., vol. 70 (1948), pp. 447–459.

[9] KAPLANSKY, I. *Locally compact rings II*. Amer. J. Math., vol. 73 (1951), pp. 20–24.

[10] LEFSCHETZ, S. Algebraic topology. Amer. Math. Soc. Colloq. Publications, vol. 27 (1942).

[11] WEIL, A. L'intégration dans les groupes topologiques et ses applications. Actualités sci. et indus., 869 (1940).

[12] WEIL, A. *Sur la théorie du corps de classes*. J. Japanese Math. Soc., vol. 3 (1951), pp 1–35.

[27] On solvable extensions of algebraic number fields

Ann. of Math., (2) 58 (1953), 548-572.

(Received January 6, 1953)

Introduction

Class field theory was, originally, a theory of abelian extensions of algebraic number fields. But it has since come to mean the theory of abelian extensions of certain kinds of fields which are not necessarily algebraic number fields, e.g. algebraic function fields with finite fields of constants. Yet the main feature of the theory still remains in that it is a theory of abelian field extensions, and it seems quite natural that considerable efforts have been made to extend class field theory so as to cover more general types of field extensions. The purpose of the present paper is also to make, we hope, some contribution in the direction of algebraic number theory. We consider, namely, solvable extensions of algebraic number fields, and our main result is stated as follows: Let k_0 be a finite algebraic number field, k the maximal abelian extension of k_0 and Σ the maximal solvable extension of k_0. Then the Galois group $G(\Sigma/k)$ of the extension Σ/k is isomorphic to some fixed totally disconnected compact group which is obtained from a free group with a countable number of free generators and is independent of the ground field k_0. Now, the factor group of the Galois group $G(\Sigma/k_0)$ of Σ/k_0 modulo $G(\Sigma/k)$ is canonically isomorphic to the Galois group $G(k/k_0)$ of k/k_0. Hence, by class field theory, it is also canonically isomorphic to the idèle class group J of k_0, and we know that the group $G(\Sigma/k_0)$ is a group extension of the group of known structure $G(\Sigma/k)$ by the group J whose structure depends only upon the ground field k_0. However, such information about $G(\Sigma/k_0)$ is of course still very far from getting a complete analogue of the present class field theory for the maximal solvable extension Σ/k_0, for it seems difficult to determine the structure of the group extension $G(\Sigma/k_0)$ of $G(\Sigma/k)$ by J and, even when that can be done somehow, it is also necessary to find a canonical description of the group $G(\Sigma/k_0)$ by means of the structures determined solely by the ground field k_0 so that we can get a generalization of the norm residue mapping in the abelian case. Nevertheless, we still think that our result might be of some interest, because there seems to exist no conjecture about the structure of the group $G(\Sigma/k_0)$, and we hope that it may give some suggestion as to further developments in the generalization of class field theory.

An outline of the paper is as follows. In the first section, we consider the so-called imbedding problem first formulated by R. Brauer [2]. The problem is, simply speaking, to imbed a given finite Galois extension K of a field k in some finite Galois extension L of k so that the Galois group $G(L/k)$ of L/k is naturally isomorphic to a given group extension of a finite group N by the Galois group $G(K/k)$ of K/k. We fix the ground field k and seek conditions on k which guarantee that every imbedding problem over k has a solution for any finite Galois extension K of k and for any finite solvable group N. An answer is given in Theorem 3 which is proved by generalizing the methods of R. Brauer [2] and E. Witt [7] and by using the cohomology theory of groups. The arguments of this section are entirely algebraic.

In the second section, we show that, if every imbedding problem over k, with solvable $G(K/k)$ and solvable N, has a solution, then the Galois group of the maximal solvable extension of k has a certain simple group-theoretical property and that this property uniquely determines the structure of the group with this property among all separable totally disconnected compact groups. On the other hand, it is proved by using arithmetic of number fields that an algebraic number field, which is an abelian extension of a finite algebraic number field and which contains all roots of unity, satisfies the sufficient conditions for the solvability of imbedding problems as stated in Theorem 3. Therefore, combining the above, the structure of the maximal solvable extension of such a field k can be completely determined and our main result then follows immediately. It is also noticed that a similar result holds for the Galois groups of the maximal solvable extensions of certain algebraic function fields.

Section I. Imbedding problems

1.1. Let k be a field and K a finite Galois extension of k. Let G be a finite group containing a normal subgroup N such that there is an isomorphic mapping φ from G/N onto the Galois group $G(K/k)$ of K/k. The problem which we are going to consider in this section is the following: *To imbed K in a finite Galois extension L of k so that there is an isomorphism ψ from G onto the Galois group $G(L/k)$ of L/k, which maps N onto the Galois group $G(L/K)$ of L/K and induces, in a natural way, the given isomorphism φ from G/N onto $G(K/k)$.* We call such a problem an *imbedding problem* $P(K/k, G/N, \varphi)$. The main purpose of the present section is to get some sufficient conditions on k in order that such a problem $P(K/k, G/N, \varphi)$ have a solution for arbitrary K, G, φ and for an arbitrary solvable normal subgroup N of G.

1.2. We shall first prove several lemmas which enable us to reduce the solution of general imbedding problems to that of simpler special cases.

LEMMA 1. *Let G be a finite group, M and N normal subgroups of G such that $M \subseteqq N$. Put $G' = G/M$, $N' = N/M$. Then, if (L', ψ') is a solution of an imbedding problem $P(K/k, G'/N', \varphi')$ and if (L, ψ) is a solution of the problem $P(L'/k, G/M, \psi')$, (L, ψ) also gives a solution of the problem $P(K/k, G/N, \varphi)$, where $\varphi = \varphi'\varphi''$ with φ'' the canonical isomorphism of G/N onto G'/N'.*

LEMMA 2. *Let $P(K/k, G/N, \varphi)$ and $P(K'/k, G'/N', \varphi')$ be imbedding problems over the same ground field k such that $K \subseteqq K'$ and such that there exists a homomorphism φ_1 of G' onto G which maps N' onto N and satisfies $\varphi\varphi_1' = \varphi_2\varphi'$, where φ_1' is the homomorphism of G'/N' onto G/N induced by φ_1 and φ_2 is the canonical homomorphism of the Galois group $G(K'/k)$ onto the Galois group $G(K/k)$. Further, let M' be the kernel of φ_1 and φ_1'' the canonical isomorphism of G onto G'/M'. Suppose that the problem $P(K'/k, G'/N', \varphi')$ has a solution (L', ψ'). Then, if we* *denote by L the subfield of L' such that $G(L'/L) = \psi'(M')$ and put $\psi = \psi''\varphi_1''$ where ψ'' is the isomorphism of G'/M' onto $G(L/k)$ induced by ψ', then (L, ψ) gives a solution of the problem $P(K/k, G/N, \varphi)$.*

The proofs of these lemmas follow immediately from the definition of imbedding problem.

LEMMA 3. *Let $P(K/k, G/N, \varphi)$ be an imbedding problem over a ground field k and let E be a finite separable extension of k. Then there is a problem $P(K'/k, G'/N', \varphi')$ such that K' contains E, N' is isomorphic to N and $P(K'/k, G'/N', \varphi')$ is related to $P(K/k, G/N, \varphi)$ as described in Lemma 2.*

PROOF. Let K' be any finite Galois extension of k containing both K and E and let G^* be the direct product of G and the Galois group $G(K'/k)$ of K'/k. We denote by φ_0, φ_2 the canonical homomorphisms from G onto G/N and from $G(K'/k)$ onto $G(K/k)$ respectively, and by G' the subgroup of G^* consisting of all (a, b) $(a \in G, b \in G(K'/k))$ such that $\varphi(\varphi_0(a)) = \varphi_2(b)$. The projections from G^* onto G and $G(K'/k)$ then induce a homomorphism φ_1 of G' onto G and an isomorphism φ' of G'/N' onto $G(K'/k)$ respectively, where N' is the normal subgroup of G' consisting of all (a, e), $a \in N$, and it is easy to see that φ_1 and φ_2 satisfy the relation $\varphi\varphi_1' = \varphi_2\varphi'$ as mentioned in the previous lemma.

The following two group-theoretical lemmas shall be also used for the reduction. Let G be a finite group containing a normal subgroup N which is elementary abelian, i.e. an abelian group of type $(p, p, \cdots, p)$, p being a prime number. We put $\Gamma = G/N$ and denote the elements of Γ by $\sigma, \tau, \cdots$. Γ operates on N by

$$u^\sigma = sus^{-1}, \qquad u \in N, \qquad \sigma \in \Gamma,$$

where s is an element in the coset σ, and we have

$$u^{\sigma\tau} = (u^\tau)^\sigma.$$

We take a basis $u_1, \cdots, u_r$ of N and put, for any σ in Γ,

$$u_j^\sigma = \prod_{i=1}^{r} u_i^{a_{ij}(\sigma)}, \qquad j = 1, \cdots, r. \tag{1}$$

The integers $a_{ij}(\sigma)$ are then uniquely determined mod p, and denoting these residue classes by $\bar{a}_{ij}(\sigma)$, we get a representation

$$\bar{A}(\sigma) = (\bar{a}_{ij}(\sigma)), \qquad i, j = 1, \cdots, r$$

of Γ in the prime field $GF(p)$ with p elements.

LEMMA 4. *Let G be a finite group as given above and let $\bar{c}(\sigma)$ be a representation of degree 1 of the group $\Gamma = G/N$ in the field $GF(p)$. Then there exists a finite group G' with the following properties:*

(i) *G' contains a normal subgroup N' which is also an elementary abelian p-group,*

(ii) *there is a homomorphism φ_1 from G' onto G which maps N' onto N and induces an isomorphism φ_1' of G'/N' onto G/N,*

(iii) *N' contains a suitable basis $u_1', \cdots, u_t'$ such that, for the representation*

$$\bar{B}'(\sigma') = \bar{c}'(\sigma')^{-1}\bar{A}'(\sigma')$$

of $\Gamma' = G'/N'$ *in* $GF(p)$, *where* $\bar{c}'(\sigma') = \bar{c}(\varphi_1'(\sigma'))$ *and* $\bar{A}'(\sigma')$ *is the representation of* Γ' *defined like* (1) *by using* u_i', *there exists a representation* $B'(\sigma')$ *of* Γ' *with integer components which gives* $\bar{B}'(\sigma')$ *when the components are reduced* mod p.

PROOF. Let $s_1, \cdots, s_m$ be a set of elements which generate the group G. We take a free group F with m free generators $x_1, \cdots, x_m$ and define a homomorphism f from F onto G by $f(x_i) = s_i$, $i = 1, \cdots, m$. Put

$$F_1 = f^{-1}(N), \qquad F_2 = f^{-1}(e), \qquad F_3 = F_1^p F_1',$$

where F_1' denotes the commutator group of F_1. Since the index $[F: F_1]$ is finite, F_1 is a free group with a finite number of generators, say, $y_1, \cdots, y_l$, and $G^* = F/F_3$ is a finite group containing an elementary abelian normal subgroup $N^* = F_1/F_3$ of order p^l, generated by the cosets u_i^* of y_i mod F_3. Moreover, since F_1/F_2 is isomorphic to N and is an elementary abelian p-group, F_3 is contained in F_2 and $G \cong F/F_2$ is a homomorphic image of $G^* = F/F_3$. On the other hand, F_1/F_1' is a free abelian group of rank l generated by the cosets of y_i mod F_1'. Therefore, for any x in F, we have

$$xy_jx^{-1} \equiv \prod_{i=1}^{l} y_i^{a_{ij}^*(x)} \qquad \text{mod } F_1',$$

and here the integers $a_{ij}^*(x)$ are uniquely determined by the coset σ^* of x mod F_1. Putting $a_{ij}^*(\sigma^*) = a_{ij}^*(x)$, we can see immediately that the group G^* and its normal subgroup N^* satisfy the conditions (i), (ii) of the lemma and that the corresponding modular representation $\bar{A}^*(\sigma^*)$ of $\Gamma^* = G^*/N^* = F/F_1$ with respect to u_i^* comes from the integer representation $A^*(\sigma^*) = (a_{ij}^*(\sigma^*))$ of Γ^*.

Now, let H be the direct product of $p - 2$ copies of N^*. For $p = 2$, this shall mean $H = e$. We take a group G' containing both G^* and H so that H is normal in G' and $G^*H = G'$, $G^* \cap H = e$. Such a group G' always exists and is uniquely determined if we define only the transforms of elements of H by elements in G^*. Here we put, for any s^* in G^* and any $a = (a_1, \cdots, a_{p-2})$ in $H (a_i \in N)$,

$$s^*as^{*-1} = (b_1, \cdots, b_{p-2}), \tag{2}$$

where b_i is the $\bar{c}(\sigma)^{-i}$th power of $s^*a_is^{*-1}$, σ being the element in Γ which corresponds to the coset σ^* of s^* in $\Gamma^* = G^*/N^*$. It is then clear that $N' = N^*H$ is an elementary abelian normal subgroup of G' and that there is a homomorphism φ_1 from G' onto G which induces an isomorphism φ_1' of $\Gamma' = G'/N'$ onto $\Gamma = G/N$.

Now, from definition (2), it follows that N' is the direct product of $p - 1$ copies of N^* and that, by a suitable choice of a basis of N', the corresponding matrix $\bar{A}'(\sigma')$ for G', N' is the Kronecker product of $\bar{A}^*(\sigma') = \bar{A}^*(\sigma^*)$ and a matrix of degree of $p - 1$ $\bar{P}'(\sigma') = (\delta_{ij}\bar{c}'(\sigma')^{1-i})$, where σ^* is the element of Γ^* corresponding to σ' in Γ' and $\bar{c}'(\sigma') = \bar{c}(\varphi_1'(\sigma'))$:

$$\bar{A}'(\sigma') = \bar{A}^*(\sigma') \otimes \bar{P}(\sigma').$$

We have then

$$\bar{B}'(\sigma') = \bar{c}'(\sigma')^{-1}\bar{A}'(\sigma') = \bar{A}^*(\sigma') \otimes \bar{c}'(\sigma')^{-1}\bar{P}(\sigma').$$

But it is easy to see that, if c is a primitive root mod p, there is an integer matrix Q of degree $p-1$ such that Q is congruent to $(\delta_{ij}c^i)$ mod p and that Q^{p-1} is the unit matrix. Therefore, there is a representation $Q(\sigma')$ of Γ' with integer components which gives the representation $\bar{c}'(\sigma')^{-1}\bar{P}(\sigma') = (\delta_{ij}\bar{c}'(\sigma')^{-i})$ when reduced mod p. Putting $A^*(\sigma') = A^*(\sigma^*)$ as above, we have then a representation

$$B'(\sigma') = A^*(\sigma') \otimes Q(\sigma')$$

of Γ' with integer components and the modular representation $\bar{B}'(\sigma')$ is obtained by reducing $B'(\sigma')$ mod p. This completes the proof of our lemma.

Let, again, G be a finite group containing a normal subgroup N which is an elementary abelian p-group. We put $\Gamma = G/N$ and denote by R the group ring of Γ over the finite field $GF(p)$. By left-multiplication in R, Γ operates on the additive group of R. Now, for any integer $n \geqq 1$, let G_n denote a finite group with the following properties: G_n has an abelian normal subgroup N_n such that G_n/N_n is isomorphic to Γ and that N_n is, considered as a Γ-group by that isomorphism, Γ-isomorphic with the direct sum of n-copies of the additive group of R. Such a group G_n always exists for any given integer $n \geqq 1$ and is uniquely determined up to isomorphism, for the extension G_n over N_n always splits.[1]

LEMMA 5. *Let G, N, Γ, G_n and N_n be as above. Suppose that the extension G of N by Γ splits. Then, for a suitable integer n, there is a homomorphism φ_1 from G_n onto G which maps N_n onto N and induces an isomorphism φ_1' from G_n/N_n onto G/N.*

PROOF. We may assume that G contains Γ so that $\Gamma N = G$, $\Gamma \cap N = e$. Similarly we have $\Gamma N_n = G_n$, $\Gamma \cap N_n = e$. Let $w_1, \cdots, w_n$ be a set of elements in N such that N is generated by the w_i and their conjugates, and let $V_1, \cdots, V_n$ be subgroups of N_n such that the V_i are Γ-isomorphic to the additive group of R and N_n is the direct sum of all these V_i's. Take an element v_i in V_i which corresponds to the unity element e in the group ring R. It is then easy to see that there exists a unique homomorphism φ_1 from G_n onto G such that $\varphi_1(\sigma) = \sigma$ and $\varphi_1(v_i^\sigma) = w_i^\sigma$ for any σ in Γ. This φ_1 satisfies the conditions of the lemma.

1.3. From now on up to the end of 1.8, we shall fix a ground field k and a prime number p which is different from the characteristic of k, and consider imbedding problems $P(K/k, G/N, \varphi)$ where K, G and φ are entirely arbitrary and N is an arbitrary p-group.

Suppose that such a problem $P(K/k, G/N, \varphi)$ is given. We can then find a sequence of normal subgroups of G:

$$N = M_1 \supset M_2 \supset \cdots \supset M_n = e$$

such that each factor group M_{i-1}/M_i is an elementary abelian p-group. By Lemma 1, we know immediately that the problem $P(K/k, G/N, \varphi)$ can be reduced to the case where N is elementary abelian. Moreover, by Lemmas 2, 3, we may also assume that K contains a primitive p^{th} root of unity.

[1] Cf. Artin and Tate [1].

Let, therefore, $P(K/k, G/N, \varphi)$ be a problem such that N is an elementary abelian group of order p^r and K contains a primitive p^{th} root of unity ζ. As before, we choose a basis $u_1, \cdots, u_r$ of N and define a modular representation $\bar{A}(\sigma)$ of $\Gamma = G/N$ by (1). We also denote by μ^σ the image of an element μ in K by the automorphism $\varphi(\sigma)$ in $G(K/k)$. Γ thus becomes an operator domain of the additive group and of the multiplicative group of the field K. In particular, we have

$$\zeta^\sigma = \zeta^{c(\sigma)}, \qquad \sigma \in \Gamma, \tag{3}$$

with a suitable integer $c(\sigma)$ which is uniquely determined mod p. The residue classes $\bar{c}(\sigma)$ of $c(\sigma)$ mod p then obviously define a representation of Γ in $GF(p)$ and, applying Lemma 4, we can see immediately that there is a problem $P(K/k, G'/N', \varphi')$ which is related to the problem $P(K/k, G/N, \varphi)$ as stated in Lemma 2 and for which the representation $\bar{B}(\sigma') = \bar{c}'(\sigma')^{-1}\bar{A}'(\sigma')$ comes from an integer representation $B'(\sigma')$ of $\Gamma' = G'/N'$.

We have thus reduced a general problem $P(K/k, G/N, \varphi)$ with an arbitrary p-group N to a special type of problem $P(K/k, G'/N', \varphi')$ as described above.

1.4. According to the above reduction, we now consider an imbedding problem $P(K/k, G/N, \varphi)$ with the following properties:

(i) N is an elementary abelian group of order p^r,

(ii) K contains a primitive p^{th} root of unity ζ,

(iii) there exists a representation $B(\sigma) = (b_{ij}(\sigma))$ of $\Gamma = G/N$ with integer components $b_{ij}(\sigma)$, which gives the modular representation $\bar{B}(\sigma) = \bar{c}(\sigma)^{-1}\bar{A}(\sigma)$ of Γ when $b_{ij}(\sigma)$ are reduced mod p. Here $\bar{A}(\sigma)$ and $\bar{c}(\sigma)$ are representations of Γ defined by (1) and (3) respectively, using a suitable basis $u_1, \cdots, u_r$ of N.

Let V_K denote the direct product of r copies of the multiplicative group of K. For any σ in Γ and $\xi = (\xi_1, \cdots, \xi_r)$ in $V_K(\xi_i \in K, \xi_i \neq 0)$, we define ξ^σ to be an element of V_K whose i^{th} component is $\prod_{j=1}^r \xi_j^{b_{ij}(\sigma)\sigma}$:

$$\xi^\sigma = (\cdots, \textstyle\prod_{j=1}^r \xi_j^{b_{ij}(\sigma)\sigma}, \cdots). \tag{4}$$

We have then

$$\xi^{\sigma\tau} = (\xi^\tau)^\sigma, \qquad \xi \in V_K, \qquad \sigma, \tau \in \Gamma,$$

and Γ becomes an operator domain of the abelian group V_K. On the other hand, since $u_1, \cdots, u_r$ is a basis of N, there exists a dual basis $\chi_1, \cdots, \chi_r$ of the character group of N such that

$$\chi_i(u_i) = \zeta, \qquad \chi_i(u_j) = 1, \qquad i \neq j.$$

For any u in N and σ in Γ, it then follows that

$$\chi_i(u^\sigma) = \textstyle\prod_{j=1}^r \chi_j(u)^{b_{ij}(\sigma)\sigma}, \qquad i = 1, \cdots, r$$

for both sides are equal to $\zeta^{a_{ij}(\sigma)}$ for $u = u_j$. Defining $\chi(u)$ in V_K by

$$\chi(u) = (\chi_1(u), \cdots, \chi_r(u)),$$

we have then

$$(5)\qquad \chi(uv) = \chi(u)\chi(v), \qquad u, v \in N,$$
$$\chi(u^{\sigma}) = \chi(u)^{\sigma}, \qquad u \in N, \sigma \in \Gamma.$$

We now take a system of representatives x_{σ} of $\Gamma = G/N$ in G and put

$$(6)\qquad x_{\sigma}x_{\tau} = u_{\sigma,\tau}x_{\sigma\tau}, \qquad u_{\sigma,\tau} \in N, \sigma, \tau \in \Gamma,$$
$$\chi_{\sigma,\tau} = \chi(u_{\sigma,\tau}).$$

The factor set $u_{\sigma,\tau}$ then obviously satisfies the relations

$$u_{\sigma,\tau}\, u_{\sigma\tau,\rho} = u_{\tau,\rho}^{\sigma}\, u_{\sigma,\tau\rho}, \qquad \sigma, \tau, \rho \in \Gamma.$$

and these relations, together with (5), imply that

$$\chi_{\sigma,\tau}\chi_{\sigma\tau,\rho} = \chi_{\tau,\rho}^{\sigma}\,\chi_{\sigma,\tau\rho}.$$

This shows that $\chi_{\sigma,\tau}$ is a factor set, or, in another terminology, a 2-cocycle of Γ operating on the group V_K.

Now, suppose that our problem $P(K/k, G/N, \varphi)$ has a solution (L, ψ). Similarly, as above, for any λ in L and s in G, we denote by λ^s the image of λ by the automorphism $\psi(s)$ in $G(L/k)$. We also denote by V_L the direct product of r copies of the multiplicative group of L and define η^s, for any $\eta = (\eta_1, \cdots, \eta_r)$ in V_L and s in G, by

$$\eta^s = (\cdots, \textstyle\prod_{j=1}^{r} \eta_j^{b_{ij}(\sigma)s}, \cdots),$$

where σ is the coset of s in G/N. G then becomes an operator domain of V_L. In a natural way, V_K can be considered as a subgroup of V_L and we have

$$\xi^s = \xi^{\sigma}, \qquad \xi \in V_K, s \in G, \sigma = sN \in \Gamma.$$

We also note that $\eta^u = (\eta_1^u, \cdots, \eta_r^u)$ for any u in N and, consequently that η is contained in the subgroup V_K if and only if $\eta^u = \eta$ for every u in N.

Now, since K contains a primitive p^{th} root of unity and L/K is an abelian extension of exponent p, we know, by Kummer's theory, that there exist r elements $\beta_1, \cdots, \beta_r$ in L such that $L = K(\beta_1, \cdots, \beta_r)$ and that

$$\beta_i^{u-1} = \chi_i(u), \qquad u \in N, i = 1, \cdots, r.$$

If we take the element $\beta = (\beta_1, \cdots, \beta_r)$ in V_L, this can be simply written as

$$(7)\qquad \beta^{u-1} = \chi(u), \qquad u \in N.$$

For any s in G, we have then, by (5),

$$(\beta^{s-1})^{u-1} = \beta^{(u-1)(s-1)} = (\beta^{s^{-1}us-1})^s(\beta^{u-1})^{-1}$$
$$= \chi(s^{-1}us)^s\chi(u)^{-1}$$
$$= 1.$$

Since u can be chosen arbitrarily in N, this shows that β^{s-1} is contained in V_K for any s in G and we put, in particular,

$$\gamma_\sigma = \beta^{x_\sigma - 1}, \qquad \gamma_\sigma \in V_K, \qquad \sigma \in \Gamma. \tag{8}$$

$\gamma_\sigma \gamma_\tau^\sigma \gamma_{\sigma\tau}^{-1}$ is then a power of β whose exponent is

$$\begin{aligned}(x_\sigma - 1) + x_\sigma(x_\tau - 1) + (x_{\sigma\tau} - 1) &= (u_{\sigma,\tau} - 1)x_{\sigma\tau} \\ &= (u_{\sigma,\tau} - 1)(x_{\sigma\tau} - 1) + (u_{\sigma,\tau} - 1).\end{aligned}$$

By (6), (7) and $\beta^{(u-1)(s-1)} = 1$, we have

$$\gamma_\sigma \gamma_\tau^\sigma \gamma_{\sigma\tau}^{-1} = \chi_{\sigma,\tau}, \tag{9}$$

showing that the 2-cocycle $\chi_{\sigma,\tau}$ of Γ in V_K is the coboundary of the 1-cochain γ_σ. Furthermore, since the components of $\chi_{\sigma,\tau}$ are p^{th} roots of unity, we have $\chi_{\sigma,\tau}^p = 1$ and (9) then implies

$$\delta_\sigma \delta_\tau^\sigma \delta_{\sigma\tau}^{-1} = 1, \qquad \delta_\sigma = \gamma_\sigma^p, \qquad \sigma \in \Gamma,$$

which shows that δ_σ is a 1-cocycle of Γ in V_K. On the other hand, if we put

$$\alpha = \beta^p = (\alpha_1, \cdots, \alpha_r), \qquad \alpha_i = \beta_i^p, \qquad i = 1, \cdots, r,$$

it follows from (7) that $\alpha^{u-1} = 1$, $u \in N$. Hence α is contained in V_K and, by (8), we have

$$\delta_\sigma = \alpha^{\sigma-1}, \qquad \sigma \in \Gamma.$$

The 1-cocycle δ_σ is, therefore, the coboundary of the 0-cochain α. Moreover, since $\beta_1, \cdots, \beta_r$ generate L over K, the components $\alpha_1, \cdots, \alpha_r$ of α are p-independent in the multiplicative group K_m of K mod K_m^p.

We have thus proved that, if our problem $P(K/k, G/N, \varphi)$ has a solution (L, ψ), the 2-cocycle $\chi_{\sigma,\tau}$ is the coboundary of a 1-cochain γ_σ, such that the 1-cocycle $\delta_\sigma = \gamma_\sigma^p$ is the coboundary of a 0-cochain $\alpha = (\alpha_1, \cdots, \alpha_r)$ whose components α_i are p-independent in K_m mod K_m^p.

We shall next show that the converse of the above statement is also true. Suppose, therefore, that cochains γ_σ and α as described above exist. Since $\alpha^{\sigma-1} = \delta_\sigma = \gamma_\sigma^p$, the subgroup of K_m generated by α_i and K_m^p is invariant under the automorphisms $\varphi(\sigma)$ of $G(K/k)$ and it contains K_m^p as a subgroup of index p^r. Therefore the field L, obtained by adjoining p^{th} roots β_i of α_i to K, is a Galois extension over k and the Galois group $G(L/K)$ of L/K is isomorphic to N. We then define a mapping ψ from G into the Galois group $G(L/k)$ as follows: For any u in N, we define $\psi(u)$ to be an automorphism of $G(L/K)$ such that the image of β by $\psi(u)$ is $\chi(u)\beta$, i.e. that (7) holds. This is possible by Kummer's theory and it defines an isomorphism from N onto $G(L/K)$. Then, we define $\psi(x_\sigma)$ to be an automorphism of L/k such that $\psi(x_\sigma)$ coincides with $\varphi(\sigma)$ on K and the image of β by $\psi(x_\sigma)$ is $\gamma_\sigma\beta$, i.e. that (8) holds. This is also permissible because the only condition which should be satisfied by an extension $\psi(x_\sigma)$ of

$\varphi(\sigma)$ in $G(L/k)$ is the validity of the relation $(\beta^{x_\sigma})^p = \alpha^\sigma$. From (5), (6), (9), it then follows that $\psi(x_\sigma)\psi(u)\psi(x_\sigma)^{-1} = \psi(u^\sigma)$, $\psi(x_\sigma)\psi(x_\tau) = \psi(u_{\sigma,\tau})\psi(x_{\sigma\tau})$. Putting $\psi(ux_\sigma) = \psi(u)\psi(x_\sigma)$ for any u in N and σ in Γ, we can see immediately that ψ is an isomorphism of G onto $G(L/k)$ and that (L, ψ) gives a solution of the problem $P(K/k, G/N, \varphi)$.

We have thus proved the following:

LEMMA 6. *Let $P(K/k, G/N, \varphi)$ be an imbedding problem satisfying the three conditions given at the beginning of* 1.4, *let V_K be a Γ-group as defined above and let $\chi_{\sigma,\tau}$ be a 2-cocycle of Γ in V_K defined by* (6). *Then, in order that the problem $P(K/k, G/N, \varphi)$ have a solution (L, ψ), it is necessary and sufficient that $\chi_{\sigma,\tau}$ is the coboundary of a 1-cochain γ_σ of Γ in V_K, such that the 1-cocycle $\delta_\sigma = \gamma_\sigma^p$ is the coboundary of a 0-cochain $\alpha = (\alpha_1, \cdots, \alpha_r)$ of Γ in V_K whose components α_i are p-independent in the multiplicative group K_m* mod K_m^p.

1.5. We first apply Lemma 6 to the following case: As in Lemma 5, let G_n denote a finite group such that G_n contains a normal subgroup N_n which is Γ-isomorphic to the direct sum of n copies of the additive group of the group ring R of $\Gamma = G_n/N_n$ over $GF(p)$, and let us consider a problem $P(K/k, G_n/N_n, \varphi)$ where K contains a primitive p^{th} root of unity ζ. The problem $P(K/k, G_n/N_n, \varphi)$ then satisfies the conditions (i), (ii) of 1.4. To see that (iii) is also satisfied, we have only to take a basis u_i of N_n which corresponds to the elements $\bar{c}(\sigma)^{-1}\sigma$ in R. The representation $\bar{B}(\sigma)$ then comes from the sum $B(\sigma)$ of n regular representations of the group Γ. We can therefore apply Lemma 6 to the present problem $P(K/k, G_n/N_n, \varphi)$. Since the extension G_n over N_n splits, we may take x_σ so that $u_{\sigma,\tau} = 1$ and it follows that

$$\chi_{\sigma,\tau} = 1, \qquad \sigma, \tau \in \Gamma.$$

Suppose, first, that the problem has a solution and take cochains γ_σ and $\alpha = (\alpha_1, \cdots, \alpha_r)$ as given in the lemma. Since $\alpha^{\sigma-1} = \delta_\sigma = \gamma_\sigma^p$ and $B(\sigma)$ is the sum of n regular representations of Γ, we can see immediately from definition (4) that there are n elements in $\alpha_1, \cdots, \alpha_r$, say $\alpha_1, \cdots, \alpha_n$ $(n \leqq r)$, such that the set $\{\alpha_1, \cdots, \alpha_r\}$ is congruent to the set $\{\alpha_i^\sigma;\ 1 \leqq i \leqq n,\ \sigma \in G(K/k)\}$ modulo the group K_m^p of pth powers of the multiplicative group K_m of K.[2] Consequently, the elements α_i^σ, $1 \leqq i \leqq n$, $\sigma \in G(K/k)$, are p-independent in K_m mod K_m^p.

Conversely, suppose that K_m contains n elements $\alpha_1, \cdots, \alpha_n$ such that α_i^σ, $1 \leqq i \leqq n$, $\sigma \in G(K/k)$, are p-independent in K_m mod K_m^p. If we then take an element α in V_K whose components are equal to α_i^σ in a suitable order, we have $\alpha^{\sigma-1} = 1$. Putting $\gamma_\sigma = \delta_\sigma = 1$, we see that γ_σ and α satisfy the relations in Lemma 6 and, consequently, that the problem $P(K/k, G_n/N_n, \varphi)$ has a solution. Therefore the following lemma is proved:

LEMMA 7. *Let K be a finite Galois extension containing a primitive p^{th} root of unity and let G_n and N_n be finite groups as given above. Then, a problem*

---

[2] For simplicity, we change the notation here and denote by σ an element of $G(K/k)$ instead of an element in Γ. This will be done also sometimes in the following.

$P(K/k, G_n/N_n, \varphi)$ has a solution if and only if the multiplicative group K_m of K contains n elements $\alpha_1, \cdots, \alpha_n$ such that α_i^σ, $1 \leqq i \leqq n$, $\sigma \in G(K/k)$, are p-independent in K_m mod K_m^p.

We now consider a problem $P(K/k, G/N, \varphi)$ where N is an elementary abelian group of order p^r and the extension G over N splits. Let $K' = K(\zeta)$ be the field obtained by adjoining a primitive p^{th} root of unity ζ to K. Then, by Lemma 3, the problem $P(K/k, G/N, \varphi)$ can be reduced to a problem $P(K'/k, G'/N', \varphi')$ where N' is isomorphic to N and, as can be seen from the proof of Lemma 3, the extension G' over N' also splits. Therefore, using Lemma 5, this problem $P(K'/k, G'/N', \varphi')$ can be again reduced to a problem $P(K'/k, G_n/N_n, \varphi'')$ where n is a sufficiently large integer and G_n and N_n are defined as before with respect to the group $\Gamma = G'/N'$. Since N' is a group of order p^r, it is sufficient to take $n = r$. From Lemma 7, the following result holds:

LEMMA 8. *Let $P(K/k, G/N, \varphi)$ be a problem such that N is an elementary abelian group of order p^r and that the extension G over N splits. In order that $P(K/k, G/N, \varphi)$ have a solution, it is sufficient that the multiplicative group K'_m of the field $K' = K(\zeta)$ contains r elements $\alpha_1, \cdots, \alpha_r$ such that α_i^σ, $1 \leqq i \leqq r$, $\sigma \in G(K'/k)$, are p-independent in K'_m* mod $K_m'^p$. *Here ζ denotes a primitive p^{th} root of unity.*

1.6. We shall now give another application of Lemma 6. Let $P(K/k, G/N, \varphi)$ be a problem such that k contains a primitive p^{th} root of unity ζ and that G is a p-group. The normal subgroup N of G is of course also a p-group, but we do not assume that N is elementary abelian. Since G is a p-group, there is a sequence of normal subgroups of G:

$$N = M_0 \supset M_1 \supset \cdots \supset M_n = e,$$

such that every factor group M_{i-1}/M_i is a group of order p. By Lemma 1, this reduces our problem to the case where N is a normal subgroup of order p.

Suppose, therefore, that N is a group of order p. It is then contained in the center of G and the representation $\bar{A}(\sigma)$ of $\Gamma = G/N$ defined by (1) is the trivial representation of degree 1. On the other hand, since ζ is contained in the ground field k, the representation $\bar{c}(\sigma)$ of Γ defined by (3) is also trivial. Hence $\bar{B}(\sigma) = \bar{c}(\sigma)^{-1}\bar{A}(\sigma)$ is again trivial and it is obviously obtained from the non-modular trivial representation $B(\sigma)$ of degree 1. The problem $P(K/k, G/N, \varphi)$ therefore satisfies the three conditions (i), (ii), (iii) of 1.4 and we can apply Lemma 6 to that problem.

In the present case, since $r = 1$ in the notation in 1.4, we can identify the Γ-group V_K with the multiplicative group K_m of K, on which the Galois group $G(K/k)$ acts as an operator domain. In general, for any finite Galois extension K of a field k, the cohomology groups of the Galois group $G(K/k)$ of K/k operating on the multiplicative group of K are called Galois cohomology groups of K/k, and their cochains and cocycles are called Galois cochains and Galois cocycles of K/k respectively. In that terminology, the 2-cocycle $\chi_{\sigma,\tau}$ of our problem $P(K/k, G/N, \varphi)$, as defined by (6), is a Galois 2-cocycle of K/k and, by Lemma 6, a necessary and sufficient condition for the existence of a solution of $P(K/k, G/N, \varphi)$ is that $\chi_{\sigma,\tau}$ is the coboundary of a Galois 1-cochain γ_σ such that $\gamma_\sigma^p = \alpha^{\sigma-1}$ for a suitable element α in K_m, not contained in K_m^p.

Now, let us assume only that $\chi_{\sigma,\tau}$ is the coboundary of a Galois 1-cochain γ_σ and see whether this condition is already sufficient for the existence of a solution of $P(K/k, G/N, \varphi)$. Since the values of $\chi_{\sigma,\tau}$ are p^{th} roots of unity, $\delta_\sigma = \gamma_\sigma^p$ is a Galois 1-cocycle. But, as is well-known, every Galois 1-cocycle is trivial and there exists an element α in K_m such that $\delta_\sigma = \alpha^{\sigma-1}$. If this α is not contained in K_m^p, then the problem $P(K/k, G/N, \varphi)$ has a solution. Suppose, therefore, that α is contained in K_m^p and put $\alpha = \beta^p$, $\beta \in K_m$. Then $\gamma'_\sigma = \gamma_\sigma \beta^{1-\sigma}$ is a Galois 1-cochain of K/k whose values are p^{th} roots of unity and $\chi_{\sigma,\tau}$ is again the coboundary of γ'_σ. Now, the group $\Gamma = G/N$ operates trivially on N and so does the Galois group $G(K/k)$ on the group W of pth roots of unity in K. Therefore, if we identify Γ with $G(K/k)$ using the isomorphism φ, N and W can be considered as operator isomorphic groups. Moreover, between N and W an isomorphism can be chosen so that $u_{\sigma,\tau}$ correspond to $\chi_{\sigma,\tau} = \chi(u_{\sigma,\tau})$, $\sigma, \tau \in \Gamma$. If we then denote by v_σ the elements of N which correspond to γ'_σ in W, we have $u_{\sigma,\tau} = v_\sigma v_\tau^\sigma v_{\sigma\tau}^{-1}$, and this shows that the extension G over N splits, namely, that G is the direct product of N and a subgroup H which is isomorphic to Γ. We have thus proved that, if G is not the direct product of N and a subgroup H, the triviality of the 2-cocycle $\chi_{\sigma,\tau}$ is sufficient for the existence of a solution of $P(K/k, G/N, \varphi)$.[3] On the other hand, if G is the direct product of N and H, our problem reduces to the problem of constructing a cyclic extension Z of degree p over k such that $K \cap Z = k$.

To state the following theorem simply, we first give a definition:

DEFINITION. *A field k is called p-trivial, when the second Galois cohomology group of K/k is trivial for any Galois extension K over k whose Galois group $G(K/k)$ is a p-group.*

THEOREM 1. *Let p be a prime number and k a field containing a primitive p^{th} root of unity. In order that every imbedding problem $P(K/k, G/N, \varphi)$ has a solution for an arbitrary p-group G, it is necessary and sufficient that k have the following properties:*

(i) *k is p-trivial,*

(ii) *the factor group k_m/k_m^p of the multiplicative group k_m of k modulo its pth power k_m^p is an infinite group.*

PROOF. We shall first prove that the above conditions are sufficient. Let $P(K/k, G/N, \varphi)$ be a problem such that G is a p-group. By the reduction given at the beginning of this section, we have only to consider the case where N is a group of order p. Then, as we have seen above, if the group G is not the direct product of N and a subgroup of G, the problem $P(K/k, G/N, \varphi)$ has a solution, because k is p-trivial and there is a Galois 1-cochain γ_σ whose coboundary is $\chi_{\sigma,\tau}$. On the other hand, if G is the direct product of N and a subgroup of G, the problem is equivalent to that of constructing a cyclic extension Z over k·

---

[3] Cf. Brauer [3].

such that $K \cap Z = k$. The existence of such Z is then assured by Kummer's theory, for the group k_m/k_m^p is infinite. This proves that the conditions (i), (ii) are sufficient for the existence of a solution of every problem $P(K/k, G/N, \varphi)$ with a p-group G.

Suppose, next, that our problem always has a solution. For an arbitrarily large integer n, let N denote an elementary abelian group of order p^n. Put $K = k$, $G = N$ and consider the problem $P(K/k, G/N, \varphi)$ where φ is the trivial mapping from G/N onto $G(K/k)$. Since such a problem has a solution (L, ψ), it follows from Kummer's theory that the order of k_m/k_m^p is at least p^n. This proves that k_m/k_m^p is an infinite group. Now, let K be a p-extension over k, i.e. a Galois extension of k whose Galois group $G(K/k)$ is a p-group, and let $\xi_{\sigma,\tau}$ $(\sigma, \tau \in G(K/k))$ be a Galois 2-cocycle of K/k. We have to show that $\xi_{\sigma,\tau}$ is trivial. Since the order of every such cocycle is a power of p, we may prove it under the assumption that $\xi_{\sigma,\tau}^p$ is trivial, i.e. that $\xi_{\sigma,\tau}^p = \eta_\sigma \eta_\tau^\sigma \eta_{\sigma\tau}^{-1}$ for a suitable 1-cochain η_σ. As k contains a primitive p^{th} root of unity, there exists a p-extension K' of k which contains K and p^{th} roots ω_σ of η_σ for all σ in $G(K/k)$. For any s, t in the Galois group $G(K'/k)$ of K'/k, we put

$$\xi'_{s,t} = \xi_{\sigma,\tau}\omega_\sigma^{-1}\omega_\tau^{-s}\omega_{\sigma\tau},$$

where σ, τ are automorphisms of $G(K/k)$, induced by s, t on the subfield K. It is then clear that the values of $\xi'_{s,t}$ are p^{th} roots of unity, but it can be also proved that $\xi'_{s,t}$ is a Galois cocycle of K'/k with the same order as $\xi_{\sigma,\tau}$.[4] Therefore, to prove that $\xi_{\sigma,\tau}$ is trivial, it suffices to show that $\xi'_{s,t}$ is trivial. Hence, changing the notation, we may assume from the beginning that the values of $\xi_{\sigma,\tau}$ are p^{th} roots of unity. $\xi_{\sigma,\tau}$ is then a factor set of $G(K/k)$ in the group W of pth roots of unity, on which $G(K/k)$ operates trivially. We put $N = W$ and denote by G the extension of N by $G(K/k)$ using the factor set $\xi_{\sigma,\tau}$. G is a p-group and N is a normal subgroup of order p contained in the center of G. We then consider an imbedding problem $P(K/k, G/N, \varphi)$ where φ is the inverse of the isomorphism from $G(K/k)$ onto G/N defined by the construction of G. It is easy to see that we can take a generator u_1 of N so that the Galois 2-cocycle $\chi_{\sigma,\tau}$ attached to this problem $P(K/k, G/N, \varphi)$ coincides with the 2-cocycle $\xi_{\sigma,\tau}$. But, since the problem $P(K/k, G/N, \varphi)$ has a solution, the cocycle $\chi_{\sigma,\tau} = \xi_{\sigma,\tau}$ must be trivial. The theorem is therefore completely proved.

1.7. As in the preceding, let p be a prime number different from the characteristic of a field k. We shall now prove that every imbedding problem $P(K/k, G/N, \varphi)$ has a solution for an arbitrary p-group N, if the ground field k has the following properties:

I_p for any finite Galois extension K of k and for any integer $n \geqq 1$, the multiplicative group K_m of K contains n elements $\alpha_1, \cdots, \alpha_n$ such that the conjugates α_i^σ of α_i, $i = 1, \cdots, n$, $\sigma \in G(K/k)$, are p-independent in K_m modulo K_m^p,

II_p every finite separable extension of k is p-trivial.

[4] Cf. Artin and Tate [1].

As we have seen in 1.3, every problem $P(K/k, G/N, \varphi)$ with a p-group N can be reduced to a special kind of problem for which a necessary and sufficient condition for the existence of a solution was given in Lemma 6. Hence we have only to show that the conditions of Lemma 6 are satisfied under the assumption that k has the properties I_p and II_p above.

To do that, we first prove the following:

LEMMA 9. *Suppose that a field k has the properties I_p above. Then every finite separable extension k' of k has the same property and the factor group k'_m/k'^p_m of the multiplicative group k'_m of k' modulo its pth power k'^p_m is an infinite group.*

PROOF. Let K' be any finite Galois extension of k' and let n be any integer $\geqq 1$. We take a finite Galois extension K of k containing K' and, by I_p, n elements $\alpha_1, \cdots, \alpha_n$ in K such that α_i^σ, $i = 1, \cdots, n$, $\sigma \in G(K/k)$, are p-independent in K_m mod K_m^p. It is then easy to see that the conjugates of the norms $\beta_i = N_{K/K'}(\alpha_i)$, $i = 1, \cdots, n$, are p-independent in K'_m mod K'^p_m. In particular, taking $K' = k'$ and arbitrarily large n, we see that k'_m/k'^p_m is an infinite group.

Now, suppose that k has both the properties I_p and II_p and consider a problem $P(K/k, G/N, \varphi)$ which satisfies the conditions (i), (ii), (iii) at the beginning of 1.4. We denote by S_p a p-Sylow group of G, by k_p the subfield of K such that $\varphi(S_p/N) = G(K/k_p)$, and by φ_p the restriction of the mapping φ on the subgroup S_p/N. By the assumption, K contains a primitive p^{th} root of unity ζ. But, since K/k_p is an extension of a p-power degree and since the degree of $k_p(\zeta)/k_p$ is a divisor of $p - 1$, ζ is already contained in k_p. Moreover, k_p is p-trivial by II_p and the factor group of the multiplicative group of k_p modulo its pth power is an infinite group by Lemma 9. Therefore, by Theorem 1, the problem

$$P(K/k_p, S_p/N, \varphi_p)$$

has a solution.

Now, let $\chi_{\sigma,\tau}$ be the 2-cocycle of $\Gamma = G/N$ defined by (6) for the problem $P(K/k, G/N, \varphi)$ and let $\chi'_{\sigma,\tau}$ be the corresponding 2-cocycle of $\Gamma_p = S_p/N$ for the problem $P(K/k_p, S_p/N, \varphi_p)$ which also satisfies the conditions (i), (ii), (iii) for the same basis $u_1, \cdots, u_r$ of N. It is then clear that $\chi'_{\sigma,\tau}$ is the restriction of $\chi_{\sigma,\tau}$ on the subgroup Γ_p. But, since $P(K/k_p, S_p/N, \varphi_p)$ has a solution, $\chi'_{\sigma,\tau}$ must be trivial by Lemma 6. We have thus shown that the restriction of the 2-cocycle $\chi_{\sigma,\tau}$ on a p-Sylow group $\Gamma_p = S_p/N$ of $\Gamma = G/N$ is trivial. On the other hand, if q is any prime number different from p, the restriction of $\chi_{\sigma,\tau}$ on a q-Sylow group Γ_q of Γ is also trivial, because the values of $\chi_{\sigma,\tau}$ are p^{th} roots of unity. Hence, by a theorem in the cohomology theory of groups,[5] the 2-cocycle $\chi_{\sigma,\tau}$ itself must be trivial, i.e. there exists a 1-cochain γ_σ of Γ such that $\chi_{\sigma,\tau} = \gamma_\sigma \gamma_\tau^\sigma \gamma_{\sigma\tau}^{-1}$. Now, since the problem $P(K/k_p, S_p/N, \varphi_p)$ has a solution, we know, by Lemma 6, not only that $\chi'_{\sigma,\tau}$ is trivial, but also that $\chi'_{\sigma,\tau}$ is the coboundary of a 1-cochain γ''_σ of Γ_p such that the cocycle γ''^p_σ is trivial. Hence, if we de-

[5] For the results in the cohomology theory of groups used here and in the following, cf. Artin and Tate [1].

note by γ'_σ the restriction of γ_σ on Γ_p and put $\omega'_\sigma = \gamma'_\sigma \gamma''^{-1}_\sigma$, then ω'_σ is a 1-cocycle of Γ_p and the cocycle $\delta'_\sigma = \gamma'^p_\sigma$ is cohomologous to ω'^p_σ: $\delta'_\sigma \sim \omega'^p_\sigma$. Taking the transfers (Verlagerung) $V(\delta'_\sigma)$, $V(\omega'_\sigma)$ of δ'_σ and ω'_σ from Γ_p to Γ respectively, we have $V(\delta'_\sigma) \sim V(\omega'_\sigma)^p$. But, since δ'_σ is obviously the restriction of the 1-cocycle $\delta_\sigma = \gamma^p_\sigma$ of Γ on the subgroup Γ_p, we know that $V(\delta'_\sigma)$ is cohomologous to δ^m_σ, where m is the index of Γ_p in Γ. It follows that $\delta^m_\sigma \sim V(\omega'_\sigma)^p$ and, since m is prime to p, this implies that

$$\delta_\sigma \sim \omega^p_\sigma$$

with a suitable 1-cocycle ω_σ of Γ. The 2-cocycle $\chi_{\sigma,\tau}$ of Γ is then also the coboundary of the 1-cochain $\gamma_\sigma \omega^{-1}_\sigma$ and the pth power of $\gamma_\sigma \omega^{-1}_\sigma$ is a trivial 1-cocycle of Γ. Denoting $\gamma_\sigma \omega^{-1}_\sigma$ again by γ_σ, we get therefore

$$\chi_{\sigma,\tau} = \gamma_\sigma \gamma^\sigma_\tau \gamma^{-1}_{\sigma,\tau}, \qquad \gamma^p_\sigma = \delta_\sigma = \alpha^{\sigma-1}, \tag{10}$$

with a suitable 0-cochain $\alpha = (\alpha_1, \cdots, \alpha_r)$.

Our next task is to show that we can choose α so that $\alpha_1, \cdots, \alpha_r$ are p-independent in K_m mod K^p_m. To do that, we start from an arbitrary solution γ_σ, $\alpha = (\alpha_1, \cdots, \alpha_r)$ of (10) as given there and put $K' = K(\sqrt[p]{\alpha_1}, \cdots, \sqrt[p]{\alpha_r})$. Since $\alpha^{\sigma-1} = \gamma^p_\sigma$, the subgroup H of K_m generated by $\alpha_1, \cdots, \alpha_r$ and K^p_m is invariant under $G(K/k)$ and it follows that K'/k is a Galois extension. We then denote by G' a finite group with the following properties: G' contains both the Galois group $G(K'/k)$ and N, and N is normal in G'; $G(K'/k)N = G'$, $G(K'/k) \cap N = e$; for any σ' in $G(K'/k)$ and u in N, we have

$$\sigma' u \sigma'^{-1} = u^\sigma,$$

where σ is an element in $\Gamma = G/N$ such that $\varphi(\sigma)$ is the restriction of σ' on K. Such a group G' surely exists and is uniquely determined by the above properties. It is then easy to see that the subgroup $G(K'/K)$ of $G(K'/k)$ is normal in G' and $G(K'/K)N$ is the direct product of $G(K'/K)$ and N. Therefore, if we identify $G(K'/K)N/G(K'/K)$ with N, the group $G^* = G'/G(K'/K)$ can be considered as an extension of N by $\Gamma = G/N \cong G(K'/k)/G(K'/K)$. By the definition of G', the extension G^* over N splits and the action of an element σ in Γ on N is the same as that by the extension G over N.

We now consider the problems $P(K'/k, G'/N, \varphi')$ and $P(K/k, G^*/N, \varphi^*)$, where φ' is the inverse of the natural mapping from $G(K'/k)$ onto G'/N and φ^* is the mapping induced by φ' in an obvious way. Since the extension G' over N splits and the ground field k has the property I_p, the problem $P(K'/k, G'/N, \varphi')$ has a solution (L', ψ') by Lemma 8 and, consequently, the subfield L^* of L' determined by the condition $G(L'/L^*) = \psi'(G(K'/K))$, together with a suitable mapping ψ^* induced by ψ', gives a solution of the problem $P(K/k, G^*/N, \varphi^*)$. Now, by the assumption, our original problem

$$P(K/k, G/N, \varphi)$$

satisfies the conditions (i), (ii), (iii) of 1.4. Therefore, since the most essential condition (iii) is concerned only with the action of $\Gamma = G/N$ on the group N and is independent of the factor set of the extension, the problem

$$P(K/k, G^*/N, \varphi^*)$$

also satisfies (i), (ii), (iii) with respect to the same basis u_i of N and the Γ-group V_K are the same for both the problems $P(K/k, G/N, \varphi)$ and $P(K/k, G^*/N, \varphi^*)$. But the extension G^* over N splits and we can put $\chi^*_{\sigma,\tau} = 1$ for a factor set $u^*_{\sigma,\tau}$ of G^*/N. It then follows from Lemma 6 that there exist a 1-cochain γ^*_σ and a 0-cochain $\alpha^* = (\alpha^*_1, \cdots, \alpha^*_r)$ of Γ in V_K such that

$$1 = \chi^*_{\sigma,\tau} = \gamma^*_\sigma \gamma^{*\sigma}_\tau \gamma^{*-1}_{\sigma\tau}, \qquad \gamma^{*p} = \alpha^{*\sigma-1},$$

and that α^*_i are p-independent in K_m mod K^p_m and p^{th} roots of α^*_i generate L^* over K. Putting

$$\bar\gamma_\sigma = \gamma_\sigma \gamma^*_\sigma, \qquad \bar\alpha = \alpha\alpha^* = (\bar\alpha_1, \cdots, \bar\alpha_r), \qquad \bar\alpha_i = \alpha_i \alpha^*_i,$$

we have again

$$\chi_{\sigma,\tau} = \bar\gamma_\sigma \bar\gamma^\sigma_\tau \bar\gamma^{-1}_{\sigma,\tau}, \qquad \bar\gamma^p_\sigma = \bar\alpha^{\sigma-1}.$$

But, since the group $G(K'/K)N$ is the direct product of $G(K'/K)$ and N, the extensions K'/K and L^*/K are independent over K and this implies that $\alpha^*_1, \cdots, \alpha^*_r$ are even p-independent modulo the subgroup H generated by $\alpha_1, \cdots, \alpha_r$ and K^p_m. Consequently, $\bar\alpha_1, \cdots, \bar\alpha_r$ are p-independent in K_m mod K^p_m and, by Lemma 6, the problem $P(K/k, G/N, \varphi)$ has a solution (L, ψ) where L is the field generated by p^{th} roots of $\bar\alpha_i$ over K.

We have thus proved that, if the ground field k has the properties I_p, II_p given at the beginning of this paragraph, then every imbedding problem

$$P(K/k, G/N, \varphi)$$

has a solution for an arbitrary p-group N.

1.8. Let us suppose again that k has the properties I_p and II_p of 1.7 and let k' be an arbitrary finite separable extension of k. Then, by Lemma 9, k' also has the property I_p. On the other hand, it is obvious that k' has the property II_p, because every finite separable extension of k' is a finite separable extension of k. Therefore, by what we have proved in 1.7, every imbedding problem $P(K/k', G/N, \varphi)$ has a solution, where k' is an arbitrary finite separable extension of k and N is an arbitrary p-group.

We shall now prove that the converse is also true, namely that, if every such problem $P(K/k', G/N, \varphi)$ has a solution, then the ground field k has the properties I_p, II_p.

Let K be an arbitrary finite Galois extension of k and n an arbitrary integer $\geqq 1$. We first assume that K contains a primitive p^{th} root of unity ζ and consider a problem $P(K/k, G_n/N_n, \varphi)$ where G_n and N_n are defined as in 1.2 with respect to $\Gamma = G(K/k)$. Since such a problem has a solution by the assumption, it follows from Lemma 7 that the multiplicative group K_m of K contains n elements $\alpha_1, \cdots, \alpha_n$ whose conjugates α_i^σ are p-independent mod K_m^p. If K does not contain ζ, we put $K' = K(\zeta)$ and apply the above result to K'. K_m' then contains n elements $\alpha_1', \cdots, \alpha_n'$ whose conjugates are p-independent mod $K_m'^p$ and the norms $\alpha_i = N_{K'/K}(\alpha_i')$, $i = 1, \cdots, n$, have the corresponding property for K/k. This proves that k has the property I_p.

We next consider an arbitrary finite separable extension k' of k and take a Galois 2-cocycle $\xi_{\sigma,\tau}$ of a p-extension K' over k'. Then the inflation $\xi'_{\sigma,\tau}$ of $\xi_{\sigma,\tau}$ in the Galois cohomology group of $K'(\zeta)/k'$ has the same order as $\xi_{\sigma,\tau}$ and hence, its order is a power of p.[6] On the other hand, the restriction $\xi''_{\sigma,\tau}$ of $\xi'_{\sigma,\tau}$ on the p-Sylow group $G(K'(\zeta)/k'(\zeta))$ of $G(K'(\zeta)/k')$ is trivial by Theorem 1, because every problem $P(K/k'(\zeta), G/N, \varphi)$ has a solution for an arbitrary p-group G, and this implies that the order of $\xi'_{\sigma,\tau}$ is prime to p. Therefore $\xi'_{\sigma,\tau}$ must be trivial and so is $\xi_{\sigma,\tau}$. This proves that k' is p-trivial and k has the property II_p.

We can now summarize our results as follows:

THEOREM 2. *Let k be a field and p a prime number which is different from the characteristic of k. Then every imbedding problem $P(K/k', G/N, \varphi)$ has a solution for an arbitrary finite separable extension k' of k and for an arbitrary p-group N, if and only if k has the following properties:*

I_p for any finite Galois extension K of k and for any integer $n \geqq 1$, the multiplicative group K_m of K contains n elements $\alpha_1, \cdots, \alpha_n$ such that the conjugates α_i^σ of α_i, $i = 1, \cdots, n$, $\sigma \in G(K/k)$, are p-independent in K_m modulo the subgroup of K_m^p of pth powers of K_m,

II_p every finite separable extension of k is p-trivial in the sense of 1.6.

We notice that the property II_p is equivalent to the following: for any finite separable extension k' of k, the order of any Galois 2-cocycle of an arbitrary finite Galois extension K' of k' is prime to p.

1.9. We now suppose that k is a field of characteristic $p > 0$ and consider imbedding problems $P(K/k, G/N, \varphi)$ where N is a p-group. By the reduction in 1.3, it is obvious that we have only to consider a problem $P(K/k, G/N, \varphi)$ where N is an elementary abelian p-group. Then, we take an arbitrary basis $u_1, \cdots, u_r$ of N and denote again by $\bar{A}(\sigma)$ the representation of $\Gamma = G/N$ defined by (1). $\bar{A}(\sigma)$ is, by the definition, a representation of Γ in the finite field $GF(p)$, but since k is a field of characteristic p, we may consider $\bar{A}(\sigma)$ as a representation of Γ in the field k. This makes unnecessary any further reduction which was done by using Lemma 4 in the previous case. Namely, this time, we put $B(\sigma) = (b_{ij}(\sigma)) = \bar{B}(\sigma) = \bar{c}(\sigma)^{-1}\bar{A}(\sigma)$ and denote by V_K the direct sum of r copies of the additive group of K. For any σ in Γ and $\xi = (\xi_1, \cdots, \xi_r)$ in V_K, we define $\sigma(\xi)$ by

$$\sigma(\xi) = (\cdots, \textstyle\sum_{j=1}^{r} b_{ij}(\sigma)\sigma(\xi_j), \cdots),$$

[6] Cf. Artin and Tate [1].

where $\sigma(\xi_j) = \xi_j^\sigma$ is the image of ξ_j in K by the automorphism $\varphi(\sigma)$ of $G(K/k)$, and V_K thus becomes a Γ-module. We then take additive characters $\chi_1, \cdots, \chi_r$ of N whose values are in the prime field of k and are such that

$$\chi_i(u_j) = \delta_{ij}, \qquad i, j = 1, \cdots, r.$$

Putting $\chi(u) = (\chi_1(u), \cdots, \chi_r(u))$ for any u in N, we have again $\chi(\sigma(u)) = \sigma(\chi(u))$ and, for a factor set $u_{\sigma,\tau}$ of the extension G over N, $\chi_{\sigma,\tau} = \chi(u_{\sigma,\tau})$ defines a 2-cocycle of Γ in V_K.

Now, suppose that the problem $P(K/k, G/N, \varphi)$ has a solution (L, ψ). We can then define a G-module V_L similarly as above and get an element $\beta = (\beta_1, \cdots, \beta_r)$ in V_L such that $L = K(\beta_1, \cdots, \beta_r)$ and $(u - 1)\beta = \chi(u)$ for any u in N. If we again denote by x_σ the representatives of Γ in G belonging to the factor set $u_{\sigma,\tau}$ and put $\gamma_\sigma = (x_\sigma - 1)\beta$, γ_σ gives a 1-cochain of Γ in V_K and its coboundary is equal to $\chi_{\sigma,\tau}$. Putting, in general, $\wp(\eta) = (\eta_1^p - \eta_1, \cdots, \eta_r^p - \eta_r)$ for any vector $\eta = (\eta_1, \cdots, \eta_r)$, we see that $\alpha = \wp(\beta)$ is a 0-cochain of Γ in V_K and its coboundary is $\delta_\sigma = \wp(\gamma_\sigma)$. Moreover, since $\beta_1, \cdots, \beta_r$ generate L over K, the components $\alpha_1, \cdots, \alpha_r$ of α are p-independent in the additive group K_a of K modulo the subgroup $\wp(K_a) = \{\lambda^p - \lambda; \lambda \in K_a\}$.

Now, the converse to the above is also true. Namely, if V_K contains a 1-cochain γ_σ and a 0-cochain $\alpha = (\alpha_1, \cdots, \alpha_r)$ such that the coboundary of γ_σ is $\chi_{\sigma,\tau}$, the coboundary of α is $\delta_\sigma = \wp(\gamma_\sigma)$ and such that $\alpha_1, \cdots, \alpha_r$ are p-independent in K_a mod $\wp(K_a)$, then the field $L = K(\beta_1, \cdots, \beta_r)$ generated over K by the solutions $\beta_1, \cdots, \beta_r$ of $\wp(\beta_i) = \alpha_i$, together with a suitable mapping ψ, gives a solution of our problem $P(K/k, G/N, \varphi)$.

We have thus shown that an analogue of Lemma 6 holds for any problem $P(K/k, G/N, \varphi)$ where k is a field of characteristic p and N is an elementary abelian p-group, but using additive groups instead of multiplicative ones. It then follows immediately that analogues of Lemmas 7, 8 and Theorems 1, 2 also hold, for the proofs of these results are essentially based upon Lemma 6. As for Theorem 1 and Theorem 2, we notice that, for any finite Galois extension K over k, the cohomology groups of $G(K/k)$ operating on the additive group of K are always trivial for any dimension $m \geqq 1$.[7] We have therefore the following theorems:

THEOREM 1′. *Let k be a field of characteristic $p > 0$. In order that every imbedding problem $P(K/k, G/N, \varphi)$ have a solution for an arbitrary p-group G, it is necessary and sufficient that the factor group $K_a/\wp(K_a)$ of the additive group K_a of K modulo the subgroup $\wp(K_a) = \{\lambda^p - \lambda\}$ is an infinite group.*

THEOREM 2′. *Let k be a field of characteristic $p > 0$. In order that every imbedding problem $P(K/k', G/N, \varphi)$ have a solution for an arbitrary finite separable extension k' of k and an arbitrary p-group N, it is necessary and sufficient that k have the following property:*

I'_p *for any finite Galois extension K of k and for any integer $n \geqq 1$, the additive*

[7] Cf. Artin and Tate [1].

group K_a *of* K *contains* n *elements* $\alpha_1, \cdots, \alpha_n$ *such that the conjugates* $\sigma(\alpha_i)$ *of* α_i, $i = 1, \cdots, n$, $\sigma \in G(K/k)$, *are* p*-independent in* K_a *modulo the subgroup* $\wp(K_a)$.

1.10. Let us suppose again that k is an arbitrary field and consider an imbedding problem $P(K/k, G/N, \varphi)$ where N is an arbitrary solvable group. We can then find a sequence of normal subgroups of G:

$$N = M_0 \supset M_1 \supset \cdots \supset M_n = e,$$

such that the order of every factor group M_{i-1}/M_i is a power of a prime number. Therefore, by Lemma 1, our problem can be reduced to a sequence of problems where the orders of the normal subgroups are powers of prime numbers. By Theorems, 2, 2′, we have then immediately the following theorem, which is the final goal of the present section:

THEOREM 3. *Let* k *be a field. In order that every imbedding problem* $P(K/k', G/N, \varphi)$ *have a solution for an arbitrary finite separable extension* k' *of* k *and an arbitrary finite solvable group* N, *it is necessary and sufficient that* k *have the properties* I_p, II_p *of Theorem 2 for every prime number* p *different from the characteristic of* k *and also the property* I'_p *of Theorem 2′ for* $p = p_0$, *when* k *is a field of characteristic* $p_0 > 0$.

We notice that k has the property II_p for any prime number p different from the characteristic of k, if and only if every Galois 2-cocycle over an arbitrary finite separable extension of k has an order prime to the characteristic of k. In case k is a field of characteristic 0, this means that there exists no non-commutative division algebra of finite rank over the ground field k.

Section II. Galois groups

2.1. We shall first define a certain class of totally disconnected compact groups which will appear in the following as Galois groups of certain Galois extensions.

Let E be a family of finite groups such that every finite group is isomorphic to one and only one group in E. In other words, E is a complete system of representatives of all different types of finite groups. In the following, we shall consider a subfamily S of E satisfying certain conditions. For simplicity, we call a finite group G an S-group when G is isomorphic to a group contained in S. Then the conditions we impose upon S are stated as follows:

(i) any subgroup of an S-group is an S-group,

(ii) any homomorphic image of an S-group is an S-group,

(iii) the direct product of two S-groups is an S-group.

We notice an immediate consequence of the definition that if a group G contains normal subgroups N_1 and N_2, and if G/N_1 and G/N_2 are both S-groups, then $G/N_1 \cap N_2$ is also an S-group.

Examples of such a subfamily S of E are given by $S = E$; the set of all solvable groups in E; the set of all nilpotent groups in E; the set of all p-groups in E for a fixed prime number p; the set of all abelian groups in E, etc.

We now generalize the definition of finite S-groups to compact groups as follows: Given a family S of finite groups as above, a compact group U is called an S-group when U is a projective limit of finite S-groups. It is then easy to see that every compact S-group U is totally disconnected and the factor group U/V of U modulo an open normal subgroup V of U is a finite S-group. Conversely, it is also clear that a compact group with these properties is a compact S-group. Moreover, the above (i), (ii), (iii) hold also for compact S-groups. We also notice that, when S is the set of all nilpotent or solvable groups in E, a group U is a compact S-group if and only if U is a totally disconnected compact nilpotent or solvable group, respectively, in the sense of the theory of topological groups.[8]

We shall now define a compact S-group which will play an important role in the following. Let F be a free group generated by a countable number of free generators $x_1, x_2, \cdots$, and let $\{M\}$ be the set of all normal subgroups of F such that each M contains all x_i except for a finite number and that F/M is a finite S-group. Taking $\{M\}$ as a fundamental system of neighborhoods of the identity, we can define a topology on F which induces a separated totally bounded topology on F/M_0, where M_0 is the intersection of all subgroups in $\{M\}$. The completion of F/M_0 is therefore compact and it is a compact S-group as can be seen readily from the construction. We shall call that group the S-completion of F and denote it by $F(S)$. If S' is another subfamily of E having the same properties as S and if S' is contained in S, there is a natural homomorphism of $F(S)$ onto $F(S')$ which maps F/M_0 onto the corresponding subgroup F/M_0' of $F(S')$. For instance, the nilpotent completion of F defined by the family of nilpotent subgroups in E is a homomorphic image of the solvable completion of F defined by the family of solvable subgroups in E. We also notice that when the intersection M_0 of all M is e, the free group F itself is imbedded in $F(S)$ as an everywhere dense subgroup and there is a one-to-one correspondence between the subgroups in $\{M\}$ and open normal subgroups of $F(S)$, the latter being the closures of the corresponding subgroups in $\{M\}$. It has been proved that $M_0 = e$ for the p-completion of F where S is the set of all p-groups in E.[9] Therefore the same is also true for the total completion $F(E)$ and for the solvable or nilpotent completion of F.

We next consider the following type of problem which is the group-theoretical counterpart of the imbedding problem considered in Section I: Let U be a totally disconnected compact group and V an open normal subgroup of U. Let G be a finite group and N a normal subgroup of G such that there is an isomorphism Φ of G/N onto U/V. *Lift the isomorphism Φ to an isomorphism Ψ of G onto a factor group U/W such that W is an open normal subgroup of U contained in V and that $\Phi\rho = \rho'\Psi$ where ρ and ρ' denote canonical homomorphisms of G onto G/N and of U/W onto U/V, respectively*. We call such a problem a *lifting problem* $P(U/V, G/N, \Phi)$ and the pair (W, Ψ) a solution of that problem.

---

[8] Cf. Iwasawa [4].

[9] Cf. Iwasawa [3].

We have then the following theorem which characterizes the compact S-group $F(S)$:

THEOREM 4. *Let S be a family of finite groups as given above and let $F(S)$ be the S-completion of a free group F generated by a countable number of free generators $x_1, x_2, \cdots$. Then $F(S)$ is a separable compact S-group and every lifting problem $P(F(S)/V, G/N, \Phi)$ has a solution for an arbitrary finite S-group G. Moreover, $F(S)$ is uniquely characterized, up to isomorphism, by these properties.*

PROOF. It is obvious that $F(S)$ is a separable group, for the family $\{M\}$ contains at most a countable number of subgroups. Let $P(F(S)/V, G/N, \Phi)$ be an arbitrary lifting problem with a finite S-group G. Since V is an open normal subgroup of $F(S)$, there is a normal subgroup M of F in $\{M\}$ such that V is the closure of M/M_0, and Φ induces an isomorphism of G/N onto F/M. We denote by Φ' the homomorphism of F onto G/N which is the product of the inverse of the above isomorphism of G/N onto F/M and the canonical homomorphism of F onto F/M. We then take a sufficiently large integer n so that M contains all x_i for $i > n$ and define a homomorphism Ψ' of F onto G as follows: For $1 \leqq i \leqq n$, we put $\Psi'(x_i) = z_i$, where z_i is any element of G contained in the coset $\Phi'(x_i)$ of G/N, and, for $i > n$, we take as $\Psi'(x_i)$ elements of N such that $\Psi'(x_i)$ generate N and are equal to e except for a finite number of them. The kernel M' of Ψ' is then contained in $\{M\}$ and the closure W of M'/M_0 is an open normal subgroup of $F(S)$. It is then clear that Ψ' induces an isomorphism Ψ of G onto $F(S)/W$ and (W, Ψ) gives a solution of the problem $P(F(S)/V, G/N, \Phi)$.

To prove the second part of the theorem, it is sufficient to show that, if U and U' are separable compact S-groups such that every problem $P(U/V, G/N, \Phi)$ or $P(U'/V', G'/N', \Phi')$ has a solution for an arbitrary finite S-group G or G', then U and U' are isomorphic groups. Let

$$U = V_1 \supseteqq V_2 \supseteqq V_3 \supseteqq \cdots$$

be a sequence of open normal subgroups of U such that the intersection of all V_i is e, and let

$$U' = V_1' \supseteqq V_2' \supseteqq V_3' \supseteqq \cdots$$

be a similar sequence for U'. Since U and U' are separable totally disconnected compact groups, such sequences do exist. By induction, we shall then define sequences

$$U = W_1 \supseteqq W_2 \supseteqq W_3 \supseteqq \cdots; \qquad U' = W_1' \supseteqq W_2' \supseteqq W_3' \supseteqq \cdots$$

of open normal subgroups of U and U' such that $W_n \subseteqq V_n$, $W_n' \subseteqq V_n'$ and such that there exist isomorphisms Φ_n of U/W_n onto U'/W_n' satisfying $\Phi_{n-1}\rho_n = \rho_{n-1}'\Phi_n$, where ρ_n is the canonical homomorphism of U/W_n onto U/W_{n-1} and ρ_n' is the canonical homomorphism of U'/W_n' onto U'/W_{n-1}'. For $n = 1$, we put of course $W_1 = U$, $W_1' = U'$ and take as Φ_1 the trivial mapping of U/W_1 onto U'/W_1'. Suppose that $n > 1$ and the sequences are already defined up to W_{n-1}, W_{n-1}'. We put $X = V_n \cap W_{n-1}$, $G = U/X$, $N = W_{n-1}/X$ and consider the problem

$P(U'/W'_{n-1}, G/N, \Phi_{n-1})$. Since G is a finite S-group, this problem has a solution (X', Φ') by the assumption on U'. X' is, then, an open normal subgroup of U' contained in W'_{n-1} and Φ' is an isomorphism of $G = U/X$ onto U'/X', which induces the isomorphism Φ_{n-1} of $G/N = U/W_{n-1}$ onto U'/W'_{n-1}. We, next, put $W'_n = V'_n \cap X'$, $G' = U'/W'_n$, $N' = X'/W'_n$ and consider the problem $P(U/X, G'/N', \Phi'^{-1})$. This problem again has a solution (W_n, Ψ) by the assumption on U. Since $W_n \subseteqq X \subseteqq W_{n-1}$, $W'_n \subseteqq X' \subseteqq W'_{n-1}$ and since Ψ is an isomorphism of $G' = U'/W'_n$ onto U/W_n which induces the isomorphism Φ'^{-1}, the groups W_n, W'_n, together with $\Phi_n = \Psi^{-1}$, give the nth terms of the sequences for U and U'. We can thus get the sequences $\{W_n\}$ and $\{W'_n\}$ as described above.

Now, since the intersections of all W_n and of all W'_n are both equal to e and since the isomorphisms Φ_n are coherent, we can see easily that Φ_n determines an isomorphism of U onto U'. The theorem is therefore proved.

By a similar argument, but much more simply, we can also prove the following lemma:

LEMMA 10. *Every separable compact S-group is a homomorphic image of the group $F(S)$.*

For instance, every totally disconnected compact solvable group is a homomorphic image of the solvable completion of F. We also notice that the property of $F(S)$ as stated in Lemma 10 does not characterize the group $F(S)$.

2.2. Let k be a field and let S again denote a family of finite groups as considered in 2.1. A finite or infinite Galois extension K of k is called an S-extension when the Galois group $G(K/k)$ of K/k is a compact S-group. It then follows immediately that a Galois extension K/k is an S-extension if and only if K is the composite of a family of finite S-extensions. In particular, if we take an algebraic closure Ω of k and denote by Σ the composite of all finite S-extensions of k in Ω, Σ is also an S-extension and every S-extension of k in Ω is contained in Σ. Therefore we call Σ the maximal S-extension of k in Ω or, simply, the S-closure of k in Ω. It is clear that the structure of the maximal S-extension Σ/k is independent of the choice of the imbedding algebraic closure Ω of k and that every S-extension is k-isomorphic to a subfield of Σ. We may therefore call Σ, simply, the S-closure of k, without referring to Ω.

Now, let us consider an imbedding problem $P(K/k, G/N, \varphi)$, where G is a finite S-group. Since K/k is an S-extension, we may suppose that K is a subfield of the S-closure Σ of k. If we then put $U = G(\Sigma/k)$, $V = G(\Sigma/K)$, V is an open normal subgroup of U and the product Φ of the canonical isomorphism of $G(K/k)$ onto U/V and the isomorphism φ of G/N onto $G(K/k)$ gives an isomorphism of G/N onto U/V. Therefore they define a lifting problem $P(U/V, G/N, \Phi)$. Moreover, if the problem $P(K/k, G/N, \varphi)$ has a solution (L, ψ) we may again consider L as a subfield of Σ and, putting $W = G(\Sigma/L)$ and denoting by Ψ the product of the canonical isomorphism of $G(L/k)$ onto U/W and ψ, (W, Ψ) gives a solution of the corresponding lifting problem $P(U/V, G/N, \Phi)$. Conversely, if a lifting problem $P(U/V, G/N, \Phi)$ is given such that $U = G(\Sigma/k)$ and G is a finite S-group, then we get immediately an imbedding problem $P(K/k, G/N, \varphi)$, where K is a subfield of Σ such that $V = G(\Sigma/K)$ and φ is the product of the canonical isomorphism of U/V onto $G(K/k)$ and the isomorphism Φ of G/N onto U/V, and a solution (W, Ψ) of the problem $P(U/V, G/N, \Phi)$ gives a solution of $P(K/k, G/N, \varphi)$ in a similar way as above. Therefore, combining with Theorem 4, we have the following theorem:

THEOREM 5. *Let k be a field and S a certain family of finite groups satisfying the conditions* (i), (ii), (iii) *of* 2.1. *Then the Galois group $G(\Sigma/k)$ of the maximal S-extension Σ of k is isomorphic to the S-completion $F(S)$ of a free group F generated by a countable number of free generators, if and only if $G(\Sigma/k)$ is separable and every imbedding problem $P(K/k, G/N, \varphi)$ has a solution for any finite S-group G.*

Applying this theorem to the case where S is the set of all solvable groups in the family E considered in 2.1, and using Theorem 3 of Section I, we get immediately the following:

COROLLARY. *Let k be a field of characteristic p_0 and let k have the properties I_p, II_p of Theorem 2 for every prime number $p \neq p_0$ and the property I'_p of Theorem 2′ for $p = p_0$, in case p_0 is different from* 0. *If the Galois group of the maximal solvable extension of k is separable, then that Galois group is isomorphic to the solvable completion of a free group F with a countable number of free generators.*

We notice that, if k has the above properties and if G is any separable totally disconnected compact solvable group, there exists, by Lemma 10, a Galois extension K of k whose Galois group is isomorphic to G. Moreover, if we take a fixed algebraic closure Ω of k, there is a one-to-one correspondence between finite solvable extensions of k contained in Ω and normal subgroups of F in the family $\{M\}$ as given in 2.1, so that F/M is isomorphic to the Galois group of the corresponding Galois extension of k in Ω.

We also notice that, using Theorem 1 or Theorem 1′, we can get a similar result for the Galois groups of maximal p-extensions.[10]

In the following, we shall prove that a certain class of algebraic number fields satisfy the assumptions of the above corollary and we shall, then, be able to determine, by that corollary, the explicit structure of the Galois groups of maximal solvable extensions of those algebraic number fields.

2.3. We first consider algebraic number fields which are abelian extensions of finite algebraic number fields. For simplicity, we shall call such fields essentially abelian algebraic number fields. It is then easy to see that a finite extension of an essentially abelian algebraic number field is also essentially abelian.

The object of the present paragraph is to prove the following lemma:

LEMMA 11. *An essentially abelian algebraic number field k has the property I_p of Theorem 2 for every prime number p; namely, given any finite Galois extension K of k and any integer $n \geqq 1$, the multiplicative group K_m of K contains n elements $\alpha_1, \cdots, \alpha_n$ such that the conjugates α_i^σ of α_i, $i = 1, \cdots, n$, $\sigma \in G(K/k)$, are p-independent in K_m modulo the subgroup K_m^p of pth powers of K_m.*

PROOF. As in the proof of Lemma 9, we can see easily that it suffices to prove

[10] For non-separable p-extensions, see a forthcoming paper of Y. Kawada. Cf. also Scholz [5].

the lemma for some finite Galois extension of k containing K. Therefore, if k_0 is a finite algebraic number field of which k is an abelian extension, we may assume that K is the composite of k and a finite Galois extension K_0 of k_0 and that K contains a primitive p^{th} root of unity ζ. Moreover, we may also put $k_0 = k \cap K_0$. K is then a Galois extension of k_0 and is an abelian extension of K_0, and the Galois group $G(K/k_0)$ is the direct product of the Galois groups $G(K/k)$ and $G(K/K_0)$.

To prove the lemma, it is sufficient to show that, given any finite group H/K_m^p of K_m/K_m^p, where $K_m^p \subseteqq H \subseteqq K_m$, there exists an element α in K_m whose conjugates α^σ, $\sigma \in G(K/k)$, are p-independent modulo H. For the proof of the above statement, we take a finite number of elements $\xi_1, \cdots, \xi_s$ in K_m which generate H over K_m^p. Replacing K_0 by $K_0(\xi_1, \cdots, \xi_s, \zeta)$, if necessary, we may assume that K_0 contains $\xi_1, \cdots, \xi_s$ and ζ. We then take a non-archimedean prime divisor P of K_0 such that the conjugates P^σ, $\alpha \in G(K/k)$, of P with respect to the Galois extension K_0/k_0 are all distinct and that P is prime to p and to all ξ_i^σ, $i = 1, \cdots, n$, $\sigma \in G(K/k)$. Such P surely exists, because there are infinitely many prime divisors of k_0 which decompose completely by the extension K_0/k_0.[11] Let E be a finite extension of K_0 contained in K such that the degree of E over K is a power of p and that P ramifies completely by the extension E/K_0. Since p is prime to the absolute norm $N(P)$ of the prime divisor P and since E/K_0 is an abelian extension, it is known that the degree of E/K_0 is a divisor of $N(P) - 1$. Hence the degree of such a field E over K_0 is bounded and we may assume that we have chosen a field E which has the above properties and has the highest degree over K_0.

Let Q be the prime divisor of E which is the unique extension of P in E. Since the conjugates P^σ of P are all different, it is easy to see that the conjugates Q^σ, $\sigma \in G(K/k)$, of Q in E are also distinct. Denoting by ν_σ the normalized exponential valuation of E corresponding to Q^σ, we can then find an element $\alpha \neq 0$ in E such that

$$\nu_e(\alpha) = 1, \qquad \nu_\sigma(\alpha) = 0, \qquad \sigma \neq e, \qquad \sigma \in G(K/k).$$

It follows, obviously, that

$$\nu_\sigma(\alpha^\sigma) = 1, \qquad \nu_\sigma(\alpha^\tau) = 0, \qquad \sigma \neq \tau. \tag{11}$$

We shall now prove that the conjugates α^σ of α are p-independent modulo H. Suppose, therefore, that a power α^z of α is contained in H, where $z = \sum m_\sigma \sigma$ is a linear combination of σ in $G(K/k)$ with integer coefficients m_σ. We have to show that all m_σ are divisible by p. Since H is generated by ξ_i and K_m^p, we have

$$\alpha^z = \xi_1^{e_1} \cdots \xi_s^{e_s} \beta^p, \qquad \beta \in K_m, \quad e_i = \text{integer}, \tag{12}$$

and since E contains α^z and ξ_i, β^p is also contained in E. On the other hand, E contains a primitive p^{th} root of unity ζ. Hence the extension $E(\beta)/E$ must be

[11] Cf. Artin and Tate [1].

either trivial or a cyclic extension of degree p and, in the latter case, the prime divisor Q does not ramify in $E(\beta)$ because of the choice of the field E. Since the automorphisms σ of $G(K/k)$ map E and $E(\beta)$ onto themselves, every Q^σ is then also unramified in $E(\beta)$. Therefore, in either case, there exists, for each σ, a normalized exponential valuation of $E(\beta)$ which takes the same values as ν_σ on E, and, for simplicity, we shall denote it again by ν_σ. It then follows from (12) that

$$\nu_\sigma(\alpha^z) = \sum_{i=1}^{s} e_i \nu_\sigma(\xi_i) + p\nu_\sigma(\beta).$$

Using (11), we have $\nu_\sigma(\alpha^z) = m_\sigma$, and, as P was prime to ξ_i^σ, we have also $\nu_\sigma(\xi_i) = 0$. It follows that

$$m_\sigma = p\nu_\sigma(\beta),$$

and this shows that α^σ, $\sigma \in G(K/k)$, are p-independent modulo H. The lemma is therefore proved.

2.4. Now, it is well-known in algebraic number theory that, if an algebraic number field k is large enough, there exists no non-trivial Galois 2-cocycle over the ground field k; it suffices that k is a totally imaginary number field and that, given any finite algebraic number field k_0 in k and any non-archimedean prime divisor P of k_0, k contains a finite extension E of k_0 such that the local degrees of the extension E/k_0 at P are arbitrarily large.[12] In particular, every algebraic number field containing all roots of unity has the above property and, hence, also the property II_p of Theorem 2 for every prime number p. Combining this with Lemma 11 and using the corollary to Theorem 5, we get immediately the following theorem:

THEOREM 6. *Let k be an algebraic number field which is an abelian extension of a finite algebraic number field and which contains all roots of unity. Then the Galois group of the maximal solvable extension of k over k is isomorphic to the solvable completion of a free group with a countable number of free generators.*

Now, let k_0 be an arbitrary finite algebraic number field, k the maximal abelian extension of k_0 and Σ the maximal solvable extension of k_0. It is then obvious that k contains all roots of unity and Σ is also the maximal solvable extension of k. Hence the structure of the Galois group $G(\Sigma/k)$ of Σ/k is given by the above theorem and we have the following final result:

THEOREM 7. *Let k_0 be an arbitrary finite algebraic number field, k the maximal abelian extension of k_0 and Σ the maximal solvable extension of k_0. Then the Galois group $G(\Sigma/k)$ of Σ/k is isomorphic to the solvable completion of a free group with a countable number of free generators.*

We notice that $G(\Sigma/k)$ is the topological commutator group of the Galois group $G(\Sigma/k_0)$ of Σ/k_0 and that the factor commutator group of $G(\Sigma/k_0)$ is canonically isomorphic to the Galois group $G(k/k_0)$ of the maximal abelian extension k of k_0, and hence also, by class field theory, to the idèle class group of the ground field k_0.

[12] Cf. Artin and Tate [1].

2.5. We now consider an algebraic function field k over an algebraically closed constant field k_0. k has then an infinite number of prime divisors and all of them are of the first degree. Moreover, by any finite extension E/k, these prime divisors of k decompose completely except for a finite number of ramifying ones. Therefore, using a similar argument to that of the proof of Lemma 11, we can see easily that k has the property I_p or I'_p of Theorem 2 or Theorem 2′ respectively, for any prime number p.

On the other hand, it is also well-known that there exists no non-trivial Galois 2-cocycle over such a function field k.[13] It follows therefore from Theorem 3 that every imbedding problem $P(K/k, G/N, \varphi)$ has a solution for an arbitrary solvable group N. Furthermore, it is easy to see that the Galois group of the maximal solvable extension of k is separable if and only if k_0 contains a countable number of elements, and again from the corollary to Theorem 5, we get the following

THEOREM 8. *Let k be an algebraic function field of one variable over an algebraically closed constant field k_0 containing a countable number of elements. Then the Galois group of the maximal solvable extension of k over k is isomorphic to the solvable completion of a free group with a countable number of free generators.*

We notice that the above theorem does not cover the classical case where k is an algebraic function field over the complex field, though every imbedding problem $P(K/k, G/N, \varphi)$ with solvable N has a solution in that case also.

MASSACHUSETTS INSTITUTE OF TECHNOLOGY

BIBLIOGRAPHY

[1] E. ARTIN and J. TATE. Algebraic numbers and algebraic functions II. Lecture notes, Princeton University (1951–1952).

[2] R. BRAUER. *Über die Konstruktion des Schiefkörper, die von endlichem Rang in bezug auf ein gegebenes Zentrum sind.* J. Reine Angew. Math., Bd. 168 (1932), 44–64.

[3] K. IWASAWA. *Einige Sätze über freie Gruppen.* Proc. Imp. Acad. Tokyo, vol. 19 (1943), 272–274.

[4] K. IWASAWA. *On Some types of topological groups.* Ann. of Math., vol. 50 (1949), 507–558.

[5] A. SCHOLZ. *Konstruktion algebraischer Zahlkörper mit beliebiger Gruppe von Primzahlpotenzordnung I.* Math. Zeit., Bd. 42 (1937), 161–188.

[6] C. TSEN. *Divisionsalgebren über Funktionenkörper.* Nachr. Ges. Wiss. Göttingen (1933), 335–339.

[7] E. WITT. *Konstruktion von galoisschen Körper der Charakteristik p zu vorgegebener Gruppe der Ordnung p^f.* J. Reine Angew. Math., Bd. 174 (1936), 237–245.

[13] Cf. Tsen [6].

[28] A note on Kummer extensions

J. Math. Soc. Japan, 5 (1953), 253-262.

(Received April 20, 1953)

1. Let k be an arbitrary field and $Z(k)$ the set of all integers $n \geqq 1$ such that k contains a primitive n-th root of unity. It is clear that, if $Z(k)$ contains m and n, it also contains the least common multiple of these two integers. Therefore the set of all rational numbers with denominators in $Z(k)$ is an additive group $R(k)$ containing the group of all integers Z, and the quotient group $\bar{R}(k)=R(k)/Z$ is isomorphic with the multiplicative group $W(k)$ of all roots of unity in k.

We now take an algebraic closure Ω of k and consider the subfield K of Ω obtained by adjoining all $\alpha^{1/n}$ to k, where α is an arbitrary element in k and n is an arbitrary integer in $Z(k)$. K is obviously the composite of all finite Kummer extensions of k contained in Ω and hence, may be called the *Kummer closure* of k in Ω. K/k is clearly an abelian extension and its structure is independent of the choice of the algebraic closure Ω of k. In particular, the structure of the Galois group $G(K/k)$ of K/k is an invariant of the field k, and we shall show in the following how we can describe it by means of groups which depend solely on the ground field k.

2. We shall first define a symbol (σ, α, r) for arbitrary σ in $G=G(K/k)$, $\alpha \neq 0$ in k and r in $R(k)$. Namely, we express r as a fraction $\frac{m}{n}$ with denominator n in $Z(k)$ and choose an elemnt a in K such that $a^n=\alpha^m$. The symbol (σ, α, r) is then defined by

$$(\sigma, \alpha, r)=a^{\sigma-1}.$$

It is easy to see that (σ, α, r) is an n-th root of unity in k and is independent of the choice of the fractional expression $\frac{m}{n}$ of r and, also, of the choice of a in K such that $a^n=\alpha^m$.

The symbol (σ, α, r), defined uniquely in this way, has the following properties which can be verified easily from the definition:

1) $(\sigma_1\sigma_2, \alpha, r)=(\sigma_1, \alpha, r)(\sigma_2, \alpha, r)$,
2) $(\sigma, \alpha_1\alpha_2, r)=(\sigma, \alpha_1, r)(\sigma, \alpha_2, r)$,
3) $(\sigma, \alpha, r_1+r_2)=(\sigma, \alpha, r_1)(\sigma, \alpha, r_2)$,
4) $(\sigma, \alpha, m)=1, \qquad m \in Z$.

From 3), 4), it follows that (σ, α, r) essentially depends upon σ, α and the residue class $\bar{r}$ or r mod. Z and that we may therefore put $(\sigma, \alpha, r)=(\sigma, \alpha, \bar{r})$. The new symbol $(\sigma, \alpha, \bar{r})$ then has properties similar to 1), 2), 3) above.

We now fix σ and α and consider a mapping $\varphi_{\sigma,\alpha}$ of $\bar{R}(k)$ into $W(k)$ defined by

$$\varphi_{\sigma,\alpha}(\bar{r})=(\sigma, \alpha, \bar{r}).$$

By 3), $\varphi_{\sigma,\alpha}$ is a homomorphism of $\bar{R}(k)$ into $W(k)$, i. e. an element of the group of homomorphisms $\mathrm{Hom}(\bar{R}(k), W(k))$. We then define, for any fixed σ, a mapping f_σ of the multiplicative group k^* of k into $\mathrm{Hom}(\bar{R}(k), W(k))$ by

$$f_\sigma(\alpha)=\varphi_{\sigma,\alpha}.$$

f_σ is again a homomorphism by 2), i. e. an element of the group of homomorphisms $\mathrm{Hom}(k^*, \mathrm{Hom}(\bar{R}(k), W(k)))$. We finally define a mapping Φ of G into $\mathrm{Hom}(k^*, \mathrm{Hom}(\bar{R}(k), W(k)))$ by

$$\Phi: \quad \sigma \rightarrow f_\sigma.$$

Φ is a homomorphism by 1) and is, in fact, an isomorphism, for, if f_σ is the identity, $(\sigma, \alpha, r)=1$ for every α in k^* and every r in $R(k)$, and it follows from the definition of (σ, α, r) that each $\alpha^{1/n}$ is invariant under σ, and that σ is, consequently, the identity of the group $G=G(K/k)$.

We now consider $W(k)$, $\bar{R}(k)$ and k^* as discrete groups and introduce the so-called compact convergence topology[1] in $\mathrm{Hom}(\bar{R}(k), W(k))$ and $\mathrm{Hom}(k^*, \mathrm{Hom}(\bar{R}(k), W(k)))$. It is then easy to see that both

1) Cf. N. Bourbaki, Topologie générale, Chap. X.

these groups become topological groups and that a fundamental system of neighborhoods of the identity in $H=\mathrm{Hom}(k^*, \mathrm{Hom}(\bar{R}(k), W(k)))$ is given by the family of subsets $U(\alpha_1, \cdots, \alpha_s; n)$, where the set $U(\alpha_1, \cdots, \alpha_s; n)$ is defined for any finite set of elements $\alpha_1, \cdots, \alpha_s$ in k^* and for any integer n in $Z(k)$, and consists of elements f in H such that $f(\alpha_i)$ in $\mathrm{Hom}(\bar{R}(k), W(k))$ maps the residue class of $\frac{1}{n}$ mod. Z to the unity element 1 in $W(k)$. Therefore, an element σ of G is contained in $\Phi^{-1}(U(\alpha_1, \cdots, \alpha_s; n))$ if and only if $\left(\sigma, \alpha_i, \frac{1}{n}\right)=1$ for $i=1, \cdots, s$, and, taking a_i in K with $a_i^n=\alpha_i$ and putting $E=k(a_1, \cdots, a_s)$, we see that $\Phi^{-1}(U(\alpha_1, \cdots, \alpha_s; n))$ coincides with the Galois group $G(K/E)$ of K/E. But, since E/k is a finite extension, $G(K/E)$ is an open subgroup of G in Krull's topology of the Galois group $G=G(K/k)$. Therefore Φ is a continuous mapping of G into H.

We shall next show that the image $\Phi(G)$ of G is everywhere dense in H. Let f be an arbitrary element of H and $U(\alpha_1, \cdots, \alpha_s; n)$ an arbitrary neighborhood of the identity as given above. We prove that there exists an element σ in G such that $f^{-1}f_\sigma$ is contained in $U(\alpha_1, \cdots, \alpha_s; n)$; namely, such that

$$f(\alpha_i)\left(\frac{1}{n}\right)^{2)}=\left(\sigma, \alpha_i, \frac{1}{n}\right), \quad i=1, \cdots, s.$$

To see this, we consider a function $\chi(\alpha)$ of k^* defined by

$$\chi(\alpha)=f(\alpha)\left(\frac{1}{n}\right).$$

Since $\chi(\alpha)$ is obviously a character of k^* and is trivial on the subgroup $(k^*)^n$, it follows from the theory of Kummer extensions that there exists a k-automorphism σ of the field K_n generated over k by all n-th roots of elements in k, such that

$$\chi(\alpha)=(\alpha^{1/n})^{\sigma-1}.$$

Denoting an extension of σ in the Galois group G of K/k again by σ, we see immediately from the definition of (σ, α, r) that

2) Here $\frac{1}{n}$ stands for the residue class of $\frac{1}{n}$ mod. Z.

$$\chi(\alpha)=\left(\sigma, \alpha, \frac{1}{n}\right),$$

which proves the assertion.

We have thus shown that Φ is a continuous isomorphism of G into H and the image $\Phi(G)$ of G is everywhere dense in H. But, since G is a compact group as the Galois group of K/k, Φ must be an isomorphism of G onto H, and we have thus obtained the following

THEOREM 1. *Let k be an arbitrary field and K the Kummer closure of k in an algebraic closure of k. Then the Galois group of K/k is canonically isomorphic with the group of homomorphisms* Hom $(k^*$, Hom $(\bar{R}(k), W(k)))$ *which is attached to the field k as described above.*

3. Now, it is easy to see that the group of roots of unity $W(k)$ of a field k is isomorphic with a subgroup of the group of ordinary roots of unity $W_0=\{e^{2\pi ir}$; $r=$rationals$\}$. Therefore, taking such an isomorphism g of $W(k)$ into W_0, every element φ of Hom $(\bar{R}(k), W(k))$ defines a character $g\circ\varphi$ of the discrete group $\bar{R}(k)$. Moreover, using the fact that $\bar{R}(k)$ is isomorphic with $W(k)$, it can be seen that every character of $\bar{R}(k)$ can be written in the form $g\circ\varphi$ with some φ in Hom $(\bar{R}(k), W(k))$ and that Hom $(\bar{R}(k), W(k))$ is consequently isomorphic with the character group $\tilde{W}(k)$ of $W(k)$, both being considered as topological groups. Hence the Galois group G of the Kummer closure K of k is isomorphic with the group of homomorphisms Hom $(k^*, \tilde{W}(k))$, and, though such a description of G is not canonical (unlike the one as given in Theorem 1), it is useful when we only consider the structure of G as a topological group and not a canonical description of it.

Let, for instance, k be a field of characteristic 0 containing all roots of unity. Every finite abelian extension of k is then a Kummer extension and the Kummer closure K of k coincides with the maximal abelian extension A over k. Moreover, in such a case, the group $W(k)$ is isomorphic with the group W_0. Hence the Galois group $G(A/k)$ of the maximal abelian extension A over k is isomorphic with the group of homomorphisms Hom $(k^*, \tilde{W}_0)$ of k^* into the character group $\tilde{W}_0$ of W_0. To determine the structure of the group $G(A/k)$

more explicitly, we have, therefore, only to study the structure of the multiplicative group k^* of k, and we shall do this in the following sections for a certain kind of algebraic number field containing all roots of unity.

4. We shall first define a special type of abelian group and give some simple properties which will be used later. Let G be an abelian group and T the torsion subgroup of G. We call G a *regular* abelian group when the factor group G/T is free abelian. The following properties of regular abelian groups are immediate consequences of the definition:

α) an abelian group G is regular if and only if it is the direct product of its torsion subgroup T and a free abelian subgroup of G,

β) if H is a subgroup of an abelian group G and if every element of H has finite order, G/H is regular if and only if G is regular,

γ) a subgroup of a regular abelian group is regular,

δ) the direct product of a finite number of regular abelian groups is regular,

ε) if $\{H_i\}$ is a finite set of subgroups of an abelian group G such that their intersection is e and if every G/H_i is regular, then G itself is also regular.[3)]

We shall now prove the following lemmas:

LEMMA 1. *Let G be a regular abelian group and let $\{\sigma\}$ be a finite set of endomorphisms of G. If H is the subgroup of G consisting of all elements of G which are invariant under all σ, then G/H is also a regular abelian group.*

PROOF. For each σ, let H_σ denote the kernel of the endomorphism $\tau(a)=\sigma(a)a^{-1}$ of G. Since G/H_σ is isomorphic with $\tau(G)$ and $\tau(G)$ is regular by γ), G/H_σ is also regular. But, as H is clearly the intersection of all H_σ, G/H is regular by ε).

LEMMA 2. *Let G be an abelian group and H a subgroup of G. If H and G/H are both regular and if the orders of elements of the torsion subgroup of G/H are bounded, then G is also regular.*

PROOF. Let U be the torsion subgroup of H. By β) above, it suffices to show that G/U is regular. We may therefore assume that $U=e$ and H is free abelian. Let V be the subgroup of G containing

3) Notice that G is isomorphic with a subgroup of the direct product of all G/H_i.

H such that V/H is the torsion subgroup of G/H. By the assumption, there is an integer m such that V^m is contained in H. If we then denote by T the kernel of the endomorphism $\sigma(a)=a^m$ of V, V/T is isomorphic with the subgroup V^m of the free abelian group H and, hence, is again free abelian. On the other hand, since G/H is regular, G/V is also free abelian. Therefore G/T is free abelian and G is regular, for T must be the torsion subgroup of G.[4)]

5. We shall now study the structure of the multiplicative groups of a certain class of algebraic number fields by applying the result of the preceding section. If there will be no risk of confusion, we shall denote, for simplicity, the multiplicative group of a field k by the same letter k, instead of k^*.

LEMMA 3. *The multiplicative group of a finite algebraic number field E is regular.*

PROOF. Since the group of ideals of E is obviously free abelian, so is the subgroup of principal ideals of E. But the latter group is isomorphic with the factor group E/U of the multiplicative group of E modulo the group of units of E, and, as U is regular by Dirichlet's theorem, E is also regular by Lemma 2.

LEMMA 4. *Let E be a finite algebraic number field and let F be a finite extension of E. Then the factor group F/E is regular.*

PROOF. Let K be a finite Galois extension of E containing F. Since F/E is a subgroup of the factor group K/E, it suffices to show that K/E is regular. But K is regular by Lemma 3 and E is the subgroup of K consisting of all elements of K which are invariant under the Galois automorphisms of the extension K/E. The group K/E is therefore regular by Lemma 1.

LEMMA 5. *Let E be a finite algebraic number field and let A be an abelian extension of E containing all roots of unity. Denote by W the group of roots of unity in A and by N the subgroup of A containing E such that N/E is the torsion subgroup of A/E. If m is the order of the finite group $E\frown W$, N^m is contained in the group EW.*

PROOF. Let p be an arbitrary prime number and let p^e $(e\geq 0)$ be the p-part of the order m. To prove the theorem, it suffices to

4) In general, an abelian group G is not regular even when H and G/H are both regular groups. Example: $G=$the additive group of rationals, $H=$the additive group of integers.

show that N contains no element of order p^{e+1} modulo EW. For the proof, we assume that there exists an element ξ of order p^{e+1} modulo EW and show that such an assumption leads to a contradiction. Let $\xi^{p^{e+1}}=\alpha\omega$ with α in E, ω in W. Taking a p^{e+1}-th root ω' of ω^{-1} and replacing ξ by $\xi\omega'$, we may assume that

$$\xi^{p^{e+1}}=\alpha, \qquad \alpha \in E.$$

Let $f(x)$ be the minimal polynomial of ξ over E and n the degree of $f(x)$. Since $f(x)$ is a divisor of $x^{p^{e+1}}-\alpha=\Pi(x-\eta\xi)$, where η runs over all p^{e+1}-th roots of unity, the constant term of $f(x)$ must be of the form $\eta_1\xi^n$ with a suitable p^{e+1}-th root of unity η_1. It follows that ξ^n is contained in EW and, consequently, that $n=p^{e+1}$, $f(x)=x^{p^{e+1}}-\alpha$. Therefore $K=E(\xi)$ is an abelian extension of degree p^{e+1} over E and it contains a primitive p^{e+1}-th root of unity ζ. Since ζ is not contained in E, while ζ^p is in E, the intermediate field $E(\eta)$ of K/E must be an extension of degree p over E, and we see, in particular, that $e\geqq 1$. The constant term of the minimal polynomial of ξ over $F=E(\zeta)$ is again of the form $\eta_2\xi^{p^e}$ with a suitable p^{e+1}-th root of unity η_2 and it follows that $\beta=\xi^{p^e}$ is in F and $F=E(\beta)$. Now, since $F=E(\zeta)=E(\beta)$ is a Kummer extension of degree p over E such that ζ^p and β^p are both contained in E, the product of β with a suitable power of ζ must be in E. But $\beta=\xi^{p^e}$ is then contained in EW and this contradicts the assumption that the order of ξ modulo EW is p^{e+1}. The lemma is therefore proved.

We now consider an arbitrary abelian extension A of a finite algebraic number field k and prove that the multiplicative group of A is regular. Let A' be the field obtained by adjoining all roots of unity to A. A' is then still abelian over k and the multiplicative group A is a subgroup of the multiplicative group A'. Hence it suffices to show that A' is regular, and we may assume from the beginning that A contains all roots of unity.

Let E and F be finite extensions of k such that $E\subseteqq F\subseteqq A$ and let W be again the group of all roots of unity in A. Since $(F\frown W)E/E$ is a finite group and F/E is regular by Lemma 4, the group $F/(F\frown W)E$ is again regular by β) above. But, since we have the isomorphism

$$FW/EW \cong F/(F \frown W)E,$$

FW/EW is also a regular abelian group. On the other hand, if we denote by N_E the subgroup of A containing E such that N_E/E is the torsion subgroup of A/E, and by N_F the corresponding group for F, the orders of elements of N_F/FW are bounded by Lemma 5. Therefore N_F/EW is regular by Lemma 2. Since N_E/EW is obviously the torsion subgroup of N_F/W, it follows that N_F/N_E is a free abelian group.

Now, let

$$k = E_0 \leqq E_1 \leqq E_2 \leqq \cdots$$

be a sequence of finite extensions of k such that A is the union of all these E_n, and let $N_n = N_{E_n}$ be the subgroup of A containing E_n such that N_n/E_n is the torsion subgroup of A/E_n. By what we have proved above, every factor group N_n/N_{n-1} $(n=1, 2, \cdots)$ is a free abelian group. Since the group A is obviously the union of all N_n, it follows immediately that A/N_0 is also free abelian. On the other hand, considering E_0W/W instead of FW/EW and using a similar argument, we can see easily that N_0 is a regular abelian group. It then follows from Lemma 2 that the group A is also regular, thus proving our previous contention. By α) above, our result can be stated as follows:

THEOREM 2. *The multiplicative group of an abelian extension A of a finite algebraic number field is the direct product of a free abelian group and the group of roots of unity in A.*

6. We now change our notation and denote by k an algebraic number field which contains all roots of unity and which is an abelian extension of a finite algebraic number field. As we have seen in § 3, the Galois group $G(A/k)$ of the maximal abelian extension A over k is isomorphic with the group of homomorphisms Hom $(k^*, \widetilde{W}_0)$. In this section, we shall determine the structure of the latter group using the result of § 5.

By Theorem 2, the multiplicative group k^* of k is the direct product of the group of roots of unity in k, $W(k)$, and a free abelian group U which has, as readily seen, a countable number of free generators $u_1, u_2, \cdots$. Let φ be any homomorphism in Hom $(k^*, \widetilde{W}_0)$.

We shall show that the character $\chi=\varphi(\omega)$ of W_0 is trivial for any root of unity ω in $W(k)$. To see this, let ζ be an arbitrary root of unity in W_0 and let $\zeta^n=1$. If we take ω_1 in $W(k)$ such that $\omega_1^n=\omega$ and put $\chi_1=\varphi(\omega_1)$, we have

$$\chi=\chi_1^n\,;$$
$$\chi(\zeta)=\chi_1^n(\zeta)=\chi_1(\zeta^n)=\chi_1(1)=1\,,$$

which proves our assertion. Hence, a homomorphism φ in Hom $(k^*, \widetilde{W}_0)$ is completely determined by its values $\varphi(u_i)$ for the basis u_i, $i=1, 2, \cdots$. On the other hand, since the u_i are free generators of the group U, there exists at least one homomorphism φ in Hom $(k^*, \widetilde{W}_0)$ satisfying $\varphi(u_i)=\chi_i$ for any given sequence of characters χ_i in $\widetilde{W}_0$. It then follows immediately that the mapping

$$\varphi\rightarrow(\varphi(u_1),\ \varphi(u_2),\ \cdots)$$

gives a topological isomorphism of Hom $(k^*, \widetilde{W}_0)$ with the direct product of a countable number of copies $\widetilde{W}_0^{(i)}$ of the group $\widetilde{W}_0$.

We have therefore the following theorem:

THEOREM 3. *Let k be an abelian extension of a finite algebraic number field containing all roots of unity and let A be the maximal abelian extension of k. Then the Galois group of A/k is isomorphic with the direct product of a countable number of groups each of which is isomorphic with the character group of the group of roots of nuity in k.*

7. We add here some remarks on the above result. Let k and A be as in Theorem 3. Let $A_0=k$, $A_1=A$ and let, in general, A_n be the maximal abelian extension of A_{n-1} $(n\geqq 2)$. We can determine the structure of the multiplicative group of A_n $(n\geqq 1)$ by a method similar to that of § 5 and, then, by using the result of § 3, we can also prove that the Galois group of the extension A_n/A_{n-1} is always isomorphic with the direct product of a countable number of copies $\widetilde{W}_0^{(i)}$ of the group $\widetilde{W}_0$. In other words, if we take the maximal solvable extension Σ of k and denote by G_n the n-th topological commutator group of the Galois group $G(\Sigma/k)$ of Σ/k $(G_0=G(\Sigma/k))$, we have

$$G_{n-1}/G_n\cong\prod_{i=1}^{\infty}\widetilde{W}_0^{(i)}\,,\qquad n=1, 2, \cdots.$$

This gives considerable information about the group $G(\Sigma/k)$, though we can determine the structure of $G(\Sigma/k)$ completely by an entirely different method.[5]

University of Tokyo
Massachusetts Institute of Technology.

5) Cf. the author's forthcoming paper "On solvable extensions of algebraic number fields".

[29] On Galois groups of local fields

Trans. Amer. Math. Soc., 80 (1955), 448-469.

Let p be a prime number, Q_p the field of p-adic numbers, and Ω an algebraic closure of Q_p. In the present paper, we take a finite extension k of Q_p in Ω as the ground field and study the structure of the Galois group $G(\Omega/k)$ of the extension Ω/k. Let V be the ramification field of Ω/k, i.e. the composite of all finite tamely ramified extensions of k in Ω, and let $G(\Omega/V)$ and $G(V/k)$ denote the Galois groups of the extensions Ω/V and V/k respectively. We shall first determine the structure of the groups $G(V/k)=G(\Omega/k)/G(\Omega/V)$, and $G(\Omega/V)$ and show that the group extension $G(\Omega/k)/G(\Omega/V)$ splits. Our main result is, then, to describe explicitly the effect of inner automorphisms of $G(\Omega/k)$ on the factor group of $G(\Omega/V)$ modulo its commutator subgroup, i.e., on the Galois group $G(V'/V)$ of the maximal abelian extension V' of V in Ω. This is, of course, not sufficient to determine the structure of the group $G(\Omega/k)$ completely; to do that, we still have to find the effect of inner automorphisms of $G(\Omega/k)$ on the normal subgroup $G(\Omega/V)$ itself. However, it gives us some insight into the structure of $G(\Omega/k)$; and we hope it will help somehow, in the future, in the study of the group $G(\Omega/k)$ as well as in that of the Galois groups of algebraic number fields.

An outline of the paper is as follows: in §1 we prove some group-theoretical lemmas which will be used later. In §2 we study the behavior of the Galois group of a certain type of finite tamely ramified Galois extension E of k acting on the multiplicative group of E. Using those results, we then prove in §3 the properties of $G(\Omega/k)$ as mentioned above(1).

1. Group-theoretical preparations

1.1. Let q be a power of a prime number $p: q=p^{f_0}$, $f_0 \geqq 1$. In sections 1.1 and 1.2, we shall denote by G a finite group of order $n=ef$ generated by two elements σ and τ satisfying the relations

$$\sigma^f = 1, \qquad \tau^e = 1, \qquad \sigma\tau\sigma^{-1} = \tau^q. \tag{1.1}$$

If such a group G exists, e and f satisfy the congruence relation

$$q^f \equiv 1 \bmod e. \tag{1.2}$$

Hence, in particular, e must be an integer prime to p. Conversely, if e and f are integers $\geqq 1$ satisfying (1.2), there exists, up to isomorphisms, a unique finite group G of order $n=ef$ as described above.

Received by the editors March 13, 1955.

(1) For the structure of Galois groups of p-extensions of k, cf. [5; 8].

Let G be such a finite group of order $n=ef$. We denote by $GF(q^f)$ the finite field with q^f elements and by η a primitive eth root of unity in $GF(q^f)$. According to (1.2), the field $GF(q^f)$ certainly contains such an η. For any integer i, we then make the additive group of $GF(q^f)$ a G-module A_i by putting

$$\sigma a = a^q, \qquad \tau a = \eta^i a$$

for any a in $GF(q^f)$. Obviously A_i is a G-module with dimension ff_0 over the prime field $GF(p)$ of $GF(q^f)$, and we denote by A_i' the G-module over the algebraic closure Ω_p of $GF(p)$ obtained from A_i by extending the scalar field $GF(p)$ to Ω_p. A_i' then contains an Ω_p-basis a_j' indexed by residue classes j mod ff_0, such that

$$\sigma a_j' = a_{j-f_0}', \qquad \tau a_j' = \eta^{ip^j} a_j' \text{(}^2\text{)}.$$

In the following, we shall denote by R_p and R_p' the group rings of G over the fields $GF(p)$ and Ω_p respectively. We make R_p and R_p' into (left) G-modules in an obvious way. We then prove the following

LEMMA 1. *Let e_0 and s be integers $\geqq 1$ such that $ee_0=(p-1)s$ and let $i_1<i_2<\cdots<i_r$ $(r=ee_0)$ be all the integers i such that $1\leqq i\leqq sp$, $(i, p)=1$. Let, furthermore, M be a G-module over $GF(p)$ containing a sequence of G-invariant submodules*

$$M = M_0 \supset M_1 \supset \cdots \supset M_r = \{0\},$$

such that M_{l-1}/M_l is isomorphic with A_{i_l} as defined above $(l=1, \cdots, r)$. Then, M is isomorphic with the direct sum of e_0f_0 copies of the G-module R_p.

Proof. Let M', M_l' denote the G-modules over Ω_p which are obtained from M, M_l by extending the scalar field $GF(p)$ to Ω_p. We have again a sequence of G-modules

$$M' = M_0' \supset M_1' \supset \cdots \supset M_r' = \{0\},$$

and M_{l-1}'/M_l' is isomorphic with A_{i_l}'. For $j=0, 1, \cdots, f_0-1$, let $a_{j,l}$ denote elements of M_{l-1}' such that the residue class of $a_{j,l}$ mod M_l' is mapped to a_j' in A_{i_l}' by the above isomorphism. Since the order e of the normal subgroup N of G generated by τ is prime to p, M' is completely reducible as an N-module; and we may choose $a_{j,l}$ so that

$$\tau a_{j,l} = \eta^{ip^j} a_{j,l}$$

with $i=i_l$. For any integer j with $j=j_0+tf_0$, $0\leqq j_0<f_0$, we then put

$$a_{j,l} = \sigma^{-t} a_{j_0,l}.$$

$a_{j,l}$ then depends only upon the residue class of j mod ff_0 and those $a_{j,l}$,

(²) Cf. Deuring [2, p. 38].

$0 \leqq j < ff_0$, represent a basis of M'_{l-1}/M'_l over Ω_p. From the definition, it also follows easily that

$$\sigma a_{j,l} = a_{j-f_0,l}, \qquad \tau a_{j,l} = \eta^{ip^j} a_{j,l}$$

with $i = i_l$.

Now, the rff_0 elements $a_{j,l}$, thus obtained for $0 \leqq j < ff_0$, $1 \leqq l \leqq r$, obviously form an Ω_p-basis of M'. They are all characteristic elements for the operator τ; and the corresponding characteristic values are given by η^t, where t runs over the residue classes ip^j mod e with $1 \leqq i \leqq sp$, $(i, p) = 1$, and $0 \leqq j < ff_0$. However, since $sp = s + ee_0$ and $p^{ff_0} \equiv 1$ mod e, we have

$$\{ip^j \bmod e;\, 1 \leqq i \leqq sp,\, 0 \leqq j < ff_0\}\text{(}^3\text{)}$$
$$= \{ip^j \bmod e;\, 1 \leqq i \leqq s,\, j\} \cup \{ip^j \bmod e;\, 1 \leqq i \leqq ee_0,\, j\},$$
$$\{ip^j \bmod e;\, 1 \leqq i \leqq sp,\, i \mid p,\, 0 \leqq j < ff_0\}$$
$$= \{ip^{j+1} \bmod e;\, 1 \leqq i \leqq s,\, 0 \leqq j < ff_0\}$$
$$= \{ip^j \bmod e;\, 1 \leqq i \leqq s,\, 0 \leqq j < ff_0\}.$$

It follows that

$$\{ip^j \bmod e;\, 1 \leqq i \leqq sp,\, (i, p) = 1,\, 0 \leqq j < ff_0\}$$
$$= \{ip^j \bmod e;\, 1 \leqq i \leqq ee_0,\, 0 \leqq j < ff_0\}$$
$$= e_0f_0f \text{ times the complete residue classes mod } e.$$

Therefore, every η^t $(0 \leqq t < e)$ appears exactly e_0f_0f times as a characteristic value of τ belonging to some $a_{j,l}$. On the other hand, the operator σ permutes the $a_{j,l}$ among themselves in ee_0f_0 cycles of length f; and if the characteristic value of τ belonging to $a_{j,l}$ is η^t, that of τ for $\sigma^{-1}a_{j,l}$ is given by η^{tq}. It then follows immediately that, as a G-module over Ω_p, M' is isomorphic with the direct sum of e_0f_0 copies of R'_p.

Now, let $e_0f_0 \times R_p$ and $e_0f_0 \times R'_p$ denote the direct sums of e_0f_0 copies of R_p and R'_p respectively. M and $e_0f_0 \times R_p$ are both G-modules over $GF(p)$ and their scalar extensions M' and $e_0f_0 \times R'_p$ are proved to be isomorphic. Therefore M and $e_0f_0 \times R_p$ are also isomorphic as G-modules over $GF(p)$, q.e.d.

1.2. Let G be a finite group as considered in 1.1. G is, namely, a group of order $n = ef$ generated by two elements σ and τ satisfying the relations (1.1) with a prime power $q = p^{f_0}$, $f_0 \geqq 1$. In the following, we shall further assume that f is divisible by p, and even by 4 if $p = 2$.

Let O_p denote the ring of p-adic integers and R the group ring of G with coefficients in O_p. We consider an R-module L with the following properties:

(i) as an O_p-module, L is the direct sum of a finite cyclic O_p-module W

(3) In these formulae, the residue classes mod e in the brackets are to be counted with multiplicities as i and j run over the domains as indicated.

generated by an element w of order p^κ, $\kappa \geqq 1$, and mn copies of the module O_p for some integer $m \geqq 1$:

$$L = W + O_p + \cdots + O_p,$$
$$W = O_p w, \qquad p^\kappa w = 0, \qquad p^{\kappa-1} w \neq 0, \qquad \kappa \geqq 1.$$

W is obviously invariant under G so that we have

$$\sigma w = g_1 w, \qquad \tau w = g_2 w \tag{1.3}$$

with suitable integers g_1, g_2, which are uniquely determined mod p^κ.

(ii) let $f = f'p$ and $\sigma_1 = \sigma^{f'}$. Then, $\sigma_1 w = w$, and L contains an element z such that $\sigma_1 z = z + w_1$, where w_1 is an element of order p in W, e.g., $p^{\kappa-1}w$.

(iii) the residue class module $\bar{L}$ of L mod pL is obviously an $(mn+1)$-dimensional G-module over the finite field $GF(p)$; and as such, it is the direct sum of m copies of the group ring R_p of G over $GF(p)$ and a one-dimensional module $\bar{L}_0$ over $GF(p)$ such that

$$\sigma \bar{a} = g_1 \bar{a}, \qquad \tau \bar{a} = g_2 \bar{a} \tag{1.4}$$

for any $\bar{a}$ in $\bar{L}_0$.

We shall now study the structure of such an R-module L. For any element c in L, let $\bar{c}$ denote the residue class of c mod pL. By (iii), we can choose elements $c_0, c_1, \cdots, c_m$ in L so that $\bar{c}_0$ is a basis of $\bar{L}_0$ and that $\bar{c}_0$ and $\rho\bar{c}_i$ ($\rho \in G$, $i = 1, \cdots, m$) together form a basis of $\bar{L}$ over $GF(p)$. We put

$$L' = Rc_1 + \cdots + Rc_m.$$

L/L' is then a cyclic O_p-module generated by the residue class of c_0 mod L'; and hence, it is either finite cyclic or isomorphic with O_p.

We first assume that L/L' is isomorphic with O_p and put

$$\sigma c_0 \equiv \zeta_1 c_0, \qquad \tau c_0 \equiv \zeta_2 c_0 \bmod L',$$

with suitable ζ_1, ζ_2 in O_p. By the assumption, ζ_1 and ζ_2 are uniquely determined by the above congruences; and, as $\sigma^f = 1$, $\tau^e = 1$, we must have $\zeta_1^f = 1$ and $\zeta_2^e = 1$. Since ζ_1 and ζ_2 are thus roots of unity in O_p, we have $\zeta_1^{p-1} = \zeta_2^{p-1} = 1$; and it also follows from (1.4) that

$$\zeta_1 \equiv g_1, \qquad \zeta_2 \equiv g_2 \bmod pO_p. \tag{1.5}$$

Let, then, $m \times R$ denote the direct sum of m copies of the group ring R considered as a G-module over O_p; and let ϕ be the R-homomorphism of $m \times R$ onto L' defined by

$$\phi(\alpha_1, \cdots, \alpha_m) = \alpha_1 c_1 + \cdots + \alpha_m c_m, \qquad \alpha_i \in R.$$

We denote by K the kernel of ϕ. As $L/L' \cong O_p$ by the assumption, the finite module W is contained in L' and it follows from (i) that L' is, as an O_p-

module, the direct sum of W and $mn-1$ copies of the module O_p. Therefore, there is an element x in $m\times R$ such that both $\phi(x)=w$ and $K=O_p p^\kappa x$ hold. Since K is clearly G-invariant, we see, as we have done for L/L', that

$$\sigma p^\kappa x = \zeta_1' p^\kappa x, \qquad \tau p^\kappa x = \zeta_2' p^\kappa x$$

with suitable roots of unity ζ_1', ζ_2' in O_p. It then follows that $\sigma x=\zeta_1' x$, $\tau x=\zeta_2' x$ and, consequently, that

$$\sigma w = \zeta_1' w, \qquad \tau w = \zeta_2' w.$$

We then have, by (1.3) and (1.5), that $\zeta_1' \equiv \zeta_1$, $\zeta_2' \equiv \zeta_2 \bmod pO_p$; and, as the ζ's are roots of unity in O_p, we get $\zeta_1' = \zeta_1$, $\zeta_2' = \zeta_2$. Let μ be an element in the center of R defined by

$$\mu = \sum_{i=0}^{f-1} \sum_{j=0}^{e-1} \zeta_1^{-i} \zeta_2^{-j} \sigma^i \tau^j \text{ (4)}.$$

Since $\sigma x=\zeta_1 x$ and $\tau x=\zeta_2 x$, it is easy to see that there is an element y of the form $y=(u_1, \cdots, u_m)$, $u_i\in O_p$, in $m\times R$, satisfying $x=\mu y$. On the other hand, as $\phi(x)=w$, there is no element y' in $m\times R$ such that $y=py'$, and, hence, some u_i is not divisible by p. Therefore, replacing $c_1, \cdots, c_m$ by their suitable linear combinations with coefficients in O_p and changing the mapping ϕ accordingly, we can make

$$y = (1, 0, \cdots, 0), \qquad x = (\mu, 0, \cdots, 0).$$

In other words, we may assume that the generators $c_1, \cdots, c_m$ of L' have a unique fundamental relation

$$p^\kappa \mu c_1 = 0.$$

We now put

$$c_1' = c_1 + \mu c_0, \qquad c_i' = c_i, \qquad \text{for } i = 0, 2, \cdots, m,$$

$$L'' = Rc_1' + \cdots + Rc_m'.$$

Using the above and

$$\sigma\mu c_0 = \zeta_1 \mu c_0, \qquad \tau\mu c_0 = \zeta_2 \mu c_0, \qquad \mu c_0 \equiv n c_0 \bmod L',$$

we can see easily that $c_1', \cdots, c_m'$ are linearly independent over R. Hence w is not contained in L'' and L/L'' is a finite cyclic O_p-module. We have thus proved that we may always assume L/L' is a finite cyclic O_p-module, for otherwise, we can take $c_0', c_1', \cdots, c_m'$ and L'' instead of the original $c_0, c_1, \cdots, c_m$ and L'.

We, now, therefore assume that L/L' is a finite cyclic O_p-module of order,

(4) To see that μ is contained in the center of R, notice $\zeta_1^e = \zeta_1$, $\zeta_2^e = \zeta_2$. μ is, up to a scalar, the unique element in R which takes the factors ζ_1, ζ_2 when multiplied by σ and τ respectively.

say, p^l, $l \geqq 0$. It then follows immediately from (i) that L' contains mn linearly independent elements over O_p, and hence that L' is isomorphic with $m \times R$ by the mapping ϕ as defined above. In particular, L' contains no element of finite order except 0 and $W \cap L' = \{0\}$. W is therefore mapped isomorphically into L/L' by the canonical homomorphism of L onto L/L'; and we see that the order of L/L' is not less than the order of W, i.e., that $l \geqq \kappa$. Suppose that $l = \kappa$. L is then the direct sum of W and L'; and, for the element z in (ii) above, we may put $z = rw + b_1$ with some integer r and some b_1 in L'. It then follows from (ii) that $w_1 = (\sigma_1 - 1)z = r(\sigma_1 - 1)w + (\sigma_1 - 1)b_1 = (\sigma_1 - 1)b_1$. But, as $(\sigma_1 - 1)b_1$ is an element in L', this contradicts the fact that $W \cap L' = \{0\}$. We must have therefore

$$l \geqq \kappa + 1. \tag{1.6}$$

Now, as L/L' is a cyclic module of order p^l, there exists an integer g such that $\sigma a \equiv ga \bmod L'$ for any element a in L. Since $\sigma^f = 1$, we obviously have $g^f \equiv 1 \bmod p^l$. It follows, in particular, that $g^f \equiv 1 \bmod p$; and as $f = f'p$, $g^{f'} \equiv 1 \bmod p$ also holds. If $p = 2$, f' is an even integer by the assumption; and we have also $g^{f'} \equiv 1 \bmod 4$. Therefore, if $g^f - 1$ were divisible by p^{l+1}, we would have $g^{f'} \equiv 1 \bmod p^l$ and

$$\sigma_1 a \equiv \sigma^{f'} a \equiv g^{f'} a \equiv a \bmod L'$$

for any a in L. Applying this to $a = z$, we see that $w_1 = (\sigma_1 - 1)z$ is in L', which is again a contradiction. We obtain therefore

$$g^f \equiv 1 \bmod p^l, \qquad g^f \not\equiv 1 \bmod p^{l+1}. \tag{1.7}$$

We next take an element a_0' in L which generates $L \bmod L'$ and satisfies $p^{l-\kappa} a_0' \equiv w \bmod L'$. As $p^\kappa w = 0$, $p^l a_0'$ is contained in $p^\kappa L'$, and so is $(g^f - 1)a_0'$. Put $(\sigma - g)a_0' = b$. b is then an element in L' such that

$$\left(\sum_{i=0}^{f-1} g^i \sigma^{f-i-1} \right) b = (\sigma^f - g^f)a_0' = (1 - g^f)a_0' \in p^\kappa L'. \tag{1.8}$$

Since, however, $L'/p^\kappa L'$ is (as a G-module) isomorphic with the direct sum of m copies of the group ring of G over $O_p/p^\kappa O_p$, it follows from (1.8) that there is an element b' in L' satisfying

$$b \equiv (\sigma - g)b' \bmod p^\kappa L'.$$

Replacing a_0' by $a_0' - b'$, if necessary, we may therefore assume that

$$(\sigma - g)a_0' \equiv 0 \bmod p^\kappa L'. \tag{1.9}$$

We now consider the action of τ on L/L'. Just as for σ, there is an integer g' such that $\tau a \equiv g'a \bmod L'$ for any a in L. Since $\tau^e = 1$, $g'^e \equiv 1 \bmod p^l$; and as $(e, p) = 1$, it follows immediately that $g' \equiv \zeta \bmod p^l O_p$ with some root of unity ζ in O_p. For any a in L, we have then $\tau a \equiv \zeta a \bmod L'$; and using $W \cap L'$

$=\{0\}$, we see in particular that

$$\tau w = \zeta w. \tag{1.10}$$

Clearly, the root of unity ζ is uniquely determined by the above equality. Using such ζ, we define an element λ in the center of R by

$$\lambda = \frac{1}{e}\sum_{j=0}^{e-1} \zeta^{-j}\tau^{j(5)}. \tag{1.11}$$

It then follows that $\lambda a_0' \equiv a_0'$ mod L', and hence we may replace a_0' by $\lambda a_0'$ and denote it again by a_0'. For the new a_0', (1.9) still holds; and furthermore we have also

$$\lambda a_0' = a_0', \qquad (\tau - \zeta)a_0' = 0. \tag{1.12}$$

Using (1.9), we put $(\sigma - g)a_0' = p^\kappa c''$ with c'' in L'. It then follows that

$$(\sigma - g)a_0' = p^\kappa c' \tag{1.13}$$

with $c' = \lambda c''$, which is also an element in L' satisfying

$$\lambda c' = c'.$$

We may hence put

$$c' = \sum_{i=1}^{m} \alpha_i \lambda c_i,$$

where the α_i $(i=1, \cdots, m)$ are linear combinations of σ^j with coefficients in O_p, i.e., elements of the group ring R_σ, over O_p, of the cyclic group generated by σ. We put

$$\alpha_i = u_i + (\sigma - g)\beta_i, \qquad u_i \in O_p, \beta_i \in R_\sigma,$$

for $i=1, \cdots, m$, and also

$$a_0 = a_0' - p^\kappa \sum_{i=1}^{m} \beta_i \lambda c_i, \qquad c = \sum_{i=1}^{m} u_i \lambda c_i.$$

We have then

$$\lambda a_0 = a_0, \qquad (\sigma - g)a_0 = p^\kappa c, \qquad (\tau - \zeta)a_0 = 0. \tag{1.14}$$

We shall next show that some u_i is not contained in pO_p. Suppose, on the contrary, that all u_i are in pO_p. We have then

$$(\sigma - g)a_0 = p^{\kappa+1}d_1$$

with some d_1 in L'. Multiplying both sides of the above by $\sum_{i=0}^{f-1} g^i\sigma^{f-i-1}$, we get

---

(5) Cf. footnote (4).

$$(1-g^f)a_0 = p^{\kappa+1}d_2, \qquad d_2 \in L'.$$

Using (1.6), (1.7), put

$$d_3 = (1-g^f)p^{-\kappa-1}a_0 - d_2.$$

We have then $p^{\kappa+1}d_3=0$. But, as $1-g^f$ is not divisible by p^{l+1}, the order of d_3 mod L' is exactly $p^{\kappa+1}$. Hence the order of d_3 itself must be also $p^{\kappa+1}$, and this contradicts the assumption (i). Therefore, some u_i, say u_1, is not contained in pO_p.

Finally, we then put

$$a_1 = \sum_{i=1}^{m} u_i c_i, \qquad a_j = c_j,\ j = 2, \cdots, m.$$

Since u_1 is a unit in O_p, we have

$$L' = Ra_1 + Ra_2 + \cdots + Ra_m;$$

and as a_0 generates L mod L', $a_0, a_1, \cdots, a_m$ obviously generate L over R. Furthermore, we also get from (1.12), (1.13), and (1.14) that

$$(1.15) \qquad (\sigma - g)a_0 = p^\kappa \lambda a_1, \qquad (\tau - \zeta)a_0 = 0.$$

From the definition of g, it follows that $\sigma w = g_1 w \equiv gw$ mod L'. But, as $W \cap L' = \{0\}$, we have $g_1 w = gw$, $\sigma w = gw$. Now, take any rational integer $g^* = g + sp^\kappa$ which is congruent to g mod p^κ. $a_0, a_1 - sa_0, a_2, \cdots, a_m$ is then clearly another system of generators of L over R and (1.15) holds for a_0, $a_1 - sa_0$, and g^* instead of a_0, a_1, and g. Hence, by such a substitution, we can let g take any given integer value satisfying $\sigma w = gw$. We have therefore proved the following lemma:

LEMMA 2. *Let L be an R-module having the properties* (i), (ii), (iii) *as given at the beginning of this section. Then, L has a system of $m+1$ generators $a_0, a_1, \cdots, a_m$ over R such that*

$$(\sigma - g)a_0 = p^\kappa \lambda a_1, \qquad (\tau - \zeta)a_0 = 0,$$

where ζ and λ are given by (1.10) *and* (1.11), *and g is a rational integer satisfying $\sigma w = gw$ which can otherwise be arbitrarily given beforehand.*

We shall next prove the following

LEMMA 3. *Let $a_0, a_1, \cdots, a_m$ be a system of generators of L over R as given in the previous lemma. Then a_0 and ρa_i ($\rho \in G$, $i = 1, \cdots, m$) form a system of generators of L over the ring O_p, and every O_p-linear relation among a_0 and the ρa_i is a consequence of the following fundamental relation:*

$$(1.16) \qquad (g^f - 1)a_0 + p^\kappa \frac{1}{e} \sum_{i=0}^{f-1} \sum_{j=0}^{e-1} g^i \zeta^j \sigma^{f-i-1} \tau^{e-j} a_1 = 0.$$

Proof. The first half of the lemma is trivial. Let L^* denote the free O_p-module which is the direct sum of O_p and m copies of the group ring R, and let ψ be the O_p-homomorphism of L^* onto L defined by

$$\psi(u, \alpha_1, \cdots, \alpha_m) = ua_0 + \sum_{i=1}^{m} \alpha_i a_i, \qquad u \in O_p,\ \alpha_i \in R.$$

Using the property (i) of L, we can then choose a basis $b_0, b_1, \cdots, b_{mn}$ of L^* over O_p such that the kernel of ψ is given by $O_p p^\kappa b_0$. Now, multiplying both sides of $(\sigma - g)a_0 = p^\kappa \lambda a_1$ by $\sum_{i=0}^{f-1} g^i \sigma^{f-i-1}$, we get immediately the relation (1.16). Therefore, if we put

$$b = \left((g^f - 1)p^{-\kappa},\ \sum_{i=0}^{f-1} g^i \sigma^{f-i-1}\lambda, 0, \cdots, 0\right),$$

$p^\kappa b$ is contained in the kernel $O_p p^\kappa b_0$; and we get

$$b = u_0 b_0$$

with some u_0 in O_p. However, b is obviously not contained in pL^*. u_0 is hence a unit in O_p and the kernel of ψ is also equal to $O_p p^\kappa b$. The lemma is therefore proved.

It follows immediately from Lemma 3 that all the R-linear relations among the generators $a_0, a_1, \cdots, a_m$ are consequences of the fundamental relations (1.15) and, hence, that the R-module L is, up to isomorphisms, uniquely determined by the properties (i), (ii), (iii) given above.

1.3. We shall next prove a lemma on a certain type of totally disconnected compact groups.

Let J be an arbitrary group. Let $\{N_\delta\}$ be the family of all normal subgroups of J such that the indices $[J:N_\delta]$ are finite, and let J_0 be the intersection of all such N_δ's. We can make the factor group $J' = J/J_0$ a totally bounded topological group by taking the family of subgroups $\{N_\delta/J_0\}$ as a system of neighborhoods of 1 in J'. The completion of J' then gives us a totally disconnected compact group $\bar{J}$ which we shall call the total completion of J.

Instead of the family $\{N_\delta\}$ of all normal subgroups of J with finite indices, we can also start with suitable subfamilies of $\{N_\delta\}$ and get various totally disconnected compact groups in a similar way. In particular, taking the family of all normal subgroups N_δ' of J such that the indices $[J:N_\delta']$ are finite and not divisible by a given prime number p, we get a compact group ${}^p\bar{J}$ which we shall call the p-complementary completion of J. The p-completion $\bar{J}^p$ of J can be obtained similarly considering those normal subgroups N_δ'' of J whose indices in J are powers of p.

Now, let G be a totally disconnected compact group and let there be a homomorphism ϕ of J into G such that the image $\phi(J)$ of J is everywhere dense in G. The group J_0 above is then contained in the kernel of ϕ so that ϕ

induces a homomorphism ϕ' of J' into G and ϕ' can be uniquely extended to a continuous homomorphism ψ of $\bar{J}$ onto G. In that sense, $\bar{J}$ may be called the universal group in the family of totally disconnected compact groups G having everywhere dense homomorphic images of the group J; and it is, up to isomorphisms, uniquely characterized by that property. We can also give similar characterizations for ${}^p\bar{J}$ and $\bar{J}^p$.

We now consider the special case where the group J is generated by two elements α and β satisfying the unique relation

$$\alpha\beta\alpha^{-1} = \beta^q,$$

where q is, as before, a power p^{f_0} of a prime number p. We shall denote by Γ the total completion of such J, and by σ and τ the elements of Γ which are the images of α and β by the canonical homomorphism of J onto J'. We have then obviously

$$\sigma\tau\sigma^{-1} = \tau^q.$$

Let Γ_0 be the closure of the cyclic subgroup of Γ generated by τ. Γ_0 is then a closed normal subgroup of Γ; and Γ/Γ_0 and Γ_0 are isomorphic with the total and p-complementary completions of an infinite cyclic group, respectively. It is also easy to see that Γ is an inverse limit of the type of finite groups as considered in 1.1 and 1.2.

LEMMA 4. *Let G be a totally disconnected compact group and N a closed normal subgroup of G such that $G/N=\Gamma$. If, furthermore, N is an inverse limit of finite p-groups, then the group extension G/N splits, i.e. there exists a closed subgroup H in G such that $G=HN$, $H\cap N=1$.*

For the proof, we first notice the following: let G_1 and G_2 be both totally disconnected compact groups. We say that G_1 and G_2 are relatively prime if the index $[G_1:N_1]$ of any open normal subgroup N_1 in G_1 is always prime to the index $[G_2:N_2]$ of any open normal subgroup N_2 in G_2. We can then prove the following

LEMMA 5. *Let G be a totally disconnected compact group and N a closed normal subgroup of G such that G/N and N are relatively prime. Then the group extension G/N splits, i.e., there is a closed subgroup H in G such that $G=HN$, $H\cap N=1$. If, furthermore, either G/N or N is solvable, then any two such closed subgroups H_1, H_2 satisfying $G=H_iN$, $H_i\cap N=1$ $(i=1, 2)$ are conjugate in G.*

It is known that Lemma 5 holds if G is a finite group and the proof for the general case can be easily reduced to that special case. We, therefore, omit the details here[6].

---

[6] For the proof of the lemma for a finite group G, cf. [9, p. 126]. To prove the existence of H in the compact case, take a minimal closed subgroup H such that $G=HN$.

Now, Lemma 4 can be proved as follows; let G' be the closed normal subgroup of G containing N such that $G'/N=\Gamma_0$. Since Γ_0 is isomorphic with the p-complementary completion of an infinite cyclic group and N is an inverse limit of p-groups, G'/N and N are relatively prime. Hence, by Lemma 5, there is a closed subgroup H_1 of G' such that $G'=H_1N$, $H_1\cap N=1$. Let a be an element of G such that the coset $a'=aG'$ generates an everywhere dense cyclic subgroup of G/G'. $H_2=aH_1a^{-1}$ then also satisfies $G'=H_2N$, $H_2\cap N=1$; and by Lemma 5, there is an element b in G' such that $H_1=bH_2b^{-1}$. Let C denote the closure of the cyclic group generated by $a_1=ba$. As G/G' is the total completion of the infinite cyclic group generated by $a'=a_1G'$, it is easy to see that $G=CG'$, $C\cap G'=1$. Since $a_1H_1a_1^{-1}=H_1$, $H=CH_1$ is then a closed subgroup of G such that $G=HN$, $H\cap N=1$.

2. Finite tamely ramified extensions

2.1. Let p be a prime number, Q_p the field of p-adic numbers, and Ω an algebraic closure of Q_p. All the fields we shall consider in the following are extensions of Q_p contained in Ω.

We take such a finite extension k of Q_p in Ω as our ground field and denote by m, e_0, and f_0 the degree, the ramification, and the residue class degree of the extension k/Q_p respectively. We have then $m=e_0f_0$. Let E be a finite extension of k in Ω and let F be the inertia field of E/k. We denote the degree, the ramification, and the residue class degree of E/k by n, e, and f, respectively. We have then clearly $e=[E:F]$, $f=[F:k]$, and $n=ef=[E:k]$. We assume that the extension E/k is tamely ramified, i.e., that e is prime to p. Furthermore, we also assume that E/k is a Galois extension and denote by G and N the Galois groups of E/k and E/F respectively. N is then a normal subgroup of G, and G/N and N are both cyclic. We say that the tamely ramified Galois extension E/k splits if the group extension G/N splits. It is easy to see that E/k splits if and only if E contains a prime element π such that π^e is a prime element of k. In fact, suppose that E contains such an element π. Since the residue class degree of the extension E/Q_p is ff_0, E contains a root of unity ξ of order $p^{ff_0}-1$; and we have

$$F=k(\xi), \qquad E=k(\xi,\pi).$$

Put $F'=k(\pi)$ and denote by σ the Frobenius automorphism of the unramified extension E/F'. If we then take any generator τ of N, the group G is generated by σ and τ; and we have

$$\begin{aligned} \xi^\sigma &= \xi^q, & \pi^\sigma &= \pi, \\ \xi^\tau &= \xi, & \pi^\tau &= \eta\pi, \end{aligned} \tag{2.1}$$

where $q=p^{f_0}$ is the number of elements in the residue class field of k and η is a suitable eth root of unity in E. It follows immediately that

$$\sigma^f=1, \qquad \tau^e=1, \qquad \sigma\tau\sigma^{-1}=\tau^q.$$

Therefore, the group extension G/N splits and G is the type of group we considered in 1.1. We can also see easily that if, conversely, E/k splits, then E contains a prime element π such that π^e is in k.

2.2. Let E/k be a splitting tamely ramified finite Galois extension of k as considered in 2.1 and let F, ξ, π, σ, τ etc. be the same as given there. We also denote by p^κ $(\kappa \geqq 0)$ the highest power of p dividing the order of any root of unity in E and by w a primitive p^κth root of unity in E. In the following, we shall assume that $\kappa \geqq 1$, i.e., that E contains at least a primitive pth root of unity, and also that the degree of the extension $E/k(w)$ is divisible by p, and even by 4 if $p=2$.

We shall now study the action of the Galois group G of E/k operating on the multiplicative group E^* of E. For any positive integer i, let U_i denote the G-invariant subgroup of E^* consisting of all units a in E such that $a \equiv 1 \mod \pi^i$. We have then obviously the following direct product decomposition:

$$E^* = \{\pi\} \times \{\xi\} \times U_1, \tag{2.2}$$

where $\{\pi\}$ denotes the infinite cyclic group generated by π and $\{\xi\}$ the finite cyclic group of order $q^f - 1$ generated by ξ. Since the action of G on the first two direct factors is given by (2.1), we simply have to study the behavior of G acting on the group U_1.

Let O_p and R denote, as in 1.2, the ring of p-adic integers and the group ring of G with coefficients in O_p respectively, and let u be any element in O_p. For any a in U_1, the element a^u in U_1 is defined as usual; and as $(a^u)^\rho = (a^\rho)^u$ for any ρ in G, we can consider the group ring R as an operator domain of the abelian group U_1 in an obvious way. We shall next show that the group U_1 with the operator domain R has the properties (i), (ii), (iii) of 1.2, though U_1 here is a multiplicative group and those conditions have to be modified accordingly.

It is well-known that (i) holds for U_1 with $m = [k:Q_p]$ and with the primitive p^κth root of unity w. We put here

$$w^\sigma = w^{a_1}, \qquad w^\tau = w^{a_2}.$$

By the assumption, $[E:k(w)]$ is divisible by p. On the other hand, as $k(w)/k$ is a Galois extension, $[k(\pi, w):k(w)]$ divides $e = [k(\pi):k]$ and, hence, is prime to p. $[E:k(\pi, w)]$ is therefore also divisible by p and $k(\pi, w)$ is contained in the fixed field of $\sigma_1 = \sigma^{f'}$ where $f = f'p$. We have then $w^{\sigma_1} = w$, and the existence of an element z in U_1 satisfying $z^{\sigma_1} = zw_1$ with a primitive pth root of unity w_1 also follows immediately from the fact that E is a cyclic unramified extension of degree p over the fixed field of σ_1 containing primitive pth roots of unity. The second property (ii) is hence verified for U_1.

To show, finally, that (iii) also holds for U_1, we first consider the action of σ and τ on the factor group U_i/U_{i+1}. By the definition of U_i, every element a in U_i satisfies a congruence relation of the form

$$a \equiv 1 + b\pi^i \bmod \pi^{i+1}$$

with some integer b in E; and the residue class b' of b mod π is uniquely determined by a. The mapping $b' = \phi(a)$ of U_i into the residue class field $GF(q^f)$ of E then induces an isomorphism of U_i/U_{i+1} onto the additive group of $GF(q^f)$, and it follows from (2.1) that the same isomorphism gives an isomorphism of the G-group U_i/U_{i+1} onto the G-module A_i as defined in 1.1.

Put $V_i = U_i U_1^p$ for $i \geqq 1$. We then get a sequence of G-invariant subgroups

$$U_1 = V_1 \supseteq V_2 \supseteq V_3 \supseteq \cdots,$$

concerning which we know the following results(7):

(1) let $ee_0 = (p-1)s$. s is then an integer; and if $i > sp$, $V_i = U_1^p$.

(2) let w_1 be a primitive pth root of unity in E. Since $sp = ee_0 + s = ee_0 + ee_0/(p-1)$, $p(1-w_1)$ is exactly divisible by π^{sp}. Hence, for any a in U_{sp}, we have

$$a \equiv 1 - bp(1 - w_1) \bmod \pi^{sp+1}$$

with some integer b in E; and a mapping $b' = \phi(a)$ is defined much as before. It can be then proved that a is in $U_{sp} \cap U_1^p$ if and only if $\phi(a)$ is in the subgroup D of the additive group of the residue class field $GF(q^f)$ formed by the elements $a'^p - a'$, $a' \in GF(q^f)$, and that ϕ induces an isomorphism of

$$V_{sp}/V_{sp+1} = V_{sp}/U_1^p$$

onto the factor group of $GF(q^f)$ mod D. Take an integer b_0 of E such that the residue class of b_0 mod π is not in D and put

$$a_0 = 1 - b_0 p(1 - w_1).$$

It then follows from the above that V_{sp}/U_1^p is a cyclic group of order p generated by the coset $a_0 U_1^p$. Now, since $w_1^\sigma = w_1^{g_1}$, we have

$$1 - w_1^\sigma = 1 - w_1^{g_1} \equiv g_1(1 - w_1) \equiv g_1^q(1 - w_1) \bmod \pi^{s+1}.$$

It then follows that

$$a_0^\sigma = 1 - b_0^\sigma p(1 - w_1^\sigma) \equiv 1 - g_1^q b_0^q p(1 - w_1) \bmod \pi^{sp+1};$$

and as the residue classes of $(g_1 b_0)^q$ and $g_1 b_0$ are congruent mod D, we obtain

$$a_0^\sigma \equiv a_0^{g_1} \bmod U_1^p.$$

Similarly, we also have

$$a_0^\tau \equiv a_0^{g_2} \bmod U_1^p.$$

(3) let i be an integer such that $1 \leqq i \leqq sp$. If p divides i, then $V_i = V_{i+1}$.

(7) Cf. [3, §15].

On the other hand, if i is prime to p, V_i/V_{i+1} is isomorphic with U_i/U_{i+1}, both being considered as G-groups. Therefore, by the remark mentioned above, the G-group V_i/V_{i+1} is isomorphic with A_i of 1.1.

Now, let $i_1<i_2< \cdots <i_r$ be all the integers i such that $1\leqq i\leqq sp$, $(i, p)=1$ and put

$$V'_{j-1} = V_{i_j}/V_{sp}, \qquad i = 1, \cdots, r,$$
$$V'_r = 1.$$

It follows from (3) that the sequence of the G-invariant subgroups

$$V'_0 \supset V'_1 \supset \cdots \supset V'_r = 1$$

satisfies the assumption of Lemma 1; and by that lemma, we see immediately that as a G-group, $V'_0 = U_1/V_{sp}$ is isomorphic with the direct sum of $m=e_0f_0$ copies of the group ring R_p of G over the finite field $GF(p)$. On the other hand, we also know by (2) that V_{sp}/U_1^p is a cyclic group of order p satisfying (1.4). It then follows easily that the G-group U_1/U_1^p has the structure as given in (iii).

We have thus verified that the group U_1 with the operator domain R has all the properties (i), (ii), (iii) in 1.2. Therefore, applying Lemmas 2, 3, we can immediately obtain the following theorem:

THEOREM 1. *Let k be a finite extension of degree m over the field of p-adic numbers Q_p and $E=k(\xi, \pi)$ a splitting tamely ramified Galois extension of degree $n=ef$ over k whose Galois group G is generated by σ and τ satisfying (2.1). Let p^κ denote the highest power of p dividing the order of any root of unity in E and w a primitive p^κth root of unity in E. Assume that $\kappa\geqq 1$ and that $[E:k(w)]$ is divisible by p, and even by 4 if $p=2$. Then, the multiplicative group U_1 consisting of all units a in E with $a\equiv 1 \bmod \pi$ contains a set of elements $a_0, a_1, \cdots, a_m$ such that*

(1) *every element a in U_1 can be written in the form*

$$a = a_0^u a_1^{\alpha_1} \cdots a_m^{\alpha_m},$$

where u is an element in the ring of p-adic integers O_p and α_i are elements of the group ring R of G with coefficients in O_p,

(2) $$a_0^{\sigma-g} = a_1^{p^\kappa\lambda}, \qquad a_0^{\tau-\zeta} = 1,$$

where g is a given rational integer satisfying $w^\sigma=w^g$, ζ is a root of unity in O_p uniquely determined by $w^\tau=w^\zeta$, and λ is an element in the center of R, defined by

$$\lambda = \frac{1}{e}\sum_{j=0}^{e-1}\zeta^{-j}\tau^j,$$

(3) $a=a_0^u a_1^{\alpha_1}\cdots a_m^{\alpha_m}$ *is* 1 *if and only if* $\alpha_i=0$ *for* $i>1$ *and*

$$u = v(g^f - 1), \qquad \alpha_1 = vp^\kappa \frac{1}{e}\sum_{i=0}^{f-1}\sum_{j=0}^{e-1} g^i\zeta^i\sigma^{f-i-1}\tau^{e-i}$$

with some v *in* O_p.

By (2.1), (2.2), and Theorem 1, the action of G on the multiplicative group E^* of E is completely determined. Here we have made certain assumptions on E/k. However, as every finite tamely ramified Galois extension E' of k can be imbedded in an extension E/k satisfying those conditions, we can also see, using the above results, the behavior of the Galois group of E'/k acting on the multiplicative group of such E'. But we omit here the details([8]).

3. The Galois group of Ω/k

3.1. Let k be, as in §2, a finite extension of degree m over the field of p-adic numbers Q_p and Ω an algebraic closure of Q_p containing k. We denote by V the composite of all finite tamely ramified extensions of k contained in Ω and call it the ramification field of Ω/k. Obviously, V contains the composite of all finite unramified extensions of k in Ω, i.e., the inertia field T of the extension Ω/k. Let $q=p^{f_0}$ be, as before, the number of elements in the residue class field of k and π_1 a prime element of k which we shall fix in the following. For any positive integer e prime to p, we choose a primtive eth root of unity ξ_e and an eth root π_e of π_1 in Ω so that if $e=e_1e_2$, then

$$\xi_{e_2} = \xi_e^{e_1}, \qquad \pi_{e_2} = \pi_e^{e_1}. \tag{3.1}$$

We put, for any integer $f\geqq 1$,

$$E_f = k(\xi_e, \pi_e), \qquad F_f = k(\xi_e),$$

where $e=e_f=q^f-1$. E_f/k is then a splitting tamely ramified Galois extension of degree ef as considered in §2 and F_f is the inertia field of E_f/k. As V and T are also the composites of all E_f and F_f ($f\geqq 1$) respectively, V is obtained by adjoining all ξ_e and π_e to k and T is obtained by adjoining all ξ_e to k.

V is obviously an infinite Galois extension of k and we denote by $G(V/k)$ the Galois group of V/k. $G(V/k)$ is then a totally disconnected compact group in Krull's topology; and as T/k is also a Galois extension, the Galois group $G(V/T)$ of V/T is a closed normal subgroup of $G(V/k)$. It is now clear from (3.1) that $G(V/k)$ contains automorphisms σ' and τ' such that

$$\begin{aligned} \xi_e^{\sigma'} &= \xi_e^{q}, & \pi_e^{\sigma'} &= \pi_e, \\ \xi_e^{\tau'} &= \xi_e, & \pi_e^{\tau'} &= \xi_e\pi_e. \end{aligned} \tag{3.2}$$

([8]) More general, but less explicit results on the action of the Galois group on the multiplicative group of a local field are given in [6; 7].

σ' and τ', then, generate an everywhere dense subgroup of $G(V/k)$ and τ' generates an everywhere dense subgroup of $G(V/T)$. It also follows from (3.2) that

$$\sigma'\tau'\sigma'^{-1} = \tau'^{q}.$$

Therefore, if we denote by J, as in 1.3, the group generated by α, β satisfying the relation $\alpha\beta\alpha^{-1}=\beta^q$, and by Γ the total completion of J, there is a continuous homomorphism ψ of Γ onto $G(V/k)$ such that $\psi(\sigma)=\sigma'$, $\psi(\tau)=\tau'$, σ and τ being the elements of Γ corresponding to α and β in J respectively. Since $G(V/k)$ contains, for each $f\geqq 1$, an open normal subgroup $G(V/E_f)$ of index ef, we can see immediately that ψ is an isomorphism of Γ onto $G(V/k)$.

3.2. We shall next consider the structure of the ramification group of Ω/k, i.e., the structure of the Galois group $G(\Omega/V)$ of the extension Ω/V. Let K be an arbitrary finite Galois extension of k and E the ramification field of the extension K/k, i.e. $E=K\cap V$. It is well-known that the Galois group $G(K/E)$ of K/E is a p-group, and it follows immediately that $G(\Omega/V)$ is an inverse limit of finite p-groups. Let V^* denote the multiplicative group of V and V^{*p} the subgroup of V^* consisting of pth powers of elements in V^*. Using the fact that V^* is the union of the multiplicative groups of E_f ($f\geqq 1$), it is easy to see that V^*/V^{*p} is an infinite group. On the other hand, given any integer $f\geqq 1$, V contains the field F_f which is an extension of degree f over k. Hence, by local class field theory, every Galois 2-cocycle over the field V splits(9). As $G(\Omega/V)$ is obviously a separable topological group, it follows from a result in [4] that the ramification group $G(\Omega/V)$ is the p-completion of a free group with a countable number of free generators. Using, then, the fact that $G(\Omega/k)/G(\Omega/V)$ is isomorphic with $G(V/k)\cong\Gamma$, we can see also by Lemma 4 that the group extension $G(\Omega/k)/G(\Omega/V)$ splits. We thus have the following

THEOREM 2. *Let k be a finite extension of the field of p-adic numbers Q_p and Ω an algebraic closure of Q_p containing k. Furthermore, let T and V be the inertia field and the ramification field of the extension Ω/k, respectively. Then:*

(i) *the Galois group $G(V/k)$ of the extension V/k is isomorphic with the total completion Γ of a group J generated by two elements α, β satisfying a unique relation $\alpha\beta\alpha^{-1}=\beta^q$, where q is the number of elements in the residue class field of k; if σ and τ denote the elements of Γ corresponding to α and β in J respectively, there exists an isomorphism ψ of Γ onto $G(V/k)$ such that $\psi(\sigma)$ induces the Frobenius automorphism of T/k and such that $\psi(\tau)$ generates an everywhere dense subgroup of $G(V/T)$, the Galois group of V/T,*

(ii) *the ramification group $G(\Omega/V)$, i.e., the Galois group of the extension Ω/V, is isomorphic with the p-completion of a free group with a countable number of free generators,*

(9) For the results in local class field theory used here and in the following, cf. [1].

(iii) *$G(\Omega/V)$ is a closed normal subgroup of the Galois group $G(\Omega/k)$ of Ω/k and the group extension $G(\Omega/k)/G(\Omega/V)$ splits, i.e., there is a closed subgroup H of $G(\Omega/k)$ such that $G(\Omega/k)=HG(\Omega/V)$, $H\cap G(\Omega/V)=1$.*

Now, Theorem 2 being proved, the structure of the Galois group $G(\Omega/k)$ can be determined completely, if we know the automorphisms $\phi_b(a)=bab^{-1}$ of the normal subgroup $G(\Omega/V)$ defined by elements b in H. However, the determination of those automorphisms ϕ_b seems to be a different problem and, in the following, we shall only describe explicitly the effect of the automorphisms ϕ_b on the factor commutator group of $G(\Omega/V)$, i.e., on the Galois group $G(V'/V)$ of the maximal abelian extension V' of V in Ω.

3.3. We shall now denote by G and N the Galois groups of V'/k and V'/V respectively. G is again a totally disconnected compact group and N a closed abelian normal subgroup of G. The factor group G/N is then canonically isomorphic with the Galois group $G(V/k)$ and hence also with the group Γ as given in Theorem 2. For simplicity, we put $G/N=G(V/k)=\Gamma$ and denote by σ and τ the elements of Γ which, when considered as Galois automorphisms of V/k, are the same as σ' and τ' in (3.2) respectively. We put, namely,

$$\xi_e^\sigma=\xi_e^q,\qquad \pi_e^\sigma=\pi_e,$$
$$\xi_e^\tau=\xi_e,\qquad \pi_e^\tau=\xi_e\pi_e, \tag{3.3}$$

where ξ_e and π_e are elements of V as given in 3.1. The group $\Gamma=G/N$ then operates on the abelian group N in a natural way and our problem now is to determine the structure of the Γ-group N.

Let E_f/k be the tamely ramified extension of degree $ef(e=q^f-1)$ as defined in 3.1, and let G_f denote the Galois group of V'/E_f. We also denote by G_f' the topological commutator group of G_f and put $N_f=N\cap G_f'$. G_f/G_f' is then clearly the Galois group of the maximal abelian extension K_f of E_f in Ω; and NG_f'/G_f' is the ramification group of K_f/E_f, for it corresponds to the ramification field $V\cap K_f$ of the extension K_f/E_f. Therefore, if we denote by $U_1(E_f)$ the group of all units a in E_f such that $a\equiv 1 \bmod \pi_e$, the norm residue mapping of E_f gives, by local class field theory, a topological isomorphism ϕ_f' of $U_1(E_f)$ onto NG_f'/G_f', the group $U_1(E_f)$ being considered as a compact group with respect to the natural topology of the local field E_f. Combining ϕ_f' with the canonical isomorphism of NG_f'/G_f' onto N/N_f, we also have a topological isomorphism ϕ_f of $U_1(E_f)$ onto N/N_f. Furthermore, $U_1(E_f)$ and N/N_f both have the group Γ as an operator domain in obvious ways, and it follows from a property of the norm residue mapping that ϕ_f defined above is a Γ-isomorphism of $U_1(E_f)$ onto N/N_f.

Now, let f' be a divisor of f. $E_{f'}$ is then a subfield of E_f and N_f is a subgroup of $N_{f'}$. We denote by $\nu_{f',f}$ the norm mapping of the field E_f to the subfield $E_{f'}$ and by $\psi_{f',f}$ the canonical homomorphism of N/N_f onto $N/N_{f'}$. $\nu_{f',f}$ then maps $U_1(E_f)$ onto $U_1(E_{f'})$ and it again follows from one of the properties of the norm residue mapping that

$$\phi_{f'} \circ \nu_{f',f} = \psi_{f',f} \circ \phi_f. \tag{3.4}$$

On the other hand, the intersection of all N_f, $f \geqq 1$, is clearly the identity and, hence, N is the inverse limit of the groups N_f, $f \geqq 1$, with the homomorphisms $\psi_{f',f}$. Therefore, we also see from (3.4) that the Γ-group N is isomorphic with the inverse limit U of the Γ-groups $U_1(E_f)$ with the homomorphisms $\nu_{f',f}$.

Now, let p^κ be the highest power of p dividing the order of any root of unity contained in V. From the definition of the ramification field V, it follows easily that there is such a finite power p^κ of p and that $\kappa \geqq 1$. We denote by w a primitive p^κth root of unity in V and put, as in Theorem 1,

$$w^\sigma = w^q, \qquad w^\tau = w^\zeta, \tag{3.5}$$

with a rational integer g prime to p and a root of unity ζ in O_p. The existence of such ζ follows from the fact that the order of the automorphism induced by τ on $k(w)$ is prime to p. Though ζ is uniquely determined by the above equality, the integer g is only determined mod p^κ by (3.5). But, we shall fix such an integer g once for all in the following considerations.

We now take a positive integer f_1 such that E_{f_1} contains w and that the degree $[E_{f_1}:k(w)]$ is divisible by p, and also by 4 if $p=2$. For any positive integer f divisible by f_1, the extension E_f/k then satisfies all the assumptions of Theorem 1, the Galois group of E_f/k being generated by the restrictions on E_f of the automorphisms σ and τ in $\Gamma = G(V/k)$ defined by (3.3). Therefore, by that theorem, the group $U_1(E_f)$ contains a system of $m+1$ elements $a_0, a_1, \cdots, a_m$, having the properties (1), (2), (3) of Theorem 1 with respect to σ, τ, g and ζ given by (3.3) and (3.5) respectively. Hence, if we denote by Θ the set of all such systems of $m+1$ elements $\theta = (a_0, a_1, \cdots, a_m)$ having the properties mentioned above, Θ is nonempty; and it is also compact when considered as a subset of the direct product of $m+1$ copies of the compact group $U_1(E_f)$ with its natural topology. We now choose a sequence of positive integers $f_2, f_3, \cdots$, such that f_i divides f_{i+1}, $i=1, 2, \cdots$, and such that every integer f is a divisor of some f_i. Then U is also the inverse limit of the sequence of Γ-groups $U_1(E_{f_i})$, $i=1, 2, \cdots$, with the homomorphisms $\nu'_{j,i} = \nu_{f_j, f_i}$ $(j \leqq i)$. For each $i \geqq 1$, let Θ_i denote the Θ-set of E_{f_i} as defined above. If, then, $j \leqq i$ and $\theta = (a_0, a_1, \cdots, a_{m_0})$ is in Θ_i, $\nu'_{j,i}\theta = (a'_0, a'_1, \cdots, a'_m)$ with $a'_t = \nu'_{j,i} a_t$ $(t=0, 1, \cdots, m)$ belongs to Θ_j; and those $\nu'_{j,i}\theta$ form a closed subset $\nu'_{j,i}\Theta_i$ of Θ_j. Since $\nu'_{1,j} \circ \nu'_{j,i} = \nu'_{1,i}$ for any $1 \leqq j \leqq i$, $\nu'_{1,i}\Theta_i$ is contained in $\nu'_{1,j}\Theta_j$ for $j \leqq i$, and the intersection of all $\nu'_{1,i}\Theta_i$, for $i=1, 2, \cdots$, is nonempty by the compactness of Θ_1. Take an element θ_1 in the intersection and let Θ'_i denote, for $i=2, 3, \cdots$, the set of all θ in Θ_i such that $\nu'_{1,i}\theta = \theta_1$. Again Θ'_i is compact and $\nu'_{j,i}\Theta'_i$ is a subset of Θ'_j for $j \leqq i$. Therefore, the intersection of all $\nu'_{2,i}\Theta'_i$, $i \geqq 2$, is nonempty and we can take an element θ_2 in the intersection. Proceeding similarly, we get a sequence of elements $\theta_1, \theta_2, \cdots$, such that θ_i is in Θ_i and that $\nu'_{j,i}\theta_i = \theta_j$ for $j \leqq i$. Put

$$\theta_i = (a_{i0}, a_{i1}, \cdots, a_{im}), \qquad a_{i,t} \in U_1(E_{f_i}).$$

We have then

$$\nu'_{j,i}a_{it} = a_{jt}, \qquad t = 0, 1, \cdots m; j \leqq i. \tag{3.6}$$

Now, if we use those elements $a_{i0}, a_{i1}, \cdots, a_{im}$ of $U_1(E_{f_i})$, the action of σ and τ on $U_1(E_{f_i})$ is given explicitly by Theorem 1. Furthermore, the systems of those elements θ_i, $i=1, 2, \cdots$, are coherent in the sense that they satisfy (3.6) for any $j \leqq i$. Therefore, taking the limits of a_{it}, we can see immediately how σ and τ act on the inverse limit U of $U_1(E_{f_i})$; and thus the structure of the Γ-group $N = G(V'/V)$ can also be determined explicitly.

3.4. The result obtained above can be stated more clearly if it is formulated, as we shall do it in the following, in terms of the character group $\tilde{N}$ of the compact abelian group N. In general, let A be a locally compact abelian group, $\tilde{A}$ the character group of A and let

$$(a, \chi), \qquad a \in A, \chi \in \tilde{A},$$

be a dual pairing of A and $\tilde{A}$. For any χ in $\tilde{A}$ and for any element ρ in a group Σ acting on A, there exists a unique character χ^ρ such that

$$(a^\rho, \chi^\rho) = (a, \chi)$$

for all a in A. Σ thus becomes also an operator domain of $\tilde{A}$ and the structure of the Σ-group A is, conversely, uniquely determined by that of the Γ-group $\tilde{A}$. In our case, we have therefore only to describe the structure of the Γ-group $\tilde{N}$ explicitly.

Now, let Q_p denote the additive group of the p-adic number field as well as the field itself and let O_p also be the additive group of p-adic integers. With respect to the p-adic topology, O_p is an open compact subgroup of Q_p so that $\overline{Q}_p = Q_p/O_p$ is discrete. On the other hand, Q_p is an O_p-module in an obvious way and, as O_p is an invariant submodule of Q_p, $\overline{Q}_p$ is also an O_p-module. We now denote by $C(\Gamma)$ the set of all continuous functions defined on the compact group Γ with values in $\overline{Q}_p$, and for any ρ in Γ and $h = h(\omega)$ in $C(\Gamma)$, we define a function ρh in $C(\Gamma)$ by

$$(\rho h)(\omega) = h(\rho^{-1}\omega), \qquad \omega \in \Gamma.$$

It is clear that Γ thus becomes an operator domain of $C(\Gamma)$. For any integer $i \geqq 1$, the set of all functions h in $C(\Gamma)$ such that $\rho h = h$ for any ρ in the Galois group of V/E_{f_i} forms a submodule of $C(\Gamma)$. We denote that submodule by $C(\Gamma_i)$, for it can be canonically identified with the set of all (continuous) functions with values in $\overline{Q}_p$ defined on the Galois group Γ_i of E_{f_i}/k. It then follows immediately from the discreteness of $\overline{Q}_p$ that $C(\Gamma)$ is the union of all such $C(\Gamma_i)$, $i=1, 2, \cdots$. Let $h=h(\omega)$ be again a function in $C(\Gamma)$ and let h be in $C(\Gamma_i)$ for some i. We put

$$m_i(h) = \frac{1}{e}\sum_{j=0}^{e-1} \zeta^{-j}h(\tau^j),$$

where $e=q^{f_i}-1$ and ζ is the root of unity in O_p given by (3.5). If h is also contained in $C(\Gamma_j)$ and $m_j(h)$ is defined in a similar way, it is easy to see that $m_i(h)=m_j(h)$. We therefore denote those common values by $m(h)$. $m(h)$ is obviously an O_p-linear function on $C(\Gamma)$ with values in $\overline{Q}_p$.

We now denote by X the direct sum of the group $\overline{Q}_p$ and m copies of the module $C(\Gamma)$:

$$X = \overline{Q}_p + C(\Gamma) + \cdots + C(\Gamma).$$

For each $i \geqq 1$, we also denote by X_i the submodule of X consisting of all $x=(\bar{y}, h_1, \cdots, h_m)$ ($\bar{y}\in\overline{Q}_p$, $h_t\in C(\Gamma)$, $t=1, \cdots, m$) such that h_t are in $C(\Gamma_i)$ for $t=1, \cdots, m$ and such that

$$(3.7) \qquad (g^f - 1)\bar{y} + p^\kappa\frac{1}{e}\sum_{i=0}^{f-1}\sum_{j=0}^{e-1} g^i\zeta^j h_1(\sigma^{f-i-1}\tau^{e-j}) = 0,$$

where $f=f_i$, $e=q^f-1$. X is then the union of those submodules X_i ($i\geqq 1$), and it is easy to see that there is a unique way of making X a Γ-module so that $\rho x = x$ if x is in X_i and ρ is an element of the Galois group of V/E_{f_i} and that

$$(3.8) \qquad \begin{aligned} \sigma x &= (\bar{y}_1, \sigma h_1, \cdots, \sigma h_m), & \bar{y}_1 &= g^{-1}(\bar{y} - p^\kappa m(\sigma h_1)),\\ \tau x &= (\bar{y}_2, \tau h_1, \cdots, \tau h_m), & \bar{y}_2 &= \zeta^{-1}\bar{y},\end{aligned}$$

for any $x=(\bar{y}, h_1, \cdots, h_m)$ in X. We shall next show that the Γ-module X thus defined is Γ-isomorphic with the group $\tilde{N}$, the character group of N.

Let χ be an additive character of Q_p with the kernel O_p, i.e. a homomorphism of Q_p into the additive group of reals mod 1 such that the kernel of χ is O_p. The bilinear function (u, y) on $Q_p\times Q_p$ defined by

$$(u, y) = \chi(uy), \qquad u, y \in Q_p,$$

then gives a dual pairing of the locally compact abelian group Q_p with itself; and since the annihilator of O_p in Q_p with respect to this pairing is O_p itself, it induces a pairing $(u, \bar{y})$ of O_p and $\overline{Q}_p = Q_p/O_p$ ($u\in O_p$, $\bar{y}\in\overline{Q}_p$). For any v in O_p, we have then

$$(vu, \bar{y}) = (u, v\bar{y}).$$

We now fix an integer $i\geqq 1$ and consider the group $U_1 = U_1(E_{f_i})$. For simplicity, we denote $f_i, a_{i0}, a_{i1}, \cdots, a_{im}$ by $f, a_0, a_1, \cdots, a_m$, respectively. Every element a in U_1 can then be written in the form

$$a = a_0^u a_1^{\alpha_1} \cdots a_m^{\alpha_m}, \tag{3.9}$$

where u is an element of O_p and α_t are elements in the group ring of the Galois group Γ_i of E_{f_i}/k over O_p. We put

$$\alpha_t = \sum_{\rho'} u_{t,\rho'}\rho'. \qquad u_{t,\rho'} \in O_p, \ \rho' \in \Gamma_i.$$

For any $x = (\bar{y}, h_1, \cdots, h_m)$ in X_i, we then define a symbol $[a, x]_i$ by

$$[a, x]_i = (u, \bar{y}) + \sum_{t\rho',} (u_{t,\rho'}, h_t(\rho)),$$

where ρ denotes any element of Γ contained in the coset ρ' of the factor group Γ_i of Γ. Though the expression (3.9) is not unique for the given element a, the value of $[a, x]_i$ is uniquely defined according to Theorem 1, (2), and (3.7). As can be verified easily, $[a, x]_i$ then gives a dual pairing of the groups $U_1(E_{f_i})$ and X_i; and it also follows from Theorem 1, (3), and (3.8) that

$$[a^\rho, \rho x]_i = [a, x]_i \tag{3.10}$$

for any ρ in Γ. Let j be another integer such that $1 \leqq j \leqq i$ and let $[a, x]_j$ be the dual pairing of $U_1(E_{f_i})$ and X_j defined in a similar way as above. X_j is then contained in X_i, and it follows again easily from the definition that

$$[a, x]_i = [\nu'_{j,i}a, x]_j$$

for any a in $U_1(E_{f_i})$ and for any x in X_j. Since the group U is the inverse limit of $U_1(E_{f_i})$ with homomorphisms $\nu'_{j,i}$ and since X is the union of all X_i, it then follows immediately that $[a, x]_i$ $(i \geqq 1)$ together define a dual pairing $[a, x]$ of the compact abelian group U and the discrete abelian group X such that

$$[a^\rho, \rho x] = [a, x],$$

for any ρ in Γ. We have thus proved that the group X is Γ-isomorphic with the character group of U, and, hence, also with the character group $\tilde{N}$ of N. We have therefore the following theorem:

THEOREM 3. *Let k, V, Ω be as in Theorem 2. Let Γ be the Galois group of V/k and N the Galois group of the maximal abelian extension V'/V of V in Ω. Let, furthermore, $C(\Gamma)$ denote the module of all continuous functions defined on Γ with values in $\bar{Q}_p = Q_p/O_p$ and X the direct sum of $\bar{Q}_p$ and m copies of the module $C(\Gamma)$, where $m = [k:Q_p]$. If we then make X a Γ-group so that* (3.8) *holds, the character group $\tilde{N}$ of N is Γ-isomorphic with the so defined Γ-module X.*

BIBLIOGRAPHY

1. E. Artin, *Algebraic numbers and algebraic functions* I, Lecture notes at Princeton University, 1950–1951.

2. M. Deuring, *Algebren*, Berlin, 1935.

3. H. Hasse, *Zahlentheorie*, Berlin, 1949.

4. K. Iwasawa, *On solvable extensions of algebraic number fields*, Ann. of Math. vol. 58 (1953) pp. 548–572.

5. Y. Kawada, *On the structure of the Galois group of some infinite extensions* I, Journal of the Faculty of Science. Imperial University of Tokyo. Section I., vol. 7 (1954) pp. 1–18.

6. M. Krasner, *Sur la représentation multiplicative dans les corps de nombres P-adiques relativement galoisiens*, C. R. Acad. Sci. Paris vol. 203 (1936) pp. 907–908.

7. ———, *Sur la représentation exponentielle dans les corps relativement galoisiens de nombres P-adiques*, Acta Arithmetica vol. 3 (1939) pp. 133–173.

8. I. R. Safarevic, *On p-extensions*, Mat. Sbornik vol. 20 (1947) pp. 351–363.

9. H. Zassenhaus, *Lehrbuch der Gruppentheorie* I, Leipzig-Berlin, 1937.

MASSACHUSETTS INSTITUTE OF TECHNOLOGY,
CAMBRIDGE, MASS.

[30] Galois groups acting on the multiplicative groups of local fields

"Proc. Int. Sympos. on Algebraic Number Theory" (Tokyo and Nikko, 1955), Science Council of Japan, Tokyo, 1956, pp. 63-64.

Let k be a finite extension of degree m over Q_p, the field of p-adic numbers, and E a finite tamely ramified Galois extension of k. Let F be the inertia field of E/k and let $[E:F]=e$, $[F:k]=f$. The Galois group $G(E/F)$ is then a normal subgroup of the Galois group $G(E/k)$, and, $G(E/F)$ and $G(E/k)/G(E/F)=G(F/k)$ are both cyclic groups. In the following, we assume that the group $G(E/k)$ contains an element σ of order f which induces the Frobenius automorphism on the unramified extension F/k. $G(E/k)$ is then generated by σ and a generator τ of $G(E/F)$, and the defining relations are given by

$$\sigma^f=1, \qquad \tau^e=1, \qquad \sigma\tau\sigma^{-1}=\tau^q,$$

where q is the number of elements in the residue class field of E. Let p^κ be the highest power of p dividing the order of any root of unity in E and let w be a primitive p^κ-th root of unity in E. We, then, further assume that $\kappa \geqq 1$ and that $[E:k(w)]$ is divisible by p, and even by 4 when $p=2$. Under these assumptions, the action of the Galois group $G(E/k)$ on the multiplicative group U_1, consisting of all units a in E with $a \equiv 1$ mod. π ($\pi=$ a prime element in E), can be explicitly described as follows.

We first notice that for any a in U_1 and for any p-adic integer α, a^α is defined as usual and is again an element of U_1. Therefore, if we denote by O_p the ring of p-adic integers and by R the group ring of $G(E/k)$ over O_p, R can be considered as an operator domain of U_1 in an obvious way, and for our purpose, it is sufficient to determine the structure of the R-group U_1. Then, as an R-group, U_1 contains $m+1$ generators $a_0, a_1, \cdots, a_m$ over R with the following fundamental relations:

$$a_0^{\sigma-g}=a_1^{p^\kappa\lambda}, \qquad a_0^{1-\sigma^f}=a_1^{p^\kappa\mu}, \qquad a_0^{\tau-\zeta}=1.$$

Here, g is a rational integer satisfying $w^\sigma=w^g$, ζ is a root of unity in O_p uniquely determined by $w^\tau=w^\zeta$, and λ and μ are elements in R defined by

$$\lambda=\frac{1}{e}\sum_{j=0}^{e-1}\zeta^{-j}\tau^{j}, \qquad \mu=\Big(\sum_{i=0}^{f-1}g^{i}\sigma^{-i-1}\Big)\lambda.$$

Now, let Ω be an algebraic closure of k and V the ramification field of the extension Ω/k, i.e. composite of all finite tamely ramified extensions of k in Ω. Furthermore, let A be the maximal abelian extension of V in Ω. As the Galois group $G(A/V)$ is an abelian normal subgroup of the Galois group $G(A/k)$ with the factor group $G(V/k)$, the latter acts on $G(A/V)$ in a natural way. Using the result mentioned above, we can, then, explicitly give the action of $G(V/k)$ on $G(A/V)$ as follows: let $\bar{Q}_p$ denote the factor group of the additive group of Q_p modulo the additive group of O_p and C the set of all continuous functions on the compact group $G(V/k)$ (in its Krull topology) with values in the discrete group $\bar{Q}_p$. C is then an O_p-module in an obvious way. But we can also make C a $G(V/k)$-module by putting

$$(\rho h)(\omega)=h(\rho^{-1}\omega),$$

for any ρ in $G(V/k)$ and any $h=h(\omega)$ in C. Let X be the direct sum of $\bar{Q}_p$ and m copies of the module C. For any ρ in $G(V/k)$ and any $x=(a, h_1, \cdots, h_m)$ in X, where $a \in \bar{Q}_p$, $h_i \in C$, we put

$$\rho x=(b, \rho h_1, \cdots, \rho h_m),$$

with a suitable b in $\bar{Q}_p$ depending on ρ, a and h_i, of which precise definition is, however, omitted here. X is thus also made a $G(V/k)$-module. Now, the compact abelian group $G(A/V)$ is, as a $G(V/k)$-group, isomorphic with the character group of the discrete $G(V/k)$-group X as defined above[1].

MASSACHUSETTS INSTITUTE OF TECHNOLOGY

1) For the details, c.f. K. IWASAWA, On Galois groups of local fields: Trans. Amer. Math. Soc., **80** (1955), pp. 448–469.

[31] A note on the group of units of an algebraic number field

J. Math. Pure and Appl., (9) 35 (1956), 189-192.

1. Let K be a finite Galois extension of a finite algebraic number field k and G the Galois group of K/k. In the present Note, we shall give some simple remarks on the cohomology group of G acting on the group of units, E, of the field K. The results are rather elementary, but they might be of some interest because of the fact that they give us some relations between E and the group of ideals of K.

As usual, we shall denote by $H^n(G, A)$, or sometimes simply by $H^n(A)$, the n-dimensional cohomology groupe of G acting on an abelian group A. Following Artin-Tate-Chevalley, we shall consider $H^n(G, A)$ also for negative integers n (1).

2. We first consider the 1-dimensional cohomology group $H^1(G, E)$. Let $\mathfrak{a}$ be an ambiguous principal ideal of K, i. e. a principal ideal of K such that

$$\mathfrak{a}^\sigma = \mathfrak{a}$$

for any σ in G. Put $\mathfrak{a} = (\alpha)$, $\alpha \in K$. It is then clear that

$$\varepsilon_\sigma = \alpha^{\sigma-1} \qquad (\sigma \in G)$$

defines a 1-cocycle of G in E. If we choose another α' in K such that $\mathfrak{a} = (\alpha')$, we have $\alpha' = \alpha\varepsilon$ with some unit ε in E and we see that ε_σ is cohomologous with $\varepsilon'_\sigma = \alpha'^{\sigma-1} = \varepsilon_\sigma \varepsilon^{\sigma-1}$. Hence the ideal $\mathfrak{a}$ uniquely determines a cohomology classe $c = c(\mathfrak{a})$ in the group $H^1(G, E)$. It is then obvious from the definition that $\mathfrak{a} \to c(\mathfrak{a})$ is a

(1) *Cf.* C. Chevalley, *Class field theory*, Nagoya Univ., 1953-1954.

homomorphism of the multiplicative group of all ambiguous principal ideals of K into $H^1(G, E)$. Let ε_σ be an arbitrary 1-cocycle of G in E. By Galois theory, there is an element α in K such that $\varepsilon_\sigma = \alpha^{\sigma-1}$. $\mathfrak{a} = (\alpha)$ is then an ambiguous principal ideal of K and ε_σ is a cocycle in the cohomology class $c(\mathfrak{a})$. Therefore, the mapping $\mathfrak{a} \to c(\mathfrak{a})$ is onto. Finally, let $\mathfrak{a}$ be in the kernel of the above homomorphism. We have then

$$\mathfrak{a} = (\alpha), \qquad \varepsilon_\sigma = \alpha^{\sigma-1} \sim 1,$$

i. e.

$$\alpha^{\sigma-1} = \varepsilon^{\sigma-1} \qquad (\sigma \in G),$$

with some ε in E. $a = \alpha\varepsilon^{-1}$ is then an element in the ground field k and $\mathfrak{a} = (a)$ is a principal ideal of k. Conversely, if $\mathfrak{a}$ is a principal of k, $c(\mathfrak{a})$ is obviously the identity of $H^1(G, E)$. By the isomorphism theorem, we then see that $H^1(G, E)$ *is canonically isomorphic with the factor group of the group of ambiguous principal ideals of* K *modulo the group of principal ideals of* k (²).

Suppose, in particular, that K/k be an unramified extension. Then ambiguous ideals of K are nothing but ideals of the ground field k. Hence it follows from the above that $H^1(G, E)$ is isomorphic with the group of ideal classes of k whose ideals become principal in K. The principal ideal theorem then tells us that if K contains the Hilbert's class field K_0 over k, $H^1(G, E)$ is isomorphic with the ideal class group of k (³). Furthermore, in this case, every 1-cocycle of G in E is cohomologous with a 1-cocycle of G in the group of units of K_0.

5. To study $H^n(G, E)$ in general, we consider the idèle group J of K on which G acts in natural way. As is well-known, there is a canonical homomorphism of J onto the group of ideals of K and we

---

(²) This result is essentially known in classical literatures, though not formulated as here. In particular, *cf.* M. Moriya, *Ueber die Klassenzahl eines relativzyklischen Zahlkörpers von Primzahlgrad* (*Jap. J. Math.*, vol. 10, 1933, p. 1-18).

(³) Conversely, if we could know the structure of the group $H^1(G, E)$ by a cohomology-theoretical method, we would have a new proof of the principal ideal theorem.

denote the kernel of the homomorphism by U. Then the intersection of U and the group of principal idèles P is obviously the group of units E, and the factor group J/PU is canonically isomorphic with the ideal class group of K. Furthermore, since those subgroups U, P, E and PU are all G-invariant, we can consider cohomology groups of G acting on those subgroups and their factor groups. The sequences of any three such groups, e. g. $J \supset PU \supset U$, then give rise to exact sequences of cohomology groups of G acting on their factor groups and, together with the canonical isomorphisms

$$PU/P \cong U/E, \qquad PU/U \cong P/E,$$

they give us various relations among the cohomology groups of G acting on those groups.

4. To be more definite, let us assume again that K/k be an unramified extension. It is then easy to see that $H^n(U)=1$ for all n, and the exact sequence for $U \supset E \supset 1$ givesus the isomorphisms

$$H^n(U/E) \cong H^{n+1}(E).$$

In the exact secquence for $J \supset PU \supset P$, we can then replace

$$H^n(PU/P) \cong H^n(U/E)$$

by $H^{n+1}(E)$ and we obtain an exact sequence

$$(1) \qquad \ldots \to H^{n+1}(E) \to H^n(\bar{J}) \to H^n(\bar{I}) \to H^{n+2}(E) \to \ldots,$$

where $\bar{J}=J/P$ denotes the idèle class group of K and $\bar{I}=J/PU$ the ideal class group of K. Here the mappings $\theta_n : H^{n+1}(E) \to H^n(\bar{J})$ and $\omega_n : H^n(\bar{I}) \to H^{n+2}(E)$ are defined as follows. Let u be an element in $H^{n+1}(E)$ and γ a cocycle in the cohomology class u. Since $H^{n+1}(U)=1$, $\gamma=\delta\beta$ with a suitable n-cochain β in U. Let $\bar{\beta}$ be the n-cocycle in $\bar{J}=J/P$ obtained by considering β modulo P. $\theta_n(u)$ is then the cohomology class of $\bar{\beta}$ in $H^n(\bar{J})$. On the other hand, let v be an element in $H^n(\bar{I})$, $\bar{\lambda}$ a cocycle in the cohomology class v and λ an n-cochain in J which gives $\bar{\lambda}$ when considered modulo PU. $\mu=\delta\lambda$ is then an $(n+1)$-cocycle in PU and we can put $\mu=\nu_1\nu_2$ where ν_1 and ν_2 are suitable $(n+1)$-cochains in P and U respectively. $\nu = \partial\nu_2 = \partial\nu_1^{-1}$ is obviously an $(n+2)$-cocycle in $E = P \cap U$, and $\omega_n(\nu)$ is the cohomology class of such an ν.

Now, by a theorem of Tate, the structure of the cohomology group $H^n(\bar{J})$ is known (4) : $H^n(\bar{J})$ is canonically isomorphic with $H^{n-2}(G, Z)$ where Z denotes the additive group of rational integers on which G acts trivially. Hence the exact sequence (1) gives us relations between the cohomology groups of G acting on E and $\bar{I}$, respectively.

Let us consider more precisely the part of the exact sequence (1) where $n = 0$:

$$(2) \qquad \ldots \to H^0(\bar{J}) \to H^0(\bar{I}) \to H^2(E) \to H^1(\bar{J}) \to \ldots$$

By the theorem of Tate, $H^1(\bar{J}) = 1$. A 0-cocycle of J is an idèle class of K invariant under G, i. e. an idèle class represented by an idèle of k. Hence, we see from (2) that $H^2(G, E)$ *is isomorphic with the factor group of the group of ambiguous ideal classes of* K *modulo the group of ideal classes represented by ideals of k.* Similarly, considering (1) for $n = 1$, we can prove that the order of $H^1(G, E)$ is equal to the order of the group of ideal classes of k which become principal in K, a less explicit result than the one we obtained in paragraph 1.

Finally, if we assume more specifically that $\bar{I} = 1$, i. e. that the class number of K is 1, we have, again from (1), the canonical isomorphisms

$$H^n(E) \cong H^{n-1}(\bar{J}) \cong H^{n-3}(Z).$$

5. In the above, we have assumed that the extension K/k is unramified. However, we can get similar results also for a ramified extension K/k, provided we know well the properties of the prime divisors of K which are ramified by the extension K/k. A typical example of such an extension is given by a cyclotomic field over the field of rational numbers, but we omit here the details.

---

(4) *Cf. loc. cit.* (1).

[32] A note on class numbers of algebraic number fields

Abh. Math. Sem. Univ. Hamburg, 20 (1956), 257-258.

In the present paper, we shall give a note on class numbers of algebraic number fields including, in particular, a proof of a theorem of WEBER and its generalization[1]).

Let k be a finite algebraic number field, K a finite Galois extension of k, and let h and H be class numbers of k and K respectively. We first prove the following:

I. *If there exists a prime divisor P of k which is fully ramified by the extension K/k, then $h|H$. In particular, for any prime number p,*

$$p|h \rightarrow p|H \quad (\text{or } p\nmid H \rightarrow p\nmid h).$$

II. *If, furthermore, K/k is a cyclic extension of p-power degree and has no ramified prime divisor other than P, then conversely*

$$p|H \rightarrow p|h \quad (\text{or } p\nmid h \rightarrow p\nmid H).$$

The proof of I is as follows[2]): Let A and A' be the maximal unramified abelian extensions of k and K respectively. By the class field theory, $[A:k] = h$ and $[A':K] = H$. Since P is fully ramified by K/k, we have $K \cap A = k$ and the Galois group of KA/k is the direct product of the Galois groups of KA/K and of KA/A. It then follows easily that KA is an unramified abelian extension of K and, hence, is contained in A'. $H = [A':K]$ is therefore divisible by $h = [A:k] = [KA:K]$.

We now prove II under the additional assumptions stated there. Let A'' be the intermediate field of K and A' such that $[A'':K]$ is a power of p and $[A':A'']$ is prime to p. A''/k is then a Galois extension and its Galois group G is a p-group. Suppose $p\,|H$. Then $A'' \neq K$ and the Galois group N of A''/K is not trivial. Since G is a p-group, we can find a normal subgroup M of G such that $N > M$ and $[N:M] = p$. N/M is then contained in the center of G/M and, as the Galois group G/N of K/k is cyclic by the assumption, G/M is an abelian group[3]).

---

[1]) Cf. H. HASSE, Über die Klassenzahl Abelscher Zahlkörper (Berlin, 1952), § 34. The author expresses here his hearty thanks to Professor H. Hasse for his kind suggestions in preparing the present paper.

[2]) This first part is a known result; cf. C. CHEVALLEY, Relation entre le nombre de classes d'un sous-corps et celui d'un sur-corps. C. R., Paris, 192, 257—258 (1931). The proof is given here only for the sake of completeness.

[3]) Cf. H. ZASSENHAUS, Lehrbuch der Gruppentheorie I (Leipzig und Berlin 1937), p. 104.

17*

Let E be the intermediate field of K and A'' corresponding to the subgroup M. E/k is then an abelian extension such that $E>K>k$, $[E:K]=p$. Since E/k is abelian, all the prime divisors of E which divide P of k have the same inertia field, say F; as P is fully ramified by K/k and E/K is an unramified extension, we must have $[F:k]=p$. Obviously, no prime divisor of k other than P is ramified by F/k. Hence F is a cyclic unramified extension of degree p over k and is contained in A. $h=[A:k]$ is therefore divisible by $p=[F:k]$, q.e.d.

We now consider an abelian extension K of the field of rational numbers R whose conductor is a prime-power p^e ($e \geqq 1$). Such K is contained in the field Z_e of p^e-th roots of unity. Assume first $p \neq 2$. It can then be readily verified that the assumptions of II are satisfied for the extension Z_e/Z_1, Z_1 being the field of p-th roots of unity. Hence, if p is a regular prime, i. e. if the class number of Z_1 is prime to p, the class number of Z_e is also prime to p by II. Applying I to Z_e/K, we find that the class number of K is again prime to p. If $p=2$, $Z_1=R$ and $Z_2=R(\sqrt{-1})$ have both the class number 1. Hence, applying II to Z_e/Z_2, the class number of Z_e ($e \geqq 2$) is known to be odd and so is the class number of K by I.

We have thus obtained the following:

III. *If p is a regular prime, the class number of an absolutely abelian field with conductor p^e is prime to p.*

In particular, an absolutely abelian field with conductor 2^e or 3^e has a class number which is prime to 2 or 3 respectively[2]).

By a similar argument, we can also prove that the class number of the cyclic extension of degree p^{e-1} over R with conductor p^e is prime to p. Here we need not assume that p is regular.

---

[2]) Cf. l. c. [1]).

Eingegangen am 1. 10. 1954

[33] On some invariants of cyclotomic fields

Amer. J. Math., 80 (1958), 773-783. Erratum, ibid., 81 (1959), 280.

Dedicated to Professor E. Artin on His Sixtieth Birthday

1. Let p be an odd prime. For each $n \geqq 0$, we denote by ζ_n a primitive p^n-th root of unity and by K_n the cyclotomic field obtained by adjoining ζ_{n+1} to the rational field Q: $K_n = Q(\zeta_{n+1})$. It is proved in [4] that if p^{e_n} is the highest power of p dividing the class number of K_n, then the exponent e_n is given, for all sufficiently large n, by a formula

$$e_n = \lambda n + \mu p^n + \nu, \qquad n \geqq n_0, \tag{1}$$

where λ, μ and ν are integers independent of n. The numbers λ and μ seem to have deep significance for the arithmetic of the fields K_n. In general, if the invariant $\mu = \mu(K/F)$ of a so-called Γ-extension K over a finite algebraic number field F is 0, then the Galois group of the maximal unramified abelian p-extension over K is, up to a finite subgroup, isomorphic with the direct sum of λ copies of the additive group of p-adic integers, where $\lambda = \lambda(K/F)$ denotes another invariant of K/F.[1] So, if $\mu = 0$, we have an analogue, for number fields, of a similar result for algebraic function fields of one variable over algebraically closed fields of constants. For this and other reasons, it seems interesting to know whether $\mu > 0$ or $\mu = 0$ for a given Γ-extension K/F, and we shall find in the present paper necessary and sufficient conditions for $\mu > 0$ when the Γ-extension is obtained from the cyclotomic fields K_n defined above, namely, when μ is given as the second coefficient in the above formula (1).

2. Let K be the union of all K_n, $n \geqq 0$. K is an abelian extension of Q and, for any element σ of the Galois group $G(K/Q)$, there exists a unique p-adic unit ξ such that

$$\sigma(\zeta_n) = \zeta_n^{\xi},$$

for every $n \geqq 0$. The mapping $\sigma \to \xi$ then defines a topological isomorphism

* Received February 6, 1958.

** Guggenheim Fellow. The present research was also supported in part by a National Science Foundation grant.

[1] Cf. [4], § 5, § 7.

of $G(K/Q)$ with the multiplicative group U of p-adic units and we denote the Galois automorphism σ corresponding to ξ by σ_ξ. Under this isomorphism, the subgroup $G(K/K_n)$ of $G(K/Q)$ corresponds to the subgroup U_n of all ξ in U satisfying $\xi \equiv 1 \bmod. p^{n+1}$. On the other hand, the group U is the direct product of U_0 and a cyclic group V of order $p-1$, the group of all roots of unity in the p-adic number field Q_p. The corresuonding subgroup Δ of $G(K/Q)$ is then canonically isomorphic with the Galois group $G(K_0/Q)$, and $G(K/Q)$ is the direct product of Δ and $\Gamma = G(K/K_0) \cong U_0$.

Now, suppose that Δ acts continuously on a compact p-primary abelian group N. Since N is p-primary, the element u^α is defined as usual for any u in N and for any p-adic integer α. For each i, $1 \leqq i \leqq p-1$, let $N^{(i)}$ be the set of all u in N such that

$$\sigma_\eta(u) = u^{\eta^i},$$

for every η in V, i.e., for every σ_η in Δ. It is then easy to see that each $N^{(i)}$ is a closed subgroup of N and N is the direct product of these $N^{(i)}$:

$$N = N^{(1)} \times \cdots \times N^{(p-1)}.$$

Let L be the maximal unramified abelian p-extension over K. Then L is a Galois extension of Q, and $G(K/Q) = \Gamma \times \Delta$ acts on the abelian normal subgroup $G(L/K)$ of $G(L/Q)$ in a natural way. For each $n \geqq 0$, let L_n denote the maximal unramified abelian p-extension of K_n. L is then the union of all L_n and, p^{e_n} being the degree of the extension L_n/K_n, the formula (1) was proved in [4] by studying the structure of the Γ-group $G(L/K)$.

We now apply the above decomposition to the Δ-group $G(L/K)$ and obtain

$$G(L/K) = G(L/K)^{(1)} \times \cdots \times G(L/K)^{(p-1)}.$$

We denote by $L^{(i)}$ $(1 \leqq i \leqq p-1)$ the subfield of L containing K such that

$$G(L/L^{(i)}) = \prod_{j \neq i} G(L/K)^{(j)}, \qquad G(L^{(i)}/K) \cong G(L/K)^{(i)}.$$

Since Δ may be considered as acting on the Galois group $G(L_n/K_n)$ in a natural way, $G(L_n/K_n)$ can be also decomposed with respect to Δ and we obtain subfields $L_n^{(i)}$ $(1 \leqq i \leqq p-1)$ of L_n containing K_n, defined similarly as the above $L^{(i)}$. Clearly, the action of Γ on $G(L/K)$ commutes with the action of Δ on $G(L/K)$. Hence $G(L^{(i)}/K)$ $(1 \leqq i \leqq p-1)$ are again Γ-groups like $G(L/K)$. Using the fact that $L^{(i)}$ is the union of all $L_n^{(i)}$, $n \geqq 0$, it can be proved in the same way as in [4] that if $p^{e_n^{(i)}}$ is the degree of the extension $L_n^{(i)}/K_n$, then the exponent $e_n^{(i)}$ is given, for all sufficiently large n, by

(2) $$e_n^{(i)} = \lambda^{(i)} n + \mu^{(i)} p^n + \nu^{(i)}, \qquad (1 \leqq i \leqq p-1),$$

where $\lambda^{(i)}$, $\mu^{(i)}$ and $\nu^{(i)}$ are again integers independent of n. Since $e_n = \sum_{i=1}^{p-1} e_n^{(i)}$, we obtain from (1) and (2) that

$$\lambda = \sum_{i=1}^{p-1} \lambda^{(i)}, \qquad \mu = \sum_{i=1}^{p-1} \mu^{(i)}, \qquad \nu = \sum_{i=1}^{p-1} \nu^{(i)}.$$

Let e_n^+ be the sum of $e_n^{(i)}$ with even i and let e_n^- be the sum of $e_n^{(i)}$ with odd i. Similarly, we define λ^+, λ^-, μ^+, μ^- and ν^+, ν^- as the sums of $\lambda^{(i)}$, $\mu^{(i)}$ and $\nu^{(i)}$ with even or odd indices respectively. Then,

$$e_n = e_n^+ + e_n^-, \quad \lambda = \lambda^+ + \lambda^-, \quad \mu = \mu^+ + \mu^-, \quad \nu = \nu^+ + \nu^-.$$

Furthermore, if L_n^+ denotes the composite of all $L_n^{(i)}$ with even i and L_n^- the composite of all $L_n^{(i)}$ with odd i, then $p^{e_n^+}$ and $p^{e_n^-}$ are the degrees of the extensions of L_n^+/K_n and L_n^-/K_n respectively, and they are also equal to the highest powers of p dividing, respectiveliy, the second and the first factors of the class number of K_n.[2]

LEMMA 1. *$\mu = 0$ if and only if $\mu^- = 0$.*

Proof. As the proof of (1) shows, $\mu = 0$ ($\mu^- = 0$) if and only if the rank of $G(L_n/K_n)$ ($G(L_n^-/K_n)$) is bounded for every $n \geqq 0$. However, it is a classical result that

$$\operatorname{rank} G(L_n^+/K_n) \leqq \operatorname{rank} G(L_n^-/K_n) \leqq \operatorname{rank} G(L_n/K_n) \leqq 2 \operatorname{rank} G(L_n^-/K_n).^3$$

The lemma then follows immediately.

Remark. By a similar method as in [7], we can obtain a more precise result:

$$\operatorname{rank} G(L_n^{(i)}/K_n) \leqq \operatorname{rank} G(L_n^{(j)}/K_n), \qquad n \geqq 0,$$

where i is even, j is odd and $i + j \equiv 1 \bmod. p-1$.

3. Assume, now, that $\mu > 0$. Then, by the above lemma, $\mu^- > 0$ and, hence, $\mu^{(i)} > 0$ for some odd index i. We fix such an index $i = a$. The Galois group $G(L^{(a)}/K)$ is then a so-called regular strictly Γ-finite compact Γ-group with the invariant $\mu^{(a)} > 0$, and it follows that, for each $n \geqq 0$, $G(L_n^{(a)}/K_n)$ contains a subgroup which is, as a $G(K_n/K_0)$-group, isomorphic with the additive group of the group ring of $G(K_n/K_0)$ over the prime field of characteristic p.[4] Let P_n be the p-Sylow group of the absolute ideal class

[2] More generally, if F is a subfield of K_0 with $[K_0 : F] = d$ and K' is the Γ-extension defined naturally over F, then the invariants of K'/F are given by the sums of $\lambda^{(i)}$ and $\mu^{(i)}$ with the indices i divisible by d.

[3] Cf. [7].

[4] Cf. [4], § 5 and a forthcoming paper of the author.

group of K_n and let $P_n^{(a)}$ be the a-th direct factor of the Δ-decomposition of P_n. By class field theory, $G(L_n^{(a)}/K_n)$ is then canonically isomorphic with $P_n^{(a)}$. Hence, by the above, there exists an ideal class C_0 of order p in $P_n^{(a)}$ such that $\sigma_\eta(C_0) = C_0^{\eta^a}$ for every η in V and that the classes $\sigma_\xi(C_0)$ are independent when ξ runs over p-adic units in U_0 mod. U_n.

Now, for any p-adic integer α, let $s_n(\alpha)$ $(n \geqq 0)$ denote the uniquely determined rational integer such that

$$\alpha \equiv s_n(\alpha) \bmod. p^{n+1}, \qquad 0 \leqq s_n(\alpha) < p^{n+1}.$$

For any ideal class C in P_n and for any ξ in U, we have then

$$\prod_\alpha \sigma_\alpha(C)^{d(\alpha)} = 1, \tag{3}$$

where the product is taken over all α in U mod. U_n and the exponent $d(\alpha)$ is given by

$$d(\alpha) = p^{-(n+1)}(s_n(\xi\alpha^{-1}) - \xi s_n(\alpha^{-1})).$$

This is a special case of a general identity which holds for the ideal classes and their conjugates of a cyclotomic field. For the field K_0, it is given in [3], § 109 and the general case can be deduced easily from the result in [5].

Applying (3) to the above ideal class C_0, we obtain

$$\sum_{\eta \in V} (s_n(\xi\eta^{-1}) - \xi s_n(\eta^{-1}))\eta^a \equiv 0 \bmod. p^{n+2}, \qquad \xi \in U,$$

or

$$\sum_\eta s_n(\xi\eta^{-1})\eta^a \equiv \xi \sum_\eta s_n(\eta^{-1})\eta^a \bmod. p^{n+2}, \qquad \xi \in U. \tag{4}$$

First, we put $\xi = 1 + p^{n+1}$ in the above. Since $s_n(\xi\eta^{-1}) = s_n(\eta^{-1})$, we have

$$\sum_\eta s_n(\eta^{-1})\eta^a \equiv (1 + p^{n+1}) \sum_\eta s_n(\eta^{-1})\eta^a \bmod. p^{n+2},$$

and, hence, also

$$\sum_\eta s_n(\eta^{-1})\eta^a \equiv 0 \bmod. p.$$

As $s_n(\eta^{-1}) \equiv \eta^{-1} \bmod. p$, it follows that $a \neq 1$.

Next, we take ξ from V, in (4). Then the left hand side is equal to $\xi^a \sum_\eta s_n(\eta^{-1})\eta^a$ and, as $a \neq 1$, $1 \leqq a \leqq p-1$, it follows that

$$\sum_\eta s_n(\eta^{-1})\eta^a \equiv 0 \bmod. p^{n+2}.$$

Hence, by (4),

$$\sum_\eta s_n(\xi\eta^{-1})\eta^a \equiv 0 \bmod. p^{n+2}, \qquad \eta \in V, \tag{5}$$

for any ξ in U. We have thus proved that the assumption $\mu > 0$ implies (5) for some odd index $a \neq 1$ and for every $n \geqq 0$. Notice that (5) does not hold for $a=1$ (and $n=0$).

To prove the converse of the above, we use the classical formula for the first factor of the class number of K_n.[5] Let Ω_p be the algebraic closure of Q_p and χ a character of U with values in Ω_p^* such that $\chi(U_n)=1$. Put

$$S_n(\chi)=p^{-(n+1)}\sum_{\xi} s_n(\xi)\chi(\xi),$$

where ξ runs over p-adic units in U mod. U_n. We notice that $S_n(\chi)$ is an element of Ω_p and that if $\chi(U_m)=1$ for $m \leqq n$, then $S_n(\chi)=S_m(\chi)$. Now, the classical formula tells us that the highest power of p dividing the first factor of the class number of K_n is equal to the highest power of p dividing, in Ω_p, the product

$$p^{n+1}\prod_{\chi} S_n(\chi), \tag{6}$$

where χ runs over all characters of U such that $\chi(U_n)=1$ and $\chi(-1)=-1$.

Suppose that (5) holds for an odd index a. Clearly, there exist p^n characters χ of U such that $\chi(U_n)=1$ and that $\chi(\eta)=\eta^{-a}$ for every η in V. Since a is odd, these characters satisfy $\chi(-1)=-1$ and the corresponding $S_n(\chi)$ appear in the product (6). It is also clear that (5) implies

$$S_n(\chi)\equiv 0 \text{ mod. } p,$$

for any such character χ. On the other hand, it is known that the product of p^{n+1} and other factors $S_n(\chi')$ in (6) is an integer of Ω_p. Hence the product (6) is divisible by p^{p^n} and we obtain that

$$e_n^- \geqq p^n.$$

Since $e_n^-=\lambda^- n+\mu^- p^n+\nu^-$, it follows that if (5) holds for every $n \geqq 0$, then $\mu^- > 0$ and, consequently, $\mu > 0$. Thus the following result is proved:

I. *The second coefficient μ in the formula* (1) *is positive if and only if there exists an odd integer a, $1 \leqq a \leqq p-1$, such that* (5) *holds for any p-adic unit ξ and for any $n \geqq 0$.*

For any p-adic integer α, let

$$\alpha=\sum_{k=0}^{\infty} t_k(\alpha)p^k, \qquad 0 \leqq t_k(\alpha) < p,$$

be the canonical p-adic expansion of α. Then,

$$s_n(\alpha)=\sum_{k=0}^{n} t_k(\alpha)p^k, \qquad s_{n+1}(\alpha)=s_n(\alpha)+t_{n+1}(\alpha)p^{n+1}.$$

[5] For the following, cf. [2], § 5, § 33.

Since $a \neq 1$ and

$$\sum_{\eta} s_{n+1}(\xi\eta^{-1})\eta^{a} \equiv \sum_{\eta} \xi\eta^{a-1} \equiv 0 \bmod. p^{n+2},$$

the condition (5) is equivalent with the following:

$$\sum_{\eta} t_{n+1}(\xi\eta^{-1})\eta^{a} \equiv 0 \bmod. p, \qquad n \geqq 0.$$

Notice that it holds trivially

$$\sum_{\eta} t_{0}(\xi\eta^{-1})\eta^{a} \equiv 0 \bmod. p.$$

4. We now consider that the roots of unity ζ_n are all contained in Ω_p and asume that $\zeta_{n+1} = \zeta_n$, $n \geqq 0$. We have then

$$p^{n+1}/(\zeta_{n+1}-1) = \sum_{i=0}^{p^{n+1}-1} i\zeta_{n+1}{}^{i} = \sum_{\alpha} s_n(\alpha)\zeta_{n+1}{}^{\alpha},$$

where α runs over all p-adic integers mod. p^{n+1}. For any η in V, it follows that

$$p^{n+1}/(\zeta_{n+1}{}^{\eta}-1) = \sum_{\alpha} s_n(\alpha)\zeta_{n+1}{}^{\alpha\eta} = \sum_{\alpha} s_n(\alpha\eta^{-1})\zeta_{n+1}{}^{\alpha}.$$

If $n \geqq 1$, we have similarly

$$p^{n+1}/(\zeta_{n}{}^{\eta}-1) = p\sum_{\beta} s_{n-1}(\beta\eta^{-1})\zeta_{n}{}^{\beta} = \sum_{\beta} s_n(p\beta\eta^{-1})\zeta_{n+1}{}^{p\beta},$$

where β runs over all p-adic integers mod. p^{n}. Therefore,

$$p^{n+1}(1/(\zeta_{n+1}{}^{\eta}-1) - 1/(\zeta_{n}{}^{\eta}-1)) = \sum_{\xi} s_n(\xi\eta^{-1})\zeta_{n+1}{}^{\xi},$$

where ξ runs over all p-adic units in U mod. U_n.

Now, for each $n \geqq 0$ and i, $1 \leqq i \leqq p-1$, put

$$T_n^{(i)} = \sum_{\eta \in V} \eta^{i}/(\zeta_{n+1}{}^{\eta}-1).$$

Then the above computation shows that, for any $n \geqq 0$,

$$(7) \qquad p^{n+1}(T_n^{(i)} - T_{n-1}^{(i)}) = \sum_{\xi} (\sum_{\eta} s_n(\xi\eta^{-1})\eta^{i})\zeta_{n+1}{}^{\xi}, \qquad \xi \text{ in } U \bmod. U_n.$$

For convenience, we put here

$$T_{-1}^{(i)} = 0, \qquad 1 \leqq i \leqq p-1.$$

Suppose that there exists an odd index a, $1 \leqq a \leqq p-1$, such that (5) holds for any p-adic unit ξ and for any $n \geqq 0$. Then, it follows from (7) that

$$T_n^{(a)} \equiv T_{n-1}^{(a)} \bmod. p, \qquad n \geqq 0,$$

and, as $T_{-1}^{(a)}=0$, we also have that

$$T_n^{(a)}\equiv 0 \text{ mod. } p, \qquad n\geqq 0. \tag{8}$$

To prove the converse, we first notice the following

LEMMA 2. *Suppose that*

$$\sum_{\xi} a_\xi \zeta_{n+1}^{\ \xi}\equiv 0 \text{ mod. } p^s, \qquad \xi \text{ in } U \text{ mod. } U_n,$$

with a_ξ *in* Q_p. *Then, for any* ω *in* U,

$$\sum_{\xi}{}' a_\xi\equiv 0 \text{ mod. } p^s,$$

where ξ *runs over all p-adic units* mod. U_n *such that* $\xi\equiv\omega$ mod. U_{n-1}.[6]

The proof is elementary and, hence, is omitted here.

Now, assume that (8) holds for an odd index a and for any $n\geqq 0$. It then follows from (7) that

$$\sum_{\xi}\Big(\sum_{\eta} s_n(\xi\eta^{-1})\eta^a\Big)\zeta_{n+1}^{\ \xi}\equiv 0 \text{ mod. } p^{n+2}.$$

By Lemma 2, we have, for any ω in U, that

$$\sum_{\xi}{}'\sum_{\eta} s_n(\xi\eta^{-1})\eta^a\equiv 0 \text{ mod. } p^{n+2}, \qquad \xi\equiv\omega \text{ mod. } U_{n-1}. \tag{9}$$

Suppose that $n\geqq 1$ and put

$$\xi\equiv\omega+jp^n \text{ mod. } p^{n+1}, \qquad 0\leqq j<p.$$

Then,

$$s_n(\xi\eta^{-1})=s_{n-1}(\omega\eta^{-1})+kp^n, \qquad 0\leqq k<p, \tag{10}$$

where k is determined by

$$k\equiv t_n(\omega\eta^{-1})+j\eta^{-1} \text{ mod. } p.$$

If ξ runs over p-adic units mod. U_n such that $\xi\equiv\omega$ mod. U_{n-1}, k runs over all integers from 0 to $p-1$, and we obtain from (9) and (10) that

$$p\sum_{\eta} s_{n-1}(\omega\eta^{-1})\eta^a+p^n\sum_{k}\sum_{\eta} k\eta^a\equiv 0 \text{ mod. } p^{n+2}.$$

Since $a\not\equiv 0$ mod. $p-1$, $\sum_{\eta}\eta^a=0$ and it follows that

$$\sum_{\eta} s_{n-1}(\omega\eta^{-1})\eta^a\equiv 0 \text{ mod. } p^{n+1}, \qquad n\geqq 1.$$

Thus, we see that (5) holds for any ξ in U and for any $n\geqq 0$, and the following result is proved:

---

[6] For $n=0$, we understand $U_{-1}=1$.

II. *The second coefficient* μ *in the formula* (1) *is positive if and only if there exists an odd integer* a, $1 \leqq a \leqq p-1$, *such that*

$$\sum_{\eta} \eta^a/(1-\zeta_{n+1}{}^{\eta}) \equiv 0 \bmod. p, \qquad \eta \in V, \tag{11}$$

for every $n \geqq 0$.

5. Let O_p be the ring of p-adic integers and $Q_p((x))$ the field of all power series in x with coefficients in Q_p. We denote by R the subring of those power series in $Q_p((x))$ which have coefficients in O_p, and by R_0 the subring of integral power series in R. Let $\alpha \neq 0$ be a non-unit integer in Ω_p. Then, for any $\phi = \phi(x)$ in R, the element $\phi(\alpha)$ in Ω_p is well defined and $\phi \rightarrow \phi(\alpha)$ gives a ring homomorphism of R into Ω_p. In particular, if $\alpha = 1-\zeta_{n+1}$, it induces a homomorphism of R_0 onto the ring of all integers in $Q_p(\zeta_{n+1})$.

Let a be an odd integer and η any element of V. The power series for $(1-x)^{\eta}$, defined in the usual way, is then contained in R_0 and we have

$$\eta^a/(1-(1-x)^{\eta}) = \eta^{a-1}/x + \phi_{\eta}(x),$$

with a power series ϕ_{η} in R_0. Putting $x = 1-\zeta_{n+1}$ in the above, it follows that

$$\sum_{\eta} \eta^a/(1-\zeta_{n+1}{}^{\eta}) = (1-\zeta_{n+1})^{-1} \cdot \sum_{\eta} \eta^{a-1} + \sum_{\eta} \phi_{\eta}(1-\zeta_{n+1}),$$

where $\phi_{\eta}(1-\zeta_{n+1})$ are integers in $Q_p(\zeta_{n+1})$.

Now, suppose that (11) holds for every $n \geqq 0$. The above equality then shows that $a \neq 1$ and that

$$\sum_{\eta} \phi_{\eta}(1-\zeta_{n+1}) \equiv 0 \bmod. p.$$

Using the fact that the elements $(1-\zeta_{n+1})^i$, $0 \leqq i < (p-1)p^n$, form a basis of the ring of integers in $Q_p(\zeta_{n+1})$ over O_p, it follows that

$$\sum_{\eta} \phi_{\eta}(x) \equiv 0 \bmod. pR_0 + x^{(p-1)p^n} R_0.$$

Since n is an arbitrary non-negative integer, we have then

$$\sum_{\eta} \phi_{\eta}(x) \equiv 0 \bmod. p,$$

and, consequently,

$$\sum_{\eta} \eta^a/(1-(1-x)^{\eta}) \equiv 0 \bmod. p.^{7}$$

Clearly, we can reverse the above argument and see that the above congruence conversely implies (11) for every $n \geqq 0$. Therefore:

---

[7] mod. p means mod. pR_0 or mod. pR.

III. *The second coefficient* μ *in the formula* (1) *is positive if and only if there exists an odd integer* a, $1 \leqq a \leqq p-1$, *such that*

$$\sum_{\eta} \eta^a/(1-(1-x)^\eta) \equiv 0 \bmod. p, \qquad \eta \in V. \tag{12}$$

Remark. III can be proved also by a different method using the classical formula for the second factor of the class number of K_n.

Now, if an element $\pi=\pi(x)$ of R_0 is of the form

$$\pi(x)=c_1x+c_2x^2+\cdots, \qquad c_1 \in U,$$

there exists an automorphism of R over O_p which maps x to $\pi(x)$. Therefore, (12) is equivalent with

$$\sum_{\eta} \eta^a/(1-(1-\pi(x))^\eta) \equiv 0 \bmod. p, \tag{13}$$

for any such π. Choosing a suitable power series for œ, we can further transform the condition (12).

Consider the formal power series $L(x)$ and $E(x)$ defined by

$$L(x)=\sum_{i=0}^{\infty} x^{p^i}/p^i, \qquad E(x)=e^{-L(x)}.\text{[8]}$$

As power series in $Q_p((x))$, they have the properties:

$$\eta L(x)=L(\eta x), \qquad E(x)^\eta=E(\eta x),$$

for any η in V. It is also known that $E(x)$ can be written in the form

$$E(x)=\prod_{m}(1-x^m)^{\mu(m)/m},$$

where $\mu(m)$ is the Mobius function and m runs over all positive rational integers prime to p. Hence $E(x)$ is contained in R_0 and it holds that

$$E(x) \equiv 1-x \bmod. x^2R_0.$$

Putting $\pi(x)=1-E(x)$ in (13), we then obtain

$$\sum_{\eta} \eta^a/(1-E(\eta x)) \equiv 0 \bmod. p.$$

Now, let $\sum_{k} a_kx^k$ be the power series for $(1-E(x))^{-1}$ and put

$$\psi_i(x)=\sum_{k\equiv i(p-1)} a_kx^k, \qquad 1 \leqq i \leqq p-1.$$

Then the left hand side of the above congruence is nothing but $(p-1)\psi_{p-a-1}(x)$ and we have immediately the following result:

---

[8] Cf. [1].

IV. *Let $\psi_i(x)$ be defined as above. Then the second coefficient μ in the formula* (1) *is positive if and only if there exists an odd integer a, $1 \leqq a \leqq p-1$, such that*

$$\psi_a(x) \equiv 0 \bmod. p.$$

The advantage of IV lies in that we can explicitly calculate $\psi_i(x)$, $1 \leqq i \leqq p-1$, by using the classical power series:

$$1/(1-e^{-z}) = 1/z + \tfrac{1}{2} - \sum_{n=1}^{\infty} (-1)^n B_n z^{2n-1}/(2n)!,$$

where B_n $(n \geqq 1)$ are the Bernoulli numbers. Put $z = L(x)$ in the above and notice that the exponents of x which appear in the power series $L(x)$ are always congruent with $1 \bmod. p-1$. We then obtain immediately that

$$\begin{aligned} &\psi_{p-1}(x) = \tfrac{1}{2}, \\ &\psi_i(x) = 0, && 2 \mid i, i \neq p-1, \\ &\psi_{p-2}(x) = L(x)^{-1} - \textstyle\sum^{(p-2)} (-1)^n B_n L(x)^{2n-1}/(2n)!, \\ &\psi_i(x) = -\textstyle\sum^{(i)} (-1)^n B_n L(x)^{2n-1}/(2n)!, && 2 \nmid i, i \neq p-2, \end{aligned}$$

where $\sum^{(i)}$ means the summation over all $n \geqq 1$ such that $2n-1 \equiv i \bmod.$ $p-1$.

Now, it is clear from the above that $\psi_{p-2}(x) \not\equiv 0 \bmod. p.$[9] For an odd $a \neq p-2$, we can then calculate the coefficients of $x^{a+k(p-1)}$, $k \geqq 0$, in $\psi_a(x)$ from the last equality and, putting them to be congruent to $0 \bmod. p$, we obtain a sequence of congruence relations on the Bernoulli numbers which altogether form an equivalent condition with $\psi_a(x) \equiv 0 \bmod. p$. For instance, we obtain from the first two coefficients the following congruences:

$$\begin{aligned} &B'(\tfrac{1}{2}(a+1)) \equiv 0 \bmod. p, \\ &B'(\tfrac{1}{2}(a+1)) \equiv B'(\tfrac{1}{2}(a+1) + \tfrac{1}{2}(p-1)) \bmod. p^2, \end{aligned}$$

where $B'(n)$ is defined by

$$B'(n) = (-1)^n B_n/n, \qquad n \geqq 1.$$

The first congruence in the above shows that if $\mu > 0$, i. e., if $\psi_a(x) \equiv 0$ mod. p for some odd a, p must be an irregular prime. But this can be otherwise proved easily without making use of the above result.

For the irregular primes $p = 37, 59, 67$, we can determine the index a for which the first congruence is satisfied, and then see that the second

---

[9] This corresponds to the fact that (5), (11) and (12) do not hold for $a = 1$.

congruence does not hold for such an a.[10] Hence, we have $\mu=0$ for these prime numbers. However, in these cases, it is also known that the order of $G(L_1/K_1)$ is p^2 and, by the structural theorem on Γ-groups,[11] we can prove more precisely that $G(L/K)$ is isomorphic with Γ and that $\lambda=1$, $\mu=0$ and $\nu=0$.

Remark. Using the fact that the coefficients of the power series for $(1-E(x))^{-1}$ are always p-adic integers, we can obtain, by a similar argument as above, various congruence relations for the Bernoulli numbers mod. p^n.

THE INSTITUTE FOR ADVANCED STUDY
AND THE MASSACHUSETTS INSTITUTE OF TECHNOLOGY.

REFERENCES.

[1] E. Artin and H. Hasse, "Die beiden Ergänzungssätze zum Reziprozitätsgesetz der l^n-ten Potenzreste im Körper der l^n-ten Einheitswurzeln," *Abhandlungen aus dem Mathematischen Seminar der Hamburg Universität*, vol. 6 (1928), pp. 146-162.

[2] H. Hasse, *Über die Klassenzahl abelscher Zahlkörper*, Berlin, 1952.

[3] D. Hilbert, "Die Theorie der algebraischen Zahlkörper," *Jahresbericht der Deutschen Mathematiker-Vereinigung*, vol. 4 (1897), pp. 175-546.

[4] K. Iwasawa, "On Γ-extensions of algebraic number fields," to appear shortly in *Bulletin of the American Mathematical Society*.

[5] R. E. MacKenzie, "Class group relations in cyclotomic fields," *American Journal of Mathematics*, vol. 74 (1952), pp. 759-763.

[6] F. Pollaczek, "Über die irregulären Kreiskörper der l-ten und l^2-ten Einheitswurzeln," *Mathematische Zeitschrift*, vol. 21 (1924), pp. 1-38.

[7] T. Takagi, "Zur Theorie der Kreiskörpers," *Journal für die reine und angewandte Mathematik*, vol. 157 (1927), pp. 230-238.

---

[10] Cf. [6].

[11] Cf. [4], § 5.

Correction to the paper "On some invariants of cyclotomic fields" by K. IWASAWA, this Journal, vol. 80 (1958), pp. 773-783.

Lemma 2 on p. 779 should be replaced by the following:

LEMMA 2. *Suppose that*

$$\sum a_\xi \zeta_{n+1}{}^\xi \equiv 0 \bmod. p^s \qquad (\xi \text{ in } U \bmod. U_n)$$

with a_ξ *in* Q_p. *Then* $a_\xi \equiv a_\omega \bmod. p^s$ *whenever* $\xi \equiv \omega \bmod. U_{n-1}$.

Using this lemma, it follows from (8), after a simple computation, that $S_n(\chi) \equiv 0 \bmod. p$ for any character χ of U satisfying $\chi(U_n) = 1$, $\chi(U_{n-1}) \neq 1$, $\chi(\eta) = \eta^{-a}$. Hence (8) implies $\mu > 0$.

[34] Sheaves for algebraic number fields

Ann. of Math., (2) 69 (1959), 408-413.

(Received April 9, 1958)

Following the analogies between algebraic number theory and algebraic geometry, we construct in the present paper a certain sheaf on the space of all non-archimedean prime divisors of an algebraic number field, and we then show that arithmetic interpretations of the cohomology groups of the space with coefficients in this sheaf lead to number-theoretical isomorphisms in a natural way. Though the results obtained in the following are rather simple, they seem to indicate further possibility of applying the method of sheaves in algebraic number theory; in particular, in the study of infinite algebraic number fields.

1. Let K be an algebraic number field, i.e., an algebraic, but not necessarily finite, extension of the rational field Q. Let X be the set of all non-archimedean prime divisors of K. For any element a in K, we denote by U_a the set of all prime divisors P in X such that a is contained in the valuation ring of P, and, for any finite extension F of Q in K and for any finite set of non-archimedean prime divisors $P_1, \cdots, P_s$ of F, we denote by $U(F; P_1, \cdots, P_s)$ the set of all P in X such that the restriction of P on F is different from $P_1, \cdots, P_s$. Then the family of all subsets U_a in X, $a \in K$, coincides with the family of all subsets $U(F; P_1, \cdots, P_s)$ in X, and we can define a topology on X by taking the family of these subsets as a base for open sets in X. It is clear that X then becomes a T_1-space and that the intersection of two non-empty open subsets in X is non-empty; the cohomology groups in any constant sheaf over X are thus trivial in positive dimensions. It is also not difficult to see that any closed subset different from X is a compact Hausdorff space and that X itself is quasi-compact.

2. Now, let U be an open set in the above topological space X and let $\mathcal{F}_U$ denote the set of all $a \neq 0$ in K such that a is a local unit in K at every P belonging to U. Then $\mathcal{F}_U$ is a subgroup of the multiplicative group K^* of K and, if V is an open subset in U, there exists a natural injective map: $\mathcal{F}_U \to \mathcal{F}_V$. Therefore these subgroups $\mathcal{F}_U$, together with the above maps, define a sheaf $\mathcal{F}$ on X. We shall next consider the cohomology groups $H^n(X, \mathcal{F})$, $n \geqq 0$.[1]

* Guggenheim Fellow. The present research was also supported in part by a National Science Foundation grant.

[1] For the theory of sheaves in general, cf. [3, Chap. I].

It is clear from the definition that the stalk $\mathfrak{F}_P$ of $\mathfrak{F}$ at P is the group of all elements in K which are local units at P. Hence $H^0(X, \mathfrak{F}) = \Gamma(X, \mathfrak{F})$ is nothing but the group E of all units in K:

$$H^0(X, \mathfrak{F}) = E\ .$$

As usual, let K^* denote the constant sheaf on X isomorphic with K^*. Then $\mathfrak{F}$ is a subsheaf of K^* and the exact sequence of sheaves

$$1 \to \mathfrak{F} \to K^* \to K^*/\mathfrak{F} \to 1$$

gives rise to the following exact sequence of cohomology groups:

$$\begin{aligned} 1 \to H^0(X, \mathfrak{F}) \to H^0(X, K^*) &\to H^0(X, K^*/\mathfrak{F}) \\ &\to H^1(X, \mathfrak{F}) \to H^1(X, K^*) \to H^1(X, K^*/\mathfrak{F})\ . \end{aligned}$$

Here we have

$$H^0(X, K^*) = K^*, \quad H^1(X, K^*) = 1\ .$$

Hence, the above exact sequence implies that

$$(1) \qquad H^1(X, \mathfrak{F}) = \mathrm{Coker}\,(K^* \to H^0(X, K^*/\mathfrak{F}))\ .$$

Let s be an element of $H^0(X, K^*/\mathfrak{F})$, i.e., a section of $K^*/\mathfrak{F}$ over X. Then there exists a finite extension F_0 of Q in K with the following property: For any finite extension F of F_0 in K and for any P on X, there is an element a in F such that $s(P)$ is the class of a mod $\mathfrak{F}_P$ and such that if P' is any other point on X having the same restriction on the subfield F, then $s(P')$ also is given by the class of a mod $\mathfrak{F}_{P'}$. Let $\tilde{P}$ be the prime ideal (=non-archimedean prime divisor) which is the restriction of P on such an F and let $\tilde{P}^e$ be the $\tilde{P}$-component of the principal ideal (a) in F. Then the product of $\tilde{P}^e$, taken over all prime ideals $\tilde{P}$ of F, defines an ideal $A(s\,; F)$ of F, uniquely determined by s.

Now, let $\{F_j\}$ be the family of all finite extensions F_j of Q contained in K and let I_j and $\bar{I}_j$ denote the ideal group and the ideal class group of each F_j, respectively. Then, if F_j is contained in F_k, there exists a natural injective map: $I_j \to I_k$ and it induces a map: $\bar{I}_j \to \bar{I}_k$. The direct limit I of the family of ideal groups $\{I_j\}$ with respect to the above maps may be called *the ideal group of* K and, similarly, the direct limit $\bar{I}$ of $\{\bar{I}_j\}$ *the ideal class group of* K. We then see that the ideals $A(s\,; F)$ associated with a section s of $K^*/\mathfrak{F}$ over X determine an element $A(s)$ of I and that the map: $s \to A(s)$ defines an isomorphism

$$H^0(X, K^*/\mathfrak{F}) \cong I .$$

Combining this with (1), we get the isomorphism

(2) $$H^1(X, \mathfrak{F}) \cong \bar{I} .$$

In general, let $\mathfrak{U}$ be any open covering of X and f any element of $C^n(\mathfrak{U}, \mathfrak{F})$, $n \geqq 0$. We can then find a finite extension F of Q in K and a finite number of non-archimedean prime divisors $P_0, P_1, \cdots, P_s$ of F with the following properties: Let $V_i (i = 0, 1, \cdots, s)$ be the subset of all P in X such that the restriction of P on F is different from $P_0, \cdots, P_{i-1}, P_{i+1}, \cdots, P_s$. Then the covering $\mathfrak{V} = \{V_i\}$ is finer than $\mathfrak{U}$ and if f' is the image of f under a map $C^n(\mathfrak{U}, \mathfrak{F}) \to C^n(\mathfrak{V}, \mathfrak{F})$, the local sections of $\mathfrak{F}$ which define f' are given by elements of F. The existence of such $\mathfrak{V}$ and f' implies immediately that

$$H^n(X, \mathfrak{F}) = 1 ,$$

for any $n \geqq 2$.[2]

3. Since the ideal class group $\bar{I}_j$ of the finite algebraic number field F_j is a finite group, the ideal class group $\bar{I}$ of K is a periodic abelian group. Hence $H^1(X, \mathfrak{F})$ is also periodic and is the direct sum of its primary components. Let p be a fixed rational prime and let $H^1(X, \mathfrak{F})^{(p)}$ denote the p-primary component of $H^1(X, \mathfrak{F})$. Clearly, $H^1(X, \mathfrak{F})^{(p)}$ is naturally isomorphic with the p-primary component $\bar{I}^{(p)}$ of $\bar{I}$. But, assuming that K contains all p^n th roots of unity for $n = 1, 2, \cdots$, we shall give another arithmetic interpretation of the group $H^1(X, \mathfrak{F})^{(p)}$.

Let L be the maximal abelian p-extension over K such that any P' in X not dividing the rational prime p is unramified in L.[3] L contains the field M which is obtained by adjoining to K all p^n th roots of all units in K for $n = 1, 2, \cdots$. L/M is obviously a Galois extension and its Galois group $G = G(L/M)$ is a p-primary compact abelian group. We shall prove that $H^1(X, \mathfrak{F})^{(p)}$, as a discrete abelian group, is dual to the compact abelian group G.

Let h be any element of $H^1(X, \mathfrak{F})^{(p)}$ and suppose that $h^{p^m} = 1$, $m \geqq 0$. Take an open covering $\mathfrak{U} = \{U_i\}$ of X and an element f of $Z^1(\mathfrak{U}, \mathfrak{F})$ representing h such that $f^{p^m} = \partial g$ with a g in $C^0(\mathfrak{U}, \mathfrak{F})$. Putting $g = \{g_i\}$ and $f = \{f_{ij}\}$, $g_i \in \Gamma(U_i, \mathfrak{F})$, $f_{ij} \in \Gamma(U_i \cap U_j, \mathfrak{F})$, we have

(3) $$f_{ij}^{p^m} = g_j g_i^{-1} .$$

Therefore, if t_i is a p^m th root of g_i, $t_j t_i^{-1}$ is equal to f_{ij} up to a factor of

[2] Cf. [3, Chap. III, § 1, Prop. 4].

[3] Cf. [1, § 6].

$p^{m\text{th}}$ root of unity and, hence, is contained in K. It follows that $K(t_i) = K(t_j)$ for any indices i and j. Furthermore, as g_i is in $\Gamma(U_i, \mathscr{F})$, this common field is contained in L; namely, every t_i is contained in L. For any σ in G, put

$$\chi(\sigma) = t_i^{\sigma-1} = t_j^{\sigma-1} .$$

It is then clear that $\chi(\sigma)$ is a continuous character of the compact abelian group G. However, we can also see that χ depends only upon h and is independent of the choice of auxiliary elements which are used to define χ. Thus we have a map

$$(4) \qquad h \rightarrow \chi$$

of $H^1(X, \mathscr{F})^{(p)}$ into the character group $\hat{G}$ of G, and it is obviously a homomorphism.

Suppose that the character χ corresponding to h is trivial. With the same notations as above, every t_i is then contained in M and it follows from the definition of M that

$$(5) \qquad g_i = t_i^{p^m} = \varepsilon_i u_i^{p^m} ,$$

with $\varepsilon_i \in E$, $u_i \in K^*$. Since ε_i is a unit in K and g_i is in $\Gamma(U_i, \mathscr{F})$, u_i is also contained in $\Gamma(U_i, \mathscr{F})$. Putting (5) in (3), we see that

$$f_{ij} = \eta_{ij} u_j u_i^{-1} ,$$

with $\eta_{ij} \in E$. But the cohomology group $H^1(X, E)$ is 1 for the constant sheaf E. Hence

$$\eta_{ij} = \eta_j \eta_i^{-1} ,$$

with $\eta_i \in E$. We thus obtain

$$f_{ij} = s_j s_i^{-1} ,$$

with $s_i = \eta_i u_i$ in $\Gamma(U_i, \mathscr{F})$, and we see that

$$f \sim 1, \quad h = 1 .$$

The map (4) is therefore injective.

Now, let χ be any continuous character of G. Take a continuous character χ' of the Galois group of L/K such that χ is the restriction of χ' on G and let K' denote the fixed field of the kernel of χ'. If p^m is the order of χ', K'/K is a cyclic extension of degree p^m and, for each P in X, there exists an element g_P of K such that g_P is a local unit at P and that K' is obtained by adjoining a $p^{m\text{th}}$ root t_P of g_P satisfying $\chi'(\sigma) = t_P^{\sigma-1}$ for every σ in the Galois group of L/K; if P does not divide the rational prime p, the existence of such g_P and t_P follows from the fact that P is unramified in K', and if P divides p, it is a consequence of the facts that K contains all p^n th roots of unity ($n \geqq 1$) and that the value group of an exponential valuation belonging to P is divisible by p. Put

$$U_P = U_{g_P} \cap U_{g_P^{-1}}, \quad f_{P,P'} = t_{P'} t_P^{-1}$$

for any P and P'. Then $\mathfrak{U} = \{U_P\}$ defines an open covering of X and $f = \{f_{P,P'}\}$ an element of $Z^1(\mathfrak{U}, \mathfrak{F})$. Furthermore, we have obviously $f^{p^m} = \partial g$ with $g = \{g_P\}$ in $C^0(\mathfrak{U}, \mathfrak{F})$. Hence f determines an element h in $H^1(X, \mathfrak{F})^{(p)}$ and we see immediately that the given character χ is the image of such an h under the map (4). Therefore (4) is also surjective.

We have thus proved the following isomorphism :

$$H^1(X, \mathfrak{F})^{(p)} \cong \hat{G} .$$

It then follows immediately from (2) that

$$\bar{I}^{(p)} \cong \hat{G} , \tag{6}$$

where $\bar{I}^{(p)}$ is the p-primary component of the ideal class group of K and $\hat{G}$ is the character group of the Galois group of L/M.

It is clear that if K contains not only all p^n th roots of unity but also all n^{th} roots of unity for $n = 1, 2, \cdots$, then we can obtain similar isomorphisms involving $H(X, \mathfrak{F})$, $\bar{I}$, and the character group of a certain Galois group like G. We also notice that once such an isomorphism (6) is known, it is not difficult to prove it directly without making use of the sheaf $\mathfrak{F}$.

4. Now, for each integer $n \geqq 0$, let K_n denote the field obtained by adjoining all p^{n+1} th roots of unity to Q, and let p^{e_n} be the highest power of p dividing the class number of K_n. Then, for all sufficiently large n, the exponent e_n is given by the formula :

$$e_n = \lambda n + \mu p^n + \nu , \tag{7}$$

where λ, μ and ν are integers independent of n, and it is also known that the second coefficient μ is 0 if and only if the rank of the p-Sylow group $\bar{I}_n^{(p)}$ of the ideal class group $\bar{I}_n$ of K_n is bounded for all $n \geqq 0$.[4]

Now, let K be the union of all K_n, $n \geqq 0$. K is obviously the smallest algebraic number field to which we may apply the result of the previous section, and the p-primary component $\bar{I}^{(p)}$ of the ideal class group $\bar{I}$ of K is the direct limit of the sequence of groups $\bar{I}_n^{(p)}$, $n \geqq 0$. However, it can be proved that the order of Ker $(\bar{I}_n^{(p)} \rightarrow \bar{I}^{(p)})$ has a fixed upper bound for

[4] Cf. [1, § 7].

all $n \geqq 0$.[5] Therefore the group $\bar{I}^{(p)}$ has a finite rank if and only if the rank of $\bar{I}_n^{(p)}$ is bounded for all $n \geqq 0$, and, by (2) and by the above remark, we see that *the second coefficient* μ *in* (7) *is* 0 *if and only if the group* $H^1(X, \mathfrak{F})^{(p)}$ *has a finite rank.*[6]

In view of such an arithmetic application, it seems quite interesting to know whether or not $H^1(X, \mathfrak{F})^{(p)}$ has a finite rank for the field K defined above, or, more generally, for any field which is obtained by adjoining all $p^{n\text{th}}$ roots of unity, $n = 1, 2, \cdots$, to a finite algebraic number field.[7]

Finally, we also notice that, though we have considered here only one sheaf $\mathfrak{F}$ for a given algebraic number field K, we can of course construct various similar sheaves for K either by applying elementary algebraic operations on $\mathfrak{F}$ or by imposing different local condition on the constant sheaf K^*; some of these sheaves are again related with certain extensions of the field K.

BIBLIOGRAPHY

1. K. IWASAWA, *On Γ-extensions of algebraic number fields*, to appear shortly in Bull. Amer. Math. Soc..
2. ———, *On some invariants of cyclotomic fields*, Amer. Jour. Math., 80 (1958), 773–783.
3. J-P. SERRE, *Faisceaux algébriques cohérents*, Ann. of Math., 61 (1955), 197-278.

[5] Cf. a forthcoming paper of the author on cyclotomic fields.

[6] Arithmetic criteria for $\mu = 0$ are given in [2]. In particular, it is shown there that $\mu = 0$ for every $p < 100$. We can also prove that $H^1(X, \mathfrak{F})^{(p)} = 1$ if and only if p is a regular prime; cf. the paper mentioned in footnote 5.

[7] In this connection, cf. [3, Chap. III, § 3, Theorem 1].

[35] On Γ-extensions of algebraic number fields

Bull. Amer. Math. Soc., 65 (1959), 183-226.

Let p be a prime number. We call a Galois extension L of a field K a Γ-extension when its Galois group is topologically isomorphic with the additive group of p-adic integers. The purpose of the present paper is to study arithmetic properties of such a Γ-extension L over a finite algebraic number field K. We consider, namely, the maximal unramified abelian p-extension M over L and study the structure of the Galois group $G(M/L)$ of the extension M/L. Using the result thus obtained for the group $G(M/L)$, we then define two invariants $l(L/K)$ and $m(L/K)$, and show that these invariants can be also determined from a simple formula which gives the exponents of the p-powers in the class numbers of the intermediate fields of K and L. Thus, giving a relation between the structure of the Galois group of M/L and the class numbers of the subfields of L, our result may be regarded, in a sense, as an analogue, for L, of the well-known theorem in classical class field theory which states that the class number of a finite algebraic number field is equal to the degree of the maximal unramified abelian extension over that field.

An outline of the paper is as follows: in §1–§5, we study the structure of what we call Γ-finite modules and find, in particular, invariants of such modules which are similar to the invariants of finite abelian groups. In §6, we give some definitions and simple results on certain extensions of (infinite) algebraic number fields, making it clear what we mean by, e.g., an unramified extension, when the ground field is an infinite algebraic number field. In the last §7, we first show that the Galois group $G(M/L)$ as considered above is a Γ-finite module, then define the invariants $l(L/K)$ and $m(L/K)$, and finally prove our main formula using the group-theoretical results obtained in previous sections.

1. Preliminaries. 1.1. Let p be a prime number. We shall first recall some definitions and elementary properties of p-primary abelian groups.[2]

An address delivered before the Summer Meeting of the Society in Seattle on August 23, 1956 under the title of *A theorem on Abelian groups and its application in algebraic number theory* by invitation of the Committee to Select Hour Speakers for Annual and Summer Meetings; received by the editors August 28, 1957.

[1] Guggenheim Fellow. The present research was also supported in part by a National Science Foundation grant.

[2] For the theory of abelian groups in general, cf. I. Kaplansky, *Infinite abelian groups*, University of Michigan Press, 1954.

A discrete group is called p-primary if it is the direct limit of a family of finite p-groups, or, what amounts to the same, if it is locally finite and if every element of the group has a finite order which is a power of p. A compact group is called p-primary if it is the inverse limit of finite p-groups. For a finite group, which is at the same time discrete and compact, both definitions coincide and the group is p-primary if and only if it is a p-group.

Now, let A be a p-primary discrete abelian (additive) group. If the orders of elements in A have a fixed upper bound, i.e. if $p^n A=0$ for some $n \geqq 0$, A is called a group of bounded order, or, simply, bounded. Let A' be the subgroup of elements a in A satisfying $pa=0$. If A' is a finite group of order p^l, A is called a group of finite rank l. A is called divisible if $pA=A$.

It is clear from the definition that the character group of a p-primary discrete abelian group is a p-primary compact abelian group and, conversely, the character group of a p-primary compact abelian group is a p-primary discrete abelian group.

Let $\{A, X\}$ be such a dual pair of a p-primary discrete abelian group A and a p-primary compact abelian group X. Obviously, $p^n A=0$ if and only if $p^n X=0$. In such a case, the compact abelian group X is also called bounded. As the subgroup A' of A defined above is dual to X/pX, A is of finite rank l if and only if X/pX is a finite group of order p^l. If this is the case, the compact abelian group X is called a group of finite rank l; for a finite abelian p-group, both definitions give the same rank l. Finally, A is divisible if and only if X is torsion-free.

1.2. We shall next give a typical example of such a dual pair of a p-primary discrete abelian group and a p-primary compact abelian group.

For every integer $n \geqq 0$, let Z_{p^n} denote a cyclic group of order p^n. Clearly, for each $n \geqq 0$, there exist an isomorphism ϕ_n of Z_{p^n} into $Z_{p^{n+1}}$ and a homomorphism ψ_n of $Z_{p^{n+1}}$ onto Z_{p^n}. Let Z_{p^∞} be the direct limit of the sequence of cyclic groups Z_{p^n} relative to ϕ_n. Z_{p^∞} is then a p-primary discrete abelian group and is isomorphic with the factor group Q_p/O_p, where Q_p denotes the additive group of p-adic numbers and O_p the subgroup of p-adic integers. On the other hand, the inverse limit of the groups Z_{p^n} relative to ψ_n gives a p-primary compact abelian group isomorphic with O_p which is a compact group with respect to its natural p-adic topology. Since the finite groups Z_{p^n} are self-dual, it follows that Z_{p^∞} and O_p, or, Q_p/O_p and O_p, form a dual pair of p-primary abelian groups.

The duality between Q_p/O_p and O_p can be seen also directly as follows. Q_p is a locally compact abelian group in its p-adic topology and there is a character χ of Q_p such that the kernel of the homomorphism $a \to \chi(a)(a \in Q_p)$ is O_p. The inner product on Q_p defined by:[3]

$$(a, b) = \chi(ab), \qquad a, b \in Q_p,$$

then gives a dual pairing of Q_p with itself such that the annihilator of O_p in Q_p coincides with O_p itself. Hence, the pairing (a, b) induces a dual pairing of Q_p/O_p and O_p.

1.3. Let G be a totally disconnected compact multiplicative group with the unity element 1. A topological additive abelian group U will be called a G-module when G acts on U so that $1 \cdot u = u$ for every u in U and that the mapping $\sigma \times u \to \sigma u$ of $G \times U$ into U is continuous.

Let $\{A, X\}$ be a dual pair as considered in 1.1 and suppose both A and X are G-modules in the sense defined above. We call A and X dual G-modules if there exists a dual pairing (a, x) of A and X such that

$$(\sigma a, \sigma x) = (a, x) \tag{1}$$

for every σ in G, a in A, and x in X.

Now, let (a, x) be any dual pairing of A and X. Suppose first that only A is given a structure of a G-module. We can then define, in a unique way, the product σx of σ in G and x in X so that X becomes a G-module satisfying (1). Thus, if A is a G-module, the dual group X can be also made into a G-module so that A and X form a pair of dual G-modules with respect to a given pairing of A and X. The structure of the G-module X defined in this way depends upon the choice of the pairing (a, x), but it is uniquely determined up to an automorphism of X. Similarly, if X is a G-module, we can define a structure of a G-module on the dual group A so that A and X form a pair of dual G-modules.

1.4. Let G be a totally disconnected compact group and $\{A, X\}$ a dual pair of a p-primary discrete abelian group A and a p-primary compact abelian group X. Defining $\sigma a = a$, $\sigma x = x$ for any a in A, x in X, and σ in G, we may consider $\{A, X\}$ as a pair of dual G-modules as defined above. Let N_α be any open normal subgroup of G and $Z(G_\alpha)$ the group ring of the finite group $G_\alpha = G/N_\alpha$ over the ring of rational integers Z. We may consider $Z(G_\alpha)$ as a G-module by defining

$$\sigma w = \sigma' w, \qquad \sigma \in G,\ w \in Z(G_\alpha),$$

where σ' is the image of σ under the canonical homomorphism $G \to G_\alpha$. Let $\mathrm{Hom}(Z(G_\alpha), A)$ be the group of all homomorphisms of $Z(G_\alpha)$ into A. Since both $Z(G_\alpha)$ and A are G-modules, $\mathrm{Hom}(Z(G_\alpha), A)$

[3] ab is the product of a and b as elements of the field Q_p.

is also made into a G-module in a natural way.[4] Furthermore, if N_β is an open normal subgroup of G contained in N_α, the canonical homomorphism $G_\beta = G/N_\beta \to G_\alpha$ induces a natural G-isomorphism $\phi_{\beta,\alpha}$ of $\mathrm{Hom}(Z(G_\alpha), A)$ into $\mathrm{Hom}(Z(G_\beta), A)$. Hence, considering the set of all groups $\mathrm{Hom}(Z(G_\alpha), A)$ attached one for each open normal subgroup N_α of G, we can form the direct limit $M_1(G, A)$ of these discrete G-modules $\mathrm{Hom}(Z(G_\alpha), A)$ relative to the homomorphisms $\phi_{\beta,\alpha}$. By the definition, $M_1(G, A)$ is a discrete G-module. Clearly, $\mathrm{Hom}\,(Z(G_\alpha), A)$ can be identified with the additive group of all functions defined on G_α taking values in A, i.e. with the additive group of those functions on G with values in A which are constant on each coset of G mod N_α. Since G is totally disconnected and A is discrete, we may then consider $M_1(G, A)$ as the G-module of all continuous functions defined on G taking values in A, where the action of G on $M_1(G, A)$ is defined by:

$$(\sigma f)(\tau) = f(\sigma^{-1}\tau), \qquad f \in M_1(G, A), \sigma \in G.$$

We now consider the tensor product $Z(G_\alpha) \otimes X$ of $Z(G_\alpha)$ and X over Z. Clearly, $Z(G_\alpha) \otimes X$ is a compact G-module. Furthermore, the canonical homomorphism $G_\beta \to G_\alpha$ again induces a continuous homomorphism $\psi_{\alpha,\beta}$ of $Z(G_\beta) \otimes X$ onto $Z(G_\alpha) \otimes X$. Hence, the inverse limit of the family of compact G-modules $Z(G_\alpha) \otimes X$ relative to the homomorphisms $\psi_{\alpha,\beta}$ gives us a compact G-module $M_2(G, X)$.

Now, by the assumption, there is a dual pairing (a, x) of A and X. Then, there also exists a unique dual pairing $(s, t)_\alpha$ of the discrete G-module $\mathrm{Hom}\,(Z(G_\alpha), A)$ and the compact G-module $Z(G_\alpha) \otimes X$ such that

$$(s, w \otimes x)_\alpha = (s(w), x)$$

for any s in $\mathrm{Hom}\,(Z(G_\alpha), A)$, w in $Z(G_\alpha)$, and x in X. Since

$$(s, \psi_{\alpha,\beta}(t'))_\alpha = (\phi_{\beta,\alpha}(s), t')_\beta, \quad s \in \mathrm{Hom}(Z(G_\alpha), A), t' \in Z(G_\beta) \otimes X,$$

for $N_\beta \subset N_\alpha$, those pairings $(s, t)_\alpha$ together define a dual pairing of $M_1(G, A)$ and $M_2(G, X)$. We can thus obtain, for each dual pair $\{A, X\}$, a pair of dual G-modules $M_1(G, A)$ and $M_2(G, X)$.

2. **Some definitions.** 2.1. Let Γ be a multiplicative topological group isomorphic with the additive group of p-adic integers O_p. We shall fix such a group Γ once and for all in the following discussions. Γ is a totally disconnected compact abelian group and, for each $n \geqq 0$, it contains an open subgroup Γ_n such that Γ/Γ_n is a cyclic group of order p^n. We have, then, a sequence of subgroups $\Gamma = \Gamma_0 \supset \Gamma_1$

[4] Cf. the definition of σf below.

$\supset\Gamma_2\supset\cdots$, and these subgroups form a fundamental system of neighborhoods of the identity 1 in Γ. Furthermore, there exists no nontrivial closed subgroup of Γ other than the Γ_n.

For convenience, we take an element γ of Γ not contained in Γ_1 and fix it once and for all in the following. For each $n\geqq 0$, put

$$\gamma_n = \gamma^{p^n}.$$

Then, each γ_n generates an infinite cyclic group which is everywhere dense in Γ_n. In particular, $\gamma=\gamma_0$ generates an everywhere dense subgroup in Γ. We also put

$$\omega_n = 1 - \gamma_n, \qquad n \geqq 0.$$

ω_n is an element of the group ring of the cyclic group generated by γ over the ring of rational integers Z.

2.2. In what follows, up to the end of §5, we shall exclusively deal with p-primary discrete or compact abelian groups which are also Γ-modules in the sense of 1.3. Therefore, if there is no risk of misunderstanding, we shall call those groups simply discrete or compact modules. Similarly, Γ-invariant subgroups, Γ-homomorphisms, etc., of those modules will be simply called submodules, homomorphisms, etc.

Let A be such a discrete module. For each $n\geqq 0$, we denote by A_n the submodule of all elements a in A such that $\sigma a=a$ for every σ in Γ_n. Since γ_n generates an everywhere dense subgroup of Γ_n, A_n is the submodule of all a in A satisfying $\gamma_n a=a$, i.e. $\omega_n a=0$. Since Γ_{n+1} is contained in Γ_n, A_n is a submodule of A_{n+1}.

LEMMA 2.1. *A is the union of the submodules A_n, $n\geqq 0$.*

PROOF. Let a be any element in A. Since $1\cdot a=a$, A is discrete and the mapping $\sigma\times a\to\sigma a$ is continuous, there exists a neighborhood Γ_n ($n\geqq 0$) of 1 in Γ such that $\sigma a=a$ for every σ in the neighborhood Γ_n. a is then contained in A_n.

We notice that, for a discrete abelian group A with operator domain Γ, the continuity of the mapping $\sigma\times a\to\sigma a$ follows, conversely, from the fact that A is the union of all A_n, $n\geqq 0$.

For each $n\geqq 0$, let A_n^* denote the submodule of A generated by elements of the form $(1-\sigma)a$ where σ and a are arbitrary elements in Γ_n and A respectively. Since γ_n generates an everywhere dense subgroup of Γ_n, A_n^* coincides with $\omega_n A=(1-\gamma_n)A$.

Now the discrete module A will be called *n-regular* if $A_n^*=A$, and A will be called *regular* if it is n-regular for all $n\geqq 0$. Clearly, a homomorphic image of an n-regular (regular) module is n-regular (regular).

In particular, any quotient module A/B of an n-regular (regular) module A is n-regular (regular). The sum of n-regular (regular) submodules of a discrete module is also n-regular (regular). Hence, every discrete module has a unique maximal n-regular (regular) submodule.

LEMMA 2.2. *Let B be a submodule of a discrete module A and let B_n and C_n be the submodules of B and $C=A/B$, respectively, defined similarly as A_n for A. Then, $B_n=A_n \cap B$, and, if B is n-regular, $C_n=(A_n+B)/B \cong A_n/B_n$.*

PROOF. It is clear from the definition that $B_n=A_n \cap B$ and that $(A_n+B)/B$ is contained in C_n. Suppose B be n-regular. Let $\bar{a}$ be any element in C_n and a an element of A in the residue class $\bar{a}$. As $\omega_n \bar{a}=0$, $\omega_n a$ is contained in B, and, as B is n-regular, $\omega_n a=\omega_n b$ for some b in B. $a'=a-b$ is then also in the same residue class $\bar{a}$ and it is contained in A_n. Hence, C_n is contained in $(A_n+B)/B$.

2.3. We now define a certain kind of discrete modules.

Let A be a discrete module and A_n ($n \geqq 0$) the submodules of A as defined in 2.2. A is called *Γ-finite* if every A_n is a group of finite rank, and A is called *strictly Γ-finite* if every A_n is a finite group.

Suppose A be strictly Γ-finite. Then, the order of the finite group A_n is a power of p. We denote the exponent of p in the order of A_n by $c(n; A)$. For given A, $c(n; A)$ then defines a nondecreasing function of the integers $n \geqq 0$, and we call it the *characteristic function* of the strictly Γ-finite discrete module A.

Clearly, a submodule B of a Γ-finite discrete module A is also Γ-finite. If A is strictly Γ-finite, so is B, and $c(n; B) \leqq c(n; A)$ for all $n \geqq 0$. The following lemma is also an immediate consequence of Lemma 2.2 and of the definition.

LEMMA 2.3. *Let B be a regular submodule of a discrete module A. Then, A is (strictly) Γ-finite if and only if both A/B and B are (strictly) Γ-finite, and, if A is strictly Γ-finite,*

$$c(n; A) = c(n; A/B) + c(n; B),$$

for all $n \geqq 0$.

2.4. We now consider compact modules, i.e. p-primary compact abelian groups which also form Γ-modules.

Let X be such a module. For each $n \geqq 0$, let X_n be the closed submodule of all a in X satisfying $\sigma a=a$ for every σ in Γ_n, and X_n^* the closure of the subgroup of X generated by all elements of the form $(1-\sigma)x$ where σ and x are arbitrary elements of Γ_n and X, respectively. As γ_n generates an everywhere dense subgroup of Γ_n, X_n is the submodule of all a in X satisfying $\omega_n a=0$, and X_n^* coincides with $\omega_n X$.

Now, as stated in general in 1.3, there exists a discrete module A such that A and X form a pair of dual Γ-modules, and such an A is, up to isomorphisms, uniquely determined by X. Let (a, x) be the dual pairing of A and X such that $(\sigma a, \sigma x)=(a, x)$ for every σ in Γ. Since

$$((1-\sigma^{-1})a, x) = (a, (1-\sigma)x), \qquad a \in A, x \in X, \sigma \in \Gamma,$$

both $\{A_n, X_n^*\}$ and $\{A_n^*, X_n\}$ are pairs of mutually orthogonal submodules of A and X, respectively, relative to the pairing (a, x). Therefore, A_n and X/X_n^* form a pair of dual Γ-modules, and so do A/A_n^* and X_n. By Lemma 2.1, A is the union of all A_n $(n \geqq 0)$. Hence, by the above, the intersection of all X_n^* $(n \geqq 0)$ is 0. It also follows that A is n-regular (i.e. $A_n^*=A$) if and only if $X_n=0$.

Now, we call a compact module X *Γ-finite* if every group X/X_n^* has a finite rank, and we call it *strictly Γ-finite* if every X/X_n^* is a finite group. In other words, a compact module X is called (strictly) Γ-finite if and only if the discrete module A dual to X is (strictly) Γ-finite. An *n-regular* (*regular*) compact module X is defined similarly, either by $X_n=0$ (for all $n \geqq 0$) or by the fact that it is dual to an n-regular (regular) discrete module.

Suppose that X be strictly Γ-finite. By the definition, X/X_n^* is a finite p-group for every $n \geqq 0$, and we denote by $c(n; X)$ the exponent of p in the order of X/X_n^*. We thus obtain a nondecreasing function $c(n; X)$ of the integers $n \geqq 0$ and call it the *characteristic function* of the strictly Γ-finite compact module X. Clearly, if A is a strictly Γ-finite discrete module dual to X, then

$$c(n; A) = c(n; X)$$

for all $n \geqq 0$.

In the following, we shall consider the structure of discrete modules and that of compact modules in parallel; by the duality between discrete and compact modules, any results on discrete (strictly) Γ-finite modules will then immediately give us corresponding results on compact (strictly) Γ-finite modules, and vice versa.

2.5. Let G be a compact group containing a closed normal subgroup X such that X is a p-primary compact abelian group and that $G/X=\Gamma$. Then Γ acts on X in an obvious way and X is thus made into a compact (Γ-)module in the sense of 2.2.[5] Furthermore, it is

[5] Of course, we understand that X is then considered as an additive group.

easy to see that the group extension G/X splits. Hence the structure of the compact group G is completely determined by the structure of the compact module X. On the other hand, given any compact module X, we can immediately find a compact group G to which X is related as stated above. Thus, there is a one-one correspondence between the set of all types of compact modules and the set of all types of group extensions of p-primary compact abelian groups by Γ.

In such a correspondence, the fact that a compact module X is Γ-finite can be interpreted for the corresponding compact group G as follows: let G_n $(n \geqq 0)$ be the closed subgroup of G such that $X \subset G_n$ and $G_n/X = \Gamma_n$. By a simple computation of commutators in G, we see that the topological commutator group $[G_n, G_n]$ of G_n is equal to the submodule $X_n^* = \omega_n X$ of X given in 2.4. Therefore X is Γ-finite if and only if $X/[G_n, G_n]$ has a finite rank for every $n \geqq 0$, or, equivalently, if and only if $G_n/[G_n, G_n]$ has a finite rank for every $n \geqq 0$.

In our later applications, Γ-finite compact modules will be obtained from compact groups such as G in the manner as described above.

3. **Modules of finite ranks.** 3.1. Clearly, every discrete or compact module of finite rank is Γ-finite.

Let A be a discrete module of finite rank. As a p-primary abelian group, A is then the direct sum of a finite group and a subgroup B isomorphic with $Z_{p^\infty}^l$, the direct sum of l copies of Z_{p^∞} $(l \geqq 0)$. B is the maximal divisible subgroup of A and is a characteristic subgroup of A. Hence B is also a (Γ-invariant) submodule of A. We shall next study the structure of such an A in the case $A = B$, i.e. in the case A is divisible.

LEMMA 3.1. *Let A be a divisible discrete module of finite rank. Then A is the direct sum of a regular submodule B and a submodule C such that $\omega_n^m C = 0$ for some $m \geqq 0$ and $n \geqq 0$. Furthermore, such a decomposition $A = B + C$ is unique for A, and B is also the unique maximal regular submodule of A.*

PROOF. Let B be the intersection of the submodules $\omega_n^m A$ for all $m \geqq 0$ and $n \geqq 0$. As a homomorphic image of divisible A, every $\omega_n^m A$ is also divisible, and, if $m' \geqq m$, $n' \geqq n$, then $\omega_{n'}^{m'} A$ is contained in $\omega_n^m A$. Since A is of finite rank, it follows that $B = \omega_n^m A$ whenever both m and n are sufficiently large, i.e. whenever $m \geqq m_0$ and $n \geqq n_0$ for some fixed $m_0 \geqq 0$, $n_0 \geqq 0$. But, then, $\omega_n B = \omega_n^{m+1} A = B$ for any $n \geqq n_0$, and B is, hence, a regular submodule of A.

Let C' be the kernel of the endomorphism of A: $a \rightarrow \omega_{n_0}^{m_0} a$. Since

$\omega_{n_0}^{m_0}A=B=\omega_{n_0}^{m_0}B$, we have $A=B+C'$, and since A and B are both divisible, we also have $A=B+p^sC'$ for any $s\geqq 0$. Now, choose s so large that $C=p^sC'$ is divisible. It then follows from the isomorphism $A/C'\cong B=\omega_{n_0}^{m_0}A$ that the rank of A is the sum of the ranks of B and C, and, hence, the sum $A=B+C$ is direct.

Next, suppose that $A=B^*+C^*$ be any direct sum decomposition of A such that B^* is regular and $\omega_n^m C^*=0$ for some m and n. Since $\omega_{n'}^{m'}C^*=0$ for any $m'\geqq m$ and $n'\geqq n$, we may assume that $m\geqq m_0$ and $n\geqq n_0$. We have then $B^*=\omega_n^m B^*=\omega_n^m A=B$. Therefore, $B=\omega_{n_0}^{m_0}A$ $=\omega_{n_0}^{m_0}B+\omega_{n_0}^{m_0}C^*=B+\omega_{n_0}^{m_0}C^*$ and, consequently, $\omega_{n_0}^{m_0}C^*=0$. Thus C^* is contained in C', and since C^* is divisible as a direct summand of A, we then see easily that $C^*=C$. The fact that B is the unique maximal regular submodule of A can be proved in a similar way.

By the duality between discrete and compact modules, we can immediately obtain the following result from the above lemma:

LEMMA 3.2. *Let X be a torsion-free compact module of finite rank. Then X is the direct sum of a regular submodule U and a submodule V such that $\omega_n^m V=0$ for some $m\geqq 0$ and $n\geqq 0$. Furthermore, such a decomposition $X=U+V$ is unique for X, and V also is the unique minimal submodule of X such that X/V is regular.*

3.2. Let X be as in Lemma 3.2 and let l be the rank of X. As a p-primary compact abelian group, X can be then identified with O_p^l, the direct sum of l copies of O_p, and the mapping $x\to\gamma x$ defines a continuous automorphism of O_p^l. Thus, there exists an $l\times l$ matrix M with entries in O_p such that the determinant of M is a p-adic unit and that

$$\gamma x = xM$$

for any $x=(x_1,\cdots,x_l)$ in O_p^l $(x_i\in O_p)$.[6] Put $M_n=M^{p^n}$ $(n\geqq 0)$ so that

$$\gamma_n x = xM_n, \qquad \omega_n x = x(I-M_n), \qquad x\in X,$$

I being the $l\times l$ identity matrix. Since the intersection of all $X_n^*=\omega_n X$ is 0, there exists an $s\geqq 0$ such that X_s^* is contained in pX. We have then

$$M_s \equiv I \bmod p, \qquad M_s = M^{p^s}. \tag{2}$$

On the other hand, if we are given any $l\times l$ integral p-adic matrix M satisfying (2) for some $s\geqq 0$, we can uniquely define a structure of a Γ-module on O_p^l so that $\gamma x=xM$ holds for any x in the compact

[6] xM is the product of the vector ($1\times l$ matrix) x and the $l\times l$ matrix M, and it is again a vector in $X=O_p^l$.

module $X=O_p^l$; the condition (2) then ensures the continuity of the action of Γ on O_p^l. Furthermore, two such matrices M_1 and M_2 define isomorphic compact modules on O_p^l if and only if $M_2=TM_1T^{-1}$ with a suitable integral p-adic matrix T whose determinant is a p-adic unit. Thus the classification problem for all torsion-free compact modules of rank l can be reduced to the problem of classifying all $l\times l$ integral p-adic matrices satisfying (2) for some $s\geqq 0$, with respect to the equivalence as stated above.

Now, let X and M be again as above. For any $n\geqq s+1$, it follows from (2) that

$$M_n \equiv I \mod p^2,$$

namely, that $M_n=I+p^2N$ with a suitable integral p-adic matrix N. By a simple computation, we then see that, for any $t\geqq 0$,

$$\sum_{i=0}^{p^t-1} M_n^i = p^t(I+pN_1),$$

with an integral p-adic matrix N_1. Since the determinant of $I+pN_1$ is a p-adic unit, it follows that

$$X\left(\sum_{i=0}^{p^t-1} M_n^i\right) = p^tX(I+pN_1) = p^tX.$$

Putting, in general,

$$\nu_{m,n} = \sum_{i=0}^{p^{m-n}-1} \gamma_n^i, \qquad m \geqq n \geqq 0, \tag{3}$$

we then get from the above that $\nu_{n+t,n}X=p^tX$ and, consequently, that

$$[X:\nu_{m,n}X] = [X:p^{m-n}X] = p^{l(m-n)},$$

for any integers $m\geqq n\geqq s+1$. Thus, the following lemma is proved:

LEMMA 3.3. *Let X be a torsion-free compact module of rank l. Then there exists an integer $n_0\geqq 0$ such that, for any integers m, n satisfying $m\geqq n\geqq n_0$, the index $[X:\nu_{m,n}X]$ is given by*

$$[X:\nu_{m,n}X] = p^{l(m-n)}.$$

3.3. Let X and M be as above. Since $X_n^*=\omega_nX=X(I-M_n)$, X/X_n^* is a finite module if and only if the determinant $|I-M_n|$ is different from 0. Therefore, X is strictly Γ-finite if and only if none of the p^nth roots of unity, for $n=0, 1, 2, \cdots$, is a characteristic root of the matrix M.

LEMMA 3.4. *A torsion-free compact module of finite rank is strictly Γ-finite if and only if it is regular.*

PROOF. As stated above, such a compact module X is strictly Γ-finite if and only if $|I-M_n|\neq 0$ for every $n\geqq 0$. But the condition $|I-M_n|\neq 0$ is equivalent to the fact that there is no $x\neq 0$ in X satisfying $x(I-M_n)=0$, i.e. $\omega_n x=0$. Thus X is strictly Γ-finite if and only if the submodule X_n of X as defined in 2.4 is 0 for every $n\geqq 0$. The lemma then follows immediately from the remark also given in 2.4.

By the duality, we see that a divisible discrete module of finite rank is strictly Γ-finite if and only if it is regular.

LEMMA 3.5. *Let X be a torsion-free compact module of rank l and let X be strictly Γ-finite. Then there exists an integer n_0 such that, for $n\geqq n_0$, the characteristic function of X is given by*

$$c(n; X) = ln + u,$$

where u is a suitable integer independent of n.

PROOF. By the previous lemma, X is regular and $X_n=0$ for every $n\geqq 0$. Hence the endomorphism $x\rightarrow\omega_n x$ of X is one-one, and we see that

$$[X:\nu_{m,n}X] = [\omega_n X:\omega_n\nu_{m,n}X] = [\omega_n X:\omega_m X],$$

for any $m\geqq n\geqq 0$. But, Lemma 3.3, $[X:\nu_{m,n}X]=p^{l(m-n)}$ when $m\geqq n\geqq n_0$. Therefore, if $n\geqq n_0$,

$$[X:\omega_n X] = [X:\omega_{n_0}X][\omega_{n_0}X:\omega_n X] = [X:\omega_{n_0}X]p^{l(n-n_0)},$$

and if we put $[X:\omega_{n_0}X]p^{-ln_0}=p^u$, then

$$c(n; X) = ln + u, \qquad n\geqq n_0.$$

Again, by the duality, we obtain the corresponding result on divisible discrete modules of finite rank l; if A is such a module, there exists an integer $n_0\geqq 0$ such that

$$c(n; A) = ln + u, \qquad n\geqq n_0,$$

with a suitable constant u.

4. **Bounded modules.** 4.1. We shall next consider Γ-finite discrete modules of bounded order. Clearly, all those modules are also strictly Γ-finite.

We shall first define an important class of such modules. Let m be any non-negative integer. Let $M_1(\Gamma, Z_{p^m})$ be, as defined in general in 2.4, the discrete Γ-module formed by all continuous functions on Γ with values in Z_{p^m}. Clearly, if a is any element in $M_1(\Gamma, Z_{p^m})$, then $p^m a=0$. Hence, $M_1(\Gamma, Z_{p^m})$ is a bounded module, and we denote it simply by $E(m)$. As noticed above, $p^m E(m)=0$, and, in particular, $E(0)=0$.

LEMMA 4.1. *Let $0 \leqq l \leqq m$. Then $p^{m-l}E(m)$ is the submodule of all a in $E(m)$ such that $p^l a=0$, and*

$$E(m)/p^{m-l}E(m) \cong p^l E(m) \cong E(m-l).$$

PROOF. The endomorphism $c \to p^l c$ of the cyclic group Z_{p^m} maps Z_{p^m} onto $p^l Z_{p^m} \cong Z_{p^{m-l}}$ and its kernel is $p^{m-l}Z_{p^m}$. Hence, the endomorphism $a \to p^l a$ of $E(m)$ induces the above isomorphisms.

LEMMA 4.2. *The discrete module $E(m)$ is bounded, regular and strictly Γ-finite, and its characteristic function is given by*

$$c(n; E(m)) = mp^n, \qquad n \geqq 0.$$

PROOF. By the definition, $E(m)$ is the union of a sequence of submodules $\mathrm{Hom}(Z(\Gamma/\Gamma_n), Z_{p^m})$, $n \geqq 0$, each considered as the submodule of all continuous functions on Γ with values in Z_{p^m} which are constant on every coset of Γ mod Γ_n. It is then clear that $\mathrm{Hom}(Z(\Gamma/\Gamma_n), Z_{p^m})$ coincides with the submodule $E(m)_n$ of all a in $E(m)$ such that $\sigma a=a$ for every σ in Γ_n. Since $\mathrm{Hom}(Z(\Gamma/\Gamma_n), Z_{p^m})$ is a group of order p^{mp^n}, $E(m)$ is strictly Γ-finite and $c(n; E(m))=mp^n$.

Now, let $n \geqq 0$ be fixed. For any element a in $E(m)$, choose an integer $s \geqq n$ so that a is in $E(m)_s$ and that $\nu_{s,n}a=0$, where $\nu_{s,n}$ is defined as (3) in 3.2. This is always possible, because if a is in $E(m)_t$, $t \geqq n$ and $s=t+m$, then $\nu_{s,n}a=p^m\nu_{t,n}a=0$. On the other hand, as a (Γ_n/Γ_s)-group, $E(m)_s=\mathrm{Hom}(Z(\Gamma/\Gamma_s), Z_{p^m})$ is isomorphic with the direct sum of p^n copies of $Z_{p^m}(\Gamma_n/\Gamma_s)=Z(\Gamma_n/\Gamma_s)\otimes Z_{p^m}$. Hence, the cohomology groups $H^i(\Gamma_n/\Gamma_s, E(m)_s)$ are 0 for all i. Since $\nu_{s,n}a=0$ and since Γ_n/Γ_s is a cyclic group generated by the coset of γ_n mod Γ_s, there then exists an element b in $E(m)_s$ such that $a=(1-\gamma_n)b=\omega_n b$. As a was an arbitrary element of $E(m)$, $E(m)$ is n-regular for every $n \geqq 0$, and the lemma is proved.

4.2. Now, let A be a discrete module such that $pA=0$. We may then consider A as a vector space over the prime field P of characteristic p and the endomorphism $a \to \omega_0 a=(1-\gamma)a$ as a linear transformation of the vector space. Since P is a field of characteristic p,

$$\omega_n a = (1-\gamma^{p^n})a = (1-\gamma)^{p^n} a = \omega_0^{p^n} a, \qquad n \geqq 0,$$

for every a in A. Hence, given any element a of A, there always exists, by Lemma 2.1, an integer $i \geqq 0$ such that $\omega_0^i a=0$. Suppose that A is

Γ-finite and that the submodule A_0 of A defined as before has a finite order p^s ($s \geqq 0$). We then see easily that the vector space A is decomposed into the direct sum of s subspaces $A^{(j)}$ such that each $A^{(j)}$ is a direct indecomposable submodule of A and has either a finite or a countable number of basis $a_i^{(j)}$, $0 \leqq i < \dim A^{(j)}$, over P, with the property $\omega_0 a_0^{(j)} = 0$ and $\omega_0 a_i^{(j)} = a_{i-1}^{(j)}$ for $i > 0$.

Now, consider the module $E(1)$. Since $pE(1) = 0$, $\dim E(1) = \infty$, and since $E(1)_0$ has order p, it follows from the above that $E(1)$ is direct indecomposable and has a basis e_i, $0 \leqq i < \infty$, such that $\omega_0 e_0 = 0$, $\omega_0 e_i = e_{i-1}$ for $i > 0$.

LEMMA 4.3. *Let A be a Γ-finite discrete module such that $pA = 0$. Then, A is the direct sum of a finite submodule and a finite number of submodules each isomorphic with $E(1)$.*

PROOF. By what was mentioned above, A is the direct sum of a finite number of submodules $A^{(j)}$ as described there. Suppose $A^{(j)}$ is infinite dimensional. Then, $A^{(j)}$ has a basis $a_i^{(j)}$, $0 \leqq i < \infty$, such that $\omega_0 a_0^{(j)} = 0$ and $\omega_0 a_i^{(j)} = a_{i-1}^{(j)}$ for $i > 0$, and it is clear that there is an isomorphism ϕ of the module $E(1)$ onto $A^{(j)}$ such that $\phi(e_i) = a_i^{(j)}$, $0 \leqq i < \infty$. Thus, every infinite dimensional $A^{(j)}$ is isomorphic with $E(1)$, and the lemma is proved.

It follows immediately from the lemma that if the rank of a Γ-finite discrete module A is infinite, A contains a submodule isomorphic with $E(1)$; for the submodule A' of all a in A satisfying $pa = 0$ is then an infinite Γ-finite discrete module with the property $pA' = 0$ and, hence, contains a submodule isomorphic with $E(1)$.

Now, let B be an infinite submodule of $E(1)$. By the above, B contains a submodule B' isomorphic with $E(1)$. Then, by Lemma 4.2, the submodules $E(1)_n$ and $B_n' = E(1)_n \cap B'$ have the same order p^n. Hence $E(1)_n = B_n'$ for all $n \geqq 0$, and $E(1) = B' = B$. Thus, there is no infinite submodule of $E(1)$ except $E(1)$ itself. It follows, in particular, that a regular submodule of $E(1)$ is either 0 or $E(1)$ itself, for a nontrivial finite submodule of $E(1)$ obviously can not be regular.

4.3. We shall now prove some lemmas on the modules $E(m)$.

LEMMA 4.4. *Let B be a regular submodule of $E(m)$, $m \geqq 0$. Then, $B = p^l E(m)$ for some l, $0 \leqq l \leqq m$.*

PROOF. If $m = 0$, the lemma is trivial. Suppose that $m > 0$ and that the lemma is proved for $m-1$. Consider the submodule $\overline{B} = (B + pE(m))/pE(m)$ of $E(m)/pE(m)$. $\overline{B}$ is a regular submodule of $E(m)/pE(m)$ and the latter is isomorphic with $E(1)$ by Lemma 4.1. Hence, by the remark in 4.2, either $\overline{B} = E(m)/pE(m)$ or $\overline{B} = 0$. In the first case, $E(m)=B+pE(m)$ and, consequently, $E(m)=B$.[7] In the second case, B is contained in $pE(m)$. But, as $pE(m)\cong E(m-1)$ by Lemma 4.1, it follows from the induction assumption that $B=p^{l-1}(pE(m))=p^lE(m)$ for some $0\leqq l\leqq m$.

LEMMA 4.5. *Let m be any positive integer. Then, $E(m)$ has a basis e_i, $0\leqq i<\infty$, such that*

(i) *every e_i has order p^m,*

(ii) *$\omega_0e_0=0$ and $\omega_0e_i=e_{i-1}$ for $i>0$.*

PROOF. For $m=1$, the lemma is already proved in 4.2. Let $m>1$. By Lemma 4.1, $E(m)/pE(m)\cong E(1)$. Hence, we take a basis $\bar{e}_i$, $0\leqq i<\infty$, of $E(m)/pE(m)$ such that $\omega_0\bar{e}_0=0$ and $\omega_0\bar{e}_i=\bar{e}_{i-1}$ for $i>0$. Let e_0' be an element of $E(m)$ such that $\bar{e}_0$ is the coset of e_0' mod $pE(m)$. Since $\omega_0\bar{e}_0=0$, ω_0e_0' is contained in $pE(m)$. But, by Lemmas 4.1, 4.2, $pE(m)$ is regular. Hence there is an element b_0 in $pE(m)$ such that $\omega_0e_0'=\omega_0b_0$. Put $e_0=e_0'-b_0$. Then, e_0 is still in the coset $\bar{e}_0$ and $\omega_0e_0=0$. Let e_1' be an element of $E(m)$ such that $\bar{e}_1$ is the coset of e_1' mod $pE(m)$. Since $\omega_0\bar{e}_1=\bar{e}_0$, $\omega_0e_1'-e_0$ is contained in $pE(m)$, and, as $pE(m)$ is regular, there is an element b_1 in $pE(m)$ such that $\omega_0e_1'-e_0=\omega_0b_1$. Put $e_1=e_1'-b_1$. Then, e_1 is again in the coset $\bar{e}_1$ and $\omega_0e_1=e_0$. Proceeding similarly, we can find elements $e_0, e_1, e_2, \cdots$ in $E(m)$, successively, so that each e_i is in the coset $\bar{e}_i$ and satisfies the condition (ii) of the lemma. As the cosets of e_i mod $pE(m)$ form a basis of $E(m)/pE(m)$, the elements e_i generate the group $E(m)$. Let n be any positive integer. Take an integer s such that all $e_0, e_1, \cdots, e_{n-1}$ are contained in $E(m)_s$. Since $e_0, e_1, \cdots, e_{n-1}$ are independent mod $pE(m)$ and, consequently, also mod $pE(m)_s$, and since $E(m)_s=\text{Hom}\,(Z(\Gamma/\Gamma_s), Z_{p^m})$ is an abelian group of type $(p^m, \cdots, p^m)$ with rank p^s, the n elements $e_0, e_1, \cdots, e_{n-1}$ generate a subgroup of order p^{mn} in $E(m)_s$. Therefore, every one of $e_0, e_1, \cdots, e_{n-1}$ has order p^m and they form a basis of the subgroup generated by themselves. Thus, the lemma is proved.

LEMMA 4.6. *Let A be a regular discrete module satisfying $p^mA=0$, $m>0$, and let B be a submodule of A containing pA such that $B/pA\cong E(1)$. Then, there exists a homomorphism ϕ of $E(m)$ into A such that $\phi(E(m))+pA=B$ and $\phi^{-1}(pA)=pE(m)$.*

PROOF. Since $B/pA\cong E(1)$, we can find a basis $\bar{a}_i$, $0\leqq i<\infty$, of B/pA such that $\omega_0\bar{a}_0=0$ and $\omega_0\bar{a}_i=\bar{a}_{i-1}$ for $i>0$. As A is regular, the

[7] In general, if B is a subgroup of a bounded p-primary abelian group A and if $A=B+pA$, then $A=B$; $A=B+pA$ implies $A=B+p(B+pA)=B+p^2A=\cdots=B+p^nA=B$. This will be often used in the following arguments.

homomorphic image pA of A is also regular. Hence, by a similar argument as in the proof of Lemma 4.5, we can find elements a_i, $0 \leqq i < \infty$, in B such that a_i is in the coset $\bar{a}_i$ and that $\omega_0 a_0 = 0$ and $\omega_0 a_i = a_{i-1}$ for $i > 0$. Now, let e_i, $0 \leqq i < \infty$, be a basis of $E(m)$ as given in Lemma 4.5. Since $p^m a_i = 0$, $0 \leqq i < \infty$, there is a homomorphism ϕ of the module $E(m)$ into B such that $\phi(e_i) = a_i$, $0 \leqq i < \infty$. It is then clear from the choice of a_i that $\phi(E(m)) + pA = B$ and that ϕ induces an isomorphism of $E(m)/pE(m)$ onto B/pA. Hence $\phi^{-1}(pA) = pE(m)$.

4.4. Let $m_1, \cdots, m_s$ be any set of non-negative integers. We denote by $E(m_1, \cdots, m_s)$ the direct sum of $E(m_1), \cdots, E(m_s)$:

$$E(m_1, \cdots, m_s) = E(m_1) + \cdots + E(m_s).$$

If $m_1 = \cdots = m_s = m$, $E(m_1, \cdots, m_s)$ will be denoted also by $E(m)^s$. Clearly, $E(m_1, \cdots, m_s)$ may be also defined as the module of all continuous functions on Γ with values in the direct sum $Z_{p^{m_1}} + \cdots + Z_{p^{m_s}}$. It follows immediately from Lemma 4.2 that $E(m_1, \cdots, m_s)$ is a bounded, regular, strictly Γ-finite, discrete module and its characteristic function is given by

$$c(n; E(m_1, \cdots, m_s)) = mp^n, \qquad n \geqq 0,$$

where $m = \sum_{i=1}^{s} m_i$.

LEMMA 4.7. *Let D be a submodule of $E(m)^s$ isomorphic with $E(1)$. Then, there exists a submodule C of $E(m)^s$ such that $C \cong E(m)$ and $p^{m-1}C = D$.*

PROOF. As stated in the proof of Lemma 4.1, the endomorphism $a \to p^{m-1}a$ of $E(m)^s$ induces an isomorphism of $E(m)^s/pE(m)^s$ onto $p^{m-1}E(m)^s$. Furthermore, by the same lemma, $p^{m-1}E(m)^s$ is the submodule of all a in $E(m)^s$ satisfying $pa = 0$. Hence, D is contained in $p^{m-1}E(m)^s$, and there exists a submodule B of $E(m)^s$ containing $pE(m)^s$ such that $p^{m-1}B = D$, $B/pE(m)^s \cong D \cong E(1)$. Since $pE(m)^s$ is regular by Lemmas 4.1, 4.2, it follows from Lemma 4.6 that there is a homomorphism ϕ of $E(m)$ onto a submodule $C = \phi(E(m))$ of $E(m)^s$ such that $C + pE(m)^s = B$ and $\phi^{-1}(pE(m)^s) = pE(m)$, inducing an isomorphism of $E(m)/pE(m)$ onto $B/pE(m)^s$. Since the endomorphism $a' \to p^{m-1}a'$ of $E(m)$ also induces an isomorphism of $E(m)/pE(m)$ onto $p^{m-1}E(m)$, ϕ maps the submodule $p^{m-1}E(m)$ of $E(m)$ isomorphically onto the submodule $D = p^{m-1}B = p^{m-1}C$ of C. Hence ϕ must be an isomorphism, and the lemma is proved.

LEMMA 4.8. *Let D be a submodule of $E(m_1, \cdots, m_s)$, isomorphic with $E(1)$. Then, $E(m_1, \cdots, m_s)/D$ is a homomorphic image of some $E(n_1, \cdots, n_t)$ where $\sum_{j=1}^{t} n_j < \sum_{i=1}^{s} m_i$.*

PROOF. We use induction on s. If $s=1$, the lemma is an immediate consequence of Lemmas 4.1, 4.4. Suppose $s>1$. We may of course assume $m_1 \geqq m_2 \geqq \cdots \geqq m_s > 0$. Put $m = m_s$ and

$$A = E(m_1, \cdots, m_s) = A' + A'',$$
$$A' = E(m_1, \cdots, m_{s-1}), \ A'' = E(m_s).$$

If D is contained in A', then $A/D \cong (A'/D) + A''$, and the lemma can be proved immediately by applying the induction assumption on $A'/D = E(m_1, \cdots, m_{s-1})/D$. Hence, we may assume that D is not contained in A'.

Now, let B denote the submodule of all a in A satisfying $p^m a = 0$, and put

$$B' = B \cap A', \qquad N' = p^{m-1}B', \qquad N'' = p^{m-1}A''.$$

Then, $B \cong E(m)^s$, $N'' = p^{m-1}E(m) \cong E(1)$, and we have also the following direct sum decompositions:

$$B = B' + A'', \qquad p^{m-1}B = N' + N''.$$

Since $p^{m-1}B$ is the submodule of all a in A satisfying $pa=0$, D is contained in $p^{m-1}B$. But, by the assumption made above, D is not contained in N'. Hence, $(N'+D)/N'$ is a nontrivial submodule of $p^{m-1}B/N' \cong N'' \cong E(1)$. As a homomorphic image of $D \cong E(1)$, $(N'+D)/N'$ is also regular. Hence, by the remark in 4.2, or by Lemma 4.4, $N'+D = p^{m-1}B$.

Now, by Lemma 4.7, B contains a submodule C such that $C \cong E(m)$ and $p^{m-1}C = D$. It then follows that

$$p^{m-1}B = N' + D = p^{m-1}B' + p^{m-1}C = p^{m-1}(B' + C).$$

As B is isomorphic with $E(m)^s$, pB is the submodule of all b in B such that the $p^{m-1}b=0$. Hence the above equality implies that $B = (B'+C) + pB$ and, consequently, that $B = B' + C$. Therefore,

$$A = A' + A'' = A' + B = A' + B' + C = A' + C,$$

though the sums are not necessarily direct. But, then A/D is a homomorphic image of the direct sum of $A' = E(m_1, \cdots, m_{s-1})$ and $C/D = C/p^{m-1}C \cong E(m-1)$. It is thus proved that A/D is a homomorphic image of $E(m_1, \cdots, m_{s-1}, m_s - 1)$, q.e.d.

4.5. A discrete module E will be called *elementary* if E is bounded, regular, and Γ-finite.

LEMMA 4.9. *A discrete module E is elementary if and only if E is a homomorphic image of a module $E(m_1, \cdots, m_s)$.*

PROOF. Let E be an elementary discrete module. As a homomorphic image of a regular module E, the submodule pE is also regular. Hence, by Lemma 2.3, the regular module E/pE is also Γ-finite, for E is Γ-finite by the definition. Then, by Lemma 4.3, E/pE is the direct sum of a finite number of submodules $\bar{E}^{(1)}, \cdots, \bar{E}^{(s)}$ each isomorphic with $E(1)$. Let $E^{(i)}$ be a submodule of E containing pE such that $\bar{E}^{(i)}=E^{(i)}/pE$. Since $p^mE=0$ for some $m\geqq 0$, there exists, by Lemma 4.6, a homomorphism ϕ_i $(1\leqq i\leqq s)$ of $E(m)$ onto a submodule $B^{(i)}$ of E such that $B^{(i)}+pE=E^{(i)}$. We have, then, $E=E^{(1)}+\cdots+E^{(s)}=B^{(1)}+\cdots+B^{(s)}+pE$, and, consequently, also $E=B^{(1)}+\cdots+B^{(s)}$. Since $B^{(i)}=\phi_i(E(m))$, it is clear that E is a homomorphic image of $E(m)^s=E(m, \cdots, m)$.

Suppose, conversely, E is a homomorphic image of a module $E(m_1, \cdots, m_s)$. Among all such $E(m_1, \cdots, m_s)$ of which E is a homomorphic image, we choose an $E(m_1, \cdots, m_s)$ for which $\sum_{i=1}^s m_i$ is minimal. Put $E=E(m_1, \cdots, m_s)/D$. Suppose, now, that D is an infinite module. Then, since D is bounded, it can not be of finite rank. Hence, by the remark in 4.2, D has a submodule D' isomorphic with $E(1)$. But, by Lemma 4.8, $E(m_1, \cdots, m_s)/D'$ is then a homomorphic image of a module $E(n_1, \cdots, n_t)$ where $\sum_{j=1}^t n_j < \sum_{i=1}^s m_i$. Hence E is also a homomorphic image of $E(n_1, \cdots, n_t)$ with $\sum_{j=1}^t n_j<\sum_{i=1}^s m_i$, and this contradicts the choice of $E(m_1, \cdots, m_s)$. It is thus proved that D is a finite module.

Now, since $E(m_1, \cdots, m_s)$ is bounded and regular, so is $E=E(m_1, \cdots, m_s)/D$. We shall next prove that E is (strictly) Γ-finite. Put $A=E(m_1, \cdots, m_s)$ and denote, as usual, by A_n the submodule of all a in A satisfying $\omega_n a=0$, and by E_n the submodule of E defined similarly for E. By the definition, $A_n=\omega_n^{-1}(0)$ and $E_n=\omega_n^{-1}(D)/D$. However, as $\omega_n A=A$, $\omega_n^{-1}(D)/\omega_n^{-1}(0)$ is isomorphic with D. Since D is finite, it then follows that the order of E_n is equal to the order of A_n. Thus, by the remark at the beginning of 4.4, we see that E_n is a finite module of order p^{mp^n} where $m=\sum_{i=1}^s m_i$. E is, therefore, Γ-finite, and the lemma is completely proved.

At the same time, the following lemma is also proved by the above argument:

LEMMA 4.10. *Let E be an elementary discrete module. Then, there is a module $E(m_1, \cdots, m_s)$ and a finite submodule D of $E(m_1, \cdots, m_s)$ such that $E\cong E(m_1, \cdots, m_s)/D$. The characteristic function of E is then given by*

$$c(n; E) = mp^n, \qquad n \geqq 0,$$

where $m=\sum_{i=1}^s m_i$.

LEMMA 4.11. *Let B be a finite submodule of an elementary discrete module E. Then*

$$c(n; E/B) = c(n; E),$$

for all $n \geqq 0$.

PROOF. Let $E \cong E(m_1, \cdots, m_s)/D$, as stated in the previous lemma. Then, there is a submodule C of $E(m_1, \cdots, m_s)$ containing D such that

$$E/B \cong E(m_1, \cdots, m_s)/C, \qquad B \cong C/D.$$

As C is also a finite module, we have, by Lemma 10,

$$c(n; E/B) = mp^n = c(n; E),$$

where $m = \sum_{i=1}^{s} m_i$.

LEMMA 4.12. *A homomorphic image of an elementary discrete module is again elementary.*

This follows immediately from Lemma 4.9.

LEMMA 4.13. *Let B be a submodule of a discrete module A. If both A/B and B are elementary, so is A.*

PROOF. Since both A/B and B are bounded, regular modules, A is also bounded and regular. By Lemma 2.3, A is also Γ-finite.

LEMMA 4.14. *Let B and C be submodules of a discrete module A. If both B and C are elementary, the sum B+C in A is also an elementary module.*

PROOF. Clearly, if both B and C are elementary, the direct sum of B and C is also elementary. Since the sum $B+C$ in A is a homomorphic image of the direct sum of B and C, $B+C$ is also elementary by Lemma 4.12.

4.6. We now consider bounded Γ-finite discrete modules in general.

LEMMA 4.15. *A bounded Γ-finite discrete module A has the unique maximal elementary submodule E in which every elementary submodule of A is contained. A/E is then a finite module.*

PROOF. Clearly, for any elementary submodule E' of A, we have $c(0; E') \leqq c(0; A)$. Hence, there exists an elementary submodule E of A such that $c(0; E') \leqq c(0; E)$ for any elementary submodule E' of A. Put $E'' = E + E'$. By Lemma 4.14, E'' is also elementary, and $E \subset E''$, $c(0; E) \leqq c(0; E'')$. Therefore $c(0; E) = c(0; E'')$ and, by Lemma 4.10, $c(n; E) = c(n; E'')$ for all $n \geqq 0$. Then, for every $n \geqq 0$, E_n and E_n''

have the same order and, consequently, $E_n = E_n''$. Hence, it follows from Lemma 2.1 that $E = E''$, $E' \subset E$, i.e. that every elementary submodule of A is contained in E.

Suppose, next, A/E be an infinite module. Then, the bounded module A/E can not be of finite rank. Since A/E is Γ-finite by Lemma 2.3, it has a submodule isomorphic with $E(1)$, by the remark in 4.2. Hence, A has a submodule B containing E such that $B/E \cong E(1)$. As both B/E and E are elementary, so is B by Lemma 4.13. However, this contradicts the fact that every elementary submodule of A is contained in E. It is, hence, proved that A/E is a finite module.

Now, as above, let A be a bounded Γ-finite discrete module and E the maximal elementary submodule of A. By Lemma 2.3, we have

$$c(n; A) = c(n; A/E) + c(n; E), \qquad n \geqq 0.$$

But, as A/E is a finite module, $c(n; A/E)$ is constant for all sufficiently large n. By Lemma 4.10, we have therefore the following

LEMMA 4.16. *Let A be a bounded Γ-finite discrete module. Then, there exists an integer $n_0 \geqq 0$ such that, for $n \geqq n_0$, the characteristic function of A is given by*

$$c(n; A) = mp^n + u,$$

where m and u are suitable non-negative integers independent of n.

Obviously, for given A, the integers m and u in the lemma are uniquely determined by the above equality, and they give us invariants of the module A. In particular, we call the invariant m the *weight* of the bounded Γ-finite discrete module A and denote it by $w(A)$. As the above proof shows, the weight of A is given by

$$w(A) = c(0; E) = \sum_{i=1}^{s} m_i,$$

if E is the maximal elementary submodule of A and if

$$E = E(m_1, \cdots, m_s)/D$$

with finite D. We also notice that $w(A) = 0$ if and only if A is a finite module.

LEMMA 4.17. *Let A be a bounded Γ-finite discrete module and B a submodule of A. Then, A/B is also Γ-finite and*

$$w(A) = w(A/B) + w(B).$$

PROOF. Let E be the maximal elementary submodule of A and E'

the maximal elementary submodule of B. Since $(E+B)/B$ is a homomorphic image of E, it is elementary by Lemma 4.11. Clearly, $A/(E+B)$ is a finite module and, hence, is Γ-finite. Therefore, by Lemma 2.3, A/B is also Γ-finite. It is then easy to see that $(E+B)/B$ is the maximal elementary submodule of A/B and that $w(A/B) = w((E+B)/B)$. As E' is an elementary submodule of A, it is contained in $E\cap B$, and $E\cap B/E'$ is a finite module, for B/E' is finite. By Lemma 4.11, we then have $w(E/E\cap B)=w(E/E')$ and, hence, $w(A/B)=w((E+B)/B)=w(E/E\cap B)=w(E/E')$. Now, by Lemma 2.3,

$$c(n; E) = c(n; E/E') + c(n; E'), \qquad n \geqq 0.$$

Putting $n=0$, we obtain

$$w(E) = w(E/E') + w(E').$$

As $w(E)=w(A)$ and $w(E')=w(B)$ by the definition, and as it is shown above that $w(E/E')=w(A/B)$, the lemma is proved.

Now, let A be again a bounded Γ-finite discrete module, E the maximal elementary submodule of A, and $E\cong E(m_1, \cdots, m_s)/D$ with finite D. We shall next show that the module $E(m_1, \cdots, m_s)$ with the property described above is uniquely determined by A.

Let i be any non-negative integer and let $A^{(i)}$ denote the submodule of all a in A satisfying $p^i a=0$. Put $E'=E\cap A^{(i)}$. Since $A^{(i)}/E'$ is a finite module, we have, by Lemma 4.17, $w(A^{(i)})=w(E')$. Now, consider the endomorphism ϕ: $b\to p^i b$ of $E(m_1, \cdots, m_s)$, and denote by B the kernel of ϕ and by C the inverse image of D under ϕ. Clearly, $E'\cong C/D$ and, hence, $w(E')=w(C/D)$. Since D is finite, C/B is also finite. By Lemma 4.17, we have then $w(C/D)=w(C)=w(B)$. Therefore, $w(A^{(i)})=w(B)$. However, by the definition, B is the submodule of all b in $E(m_1, \cdots, m_s)$ satisfying $p^i b=0$. Hence, obviously, $B\cong E(n_1, \cdots, n_s)$ where $n_j=\min\,(m_j, i)$, $j=1, \cdots, s$. As $w(B) = \sum_{j=1}^{s} n_j$, we have

$$w(A^{(i)}) = \sum_{j=1}^{s} \min\,(m_j, i).$$

Since i is an arbitrary non-negative integer and $A^{(i)}$ is a module defined uniquely by A and i, the above equality shows that the nonzero integers in $m_1, \cdots, m_s$ are uniquely determined by A. The module $E(m_1, \cdots, m_s)$ is therefore also uniquely determined by A.

As shown above, the nonzero integers in $m_1, \cdots, m_s$ are invariants of the bounded Γ-finite discrete module A and those invariants determine the structure of A up to finite modules. Furthermore, they have, in various respects, similar properties as the invariants of finite abelian groups. For instance, we can prove, by a similar argument as above, that

$$w(p^i A) = \sum_{j=1}^{s} \max(m_j - i, 0),$$

for any integer $i \geqq 0$. Using Lemma 4.17, it then follows, in particular, that

$$w(A^{(1)}) = w(A/pA). \tag{4}$$

4.7. We shall now briefly state, without proofs, the results on the structure of bounded Γ-finite compact modules which correspond, by the duality between discrete and compact modules, to what we have proved above for bounded Γ-finite discrete modules.

For any non-negative integer m, we denote by $Y(m)$ the compact module $M_2(\Gamma, Z_{p^m})$ as defined in general in 1.4. Since the finite group Z_{p^m} is self-dual, $E(m)$ and $Y(m)$ form a pair of dual Γ-modules, and it follows that $Y(m)$ is a bounded Γ-finite compact module. More generally, for any non-negative integers $m_1, \cdots, m_s$, we denote by $Y(m_1, \cdots, m_s)$ the direct sum of $Y(m_i)$, $i=1, \cdots, s$. $Y(m_1, \cdots, m_s)$ is again a bounded Γ-finite compact module and it is dual to the discrete module $E(m_1, \cdots, m_s)$.

A compact module Y is called *elementary* if it is bounded, regular and Γ-finite, i.e. if Y is dual to an elementary discrete module. By Lemma 4.9, a compact module Y is elementary if and only if it is isomorphic with a submodule W of some $Y(m_1, \cdots, m_s)$. In fact, if Y is elementary, we can find $Y(m_1, \cdots, m_s)$ and a submodule W of $Y(m_1, \cdots, m_s)$ isomorphic with Y such that $Y(m_1, \cdots, m_s)/W$ is finite.

In general, a bounded Γ-finite compact module X has the unique minimal submodule U such that X/U is elementary. U is a finite module and, by the above, X/U is isomorphic with a submodule of a module $Y(m_1, \cdots, m_s)$ having a finite index in $Y(m_1, \cdots, m_s)$. The nonzero integers in $m_1, \cdots, m_s$ are, then, invariants of X and they determine the structure of X up to finite modules. The sum $m = \sum_{i=1}^{s} m_i$ is again called the *weight* of X and is denoted by $w(X)$. For the characteristic function $c(n; X)$ of X, we also have the result corresponding to Lemma 4.16, for $c(n; X) = c(n; A)$ if A is a bounded Γ-finite discrete module dual to X.

5. Γ-finite modules in general. 5.1. We now consider the Γ-module $M_1(\Gamma, Z_{p^\infty})$ defined in 1.4 and denote it by $E(\infty)$. By the definition,

$E(\infty)$ consists of all continuous functions on Γ with values in Z_{p^∞}. Since Γ is totally disconnected, every continuous function in $E(\infty)$ takes only a finite number of distinct values in the discrete group Z_{p^∞}, which, on the other hand, may be considered as the union of finite cyclic groups Z_{p^l}, $l \geqq 0$. Hence, if, for each $l \geqq 0$, $E(\infty)^{(l)}$ denotes the submodule of all a in $E(\infty)$ satisfying $p^l a = 0$, $E(\infty)^{(l)}$ is naturally isomorphic with $E(l)$, and $E(\infty)$ is the union of all those $E(\infty)^{(l)}$, $l \geqq 0$. It then follows immediately that $E(\infty)$ is a discrete module in the sense of 2.2 and that it is also regular and Γ-finite, though not bounded.

More generally, for any $m_1, \cdots, m_s$ which are either non-negative integers or ∞, we denote by $E(m_1, \cdots, m_s)$ the direct sum of $E(m_i)$, $i = 1, \cdots, s$. Clearly, $E(m_1, \cdots, m_s)$ is again a regular Γ-finite discrete module.

Now, let A be a discrete module. For any integer $l \geqq 0$, let $A^{(l)}$ denote the submodule of all a in A satisfying $p^l a = 0$.

Lemma 5.1. *Let $E(\infty)^s$ be the direct sum of s copies of $E(\infty)$, $s \geqq 0$. A discrete module A is isomorphic with $E(\infty)^s$ if and only if A is divisible and $A^{(1)} \cong E(1)^s$.*

Proof. The lemma is trivial for $s = 0$. Using the remark on $E(\infty)^{(l)}$ mentioned at the beginning, it is also easy to see that $E(\infty)^s$ has the properties stated above.

Now, let $s \geqq 1$ and let A be any discrete module having the properties given in the lemma. We first prove the existence of a set of elements $a_{ij}^{(k)}$ in A, $1 \leqq i \leqq s$, $0 \leqq j < \infty$, $1 \leqq k < \infty$, such that the elements $a_{ij}^{(k)}$, $1 \leqq i \leqq s$, $0 \leqq j < \infty$, form a basis of $A^{(k)}$, such that $pa_{ij}^{(k)} = a_{ij}^{(k-1)}$ for $k > 1$ and that $\omega_0 a_{i0}^{(k)} = 0$ and $\omega_0 a_{ij}^{(k)} = a_{i,j-1}^{(k)}$ for $j > 0$. We use induction on the upper index k. Since $A^{(1)} \cong E(1)^s$, it is clear from Lemma 4.5 that there exist elements $a_{ij}^{(1)}$, $1 \leqq i \leqq s$, $0 \leqq j < \infty$, satisfying the above conditions for $k = 1$. Suppose we have found such elements $a_{ij}^{(k)}$ in A for $1 \leqq i \leqq s$, $0 \leqq j < \infty$, $1 \leqq k \leqq l$. Let i be fixed. Since A is divisible and $pA^{(l+1)} = A^{(l)}$, there is an element a in $A^{(l+1)}$ such that $pa = a_{i0}^{(l)}$. We have then $p(\omega_0 a) = \omega_0 a_{i0}^{(l)} = 0$. Hence, $\omega_0 a$ is an element of $A^{(1)} \cong E(1)^s$, and there is an element b in $A^{(1)}$ such that $\omega_0 a = \omega_0 b$. Put $a_{i0}^{(l+1)} = a - b$. Then, $pa_{i0}^{(l+1)} = a_{i0}^{(l)}$ and $\omega_0 a_{i0}^{(l+1)} = 0$. Next we take an element a' in $A^{(l+1)}$ such that $pa' = a_{i1}^{(l)}$. We have then $p(\omega_0 a') = \omega_0 a_{i1}^{(l)} = a_{i0}^{(l)} = pa_{i0}^{(l+1)}$. Hence, $\omega_0 a' - a_{i0}^{(l+1)}$ is contained in $A^{(1)}$, and there is an element b' in $A^{(1)}$ such that $\omega_0 a' - a_{i0}^{(l+1)} = \omega_0 b'$. Put $a_{i1}^{(l+1)} = a' - b'$. Then, $pa_{i1}^{(l+1)} = a_{i1}^{(l)}$ and $\omega_0 a_{i1}^{(l+1)} = a_{i0}^{(l+1)}$. Proceeding similarly, we can obtain elements $a_{ij}^{(l+1)}$, $1 \leqq i \leqq s$, $0 \leqq j < \infty$, in $A^{(l+1)}$ such that $pa_{ij}^{(l+1)} = a_{ij}^{(l)}$, $\omega_0 a_{i0}^{(l+1)} = 0$ and $\omega_0 a_{ij}^{(l+1)} = a_{i,j-1}^{l+1}$ for

$j>0$. Let B be the subgroup of $A^{(l+1)}$ generated by those $a_{ij}^{(l+1)}$, $1 \leqq i \leqq s$, $0 \leqq j < \infty$. Since the elements $a_{ij}^{(l)}$ form a basis of $A^{(l)}$, we have $pB = A^{(l)} = pA^{(l+1)}$. But, as $A^{(1)}$ is contained in $A^{(l)}$, and, hence, also in B, we see immediately that $B = A^{(l+1)}$. From $pa_{ij}^{(l+1)} = a_{ij}^{(l)}$, it then follows that the elements $a_{ij}^{(l+1)}$ form a basis of $A^{(l+1)}$. Thus, by induction, the existence of $a_{ij}^{(k)}$, $1 \leqq i \leqq s$, $0 \leqq j < \infty$, $1 \leqq k < \infty$, is proved. We notice that every $a_{ij}^{(k)}$ has order p^k.

Now, let $\bar{A}$ be another divisible discrete module such that $\bar{A}^{(1)} \cong E(1)^s$. Then, $\bar{A}$ also contains a set of elements $\bar{a}_{ij}^{(k)}$, $1 \leqq i \leqq s$, $0 \leqq j < \infty$, $1 \leqq k < \infty$, having similar properties as $a_{ij}^{(k)}$. But, it is then clear that there is an isomorphism ϕ of the module A onto the module $\bar{A}$ such that $\phi(a_{ij}^{(k)}) = \bar{a}_{ij}^{(k)}$. Thus, any two discrete modules having the properties stated in the lemma are isomorphic with each other. Since $E(\infty)^s$ has these properties, the lemma is proved.

Lemma 5.2. *Let A be a discrete module and B a submodule of A isomorphic with $E(\infty)^s$, $s \geqq 0$. Then, A is the direct sum of B and a suitable submodule C: $A = B + C$.*

Proof. Let D be a submodule of A such that $B \cap D = 0$. Suppose $A \neq B + D$. Then, there exists an element a in A such that a is not in $B + D$ but both pa and $\omega_0 a$ are in $B + D$. Put $pa = b + d$, $b \in B$, $d \in D$. Since $B \cong E(\infty)^s$, there is an element b_0 in B such that $b = pb_0$. Put $a' = a - b_0$ so that $pa' = d$. Put also $\omega_0 a' = b_1 + d_1$, $b_1 \in B$, $d_1 \in D$. Then, $pb_1 + pd_1 = \omega_0 pa' = \omega_0 d$, and, since $B \cap D = 0$, it follows that $pb_1 = 0$. As $B \cong E(\infty)^s$, there then exists an element b_2 in B such that $pb_2 = 0$, $\omega_0 b_2 = b_1$. Put $a'' = a' - b_2$. Then, $pa'' = d$ and $\omega_0 a'' = d_1$. Hence, if D^* denotes the subgroup of A generated by D and a'', D^* contains D as a subgroup of index p, $B \cap D^* = 0$ and D^* is Γ-invariant, i.e. a submodule of A. Now, take a maximal submodule C of A such that $B \cap C = 0$. By the above, we have then $A = B + C$, and the lemma is proved.

5.2. To study the structure of Γ-finite discrete modules in general, we shall first prove the following

Lemma 5.3. *Let A be a Γ-finite discrete module, A' the submodule of all a in A satisfying $pa = 0$. Let B be an elementary submodule of A containing A', and C a submodule of A such that pC is contained in B. Suppose $w(B/pB) \leqq w(C/B)$. Then $pC = B$, and C/B is isomorphic with B/pB.*

Proof. The endomorphism $c \to pc$ of C obviously induces a homomorphism ϕ of C/B into B/pB. Suppose that pc ($c \in C$) is contained in pB. Then $pc = pb$ for some b in B, and $c - b$ is contained in A', and, hence, in B. Thus, c is also contained in B, and we see that ϕ is an isomorphism of C/B into B/pB. It then follows from Lemma 4.17, that $w(C/B) \leqq w(B/pB)$ and, consequently, by the assumption, that $w(C/B)=w(B/pB)$.

Now, since B is elementary, so is B/pB by Lemma 4.12. Hence, by Lemma 4.3, $B/pB \cong E(1)^t$ where $t=w(E(1)^t)=w(B/pB)$. However, it is easy to see that no submodule of $E(1)^t$ has weight t unless it coincides with $E(1)^t$ itself. Therefore, $\phi(C/B)=B/pB$ and ϕ is an isomorphism of C/B onto B/pB. It then follows that $pC+pB=B$ and, hence, that $pC=B$.

Now, let A be any Γ-finite discrete module. By Lemma 4.14, the union of all elementary submodules in A is a submodule of A. We shall first study the structure of A in the case where A itself is the union of all elementary submodules of A.

Let C be any bounded submodule of such a discrete module A. Then the maximal elementary submodule E of C has a finite index in C, and we can find a finite number of elements a_i which generate $C \bmod E$. Since, by the assumption, every a_i is contained in some elementary submodule of A, it follows from Lemma 4.14, that C is also contained in an elementary submodule of A.

Now, for any elementary submodule E of A, let E^* denote the submodule of all a in A such that pa is contained in E. Then, among all elementary submodules of A, we choose an E for which the weight $w(E^*/E)$ attains the minimum. Since E^* is obviously a bounded module, there exists, by the above remark, an elementary submodule B of A containing E^*. Put $\bar{A}=A/E$, $\bar{B}=B/E$. $\bar{E}^*=E^*/E$ is then the submodule of all $\bar{a}$ in $\bar{A}$ satisfying $p\bar{a}=0$. But, as $\bar{E}^*$ is contained in $\bar{B}$, $\bar{E}^*$ is also the submodule of all $\bar{a}$ in $\bar{B}$ satisfying $p\bar{a}=0$. Hence, by (4) in 4.6, $\bar{E}^*$ has the same weight as $\bar{B}/p\bar{B}$: $w(\bar{E}^*)=w(\bar{B}/p\bar{B})$. Let B^* be the submodule of all b in A such that pb is contained in B. Then, $\bar{B}^*=B^*/E$ is the submodule of all $\bar{b}$ in $\bar{A}$ such that $p\bar{b}$ is contained in $\bar{B}$, and, by the choice of E, $w(\bar{B}/p\bar{B})=w(\bar{E}^*)=w(E^*/E) \leqq w(B^*/B) =w(\bar{B}^*/\bar{B})$. Hence, applying Lemma 5.3 to $\bar{B}$ and $\bar{B}^*$, we see that $p\bar{B}^*=\bar{B}$ and $\bar{B}^*/\bar{B} \cong \bar{B}/p\bar{B}$.

Put $\tilde{A}=A/B$ and $\tilde{A}'=B^*/B$. $\tilde{A}'$ is then the submodule of all $\tilde{a}$ in $\tilde{A}$ satisfying $p\tilde{a}=0$ and $\tilde{A}' \cong \bar{B}^*/\bar{B} \cong \bar{B}/p\bar{B}$. As B is elementary, $\bar{B}$ and $\bar{B}/p\bar{B}$ are also elementary. Hence $\tilde{A}'$ is an elementary module with $p\tilde{A}'=0$, and it is therefore isomorphic with $E(1)^t$, $t \geqq 0$. Let a be an arbitrary element in A. By the remark mentioned above, there then exists an elementary submodule C of A containing both a and E^*. Applying the above argument for $\bar{C}=C/E$ instead of $\bar{B}=B/E$, we see that there is a submodule $\bar{C}^*$ of $\bar{A}=A/E$ such that $p\bar{C}^*=\bar{C}$.

Therefore, there also exists an element c in A such that $pc \equiv a \bmod E$, and, hence, such that $pc \equiv a \bmod B$. Thus the module $\tilde{A} = A/B$ is divisible and, by Lemma 5.1, it is isomorphic with $E(\infty)^t$.

Now, since B is bounded, $p^m B = 0$ for some $m \geqq 0$. Consider, then, the endomorphism $a \to p^m a$ of A and denote its kernel by $A^{(m)}$. As A/B is isomorphic with $E(\infty)^t$ and B is contained in $A^{(m)}$, the endomorphism induces an isomorphism $p^m A \cong E(\infty)^t/K$ where K is a bounded submodule of $E(\infty)^t$ isomorphic with $A^{(m)}/B$. On the other hand, $A/B \cong E(\infty)^t$ is divisible by Lemma 5.1, and we have $p^m A + B = A$. Applying Lemma 4.9 to the elementary module B, it then follows immediately that if a Γ-finite discrete module A is the union of its elementary submodules, A is isomorphic with a module $E(m_1, \cdots, m_s)/D$ where $m_1, \cdots, m_s$ are either non-negative integers or ∞ and D is a suitable bounded submodule of $E(m_1, \cdots, m_s)$.

5.3. Let A be as above. Among all modules $E(m_1, \cdots, m_s)$, $0 \leqq m_i \leqq \infty$, such that $A \cong E(m_1, \cdots, m_s)/D$ with bounded D, we choose an $E(m_1, \cdots, m_s)$ for which the weight $w(D)$ is minimal. We shall next show that $w(D)$ is then 0, i.e. that D is a finite module.

Suppose D be infinite. By the remark in 4.2, the bounded module D then contains a submodule D' isomorphic with $E(1)$. Following the proof of Lemma 4.8, we may assume that $m_1 \geqq \cdots \geqq m_s > 0$ and that D' is not contained in the direct summand $E(m_1, \cdots, m_{s-1})$ of $E(m_1, \cdots, m_s)$. Suppose, first, that $m = m_s$ is not ∞. By the same argument as in the proof of Lemma 4.8, we can then see that $E(m_1, \cdots, m_s)$ has a submodule C such that $C = E(m)$, $p^{m-1}C = D'$ and $E(m_1, \cdots, m_s) = E(m_1, \cdots, m_{s-1}) + C$, though the sum is not necessarily direct. The intersection of $E(m_1, \cdots, m_{s-1})$ and C is finite, for, otherwise, it would contain $D' = p^{m-1}C \cong E(1)$. A homomorphism of $E(m_1, \cdots, m_s) = E(m_1, \cdots, m_{s-1}) + E(m_s)$ onto $E(m_1, \cdots, m_s) = E(m_1, \cdots, m_{s-1}) + C$, mapping $E(m_1, \cdots, m_{s-1})$ and $E(m_s)$ isomorphically onto $E(m_1, \cdots, m_{s-1})$ and C respectively, induces a homomorphism ϕ of $E(m_1, \cdots, m_{s-1}, m_s - 1)$ onto $E(m_1, \cdots, m_s)/D'$ whose kernel K is a finite module. Let ψ denote the homomorphism of

$$E(m_1, \cdots, m_{s-1}, m_s - 1) \text{ onto } E(m_1, \cdots, m_s)/D$$

which is the product of ϕ and the canonical homomorphism $E(m_1, \cdots, m_s)/D' \to E(m_1, \cdots, m_s)/D$. The kernel D^* of ψ then satisfies $D^*/K \cong D/D'$. Since K is finite, we have, by Lemma 4.17, $w(D^*) = w(D^*/K) = w(D/D') = w(D) - w(D') = w(D) - 1$. As A is obviously isomorphic with $E(m_1, \cdots, m_{s-1}, m_s - 1)/D^*$, this contradicts the choice of D.

Assume, next, $m_s=\infty$ so that $m_1=\cdots=m_s=\infty$, $E(m_1,\cdots,m_s)=E(\infty)^s$. Let E' be the submodule of all a in $E(\infty)^s$ satisfying $pa=0$. Since $E'\cong E(1)^s$ and D' is a submodule of E' isomorphic with $E(1)$, it is easily seen that E' has a basis $a_{ij}^{(1)}$, $1\leqq i\leqq s$, $0\leqq j<\infty$ such that $\omega_0 a_{i0}^{(1)}=0$, $\omega_0 a_{ij}^{(1)}=a_{i,j-1}^{(1)}$ for $j>0$ and that $a_{sj}^{(1)}$, $0\leqq j<\infty$, form a basis of D'. As shown in the proof of Lemma 5.1, we can then find elements $a_{ij}^{(k)}$ in $E(\infty)^s$, $1\leqq i\leqq s$, $0\leqq j<\infty$, $1\leqq k<\infty$, which include those $a_{ij}^{(1)}$ above and satisfy the conditions stated there. Let B be the submodule of $E(\infty)^s$ generated by $a_{ij}^{(k)}$, $1\leqq i\leqq s-1$, $0\leqq j<\infty$, $1\leqq k<\infty$, and C the submodule of $E(\infty)^s$ generated by $a_{sj}^{(k)}$, $0\leqq j<\infty$, $1\leqq k<\infty$. $E(\infty)^s$ is then the direct sum of B and C, and D' is the submodule of all a in C satisfying $pa=0$. Clearly, $B\cong E(\infty)^{s-1}$, $C\cong E(\infty)$, and the endomorphism $c\rightarrow pc$ of C induces an isomorphism of C/D' onto $C\cong E(\infty)$. These isomorphisms then define an isomorphism ϕ of $E(\infty)^s$ onto $E(\infty)^s/D'$ in an obvious way and we denote by ψ the homomorphism of $E(\infty)^s$ onto $E(\infty)^s/D$, which is the product of ϕ and the canonical homomorphism $E(\infty)^s/D'\rightarrow E(\infty)^s/D$. Denoting the kernel of ψ by D^*, we have thus $E(\infty)^s/D\cong E(\infty)^s/D^*$ and $D^*\cong D/D'$. However, it then follows from Lemma 4.17 that $w(D^*)=w(D)-w(D')=w(D)-1$, and this again contradicts the choice of D.

We have thus proved the following.

LEMMA 5.4. *Let A be a Γ-finite discrete module. Suppose that A is the union of all elementary submodules of A. Then, there exists a module $E(m_1,\cdots,m_s)$, $0\leqq m_i\leqq\infty$, and a finite submodule D of $E(m_1,\cdots,m_s)$ such that A is isomorphic with $E(m_1,\cdots,m_s)/D$.*

5.4. We now consider an arbitrary Γ-finite discrete module A. Let C denote the union of all elementary submodules of A. By Lemma 5.4, $C\cong E(m_1,\cdots,m_s)/D$, $0\leqq m_i\leqq\infty$, with a finite submodule D of $E(m_1,\cdots,m_s)$. As can be seen from the proof of Lemma 5.4 (or, directly from the fact $C\cong E(m_1,\cdots,m_s)/D$), C has an elementary submodule E such that $C/E\cong E(\infty)^t$ for some $t\geqq 0$. Then, applying Lemma 5.2 to A/E and C/E, we see that A has a submodule B^* containing E such that A/E is the direct sum of B^*/E and C/E. By Lemma 2.3, B^*/E is Γ-finite. Hence, if B^*/E were not of finite rank, B^* would contain, by the remark in 4.2, a submodule E^* such that $E^*/E\cong E(1)$. But, then, E^* would be also elementary by Lemma 4.13, and were not contained in C, a contradiction to the definition of C. Therefore, B^*/E must be a module of finite rank. Let B' be the submodule of B^* containing E such that B^*/B' is finite and B'/E is divisible. As an abelian group, B'/E is then isomorphic with

$Z_{p^\infty}^l$ for some $l \geqq 0$. Now, since E is elementary, $p^nE=0$ for some $n \geqq 0$. Put, then, $B=p^nB'$. Considering the endomorphism $b \to p^nb$ of B', we see that B is isomorphic with a quotient module of B'/E modulo a finite submodule. Hence, as an abelian group, B is again isomorphic with $Z_{p^\infty}^l$. Furthermore, since B'/E is divisible, it holds that $B'=B+E$, and $B+C=B'+C$ has a finite index in $A=B^*+C$. On the other hand, since B is of finite rank and E is bounded, $B \cap C = B \cap E$ is a finite module. Therefore, the following theorem is proved:

THEOREM 1. *Let A be a Γ-finite discrete module. Then, A has submodules B and C with the following properties*:

(i) *both $A/(B+C)$ and $B \cap C$ are finite modules,*

(ii) *B is divisible and of finite rank, i.e., B is, as a discrete abelian group, isomorphic with $Z_{p^\infty}^l$ for some $l \geqq 0$,*

(iii) *C is isomorphic with $E(m_1, \cdots, m_s)/D$ for some $m_1, \cdots, m_s$, $0 \leqq m_i \leqq \infty$, and for a finite submodule D of $E(m_1, \cdots, m_s)$.*

We notice that the structure of a discrete module like B was studied in §3.

We now consider the uniqueness of submodules given in the theorem. Let B and C be any submodules of A having the properties (i), (ii), (iii) above; B and C need not be those which were considered in the above proof of Theorem 1. Then, A/C is of finite rank and $(B+C)/C$ is divisible. Take any elementary submodule E of A. By Lemma 4.12, $(E+C)/C$ is elementary and, hence, is bounded. But, as A/C is of finite rank, $(E+C)/C$ is finite. Since $(E+C)/C$ is also regular, we have $(E+C)/C=0$, i.e. $E+C=C$. Thus, every elementary submodule of A is contained in C. On the other hand, it follows easily from $C \cong E(m_1, \cdots, m_s)/D$ that C is the union of its elementary submodules. Therefore, C is also the union of all elementary submodules of A and is thus uniquely determined by A.

Now, let A^* be any submodule with a finite index in A. By a similar argument as above, we see that every elementary submodule of A is contained in A^*. Hence, C is also a submodule of A^*. On the other hand, if the index of A^* in A is p^n, p^nA is contained in A^*. Hence $B=p^nB$ is contained in A^*, and so is $B+C$. Thus $B+C$ is the minimal submodule of A having a finite index in A and is uniquely characterized by this property.

A simple example shows that the module B satisfying (i), (ii), (iii) together with C is not unique for given A. However, the rank l of B is uniquely determined by A, for it is equal to the rank of $(B+C)/C$.

Finally, for any integer $i \geqq 0$, let $A^{(i)}$ be the submodule of all a in A

satisfying $p^i a=0$. Since A/C is of finite rank, $A^{(i)}\cap C$ has a finite index in $A^{(i)}$. Hence, by a similar argument as in 4.6, we see from $C\cong E(m_1, \cdots, m_s)/D$ that

$$w(A^{(i)}) = \sum_{j=1}^{s} \min(m_j, i), \qquad i \geqq 0.$$

Therefore, nonzero m_j's in $m_1, \cdots, m_s$ are uniquely determined by A and they give us a set of invariants of the module A. Clearly, the module $E(m_1, \cdots, m_s)$ is then also uniquely determined by A, and so is the sum $m=\sum_{j=1}^{s} m_j$. Here, m is meant to be ∞ if one of m_j is ∞.

We have thus obtained the following corollary to Theorem 1.

COROLLARY. *Let B and C be any submodules of a Γ-finite discrete module A, having the properties* (i), (ii), (iii) *stated in Thorem* 1. *Then, the modules C, $B+C$ and $E(m_1, \cdots, m_s)$ are uniquely determined by A with these properties, and so are the rank l of the module B and the sum $m=\sum_{j=1}^{s} m_j$.*

In the following, we shall denote the invariants l and m of A by $l(A)$ and $m(A)$ respectively. $l(A)$ is a non-negative integer and $m(A)$ is either a non-negative integer or ∞. As can be seen easily, $m(A)$ is the supremum of the weights $w(D)$ of bounded submodules D in A. In particular, if A itself is bounded, $m(A)=w(A)$.

We now apply Lemma 3.1 to the submodule B in Theorem 1. B is then the direct sum of a regular submodule B' and a submodule B'' such that $\omega_{u'}^{v'}B''=0$ for some u' and v'. Put $M=B'+C$. Since B' and C are both regular, so is M. Furthermore, since $\omega_{u'}^{v'}B''=0$ and $A/(B+C)$ is finite, $\omega_u^v(A/M)=0$ for some u and v. It follows that $\omega_u^v A=M$, and we see, as in the proof of Lemma 3.1, that M is the unique maximal regular submodule of A in which every regular submodule of A is contained.[8] Thus the following theorem is proved:

THEOREM 2. *Let M be the unique maximal regular submodule of a Γ-finite discrete module A. Then, M is the sum of a divisible regular submodule B' of finite rank and the characteristic submodule C of A given in Theorem* 1. *A/M is a module of finite rank and $\omega_u^v(A/M)=0$ for some $u\geqq 0$ and $v\geqq 0$.*

5.5. We now consider a special kind of Γ-finite discrete modules.

LEMMA 5.5. *For a Γ-finite discrete module A, the following conditions are mutually equivalent*:

[8] The existence of such a unique maximal regular submodule of A was already noticed in 2.2.

(i) *the invariant* $m(A)$ *is finite,*

(ii) *the maximal regular submodule* M *of* A *is strictly* Γ*-finite,*

(iii) *if* A_n' *denotes the maximal divisible submodule of* A_n $(n \geqq 0)$, *the rank of* A_n' *has a fixed upper bound for all* $n \geqq 0$.

PROOF. Let $M = B' + C$ as in the above. By Lemma 3.4, B' is strictly Γ-finite. Hence M is strictly Γ-finite if and only if C is so. But it is easy to see that $C \cong E(m_1, \cdots, m_s)/D$ is strictly Γ-finite if and only if $m(A) = \sum_{i=1}^{s} m_i$ is finite. (i) and (ii) are therefore equivalent.

Now, by Lemma 2.2, $(A/M)_n \cong A_n/M_n$. If M is strictly Γ-finite, M_n is a finite module and it follows from the above isomorphism that the rank of A_n' is at most equal to the rank of $(A/M)_n$. The rank of A_n' is, hence, not greater than the rank of A/M and we see that (ii) implies (iii). On the other hand, if $m(A)$ is infinite, then at least one of $m_1, \cdots, m_s$ in $E(m_1, \cdots, m_s)/D$ is infinite and the rank of $A_n' \cap C$ is at least p^n, as can be seen readily from the definition of $E(\infty)$. Therefore (iii) implies (i), and the lemma is proved.

Now, assume that $m(A)$ is finite. C is then elementary; in fact, it is the maximal elementary submodule of A, having the weight $w(C) = m(A)$. Therefore, $p^m C = 0$ for $m = m(A)$. On the other hand, as $A/(B+C)$ is finite, $p^n A$ is contained in $B + C$ for all sufficiently large $n \geqq 0$. Hence $p^{m+n}A$ is a submodule of $p^m B + p^m C = p^m B$. However, as B is divisible, $p^m B = B = p^{m+n} B$. Therefore, $p^{m+n} A = B$, and it follows that B is the intersection of all $p^n A$, $n \geqq 0$, and is the unique maximal divisible submodule of A.

We summarize our results in the following

THEOREM 3. *Let* A *be a* Γ*-finite discrete module such that the invariant* $m(A)$ *is finite. Then the submodules* B *and* C *in Theorem* 1 *are both uniquely determined by* A; B *is the unique maximal divisible submodule of* A *and* C *is the unique maximal elementary submodule of* A. *The invariant* $l(A)$ *is the rank of* B *as well as that of the maximal divisible submodule of* A/C, *and the invariant* $m(A)$ *is the weight of* C *as well as that of the bounded* Γ*-finite discrete module* A/B. *Furthermore, the maximal regular submodule* M *of* A *is strictly* Γ*-finite.*

We consider next a strictly Γ-finite module A. The rank of the module A_n' in Lemma 5.5 is then 0, and we know that $m(A)$ is finite. By Lemma 3.4, the maximal divisible submodule B of A is regular. Hence $B + C$ is also regular and it coincides with the maximal regular submodule M. Now, by Lemma 2.3, we have

$$c(n; A) = c(n; A/M) + c(n; M), \qquad n \geqq 0.$$

Let B' and C' be discrete modules such that $B'\cong B$ and $C'\cong C$, and let A' be the direct sum of B' and C'. It is clear that there is a homomorphism ϕ of A' onto $M=B+C$, mapping B' and C' isomorphically onto B and C, respectively. The kernel D' of ϕ is then a finite module isomorphic with $B\cap C$. Since $A'=B'+C'$ is regular and $\omega_n(A')=A'$ for every $n\geqq 0$, $\omega_n^{-1}(D')/\omega_n^{-1}(0)$ has the same finite order as D'. Hence $(A'/D')_n=\omega_n^{-1}(D')/D'$ also has the same order as $A'_n=\omega_n^{-1}(0)$, and we have $c(n;M)=c(n;A'/D')=c(n;A')=c(n;B')+c(n;C')=c(n;B)+c(n;C)$, for all $n\geqq 0$. Therefore,

$$c(n;A)=c(n;A/M)+c(n;B)+c(n;C), \qquad n\geqq 0.$$

However, as A/M is finite, $c(n;A/M)$ is constant for all sufficiently large n. Hence, by Lemmas 3.5, 4.10, we immediately obtain the following

THEOREM 4. *Let A be a strictly Γ-finite discrete module. Then the invariant $m(A)$ is finite and $B+C=M$ for the submodules B, C and M in Theorem 3. Furthermore, there exists a non-negative integer n_0 such that, for $n\geqq n_0$, the characteristic function of A is given by*

$$c(n;A)=l(A)n+m(A)p^n+u,$$

with a suitable integer u independent of n.

Now, for any function of the form $f(n)=ln+mp^n+u$, the coefficients l, m and u are uniquely determined by f. Hence, if A is a strictly Γ-finite discrete module, the invariants $l(A)$ and $m(A)$ are uniquely determined from the characteristic function $c(n;A)$ of A by the above formula. On the other hand, such an A is bounded if and only if $l(A)=0$ and it is of finite rank if and only if $m(A)=0$. Therefore, we can see from the characteristic function of A whether or not A is bounded or of finite rank.

5.6. Now, let O_p be, as before, the additive group of p-adic integers and let $Y(\infty)$ denote the module $M_2(\Gamma, O_p)$ defined in 1.4. Since O_p is a compact abelian group dual to the discrete abelian group Z_{p^∞}, $Y(\infty)=M_2(\Gamma, O_p)$ is a compact module in the sense of 2.2 and is dual to the discrete module $E(\infty)=M_1(\Gamma, Z_{p^\infty})$. More generally, for any $m_1, \cdots, m_s$, $0\leqq m_i\leqq\infty$, we denote by $Y(m_1, \cdots, m_s)$ the direct sum of $Y(m_i)$, $i=1, \cdots, s$. Clearly, $Y(m_1, \cdots, m_s)$ is a regular Γ-finite compact module dual to the regular Γ-finite discrete module $E(m_1, \cdots, m_s)$.

By the duality between discrete and compact modules, we can then immediately obtain theorems on Γ-finite compact modules which correspond to the above results on Γ-finite discrete modules. We state here only some of them.

THEOREM 5. *Let X be a Γ-finite compact module. Then, X has submodules U and V with the following properties:*

(i) *both $X/(U+V)$ and $U\cap V$ are finite modules,*

(ii) *X/U is torsion-free and of finite rank, i.e., X/U is, as a compact abelian group, isomorphic with O_p^l for some $l\geqq 0$,*

(iii) *X/V is isomorphic with a submodule of finite index in $Y(m_1, \cdots, m_s)$ for some $m_1, \cdots, m_s$, $0\leqq m_i\leqq\infty$.*

COROLLARY. *Let U and V be any submodules of a Γ-finite compact module A with the properties* (i), (ii), (iii) *in the above theorem. Then, the modules V, $U\cap V$ and $Y(m_1, \cdots, m_s)$ are uniquely determined by X, and so are the rank l of X/U and the sum $m=\sum_{i=1}^{s} m_i$.*

Thus, we have again invariants $l=l(X)$ and $m=m(X)$ for any Γ-finite compact module X. Clearly, if A is a discrete module dual to X, then $l(X)=l(A)$, $m(X)=m(A)$. We also notice that the structure of a module like X/U was studied in §3.

THEOREM 6. *Let X be a Γ-finite compact module such that $m(X)$ is finite. Then the submodules U and V in Theorem 5 are both uniquely determined by X; U is the torsion submodule of X and V is the unique minimal submodule of X such that X/V is elementary. The invariant $l(X)$ is the rank of X/U as well as that of the factor torsion module of V, and the invariant $m(X)$ is the weight of X/V as well as that of the bounded Γ-finite compact module U. Furthermore, if S is the unique minimal submodule of X such that X/S is regular, then X/S is also strictly Γ-finite.*

6. Unramified extensions. 6.1. Let Ω be the field of all algebraic numbers. In what follows, we shall always consider the structure of various algebraic number fields, i.e. the structure of various subfields of Ω. So, if there is no risk of misunderstanding, we shall call those algebraic number fields simply fields. By the definition, our fields are algebraic extensions of the field of rational numbers Q, but they need not be finite extensions of Q; in other words, our fields are not necessarily finite algebraic number fields.

If both E and F are such fields and if E is a Galois extension of F, we denote the Galois group of the extension E/F by $G(E/F)$. $G(E/F)$ is a totally disconnected compact group in Krull's topology. For any prime divisor of E, archimedean or non-archimedean, the decomposition group and the inertia group of the prime divisor for the extension E/F can be defined just as in the case of finite algebraic number fields.[9] They are closed subgroups of $G(E/F)$ and have sim-

[9] For an archimedean prime divisor, the inertia group is defined to be the same as the decomposition group.

ilar properties as those defined for finite algebraic number fields.

Let $\mathfrak{p}$ be a prime divisor of Ω and T the inertia group of $\mathfrak{p}$ for the extension Ω/Q. Let K and L be fields such that $K\subset L$. Then $T\cap G(\Omega/K)$ and $T\cap G(\Omega/L)$ are the inertia groups of $\mathfrak{p}$ for Ω/K and Ω/L, respectively, and the latter is a subgroup of the former. Now, the prime divisor $\mathfrak{p}$ is said to be ramified for the extension L/K if $T\cap G(\Omega/K) \neq T\cap G(\Omega/L)$, and it is said to be unramified for L/K if $T\cap G(\Omega/K) = T\cap G(\Omega/L)$. Obviously, $\mathfrak{p}$ is unramified if and only if $T\cap G(\Omega/K)$ is contained in $G(\Omega/L)$.

A prime divisor $\mathfrak{p}'$ of K (or a prime divisor $\mathfrak{p}''$ of L) is said to be unramified for L/K if and only if every extension $\mathfrak{p}$ of $\mathfrak{p}'$ (of $\mathfrak{p}''$) on Ω is unramified for L/K. Otherwise $\mathfrak{p}'(\mathfrak{p}'')$ is said to be ramified for L/K. If L/K is a Galois extension and T is the inertia group, for Ω/Q, of an extension $\mathfrak{p}$ of $\mathfrak{p}''$ on Ω, then the inertia group of $\mathfrak{p}''$ for the extension L/K is given by the image of $T\cap G(\Omega/K)$ under the canonical homomorphism $G(\Omega/K)\rightarrow G(L/K) = G(\Omega/K)/G(\Omega/L)$. Hence, $\mathfrak{p}''$ is unramified for L/K if and only if the inertia group of $\mathfrak{p}''$ for L/K is trivial. We also notice that our definition of ramified or unramified prime divisors coincides with the usual one when both L and K are finite algebraic number fields.

Now, an extension L/K is called an unramified extension if and only if every prime divisor of K, or, equivalently, every prime divisor of L, is unramified for the extension L/K. The following properties of unramified extensions are immediate consequences of the definition: if L/K is unramified and L'/K is conjugate with L/K in Ω, then L'/K is also unramified; if $F\subset K\subset L$, then L/F is unramified if and only if both L/K and K/F are unramified; if $F\subset E$, $F\subset K$ and E/F is unramified, then the composite $L=EK$ of E and K in Ω is unramified over K; if L/K is the composite, in Ω, of a family of unramified extensions L_α/K, then L/K is also unramified. From these, it follows in particular that every field K has a unique maximal unramified extension L in Ω which contains every unramified extension of K in Ω; L is a Galois extension of K. Similarly, there also exists a unique maximal unramified abelian extension A of K in Ω which contains every unramified abelian extension of K in Ω. If K is a finite algebraic number field, A is nothing but the Hilbert's class field over K, and it is well-known that A/K is a finite extension with degree equal to the class number of K. If the degree of K/Q is infinite, A/K is not necessarily a finite extension,[10] but we shall still call A the Hilbert's class field over K.

---

[10] Cf. 7.7 below.

6.2. We shall next show that every unramified extension can be obtained by composing, in a suitable manner, finite unramified extensions of finite algebraic number fields.

LEMMA 6.1. *Given any finite unramified extension L/K, there exist finite algebraic number fields E and F such that*

(i) *E is an unramified extension of F, and*

(ii) *$F \subset K$, $E \subset L$ and $L = EK$.*

If L/K is furthermore a Galois extension, then the fields E and F can be chosen so that E/F is also a Galois extension and its Galois group is isomorphic with the Galois group of L/K.

PROOF. Let S be the space of all nonempty closed subsets of the compact group $G(\Omega/Q)$. We may regard $G(\Omega/Q)$ as a subspace of S by identifying each element of $G(\Omega/Q)$ with the subset consisting of that single element. Now, as a separable compact topological group, $G(\Omega/Q)$ can be topologized by a metric $\rho(a, b)(a, b \in G(\Omega/Q))$, and, as is known, this metric can be extended to a metric $\rho(A, B)$ $(A, B \in S)$ on S so that S becomes a compact metric space.[11] Define a function $\rho'(A, B)$ on $S \times S$ by

$$\rho'(A, B) = \inf \rho(a, b), \qquad a \in A, b \in B.$$

Then, $\rho'(A, B)$ is a non-negative continuous function on $S \times S$ and $\rho'(A, B) > 0$ if and only if A and B are disjoint.

In proving the lemma, we may of course assume that both K and L are infinite extensions of Q and that $K \neq L$. Since L/K is a finite extension, there exist finite algebraic number fields K_0 and L_0 such that $K_0 \subset K$, $L_0 \subset L$, $K_0 \subset L_0$, $L = KL_0$ and $[L_0: K_0] = [L: K]$. We then choose a sequence of fields, $K_0 \subset K_1 \subset K_2 \subset \cdots$, so that each K_n is a finite extension of Q and that K is the union of all K_n. Put $L_n = K_n L_0$ for $n \geqq 1$. Then we have again a sequence of fields, $L_0 \subset L_1 \subset L_2 \subset \cdots$, such that each L_n is finite over Q, that L is the union of all L_n and that $[L_n: K_n] = [L: K]$ for every $n \geqq 0$. Put

$$G = G(\Omega/K), \qquad G_n = G(\Omega/K_n), \quad n = 0, 1, 2, \cdots$$
$$H = G(\Omega/L), \qquad H_n = G(\Omega/L_n), \quad n = 0, 1, 2, \cdots.$$

As K is the union of all K_n, G is the intersection of all G_n, and as L is the union of all L_n, H is the intersection of all H_n. Furthermore, since $L = KL_n$, $H = G \cap H_n$ for every $n \geqq 0$. Let C be the set-theoretical complement of H in G. As $K \neq L$ and $[G: H] = [L: K]$ is finite, C is

[11] Cf. D. Montgomery and L. Zippin, *Topological transformation groups*, New York, Interscience Publishers, 1955, p. 17.

a nonempty compact subset of $G(\Omega/Q)$. From the fact that $[G_n: H_n] = [L_n: K_n] = [L: K] = [G: H]$, it also follows easily that G_n is the disjoint union of H_n and CH_n.

Now, let $\mathfrak{p}_1', \cdots, \mathfrak{p}_s'$ be all the prime divisors of the finite algebraic number field L_0 which are ramified for the extension L_0/K_0. For each j $(1 \leqq j \leqq s)$, let T_j be the inertia group, for Ω/Q, of an extension $\mathfrak{p}_j$ of $\mathfrak{p}_j'$ on Ω, and let ϕ_j be a continuous function on $G(\Omega/Q) \times S$ defined by

$$\phi_j(\sigma, A) = \rho'(\sigma T_j \sigma^{-1}, CA), \qquad \sigma \in G(\Omega/Q), A \in S.$$

By the definition of C, we have $CH = C$. Hence $\sigma T_j \sigma^{-1} \cap CH = (\sigma T_j \sigma^{-1} \cap G) \cap C$. But, since L/K is unramified and $\sigma T_j \sigma^{-1}$ is the inertia group of the prime divisor $\mathfrak{p}_j^\sigma$ for Ω/Q, $\sigma T_j \sigma^{-1} \cap G$ is contained in H and the intersection $(\sigma T_j \sigma^{-1} \cap G) \cap C$ is empty. Therefore, $\phi_j(\sigma, H) > 0$ for every σ in $G(\Omega/Q)$ and for every j, $1 \leqq j \leqq s$. On the other hand, since H is the intersection of all H_n, H is the limit of the sequence, $H_0, H_1, H_2, \cdots$, in S. Using the compactness of $G(\Omega/Q)$, it then follows that there exists an integer $n_0 \geqq 0$ such that $\phi_j(\sigma, H_{n_0}) > 0$ for every σ in $G(\Omega/Q)$ and for every j, $1 \leqq j \leqq s$. But, then, $\sigma T_j \sigma^{-1} \cap CH_{n_0}$ is empty by the definition of ϕ_j and, as G_{n_0} is the disjoint union of H_{n_0} and CH_{n_0}, $\sigma T_j \sigma^{-1} \cap G_{n_0}$ must be contained in H_{n_0}. Thus, for every σ in $G(\Omega/Q)$ and for every j, $1 \leqq j \leqq s$, the prime divisor $\mathfrak{p}_j^\sigma$ of Ω is unramified for the extension L_{n_0}/K_{n_0}.

Now, let $\mathfrak{p}$ be any prime divisor of Ω different from $\mathfrak{p}_j^\sigma$, $\sigma \in G(\Omega/Q)$, $1 \leqq j \leqq s$. The restriction $\mathfrak{p}'$ of $\mathfrak{p}$ on L_0 is then different from $\mathfrak{p}_1', \cdots, \mathfrak{p}_s'$ and it is unramified for L_0/K_0. $\mathfrak{p}$ is therefore unramified for L_0/K_0 and, hence, also for L_{n_0}/K_{n_0}. This, combined with the above, shows that L_{n_0}/K_{n_0} is an unramified extension. Putting $F = K_{n_0}$, $E = L_{n_0}$, the conditions (i), (ii) of the lemma are then satisfied.

If L/K is a Galois extension, we can choose K_0 and L_0 in the above so that L_0/K_0 is also a Galois extension. Every L_n/K_n is then also a Galois extension and its Galois group $G(L_n/K_n)$ is canonically isomorphic with $G(L/K)$. Thus, in particular, $G(E/F)$ is isomorphic with $G(L/K)$ and the lemma is completely proved.

Theorem 7. *An extension L/K $(K \subset L \subset \Omega)$ is an unramified extension if and only if there exists a family of extensions $\{L_\alpha/K_\alpha\}$ of finite algebraic number fields K_α and L_α such that*

(i) *every L_α/K_α is an unramified extension, and that*

(ii) *K is the composite of all K_α and L is the composite of all L_α.*

Proof. Suppose first that there exists such a family of extensions $\{L_\alpha/K_\alpha\}$. Put $L_\alpha' = KL_\alpha$. Since L_α/K_α is unramified, L_α'/K is also unramified. Since L is the composite of all L_α', L/K is again unramified.

Suppose, conversely, that L/K is unramified. Let $\{L_i/K\}$ be a family of finite extensions such that L is the composite of all L_i, and let $\{K_j/Q\}$ be a family of finite extensions such that K is the composite of all K_j. By Lemma 6.1, there exist, for each i, finite algebraic number fields E_i and F_i such that E_i/F_i is unramified and $L_i = KE_i$. Put $K_{i,j} = F_iK_j$, $L_{i,j} = E_iK_j$. Then the family of extensions $\{L_{i,j}/K_{i,j}\}$ has the properties (i), (ii) stated in the theorem.

COROLLARY. *Let K be a field and $\{K_\alpha\}$ a family of finite algebraic number fields such that K is the union of all K_α. Then the maximal unramified extension of K in Ω is the composite of all finite unramified extensions of all fields K_α in the family.*

We next consider abelian extensions. If the extension L/K in Lemma 6.1 is abelian, we may take also an abelian extension for E/F in the lemma. By a similar argument as in the proof of Theorem 7, we can then immediately obtain the following

THEOREM 8. *An extension L/K ($K \subset L \subset \Omega$) is an unramified abelian extension if and only if there exists a family of extensions $\{L_\alpha/K_\alpha\}$ of finite algebraic number fields K_α and L_α such that*

(i) *every L_α/K_α is an unramified abelian extension, and that*

(ii) *K is the composite of all K_α and L is the composite of all L_α.*

COROLLARY. *Let K be a field and $\{K_\alpha\}$ a family of finite algebraic number fields such that K is the union of all K_α. For each α, let L_α denote the Hilbert's class field over K_α. Then the Hilbert's class field over K is the composite of all such L_α.*

6.3. As before, let p denote a prime number. An extension L/K is called a p-extension if L/K is a Galois extension and the Galois group $G(L/K)$ is a p-primary compact group. Every field K has a unique maximal (abelian) p-extension in Ω which contains every (abelian) p-extension of K in Ω. By the properties of unramified extensions stated in 6.1, K has also a unique maximal unramified p-extension and a unique maximal unramified abelian p-extension in Ω. The latter is nothing but the p-part of the Hilbert's class field over K, and, if K/Q is finite, its degree over K is equal to the highest power of p dividing the class number of K.

Using again Lemma 6.1, we get immediately such results on unramified (abelian) p-extensions which are similar to those on unramified abelian extensions given in Theorem 8 and its corollary. We state here only the following

THEOREM 9. *Let K be a field and $\{K_\alpha\}$ a family of finite algebraic number fields such that K is the union of all K_α. For each α, let L_α denote* *the maximal unramified abelian p-extension of* K_α *in* Ω. *Then the maximal unramified abelian p-extension of* K *in* Ω *is the composite of all such* L_α.

7. **Γ-extensions.** 7.1. Let p be a prime number and let Γ be, as in previous sections, a fixed p-primary compact abelian group isomorphic with the additive group of p-adic integers. An extension L of a field K is called a Γ-*extension* of K if L/K is a Galois extension and if the Galois group $G(L/K)$ is isomorphic with Γ.

Let L/K be such a Γ-extension and let, for simplicity, $G(L/K)$ be identified with Γ. For each $n \geqq 0$, we denote by K_n the intermediate field of K and L such that $G(L/K_n) = \Gamma_n$. We have then a sequence of fields:

$$K = K_0 \subset K_1 \subset K_2 \subset \cdots \subset L,$$

such that K_n is a cyclic extension of degree p^n over K and L is the union of all K_n, $n \geqq 0$.

LEMMA 7.1. *Let* K *be a finite algebraic number field and* L *a* Γ-*extension of* K. *Then a prime divisor* $\mathfrak{p}$ *of* K *is ramified for* L/K *only when* $\mathfrak{p}$ *is a non-archimedean prime divisor dividing the rational prime* p.

PROOF. Assume that $\mathfrak{p}$ is ramified for L/K and denote by T the inertia group of $\mathfrak{p}$ for the abelian extension L/K. Since T is a non-trivial closed subgroup of Γ, T must be equal to Γ_n for some $n \geqq 0$. Hence T is an infinite group and it follows immediately that $\mathfrak{p}$ is non-archimedean. Let $\mathfrak{p}'$ be an extension of $\mathfrak{p}$ on K_n. For any integer $m \geqq n$, $\mathfrak{p}'$ is then completely ramified for the extension K_m/K_n and its ramification is p^{m-n}. Therefore, if $\mathfrak{p}$ and, hence, $\mathfrak{p}'$ did not divide p, we would have

$$N(\mathfrak{p}') \equiv 1 \mod p^{m-n},$$

where $N(\mathfrak{p}')$ denotes the absolute norm of the prime divisor $\mathfrak{p}'$. But, since m can be taken arbitrarily large, this is obviously a contradiction, and the lemma is proved.

From the lemma, it follows in particular that there exist only a finite number of prime divisors of K which are ramified for L/K. On the other hand, since an unramified abelian extension of K is always a finite extension, there exists at least one prime divisor which is ramified for L/K.

LEMMA 7.2. *Let* K *and* L *be as in Lemma* 7.1 *and let* s *be the number of prime divisors of* K *which are ramified for* L/K. *Furthermore, let* L' *be an unramified p-extension of* L *such that* L'/K *is an abelian exten-*

sion. Then the Galois group $G(L'/K)$ is a p-primary compact abelian group of finite rank and the rank of the factor torsion group of $G(L'/K)$ is at most s.

PROOF. Put $G=G(L'/K)$ and $N=G(L'/L)$. It is clear that G is a p-primary compact group, for both N and $G/N=G(L/K)=\Gamma$ are such groups. Let $\mathfrak{p}_1, \cdots, \mathfrak{p}_s$ be all the prime divisors of K which are ramified for L/K and let $T_1, \cdots, T_s$ denote the inertia groups of $\mathfrak{p}_1, \cdots, \mathfrak{p}_s$, respectively, for the abelian extension L'/K. Since L'/L is an unramified extension, the intersection $T_i \cap N$ is 1, and we have $T_i \cong T_iN/N$. T_i is thus isomorphic with a nontrivial subgroup of $\Gamma=G/N$ and, hence, also isomorphic with Γ itself. Let T be the product of the subgroups $T_1, \cdots, T_s$ in G and E the intermediate field of K and L' such that $T=G(L'/E)$. As L'/L is an unramified extension, no prime divisor of K, different from $\mathfrak{p}_1, \cdots, \mathfrak{p}_s$, is ramified for L'/K. It then follows from the definition of T that E/K is an unramified abelian extension. Therefore, E/K is a finite extension and G/T is a finite group. From this and from the fact that T is the product of $T_1, \cdots, T_s$, each isomorphic with Γ, the lemma follows immediately by a simple group-theoretical consideration.

We notice that the rank of the factor torsion group of $G(L'/L)$ is at most $s-1$. Hence, if $s=1$, $G(L'/L)$ is a finite group.

7.2. As before, let L be a Γ-extension of a finite algebraic number field K. Let M be an unramified abelian p-extension of L such that M/K is also a Galois extension. We put $G=G(M/K)$, $X=G(M/L)$. Then X is a closed normal subgroup of G and $G/X=G(L/K)=\Gamma$. As X is a p-primary compact abelian group, G is such a compact group as we considered in 2.5, and X is thus made into a compact Γ-module in a natural way. We shall next show that this compact Γ-module X is Γ-finite.

For each $n \geqq 0$, put $G_n=G(M/K_n)$. Then X is contained in G_n and $G_n/X=\Gamma_n$. By 2.5, we have only to prove that, if $[G_n, G_n]$ is the topological commutator group of G_n, then $G_n/[G_n, G_n]$ has a finite rank for every $n \geqq 0$. Let L_n denote the intermediate field of K and M such that $G(M/L_n)=[G_n, G_n]$. By the definition of $[G_n, G_n]$, L_n is the maximal abelian extension of K_n contained in M. Hence, in particular, L is contained in L_n. On the other hand, as $G(L/K_n)=\Gamma_n \cong \Gamma$, L/K_n is a Γ-extension. Therefore, we may apply Lemma 7.2 to K_n, L and L_n, and we see that $G(L_n/K_n)=G_n/[G_n, G_n]$ has a finite rank. Our assertion is thus proved.

Now, let $\mathfrak{p}_1, \cdots, \mathfrak{p}_s$ be all the prime divisors of K which are ramified for L/K. The inertia groups of $\mathfrak{p}_1, \cdots, \mathfrak{p}_s$ for L/K are then nontrivial subgroups of Γ and we can find a suitable integer $n_0 \geqq 0$ such that Γ_{n_0} is contained in all these inertia groups. Then a prime divisor of K_{n_0}, which is ramified for L/K_{n_0}, is not decomposed for any extension K_n/K_{n_0}, $n \geqq n_0$, and the number of the prime divisors of K_n which are ramified for L/K_n remains the same for all K_n, $n \geqq n_0$. Thus, it follows from Lemma 7.2 that the rank of the factor torsion group of $G_n/[G_n, G_n]$ has a fixed upper bound for all $n \geqq 0$. But, since $[G_n, G_n] = \omega_n X = X_n^*$, this implies that the rank of the factor torsion module of X/X_n^* also has a fixed upper bound for all $n \geqq 0$. Now, let A be a discrete Γ-module dual to the compact Γ-module X and let A_n $(n \geqq 0)$ be the submodules of A as defined before. Then the duality between A and X implies a duality between A_n and X/X_n^*, and also a duality between the maximal divisible submodule of A_n and the factor torsion module of X/X_n^*. Hence, by above, the maximal divisible submodule of A_n has a fixed upper bound for all $n \geqq 0$. It then follows from Lemma 5.5 that the invariant $m(A)$ of A is finite, and, as the invariant $m(X)$ of X is equal to $m(A)$, the following theorem is proved:

THEOREM 10. *Let L be a Γ-extension of a finite algebraic number field K and M an unramified abelian p-extension of L such that M/K is also a Galois extension. Then the Galois group $G(M/L)$ is a Γ-finite compact Γ-module with respect to $\Gamma = G(L/K)$, and the invariant $m(X)$ of the compact Γ-module $X = G(M/L)$ is finite. The structure of $G(M/L)$ as a p-primary compact abelian group with operator domain $G(L/K)$ is thus given by Theorem* 6.

7.3. Let K and L be as above. We now take as M the maximal unramified abelian p-extension of L in Ω, i.e. the p-part of the Hilbert's class field over L; M/K is then obviously a Galois extension. By Theorem 10, the Galois group $G(M/L)$ is a Γ-finite compact Γ-module with respect to $\Gamma = G(L/K)$ and we denote the invariants $l(X)$ and $m(X)$ of the Γ-module $X = G(M/L)$ by $l(L/K)$ and $m(L/K)$, respectively. By the above theorem, not only $l(L/K)$ but also $m(L/K)$ are non-negative integers, and they give us information on the structure of the Galois group $G(M/L)$ of the maximal unramified abelian p-extension M over L. For instance, $G(M/L)$ is of bounded order if and only if $l(L/K) = 0$ and it is of finite rank if and only if $m(L/K) = 0$.

Actually, $l(L/K)$ depends only upon L (and M), but not upon K, for it is an invariant of the Galois group $G(M/L)$ considered merely as an abelian group. On the other hand, the invariant $m(L/K)$ is defined by means of the Γ-structure of $G(M/L)$ and, hence, essentially depends upon the ground field K. In fact, if we consider the Γ-extension L/K_n $(n \geqq 0)$ instead of L/K, then we see easily that

$$m(L/K_n) = p^n m(L/K).$$

Now, let K' be any finite extension of K and L' the composite of K' and L in $\Omega: L' = K'L$. Then $K' \cap L = K_n$ for some $n \geqq 0$ and $G(L'/K') \cong G(L/K_n) \cong \Gamma$. Hence L'/K' is also a Γ-extension. If M is, as before, the maximal unramified abelian p-extension of L in Ω, the composite ML' of M and L' in Ω is contained in the maximal unramified abelian p-extension M' of L' in Ω, and $G(ML'/L')$ is a factor group of $G(M'/L')$. On the other hand, $G(ML'/L')$ is canonically isomorphic with $G(M/M \cap L')$ which is a subgroup of finite index $[M \cap L' : L]$ in $G(M/L)$. Thus the Galois group $G(M/L)$ is, up to a finite factor group, isomorphic with a factor group of the Galois group $G(M'/L')$, and this is so even when both groups are considered as modules over the same operator domain $\Gamma = G(L'/K') = G(L/K_n)$. It then follows immediately that

$$l(L/K_n) \leqq l(L'/K'), \quad m(L/K_n) \leqq m(L'/K'),$$

or, by the above, that

$$l(L/K) \leqq l(L'/K'), \quad m(L/K) \leqq p^{-n} m(L'/K').$$

7.4. We shall next give another arithmetic characterization of the invariants $l(L/K)$ and $m(L/K)$. Let K, L and M be as in 7.3 and put $G = G(M/K)$, $X = G(M/L)$, $\Gamma = G/X = G(L/K)$, $G_n = G(L/K_n)$, $n \geqq 0$. As in 7.2, we choose an integer n_0 such that, for any $n \geqq n_0$, the field K_n has the same number of prime divisors which are ramified for L/K_n. We then denote by $\mathfrak{p}_1, \cdots, \mathfrak{p}_s$ all the prime divisors of K_{n_0} which are ramified for L/K_{n_0}. Let $\mathfrak{p}_i^*$ $(1 \leqq i \leqq s)$ be an extension of $\mathfrak{p}_i$ on M and T_i the inertia group of $\mathfrak{p}_i^*$ for the extension M/K_{n_0}. Since M/L is unramified, $T_i \cap X = 1$, and since $\mathfrak{p}_i$ is completely ramified for K_n/K_{n_0} for any $n \geqq n_0$, $T_i X = G_{n_0}$. Hence T_i is naturally isomorphic with $\Gamma_{n_0} = G_{n_0}/X$ and it contains an element σ_i such that the coset of σ_i mod X is the element $\gamma_{n_0}^{-1}$ of Γ_{n_0}. Put $\sigma = \sigma_1$ and $\sigma_i = \sigma x_i$, $1 \leqq i \leqq s$, with x_i in X. For any $t \geqq 0$, we have then

$$\sigma_i^{p^t} = \sigma^{p^t} x_i^{\nu}, \tag{5}$$

where $\nu = \nu_{n_0+t, n_0}$ is defined as (3) in 3.2.

We now fix an integer $n \geqq n_0$ and put $G' = G_n/[G_n, G_n] = G(L_n/K_n)$ and $X' = X/[G_n, G_n] = G(L_n/L)$, where L_n denotes, as before, the maximal abelian extension of K_n contained in M. Let $\mathfrak{p}_i'$ $(1 \leqq i \leqq s)$ be the unique extension of $\mathfrak{p}_i$ on K_n and let T_i' denote the inertia group of $\mathfrak{p}_i'$ for the abelian extension L_n/K_n. Since the inertia group of $\mathfrak{p}_i^*$ for the extension M/K_n is $T_i \cap G_n = T_i^{p^t}$, where $t = n - n_0$, T_i' is the image of $T_i^{p^t}$ under the canonical homomorphism $G_n \rightarrow G'$ and, hence, is the closure of the cyclic subgroup of G' generated by the coset of $\sigma_i^{p^t}$ mod $[G_n, G_n]$. On the other hand, as $G_n = T_1^{p^t} X$, G' is the direct product of T_1' and X': $G' = T_1' \times X'$. Therefore, if we denote by T' the product of the subgroups $T_1', \cdots, T_s'$ in G' and if we put

$$T' = T_1' \times Y', \qquad Y' \subset X',$$

it follows from (5) that Y' is the closure of the subgroup of G' generated by the cosets of x_i^ν mod $[G_n, G_n]$ for $1 \leqq i \leqq s$.

Now, let E_n be the intermediate field of K_n and L_n such that $G(L_n/E_n) = T'$. From the definition of T', it follows that E_n is the maximal unramified extension of K_n contained in L_n, and, hence, is the maximal unramified abelian extension of K_n contained in M. As M was the maximal unramified abelian p-extension of L in Ω, it is easy to see from Theorem 9 that E_n is the maximal unramified abelian p-extension of K_n in Ω, i.e., the p-part of the Hilbert's class field over K_n. E_n/K_n is therefore a finite extension and the degree $[E_n : K_n]$ is equal to the highest power of p dividing the class number of K_n. By the above, $G(E_n/K_n) = G'/T' \cong X'/Y'$. Let Y_n be the closure of the subgroup of X generated by $[G_n, G_n]$ and x_i^ν, $1 \leqq i \leqq s$, where $\nu = \nu_{n,n_0}$. Then, $Y' = Y_n/[G_n, G_n]$ and we have $G(E_n/K_n) \cong X/Y_n$.

We now consider the group Y_n for every $n \geqq n_0$. For simplicity, put $Y = Y_{n_0}$. If we use the additive notation for the group X, Y_n is the closure of the subgroup of X generated by $X_n^* = [G_n, G_n]$ and $\nu_{n,n_0} x_i$, $1 \leqq i \leqq s$. As $X_n^* = \omega_n X = \nu_{n,n_0} \omega_{n_0} X = \nu_{n,n_0} X_{n_0}^*$ and $\nu_{n_0,n_0} x_i = x_i$, we then see that $Y_n = \nu_{n,n_0} Y$, and the isomorphism $G(E_n/K_n) \cong X/Y_n$ immediately implies that

$$[E_n : K_n] = [X : \nu_{n,n_0} Y],$$

for any $n \geqq n_0$.

7.5. We shall next compute the group index on the right hand side of the above equality. As in Theorem 6, let V be the minimal submodule of X such that $\overline{X} = X/V$ is elementary. V is then a module of finite rank, and if we denote by W the finite torsion submodule of V, $\overline{V} = V/W$ is a torsion-free compact module of finite rank $l = l(X)$. Therefore, if n_0 is large enough, then, by Lemma 3.3,

$$(6) \qquad [\overline{V} : \nu_{n,n_0} \overline{V}] = p^{l(n-n_0)},$$

for any $n \geqq n_0$. For large n_0, we also have that $\omega_{n_0} W = 0$ and $\nu_{n,n_0} W = p^t W$ with $t = n - n_0$. In the following, we shall assume, as we can, that n_0 is chosen so large that all these conditions are satisfied, together with what is mentioned at the beginning of 7.4.

Now, we have

$$[X:\nu Y] = [X:\nu X][\nu X:\nu Y], \qquad \nu = \nu_{n,n_0},$$

and

$$[X:\nu X] = [X:\nu X + V][\nu X + V:\nu X]$$
$$= [\overline{X}:\nu\overline{X}][V:\nu X \cap V].$$

Let x be an element of X such that νx is contained in V. Then $\omega_n x = \omega_{n_0}\nu_{n,n_0}x$ is also in V. But, as $\overline{X}=X/V$ is regular and $\overline{X}_n=0$, x itself must be in V. Therefore $\nu X\cap V=\nu V$, and it follows that

$$[X:\nu X] = [\overline{X}:\nu\overline{X}][V:\nu V].$$

Now, we can see from the proof of Lemma 3.3 that $\nu\bar{v}=0$, $\bar{v}\in\overline{V}$, implies $\bar{v}=0$.[12] Hence, by a similar argument as above, we obtain that $\nu V\cap W=\nu W$ and $[V:\nu V]=[\overline{V}:\nu\overline{V}][W:\nu W]$. On the other hand, since $\overline{X}$ is a regular module and $\overline{X}_{n_0}=0$, the endomorphism $\bar{x}\to\omega_{n_0}\bar{x}$ of $\overline{X}$ is one-one. Therefore, using $\omega_n=\omega_{n_0}\nu_{n,n_0}$, we also have

$$[\overline{X}:\nu\overline{X}] = [\omega_{n_0}\overline{X}:\omega_n\overline{X}] = [\overline{X}:\overline{X}_n^*][\overline{X}:\overline{X}_{n_0}^*]^{-1}.$$

Thus, we obtain

$$[X:\nu X] = [\overline{X}:\overline{X}_n^*][\overline{X}:\overline{X}_{n_0}^*]^{-1}[\overline{V}:\nu\overline{V}][W:\nu W].$$

We next compute $[\nu X:\nu Y]$. As is readily seen,

$$[\nu X:\nu Y] = [X:Y][X^0:Y^0]^{-1},$$

where X^0 is the submodule of all x in X satisfying $\nu x=0$ and $Y^0=Y\cap X^0$. By the above argument, we know that X^0 is contained in W. However, as W is a finite module, $\nu W=p^tW=0$ whenever $t=n-n_0$ is sufficiently large. Hence, there exists an integer $n_1\geqq n_0$ such that, if $n\geqq n_1$, then $\nu W=0$, $X^0=W$, $Y^0=Y\cap W$ and, consequently, $[\nu X:\nu Y]=[X:Y][W:Y\cap W]^{-1}$.

Putting all these together, we then see that, for any $n\geqq n_1$,

$$[X:\nu_{n,n_0}Y] = p^u[\overline{X}:\overline{X}_n^*][\overline{V}:\nu_{n,n_0}\overline{V}],$$

with an integer u independent of n. Here the factor $[\overline{V}:\nu_{n,n_0}\overline{V}]$ on the right hand side is given by (6), and the other factor $[X:X_n^*]$ is a power of p whose exponent is given by

$$c(n;\overline{X}) = m(X)p^n,$$

for $\overline{X}=X/V$ is an elementary compact Γ-module of weight $w(\overline{X})$

[12] The endomorphism $\bar{v}\to\nu\bar{v}$ of $\overline{V}$ is the product of the endomorphism $\bar{v}\to p^t\bar{v}$ and an automorphism of $\overline{V}$.

$=m(X)$. It then follows immediately that, for any $n \geqq n_1$, $[E_n : K_n] = [X : \nu_{n,n_0} Y]$ is a power of p whose exponent is equal to

$$l(X)n + m(X)p^n + u',$$

where u' is an integer independent of n.

Changing the notation slightly, we can now state our result as follows:

THEOREM 11. *Let K be a finite algebraic number field and L a Γ-extension of K. For each $n \geqq 0$, let K_n be the intermediate field of K and L with degree p^n over K, and let p^{e_n} be the highest power of p dividing the class number of K_n. Then, there exist an integer $n_0 \geqq 0$ and an integer c such that, for any $n \geqq n_0$,*

$$e_n = ln + mp^n + c,$$

where $l = l(L/K)$ and $m = m(L/K)$ are the invariants of L/K as defined in 7.3.

The theorem shows, in particular, that the invariants $l(L/K)$ and $m(L/K)$ are uniquely determined by e_n $(n \geqq 0)$, and, hence, also by the extension L/K, without knowing the structure of the extension M over L.

7.6. In some special cases, the result of Theorem 11 can be obtained more simply in the following manner: suppose, namely, that the field K has only one prime divisor $\mathfrak{p}$ which is ramified for the extension L/K and suppose also that the inertia group of $\mathfrak{p}$ for L/K coincides with $\Gamma = G(L/K)$. Then, for each $n \geqq 0$, the field K_n also has exactly one prime divisor $\mathfrak{p}_n$ which is ramified for L/K_n, namely, the unique extension of $\mathfrak{p}$ on K_n. Applying the same argument as in 7.4 for the case $s=1$, we then see that the Galois group $G(L_n/K_n) = G_n/[G_n, G_n]$ is the direct product of $X/[G_n, G_n]$ and the inertia group of $\mathfrak{p}_n$ for the abelian extension L_n/K_n, and, consequently, that the Galois group $G(E_n/K_n)$ of the maximal unramified abelian p-extension E_n over K_n is isomorphic with X/X_n^*. Therefore, the compact Γ-module X is strictly Γ-finite and, for every $n \geqq 0$, the degree $p^{e_n} = [E_n : K_n]$ is equal to the order of X/X_n^*. We have thus:

$$e_n = c(n; X), \qquad n \geqq 0,$$

with the characteristic function $c(n; X)$ of the strictly Γ-finite compact Γ-module X, and the result in Theorem 11 then follows immediately from the dual of Theorem 4.

7.7. We finally give here some examples of Γ-extensions of finite algebraic number fields, illustrating the results obtained above.

For each $n \geqq 0$, let C_n denote the field obtained by adjoining all p^nth roots of unity to the field of rational numbers Q. If $p \neq 2$, C_{n+1} is a cyclic extension of degree $(p-1)p^n$ over Q and we denote by F_n the cyclic subextension of C_{n+1}/Q with degree p^n over Q. On the other hand, if $p=2$, we denote by F_n the maximal real subfield of C_{n+2}; F_n is again a cyclic extension of degree p^n $(=2^n)$ over Q. In both cases, we have then a sequence of fields:

$$Q = F_0 \subset F_1 \subset F_2 \subset \cdots,$$

and the union of all these F_n $(n \geqq 0)$ obviously gives us a Γ-extension E of Q. Suppose next that E' be any Γ-extension of Q and denote by F_n' the subfield of E' such that $[F_n' : Q] = p^n$. By Lemma 7.1, the conductor of the cyclic extension F_n'/Q is a power of p, and it follows immediately that $F_n' = F_n$ for every $n \geqq 0$. Therefore, E' must coincide with E and we know that E is the unique Γ-extension of the field of rational numbers Q. Now, it can be proved that, for every $n \geqq 0$, the class number of F_n is prime to p.[13] Hence, we see from Theorem 9 that the maximal unramified abelian p-extension of E just coincides with E itself and, consequently, that

$$l(E/Q) = m(E/Q) = 0.$$

More generally, for any finite algebraic number field K, the composite of K and E in Ω always defines a Γ-extension L over K, though this is not necessarily the unique Γ-extension of K.

Consider, in particular, the case where $p \neq 2$ and K is the cyclotomic field of pth roots of unity. Then the subfield K_n of $L = KE$ with degree p^n over K is nothing but C_{n+1}, and the assumptions stated in 7.6 are satisfied for the Γ-extension L/K. Therefore, by Theorem 11 or by the remark in 7.6, we immediately obtain the following

THEOREM 12. *Let $p \neq 2$ and let C_n denote the cyclotomic field of p^nth roots of unity $(n \geqq 0)$. Furthermore, let $K = C_1$ and let L be the union of all C_n, $n \geqq 0$. Then there exists an integer $n_0 \geqq 0$ such that, for $n \geqq n_0$, the exponent e_n of the highest power of p dividing the class number of C_{n+1} is given by*

$$e_n = ln + mp^n + c,$$

where $l = l(L/K)$ and $m = m(L/K)$ are the invariants of the Γ-extension L/K as defined in 7.3, and c is a suitable integer independent of n.

[13] Cf. K. Iwasawa, *A note on class numbers of algebraic number fields*, Abh. Math. Sem. Univ. Hamburg vol. 20 (1956) pp. 257–258.

Of course, a similar result can be obtained for $p=2$, if we only put $K=C_2$ and denote by e_n the highest power of 2 dividing the class number of C_{n+2}. However, for any regular prime p including $p=2$, we know a more precise result that $e_n=0$ for all $n\geqq 0$.[14] Thus, in such a case, the maximal unramified abelian p-extension M of L coincides with L itself and both invariants $l(L/K)$ and $m(L/K)$ are 0. On the other hand, if p is irregular, it can be shown that at least one of $l(L/K)$ and $m(L/K)$ is different from 0; in such a case, M is therefore an infinite extension of L.

Further arithmetic properties of our invariants will be studied in our forthcoming papers.

MASSACHUSETTS INSTITUTE OF TECHNOLOGY

[14] Cf. (12) above.

[36] On some properties of Γ-finite modules

Ann. of Math., (2) 70 (1959), 291-312.

(Received March 9, 1959)

In a previous publication, we have introduced the notion of Γ-finite modules to study the structure of certain Galois extensions of algebraic number fields.[1] In the present paper, we shall further investigate such Γ-finite modules and prove, in particular, those results which we need in the study of cyclotomic fields.[2] Our previous results on the structure of Γ-finite modules will be briefly reviewed in § 1 for the convenience of the reader.

1. Γ-finite modules[3]

1.1. *Γ-modules.* Let p be a prime number and Γ a multiplicative compact topological group isomorphic with the additive group of p-adic integers. We shall fix such a group Γ once and for all in the following. For each $n \geqq 0$, Γ has a unique closed subgroup Γ_n such that Γ/Γ_n is a cyclic group of order p^n, and the groups Γ_n, $n \geqq 0$, give a decreasing sequence of closed subgroups of Γ with intersection 1. For convenience, we fix once and for all an element γ in Γ which is not contained in Γ_1 and put

$$\gamma_n = \gamma^{p^n}, \qquad n \geqq 0.$$

In the group ring of Γ over the ring of rational integers, we then define elements ω_n and $\nu_{n,m}$ by

$$\omega_n = 1 - \gamma_n, \qquad n \geqq 0,$$

$$\nu_{n,m} = \sum_{i=0}^{p^{m-n}-1} \gamma_n^i, \qquad m \geqq n \geqq 0.$$

Now, let A be a discrete p-primary additive abelian group such that Γ acts on A unitarily and continuously, namely, such that $1 \cdot a = a$, $a \in A$, and $\sigma \times a \to \sigma a$ ($\sigma \in \Gamma$, $a \in A$) is a continuous mapping of $\Gamma \times A$ into A. In what follows, such an A will be called a (discrete) *Γ-module*, or, simply,

---

* The present research was supported in part by a National Science Foundation grant.

[1] Cf., K. Iwasawa, *On Γ-extensions of algebraic number fields*, Bull. Amer. Math. Soc., 65 (1959), 183-226. In what follows, this paper will be always referred to as [a].

[2] Cf., K. Iwasawa, On the theory of cyclotomic fields, to appear in these Annals.

[3] For the results given in this section without any proofs, cf. [a], § 1 — § 5. Cf., also the Bourbaki Seminar note (174) by J-P. Serre where the elementary proofs given in [a] are simplified by means of the theory of local rings.

a (*discrete*) *module*. Similarly, Γ-invariant subgroups, Γ-isomorphisms, etc., of such a module will be simply called submodules, isomorphisms, etc.

Let A be a Γ-module. For each $n \geqq 0$, we denote by A_n the submodule of all a in A such that $\sigma a = a$ for every σ in Γ_n. The modules A_n, $n \geqq 0$, then give an increasing sequence of submodules with the union A. Unless otherwise specified, A_n (B_n, C_n, etc.) will always denote in the following the submodule of a given Γ-module A (B, C, etc.) defined as in the above.

1.2. *Γ-finite modules.* A Γ-module A is called *Γ-finite* if every submodule A_n, $n \geqq 0$, has finite rank; and A is called *strictly Γ-finite* if every submodule A_n, $n \geqq 0$, has finite order. Suppose that A is strictly Γ-finite and let $c(n; A)$ denote the exponent of p in the order of the finite group A_n, $n \geqq 0$. $c(n; A)$ then defines a non-decreasing integral-valued function of the integers $n \geqq 0$, and we call it the *characteristic function* of A.

For $\mu = 0, 1, 2, \cdots, \infty$, let $Z(p^\mu)$ denote a p-primary abelian group of rank 1 with order p^μ; if μ is finite, $Z(p^\mu)$ is a finite cyclic group of order p^μ; and if $\mu = \infty$, $Z(p^\mu)$ is divisible. For any μ, $0 \leqq \mu \leqq \infty$, we consider $Z(p^\mu)$ as a discrete group and denote by $E(\mu)$ the additive group of all continuous functions on the compact group Γ with values in $Z(p^\mu)$. If $f = f(\tau)$ is an element of $E(\mu)$, we define σf ($\sigma \in \Gamma$) by $(\sigma f)(\tau) = f(\sigma^{-1}\tau)$. $E(\mu)$ then becomes a Γ-finite module and, if $\mu \neq \infty$, it is even strictly Γ-finite and has the characteristic function $c(n; E(\mu)) = \mu p^n$. For any $\mu_1, \cdots, \mu_s$, $0 \leqq \mu_i \leqq \infty$, we denote by $E(\mu_1, \cdots, \mu_s)$ the direct sum of $E(\mu_i)$, $1 \leqq i \leqq s$. $E(\mu_1, \cdots, \mu_s)$ is again a Γ-finite module and is even strictly Γ-finite if none of μ_i is ∞.

1.3. *Elementary and quasi-elementary modules.* A Γ-module A is called *regular* if, for any $n \geqq 0$, every element a in A can be written in the form $a = (1 - \sigma)a'$ with some σ in Γ_n and some a' in A. This condition is equivalent with the fact that $A = \omega_n A$ for every $n \geqq 0$. If A is regular and finite, then $A = 0$.

A regular Γ-finite module of bounded order is called an *elementary* module. A Γ-module A is elementary if and only if $A = E(\mu_1, \cdots, \mu_s)/D$ with some $\mu_1, \cdots, \mu_s$, $0 \leqq \mu_i < \infty$, and with a suitable finite submodule D of $E(\mu_1, \cdots, \mu_s)$. A homomorphic image of an elementary module is again elementary.

A Γ-module A is called *quasi-elementary* if A is Γ-finite and if every element of A is contained in some elementary submodule of A.[4] A Γ-module

[4] Quasi-elementary modules were not defined in [a].

A is quasi-elementary if and only if $A = E(\mu_1, \cdots, \mu_s)/D$ with some $\mu_1, \cdots, \mu_s$, $0 \leqq \mu_i \leqq \infty$, and with a suitable finite submodule D of $E(\mu_1, \cdots, \mu_s)$. Hence, a Γ-module A is elementary if and only if A is quasi-elementary and bounded. A quasi-elementary module is regular.

1.4. *Structure of Γ-finite modules.* In general, if A is a p-primary abelian group of finite rank, we shall call, for simplicity, the rank of the maximal divisible subgroup of A the *essential rank* of A. If B is a subgroup of A, the essential rank of A is the sum of the essential rank of A/B and that of B.

Now, let A be an arbitrary Γ-finite module. Then A has a unique maximal quasi-elementary submodule A' which contains every quasi-elementary submodule of A. A/A' has a finite rank and if the essential rank of A/A' is λ, A also contains a divisible submodule A'' of rank λ such that both $A/(A' + A'')$ and $A' \cap A''$ are finite modules. Since A' is quasi-elementary, put $A' = E(\mu_1, \cdots, \mu_s)/D$ with a finite submodule D. Then the non-zero numbers (including ∞) in $\mu_1, \cdots, \mu_s$ are uniquely determined by A. Hence, $\lambda(A) = \lambda$ and $\mu(A) = \sum_{i=1}^{s} \mu_i$ are both invariants of the Γ-module A, and so are the numbers $\mu_1, \cdots, \mu_s$ when zeros are omitted.

The invariant $\mu(A)$ is either a non-negative integer or ∞ and if $\mu(A) \neq \infty$, then the submodule A'' in the above is also uniquely determined by A as the maximal divisible submodule of A. In particular, if A is strictly Γ-finite, then $\mu(A) \neq \infty$ and the characteristic function of A is given, for all sufficiently large n, in the form:

$$c(n; A) = \lambda n + \mu p^n + \nu ,$$

where $\lambda = \lambda(A)$, $\mu = \mu(A)$ and ν is also an integer independent of n.

A Γ-finite module A has also a unique maximal regular submodule R in which every regular submodule of A is contained. A/R has a finite rank and $\omega_n^m(A/R) = 0$ for some $m \geqq 0$, $n \geqq 0$. If A is strictly Γ-finite, A/R is even finite.

1.5. *Dual modules.* Let A be a discrete p-primary abelian group and X a compact p-primary abelian group dual to A. Let G be a topological group acting unitarily and continuously on both A and X so that A is a left G-module and X is a right G-module. Then a dual pairing (a, x) of A and X will be called a G-pairing if it satisfies

$$(\sigma a, x) = (a, x\sigma) , \qquad a \in A, x \in X,$$

for every σ in G.[5] When such a G-pairing exists, A and X are called a

[5] This definition of a G-pairing is slightly different from the one given in [a].

pair of dual G-modules. Given any discrete G-module A, there exists, up to isomorphisms, a unique compact G-module X dual to A, and *vice versa*. If the group G is abelian, the compact module X dual to the discrete left Γ-module A may be considered also as a left Γ-module and a G-pairing (a, x) then satisfies $(\sigma a, x) = (a, \sigma x)$.

Now, consider the case where $G = \Gamma$. Let X be any compact Γ-module and A a discrete Γ-module dual to X. We call X (strictly) Γ-finite if and only if A is (strictly) Γ-finite. Then, using the duality, our previous results on discrete Γ-finite modules can be immediately translated into the corresponding results on compact Γ-finite Γ-modules. But we omit here the details.

Since we shall not often deal with compact Γ-modules in the following, the words Γ-modules or modules will be reserved to mean only discrete Γ-modules.

2. Submodules and factor modules

2.1. *Characteristic roots.* We shall first introduce some more invariants of Γ-finite modules.

Let Q_p be the p-adic number field and O_p the ring of all p-adic integers. We shall use the same letters Q_p and O_p to denote also the additive group of the field Q_p and that of the ring O_p, respectively.

Now, let A be a Γ-finite module and let A be divisible and have finite rank λ. Then the underlying additive group of A can be identified with Q_p^λ/O_p^λ [6] and, for any σ in Γ, there is a $\lambda \times \lambda$ integral p-adic matrix $M(\sigma)$ such that if a ($a \in A$) is the residue class of a vector $\alpha = (\alpha_1, \cdots, \alpha_\lambda)$ in Q_p^λ mod O_p^λ, then σa is given by the residue class of the vector $\alpha M(\sigma) = (\alpha_1, \cdots, \alpha_\lambda)M(\sigma)$ mod O_p^λ. $\sigma \to M(\sigma)$ is obviously a continuous representation of Γ by $\lambda \times \lambda$ integral p-adic matrices. For a different identification of A with Q_p^λ/O_p^λ, $M(\sigma)$ is replaced by $T^{-1}M(\sigma)T$ where T is a $\lambda \times \lambda$ integral p-adic matrix with determinant a p-adic unit. Hence the representation class of the integral p-adic representation of Γ: $\sigma \to M(\sigma)$ is uniquely determined by the Γ-module A. Let $\chi_1(\sigma), \cdots, \chi_\lambda(\sigma)$ be the absolutely irreducible (one-dimensional) representations of Γ in a fixed algebraic closure of Q_p derived from the above representation $M(\sigma)$ in Q_p. Then these representations are invariantly defined by the Γ-module A and we call them the characteristic roots of A.

LEMMA 2.1. *Let A be as in the above and let B be a finite submodule of A. Then A and A/B have the same characteristic roots.*

PROOF. We first notice that A/B also is divisible and has the rank λ.

[6] In general, we denote by A^λ the direct sum of λ copies of any additive group A.

Let $A = Q_p^\lambda/O_p^\lambda$ and $B = O'/O_p$, $O_p^\lambda \subset O' \subset Q_p^\lambda$. Since B is finite, there exists a $\lambda \times \lambda$ integral p-adic matrix T with determinant the order of B such that the linear transformation of Q_p^λ defined by T maps O' onto O_p^λ. Then the same linear transformation induces an isomorphism of $A/B = Q_p^\lambda/O'$ onto Q_p^λ/O_p^λ and it follows immediately that if $M(\sigma)$ is the representation of Γ defining the Γ-structure of A on Q_p^λ/O_p^λ, then the representation defining the Γ-structure of A/B on Q_p^λ/O_p^λ is given by $T^{-1}M(\sigma)T$. Since the representations $M(\sigma)$ and $T^{-1}M(\sigma)T$ have the same absolutely irreducible components, the lemma is proved.

We now consider an arbitrary Γ-finite module A and denote by A' the maximal quasi-elementary submodule of A. Then A/A' has a finite rank, and we denote by B the submodule of A containing A' such that B/A' is the maximal divisible submodule of A/A'. B/A' is a divisible module of rank $\lambda(A)$, and we shall call the characteristic roots of B/A' the *characteristic roots* of A. They are obviously invariants of the module A, and we denote the set of these characteristic roots of A by $\chi(A)$.

Let A'' be a divisible submodule of A with rank $\lambda(A)$ such that $A/(A'+A'')$ and $A' \cap A''$ are finite. It is then easy to see that $B = A' + A''$ and, hence, that $B/A' \cong A''/(A' \cap A'')$. Since $A' \cap A''$ is finite, it follows from Lemma 2.1 that B/A' and A'' have the same characteristic roots. Therefore the characteristic roots of A are the same as those of any submodule A'' of A as considered above.

LEMMA 2.2. *Let A be a Γ-finite module of finite rank and B a submodule of A. Then $\chi(A)$ is the union of $\chi(A/B)$ and $\chi(B)$.*

PROOF. We first notice that A/B has a finite rank and is also a Γ-finite module. Let A'' be the maximal divisible submodule of A and B'' the maximal divisible submodule of B. Then B'' is a submodule of finite index in $A'' \cap B$, and $(A'' + B)/B$ is the maximal divisible submodule of A/B. Since $(A'' + B)/B \cong A''/(A'' \cap B)$ and since $(A'' \cap B)/B''$ is finite, it follows from Lemma 2.1 that A/B and A''/B'' have the same characteristic roots. Let λ be the rank of A'' and λ' the rank of B''. Then we can identify the additive group of A'' with Q_p^λ/O_p^λ so that the elements of B'' are represented by the vectors whose first $\lambda - \lambda'$ coordinates are 0. The corresponding representation $M(\sigma)$ for A'' then contains a $\lambda' \times (\lambda - \lambda')$ zero matrix on the left low corner, and the lemma follows immediately.

2.2. *Lemmas on quasi-elementary modules.*

LEMMA 2.3. *Let A be a Γ-finite module and A' a submodule of A. If A' is quasi-elementary and A/A' has finite rank, then A' is the maximal quasi-elementary submodule of A.*

PROOF. Let a be any element in the maximal quasi-elementary submodule of A and let E be an elementary submodule of A containing a. Then, as a homomorphic image of E, $(E + A')/A'$ is also elementary and, hence, is bounded. However, as a submodule of A/A', $(E + A')/A'$ has finite rank. Therefore $(E + A')/A'$ is finite and, as a regular module, it must be 0. Thus E is contained in A', and a is an element in A'. Since a was an arbitrary element in the maximal quasi-elementary submodule of A, the lemma is proved.

LEMMA 2.4. *Let φ be a homomorphism of a Γ-finite module A onto a Γ-finite module B. If A' is the maximal quasi-elementary submodule of A, $\varphi(A')$ is the maximal quasi-elementary submodule of B. In particular, if A is quasi-elementary, so is the homomorphic image B.*

PROOF. Let a be any element in A' and let E be an elementary submodule of A containing a. Then $\varphi(E)$ is an elementary submodule of B and $\varphi(a)$ is contained in $\varphi(E)$. It follows that $\varphi(A')$ is quasi-elementary. On the other hand, A/A' has finite rank and $B/\varphi(A')$ is a homomorphic image of A/A'. Hence, $B/\varphi(A')$ also has finite rank and, by Lemma 2.3, $\varphi(A')$ is the maximal quasi-elementary submodule of B.

LEMMA 2.5. *Let A be a quasi-elementary module and B a submodule of A. If A/B is elementary and B has a finite rank, then A itself is elementary and B is finite.*

PROOF. Since A/B is bounded, p^nA is contained in B for some $n \geqq 0$. By Lemma 2.4, p^nA is quasi-elementary. But, as B has a finite rank, the maximal quasi-elementary submodule of B is 0, by Lemma 2.3. Hence $p^nA = 0$ and A is bounded and elementary. Since B has a finite rank and $p^nB = 0$, B is finite.

2.3. *Invariants of submodules and factor modules.*

THEOREM 2.6. *Let A be a Γ-module and B a submodule of A. Then A is (strictly) Γ-finite if and only if both A/B and B are (strictly) Γ-finite.*

PROOF. Let A_n, B_n and $(A/B)_n$ $(n \geqq 0)$ be the submodules of A, B and A/B, respectively, as defined in 1.1. Then $(A_n + B)/B$ is contained in $(A/B)_n$, and $B_n = A_n \cap B$, $(A_n + B)/B \cong A_n/B_n$.

Suppose first that both A/B and B are (strictly) Γ-finite, i.e., that $(A/B)_n$ and B_n have finite ranks (finite orders) for every $n \geqq 0$. Then, by the above remark, A_n also has a finite rank (finite order) for every $n \geqq 0$ and, hence, A is (strictly) Γ-finite.

Next, suppose that A is (strictly) Γ-finite. It is trivial that B is then also (strictly) Γ-finite. Let C be the maximal regular submodule of B. Then B/C has a finite rank (finite order); and, as C is regular, $(B/C)_n$ is isomorphic with B_n/C_n, and B/C is also (strictly) Γ-finite.[7] Hence, to prove that A/B is (strictly) Γ-finite, we may assume that B has finite rank (finite order). For each $n \geqq 0$, let A'_n denote the submodule of all a in A such that $\omega_n a$ is contained in B. Then $(A/B)_n = A'_n/B$ and $A'_n/A_n \cong \omega_n A'_n \subset B$. Since both A_n and B have finite ranks (finite orders), so does the module A'_n. Hence $(A/B)_n$ also has finite rank (finite order), and A/B is (strictly) Γ-finite.

THEOREM 2.7. *Let A be a Γ-finite module and B a submodule of A. Then,*

$$\lambda(A) \leqq \lambda(A/B) + \lambda(B)\,, \qquad \mu(A) = \mu(A/B) + \mu(B)\,.$$

Furthermore, if $\mu(A/B)$ is finite, then the equality holds for the λ-invariants and the set $\chi(A)$ of the characteristic roots of A is the union of $\chi(A/B)$ and $\chi(B)$.

PROOF. Let A' and B' denote the maximal quasi-elementary submodules of A and B, respectively. Clearly, B' is contained in $A' \cap B$. By Lemma 2.4, $(A'+B)/B$ is the maximal quasi-elementary submodule of A/B and, by the definition, $\lambda(A/B)$ is the essential rank of $A/(A'+B)$. Hence $\lambda(A)$, i.e., the essential rank of A/A', is the sum of $\lambda(A/B)$ and the essential rank of $(A'+B)/A'$. But, as $(A'+B)/A' \cong B/(A' \cap B)$ and $B' \subset A' \cap B$, the essential rank of $(A'+B)/A'$ is not greater than the essential rank of B/B', namely, $\lambda(B)$. Therefore, $\lambda(A) \leqq \lambda(A/B) + \lambda(B)$.

Next, suppose first that $\mu(A/B) = \infty$. Then $(A'+B)/B$ is unbounded, and so is $A'/(A' \cap B)$. Therefore A' is also unbounded, and $\mu(A) = \infty$. Thus, $\mu(A) = \mu(A/B) + \mu(B)$. Suppose now that $\mu(A/B) \neq \infty$. Then $(A'+B)/B$ is elementary. Hence $A'/(A' \cap B)$ is also elementary. But A'/B' is quasi-elementary by Lemma 2.4, and $(A' \cap B)/B'$ has finite rank as a submodule of B/B'. Therefore, by Lemma 2.5, A'/B' is also elementary and $(A' \cap B)/B'$ is finite. If $\mu(B) = \infty$, then B' is unbounded and so is A'. Hence $\mu(A) = \infty = \mu(A/B) + \mu(B)$. If $\mu(B) \neq \infty$, then B' is elementary and so is A'.[8] Since $(A' \cap B)/B'$ is finite and $\mu(B') = \mu(A' \cap B)$, we have

$$\begin{aligned}\mu(A) = \mu(A') = \mu(A'/B') + \mu(B') &= \mu(A'/(A' \cap B)) + \mu(B') \\ = \mu((A'+B)/B) + \mu(B') &= \mu(A/B) + \mu(B)\,.\end{aligned}$$ [9]

[7] Cf., [a], Lemma 2.2.
[8] Cf., [a], Lemma 4.13.
[9] Cf., [a], Lemma 4.17.

As shown in the above, $(A' \cap B)/B'$ is finite if $\mu(A/B) \neq \infty$. Hence the essential rank of $B/(A' \cap B)$ is equal to the essential rank of B/B', and we see that $\lambda(A) = \lambda(A/B) + \lambda(B)$. Furthermore, in such a case, $\chi(B) = \chi(B/B') = \chi(B/(A' \cap B))$ because $(A' \cap B)/B'$ is finite. Since $(A' + B)/A' \cong B/(A' \cap B)$, we then have that $\chi(B) = \chi((A' + B)/A')$. However, by Lemma 2.2, $\chi(A) = \chi(A/B')$ is the union of $\chi(A/(A' + B))$ and $\chi((A' + B)/A')$; and $\chi(A/(A' + B)) = \chi(A/B)$ because $(A' + B)/B$ is the maximal quasi-elementary submodule of A/B. Hence $\chi(A)$ is the union of $\chi(A/B)$ and $\chi(B)$.

REMARK. The assumption $\mu(A/B) \neq \infty$ in the second half of the theorem is indispensable, as one sees by the example where $A = E(\infty)$ and $B = A_0$.

3. Cohomology of Γ-modules

3.1. *Regular modules.* Let A be a Γ-module. For any $m \geqq n \geqq 0$, Γ_n/Γ_m is a cyclic group of order p^{m-n} generated by the coset of γ_n mod Γ_m and it acts on A_m in a natural way. Hence the cohomology groups $H^i(\Gamma_n/\Gamma_m, A_m)$ are defined. Since Γ_n/Γ_m is a finite cyclic group, we need to consider these cohomology groups only for the dimensions $i = 1, 2$.

THEOREM 3.1. *A Γ-module A is regular if and only if $H^i(\Gamma_n/\Gamma_m, A_m)=0$ for all i and $m \geqq n \geqq 0$.*

PROOF. Let ω_n and $\nu_{n,m}$ be the elements of the group ring of Γ defined in 1.1 and, for simplicity, let ω_n and $\nu_{n,m}$ also denote the endomorphisms of A defined by $a \to \omega_n a$ and $a \to \nu_{n,m} a$, respectively. Then A_n is nothing but the kernel of ω_n: $A_n = \omega_n^{-1}(0)$.

Suppose first that A is regular, i.e., that $\omega_n A = A$ for every $n \geqq 0$. Since $\omega_m = \omega_n \nu_{n,m} = \nu_{n,m}\omega_n$, it follows that $\nu_{n,m}A = A$. As $\omega_m^{-1}(0) = \omega_n^{-1}(\nu_{n,m}^{-1}(0))$ and $\omega_n A = A$, we see that $\omega_n(\omega_m^{-1}(0)) = \nu_{n,m}^{-1}(0)$; namely, that $H^1(\Gamma_n/\Gamma_m, A_m) = 0$. Similarly, it follows from $\omega_m^{-1}(0) = \nu_{n,m}^{-1}(\omega_n^{-1}(0))$ and $\nu_{n,m}A = A$ that $\nu_{n,m}(\omega_m^{-1}(0)) = \omega_n^{-1}(0)$; namely, that $H^2(\Gamma_n/\Gamma_m, A_m) = 0$.

Suppose next that $H^i(\Gamma_n/\Gamma_m, A_m) = 0$ for all i and $m \geqq n \geqq 0$. Let a be any element in A and let $p^t a = 0$. Let n be any sufficiently large integer such that a is contained in A_n. Then $\nu_{n,m}a = 0$ with $m = n + t$. Since $H^1(\Gamma_n/\Gamma_m, A_m) = 0$, $a = \omega_n b$ with some b in A_m. Thus a is contained in $\omega_n A$ for all sufficiently large n and, hence, for all $n \geqq 0$. Since a was an arbitrary element in A, $\omega_n A = A$ for every $n \geqq 0$, and A is regular.

REMARK. As the above proof shows, a Γ-module A is regular if only $H^1(\Gamma_n/\Gamma_m, A_m) = 0$ for all sufficiently large m and n ($m \geqq n$).

3.2. *Modules with finite ranks.* Let A be a Γ-module with finite rank and B the maximal divisible submodule of A. A is obviously Γ-finite and the rank of B is $\lambda = \lambda(A)$. For each $n \geqq 0$, let $B^{(n)}$ denote the submodule of all b in B such that $p^n b = 0$. $B^{(n)}$ is a finite abelian group of type $(p^n, \cdots, p^n)$ with rank λ: $B^{(n)} \cong Z(p^n)^\lambda$. Since the modules $\omega_n B$, $n \geqq 0$, form a decreasing sequence of divisible submodules of B, there is an integer n_1 such that $\omega_n B = \omega_{n_1} B$ for $n \geqq n_1$. Let $\lambda - \lambda_0$ $(0 \leqq \lambda_0 \leqq \lambda)$ be the rank of $B' = \omega_{n_1} B$.

On the other hand, since A/B is finite, there is a finite submodule C of A such that $A = B + C$. Suppose that C is contained in A_{n_2} $(n_2 \geqq 0)$. Then $A_n = B_n + C$ and $B_n \cap C = B \cap C$ for every $n \geqq n_2$, and $A_n/B_n \cong C/(B_n \cap C) = C/(B \cap C) \cong A/B$. Furthermore, if $p^t C = 0$ $(t \geqq 0)$, then $\nu_{n,m} C = 0$ and $\nu_{n,m} A = \nu_{n,m} B$ for $n \geqq n_2$, $m \geqq n + t$.

Finally, let ω'_n and $\nu'_{n,m}$ denote the endomorphisms of B induced by the endomorphisms ω_n and $\nu_{n,m}$ of A, respectively. Then there also exists an integer $n_3 \geqq 0$ such that $\nu'^{-1}_{n,m}(0) = B^{(m-n)}$ for $m \geqq n \geqq n_3$.[10] It follows that, as abelian groups, $\nu'_{n,m} B \cong B/B^{(m-n)} \cong B$, and hence that $\nu'_{n,m} B = B$, for $\nu'_{n,m} B$ is a divisible submodule of rank λ in B.

Now, let n_0 be the maximum of n_1, n_2 and n_3, and let $n \geqq n_0$, $m \geqq n + t$. Since $\nu'_{n,m} B = B$, we see by the same argument as in the proof of Theorem 3.1 that $\nu'_{n,m} B_m = B_n$. Hence $\nu_{n,m} A_m = \nu_{n,m}(B_m + C) = B_n$, and we have

$$H^2(\Gamma_n/\Gamma_m, A_m) = A_n/B_n \cong A/B .$$

Since $p^{m-n} C = 0$, $B^{(m-n)} + C$ is, as an abelian group, the direct sum of $B^{(m-n)}$ and a subgroup C': $B^{(m-n)} + C = B^{(m-n)} + C'$. Here

$$C' \cong (B^{(m-n)} + C)/B^{(m-n)} \cong C/(B^{(m-n)} \cap C) = C/(B \cap C) \cong A/B .$$

But, $\nu^{-1}_{n,m}(0) = \nu'^{-1}_{n,m}(0) + C = B^{(m-n)} + C = B^{(m-n)} + C'$ and $\omega_n A_m = \omega_n B_m = \omega_n \omega'^{-1}_n \nu'^{-1}_{n,m}(0) = \omega_n \omega'^{-1}_n B^{(m-n)} = \omega_n B \cap B^{(m-n)} = B' \cap B^{(m-n)}$. Since B' is a divisible submodule of rank $\lambda - \lambda_0$ in B, $B' \cap B^{(m-n)}$ is an abelian group isomorphic with $Z(p^{m-n})^{\lambda-\lambda_0}$. Hence $B^{(m-n)}/(B' \cap B^{(m-n)}) \cong Z(p^{m-n})^{\lambda_0}$, and it follows that

$$H^1(\Gamma_n/\Gamma_m, A_m) = (B^{(m-n)} + C)/(B' \cap B^{(m-n)}) \cong A/B + Z(p^{m-n})^{\lambda_0} .$$

Therefore the following lemma is proved.

LEMMA 3.2. *Let A be a Γ-module of finite rank and B the maximal divisible submodule of A. If both n and $m - n$ are sufficiently large, then*

$$H^1(\Gamma_n/\Gamma_m, A_m) \cong A/B + Z(p^{m-n})^{\lambda_0} \qquad (\textit{direct sum}),$$
$$H^2(\Gamma_n/\Gamma_m, A_m) \cong A/B ,$$

[10] Cf., the dual statement contained in the proof of [a], Lemma 3.3.

where λ_0 *is a fixed integer with* $0 \leqq \lambda_0 \leqq \lambda(A) = \lambda(B)$.

We notice that the second isomorphism is even a Γ-isomorphism if $H^2(\Gamma_n/\Gamma_m, A_m)$ is considered as a Γ-module in a natural way.

3.3. *Γ-finite modules in general.* Let A be an arbitrary Γ-finite module. Let R be the maximal regular submodule of A and S the submodule of A containing R such that S/R is the maximal divisible submodule of A/R. S is then also characterized as the minimal submodule of A having finite index in A.[11] Since R is regular, we have an exact sequence $0 \to R_m \to A_m \to (A/R)_m \to 0$ for every $m \geqq 0$.[12] But it follows from Theorem 3.1 that $H^i(\Gamma_n/\Gamma_m, R_m) = 0$ for all i and $m \geqq n \geqq 0$. Therefore we see immediately that

$$H^i(\Gamma_n/\Gamma_m, A_m) \cong H^i(\Gamma_n/\Gamma_m, (A/R)_m)\ .$$

Since A/R has finite rank, we may apply Lemma 3.2 to A/R and obtain the following:

THEOREM 3.3. *Let A be an arbitrary Γ-finite module and S the minimal submodule of A having finite index in A. If both n and $m-n$ are sufficiently large, then*

$$H^1(\Gamma_n/\Gamma_m, A_m) \cong A/S + Z(p^{m-n})^{\lambda_0}\ , \qquad (\textit{direct sum}),$$
$$H^2(\Gamma_n/\Gamma_m, A_m) \cong A/S\ ,$$

where λ_0 *is a fixed integer with* $0 \leqq \lambda_0 \leqq \lambda(A)$.

Now, it is known that $\omega_u^v(S/R) = 0$ for some $u, v \geqq 0$. Hence, if $S/R \neq 0$, then $\omega_u(S/R) \neq S/R$ and, as can be seen from the proof of Lemma 3.2, the integer λ_0 in the above theorem is positive. Therefore the order of $H^1(\Gamma_n/\Gamma_m, A_m)$ is unbounded.

On the other hand, if $S = R$, then $\lambda_0 = 0$ and it follows from the above that the order of $H^i(\Gamma_n/\Gamma_m, A_m)$ does not exceed the order of the finite module A/R. Since S coincides with R whenever A is strictly Γ-finite,[13] we obtain the following:

THEOREM 3.4. *Let A be a strictly Γ-finite module and R the maximal regular submodule of A. Then the order of any cohomology group $H^i(\Gamma_n/\Gamma_m, A_m)$ does not exceed the order of the finite module A/R and it is isomorphic with A/R whenever both n and $m-n$ are sufficiently large.*

[11] Cf., [a], § 5.
[12] Cf., l. c. 7.
[13] Cf., [a]. § 5.

REMARK. In the theory of Galois cohomology,[14] it is known that the limit of $H(\Gamma_n/\Gamma_m, A_m)$ for $m \to \infty$ is the cohomology group $H'(\Gamma_n, A)$ defined by cocycles of finite type. From the above, we can therefore see the structure of such groups $H'(\Gamma_n, A)$; for instance, $H'(\Gamma_n, A) = 0$ for every $n \geqq 0$ if A is regular, and conversely.

4. κ-isomorphisms

4.1. *Definition.* We first notice that for any p-adic integer ξ in O_p and for any element a in a p-primary abelian group A, the element ξa of A is defined as usual so that such an abelian group A can always be considered as an O_p-module. Obviously, the action of O_p on A commutes with any endomorphism of A.

Now, let U be the multiplicative group of the ring O_p, namely, the group of all p-adic units in the p-adic number field Q_p. U is a compact topological group in its natural p-adic topology. Let κ be a homomorphism of a topological group G into U and let both A and B be discrete G-modules as considered in 1.5. Then an isomorphism φ of the underlying additive group of A into that of B is called a *κ-isomorphism* of the G-module A into the G-module B if

$$\varphi(\sigma a) = \kappa(\sigma)\sigma^{-1}\varphi(a)$$

holds for any σ in G and any a in A. A κ-homomorphism of A into B and also κ-isomorphisms and κ-homomorphisms of compact G-modules are defined similarly.

Let φ be a κ-isomorphism of A into B. Then the image of a G-submodule of A under φ is again a G-submodule of B. If φ is surjective, φ^{-1} is also a κ-isomorphism of B onto A. If ψ is a κ-isomorphism of B into another G-module C, then $\psi \circ \varphi$ is a G-isomorphism of A into C in the usual sense. It follows that a κ-isomorphic image of a G-module A is uniquely determined by A up to G-isomorphisms.

4.2. *κ-isomorphic Γ-finite modules.* We now consider κ-isomorphisms of Γ-modules. We first notice that if κ is a homomorphism of Γ into U, then $\kappa(\sigma) \equiv 1 \bmod p$ for every σ in Γ. It then follows that $\kappa(\sigma) \equiv 1 \bmod p^{n+1}$ for every σ in Γ_n ($n \geqq 0$). Let φ be a κ-isomorphism of a Γ-module A onto a Γ-module B. For each $i \geqq 0$, let $A^{(i)}$ denote the submodule of all a in A satisfying $p^i a = 0$ and let $B^{(i)}$ be the submodule of B defined similarly. It is clear that φ maps $A^{(i)}$ onto $B^{(i)}$. But, by the above remark, if $n \geqq i-1$ and σ is in Γ_n, then $\varphi(\sigma a) = \sigma^{-1}\varphi(a)$ for any a

[14] Cf., S. Lang and J. Tate, *Principal homogeneous spaces over abelian varieties*. Amer. J. Math., 80 (1958), 659-684.

in $A^{(i)}$. It follows that if $n \geqq i-1 \geqq 0$, then φ maps $A_n^{(i)} = A^{(i)} \cap A_n$ into $B_n^{(i)} = B^{(i)} \cap B_n$. Since φ^{-1} is a κ-isomorphism of B onto A, we have

$$\varphi(A_n^{(i)}) = B_n^{(i)}, \qquad n \geqq i-1 \geqq 0.$$

LEMMA 4.1. *Let A be a Γ-module of finite rank, and let there be a κ-isomorphism of A onto a Γ-module B. If $\chi_1, \cdots, \chi_\lambda$ are the characteristic roots of A, then the characteristic roots of B are given by $\kappa\chi_1^{-1}, \cdots, \kappa\chi_\lambda^{-1}$.*

PROOF. Obviously, φ induces a κ-isomorphism of the maximal divisible submodule of A onto the maximal divisible submodule of B. Hence, replacing A and B by their maximal divisible submodules, we may assume that both A and B are divisible. Let λ be the rank of A and B. Then we can identify the underlying additive groups of A and B with Q_p^λ/O_p^λ so that φ becomes the identity map on Q_p^λ/O_p^λ. Let $M(\sigma)$ and $N(\sigma)$ denote the representations of Γ defining, respectively, the Γ-structures of A and B on Q_p^λ/O_p^λ with respect to such identifications. For any vector α in Q_p^λ, we then see that

$$\alpha M(\sigma) \equiv \alpha(\kappa(\sigma)N(\sigma)^{-1}) \qquad \text{mod } O_p^\lambda,$$

and hence that

$$M(\sigma) = \kappa(\sigma)N(\sigma)^{-1}.$$

The lemma then follows immediately.

LEMMA 4.2. *Let A be an elementary module, and let there be a κ-isomorphism of A onto a Γ-module B. Then B is also elementary, and $\mu(A) = \mu(B)$.*

PROOF. Since A is bounded, $A = A^{(s)}$ for some $s \geqq 0$. Hence, if $n \geqq s-1$ and σ is in Γ_n, then $\varphi(\sigma a) = \sigma^{-1}\varphi(a)$ for every a in A, and φ maps A_n onto B_n. It follows that B is strictly Γ-finite. On the other hand, we also see immediately that $H^i(\Gamma_n/\Gamma_m, A_m) \cong H^i(\Gamma_n/\Gamma_m, B_m)$ for $m \geqq n \geqq s-1$. But, as A is regular, $H^i(\Gamma_n/\Gamma_m, A_m) = 0$ by Theorem 3.1. Hence $H^i(\Gamma_n/\Gamma_m, B_m)$ is also 0 for $m \geqq n \geqq s-1$ and, by Theorem 3.4 or by the remark after Theorem 3.1, B is regular. Since B is obviously bounded, it is therefore an elementary module. It then follows from $\varphi(A_n) = B_n$, $n \geqq s-1$, that $c(n; A) = c(n; B)$ for $n \geqq s-1$ and hence that $\mu(A) = \mu(B)$.

LEMMA 4.3. *Let A be a quasi-elementary module and let $A \cong E(\mu_1, \cdots, \mu_s)/D$ with a finite submodule D of $E(\mu_1, \cdots, \mu_s)$. Suppose that there is a κ-isomorphism φ of A onto a Γ-module B. Then B is also quasi-elementary and $B \cong E(\mu_1, \cdots, \mu_s)/D'$ with a finite submodule D' of $E(\mu_1, \cdots, \mu_s)$.*

PROOF. For each $i \geqq 0$, let $E^{(i)}$ denote the maximal elementary submodule of the submodule $A^{(i)}$ of A defined in the above. Since $A^{(i)}/E^{(i)}$ is finite and $\varphi(A^{(i)}) = B^{(i)}$, $B^{(i)}/\varphi(E^{(i)})$ is also finite. But $\varphi(E^{(i)})$ is elementary by Lemma 4.2. Therefore $\varphi(E^{(i)})$ is the maximal elementary submodule of $B^{(i)}$, and it follows from the same lemma that $\mu(A^{(i)}) = \mu(E^{(i)}) = \mu(\varphi(E^{(i)})) = \mu(B^{(i)})$.

Now, since $\varphi(E^{(1)})$ is elementary and $B^{(1)}/\varphi(E^{(1)})$ is finite, $B^{(1)}$ is strictly Γ-finite and, consequently, B is Γ-finite. Since A is quasi-elementary and is the union of all $E^{(i)}$, $i \geqq 0$, B is also the union of the elementary submodules $\varphi(E^{(i)})$, $i \geqq 0$. Hence B is also quasi-elementary. Finally, the invariants $\mu_1, \cdots, \mu_s$ of A are determined by the invariants $\mu(A^{(i)})$ of the submodules $A^{(i)}$, $i \geqq 0$, and the same is true for the invariants $\mu'_1, \cdots, \mu'_t$ of B.[15] But, since $\mu(A^{(i)}) = \mu(B^{(i)})$, $i \geqq 0$, we have $\{\mu_1, \cdots, \mu_s\} = \{\mu'_1, \cdots, \mu'_t\}$.

THEOREM 4.4. *Let A be a Γ-finite module and let there be a κ-isomorphism φ of A onto a Γ-module B. Then B is also Γ-finite, and A and B have the same invariants λ, μ, $\mu_1, \cdots, \mu_s$. Furthermore, if $\chi_1, \cdots, \chi_\lambda$ are the characteristic roots of A, then the characteristic roots of B are given by $\kappa\chi_1^{-1}, \cdots, \kappa\chi_\lambda^{-1}$.*

PROOF. Let A' be the maximal quasi-elementary submodule of A. Then, by the previous lemma, $B' = \varphi(A')$ is also quasi-elementary. But, since A/A' has finite rank, so does B/B'. Hence, by Theorem 2.6, B is Γ-finite and, by Lemma 2.3, B' is the maximal quasi-elementary submodule of B. It then follows from Lemma 4.3 that A and B have the same invariants $\mu_1, \cdots, \mu_s$ and $\mu = \sum_{i=1}^{s} \mu_i$. Let S be the submodule of A containing A' such that S/A' is the maximal divisible submodule of A/A'. Then $\varphi(S)$ is obviously the submodule of B containing B' such that $\varphi(S)/B'$ is the maximal divisible submodule of B/B'. Since φ induces a κ-isomorphism of S/A' onto $\varphi(S)/B'$, A and B have the same invariant λ; and it follows from Lemma 4.1 that B has the characteristic roots as stated in the theorem.

The following theorem will be used in the next section.

THEOREM 4.5. *Let A and B be as in the above. Suppose that A is regular and B is strictly Γ-finite. Then A is also strictly Γ-finite and B is also regular.*

PROOF. As shown in the above, the image B' of the maximal quasi-elementary submodule A' of A is the maximal quasi-elementary submodule of B. Since A/A' is a regular module of finite rank, it is divisible and is

[15] Cf., [a], § 5.

strictly Γ-finite.[16] Since B is strictly Γ-finite, B' is elementary. Hence, by Lemma 4.2, A' is also elementary and it follows from Theorem 2.6 that A is strictly Γ-finite. Since A/A' is divisible, so is B/B'. Let R be the maximal regular submodule of B. Since B is strictly Γ-finite, B/R is finite. But B' is obviously contained in R and B/R is, as a factor module of B/B', also divisible. Hence $B = R$ and B is regular.

5. The adjoints of strictly Γ-finite modules

5.1. *Definition.* Let A be a strictly Γ-finite module. For each $n \geqq 0$, let B'_n denote a dual of A_n and let (a_n, b'_n), $a_n \in A_n$, $b'_n \in B'_n$, be a Γ-pairing of A_n and B'_n. Since A_n is finite, so is the dual B'_n; and they have the same finite order, a power of p. For $m \geqq n \geqq 0$, let $\varphi_{m,n}$ denote the dual of the homomorphism $a_m \to \nu_{n,m} a_m$ of A_m into A_n; $\varphi_{m,n}$ is the unique homomorphism of B'_n into B'_m satisfying

$$(\nu_{n,m} a_m, b'_n)_n = (a_m, \varphi_{m,n} b'_n)_m, \qquad a_m \in A_m,\ b'_n \in B'_n.$$

Then we have $\varphi_{l,m} \circ \varphi_{m,n} = \varphi_{l,n}$ for $l \geqq m \geqq n$, and these homomorphisms $\varphi_{m,n}$ define a direct limit B of the modules B'_n $(n \geqq 0)$. B will be considered as a discrete group; as the modules B'_n are p-groups, B is a discrete p-primary abelian group. The action of Γ on B can be also defined in an obvious way, and Γ then acts on B unitarily. For each $n \geqq 0$, let φ_n denote the canonical homomorphism of B'_n into B. Then $\varphi_n(B'_n)$ is clearly contained in B_n, the submodule of all b in B satisfying $\omega_n b = 0$. Since B is the union of $\varphi_n(B'_n)$, it is also the union of B_n $(n \geqq 0)$, and it follows that Γ acts on B also continuously. Therefore, B is a Γ-module in the sense of 1.1. Since the structure of the Γ-module B is uniquely determined by A up to isomorphism, we shall call B the *adjoint* of A.

The adjoint B of A can be defined also as follows: let X be a compact Γ-module dual to A and let (a, x) be a Γ-pairing of A and X. For each $n \geqq 0$, let X_n denote the annihilator of A_n in X with respect to the pairing (a, x). Then $X_n = \omega_n X$ and the modules X_n $(n \geqq 0)$ form a decreasing sequence of closed submodules of X with intersection 0. Since $\omega_m = \nu_{n,m}\omega_n$ $(m \geqq n)$, the endomorphism $x \to \nu_{n,m} x$ of X induces a homomorphism $\varphi'_{m,n}$ of X/X_n into X/X_m. The adjoint B is then nothing but the direct limit of X/X_n $(n \geqq 0)$ relative to these homomorphisms $\varphi'_{m,n}$.

5.2. *The structure of adjoints.* We continue to use the same notations as in the above. Let R be the maximal regular submodule of A. For each $n \geqq 0$, let S'_n denote a dual of $R_n = A_n \cap R$, and let $R_n \to A_n$ be the natural injection of the submodule R_n into A_n. Then we have the following com-

[16] Cf., [a], § 3 or Theorem 3.1, Lemma 3.2 of the present paper.

mutative diagrams of homomorphisms which are dual to each other $(m \geqq n)$:

$$\begin{array}{ccc} A_m & \xrightarrow{\nu_{n,m}} & A_n \\ \uparrow & & \uparrow \\ R_m & \xrightarrow{\nu_{n,m}} & R_n \end{array}, \qquad \begin{array}{ccc} B'_m & \xleftarrow{\varphi_{m,n}} & B'_n \\ \downarrow & & \downarrow \\ S'_m & \xleftarrow{\psi_{m,n}} & S'_n \end{array}.$$

Since the adjoint S of R is the direct limit of S'_n relative to $\psi_{m,n}$, the homomorphisms $B'_n \to S'_n$ $(n \geqq 0)$ define a homomorphism f of B into S. Since $R_n \to A_n$ is injective, $B'_n \to S'_n$ is surjective. Hence f is also surjective. Suppose that $f(b) = 0$ for an element b in B. Then, for any sufficiently large n, B'_n contains an element b'_n such that b'_n is mapped to 0 under $B'_n \to S'_n$ and that $b = \varphi_n(b'_n)$. On the other hand, since A/R is finite, $A = A_{n_0} + R$ for some $n_0 \geqq 0$. Hence we have $\nu_{n,m}A_m = \nu_{n,m}R_m = R_n$ whenever both n and $m - n$ are sufficiently large. For such m and n, we may then adjoin to the above diagrams arrows from A_m to R_n and from S'_n to B'_m, preserving the commutativity of the diagrams. It then follows that $\varphi_{m,n}(b'_n) = 0$ and hence that $b = \varphi_n(b'_n) = \varphi_m \circ \varphi_{m,n}(b'_n) = 0$. It is therefore proved that f is an isomorphism of B onto S.

By what has been proved in the above, we shall now assume that A is regular. Then $\nu_{n,m}A_m = A_n$ for any $m \geqq n \geqq 0$. Hence the homomorphism $\varphi_{m,n}$ is injective and so is every φ_n, $n \geqq 0$. For each $n \geqq 0$, put $B''_n = \varphi_n(B'_n)$, and let $[a_n, b''_n]_n$, $a_n \in A_n$, $b''_n \in B''_n$, denote the Γ-pairing of A_n and B''_n such that $(a_n, b'_n)_n = [a_n, \varphi_n(b'_n)]_n$, $a_n \in A_n$, $b'_n \in B'_n$. Since $\varphi_n = \varphi_m \circ \varphi_{m,n}$, we have then:

$$(1) \qquad [\nu_{n,m}a_m, b''_n]_n = [a_m, b''_n]_m, \qquad a_m \in A_m,\ b''_n \in B''_n,\ m \geqq n.$$

Consider the following diagrams:

$$A_m \xrightarrow{\nu_{n,m}} A_n \xrightarrow{\alpha} A_m, \qquad B''_m \xleftarrow{\beta} B''_n \xleftarrow{\alpha^*} B''_m,$$

where α and β are natural injections and α^* is the dual of α. By (1), β is dual to $\nu_{n,m}\colon A_m \to A_n$. Hence $\beta \circ \alpha^*$ is dual to $\nu_{n,m}\colon A_m \to A_m$, and it therefore coincides with the endomorphism $\nu_{n,m}\colon B''_m \to B''_m$. It follows that $\nu_{n,m}B''_m$ is contained in B''_n and that α^* is nothing but the homomorphism $\nu_{n,m}\colon B''_m \to B''_n$. Since α^* is the dual of the injection α, we have

$$(2) \qquad [a_n, \nu_{n,m}b''_m]_n = [a_n, b''_m]_m, \qquad a_n \in A_n,\ b''_m \in B''_m,\ m \geqq n.$$

Let a_n be an element of A_n such that $[a_n, \nu_{n,m}B''_m]_n = 0$. Then, by the above, $[a_n, B''_m]_m = 0$, and hence $a_n = 0$. It follows that $B''_n = \nu_{n,m}B''_m$ for any $m \geqq n \geqq 0$. On the other hand, by Theorem 3.1, $H^1(\Gamma_n/\Gamma_m, A_m) = 0$

for every i and $m \geqq n \geqq 0$. Since B''_m is dual to A_m, we see readily that $H^i(\Gamma_n/\Gamma_m, B''_m) = 0$ for every i and $m \geqq n \geqq 0$. Therefore, $B_n \cap B''_m = \nu_{n,m}B''_m = B''_n$ for $m \geqq n \geqq 0$. Since B is the union of all B''_m, we then have $B_n = B''_n$ for every $n \geqq 0$.

Now, as A_n and the dual $B_n = B''_n$ have the same finite order, the adjoint B of A is strictly Γ-finite and has the same characteristic function as that of A. Hence, in particular, $\lambda(B) = \lambda(A)$, $\mu(B) = \mu(A)$. Since $\nu_{n,m}B_m = B_n$ for $m \geqq n$, B is regular and it also follows immediately from (2) that the adjoint of B is A. Therefore, we can summarize our results as follows:

THEOREM 5.1. *Let A be a strictly Γ-finite module and B the adjoint of A. Then B is a regular strictly Γ-finite module with the invariants $\lambda(B)=\lambda(A)$, $\mu(B) = \mu(A)$, and it is also the adjoint of the maximal regular submodule of A. Furthermore, if A is regular, A and B have the same characteristic function and A is, conversely, the adjoint of B.*

5.3. *Adjoints of submodules and factor modules.* Let both A and B be regular strictly Γ-finite modules. Suppose first that B is the adjoint of A. Then, as shown in 5.2, B is also the adjoint of A and there exists, for each $n \geqq 0$, a Γ-pairing $(a_n, b_n)_n$ of A_n and B_n satisfying

$$(3) \qquad (\nu_{n,m}a_m, b_n)_n = (a_m, b_n)_m\,, \quad (a_n, \nu_{n,m}b_m)_n = (a_n, b_m)_m\,,$$

for any a_n in A_n, a_m in A_m, b_n in B_n, and b_m in B_m $(m \geqq n)$. Conversely, if there exists such a system of Γ-pairings $(a_n, b_n)_n$ between the submodules A_n and B_n $(n \geqq 0)$, then A and B are the adjoints of each other. Hence we shall call such a set of Γ-pairings $(a_n, b_n)_n$ a *system of adjoint pairings* for A and B.

We now assume that A and B have a system of adjoint pairings $(a_n, b_n)_n$ and are the adjoints of each other. Let C be a regular submodule of A. For each $n \geqq 0$, we denote by D_n the annihilator of C_n in B_n with respect to the pairing $(a_n, b_n)_n$ of A_n and B_n. Let $m \geqq n$ and let b_n be an element of B_n. b_n is contained in D_m if and only if $[C_m, b_n]_m=0$, namely, by (3), if and only if $[\nu_{n,m}C_m, b_n]_n=0$. But, as C is regular, $\nu_{n,m}C_m = C_n$. Therefore b_n is in D_m if and only if $[C_n, b_n]_n = 0$, namely, if and only if b_n is in D_n. Hence $D_n = B_n \cap D_m$ for any $m \geqq n \geqq 0$. In particular, the modules D_n $(n \geqq 0)$ form an increasing sequence of submodules in B, and we denote the union of all D_n by D. Then $D_n = B_n \cap D$ for every $n \geqq 0$ and, as the notation shows, D_n is the submodule of all d in D with $\omega_n d = 0$. Next, let a_n be an element in the annihilator of $\nu_{n,m}D_m$ in A_n with respect to the pairing $(a_n, b_n)_n$. Then, by (3), $[a_n, D_m]_m = [a_n, \nu_{n,m}D_m]_n=0$, and hence a_n is contained in $C_n = A_n \cap C_m$. Conversely, if a_n is in C_n, then $[a_n, \nu_{n,m}D_m]_n=0$. However, the annihilator of C_n in B_n is D_n. Therefore $D_n = \nu_{n,m}D_m$ for any $m \geqq n \geqq 0$, and we see readily that the module D is also regular.

Now, since D is regular, $(B/D)_n \cong B_n/D_n$ canonically. Hence the pairing $(a_n, b_n)_n$ of A_n and B_n induces a Γ-pairing of C_n and $(B/D)_n$ in a natural way, and these pairings, defined for all $n \geqq 0$, obviously form a system of adjoint pairings for C and B/D. Therefore, C and B/D are the adjoints of each other. Similarly, A/C and D have a system of adjoint pairings induced by $(a_n, b_n)_n$, and they form a pair of mutually adjoint modules.

We have thus proved the following:

THEOREM 5.2. *Let both A and B be regular strictly Γ-finite modules, and let there be a system of adjoint pairings $(a_n, b_n)_n$ for A and B so that A and B form a pair of mutually adjoint modules. Let C be any regular submodule of A. Then there exists a unique regular submodule D of B such that, for each $n \geqq 0$, C_n and D_n form a pair of annihilators in A_n and B_n, respectively, relative to the pairing $(a_n, b_n)_n$. B/D is then the adjoint of C, and A/C is the adjoint of D. The mapping $C \to D$ defines a one-one correspondence between the regular submodules of A and B reversing the inclusion relation.*

5.4. *Adjoints of modules with finite ranks.*

LEMMA 5.3. *Let A be a divisible strictly Γ-finite module with a finite rank λ, and let the Γ-structure of A be defined on the additive group Q_p^λ/O_p^λ by a representation $\sigma \to M(\sigma)$. Then the adjoint B of A is also a divisible Γ-module of rank λ, and the Γ-structure of B is defined on Q_p^λ/O_p^λ by the representation $\sigma \to {}^tM(\sigma)$, the transpose of $M(\sigma)$. In particular, A and the adjoint B have the same characteristic roots: $\chi(A) = \chi(B)$.*

PROOF. Let B be the divisible Γ-module defined by ${}^tM(\sigma)$ on the additive group Q_p^λ/O_p^λ. For $n \geqq 0$, put $M_n = M^{p^n}$, $M = M(\gamma)$. Since A is strictly Γ-finite, the matrix $I - M_n$ is non-singular for every $n \geqq 0$.[17] Hence $I - {}^tM_n$ is also non-singular for every $n \geqq 0$, and the module B is strictly Γ-finite. Thus both A and B are regular modules. We shall show that A and B have a system of adjoint pairings.

Let X be a character of the locally compact group Q_p such that $X(\xi) = 0$ if and only if ξ is in the subgroup O_p. For vectors $\alpha = (\alpha_1, \cdots, \alpha_\lambda)$ and $\beta = (\beta_1, \cdots, \beta_\lambda)$ in Q_p^λ, put

$$(\alpha, \beta) = \sum_{i=1}^{\lambda} X(\alpha_i \beta_i) .$$

Then (α, β) defines a dual pairing of Q_p^λ with itself, and the annihilator of O_p^λ in Q_p^λ coincides with O_p^λ itself. Clearly,

[17] For the argument here, cf., [a], § 3.

$$(\alpha M(\sigma), \beta) = (\alpha, \beta\, {}^tM(\sigma)), \qquad \sigma \in \Gamma,\ \alpha, \beta \in Q_p^\lambda.$$

Now, for each $n \geqq 0$, we have $A_n = A'_n/O_p^\lambda$, $B_n = B'_n/O_p^\lambda$ with $A'_n = O_p^\lambda(I - M_n)^{-1}$ and $B'_n = O_p^\lambda(I - {}^tM_n)^{-1}$. Let a_n and b_n be arbitrary elements in A_n and B_n, respectively, and let α and β be vectors of A'_n and B'_n, respectively, representing a_n and b_n mod O_p^λ. We then put

$$(a_n, b_n)_n = (\alpha(I - M_n), \beta) = (\alpha, \beta(I - {}^tM_n)).$$

It is easy to see that $(a_n, b_n)_n$ actually depends only upon a_n and b_n and that it defines a Γ-pairing of A_n and B_n. Furthermore, if α' is a vector in A'_m representing an element a_m in A_m and if β is a vector in B''_n representing an element b_n in B_n $(m \geqq n)$, then $\nu_{n,m}a_m$ is represented by the vector $\alpha'(\sum_i M_n^i)$ $(i = 0, \cdots, p^{m-n} - 1)$. Hence we have

$$(\nu_{n,m}a_m, b_n)_n = (\alpha'(\textstyle\sum_i M_n^i)(I - M_n), \beta) = (\alpha'(I - M_m), \beta) = (a_m, b_n)_m .$$

Similarly, $(a_n, \nu_{n,m}b_m)_n = (a_n, b_m)_m$ for a_n in A_n and b_m in B_m. Hence the pairings $(a_n, b_n)_n$, $n \geqq 0$, give a system of adjoint pairings for A and B, and the lemma is proved.

5.5. *Invariants of adjoints.*

LEMMA 5.4. *Let A be an elementary module and B the adjoint of A. If $A = E(\mu_1, \cdots, \mu_s)/D$ $(0 \leqq \mu_i < \infty)$ with a finite submodule D of $E(\mu_1, \cdots, \mu_s)$, then $B = E(\mu_1, \cdots, \mu_s)/D'$ with a finite submodule D' of $E(\mu_1, \cdots, \mu_s)$.*

PROOF. By Theorem 5.1, B is a regular strictly Γ-finite module and $\lambda(B) = \lambda(A) = 0$. Hence B is elementary. For each $i \geqq 0$, let $A^{(i)}$ denote the submodule of all a in A satisfying $p^ia = 0$ and let $B^{(i)}$ be the submodule of B defined similarly. As shown in 5.2, A_n and B_n form a pair of dual finite modules and, hence, are isomorphic to each other as abelian groups. Therefore, $A_n^{(i)} = A_n \cap A^{(i)}$ and $B_n^{(i)} = B_n \cap B^{(i)}$ are also isomorphic to each other as abelian groups, and they have the same finite order. It follows that the characteristic functions of $A^{(i)}$ and $B^{(i)}$ are the same and, consequently, that $\mu(A^{(i)}) = \mu(B^{(i)})$. However, the invariants $\mu_1, \cdots, \mu_s$ of the elementary module A are determined by the invariants $\mu(A^{(i)})$ of the submodules $A^{(i)}$, $i \geqq 0$; and the same is true for the elementary module B.[18] Hence A and B have the same invariants $\mu_1, \cdots, \mu_s$, and the lemma is proved.

Now, let A be an arbitrary strictly Γ-finite module and B the adjoint of A. We compare the numerical invariants $\lambda, \mu, \mu_1, \cdots, \mu_s$ of A and the characteristic roots of A with those of B. Let R be the maximal

[18] Cf., l. c. 15.

regular submodule of A. Since the invariants and the characteristic roots of A are the same as those of R and since B is also the adjoint of R by Theorem 5.1, we may replace A by R and assume that A is regular. So, let A be regular and let A' be the maximal elementary submodule of A and A'' the maximal divisible submodule of A. Then both A' and A'' are regular submodules of A. Fixing a system of adjoint pairings for A and B, let B' and B'' denote the regular submodules of B corresponding to A'' and A' respectively in the sense of Theorem 5.2. Then B'' is the adjoint of A/A' and, since A/A' is a strictly Γ-finite divisible module of rank $\lambda = \lambda(A)$, B'' is also a divisible module of the same rank λ by Lemma 5.3. However, $\lambda(A) = \lambda(B)$ by Theorem 5.1. Hence B'' must be the maximal divisible submodule of B and, again by Lemma 5.3, we have $\chi(A) = \chi(A/A') = \chi(B'') = \chi(B)$. On the other hand, B' is the adjoint of A/A''. But, since $A = A' + A''$, $A/A'' \cong A'/(A' \cap A'')$ and $A' \cap A''$ is finite, A/A'' is an elementary module having the same invariants $\mu_1, \cdots, \mu_s$ as those of A'. By Lemma 5.4, B' is then also elementary and has the same invariants $\mu_1, \cdots, \mu_s$. It follows, in particular, that $\mu(B') = \mu(A)$. But we know that $\mu(A) = \mu(B)$ by Theorem 5.1. Hence B' must be the maximal elementary submodule of B, and consequently A and B have the same invariants $\mu_1, \cdots, \mu_s$. Thus the following theorem is proved.

THEOREM 5.5. *Let A be a strictly Γ-finite module and B the adjoint of A. Then, not only $\lambda(A) = \lambda(B)$, $\mu(A) = \mu(B)$, but also A and B have the same invariants $\mu_1, \cdots, \mu_s$ and the same set of characteristic roots $\chi(A) = \chi(B)$. If A is regular and if A' and A'' denote the maximal elementary submodule of A and the maximal divisible submodule of A, respectively, then the regular submodules B' and B'' in B which correspond to A'' and A' in the sense of Theorem* 5.2 *are the maximal elementary submodule of B and the maximal divisible submodule of B, respectively.*

5.6. *κ-adjoints.* Let A be a regular strictly Γ-finite module and X a compact Γ-module dual to A. Then X is a regular strictly Γ-finite compact module, and the endomorphism of X defined by $x \rightarrow \omega_n x$ is injective for every $n \geqq 0$.

LEMMA 5.6. *For the module X, there exists a locally compact abelian group Λ with the following three properties:*

(i) *Γ acts on Λ unitarily and continuously,*

(ii) *X is contained in Λ as an open compact submodule so that Λ/X is a discrete Γ-module,*

(iii) *for every $n \geqq 0$, $u \rightarrow \omega_n u$ defines an automorphism of Λ. Furtheremore, by these properties, Λ is uniquely determined up to* *isomorphisms which leave every element of X fixed.*

PROOF. For $n \geqq 0$, put $X^{(n)} = X$, and denote by Λ the direct limit of $X^{(n)}$ with respect to the maps ω_n: $X^{(n)} \to X^{(n+1)}$, defined by $x \to \omega_n x (x \in X = X^{(n)})$. Let φ_n be the natural homomorphism of $X^{(n)}$ into Λ, and put $\Lambda^{(n)} = \varphi_n(X^{(n)})$, $n \geqq 0$. Since every ω_n is injective, so is every φ_n. Hence, in particular, we may identify $X = X^{(0)}$ with $\Lambda^{(0)}$, using φ_0. Then we have

$$X = \Lambda^{(0)} \subset \Lambda^{(1)} \subset \Lambda^{(2)} \subset \cdots \subset \Lambda = \bigcup_n \Lambda^{(n)} .$$

Since every ω_n is a Γ-isomorphism, we can define the action of Γ on Λ in an obvious way. We then see immediately from the definition that $\omega_n \Lambda^{(n+1)} = \Lambda^{(n)}$ and $\Lambda^{(n+1)}/\Lambda^{(n)} \cong X/\omega_n X$. Since X is strictly Γ-finite, $X/\omega_n X$ is a finite p-group, and so is $\Lambda^{(n+1)}/\Lambda^{(n)}$. Therefore $\Lambda^{(n)}/X$ is a finite p-group, and it follows that $\omega_m \Lambda^{(n)}$ is contained in X if m is large enough.

We now define a topology on Λ so that Λ becomes a topological group and the compact group X is an open subgroup of Λ. This is possible, and we see from the above that the locally compact abelian group Λ has the properties (i) and (ii).

For every $m \geqq 0$, $\Lambda^{(m)}/X$ is finite, and hence $\Lambda^{(m)}$ is an open compact submodule of Λ isomorphic with X. Therefore the map $u \to \omega_n u$ induces a continuous injective endomorphism of $\Lambda^{(m)}$ into itself. However, since $\Lambda^{(m)} = \omega_m \Lambda^{(m+1)} = \omega_n(\nu_{n,m} \Lambda^{(m+1)})$ for $m \geqq n$, the map $u \to \omega_n u$ is also surjective on Λ. It is then easy to see that the map $\omega_n u \to u$ is also continuous. Hence Λ has the property (iii) and the existence of a locally compact abelian group Λ is proved.

To prove the uniqueness, let Λ' be any other locally compact abelian group satisfying (i), (ii), (iii). Let u be any element in Λ. Since Λ/X is a discrete Γ-module, $\omega_n u$ is contained in X for some $n \geqq 0$. But, by (iii), Λ' contains an element u' such that $\omega_n u = \omega_n u'$, and we see immediately that such u' is uniquely determined by u and that $u \to u'$ is a Γ-isomorphism of the locally compact Γ-group Λ onto the locally compact Γ-group Λ' leaving every element of X fixed. Therefore the lemma is proved.

We now consider the discrete module $B = \Lambda/X$. Clearly $B_n = \omega_n^{-1}X/X$ for every $n \geqq 0$, and we have the following commutative diagram:

$$\begin{array}{ccc} B_n & \longrightarrow & B_m \\ \downarrow{\scriptstyle \omega_n} & & \downarrow{\scriptstyle \omega_m} \\ X/\omega_n X & \xrightarrow{\nu_{n,m}} & X/\omega_m X \end{array}, \qquad m \geqq n \geqq 0,$$

where $B_n \to B_m$ is the natural injection of B_n into B_m. It then follows that $B = \Lambda/X$ is nothing but the adjoint of the regular strictly Γ-finite module A. Such a construction of the adjoint B of A has the advantage that we need not use the submodules A_n of A to define the adjoint.

Now, let κ be a homomorphism of Γ into the group U of p-adic units as considered in 4.1. Let A' be a κ-isomorphic image of the regular strictly Γ-finite module A and B' a κ-isomorphic image of the adjoint B of A. By Theorem 4.4, both A' and B' are Γ-finite modules. Using the above construction of B, we shall prove the following

THEOREM 5.7. *Let A, B, A' and B' be as in the above. If A' is strictly Γ-finite, then B' is also strictly Γ-finite, and A' and B' are the adjoints of each other. In particular, both A' and B' are also regular modules.*

PROOF. Let Λ/X as in the above and let M be a locally compact abelian group dual to Λ. We denote by (u, v) a dual pairing of Λ and M. As in 1.5, we can define the action of Γ on M so that $(\sigma u, v) = (u, \sigma v)$ for any σ in Γ, u in Λ and v in M. Let Y be the annihilator of X in M with respect to the pairing (u, v). Since X is an open compact submodule of Λ, Y is an open compact submodule of M. Furthermore, X is dual to M/Y, and Λ/X is dual to Y. Hence we may put $A = \mathrm{M}/Y$, and we also see easily that M is a locally compact Γ-group associated with $B = \Lambda/X$ in the sense of Lemma 5.6.

We now define a new Γ-structure on the locally compact abelian group Λ. Namely, we define a new product $\sigma \circ u$ of σ in Γ and u in Λ by

$$\sigma \circ u = \kappa(\sigma)\sigma^{-1}u \ .$$

Let Λ' denote the Γ-group defined in this manner on the abelian group Λ and let X' be the submodule of Λ' corresponding to X. Then the mapping $u \to u$ of Λ onto Λ' induces a κ-isomorphism of Λ/X onto Λ'/X', and hence we may put $B' = \Lambda'/X'$. Similarly, Γ-groups M' and Y' are defined from M and Y, and we have $A' = \mathrm{M}'/Y'$. The pairing (u, v) of Λ and M then also gives a dual pairing of Λ' and M' such that

$$(\sigma \circ u, v) = (u, \sigma \circ v) \ , \qquad \sigma \in \Gamma.$$

It follows in particular that X' is dual to A' and Y' is dual to B'. By the assumption, A' is strictly Γ-finite. Hence, by Theorem 4.5, A' is also regular. To prove the theorem, it is therefore sufficient to show that Λ' is a locally compact abelian group associated with the regular strictly Γ-finite module A' in the sense of Lemma 5.6, namely, that Λ' has the properties (i), (ii), (iii) in the lemma with respect to the submodule X'. Since (i) and (ii) are easily verified for Λ', we shall prove that $u \to \omega_n \circ u$ is an automorphism of Λ' for every $n \geqq 0$. Let u be an element of Λ' such that $\omega_n \circ u = 0$. Since u is an element in Λ, $\omega_m u$ is contained in X for a sufficiently large

m. Then $\omega_n \circ (\omega_m u) = \omega_m(\omega_n \circ u) = 0$. But, since X' is regular and $\omega_m u$ is an element of X', $\omega_n \circ (\omega_m u) = 0$ implies $\omega_m u = 0$. It then follows from the property (iii) for Λ that $u = 0$. Therefore $u \to \omega_n \circ u$ is injective. On the other hand, since X' is strictly Γ-finite, $X'/(\omega_n \circ X')$ is a finite module. As a subset of Λ, $\omega_n \circ X' = \omega_n \circ X$ is a closed submodule of Λ and, since $X/(\omega_n \circ X)$ is finite, $\Lambda/(\omega_n \circ X)$ is a discrete Γ-finite factor module of Λ. Let u be any element of Λ'. Considering u as an element in Λ, it follows from the above that $\omega_m u$ is contained in $\omega_n \circ X$ if m is sufficiently large. Hence $\omega_m u = \omega_n \circ x$ with x in X and, consequently, $u = \omega_n \circ (\omega_m^{-1} x)$ with $\omega_m^{-1} x$ in Λ'. Hence $u \to \omega_n \circ u$ is surjective. Finally, it is easy to see that $u \to \omega_n \circ u$ is a homeomorphism of Λ'. Thus the property (iii) is also verified for Λ', and the lemma is completely proved.

REMARK. The locally compact abelian groups Λ and M constructed in the above may be used to obtain some of our previous results as well as further information on adjoints of regular strictly Γ-finite modules. But we do not discuss this approach here.

Now, if a κ-isomorphic image A' of a regular strictly Γ-finite module A is strictly Γ-finite, we call the adjoint B' of A' a κ-adjoint of A. If A has such a κ-adjoint B', B' is uniquely determined by A up to isomorphism and, by Theorem 5.7, it is also a κ-isomorphic image of the adjoint of A. B' is a regular strictly Γ-finite module and A is, conversely, a κ-adjoint of B'.

THEOREM 5.8. *Let A be a regular strictly Γ-finite module with characteristic roots $\chi_1, \cdots, \chi_\lambda$, and let A have a κ-adjoint B'. Then A and B' have the same invariants $\lambda, \mu, \mu_1, \cdots, \mu_s$, and the characteristic roots of B' are given by $\kappa\chi_1^{-1}, \cdots, \kappa\chi_\lambda^{-1}$.*

This theorem is an immediate consequence of Theorems 4.4 and 5.5. Actually, we can obtain more precise information on the maximal elementary submodules and the maximal divisible submodules of A and B', but we omit the details here.

MASSACHUSETTS INSTITUTE OF TECHNOLOGY

[37] On the theory of cyclotomic fields

Ann. of Math., (2) 70 (1959), 530-561.

(Received March 9, 1959)

The theory of cyclotomic fields is in a unique position in algebraic number theory. On the one hand, it has provided us a typical example of algebraic number fields from which we have been able to develop the theory of algebraic number fields in general; and on the other hand, it has also revealed to us many beautiful properties of the cyclotomic fields which are proper to these fields and which give us deep insight into important arithmetic results in elementary number theory.

Such a twofold character of the theory of cyclotomic fields will be seen also in the present paper. In a previous publication, we have introduced the notion of Γ-extensions of algebraic number fields.[1] In the present paper, we shall consider such a Γ-extension in the special case where the extension is obtained by a sequence of cyclotomic fields; and using the special features of these fields, we shall study in detail those arithmetic properties of Γ-extensions which are proper to such a cyclotomic case, hoping at the same time that our study will also eventually lead us to obtain deeper results on the structure of Γ-extensions in general.

An outline of the paper is as follows. We shall first give some preliminaries in § 1 – § 4. We then introduce fundamental field extensions L/F, K/F, and M/F in § 5, § 6, and § 7, respectively. We shall prove our main results in § 8 – § 10; and finally, in § 11, we discuss briefly the local cyclotomic fields. Throughout the paper, our proofs will be based upon classical class field theory and the theory of Γ-finite modules developed in our previous papers.

1. Galois groups

In general, for any Galois extension K of a field F, we shall denote the Galois group of K/F by $G(K/F)$. $G(K/F)$ will always be considered as a compact topological group with respect to its Krull topology. If E is any field between F and K, $G(K/E)$ is a closed subgroup of $G(K/F)$; if, in particular, E/F is also a Galois extension, then $G(K/E)$ is normal in $G(K/F)$ and the Galois group $G(E/F)$ can be canonically identified with the factor group $G(K/F)/G(K/E)$. If $G(K/E)$ is furthermore an abelian group, then every inner automorphism $x \to s^{-1}xs$ of $G(K/F)$ induces an automorphism

* The present research was supported in part by a National Science Foundation grant.

[1] Cf. [6]. For the bibliography, see the end of the paper.

$x \to x^{\sigma}$ of the normal subgroup $G(K/E)$ which depends only upon the coset σ of s mod $G(K/E)$, and the abelian group $G(K/E)$ becomes a $G(E/F)$-group on which $G(E/F)$ acts unitarily and continuously.[2]

In case K/F is an abelian extension, $G(K/F)$ is a compact abelian group and we shall denote the character group of $G(K/F)$ by $A(K/F)$; $A(K/F)$ is then a discrete abelian group and, in most cases, we shall consider it as an additive group. Since $G(E/F) = G(K/F)/G(K/E)$, $A(E/F)$ can be naturally imbedded as a subgroup in $A(K/F)$, and $A(K/E)$ can be identified with $A(K/F)/A(E/F)$. By Galois theory, $E \to A(E/F)$ then defines a one-one correspondence between subextensions of K/F and subgroups of $A(K/F)$.

We now consider the case where $G(K/E)$ is an abelian normal subgroup of $G(K/F)$ but $G(K/F)$ itself is not necessarily abelian. As stated in the above, $G(K/E)$ is then a compact abelian $G(E/F)$-group. For any σ in $G(E/F)$ and for any character $a = a(x)$ in $A(K/E)$ ($x \in G(K/E)$), we define σa in $A(K/E)$ by

$$(\sigma a)(x) = a(x\sigma) = a(s^{-1}xs)\,, \qquad x \in G(K/E),$$

where s is any element of $G(K/F)$ such that σ is the coset of s mod $G(K/E)$. Then $A(K/E)$ becomes a discrete $G(E/F)$-module on which $G(E/F)$ acts unitarily and continuously, and

$$(a, x) = a(x)\,, \quad a \in A(K/E),\, x \in G(K/E),$$

defines a so-called $G(E/F)$-pairing of $A(K/E)$ and $G(K/E)$.[3]

2. Cyclotomic extensions

Let p be an odd prime number which we shall fix once and for all in the following. For each $n \geqq 0$, let ζ_n denote a primitive $p^{n\text{th}}$ root of unity and let F_n be the cyclotomic field obtained by adjoining ζ_{n+1} to the rational field Q: $F_n = Q(\zeta_{n+1})$. In the following, these fields $F_n (n \geqq 0)$ and all other algebraic extensions of the rational field Q will be supposed to be contained in a fixed algebraic closure of Q. Since every F_n/Q is a cyclic extension, the union F of the increasing sequence of extensions F_n/Q, $n \geqq 0$, gives an abelian extension of Q and we denote the Galois group of the extension F/Q by G: $G = G(F/Q)$.

Let Ξ denote the multiplicative group of all p-adic units, and let Ξ_n ($n \geqq 0$) be the subgroup of all ξ in Ξ such that $\xi \equiv 1 \bmod p^{n+1}$. Then

---

[2] We write $x\sigma$ instead of x^{σ} when $G(K/E)$ is considered as an additive group. Notice that $x^{\sigma\tau} = (x^{\sigma})^{\tau}$.

[3] Cf. [7], 1.5.

there exists a canonical topological isomorphism κ of G onto Ξ such that

$$\zeta_n^\sigma = \zeta_n^{\kappa(\sigma)}, \qquad \sigma \in G,$$

for every $n \geqq 0$. It is clear from the above that κ maps the Galois group $G(F/F_n)$ onto the subgroup Ξ_n ($n \geqq 0$). In particular, $G(F/F_0)$ is isomorphic with Ξ_0 and, hence, with the compact group Γ considered in [6], [7]. Therefore, we simply put

$$\Gamma = G(F/F_0) .$$

Using the same notation as in [6], [7], let Γ_n ($n \geqq 0$) denote the unique closed subgroup of index p^n in Γ. Then we have

$$\Gamma_n = G(F/F_n) = \kappa^{-1}(\Xi_n) , \qquad n \geqq 0.$$

We also take an element γ in Γ which is not in Γ_1, and put

$$\omega_n = 1 - \gamma^{p^n}, \quad \nu_{n,m} = \sum_{i=0}^{p^{m-n}-1} \gamma^{ip^n} , \qquad m \geqq n \geqq 0.$$

Here ω_n and $\nu_{n,m}$ are elements of the group ring of Γ over the ring of rational integers.

Now, the group Ξ is the direct product of Ξ_0 and a cyclic group of order $p-1$, namely, the group H of all roots of unity in the p-adic number field Q_p. Hence, if we put $\Delta = \kappa^{-1}(\mathrm{H})$, then

$$G = \Gamma \times \Delta .$$

Clearly, Δ is canonically isomorphic with $G(F_0/Q)$; and, for any η in H, we denote the element $\kappa^{-1}(\eta)$ of Δ by δ_η: $\kappa(\delta_\eta) = \eta$, $\eta \in \mathrm{H}$.

3. Δ-decompositions

Let A be an additive discrete p-primary abelian group and suppose that $G = \Gamma \times \Delta$ acts on A unitarily and continuously. Since A is a discrete p-primary abelian group, the product ξa of a p-adic integer ξ and an element a in A is defined as usual, and the ring O_p of p-adic integers may be considered as acting on A. As the action of G on A obviously commutes with the action of O_p on A, we may also consider the group ring of G over O_p as an operator domain of A.

For any integer i, $0 \leqq i \leqq p-2$, let ε_i denote an element of the group ring of Δ over O_p defined by

$$\varepsilon_i = \frac{1}{p-1} \sum_\eta \eta^{-i} \delta_\eta , \qquad \eta \in \mathrm{H}.$$

Then

$$\sum_{i=0}^{p-2} \varepsilon_i = 1, \qquad \varepsilon_i^2 = \varepsilon_i, \qquad \varepsilon_i\varepsilon_j = 0 \ (i \neq j) .$$

Hence, if we put ${}^iA = \varepsilon_i A$, $0 \leqq i \leqq p-2$, then we have the direct decomposition:

$$A = {}^0A + \cdots + {}^{(p-1)}A .$$

Here each ${}^iA$ is also characterized as the submodule of all a in A satisfying $\delta_\eta a = \eta^i a$ for every δ_η in Δ. We shall call such a direct decomposition of A the *Δ-decomposition* of the G-module A.

If B is any G-submodule of A and if $B = \sum_{i=0}^{p-2} {}^iB$ and $A/B = \sum_{i=0}^{p-2} {}^i(A/B)$ are the Δ-decompositions of B and A/B respectively, then we see easily that

$${}^iB = {}^iA \cap B, \quad {}^i(A/B) = ({}^iA + B)/B \cong {}^iA/{}^iB , \qquad 0 \leqq i \leqq p-2.$$

Now, the group A and each direct summand ${}^iA$ of the Δ-decomposition of A are also Γ-modules as considered in [6], [7], and A is (strictly) Γ-finite if and only if every ${}^iA$ $(0 \leqq i \leqq p-2)$ is (strictly) Γ-finite. If A is Γ-finite, it is clear that

$$\lambda(A) = \sum_{i=0}^{p-2} \lambda({}^iA), \qquad \mu(A) = \sum_{i=0}^{p-2} \mu({}^iA) ,$$

for the invariants of A and ${}^iA$. Furthermore, the set of the numerical invariants μ_j for A is the union of the sets of such invariants for ${}^iA$ $(0 \leqq i \leqq p-2)$, and the same is also true for the sets of characteristic roots of A and ${}^iA$.

Let X be an additive compact p-primary abelian group and suppose that G acts on X unitarily and continuously. Then we have again a direct decomposition of X:

$$X = {}^0X + \cdots + {}^{(p-2)}X ,$$

where ${}^iX = X\varepsilon_i$ is the submodule of all x in X satisfying $x\delta_\eta = \eta^i x$ for every δ_η in Δ. Such a decomposition will also be called the *Δ-decomposition* of X.

Suppose that X is dual to a discrete G-module A and that (a, x), $a \in A$, $x \in X$ is a G-pairing of A and X, namely, a dual pairing of A and X satisfying $(\sigma a, x) = (a, x\sigma)$ for any σ in G. We have then $({}^iA, {}^jX) = 0$ for $i \neq j$ and it follows that ${}^iA$ and ${}^iX$ $(0 \leqq i \leqq p-2)$ are dual to each other and that (a, x) induces a G-pairing of these G-modules.

4. A Galois extension Φ/Q

Let Φ be a Galois extension of Q containing F such that Φ is a p-primary abelian extension of F. Then $G(\Phi/F)$ is a closed normal subgroup of

$G(\Phi/Q)$ and $G(\Phi/Q)/G(\Phi/F) = G$. As explained in general in § 1, the Galois group $G(\Phi/F)$ and its character group $A(\Phi/F)$ can be then considered as G-modules and $(a, x) = a(x)$ gives a G-pairing of these modules.

For each $n \geqq 0$, let Φ_n denote the maximal abelian extension of F_n contained in Φ. Clearly, F is contained in Φ_n: $F \subset \Phi_n \subset \Phi$. By the definition of Φ_n, $G(\Phi/\Phi_n)$ is the topological commutator group of $G(\Phi/F_n)$. Using the fact that γ^{p^n} generates a subgroup everywhere dense in $\Gamma_n = G(F/F_n)$, we then see that $G(\Phi/\Phi_n)$ is equal to $G(\Phi/F)^{\omega_n}$ and that $A(\Phi_n/F)$ coincides with the submodule $A(\Phi/F)_n = \omega_n A(\Phi/F)$ of $A(\Phi/F)$:

$$A(\Phi_n/F) = A(\Phi/F)_n\,, \qquad n \geqq 0.$$

Since Φ/Q and F_n/Q are both Galois extensions, Φ_n/Q is also a Galois extension and $G(\Phi_n/F_n)$ is an abelian normal subgroup of $G(\Phi_n/Q)$. Hence $G(F_n/Q) = G/\Gamma_n$ may be considered as acting on both $G(\Phi_n/F_n)$ and $A(\Phi_n/F_n)$, and we may consider these groups also as G-modules in a natural way. Since $G(\Phi/F_0)$ is a p-primary compact group, $G(\Phi_n/F_n)$ and $A(\Phi_n/F_n)$ are p-primary abelian groups.

Let

$$A(\Phi/F) = \sum_{i=0}^{p-2} {}^iA(\Phi/F)$$

be the Δ-decomposition of the G-module $A(\Phi/F)$. Since $A(\Phi_n/F) = A(\Phi/F)_n$, the ith component ${}^iA(\Phi_n/F)$ of the Δ-decomposition of $A(\Phi_n/F)$ is then given by

$${}^iA(\Phi_n/F) = {}^iA(\Phi/F)_n = {}^iA(\Phi/F) \cap A(\Phi/F)_n\,.$$

Let

$$A(\Phi_n/F_n) = \sum_{i=0}^{p-2} {}^iA(\Phi_n/F_n)$$

be the Δ-decomposition of the G-module $A(\Phi_n/F_n)$ and let ${}^i\Phi_n$ be the field between F_n and Φ_n such that

$$A({}^i\Phi_n/F_n) = {}^iA(\Phi_n/F_n)\,, \qquad 0 \leqq i \leqq p-2.$$

Since $A(\Phi_n/F) = A(\Phi_n/F_n)/A(F/F_n)$, it follows from § 3 that ${}^iA(\Phi_n/F)$ is naturally isomorphic with ${}^iA(\Phi_n/F_n)/{}^iA(F/F_n) = A({}^i\Phi_n/F_n)/{}^iA(F/F_n)$. However, since G acts trivially on $G(F/F_n)$, $A(F/F_n)$ coincides with ${}^0A(F/F_n)$. Hence F is contained in ${}^0\Phi_n$ and $F \cap {}^i\Phi_n = F_n$ for $i \neq 0$. We then see that

$${}^0A(\Phi/F)_n = {}^0A(\Phi_n/F) = A({}^0\Phi_n/F) = A({}^0\Phi_n/F_n)/A(F/F_n) = {}^0A(\Phi_n/F_n)/A(F/F_n)\,,$$

$${}^iA(\Phi/F)_n = {}^iA(\Phi_n/F) = A({}^i\Phi_n F/F) \cong A({}^i\Phi_n/F_n) = {}^iA(\Phi_n/F_n)\,, \qquad i \neq 0.$$

As $A(F/F_n)$ is a p-primary abelian group of rank 1, it follows in particular that the module ${}^iA(\Phi/F)$ $(0 \leqq i \leqq p-2)$ is Γ-finite if and only if ${}^iA(\Phi_n/F_n)$ has finite rank for every $n \geqq 0$ and, consequently, that $A(\Phi/F)$ is Γ-finite if and only if $A(\Phi_n/F_n)$ has finite rank for every $n \geqq 0$.

5. The extension L/F

Let L be the maximal unramified p-primary abelian extension of F.[4] It is clear from the definition that L is a Galois extension of Q, and we may apply the above result to the extension L/Q. Let L_n $(n \geqq 0)$ be the maximal abelian extension of F_n contained in L and let L'_n be the maximal unramified p-primary abelian extension of F_n. Then L'_n is obviously contained in L_n. Let $\mathfrak{p}_n$ be the unique prime divisor of F_n dividing the rational prime p. Since no prime divisor of F_n different from $\mathfrak{p}_n$ is ramified in L_n, L'_n coincides with the inertia field of $\mathfrak{p}_n$ in L_n (i. e., the inertia field of any extension of $\mathfrak{p}_n$ in the abelian extension L_n of F_n). Using the fact that $\mathfrak{p}_n$ is fully ramified in F_m for any $m \geqq n$, it then follows that

$$L_n = FL'_n, \quad F_n = F \cap L'_n, \quad G(L_n/F) \cong G(L'_n/F_n) .$$

By class field theory, $G(L'_n/F_n)$ is a finite group isomorphic with the Sylow p-group of the ideal class group of F_n.[5] Hence $G(L_n/F)$ and $A(L_n/F) = A(L/F)_n$ are also finite for every $n \geqq 0$, and we see that $A(L/F)$ is strictly Γ-finite.

Let

$$A(L/F) = \sum_{i=0}^{p-2} {}^iA(L/F)$$

be the Δ-decomposition of $A(L/F)$. Then, by the above, every ${}^iA(L/F)$ is strictly Γ-finite. For simplicity, we put

$$\lambda = \lambda(A(L/F)) , \qquad \mu = \mu(A(L/F)) ,$$
$${}^i\lambda = \lambda({}^iA(L/F)) , \qquad {}^i\mu = \mu({}^iA(L/F)) , \qquad 0 \leqq i \leqq p-2.$$

Then we have

$$\lambda = \sum_{i=0}^{p-2} {}^i\lambda , \qquad \mu = \sum_{i=0}^{p-2} {}^i\mu .$$

We shall next give arithmetic interpretations for the invariants ${}^i\lambda$ and ${}^i\mu$. Since L'_n/Q is a Galois extension, $G(F_n/Q) = G/\Gamma_n$ acts on the finite abelian group $G(L'_n/F_n)$ as explained in § 1. Hence $G(L'_n/F_n)$ may be considered also as a G-module in a natural way. Then the isomorphism

[4] Cf. [6], § 6.

[5] For class field theory used here and in the following, cf. [2], [3].

$G(L_n/F) \cong G(L'_n/F_n)$ becomes a G-isomorphism, and it induces a G-isomorphism on their character groups:

$$A(L_n/F) \cong A(L'_n/F_n) .$$

Let

$$A(L'_n/F_n) = \sum_{i=0}^{p-2} {}^iA(L'_n/F_n)$$

be the Δ-decomposition of $A(L'_n/F_n)$. Then the above isomorphism induces the isomorphisms:

$${}^iA(L_n/F) \cong {}^iA(L'_n/F_n) , \qquad 0 \leqq i \leqq p-2.$$

Let $S(F_n)$ denote the Sylow p-group of the ideal-class group of F_n $(n \geqq 0)$. The Galois group $G(F_n/Q) = G/\Gamma_n$ acts on $S(F_n)$ in an obvious way, and we may consider $S(F_n)$ also as a G-group. Let

$$S(F_n) = \prod_{i=0}^{p-2} {}^iS(F_n)$$

be the Δ-decomposition of $S(F_n)$. The canonical isomorphism $S(F_n) \cong G(L'_n/F_n)$ then induces isomorphisms between the corresponding direct factors of the Δ-decompositions of these groups, and it follows from the above that ${}^iS(F_n)$ is dual to ${}^iA(L_n/F) = {}^iA(L/F)_n$ in a natural way. In particular, ${}^iS(F_n)$ has the same order as ${}^iA(L/F)_n$. But it is known that, for every sufficiently large n, the exponent of p in the order of ${}^iA(L/F)_n$ is given by a formula:

$${}^i\lambda n + {}^i\mu p^n + {}^i\nu ,$$

where ${}^i\lambda$ and ${}^i\mu$ are the invariants of ${}^iA(L/F)$ as defined in the above and ${}^i\nu$ is also an integer independent of n.[6] Hence we have the following:

THEOREM 1. *Let ${}^iS(F_n)$ $(0 \leqq i \leqq p-2)$ be the ith component of the Δ-decomposition of the Sylow p-group $S(F_n)$ of the ideal-class group of F_n $(n \geqq 0)$. Then, for every sufficiently large n, the exponent of p in the order of ${}^iS(F_n)$ is given by a formula:*

$${}^i\lambda n + {}^i\mu p^n + {}^i\nu ,$$

where ${}^i\lambda$ and ${}^i\mu$ are the invariants as defined in the above and ${}^i\nu$ is also an integer independent of n.

Now, let d be a divisor of $p-1$ and let $\Delta^{(d)}$ be the subgroup of Δ with order d. Let $F^{(d)}$ be the fixed field of $\Delta^{(d)}$ in F, and put $F_n^{(d)} = F_n \cap F^{(d)}$ for every $n \geqq 0$. Then $[F : F^{(d)}] = [F_n : F_n^{(d)}] = d$, and $F^{(d)}$ is the union of the increasing sequence of subfields $F_n^{(d)}$, $n \geqq 0$. Since $G(F^{(d)}/F_0^{(d)}) =$

[6] Cf. [6], § 5, Theorem 4.

$(\Gamma \times \Delta^{(d)})/\Delta^{(d)} \cong \Gamma$, $F^{(d)}$ is a so-called Γ-extension of the subfield $F_0^{(d)}$. For each $n \geqq 0$, let L_n'' denote the maximal unramified p-primary abelian extension of $F_n^{(d)}$. It is then easy to see that $F_n \cap L_n'' = F_n^{(d)}$ and that

$$A(L_n''/F_n^{(d)}) \cong A(F_n L_n''/F_n) = \sum_{d|i} {}^iA(L_n'/F_n) .$$

Since the order of $A(L_n''/F_n^{(d)})$ is equal to the highest power of p dividing the class number of $F_n^{(d)}$, we obtain from the above the following:

THEOREM 2. *Let d be any divisor of $p-1$ and let $F_n^{(d)}$ be the unique subfield of F_n with $[F_n : F_n^{(d)}] = d$ $(n \geqq 0)$. Then, for every sufficiently large n, the exponent of the highest power of p dividing the class number of $F_n^{(d)}$ is given by a formula:*

$$\lambda^{(d)} n + \mu^{(d)} p^n + \nu^{(d)} ,$$

where $\lambda^{(d)} = \lambda(F^{(d)}/F_0^{(d)})$ and $\mu^{(d)} = \mu(F^{(d)}/F_0^{(d)})$ are the invariants of the Γ-extension $F^{(d)}/F_0^{(d)}$ and are given by

$$\lambda^{(d)} = \sum_{d|i} {}^i\lambda , \qquad \mu^{(d)} = \sum_{d|i} {}^i\mu ,$$

and where $\nu^{(d)}$ is also an integer independent of n.

An important special case occurs when $d = 2$. Put $F^+ = F^{(2)}$, $F_n^+ = F_n^{(2)}$ and $\lambda^+ = \lambda^{(2)}$, $\mu^+ = \mu^{(2)}$. Then F^+ and F_n^+ are the maximal real subfields of F and F_n, respectively, and

$$\lambda^+ n + \mu^+ p^n + \nu^+$$

gives, for sufficiently large n, the exponent of the highest power of p dividing the class number of F_n^+, namely, the second factor of the class number of F_n.

6. The extension K/F

Let $\mathfrak{p}$ be the unique prime divisor of F dividing the rational prime p and let K be the maximal p-primary abelian extension of F such that no prime divisor of F different from $\mathfrak{p}$ is ramified in K.[7] Since $\mathfrak{p}$ is invariant under every automorphism of $G = G(F/Q)$, K/Q is a Galois extension, and the result in § 4 may be applied to the extension K/Q. Clearly, the unramified extension L of F is contained in K so that $A(L/F)$ is a G-submodule of $A(K/F)$. For each $n \geqq 0$, let K_n denote the maximal abelian extension of F_n contained in K. Then K_n is the maximal p-primary abelian extension of F_n such that no prime divisor of F_n different from $\mathfrak{p}_n$ is ramified in K_n. We shall next show that $A(K/F)$ is Γ-finite, proving that $A(K_n/F_n)$ has finite rank for every $n \geqq 0$.

---

[7] The existence of such a field K can be proved in a way similar to the proof of the existence of the field L in § 5. Cf. [6], § 6.

For $m \geqq 0$, let $K_{n,m}$ denote the intersection of K_n with the "Strahl" class field mod $\mathfrak{p}_n^{m+1}$ over F_n. Then $L'_n = K_{n,0}$, and K_n is the union of the increasing sequence of subfields $K_{n,m}$, $m \geqq 0$. By class field theory, $G(K_{n,m}/L'_n)$ is canonically isomorphic with the factor group of the group of principal ideals of F_n modulo the subgroup of principal ideals (a) such that $a \equiv 1 \bmod \mathfrak{p}_n^{m+1}$.[8] It follows that $G(K_n/L'_n)$ is canonically isomorphic with the inverse limit of such factor groups for $m \to \infty$, the limit being taken in an obvious manner. Let E_n be the group of units in F_n and let U_n be the group of $\mathfrak{p}_n$-adic units in the $\mathfrak{p}_n$-completion of F_n; U_n is a compact topological group in its $\mathfrak{p}_n$-adic topology, and we may consider E_n as a subgroup of U_n. The above inverse limit is then naturally isomorphic with the factor group of U_n modulo the closure $\bar{E}_n$ of E_n in U_n, and we have a canonical isomorphism[9]:

$$G(K_n/L'_n) \cong U_n/\bar{E}_n .$$

For $m \geqq 0$, let $U_{n,m}$ denote the subgroup of all ξ in U_n such that $\xi \equiv 1 \bmod \mathfrak{p}_n^{m+1}$, and let $E_{n,m} = E_n \cap U_{n,m}$. If r is any rational integer prime to p, then $\varepsilon = (1 - \zeta_{n+1}^r)/(1 - \zeta_{n+1})$ is a unit of F_n and $\varepsilon \equiv r \bmod \mathfrak{p}_n$. It follows that $U_n = E_n U_{n,0}$ and $G(K_n/L'_n) \cong U_n/\bar{E}_n \cong U_{n,0}/\bar{E}_{n,0}$, where $\bar{E}_{n,0}$ is the closure of $E_{n,0}$ in U_n. However, it is well known that $U_{n,0}$ is a compact p-primary abelian group of rank $p^n(p-1)+1$. Hence $G(K_n/L'_n)$ has finite rank. As $G(L'_n/F_n)$ is a finite group, $G(K_n/F_n)$ is then also a compact p-primary abelian group of finite rank. Therefore $A(K_n/F_n)$ has finite rank for every $n \geqq 0$, and the module $A(K/F)$ is Γ-finite.

Now, the Galois group of the $\mathfrak{p}_n$-adic completion of F_n over the p-adic number field Q_p may be canonically identified with $G(F_n/Q) = G/F_n$. Therefore we may consider G/F_n, and hence G, as acting on the group U_n. $U_{n,m}$, E_n, and $E_{n,m}$ are then invariant subgroups of U_n, and it follows from the above proof that we have a G-isomorphism:

$$G(K_n/L'_n) \cong U_{n,0}/\bar{E}_{n,0} .$$

We have then also Γ-isomorphisms between the direct factors of the Δ-decompositions of these groups:

$${}^iG(K_n/L'_n) \cong {}^iU_{n,0}/{}^i\bar{E}_{n,0} , \qquad 0 \leqq i \leqq p-2.$$

Let W_n denote the cyclic group of order p^{n+1} generated by ζ_{n+1}. Since

$$\zeta_{n+1}^{\delta_\eta} = \zeta_{n+1}^{\eta} , \qquad \eta \in \mathrm{H},$$

[8] Cf. [3]. Notice that every residue class of integers mod $\mathfrak{p}_n$ contains a unit. See the argument below.

[9] Of course, this can be seen also by considering the idèle group of F_n.

W_n is contained in ${}^1U_{n,0}$. It is then not difficult to prove that, as compact abelian groups,

$$ {}^1U_{n,0} \cong W_n \times \Gamma \times \cdots \times \Gamma, \qquad {}^iU_{n,0} \cong \Gamma \times \cdots \times \Gamma, \qquad i \neq 1,$$

where $\Gamma \times \cdots \times \Gamma$ is the direct product of p^n copies of Γ.[10] On the other hand, since E_n is the direct product of W_n and the group of units E_n^+ of the maximal real subfield F_n^+ of F_n, and since the complex conjugation δ_{-1} acts trivially on E_n^+, we see that

$$ {}^1\bar{E}_{n,0} = W_n , \qquad {}^j\bar{E}_{n,0} = 1,$$

for every odd index $j \neq 1$. Hence we have

$$ {}^jG(K_n/L'_n) \cong \Gamma \times \cdots \times \Gamma , \qquad (p^n \text{ copies of } \Gamma),$$

for any odd index j.

7. The extension M/F

As in the above, let E_n denote the group of units of the cyclotomic field F_n $(n \geqq 0)$. Then E_n is contained in E_{n+1}, and the union E of all E_n $(n \geqq 0)$ is the group of all units in the field F. The Galois group $G = G(F/Q)$ acts on E in an obvious way, and E_n is the subgroup of all ε in E such that $\varepsilon^\sigma = \varepsilon$ for every σ in Γ_n.

Now, let M be the extension of F obtained by adjoining all $p^{n\text{th}}$ roots, $n = 1, 2, \cdots$, of all units in E. Let ε be any element in E and let n be any integer. Choose an integer $m \geqq n-1$ such that ε is contained in E_m. Then $F_m(\varepsilon^{p^{-n}})$ is a cyclic extension of F_n in which no prime divisor of F_m different from $\mathfrak{p}_m$ is ramified. Therefore the extension is contained in K_m and, hence, $\varepsilon^{p^{-n}}$ is contained in K. It follows that M is a subfield of K. It is clear that M is a Galois extension of Q. Therefore we are again in the situation that we have considered in § 4. Since $A(M/F)$ is a submodule of the Γ-finite module $A(K/F)$, it is also Γ-finite.

For each $n \geqq 0$, let M_n be the maximal abelian extension of F_n contained in M, and let M'_n be the subextension of M_n such that $A(M'_n/F_n)$ is the submodule of all u in $A(M_n/F_n)$ satisfying $p^n u = 0$. Then, both M_n/Q and M'_n/Q are Galois extensions, and since F_n contains a primitive $p^{n\text{th}}$ root of unity, M'_n/F_n is a Kummer extension. With n fixed, let T denote the multiplicative group of all non-zero elements in F_n whose $p^{n\text{th}}$ roots are contained in M'_n. Let a be any element in T and α a $p^{n\text{th}}$ root of a in M'_n. For any x in $G(M'_n/F_n)$, put

[10] Cf., § 11 below.

$$\chi_a(x) = \alpha^{1-x}\,.$$

By the theory of Kummer extensions, χ_a then defines a character of $G(M'_n/F_n)$ depending only upon a; and the mapping φ: $a \to \chi_a$ gives a homomorphism of T onto $A(M'_n/F_n)$ whose kernel N is the group of all p^{n}th powers of non-zero elements in F_n. For any σ in G, choose an element s in $G(M'_n/Q)$ which induces on F_n the same automorphism as σ does. We have then:

$$\chi_{a^\sigma}(x) = (\alpha^s)^{1-x} = (\alpha^{1-sxs^{-1}})^s = \chi_a(sxs^{-1})^s = \chi_a^{\sigma^{-1}}(x)^\sigma = (\chi_a^{\sigma^{-1}}(x))^{\kappa(\sigma)}\,,$$

namely,

$$\varphi(a^\sigma) = \varphi(a)^{\kappa(\sigma)\sigma^{-1}} \qquad \sigma \in G.$$

Hence φ induces a so-called *κ-isomorphism* of T/N onto $A(M'_n/F_n)$ when these groups are considered as G-groups in an obvious way.[11] Since $\kappa(\delta_\eta)=\eta$, the above κ-isomorphism maps each ${}^i(T/N)$ onto ${}^jA(M'_n/F_n)$ where $i+j\equiv 1$ mod $p-1$. Here ${}^i(T/N)$ and ${}^jA(M'_n/F_n)$ are of course the components of the Δ-decompositions of (T/N) and $A(M'_n/F_n)$, respectively. We shall next consider the structure of ${}^i(T/N)$.

Let a and α be as in the above. Since $F_n(\alpha)$ is contained in M'_n (and, hence, in K_n), no prime ideal of F_n different from $\mathfrak{p}_n$ is ramified in $F_n(\alpha)$. Hence the principal ideal (a) can be uniquely written in the form:

$$(a) = \mathfrak{p}_n^t \mathfrak{a}^{p^n}\,,$$

where $\mathfrak{a}$ is an ideal of F_n prime to $\mathfrak{p}_n$. Let $c(\mathfrak{a})$ denote the ideal-class of $\mathfrak{a}$. Then the map: $a \to c(\mathfrak{a})$ is a homomorphism of T into the ideal-class group of F_n, and we denote the image by C and the kernel by T'. N is then contained in T' and $T/T' \cong C$.

Now, $F(\alpha)/F$ is a subextension of M/F and has degree dividing p^n. Hence, by the definition of M, there exists a unit ε in E such that $F(\alpha) = F(\varepsilon^{p^{-n}})$ and that $\alpha = \varepsilon^{p^{-n}}b$ with b in F. Let $m \geqq 2n$ and let F_m contain both ε and b. Then $a = \varepsilon b^{p^n}$, and hence $\mathfrak{p}_n^t\mathfrak{a}^{p^n} = (a) = (b)^{p^n}$. Since $\mathfrak{p}_n = \mathfrak{p}_m^{p^{m-n}}$ and $\mathfrak{p}_m$ is a principal ideal of F_m, we see that $\mathfrak{a}$ becomes a principal ideal in F_m. Since $p^nA(M'_n/F_n) = 0$ and $T/N \cong A(M'_n/F_n)$, we also have $c(\mathfrak{a})^{p^n} = 1$. Suppose, conversely, that c is an ideal-class of F_n such that $c^{p^n} = 1$ and such that an ideal $\mathfrak{a}$ of the class c (hence, every ideal of the class c) becomes a principal ideal in some F_m, $m \geqq n$. Since $\mathfrak{p}_n = (\pi_n)$, $\pi_n = 1 - \zeta_{n+1}$, we may assume that $\mathfrak{a}$ is prime to $\mathfrak{p}_n$. Let $\mathfrak{a}^{p^n} = (a)$, $a \in F_n$ and let $\mathfrak{a} = (b)$, $b \in F_m$. Then $a = \varepsilon b^{p^n}$ with a unit ε in F_m, and $\alpha = \varepsilon^{p^{-n}} b$

[11] Cf. [7], § 4.

is a $p^{n\text{th}}$ root of a. It follows that $F(\alpha)=F(\varepsilon^{p^{-n}})$ is a subextension of M/F, that $F_n(\alpha)$ is contained in M'_n, and that $c=c(a)$ is the ideal-class of F_n associated with the element a in T under the above homomorphism. It is thus proved that C is the group of all ideal-classes c of F_n such that $c^{p^n}=1$ and such that every ideal of the class c becomes principal in some F_m, $m \geqq n$.

Assume next that a is an element of T'. Then, by the definition, $(a)=\mathfrak{p}_n^i(b)^{p^n}$ with b in F_n; and

$$a=\varepsilon\pi_n^i b^{p^n},$$

with a unit ε in E_n. Let Π' denote the multiplicative group generated by π_n and E_n. Since $\pi_n^{1-\sigma}$ is a unit in E_n for any σ in G, Π' is invariant under G. Since $\pi_n\pi_{2n}^{-p^n}$ is also a unit in E_{2n}, $p^{n\text{th}}$ roots of π_n are contained in M'_n. It follows that π_n is in T', and Π' is a G-subgroup of T'. By the above, $T'=\Pi'N$. Hence, we have

$$T'/N \cong \Pi'/N',$$

where $N'=\Pi'\cap N$. It is easy to see that Π' is the direct product of E_n and the cyclic group $\{\pi_n\}$ generated by π_n and that N' is the group of all $p^{n\text{th}}$ powers of elements in Π'.

As before, let F_n^+ be the maximal real subfield of F_n and let E_n^+ be the group of units of F_n^+. Then E_n^+ is the direct product of a free abelian group of rank $((p-1)p^n)/2-1$ and a cyclic group of order 2 generated by -1, and E_n is the direct product of E_n^+ and the cyclic group W_n of order p^{n+1} generated by ζ_{n+1}. Hence $\Pi'=\{\pi_n\}\times E_n^+\times W_n$, and it follows that Π'/N' is an abelian group of type $(p^n,\cdots,p^n)$ with rank $((p-1)p^n)/2+1$. Let $\pi_n^+=\pi_n\bar{\pi}_n$, where $\bar{\pi}_n$ is the complex conjugate of π_n, and let Π denote the subgroup of Π' generated by the conjugates $(\pi_n^+)^\sigma$ of π_n^+ $(\sigma\in G)$. It is known that $\Pi\cap E_n^+$ has finite index in E_n^+.[12] Hence Π contains a free abelian group of rank $((p-1)p^n)/2$. Let R be the group ring of $G(F_n^+/Q)$ over the ring of rational integers. Since the additive group of R has rank $((p-1)p^n)/2$, it follows from the above that the map: $r\to(\pi_n^+)^r$, $r\in R$, defines a $G(F_n^+/Q)$-isomorphism of the additive group of R onto Π. Since Π' is the direct product of a finite group and a free abelian group of rank $((p-1)p^n)/2$, we then see that Π'/Π is finite and, hence, that $\Pi\cap(\Pi')^{p^m}$ is contained in Π^p for some $m\geqq n$, Π^p and $(\Pi')^{p^m}$ being here the groups of pth or $p^{m\text{th}}$ powers of elements in Π or in Π', respectively.[13] Let $\Lambda_1=\Pi\Pi'^{p^m}$, $\Lambda_2=\Pi^p\Pi'^{p^m}$. Then $\Lambda_1/\Lambda_2\cong\Pi/\Pi^p\cong R/pR$ as G-groups. We consider the components of the Δ-decompositions of various

[12] Cf., the well-known class number formula for F_n.

[13] Thus, Π^p, for instance, is not the direct product of p copies of Π.

G-groups. First, it is easy to see that ${}^{i}(R/pR)$ is an abelian group of type $(p, \cdots, p)$ with rank p^n for every even index i. Hence the same is true for ${}^{i}(\Lambda_1/\Lambda_2)$. Since $\Pi'^{p^m} \subset \Lambda_2 \subset \Lambda_1 \subset \Pi'$, it follows that the rank of ${}^{i}(\Pi'/\Pi'^{p^m})$ is at least p^n. However, Π'^{p^m} is contained in $N' = \Pi'^{p^n}$, and ${}^{i}(\Pi'/N') \cong {}^{i}(\Pi'/\Pi'^{p^m})/{}^{i}(\Pi'/\Pi'^{p^m})^{p^n}$. Therefore the rank of ${}^{i}(\Pi'/N')$ is also at least p^n for every even i. On the other hand, the existence of the subgroup W_n implies that the rank of ${}^{1}(\Pi'/N')$ is at least 1. Since Π'/N' is an abelian group of type $(p^n, \cdots, p^n)$ with rank $((p-1)p^n)/2+1$, it follows that ${}^{i}(\Pi'/N')$ is an abelian group of type $(p^n, \cdots, p^n)$ with rank p^n for every even index i, that ${}^{1}(\Pi'/N')$ is a cyclic group of order p^n, and that ${}^{i}(\Pi'/N')$ is trivial for every odd index $i \neq 1$. As T'/N is naturally isomorphic with Π'/N', we also obtain a similar result for ${}^{i}(T'/N)$. It follows in particular that the rank of ${}^{i}(T/N)$ is at least p^n for every even index p^n, and hence that the rank of ${}^{j}A(M'_n/F_n)$ is also at least p^n for every odd index j.

We now consider the structure of $A(M/F)$. Let χ be any character in $A(M/F)$ and let $\chi^{p^m} = 1$, $m \geqq 0$. Since M/F is a Kummer extension, there exists an element ε in E and a $p^{m\text{ th}}$ root of ε, say α, in M such that

$$\chi(x) = \alpha^{1-x} ,$$

for every x in $G(M/K)$. If χ' is the character in $A(M/F)$ defined by a $p^{m+1\text{ th}}$ root of ε in the same way as χ is defined by α, then we have $\chi'^{p} = \chi$. Therefore $A(M/K)$ is a divisible p-primary abelian group.

Since the group E is the direct product of the union E^+ of all E_n^+ $(n \geqq 0)$ and the union W of all W_n $(n \geqq 0)$, the above ε can be written in the form $\varepsilon = \varepsilon_0\zeta$ where $\varepsilon_0 \in E^+$, $\zeta \in W$. But ζ has a $p^{m\text{ th}}$ root in W. Hence the above character χ can be defined also by a $p^{m\text{ th}}$ root of ε_0. Let σ be any element in G. Then, by the same argument as in the above, we see that the character in $A(M/F)$ defined by a $p^{m\text{ th}}$ root of ε_0^{σ} is $\chi^{\kappa(\sigma)\sigma^{-1}}$. Since ε_0 is in E^+ and is invariant under the complex conjugation δ_{-1}, it follows that $\chi^{\delta_{-1}} = \chi^{-1}$. As χ was an arbitrary element in $A(M/F)$, we see that

$${}^{i}A(M/F) = 0 ,$$

for every even index i.

We now fix an odd index j and consider the structure of ${}^{j}A(M/F)$. We know already that ${}^{j}A(M/F)$ is a divisible Γ-finite Γ-module. Let A be the maximal quasi-elementary submodule of ${}^{j}A(M/F)$. Since ${}^{j}A(M/F)$ is divisible, A is also divisible and

$$A \cong E(\infty)^s/D ,$$

where $E(\infty)^s$ is the direct sum of s copies of the module $E(\infty)$ and D is a suitable finite submodule of $E(\infty)^s$. As proved in the above, the rank of ${}^jA(M'_n/F_n)$ is at least p^n. Hence the rank of ${}^jA(M_n/F_n)$ is also at least p^n. But, by § 4, ${}^jA(M/F)_n \cong {}^jA(M_n/F_n)$. Therefore the rank of ${}^jA(M/F)_n$ is also at least p^n. Since ${}^jA(M/F)/A$ is a module of finite rank, it follows that $A \neq 0$ and $s \geqq 1$. On the other hand, the isomorphism $A \cong E(\infty)^s/D$ implies that ${}^jA(M/F)_n$ contains a submodule isomorphic with $(E(\infty)^s_n+D)/D$, a divisible module of rank sp^n. Hence ${}^jA(M_n/F_n)$ also contains such a submodule. Since ${}^jA(L'_n/F_n)$ is finite, it then follows that $({}^jA(L'_n/F) + {}^jA(M_n/F_n))/{}^jA(L'_n/F_n)$ also contains a divisible submodule of rank sp^n, $s \geqq 1$. However, it is proved in § 6 that ${}^jG(K_n/L'_n)$ is the direct product of p^n copies of Γ. Hence ${}^jA(K_n/L'_n) = {}^jA(K_n/F_n)/{}^jA(L'_n/F_n)$ is a divisible module of rank p^n. Since ${}^jA(L'_n/F_n) + {}^jA(M_n/F_n)$ is contained in ${}^jA(K_n/F_n)$, we see immediately that $s = 1$ and

$$ {}^jA(K_n/F_n) = {}^jA(L'_n/F_n) + {}^jA(M_n/F_n) . $$

It then follows that

$$ {}^jA(K/F)_n = {}^jA(L/F)_n + {}^jA(M/F)_n , \qquad n \geqq 0, $$

and, consequently, that

$$ {}^jA(K/F) = {}^jA(L/F) + {}^jA(M/F) , $$

for every odd index j.

8. Fundamental isomorphisms

Let $I(F_n)$ and $C(F_n)$ denote the ideal group of F_n and the ideal-class group of F_n, respectively, and let $S(F_n)$ be, as in § 5, the Sylow p-group of $C(F_n)$. Since $G(F_n/Q) = G/\Gamma_n$ acts on these groups in a natural manner, they may be considered also as G-groups. If $m \geqq n \geqq 0$, then there is a natural injective mapping $I(F_n) \rightarrow I(F_m)$, and it induces G-homomorphisms $C(F_n) \rightarrow C(F_m)$ and $S(F_n) \rightarrow S(F_m)$. We denote by $C(F)$ the direct limit of the groups $C(F_n)$ ($n \geqq 0$) with respect to the above homomorphisms, and we call it *the ideal-class group of* F. Since every $C(F_n)$ is a finite group, $C(F)$ is a periodic abelian group, and its Sylow p-group is nothing but the direct limit $S(F)$ of the groups $S(F_n)$ ($n \geqq 0$). As $C(F_n)$ and $S(F_n)$ are G-groups, so are the direct limits $C(F)$ and $S(F)$.

Now, put

$$ X = G(L/F) , \qquad X_n = G(L/L_n) , \qquad n \geqq 0. $$

Then X_n is the annihilator of $A(L/F)_n = A(L_n/F)$ in X with respect to the canonical G-pairing of $A(L/F)$ and $G(L/F)$. But, as explained in § 5, $X/X_n = G(L_n/F)$ is isomorphic with $G(L'_n/F)$, and the latter is also isomorphic with $S(F_n)$. Hence, for each $n \geqq 0$, we have a natural G-isomorphism

$$X/X_n \to S(F_n)\,, \qquad n \geqq 0.$$

Let $m \geqq n \geqq 0$. Then the canonical mapping $S(F_n) \to S(F_m)$ considered above defines a corresponding homomorphism φ of $G(L'_n/F_n)$ into $G(L'_m/F_m)$. In class field theory, φ is given by the so-called transfer map[14]; in the present case, it can be described as follows: Since $F_m \cap L'_n = F_n$, $G(L'_n/F_n)$ is naturally isomorphic with $G(F_mL'_n/F_m) = G(L'_m/F_m)/G(L'_m/F_mL'_n)$. For any u in $G(L'_n/F_n)$, choose an element v in $G(L'_m/F_m)$ such that u is mapped to the coset of v mod $G(L'_m/F_mL'_n)$ under the isomorphism $G(L'_n/F_n) \to G(F_mL'_n/F_m)$. Then the homomorphism φ is given by

$$\varphi(u) = v^{\nu_{n,m}}\,,$$

where $\nu_{n,m}$ is defined as in § 2.

It follows that we have a commutative diagram of G-homomorphisms:

$$\begin{array}{ccc} X/X_n & \longrightarrow & S(F_n) \\ \downarrow & & \downarrow \\ X/X_m & \longrightarrow & S(F_m)\,, \end{array}$$

where $X/X_n \to X/X_m$ is the mapping induced by the endomorphism of X: $x \to x^{\nu_{n,m}}$; and we see that the direct limit of X/X_n is G-isomorphic with the direct limit $S(F)$ of $S(F_n)$. But, as $X = G(L/F)$ is the dual of the strictly Γ-finite module $A(L/F)$, the direct limit of X/X_n is, by definition, *the adjoint of* $A(L/F)$ and is a regular strictly Γ-finite module.[15] Thus we have the following:

THEOREM 3. *Let $S(F)$ be the Sylow p-group of the ideal-class group of F. Then there is a cononical G-isomorphism of $S(F)$ onto the adjoint $A^*(L/F)$ of the strictly Γ-finite module $A(L/F)$. In particular, $S(F)$ is a regular strictly Γ-finite Γ-group.*

If

$$S(F) = \prod_{i=0}^{p-2} {}^{\iota}S(F)\,, \quad A^*(L/F) = \sum_{i=0}^{p-2} {}^{\iota}A^*(L/F)$$

are the Δ-decompositions of $S(F)$ and $A^*(L/F)$ respectively, then the above isomorphism obviously induces a G-isomorphism

$${}^{\iota}S(F) \cong {}^{\iota}A^*(L/F)\,,$$

---

[14] Cf. [3], II.

[15] Cf. [7], § 5.

for every i. Notice that ${}^iA^*(L/F)$ may be also identified with the adjoint of ${}^iA(L/F)$.

Now, let c be any element of the group $S(F)$. Let $\mathfrak{a}$ be an ideal of an F_n such that the ideal-class of $\mathfrak{a}$ is mapped to c under the homomorphism $S(F_n) \to S(F)$. Since the ideal-class of $\mathfrak{a}$ is in the Sylow p-group $S(F_n)$ of $C(F_n)$, there is an integer $m \geqq 0$ such that $\mathfrak{a}^{p^m} = (a)$ is a principal ideal of F_n. Let α be a $p^{m\text{ th}}$ root of a. Then α is contained in K, and the function $\chi(x)$ on $G(K/M)$ defined by

$$\chi(x) = \alpha^{1-x}, \qquad x \in G(K/M),$$

gives a character of the group $G(K/M)$ with values in W. Furthermore, it is known that such a character χ depends only upon the element c in $S(F)$ and that the mapping $c \to \chi$ defines an isomorphism of $S(F)$ onto the character group $A(K/M)$ of $G(K/M)$.[16] Let σ be any element of G and let s be an element of $G(K/Q)$ which induces the automorphism σ on F. Then $(\mathfrak{a}^\sigma)^{p^m} = (a^\sigma)$ and α^s is a $p^{m\text{ th}}$ root of a^σ in K. Since the ideal-class of $\mathfrak{a}^\sigma$ is mapped to c^σ under the homomorphism $S(F_n) \to S(F)$, the character χ' corresponding to c^σ is given by

$$\chi'(x) = (\alpha^s)^{1-x}, \qquad x \in G(K/M).$$

From an argument similar to that in § 7, it follows that

$$\chi' = \chi^{\kappa(\sigma)\sigma^{-1}}.$$

Thus, we have the following :

THEOREM 4. *Let $S(F)$ be as in Theorem 3. Then there exists a canonical κ-isomorphism of the G-group $S(F)$ onto the G-module $A(K/M) = A(K/F)/A(M/F)$.*

The above κ-isomorphism induces κ-isomorphisms between the components of the Δ-decompositions of $S(F)$ and $A(K/M)$; for each i, it induces a κ-isomorphism of ${}^iS(F)$ onto ${}^jA(K/M) = {}^jA(K/F)/{}^jA(M/F)$, where $i + j \equiv 1 \bmod p - 1$.

Combining Theorems 3 and 4, we also obtain the following:

THEOREM 5. *There exists a canonical κ-isomorphism of the adjoint $A^*(L/F)$ of the G-module $A(L/F)$ onto the G-module $A(K/M) = A(K/F)/A(M/F)$.*

Again there also exists, for each i, a κ-isomorphism of ${}^iA^*(L/F)$ onto ${}^jA(K/M) = {}^jA(K/F)/{}^jA(M/F)$, where $i + j \equiv 1 \bmod p - 1$.

Now, let R be the maximal regular submodule of the strictly Γ-finite module $A(L/F)$. Since R is a G-submodule of $A(L/F)$, we have the

[16] Cf. [5].

Δ-decomposition:

$$R = \sum_{i=0}^{p-2} {}^iR .$$

${}^iR$ is then the maximal regular submodule of ${}^iA(L/F)$. Hence ${}^iA(L/F)$ and ${}^iR$ have the same adjoint $({}^iA(L/F))^* = ({}^iR)^*$, and it may be identified with ${}^iA^*(L/F) = {}^i(R^*)$. Thus, by Theorem 3,

$$({}^iR)^* \cong {}^iS(F) ,$$

for every index i.

We now fix an even index i and an odd index j such that $i + j \equiv 1 \bmod p - 1$. By the last result in § 7, we have

$${}^jA(K/F)/{}^jA(M/F) \cong {}^jA(L/F)/{}^jA(L/F) \cap {}^jA(M/F) .$$

Since ${}^jA(L/F)$ is strictly Γ-finite, so is ${}^jA(L/F)/{}^jA(L/F) \cap {}^jA(M/F)$. But ${}^jA(K/F)/{}^jA(M/F)$ is the κ-isomorphic image of the regular strictly Γ-finite module ${}^iS(F)$. It follows that ${}^jA(L/F)/{}^jA(L/F) \cap {}^jA(M/F)$ is also regular.[17] By the definition of *κ-adjoints*,[18] we then have immediately the following:

THEOREM 6. *Let R be the maximal regular submodule of $A(L/F)$ and let ${}^iR$ be the maximal regular submodule of ${}^iA(L/F)$ so that $R = \sum_{i=0}^{p-2} {}^iR$ is the Δ-decomposition of R. Let i be an even index and j an odd index such that $i + j \equiv 1 \bmod p - 1$. Then both ${}^iR$ and ${}^jA(L/F)/{}^jA(L/F) \cap {}^jA(M/F)$ are regular strictly Γ-finite modules, and they are the κ-adjoints of each other.*

The above theorem shows that *there exist close relations between the structures of ${}^iR$ and ${}^jA(L/F)/{}^jA(L/F) \cap {}^jA(M/F)$ and, hence, between those of ${}^iA(L/F)$ and ${}^jA(L/F)$.* Such relations are discussed in detail in [7, § 5]. In particular, we have

$$\lambda({}^iA(L/F)) = \lambda({}^iR) = \lambda({}^jA(L/F)/{}^jA(L/F) \cap {}^jA(M/F)) \leqq \lambda({}^jA(L/F)) ,$$

for the λ-invariants, and similarly for the μ-invariants.[19] Thus:

THEOREM 7. *Let ${}^i\lambda = \lambda({}^iA(L/F))$ and ${}^i\mu = \mu({}^iA(L/F))$ as in* § 5. *Let i be an even index and j an odd index such that $i + j \equiv 1 \bmod p - 1$. Then*

$${}^i\lambda \leqq {}^j\lambda , \qquad {}^i\mu \leqq {}^j\mu .$$

It follows in particular that

$$\lambda^+ \leqq \lambda^- , \qquad \mu^+ \leqq \mu^- ,$$

[17] Cf. [7], Theorem 4.5.

[18] Cf. [7], 5.6.

[19] Cf. [7], Theorem 2.7, Theorem 5.8.

where $\lambda^+ = \lambda(F^+/F_0^+)$ and $\mu^+ = \mu(F^+/F_0^+)$ are defined as in § 5, and where λ^- and μ^- are defined by

$$\lambda^- = \sum_{2\nmid i} {}^i\lambda\,, \qquad \mu^- = \sum_{2\nmid i} {}^i\mu\,.$$

Since $\lambda = \lambda^+ + \lambda^-$ and $\mu = \mu^+ + \mu^-$, we see that *$\lambda = 0$ or $\mu = 0$ if and only if $\lambda^- = 0$ or $\mu^- = 0$, respectively.*

Now, it is shown in § 7 that ${}^iA(M/F) = 0$ for any even index i. Hence we obtain also the following theorem from Theorem 5.

THEOREM 8. *Let i be an even index and j an odd index such that $i + j \equiv 1 \bmod p - 1$. Then ${}^iA(K/F)$ is the κ-isomorphic image of the adjoint ${}^jS(F)$ of ${}^jA(L/F)$.*

It follows, in particular, that if ${}^jA(L/F) = 0$, then ${}^iA(K/F) = 0$ and, hence, also ${}^iA(L/F) = 0$.

Let i and j be fixed as in the above. It was proved in § 7 that

$${}^jA(K_n/F_n) = {}^jA(L'_n/F_n) + {}^jA(M_n/F_n)\,, \qquad n \geqq 0,$$

for the odd index j. Put $D = {}^jA(L'_n/F_n) \cap {}^jA(M_n/F_n) = {}^jA((L'_n \cap M_n)/F_n)$. It then follows from the above that ${}^jA(K_n/F_n)/D$ is the direct sum of ${}^jA(L'_n/F_n)/D$ and ${}^jA(M_n/F_n)/D$. Hence, for any subfield $K_{n,m}$ $(m \geqq 0)$ of K_n defined in § 6, ${}^jA(K_{n,m}/F_n)/D$ is also the direct sum of ${}^jA(L'_n/F_n)/D$ and $({}^jA(K_{n,m}/F_n) \cap {}^jA(M_n/F_n))/D$; namely, ${}^jA(K_{n,m}/(L'_n \cap M_n))$ is the direct sum of ${}^jA(L'_n/(L'_n \cap M_n))$ and ${}^iA((K_{n,m} \cap M_n)/(L'_n \cap M_n))$.

Let $P(F_n)$ denote the group of principal ideals of F_n and $P_m(F_n)$ the subgroup of all principal ideals (a) in $P(F_n)$ such that $a \equiv 1 \bmod \mathfrak{p}_n^{m+1}$. Let J be the subgroup of $I(F_n)$ containing $P(F_n)$ such that $J/P(F_n)$ is the Sylow p-group $S(F_n)$ of $C(F_n) = I(F_n)/P(F_n)$. Then, by class field theory, there is a canonical isomorphism of $G(K_{n,m}/F_n)$ onto $J/P_m(F_n)$[20]. Let $S'(F_n)$ be the subgroup of $S(F_n)$ corresponding to the Galois group $G(L'_n/(L'_n \cap M_n))$ under the canonical isomorphism of $S(F_n)$ onto $G(L'_n/F_n)$. If J' is the subgroup of J containing $P(F_n)$ such that $S'(F_n) = J'/P(F_n)$, then $G(K_{n,m}/(L'_n \cap M_n))$ is mapped to $J'/P_m(F_n)$ under the above isomorphism: $G(K_{n,m}/F_n) \to J/P_m(F_n)$. Since ${}^jA(K_{n,m}/(L'_n \cap M_n))$ is the direct sum of ${}^jA(L'_n/(L'_n \cap M_n))$ and ${}^jA((K_{n,m} \cap M_n)/(L'_n \cap M_n))$, and since ${}^jG(K_{n,m}/(L'_n \cap M_n))$ corresponds to ${}^j(J'/P_m(F_n))$, the latter is also the direct product of ${}^j(P(F_n)/P_m(F_n))$ and a subgroup naturally isomorphic with ${}^jS'(F_n) = {}^j(J'/P(F_n))$. It follows that if c is any ideal-class in ${}^jS'(F_n)$ with order p^s, then c contains an ideal $\mathfrak{a}$ such that $\mathfrak{a}^{p^s}$ is in $P_m(F_n)$, namely, such that $\mathfrak{a}^{p^s} = (a)$ with $a \equiv 1 \bmod \mathfrak{p}_n^{m+1}$. Notice that m can be any large integer given beforehand.

[20] Cf., l.c. 8.

Now, it will be proved below that, for the odd index j, the canonical homomorphism ${}^jS(F_n) \to {}^jS(F)$ is injective for every $n \geqq 0$, that the image of ${}^jS(F_n)$ coincides with the submodule ${}^jS(F)_n$ of those elements in ${}^jS(F)$ which are fixed under every σ in Γ_n, and that both ${}^iA(L/F)$ and ${}^iA(L/F) \cap {}^iA(M/F)$ are regular Γ-modules.[21] The canonical isomorphisms ${}^jS(F)_n \cong {}^jS(F_n) \cong {}^jG(L'_n/F_n) \cong {}^jG(L_n/F)$ then define a natural pairing of ${}^iA(L/F)_n = {}^iA(L_n/F)$ and ${}^jS(F)_n$ for each $n \geqq 0$; and these pairings, defined for every $n \geqq 0$, give a so-called system of adjoint pairings for the regular strictly Γ-finite modules ${}^iA(L/F)$ and ${}^jS(F)$. Let B be the annihilator of ${}^iA(L/F) \cap {}^iA(M/F)$ in ${}^jS(F)$ with respect to the above pairings.[22] Then B is a regular submodule of ${}^jS(F)$, and ${}^iA(L/F) \cap {}^iA(M/F)$ is the annihilator of B in ${}^iA(L/F)$. Furthermore, ${}^iA(L/F)/{}^iA(L/F) \cap {}^iA(M/F)$ and B are the adjoints of each other; and so are ${}^iA(L/F) \cap {}^iA(M/F)$ and ${}^jS(F)/B$. Let b be an element of ${}^jS(F)_n$ and let c be the corresponding element in ${}^jS(F_n)$. By the definition of B, b is contained in $B_n = B \cap {}^jS(F)_n$ if and only if b is orthogonal to every element of $({}^iA(L/F) \cap {}^iA(M/F))_n = {}^iA(L/F)_n \cap {}^iA(M/F)_n = {}^iA(L_n/F) \cap {}^iA(M_n/F)$ with respect to the above pairing of ${}^iA(L/F)_n$ and ${}^jS(F)_n$. But ${}^iA(L_n/F) \cap {}^iA(M_n/F)$ is the submodule of ${}^iA(L_n/F)$ corresponding to ${}^iA(L'_n/F_n) \cap {}^iA(M_m/F_m) = {}^iA((L'_n \cap M_n)/F_n)$ under the natural isomorphism of ${}^iA(L'_n/F_n)$ onto ${}^iA(L_n/F)$. Hence b is in B_n if and only if c is orthogonal to every element of ${}^iA((L'_n \cap M_n)/F_n)$ with respect to the canonical pairing of ${}^iA(L'_n/F_n)$ and ${}^jS(F_n) \cong {}^jG(L'_n/F_n)$; namely, if and only if c is contained in ${}^jS'(F_n)$. Therefore B_n is the image of ${}^jS'(F_n)$ under the injective map ${}^jS(F_n) \to {}^jS(F)$, and B is nothing but the direct limit of ${}^jS'(F_n)$ for $n \to \infty$. Hence we shall denote B also by ${}^jS'(F)$.

Let b and c be as in the above and let p^s be the order of these elements. Suppose that b is in $B_n = {}^jS'(F)_n$ and, hence, that c is in ${}^jS'(F_n)$. For any integer $m \geqq 0$, the ideal class c then contains an ideal $\mathfrak{a}$ of F_n such that $\mathfrak{a}^{p^s} = (a)$ with an element a in F_n satisfying $a \equiv 1 \bmod \mathfrak{p}_n^{m+1}$. But, if m is sufficiently large, any such element a with $a \equiv 1 \bmod \mathfrak{p}_n^{m+1}$ is a p^{s} th power of an element in the $\mathfrak{p}_n$-completion of the field F_n. Hence, for such an m, the extension $F(\alpha)$ generated by a p^{s} th root α of a over F is an unramified extension of F; and as b is in ${}^jS(F)$, the character $\chi(x) = \alpha^{1-x}$ associated with b under the κ-isomorphism: ${}^jS(F) \to {}^iA(K/F)$ in Theorem 8 is contained in the submodule ${}^iA(L/F)$ of ${}^iA(K/F)$. It follows that the κ-isomorphic image B' of $B = {}^jS'(F)$ under ${}^jS(F) \to {}^iA(K/F)$ is contained in ${}^iA(L/F)$. However, B is a regular strictly Γ-finite module, and B' is

[21] Cf., Theorems 15, 16 in § 10.

[22] Cf. [7], 5.3.

also strictly Γ-finite as a submodule of ${}^iA(L/F)$. Therefore B' is also regular,[23] and we see that ${}^jA(L/F)/{}^jA(L/F) \cap {}^jA(M/F)$ and B' are the κ-adjoints of each other. But, by Theorem 6, ${}^jA(L/F)/{}^jA(L/F) \cap {}^jA(M/F)$ is also the κ-adjoint of the maximal regular submodule ${}^iR$ of ${}^iA(L/F)$. It follows that B' is isomorphic with ${}^iR$ and that $\lambda(B') = \lambda({}^iR)$, $\mu(B') = \mu({}^iR)$. Since the regular submodule B' of ${}^iA(L/F)$ is contained in ${}^iR$, this implies immediately that ${}^iR = B'$. Thus the following theorem is proved.

THEOREM 9. *Let i be an even index and j an odd index such that $i + j \equiv 1 \bmod p - 1$. Then the annihilator of ${}^jA(L/F) \cap {}^jA(M/F)$ in ${}^jS(F)$ with respect to the system of natural adjoint pairings for ${}^jA(L/F)$ and ${}^jS(F)$ is the image ${}^jS'(F)$ of the maximal regular submodule ${}^iR$ of ${}^iA(L/F)$ under the κ-isomorphism of ${}^iA(K/F)$ onto ${}^jS(F)$ given in Theorem* 8.

The results in Theorems 6, 8, and 9 can be simply expressed by the following diagram where the vertical arrows denote natural surjective and injective homomorphisms, respectively:

$$\begin{array}{ccccccccc}
{}^iR & \overset{\text{adj}}{\longleftrightarrow} & {}^iS(F) & \overset{\kappa}{\longleftrightarrow} & {}^jA(L/F)/{}^jA(L/F) \cap {}^jA(M/F) & \overset{\text{adj}}{\longleftrightarrow} & {}^jS'(F) & \overset{\kappa}{\longleftrightarrow} & {}^iR \\
 & & & & \uparrow & & \downarrow & & \downarrow \\
 & & & & {}^jA(L/F) & \overset{\text{adj}}{\longleftrightarrow} & {}^jS(F) & \overset{\kappa}{\longleftrightarrow} & {}^iA(K/F).
\end{array}$$

Here the square on the left represents the situation stated in [7, Theorem 5.2]. It is also directly shown here that ${}^jA(L/F)/{}^jA(L/F) \cap {}^jA(M/F)$ is the κ-isomorphic image of the adjoint of ${}^iR$ as well as the adjoint of the κ-isomorphic image of ${}^iR$.[24]

9. Cohomology groups

Let $m \geqq n \geqq 0$. We shall consider the cohomology group $H(G(F_m/F_n), C(F_m))$, where $C(F_m)$ denotes, as before, the ideal-class group of F_m. Since $G(F_m/F_n) = \Gamma_n/\Gamma_m$ is a cyclic group of order p^{m-n}, we have

$$H(G(F_m/F_n), C(F_m)) = H(\Gamma_n/\Gamma_m, S(F_m)) ,$$

for the Sylow p-group $S(F_m)$ of $C(F_m)$. But, as we have seen in § 5, there exists a natural dual pairing of $S(F_m)$ and $A(L_m/F) = A(L/F)_m$, and we obtain easily the following:

THEOREM 10. *For any $m \geqq n \geqq 0$, $H^1(G(F_m/F_n), C(F_m))$ is dual to $H^2(\Gamma_n/\Gamma_m, A(L/F)_m)$ and $H^2(G(F_m/F_n), C(F_m))$ is dual to $H^1(\Gamma_n/\Gamma_m, A(L/F)_m)$ in a natural way.*

Now, let $N = N_{n,m}$ denote the norm map from F_m to F_n. By class field

[23] Cf., l.c. 17.

[24] Cf. [7], 5.6.

theory, the index of $N(C(F_m))$ in $C(F_n)$ is equal to the degree of the maximal unramified extension of F_n contained in F_m. But the prime ideal $\mathfrak{p}_n$ of F_n is fully ramified in F_m; namely, $\mathfrak{p}_n = \mathfrak{p}_m^{p^{m-n}}$. Hence there is no non-trivial unramified extension of F_n contained in F_m, and we have

$$N_{n,m}(C(F_m)) = C(F_n), \qquad m \geqq n \geqq 0.$$

Let $\omega_n = 1 - \gamma^{p^n}$ as in § 2. Then $C(F_m)^{\omega_n}$ is clearly contained in the kernel of N: $C(F_m) \to C(F_n)$. Let $\mathfrak{A}$ be an ideal of F_m such that $N(\mathfrak{A})$ is a principal ideal of F_n: $N(\mathfrak{A}) = (a)$, $a \in F_n$. If $\mathfrak{q}$ is any prime divisor of F_n different from $\mathfrak{p}_n$, then $\mathfrak{q}$ is unramified in F_m and a is a local norm at $\mathfrak{q}$ for the extension F_m/F_n. By the product formula for the norm residue symbols, a is then also a local norm at $\mathfrak{p}_n$ and hence is a global norm for the extension F_m/F_n.[25] Therefore we have $a = N(\alpha)$ for an element α in F_m. But, then $N(\alpha^{-1}\mathfrak{A}) = 1$ and $\alpha^{-1}\mathfrak{A} = \mathfrak{B}^{\omega_n}$ with an ideal $\mathfrak{B}$ of F_m. Hence we see that the kernel of N: $C(F_m) \to C(F_n)$ is $C(F_m)^{\omega_n}$.

Let ψ be the product of the norm map N: $C(F_m) \to C(F_n)$ and the homomorphism $\varphi = \varphi_{m,n}$: $C(F_n) \to C(F_m)$ induced by the natural injective map $I(F_n) \to I(F_m)$. Since N is surjective, we have

$$\mathrm{Ker}(\psi)/\mathrm{Ker}(N) \cong \mathrm{Ker}(\varphi) .$$

Since $\mathrm{Ker}(N) = C(F_m)^{\omega_n}$, the left hand side of the above is $H^1(G(F_m/F_n), C(F_m))$. Hence the following theorem is proved.

Theorem 11. *For $m \geqq n \geqq 0$,*

$$H^1(G(F_m/F_n), C(F_m)) \cong \mathrm{Ker}(C(F_n) \to C(F_m)) .$$

We notice that

$$\mathrm{Ker}(C(F_n) \to C(F_m)) = \mathrm{Ker}(S(F_n) \to S(F_m)) .$$

This follows either from the above theorem or from the fact that, if $\mathfrak{a}$ is an ideal of F_n and $\mathfrak{a} = (\alpha)$, $\alpha \in F_m$, then $\mathfrak{a}^{p^{m-n}} = N(\mathfrak{a}) = (N(\alpha))$. It is also clear that, for every sufficiently large m, we have

$$\mathrm{Ker}(C(F_n) \to C(F_m)) = \mathrm{Ker}(C(F_n) \to C(F)) .$$

Now, by Theorems 10 and 11, $\mathrm{Ker}(C(F_n) \to C(F_m))$ is dual to $H^2(\Gamma_n/\Gamma_m, A(L/F)_m)$. Hence, using a theorem on the cohomology of strictly Γ-finite modules,[26] we obtain the following:

Theorem 12. *Let R be, as before, the maximal regular submodule of $A(L/F)$. Then, for every $n \geqq 0$, the order of* $\mathrm{Ker}(C(F_n) \to C(F))$ *does*

[25] Cf. [3], II.

[26] Cf. [7], Theorem 3.4.

not exceed the order of the finite module $A(L/F)/R$. Furthermore, if n is sufficiently large, then $\mathrm{Ker}(C(F_n) \to C(F))$ *is dual to* $A(L/F)/R$ *in a natural way.*

We shall next consider the cohomology group $H(G(F_m/F_n), E_m)$, where E_m denotes, as in § 6, the group of units of the field F_m.

THEOREM 13. *For* $m \geqq n \geqq 0$,

$$H^1(G(F_m/F_n), E_m) \cong H^1(G(F_m/F_n), C(F_m)) \times Z(p^{m-n}),$$
$$H^2(G(F_m/F_n), E_m) \cong H^2(G(F_m/F_n), C(F_m)),$$

where $Z(p^{m-n})$ *denotes a cyclic group of order* p^{m-n}.

For the proof, let I' denote the group of all principal ideals of F_m which are invariant under the Galois group $G(F_m/F_n)$. It is then easy to see that I' is the direct product of the cyclic group generated by $\mathfrak{p}_m$ and the subgroup consisting of all such ideals of F_n which are prime to $\mathfrak{p}_n$ and which become principal in F_m. Clearly $P(F_n)$ is contained in I'; and by the above, $I'/P(F_n)$ is the direct product of $\mathrm{Ker}(C(F_n) \to C(F_m))$ and cyclic group $Z(p^{m-n})$ generated by the coset of $\mathfrak{p}_m$ mod $P(F_n)$. However, it is known that $H^1(G(F_m/F_n), E_m)$ is naturally isomorphic with the factor group $I'/P(F_n)$.[27] Hence the first isomorphism of the theorem follows immediately from Theorem 11.

To prove the second isomorphism, let c be any ideal-class of F_m invariant under the Galois group $G(F_m/F_n)$ and let $\mathfrak{A}$ be an ideal in the class c. Then $\mathfrak{A}^{\omega_n} = (\alpha)$, $\alpha \in F_m$, and $N(\alpha) = \varepsilon$ is a unit of F_n. The cohomology class of $H^2(G(F_m/F_n), E_m)$ defined by ε then depends only upon the ideal-class c and not upon the choice of $\mathfrak{A}$ and α, and we have a homomorphism of the group of invariant classes of $C(F_m)$ into the group $H^2(G(F_m/F_n), E_m)$ mapping c to the cohomology class of ε. Now, suppose, in the above, that the cohomology class of ε is trivial. Then $\varepsilon = N(\varepsilon')$ for a unit ε' in F_m, and we have that $N(\varepsilon'^{-1}\alpha) = 1$ and $\varepsilon'^{-1}\alpha = \beta^{\omega_n}$, $\beta \in F_m$. It follows that $\mathfrak{A}^{\omega_n} = (\beta)^{\omega_n}$ and that the ideal-class c contains an ideal $\mathfrak{A}' = \beta^{-1}\mathfrak{A}$ which is invariant under $G(F_m/F_n)$. Multiplying $\mathfrak{A}'$ by a suitable power of $\mathfrak{p}_m$, we then see that c contains also an ideal of F_n and, hence, is contained in $\psi(C(F_m))$, ψ being the map defined in the proof of Theorem 11. Thus the above homomorphism, which maps c to the cohomology class of ε, induces an isomorphism of $H^2(G(F_m/F_n), C(F_m))$ into $H^2(G(F_m/F_n), E_m)$. However, as in the proof of Theorem 11, we can prove that every unit ε in F_n is the norm of an element α in F_m: $\varepsilon = N(\alpha)$. Then $(\alpha) = \mathfrak{A}^{\omega_n}$ for an ideal $\mathfrak{A}$ of F_m, and we see that $H^2(G(F_m/F_n), C(F_m)) \to H^2(G(F_m/F_n), E_m)$ is surjective. Therefore the second isomorphism of the theorem is also proved.

[27] Cf. [4].

For any $m \geqq n \geqq 0$, $N_{n,m+1}(E_{m+1})$ is clearly contained in $N_{n,m}(E_m)$ $(m \geqq n)$; and we denote by $N_{n,\infty}(E)$ the intersection of $N_{n,m}(E_m)$ for all $m \geqq n$. By Theorems 10 and 13, $E_n/N_{n,m}(E_m) = H^2(G(F_m/F_n), E_m)$ is dual to $H^1(\Gamma_n/\Gamma_m, A(L/F)_m)$. Hence, by a theorem on the cohomology of strictly Γ-finite modules,[28] we obtain the following:

THEOREM 14. *The group $N_{n,\infty}(E)$ defined in the above coincides with $N_{n,m}(E_m)$ whenever m is sufficiently large, and for every $n \geqq 0$, the order of $E_n/N_{n,\infty}(E)$ does not exceed the order of the module $A(L/F)/R$. Furthermore, if n is sufficiently large, then $E_n/N_{n,\infty}(E)$ is dual to $A(L/F)/R$ in a natural way.*

Now, the Galois group $G(F_m/Q) = G/\Gamma_m$ and, hence, also the Galois group $G(F/Q) = G$ act on the groups $C(F_m)$, $\mathrm{Ker}(C(F_n) \to C(F_m))$, E_m, etc., in an obvious way. Therefore the cohomology groups $H(G(F_m/F_n), C(F_m))$, $H(G(F_m/F_n), E_m)$, etc., can be considered also as G-groups in the usual manner; and as all these cohomology groups are finite p-groups, we may also consider their Δ-decompositions. But, a re-examination of the above proofs shows that all the isomorphisms and dualisms obtained in the above are G-isomorphisms and G-dualisms and that they induce G-isomorphisms and G-dualisms between the corresponding components of the Δ-decompositions of respective groups. For instance, we obtain from Theorem 11 that

$${}^iH^1(G(F_m/F_n), C(F_m)) \cong {}^i\mathrm{Ker}(C(F_n) \to C(F_m)) , \qquad 0 \leqq i \leqq p-2,$$

namely, that

$$H^1(G(F_m/F_n), {}^iS(F_m)) \cong \mathrm{Ker}({}^iS(F_n) \to {}^iS(F_m)) , \qquad 0 \leqq i \leqq p-2.$$

We do not write down here all such isomorphisms and dualisms, but we notice that the cyclic group $Z(p^{m-n})$ in Theorem 13 is trivially acted on by G and, hence, that $Z(p^{m-n}) = {}^0Z(p^{m-n})$.

10. The regularity of modules

As stated before, the group E_n of all units in F_n is the direct product of the group E_n^+ of all units in the maximal real subfield F_n^+ of F_n and the cyclic group W_n of order p^{n+1} generated by ζ_{n+1}. Since $N_{n,m}(W_m) = W_n$ $(m \geqq n)$, $E_n/N_{n,m}(E_m)$ is naturally isomorphic with $E_n^+/N_{n,m}(E_m^+)$; and since the complex conjugation δ_{-1} acts trivially on E_n^+, we see that

$${}^jH^2(G(F_m/F_n), E_m) = {}^j(N_n/N_{n,m}(E_m)) = 1 ,$$

for every odd index j. By Theorem 13, we then have

[28] Cf., l.c. 26.

$$^jH^2(G(F_m/F_n),\ C(F_m)) = H^2(G(F_m/F_n),\ ^jS(F_m)) = 1\,.$$

Since $^jS(F_m)$ is a finite group, $H^1(G(F_m/F_n),\ ^jS(F_m))$ is then also trivial. Hence, by the remark at the end of § 9,

$$\mathrm{Ker}(^jS(F_n) \to {}^jS(F_m)) = 1\,.$$

It follows that

$$\mathrm{Ker}(^jS(F_n) \to {}^jS(F)) = 1\,;$$

namely, that the map $^jS(F_n) \to {}^jS(F)$ is injective for every odd j.

By Theorem 12 (and by the remark at the end of § 9), we then also see that $^j(A(L/F)/R) = {}^jA(L/F)/{}^jR = 1$, namely, that $^jA(L/F)$ is regular for every odd j. Since $^jA(L/F)$ is regular and $^jS(F)$ is the adjoint of $^jA(L/F)$, $^jA(L/F)_n$ and $^jS(F)_n$ are dual to each other and have the same finite order. Clearly, the image of $^jS(F_n)$ in $^jS(F)$ is contained in $^jS(F)_n$. But, since $^jA(L/F)_n = {}^jA(L_n/F) \cong {}^jA(L'_n/F_n)$, $^jS(F_n)$ is dual to $^jA(L/F)_n$ and has the same finite order. It follows that the image of $^jS(F_n)$ in $^jS(F)$ coincides with $^jS(F)_n$, and the following theorem is proved.

THEOREM 15. *Let j be an odd index. Then the canonical homomorphism $^jS(F_n) \to {}^jS(F)$ is injective for every $n \geqq 0$, and the image of $^jS(F_n)$ in $^jS(F)$ is $^jS(F)_n$.*[29]

Now, it is clear that if $A(L/F) = 0$, then the invariants λ and μ of the Γ-finite Γ-module $A(L/F)$ are both 0. Suppose, conversely, that $\lambda = \mu = 0$. Then $^j\lambda = {}^j\mu = 0$ for every odd index j. Since $^j\lambda$ and $^j\mu$ are the invariants of $^jA(L/F)$ and since $^jA(L/F)$ is regular as shown in the above, it follows that $^jA(L/F) = 0$. By the remark after Theorem 8, we then also have $^iA(L/F) = 0$ for the even index i with $i + j \equiv 1 \bmod p - 1$. Therefore $\lambda = \mu = 0$ implies $A(L/F) = 0$. However, it is easy to see that $A(L/F)$ is 0 if and only if $A(L/F)_0 = 0$. Since $A(L/F)_0 = A(L_0/F)$ is isomorphic with $A(L'_0/F_0)$ and since the latter is 0 if and only if the class number of F_0 is not divisible by p, it follows that *p is a regular prime if and only if $\lambda = \mu = 0$.*

THEOREM 16. *Let j be an odd index. Then the Γ-module $^jA(L/F)$, $^jA(M/F)$, $^jA(K/F)$ and $^jA(L/F) \cap {}^jA(M/F)$ are all regular.*

That $^jA(L/F)$ is regular is already proved. Suppose that $^jA(M/F)$ is also regular. It is proved in § 7 that

$$^jA(K/F)_n = {}^jA(L/F)_n + {}^jA(M/F)_n\,, \quad {}^jA(K/F) = {}^jA(L/F) + {}^jA(M/F).$$

Hence $^jA(K/F)$ is also regular. Let a be any element of $^jA(L/F) \cap {}^jA(M/F)$

[29] The fact that $^jS(F_0) \to {}^jS(F_1)$ is injective is already proved in [8], where one may also trace the germ of other results proved in the present paper.

and let n be any large integer. Choose $m \geqq n$ so that a is contained in $({}^jA(L/F) \cap {}^jA(M/F))_m = {}^jA(L/F)_m \cap {}^jA(M/F)_m$. Since ${}^jA(L/F)$ is regular, there is an element b in ${}^jA(L/F)$ such that $\omega_m b = a$; similarly, $\omega_m c = a$ for some c in ${}^jA(M/F)$. But then $\omega_m(b-c) = 0$, and $b - c$ is contained in ${}^jA(K/F)_m$. Since ${}^jA(K/F)_m = {}^jA(L/F)_m + {}^jA(M/F)_m$, we have $b - c = b' - c'$ with b' in ${}^jA(L/F)_m$ and c' in ${}^jA(M/F)_m$. Put $a' = b - b' = c - c'$. Then a' is contained in ${}^jA(L/F) \cap {}^jA(M/F)$ and $\omega_m a' = \omega_m b - \omega_m b' = \omega_m b = a$ because b' is in ${}^jA(L/F)_m$. Hence a is contained in $\omega_m({}^jA(L/F) \cap {}^jA(M/F))$ and, consequently, also in $\omega_n({}^jA(L/F) \cap {}^jA(M/F))$. This proves that the Γ-module ${}^jA(L/F) \cap {}^jA(M/F)$ is regular. Thus, the above theorem will be completely proved if we can show that ${}^jA(M/F)$ is regular.

Let n be fixed and let M_n, M'_n, T, T' and N be defined as in § 7. Then $A(M'_n/F_n)$ is κ-isomorphic with T/N and the factor group T/T' of T/N is naturally isomorphic with the group of all ideal-classes c in $\mathrm{Ker}(C(F_n) \to C(F))$ such that $c^{p^n} = 1$. But $\mathrm{Ker}(C(F_n) \to C(F)) = \mathrm{Ker}(S(F_n) \to S(F))$; and the order of $\mathrm{Ker}(C(F_n) \to C(F))$ is bounded for all $n \geqq 0$, by Theorem 12. Hence we have

$$T/T' \cong \mathrm{Ker}(C(F_n) \to C(F))$$

whenever n is sufficiently large.

Let ${}^jA(M/F)'_n$ denote the submodule of all a in ${}^jA(M/F)_n$ such that $p^n a = 0$. Then ${}^jA(M/F)'_n$ is canonically isomorphic with ${}^jA(M'_n/F_n)$ and, hence, is κ-isomorphic with ${}^i(T/N)$, where $i + j \equiv 1 \bmod p - 1$. Let H_n be the inverse image of ${}^i(T'/N)$ in ${}^jA(M/F)'_n$. By the above, ${}^jA(M/F)'_n/H_n$ is κ-isomorphic with ${}^i\mathrm{Ker}(C(F_n) \to C(F))$. Hence, by Theorem 12, it is also κ-isomorphic to the dual of ${}^iA(L/F)/{}^iR$ whenever n is sufficiently large. On the other hand, it is proved in § 7 that ${}^i(T'/N)$ is an abelian group of type $(p^n, \cdots, p^n)$ with rank p^n. Therefore, H_n is also an abelian group of the same structure. Since we know that ${}^jA(M/F)$ is a divisible module, its structure can be determined by the following lemma.

Lemma. *Let A be a divisible Γ-finite Γ-module. For each $n \geqq 0$, let A'_n be the submodule of all a in A_n such that $p^n a = 0$. Suppose that, for every sufficiently large n, A'_n contains a submodule H_n such that H_n is an abelian group of type $(p^n, \cdots, p^n)$ with rank p^n and such that A'_n/H_n is isomorphic with a fixed finite module D'. Then A is isomorphic with $E(\infty)/D$, where D is a finite submodule of $E(\infty)$ isomorphic with D'. In particular, A is a regular Γ-finite Γ-module.*

For the proof, let B denote the maximal quasi-elementary submodule of A and let $B = E(\mu'_1, \cdots, \mu'_s)/D$ with a finite submodule D of $E(\mu'_1, \cdots, \mu'_s)$. Since A is divisible and since $p^n \leqq \mathrm{rank}(A_n) \leqq p^n + \mathrm{rank}(D')$ for every sufficiently large n, it is easy to see that $s = 1$, $\mu'_1 = \infty$, and $B = E(\infty)/D$. Let n be large enough so that D is contained in $E(\infty)_n$, and let H'_n denote the intersection of A'_n and the submodule $E(\infty)_n/D$ of B_n. Since $E(\infty)_n/D$ is a divisible module of rank p^n, H'_n is an abelian group of type $(p^n, \cdots, p^n)$ with rank p^n; and it is contained in $A'_n \cap B_n$. Therefore the order of $A'_n/(A'_n \cap B_n)$ does not exceed the order of D', and the same is true for the order of $(A'_n + B_n)/B_n$ and also for the order of $(A'_n + B)/B$. Since A is the union of all A'_n, it follows that A/B is a finite module. As A is divisible, we then see that $A = B = E(\infty)/D$.

Let n be again sufficiently large, and let $\omega_n^{-1}(D)$ denote the inverse image of D under the endomorphism $u \to \omega_n u$ of $E(\infty)$. Then $A_n = \omega_n^{-1}(D)/D$; and as an abelian group, A_n is the direct sum of $E(\infty)_n/D$ and a subgroup D'' which is isomorphic with $\omega_n^{-1}(D)/E(\infty)_n \cong D$. Suppose that $p^tD = p^tD' = p^tD'' = 0$, $t \geqq 0$ and $n \geqq 2t$. Then D'' is contained in A'_n, and hence $A_n = A'_n + (E(\infty)_n/D)$, $A'_n/H'_n \cong \omega_n^{-1}(D)/E(\infty)_n \cong D$ as Γ-modules. Let A''_n be the submodule of all a in A'_n such that $p^ta = 0$. Since D'' is contained in A''_n, we see easily that $A'_n = A''_n + H'_n$ and $A'_n/H'_n \cong A''_n/(A''_n \cap H'_n)$. But H'_n is an abelian group of type $(p^n, \cdots, p^n)$ and $p^{n-t}A''_n = 0$. Hence $A''_n \cap H'_n = p^{n-t}H'_n = p^{n-t}A'_n$, and we have the following Γ-isomorphisms:

$$D \cong A'_n/H'_n \cong A''_n/p^{n-t}A'_n .$$

However, since $p^nA'_n = 0$ and since H_n is an abelian group of type $(p^n, \cdots, p^n)$, H_n is a direct summand of the abelian group A'_n. By an argument similar to that above, we then also see that

$$D' \cong A'_n/H_n \cong A''_n/p^{n-t}A'_n .$$

Hence D is isomorphic with D', and the lemma is proved.

We now apply the above lemma to $A = {}^jA(M/F)$ and see immediately the structure of the module ${}^jA(M/F)$. Since ${}^iA(M/F) = 0$ for an even index i, we have the following theorem, which also completes the proof of Theorem 16.

THEOREM 17. *The Γ-finite module $A(M/F)$ is regular. If i is an even index and j an odd index such that $i + j \equiv 1 \bmod p-1$, then ${}^iA(M/F) = 0$ and ${}^jA(M/F) \cong E(\infty)/{}^jD$, where ${}^jD$ is a finite submodule of $E(\infty)$, κ-isomorphic with the dual of ${}^iA(L/F)/{}^iR$.*

We shall next consider the regularity of the modules ${}^iA(K/F)$ and ${}^iA(L/F)$ for an even index i.

By Theorem 8, ${}^iA(K/F)$ is the κ-isomorphic image of the regular strictly Γ-finite module ${}^jS(F)$ $(i + j \equiv 1 \bmod p - 1)$. Hence ${}^iA(K/F)$ is regular if and only if it is strictly Γ-finite;[30] namely, if and only if ${}^iA(K/F)_n$ is finite for every $n \geqq 0$. Let $i \neq 0$. Then ${}^iA(K/F)_n \cong {}^iA(K_n/F_n)$. But ${}^iA(L'_n/F_n)$ is always finite and, by § 6, ${}^iG(K_n/L'_n) \cong {}^iU_{n,0}/{}^i\bar{E}_{n,0}$. Hence ${}^iA(K/F)_n$ is finite if and only if ${}^iU_{n,0}/{}^i\bar{E}_{n,0}$ is finite; namely, if and only if the compact p-primary abelian group ${}^i\bar{E}_{n,0}$ has rank p^n. Similarly, for $i = 0$, ${}^0A(K/F)_n$ is finite if and only if ${}^0\bar{E}_{n,0}$ has rank $p^n - 1$. It follows that ${}^iA(K/F)$, $i \neq 0$, (or ${}^0A(K/F)$) is regular if and only if ${}^i\bar{E}_{n,0}$ has rank p^n (or $p^n - 1$) for every $n \geqq 0$. Since ${}^jA(K/F)$ is regular for every odd index j, we also see that *$A(K/F)$ is regular if and only if the closure $\bar{E}_{n,0}$ of $E_{n,0}$ in $U_{n,0}$ has rank $((p-1)p^n)/2 - 1$ for every $n \geqq 0$.* However, this last statement, on the rank of $\bar{E}_{n,0}$, is yet unproved, though it looks quite likely to be true. We notice that the strict Γ-finiteness of ${}^iA(K/F)$ can be expressed also as a property of the characteristic roots of the Γ-module ${}^iA(K/F)$ and, hence, those of ${}^jA(L/F)$, for ${}^iA(K/F)$ and ${}^jA(L/F)$ are the κ-adjoints of each other by Theorem 8.

It is also yet unknown whether or not ${}^iA(L/F)$ is regular for every even index i; namely, whether or not the module $A(L/F)$ is regular. But, by Theorems 11–14, we see immediately that *the regularity of $A(L/F)$ is equivalent to each one of the following arithmetic properties of the cyclotomic fields F_n ($n \geqq 0$):*

(i) *for any $m \geqq n \geqq 0$, no non-principal ideal of F_n becomes a principal ideal in F_m,*

(ii) *for any $m \geqq n \geqq 0$, every unit of F_n is the norm of a unit in F_m.*

Thus the regularity of the modules $A(K/F)$ and $A(L/F)$ (and also many important arithmetic properties of the cyclotomic fields F_n ($n \geqq 0$)) essentially depends upon the structure of the G-group E, the group of units of the field F. But we leave the study of the structure of E to a future publication and mention here only the following fact: Let E^+ denote as before, the group of real units in E and E' the subgroup of E^+ generated by the so-called circular units. Let S be the Sylow p-group of the periodic abelian group E^+/E'. Then, as a Γ-group, S is strictly Γ-finite and

$$\lambda(S) = \lambda^+ , \qquad \mu(S) = \mu^+ ,$$

where λ^+ and μ^+ are the invariants of the Γ-extension F^+/F_0^+ as defined in § 5.

Finally, we also notice that *if the class number of the maximal real subfield F_0^+ of F_0 is prime to p, then ${}^iA(L/F) = 0$ for every even index i and $A(L/F)$ is regular.*

30 Cf., l.c. 17.

11. Local cyclotomic fields

In this last section, we shall briefly study local cyclotomic fields, where the situation is much simpler than the global case. To avoid the bulk of new notations, we shall freely use here the same letters as before, such as F, G, etc., to denote different fields and groups.

For each $n \geqq 0$, let F_n denote, in this section, the local cyclotomic field generated by $p^{n+1\text{th}}$ roots of unity over the p-adic number field Q_p, and let F be the union of all F_n ($n \geqq 0$). Then F is an abelian extension of Q_p, and the Galois group $G = G(F/Q_p)$ has the same structure as the Galois group denoted by the same letter G in § 2. In particular, we have

$$G = \Gamma \times \Delta ,$$

where $\Gamma = G(F/F_0)$ and Δ is canonically isomorphic with $G(F_0/Q_p)$. Let K be the maximal p-primary abelian extension of F. Since K is a Galois extension over Q_p, the group G acts on the Galois group $G(K/F)$ and on its character group $A(K/F)$ in a natural way. We shall next consider the structure of the G-module $A(K/F)$.

For each $n \geqq 0$, let K_n denote the maximal abelian extension of F_n contained in K. Then K_n contains F and is the maximal p-primary abelian extension of F_n. Furthermore, as in § 4, we see that

$$A(K/F)_n = A(K_n/F) , \qquad n \geqq 0.$$

Clearly, the group G also acts on $A(K_n/F_n)$ in a natural way. Let

$$A(K_n/F_n) = \textstyle\sum_{i=0}^{p-2} {}^iA(K_n/F_n)$$

be the Δ-decomposition of $A(K_n/F_n)$, and let ${}^iK_n$ ($0 \leqq i \leqq p-2$) be the subfield of K_n containing F_n such that ${}^iA(K_n/F_n) = A({}^iK_n/F_n)$. Then we have

$$F \subset {}^0K_n , \quad F \cap {}^iK_n = F_n , \qquad i \neq 0.$$

Let

$$A(K/F) = \textstyle\sum_{i=0}^{p-2} {}^iA(K/F)$$

be the Δ-decomposition of $A(K/F)$, and let ${}^iA(K/F)_n = {}^iA(K/F) \cap A(K/F)_n$. It then follows from the above that

$$\begin{aligned} {}^0A(K/F)_n &= {}^0A(K_n/F) = A({}^0K_n/F) = A({}^0K_n/F_n)/A(F/F_n) \\ &= {}^0A(K_n/F_n)/A(F/F_n) \\ {}^iA(K/F)_n &= {}^iA(K_n/F) = A(F\,{}^iK_n/F) \cong A({}^iK_n/F_n) = {}^iA(K_n/F_n) , \qquad i \neq 0. \end{aligned}$$

Let $\mathfrak{p}_n$ be the unique prime ideal (prime divisor) of the local field F_n, and let F_n^* be the multiplicative group of F_n. F_n^* is a locally compact group in its $\mathfrak{p}_n$-adic topology. Furthermore, let U_n be the subgroup of $\mathfrak{p}_n$-adic units in F_n^*, and let $U_{n,m}$ $(m \geqq 0)$ be the subgroup of all ξ in U_n such that $\xi \equiv 1 \bmod \mathfrak{p}_n^{m+1}$. Then U_n is the direct product of $U_{n,0}$ and a cyclic group H of order $p-1$, the group of all $(p-1)^{\text{th}}$ roots of unity in F_n. U_n and $U_{n,m}$ are all open compact subgroups of F_n^*, and F_n^* is the direct product of $U_n = U_{n,0} \times \mathrm{H}$ and an infinite cyclic group generated by $\pi_n = 1 - \zeta_{n+1}$, where ζ_{n+1} denotes a primitive $p^{n+1\,\text{th}}$ root of unity in F_n.

Now, let K_n' be the maximal abelian extension of F_n. By local class field theory,[31] the Galois group $G(K_n'/F_n)$ is then canonically isomorphic with a certain compact completion $\bar{F}_n^*$ of the locally compact group F_n^*, and it induces an isomorphism of $G(K_n/F_n)$ onto the Sylow p-group S_n of the totally disconnected compact abelian group $\bar{F}_n^*$. Since $G(F_n/Q_p) = G/\Gamma_n$ acts on F_n^* in an obvious manner, we may consider F_n^* also as a G-group. But, by continuity, the action of G can be uniquely extended on the completion $\bar{F}_n^*$ and we then see that the canonical isomorphisms considered above are G-isomorphisms. It follows in particular that if

$$S_n = \prod_{i=0}^{p-2} {}^iS_n$$

is the Δ-decomposition of S_n, then ${}^iS_n$ is isomorphic with ${}^iG(K_n/F_n)$ and hence is dual to ${}^iA(K_n/F_n)$.

As follows from the definition of the completion $\bar{F}_n$, the compact group $U_{n,0}$ is a G-subgroup of S_n such that $S_n/U_{n,0}$ is isomorphic with Γ and is trivially acted on by G. Therefore, if

$$U_{n,0} = \prod_{i=0}^{p-2} {}^iU_{n,0}$$

is the Δ-decomposition of $U_{n,0}$, then

$$ {}^0S_n/{}^0U_{n,0} \cong \Gamma\,, \quad {}^iS_n = {}^iU_{n,0}\,, \qquad i \neq 0,$$

and G acts trivially on ${}^0S_n/{}^0U_{n,0}$. However, it is known that, as an abelian group, $U_{n,0}$ is the direct product of the cyclic group W_n of order p^{n+1} generated by ζ_{n+1} and $(p-1)p^n$ copies of the group Γ and, also, that $U_{n,0}$ contains a subgroup which is, as a $G(F_n/Q_p)$-group, isomorphic with the additive group of the group ring of $G(F_n/Q_p)$ over the ring O_p of p-adic integers.[32] Since W_n is contained in ${}^1U_{n,0}$, it follows that ${}^iU_{n,0}$ $(i \neq 1)$ is, as an abelian group, the direct product of p^n copies of Γ and that ${}^1U_{n,0}$ is the direct product of W_n and p^n copies of Γ. In general, let $Z(p^m)$ denote

[31] For local class field theory used here and in the following, cf. [1].
[32] Cf. [2], § 11.

a discrete divisible p-primary abelian group of rank 1 with order p^m, $0 \leqq m \leqq \infty$; if m is finite, $Z(p^m)$ is a cyclic group of order p^m and, for $m = \infty$, $Z(p^\infty)$ is the discrete group dual to the compact abelian group Γ. From the above, we then have the following direct sum decompositions:

$$^0A(K_n/F_n) \cong Z(p^\infty) + Z(p^\infty)^{p^n}, \qquad ^1A(K_n/F_n) \cong Z(p^{n+1}) + Z(p^\infty)^{p^n},$$
$$^iA(K_n/F_n) \cong Z(p^\infty)^{p^n}, \qquad i \neq 0, 1,$$

where $Z(p^\infty)^{p^n}$ denotes the direct sum of p^n copies of $Z(p^\infty)$. Since $A(F/F_n) \cong Z(p^\infty)$, it also follows that

$$^1A(K/F)_n \cong Z(p^{n+1}) + Z(p^\infty)^{p^n}, \qquad ^iA(K/F)_n \cong Z(p^\infty)^{p^n}, \qquad i \neq 1.$$

Now, it is easy to prove that if A is a discrete Γ-module such that $A_n \cong Z(p^\infty)^{p^n}$ for every $n \geqq 0$, then A is isomorphic with the module $E(\infty)$.[33] Hence, by the above, we have $^iA(K/F) \cong E(\infty)$ for any $i \neq 1$. Since W_n is a G-subgroup of $^1U_{n,0}$ and since $^1U_{n,0} \cap W_{n+1} = W_n$, $^1A(K/F)$ contains a submodule B such that $B_n = {}^1A(K/F)_n \cap B \cong Z(p^\infty)^{p^n}$ and such that $^1A(K/F)_n/B_n$ is, as a Γ-module, dual to W_n and hence isomorphic to W_n. It follows that $B \cong E(\infty)$ and that $^1A(K/F)$ is the direct sum of B and a submodule isomorphic with $^1A(K/F)/B$. Since B is regular, we have $(^1A(K/F)/B)_n \cong {}^1A(K/F)_n/B_n \cong W_n$, and the following theorem is proved.

Theorem 18. *Let K be the maximal p-primary abelian extension of the local field F and let*

$$A(K/F) = \sum_{i=0}^{p-2} {}^iA(K/F)$$

be the Δ-decomposition of the character group $A(K/F)$ of the Galois group $G(K/F)$. Then, as a Γ-module, $^iA(K/F)$ is isomorphic with $E(\infty)$ for any $i \neq 1$, and $^1A(K/F)$ is isomorphic with the direct sum of $E(\infty)$ and the Γ-group W, the group of all $p^{n\text{th}}$ roots of unity, $n \geqq 0$, contained in F.

By the above theorem, we can see how $G = G(F/Q_p)$ acts on the normal subgroup $G(K/F)$ of $G(K/Q_p)$. Since it is easy to prove that the group extension $G(K/Q_p)/G(K/F)$ splits, the structure of the Galois group is thus completely determined. We also notice that since $G(K/F)$ is canonically isomorphic with the Sylow p-group of the Galois group of the maximal abelian extension K' of F, we can even determine without much difficulty the structure of the Galois group $G(K'/Q_p)$, though the group extension $G(K'/Q_p)/G(K'/F)$ does not split in this case.

We now fix an integer $n \geqq 0$ and consider the structure of the G-group

[33] Cf. [6], Lemma 5.1.

$U_{n,0}$. If $i \neq 0$, then ${}^iU_{n,0} = {}^iS_n$, and it is dual to ${}^iA(K/F)_n \cong {}^iA(K_n/F_n)$. But ${}^iA(K/F)_n \cong E(\infty)_n$ and ${}^1A(K/F)_n \cong W_n + E(\infty)_n$ by the above theorem. Hence the structure of ${}^iU_{n,0}$ $(i \neq 0)$ is determined.

Let Ξ_0 denote the group of all p-adic units ξ in Q_p such that $\xi \equiv 1 \bmod p$. Since Ξ_0 is a closed subgroup of ${}^0U_{n,0}$ and hence also a closed subgroup of ${}^0S_n$, we denote by L_n the subfield of ${}^0K_n$ containing F_n such that $G({}^0K_n/L_n)$ is mapped to Ξ_0 under the canonical isomorphism of $G({}^0K_n/F_n) = {}^0G(K_n/F_n)$ onto ${}^0S_n$. We shall next prove that $F \cap L_n = F_n$. Suppose that this is not true. Then F_{n+1} is contained in L_n. In general, we shall denote by N_m the norm map from F_m to Q_p and by $N_{n,m}$ the norm map from F_m to F_n $(m \geqq n \geqq 0)$. Then, since F_{n+1} is contained in L_n, by local class field theory, Ξ_0 is contained in $N_{n,n+1}(F^*_{n+1})$. It follows that $N_n(\Xi_0) = \Xi_0^{(p-1)p^n} = \Xi_0^{p^n}$ is contained in $N_n(N_{n,n+1}(F^*_{n+1})) = N_{n+1}(F^*_{n+1})$. But, again by local class field theory, we know that $N_{n+1}(F^*_{n+1}) \cap \Xi_0 = \Xi_0^{p^{n+1}}$. Hence we have a contradiction, and our assertion is proved.

Now, since Ξ_0 is G-invariant, $A(L_n/F_n)$ is a G-submodule of ${}^0A(K_n/F_n) = A({}^0K_n/F_n)$; and by the above, $A(F/F_n) \cap A(L_n/F_n) = 0$. But, since $A(F/F_n) \cong Z(p^\infty)$ and since $A({}^0K_n/L_n) = A({}^0K_n/F_n)/A(L_n/F_n)$ is dual to Ξ_0 and is also isomorphic with $Z(p^\infty)$, it follows that ${}^0A(K_n/F_n)$ is the direct sum of $A(F/F_n)$ and $A(L_n/F_n)$. We then have that $A(L_n/F_n) \cong {}^0A(K_n/F_n)/A(F/F_n) = A({}^0K_n/F) = {}^0A(K/F)_n \cong E(\infty)_n$ as Γ-modules.

Let M_n be the subfield of ${}^0K_n$ such that $G({}^0K_n/M_n)$ is mapped to ${}^0U_{n,0}$ under the canonical isomorphism of $G({}^0K_n/F_n)$ onto ${}^0S_n$. Since Ξ_0 is contained in ${}^0U_{n,0}$, M_n is a subfield of L_n and $A(M_n/F_n)$ is a submodule of $A(L_n/F_n)$. Furthermore, $G(M_n/F_n)$ is isomorphic with ${}^0S_n/{}^0U_{n,0} \cong \Gamma$ and is trivially acted on by G. Hence $A(M_n/F_n) \cong Z(p^\infty)$ and is trivially acted on by G. Since $A(L_n/F_n) \cong E(\infty)_n$, it follows that $A(M_n/F_n)$ corresponds to $E(\infty)_0$ under this isomorphism and hence that $A(L_n/M_n) \cong E(\infty)_n/E(\infty)_0$. As ${}^0U_{n,0}$ is dual to $A({}^0K_n/M_n)$ and the latter is the direct sum of $A(L_n/M_n)$ and a submodule naturally isomorphic with $A(F/F_n)$, we obtain the following:

THEOREM 19. *Let $U_{n,0}$ be the compact multiplicative group of all units ξ in the local cyclotomic field F_n such that $\xi \equiv 1 \bmod \mathfrak{p}_n$ $(n \geqq 0)$, and let $U_{n,0} = \prod_{i=0}^{p-2} {}^iU_{n,0}$ be the Δ-decomposition of $U_{n,0}$. Then, as Γ-modules, ${}^0U_{n,0}$ is dual to the direct sum of $E(\infty)_0$ and $E(\infty)_n/E(\infty)_0$, ${}^1U_{n,0}$ is dual to the direct sum of W_n and $E(\infty)_n$, and ${}^iU_{n,0}$ is dual to $E(\infty)_n$ for $i \neq 0, 1$. Here W_n is the group of all $p^{n+1\text{th}}$ roots of unity in F_n.*

Now, since $U = U_{n,0} \times \mathrm{H}$ and G acts trivially on H, the action of G on U is completely determined by the above theorem. Using this, we can also determine the structure of the G-group F_n^*, but we do not discuss this question here.

MASSACHUSETTS INSTITUTE OF TECHNOLOGY

BIBLIOGRAPHY

1. E. ARTIN, Algebraic numbers and algebraic functions. Lecture notes at Princeton University, 1950-1951.
2. C. CHEVALLEY, Class field theory. Lecture notes at Nagoya University, 1953–1954.
3. H. HASSE, Bericht über die neuere Untersuchungen und Probleme aus der Theorie der algebraischen Zahlkörper, I, Ia, II. Jahresber. der Deutsch. Math. Ver., (1926, 1927, 1930).
4. K. IWASAWA, *A note on the group of units of an algebraic number field.* Jour. math. pures et appl., 35 (1956), 189-192.
5. ———, Sheaves for algebraic number fields. Ann. of Math., 69 (1959), 408–413.
6. ———, On Γ-extensions of algebraic number fields. Bull. Amer. Math. Soc., vol. 65 (1959), 183-226.
7. ———, On some properties of Γ-finite modules. Ann. of Math., 70 (1959), 291–312.
8. F. POLLACZEK, *Über die irregulären Kreiskörper der l-ten und l²-ten Einheitswurzeln.* Math. Zeit., 21 (1924), 1-38.

[38] On local cyclotomic fields

J. Math. Soc. Japan, 12 (1960), 16-21.

Dedicated to Professor Z. Suetuna.

(Received May 4, 1959)

Introduction.

Let p be an odd prime, Q_p the p-adic number field, and Ω an algebraic closure of Q_p. For each $n \geqq 0$, we denote by F_n the extension field of Q_p generated by the set W_n of all p^{n+1}-th roots of unity in Ω. The local cyclotomic field F_n is then a cyclic extension of degree $p^n(p-1)$ over Q_p. Let W be the union of the increasing sequence of groups W_n ($n \geqq 0$) and let F be the union of the increasing sequence of fields F_n ($n \geqq 0$). Then $F = Q_p(W)$, and it is an infinite abelian extension of Q_p. Let M be the maximal abelian extension of F in Ω; M is clearly a Galois extension of Q_p.

We now consider the following problems on the local fields F_n and M: To determine the structure of the multiplicative group of the field F_n acted on by the Galois group $G(F_n/Q_p)$, and to describe explicitly the structure of the Galois group of the extension M/Q_p. In the present paper, we shall give a solution to these problems by using the result of a previous paper, in which we studied some arithmetic properties of local cyclotomic fields in applying the theory of Γ-finite modules.[1] We hope that the result of the present paper, combined with our previous results on Galois groups of local fields,[2] will give us further insight into the structure of the Galois group of the extension Ω/Q_p.

1. The structure of the multiplicative group of F_n.

Let U be the group of all p-adic units in Q_p and U^0 the subgroup of all a in U such that $a \equiv 1 \bmod p$. Then U is the direct product of U^0 and a cyclic subgroup V of order $p-1$ consisting of all roots of unity in Q_p:

$$U = U^0 \times V .$$

1) Cf. K. Iwasawa, On the theory of cyclotomic fields, Ann. of Math., 70 (1959), 530-561.

2) Cf. K. Iwasawa, On Galois groups of local fields, Trans. Amer. Math. Soc., 80 (1955), 448-469.

By local class field theory, there exists a topological isomorphism κ of $G=G(F/Q_p)$ onto U such that

$$\zeta^\sigma=\zeta^{\kappa(\sigma)}, \quad \sigma\in G,$$

for every ζ in W. Then, for any σ in G, there exists a unique element η_σ in V such that

$$\kappa(\sigma)\equiv\eta_\sigma \mod p,$$

and the mapping $\sigma\to\eta_\sigma$ defines a homomorphism of G onto V with kernel $G(F/F_0)$.

Let n ($\geqq 0$) be fixed. Let $\mathfrak{p}_n$ be the unique prime ideal of F_n dividing the rational prime p, and let B_n and $B_n{}^0$ denote, respectively, the group of all $\mathfrak{p}_n$-adic units in F_n and the subgroup of all β in B_n such that $\beta\equiv 1 \mod \mathfrak{p}_n$. Then B_n is the direct product of $B_n{}^0$ and V:

$$B_n=B_n{}^0\times V.$$

The groups B_n, $B_n{}^0$, and V are invariant under the Galois group $G_n=G(F_n/Q_p)$. The action of G_n on V is obviously trivial. But the action of G_n on $B_n{}^0$ is given as follows[3]: Let R_n be the group ring of G_n over the ring O_p of p-adic integers, and let I_n be the ideal of R_n consisting of all elements of the form $\sum_\sigma a_\sigma\sigma$ ($a_\sigma\in O_p$) with $\sum_\sigma a_\sigma=0$. Since $B_n{}^0$ is a p-primary compact abelian group, we may consider O_p as an operator domain of $B_n{}^0$. Hence we may also consider R_n as acting on $B_n{}^0$. As an R_n-group, $B_n{}^0$ is then the direct product of U^0, W_n, and a subgroup C_n isomorphic with the R_n-module I_n:

$$B_n{}^0=U^0\times W_n\times C_n.$$

Since $U=U^0\times V$, we also have

$$B_n=U\times W_n\times C_n, \qquad C_n\cong I_n.$$

Now, let A_n denote the multiplicative group of the field F_n and let π_n be any prime element of F_n. Then A_n/B_n is an infinite cyclic group generated by the coset of $\pi_n \mod B_n$, and the Galois group G_n acts trivially on A_n/B_n. Therefore $\pi_n{}^{\sigma-1}$ is contained in B_n for any σ in G_n. For such a σ, we also put

$$\eta_\sigma=\eta_{\sigma'},$$

where σ' is any element of $G=G(F/Q_p)$ inducing σ on F_n. We then have the following

LEMMA. *For any prime element π_n of F_n and for any σ in G_n,*

$$\pi_n{}^{\sigma-1}\equiv\eta_\sigma \mod \mathfrak{p}_n.$$

---

3) Cf. l. c. 1), Theorem 19.

PROOF. Let π_n' be any other prime element of F_n. Then $\pi_n' = \beta\pi_n$, with β in B_n; and since G_n acts trivially on V, $\beta^{\sigma-1} \equiv 1 \bmod \mathfrak{p}_n$. Hence $\pi_n'^{\sigma-1} \equiv \pi_n^{\sigma-1} \bmod \mathfrak{p}_n$, and we see that it is sufficient to prove the lemma for one particular π_n. Let ζ_{n+1} be a primitive p^{n+1}-th root of unity in F_n. Then $\pi_n = 1-\zeta_{n+1}$ is a prime element of F_n, and

$$\pi_n^{\sigma} \equiv \pi_n^{\sigma'} \equiv 1-\zeta_{n+1}^{\kappa(\sigma')} \equiv 1-(1-\pi_n)^{\kappa(\sigma')}$$
$$\equiv \kappa(\sigma')\pi_n \equiv \eta_{\sigma'}\pi_n \equiv \eta_\sigma\pi_n \qquad \bmod \mathfrak{p}_n^2 .$$

Threfore $\pi_n^{\sigma-1} \equiv \eta_\sigma \bmod \mathfrak{p}_n$, q. e. d.

Let π_n be again any prime element of F_n. By the above lemma, we put

$$\pi_n^{\sigma-1} = \beta_\sigma\eta_\sigma , \qquad \sigma \in G_n ,$$

with β_σ in B_n^0. We then denote by $D(\pi_n)$ the closure of the subgroup of the compact group B_n^0 generated by these β_σ ($\sigma \in G_n$); $D(\pi_n)$ consists of all elements of the form

$$\prod_\sigma \beta_\sigma^{a_\sigma}$$

with arbitrary p-adic integers a_σ. Since the elements β_σ ($\sigma \in G_n$) define a 1-cocycle of G_n in B_n^0 and satisfy the relations $\beta_{\tau\sigma} = \beta_\sigma\beta_\tau^\sigma$ ($\sigma, \tau \in G_n$), $D(\pi_n)$ is an R_n-subgroup of B_n^0.

THEOREM 1. *There exists a prime element π_n of F_n such that*

$$B_n = U \times W_n \times D(\pi_n) .$$

The R_n-group $D(\pi_n)$ is then isomorphic with the R_n-module I_n under an isomorphism φ such that $\varphi(\beta_\sigma) = \sigma - 1$ ($\sigma \in G_n$).

PROOF. Let $B_n = U \times W_n \times C_n$ as in the above, and let g be the projection from B_n on the factor C_n. For any ξ in A_n, $\xi^{\sigma-1}$ ($\sigma \in G_n$) is always contained in B_n. Hence we put

$$\xi_\sigma = g(\xi^{\sigma-1}) , \qquad \sigma \in G_n .$$

Then $\{\xi_\sigma\}$ defines a 1-cocycle of G_n in C_n; and since $H^1(G_n; A_n) = 1$, the mapping $\xi \to \{\xi_\sigma\}$ induces a homomorphism of A_n/B_n onto the cohomology group $H^1(G_n; C_n)$. Let f be an R_n-isomorphism of C_n onto I_n, and let ω_σ ($\sigma \in G_n$) be the elements of C_n such that $f(\omega_\sigma) = \sigma - 1$. It is then easy to see that $H^1(G_n; C_n)$ is a cyclic group of order p^n generated by the cohomology class of $\{\omega_\sigma\}$. Take a prime element π_n of F_n. Since A_n/B_n is an infinite cyclic group generated by the coset of $\pi_n \bmod B_n$, the 1-cocycle $\{g(\pi_n^{\sigma-1})\}$ also generates $H^1(G_n; C_n)$. Therefore there is an integer m, prime to p, such that

$$g(\pi_n^{\sigma-1}) = \omega_\sigma^m\gamma^{\sigma-1} , \qquad \sigma \in G_n ,$$

with an element γ in C_n. Since $\pi_n\gamma^{-1}$ is also a prime element of F_n, we replace π_n by $\pi_n\gamma^{-1}$ and denote the latter again by π_n. Then we have

$$g(\pi_n{}^{\sigma-1}) = \omega_\sigma{}^m, \qquad \sigma \in G_n.$$

As in the above, let $\pi_n{}^{\sigma-1} = \beta_\sigma \eta_\sigma$. Then $g(\beta_\sigma) = g(\pi_n{}^{\sigma-1}) = \omega_\sigma{}^m$ ($\sigma \in G_n$) and g induces an O_p-homomorphism of $D(\pi_n)$ into C_n. Therefore, if h is the O_p-homomorphism of I_n onto $D(\pi_n)$ such that $h(\sigma-1) = \beta_\sigma$, then

$$f \circ g \circ h(\sigma-1) = m(\sigma-1), \qquad \sigma \in G_n.$$

Since m is prime to p, $f \circ g \circ h$ is an automorphism of I_n. It follows that g induces an isomorphism of $D(\pi_n)$ onto C_n, and we have

$$B_n = U \times W \times D(\pi_n).$$

Suppose next that π_n is any prime element of F_n satisfying $B_n = U \times W \times D(\pi_n)$; π_n need not be the particular prime element obtained in the above argument. Clearly, there is an O_p-homomorphism ψ of I_n onto $D(\pi_n)$ such that $\psi(\sigma-1) = \beta_\sigma$. Since $\beta_{\tau\sigma} = \beta_\sigma \beta_\tau{}^\sigma$, ψ is then also an R_n-homomorphism. However, it follows from $B_n = U \times W_n \times C_n$ that $I_n \cong C_n \cong D(\pi_n)$. In particular, as compact abelian groups, both I_n and $D(\pi_n)$ are isomorphic with the direct sum of $p^n(p-1)-1$ copies of O_p. Hence ψ must be one-one, and $\varphi = \psi^{-1}$ is an R_n-isomorphism of $D(\pi_n)$ onto I_n such that $\varphi(\beta_\sigma) = \sigma - 1$. Thus the theorem is completely proved.

Since A_n/B_n is an infinite cyclic group generated by the coset of π_n mod B_n and since the action of G_n on $U \times W_n$ is well-known, *the structure of the G_n-group A_n, the multiplicative group of F_n, is completely determined by* Theorem 1.

2. The structure of the Galois group $G(M/Q_p)$.

Let E be the maximal unramified extension of Q_p in Ω. It is known that E is an abelian extension of Q_p generated by all roots of unity in Ω whose orders are prime to p, and also that the Galois group $G(E/Q_p)$ is isomorphic with the so-called total completion $\bar{Z}$ of the additive group Z of rational integers.[4] It follows that the Galois group $G(E'/Q_p)$ of the maximal p-complementary unramified extension E' of Q_p is isomorphic with the p-complementary completion ${}^p\bar{Z}$ of Z. Furthermore, for each $n \geq 0$, EF_n is the maximal unramified extension of F_n in Ω, and $E'F_n$ is the maximal p-complementary unramified extension of F_n in Ω. Let L_n be the maximal p-complementary abelian extension of F_n in Ω. Then $E'F_n$ is contained in L_n and,

4) For compact completions of (discrete) groups, cf. l. c. 2), 1.3. We also notice that a compact topological group is called p-primary (p-complementary) if and only if it is the inverse limit of a family of finite groups whose orders are powers of p (prime to p).

by local class field theory, $G(L_n/E'F_n)$ is naturally isomorphic with $B_n/B_n^0 \cong V$. Since $F_n \cap L_0 = F_0$, $G(F_nL_0/F_n) \cong G(L_0/F_0)$, F_nL_0 is clearly contained in L_n. But, since F_nL_0 contains both $E'F_n$ and a ramified extension of degree $p-1$ over F_n, it follows that

$$F_nL_0 = L_n, \qquad n \geqq 0.$$

If F_n' denotes the unique subfield of F_n with degree p^n over Q_p, then we also have

$$F_n'L_0 = L_n, \quad F_n' \cap L_0 = Q_p, \qquad n \geqq 0.$$

Let F' be the union of the increasing sequence of subfields F_n' in Ω. Then F' is a subfield of F such that $\kappa(G(F/F')) = V$, and we have

$$G(F'/Q_p) \cong U^0.$$

On the other hand, the union L of the increasing sequence of subfields L_n in Ω is, as one sees easily, the maximal p-complementary abelian extension of F in Ω. We then prove the following

THEOREM 2. *Let F' be the subfield of F such that $\kappa(G(F/F')) = V$ and let L_0 and L be the maximal p-complementary abelian extensions of F_0 and F in Ω, respectively. Then*

$$F'L_0 = L, \quad F' \cap L_0 = Q_p,$$

$$G(L/Q_p) = G(L/F') \times G(L/L_0),$$

$$G(L/F') \cong G(L_0/Q_p), \quad G(L/L_0) \cong G(F'/Q_p) \cong U^0.$$

Furthermore, $G(L_0/Q_p)$ is the p-complementary completion of a group generated by two elements σ and τ satisfying the only relations

$$\sigma\tau\sigma^{-1} = \tau^p, \quad \tau^{(p-1)^2} = 1;$$

σ is a Frobenius automorphism for L_0/Q_p and τ is a generator of the inertia group for L_0/Q_p.

PROOF. The first half of the theorem is an immediate consequence of what is stated in the above; one has only to notice that L_0 is a Galois extension of Q_p.

The field E' defined in the above is obviously the inertia field for the tamely ramified extension L_0/Q_p. Since $[L_0 : E'F_0] = [F_0 : Q_p] = p-1$ and $E' \cap F_0 = Q_p$, we see that $[L_0 : E'] = (p-1)^2$. The second half of the theorem is then an easy consequence of a result on the structure of the Galois group for the maximal tamely ramified extension of a local field.[5)]

If we are merely interested in the purely group-theoretical structure of the group $G(L/Q_p)$, we have the following corollary, which is an immediate consequence of the above theorem:

5) Cf. l. c. 2), 3.1.

COROLLARY. *The Galois group $G(L/Q_p)$ is the total completion of a group generated by two element λ and μ satisfying the only relations*

$$\lambda\mu\lambda^{-1}=\mu^p,\quad \mu^{(p-1)^s}=1.$$

THEOREM 3. *Let L and M be as in the above and let K be the maximal p-primary abelian extension of F in Ω so that $KL=M$, $K\cap L=F$. Then:*

i) *$G(M/L)$ is a closed normal subgroup of $G(M/Q_p)$ such that $G(M/Q_p)/G(M/L)=G(L/Q_p)$, and the group extension $G(M/Q_p)/G(M/L)$ splits,*

ii) *$G(L/F)$ acts trivially on $G(M/L)$ so that $G(M/L)$ can be considered as a G-group $(G=G(F/Q_p)=G(L/Q_p)/G(L/F))$, and as such, $G(M/L)$ is naturally isomorphic with $G(K/F)$.*

PROOF. Let

$$X=G(M/Q_p),\quad P=G(M/L_0),\quad N=G(M/L).$$

Then P is a closed p-primary normal subgroup of X, and $X/P=G(L_0/Q_p)$ is a p-complementary compact group. Hence the group extension X/P splits and there exists a closed subgroups H of X such that

$$HP=X,\quad H\cap P=1,\quad H\cong X/P.$$ [6]

Such a group H also satisfies $HN=G(M/F')$. On the other hand, since $P/N=G(L/L_0)\cong U^0$, there is an element σ in P such that $N\sigma$ generates a cyclic group which is everywhere dense in P/N. Let S be the closure of the cyclic subgroup of P generated by σ. Using $P/N\cong U^0$, we then see easily that

$$NS=P,\quad N\cap S=1.$$

Since both N and $HN=G(M/F')$ are normal in X, we have $(\sigma H\sigma^{-1})N=\sigma(HN)\sigma^{-1}=HN$, and $\sigma H\sigma^{-1}\cap N=\sigma(H\cap N)\sigma^{-1}=1$. Hence there is an element τ in N such that $\tau\sigma H\sigma^{-1}\tau^{-1}=H$.[7] Let $\sigma'=\tau\sigma$. Then $N\sigma=N\sigma'$, and the closure S' of the cyclic subgroup of P generated by σ' also satisfies $NS'=P$ and $N\cap S'=1$. Furthermore, since $\sigma'H\sigma'^{-1}=H$, S' is contained in the normalizer of H in X. Therefore $T=HS'$ is a closed subgroup of X, and it is easy to see that $NT=X$, $N\cap T=1$. Thus the first part of the theorem is proved.

The second part is an immediate consequence of the fact that $KL=M$, $K\cap L=F$, and $G(M/F)=G(M/K)\times G(M/L)$.

Now, the action of $G=G(F/Q_p)$ on $G(K/F)$ is explicitly known.[8] Therefore, combining that with the above Theorems 2, 3, we see that *the structure of the Galois group $G(M/Q_p)$ is thus completely determined.*

Massachusetts Institute of Technology.

6) Cf. l. c. 2), Lemma 5.
7) Cf. l. c. 6).
8) Cf. l. c. 1), Theorem 18.

[39] A class number formula for cyclotomic fields

Ann. of Math., (2) 76 (1962), 171-179.

(Received December 11, 1961)

Let p be an odd prime. Let F_n ($n \geqq 0$) denote the cyclotomic field generated by $p^{n+1\text{th}}$ roots of unity over the rational field $\mathbf{Q}$. As is well known, the class number h_n of F_n is the product of two integers h_n^- and h_n^+: $h_n = h_n^- h_n^+$. The so-called first factor h_n^- is given by the formula:

$$(1) \qquad h_n^- = 2p^{n+1} \prod_\chi \left(-\frac{1}{2p^{n+1}} \sum_m m\chi^{-1}(m) \right),$$

where m ranges over all integers satisfying $0 \leqq m < p^{n+1}$, $(m, p) = 1$, and χ over all characters of the multiplicative group of integers mod p^{n+1} with $\chi(-1) = -1$; and the second factor h_n^+, which is the class number of the maximal real subfield F_n^+ of F_n, is given by a group index:

$$(2) \qquad h_n^+ = [E_n^+ : T_n],$$

where E_n^+ is the group of units in F_n^+ and T_n is the subgroup of circular units in E_n^+.[1]

Let G_n denote the Galois group of F_n over $\mathbf{Q}$ and let $\mathfrak{R} = \mathbf{Z}[G_n]$ be the group ring of G_n over the ring of rational integers $\mathbf{Z}$. In the present paper, we shall first transform the formula (1) and show that the first factor h_n^- also can be expressed as a group index of certain additive groups in $\mathfrak{R}$. In general, any such class number formula suggests the existence of deeper group-theoretical relations between the ideal class group and the factor group by the order of which the class number is expressed.[2] In the rest of the paper, we shall study such relations between the two groups involved in our formula for h_n^-.

1. For simplicity, let $q = p^{n+1}$, $N = \frac{1}{2}(p-1)p^n$, and $G = G_n$. For any integer a prime to p, let σ_a denote the element of G such that $\sigma_a(\zeta) = \zeta^a$ for any q^{th} root of unity ζ in F_n. Then the mapping $a \to \sigma_a$ induces in the obvious way a natural isomorphism of G onto the multiplicative group of rational integers mod q. Let $\mathfrak{S} = \mathbf{Q}[G]$ be the group ring of G over $\mathbf{Q}$ and let

* The present research was supported in part by the National Science Foundation grants NSF-G13985 and NSF-G19992.

[1] Cf. [1]. The formula in [1] is slightly different from (1), but it can be easily transformed into (1).

[2] For the formula (2), see a remark [3, p. 556].

$$\omega_0 = \sum m\sigma_m^{-1}\,, \qquad \omega_0 \in \mathfrak{R}\,,$$

$$\omega = \frac{1}{q}\omega_0\,, \qquad \omega \in \mathfrak{S}\,,$$

where the sum for ω_0 is taken over all integers m such that $0 \leqq m < q$, $(m, p) = 1$. For any character χ of the abelian group G, let

$$\chi(\omega_0) = \sum_m m\chi(\sigma_m^{-1})\,.$$

Then formula (1) can be written in the form

$$h_n^- = 2q \prod_\chi \left(-\frac{1}{2q}\chi(\omega_0)\right),$$

where χ now ranges over all characters of G such that $\chi(\sigma_{-1}) = -1$.

Put

$$\mathfrak{J} = \mathfrak{R} \cap \mathfrak{R}\omega\,.$$

Let $\tau = \sigma_{-1}$ denote the complex conjugation of the imaginary field F_n. Let $\mathfrak{R}^-$ be the ideal of all elements α in $\mathfrak{R}$ such that $(1+\tau)\alpha = 0$, and let

$$\mathfrak{J}^- = \mathfrak{R}^- \cap \mathfrak{J}\,.$$

Then our class number formula says:

THEOREM 1. $h_n^- = [\mathfrak{R}^- : \mathfrak{J}^-]$.

For the proof, let

$$\varepsilon^+ = \frac{1}{2}(1+\tau)\,, \qquad \varepsilon^- = \frac{1}{2}(1-\tau)\,, \qquad \varepsilon^+ + \varepsilon^- = 1\,.$$

Then

$$\mathfrak{S} = \mathfrak{S}^+ \oplus \mathfrak{S}^-\,,$$

with $\mathfrak{S}^+ = \varepsilon^+\mathfrak{S}$, $\mathfrak{S}^- = \varepsilon^-\mathfrak{S}$. $\mathfrak{S}^-$ is a semi-simple algebra over $\mathbf{Q}$ with identity ε^-, and its absolutely irreducible (one-dimensional) representations φ are obtained from the characters χ of G with $\chi(\tau) = -1$ in the obvious manner. Hence the determinant of the matrix for $\varepsilon^-\omega_0$ in a regular representation of $\mathfrak{S}^-$ is given by

$$(3) \qquad \prod_\varphi \varphi(\varepsilon^-\omega_0) = \prod_\chi \chi(\omega_0) = \pm(2q)^{N-1}h_n^- \neq 0\,.$$

It follows that $\varepsilon^-\omega_0$ is a regular element of the algebra $\mathfrak{S}^-$.

Let $\mathfrak{A}$ denote the additive group in $\mathfrak{R}$ generated by all $\sigma_a - a$ with $(a, p) = 1$. It is easy to see that $(\sigma_a - a)\omega$ is contained in $\mathfrak{R}$ for any $\sigma_a - a$ in $\mathfrak{A}$. It follows that

$$q\mathfrak{J} = \mathfrak{A}\omega_0\,, \qquad q\mathfrak{J}^- = \mathfrak{A}\omega_0 \cap \mathfrak{S}^-\,.$$

For any $\alpha = \sum_\sigma a_\sigma \sigma$ $(a_\sigma \in \mathbf{Q})$ in $\mathfrak{S}$, let $\lambda(\alpha) = \sum_\sigma a_\sigma$. Since

$$\varepsilon^+ \omega_0 = q \sum_\sigma \sigma ,$$

we have

$$\varepsilon^+(\alpha\omega_0) = \alpha(\varepsilon^+\omega_0) = q\alpha(\textstyle\sum_\sigma \sigma) = q\lambda(\alpha)(\textstyle\sum_\sigma \sigma) .$$

Therefore $\alpha\omega_0$ is contained in $\mathfrak{S}^-$ if and only if $\lambda(\alpha) = 0$. It follows that $q\mathfrak{J}^-$ is the set of all elements of the form $\alpha\omega_0(= \varepsilon^-(\alpha\omega_0))$ with α in $\mathfrak{A}$ satisfying $\lambda(\alpha) = 0$. Let $\mathfrak{B}$ be the additive group in $\varepsilon^-\mathfrak{R}$ consisting of all elements of the form $\varepsilon^-\alpha$ with α in $\mathfrak{A}$ satisfying $\lambda(\alpha) = 0$. Then, by the above, $q\mathfrak{J}^- = \mathfrak{B}\omega_0 = \mathfrak{B}\varepsilon^-\omega_0$. Since $\varepsilon^-\omega_0$ is a regular element in $\varepsilon^-\mathfrak{R}$, it follows from (3) that

$$\begin{aligned}[\varepsilon^-\mathfrak{R} : q\mathfrak{J}^-] &= [\varepsilon^-\mathfrak{R} : \varepsilon^-\mathfrak{R}\varepsilon^-\omega_0][\varepsilon^-\mathfrak{R}\varepsilon^-\omega_0 : \mathfrak{B}\varepsilon^-\omega_0] \\ &= (2q)^{N-1}h_n^-[\varepsilon^-\mathfrak{R} : \mathfrak{B}] .\end{aligned}$$

Now, since $q = -(\sigma_{1+q} - (1+q))$ and $2\varepsilon^+ = \sigma_{-1} - (-1)$, we see easily that q, $2\varepsilon^+$, $\sigma_a - a$, and $\sigma_{-a} + a$, with $1 < a < q/2$, $(a, p) = 1$, form a basis of $\mathfrak{A}$ over $\mathbf{Z}$. Let α be any element in $\mathfrak{A}$ and let

$$\alpha = sq + t(2\varepsilon^+) + \textstyle\sum_a' \big(s_a(\sigma_a - a) + s_{-a}(\sigma_{-a} + a)\big) ,$$

where s, t, s_a and s_{-a} are all in $\mathbf{Z}$ and the sum is taken over all integers a such that $1 < a < q/2$, $(a, p) = 1$. Then

$$\lambda(\alpha) = sq + 2t + \textstyle\sum_a' \big(s_a(1 - a) + s_{-a}(1 + a)\big) .$$

Let $\tau_a = \varepsilon^-\sigma_a = \frac{1}{2}(\sigma_a - \sigma_{-a})$. Then $\tau_{-a} = -\tau_a$, and these elements τ_a, with $0 \leqq a < q/2$, $(a, p) = 1$, form a basis of $\varepsilon^-\mathfrak{R}$ over $\mathbf{Z}$. For the above α, we then have

$$\varepsilon^-\alpha = \textstyle\sum_a u_a\tau_a , \qquad 1 \leqq a < \frac{q}{2}, (a, p) = 1 ,$$

where the integers u_a are given by

$$\begin{aligned} u_1 &= sq + \textstyle\sum_a' (-as_a + as_{-a}) , \\ u_a &= s_a - s_{-a} , \qquad\qquad a \neq 1 . \end{aligned}$$

It follows that

$$\begin{aligned} \textstyle\sum_a a u_a &= sq \equiv 0 \qquad \bmod q , \\ \textstyle\sum_a u_a &\equiv \lambda(\alpha) \qquad \bmod 2 . \end{aligned}$$

Hence, if $\lambda(\alpha) = 0$, then $\sum_a u_a \equiv 0 \bmod 2$.

Suppose conversely that the coefficients u_a of an element $\beta = \sum_a u_a\tau_a$ in $\varepsilon^-\mathfrak{R}$ satisfy the congruences

$$(4) \qquad \sum_a au_a \equiv 0 \mod q, \qquad \sum_a u_a \equiv 0 \mod 2.$$

Let

$$\sum_a au_a = sq, \qquad \sum_a u_a = -2t,$$

and put

$$\alpha = sq + t(2\varepsilon^+) + \sum_a{}' u_a(\sigma_a - a), \qquad 1 < a < \frac{q}{2}, (a, p) = 1.$$

Then α is an element of $\mathfrak{A}$ and satisfies $\lambda(\alpha) = 0$, $\varepsilon^-\alpha = \beta$.

We now see that $\mathfrak{B}$ is the subgroup of all elements $\sum_a u_a\tau_a$ in $\varepsilon^-\mathfrak{R}$ with coefficients u_a satisfying the congruences (4). It then follows easily that

$$[\varepsilon^-\mathfrak{R} : \mathfrak{B}] = 2q,$$

and consequently that

$$[\varepsilon^-\mathfrak{R} : q\mathfrak{I}^-] = (2q)^N h_n^-.$$

This shows in particular that $q\mathfrak{I}^-$ is a free abelian group of the same rank N as $\varepsilon^-\mathfrak{R}$. Hence

$$[\mathfrak{I}^- : q\mathfrak{I}^-] = q^N.$$

On the other hand, since $\mathfrak{R}^- = 2(\varepsilon^-\mathfrak{R})$, we have

$$[\varepsilon^-\mathfrak{R} : \mathfrak{R}^-] = 2^N.$$

It follows that

$$[\mathfrak{R}^- : \mathfrak{I}^-] = h_n^-, \qquad \text{q.e.d.}$$

2. Let $\mathbf{Q}_p$ denote the p-adic number field and $\mathbf{Z}_p$ the subring of p-adic integers. Let $\mathfrak{R}_p = \mathbf{Z}_p[G] = \mathfrak{R} \otimes \mathbf{Z}_p$ (over $\mathbf{Z}$), $\mathfrak{S}_p = \mathbf{Q}_p[G] = \mathfrak{S} \otimes \mathbf{Q}_p$ (over $\mathbf{Q}$), and $\mathfrak{S}_p^- = \varepsilon^-\mathfrak{S}_p$. Furthermore, put

$$\mathfrak{I}_p = \mathfrak{R}_p \cap \mathfrak{R}_p\omega, \qquad \mathfrak{R}_p^- = \mathfrak{R}_p \cap \mathfrak{S}_p^-, \qquad \mathfrak{I}_p^- = \mathfrak{I}_p \cap \mathfrak{S}_p^-.$$

Since we have for $\mathfrak{I}_p$ and $\mathfrak{I}_p^-$ formulas similar to $q\mathfrak{I} = \mathfrak{A}\omega_0$ and $q\mathfrak{I}^- = \mathfrak{B}\omega_0$, we see that $\mathfrak{R}_p^-$ and $\mathfrak{I}_p^-$ are respectively the closures of $\mathfrak{R}^-$ and $\mathfrak{I}^-$ in $\mathfrak{S}_p$ with respect to the natural p-adic topology in $\mathfrak{S}_p$. Hence $\mathfrak{R}_p^- = \mathfrak{R}^- \otimes \mathbf{Z}_p$, $\mathfrak{I}_p^- = \mathfrak{I}^- \otimes \mathbf{Z}_p$ (over $\mathbf{Z}$), and $[\mathfrak{R}_p^- : \mathfrak{I}_p^-]$ is equal to the p-part of the index $[\mathfrak{R}^- : \mathfrak{I}^-]$. It follows from Theorem 1 that

$$(5) \qquad [\mathfrak{R}_p^- : \mathfrak{I}_p^-] = \text{the } p\text{-part of } h_n^-.^3$$

Let V denote the multiplicative group of all $(p-1)^{\text{st}}$ roots of unity in $\mathbf{Q}_p$. For any integer i, $0 \leqq i \leqq p-2$, let

[3] A similar formula holds for any prime number instead of p.

$$\varepsilon_i = \frac{1}{p-1} \sum_{\eta \in V} \eta^{-i} \sigma_\eta \,,$$

where σ_η denotes the element in the Galois group G such that $\sigma_\eta(\zeta) = \zeta^\eta$ for any q^{th} root of unity ζ in F_n. These ε_i are orthogonal idempotents in $\mathfrak{R}_p$ with $\sum_i \varepsilon_i = 1$. Let

$$\mathfrak{R}_p^{(i)} = \varepsilon_i \mathfrak{R}_p \,, \qquad \mathfrak{J}_p^{(i)} = \varepsilon_i \mathfrak{J}_p \,, \qquad 0 \leqq i \leqq p-2 \,.$$

Then we have the direct sum decompositions:

$$\mathfrak{R}_p^- = \sum\nolimits_i' \mathfrak{R}_p^{(i)} \,, \qquad \mathfrak{J}_p^- = \sum\nolimits_i' \mathfrak{J}_p^{(i)} \,,$$

where i ranges over all odd integers with $0 \leqq i \leqq p-2$. It follows that

$$[\mathfrak{R}_p^- : \mathfrak{J}_p^-] = \prod\nolimits_i' [\mathfrak{R}_p^{(i)} : \mathfrak{J}_p^{(i)}] \,. \tag{6}$$

We now suppose that the characters χ of the abelian group G take values in an algebraic closure of $\mathbf{Q}_p$ and consider the product

$$\prod\nolimits_\chi^{(i)} \left(-\frac{1}{2q} \chi(\omega_0) \right),$$

where i is any fixed odd integer with $0 \leqq i \leqq p-2$ and χ ranges over all characters such that $\chi(\sigma_\eta) = \eta^i$ for every η in V. It is easy to see that the above product is contained in $\mathbf{Q}_p$. Let $q^{(i)}$ denote the exact power of p contained in that product in $\mathbf{Q}_p$. It then follows from (1) and (5) that

$$[\mathfrak{R}_p^- : \mathfrak{J}_p^-] = q q^{(1)} q^{(3)} \cdots q^{(p-2)} \,.$$

Therefore, by (6),

$$\prod\nolimits_i' [\mathfrak{R}_p^{(i)} : \mathfrak{J}_p^{(i)}] = q \prod\nolimits_i' q^{(i)} \,. \tag{7}$$

However, we can prove the following more precise result:

LEMMA.
$$[\mathfrak{R}_p^{(1)} : \mathfrak{J}_p^{(1)}] = q q^{(1)} = 1,$$
$$[\mathfrak{R}_p^{(i)} : \mathfrak{J}_p^{(i)}] = q^{(i)}, \qquad 1 < i \leqq p-2, \ (i, 2) = 1.$$

Suppose first that $i \neq 1$. Then

$$\varepsilon_i = \frac{1}{p-1} \sum_{\eta \in V} \eta^{-i} (\sigma_\eta - \eta) \,.$$

Hence $\varepsilon_i \omega$ is contained in $\mathfrak{R}_p$, and

$$\mathfrak{J}_p^{(i)} = \varepsilon_i \mathfrak{R}_p \cap \mathfrak{R}_p \varepsilon_i \omega = \mathfrak{R}_p^{(i)} \varepsilon_i \omega \,.$$

It follows that $[\mathfrak{R}_p^{(i)} : \mathfrak{J}_p^{(i)}]$ is equal to the highest power of p dividing in $\mathbf{Q}_p$ the determinant of the matrix for $\varepsilon_i \omega$ in a regular representation of the semi-simple algebra $\varepsilon_i \mathfrak{S}_p$ over $\mathbf{Q}_p$. Since absolutely irreducible representations φ of $\varepsilon_i \mathfrak{S}_p$ are obtained from characters of G such that

$\chi(\sigma_\eta) = \eta^i$ ($\eta \in V$) and since

$$\varphi(\varepsilon_i) = 1, \qquad \varphi(\varepsilon_i \omega) = \frac{1}{q}\chi(\omega_0),$$

we have $q^{(i)} = [\Re_p^{(i)} : \Im_p^{(i)}]$ for $i \neq 1$.

Now, $[\Re_p^{(1)} : \Im_p^{(1)}] = qq^{(1)}$ follows immediately from (7). Let χ be any character of G such that $\chi(\sigma_\eta) = \eta$ for any η in V. Then

$$\begin{aligned}\chi(\omega_0) &= \sum_m m\chi(\sigma_m^{-1}) \\ &\equiv \sum \eta(1+p)^i\eta^{-1}\zeta'^i \qquad \bmod q, \qquad \eta \in V, 0 \leqq i < p^n,\end{aligned}$$

where $\zeta' = \chi(\sigma_{1+p}^{-1})$ is a $p^{n\text{th}}$ root of unity. Hence

$$\chi(\omega_0) \equiv (p-1)\frac{1-(1+p)^{p^n}}{1-(1+p)\zeta'} \qquad \bmod q.$$

Since the product $\prod(1-(1+p)\zeta')$ is divisible exactly by q when ζ' runs over all $p^{n\text{th}}$ roots of unity, we see immediately from the above that $qq^{(1)} = 1$.

3. Let C_n denote the ideal class group of F_n. Since the Galois group G of $F_n/\mathbf{Q}$ acts on C_n in the natural way, so does the group ring $\Re = \mathbf{Z}[G]$. Let C_n^- be the subgroup of C_n consisting of all ideal classes c such that $c^{1+\tau} = 1$. Let $\mathfrak{a}$ be any ideal in such a class c. Then $\mathfrak{a}^{1+\tau}$ is a principal ideal of F_n. However $\mathfrak{a}^{1+\tau}$ is an ideal of the subfield F_n^+ of F_n, and it is known that an ideal of F_n^+ is principal in F_n only when it is already principal in F_n^+.[4] Therefore $\mathfrak{a}^{1+\tau}$ is also a principal ideal of F_n^+. It follows that C_n^- is the kernel of the norm map from C_n to the ideal class group of F_n^+ and consequently[5] that h_n^- is equal to the order of C_n^-. Hence our class number formula (Theorem 1) can be also written as

$$[C_n^- : 1] = [\Re^- : \Im^-].$$

On the other hand, it is a classical result[6] that every element in the ideal $\Im$ of $\Re$ acts trivially on C_n:

$$c^\alpha = 1, \qquad c \in C_n, \alpha \in \Im. \tag{8}$$

These facts suggest, as we have mentioned in the introduction, the existence of deeper group-theoretical relations between the groups C_n^- and $\Re^-/\Im^-$.

Suppose first that C_n^- and $\Re^-/\Im^-$ are isomorphic as $\Re$-groups. Since

4 Cf. [1, Satz 28].
5 Cf. [1, Satz 11].
6 For $n = 0$, cf. [2, §109].

$\mathfrak{R}^- = \mathfrak{R}(1-\tau)$, C_n^- is then generated by a single element over $\mathfrak{R}$. Suppose conversely that there exists a class c in C_n^- which generates C_n^- over $\mathfrak{R}$. Let $\alpha(1-\tau) = 0$ for an element α in $\mathfrak{R}$. Then $\alpha = \beta(1+\tau)$ for some β in $\mathfrak{R}$. Since $c^{1+\tau} = 1$, we then have $c^\alpha = 1$. Therefore a mapping $f : \mathfrak{R}^- = \mathfrak{R}(1-\tau) \to C_n^-$ is well defined by

$$f(\alpha(1-\tau)) = c^\alpha , \qquad \alpha \in \mathfrak{R} ,$$

and f is obviously a surjective $\mathfrak{R}$-homomorphism. Let $\mathfrak{J}'$ be the kernel of f. Then the isomorphism $\mathfrak{R}^-/\mathfrak{J}' \cong C_n^-$ implies that $[\mathfrak{R}^- : \mathfrak{J}'] = [C_n^- : 1] = [\mathfrak{R}^- : \mathfrak{J}^-]$. However, it follows from (8) that $\mathfrak{J}^-$ is contained in $\mathfrak{J}'$. Hence $\mathfrak{J}' = \mathfrak{J}^-$, and we obtain an $\mathfrak{R}$-isomorphism:

$$C_n^- \cong \mathfrak{R}^-/\mathfrak{J}^- . \tag{9}$$

At the present, we know little about the existence of such an ideal class c as described above, and hence also about the existence of an $\mathfrak{R}$-isomorphism (9). However, for the p-Sylow subgroups of C_n^- and $\mathfrak{R}^-/\mathfrak{J}^-$, we can obtain more definite results as we shall show in the following.

4. Let S_n denote the p-Sylow subgroup of the ideal class group C_n. Since S_n is a p-group, the group ring $\mathfrak{R}_p = \mathbf{Z}_p[G]$ acts on S_n in the natural way. Using the fact that p is odd, we then have

$$S_n = S_n^+ \times S_n^- ,$$

with $S_n^+ = S_n^{\varepsilon^+} = S_n^{1+\tau}$ and $S_n^- = S_n^{\varepsilon^-} = S_n^{1-\tau}$. Since S_n^- is the p-Sylow subgroup of C_n^-, the order of S_n^- is equal to the p-part of $h_n^- = [C_n^- : 1]$. It then follows from (5) that

$$[S_n^- : 1] = [\mathfrak{R}_p^- : \mathfrak{J}_p^-] .$$

Therefore, by an argument similar to that in the previous section, we see immediately that S_n^- is $\mathfrak{R}_p$-isomorphic to $\mathfrak{R}_p^-/\mathfrak{J}_p^-$ if and only if S_n^- is generated over $\mathfrak{R}_p$ by a single element.

We now prove the following

THEOREM 2. *Assume that h_0^+, the class number of the maximal real subfield of the cyclotomic field of pth roots of unity, is prime to p.*[7] *Then there exists an $\mathfrak{R}_p$-isomorphism:*

$$S_n \cong \mathfrak{R}_p^-/\mathfrak{J}_p^- .$$

It is known that if h_0^+ is prime to p, then h_n^+ is also prime to p.[8] Hence $S_n^+ = 1$, $S_n = S_n^-$. By the remark stated before the theorem, it is there-

[7] This assumption is verified by numerical computation for many prime numbers, and no counter example has yet been found.

[8] Cf. [4].

fore sufficient to prove that S_n $(=S_n^-)$ is generated over $\mathfrak{R}_p$ by a single element. Let L_n and M_n be the abelian p-extensions of the field F_n as defined in § 5 and § 7 of [3] respectively. The Galois group G of the extension $F_n/\mathbf{Q}$ acts in the natural way on the Galois groups $G(L_n/F_n)$ and $G(M_n/F_n)$ of the extensions L_n/F_n and M_n/F_n respectively. Since both $G(L_n/F_n)$ and $G(M_n/F_n)$ are p-primary compact abelian groups, the group ring $\mathfrak{R}_p = \mathbf{Z}_p[G]$ also acts on $G(L_n/F_n)$ and $G(M_n/F_n)$. By class field theory, S_n is then $\mathfrak{R}_p$-isomorphic to $G(L_n/F_n)$. Hence it is sufficient to show that $G(L_n/F_n)$ is generated over $\mathfrak{R}_p$ by a single element. Now,

$$G(L_n/F_n) = G(L_n/F_n)^+ \times G(L_n/F_n)^- ,$$
$$G(M_n/F_n) = G(M_n/F_n)^+ \times G(M_n/F_n)^- ,$$

where $G(L_n/F_n)^\pm = G(L_n/F_n)^{1\pm\tau}$ and similarly for $G(M_n/F_n)^\pm$. Since $S_n^+ = 1$, $S_n = S_n^-$, we have $G(L_n/F_n)^+ = 1$, $G(L_n/F_n) = G(L_n/F_n)^-$. Since $G(L_n/F_n)^+ = 1$, it follows from [3, Theorem 17], that $G(M_n/F_n)^-$ is $\mathfrak{R}_p$-isomorphic to $\mathfrak{R}_p^- = \mathfrak{R}_p(1-\tau)$. It also follows from [3, Theorem 6], that $G(L_n/F_n) = G(L_n/F_n)^-$ is a factor group of $G(M_n/F_n)^-$. Therefore $G(L_n/F_n)$, as well as $G(M_n/F_n)^-$, is generated by a single element over $\mathfrak{R}_p$, q.e.d.

We can prove the theorem also directly without referring to the general results in [3] (although it essentially amounts to the same as above). Let K be the extension of F_n generated over F_n by all p^{th} roots of real units in F_n. Then it can be shown that the Galois group of K/F_n is generated over $\mathfrak{R}_p$ by a single element. It follows easily that $G(L_n/F_n)$ also has the same property.

The above theorem has various applications in the theory of cyclotomic fields. We state here only the following result which is an immediate consequence of the theorem and the lemma in § 2.

THEOREM 3. *Let $q^{(i)}$ $(1 < i \leqq p-2, (i, 2) = 1)$ denote, as before, the highest power of p dividing (in $\mathbf{Q}_p$) the product:*

$$\Pi_\chi^{(i)} \left(-\frac{1}{2p^{n+1}}\chi(\omega_0)\right) ,$$

where χ ranges over all characters of G_n such that $\chi(\sigma_\eta) = \eta^i$, $\eta \in V$. Let

$$S_n = S_n^{(0)} \times S_n^{(1)} \times \cdots \times S_n^{(p-2)}$$

be the Δ-decomposition[9] of the p-Sylow subgroup S_n of the ideal class group of F_n. Assume that h_0^+ is prime to p. Then

[9] Cf. [3, §3].

$$[S_n^{(i)} : 1] = \begin{cases} 1\ , & \textit{if } i \textit{ is even or } i = 1\ , \\ q^{(i)}\ , & \textit{if } i \textit{ is odd and } i \neq 1\ . \end{cases}$$

MASSACHUSETTS INSTITUTE OF TECHNOLOGY

BIBLIOGRAPHY

1. H. HASSE, Über die Klassenzahl abelscher Zahlkörper, Berlin, 1952.
2. D. HILBERT, *Die Theorie der algebraischen Zahlkörper*, Jahresber. der Deutsch. Math. Ver., 4 (1897), 175-546.
3. K. IWASAWA, *On the theory of cyclotomic fields*, Ann. of Math., 70 (1959), 530-561.
4. ———, *A note on class numbers of algebraic number fields*, Abh. Math. Sem. Univ. Hamburg, 20 (1956), 257-258.

[40] On a certain analogy between algebraic number fields and function fields

Sûgaku, 15-2 (10/1963), 65-67

Abstract

This is a short summary of Iwasawa's theory of $\mathbf{Z}_p$-extensions, beautifully explained following an analogy between the ideal class group of a number field and the Jacobian variety of a function field. More precisely, for a number field E, the author considers the field F obtained by adjoining all p-power roots of unity to E and discusses the action of $\mathrm{Gal}(F/E)$ on the p-primary subgroup A of the ideal class group of F. He states his famous class number formula

$$e_n = \lambda n + \mu p^n + \nu, \qquad n \gg 0$$

for the exponent of p in the class numbers of the intermediate fields of $F/E(\zeta_p)$ and explains the meaning and properties of the λ- and μ-invariants. In particular, he suggests that A is a good analogue of the p-power torsion points of the Jacobian variety of an algebraic curve over an algebraically closed field, if $\mu = 0$ in the above formula. The author also emphasizes the importance of the study of zeros of the characteristic polynomial of the action of the Galois group $\mathrm{Gal}(F/E(\zeta_p))$ on A.

The content of this article is essentially contained in [47] (1969). It would be worth noting that in this article no mention is made about the connection of his theory to the p-adic L-functions of Kubota-Leopoldt, while in [47] this remarkable relationship (now called Iwasawa's Main Conjecture) is explicitly mentioned and proved in the case $E = \mathbf{Q}$ by assuming Vandiver's conjecture.

論　説

代数体と函数体とのある類似について

M. I. T.　岩沢健吉

1. 代数体と函数体——くわしくいえば有限次代数体と一変数の代数函数体——との間に密接な類似があって，この類似を追うことによって，これらの理論がしばしば大きな発展を成しとげてきたことはよく知られている．一方一変数の代数函数体の理論，すなわち，幾何学的にいえば代数曲線の理論において，与えられた代数曲線の Jacobi 多様体が近来いかに大きな役割を果してきたかということも周知のとおりである．そこでなんらかの意味で Jacobi 多様体に対応するものを代数体に対して構成することができれば，代数的整数論の将来の展開に対して寄与するところがあろうということは誰しも思いつくところであって，以下ここに述べようとするのもそういった試みの一つである．

2. さて C を代数的閉体 k 上で定義された complete, non-singular な代数曲線とし，J をその Jacobi 多様体としよう．C の種数を g とすれば J は g 次元の代数的多様体であるが，それはいわゆる Abel 多様体であるから同時に Abel 群を成している．高次元の代数的多様体としての J に相当するものを代数体の理論の中で見出すことは今のところ難しいから，Abel 群としての J の構造に注目することにしよう．

今 k の標数と異なる任意の素数 p を一つ定めて，J の元で位数が p の巾であるようなものの全体のなす部分群を J_p とすると，この部分群 J_p は次のような簡単な群論的構造をもっている．すなわち p 進数体の加法群を $\boldsymbol{Q}_p$，p 進整数全体の成す部分群を $\boldsymbol{Z}_p$ とし，$\bar{\boldsymbol{Q}}_p=\boldsymbol{Q}_p/\boldsymbol{Z}_p$ とおけば，J_p は $2g$ 個の $\bar{\boldsymbol{Q}}_p$ の直和となる：

(1)　$J_p=\bar{\boldsymbol{Q}}_p\oplus\cdots\oplus\bar{\boldsymbol{Q}}_p$　($2g$ 個).

集合論的にいえば，J_p は加算集合であるから J のほんの一部分にすぎないが，代数幾何学的には J の中で稠密な部分集合となっているので代数曲線 C の理論においてしばしば重要な役を演ずる．たとえば，C の任意の代数的対応は Abel 群 J の自己準同型，従ってまたその特性部分群 J_p の自己準同型をひきおこすが，(1) を用いればこのような自己準同型を $\boldsymbol{Z}_p$ の元を成分とする $2g\times 2g$ の行列で表わすことができる．このようにして C の虚数乗法環を $\boldsymbol{Z}_p$ 上の行列によって同型に表現することができるのである．特に C が既に有限体の上で定義されている場合には Frobenius 置換から生ずる C の代数的対応があって，それを表現する行列の特性方程式が C の ζ 函数の主要部分を与える (A. Weil) ことを注意しておこう．

次に k 上で定義される C の上の有理函数全体のなす体を $K=k(C)$ とする．それは k を係数体とする一変数の代数函数体である．Jacobi 多様体を K に関して代数的な言葉でいい表わすには，K の次数 0 の因子類群を考えればよい．それは J に含まれる k-有理点全体の成す部分群 $J^{(k)}$ にほかならないからである．先の部分群 J_p は $J^{(k)}$ に含まれるから，K の因子類で位数が p の巾となるものの全体として，J_p を純代数的に定義することができる．

しかし，一方また次のように考えてもよい．K を含む代数的閉体を一つ定めてそれに含まれる K の不分岐 Abel 拡大 K' で次数 $[K':K]$ が p の巾となるようなものを考え，このようなすべての K' の合成体を A_K としよう．A_K は K 上の Abel 拡大体であって，これを K の最大不分岐 Abel p-拡大とよび，A_K/K の Galois 群を G_K と記すことにする．G_K はコンパクトな位相 Abel 群であるからその dual (指標群) を $\hat{G}_K$ とすると，Kummer 拡大の理論を用いて容易に上の群 J_p がこの $\hat{G}_K$ に同型になることが証明される．このように J_p は Galois 群 G_K の dual としても定義される．われわれはこの J_p の analogue を代数体に対して構成したいと思うのである.

3. さて任意の有限次代数体Fをとり，Fと函数体Kとの対応において因子類群$J^{(k)}$，あるいはJ_pに相当するものをFの方に求めれば，それがFの ideal 類群あるいはその p-Sylow 群となることは明らかであろう．特にJ_pについていえば，それを Galois 群の dual と考えて類似を追っても同型な群が得られる．すなわち，Fの最大不分岐 Abel p-拡大体 A_F を A_K と同様に定義し，A_F/F の Galois 群を G_F，その dual を $\hat{G}_F$ とすれば類体論によってG_F, 従って$\hat{G}_F$はFの ideal 類群の p-Sylow 群に同型となるからである．しかしこのp-Sylow 群あるいは，$\hat{G}_F$はもちろんともに有限群であってその構造は J_p のそれと一見全く相異なるばかりでなく，それはまた与えられた代数体Fの整数論的性質に深く依存していて，J_pの構造(1)に見られるような簡単な規則性がここには少しも見出されない．それゆえ，J_p の有用性がその簡単な構造 (1) に因っていることを思えば，$\hat{G}_F$ ないし ideal 類群の p-Sylow 群はJ_pの analogue としてあまり適当でないといえよう．なぜこのように $\hat{G}_F$ と J_p との間に差異が生じたのであろうか．それには幾つかの理由があげられるであろうが，ここでは次のように考えてみたい．上に J_p が $J^{(k)}$ に含まれ，$\hat{G}_K$ と同型であるといったが，実はこのことを証明するためには K の係数体 k が代数的に閉じているということ，少なくともkが十分多くの元を含んでいるということが極めて essential なのである——実際たとえばkが有限体であるような場合には$\hat{G}_K$は一般にJ_pと同型にならないでその構造は $\hat{G}_F$ のそれと同様に不規則である．ところで代数体と函数体との類似において函数体の係数体に対応するものは代数体に含まれる1の巾根にほかならない．それゆえ$\hat{G}_F \cong J_p$ となる函数体に対応する代数体，いいかえれば $\hat{G}_F$ が J_p に近い構造をもつような代数体を見つけるには1の巾根を十分多く含んでいるものを考えた方がよいことがわかるであろう．しかるに有限次代数体Fに含まれる1の巾根は有限個しかない．このことが上に述べた $\hat{G}_F$ と J_p との間の差異を生じた有力な原因の一つであるように思われるのである．

4. 以上の考察からわれわれは有限次代数体を捨てて，基礎体Fとして次のような1の巾根をもっと沢山含む（無限次）代数体をとることにする．すなわちEを任意の有限次代数体とし，pを定めてEにすべての1の p^n 乗根 ($n=1,2,3,\cdots$) を添加して得られる体を F とする．なぜ p^n 乗根だけでなく1のすべての巾根をEに添加しないかということその理由もいろいろあるが，ここでは省略しよう．とも角上のようなFを基礎体として前と全く同様に A_F, G_F および $\hat{G}_F$ を定義する．そうするとこの群 $\hat{G}_F$ の構造について次のことを証明することができる．今位数 p^μ ($\mu \geqq 0$) の環状群をZ_μ とし，Z_μ の可算無限個の直和を Z_μ^∞ とする．また任意の $\mu_1, \cdots, \mu_s \geqq 0$ に対し

$$E(\mu_1, \cdots, \mu_s) = Z_{\mu_1}^\infty \oplus \cdots \oplus Z_{\mu_s}^\infty$$

とおく．そうすると $A = \hat{G}_F$ は次のような部分群B, Cを含む：

1) $A/(B+C)$ および $B \cap C$ はともに有限群，
2) $B = \bar{Q}_p \oplus \cdots \oplus \bar{Q}_p$, λ 個の $\bar{Q}_p$,
3) $C = E(\mu_1, \cdots, \mu_s)/D$.

ただしここに $\lambda, \mu_1, \cdots, \mu_s$ は適当な整数 ($\geqq 0$)，Dは $E(\mu_1, \cdots, \mu_s)$ の適当な有限部分群である．1)は$A = \hat{G}_F$ がほとんど（すなわち有限群を除いて）BとCとの直和に等しいことを表わし，2)はBがJ_p と同じ型の群であることを示し，また3)からはCの各元の位数が有界であることがわかる．従って特にnを十分大きく取れば

$$p^n A = B = \bar{Q}_p \oplus \cdots \oplus \bar{Q}_p$$

となる．これらのことから $A = \hat{G}_F$ の構造が J_p のそれによほど近いものであることが見られるであろう．

さてEに1のp乗限を添加して得られるFの部分体をF_0とすれば，F/F_0は Abel 拡大でその Galois 群 Γ は Z_p に同型となる．そして Galois の理論から直ちにΓはG_Fおよび$\hat{G}_F$の上に自然に作用していることがわかる．すなわち $A = \hat{G}_F$ は Γ-加群であって，上に述べたAの構造に関する結果はこのことを用いて証明されたのである[1)]．しかもB, C等はAの単なる部分群ではなくて，Γ-部分群であり，また $\lambda, \mu_1, \cdots, \mu_s$，従って$\mu = \sum_{i=1}^{s} \mu_i$ が Γ-加群 A の invariant であること

も同時に示される．E と p とを与えれば，$F, A_F, G_F, \hat{G}_F$ が定まり，従って λ, μ が一意的に定まるからこれらを $\lambda(E,p), \mu(E,p)$ と記すことにすれば，このようにして任意の有限次代数体 E と任意の素数 p とに対し二つの invariant $\lambda(E,p), \mu(E,p)$ が定義される．Γ は $\boldsymbol{Z}_p$ に同型であるから任意の整数 $n \geqq 0$ に対し $[\Gamma:\Gamma_n]=p^n$ となる Γ の閉部分群 Γ_n がただ一つ存在する．この Γ_n に対応する F_0 と F との中間体を F_n とすれば F_n は有限次代数体であって，その類数 h_n の p 成分を p^{e_n} $(e_n \geqq 0)$ と書くとき，十分大きなすべての n に対して次の公式が成り立つ：

$$e_n=\lambda n+\mu p^n+\nu, \quad n \gg 0. \tag{2}$$

ここに $\lambda=\lambda(E,p), \mu=\mu(E,p)$ であって，また ν も n に無関係な整数である．これから invariant $\lambda(E,p), \mu(E,p)$ の整数論的意味がわかる．

5. われわれは Jacobi 多様体 J の稠密部分群 J_p の類似を求めて，Γ-加群 $A=\hat{G}_F$ に到達した．そして Abel 群として A が J_p に比較的近い構造をもっていることは上に見たとおりであるが，この Γ-加群 A がはたして J や J_p が函数体の理論においてなしてきたと同様な大きな役割を代数体の理論においてなしうるかどうかは将来の研究の進展にまつほかはない．ただ (2) からも見えるように，Γ-加群 A の構造，特に invariant λ, μ が(有限次) 代数体の整数論的性質に深く関連していて，それらがとも角今後の研究の対象として価値のないものでないことだけは確かのように思われる．今 λ, μ について直ちに思いつく二，三の問題をあげてみれば i) p を定め，相異なる有限次代数体 E に対して $\lambda(E,p), \mu(E,p)$ の関係を調べる (たとえば $E \subset E'$ ならば $\lambda(E,p) \leqq \lambda(E',p)$, $\mu(E,p) \leqq \mu(E',p)$ は容易に証明される)，ii) 今度は E を定めて，相異なる素数 p に対し $\lambda(E,p), \mu(E,p)$ の集合を調べる，iii) $\mu(E,p)=0$ となる条件を求める，等々．特に iii) についていえば，$\mu(E,p)=0$ は A の部分群 C が 0 になることと同値で，いいかえれば A が有限群を除いて $B=\bar{\boldsymbol{Q}}_p \oplus \cdots \oplus \bar{\boldsymbol{Q}}_p$ と一致することと同値である．これは Γ-加群 $A=\hat{G}_F$ が J_p の非常によい analogue を与える場合で，そういう意味で興味がある．

先に函数体の場合には J_p を用いて K (あるいは C) の代数的対応を $\boldsymbol{Z}_p$ 上の行列によって表現することができることを注意したが，同様なことが A を用いて代数体の場合にも考えられる：今特に E が有理数体 $\boldsymbol{Q}$ (あるいは $\boldsymbol{Q}$ と E とのある中間体) の上の Galois 拡大体であるとすると，$F/\boldsymbol{Q}$ もまた Galois 拡大となりその Galois 群 G は G_F および $A=\hat{G}_F$ の上に自然に作用し，しかも B は A の G-不変な部分群となる．よって B が J_p と同様な構造をもっていることから，$\boldsymbol{Z}_p$ 上の $\lambda \times \lambda$ の行列による Galois 群の表現が得られる．F/F_0 の Galois 群 Γ は G の部分群であるから，このようにして特に Γ の表現も得られるわけであるが，これらの表現の性質を調べることはまた興味ある問題と思う．

上に述べてきた一般論において，特に $E=\boldsymbol{Q}$ とおけば最も簡単な，しかしおそらくまた一番典型的な例が得られる．この場合 F_n は円の p^{n+1}-分体となるから，円分体論を用いて $A=\hat{G}_F$ や λ, μ をいくらか詳しく調べることができる[2]．たとえばすべての $p<100$ に対し $\lambda(\boldsymbol{Q},p)=0$ あるいは 1, $\mu(\boldsymbol{Q},p)=0$ であって，$\lambda(\boldsymbol{Q},p)=1$ のときには $A=B=\bar{\boldsymbol{Q}}_p$ となることなどが容易に知られる．この円分体の場合に上の G および Γ の表現を決めること，表現行列の特性方程式の根を求めることは特に重要であると思われる．

注

1) K. Iwasawa, On Γ-extensions of algebraic number fields, Bull. Amer. Math. Soc., 65 (1959), 183-226.

2) K. Iwasawa, On the theory of cyclotomic fields, Ann. of Math., 70 (1959), 530-561.

[41] On some modules in the theory of cyclotomic fields

J. Math. Soc. Japan, 16 (1964), 42-82.

(Received Oct. 3, 1963)

Let p be a fixed odd prime, and let $q_n = p^{n+1}$ for any integer $n \geqq 0$. Let F_n denote the cyclotomic field of q_n-th roots of unity over the rational field; and Φ_n, the local cyclotomic field of q_n-th roots of unity over the p-adic number field. The main purpose of the present paper is to introduce three compact modules $\mathfrak{X}$, $\mathfrak{Y}$, and $\mathfrak{Z}$ into the theory of cyclotomic fields F_n and Φ_n, $n \geqq 0$. They are defined as inverse limits of certain subgroups $\mathfrak{X}_n$, $\mathfrak{Y}_n$, and $\mathfrak{Z}_n$ respectively, of the additive group of Φ_n, $n \geqq 0$, and $\mathfrak{Y}$ is a submodule of $\mathfrak{X}$; $\mathfrak{Z}$, a submodule of $\mathfrak{Y}$. We shall determine the algebraic structure of $\mathfrak{X}/\mathfrak{Z}$ by direct computation, and we shall also show by class field theory how $\mathfrak{X}/\mathfrak{Y}$ and $\mathfrak{Y}/\mathfrak{Z}$ are related respectively to the ideal class groups and the unit groups of the fields F_n, $n \geqq 0$. Using these results, we shall then study the group-theoretical meaning of the classical class number formula for F_n as noted in a previous paper [10].

§ 1.

1.1. Let $\boldsymbol{Z}$, $\boldsymbol{Z}_p$, $\boldsymbol{Q}$ and $\boldsymbol{Q}_p$ denote the ring of rational integers, the ring of p-adic integers, the rational field, and the p-adic number field respectively. We shall fix an algebraic closure Ω of $\boldsymbol{Q}_p$, and consider all algebraic extensions of $\boldsymbol{Q}$ and $\boldsymbol{Q}_p$ as subfields of Ω.

Let F denote the union of all F_n, $n \geqq 0$; and Φ, the union of all Φ_n, $n \geqq 0$. Then both $F/\boldsymbol{Q}$ and $\Phi/\boldsymbol{Q}_p$ are abelian extensions, and their Galois groups are identified in a natural way. Put

$$G = G(F/\boldsymbol{Q}) = G(\Phi/\boldsymbol{Q}_p)^{1)},$$

$$G_n = G(F_n/\boldsymbol{Q}) = G(\Phi_n/\boldsymbol{Q}_p),$$

$$\Gamma = G(F/F_0) = G(\Phi/\Phi_0),$$

$$\Gamma_n = G(F/F_n) = G(\Phi/\Phi_n), \qquad n \geqq 0,$$

* The present research was supported in part by the National Science Foundation grant NSF-GP-379.

1) $G(\ /\ )$ shall denote the Galois group of the Galois extension in the parenthesis.

so that $\Gamma=\Gamma_0, G_n=G/\Gamma_n$.

Let W_n ($n\geqq 0$) denote the group of all q_n-th roots of unity in F_n (in Φ_n), and let W be the union of all $W_n, n\geqq 0$. Let $\boldsymbol{U}$ be the multiplicative group of all p-adic units in $\boldsymbol{Q}_p$. Then there exists an isomorphism

$$\kappa: G\rightarrow \boldsymbol{U}$$

such that

$$\zeta^\sigma=\zeta^{\kappa(\sigma)},$$

for any σ in G and ζ in W. The image of Γ under κ is the subgroup $\boldsymbol{U}_0$ of all u in $\boldsymbol{U}$ such that $u\equiv 1 \bmod p$. Let $\boldsymbol{V}$ denote the subgroup of all $(p-1)$-st roots of unity in $\boldsymbol{U}$; $\boldsymbol{V}$ is a cyclic group of order $p-1$, and

$$\boldsymbol{U}=\boldsymbol{U}_0\times \boldsymbol{V}.$$

Let Δ denote the subgroup of G, corresponding to $\boldsymbol{V}$ under κ. Then

$$G=\Gamma\times\Delta,$$

and Δ is canonically isomorphic to G_0.

For any u in $\boldsymbol{U}$, let $\sigma(u)$ denote the inverse image of u under κ; $\sigma(u)$ is the element of G such that $\zeta^{\sigma(u)}=\zeta^u$ for any ζ in W. We denote by $\sigma(u)_n$ the image of $\sigma(u)$ under the natural homomorphism $G\rightarrow G_n$ ($n\geqq 0$). Clearly $\sigma(u)_n$ depends only upon the residue class of u mod q_n.

In general, for any group $\mathfrak{G}$ and any additive abelian group A, let $A[\mathfrak{G}]$ denote the set of all maps $f:\mathfrak{G}\rightarrow A$ such that $f(x)=0$ except for a finite number of x's in $\mathfrak{G}$. Defining $f+g$ by

$$(f+g)(x)=f(x)+g(x),$$

we make $A[\mathfrak{G}]$ into a module. If A is a ring, $A[\mathfrak{G}]$ is nothing but the additive group of the group ring of $\mathfrak{G}$ over A. If there is a homomorphism $G\rightarrow\mathfrak{G}$, we can also make $A[\mathfrak{G}]$ into a G-module, by defining σf as

$$(\sigma f)(x)=f(s^{-1}x), \qquad x\in\mathfrak{G},$$

where s denotes the image of σ under $G\rightarrow\mathfrak{G}$.

For any integer i, we define an element ${}^i\varepsilon$ of the group ring $\boldsymbol{Z}_p[G]$ by

$${}^i\varepsilon=(p-1)^{-1}\sum_v v^{-i}\sigma(v), \qquad v\in \boldsymbol{V}.$$

The elements ${}^0\varepsilon, {}^1\varepsilon, \cdots, {}^{p-2}\varepsilon$ form a system of orthogonal idempotents in $\boldsymbol{Z}_p[G]$ with $\sum_i {}^i\varepsilon=1$. We also put

$${}^+\varepsilon=\sum_{i \text{ even}} {}^i\varepsilon, \qquad {}^-\varepsilon=\sum_{i \text{ odd}} {}^i\varepsilon.$$

Then ${}^+\varepsilon+{}^-\varepsilon=1$, ${}^+\varepsilon{}^-\varepsilon=0$. If A is a $\boldsymbol{Z}_p[G]$-module, we define submodules of A by

$${}^+A={}^+\varepsilon A, \qquad {}^-A={}^-\varepsilon A, \qquad {}^iA={}^i\varepsilon A, \qquad 0\leqq i\leqq p-2.$$

Then we obtain the direct decompositions

$$A={}^{+}A\oplus{}^{-}A={}^{0}A\oplus{}^{1}A\oplus\cdots\oplus{}^{p-2}A.$$

1.2. Let

$$\mathfrak{R}_n=\boldsymbol{Z}_p[G_n],\qquad \mathfrak{S}_n=\boldsymbol{Q}_p[G_n],\qquad n\geqq 0.$$

By means of the natural homomorphism $G\to G_n$, both $\mathfrak{R}_n$ and $\mathfrak{S}_n$ become G-modules and hence, also $\boldsymbol{Z}_p[G]$-modules. Let $\mathfrak{R}_n^0$ denote the submodule of all $\sum_\rho a_\rho\rho$ ($\rho\in G_n$, $a_\rho\in\boldsymbol{Z}_p$) in $\mathfrak{R}_n$ such that $\sum_\rho a_\rho=0$, and let

$$\mathfrak{A}_n=\mathfrak{B}_n+\mathfrak{R}_n^0,\qquad \mathfrak{B}_n=\mathfrak{R}_n\xi_n,$$

where

$$\xi_n=q_n^{-1}\sum_a\left(a-\frac{q_n-p}{2}\right)\sigma(a)_n,\qquad 0\leqq a<q_n, (a,p)=1.$$

Clearly $\mathfrak{A}_n$ and $\mathfrak{B}_n$ are $\boldsymbol{Z}_p[G]$-submodules of $\mathfrak{S}_n$.

Let $m\geqq n\geqq 0$. Then the natural homomorphism $G_m\to G_n$ defines a $\boldsymbol{Z}_p[G]$-homomorphism

$$t_{n,m}:\mathfrak{S}_m\to\mathfrak{S}_n$$

in the obvious manner. There also exists an injective $\boldsymbol{Z}_p[G]$-homomorphism

$$t'_{m,n}:\mathfrak{S}_n\to\mathfrak{S}_m$$

such that

$$t'_{m,n}\circ t_{n,m}(\alpha)=\nu_{n,m}\alpha,\qquad \alpha\in\mathfrak{S}_m,$$

where

$$\nu_{n,m}=\sum_{i=0}^{p^{m-n}-1}\sigma(1+p)^{ip^n}.$$

Since $t_{n,m}(\sigma(a)_m)=\sigma(a)_n$, we see easily that $t_{n,m}(\mathfrak{R}_m)=\mathfrak{R}_n$, $t_{n,m}(\mathfrak{R}_m^0)=\mathfrak{R}_n^0$, and

$$t_{n,m}(\xi_m)=\xi_n.$$

Hence

$$t_{n,m}(\mathfrak{A}_m)=\mathfrak{A}_n,\qquad t_{n,m}(\mathfrak{B}_m)=\mathfrak{B}_n.$$

Let $\mathfrak{A}$ and $\mathfrak{B}$ denote the inverse limits of $\mathfrak{A}_n$ and $\mathfrak{B}_n$, $n\geqq 0$, respectively, relative to the maps $t_{n,m}$, $m\geqq n$:

$$\mathfrak{A}=\lim\mathfrak{A}_n,\qquad \mathfrak{B}=\lim\mathfrak{B}_n.$$

$\mathfrak{A}_n$ is a compact submodule of $\mathfrak{S}_n$ in the natural topology of $\mathfrak{S}_n$ defined by the p-adic topology of $\boldsymbol{Q}_p$. Hence $\mathfrak{A}$ is a compact $\boldsymbol{Z}_p[G]$-module, and $\mathfrak{B}$ is a closed submodule of $\mathfrak{A}$. Since $\mathfrak{B}_n$, $n\geqq 0$, is compact, the compact $\boldsymbol{Z}_p[G]$-module $\mathfrak{A}/\mathfrak{B}$ is the inverse limit of $\mathfrak{A}_n/\mathfrak{B}_n$, $n\geqq 0$:

$$\mathfrak{A}/\mathfrak{B}=\lim\mathfrak{A}_n/\mathfrak{B}_n.$$

Hence

$${}^{\pm}(\mathfrak{A}/\mathfrak{B})=\lim{}^{\pm}\mathfrak{A}_n/{}^{\pm}\mathfrak{B}_n,\qquad {}^{i}(\mathfrak{A}/\mathfrak{B})=\lim{}^{i}\mathfrak{A}_n/{}^{i}\mathfrak{B}_n,\qquad 0\leqq i\leqq p-2.$$

where ${}^{\pm}(\mathfrak{A}/\mathfrak{B})$ etc. denote the components of the decompositions of respective modules defined in 1.1.

Let

$$ {}^{\pm}\xi_n = {}^{\pm}\varepsilon\xi_n, \qquad {}^{i}\xi_n = {}^{i}\varepsilon\xi_n, \qquad 0 \leqq i \leqq p-2. $$

Then

$$ (1) \qquad {}^{+}\xi_n = (2p^n)^{-1}\sum_{\rho} \rho, \qquad \rho \in G_n, $$
$$ {}^{-}\xi_n = (2q_n)^{-1}\sum_{a} (2a-q_n)\sigma(a)_n, \qquad 0 \leqq a < q_n, (a,p)=1, $$

and

$$ (2) \qquad \kappa(\sigma)\sigma^{-}\xi_n \equiv {}^{-}\xi_n \bmod \mathfrak{R}_n^0, \qquad \sigma \in G. $$

PROPOSITION 1.

$$ {}^{0}\mathfrak{A}_n = {}^{0}\mathfrak{B}_n \oplus {}^{0}\mathfrak{R}_n^0, \qquad {}^{0}\mathfrak{B}_n = \boldsymbol{Z}_p{}^{+}\xi_n, $$
$$ {}^{i}\mathfrak{A}_n = {}^{i}\mathfrak{R}_n, \qquad {}^{i}\mathfrak{B}_n = 0, \qquad i \text{ even}, i \neq 0, $$
$$ {}^{i}\mathfrak{A}_n = {}^{i}\mathfrak{R}_n, \qquad {}^{i}\mathfrak{B}_n = \mathfrak{R}_n{}^{i}\xi_n = {}^{i}\mathfrak{R}_n{}^{i}\xi_n, \qquad i \text{ odd}, i \neq p-2, $$
$$ {}^{p-2}\mathfrak{A}_n = {}^{p-2}\mathfrak{B}_n. $$

PROOF. It is clear that ${}^{i}\mathfrak{A}_n = {}^{i}\mathfrak{B}_n + {}^{i}\mathfrak{R}_n^0$ and ${}^{i}\mathfrak{B}_n = \mathfrak{R}_n{}^{i}\xi_n = {}^{i}\mathfrak{R}_n{}^{i}\xi_n$ for every i. Let i be even. It follows from (1) that ${}^{i}\xi_n = {}^{i}\varepsilon^{+}\xi_n = {}^{+}\xi_n$ or ${}^{i}\xi_n = 0$ according as $i=0$ or $i \neq 0$. Since ${}^{0}\mathfrak{B}_n = \mathfrak{R}_n{}^{+}\xi_n = \boldsymbol{Z}_p{}^{+}\xi_n$ and $\boldsymbol{Z}_p{}^{+}\xi_n \cap {}^{0}\mathfrak{R}_n^0 = 0$, we obtain ${}^{0}\mathfrak{A}_n = {}^{0}\mathfrak{B}_n \oplus {}^{0}\mathfrak{R}_n^0$. We also see from the above that ${}^{i}\mathfrak{B}_n = \mathfrak{R}_n{}^{i}\xi_n = 0$ and ${}^{i}\mathfrak{A}_n = {}^{i}\mathfrak{R}_n^0 = {}^{i}\mathfrak{R}_n$ for $i \neq 0$.

Next, let i be odd and $i \neq p-2$. It then follows from (2) that ${}^{i}\xi_n = {}^{i}\varepsilon^{-}\xi_n$ is contained in ${}^{i}\mathfrak{R}_n^0 = {}^{i}\varepsilon\mathfrak{R}_n^0$. Hence ${}^{i}\mathfrak{A}_n = \mathfrak{R}_n{}^{i}\xi_n + {}^{i}\mathfrak{R}_n^0 = {}^{i}\mathfrak{R}_n^0 = {}^{i}\mathfrak{R}_n$.

To prove the last equality, we consider cohomology groups of Γ/Γ_n. It follows from (2) that

$$ \kappa(\sigma)\sigma^{p-2}\xi_n \equiv {}^{p-2}\xi_n \quad \bmod {}^{p-2}\mathfrak{R}_n^0 = {}^{p-2}\mathfrak{R}_n, \qquad \sigma \in G, $$

and that ${}^{p-2}\mathfrak{A}_n/{}^{p-2}\mathfrak{R}_n$ is a cyclic group of order q_n generated by the coset of ${}^{p-2}\xi_n$. The above congruence then shows that $H^k(\Gamma/\Gamma_n, {}^{p-2}\mathfrak{A}_n/{}^{p-2}\mathfrak{R}_n) = 0$ for every k. Since ${}^{p-2}\mathfrak{R}_n \cong \boldsymbol{Z}_p[\Gamma/\Gamma_n]$ as (Γ/Γ_n)-modules, we also have $H^k(\Gamma/\Gamma_n, {}^{p-2}\mathfrak{R}_n) = 0$. Hence $H^k(\Gamma/\Gamma_n, {}^{p-2}\mathfrak{A}_n) = 0$. However, as ${}^{p-2}\mathfrak{A}_n$ is a free $\boldsymbol{Z}_p$-module of rank p^n, ${}^{p-2}\mathfrak{A}_n$ is isomorphic to $p({}^{p-2}\mathfrak{A}_n)$, and $H^k(\Gamma/\Gamma_n, p({}^{p-2}\mathfrak{A}_n)) = 0$. Therefore, $H^k(\Gamma/\Gamma_n, {}^{p-2}\mathfrak{A}_n/p({}^{p-2}\mathfrak{A}_n)) = 0$ for every k, and we see that ${}^{p-2}\mathfrak{A}_n/p({}^{p-2}\mathfrak{A}_n) \cong (\boldsymbol{Z}/p\boldsymbol{Z})[\Gamma/\Gamma_n]$ as (Γ/Γ_n)-modules. On the other hand, since ${}^{p-2}\xi_n$ generates ${}^{p-2}\mathfrak{A}_n/{}^{p-2}\mathfrak{R}_n$, it is not contained in the submodule $p({}^{p-2}\mathfrak{A}_n) + {}^{p-2}\mathfrak{R}_n$ of index p in ${}^{p-2}\mathfrak{A}_n$. Hence the cosets of $\rho^{p-2}\xi_n$, $\rho \in \Gamma/\Gamma_n$, form a basis of the abelian group ${}^{p-2}\mathfrak{A}_n/p({}^{p-2}\mathfrak{A}_n)$. It follows that ${}^{p-2}\mathfrak{A}_n = {}^{p-2}\mathfrak{B}_n + p({}^{p-2}\mathfrak{A}_n)$, and hence that ${}^{p-2}\mathfrak{A}_n = {}^{p-2}\mathfrak{B}_n$.

It also follows from the above proof that

$$\mathfrak{A}_n \cap \mathfrak{R}_n = p^n({}^0\mathfrak{B}_n) \oplus \mathfrak{R}_n^0,$$

$$\mathfrak{A}_n/(\mathfrak{A}_n \cap \mathfrak{R}_n) = ({}^0\mathfrak{B}_n/p^n({}^0\mathfrak{B}_n)) \oplus ({}^{p-2}\mathfrak{A}_n/{}^{p-2}\mathfrak{R}_n).$$

Since $[{}^{p-2}\mathfrak{A}_n : {}^{p-2}\mathfrak{R}_n] = q_n$, we obtain

$$[\mathfrak{R}_n : \mathfrak{A}_n \cap \mathfrak{R}_n] = p^n,$$
$$(3) \qquad [\mathfrak{A}_n : \mathfrak{A}_n \cap \mathfrak{R}_n] = p^{2n+1}.$$

Now, let $\mathfrak{R}$ be the inverse limit of $\mathfrak{R}_n$, $n \geqq 0$, relative to $t_{n,m} : \mathfrak{R}_m \to \mathfrak{R}_n$, $m \geqq n$, and let $\mathfrak{R}^0$ be defined similarly. $\mathfrak{R}$ is a compact $\boldsymbol{Z}_p[G]$-module, and $\mathfrak{R}^0$ is a closed submodule of $\mathfrak{R}$. Since $t_{n,m}({}^i\xi_m) = {}^i\xi_n$, $m \geqq n$, the elements ${}^i\xi_n$, $n \geqq 0$, determine an element ${}^i\xi$ in ${}^i\mathfrak{A}$. It then follows immediately from the above proposition that

$$^0(\mathfrak{A}/\mathfrak{B}) = {}^0\mathfrak{R}^0,$$
$$^i(\mathfrak{A}/\mathfrak{B}) = {}^i\mathfrak{R}, \qquad i \text{ even}, i \neq 0,$$
$$^i(\mathfrak{A}/\mathfrak{B}) = {}^i\mathfrak{R}/{}^i\mathfrak{R}\,{}^i\xi, \qquad i \text{ odd}, i \neq p-2,$$
$$^{p-2}(\mathfrak{A}/\mathfrak{B}) = 0.$$

In general, for any compact $\boldsymbol{Z}_p[\Gamma]$-module A, we put

$$A^{(n)} = A/\omega_n A, \qquad n \geqq 0,$$

where

$$\omega_n = 1 - \sigma(1+p)^{p^n}, \qquad n \geqq 0.$$

PROPOSITION 2. *The natural homomorphisms* ${}^-(\mathfrak{A}/\mathfrak{B}) \to {}^-\mathfrak{A}_n/{}^-\mathfrak{B}_n$ *and* ${}^i(\mathfrak{A}/\mathfrak{B}) \to {}^i\mathfrak{A}_n/{}^i\mathfrak{B}_n$, $i \neq 0$, *induce isomorphisms*

$$^-(\mathfrak{A}/\mathfrak{B})^{(n)} \cong {}^-\mathfrak{A}_n/{}^-\mathfrak{B}_n,$$
$$^i(\mathfrak{A}/\mathfrak{B})^{(n)} \cong {}^i\mathfrak{A}_n/{}^i\mathfrak{B}_n, \qquad i \neq 0, n \geqq 0.$$

PROOF. Let $i \neq 0$ and $m \geqq n \geqq 0$. We first notice that $H^k(\Gamma_n/\Gamma_m, {}^i\mathfrak{A}_m) = 0$ for every k. For $i = p-2$, this can be shown similarly as in the proof of the previous proposition. For $i \neq p-2$, it follows from ${}^i\mathfrak{A}_m = {}^i\mathfrak{R}_m$.

Since $t'_{m,n} \circ t_{n,m}(\alpha) = \nu_{n,m}\alpha$, $\alpha \in {}^i\mathfrak{A}_m$, and since $t'_{m,n}$ is injective, we see from $H^1(\Gamma_n/\Gamma_m, {}^i\mathfrak{A}_m) = 0$ that the kernel of $t_{n,m} : {}^i\mathfrak{A}_m \to {}^i\mathfrak{A}_n$ is $\omega_n {}^i\mathfrak{A}_m$. Therefore, the kernel of ${}^i\mathfrak{A}_m/{}^i\mathfrak{B}_m \to {}^i\mathfrak{A}_n/{}^i\mathfrak{B}_n$ is $\omega_n({}^i\mathfrak{A}_m/{}^i\mathfrak{B}_m)$. Since this holds for every $m \geqq n$, the kernel of ${}^i(\mathfrak{A}/\mathfrak{B}) \to {}^i\mathfrak{A}_n/{}^i\mathfrak{B}_n$ is $\omega_n {}^i(\mathfrak{A}/\mathfrak{B})$. As ${}^i(\mathfrak{A}/\mathfrak{B}) \to {}^i\mathfrak{A}_n/{}^i\mathfrak{B}_n$ is surjective, we obtain an isomorphism ${}^i(\mathfrak{A}/\mathfrak{B})^{(n)} \to {}^i\mathfrak{A}_n/{}^i\mathfrak{B}_n$. The isomorphism ${}^-(\mathfrak{A}/\mathfrak{B})^{(n)} \cong {}^-\mathfrak{A}_n/{}^-\mathfrak{B}_n$ is a consequence of what has just been proved.

1.3. Let Λ denote the local ring of formal power series in an indeterminate T with coefficients in $\boldsymbol{Z}_p$:

$$\Lambda = \boldsymbol{Z}_p[[T]].$$

Since ${}^i\mathfrak{R}_n={}^i\varepsilon\mathfrak{R}_n$ is isomorphic to $\boldsymbol{Z}_p[\Gamma/\Gamma_n]$ as algebras over $\boldsymbol{Z}_p$, there exists a ring isomorphism over $\boldsymbol{Z}_p$:

$$ {}^i\mathfrak{R}_n\to\Lambda/(1-(1+T)^{p^n}), $$

which maps ${}^i\varepsilon\sigma(1+p)_n$ to the coset of $1+T \bmod (1-(1+T)^{p^n})$; here $(1-(1+T)^{p^n})$ denotes the principal ideal of Λ generated by $1-(1+T)^{p^n}$. These isomorphisms, for $n\geqq 0$, then define a topological ring isomorphism over $\boldsymbol{Z}_p$:

$$ {}^i\mathfrak{R}\to\Lambda, $$

mapping ${}^i\varepsilon\sigma(1+p)$ to $1+T$. We see easily that ${}^i\mathfrak{R}_n^0$ corresponds to $(T)/(1-(1+T)^{p^n})$ so that ${}^i\mathfrak{R}^0\to(T)$ under the above isomorphism. For odd $i\neq p-2$, let

$$ {}^i\xi\to{}^ig $$

under the same isomorphism. We shall next describe the power series ${}^ig$.

For any a in $\boldsymbol{Q}_p$, there exist a rational integer b and a power of p, p^m $(m\geqq 0)$, such that $p^m a\equiv b \mod p^m, 0\leqq b<p^m$. The rational number b/p^m is then uniquely determined by a, and hence will be denoted by $\langle a\rangle$. If a is a rational number with its denominator a power of p, then $\langle a\rangle$ is nothing but the fractional part of a: $\langle a\rangle=a-[a]$. With this notation, we may write ${}^-\xi_n$ in the form

$$ {}^-\xi_n={}^-\varepsilon\Big(\sum_a\Big\langle\frac{a}{q_n}\Big\rangle\sigma(a)_n\Big), $$

where a ranges over any system of representatives of $U/(1+q_n\boldsymbol{Z}_p)$. Since

$$ v(1+p)^e, \qquad 0\leqq e<p^n, v\in\boldsymbol{V}, $$

form such a system of representatives, we obtain

$$ \begin{aligned} {}^i\xi_n={}^i\varepsilon{}^-\xi_n&={}^i\varepsilon\Big(\sum_e\sum_v\Big\langle\frac{v(1+p)^e}{q_n}\Big\rangle\sigma(v)_n\sigma(1+p)_n^e\Big)\\ &={}^i\varepsilon\Big(\sum_e\sum_v\Big\langle\frac{v(1+p)^e}{q_n}\Big\rangle v^i\sigma(1+p)_n^e\Big), \end{aligned} $$

for any odd $i\neq p-2$. Hence we see that

$$ {}^ig(T)=\lim_{n\to\infty}{}^ig_n(T), $$

where ${}^ig_n(T)$ denotes the polynomial of degree at most p^n defined by

$$ {}^ig_n(T)=\sum_{e=0}^{p^n-1}\sum_v\Big\langle\frac{v(1+p)^e}{q_n}\Big\rangle v^i(1+T)^e. $$

We notice that

$$ {}^ig(T)\equiv{}^ig_m(T)\equiv{}^ig_n(T) \qquad \bmod (1-(1+T)^{p^n}), \qquad m\geqq n. $$

Hence

$$ {}^ig(0)={}^ig_0(0)=\frac{1}{p}\sum_{a=1}^{p-1}av(a)^i, $$

where $v(a)$ denotes the element of V such that $v(a)\equiv a \bmod p$.

Now the following proposition is an immediate consequence of what is stated in the above:

PROPOSITION 3. *Let $\Lambda = Z_p[[T]]$ be made into a $Z_p[\Gamma]$-module so that*

$$\sigma(1+p)x = (1+T)x$$

for any x in Λ. Then, as $Z_p[\Gamma]$-modules,

$$^0(\mathfrak{A}/\mathfrak{B}) \cong (T) = T\Lambda \ (\cong \Lambda),$$

$$^i(\mathfrak{A}/\mathfrak{B}) \cong \Lambda, \qquad i \text{ even}, \ i \neq 0,$$

$$^i(\mathfrak{A}/\mathfrak{B}) \cong \Lambda/(^i g), \qquad i \text{ odd}, \ i \neq p-2,$$

$$^{p-2}(\mathfrak{A}/\mathfrak{B}) = 0,$$

where $(^i g)$ denotes the principal ideal of Λ generated by the $^i g$ defined in the above.

Let χ be any character of $U/(1+q_n Z_p)$ with values in Ω. Then there exists an integer i such that $\chi(v) = v^i$ for every v in V. With i fixed, let

$$^iD_n = \prod_\chi \left(\frac{1}{q_n} \sum_a a\chi(a) \right),$$

where $0 \leqq a < q_n, (a, p) = 1$, and where χ ranges over all characters of $U/(1+q_n Z_p)$ satisfying $\chi(v) = v^i, v \in V$. Then the classical formula for the first factor ^-h_n of the class number of F_n states that

$$\text{(I)} \qquad {}^-h_n = 2q_n \prod_i \left(-\frac{1}{2}\, {}^iD_n \right), \qquad 0 \leqq i \leqq p-2, (i, 2) = 1^{2)}.$$

Hence $^iD_n \neq 0$ for every odd index i $(0 \leqq i \leqq p-2)$. Furthermore, it is easy to see that ^iD_n is a p-adic integer for $i \neq p-2$ and $q_n{}^{p-2}D_n$ is a p-adic unit[3].

PROPOSITION 4. *The compact Γ-module $^-(\mathfrak{A}/\mathfrak{B})$ is strictly Γ-finite*[4]*, and the order of $^-(\mathfrak{A}/\mathfrak{B})^{(n)} = {}^-\mathfrak{A}_n/{}^-\mathfrak{B}_n$ is equal to the exact power of p dividing the first factor ^-h_n of the class number of F_n $(n \geqq 0)$. For any odd i, the Γ-module $^i(\mathfrak{A}/\mathfrak{B})$ is also strictly Γ-finite, and if $i \neq p-2$, the order of $^i(\mathfrak{A}/\mathfrak{B})^{(n)} = {}^i\mathfrak{A}_n/{}^i\mathfrak{B}_n$ is equal to the exact power of p dividing the p-adic integer $^iD_n \neq 0$.*

PROOF. We may write ^iD_n in the form:

$$^iD_n = \prod_\zeta \left(\sum_{e=0}^{p^n-1} \sum_v \left\langle \frac{v(1+p)^e}{q_n} \right\rangle v^i \zeta^e \right),$$

where ζ ranges over all p^n-th roots of unity in Ω. Hence we see that ^iD_n is equal to the determinant of the circulant matrix whose first row is

2) See [5], §5.

3) See [10], §2.

4) For the theory of Γ-finite modules which will be used throughout the paper, see [8] and [11].

$\left(\cdots, \sum_{v}\left\langle\frac{v(1+p)^{e}}{q_n}\right\rangle v^{i}, \cdots\right)$, $e=0, 1, \cdots, p^{n}-1$. On the other hand, it follows from Proposition 3 that for odd $i \neq p-2$, we have

$$\begin{aligned} {}^{i}(\mathfrak{A}/\mathfrak{B})^{(n)} &\cong \Lambda/({}^{i}g, 1-(1+T)^{p^n}) \\ &= \Lambda^{(n)}/({}^{i}g_n), \qquad n \geqq 0, \end{aligned}$$

where

$$\Lambda^{(n)} = \Lambda/\omega_n\Lambda = \sum_{e=0}^{p^n-1} \boldsymbol{Z}_p(1+T)^{e}.$$

Since $(1+T)^{p^n}=1$ on $\Lambda^{(n)}$, we see from the above that ${}^{i}(\mathfrak{A}/\mathfrak{B})^{(n)}$ is finite, and its order is equal to the exact power of p dividing ${}^{i}D_n \neq 0$. On the other hand, we know that $q_n{}^{p-2}D_n$ is a p-adic unit, that ${}^{-}(\mathfrak{A}/\mathfrak{B})^{(n)}$ is the direct sum of ${}^{i}(\mathfrak{A}/\mathfrak{B})^{(n)}$ with odd i, and that ${}^{p-2}(\mathfrak{A}/\mathfrak{B})^{(n)} = {}^{p-2}\mathfrak{A}_n/{}^{p-2}\mathfrak{B}_n = 0$. Hence it follows from (I) that the order of ${}^{-}(\mathfrak{A}/\mathfrak{B})^{(n)}$ is equal to the exact power of p dividing ${}^{-}h_n$.

For odd $i \neq p-2$, let $p^{c(n;i)}$ denote the order of ${}^{i}(\mathfrak{A}/\mathfrak{B})^{(n)}$. Using ${}^{i}(\mathfrak{A}/\mathfrak{B}) \cong \Lambda/({}^{i}g)$, we may express the characteristic function $c(n;i)$ of the Γ-module ${}^{i}(\mathfrak{A}/\mathfrak{B})$ as follows[5]: By Weierstrass' preparation theorem, there exist an integer $e_i \geqq 0$, a unit ${}^{i}u(T)$ in $\Lambda = \boldsymbol{Z}_p[[T]]$, and a polynomial ${}^{i}m(T)$ of the form:

$${}^{i}m(T) = {}^{i}a_0 + \cdots + {}^{i}a_{d_i-1}T^{d_i-1} + T^{d_i}, \qquad {}^{i}a_k \in p\boldsymbol{Z}_p.$$

such that

$${}^{i}g(T) = p^{e_i}\,{}^{i}u(T)\,{}^{i}m(T).$$

The integers $d_i \geqq 0$ and $e_i \geqq 0$ then give the *invariants* of ${}^{i}(\mathfrak{A}/\mathfrak{B})$; namely, for all sufficiently large n, we have

$$c(n;i) = d_i n + e_i p^n + r_i,$$

with an integer r_i independent of n. It follows in particular that the invariant e_i is positive if and only if ${}^{i}g(T)$ is divisible by p in $\Lambda = \boldsymbol{Z}_p[[T]]$, namely, if and only if for every $n \geqq 0$,

$$\sum_{v}\left\langle\frac{v(1+p)^{e}}{q_n}\right\rangle v^{i} \equiv 0 \mod p, \qquad 0 \leqq e < p^n,$$

or equivalently

$$\sum\left\langle\frac{av}{q_n}\right\rangle v^{i} \equiv 0 \mod p,$$

for any a in U[6].

1.4. The algebra $\mathfrak{S}_n = \boldsymbol{Q}_p[G_n]$ has an involution $\alpha \to \alpha^*$ such that $\rho^* = \rho^{-1}$ for any ρ in G_n. We see easily that $\mathfrak{R}_n^* = \mathfrak{R}_n$, $(\mathfrak{R}_n^0)^* = \mathfrak{R}_n^0$, and

5) See [11]. The ring Λ was first introduced by Serre into the theory of Γ-finite modules.

6) See [7].

$$\mathfrak{A}_n^* = \mathfrak{B}_n^* + \mathfrak{N}_n^0, \quad \mathfrak{B}_n^* = \mathfrak{N}_n \xi_n^*,$$

where

$$\xi_n^* = q_n^{-1} \sum_a \left(a - \frac{q_n - p}{2}\right) \sigma(a)_n^{-1}, \qquad 0 \leqq a < q_n, (a, p) = 1.$$

The homomorphism $t_{n,m}: \mathfrak{S}_m \to \mathfrak{S}_n$, $m \geqq n$, commutes with the involutions on $\mathfrak{S}_m$ and $\mathfrak{S}_n$, and maps $\mathfrak{A}_m^*$ onto $\mathfrak{A}_n^*$. Hence if we denote by $\mathfrak{A}^*$ the inverse limit of $\mathfrak{A}_n^*$, $n \geqq 0$, relative to $t_{n,m}$, then the maps $\mathfrak{A}_n \to \mathfrak{A}_n^*$, $n \geqq 0$, define a topological $\boldsymbol{Z}_p$-isomorphism

$$\mathfrak{A} \to \mathfrak{A}^*$$

such that

$$(\sigma\alpha)^* = \sigma^{-1}\alpha^*, \qquad \sigma \in G.$$

The inverse limit of $\mathfrak{B}_n^*$, $n \geqq 0$, gives a closed submodule $\mathfrak{B}^*$ of $\mathfrak{A}^*$, and the above isomorphism induces similar isomorphisms $\mathfrak{B} \to \mathfrak{B}^*$ and $\mathfrak{A}/\mathfrak{B} \to \mathfrak{A}^*/\mathfrak{B}^*$. Since

$$({}^{\pm}\varepsilon)^* = {}^{\mp}\varepsilon, \quad ({}^{j}\varepsilon)^* = {}^{i}\varepsilon, \qquad 0 \leqq i \leqq p-2, i+j = p-1,$$

we have

$${}^{\pm}(\mathfrak{A}^*) = ({}^{\mp}\mathfrak{A})^*, \quad {}^{i}(\mathfrak{A}^*) = ({}^{j}\mathfrak{A})^*, \qquad i+j = p-1.$$

These modules will be denoted simply by ${}^{\pm}\mathfrak{A}^*$ and ${}^{i}\mathfrak{A}^*$ respectively, and similarly for the submodules of $\mathfrak{B}^*$ and $\mathfrak{A}^*/\mathfrak{B}^*$.

Let

$$\mathrm{Res}: \boldsymbol{Q}_p \to \boldsymbol{Q}_p/\boldsymbol{Z}_p$$

be the natural homomorphism of the additive group $\boldsymbol{Q}_p$ so that $\mathrm{Res}(a)$, $a \in \boldsymbol{Q}_p$, denotes the residue class of $a \bmod \boldsymbol{Z}_p$. For each $n \geqq 0$, we define a non-degenerate symmetric pairing $\mathfrak{S}_n \times \mathfrak{S}_n \to \boldsymbol{Q}_p/\boldsymbol{Z}_p$ by

$$(\alpha, \beta)_n = \mathrm{Res}(\sum_\rho a_\rho b_\rho),$$

for any $\alpha = \sum_\rho a_\rho \rho$ and $\beta = \sum_\rho b_\rho \rho$ $(\rho \in G_n)$ in $\mathfrak{S}_n$. Then we see easily that

$$(\sigma\alpha, \sigma\beta)_n = (\alpha, \beta)_n, \qquad \alpha, \beta \in \mathfrak{S}_n, \sigma \in G,$$

$$(\alpha\gamma, \beta)_n = (\alpha, \beta\gamma^*)_n, \qquad \alpha, \beta, \gamma \in \mathfrak{S}_n,$$

$$(t_{n,m}(\alpha), \beta)_n = (\alpha, t'_{m,n}(\beta))_m, \qquad \alpha \in \mathfrak{S}_m, \beta \in \mathfrak{S}_n, m \geqq n.$$

PROPOSITION 5. *For each $n \geqq 0$, there exists a non-degenerate pairing $(x, y^*)_n$:*

$${}^{-}(\mathfrak{A}_n/\mathfrak{B}_n) \times {}^{-}(\mathfrak{A}_n^*/\mathfrak{B}_n^*) \to \boldsymbol{Q}_p/\boldsymbol{Z}_p$$

such that

$$(\sigma x, \sigma y^*)_n = (x, y^*)_n, \qquad \sigma \in G,$$

$$(t_{n,m}(x), y^*)_n = (x, t'_{m,n}(y^*))_m, \qquad m \geqq n,$$

for x in ${}^{-}(\mathfrak{A}_n/\mathfrak{B}_n)$ and y^* in ${}^{-}(\mathfrak{A}_n^*/\mathfrak{B}_n^*)$ or in ${}^{-}(\mathfrak{A}_m^*/\mathfrak{B}_m^*)$.

PROOF. We first notice that since $t'_{m,n}(\mathfrak{A}^*_n)=t'_{m,n}\circ t_{n,m}(\mathfrak{A}^*_m)=\nu_{n,m}\mathfrak{A}^*_m$ and similarly $t'_{m,n}(\mathfrak{B}^*_n)=\nu_{n,m}\mathfrak{B}^*_m$, $t'_{m,n}$ induces homomorphisms $\mathfrak{A}^*_n/\mathfrak{B}^*_n\to\mathfrak{A}^*_m/\mathfrak{B}^*_m$ and ${}^-\mathfrak{A}^*_n/{}^-\mathfrak{B}^*_n\to{}^-\mathfrak{A}^*_m/{}^-\mathfrak{B}^*_m$, $m\geqq n$.

Now the pairing $(\alpha,\beta)_n$ on $\mathfrak{S}_n$ induces a non-degenerate pairing ${}^-\mathfrak{S}_n\times{}^-\mathfrak{S}_n\to\boldsymbol{Q}_p/\boldsymbol{Z}_p$. For any subgroup A of ${}^-\mathfrak{S}_n$, let $A^\perp$ denote the annihilator of A in ${}^-\mathfrak{S}_n$. Since ${}^-\mathfrak{A}_n/{}^-\mathfrak{B}_n$ is finite, it follows from ${}^-\mathfrak{B}_n={}^-\mathfrak{R}_n{}^-\xi_n$ that ${}^-\xi_n$ has an inverse in the algebra ${}^-\mathfrak{S}_n$. Therefore, ${}^-\xi^*_n=({}^-\xi_n)^*$ also has an inverse in ${}^-\mathfrak{S}_n$. As $({}^-\mathfrak{R}_n)^\perp={}^-\mathfrak{R}_n$, it follows from $({}^-\mathfrak{R}_n{}^-\xi_n,\beta)_n=({}^-\mathfrak{R}_n,\beta({}^-\xi_n)^*)_n$ that

$$({}^-\mathfrak{B}_n)^\perp=({}^-\mathfrak{R}_n)^\perp({}^-\xi^*_n)^{-1}={}^-\mathfrak{R}_n({}^-\xi^*_n)^{-1}.$$

Since ${}^-\mathfrak{A}_n={}^-\mathfrak{B}_n+{}^-\mathfrak{R}^0_n={}^-\mathfrak{B}_n+{}^-\mathfrak{R}_n$, we have

$$({}^-\mathfrak{A}_n)^\perp=({}^-\mathfrak{B}_n)^\perp\cap({}^-\mathfrak{R}_n)^\perp={}^-\mathfrak{R}_n({}^-\xi^*_n)^{-1}\cap{}^-\mathfrak{R}_n,$$

and

$$\begin{aligned}({}^-\mathfrak{B}_n)^\perp/({}^-\mathfrak{A}_n)^\perp&={}^-\mathfrak{R}_n({}^-\xi^*_n)^{-1}/({}^-\mathfrak{R}_n({}^-\xi^*_n)^{-1}\cap{}^-\mathfrak{R}_n)\\&\cong{}^-\mathfrak{R}_n/({}^-\mathfrak{R}_n\cap{}^-\mathfrak{R}_n{}^-\xi^*_n)\\&\cong({}^-\mathfrak{R}_n{}^-\xi^*_n+{}^-\mathfrak{R}_n)/{}^-\mathfrak{R}_n{}^-\xi^*_n\\&={}^-\mathfrak{A}^*_n/{}^-\mathfrak{B}^*_n.\end{aligned}$$

Clearly the pairing on ${}^-\mathfrak{S}_n$ induces a non-degenerate pairing of ${}^-\mathfrak{A}_n/{}^-\mathfrak{B}_n$ and $({}^-\mathfrak{B}_n)^\perp/({}^-\mathfrak{A}_n)^\perp$ into $\boldsymbol{Q}_p/\boldsymbol{Z}_p$. Using the above isomorphisms, we then obtain a pairing $(x,y^*)_n:({}^-\mathfrak{A}_n/{}^-\mathfrak{B}_n)\times({}^-\mathfrak{A}^*_n/{}^-\mathfrak{B}^*_n)\to\boldsymbol{Q}_p/\boldsymbol{Z}_p$.

Let α be an element of ${}^-\mathfrak{A}_n$ representing x in ${}^-\mathfrak{A}_n/{}^-\mathfrak{B}_n$, and let β be an element of ${}^-\mathfrak{R}_n$ representing y^* in ${}^-\mathfrak{A}^*_n/{}^-\mathfrak{B}^*_n$, where ${}^-\mathfrak{A}^*_n={}^-\mathfrak{B}^*_n+{}^-\mathfrak{R}_n$. Then, by the definition,

$$(x,y^*)_n=(\alpha,\beta({}^-\xi^*_n)^{-1})_n=(\alpha{}^-\xi_n^{-1},\beta)_n.$$

Hence it is clear that $(\sigma x,\sigma y^*)_n=(x,y^*)_n$ for any σ in G. Next, let x be in ${}^-\mathfrak{A}_m/{}^-\mathfrak{B}_m$, and let α be an element of ${}^-\mathfrak{A}_m$ representing x. Using the fact that $t_{n,m}$ is a ring homomorphism mapping ${}^-\xi_m$ to ${}^-\xi_n$, $m\geqq n$, we have

$$\begin{aligned}(t_{n,m}(x),y^*)_n&=(t_{n,m}(\alpha){}^-\xi_n^{-1},\beta)_n=(t_{n,m}(\alpha{}^-\xi_m^{-1}),\beta)_n\\&=(\alpha{}^-\xi_m^{-1},t'_{m,n}(\beta))_m\\&=(x,t'_{m,n}(y^*))_m.\end{aligned}$$

Hence the proposition is proved.

PROPOSITION 6. *The Γ-modules ${}^-(\mathfrak{A}/\mathfrak{B})$ and ${}^i(\mathfrak{A}/\mathfrak{B})$, with odd i, are regular and self-adjoint*[7].

PROOF. For each $n\geqq 0$, we define a pairing ${}^-(\mathfrak{A}_n/\mathfrak{B}_n)\times{}^-(\mathfrak{A}_n/\mathfrak{B}_n)\to\boldsymbol{Q}_p/\boldsymbol{Z}_p$ by

$$(x,y)_n=(x,y^*)_n,\qquad x,y\in{}^-(\mathfrak{A}_n/\mathfrak{B}_n),$$

where the right-hand side denotes the pairing of ${}^-(\mathfrak{A}_n/\mathfrak{B}_n)$ and ${}^-(\mathfrak{A}^*_n/\mathfrak{B}^*_n)$ given

7) See [8], § 5.

in Proposition 5. It is clear that $(x, y)_n$ is non-degenerate and satisfies

$$(\sigma x, y)_n = (x, \sigma y)_n, \qquad \sigma \in G,$$
$$(t_{n,m}(x), y)_n = (x, t'_{m,n}(y))_m, \qquad m \geqq n,$$

for x in $^-(\mathfrak{A}_n/\mathfrak{B}_n)$ and y in $^-(\mathfrak{A}_n/\mathfrak{B}_n)$ or in $^-(\mathfrak{A}_m/\mathfrak{B}_m)$. Since $^-(\mathfrak{A}/\mathfrak{B})^{(n)} = {}^-(\mathfrak{A}_n/\mathfrak{B}_n)$, the existence of $(x, y)_n$ shows that $^-(\mathfrak{A}/\mathfrak{B})$ is self-adjoint. It is easy to see that for each odd i, $(x, y)_n$ induces a pairing $^i(\mathfrak{A}_n/\mathfrak{B}_n) \times {}^i(\mathfrak{A}_n/\mathfrak{B}_n) \to \boldsymbol{Q}_p/\boldsymbol{Z}_p$ with similar properties. Hence $^i(\mathfrak{A}/\mathfrak{B})$ is also self-adjoint. Since every self-adjoint Γ-module is regular, the proposition is completely proved.

We note that we can obtain a similar result for any strictly Γ-finite Γ-module of the type $\Lambda/(g), g \in \Lambda$.

1.5. The group W, which is the union of all W_n, $n \geqq 0$, the group of q_n-th roots of unity in F_n, is isomorphic to the additive group $\boldsymbol{Q}_p/\boldsymbol{Z}_p$. In the following, we shall fix an isomorphism

$$\iota : W \to \boldsymbol{Q}_p/\boldsymbol{Z}_p,$$

and denote by ζ_n the q_n-th root of unity in W_n such that

$$\iota(\zeta_n) = \operatorname{Res}\left(\frac{1}{q_n}\right), \qquad n \geqq 0.$$

Let $\mathfrak{o}_n$ denote the ring of all integers in the local cyclotomic field Φ_n, and let $\mathfrak{p}_n$ be the unique prime ideal of $\mathfrak{o}_n$; $\mathfrak{p}_n$ is the principal ideal generated by

$$\pi_n = 1 - \zeta_n.$$

The additive group of the field Φ_n, which we shall denote again by Φ_n, is a locally compact abelian group in its natural $\mathfrak{p}_n$-adic topology. Let T_n and $T_{n,m}$ $(m \geqq n \geqq 0)$ denote the trace map from Φ_n to $\boldsymbol{Q}_p$ and that from Φ_m to Φ_n respectively, and let

$$\langle \alpha, \beta \rangle_n = \operatorname{Res}(T_n(\alpha\beta))$$

for any α, β in Φ_n. Then we have a non-degenerate, symmetric pairing $\Phi_n \times \Phi_n \to \boldsymbol{Q}_p/\boldsymbol{Z}_p$ such that

$$\langle \alpha, \beta \rangle_n = \langle \alpha^\sigma, \beta^\sigma \rangle_n, \qquad \alpha, \beta \in \Phi_n, \sigma \in G,$$
$$\langle \alpha, \beta \rangle_m = \langle T_{n,m}(\alpha), \beta \rangle_n, \qquad \alpha \in \Phi_m, \beta \in \Phi_n, m \geqq n.$$

For any closed subgroup A of Φ_n, we denote by $A^\perp$ the annihilator of A in Φ_n relative to this pairing. Then $A^\perp$ is a closed subgroup of Φ_n such that $(A^\perp)^\perp = A$, and A and $\Phi_n/A^\perp$, as well as $A^\perp$ and Φ_n/A, form a pair of mutually dual locally compact abelian groups in the sense of Pontrjagin. If A is in particular a non-zero ideal of Φ_n, then $A^\perp = A^{-1}\mathfrak{d}_n^{-1}$, where

$$\mathfrak{d}_n = q_n \mathfrak{p}_n^{-p^n}$$

is the different of $\Phi_n/\boldsymbol{Q}_p$. In the following, any pairing of locally compact abelian groups with similar properties will be simply called a dual pairing.

Let Φ_n^* denote the multiplicative group of Φ_n; Φ_n^* is also a locally compact group in the $\mathfrak{p}_n$-adic topology. Let U_n denote the subgroup of all local units in Φ_n^*, and let

$$U_{n,0}=1+\mathfrak{p}_n .$$

Then U_n is an open, compact subgroup of Φ_n^*, and

$$\Phi_n^*=\Pi_n\times U_n, \qquad U_n=U_{n,0}\times \boldsymbol{V},$$

where Π_n denotes the cyclic subgroup of Φ_n^* generated by π_n.

Let $\mathfrak{L}_n$ be the set of all elements of the form $\log\alpha$ with α in $U_{n,0}$:

$$\mathfrak{L}_n=\log U_{n,0} .$$

$\mathfrak{L}_n$ is an open, compact $\boldsymbol{Z}_p$-submodule of Φ_n, invariant under G. Let

$$\mathfrak{X}_n=\mathfrak{L}_n^{\perp} ;$$

$\mathfrak{X}_n$ is the set of all α in Φ_n such that $T_n(\alpha\log\beta)$ is contained in $\boldsymbol{Z}_p$ for any β in $U_{n,0}$. By the duality, $\mathfrak{X}_n$ is also an open, compact $\boldsymbol{Z}_p$-submodule of Φ_n, invariant under G.

Let

$$\theta_n=q_n^{-1}\sum_{s=0}^{n}\zeta_s^{-1}=q_n^{-1}\sum_{s=0}^{n}\zeta_n^{-p^s}, \qquad n\geqq 0 .$$

PROPOSITION 7. *The $(p-1)p^n$ elements θ_n^ρ, $\rho\in G_n$, form a normal basis of $\Phi_n/\boldsymbol{Q}_p$ (as well as that of $F_n/\boldsymbol{Q}$), and they generate over $\boldsymbol{Z}_p$ a submodule $\mathfrak{M}_n$ of index p^{p^n-1} in $q_n^{-1}\mathfrak{o}_n$:*

$$[q_n^{-1}\mathfrak{o}_n:\mathfrak{M}_n]=p^{p^n-1} .$$

PROOF. We use induction on n. Since $\theta_0=p^{-1}\zeta_0^{-1}$, the lemma is obvious for $n=0$. Let $n>0$. It follows from the induction assumption that $q_{n-1}\mathfrak{M}_{n-1}$ is a free $\boldsymbol{Z}_p$-module of rank $(p-1)p^{n-1}$, with index $p^{p^{n-1}-1}$ in $\mathfrak{o}_{n-1}$. On the other hand, it is easy to see that $\mathfrak{o}_n=\mathfrak{o}_{n-1}+q_n\mathfrak{M}_n$ and that $q_n\mathfrak{M}_{n-1}$ is contained in $\mathfrak{o}_{n-1}\cap q_n\mathfrak{M}_n$. Hence we obtain from $\mathfrak{o}_n/q_n\mathfrak{M}_n=\mathfrak{o}_{n-1}/(\mathfrak{o}_{n-1}\cap q_n\mathfrak{M}_n)$ that

$$\begin{aligned}[\mathfrak{o}_n:q_n\mathfrak{M}_n]&\leqq[\mathfrak{o}_{n-1}:q_n\mathfrak{M}_{n-1}]\\&=[\mathfrak{o}_{n-1}:q_{n-1}\mathfrak{M}_{n-1}][q_{n-1}\mathfrak{M}_{n-1}:q_n\mathfrak{M}_{n-1}]\\&=p^{p^{n-1}-1}p^{(p-1)p^{n-1}}\\&=p^{p^n-1} .\end{aligned}$$

It follows that $(p-1)p^n$ elements θ_n^ρ are linearly independent over $\boldsymbol{Q}_p$ and form a normal basis of $\Phi_n/\boldsymbol{Q}_p$. Considering the action of the Galois group of Φ_n/Φ_{n-1} on $q_n\mathfrak{M}_n$, we then see that $q_n\mathfrak{M}_{n-1}=\mathfrak{o}_{n-1}\cap q_n\mathfrak{M}_n$. Hence we obtain from the above that $[\mathfrak{o}_n : q_n\mathfrak{M}_n]=p^{p^{n-1}}$, q. e. d.

LEMMA. *Let*

$$\alpha=\sum_\rho c_\rho\theta_n^\rho, \qquad \rho\in G_n,$$

with c_ρ in Z_p satisfying $\sum_\rho c_\rho=0$. Then α is contained in $\mathfrak{X}_n$.

PROOF. Let $\mu(m)$ $(m\geqq 1)$ be the Möbius' function, and let

$$\tau_n^{(a)}=\prod_{\substack{m=1\\(m,p)=1}}^{\infty}(1-\pi_n^{am})^{\frac{\mu(m)}{m}}.$$

for any integer $a\geqq 1$[8]. Then these elements $\tau_n^{(a)}$ generate the group $U_{n,0}=1+\mathfrak{p}_n$ topologically. Hence an element α in Φ_n belongs to $\mathfrak{X}_n$ if and only if $\langle\alpha, \log\tau_n^{(a)}\rangle_n=0$ for every $a\geqq 1$.

Since

$$T_n(\zeta_n^j)=\begin{cases}(p-1)p^n, & \text{if} \quad p^{n+1}\mid j,\\ -p^n, & \text{if} \quad p^{n+1}\nmid j,\ p^n\mid j,\\ 0, & \text{if} \quad p^n\nmid j,\end{cases}$$

we have

$$T_n(\zeta_n^{-ip^s}\pi_n^{ap^e})=\sum_{t=0}^{ap^e}(-1)^t\binom{ap^e}{t}T_n(\zeta_n^{t-ip^s})$$

$$=p^{n+1}\sum_{\substack{t\equiv ip^s(p^{n+1})\\0\leqq t\leqq ap^e}}(-1)^t\binom{ap^e}{t}-p^n\sum_{\substack{t\equiv ip^s(p^n)\\0\leqq t\leqq ap^e}}(-1)^t\binom{ap^e}{t},$$

for any i prime to p. Using

$$\log\tau_n^{(a)}=-\sum_{e=0}^{\infty}p^{-e}\pi_n^{ap^e},$$

we obtain

$$T_n(q_n^{-1}\zeta_n^{-ip^s}\log\tau_n^{(a)})=-\sum_{e=0}^{\infty}p^{-e}\Big\{\sum_{\substack{t\equiv ip^s(p^{n+1})\\0\leqq t\leqq ap^e}}(-1)^t\binom{ap^e}{t} \tag{4}$$
$$-p^{-1}\sum_{\substack{t\equiv ip^s(p^n)\\0\leqq t\leqq ap^e}}(-1)^t\binom{ap^e}{t}\Big\}.$$

The series

$$\sum_{e=0}^{\infty}p^{-e}\Big\{\sum_{\substack{t\equiv ip^s(p^m)\\0\leqq t\leqq ap^e}}(-1)^t\binom{ap^e}{t}\Big\} \tag{5}$$

converges for $m=0$, because $\sum_{0\leqq t\leqq ap^e}(-1)^t\binom{ap^e}{t}=(1-1)^{ap^e}=0$. Since the series on the right-hand side of (4) converges, we see by induction on m that the series (5) converges for every $m\geqq 0$. Hence

8) See [2], § 7.

$$T_n(q_n^{-1}\zeta_n^{-ip^s}\log\tau_n^{(a)})=-\sum_{e=0}^{\infty}p^{-e}\Big\{\sum_{\substack{t\equiv ip^s(p^{n+1})\\0\leqq t\leqq ap^e}}(-1)^t\binom{ap^e}{t}\Big\}$$
$$+\sum_{e=0}^{\infty}p^{-(e+1)}\Big\{\sum_{\substack{t\equiv ip^s(p^n)\\0\leqq t\leqq ap^e}}(-1)^t\binom{ap^e}{t}\Big\}.$$

Using

$$\binom{ap^e}{t}\equiv\binom{ap^{e+1}}{tp}\qquad \mathrm{mod}\ p^{2e+2}\ ^{9)},$$

we have

$$\sum_{\substack{t\equiv ip^s(p^n)\\0\leqq t\leqq ap^e}}(-1)^t\binom{ap^e}{t}=\sum_{\substack{tp\equiv ip^{s+1}(p^{n+1})\\0\leqq tp\leqq ap^{e+1}}}(-1)^t\Big[\binom{ap^{e+1}}{tp}+\Big\{\binom{ap^e}{tp}-\binom{ap^{e+1}}{tp}\Big\}\Big]$$
$$=\sum_{\substack{t\equiv ip^{s+1}(p^{n+1})\\0\leqq t\leqq ap^{e+1}}}(-1)^t\binom{ap^{e+1}}{t}+r_e,$$

where $r_e\equiv 0$ mod p^{2e+2}. It follows that

$$T_n(q_n^{-1}\zeta_n^{-ip^s}\log\tau_n^{(a)})=-\sum_{e=0}^{\infty}p^{-e}\Big\{\sum_{\substack{t\equiv ip^s(p^{n+1})\\0\leqq t\leqq ap^e}}(-1)^t\binom{ap^e}{t}\Big\}$$
$$+\sum_{e=0}^{\infty}p^{-(e+1)}\Big\{\sum_{\substack{t\equiv ip^{s+1}(p^{n+1})\\0\leqq t\leqq ap^{e+1}}}(-1)^t\binom{ap^{e+1}}{t}\Big\}$$
$$+\sum_{e=0}^{\infty}p^{-(e+1)}r_e$$
$$\equiv-\sum_{e=0}^{\infty}p^{-e}\Big\{\sum_{\substack{t\equiv ip^s(p^{n+1})\\0\leqq t\leqq ap^e}}(-1)^t\binom{ap^e}{t}\Big\}$$
$$+\sum_{e=0}^{\infty}p^{-e}\Big\{\sum_{\substack{t\equiv ip^{s+1}(p^{n+1})\\0\leqq t\leqq ap^e}}(-1)^t\binom{ap^e}{t}\Big\}\quad \mathrm{mod}\ \boldsymbol{Z}_p.$$

Taking the sum over $s=0,1,\cdots,n$, we obtain for $\rho=\sigma(i)_n$ that

$$T_n(\theta_n^{\rho}\log\tau_n^{(a)})\equiv-\sum_{e=0}^{\infty}p^{-e}\Big\{\sum_{\substack{t\equiv i(p^{n+1})\\0\leqq t\leqq ap^e}}(-1)^t\binom{ap^e}{t}\Big\}$$
$$+\sum_{e=0}^{\infty}p^{-e}\Big\{\sum_{\substack{t\equiv 0(p^{n+1})\\0\leqq t\leqq ap^e}}(-1)^t\binom{ap^e}{t}\Big\}\quad \mathrm{mod}\ \boldsymbol{Z}_p.$$

However, if $t\equiv i$ mod p^{n+1}, then t is prime to p, and

$$\binom{ap^e}{t}=\frac{ap^e}{t}\binom{ap^e-1}{t-1}\equiv 0\quad \mathrm{mod}\ p^e.$$

9) See [2], Hilfssatz 2.

Hence the first sum on the right-hand side of the above is contained in $\boldsymbol{Z}_p$. Therefore,

$$T_n(\theta_n^\rho \log \tau_n^{(a)}) \equiv u_n^{(a)} \quad \bmod \boldsymbol{Z}_p,$$

where

$$u_n^{(a)} = \sum_{e=0}^{\infty} p^{-e}\Big\{\sum_{\substack{t\equiv 0 (p^{n+1})\\ 0\leqq t\leqq ap^e}} (-1)^t \binom{ap^e}{t}\Big\}.$$

It then follows that

$$T_n(\alpha \log \tau_n^{(a)}) \equiv (\sum_\rho c_\rho) u_n^{(a)} \equiv 0 \qquad \bmod \boldsymbol{Z}_p.$$

Hence α is contained in $\mathfrak{X}_n$.

1.6. Now, by Proposition 7, there exists a $\boldsymbol{Z}_p[G]$-isomorphism

$$\varphi_n : \mathfrak{S}_n = \boldsymbol{Q}_p[G_n] \to \Phi_n \qquad n \geqq 0,$$

such that

$$\varphi_n(\rho) = \theta_n^\rho$$

for any ρ in G_n. Since

$$\sum_\rho \theta_n^\rho = T_n(\theta) = -\frac{1}{p}, \qquad \rho \in G_n,$$

we have

$$\varphi_n(-p\sum_\rho \rho) = 1, \qquad \rho \in G_n.$$

Let

$$\mu_n = \frac{1}{q_n}\frac{\zeta_n}{\pi_n} \qquad n \geqq 0.$$

Then

$$\begin{aligned}
\sum_{\substack{0\leqq a<q_n\\ (a,p)=1}} a\theta_n^{\sigma(a)} &= q_n^{-1} \sum_{s=0}^{n} \sum_a a\zeta_s^{-a} \\
&= q_n^{-1} \sum_{s=0}^{n} \sum_{\substack{0\leqq a<q_s\\ (a,p)=1}} \Big(\sum_{k=0}^{p^{n-s}-1} (a+kq_s)\Big)\zeta_s^{-a} \\
&= q_n^{-1} \sum_{s=0}^{n} \sum_{\substack{0\leqq a<q_s\\ (a,p)=1}} p^{n-s} a \zeta_n^{-p^{n-s}a} \\
&\quad + q_n^{-1} \sum_{s=0}^{n} \frac{(p^{n-s}-1)p^{n-s}}{2} q_s \sum_{\substack{0\leqq a<q_s\\ (a,p)=1}} \zeta_s^{-a} \\
&= q_n^{-1} \sum_{a=0}^{q_n-1} a\zeta_n^{-a} - q_n^{-1}\frac{(p^n-1)p^{n+1}}{2} \\
&= \frac{\zeta_n}{\pi_n} - \frac{p^n-1}{2}.
\end{aligned}$$

Hence

$$\mu_n = q_n^{-1}\sum_a \left(a - \frac{q_n - p}{2}\right)\theta_n^{\sigma(a)}, \qquad 0 \leqq a < q_n, (a, p) = 1,$$

and we have

$$\varphi_n(\xi_n) = \mu_n .$$

Since

$$T_{n,m}(\theta_m) = \theta_n . \qquad m \geqq n \geqq 0,$$

we also see that the following diagram is commutative:

$$\begin{array}{ccc} \mathfrak{S}_m & \xrightarrow{\varphi_m} & \Phi_m \\ \downarrow t_{n,m} & & \downarrow T_{n,m} \\ \mathfrak{S}_n & \xrightarrow{\varphi_n} & \Phi_n \end{array} .$$

THEOREM 1. *The map $\varphi_n : \mathfrak{S}_n \to \Phi_n$ defines a $\boldsymbol{Z}_p[G]$-isomorphism*

$$\mathfrak{A}_n \to \mathfrak{X}_n , \qquad n \geqq 0 .$$

PROOF. It follows from the lemma in 1.5 that $\varphi_n(\mathfrak{R}_n^0)$ is contained in $\mathfrak{X}_n$. It also follows from Artin-Hasse's explicit formula for Hilbert's norm residue symbol in Φ_n that $\mu_n = \varphi_n(\xi_n)$ is contained in $\mathfrak{X}_n$[10]. Hence $\varphi_n(\mathfrak{A}_n) = \varphi_n(\mathfrak{R}_n \xi_n + \mathfrak{R}_n^0)$ is a submodule of $\mathfrak{X}_n$.

The annihilator of $\mathfrak{p}_n^{p^{n+1}} = \log(1 + \mathfrak{p}_n^{p^{n+1}})$ relative to the pairing $\langle \alpha, \beta \rangle_n$ is $\mathfrak{p}_n^{-p^{n-1}}\mathfrak{d}_n^{-1} = q_n^{-1}\mathfrak{p}_n^{-1}$. Hence $\mathfrak{X}_n = (\log(1+\mathfrak{p}_n))^{\perp}$ is contained in $q_n^{-1}\mathfrak{p}_n^{-1}$, and $[q_n^{-1}\mathfrak{p}_n^{-1} : \mathfrak{X}_n] = [\log(1+\mathfrak{p}_n) : \log(1+\mathfrak{p}_n^{p^{n+1}})]$. However, the kernel of the log map: $1+\mathfrak{p}_n \to \log(1+\mathfrak{p}_n)$ is W_n, and $W_n \cap (1+\mathfrak{p}_n^{p^{n+1}}) = 1$. Hence $[\log(1+\mathfrak{p}_n) : \log(1+\mathfrak{p}_n^{p^{n+1}})] = q_n^{-1}[1+\mathfrak{p}_n : 1+\mathfrak{p}_n^{p^{n+1}}] = q_n^{-1}p^{p^n}$, and

$$[q_n^{-1}\mathfrak{p}_n^{-1} : \mathfrak{X}_n] = p^{p^n - n - 1} .$$

On the other hand, it follows from (3) that $[\varphi_n(\mathfrak{R}_n) : \varphi_n(\mathfrak{A}_n \cap \mathfrak{R}_n)] = p^n$, $[\varphi_n(\mathfrak{A}_n) : \varphi_n(\mathfrak{A}_n \cap \mathfrak{R}_n)] = p^{2n+1}$. By Proposition 7, $\mathfrak{M}_n = \varphi_n(\mathfrak{R}_n)$ has index $p^{p^n - 1}$ in $q_n^{-1}\mathfrak{o}_n$. Hence $[q_n^{-1}\mathfrak{p}_n^{-1} : \varphi_n(\mathfrak{R}_n)] = p^{p^n}$, and we obtain from the above that

$$[q_n^{-1}\mathfrak{p}_n^{-1} : \varphi_n(\mathfrak{A}_n)] = p^{p^n - n - 1} .$$

Since $\varphi_n(\mathfrak{A}_n)$ is contained in $\mathfrak{X}_n$, we see that $\varphi_n(\mathfrak{A}_n) = \mathfrak{X}_n$, q. e. d.

Let

$$\mathfrak{Z}_n = \varphi_n(\mathfrak{B}_n), \qquad n \geqq 0 .$$

$\mathfrak{Z}_n$ is the submodule of $\mathfrak{X}_n$ generated over $\boldsymbol{Z}_p$ by the conjugates μ_n^ρ, $\rho \in G_n$, of μ_n. Clearly we have $\boldsymbol{Z}_p[G]$-isomorphisms

$$\mathfrak{B}_n \to \mathfrak{Z}_n , \qquad \mathfrak{A}_n/\mathfrak{B}_n \to \mathfrak{X}_n/\mathfrak{Z}_n .$$

Since the diagram

10) See [2] and 3.1 below.

$$\begin{array}{ccc} \mathfrak{A}_m & \xrightarrow{\varphi_m} & \mathfrak{X}_m \\ \downarrow t_{n,m} & & \downarrow T_{n,m} \\ \mathfrak{A}_n & \xrightarrow{\varphi_n} & \mathfrak{X}_n \end{array}, \qquad m \geqq n \geqq 0,$$

is commutative, we obtain $T_{n,m}(\mathfrak{X}_m)=\mathfrak{X}_n$, and similarly $T_{n,m}(\mathfrak{Z}_m)=\mathfrak{Z}_n$. Hence, let $\mathfrak{X}$ and $\mathfrak{Z}$ denote the inverse limits of $\mathfrak{X}_n$ and $\mathfrak{Z}_n$, $n \geqq 0$, relative to $T_{n,m}: \mathfrak{X}_m \to \mathfrak{X}_n$ and $T_{n,m}: \mathfrak{Z}_m \to \mathfrak{Z}_n$, $m \geqq n$, respectively:

$$\mathfrak{X} = \lim \mathfrak{X}_n, \qquad \mathfrak{Z} = \lim \mathfrak{Z}_n .$$

$\mathfrak{X}$ is a compact $\boldsymbol{Z}_p[G]$-module, and $\mathfrak{Z}$ is a closed submodule of $\mathfrak{X}$. From the above, we obtain immediately the following

THEOREM 2. *The maps* $\varphi_n: \mathfrak{S}_n \to \Phi_n$, $n \geqq 0$, *induce* $\boldsymbol{Z}_p[G]$-*isomorphisms:*

$$\mathfrak{A} \to \mathfrak{X}, \qquad \mathfrak{B} \to \mathfrak{Z}, \qquad \mathfrak{A}/\mathfrak{B} \to \mathfrak{X}/\mathfrak{Z}.$$

We now see the structure of the compact $\boldsymbol{Z}_p[G]$-modules $\mathfrak{X}$, $\mathfrak{Z}$ and $\mathfrak{X}/\mathfrak{Z}$ from the propositions in 1.2–1.4. We note in particular that $^-(\mathfrak{X}/\mathfrak{Z})$ is a regular strictly Γ-finite Γ-module such that

$$^-(\mathfrak{X}/\mathfrak{Z})^{(n)} = {}^-\mathfrak{X}_n/{}^-\mathfrak{Z}_n \cong {}^-\mathfrak{A}_n/{}^-\mathfrak{B}_n, \qquad n \geqq 0,$$

and that the order of $^-(\mathfrak{X}/\mathfrak{Z})^{(n)}$ is equal to the exact power of p dividing the first factor ^-h_n of the class number of F_n.

§ 2.

2.1. For $m \geqq n \geqq 0$, let N_n and $N_{n,m}$ denote the norm map from Φ_n to $\boldsymbol{Q}_p$ and that from Φ_m to Φ_n respectively. The restriction of N_n on F_n obviously gives the norm map from F_n to $\boldsymbol{Q}$, and similarly for the restriction of $N_{n,m}$ on F_m.

Let P_n be the set of all α in F_n such that the principal ideal (α) in F_n is a power of the prime ideal (π_n). P_n is a subgroup of the multiplicative group F_n^* of the field F_n, and

$$P_n = \Pi_n \times E_n,$$

where E_n is the group of all units of F_n.

Let ^+F_n denote the maximal real subfield of F_n, and let $^+F_n^*$ be its multiplicative group. Put

$$^+P_n = {}^+F_n^* \cap P_n, \qquad {}^+E_n = {}^+F_n^* \cap E_n .$$

Then ^+E_n is the group of all units of ^+F_n, and

$$^+P_n = {}^+\Pi_n \times {}^+E_n, \qquad E_n = {}^+E_n \times W_n,$$

where $^+\Pi_n$ denotes the cyclic subgroup of $^+F_n^*$ generated by $^+\pi_n = \pi_n^{1+\delta}$, $\delta = \sigma(-1)$. It follows immediately from the definition that

$$P_n = F_n^* \cap P_m, \qquad m \geqq n \geqq 0,$$

and similarly for ${}^+P_n$, E_n, and ${}^+E_n$.

For any $m \geqq n$, $N_{n,m}(P_m)$ is a subgroup of P_n. Let

$$P'_n = \bigcap_{m \geqq n} N_{n,m}(P_m).$$

Since $N_{n,m}(\pi_m) = \pi_n$, π_n is contained in P'_n, and

$$P'_n = \Pi_n \times E'_n,$$

with

$$E'_n = P'_n \cap E_n = \bigcap_{m \geqq n} N_{n,m}(E_m).$$

Since $N_{n,m}({}^+\pi_m) = {}^+\pi_n$, $N_{n,m}(\zeta_m) = \zeta_n$, we also have

$${}^+P'_n = {}^+\Pi_n \times {}^+E'_n, \qquad E'_n = {}^+E'_n \times W_n,$$

where

$${}^+P'_n = {}^+F_n^* \cap P'_n = \bigcap_{m \geqq n} N_{n,m}({}^+P_m), \qquad {}^+E'_n = {}^+F_n^* \cap E'_n = \bigcap_{m \geqq n} N_{n,m}({}^+E_m).$$

It follows that

$$P_n/P'_n = {}^+P_n/{}^+P'_n = E_n/E'_n = {}^+E_n/{}^+E'_n, \qquad n \geqq 0.$$

Let Q_n denote the subgroup of F_n^* generated by π_n and its conjugates π_n^ρ, $\rho \in G_n$. Since P'_n is invariant under G, Q_n is contained in P'_n. Let

$$C_n = E_n \cap Q_n = E'_n \cap Q_n.$$

Then

$$Q_n = \Pi_n \times C_n.$$

Since $\pi_n^{1-\delta} = -\zeta_n$, $\pm W_n$ is contained in Q_n. It follows that

$$C_n = {}^+C_n \times W_n, \qquad {}^+Q_n = {}^+\Pi_n \times {}^+C_n,$$

where ${}^+Q_n = Q_n \cap {}^+F_n^*$ and ${}^+C_n = C_n \cap {}^+F_n^*$. Hence

$$P_n/Q_n = {}^+P_n/{}^+Q_n = E_n/C_n = {}^+E_n/{}^+C_n,$$

$$P'_n/Q_n = {}^+P'_n/{}^+Q_n = E'_n/C_n = {}^+E'_n/{}^+C_n, \qquad n \geqq 0.$$

Let r be a fixed primitive root mod p^2, and let

$$\varepsilon_n = \zeta_n^{-\frac{r-1}{2}} \frac{1-\zeta_n^r}{1-\zeta_n} = \frac{\zeta_n^{\frac{r}{2}} - \zeta_n^{-\frac{r}{2}}}{\zeta_n - \zeta_n^{-1}}, \qquad n \geqq 0.$$

ε_n is a so-called circular unit of the cyclotomic field F_n, and we see readily that ${}^+C_n$ is generated by $\pm\varepsilon_n^\rho$, $\rho \in G_n$. Hence the classical formula for the second factor ${}^+h_n$ of the class number of F_n states that

(II) $$ {}^+h_n = [{}^+E_n : {}^+C_n]^{11)}.$$

11) See [5], § 11.

It follows in particular that ${}^+E_n/{}^+C_n$ is a finite group.

Let ${}^+G_n$ denote the Galois group of ${}^+F_n/Q$. Since ${}^+P_n/{}^+Q_n$ is finite and since

$$\varepsilon_n^2=\varepsilon_n^{1+\delta}=\frac{(1-\zeta_n^r)(1-\zeta_n^{-r})}{(1-\zeta_n)(1-\zeta_n^{-1})}={}^+\pi_n^{\sigma-1}, \qquad \sigma=\sigma(r)_n,$$

we see that the subgroup of ${}^+Q_n$ generated by ${}^+\pi_n^\rho, \rho\in{}^+G_n$, is G-isomorphic to $\boldsymbol{Z}[{}^+G_n]$ and has a finite index in ${}^+Q_n$ which is a power of 2. Hence the cohomology groups

$$H^k(\Gamma_n/\Gamma_m, {}^+Q_m)=0,$$

for any k and $m\geqq n\geqq 0$.

PROPOSITION 8. *The group ${}^+P'_n$ has a G-subgroup P''_n such that $P''_n\cong\boldsymbol{Z}[{}^+G_n]$ as G-groups and such that the index $[{}^+P'_n : P''_n]$ is finite and prime to p.*

PROOF. For any $m\geqq 0$, let M'_m denote the extension of F_m obtained by adjoining all q_m-th roots of elements in ${}^+P_m$, and let M be the union of the increasing sequence of subfields M'_m, $m\geqq 0$, in Ω. Since $E_m={}^+E_m\times W_m$ and since ${}^+\pi_l^{p^{l-m}}/{}^+\pi_m$ belongs to ${}^+E_l$ for any $l\geqq m$, M is nothing but the abelian extension of F obtained by adjoining p^l-th roots of elements in E_m for all l and $m\geqq 0$. Let M_m be the maximal abelian extension of F_m contained in M. Clearly F is a subfield of M_m. Let A_m denote the character group of the compact abelian Galois group $G(M_m/F)$. For any χ in A_m and σ in G, we define χ^σ by

$$\chi^\sigma(u)=\chi(u^{\sigma^{-1}}), \qquad u\in G(M_m/F),$$

where $u^{\sigma^{-1}}=s^{-1}us$ with an element s in $G(M_m/Q)$ which induces σ on F. We then have

$$\chi^{\sigma\tau}=(\chi^\tau)^\sigma, \qquad \chi\in A_m, \sigma,\tau\in G,$$

and A_m becomes a $\boldsymbol{Z}_p[G]$-module. For the structure of A_m, we know the following[12]: Let $J_m={}^-((\boldsymbol{Q}_p/\boldsymbol{Z}_p)[G_m])$. Then there exist an integer $n_1\geqq 0$ and a fixed finite p-primary $\boldsymbol{Z}_p[G]$-module D such that whenever $m\geqq n_1$, the $\boldsymbol{Z}_p[G]$-modules A_m and J_m have submodules B_m and D_m respectively, with the property

$$A_m/B_m\cong D, \qquad B_m\cong J_m/D_m, \qquad D_m\cong D.$$

Since D is finite, $p^eD=0$ for some $e\geqq 0$.

Now, given $n\geqq 0$, we choose an m such that $m\geqq n, n_1, 2e$. It is easy to see that $F\cap M'_m=F_m$ and $G(FM'_m/F)=G(M'_m/F_m)$. Let A'_m be the submodule of A_m corresponding to the sub-extension FM'_m/F of M/F. Then by the theory of Kummer extensions, there exists an isomorphism $f: {}^+P_m/({}^+P_m)^{q_m}\to A'_m$ such that

12) See [9], § 10, Theorem 17.

$$f(x^\sigma)=\kappa(\sigma)f(x)^\sigma, \qquad \sigma\in G,$$

for any x in ${}^+P_m/({}^+P_m)^{q_m}$. Such an isomorphism or homomorphism of $Z_p[G]$-modules will be called in general a κ-isomorphism or κ-homomorphism[13].

As an abelian group, ${}^+P_m/\{\pm 1\}$ is isomorphic to Z^s, $s=\frac{1}{2}(p-1)p^m$. Hence $A'_m\cong {}^+P_m/({}^+P_m)^{q_m}\cong (Z/q_mZ)^s$. Let $A''_m=p^eA'_m$. Since $A_m/B_m\cong D$ and $p^eD=0$, A''_m is a submodule of B_m, isomorphic to $(Z/q_{m-e}Z)^s$. Let J'_m be the submodule of all x in J_m such that $q_{m-e}x=0$. Since $p^eD_m=0$ and $m\geqq 2e$, D_m is contained in pJ'_m. Identifying B_m with J_m/D_m, put $B'_m=J'_m/D_m$. Then B'_m is contained in A''_m, and $A''_m/B'_m\cong D, B'_m/pB'_m\cong J'_m/pJ'_m\cong {}^-((Z/pZ)[G_m])$. Clearly f induces a κ-isomorphism $g: {}^+P_m/({}^+P_m)^{q_{m-e}}\to A''_m=p^eA'_m$. Let S_m be the subgroup of ${}^+P_m$ containing $({}^+P_m)^{q_{m-e}}$ such that $S_m/({}^+P_m)^{q_{m-e}}\to B'_m$ under g. Then g defines κ-isomorphisms ${}^+P_m/S_m\to D$ and $S_m/S_m^p\to {}^-((Z/pZ)[G_m])$. It follows that S_m/S_m^p is $Z_p[G]$-isomorphic to ${}^+((Z/pZ)[G_m])=(Z/pZ)[{}^+G_m]$. Let α'_m be an element of S_m such that the cosets of $(\alpha'_m)^\rho$, $\rho\in {}^+G_m$, form a basis of S_m/S_m^p, and let S'_m be the subgroup of S_m generated by these $(\alpha'_m)^\rho$. We then see easily that $S'_m\cong Z[{}^+G_m]$ as G-groups and that the index of S'_m in S_m is finite and prime to p. It follows that $H^k(\Gamma_n/\Gamma_m, S_m)=0$ for every k. Hence $N_{n,m}(S_m)=F_n\cap S_m={}^+P_n\cap S_m$.

Since ${}^+P_m/S_m$ is κ-isomorphic to a fixed finite module D, $N_{n,m}({}^+P_m)$ is contained in S_m whenever m is sufficiently large. $N_{n,m}({}^+P_m)$ is then contained in $F_n\cap S_m=N_{n,m}(S_m)$, and we see that $N_{n,m}({}^+P_m)=F_n\cap S_m={}^+P_n\cap S_m=N_{n,m}(S_m)$. Hence ${}^+P_n/N_{n,m}({}^+P_m)={}^+P_nS_m/S_m$ and $[{}^+P_n:N_{n,m}({}^+P_m)]\leqq[{}^+P_m:S_m]=[D:0]$. Since $N_{n,m}({}^+P_m)$ decreases as m increases, it follows that

$${}^+P'_n=N_{n,m}({}^+P_m)=N_{n,m}(S_m),$$

whenever m is sufficiently large. We fix such an m and put

$$P''_n=N_{n,m}(S'_m).$$

It then follows from $S'_n\cong Z[{}^+G_m]$ that $P''_n\cong Z[{}^+G_n]$ as G-groups. Since $[S_m:S'_m]$ is finite and prime to p, so is $[N_{n,m}(S_m):N_{n,m}(S'_m)]=[{}^+P'_n:P''_n]$.

COROLLARY.

i) *For any n and $s\geqq 0$, there exists a $Z_p[G]$-isomorphism*

$${}^+P'_n/({}^+P'_n)^{p^s}\cong (Z/p^sZ)[{}^+G_n].$$

ii) *For any k and $m\geqq n\geqq 0$,*

$$H^k(\Gamma_n/\Gamma_m, {}^+P'_m)=0.$$

iii) *For $m\geqq n\geqq 0$,*

---

13) The definition of κ-isomorphisms and κ-homomorphisms given here, as well as that of u^σ and χ^σ in the above, differs slightly from those given in [8], [9]. We feel that the present definitions are more adequate.

$$^{+}P'_n = N_{n,m}(^{+}P'_m) = {}^{+}P_n \cap {}^{+}P'_m = F_n \cap {}^{+}P'_m ,$$

so that the natural homomorphism $^{+}P_n/^{+}P'_n \to {}^{+}P_m/^{+}P'_m$ *is injective. If m is sufficiently larger than n, then*

$$^{+}P'_n = N_{n,m}(^{+}P_m).$$

iv) *If n is sufficiently large,* $n \geqq n_0$, *then*

$$^{+}P_m = {}^{+}P_n {}^{+}P'_m , \qquad m \geqq n ,$$

so that $^{+}P_n/^{+}P'_n \to {}^{+}P_m/^{+}P'_m$ *is also surjective, and*

$$^{+}P_n/^{+}P'_n \cong {}^{+}P_m/^{+}P'_m , \qquad m \geqq n \geqq n_0 .$$

The groups $^{+}P_n/^{+}P'_n$, $n \geqq n_0$, *are* κ*-isomorphic to the finite p-primary* $\boldsymbol{Z}_p[G]$*-module D defined in the above.*

PROOF. i) Since $[^{+}P'_n : {}^{+}P''_n]$ is prime to p, we have

$$^{+}P'_n/(^{+}P'_n)^{p^s} \cong {}^{+}P''_n/(^{+}P''_n)^{p^s} \cong (\boldsymbol{Z}/p^s\boldsymbol{Z})[^{+}G_n].$$

ii) This follows immediately from $H^k(\Gamma_n/\Gamma_m, {}^{+}P'_m/^{+}P''_m) = H^k(\Gamma_n/\Gamma_m, {}^{+}P''_m) = 0$.

iii) It has already been proved in the above that $^{+}P'_n = N_{n,m}(^{+}P_m)$ for $m \gg n$. Given any $m \geqq n$, choose a large $l \geqq m$ so that $^{+}P'_m = N_{m,l}(^{+}P_l)$ and $^{+}P'_n = N_{n,l}(^{+}P_l)$. Since $N_{n,l} = N_{n,m} \circ N_{m,l}$, we obtain $^{+}P'_n = N_{n,m}(^{+}P'_m)$. It then follows from $H^0(\Gamma_n/\Gamma_m, {}^{+}P'_m) = 0$ that $^{+}P'_n = N_{n,m}(^{+}P'_m) = F_n \cap {}^{+}P'_m = {}^{+}P_n \cap {}^{+}P'_m$.

iv) Since D is a finite p-primary $\boldsymbol{Z}_p[G]$-module, there exists an integer $n_0 \geqq 0$ such that Γ_{n_0} acts trivially on D. We may assume that $p^{n_0}D = 0$. Let $n \geqq n_0$ and let m be sufficiently larger than n so that $^{+}P'_n = N_{n,m}(^{+}P_m)$. Since $^{+}P_m/S_m$ is κ-isomorphic to D, Γ_n acts trivially on $^{+}P_m/S_m$. Hence $(^{+}P_m)^{\omega_n}$ is a submodule of S_m, and $H^1(\Gamma_n/\Gamma_m, S_m) = 0$ implies $(^{+}P_m)^{\omega_n} = S_m^{\omega_n}$. It follows that $^{+}P_m = (^{+}P_m \cap F_n)S_m = {}^{+}P_n S_m$ and $^{+}P_m/S_m = {}^{+}P_n S_m/S_m = {}^{+}P_n/N_{n,m}(P_m^{+}) = {}^{+}P_n/^{+}P'_n$. Therefore $^{+}P_n/^{+}P'_n$, $n \geqq n_0$, is κ-isomorphic to D. For any $m \geqq n \geqq n_0$, we then see that $^{+}P_n/^{+}P'_n \to {}^{+}P_m/^{+}P'_m$ is surjective and $^{+}P_m = {}^{+}P_n {}^{+}P'_m$.

From the above, we can obtain similar results for P'_n, E'_n, and $^{+}E'_n$. For example:

$$N_{n,m}(^{+}E'_m) = {}^{+}E'_n , \qquad m \geqq n \geqq 0 .$$

Some of these results were already obtained in [9].

2.2. Let Φ_n^* be as before the multiplicative group of the local field Φ_n. For any $m \geqq n$, $N_{n,m}(\Phi_m^*)$ is a subgroup of Φ_n^*. Let

$$\Phi'_n = \bigcap_{m \geqq n} N_{n,m}(\Phi_n^*) .$$

and let

$$U'_n = U_n \cap \Phi'_n , \qquad U'_{n,0} = U_{n,0} \cap \Phi'_n .$$

Clearly Φ'_n, U'_n, and $U'_{n,0}$ are subgroups of Φ_n^*. Since $N_{n,m}(\pi_m) = \pi_n$ and

$N_{n,m}(v)=v$ for any v in $\boldsymbol{V}$, $m\geqq n$, we see immediately that

$$\Phi'_n=\Pi_n\times U'_n,\qquad U'_n=U'_{n,0}\times \boldsymbol{V}.$$

PROPOSITION 9.

i) *Φ'_n is a closed subgroup of Φ^*_n, consisting of all α in Φ_n such that $N_n(\alpha)$ is a power of p, and U'_n is a compact subgroup of U_n, consisting of all β in U_n such that $N_n(\beta)=1$,*

ii) $\Phi^*_n=\Phi'_n\times \boldsymbol{U}_0$, $U_n=U'_n\times \boldsymbol{U}_0=U'_{n,0}\times \boldsymbol{U}$.

iii) $N_{n,m}(\Phi'_m)=\Phi'_n$, $N_{n,m}(U'_m)=U'_n$, $N_{n,m}(U'_{m,0})=U'_{n,0}$, $\quad m\geqq n$.

PROOF. i) By local class field theory[14], $\boldsymbol{Q}^*_p/N_m(\Phi^*_m)$ is a cyclic group of order $(p-1)p^m$. Since $p=N_m(\pi_m)$ is contained in $N_m(\Phi^*_m)$, we see from $\boldsymbol{Q}^*_p=\{p\}\times\boldsymbol{V}\times\boldsymbol{U}_0$ that $N_m(\Phi^*_m)=\{p\}\times\boldsymbol{U}_0^{p^m}$; here we denote by $\{p\}$ the multiplicative group generated by p. Again by local class field theory, an element α in Φ_n is contained in $N_{n,m}(\Phi^*_m)$, $m\geqq n$, if and only if $N_n(\alpha)$ is contained in $N_m(\Phi^*_m)$. Hence α is in Φ'_n if and only if $N_n(\alpha)$ is contained in the intersection of $N_m(\Phi^*_m)=\{p\}\times\boldsymbol{U}_0^{p^m}$ for all $m\geqq n$, namely, in $\{p\}$. It then follows immediately that U'_n consists of all β in U_n such that $N_n(\beta)=1$. It also follows that Φ'_n is a closed subgroup of Φ^*_n, and that U'_n is a compact subgroup of U_n.

ii) Let α be any element of Φ^*_n, and let $N_n(\alpha)=p^t a$ with a in $\boldsymbol{U}_0^{p^n}=\boldsymbol{U}_0^{(p-1)p^n}$. Then $a=b^{(p-1)p^n}=N_n(b)$ for some b in $\boldsymbol{U}_0$, and $N_n(\alpha b^{-1})=p^t$ so that αb^{-1} is contained in Φ'_n. Hence $\Phi^*_n=\Phi'_n\boldsymbol{U}_0$. Since $N_n(b)=b^{(p-1)p^n}\neq 1$ for any $b\neq 1$ in $\boldsymbol{U}_0$, we have $\Phi'_n\cap\boldsymbol{U}_0=1$. Therefore $\Phi^*_n=\Phi'_n\times\boldsymbol{U}_0$. It then follows immediately that $U_n=U'_n\times\boldsymbol{U}_0=U'_{n,0}\times\boldsymbol{U}$.

iii) It is obvious that $N_{n,m}(\Phi'_m)$ is contained in Φ'_n, $m\geqq n$. Let α be any element of Φ'_n. Then $\alpha=N_{n,m}(\alpha')$ with α' in Φ^*_m. By ii), $N_n(\alpha)=N_m(\alpha')$ is a power of p. Hence α' is in Φ'_m, again by ii). Therefore α is contained in $N_{n,m}(\Phi'_m)$. Thus $\Phi'_n=N_{n,m}(\Phi'_m)$. It is then clear that $U'_n=N_{n,m}(U'_m)$, $U'_{n,0}=N_{n,m}(U'_{m,0})$.

Now, let X_n denote the inverse limit of the sequence of finite groups $\Phi'_n/\Phi'^{p^s}_n$, $s\geqq 0$, relative to the natural homomorphisms $\Phi'_n/\Phi'^{p^t}_n\to\Phi'_n/\Phi'^{p^s}_n$, $t\geqq s$:

$$X_n=\lim \Phi'_n/\Phi'^{p^s}_n.$$

X_n is a p-primary compact abelian group, and for any a in $\boldsymbol{Z}_p$ and x in X_n, x^a is defined as usual. Since $\boldsymbol{V}$ is the intersection of all $\Phi'^{p^s}_n$, $s\geqq 0$, the natural map $\Phi'_n\to X_n$ imbeds $\Phi'_n/\boldsymbol{V}=\Pi_n\times U'_{n,0}$ in X_n as a dense subgroup of X_n, and we see that

$$X_n=\overline{\Pi}_n\times U'_{n,0},$$

where $\overline{\Pi}_n$ denotes the closure of Π_n in X_n, consisting of all elements of the form π^a_n with a in $\boldsymbol{Z}_p$.

---

14) For local and global class field theory used here and in the following, see [1], [3], and [4].

Clearly the surjective homomorphism $N_{n,m}: \Phi'_m \to \Phi'_n$ induces a continuous surjective homomorphism $N_{n,m}: X_m \to X_n$, $m \geqq n$. Let X be the inverse limit of X_n, $n \geqq 0$, relative to such homomorphisms:

$$X = \lim X_n .$$

Then X is again a p-primary compact abelian group, and x^a is defined for any a in Z_p and x in X. Furthermore, since Φ'_n is invariant under the action of the Galois group G, we may extend the action of G on X_n, $n \geqq 0$, and on X in the natural way. Thus X_n, $n \geqq 0$, and X are compact $Z_p[G]$-groups.

Let Ψ_n be the maximal p-primary abelian extension of Φ_n in Ω. Clearly Φ is contained in Ψ_n, and these Ψ_n, $n \geqq 0$, form an increasing sequence of subfields of Ω. Let Ψ be the union of all Ψ_n, $n \geqq 0$; Ψ is the maximal p-primary abelian extension of Φ contained in Ω. Since Ψ/Q_p is a Galois extension, $G = G(\Phi/Q_p)$ acts on the abelian normal subgroup $G(\Psi/\Phi)$ of $G(\Psi/Q_p)$ in the obvious manner.

PROPOSITION 10. *There exists a canonical G-isomorphism*

$$X \to G(\Psi/\Phi)$$

which induces isomorphisms

$$X^{(n)} \cong X_n \cong G(\Psi_n/\Phi), \qquad n \geqq 0 .$$

PROOF. Let $G(\Psi_m/\Phi) \to G(\Psi_n/\Phi)$, $m \geqq n$, be the natural homomorphism of Galois groups. Then by local class field theory, there exists a canonical isomorphism $X_n \to G(\Psi_n/\Phi)$ for each $n \geqq 0$ such that the following diagram is commutative:

$$\begin{array}{ccc} X_m & \longrightarrow & G(\Psi_m/\Phi) \\ \downarrow & & \downarrow \\ X_n & \longrightarrow & G(\Psi_n/\Phi), \end{array} \qquad m \geqq n .$$

Hence we have a G-isomorphism of X onto $G(\Psi/\Phi)$ such that

$$\begin{array}{ccc} X & \longrightarrow & G(\Psi/\Phi) \\ \downarrow & & \downarrow \\ X_n & \longrightarrow & G(\Psi_n/\Phi) \end{array}$$

is commutative. Since $\sigma(1+p)^{p^n}$ generates $\Gamma_n = G(\Phi/\Phi_n)$ topologically, and since Ψ_n is the maximal abelian extension of Φ_n contained in Ψ, we see that $G(\Psi/\Psi_n) = G(\Psi/\Phi)^{\omega_n}$. However, $G(\Psi/\Psi_n)$ is the kernel of the surjective homomorphism $G(\Psi/\Phi) \to G(\Psi_n/\Phi)$. Hence it follows from the above diagram that $X \to X_n$ is also surjective, and its kernel is X^{ω_n}, thus proving $X^{(n)} = X/X^{\omega_n} \cong X_n$.

2.3. Let A be any subgroup of Φ'_n. Then the closure of the image of A under $\Phi'_n \to X_n$ defines a closed subgroup of X_n. Since the norm, from F_n

to $\boldsymbol{Q}$, of any element in P_n is a power of p, it follows from Proposition 9 that P_n is contained in Φ'_n. Let Y_n be the closed subgroup of X_n determined by P_n as stated in the above, and let Y'_n and Z_n denote the closed subgroups of X_n defined similarly by P'_n and Q_n, respectively.

Let

$$E_{n,0}=E_n\cap U_{n,0}, \qquad E'_{n,0}=E'_n\cap U_{n,0}, \qquad C_{n,0}=C_n\cap U_{n,0},$$

and let $\bar{E}_n$, $\bar{E}_{n,0}$, etc. denote the closures of the respective groups in U_n in the $\mathfrak{p}_n$-adic topology of Φ_n. Since

$$\pi_n^{\sigma-1}\equiv\kappa(\sigma) \qquad \mathrm{mod}\,\mathfrak{p}_n,$$

we see that

$$\bar{E}_n=E_{n,0}\times \boldsymbol{V}, \qquad \bar{E}'_n=\bar{E}'_{n,0}\times \boldsymbol{V}, \qquad \bar{C}_n=\bar{C}_{n,0}\times \boldsymbol{V}.$$

Since $[P_n:\Pi_n\times E_{n,0}]=[E_n:E_{n,0}]$ is prime to p, we also have

$$Y_n=\overline{\Pi}_n\times\bar{E}_{n,0}$$

in $X_n=\overline{\Pi}_n\times U'_{n,0}$. Similarly

$$Y'_n=\overline{\Pi}_n\times\bar{E}'_{n,0}, \qquad Z_n=\overline{\Pi}_n\times\bar{C}_{n,0}.$$

Hence it follows that

$$\begin{aligned} X_n/Y_n&=U'_{n,0}/\bar{E}_{n,0}=U'_n/\bar{E}_n,\\ X_n/Y'_n&=U'_{n,0}/\bar{E}'_{n,0}=U'_n/\bar{E}'_n,\\ X_n/Z_n&=U'_{n,0}/\bar{C}_{n,0}=U'_n/\bar{C}_n,\\ Y'_n/Z_n&=\bar{E}'_{n,0}/\bar{C}_{n,0}=\bar{E}'_n/\bar{C}_n, \qquad n\geqq 0. \end{aligned}$$

It is clear that for any $m\geqq n$, $N_{n,m}:X_m\to X_n$ maps Y_m, Y'_m, and Z_m into Y_n, Y'_n and Z_n respectively. Hence we may define the inverse limits:

$$Y=\lim Y_n, \qquad Y'=\lim Y'_n, \qquad Z=\lim Z_n,$$

relative to the homomorphisms $N_{n,m}$, $m\geqq n$, and we may consider Y, Y' and Z as closed $\boldsymbol{Z}_p[G]$-subgroups of X. However, by Proposition 8, $N_{n,m}(P_m)=P'_n$ whenever m is sufficiently larger than n. Hence also $N_{n,m}(Y_m)=Y'_n$ for $m\gg n$, and we see that

$$Y=Y'.$$

PROPOSITION 11. *X/Y is the inverse limit of $U'_n/\bar{E}'_n$, $n\geqq 0$, and the natural homomorphism $X/Y\to U'_n/\bar{E}'_n$ induces an isomorphism $(X/Y)^{(n)}\to U'_n/\bar{E}'_n$:*

$$X/Y=\lim U'_n/\bar{E}'_n, \qquad (X/Y)^{(n)}=U'_n/\bar{E}'_n.$$

Similarly,

$$X/Z=\lim U'_n/\bar{C}_n, \qquad (X/Z)^{(n)}=U'_n/\bar{C}_n.$$

PROOF. Since $N_{n,m}(P'_m)=P'_n$, $m\geqq n$, we have $N_{n,m}(Y'_m)=Y'_n$, $m\geqq n$.

Hence the natural homomorphism $X \to X_n$ maps Y' onto Y'_n. Since $Y = Y'$, $X/Y = \lim X_n/Y'_n = \lim U'_n/\bar{E}'_n$. By Proposition 10, the kernel of $X \to X_n$ is X^{ω_n}. Hence the kernel of $X \to X_n/Y'_n$ is $Y'X^{\omega_n}$. Since $X \to X_n/Y'_n$ is surjective, we have $X_n/Y'_n \cong X/Y'X^{\omega_n} = (X/Y')^{\omega_n}$. The proof is similar for X/Z.

For each $n \geqq 0$, let L'_n denote the maximal unramified p-primary abelian extension of F_n contained in Ω, and let $L_n = FL'_n$. Then the union of the increasing sequence of subfields $L_n, n \geqq 0$, in Ω defines the maximal unramified p-primary abelian extension L over F in Ω. Let K_n be the maximal p-primary abelian extension of F_n, in Ω, such that no prime ideal of F_n different from (π_n) is ramified in that extension. The union of the increasing sequence of subfields $K_n, n \geqq 0$, in Ω again defines a p-primary abelian extension K of F, containing the above L. Since K is obviously a Galois extension of $\boldsymbol{Q}$, the Galois group G of $F/\boldsymbol{Q}$ acts on the abelian normal subgroup $G(K/F)$ of $G(K/\boldsymbol{Q})$. Similarly G also acts on $G(L/F)$ and $G(K/L)$[15].

PROPOSITION 12. *There exists a G-isomorphism:*

$$G(K/L) \to X/Y.$$

PROOF. We know that

$$X/Y = \lim X_n/Y_n = \lim U'_n/\bar{E}_n.$$

By class field theory, there exists a canonical G-isomorphism $G(K_n/L'_n) \to U_n/\bar{E}_n$, inducing a G-isomorphism $G(K_n/L_n) \to U'_n/\bar{E}_n$[16]. Let $G(K_m/L_m) \to G(K_n/L_n)$ be the homomorphism obtained by restricting the action of $G(K_m/L_m)$ on K_n. Then the following diagram is commutative:

$$\begin{array}{ccc} G(K_m/L_m) & \longrightarrow & U'_m/\bar{E}_m \\ \downarrow & & \downarrow \\ G(K_n/L_n) & \longrightarrow & U'_n/\bar{E}_n, \end{array} \qquad m \geqq n.$$

Since $G(K/L)$ is the inverse limit of $G(K_n/L_n), n \geqq 0$, relative to $G(K_m/L_m) \to G(K_n/L_n)$, $m \geqq n$, we obtain from the above a G-isomorphism $G(K/L) \to X/Y$.

2.4. As in the proof of Proposition 11, we see that

$$Y/Z = Y'/Z = \lim Y'_n/Z_n = \lim \bar{E}'_n/\bar{C}_n.$$

Let $^+\Phi_n$ be the closure of ^+F_n in Φ_n, and let

$$^+\bar{E}'_n = {}^+\Phi_n \cap \bar{E}'_n, \qquad ^+\bar{C}_n = {}^+\Phi_n \cap \bar{C}_n, \qquad n \geqq 0.$$

15) For the extensions L/F and K/F, see [9], § 5, § 6.
16) See [9], § 6.

Then ${}^+\bar{E}'_n$ and ${}^+\bar{C}_n$ are also closures of ${}^+E'_n$ and ${}^+C_n$ in U_n, and $\bar{E}'_n={}^+\bar{E}'_n\times W_n$, $\bar{C}_n={}^+\bar{C}_n\times W_n$ so that

$$\bar{E}'_n/\bar{C}_n={}^+\bar{E}'_n/{}^+\bar{C}_n={}^+(\bar{E}'_n/\bar{C}_n).$$

Hence

$$Y/Z={}^+(Y/Z).$$

In general, for any finite abelian group $\mathfrak{G}$, let $(\mathfrak{G})_p$ denote the Sylow p-subgroup of $\mathfrak{G}$. Since ${}^+E_n/{}^+C_n$ is a finite group, so is ${}^+E'_n/{}^+C_n$. Let A denote the inverse limit of the p-groups $({}^+E'_n/{}^+C_n)_p$, $n\geqq 0$, relative to the homomorphisms $N_{n,m}:({}^+E'_m/{}^+C_m)_p\to({}^+E'_n/{}^+C_n)_p$, $m\geqq n$. Clearly A is a compact p-primary $\boldsymbol{Z}_p[G]$-group. Since $N_{n,m}({}^+E'_m)={}^+E'_n$, $m\geqq n$, $N_{n,m}:({}^+E'_m/{}^+C_m)_p\to({}^+E'_n/{}^+C_n)_p$ is surjective. Hence $A\to({}^+E'_n/{}^+C_n)_p$ is also surjective for every $n\geqq 0$. On the other hand, as $H^k(\Gamma_n/\Gamma_m,{}^+P'_m)=H^k(\Gamma_n/\Gamma_m,{}^+Q_m)=0$ for any k and $m\geqq n\geqq 0$, we see that the kernel of $N_{n,m}:{}^+E'_m/{}^+C_m={}^+P'_m/{}^+Q_m\to{}^+E'_n/{}^+C_n={}^+P'_n/{}^+Q_n$ is $({}^+E'_m/{}^+C_m)^{\omega_n}$. Therefore, the kernel of $N_{n,m}:({}^+E'_m/{}^+C_m)_p\to({}^+E'_n/{}^+C_n)_p$ is $({}^+E'_m/{}^+C_m)_p^{\omega_n}$, and hence the kernel of $A\to({}^+E'_n/{}^+C_n)_p$ is A^{ω_n}. Thus we obtain a $\boldsymbol{Z}_p[G]$-isomorphism

$$A^{(n)}=A/A^{\omega_n}\to({}^+E'_n/{}^+C_n)_p\,,\qquad n\geqq 0\,.$$

Since ${}^+E'_n/{}^+C_n$ is finite, ${}^+E'_n{}^+\bar{C}_n/{}^+\bar{C}_n$ is also finite. Hence ${}^+E'_n{}^+\bar{C}_n$ is closed in U_n, and we see that ${}^+\bar{E}'_n={}^+E'_n{}^+\bar{C}_n$. Therefore, the injection ${}^+E'_n\to{}^+\bar{E}'_n$ induces a surjective homomorphism ${}^+E'_n/{}^+C_n\to{}^+\bar{E}'_n/{}^+\bar{C}_n$. As ${}^+\bar{E}'_n/{}^+\bar{C}_n=Y'_n/Z_n$ is a finite p-group, $({}^+E'_n/{}^+C_n)_p\to{}^+\bar{E}'_n/{}^+\bar{C}_n=\bar{E}'_n/\bar{C}_n$ is also surjective. Hence we obtain a surjective $\boldsymbol{Z}_p[G]$-homomorphism

$$A\to Y/Z\,.$$

We shall next show that it is injective.

Suppose that there exists an element $a\neq 1$ in the kernel of $A\to Y/Z$. Since A is a p-primary compact group, there is an integer $d\geqq 1$ such that a is not contained in A^{p^d}. For each $n\geqq 0$, let a_n denote the image of a under $A\to({}^+E'_n/{}^+C_n)_p$, and let α_n be an element of ${}^+E'_n$ representing a_n mod ${}^+C_n$. Since a is in the kernel of $A\to Y/Z$, a_n is mapped to the identity under $({}^+E'_n/{}^+C_n)_p\to{}^+\bar{E}'_n/{}^+\bar{C}_n$. Hence α_n is contained in ${}^+\bar{C}_n$, and there exist an element β_n in ${}^+E'_n\cap U_n^{q_n}$ and an element γ_n in ${}^+C_n$ such that $\alpha_n=\beta_n\gamma_n$. As $N_{n,m}(a_m)=a_n$, $m\geqq n$, we have $\alpha_n{}^+C_n=N_{n,m}(\alpha_m){}^+C_n=N_{n,m}(\beta_m){}^+C_n$. Since a is not contained in A^{p^d}, α_n does not belong to $({}^+E'_n)^{p^d}{}^+C_n$ whenever n is sufficiently large. Hence $N_{n,m}(\beta_m)$, $m\geqq n$, is not contained in $({}^+E'_n)^{p^d}$, if n is large. On the other hand, it follows from Corollary iii), iv) of Proposition 8 that there exists an integer $e\geqq 0$ such that $({}^+P_n/{}^+P'_n)^{p^e}=1$ for every $n\geqq 0$. In the following, we shall fix an n such that $n\geqq d+e$ and that $N_{n,m}(\beta_m)$ is not contained in $({}^+E'_n)^{p^d}$ for any $m\geqq n$.

For each $m\geqq n$, let B_m denote the multiplicative group generated by

$\beta_m^\sigma, \sigma \in G$. Then B_m is a G-subgroup of ${}^+E'_m \cap U_m^{q_m}$, and $N_{n,m}(B_m)$ is not contained in $({}^+E'_n)^{p^d}$. Let D_m, D'_m, and D''_m denote the images of B_m under the natural homomorphisms ${}^+P'_m \to {}^+P'_m/({}^+P'_m)^{p^d}$, ${}^+P'_m \to {}^+P'_m/({}^+P'_m)^{q_m}$, and ${}^+P_m \to {}^+P_m/({}^+P_m)^{q_m}$ respectively. Let r_m be the rank of D_m. As an abelian group, ${}^+P'_m/\{\pm 1\}$ is isomorphic to $\boldsymbol{Z}^s$, where $s=\frac{1}{2}(p-1)p^m$. Hence ${}^+P'_m/({}^+P'_m)^{q_m} \cong (\boldsymbol{Z}/q_m\boldsymbol{Z})^s$ and $({}^+P'_m)^{p^d}/({}^+P'_m)^{q_m} \cong (\boldsymbol{Z}/q_{m-d}\boldsymbol{Z})^s$. We then see easily that D'_m contains a subgroup isomorphic to $(\boldsymbol{Z}/q_{m-d}\boldsymbol{Z})^{r_m}$. Now the kernel of the natural homomorphism ${}^+P'_m/({}^+P'_m)^{q_m} \to {}^+P_m/({}^+P_m)^{q_m}$ is a subgroup of $({}^+P_m)^{q_m}/({}^+P'_m)^{q_m} \cong {}^+P_m/{}^+P'_m$, and D'_m is mapped onto D''_m under that homomorphism. Since $({}^+P_m/{}^+P'_m)^{p^e}=1$, it follows from the above that D''_m contains a subgroup isomorphic to $(\boldsymbol{Z}/q_{m-d-e}\boldsymbol{Z})^{r_m}$.

Let L''_m denote the abelian extension of F_m generated by the q_m-th roots of elements in B_m. Since ${}^+P_m \cap (F_m^*)^{q_m} = {}^+P_m^{q_m}$, the Galois group $G(L''_m/F_m)$ is isomorphic to $D''_m = B_m({}^+P_m)^{q_m}/({}^+P_m)^{q_m}$ as abelian groups. On the other hand, since B_m is contained in ${}^+E'_m \cap U_m^{q_m}$, L''_m/F_m is unramified. Hence L''_m is contained in the maximal unramified abelian p-extension L'_m over F_m. It then follows from the above that the Galois group $G(L'_m/F_m)$ contains a subgroup isomorphic to $(\boldsymbol{Z}/q_{m-d-e}\boldsymbol{Z})^{r_m}$. However, since $G(L/F)$ is a strictly Γ-finite Γ-group, there exists an integer $f \geqq 0$ such that the rank of $G(L'_m/F_m)^{p^f}$ does not exceed a fixed integer $r \geqq 0$ for every $m \geqq 0$[17]. Hence we see that $r_m \leqq r$ whenever $m \geqq d+e+f$.

Now, since D_m is a subgroup of ${}^+P'_m/({}^+P'_m)^{p^d}$ with rank r_m, it follows from the above that the order of D_m does not exceed p^{dr} for any $m \geqq n, d+e+f$. Hence we can find a large m such that $N_{n,m}(D_m)=1$. As $D_m = B_m({}^+P'_m)/({}^+P'_m)^{p^d}$, $N_{n,m}(B_m)$ is then contained in $({}^+P'_n)^{p^d}$. On the other hand, Corollary iii) of Proposition 8 implies that $({}^+P'_n)^{p^d} \cap F_n = ({}^+P'_m \cap F_n)^{p^d} = ({}^+P'_n)^{p^d}$. It also follows from ${}^+P'_n = {}^+\Pi_m \times {}^+E'_n$ that $({}^+P'_n)^{p^d} \cap E_n = ({}^+E'_n)^{p^d}$. Hence the group $N_{n,m}(B_m)$, which is obviously a subgroup of E_n, must be contained in $({}^+E'_n)^{p^d}$. However we have chosen n so that $N_{n,m}(B_m)$ is not contained in $({}^+E'_n)^{p^d}$ for any $m \geqq n$. Hence we have a contradiction, and we see that there exists no $a \neq 1$ in the kernel of $A \to Y/Z$. Thus the homomorphism $A \to Y/Z$ is injective, and the following proposition is proved:

PROPOSITION 13. *Y/Z is the inverse limit of $({}^+E'_n/{}^+C_n)_p$, $n \geqq 0$ relative to $N_{n,m}: ({}^+E'_m/{}^+C_m)_p \to ({}^+E'_n/{}^+C_n)_p$, $m \geqq n$, and the natural homomorphism $Y/Z \to ({}^+E'_n/{}^+C_n)_p$ induces an isomorphism $(Y/Z)^{(n)} \to ({}^+E'_n/{}^+C_n)_p$:*

$$Y/Z = \lim ({}^+E'_n/{}^+C_n)_p, \qquad (Y/Z)^{(n)} = ({}^+E'_n/{}^+C_n)_p, \qquad n \geqq 0$$

---

17) See [9], § 5 and [8], 1.4.

§ 3

3.1. For any α, β in Φ_n^*, let

$$(\alpha, \beta)_n$$

denote Hilbert's norm residue symbol for the power q_n in the local field Φ_n. The symbol $(\alpha, \beta)_n$ defines a pairing

$$\Phi_n^* \times \Phi_n^* \to W_n$$

with the following properties[18]:

i) $(\alpha, \beta)_n = (\beta, \alpha)_n^{-1}$, $(\alpha^\sigma, \beta^\sigma)_n = (\alpha, \beta)_n^{\kappa(\sigma)}$, $\qquad \sigma \in G$,

ii) $(\alpha, \beta)_m^{p^{m-n}} = (N_{n,m}(\alpha), \beta)_n$, $\qquad \alpha \in \Phi_m^*, \beta \in \Phi_n^*, m \geqq n$,

iii) $(\alpha, \beta)_n = 1$ if and only if α is the norm of an element in $\Phi_n(\beta^{q_n^{-1}})$,

iv) $(\alpha, \Phi_n^*) = 1$ (or $(\Phi_n^*, \alpha) = 1$) if and only if α belongs to $(\Phi_n^*)^{q_n}$; hence $\alpha, \beta)_n$ induces a non-degenerate pairing

$$(\Phi_n^* / (\Phi_n^*)^{q_n}) \times (\Phi_n^* / (\Phi_n^*)^{q_n}) \to W_n .$$

v) (Artin-Hasse's explicit formula) For any β in $U_{n,0}$, both $q_n^{-1} T_n(\log \beta)$ and $q_n^{-1} T_n(\zeta_n \pi_n^{-1} \log \beta)$ are contained in $\boldsymbol{Z}_p$, and

$$(\zeta_n, \beta)_n = \zeta_n^{-q_n^{-1} T_n(\log \beta)},$$

$$(\pi_n, \beta)_n = \zeta_n^{q_n^{-1} T_n(\zeta_n \pi_n^{-1} \log \beta)}.$$

Let β be an element of $U_{n,0}$; and ξ, an element of $\mathfrak{X}_n$. Since $T_n(q_n \mathfrak{X}_n \log \beta) \equiv 0$ mod $q_n \boldsymbol{Z}_p$, $\zeta_n^{T_n(\xi \log \beta)}$ depends only upon the coset ξ' of ξ mod $q_n \mathfrak{X}_n$. Hence we may write $\zeta_n^{T_n(\xi' \log \beta)}$ for $\zeta_n^{T_n(\xi \log \beta)}$.

PROPOSITION 14. *There exists a unique map*

$$\psi_n : \Phi'_n \to \mathfrak{X}_n / q_n \mathfrak{X}_n$$

such that

$$(\alpha, \beta)_n = \zeta_n^{T_n(\psi_n(\alpha) \log \beta)}, \qquad \alpha \in \Phi'_n, \beta \in U_{n,0}.$$

ψ_n *is a surjective* κ*-homomorphism:*

$$\psi_n(\alpha^\sigma) = \kappa(\sigma) \psi_n(\alpha)^\sigma, \qquad \alpha \in \Phi'_n, \sigma \in G.$$

PROOF. Let α be fixed in Φ'_n. Since $N_{n,2n+1}(W_{2n+1}) = W_n$, $(\alpha, W_n)_n = 1$ by i) and iii). As W_n is the kernel of the log map: $U_{n,0} \to \mathfrak{L}_n$, $(\alpha, \beta)_n$ depends only upon $\log \beta$ for any β in $U_{n,0}$. Hence $\log \beta \to \iota((\alpha, \beta)_n)$ defines a homomorphism $\mathfrak{L}_n \to \boldsymbol{Q}_n / \boldsymbol{Z}_p$, $\iota : W \to \boldsymbol{Q}_p / \boldsymbol{Z}_p$ being the homomorphism fixed in 1.5. Since $\Phi_n / \mathfrak{X}_n$ is dual to $\mathfrak{L}_n$, there exists an element α'' in Φ_n such that $\iota((\alpha, \beta)_n) = \langle \alpha'', \log \beta \rangle_n$ for any β in $U_{n,0}$. We then have

$$(\alpha, \beta)_n = \zeta_n^{T_n(\alpha' \log \beta)},$$

18) See [4], II, § 11, § 19 and [2].

with $\alpha' = q_n\alpha''$. Since $T_n(\alpha' \log \beta)$ is in $\boldsymbol{Z}_p$ for any β in $U_{n,0}$, α' is contained in $\mathfrak{X}_n$. Let $\psi_n(\alpha)$ denote the coset of α' mod $q_n\mathfrak{X}_n$. Then we have

$$(\alpha, \beta)_n = \zeta_n^{T_n(\psi_n(\alpha)\log \beta)}, \qquad \beta \in U_{n,0}.$$

It is clear from the definition of $\mathfrak{X}_n$ that $\psi_n(\alpha)$ is uniquely determined for α by the above equality. Since $(\alpha, \beta)_n$ is multiplicative in α, it follows in particular that $\psi_n : \Phi'_n \to \mathfrak{X}_n/q_n\mathfrak{X}_n$ is a homomorphism. It also follows from i) that

$$(\alpha^\sigma, \beta^\sigma)_n = (\alpha, \beta)_n^{\kappa(\sigma)} = \zeta_n^{\kappa(\sigma)T_n(\psi_n(\alpha)\log \beta)}$$
$$= \zeta_n^{T_n(\kappa(\sigma)\psi_n(\alpha)^\sigma \log \beta^\sigma)}, \qquad \sigma \in G, \beta^\sigma \in U_{n,0}.$$

Hence by the uniqueness mentioned above, we have

$$\psi_n(\alpha^\sigma) = \kappa(\sigma)\psi_n(\alpha)^\sigma, \qquad \sigma \in G, \alpha \in \Phi'_n.$$

To show that ψ_n is surjective, let α' be any element of $\mathfrak{X}_n$. Let $\psi : \Phi_n^* \to W_n$ be an extension of the homomorphism $U_{n,0} \to W_n$ defined by $\beta \to \zeta_n^{T_n(\alpha' \log \beta)}$; such an extension exists because $\Phi_n^* = \Pi_n \times U_{n,0} \times V$. Since $W_n^{q_n} = 1$, ψ is trivial on $(\Phi_n^*)^{q_n}$. Hence it follows from iv) that there exists an α in Φ_n^* such that

$$(\alpha, \beta)_n = \zeta_n^{T_n(\alpha' \log \beta)}, \qquad \beta \in U_{n,0}.$$

Since $(\alpha, \zeta_n)_n = \zeta_n^{T_n(\alpha' \log \zeta_n)} = 1$, α is contained in $N_{n,2n+1}(\Phi_{2n+1}^*) = N_{n,2n+1}(\Phi'_{2n+1} \times U_0)$ $= \Phi'_n \times U_0^{q_n}$. As $(\alpha, \beta)_n$ is unchanged when α is replaced by any element of $\alpha U_0^{q_n}$, we may assume that α is contained in Φ'_n. Then we have $\alpha' = \psi_n(\alpha)$, and we see that ψ_n is surjective.

Let $f : X \to \mathfrak{X}$ be any κ-homomorphism. Then

$$f(x^{\omega_n}) \equiv \omega_n f(x) \mod q_n\mathfrak{X}, \qquad x \in X.$$

Hence f maps X^{ω_n} into $\omega_n\mathfrak{X} + q_n\mathfrak{X}$. Since $X_n = X^{(n)} = X/X^{\omega_n}$ by Proposition 10 and since $\omega_n\mathfrak{X} + q_n\mathfrak{X}$ is contained in the kernel of $\mathfrak{X} \to \mathfrak{X}_n/q_n\mathfrak{X}_n$, we see that f induces a homomorphism $X_n \to \mathfrak{X}_n/q_n\mathfrak{X}_n$, and hence also a homomorphism $\Phi'_n \to \mathfrak{X}_n/q_n\mathfrak{X}_n$.

THEOREM 4. *There exists a unique κ-isomorphism*

$$f : X \to \mathfrak{X}$$

which induces $\psi_n : \Phi'_n \to \mathfrak{X}_n/q_n\mathfrak{X}_n$ for every $n \geqq 0$.

PROOF. Let $m \geqq n$. For any α in Φ'_m and β in $U_{n,0}$, we have

$$(N_{n,m}(\alpha), \beta)_n = (\alpha, \beta)_m^{p^{m-n}} = \zeta_n^{T_m(\psi_m(\alpha)\log \beta)}$$
$$= \zeta_n^{T_n(T_{n,m}(\psi_m(\alpha))\log \beta)}.$$

This shows that

$$\psi_n(N_{n,m}(\alpha)) = T_{n,m}(\psi_m(\alpha)), \qquad \alpha \in \Phi'_m,$$

namely that the following diagram is commutative:

$$\begin{array}{ccc} \Phi'_m & \longrightarrow & \mathfrak{X}_m/q_m\mathfrak{X}_m \\ \downarrow N_{n,m} & & \downarrow T_{n,m} \\ \Phi'_n & \longrightarrow & \mathfrak{X}_n/q_n\mathfrak{X}_n \end{array} \qquad m \geqq n.$$

Since $(\Phi'_n)^{q_n}$ is contained in the kernel of ψ_n, ψ_n induces a homomorphism $f_n: X_n \to \mathfrak{X}_n/q_n\mathfrak{X}_n$. For these f_n, $n \geqq 0$, we then have commutative diagrams similar to the one in the above. Since $\mathfrak{X}$ is also the inverse limit of $\mathfrak{X}_n/q_n\mathfrak{X}_n$, $n \geqq 0$, we obtain a continuous homomorphism $f: X \to \mathfrak{X}$ which induces f_n and ψ_n for every $n \geqq 0$. By Proposition 14, ψ_n is a surjective κ-homomorphism. Hence f_n and f are also surjective κ-homomorphisms.

Suppose that $f(x)=0$ for an element x in X. Let x_n be the image of x under $X \to X_n$, and let α_n be an element of Φ'_n representing $x_n \bmod (\Phi'_n)^{q_n}$. Then $f(x)=0$ implies $f_n(x_n)=0$ and $\psi_n(\alpha_n)=0$. It follows that $(\alpha_n, \beta)_n=1$ for every β in $U_{n,0}$, and hence also for every β in $U_n=U_{n,0}\times V$, $n \geqq 0$. Since $x_n=N_{n,2n+1}(x_{2n+1})$, we have $\alpha_n \equiv N_{n,2n+1}(\alpha_{2n+1}) \bmod (\Phi'_n)^{q_n}$. As $\pi_n=\pi_{2n+1}^{q_n}\beta$ with β in U_{2n+1}, we see that

$$\begin{aligned}(\alpha_n, \pi_n)_n &= (N_{n,2n+1}(\alpha_{2n+1}), \pi_n)_n = (\alpha_{2n+1}, \pi_n)_{2n+1}^{q_n} \\ &= (\alpha_{2n+1}, \pi_{2n+1})_{2n+1}^{q_{2n+1}}(\alpha_{2n+1}, \beta)_{2n+1} \\ &= 1.\end{aligned}$$

It then follows from $\Phi_n^*=\Pi_n \times U_n$ that $(\alpha_n, \Phi_n^*)_n=1$. Therefore, by iv), α_n is contained in $(\Phi_n^*)^{q_n}$, and x_n belongs to $X_n^{q_n}$ for every $n \geqq 0$. Hence $x=1$, and f is injective. As X is compact, f is then a topological isomorphism. The uniqueness is obvious.

The definition of ψ_n, and hence also the definition of f depend upon the choice of the sequence of roots of unity ζ_n, $n \geqq 0$, namely, the choice of the isomorphism $\iota: W \to \boldsymbol{Q}_p/\boldsymbol{Z}_p$. Nevertheless, f is essentially unique in the following sense; let $\iota': W \to \boldsymbol{Q}_p/\boldsymbol{Z}_p$ be any other isomorphism and let $f': X \to \mathfrak{X}$ be the κ-isomorphism defined by ι'. Then there exists a p-adic unit u such that $\iota(\zeta)=u\iota'(\zeta)$ for every ζ in W, and such that

$$f'(x)=uf(x), \qquad x \in X.$$

Let $\mathfrak{Z}$ be the submodule of $\mathfrak{X}$ defined in 1.6, and let Z be the subgroup of X defined in 2.3. Then:

PROPOSITION 15.

$$f(Z)=\mathfrak{Z}.$$

PROOF. Let $f(Z)=\mathfrak{Z}'$, and let $\mathfrak{Z}'_n$ be the image of $\mathfrak{Z}'$ under $\mathfrak{X} \to \mathfrak{X}_n$. It follows from Artin-Hasse's explicit formula that

$$\psi_n(\pi_n)=\mu_n+q_n\mathfrak{X}_n.$$

Hence

$$\psi_n(\pi_n^\sigma)=\kappa(\sigma)\mu_n^\sigma+q_n\mathfrak{X}_n \qquad \sigma\in G\,,$$

by Proposition 14. Therefore $\psi_n(Q_n)=(\mathfrak{Z}_n+q_n\mathfrak{X}_n)/q_n\mathfrak{X}_n$, and consequently $f_n(Z_n)=(\mathfrak{Z}_n+q_n\mathfrak{X}_n)/q_n\mathfrak{X}_n$. However, since

$$\begin{array}{ccc} X & \xrightarrow{\ f\ } & \mathfrak{X} \\ \downarrow & & \downarrow \\ X_n & \xrightarrow{\ f_n\ } & \mathfrak{X}_n/q_n\mathfrak{X}_n \end{array}$$

is commutative, we see that $f_n(Z_n)=(\mathfrak{Z}'_n+q_n\mathfrak{X}_n)/q_n\mathfrak{X}_n$. Hence $\mathfrak{Z}_n+q_n\mathfrak{X}_n=\mathfrak{Z}'_n+q_n\mathfrak{X}_n$. We then obtain $\mathfrak{Z}=\mathfrak{Z}'$, q. e. d.

3.2. Let

$$\mathfrak{Y}=f(Y)\,,$$

and let $\mathfrak{Y}_n$ be the image of $\mathfrak{Y}$ under $\mathfrak{X}\to\mathfrak{X}_n$ $(n\geqq 0)$. $\mathfrak{Y}$ is a closed $\boldsymbol{Z}_p[G]$-submodule of $\mathfrak{X}$ containing $\mathfrak{Z}$, and

$$\mathfrak{X}/\mathfrak{Y}=\lim \mathfrak{X}_n/\mathfrak{Y}_n\,,$$

$$\mathfrak{Y}/\mathfrak{Z}=\lim \mathfrak{Y}_n/\mathfrak{Z}_n\,,$$

in the obvious manner. Similarly for $^{\pm}(\mathfrak{X}/\mathfrak{Y})$, etc.

It follows from Corollary i) of Proposition 8 that for each $n\geqq 0$, there exists an element α_n in $^+P'_n$ such that $^+P'_n/(^+P'_n)^p$ is generated by the cosets of α_n^ρ, $\rho\in{}^+G_n$. Using the same corollary, we also see readily that we may choose α_n so that $N_{n,n+1}(\alpha_{n+1})=\alpha_n$, $n\geqq 0$. Let y_n denote the image of α_n under $^+P'_n\to{}^+Y'_n=(Y'_n)^{+\varepsilon}$. Then $^+Y'_n$ is generated by y_n over $\boldsymbol{Z}_p[G]$, and $N_{n,m}(y_m)=y_n$ for any $m\geqq n\geqq 0$. Let y be the element of ^+Y determined by these y_n, $n\geqq 0$. It follows from the above that the elements y^σ, $\sigma\in G$, generate a dense subgroup of ^+Y. Hence $f(y)^\sigma$, $\sigma\in G$, also generate a dense subgroup of $^-\mathfrak{Y}=f(^+Y)$.

Let i be odd and $i\neq p-2$. By Theorem 2 and Proposition 1, we have $\boldsymbol{Z}_p[G]$-isomorphisms $^i\mathfrak{X}\to{}^i\mathfrak{R}\to\Lambda=\boldsymbol{Z}_p[[T]]$. Let ^ih be the element of Λ corresponding to $^i\varepsilon(f(y))$ in $^i\mathfrak{X}$. Then we have isomorphisms of compact Γ-modules

$$^i\mathfrak{Y}\to(^ih)=\Lambda^ih\,,\qquad {}^i\mathfrak{X}/{}^i\mathfrak{Y}\to\Lambda/(^ih)\,.$$

Since $^i\mathfrak{Z}\to(^ig)$ under the same map $^i\mathfrak{X}\to\Lambda$, we see that the ideal (^ig) is contained in (^ih), namely, that

$$^ig={}^ih{}^ik\,,$$

for some ^ik in Λ. It follows that

$$^i(\mathfrak{Y}/\mathfrak{Z})={}^i\mathfrak{Y}/{}^i\mathfrak{Z}\cong\Lambda/(^ik)\,.$$

Since $^{p-2}(\mathfrak{X}/\mathfrak{Z})\cong{}^{p-2}(\mathfrak{A}/\mathfrak{B})=0$, we have

$$^{p-2}(\mathfrak{X}/\mathfrak{Y})={}^{p-2}(\mathfrak{Y}/\mathfrak{Z})=0\,.$$

Hence the above result also holds for $i=p-2$, if we put simply ${}^{p-2}g={}^{p-2}h={}^{p-2}k=1$.

PROPOSITION 16. *Both* ${}^{-}(\mathfrak{X}/\mathfrak{Y})$ *and* ${}^{-}(\mathfrak{Y}/\mathfrak{Z})=\mathfrak{Y}/\mathfrak{Z}$ *are regular strictly* Γ*-finite* Γ*-modules, and*

$$
\begin{aligned}
{}^{-}(\mathfrak{X}/\mathfrak{Y})^{(n)} &= {}^{-}\mathfrak{X}_n/{}^{-}\mathfrak{Y}_n\,,\\
{}^{-}(\mathfrak{Y}/\mathfrak{Z})^{(n)} &= {}^{-}\mathfrak{Y}_n/{}^{-}\mathfrak{Z}_n\,, \qquad n\geqq 0\,.
\end{aligned}
$$

PROOF. We first notice that since ${}^{+}(Y/Z)=Y/Z$ and Y/Z is κ-isomorphic to $\mathfrak{Y}/\mathfrak{Z}$, we have ${}^{-}(\mathfrak{Y}/\mathfrak{Z})=\mathfrak{Y}/\mathfrak{Z}$. We also know by Theorem 2 that ${}^{-}(\mathfrak{X}/\mathfrak{Z})$ is strictly Γ-finite. Hence both ${}^{-}(\mathfrak{X}/\mathfrak{Y})$ and ${}^{-}(\mathfrak{Y}/\mathfrak{Z})$ are strictly Γ-finite. Let i be any odd index. Then ${}^{i}(\mathfrak{X}/\mathfrak{Y})$ is strictly Γ-finite, and ${}^{i}(\mathfrak{X}/\mathfrak{Y})\cong \Lambda/({}^{i}h)$. As we have noticed in 1.4, a strictly Γ-finite Γ-module of the type $\Lambda/(g)$ is always regular. Hence ${}^{i}(\mathfrak{X}/\mathfrak{Y})$ is regular. Therefore ${}^{-}(\mathfrak{X}/\mathfrak{Y})$, the direct sum of such ${}^{i}(\mathfrak{X}/\mathfrak{Y})$, is also regular. Similarly ${}^{-}(\mathfrak{Y}/\mathfrak{Z})$ is regular.

Since ${}^{-}(\mathfrak{X}/\mathfrak{Z})^{(n)}={}^{-}\mathfrak{X}_n/{}^{-}\mathfrak{Z}_n$ by Theorem 2, we see immediately that ${}^{-}(\mathfrak{X}/\mathfrak{Y})^{(n)}={}^{-}\mathfrak{X}_n/{}^{-}\mathfrak{Y}_n$, $n\geqq 0$. On the other hand, as ${}^{-}(\mathfrak{X}/\mathfrak{Y})$ is regular, we have ${}^{-}(\mathfrak{X}/\mathfrak{Y})^{(n)}={}^{-}(\mathfrak{X}/\mathfrak{Z})^{(n)}/{}^{-}(\mathfrak{Y}/\mathfrak{Z})^{(n)}$. Hence ${}^{-}(\mathfrak{Y}/\mathfrak{Z})^{(n)}$ and ${}^{-}\mathfrak{Y}_n/{}^{-}\mathfrak{Z}_n$ have the same finite order. However, it is clear that $\omega_n({}^{-}(\mathfrak{Y}/\mathfrak{Z}))$ is contained in the kernel of the natural homomorphism ${}^{-}(\mathfrak{Y}/\mathfrak{Z})\rightarrow{}^{-}\mathfrak{Y}_n/{}^{-}\mathfrak{Z}_n$. Hence we see from the above that ${}^{-}(\mathfrak{Y}/\mathfrak{Z})^{(n)}={}^{-}(\mathfrak{Y}/\mathfrak{Z})/\omega_n({}^{-}(\mathfrak{Y}/\mathfrak{Z}))\rightarrow{}^{-}\mathfrak{Y}_n/{}^{-}\mathfrak{Z}_n$ is an isomorphism.

Now, the map $f: X\rightarrow\mathfrak{X}$ induces κ-isomorphisms

$$
\begin{aligned}
X/Y &\rightarrow \mathfrak{X}/\mathfrak{Y}\,,\\
Y/Z &\rightarrow \mathfrak{Y}/\mathfrak{Z}\,,\\
X/Z &\rightarrow \mathfrak{X}/\mathfrak{Z}\,.
\end{aligned}
$$

Of these six $Z_p[G]$-groups, the arithmetic properties of X/Y and Y/Z are given by Proposition 12 and Proposition 13 respectively, and the algebraic structure of $\mathfrak{X}/\mathfrak{Z}$ is described in Theorem 2. Using these facts, we shall next study some arithmetic consequences of the above κ-isomorphisms.

3.3. Let S_n denote the Sylow p-subgroup of the ideal class group of F_n, $n\geqq 0$. The injection of the ideal group of F_n into the ideal group of F_m, $m\geqq n$, induces a homomorphism $S_n\rightarrow S_m$. Let S denote the direct limit of S_n, $n\geqq 0$, relative to these maps $S_n\rightarrow S_m$, $m\geqq n$. Clearly S_n, $n\geqq 0$, and S are $\boldsymbol{Z}_p[G]$-groups in the natural manner so that ${}^{\pm}S_n$, ${}^{\pm}S$, etc. are defined. It is known that ${}^{-}S_n\rightarrow{}^{-}S_m$ is injective for any $m\geqq n$[19]. Hence we may consider ${}^{-}S$ simply as the union of all ${}^{-}S_n$, $n\geqq 0$.

Let L/F and K/F be the abelian extensions defined in 2.3. Let c be an ideal class in ${}^{-}S$ and let $\mathfrak{a}$ be an ideal of an F_n representing the class c.

19) See [9], § 10. Theorem 15.

Then $\mathfrak{a}^{p^m}=(a)$ for some integer $m\geqq 0$ and for some a in F_n^*. Let α be a p^m-th root of a in Ω. It follows from the definition of K that α is contained in K. For any g in ${}^+G(K/F)$, we put

$$(g, c)=\iota(\alpha^{g-1});$$

since α^{g-1} is a p^m-th root of unity in W, the right-hand side is an element of $\boldsymbol{Q}_p/\boldsymbol{Z}_p$. We can then show that (g, c) depends only upon g and c, and defines a dual pairing of the compact abelian group ${}^+G(K/F)$ and the discrete abelian group ${}^-S$ into $\boldsymbol{Q}_p/\boldsymbol{Z}_p$:

$${}^+G(K/F)\times{}^-S\rightarrow\boldsymbol{Q}_p/\boldsymbol{Z}_p,$$

such that

$$(g^\sigma, c^\sigma)=\kappa(\sigma)(g, c), \qquad \sigma\in G^{20)}$$

Let L'/F denote the maximal sub-extension of L/F such that the Galois group $G(L'/F)$ is a regular Γ-group. It is known that L/L' is a finite extension and ${}^-G(L/F)={}^-G(L'/F)$[21]. Let ${}^-S'$ and ${}^-S''$ denote the annihilators of ${}^+G(K/L')$ and ${}^+G(K/L)$ in ${}^-S$ respectively, with respect to the above pairing. Then we have similar pairings

$${}^+G(K/L')\times({}^-S/{}^-S')\rightarrow\boldsymbol{Q}_p/\boldsymbol{Z}_p,$$

$${}^+G(K/L)\times({}^-S/{}^-S'')\rightarrow\boldsymbol{Q}_p/\boldsymbol{Z}_p,$$

$${}^+G(L'/F)\times{}^-S'\rightarrow\boldsymbol{Q}_p/\boldsymbol{Z}_p,$$

$${}^+G(L/F)\times{}^-S''\rightarrow\boldsymbol{Q}_p/\boldsymbol{Z}_p.$$

Clearly ${}^-S'$ is contained in ${}^-S''$, and ${}^-S''/{}^-S'$ is dual to the finite group ${}^+G(L/L')$. Since ${}^+G(L'/F)$ is a regular Γ-group, ${}^-S'$ is also regular. It follows that ${}^-S'$ is the maximal regular subgroup of the discrete Γ-finite Γ-group ${}^-S''$.

PROPOSITION 17. *Let c be an element of order p^m, $m\geqq 0$, in ${}^-S_n$. Then the following properties for c are equivalent:*

i) *c is contained in ${}^-S_n''={}^-S''\cap{}^-S_n$,*

ii) *There exist an integer $s\geqq m$ and an ideal $\mathfrak{a}$ in the class c such that $\mathfrak{a}^{p^s}=(a)$ with an element a in $F_n^*\cap\Phi_n^{p^s}$,*

iii) *For any integer $s\geqq m$ and for any ideal $\mathfrak{a}$ in the class c, there exists an element a in $F_n^*\cap\Phi_n^{p^s}$ such that $\mathfrak{a}^{p^s}=(a)$.*

PROOF. Clearly iii) implies ii). Let s, $\mathfrak{a}$, and a be as stated in ii). Let $l\geqq n$, s, and let α be a p^s-th root of a in Ω. By the assumption on a, $F_l(\alpha)/F_l$ is an unramified abelian p-extension so that α is contained in L_l', and hence in L. Consequently $\alpha^{g-1}=1$ for every g in $G(K/L)$, and $({}^+G(K/L), c)=0$. Therefore, c is contained in ${}^-S''$, and we see that ii) implies i).

20) See [9], § 8.
21) See [9], § 10. Theorem 16.

Suppose next that c belongs to $^{-}S''$. Let s and $\mathfrak{a}$ be as stated in iii). Since c is an element of $^{-}S_n$, we have $c^{1+\delta}=1$ for $\delta=\sigma(-1)$. Let $c_1=c^{1/2}$. Then $c=c_1^{1-\delta}$. Let $\mathfrak{a}_1$ be an ideal of c_1, prime to (π_n). Since $c_1^{p^m}=1$, we have $\mathfrak{a}_1^{p^s}=(a_1)$ with some a_1 in $F_n^*\cap U_n$. Let $\mathfrak{a}_2=\mathfrak{a}_1^{1-\delta}$, $a_2=a_1^{1-\delta}$ so that $\mathfrak{a}_2$ is an ideal of the class c satisfying $\mathfrak{a}_2^{p^s}=(a_2)$, $a_2^{1+\delta}=1$. Let α be a p^s-th root of a_2 in K. Since $({}^{+}G(K/L),c)=1$, it follows from the definition of the pairing that $\alpha^{g-1}=1$ for every g in ${}^{+}G(K/L)$. On the other hand, we see from $a_2^{1+\delta}=1$ that $\alpha^{g-1}=1$ for every g in ${}^{-}G(K/L)$. Hence $\alpha^{g-1}=1$ for any g in $G(K/L)$, and α must be an element of L. Let $l\geqq n,s$. Then $F_l(\alpha)/F_l$ is an unramified cyclic extension, and the principal prime ideal (π_l) is completely decomposed in $F_l(\alpha)$. Hence a_2 is contained in $\Phi_l^{p^s}$, and consequently in ${}^{-}U_l^{p^s}$. However ${}^{-}U_l$ is the direct product of W_l and a subgroup which is G-isomorphic to ${}^{-}(\boldsymbol{Z}_p[G_l])$[22]. It follows that $a_2=\zeta_n^i a_2'$ where i is a suitable integer and a_2' is an element of ${}^{-}U_n^{p^s}$. Thus $\mathfrak{a}_2^{p^s}=(a_2')$ with a' in $F_n^*\cap\Phi_n^{p^s}$. Since $\mathfrak{a}$ and $\mathfrak{a}_2$ belong to the same class c, there exists an element b in F_n^* such that $\mathfrak{a}=b\mathfrak{a}_2$. Then $\mathfrak{a}^{p^s}=(a)$ with $a=b^{p^s}a'$ in $F_n^*\cap\Phi_n^{p^s}$. Hence i) implies iii).

THEOREM 5. *There exists a dual pairing of the compact abelian group* ${}^{-}(\mathfrak{X}/\mathfrak{Y})$ *and the discrete abelian group* ${}^{-}S/{}^{-}S''$ *into* $\boldsymbol{Q}_p/\boldsymbol{Z}_p$:

$$ {}^{-}(\mathfrak{X}/\mathfrak{Y})\times({}^{-}S/{}^{-}S'')\rightarrow\boldsymbol{Q}_p/\boldsymbol{Z}_p\,, $$

such that

$$ (\tilde{x}^{\sigma},\tilde{c}^{\sigma})=(\tilde{x},\tilde{c})\,,\qquad\qquad \sigma\in G\,, $$

for any $\tilde{x}$ *in* ${}^{-}(\mathfrak{X}/\mathfrak{Y})$ *and* $\tilde{c}$ *in* ${}^{-}S/{}^{-}S''$.

PROOF. By Proposition 12, ${}^{+}G(K/L)={}^{+}(X/Y)$. Hence the map: $X/Y\rightarrow\mathfrak{X}/\mathfrak{Y}$ induces a κ-isomorphism ${}^{+}G(K/L)\rightarrow{}^{-}(\mathfrak{X}/\mathfrak{Y})$. Combining this with the pairing ${}^{+}G(K/L)\times({}^{-}S/{}^{-}S'')\rightarrow\boldsymbol{Q}_p/\boldsymbol{Z}_p$, we obtain a pairing ${}^{-}(\mathfrak{X}/\mathfrak{Y})\times({}^{-}S/{}^{-}S'')\rightarrow\boldsymbol{Q}_p/\boldsymbol{Z}_p$ as stated in the theorem.

Suppose that $\iota: W\rightarrow\boldsymbol{Q}_p/\boldsymbol{Z}_p$ is replaced by $\iota': W\rightarrow\boldsymbol{Q}_p/\boldsymbol{Z}_p$. Then $\iota=u\iota'$ with u in $\boldsymbol{U}$, and f is replaced by $f'=uf$. Hence $(\tilde{x},\tilde{c})$ is unchanged, and we see that the pairing is canonically defined. More precisely, the value of $(\tilde{x},\tilde{c})$ can be computed explicitly as follows: Let c be an element of ${}^{-}S_n$ $(n\geqq 0)$ representing $\tilde{c}$, and let $\mathfrak{a}$ be any ideal of the class c, prime to (π_n). Let $\mathfrak{a}^{p^m}=(a)$, $m\geqq 0$, with an element a in F_n^*. We may assume that $a\equiv 1 \bmod \pi_n$, because every non-zero residue class mod π_n contains a unit of F_n. Let x be an element of ${}^{-}\mathfrak{X}$ representing $\tilde{x}$, and let x_n be the image of x under ${}^{-}\mathfrak{X}\rightarrow{}^{-}\mathfrak{X}_n$. Then:

PROPOSITION 18.

$$ (\tilde{x},\tilde{c})=\mathrm{Res}\,(p^{-m}T_n(x_n\log a))\,. $$

PROOF. We fix an $s\geqq m,n$. Let α be a p^m-th root of a in K. Let g be

22) See [9], § 11, Theorem 19.

the element of ${}^+G(K/L)$ corresponding to $\tilde{x}$ in ${}^-(\mathfrak{X}/\mathfrak{Y})$, and let g' be the restriction of g on K_s. We choose an element b in U'_s such that g' is mapped to b mod $\bar{E}_s$ under the isomorphism $G(K_s/L_s)\to U'_s/\bar{E}_s=X_s/Y_s$. Then $\psi_s(b)=f_s(b)=x_s$ mod $q_s\mathfrak{X}_s$, where x_s denotes the image of x under ${}^-\mathfrak{X}\to{}^-\mathfrak{X}_s$, and ψ_s and f_s are the maps defined in 3.1. Hence it follows from the definition of $(\tilde{x},\tilde{c})$ and $(\ ,\ )_s$ that

$$\begin{aligned}(\tilde{x},\tilde{c})=(g,c)&=\iota(\alpha^{g-1})=\iota(\alpha^{g'-1})\\&=\iota((b,a^{p^{s+1-m}})_s)=p^{s+1-m}\iota((b,a)_s)\\&=p^{s+1-m}\iota(\zeta_s^{T_s(x_s\log a)})\\&=p^{s+1-m}\operatorname{Res}(q_s^{-1}T_s(x_s\log a))\\&=\operatorname{Res}(p^{-m}T_n(T_{n,s}(x_s)\log a))\\&=\operatorname{Res}(p^{-m}T_n(x_n\log a)),\qquad\text{q. e. d.}\end{aligned}$$

The formula indicates clearly that the pairing is independent of the choice of $\iota: W\to\boldsymbol{Q}_p/\boldsymbol{Z}_p$ and that $(\tilde{x}^\sigma,\tilde{c}^\sigma)=(\tilde{x},\tilde{c})$ for any σ in G. It is to be noted that we can also define $(\tilde{x},\tilde{c})$ by the above formula and then prove directly that it gives a dual pairing of ${}^-(\mathfrak{X}/\mathfrak{Y})$ and ${}^-S/{}^-S''$ into $\boldsymbol{Q}_p/\boldsymbol{Z}_p$.

3.4. It is known that the discrete Γ-group ${}^-S$ is regular and that ${}^-S_n$ consists of all those elements in ${}^-S$ which are invariant under the automorphisms of $\Gamma_n=G(F/F_n)$[23]. Hence $N_{n,m}:{}^-S_m\to{}^-S_n$ induces an isomorphism ${}^-S_m^{(n)}={}^-S_m/{}^-S_m^{\omega_n}\to{}^-S_n$, $m\geqq n\geqq 0$.

Let c_0 be any element of ${}^-S_0$. Since $N_{n,m}:{}^-S_m\to{}^-S_n$, $m\geqq n\geqq 0$, is surjective, we can find c_n, $n\geqq 1$, so that $N_{n,m}(c_m)=c_n$ for any $m\geqq n\geqq 0$. For each $n\geqq 0$, we then define a $\boldsymbol{Z}_p[G]$-homomorphism

$$\rho_n:\mathfrak{R}_n\to{}^-S_n,$$

by

$$\rho_n(\alpha)=c_n^\alpha,\qquad \alpha\in\mathfrak{R}_n.$$

Since $c_n^{-\varepsilon}=c_n$, we have $\rho_n(\mathfrak{R}_n)=\rho_n({}^-\mathfrak{R}_n)$. We also know by a classical result on cyclotomic fields[24] that $\rho_n(\alpha)=c_n^\alpha=1$ for any α in $\mathfrak{R}_n\cap\mathfrak{R}_n\xi_n^*$. Let $\mathfrak{A}_n^*$ and $\mathfrak{B}_n^*$ be the $\boldsymbol{Z}_p[G]$-modules defined in 1.4. Then ${}^-\mathfrak{A}_n^*={}^-\mathfrak{B}_n^*+{}^-\mathfrak{R}_n$ and ${}^-\mathfrak{B}_n^*=\mathfrak{R}_n{}^-\xi_n^*$ so that

$${}^-\mathfrak{A}_n^*/{}^-\mathfrak{B}_n^*={}^-\mathfrak{R}_n/({}^-\mathfrak{R}_n\cap\mathfrak{R}_n{}^-\xi_n^*)={}^-\mathfrak{R}_n/({}^-\mathfrak{R}_n\cap\mathfrak{R}_n\xi_n^*).$$

Hence it follows from the above that ρ_n induces a $\boldsymbol{Z}_p[G]$-homomorphism

$$\rho'_n:{}^-\mathfrak{A}_n^*/{}^-\mathfrak{B}_n^*\to{}^-S_n.$$

Since $N_{n,m}(c_m)=c_n$, $m\geqq n\geqq 0$, we have

23) See [9], § 10.
24) See [10], § 3.

$$\rho'_n = \rho'_m \circ t'_{m,n}, \qquad m \geqq n \geqq 0,$$

for the homomorphism $t'_{m,n}: {}^-\mathfrak{A}^*_n/{}^-\mathfrak{B}^*_n \to {}^-\mathfrak{A}^*_m/{}^-\mathfrak{B}^*_m$ defined in 1.2. It then follows that

$$\rho'_n \circ t_{n,m} = \rho'_m \circ \nu_{n,m} = N_{n,m} \circ \rho'_m, \qquad m \geqq n \geqq 0.$$

Hence the maps ρ'_n, $n \geqq 0$, define a homomorphism from the inverse limit ${}^-(\mathfrak{A}^*/\mathfrak{B}^*)$ of ${}^-\mathfrak{A}^*_n/{}^-\mathfrak{B}^*_n$, $n \geqq 0$, into the inverse limit of ${}^-S_n$, $n \geqq 0$, relative to the homomorphisms $N_{n,m}: {}^-S_m \to {}^-S_n$. However, we can see by class field theory that the inverse limit of ${}^-S_n$, $n \geqq 0$, is canonically isomorphic to the inverse limit of ${}^-G(L'_n/F_n)$, $n \geqq 0$, relative to the natural maps ${}^-G(L'_m/F_m) \to {}^-G(L'_n/F_n)$, $m \geqq n \geqq 0$, namely, to the Galois group ${}^-G(L/F)$. Thus we obtain a $\boldsymbol{Z}_p[G]$-homomorphism

$$\rho: {}^-(\mathfrak{A}^*/\mathfrak{B}^*) \to {}^-G(L/F),$$

depending upon the sequence of elements c_n, $n \geqq 0$.

Suppose that $c_0^{(1)}, \cdots, c_0^{(r)}$ generate ${}^-S_0$ over $\boldsymbol{Z}_p[G]$. Let $\rho^{(j)}: {}^-(\mathfrak{A}^*/\mathfrak{B}^*) \to {}^-G(L/F)$ be the homomorphism defined by a sequence of elements $c_n^{(j)}$, $n \geqq 0$, starting with $c_0^{(j)}$ $(1 \leqq j \leqq r)$. We then see easily that the homomorphism $\rho: {}^-(\mathfrak{A}^*/\mathfrak{B}^*)^r \to {}^-G(L/F)$ defined by these $\rho^{(1)}, \cdots, \rho^{(r)}$ is surjective. Hence ${}^-G(L/F)$ is always a homomorphic image of the $\boldsymbol{Z}_p[G]$-module ${}^-(\mathfrak{A}^*/\mathfrak{B}^*)^r$ for some integer $r \geqq 1$. We shall next consider the case $r=1$.

In general, let A be a p-primary compact G-module on which G of course acts continuously. For simplicity, we call A G-cyclic when A contains an element a such that the elements σa, $\sigma \in G$, generate a dense subgroup of A. For example, both ${}^-\mathfrak{Y}$ and ${}^-\mathfrak{Z}$ are G-cyclic modules.

Let $A = {}^+A \oplus {}^-A = {}^0A \oplus \cdots \oplus {}^{p-2}A$ be the decompositions of A defined as usual. Then we see easily that each of the following conditions is necessary and sufficient for A to be G-cyclic:

1) $A^{(0)} = A/A^{\omega_0}$ is G-cyclic,

2) Both ${}^+A$ and ${}^-A$ are G-cyclic,

3) Every ${}^iA$, $0 \leqq i \leqq p-2$, is a Γ-module of the type $\Lambda/(g)$, $g \in \Lambda$,

4) Every ${}^iA^{(0)} = {}^iA/{}^iA^{\omega_0}$, $0 \leqq i \leqq p-2$, is cyclic over $\boldsymbol{Z}_p$, namely, every ${}^iA^{(0)}$ is either isomorphic to $\boldsymbol{Z}_p$ or to a finite cyclic group with order a power of p.

For the Galois group $G(L/F)$, we have natural isomorphisms $G(L/F)^{(0)} \to G(L'_0/F_0) \to S_0$. Hence $G(L/F)$ is G-cyclic if and only if the finite group S_0 is G-cyclic.

PROPOSITION 19. *The following properties for S_0 are equivalent:*

i) *S_0 is G-cyclic,*

ii) *${}^-S_0$ is G-cyclic,*

iii) *${}^iS_0$ is cyclic for every index i,*

iv) *${}^iS_0$ is cyclic for every odd i.*

PROOF. It is sufficient to show that iv) implies iii). Let ${}^{-}S_0^{(p)}$ denote the subgroup of all c in ${}^{-}S_0$ such that $c^p=1$. Then the pairing ${}^{+}G(K/F)\times{}^{-}S\to \boldsymbol{Q}_p/\boldsymbol{Z}_p$ in 3.3 induces a non-degenerate pairing

$$({}^{+}G(K_0/F_0)/{}^{+}G(K_0/F_0)^p)\times{}^{-}S_0^{(p)}\to \boldsymbol{Q}_p/\boldsymbol{Z}_p\,,$$

such that

$$(g^\sigma, c^\sigma)=\kappa(\sigma)(g, c) \qquad \sigma\in\Delta\,.$$

Hence we see that

$$\operatorname{rank}{}^{i}G(K_0/F_0)=\operatorname{rank}{}^{j}S_0\,,$$

for any even i and odd j such that $i+j\equiv 1 \bmod p-1$. On the other hand, since ${}^{i}G(K_0/L_0)\cong{}^{i}U_0'/{}^{i}\bar{E}_0={}^{i}U_{0,0}'/{}^{i}\bar{E}_{0,0}$ as stated in the proof of Proposition 12, and since $\operatorname{rank}{}^{i}U_{0,0}'=0$ or 1 according as $i=0$ or $i\neq 0$, we know that $\operatorname{rank}{}^{i}G(K_0/L_0)\leqq 1$ for any even index i. Hence $\operatorname{rank}{}^{i}G(K_0/F_0)$ is either equal to $\operatorname{rank}{}^{i}G(L_0/F_0)$ or equal to $1+\operatorname{rank}{}^{i}G(L_0/F_0)$. However, by class field theory, ${}^{i}G(L_0/F_0)$ is isomorphic to ${}^{i}S_0$. Therefore, it follows from the above that

$$\operatorname{rank}{}^{i}S_0\leqq\operatorname{rank}{}^{j}S_0\leqq 1+\operatorname{rank}{}^{i}S_0\,, \tag{6}$$

for even i and odd j satisfying $i+j\equiv 1 \bmod p-1$. Hence if ${}^{j}S_0$ is cyclic, so is ${}^{i}S_0$, and we see that iii) follows from iv).

PROPOSITION 20. *Suppose that* S_0 *is* G*-cyclic. Then there exist* $\boldsymbol{Z}_p[G]$*-isomorphisms*

$$ {}^{+}G(L'/F)\to{}^{+}(Y/Z)=Y/Z\,,$$

$$ {}^{-}G(L'/F)={}^{-}G(L/F)\to{}^{-}(\mathfrak{A}^*/\mathfrak{B}^*)\,.$$

There also exist dual pairings

$$ {}^{-}(\mathfrak{X}/\mathfrak{Y})\times({}^{-}S/{}^{-}S')\to\boldsymbol{Q}_p/\boldsymbol{Z}_p\,,$$

$$ {}^{-}(\mathfrak{Y}/\mathfrak{Z})\times{}^{-}S'\to\boldsymbol{Q}_p/\boldsymbol{Z}_p\,,$$

$$ {}^{-}(\mathfrak{X}/\mathfrak{Z})\times{}^{-}S\to\boldsymbol{Q}_p/\boldsymbol{Z}_p\,,$$

such that

$$[x^\sigma, c^\sigma]=[x, c]\,, \qquad \sigma\in G\,.$$

PROOF. Since S_0 is G-cyclic, so is ${}^{-}S_0$ by Proposition 19. Hence ${}^{-}S_0$ has an element c_0 whose conjugates $c_0^\sigma, \sigma\in G$, generate ${}^{-}S_0$. Let $c_n, n\geqq 1$, be chosen from ${}^{-}S_n$ as stated in the above. Since $N_{0,n}(c_n)=c_0$ and since $N_{0,n}:{}^{-}S_n\to{}^{-}S_0$ induces an isomorphism of ${}^{-}S_n/{}^{-}S_n^{\omega_0}$ onto ${}^{-}S_0$, ${}^{-}S_n/{}^{-}S_n^{\omega_0}$ is generated by the cosets of $c_n^\sigma, \sigma\in G$. We then see that ${}^{-}S_n$ itself is generated by $c_n^\sigma, \sigma\in G$, and the map $\rho_n:\mathfrak{R}_n\to{}^{-}S_n$ is surjective. Since $\rho_n({}^{-}\mathfrak{R}_n)=\rho_n(\mathfrak{R}_n)$, $\rho_n':{}^{-}\mathfrak{A}_n^*/{}^{-}\mathfrak{B}_n^*={}^{-}\mathfrak{R}_n/({}^{-}\mathfrak{R}_n\cap\mathfrak{R}_n\xi_n^*)\to{}^{-}S_n$ is also surjective. However, ${}^{-}\mathfrak{A}_n^*/{}^{-}\mathfrak{B}_n^*$ is isomorphic to ${}^{-}\mathfrak{A}_n/{}^{-}\mathfrak{B}_n$ as abelian groups so that the order of ${}^{-}\mathfrak{A}_n^*/{}^{-}\mathfrak{B}_n^*$ is equal to the exact power of p dividing the first factor ${}^{-}h_n$ of the class number of F_n, namely, to the order of ${}^{-}S_n$. Hence ρ_n' must be an isomorphism. Therefore

$\rho: {}^-(\mathfrak{A}^*/\mathfrak{B}^*) \to {}^-G(L/F)$ is also an isomorphism.

By Proposition 5, we have a non-degenerate pairing $({}^-\mathfrak{A}_n/{}^-\mathfrak{B}_n) \times ({}^-\mathfrak{A}_n^*/{}^-\mathfrak{B}_n^*) \to \boldsymbol{Q}_p/\boldsymbol{Z}_p$ for each $n \geqq 0$. Using the isomorphisms $\varphi_n: {}^-(\mathfrak{X}/\mathfrak{Z})^{(n)} = {}^-\mathfrak{X}_n/{}^-\mathfrak{Z}_n \to {}^-\mathfrak{A}_n/{}^-\mathfrak{B}_n$ and $\rho_n': {}^-\mathfrak{A}_n^*/{}^-\mathfrak{B}_n^* \to {}^-S_n$, we obtain a non-degenerate pairing $[x, c]_n: {}^-(\mathfrak{X}/\mathfrak{Z})^{(n)} \times {}^-S_n \to \boldsymbol{Q}_p/\boldsymbol{Z}_p$ such that

$$[x^\sigma, c^\sigma]_n = [x, c]_n, \qquad \sigma \in G,$$

$$[N_{n,m}(x), c]_n = [x, c]_m, \qquad m \geqq n \geqq 0,$$

where the element x in the second equality stands for an arbitrary element in ${}^-(\mathfrak{X}/\mathfrak{Z})^{(m)}$. Since ${}^-(\mathfrak{X}/\mathfrak{Z})$ is the inverse limit of ${}^-(\mathfrak{X}/\mathfrak{Z})^{(n)}$, $n \geqq 0$, and ${}^-S$ is the direct limit of ${}^-S_n$, $n \geqq 0$, it is clear that the above pairings $[x, c]_n$, $n \geqq 0$, define a dual pairing

$${}^-(\mathfrak{X}/\mathfrak{Z}) \times {}^-S \to \boldsymbol{Q}_p/\boldsymbol{Z}_p,$$

such that $[x^\sigma, c^\sigma] = [x, c]$ for any σ in G.

Now, let A and B denote the annihilators of ${}^-S'$ and ${}^-S''$ in ${}^-(\mathfrak{X}/\mathfrak{Z})$ respectively, with regard to the above pairing of ${}^-(\mathfrak{X}/\mathfrak{Z})$ and ${}^-S$. Then we have similar pairings

$$({}^-(\mathfrak{X}/\mathfrak{Z})/A) \times {}^-S' \to \boldsymbol{Q}_p/\boldsymbol{Z}_p,$$

$$B \times ({}^-S/{}^-S'') \to \boldsymbol{Q}_p/\boldsymbol{Z}_p.$$

Hence it follows from Theorem 5 that B is $\boldsymbol{Z}_p[G]$-isomorphic to ${}^-(\mathfrak{X}/\mathfrak{Y})$ and consequently that ${}^iB \cong {}^i(\mathfrak{X}/\mathfrak{Y}) \cong \Lambda/({}^ih)$ for every odd i; here $\Lambda = \boldsymbol{Z}_p[[T]]$, and ${}^ih$ is an element of Λ as defined in 3.2. On the other hand, we know that there exists a $\boldsymbol{Z}_p[G]$-isomorphism ${}^i\lambda: {}^i(\mathfrak{X}/\mathfrak{Z}) \to \Lambda/({}^ig)$ for each odd i. Let ${}^ib$ be the image of $1 \bmod ({}^ih)$ under an isomorphism $\Lambda/({}^ih) \to {}^iB$, and let ${}^if$ be an element of Λ such that ${}^i\lambda({}^ib) = {}^if \bmod ({}^ig)$. Then ${}^i\lambda({}^iB) = ({}^if, {}^ig)/({}^ig)$, and we see from the isomorphisms $\Lambda/({}^ih) \to {}^iB \to {}^i\lambda({}^iB)$ that an element u of Λ belongs to $({}^ih)$ if and only if $u{}^if$ is contained in $({}^ig)$. Since ${}^ig = {}^ih{}^ik$, it follows that ${}^if$ is divisible by ${}^ik$ in Λ: ${}^if = {}^ik'{}^ik$, and that ${}^ik'$ is not a zero-divisor $\bmod ({}^ih)$. As ${}^ih \neq 0$, we then see that $({}^ik', {}^ih)$ is a primary ideal of Λ, belonging to the maximal ideal of the 2-dimensional local ring Λ. Hence $\Lambda/({}^ik', {}^ih)$ is a finite module. Let ${}^iA'$ be the inverse image of $({}^ik)/({}^ig)$ under ${}^i\lambda: {}^i(\mathfrak{X}/\mathfrak{Z}) \to \Lambda/({}^ig)$. Since $({}^if, {}^ig) = ({}^ik', {}^ih)({}^ik)$, we have ${}^iA'/{}^iB \cong ({}^ik)/({}^if, {}^ig) = \Lambda/({}^ik', {}^ih)$ and ${}^i(\mathfrak{X}/\mathfrak{Z})/{}^iA' \cong \Lambda/({}^ik) \cong {}^i(\mathfrak{Y}/\mathfrak{Z})$. Hence ${}^iA'/{}^iB$ is finite and ${}^i(\mathfrak{X}/\mathfrak{Z})/{}^iA'$ is a regular Γ-module. On the other hand, as A/B is dual to ${}^-S''/{}^-S'$, A/B is finite, and so is ${}^iA/{}^iB$. Since ${}^i(\mathfrak{X}/\mathfrak{Z})/{}^iA$ is dual to the regular Γ-module ${}^{-i}S'$, ${}^i(\mathfrak{X}/\mathfrak{Z})/{}^iA$ is also regular. Therefore both ${}^iA'/{}^iB$ and ${}^iA/{}^iB$ are finite submodules of the Γ-module ${}^i(\mathfrak{X}/\mathfrak{Z})/{}^iB$ with regular factor modules, and we see that ${}^iA' = {}^iA$[25]. It follows

25) A discrete Γ-finite module has a unique maximal regular submodule. See [8], 1.4.

that ${}^{i}(\mathfrak{X}/\mathfrak{Z})/{}^{i}A \cong {}^{i}(\mathfrak{Y}/\mathfrak{Z})$ for every odd i, and we have a $\boldsymbol{Z}_p[G]$-isomorphism

$$^{-}(\mathfrak{X}/\mathfrak{Z})/A \to {}^{-}(\mathfrak{Y}/\mathfrak{Z}).$$

Combining it with $({}^{-}(\mathfrak{X}/\mathfrak{Z})/A) \times {}^{-}S' \to \boldsymbol{Q}_p/\boldsymbol{Z}_p$, we obtain a dual pairing ${}^{-}(\mathfrak{Y}/\mathfrak{Z}) \times {}^{-}S' \to \boldsymbol{Q}_p/\boldsymbol{Z}_p$.

The pairing ${}^{-}(\mathfrak{X}/\mathfrak{Z}) \times {}^{-}S \to \boldsymbol{Q}_p/\boldsymbol{Z}_p$ also induces a similar pairing $A \times ({}^{-}S/{}^{-}S') \to \boldsymbol{Q}_p/\boldsymbol{Z}_p$. However, it has been proved in the above that ${}^{i}A \cong ({}^{i}k)/({}^{i}g) = \Lambda/({}^{i}h) \cong {}^{i}(\mathfrak{X}/\mathfrak{Y})$. Hence $A \cong {}^{-}(\mathfrak{X}/\mathfrak{Y})$, and we have a pairing ${}^{-}(\mathfrak{X}/\mathfrak{Y}) \times ({}^{-}S/{}^{-}S') \to \boldsymbol{Q}_p/\boldsymbol{Z}_p$.

Finally, combining the pairing ${}^{-}(\mathfrak{Y}/\mathfrak{Z}) \times {}^{-}S' \to \boldsymbol{Q}_p/\boldsymbol{Z}_p$ with the κ-isomorphism ${}^{+}(Y/Z) \to {}^{-}(\mathfrak{Y}/\mathfrak{Z})$, we obtain a dual pairing ${}^{+}(Y/Z) \times {}^{-}S' \to \boldsymbol{Q}_p/\boldsymbol{Z}_p$ such that $[y^{\sigma}, c^{\sigma}] = \kappa(\sigma)[y, c]$ for any σ in G. As stated in 3.3, there also exists a dual pairing ${}^{+}G(L'/F) \times {}^{-}S' \to \boldsymbol{Q}_p/\boldsymbol{Z}_p$ such that $(g^{\sigma}, c^{\sigma}) = \kappa(\sigma)(g, c)$ for any σ in G. It follows that ${}^{+}(Y/Z) = Y/Z$ is $\boldsymbol{Z}_p[G]$-isomorphic to ${}^{+}G(L'/F)$.

Since ${}^{-}G(L/F)^{(n)} = {}^{-}G(L'_n/F_n) \cong {}^{-}S_n$, $n \geqq 0$, the isomorphism ${}^{-}G(L/F) \cong {}^{-}(\mathfrak{A}^*/\mathfrak{B}^*)$ implies

$$^{-}S_n \cong {}^{-}\mathfrak{A}_n^* / {}^{-}\mathfrak{B}_n^*, \qquad n \geqq 0.$$

As stated in [10], this gives us a group-theoretical interpretation of the p-part of the classical class number formula (I) for the first factor ${}^{-}h_n$ of the class number of F_n. We also notice that the existence of a $\boldsymbol{Z}_p[G]$-isomorphism ${}^{-}G(L/F) \to {}^{-}(\mathfrak{A}^*/\mathfrak{B}^*)$ implies conversely that S_0 is G-cyclic, because ${}^{-}S_0 \cong {}^{-}\mathfrak{A}_0^*/{}^{-}\mathfrak{B}_0^* = {}^{-}\mathfrak{R}_0/({}^{-}\mathfrak{R}_0 \cap \mathfrak{R}_0 \xi_0^*)$ and ${}^{-}\mathfrak{R}_0$ is G-cyclic.

To obtain a similar result from ${}^{+}G(L'/F) \cong Y/Z$, we assume that $G(L/F)$ is a regular Γ-groups. Then $L' = L$, and ${}^{+}G(L'/F)^{(n)} = {}^{+}G(L/F)^{(n)} \cong {}^{+}S_n$, $n \geqq 0$. On the other hand, the regularity of $G(L/F)$ also implies $E'_n = E_n$, $n \geqq 0$[26] so that $(Y/Z)^{(n)} = ({}^{+}E_n/{}^{+}C_n)_p$ by Proposition 13. Hence we see from ${}^{+}G(L'/F) \cong Y/Z$ that

$$^{+}S_n \cong ({}^{+}E_n/{}^{+}C_n)_p, \qquad n \geqq 0,$$

as $\boldsymbol{Z}_p[G]$-groups. This may be considered as a group-theoretical interpretation of the p-part of the classical class number formula (II) for the second factor ${}^{+}h_n$ of the class number of F_n, because the p-part of (II) simply states that the order of ${}^{+}S_n$ is equal to the order of $({}^{+}E_n/{}^{+}C_n)_p$. Without assuming that $G(L/F)$ is regular, we can still obtain a certain group-theoretical relation between ${}^{+}S_n$ and $({}^{+}E_n/{}^{+}C_n)_p$ by using Corollary iv) of Proposition 8. But we omit the detail here.

Finally, we notice that unlike the pairing given in Theorem 5, the isomorphisms and the pairings in the preceding proposition are not canonical. We feel that some essential link is still missing in the relation between ${}^{-}(\mathfrak{X}/\mathfrak{Z})$ and ${}^{-}S$.

---

26) See [9], § 9, Theorem 14.

3.5. Suppose now that the prime number p is properly irregular; namely, that the second factor ${}^+h_0$ of the class number of F_0 is prime to p. Then ${}^+S_0=1$, and it follows from (6) that ${}^jS_0$ is cyclic for every odd j. Hence S_0 is G-cyclic by Proposition 19, and we see from Proposition 20 that ${}^-S_n\cong{}^-\mathfrak{A}_n^*/{}^-\mathfrak{B}_n^*$, $n\geqq 0$. However, in this case, we can also proceed as follows, without referring to the result of 3.4.

It is known that the assumption on ${}^+h_0$ implies that ${}^+h_n$ is not divisible by p for any $n\geqq 0$[27]. Hence the class number formula (II) shows that $({}^+E_n/{}^+C_n)_p=1$. It then follows from Proposition 13 that $Y/Z=1$. Therefore ${}^-(\mathfrak{Y}/\mathfrak{Z})=\mathfrak{Y}/\mathfrak{Z}=0$, and ${}^-(\mathfrak{X}/\mathfrak{Y})={}^-(\mathfrak{X}/\mathfrak{Z})$. On the other hand, the fact that ${}^+h_n$ is prime to p also implies that ${}^+S_n=1$ and ${}^+G(L_n'/F_n)=1$, $n\geqq 0$. Hence ${}^+S=1$, ${}^-S=S$, ${}^+G(L/F)=1$, and consequently ${}^-S''=1$, because ${}^-S''$ is dual to ${}^+G(L/F)$ as explained in 3.3. We then see from Theorem 5 that there exists a canonical dual pairing

$$^-(\mathfrak{X}/\mathfrak{Z})\times S\rightarrow \boldsymbol{Q}_p/\boldsymbol{Z}_p$$

such that $[x^\sigma, c^\sigma]=[x, c]$ for any σ in G. Hence we also have, for each $n\geqq 0$, a similar non-degenerate pairing

$$^-\mathfrak{X}_n/{}^-\mathfrak{Z}_n\times S_n\rightarrow \boldsymbol{Q}_p/\boldsymbol{Z}_p\,.$$

It follows in particular that ${}^-\mathfrak{X}_n/{}^-\mathfrak{Z}_n$ is finite. Hence ${}^-\mathfrak{A}_n/{}^-\mathfrak{B}_n$ is also finite by Theorem 2. The proof of Proposition 4 then shows that ${}^iD_n\neq 0$ for every odd i and that the order of S_n, which equals the order of ${}^-\mathfrak{X}_n/{}^-\mathfrak{Z}_n$, is equal to the exact power of p dividing $2q_n\prod\left(-\frac{1}{2}{}^iD_n\right)$. Since $S_n={}^-S_n$, this is nothing but the p-part of the class number formula (I). We also see from Proposition 5 that S_n is $\boldsymbol{Z}_p[G]$-isomorphic to ${}^-\mathfrak{A}_n^*/{}^-\mathfrak{B}_n^*$.

There exists an essential difference between the above proof and the one which uses Proposition 20. Whereas the isomorphism $S_n={}^-S_n\cong{}^-\mathfrak{A}_n^*/{}^-\mathfrak{B}_n^*$ obtained from Proposition 20 is not canonical, the pairing ${}^-\mathfrak{X}_n/{}^-\mathfrak{Z}_n\times S_n\rightarrow\boldsymbol{Q}_p/\boldsymbol{Z}_p$ in the above is canonical and is explicitly given by Proposition 18. Furthermore, in the proof of Proposition 20, we had to use the class number formula (I) in the form that ${}^-S_n$ and ${}^-\mathfrak{A}_n^*/{}^-\mathfrak{B}_n^*$ have the same order. However, in the above proof, in addition to the assumption that ${}^+h_0$ is not divisible by p, we have used only the class number formula (II) to the effect that $[{}^+E_n:{}^+C_n]$ is finite and prime to p, and have proved the p-part of the class number formula (I) by purely algebraic deduction.

Massachusetts Institute of Technology

27) See [6].

Bibliography

[1] E. Artin, Algebraic numbers and algebraic functions, Lecture notes at Princeton Univ., 1950–1951.

[2] E. Artin und H. Hasse, Die beiden Ergänzungssätze zum Reziprozitätsgesetz der l^n-ten Potenzreste im Körper der l^n-ten Einheitswurzeln, Abh. Math. Sem. Univ. Hamburg, 6 (1928), 146–162.

[3] E. Artin and J. Tate, Class field theory, Lecture notes at Princeton Univ., 1951–1952.

[4] H. Hasse, Bericht über die neuere Untersuchungen und Probleme aus der Theorie der algebraischen Zahlkörper, I, Ia, II, Leipzig und Berlin, 1930.

[5] H. Hasse, Über die Klassenzahl abelscher Zahlkörper, Berlin, 1952.

[6] K. Iwasawa, A note on class numbers of algebraic number fields, Abh. Math. Sem. Univ. Hamburg, 20 (1956), 257–258.

[7] K. Iwasawa, On some invariants of cyclotomic fields, Amer. J. Math., 80 (1958), 773–783.

[8] K. Iwasawa, On some properties of Γ-finite modules, Ann. of Math., 70 (1959), 291–312.

[9] K. Iwasawa, On the theory of cyclotomic fields. Ann. of Math., 70 (1959), 530–561.

[10] K. Iwasawa, A class number formula for cyclotomic fields. Ann. of Math., 76 (1962), 171–179.

[11] J.-P. Serre, Classes des corps cyclotomiques, Seminaire Bourbaki, Exposé 174 (1958/1959).

[42] Some results in the theory of cyclotomic fields

Proc. Sympos. Pure Math., Vol. VIII, Amer. Math. Soc., 1965, pp. 66-69.

1. Let q be a positive integer, and F, a local field containing primitive qth roots of unity. Let α and β be any nonzero elements of F, and let $K=F(\beta^{1/q})$. Then by local class field theory, α is canonically associated with a Galois automorphism σ of the abelian extension K/F: $\sigma=(\alpha, K/F)$. Let (α, β) denote the qth root of unity $(\beta^{1/q})^{\sigma-1}$ in F. The symbol (α, β) defined in this manner is called Hilbert's norm residue symbol for the power q in the local field F. [2] Various formulas are known which express (α, β) as a function of α and β. We shall next present such a formula for a certain type of local cyclotomic field F.[3]

2. Let p be an odd prime, and let $q_n=p^{n+1}$ for any integer $n\geqq 0$. Let F_n be the local cyclotomic field of q_nth roots of unity over the p-adic number field $\mathbf{Q}_p$. We shall consider Hilbert's norm residue symbol $(\alpha, \beta)_n$ for the power q_n in F_n.

For each $n\geqq 0$, we fix a primitive q_nth root of unity ζ_n in F_n so that $\zeta_{n+1}^p=\zeta_n$. Let $\mathfrak{o}_n$ denote the ring of all local integers in F_n, and $\mathfrak{p}_n$, the unique prime ideal of $\mathfrak{o}_n$. Put $\pi_n=1-\zeta_n$. Then each unit ξ in F_n can be written in the form

$$\xi = \sum_{i=0}^{\infty} c_i \pi_n^i,$$

with coefficients c_i in the ring of p-adic integers $\mathbf{Z}_p$. Define the "derivative" of ξ with respect to π_n by

$$D\xi = \sum_{i=0}^{\infty} i c_i \pi_n^{i-1}.$$

Since the power series expansion of ξ such as in the above is not unique, $D\xi$ is not uniquely determined by ξ . However, it can be shown that for fixed π_n, the values of $D\xi$ corresponding to various power series expansions of ξ just fulfil a residue class of $\mathfrak{o}_n$ modulo the different $\mathfrak{d}_n$ of $F_n/\mathbf{Q}_p$.[4] In the

[1] The present research was supported in part by the National Science Foundation Grant NSF-GP-379.

[2] For the symbol (α, β), see E. Artin and J. Tate, *Class field theory*, Lecture notes at Princeton University, 1951-1952, Chapter 12.

[3] The proof of the results stated below will be published elsewhere.

[4] See Artin-Tate, p. 151.

following, we shall consider $D\xi$ as a multi-valued function of ξ, representing any such value as defined in the above. Let

$$D\log\xi = D\xi/\xi,$$

for any unit ξ in F_n. Then $D\log\xi$ is also a multi-valued function of ξ, similar to $D\xi$.

Now, let

$$(\alpha, \beta)_n = \zeta_n^{[\alpha,\beta]_n},$$

for any nonzero α and β in F_n; $[\alpha, \beta]_n$ is of course determined only modulo q_n. Let T_n denote the trace map from F_n to $\mathbf{Q}_p$. Then we have the following beautiful formulas for $[\alpha, \beta]_n$:

I. Let $n=0$, and let $\alpha\equiv 1 \bmod \mathfrak{p}_0^2$ and $\beta\equiv 1 \bmod \mathfrak{p}_0$. Then

$$[\alpha, \beta]_0 = q_0^{-1}\, T_0(\zeta_0 \log\alpha\, D\log\beta).$$[5]

II. Let $n\geqq 0$, and let $\alpha\equiv 1 \bmod \mathfrak{p}_n$. Then

$$[\alpha, \zeta_n]_n = q_n^{-1}\, T_n(-\log\alpha),$$
$$[\alpha, \pi_n]_n = q_n^{-1}\, T_n(\zeta_n \pi_n^{-1} \log\alpha).$$[6]

Since $\zeta_n = 1-\pi_n$, we see that

$$\zeta_n D\log\zeta_n = -1.$$

Similarly

$$\zeta_n D\log\pi_n = \zeta_n\pi_n^{-1},$$

if we define $D\log\pi_n$ formally in the same way as in the above. Hence II may be written also as

$$(1)\qquad \begin{aligned}[\alpha, \zeta_n]_n &= q_n^{-1}\, T_n(\zeta_n \log\alpha\, D\log\zeta_n),\\ [\alpha, \pi_n]_n &= q_n^{-1}\, T_n(\zeta_n \log\alpha\, D\log\pi_n).\end{aligned}$$

These equalities suggest the existence of a formula for $[\alpha, \beta]_n$, $n\geqq 0$, which generalizes both I and II. One difficulty in finding such a formula lies in the fact that $D\log\xi$ is a multi-valued function and that the formula (1) holds only for some special values of $D\log\zeta_n$ and $D\log\pi_n$. We shall next present a formula for $[\alpha, \beta]_n$, $n\geqq 0$, which is similar to I, and which explains why the particular values of $D\log\zeta_n$ and $D\log\pi_n$ appear in the formula II.

---

[5] See Artin-Tate, p. 164, and H. Hasse, Zum expliziten Reziprozitätsgesetz. Arch. der Math., 13(1962), 479-485.

[6] E. Artin and H. Hasse, *Die beiden Ergänzungssätze zum Reziprozitätsgesetz der l^n-ten Potenzreste im Körper der l^n-ten Einheitswurzeln*, Abh. Math. Sem. Univ. Hamburg 6(1928), 146-162.

3. We first generalize the definition of $D \log \xi$. Let ξ be any element of $\mathfrak{o}_n$, $\xi \neq 0$, and let s be the order of ξ: $(\xi) = \mathfrak{p}_n^s$, $s \geqq 0$. Then ξ can be written in the form

$$\xi = \sum_{i=s}^{\infty} c_i \pi_n^i, \qquad c_i \in \mathbf{Z}_p.$$

Define $D\xi$ and $D \log \xi$ as before by using such a power series expansion for ξ. Then we can see easily that $D \log \xi$ is a multi-valued function of ξ and that for a given ξ, the values of $D \log \xi$ fulfil a residue class of $\mathfrak{p}_n^{-1}$ modulo $\mathfrak{d}_n$. Let ξ now be an arbitrary nonzero element of F_n. We write ξ in the form $\xi = \xi_1/\xi_2$ with ξ_1, ξ_2 in $\mathfrak{o}_n$, $\xi_1, \xi_2 \neq 0$, and put

$$D \log \xi = D \log \xi_1 - D \log \xi_2.$$

Then $D \log \xi$ is again a multi-valued function of ξ, and it defines a homomorphism of the multiplicative group F_n^* of the field F_n into the additive group $\mathfrak{p}_n^{-1}/\mathfrak{d}_n$.

Now, let N_n and $N_{n,m}$, $m \geqq n \geqq 0$, denote the norm map from F_n to $\mathbf{Q}_p$ and that from F_m to F_n respectively. Let F_n' be the intersection of $N_{n,m}(F_m^*)$ for all $m \geqq n$. F_n' is a subgroup of the multiplicative group F_n^*, and it can be proved that F_n' consists of all ξ in F_n such that $N_n(\xi)$ is a power of p, that F_n^* is the direct product of F_n' and the group of all a in $\mathbf{Z}_p$ with $a \equiv 1 \bmod p$, and that $N_{n,m}(F_m') = F_n'$ for anv $m \geqq n$.

For any ξ in F_n', let $N_{n,m}^{-1}(\xi)$ denote any element η in F_m', $m \geqq n$, such that $N_{n,m}(\eta) = \xi$. Then:

THEOREM 1. *Let α be any element of F_n such that $\alpha \equiv 1 \bmod \mathfrak{p}_n$, and let β be any element of F_n'.* Then

$$[\alpha, \beta]_n = q_m^{-1} T_m(\zeta_m \log \alpha\, D \log N_{n,m}^{-1}(\beta)), \tag{2}$$

for any $m \geqq 2n+1$.

More precisely, the theorem states that for any choice of the value of $D \log N_{n,m}^{-1}(\beta)$, the right-hand side of (2) is a p-adic integer, uniquely determined modulo q_n, and is equal to $[\alpha, \beta]_n$.

Let $\beta = \pi_n$ in (2). Then $D \log N_{n,m}^{-1}(\beta) = D \log \pi_m = \pi_m^{-1}$. Since the trace of $\zeta_m \pi_m^{-1}$ from F_m to F_n is $p^{m-n}\zeta_n \pi_n^{-1}$, we obtain the second formula in II. Similarly, we have the first formula in II by putting $\beta = \zeta_n$ in (2).

If $\alpha - 1$ is divisible by a higher power of $\mathfrak{p}_n$, then the formula (2) holds with a smaller m. For example, we may let $m = n+1$ in (2) if $\alpha \equiv 1 \bmod \mathfrak{p}_n^{p^n}$, and $m = n$ if $\alpha \equiv 1 \bmod \mathfrak{p}_n^{2p^n}$. Using the fact that $(\alpha, a)_n = 1$ for $\alpha \equiv 1 \bmod \mathfrak{p}_n^{2p^n}$ and $a \equiv 1 \bmod p$, $a \in \mathbf{Z}_p$, we then have the following generalization of I:

THEOREM 2. *Let α be any element of F_n such that $\alpha \equiv 1 \bmod \mathfrak{p}_n^{2p^n}$. Then*

$$[\alpha, \beta]_n = q_n^{-1} T_n(\zeta_n \log \alpha\, D \log \beta),$$

for any nonzero β in F_n.

4. Let $L_n = \log(1+\mathfrak{p}_n)$; L_n is an open, compact subgroup of the additive group of F_n. Let X_n be the set of all ξ in F_n such that $T_n(\lambda\xi)$ is contained in $\mathbf{Z}_p$ for any λ in L_n. Then X_n is again an open, compact subgroup of the additive group of F_n. Let G_n denote the Galois group of the cyclic extension $F_n/\mathbf{Q}_p$, and let R_n be the group ring of G_n over $\mathbf{Z}_p$: $R_n = \mathbf{Z}_p[G_n]$. We see easily that X_n is invariant under the action of G_n, and may be considered as an R_n-module in the natural way.

Let $m \geqq 2n+1$, and let

$$\varphi_{n,m}(\beta) = q_m^{-1} T_{n,m}(\zeta_m D \log N_{n,m}^{-1}(\beta)),$$

for any β in F'_n. Then $\varphi_{n,m}(\beta)$ is contained in X_n by Theorem 1, and the multi-valued map $\varphi_{n,m}$: $F'_n \to X_n$ induces a surjective homomorphism

$$F'_n \to X_n/q_n X_n.$$

The homomorphism does not depend upon the choice of $m \geqq 2n+1$, and it maps π_n to the coset of $\mu_n = q_n^{-1}\zeta_n\pi_n^{-1}$. We denote by Y_n the R_n-submodule of X_n generated over R_n by μ_n: $Y_n = R_n\mu_n$.

The R_n-modules X_n, Y_n, and X_n/Y_n, which are defined here for the local field F_n, also occur in the theory of global cyclotomic fields, and it seems important to study the structure of these modules. Let

$$\theta_n = q_n^{-1} \sum_{s=0}^{n} \zeta_s^{-1}, \qquad n \geqq 0.$$

Then we can prove the following result on X_n:

THEOREM 3. *X_n consists of all elements of the form*

$$\sum_{\rho} (c_\rho \mu_n^\rho + d_\rho \theta_n^\rho), \qquad \rho \in G_n,$$

with c_ρ and d_ρ in $\mathbf{Z}_p$ satisfying $\sum_\rho d_\rho = 0$.

Let K_n denote the global cyclotomic field of q_nth roots of unity over the rational field, and let S_n be the Sylow p-subgroup of the ideal class group of K_n, $n \geqq 0$. The above theorem is used to investigate the structure of S_n by means of the homomorphism $F'_n \to X_n/q_n X_n$.

[43] (with C. C. Sims) Computation of invariants in the theory of cyclotomic fields

J. Math. Soc. Japan, 18 (1966), 86-96.

(Received July 26, 1965)

1. Let a prime number p be fixed, and let F_n, $n \geqq 0$, denote the cyclotomic field of p^{n+1}-th roots of unity over the rational field Q. Let $p^{c(n)}$ be the highest power of p dividing the class number h_n of F_n. Then there exist integers λ_p, μ_p, and ν_p ($\lambda_p, \mu_p \geqq 0$), depending only upon p, such that

$$c(n) = \lambda_p n + \mu_p p^n + \nu_p ,$$

for every sufficiently large integer n[1]. In the present paper, we shall determine, by the help of a computer, the coefficients λ_p, μ_p, and ν_p in the above formula for all prime numbers $p \leqq 4001$. We shall see in particular that $\mu_p = 0$ for every $p \leqq 4001$. Let S_n denote the Sylow p-subgroup of the ideal class group of F_n. For the above primes, we shall determine not only the order $p^{c(n)}$ of S_n but also the structure of the abelian group S_n for every $n \geqq 0$.

Let $p=2$. Then we know by Weber's theorem that $c(n)=0$, $S_n=1$ for any $n \geqq 0$ so that $\lambda_2 = \mu_2 = \nu_2 = 0$. Therefore, we shall assume throughout the following that p is an odd prime, $p>2$.

2. Let Q_p and Z_p denote the field of p-adic numbers and the ring of p-adic integers, respectively. Let F be the union of all fields F_n, $n \geqq 0$. Then F is an abelian extension of Q, and we denote the Galois group of F/Q by G. For each p-adic unit u in Q_p, there is a unique automorphism σ_u of F such that $\sigma_u(\zeta) = \zeta^u$ for any root of unity ζ in F with order a power of p. The mapping $u \to \sigma_u$ then defines a topological isomorphism of the group of p-adic units in Q_p onto the compact abelian group G. Let Γ and Δ denote the subgroups of G corresponding to the group of 1-units in Q_p and the group V of all $(p-1)$-st roots of unity in Q_p, respectively. Then we have

$$G = \Gamma \times \Delta ;$$

---

* The work of this author was supported in part by the National Science Foundation grant GP-2496.

1) For the results on cyclotomic fields used in the present paper, see K. Iwasawa, On the theory of cyclotomic fields, Ann. of Math., 70 (1959), 530-561; K. Iwasawa, On some modules in the theory of cyclotomic fields, J. Math. Soc. Japan, 16 (1964), 42-82.

Γ is the Galois group of F/F_0, and Δ is a cyclic group of order $p-1$, canonically isomorphic to the Galois group of F_0/Q.

For any $m \geqq n \geqq 0$, the injective homomorphism of the ideal group of F_n into that of F_m induces a natural homomorphism $S_n \to S_m$. Let S be the direct limit of S_n, $n \geqq 0$, relative to these homomorphisms. The Galois group G acts on S in the obvious manner. For each integer i, $0 \leqq i < p-1$, let ${}^iS$ denote the subgroup of all elements s in S such that $\sigma_v(s) = s^{v^i}$ for every v in V. Then S is the direct product of the G-subgroups ${}^iS$:

$$S = \prod_{i=0}^{p-2} {}^iS .$$

We have a similar decomposition for each S_n, $n \geqq 0$, and ${}^iS$ is the direct limit of the subgroups ${}^iS_n$, $n \geqq 0$.

3. Let Λ denote the ring of formal power series in an indeterminate T with coefficients in Z_p: $\Lambda = Z_p[[T]]$. Then there is an injective homomorphism of Γ into the multiplicative group of Λ such that $\sigma_{1+p} \to 1+T$. Therefore, if M is any Λ-module, we can make it into a Γ-module so that $\sigma_{1+p}(x) = (1+T)x$ for every x in M.

For any a in Q_p, there exist a rational integer b and a power of p, p^m $(m \geqq 0)$, such that $p^m a \equiv b \bmod p^m$, $0 \leqq b < p^m$. The rational number b/p^m is then uniquely determined by a so that we denote it by $\langle a \rangle$.

For each odd integer i, $0 \leqq i < p-1$, we shall next define a power series ${}^ig(T)$ in Λ. First, we put ${}^{p-2}g(T) = 1$. Let $i \neq p-2$. For each $n \geqq 0$, let

$${}^ig_n(T) = \sum_m \sum_v \langle v(1+p)^m / p^{n+1} \rangle v^i (1+T)^m ,$$

where $0 \leqq m < p^n$, $v \in V$. Then ${}^ig_n(T)$ is a polynomial in T with coefficients in Z_p, and when n tends to infinity, ${}^ig_n(T)$ converges, on each coefficient of T^m, $m \geqq 0$, to a power series in Λ, which we denote by ${}^ig(T)$:

$${}^ig(T) = \lim_{n \to \infty} {}^ig_n(T) .$$

We see easily that

$$(1) \qquad {}^ig(T) \equiv {}^ig_n(T) \qquad \bmod (1-(1+T)^{p^n})\Lambda , \quad n \geqq 0 .$$

For each odd i, there exist, by Weierstrass' preparation theorem, an integer $e_i \geqq 0$, a unit ${}^iu(T)$ of the ring Λ, and a polynomial ${}^im(T)$ of the form

$${}^im(T) = {}^ia_0 + \cdots + {}^ia_{d_i-1}T^{d_i-1} + T^{d_i} , \qquad {}^ia_k \in pZ_p ,$$

such that

$${}^ig(T) = p^{e_i}\, {}^iu(T)\, {}^im(T) .$$

Now, let

$${}^iM = \Lambda / {}^ig(T)\Lambda , \quad 0 \leqq i < p-1 , \quad (i, 2) = 1 .$$

As noted in the above, we may consider ${}^iM$ as Γ-modules. These Γ-modules are fundamental in the theory of cyclotomic fields, and we shall next consider some special cases in which the structure of ${}^iM$ can be easily determined.

4. For each odd index i, let

$${}^ig(T) = {}^i\alpha + {}^i\beta T + {}^i\gamma T^2 + \cdots,$$

with ${}^i\alpha$, ${}^i\beta$, ${}^i\gamma$, etc. in Z_p. Then it is clear that ${}^iM = 0$ if and only if ${}^i\alpha$ is a p-adic unit, namely, if and only if $d_i = e_i = 0$. Since ${}^{p-2}g(T) = 1$, we immediately have ${}^{p-2}M = 0$.

For any integer a, $1 \leqq a \leqq p-1$, let v_a denote the element of V such that

$$v_a \equiv a \mod p.$$

Let

$$A(p, i) = \sum_{a=1}^{p-1} a v_a^i, \qquad 0 \leqq i < p-1, \qquad (i, 2) = 1.$$

Then $A(p, i) \equiv 0 \bmod p$ for $i \neq p-2$, and $A(p, p-2) \equiv -1 \bmod p$. Using $v_a \equiv a^p \bmod p^2$, we see easily that $A(p, i) \equiv 0 \bmod p^2$ if and only if the Bernoulli number $B_{(i+1)/2}$ is divisible by p.

Suppose that $i \neq p-2$. It follows from (1), with $n = 0$, that

$${}^i\alpha = {}^ig_0(0) = \sum_{a=1}^{p-1} \langle v_a/p \rangle v_a^i = \frac{1}{p} \sum_{a=1}^{p-1} a v_a^i = \frac{1}{p} A(p, i).$$

Hence we obtain the following result (including $i = p-2$):

I. $M_i = 0$ if and only if

$$A(p, i) \not\equiv 0 \mod p^2,$$

namely, if and only if

$$B_{(i+1)/2} \not\equiv 0 \mod p.$$

For each odd index i, $0 \leqq i < p-1$, let

$$B(p, i) = \sum_{a,b=1}^{p-1} C_{a,b} b v_a^i,$$

where $C_{a,b}$ denotes the integer defined by

$$C_{a,b} \equiv \frac{1}{p}(v_a - a) + ab \mod p, \qquad 0 \leqq C_{a,b} < p$$

It follows from (1), with $n = 1$, that

$${}^ig(T) \equiv {}^ig_1(T) \mod (pT, T^2) \qquad i \neq p-2.$$

Hence we obtain

$${}^i\beta \equiv \sum_{m=0}^{p-1} \sum_{v} \langle v(1+mp)/p^2 \rangle v^i m$$

$$\equiv \sum_{a,b=1}^{p-1} \langle (v_a + v_a bp)/p^2 \rangle b v_a^i \mod p.$$

However,

$$v_a+v_abp\equiv a+\frac{1}{p}(v_a-a)p+abp\equiv a+C_{a,b}p \mod p^2$$

$$0\leqq a+C_{a,b}p\leqq(p-1)+(p-1)p<p^2$$

so that

$$\langle(v_a+v_abp)/p^2\rangle=\frac{1}{p^2}(a+C_{a,b}p).$$

Therefore,

$$^i\beta\equiv\frac{1}{p^2}\sum_{a,b=1}^{p-1}(a+C_{a,b}p)bv_a^i$$

$$\equiv\frac{p-1}{2p}A(p,i)+\frac{1}{p}B(p,i) \mod p.$$

It follows in particular that $B(p,i)\equiv 0 \mod p$ for $i\neq p-2$.

Now, suppose that $A(p,i)\equiv 0 \mod p^2$ and $B(p,i)\not\equiv 0 \mod p^2$ $(i\neq p-2)$. We see from the above that ${}^i\alpha\equiv 0 \mod p$, ${}^i\beta\not\equiv 0 \mod p$ so that $d_i=1$, $e_i=0$. Let

$$^iM(T)=T-{}^i\omega, \qquad {}^i\omega\in p\boldsymbol{Z}_p.$$

Then ${}^ig(T)={}^iu(T)(T-{}^i\omega)$ and ${}^iM=\Lambda/{}^ig(T)\Lambda=\Lambda/(T-{}^i\omega)\Lambda$. Hence we obtain the following result:

II. Suppose that

$$A(p,i)\equiv 0 \mod p^2, \qquad B(p,i)\not\equiv 0 \mod p^2.$$

Then ${}^ig(T)=0$ has a unique solution $T={}^i\omega$ in $p\boldsymbol{Z}_p$, and there is a Γ-isomorphism

$$^iM\cong\boldsymbol{Z}_p,$$

where the action of Γ on $\boldsymbol{Z}_p$ is defined by

$$\sigma_{1+p}(y)=(1+{}^i\omega)y, \qquad y\in\boldsymbol{Z}_p.$$

Let p^f, $f\geqq 1$, be the highest power of p dividing ${}^i\omega$. Then, for each $n\geqq 0$, the above isomorphism induces a Γ-isomorphism

$$^iM/(\sigma_{1+p}^{p^n}-1){}^iM\cong\boldsymbol{Z}_p/p^{n+f}\boldsymbol{Z}_p.$$

It follows in particular that ${}^iM/(\sigma_{1+p}^{p^n}-1){}^iM$ is a cyclic group of order p^{n+f}. We also note that ${}^ig({}^i\omega)=0$ implies

(2) $$^i\omega\equiv-{}^i\alpha/{}^i\beta\equiv-A(p,i)/B(p,i) \mod p^2.$$

Therefore, $f=1$ if and only if

$$A(p,i)\not\equiv 0 \mod p^3.$$

5. We shall now explain the arithmetic meaning of the modules ${}^iM$.

It is well known that the class number h_n of F_n is the product of two integers, the so-called first and the second factor of h_n:

$$h_n = {}^-h_n {}^+h_n .$$

Let $p^{c(n)'}$ denote the highest power of p dividing the first factor ${}^-h_n$ of h_n. Then there exist again integers λ'_p, μ'_p, and ν'_p ($\lambda'_p, \mu'_p \geqq 0$) such that

$$c(n)' = \lambda'_p n + \mu'_p p^n + \nu'_p ,$$

for every sufficiently large n. For the coefficients λ'_p and μ'_p, we then have the following formula:

$$\lambda'_p = \sum_i d_i , \qquad \mu'_p = \sum_i e_i , \qquad 0 \leqq i < p-1, (i, 2) = 1 .$$

Therefore, the integers λ'_p and μ'_p can be obtained by computing d_i and e_i from the power series ${}^ig(T)$.

A prime number p is called regular if the class number h_0 is prime to p. In the following, we shall make an assumption on p which is weaker than the regularity. Namely, we assume that the second factor ${}^+h_0$ of h_0 is prime to p:

(A) $$({}^+h_0, p) = 1 .$$

Under this assumption, we have the following results on F_n:

i) For each $n \geqq 0$, the second factor ${}^+h_n$ of h_n is also prime to p so that $c(n) = c(n)'$. Hence

$$\lambda_p = \lambda'_p , \qquad \mu_p = \mu'_p , \qquad \nu_p = \nu'_p .$$

ii) For every even index i and for every $n \geqq 0$,

$${}^iS = {}^iS_n = 1 .$$

iii) For any $m \geqq n \geqq 0$, the homomorphism $S_n \rightarrow S_m$ is injective so that S may be simply regarded as the union of all S_n, $n \geqq 0$. S_n is then the subgroup of S consisting of all s in S such that $\sigma_{1+p}^{p^n}(s) = s$. For each i, a similar result holds also for ${}^iS$ and ${}^iS_n$, $n \geqq 0$.

iv) Let i and j be odd indices such that $i+j = p-1$. Then there exist a non-degenerate pairing of ${}^iM$ and ${}^jS$ into the additive group $\boldsymbol{Q}_p/\boldsymbol{Z}_p$ such that

$$[\sigma(x), \sigma(s)] = [x, s] , \qquad x \in {}^iM, \ s \in {}^jS ,$$

for any σ in Γ.

v) It follows from iii) that for each $n \geqq 0$, the above pairing induces a similar pairing of ${}^iM/(\sigma_{1+p}^{p^n} - 1)^iM$ and ${}^jS_n$. Hence these two are isomorphic finite abelian groups.

It is now clear that we can obtain the following results from I and II in the above:

III. Under the assumption (A), suppose that

$$A(p, i) \not\equiv 0 \mod p^2 ,$$

namely,

$$B_{(i+1)}/2 \not\equiv 0 \mod p ,$$

for an odd index i, $0 \leqq i < p-1$. Then, for the odd index $j = p-1-i$ and for every $n \geqq 0$,

$$ {}^jS = {}^jS_n = 1 . $$

IV. Under the same assumption (A), suppose that

$$ A(p, i) \equiv 0 \mod p^2, \qquad B(p, i) \not\equiv 0 \mod p^2, $$

for an odd index i. Let ${}^i\omega$ and f be defined as in II, and let $j = p-1-i$. Then there is a Γ-isomorphism

$$ {}^jS \cong \boldsymbol{Q}_p/\boldsymbol{Z}_p , $$

where the action of Γ on $\boldsymbol{Q}_p/\boldsymbol{Z}_p$ is defined by

$$ \sigma_{1+p}(z) = (1 + {}^i\omega)^{-1} z , \qquad z \in \boldsymbol{Q}_p/\boldsymbol{Z}_p . $$

For each $n \geqq 0$, it induces a Γ-isomorphism

$$ {}^jS_n \cong p^{-(n+f)}\boldsymbol{Z}_p/\boldsymbol{Z}_p , $$

so that ${}^jS_n$ is a cyclic group of order p^{n+f}. Furthermore, if

$$ A(p, i) \not\equiv 0 \mod p^3 , $$

then the above integer f is equal to $1 : f = 1$.

Suppose that p is a regular prime ($p > 2$) so that (A) is satisfied for p. Then, by a theorem of Kummer, the Bernoulli numbers B_k, $1 \leqq k \leqq (p-1)/2$, are not divisible by p. Hence it follows from ii) and III that ${}^iS_n = 1$ for any i and n, namely, that $S_n = 1$ for every $n \geqq 0$. Therefore $c(n) = 0$ for $n \geqq 0$, and, consequently, $\lambda_p = \mu_p = \nu_p = 0$. We note that this result can be proved also by a direct method without referring to the modules ${}^iM$.

6. In a sequence of papers[2)], Vandiver and others verified that our assumption (A) is satisfied for all prime numbers $p \leqq 4001$. For such a prime p, they also determined all integers k, $1 \leqq k \leqq (p-1)/2$, such that B_k is divisible by p. Putting

$$ i = 2k - 1 , $$

we then obtain all odd indices i for p such that

$$ A(p, i) \equiv 0 \mod p^2 . $$

Let $\{p, i\}$ be such a pair, $p \leqq 4001$, and let

---

2) D. H. Lehmer, Emma Lehmer, and H. S. Vandiver, An application of high-speed computing to Fermat's last theorem, Proc. Nat. Acad. Sci. USA, **40** (1954), 25–33; H. S. Vandiver, Examination of methods of attack on the second case of Fermat's last theorem, Ibid., **40** (1954), 732–735; J. L. Selfridge, C. A. Nicol, and H. S. Vandiver, Proof of Fermat's last theorem for all prime exponents less than 4002, Ibid., **41** (1955), 970–973.

$$A(p, i)=\sum_{n=0}^{\infty} a_n p^n, \quad B(p, i)=\sum_{n=0}^{\infty} b_n p^n, \quad 0 \leqq a_n, b_n < p,$$

be the p-adic expansions of the p-adic integers $A(p, i)$ and $B(p, i)$ respectively. We know from the above that

$$a_0 = a_1 = b_0 = 0.$$

By using a computer, we have computed the next coefficients a_2 and b_1, and found that

(3) $$a_2 \neq 0, \quad b_1 \neq 0$$

for every such pair $\{p, i\}$. A part of the results of these computations will be given at the end of the paper.

Now, it follows from (3) that

$$A(p, i) \not\equiv 0 \mod p^3, \quad B(p, i) \not\equiv 0 \mod p^2.$$

Therefore the following result is obtained from III and IV above:

Let $p \leqq 4001$ and let δ_p denote the number of those Bernoulli numbers B_k, $1 \leqq k \leqq (p-1)/2$, which are divisible by p. Then, for each $n \geqq 0$, the Sylow p-subgroup S_n of the ideal class group of F_n is the direct product of δ_p cyclic groups of order p^{n+1}. Hence

$$c(n) = (n+1)\delta_p,$$

for every $n \geqq 0$, and consequently

$$\lambda_p = \nu_p = \delta_p, \quad \mu_p = 0.$$

Since the values of δ_p are known for $p \leqq 4001$[3)], the structure of S_n is completely determined for such primes.

Actually, III and IV provide us more information on the structure of the G-groups $S = \prod {}^iS$ and $S_n = \prod {}^iS_n$, $n \geqq 0$: if i is an odd index such that $A(p, i) \equiv 0 \mod p^2$, $p \leqq 4001$, then ${}^jS$, $j = p-1-i$, is isomorphic to the Γ-module $\boldsymbol{Q}_p/\boldsymbol{Z}_p$ as described in IV.

Now, our computations of a_2 and b_1 show that

$$a_2 \neq b_1$$

for every pair $\{p, i\}$ as stated above. Hence it follows from (2) that

$${}^i\omega \not\equiv -p \mod p^2.$$

Therefore, if z is an element of $\boldsymbol{Q}_p/\boldsymbol{Z}_p$ such that $\sigma_{1+p}^{p^n}(z) = (1+p)^{p^n} z$ for some $n \geqq 0$, then $p^{n+1}z = 0$. Since ${}^jS \cong \boldsymbol{Q}_p/\boldsymbol{Z}_p$, the Γ-group ${}^jS$ has the same property. By the theory of cyclotomic fields, we can then obtain the following result:

---

3) See the tables in the papers of the footnote 2). For example, $\delta_p = 1$ for $p = 37$, 59, 67, $\delta_p = 2$ for $p = 157$, and $\delta_p = 3$ for $p = 491$.

Let $p \leqq 4001$. *Let* Φ_n, $n \geqq 0$, *be the local cyclotomic field of* p^{n+1}*-th roots of unity over* Q_p. *Then the group of* 1*-units in the local field* Φ_n *contains* $\frac{1}{2}(p-1)p^n-1$ *global units in* F_n *which are multiplicatively independent over the ring of p-adic integers* Z_p.

7. The computations of a_2 and b_1 for those pairs $\{p, i\}$ such that $A(p, i) \equiv 0 \bmod p^2$ were carried out on an IBM 7094 computer[4]. During the preparation of the program it became clear that b_1 presented by far the greater difficulty. As defined,

$$B(p, i) = \sum_{a,b=1}^{p-1} C_{a,b} b v_a^i .$$

For $p=4001$, the largest value we were considering, this sum has 16×10^6 terms. No more than about 10^4 terms could be computed per second, and so it seemed that for the larger values of p the computation time might be 30 minutes or more for each case. With 278 pairs to be run, this would have required more computer time than could be justified.

The problem was solved by finding a more efficient method of computing

$$\sum_{b=1}^{p-1} C_{a,b} b .$$

If $\frac{1}{p}(v_a-a) \equiv m \bmod p$, $0 \leqq m < p$, then

$$C_{a,b} = m+ab-p\left[\frac{m+ab}{p}\right].$$

(Here and throughout this section $[x]$ denotes the greatest integer less than or equal to x.) Thus

$$\begin{aligned}\sum_{b=1}^{p-1} C_{a,b} b &= \sum_{b=1}^{p-1} b\left(m+ab-p\left[\frac{m+ab}{p}\right]\right)\\ &= \frac{mp(p-1)}{2}+\frac{ap(p-1)(2p-1)}{6}-p\sum_{b=1}^{p-1} b\left[\frac{m+ab}{p}\right].\end{aligned}$$

For any integers m, a, r, and s with $s>0$, $r>m\geqq 0$, and $a\geqq 0$, define

$$F(m, a, r, s) = \sum_{b=1}^{s}\left[\frac{m+ab}{r}\right],$$

$$G(m, a, r, s) = \sum_{b=1}^{s} b\left[\frac{m+ab}{r}\right],$$

$$H(m, a, r, s) = \sum_{b=1}^{s}\left[\frac{m+ab}{r}\right]^2 .$$

We have

---

4) The computation was done at the M.I.T. Computation Center, Cambridge, Massachusetts.

$$\sum_{b=1}^{p-1} C_{a,b}b = \frac{mp(p-1)}{2} + \frac{ap(p-1)(2p-1)}{6} - pG(m, a, p, p-1).$$

F, G, and H satisfy certain recursion relations. Let $r = ua + v$, $m+1 = xa - y$, $0 \leqq v$, $y < a$. Also let

$$z = \left[\frac{m+as}{r}\right].$$

If $z = 0$, then $F(m, a, r, s) = G(m, a, r, s) = H(m, a, r, s) = 0$. If $z > 0$, then $a > 0$ and

$$F(m, a, r, s) = z(s+x) - \frac{uz(z+1)}{2} - F(y, v, a, z),$$

$$2G(m, a, r, s) = zs(s+1) - zx(x-1) - \frac{u^2 z(z+1)(2z+1)}{6}$$
$$- \frac{u(1-2x)z(z+1)}{2} - 2uG(y, v, a, z)$$
$$-(1-2x)F(y, v, a, z) - H(y, v, a, z),$$

$$H(m, a, r, s) = sz^2 - \frac{uz(z+1)(2z+1)}{3} + \frac{(2x+u)z(z+1)}{2} - xz$$
$$-2G(y, v, a, z) + F(y, v, a, z).$$

The proofs of these formulas are similar and we give only the proof of the first. We may assume $z > 0$ and therefore $a > 0$. For any positive integer t,

$$\left[\frac{m+ab}{r}\right] = t$$

for $k_t + 1 \leqq b \leqq k_{t+1}$, where

$$k_t = \left[\frac{tr-1-m}{a}\right].$$

Since $r > m$ and $z > 0$, we have $0 \leqq k_1 < s$. If we redefine k_{z+1} to be s, then

$$F(m, a, r, s) = \sum_{b=1}^{s} \left[\frac{m+ab}{r}\right] = \sum_{t=1}^{z} \sum_{k_t+1}^{k_{t+1}} t = sz - \sum_{t=1}^{z} k_t.$$

If $r = ua + v$ and $m + 1 = xa - y$, $0 \leqq v$, $y < a$, then

$$k_t = tu - x + \left[\frac{y+tv}{a}\right].$$

Thus

$$\sum_{t=1}^{z} k_t = \frac{uz(z+1)}{2} - xz + F(y, v, a, z)$$

and

$$F(m, a, r, s) = z(s+x) - \frac{uz(z+1)}{2} - F(y, v, a, z).$$

If these formulas are used to compute $G(m, a, p, p-1)$, the computation time for b_1 becomes proportional to $p \log p$ and for $p = 4001$ is under two minutes.

8. We possess a complete table of a_2 and b_1, computed for all pairs $\{p, i\}$, $p \leqq 4001$, satisfying $A(p, i) \equiv 0 \bmod p^2$. However, we produce here only the part of the table where $1 < p < 400$ or $3600 < p \leqq 4001$. Since the root ${}^i\omega$ of ${}^ig(T)=0$ seems to have an important meaning in the theory of cyclotomic fields, we also indicate in the last column the values of the integer c such that $c \equiv -a_2/b_1 \bmod p$, $0 \leqq c < p$, namely, such that

$$ {}^i\omega \equiv cp \mod p^2, \qquad 0 \leqq c < p. $$

p	i	a_2	b_1	c
37	31	23	16	24
59	43	20	33	28
67	57	34	46	8
101	67	16	59	10
103	23	1	49	21
131	21	34	106	59
149	129	24	70	55
157	61	66	109	21
157	109	109	106	36
233	83	3	101	143
257	163	124	69	28
263	99	66	176	164
271	83	141	92	78
283	19	272	268	37
293	155	57	200	218
307	87	108	102	17
311	291	152	34	87
347	279	246	241	166
353	185	260	52	348
353	299	289	192	118
379	99	327	103	236
379	173	256	297	188
389	199	340	341	234
3607	1975	3279	2832	2578
3613	2081	1991	1798	2147
3617	15	2574	1314	989
3617	2855	57	667	2733
3631	1103	3591	3510	1200
3637	2525	2894	1313	2139
3637	3201	1685	1504	3174
3671	1579	3619	555	3261
3677	2237	31	3273	2594
3697	1883	3575	1905	1638
3779	2361	2454	2855	1794
3797	1255	3066	1548	3692
3821	3295	2776	1320	160

3833	1839	156	886	95
3833	1997	2944	328	178
3833	3285	1307	1329	547
3851	215	2297	1909	606
3851	403	1828	2438	2555
3853	747	2331	2270	1844
3881	1685	3189	252	1050
3881	2137	2674	692	1645
3917	1489	1658	3382	889
3967	105	2505	1543	2883
3989	1935	679	3616	2130
4001	533	3054	3587	1515

Massachusetts Institute of Technology

[44] A note on ideal class groups

Nagoya Math J., 27 (1966), 239-247.

To the memory of TADASI NAKAYAMA

In the first part of the present paper, we shall make some simple observations on the ideal class groups of algebraic number fields, following the group-theoretical method of Tschebotarew[1]. The applications on cyclotomic fields (Theorems 5, 6) may be of some interest. In the last section, we shall give a proof to a theorem of Kummer on the ideal class group of a cyclotomic field.

1. For any prime numbers p and q, let

$$\begin{aligned} d(q, p) &= 2, & \text{for } p = q, \\ &= \text{the order of } p \bmod q, & \text{for } p \neq q. \end{aligned}$$

For any integer $n \geq 1$, we then define

$$d(n, p) = \text{the minimum of } d(q, p) \text{ for all prime factors } q \text{ of } n.$$

THEOREM 1. *Let G be a finite group of order n. Let M be a G-module over the prime field P with p elements, and let d be the dimension of M over P. Suppose that the action of G on M is non-trivial. Then*

$$d \geqq d(n, p).$$

Proof. Let σ be an element with minimal order in G such that the action of σ on M is non-trivial. Let q be a prime dividing the order of σ. Put $H = G_1/G_2$, where G_1 and G_2 denote the subgroups of G generated by σ and σ^q respectively. Then M is also an H-module over P, and the action of H on M is non-trivial. If $q = p$, we see immediately that $d \geqq 2 = d(p, p)$. Suppose that $q \neq p$. Then M is completely reducible, and it has an irreducible submodule on

Received June 7, 1965.

* The present research was supported in part by the National Science Foundation grant GP-4361.

[1] N. Tschebotarew, Zur Gruppentheorie des Klassenkörpers, J. reine u. angew. Math., **161** (1929), pp. 179-183.

which the action of H is again non-trivial. As is well-known, such an irreducible submodule is obtained by decomposing $P[H]$, the group ring of H over P. Let $\mathfrak{o}$ denote the maximal order of the cyclotomic field of q-th roots of unity. Identifying H with the group of q-th roots of unity in $\mathfrak{o}$, we may consider the $\mathfrak{o}$-module $\mathfrak{o}/p\mathfrak{o}$ as an H-module over P. We then see easily that

$$P[H] \cong P \oplus (\mathfrak{o}/p\mathfrak{o})$$
$$\cong P \oplus \sum_{i=1}^{g} (\mathfrak{o}/\mathfrak{p}_i).$$

Here P denotes the 1-dimensional trivial H-module, and $\mathfrak{p}_1, \ldots, \mathfrak{p}_g$ are the prime ideals of $\mathfrak{o}$ containing p. Since $\mathfrak{o}/\mathfrak{p}_i$ is a field, it is an irreducible $\mathfrak{o}$-module. Hence it is also irreducible as an H-module over P, and the action of H on it is non-trivial. It is known that the dimension of $\mathfrak{o}/\mathfrak{p}_i$ over P, namely, the degree of the extension $\mathfrak{o}/\mathfrak{p}_i$ over P, is equal to $d(q, p)$, the order of p mod q. Since M contains such a submodule $\mathfrak{o}/\mathfrak{p}_i$, we have $d \geqq d(q, p)$, *q.e.d.*

We note that if $p = 2$, then $d(q, p) \geqq 2$ for every prime q so that $d(n, p) \geqq 2$ for any integer $n \geqq 1$. Hence we also have $d \geqq 2$ in Theorem 1.

2. Let k be a finite algebraic number field, and let $\mathfrak{m}$ be an integral divisor of k, namely, a product of a finite number of prime divisors of k, archimedean or non-archimedean[2]. Let $I_{\mathfrak{m}}(k)$ denote the group of all ideals of k which are prime to $\mathfrak{m}$, and let $H_{\mathfrak{m}}(k)$ be the subgroup of all principal ideals (α) with $\alpha \equiv 1$ mod $\mathfrak{m}$. We put $C_{\mathfrak{m}}(k) = I_{\mathfrak{m}}(k)/H_{\mathfrak{m}}(k)$, and denote the order of $C_{\mathfrak{m}}(k)$ by $h_{\mathfrak{m}}(k)$. For $\mathfrak{m} = 1$, $C(k) = C_1(k)$ is the ideal class group of k, and $h(k) = h_1(k)$ is the class number of k.

Let M be a factor group of $C_{\mathfrak{m}}(k)$: $M = I_{\mathfrak{m}}(k)/H$, $H_{\mathfrak{m}}(k) \subset H \subset I_{\mathfrak{m}}(k)$. Let G be a group of automorphisms of k. If both $I_{\mathfrak{m}}(k)$ and H are invariant under the action of G, we may consider M as a G-group (or G-module). In such a case, we shall simply say that M is G-invariant.

3. Throughout this section, F will denote a finite algebraic number field, K a finite Galois extension of F with degree n, and G the Galois group of K/F.

THEOREM 2. *Let $\mathfrak{m}$ be an integral divisor of F, and let p be a prime number such that $(p, n) = (p, h_{\mathfrak{m}}(F)) = 1$. Let M be a G-invariant factor group of $C_{\mathfrak{m}}(K)$*

[2] For the classical class field theory used here and in the following, see H. Hasse's "Klassenkörperbericht" in Jahresbericht D. M.-V., **1926**, **1927**, **1930**.

with order a power of p, and let $M \neq 1$. Then the rank r of the finite abelian group M is at least equal to $d(n, p)$:

$$r \geqq d(n, p).$$

Proof. We first note that $\mathfrak{m}$ may be considered also as a divisor of K in the obvious manner so that the group $C_{\mathfrak{m}}(K)$ is well defined. Let $N = M/M^p$. Then $N \neq 1$, and it has the same rank as M. Hence, replacing M by N if necessary, we may assume that $M^p = 1$. Since M is a G-invariant factor group of $C_{\mathfrak{m}}(K)$, we may then consider M as a G-module over P. By Theorem 1, it is sufficient to show that the action of G on M is non-trivial.

Suppose that G acts trivially on M. Let L be the abelian extension of K which corresponds by class field theory to the ideal class group M. Since M is G-invariant, L/F is a Galois extension. Let A and B denote the Galois groups of L/F and L/K respectively. Then $A/B = G$, and B is canonically isomorphic to M so that G also acts trivially on B. Since the order of B is a power of p and is prime to the order n of G, the group extension A/B splits, and we have $A = B \times C$, $C \cong G$. Let E be the intermediate field of F and L such that the Galois group of L/E is C. Then E is an abelian extension of F with Galois group $A/C \cong B \cong M$. Let $\mathfrak{P}$ be a prime divisor of L, prime to $\mathfrak{m}$, and let T be the inertia group of $\mathfrak{P}$ for the extension L/F. Since L/K is the abelian extension corresponding to the factor group M of $C_{\mathfrak{m}}(K)$, $\mathfrak{P}$ is unramified by the extension L/K so that $T \cap B = 1$. Since the orders of B and C are prime to each other, it follows that T is contained in C. Therefore, if $\mathfrak{p}$ is any prime divisor of F, prime to $\mathfrak{m}$, then $\mathfrak{p}$ is unramified in K. By class field theory, the abelian extension E/F then corresponds to a factor group of $C_{\mathfrak{m}}(F)$, isomorphic to the Galois group $A/C \cong M$. Since $M \neq 1$, this implies that the order of $C_{\mathfrak{m}}(F)$ is divisible by p, and it contradicts the assumption $(p, h_{\mathfrak{m}}(F)) = 1$. Therefore the action of G on M is not trivial, and the theorem is proved.

Corollary. *Let p be a prime number such that $(p, n) = (p, h(F)) = 1$. Let M be a G-invariant factor group of $C(K)$ with order a power of p, and let $M \neq 1$. Then the rank r of M is at least equal to $d(n, p)$:*

$$r \geqq d(n, p).$$

In Theorem 2, suppose further that $(p-1, n) = 1$. For any prime factor q of n, we then have $p \not\equiv 1 \bmod q$ so that $d(q, p) \geqq 2$. Hence $d(n, p) \geqq 2$, and it follows from Theorem 2 that M is a non-cyclic group. Note that for $p=2$, the above assumption is always satisfied.

THEOREM 3. *Let* $\mathfrak{m}$ *and* p *be as stated in Theorem* 2: $(p, n) = (p, h_{\mathfrak{m}}(F)) = 1$. *If* p *divides* $h_{\mathfrak{m}}(K)$, *then the rank of the Sylow* p*-subgroup of* $C_{\mathfrak{m}}(K)$ *is at least equal to* $d(n, p)$.

Proof. This follows immediately from Theorem 2, because $C_{\mathfrak{m}}(K)$ has a G-invariant factor group isomorphic to its Sylow p-subgroup.

COROLLARY. *Let* p *be a prime number such that* $(p, n) = (p, h(F)) = 1$. *If* p *divides the class number* $h(K)$, *then the rank of the Sylow* p*-subgroup of the ideal class group* $C(K)$ *is at least equal to* $d(n, p)$.

Under the additional assumption $(p-1, n)=1$, we see that the Sylow p-subgroup in Theorem 3 and its corollary is non-cyclic. In particular, if $n = [K:F]$ is odd, $h(F)$ is odd, but $h(K)$ is even, then the Sylow 2-subgroup of $C(K)$ is a non-cyclic group. We can also prove by using the corollary of Theorem 2 that under the same assumption, if $h(K)$ is exactly divisible by an odd power of 2, then the rank of the Sylow 2-subgroup is at least equal to 3. For example, if $h(K)$ is exactly divisible by $8=2^3$, then the Sylow 2-subgroup is an abelian group of type (2,2,2).

4. Since $h(\mathbf{Q})=1$ for the rational field $\mathbf{Q}$, we obtain the following result from the corollary of Theorem 3:

THEOREM 4. *Let* K *be a finite Galois extension of* $\mathbf{Q}$ *with degree* n, *and let* p *be a prime number, prime to* n. *Suppose that the class number* $h(K)$ *is divisible by* p. *Then the rank of the Sylow* p*-subgroup of the ideal class group* $C(K)$ *is at least equal to* $d(n, p)$.

COROLLARY. *Let* K *be a finite Galois extension of* $\mathbf{Q}$ *with an odd degree* n. *Suppose that the class number* $h(K)$ *is even. Then the Sylow* 2*-subgroup of the ideal class group* $C(K)$ *is non-cyclic, and its rank is at least equal to* $d(n, 2)$.

The assumption $(p, n)=1$ in Theorem 4 can be replaced by various other conditions on K. As a typical example, we consider the following case of cyclotomic fields.

THEOREM 5. *Let* l *be a prime number and let* K *be the cyclotomic field of* l^e*-th roots of unity* $(e \geqq 1)$. *Suppose that the class number* $h(K)$ *is divisible by* *a prime number p, and let*

$$(l-1)l^{e-1}=p^a n, \quad (p, n)=1, \ a\geqq 0.$$

Then the rank of the Sylow p-subgroup of the ideal class group $C(K)$ is at least equal to $d(n, p)$.

Proof. Let F be the intermediate field of $\mathbf{Q}$ and K such that $[K:F]=n$, $[F:\mathbf{Q}]=p^a$. Since we know that $h(F)$ is not divisible by p[3], the theorem follows from the corollary of Theorem 3.

COROLLARY. *Let K be as in Theorem 5. Suppose that the class number $h(K)$ is even, and let*

$$(l-1)l^{e-1}=2^a n, \quad (2, n)=1, \ a\geqq 0.$$

Then the Sylow 2-subgroup of the ideal class group $C(K)$ is non-cyclic, and its rank is at least equal to $d(n, 2)$.

Remark. By a theorem of Weber, the class number $h(K)$ is odd for $l=2$.

The above corollary can be further refined as follows. Let J denote the automorphism of the cyclotomic field K, mapping each element in K to its complex-conjugate. Clearly J induces an automorphism of $C=C(K)$, $J: C\to C$. Let C^+ and C^- denote the kernels of the endomorphisms $1-J: C\to C$ and $1+J: C\to C$, respectively, so that we have

$$C/C^+\cong C^{1-J}\subset C^-, \quad C/C^-\cong C^{1+J}\subset C^+.$$

It follows that the class number $h(K)$ is the product of the order $h'(K)$ of C^- and the order $h''(K)$ of C^{1+J}. $h'(K)$ is called the first factor of $h(K)$, and $h''(K)$ the second factor of $h(K)$.

Let $S_2=S_2(K)$ denote the Sylow 2-subgroup of $C=C(K)$. Then $S_2^+=S_2\cap C^+$ and $S_2^-=S_2\cap C^-$ are the Sylow 2-subgroups of C^+ and C^- respectively. We see immediately from the definition that $S_2^+\cap S_2^-$ is the group of all x in S_2^+ satisfying $x^2=1$, and that it is also the group of all y in S_2^- satisfying $y^2=1$. Hence S_2^+ and S_2^- have the same rank. It follows in particular that $S_2^+=1$ if and only if $S_2^-=1$. Suppose that $S_2^+=S_2^-=1$. Then we see from $S_2/S_2^+\cong S_2^{1-J}\subset S_2^-$ that $S_2=1$. Therefore the three conditions $S_2=1$, $S_2^+=1$, and $S_2^-=1$ are all equi-

---

[3] K. Iwasawa, A note on class numbers of algebraic number fields, Abh. Math. Sem. Univ. Hamburg, **20** (1956), pp. 257-258.

valent to each other. This result was first obtained by Kummer in the form that the class number $h(K)$ is odd if and only if its first factor $h'(K)$ is odd[4].

THEOREM 6. *Let K be as in Theorem* 5 *and suppose that the class number* $h(K)$ *is even. Then the groups* S_2^+ *and* S_2^- *are both non-cyclic, and they have the same rank which is at least equal to* $d(n, 2)$, *n being the same as in the corollary of Theorem* 5.

Proof. We have already noted that S_2^+ and S_2^- have the same rank. If $S_2^+ = S_2$, then the theorem follows immediately from the corollary of Theorem 5. Suppose that $S_2^+ \neq S_2$. Let F be the intermediate field of $\mathbf{Q}$ and K such that $[K : F] = n$, and let G denote the Galois group of K/F. Then the ideal class group $C(K)$ has a G-invariant factor group isomorphic to S_2/S_2^+. Since $h(F)$ is odd[5], it follows from the corollary of Theorem 2 that the rank of S_2/S_2^+ is at least equal to $d(n, 2) \geqq 2$. Since S_2^- contains the subgroup S_2^{1-J} which is isomorphic to S_2/S_2^+, the theorem is proved also in the case $S_2^+ \neq S_2$.

EXAMPLE. Let K be the cyclotomic field of 29-th roots of unity. It is known that C^- is a group of order 8 so that $C^- = S_2^-$[6]. Since $28 = 2^2\ 7$, $d(7,2) = 3$, we see immediately from the above that C^- is an abelian group of type $(2,2,2)$.

5. Let K be the cyclotomic field of 41-st roots of unity. We know that the class number $h(K)$ is then divisible by $121 = 11^2$.[7] However, since $d(40, 11) = d(5, 11) = 1$, we cannot see from Theorem 5 whether the Sylow 11-subgroup of $C(K)$ is cyclic or non-cyclic. In a paper of 1853, Kummer proved an interesting theorem on cyclotomic fields by which we can settle in certain cases such as above whether or not the subgroup $C^-(K)$ of $C(K)$ is cyclic[8]. However, in his paper, Kummer worked with logarithums of ideals, not of ordinary numbers, and it seems that his proof needs some further explanation[9]. Therefore, we shall show in the following how Kummer's result can be justified from our point of view.

---

4) For a more complete result in this direction, see H. Hasse, *Über die Klassenzahl abelscher Zahlkörper*, Berlin, **1952**, § 37.

5) Iwasawa, op. cit.

6) Hasse, *Klassenzahl*, p. 150.

7) Hasse, *Klassenzahl*, p. 152.

8) E. Kummer, Über die Irregularität der Determinanten, Monatsber. Akad. d. Wissensch., Berlin, **1853**, pp. 194-200.

9) Hasse, *Klassenzahl*, p. 99.

Let l be an odd prime, and let K denote as before the cyclotomic field of l^e-th roots of unity ($e \geqq 1$). The Galois group G of $K/\mathbf{Q}$ is a cyclic group of order $m = (l-1)l^{e-1}$, and it is isomorphic to the multiplicative group of integers mod l^e, a canonical isomorphism being given by $\sigma_a \to a$ mod l^e, where σ_a denotes the automorphism of K mapping each l^e-th root of unity ζ to ζ^a : $\sigma_a(\zeta) = \zeta^a$. Let $R = \mathbf{Z}[G]$ be the group ring of G over the ring of rational integers $\mathbf{Z}$. Let ω be an element of the group ring $\mathbf{Q}[G]$ defined by

$$\omega = l^{-e} \sum_a a\sigma_a^{-1}, \quad 0 \leqq a < l^e,\ (a, l) = 1,$$

and put

$$I = \omega R \cap R.$$

Further, let R^- denote the set of all α in R such that $(1+J)\alpha = 0$, and let $I^- = I \cap R^-$. Then both R^- and I^- are ideals of R, and we have

$$h'(K) = [R^- : I^-].$$ [10]

We shall next consider the exponent of the finite abelian group R^-/I^-.

Let F denote the cyclotomic field of m-th roots of unity. For each character χ of the multiplicative group of integers mod l^e, we define an element ε_χ of the group ring $F[G]$ by

$$\varepsilon_\chi = m^{-1} \sum_a \chi(a) \sigma_a^{-1}, \quad 0 \leqq a < l^e,\ (a, l) = 1.$$

Then the elements ε_χ form a set of orthogonal idempotents in $F[G]$ such that

$$F[G] = \sum_\chi F\varepsilon_\chi, \quad 1 = \sum_\chi \varepsilon_\chi,$$

$$\omega\varepsilon_\chi = h_\chi \varepsilon_\chi,$$

with

$$h_\chi = l^{-e} \sum_a a\chi(a)^{-1}, \quad h_\chi \in F.$$

By the classical class number formula,

$$h'(K) = 2l^e \prod_\chi{}' \left(-\frac{1}{2} h_\chi\right),$$

where χ ranges over all characters mod l^e such that $\chi(-1) = -1$. Therefore $h_\chi \neq 0$ for $\chi(-1) = -1$.

---

[10] K. Iwasawa, A class number formula for cyclotomic fields, Ann. of Math., **76** (1962), pp. 171–179.

THEOREM 7. *Let t be the exponent of the finite abelian group R^-/I^-, and let N denote the least positive rational integer such that N/h_χ is an algebraic integer for every character χ with $\chi(-1) = -1$. Then N is a factor of $2t$, and t is a factor of $\frac{1}{2}mN$.*

Proof. For each character χ with $\chi(-1) = -1$, let

$$N = g_\chi h_\chi.$$

Then g_χ is an algebraic integer in F. Since $1-J$ is an element of R^-, $t(1-J)$ is contained in I^-. Hence $t(1-J) = \omega\alpha$ with some α in R. Let $\alpha\varepsilon_\chi = a_\chi\varepsilon_\chi$ with a_χ in F. Since α is in R, a_χ is an algebraic integer in F. On the other hand, $(1-J)\varepsilon_\chi = 2\varepsilon_\chi$ for $\chi(-1) = -1$. Therefore $2t\varepsilon_\chi = t(1-J)\varepsilon_\chi = \omega\alpha\varepsilon_\chi = \omega\varepsilon_\chi\alpha\varepsilon_\chi = h_\chi\varepsilon_\chi a_\chi\varepsilon_\chi = a_\chi h_\chi\varepsilon_\chi$, and we have

$$2t = a_\chi h_\chi, \qquad \chi(-1) = -1.$$

Therefore $2t/h_\chi$ is an algebraic integer for every character χ with $\chi(-1) = -1$, and we see from the definition of N that N is a factor of $2t$.

Let

$$\xi = m\sum_\chi{}' g_\chi\varepsilon_\chi,$$

where χ ranges over all characters with $\chi(-1) = -1$. It is clear that ξ is a linear combination of the elements of G with all coefficients algebraic integers in F. We also see easily that these coefficients are invariant under the Galois automorphisms of $F/\mathbf{Q}$. Therefore ξ is contained in $R = \mathbf{Z}[G]$, and hence in R^-. Since

$$1-J = 2\sum_\chi{}' \varepsilon_\chi, \qquad \chi(-1) = -1,$$

we obtain

$$\frac{1}{2}mN(1-J) = m\sum_\chi{}' g_\chi h_\chi\varepsilon_\chi = m\sum_\chi{}' g_\chi\omega\varepsilon_\chi = \omega\xi.$$

Therefore $\frac{1}{2}mN(1-J)$ is contained in $R^- \cap \omega R = I^-$. Since $R^- = (1-J)R$, $\frac{1}{2}mNR^-$ is then contained in I^-, and we see that t is a factor of $\frac{1}{2}mN$.

We now prove the following theorem of Kummer mentioned in the above.

THEOREM 8. *The exponent of the group $C^- = C^-(K)$ is a factor of mN, and the exponent of C^{1-J} is a factor of $\frac{1}{2}mN$.*

Proof. The group ring $R = \mathbf{Z}[G]$ may be considered as an operator domain on C in the obvious manner. It is well known that $x^{\alpha} = 1$ for any x in C and for any α in I. Since $\frac{1}{2}mN(1-J)$ is contained in I^- by Theorem 7, we have

$$x^{\frac{1}{2}mN(1-J)} = 1$$

for any x in C. Therefore the exponent of C^{1-J} is a factor of $\frac{1}{2}mN$.

Now, let y be any element of C^-. Since $y^{1+J} = 1$, we have $y^{1-J} = y^2$. Hence it follows from the above that $y^{mN} = 1$. Therefore the exponent of C^- is a factor of mN.

COROLLARY. *Suppose that $C^-(K)$ is a cyclic group. Then $h'(K)$ is a factor of mN.*

Proof. This is obvious, because $h'(K)$ is the order of $C^-(K)$.

Let p be a prime number. For any rational integer $a \geqq 1$, let $(a)_p$ denote the highest power of p dividing a. Then it follows from Theorem 7 that $(t)_p = (N)_p$ for any p with $(p, m) = 1$. We also see from Theorem 8 and from its corollary that for any prime number p, the exponent of the Sylow p-subgroup of C^- is a factor of $(mN)_p$, and that if the Sylow p-subgroup is cyclic, then $(h'(K))_p$ must be a factor of $(mN)_p$. By using this fact and by computing $h'(K)$ and mN, Kummer was able to see that the Sylow 11-subgroup of C^- for the cyclotomic field of 41-st roots of unity is non-cyclic. He also verified that the group C^- is cyclic for every prime $p < 100$, $p \neq 29, 41$.[11)]

Massachusetts Institute of Technology

---

11) Kummer, op. cit.

[45] Some modules in local cyclotomic fields

"Les Tendances Géom. en Algèbre et Théorie des Nombres",
Éditions du Centre National de la Recherche Scientifique, Paris,
1966, pp. 87-96.

(Massachussetts Institute of Technology)

Let p be an odd prime, and let $\mathbf{Q}$ and $\mathbf{Q}_p$ denote the rational field and the p-adic number field respectively. In the following, we shall fix an algebraic closure Ω of $\mathbf{Q}_p$ and consider algebraic extensions of $\mathbf{Q}$ and $\mathbf{Q}_p$ as subfields of Ω.

For each integer $n \geqq 0$, we put $q_n = p^{n+1}$, and fix a primitive q_n-th root of unity ζ_n in Ω so that $\zeta_{n+1}^p = \zeta_n$. Let $F_n = \mathbf{Q}(\zeta_n)$, $\Phi_n = \mathbf{Q}_p(\zeta_n)$, and let F and Φ denote the union of all F_n, $n \geqq 0$, and the union of all Φ_n, $n \geqq 0$, respectively. We identify the Galois group of $F/\mathbf{Q}$ with the Galois group of $\Phi/\mathbf{Q}_p$ in the obvious manner, and denote it by G. We also denote by R the group ring of G over the ring $\mathbf{Z}_p$ of p-adic integers :

$$R = \mathbf{Z}_p [G].$$

For any $m \geqq n \geqq 0$, let T_n and $T_{n,m}$ denote the trace map from Φ_n to $\mathbf{Q}_p$ and that from Φ_m to Φ_n respectively, and let N_n and $N_{n,m}$ denote the corresponding norm maps. Let Φ'_n be the intersection of all $N_{n,m}(\Phi^*_m)$, $m \geqq n$, Φ^*_m being the multiplicative group of Φ_m. Then Φ'_n is a subgroup of Φ^*_n, and $N_{n,m}(\Phi'_m) = \Phi'_n$ for $m \geqq n$. Let $\mathfrak{p}_n$ be the principal ideal of F_n generated by $\pi_n = 1 - \zeta_n$; $\mathfrak{p}_n$ is the unique prime ideal of F_n which divides the rational prime p. For simplicity, we shall use the same $\mathfrak{p}_n$ to denote the unique prime ideal in the local field Φ_n. Let P_n denote the group of p-units in F_n, namely, the multiplicative group of all α in F_n such that the principal ideal (α) is a power of $\mathfrak{p}_n$.

88

Then P_n is contained in Φ'_n. Let Q_n be the subgroup of P_n generated by π_n^ρ, $\rho \in G$, the conjugates of π_n. Since $N_{n,m}(\pi_m) = \pi_n$, we have

$$N_{n,m}(Q_m) = Q_n, \; m \geqq n.$$

However, it is not known whether $N_{n,m}(P_m) = P_n$, $m \geqq n$. In the following, we shall assume that $N_{n,m}(P_m) = P_n$ for any $m \geqq n \geqq 0$. Let E_n be the group of all units in F_n. Then our assumption is equivalent to the single equality

$$N_{0,1}(E_1) = E_0. \tag{A}$$

Without any such assumption, most of the results which we shall state below still hold with minor modifications. However, we shall assume (A) throughout the following because it will simplify the statement of our results considerably.

Now, for each $n \geqq 0$, let X_n be the inverse limit of $\Phi'_n/\Phi'^{p^s}_n$, $s \geqq 0$, relative to the natural homomorphisms

$$\Phi'_n/\Phi'^{p^t}_n \longrightarrow \Phi'_n/\Phi'^{p^s}_n, \qquad t \geqq s.$$

X_n is a compact abelian group, and $N_{n,m} : \Phi'_m \to \Phi'_n$ induces a homomorphism $X_m \to X_n$ for any $m \geqq n$. Let X denote the inverse limit of X_n, $n \geqq 0$, relative to those homomorphisms. Then X is again a compact abelian group, and it may be considered as an R-group in a natural way. Let Y_n and Z_n denote the closures of the images of P_n and Q_n respectively, under the canonical map $\Phi'_n \to X_n$. Let Y and Z be the subgroups of X determined by Y_n and Z_n, $n \geqq 0$, in the obvious manner. Both Y and Z are closed R-subgroups of X, and Z is contained in Y.

For each $n \geqq 0$, let A_n denote the set of all α in Φ_n such that $T_n(\alpha \log \beta)$ is contained in $\mathbf{Z}_p$ for any β in Φ_n with $\beta \equiv 1 \bmod \mathfrak{p}_n$. A_n is an open, compact subgroup of the additive group of Φ_n, and is invariant under the action of G. Hence it is an R-module in the natural way. Let $(\alpha, \beta)_n$ denote Hilbert's norm residue symbol for the power q_n in the local field Φ_n. Then there exists a map $\varphi'_n : \Phi'_n \to A_n$ such that

$$(\alpha, \beta)_n = \zeta_n^{T_n(\varphi'_n(\alpha) \log \beta)}$$

for any α in Φ'_n and β in Φ_n with $\beta \equiv 1 \bmod \mathfrak{p}_n$. φ'_n induces a surjective $\mathbf{Z}_p$-homomorphism $\varphi_n : \Phi'_n \to A_n/q_n A_n$ which satisfies

$$\varphi_n(\sigma\alpha) = \varkappa(\sigma)\,\sigma\,\varphi_n(\alpha), \qquad \sigma \in G, \qquad \alpha \in \Phi'_n,$$

where $\varkappa(\sigma)$ denotes the unit of $\mathbf{Q}_p$ such that $\sigma(\zeta_m) = \zeta_m^{\varkappa(\sigma)}$ for every $m \geqq 0$. A $\mathbf{Z}_p$-homomorphism between R-modules such as in the above will be simply called a $\varkappa$-homomorphism. We also note that by a formula due to Artin-Hasse, $\varphi_n(\pi_n)$ is the coset of

$$\mu_n = q_n^{-1} \zeta_n \pi_n^{-1} \qquad \mathrm{mod}\ q_n A_n.$$

We see easily that $T_{n,m}(A_m) = A_n$, $m \geqq n \geqq 0$. Hence we may define the inverse limit A of A_n, $n \geqq 0$, relative to $T_{n,m} : A_m \to A_n$, $m \geqq n$; A is a compact R-module. We can then prove that there exists a unique $\varkappa$-isomorphism

$$\varphi : X \rightarrow A,$$

which induces $\varphi_n : \Phi'_n \to A_n/q_n A_n$ for each $n \geqq 0$ in a natural way. Such an isomorphism φ, as well as φ_n, $n \geqq 0$, depends upon the choice of the sequence of the roots of unity ζ_n, $n \geqq 0$, which we have fixed in the above. However, a different choice of ζ_n, $n \geqq 0$, simply replaces φ by $u\varphi$ where u is a certain unit of $\mathbf{Q}_p$. So φ is canonically defined up to a factor which is a unit of $\mathbf{Q}_p$.

Let $B = \varphi(Y)$ and $C = \varphi(Z)$. As follows from the above remark, the R-submodules B and C in A are canonically determined. Obviously φ induces $\varkappa$-isomorphims $X/Y \to A/B$, $Y/Z \to B/C$, and $X/Z \to A/C$. Using the first two $\varkappa$-isomorphisms, we obtain the following results on A/B and B/C :

i) Let L be the maximal unramified abelian p-extension of F in Ω. Let $\mathfrak{p}$ denote the unique prime divisor of F which divides the rational prime p, and let K be the maximal abelian p-extension of F in Ω such that no prime divisor of F different from $\mathfrak{p}$ is ramified in K. Then there exists a $\varkappa$-isomorphism

$$G(K/L) \rightarrow A/B,$$

which is canonically defined up to a factor u, a unit of $\mathbf{Q}_p$.

ii) Let $E'_n = E_n \cap Q_n$, $n \geqq 0$. E'_n is the direct product of the group of q_n-th roots of unity in F_n and the group of the so-called circular units in the cyclotomic field F_n. Hence by the classical class number formula for F_n, E_n/E'_n is a finite group with order equal to the class number $h_n^{(2)}$ of the maximal real subfield of F_n. Let $(E_n/E'_n)_p$ denote the Sylow p-subgroup of E_n/E'_n, and let E^* be the compact R-group which is the inverse limit of $(E_n/E'_n)_p$, $n \geqq 0$, relative to

$$N_{n,m} : (E_m/E'_m)_p \rightarrow (E_n/E'_n)_p, \qquad m \geqq n.$$

Then there exists a $\varkappa$-isomorphism

$$E^* \rightarrow B/C,$$

which is canonically defined up to a factor u, a unit of $\mathbf{Q}_p$.

iii) Let δ denote the Galois automorphism in G which maps each element in F to its complex conjugate. For any R-module M, we put

$$M^+ = (1 + \delta) M, \qquad M^- = (1 - \delta) M;$$

M is the direct sum of the submodules M^+ and M^-.

For each $n \geqq 0$, let S_n denote the Sylow p-subgroup of the ideal class group of F_n, and let S be the direct limit of these S_n, $n \geqq 0$, relative to the natural homomorphisms $S_n \rightarrow S_m$, $m \geqq n$. Under the assumption (A), the maps $S_n \rightarrow S_m$, $m \geqq n$, are always injective. Hence S may be simply considered as the union of all S_n, $n \geqq 0$. It is a discrete R-group. Now, we call an ideal $\mathfrak{a}$ of F_n singular if $\mathfrak{a}^{p^m} = (\alpha)$ for some $m \geqq 0$ and for some α in F_n which is a p^m-th power in Φ_n. The set S'_n of all ideal classes in S_n^- which contain singular ideals forms an R-subgroup of S_n^-. Let S' denote the R-subgroup of S^- determined by these S'_n, $n \geqq 0$. For any f in Hom $(S^-/S', \mathbf{Q}_p/\mathbf{Z}_p)$ or Hom $((A/B)^-, \mathbf{Q}_p/\mathbf{Z}_p)$, we define σf, $\sigma \in G$, by $(\sigma f)(x) = f(\sigma^{-1} x)$ so that these groups become R-modules. Then there exist canonical R-isomorphisms

$$\mathrm{Hom}\,(S^-/S', \mathbf{Q}_p/\mathbf{Z}_p) \rightarrow (A/B)^-,$$

$$\mathrm{Hom}\,[(A/B)^-, \mathbf{Q}_p/\mathbf{Z}_p] \rightarrow S^-/S'.$$

Now, we see from the above *i*), *ii*) and *iii*), that if we know the structure of the R-modules A/B and B/C, then we also have various information on the arithmetic structure of the cyclotomic fields F_n, $n \geqq 0$. Although it seems quite difficult to study A/B and B/C separately, the structure of A/C can be investigated in the following manner. Let C_n denote the image of C under the natural homomorphism $A \rightarrow A_n$, $n \geqq 0$. Then A/C is the inverse limit of A_n/C_n, $n \geqq 0$, in the obvious manner, and the R-modules A_n and C_n are determined as follows : Let G_n denote the Galois group of $F_n/\mathbf{Q}$ as well as that of $\Phi_n/\mathbf{Q}_p$, and let

$$\theta_n = q_n^{-1} \sum_{s=0}^{n} \zeta_s^{-1} \qquad (n \geqq 0).$$

Then the module A_n consists of all elements of the form

$$\sum_\rho c'_\rho \mu_n^\rho + \sum_\rho c_\rho \theta_n^\rho,$$

where ρ ranges over G_n, and c'_ρ and c_ρ are arbitrary p-adic integers satisfying

$$\sum c_\rho = 0.$$

The submodule C_n consists of all elements of the form

$$\sum_\rho c'_\rho \mu_n^\rho,$$

with arbitrary c'_ρ, $\rho \in G_n$, in $\mathbf{Z}_p$; namely, C_n is the submodule of A_n generated over R by μ_n.

The elements θ_n^ρ, $\rho \in G_n$, in the above, form a normal basis of the Galois extension $\Phi_n/\mathbf{Q}_p$ (as well as that of $F_n/\mathbf{Q}$). Hence there exists a $\mathbf{Q}_p$-linear map ψ_n of Φ_n onto the group ring $\mathbf{Q}_p[G_n]$ such that $\psi_n(\theta_n^\rho) = \rho$ for any ρ in G_n. If we consider both Φ_n and $\mathbf{Q}_p[G_n]$ as R-modules in the natural way, then ψ_n gives an R-isomorphism. Let

$$A'_n = \psi_n(A_n), \qquad C'_n = \psi_n(C_n), \qquad \text{and} \quad \omega_n = \psi_n(\mu_n).$$

Then

$$\omega_n = (2q_n)^{-1} \sum_a (2a - q_n + p)\, \sigma_a^{(n)}, \qquad 0 \leqq a < q_n, \qquad (a, p) = 1,$$

where $\sigma_a^{(n)}$ denotes the element of G_n which maps ζ_n to ζ_n^a. Let $R_n = \mathbf{Z}_p[G_n]$, and let R_n^0 denote the submodule of all $\Sigma c_\rho \rho$ ($\rho \in G_n$, $c_\rho \in \mathbf{Z}_p$) in R_n such that $\Sigma c_\rho = 0$. Then it follows from the above that

$$A'_n = R_n^0 + C'_n, \qquad C'_n = R_n \omega_n.$$

Since A_n/C_n is R-isomorphic to A'_n/C'_n, we can see the algebraic structure of A_n/C_n by means of its « algebraic model » A'_n/C'_n.

Although such information on A/C does not determine the structure of A/B and B/C, nevertheless it seems to provide us with various results on the arithmetic of the cyclotomic fields F_n, $n \geqq 0$, through *i*), *ii*), and *iii*). Such consequences are not yet fully investigated. Here we shall state one of the immediate consequences.

Suppose that the prime p is either regular or properly irregular, namely, that the class number $h_0^{(2)}$ of the maximal real subfield of F_0 is prime to p. In this case, the assumption (A) is certainly satisfied, and furthermore $h_n^{(2)}$ is prime to p for every $n \geqq 0$. Hence $(E_n/E'_n)_p = 1$, $n \geqq 0$, and $E^* = 1$. It then follows from *ii*) that $B = C$. On the other hand, it also follows from the assumption that $S = S^-$ and $S' = 1$. Hence we see from *iii*) that S is canonically R-isomorphic to $\mathrm{Hom}\,[(A/C)^-, \mathbf{Q}_p/\mathbf{Z}_p]$, and consequently that there is a canonical R-isomorphism

$$S_n \rightarrow \mathrm{Hom}\,[(A_n/C_n)^-, \mathbf{Q}_p/\mathbf{Z}_p],$$

for each $n \geqq 0$. The group on the right hand side can be studied by means of the isomorphism $(A_n/C_n)^- \cong (A'_n/C'_n)^-$. Let

$$d_n = 2\, q_n \prod_{\chi} \left(-(2\, q_n)^{-1} \sum_{a} a\, \chi\,(a)^{-1} \right),$$

where $0 \leqq a < q_n$, $(a, p) = 1$, and where χ ranges over all characters mod q_n such that $\chi(-1) = -1$. It then follows from the above that the order of S_n is equal to the exact power of p dividing d_n. This is the p-part of the classical class number formula for the first factor $h_n^{(1)}$ of the class number of F_n which states that $h_n^{(1)} = d_n$. We note that in proving the above result, we have used the class number formula for $h_n^{(2)}$ to the effect that $(E_n/E'_n)_p = 1$ if $h_n^{(2)}$ is prime to p; the proof, however, is otherwise entirely arithmetic.

RÉSUMÉ

Soit p un premier rationnel, et soient Q, $Q_p \supset Q$, Z_p, Ω respectivement corps rationnel, corps p-adique rationnel, son anneau de valuation et une clôture algébrique valuée de Q_p. Si $q_n = p^{n+1}$ $(n \geqslant 0)$, fixons, d'une manière cohérente, pour tout n, une racine primitive q_n-ième de l'unité ζ_n (autrement dit, on a, pour tout $n \geqslant 0$, $\zeta_{n+1}^p = \zeta_n$). Soient

$$F_n = Q(\zeta_n), \qquad \Phi_n = Q_p(\zeta_n),$$

$$F = \bigcup_{n \geqslant 0} F_n, \qquad \Phi = \bigcup_{n \geqslant 0} \Phi_n,$$

G le groupe de Galois de F/Q et de Φ/Q_p (en identifiant tout automorphisme de F/Q avec son prolongement continu à Φ par rapport à la valuation de Ω), R l'anneau de groupe $Z_p[G]$. Soient

$$T_n : \Phi_n \to Q_p, \quad T_{n,m} : \Phi_m \to \Phi_n,$$

$$N_n : \Phi_n \to Q_p, \quad N_{n,m} : \Phi_m \to \Phi_n \quad (m \geqslant n)$$

les applications traciques et normiques avec les départs et les arrivées indiqués. Soient Φ^*_m le groupe multiplicatif de Φ_m,

$$\Phi'_n = \bigcap_{m \geqslant n} \Phi^*_m, \qquad \pi_n = 1 - \zeta_n,$$

$\mathfrak{p}_n$ l'idéal (π_n) de F_n, V_n le groupe multiplicatif des $\alpha \in F_n$ tels que (α) soit une puissance de $\mathfrak{p}_n$, W_n le sous-groupe de V_n engendré par les π_n^σ, $\sigma \in G$, E_n le groupe des unités de F_n. Si $m \geqslant n$, on a

$$N_{n,m} . \Phi'_m = \Phi'_n \quad \text{et} \quad N_{n,m} . W_m = W_n,$$

mais on ne sait pas si $N_{n,m} . V_m = V_n$. Toutefois, cela a lieu si

$$N_{0,1}(E_1) = E_0, \tag{A}$$

et l'on supposera que cette condition (A) satisfaite dans la suite du travail. Soit X_n la limite projective des $\Phi'_n/\Phi'^{p^s}_n$ $(s \geqslant 0)$ par rapport aux homomorphismes naturels

$$\Phi'_n/\Phi'^{p^t}_n \to \Phi'_n/\Phi'^{p^s}_n \quad (t \geqslant s)$$

et X celui des X_n par rapport aux applications normiques $N_{n,m} : X_m \to X_n$, induites par les $N_{n,m} : \Phi'_m \to \Phi'_n$ $(m \geqslant n)$. Les X_n et X sont des groupes abéliens compacts, et soient Y_n, Z_n les fermetures des V_n, W_n dans X_n et $Y \subseteq X$, $Z \subseteq X$ les limites projectives des Y_n respectivement des Z_n par rapport aux $N_{n,m}$.

Soit A_n l'ensemble des $\alpha \in \Phi_n$ tels que, pour tout $\beta \in \Phi_n$,

$$\beta \equiv 1 \pmod{\mathfrak{p}_n} \quad \text{implique} \quad T_n . \alpha \log \beta \in Z_p .$$

A_n est un sous-groupe additif ouvert et compact de Φ_n, qu'on peut considérer comme un R-module. Si $(\alpha, \beta)_n$ est le q_n-symbole des restes normiques de Hilbert dans Φ_n, il existe une application $\varphi'_n : \Phi'_n \to A_n$ telle que, pour tout $\alpha \in \Phi'_n$ et pour tout $\beta \in \Phi_n$ tel que $\beta \equiv 1 \pmod{\mathfrak{p}_n}$, on ait

$$(\alpha, \beta)_n = \zeta_n^{T_n(\varphi'_n \alpha) \log \beta},$$

94

et φ'_n induit un Z_p-épimorohisme $\varphi_n : \Phi'_n \longrightarrow A_n/q_n A_n$, qui satisfait à la condition supplémentaire

$$\varphi_n.(\sigma.\alpha) = \varkappa(\sigma)[\sigma.(\varphi_n.\alpha)] \qquad (\sigma \in G, \ \alpha \in \Phi'_n),$$

où $\varkappa(\sigma)$ est l'unité de Q_p telle que

$$\sigma.\zeta_m = \zeta_m^{\varkappa(\sigma)} \qquad \text{pour tout} \quad m \geqslant 0.$$

Un tel Z_p-homomorphisme sera appelé un $\varkappa$-homomorphisme.

On a $T_{n,m}.A_m = A_n$ $(m \geqslant n \geqslant 0)$. Soit A la limite projective des A_n par rapport aux $T_{n,m}$, qui est un R-module compact. Il existe l'unique $\varkappa$-isomorphisme $\varphi : X \longrightarrow A$, qui induit les $\varphi_n : \Phi'_n \longrightarrow A_n/q_n A_n$, et il est défini à un multiplicateur u près, qui est une unité arbitraire de Q_p, quand on varie arbitrairement le choix cohérent des ζ_n. Soient

$$B = \varphi.Y \qquad \text{et} \qquad C = \varphi.Z.$$

Le $\varkappa$-isomorphisme φ induit les $\varkappa$-isomorphismes $X/Y \longrightarrow A/B$, $Y/Z \longrightarrow A/C$. On obtient, en se servant de deux premiers isomorphismes, les résultats suivants :

(I) Soient $L \subset \Omega$ la p-extension abélienne non ramifiée maximale de F, K sa p-extension abélienne maximale telle que le seul idéal $\mathfrak{p} \mid (p)$ puisse y être ramifié, $G_{K/L}$ le groupe de Galois de K/L. Alors, *il existe un* $\varkappa$*-isomorphisme* $G_{K/L} \longrightarrow A/B$, *defini canoniquement à un multiplicateur unité de* Q_p *près.*

(II) Soient $E'_n = E_n \cap W_n$ $(n \geqslant 0)$ [donc $(E_n : E'_n) = h_n^{(2)}$ est le nombre des classes d'idéaux du sous-corps réel maximal de F_n], $(E_n/E'_n)_p$ le p-groupe de Sylow de E_n/E'_n et E^* la limite projective des $(E_n/E'_n)_p$ par rapport aux $N_{n,m}$: $(E_m/E'_m)_p \to (E_n/E'_n)_p$ $(m \geqslant n)$. Alors, *il existe un* $\varkappa$*-isomorphisme canonique* $E^* \to B/C$, *defini à un multiplicateur* u *près, qui est une unité arbitraire de* Q_p.

(III) Soit δ l'élément de G, qui applique tout élément de F sur son conjugué complexe, et soient, pour tout R-module M, $M^+ = (1+\delta).M$, $M^- = (1-\delta).M$ (donc, $M = M^+ \oplus M^-$). Soient S_n $(n \geqslant 0)$ le p-groupe de Sylow du groupe des classes d'idéaux de F_n et S la limite inductive des S_n par rapport aux applications naturelles $S_n \to S_m$ $(m < n)$. Sous l'hypothèse (A), $S_n \to S_m$ est une injection et S est la limite injective des S_n, qui est un R-groupe discret. Un idéal $\mathfrak{A}$ de F_n est dit *singulier* si, pour quelque $s \neq 0$, $\mathfrak{A}^{p^s} = (\alpha)$, où $\alpha \in \Phi_n{}^{p^s}$. Soient S'_n l'ensemble des classes d'idéaux $\in S_n^-$, auxquelles appartiennent des idéaux singuliers, et S' le sous-groupe de S^- que les S'_n engendrent par l'injection canonique $S_n \to S$ (ce sont des R-groupes). On organise Hom $(S^-/S', Q_p/Z_p)$ et Hom $[(A/B)^-, Q_p/Z_p]$ en R-modules en posant, pour tout $\sigma \in G$,

$$(\sigma.f).x = f.(\sigma^{-1}.x).$$

Alors, *il existe des R-isomorphismes canoniques* Hom $(S^-/S', Q_p/Z_p) \to (A/B)^-$ *et* Hom $[(A/B)^-, Q_p/Z_p] \to S^-/S$.

En vertu des (I), (II) et (III), la connaissance de la structure des R-modules A/B et B/C donnerait des informations importantes sur la structure arithmétique des F_n. Mais il semble difficile d'étudier A/B et B/C séparément, tandis que la structure de A/C peut l'être comme suit. Soit G_n le groupe de Galois de F_n/Q (et de Φ_n/Q_p), et soit

$$\theta_n = q_n^{-1} \sum_{s=0}^{n} \zeta_s^{-1}.$$

Alors, A_n est l'ensemble des éléments de Φ_n de la forme

$$\sum_\rho c'_\rho \mu_n^\rho + \sum_\rho c'_\rho \theta_n^\rho, \qquad \text{où} \quad \mu_n = q_n^{-1} \zeta_n \pi_n^{-1},$$

ρ parcourt G_n et c'_ρ et les c_ρ parcourent les éléments quelconques de Z_p satisfaisant à la seule condition $\sum_\rho c_\rho = 0$. L'image inverse C_n de C par l'application $A_n \to A$ est $\sum Z_p \mu_n^\rho$, autrement dit, le R-sous-module de A_n engendré par μ_n. Et A/C est la limite projective des A_n/C_n.

Les θ_n^ρ $(\rho \in G_n)$ forment une base normale de Φ_n/Q_p (et de F_n/Q), qui induit une bijection Q_p-linéaire

$$\Psi_n : \Phi_n \to Q_p [G_n] \qquad \text{en posant} \qquad \Psi_n . \theta_n^\rho = \rho,$$

et Ψ_n peut être considérée, d'une manière naturelle, comme un R-isomorphisme. Soient

$$A'_n = \Psi_n . A_n, \qquad C'_n = \Psi_n . C_n$$

et

$$\omega_n = \Psi_n . \mu_n = (2\, q_n)^{-1} \sum_a \; (2\, a - q_n + p)\, \sigma_n^{(a)}$$

$$[0 < a < q_n;\ (a, p) = 1],$$

où $\sigma_n^{(a)} \in G_n$ est tel que

$$\sigma_n^{(a)} . \zeta_n = \zeta_n^a.$$

Si $R_n = Z_p [G_n]$, et R_n^0 est le sous-module de R_n, formé par les

$$\sum c_\rho\, \rho \ (c_\rho \in Z_p,\ \rho \in G_n)$$

96

tels que $\sum c_\rho = 0$, on a $A'_n = R^0_n + C'_n$, $C'_n = R_n\,\omega_n$, ce qui, en vertu du R-isomorphisme $A_n/C_n \cong A'_n/C'_n$, donne la description algébrique de

$$A/C = \text{lim proj. } A_n/C_n.$$

Bien que cela ne détermine pas celle des A/B et B/C, la connaissance de la structure algébrique de A/C permet, grâce aux (I), (II) et (III), de tirer des conséquences variées (et pas encore entièrement étudiées) au sujet des arithmétiques des F_n. Ainsi, sous l'hypothèse que $h_n^{(2)}$ soit premier à p [auquel cas, la condition (A) est automatiquement satisfaite,] je donne une démonstration simple et purement arithmétique que l'ordre de S_n est égal à la contribution de p à

$$d_n = 2\,q_n \prod_{\chi} \left[-(2\,q_n)^{-1} \sum_{a} a\,\chi(a)^{-1} \right]$$

(où χ parcourt les caractères (mod q_n) tels que $\chi(-1) = -1$, et où a parcourt les entiers rationnels > 0 et $< q_n$ premiers à p).

[46] On explicit formulas for the norm residue symbol

J. Math. Soc. Japan, 20 (1968), 151-165.

(Received Dec. 1, 1966)

Let p be an odd prime and let $\boldsymbol{Z}_p$ and $\boldsymbol{Q}_p$ denote the ring of p-adic integers and the field of p-adic numbers respectively. For each integer $n \geqq 0$, let $q_n = p^{n+1}$ and let Φ_n denote the local cyclotomic field of q_n-th roots of unity over $\boldsymbol{Q}_p$. We fix a primitive q_n-th root of unity ζ_n in Φ_n so that $\zeta_{n+1}^p = \zeta_n$, and put $\pi_n = 1-\zeta_n$; π_n generates the unique prime ideal $\mathfrak{p}_n$ in the ring $\mathfrak{o}_n$ of local integers in Φ_n. Let $(\alpha, \beta)_n$ denote Hilbert's norm residue symbol in Φ_n for the power q_n and let

$$(\alpha, \beta)_n = \zeta_n^{[\alpha, \beta]_n}$$

with $[\alpha, \beta]_n$ in $\boldsymbol{Z}_p$, well determined mod q_n. The classical formulas for $[\alpha, \beta]_n$ state that

$$[\zeta_n, \beta]_n = q_n^{-1} T_n(\log \beta), \qquad \beta \in 1+\mathfrak{p}_n, \quad n \geqq 0,$$

$$[\pi_n, \beta]_n = -q_n^{-1} T_n(\zeta_n \pi_n^{-1} \log \beta), \qquad \beta \in 1+\mathfrak{p}_n, \quad n \geqq 0,$$

$$[\alpha, \beta]_0 = -q_0^{-1} T_0\left(\zeta_0 \alpha^{-1} \frac{d\alpha}{d\pi_0} \log \beta\right), \qquad \alpha \in 1+\mathfrak{p}_0, \quad \beta \in 1+\mathfrak{p}_0^2,$$

where T_n denotes the trace from Φ_n to $\boldsymbol{Q}_p$[1].

In a previous note [7], we have announced formulas for $[\alpha, \beta]_n$ which generalize the above formulas of Artin-Hasse. In the present paper, we shall prove those formulas and then discuss some related results.

As in the above, we retain most of the notations introduced in our earlier paper [6]. In particular, we denote by N_n the norm from Φ_n to $\boldsymbol{Q}_p$, and by $T_{n,m}$ and $N_{n,m}$ the trace and the norm, respectively, from Φ_m to Φ_n, $m \geqq n \geqq 0$; for an automorphism σ of the union Φ of all Φ_n, $n \geqq 0$, we denote by $\kappa(\sigma)$ the

---

*) The present research was supported in part by the National Science Foundation grant NSF-GP-4361.

1) See [1], [2], [5]. For the general theory of the norm residue symbol needed in the following, see [2], Chap. 12 or [4], II, §11, §19. It is to be noted that the symbol $(\alpha, \beta)_n$ in [2] is the inverse of the same symbol in [4]. Here we follow the definition of $(\alpha, \beta)_n$ in [2] as we did in [7].

unique p-adic unit such that $\zeta_n^{\sigma}=\zeta_n^{\kappa(\sigma)}$ for every $n \geqq 0$[2].

1. We shall first prove several elementary lemmas. For $n \geqq 0$, we denote by ν_n the normalized valuation on Φ_n such that $\nu_n(\pi_n)=1$.

LEMMA 1. *Let $\mathfrak{d}_n$ be the different of Φ_n/Q_p: $\mathfrak{d}_n=q_n\mathfrak{p}_n^{-p^n}$. Then*

$$T_n(\mathfrak{d}_n \log(1+\mathfrak{p}_n)) \equiv 0 \quad \bmod q_n .$$

PROOF. Since the multiplicative group $(1+\mathfrak{p}_n)/(1+\mathfrak{p}_n^2)$ is generated by the coset of ζ_n, we have $\log(1+\mathfrak{p}_n)=\log(1+\mathfrak{p}_n^2)$. We shall show that $T_n(\mathfrak{d}_n \log \alpha)\equiv 0 \bmod q_n$, or, equivalently, $\nu_n(q_n \log \alpha) \geqq 2p^n$, for any α in $1+\mathfrak{p}_n^2$. Let $\alpha=1-\beta$ with β in $\mathfrak{p}_n^2$ so that $\log \alpha=-\sum_{i=1}^{\infty} i^{-1}\beta^i$. It is sufficient to show $\nu_n(q_n i^{-1}\beta^i) \geqq 2p^n$ for $i \geqq 1$. Let p^e $(e \geqq 0)$ be the exact power of p dividing i. Then $\nu_n(q_n i^{-1}\beta^i) \geqq (n+1-e)(p-1)p^n+2i \geqq (n+1-e)(p-1)p^n+2p^e$. If $e \leqq n$, then clearly $(n+1-e)(p-1)p^n+2p^e \geqq (p-1)p^n \geqq 2p^n$. If $e>n$, then $(n+1-e)(p-1)p^n+2p^e=(2+2(p^{e-n}-1)-(e-n-1)(p-1))p^n \geqq 2p^n$ because $p^{e-n}-1=(1+p-1)^{e-n}-1 \geqq (e-n)(p-1)$. Thus $\nu_n(q_n i^{-1}\beta^i) \geqq 2p^n$ in either case.

LEMMA 2.

i) $$q_m^{-1}T_m(\mathfrak{d}_m \log(1+\mathfrak{p}_n)) \equiv 0 \quad \bmod q_n, \qquad m \geqq 2n+1,$$

ii) $$q_m^{-1}T_m(\mathfrak{d}_m \log(1+\mathfrak{p}_n^{p^n})) \equiv 0 \quad \bmod q_n, \qquad m \geqq n+1,$$

iii) $$q_m^{-1}T_m(\mathfrak{d}_m \log(1+\mathfrak{p}_n^{2p^n})) \equiv 0 \quad \bmod q_n, \qquad m \geqq n.$$

PROOF. Clearly $T_{n,m}(\mathfrak{d}_m)=T_{n,m}(q_m\mathfrak{p}_0^{-1})$ is contained in $p^{m-n}q_m\mathfrak{p}_0^{-1}$, and if $m \geqq 2n+1$, the latter is contained in $p^{n+1}q_m\mathfrak{p}_0^{-1}=q_m\mathfrak{d}_n$. Hence, for $m \geqq 2n+1$, $q_m^{-1}T_m(\mathfrak{d}_m \log(1+\mathfrak{p}_n))=q_m^{-1}T_n(T_{n,m}(\mathfrak{d}_m)\log(1+\mathfrak{p}_n))$ is contained in $T_n(\mathfrak{d}_n \log(1+\mathfrak{p}_n))$, and i) follows from Lemma 1.

Since $(1+\mathfrak{p}_n^{p^n})/(1+\mathfrak{p}_n^{p^n+1})$ is generated by the coset of ζ_0, we have $\log(1+\mathfrak{p}_n^{p^n})=\log(1+\mathfrak{p}_n^{p^n+1})=\mathfrak{p}_n^{p^n+1}$. Hence $q_m^{-1}T_m(\mathfrak{d}_m \log(1+\mathfrak{p}_n^{p^n}))=T_m(\mathfrak{p}_n)\equiv 0 \bmod p^m$, and ii) holds for $m \geqq n+1$. iii) can be proved similarly by using $\log(1+\mathfrak{p}_n^{2p^n})=\mathfrak{p}_n^{2p^n}=\mathfrak{p}_0^2$ and $T_m(\mathfrak{p}_0)\equiv 0 \bmod q_m$.

Let Λ denote the ring of formal power series in T with coefficients in Z_p: $\Lambda=Z_p[[T]]$. For any nonzero element ξ in $\mathfrak{o}_n$, there exists a power series $f(T)$ in Λ such that

$$\xi=f(\pi_n),$$

$$f(T)=\sum_{i=s}^{\infty} a_i T^i, \qquad a_i \in Z_p,\ s=\nu_n(\xi) \geqq 0 .$$

Such $f(T)$ will be simply called a power series for ξ. Let

2) However, $(\alpha, \beta)_n$ in the formulas of [6], § 3, is the inverse of the same symbol in the present paper. See the footnote 1).

$$\frac{d\xi}{d\pi_n}=f'(\pi_n),$$

$$\delta_n(\xi)=\frac{\zeta_n}{\xi}\frac{d\xi}{d\pi_n},$$

where $f'(T)$ denotes the formal derivative of $f(T)$ with respect to T. Since a power series for ξ such as $f(T)$ is not unique for ξ, $d\xi/d\pi_n$ and $\delta_n(\xi)$ are not uniquely determined by ξ. However we can prove the following lemma on the values taken by $\delta_n(\xi)$.

LEMMA 3. *The values of $\delta_n(\xi)$ which are obtained from all possible power series for ξ fulfil a residue class of $\mathfrak{p}_n^{-1}$ mod $\mathfrak{d}_n$.*

PROOF. It is clear from the definition that $\delta_n(\xi_n)$ always belongs to $\mathfrak{p}_n^{-1}$. In general, let $f(T)$ and $g(T)$ be power series in Λ such that $f(\pi_n)=g(\pi_n)$. Then

$$f(T)=g(T)+u(T)d(T),$$

where $u(T)$ is a certain power series in Λ and $d(T)=\sum_{i=0}^{p-1}(1-T)^{ip^n}$ is the minimal polynomial of $\pi_n=1-\zeta_n$ over Q_p[3]. Differentiating the both sides, we obtain

$$f'(\pi_n)=g'(\pi_n)+u(\pi_n)d'(\pi_n).$$

Since $\mathfrak{d}_n=(d'(\pi_n))$, it follows that

$$f'(\pi_n)\equiv g'(\pi_n) \qquad \text{mod } u(\pi_n)\mathfrak{d}_n.$$

Now, suppose that both $f(T)$ and $g(T)$ are power series for ξ. Then the coefficient of T^i vanishes for $0\leqq i<s=\nu_n(\xi)$ in both $f(T)$ and $g(T)$. Since $d(0)=p\neq 0$, the same holds for the coefficient of T^i in $u(T)$ for $0\leqq i<s$. Hence $\nu_n(u(\pi_n))\geqq s$ and it follows from the above congruence that

$$\zeta_n\xi^{-1}f'(\pi_n)\equiv\zeta_n\xi^{-1}g'(\pi_n) \qquad \text{mod } \mathfrak{d}_n.$$

On the other hand, let η be any element of $\mathfrak{d}_n=(d'(\pi_n))$. Then there is a power series $v(T)$ in Λ such that $\eta=\zeta_n v(\pi_n)d'(\pi_n)$. Let

$$h(T)=(1+v(T)d(T))f(T).$$

Then $h(T)$ is also a power series for ξ and $\zeta_n\xi^{-1}h'(\pi_n)=\zeta_n\xi^{-1}f'(\pi_n)+\eta$. Therefore the lemma is proved.

In the following, we regard $\delta_n(\xi)$ as a multi-valued function of $\xi\neq 0$ in $\mathfrak{o}_n$, representing any such value as described above.

LEMMA 4.

i) *For $\xi\neq 0$ and $\eta\neq 0$ in $\mathfrak{o}_0$,*

$$\delta_n(\xi\eta)\equiv\delta_n(\xi)+\delta_n(\eta) \qquad \text{mod } \mathfrak{d}_n.$$

ii) *For any $\xi\neq 0$ in $\mathfrak{o}_n$ and for any automorphism σ of Φ,*

3) See [2], p. 151.

$$\delta_n(\xi^\sigma)\equiv\kappa(\sigma)\delta_n(\xi)^\sigma \quad \bmod \mathfrak{d}_n .$$

iii) *For $m\geqq n$ and for any unit ξ in $\mathfrak{o}_n$,*

$$\delta_m(\xi)\equiv p^{m-n}\delta_n(\xi) \quad \bmod \mathfrak{d}_m .$$

PROOF. i) follows immediately from the fact that if $f(T)$ is a power series for ξ and $g(T)$ a power series for η, then $f(T)g(T)$ is a power series for $\xi\eta$. To prove ii), let

$$u(T)=1-(1-T)^{\kappa(\sigma)}=\kappa(\sigma)T+\cdots .$$

Then $u(T)$ is a power series for π_n^σ and

$$u'(\pi_n)=\kappa(\sigma)(1-\pi_n)^{\kappa(\sigma)-1}=\kappa(\sigma)\zeta_n^{\sigma-1} .$$

Let $f(T)$ be a power series for ξ. Then $f(u(T))$ is a power series for ξ^σ, and

$$\delta_n(\xi^\sigma)=\zeta_n(\xi^\sigma)^{-1}f'(u(\pi_n))u'(\pi_n)=\zeta_n(\xi^{-1})^\sigma f'(\pi_n)^\sigma\kappa(\sigma)\zeta_n^{\sigma-1}$$
$$=\kappa(\sigma)(\zeta_n\xi^{-1}f'(\pi_n))^\sigma=\kappa(\sigma)\delta_n(\xi)^\sigma .$$

Finally, let ξ be a unit in $\mathfrak{o}_n$. Then ξ is also a unit in $\mathfrak{o}_m$ and every power series $g(T)$ such that $g(\pi_m)=\xi$ is a power series for ξ in $\mathfrak{o}_m$. Let $f(T)$ be a power series for ξ in $\mathfrak{o}_n$. Since $f(\pi_n)=\xi$, $\pi_n=1-(1-\pi_m)^{p^{m-n}}$, $f(1-(1-T)^{p^{m-n}})$ is a power series for ξ in $\mathfrak{o}_m$. Computing $d\xi/d\pi_m$ by means of this power series, we obtain immediately the formula in iii). Note that the both sides of the congruence in iii) are well determined mod $\mathfrak{d}_m$ because $\mathfrak{d}_m=p^{m-n}\mathfrak{d}_n$.

Let ξ now be an arbitrary element of the multiplicative group Φ_n^* of Φ_n, i.e., an arbitrary nonzero element of Φ_n. We write ξ in the form $\xi=\xi_1/\xi_2$ with $\xi_1\neq 0$, $\xi_2\neq 0$ in $\mathfrak{o}_n$, and define

$$\delta_n(\xi)=\delta_n(\xi_1)-\delta_n(\xi_2) .$$

By Lemma 4, i), we see that the values of $\delta_n(\xi)$ again fulfil a residue class of $\mathfrak{p}_n^{-1}$ mod $\mathfrak{d}_n$. Furthermore, i), ii) of Lemma 4 now hold for any ξ and η in Φ_n^*. Hence δ_n defines a so-called κ-homomorphism

$$\delta_n:\Phi_n^*\rightarrow\mathfrak{p}_n^{-1}/\mathfrak{d}_n .$$

δ_n is continuous in the sense that if $\xi\equiv 1 \bmod \mathfrak{p}_n^k$, $k\gg 0$, then $\delta_n(\xi)\equiv 0 \bmod \mathfrak{d}_n$.

LEMMA 5. *For $m\geqq n$ and for ξ in Φ_m^*,*

$$\delta_n(N_{n,m}(\xi))\equiv p^{-(m-n)}T_{n,m}(\delta_m(\xi)) \quad \bmod \mathfrak{d}_n .$$

PROOF. Since $\mathfrak{d}_m=p^{m-n}\mathfrak{d}_n$, the both sides of the above are well determined mod $\mathfrak{d}_n$. By Lemma 4, i), it is sufficient to prove the lemma for $\xi\equiv\pi_m$ and for a unit ξ in $\mathfrak{o}_m$. Since

$$N_{n,m}(\pi_m)=\pi_n , \qquad T_{n,m}(\zeta_m\pi_m^{-1})=p^{m-n}\zeta_n\pi_n^{-1} ,$$
$$\delta_n(\pi_n)=\zeta_n\pi_n^{-1} , \qquad \delta_m(\pi_m)=\zeta_m\pi_m^{-1} ,$$

the lemma holds for $\xi=\pi_m$. Let ξ be a unit in $\mathfrak{o}_m$. When σ ranges over all Galois automorphisms of Φ_m/Φ_n, we obtain from Lemma 4, ii), iii) that

$$p^{m-n}\delta_n(N_{n,m}(\xi))\equiv\delta_m(N_{n,m}(\xi))\equiv\sum_\sigma\delta_m(\xi^\sigma)$$
$$\equiv\sum_\sigma\kappa(\sigma)\delta_m(\xi)^\sigma \qquad \mathrm{mod}\ \mathfrak{d}_m\,.$$

Apply $p^{-(m-n)}T_{n,m}$ to the above. Since

$$T_{n,m}(\mathfrak{d}_m)=T_{n,m}(q_m\mathfrak{p}_0^{-1})=p^{m-n}q_m\mathfrak{p}_0^{-1}=p^{m-n}\mathfrak{d}_m\,,$$

we have

$$p^{m-n}\delta_n(N_{n,m}(\xi))\equiv p^{-(m-n)}\sum_\sigma\kappa(\sigma)T_{n,m}(\delta_m(\xi)) \qquad \mathrm{mod}\ \mathfrak{d}_m\,.$$

Here

$$\sum_\sigma\kappa(\sigma)\equiv\sum_{i=0}^{p^{m-n}-1}(1+iq_n)\equiv p^{m-n} \qquad \mathrm{mod}\ q_m\,.$$

Since ξ is a unit, $\delta_m(\xi)$ belongs to $\mathfrak{o}_m$. Hence $p^{-(m-n)}T_{n,m}(\delta_m(\xi))$ is in $\mathfrak{o}_n$, and we obtain from the above

$$p^{m-n}\delta_n(N_{n,m}(\xi))\equiv T_{n,m}(\delta_m(\xi)) \qquad \mathrm{mod}\ \mathfrak{d}_m\,,$$

which implies the congruence of the lemma.

2. For α in Φ_n^* and β in $1+\mathfrak{p}_n$, $n\geqq 0$, let

$$\langle\alpha,\beta\rangle_n=-q_n^{-1}T_n(\delta_n(\alpha)\log\beta)^{4)}\,.$$

Although the value of $\langle\alpha,\beta\rangle_n$ depends upon the choice of the value of $\delta_n(\alpha)$, it is, in certain cases, well determined modulo a high power of p.

LEMMA 6. *Let $m\geqq n$ and let α be an arbitrary element of Φ_m^*. In each one of the following three cases, the value of $\langle\alpha,\beta\rangle_m$ is well determined* mod q_n:

i) $m\geqq 2n+1$ *and* β *in* $1+\mathfrak{p}_n$,

ii) $m\geqq n+1$ *and* β *in* $1+\mathfrak{p}_n^{p^n}$,

iii) $m\geqq n$ *and* β *in* $1+\mathfrak{p}_n^{2p^n}$.

PROOF. Since $\delta_m(\alpha)$ is well determined mod $\mathfrak{d}_m$, the lemma follows immediately from Lemma 2.

In these three cases, we have

$$\langle\alpha_1\alpha_2,\beta\rangle_m\equiv\langle\alpha_1,\beta\rangle_m+\langle\alpha_2,\beta\rangle_m \qquad \mathrm{mod}\ q_n\,,$$
$$\langle\alpha,\beta_1\beta_2\rangle_m\equiv\langle\alpha,\beta_1\rangle_m+\langle\alpha,\beta_2\rangle_m \qquad \mathrm{mod}\ q_n\,,$$

for α,α_1,α_2 in Φ_m^* and β,β_1,β_2 in Φ_n satisfying respective conditions.

LEMMA 7. *Let $l\geqq m\geqq n$. Let α be an element of Φ_l^*, and β an element*

---

4) The symbol $\langle\alpha,\beta\rangle_n$ here is completely different from the same symbol defined in [6], §1, p. 52.

of $1+\mathfrak{p}_n$. *Suppose that* β *and* m *satisfy one of the three conditions in Lemma* 6. *Then*

$$\langle N_{m,l}(\alpha), \beta\rangle_m \equiv \langle \alpha, \beta\rangle_l \quad \mod q_n.$$

PROOF. By Lemma 5,

$$\delta_m(N_{m,l}(\alpha)) \equiv p^{-(l-m)}T_{m,l}(\delta_l(\alpha)) \quad \mod \mathfrak{d}_m.$$

Multiply the both sides of the above by $-\log\beta$ and apply $q_m^{-1}T_m$. The congruence of the lemma then follows immediately from Lemma 2.

For $n\geqq 0$, $i\geqq 1$, let

$$\eta_i^{(n)} = 1-\pi_n^i.$$

For any $k\geqq 0$, the elements $\eta_i^{(n)}$, $i\geqq k$, generate the compact multiplicative group $1+\mathfrak{p}_n^k$ topologically. We shall next prove a key lemma for such $\eta_i^{(n)}$.

LEMMA 8. *For* $i\geqq 1$, $j\geqq((n+2)(p-1)+1)p^{n-1}$,

$$[\eta_i^{(n)}, \eta_j^{(n)}]_n = \langle \eta_i^{(n)}, \eta_j^{(n)}\rangle_n.$$

PROOF. Since n is fixed throughout the following proof, we shall write η_i for $\eta_i^{(n)}$, $i\geqq 1$. By Lemma 6, iii) $\langle\eta_i, \eta_j\rangle_n$ is well determined mod q_n. Hence it is sufficient to prove the lemma for the following particular value of $\delta_n(\eta_i)$:

$$\delta_n(\eta_i) = -\zeta_n \sum_{r=1}^{\infty} i\pi_n^{ir-1}.$$

Since

$$\log\eta_j = -\sum_{s=1}^{\infty} s^{-1}\pi_n^{js},$$

we have

$$\langle\eta_i, \eta_j\rangle_n = -q_n^{-1}\sum_{r,s\geqq 1} is^{-1}T_n(\zeta_n\pi_n^{ir+js-1})$$

$$= -q_n^{-1}\sum_{\substack{r,s\geqq 1\\(r,s)=1}}\sum_{u=1}^{\infty} i(us)^{-1}T_n(\zeta_n\pi_n^{u(ir+js)-1}).$$

On the other hand, it is known that

$$[\eta_i, \eta_j]_n = \sum_{\substack{r,s\geqq 1\\(r,s)=1}} (ir'+js')[\pi_n, \eta_{ir+js}]_n,$$

where for each pair r, s, the integers r', s' are chosen so that $rs'-sr'=1$[5]. Using Artin-Hasse's formula

$$[\pi_n, \beta]_n = -q_n^{-1}T_n(\zeta_n\pi_n^{-1}\log\beta), \qquad \beta\in 1+\mathfrak{p}_n,$$

we obtain

---

5) See [3].

$$[\eta_i, \eta_j]_n = -q_n^{-1} \sum_{\substack{r,s\geq 1\\(r,s)=1}} (ir'+js')T_n(\zeta_n\pi_n^{-1}\log\eta_{ir+js})$$

$$= q_n^{-1} \sum_{\substack{r,s\geq 1\\(r,s)=1}} \sum_{u=1}^{\infty} (ir'+js')u^{-1}T_n(\zeta_n\pi_n^{u(ir+js)-1}).$$

Since $rs'-sr'=1$, we see that

$$[\eta_i, \eta_j]_n - \langle \eta_i, \eta_j\rangle_n = q_n^{-1} \sum_{\substack{r,s\geq 1\\(r,s)=1}} \sum_{u=1}^{\infty} s'(ir+js)(us)^{-1}T_n(\zeta_n\pi_n^{u(ir+js)-1}).$$

However

$$q_n^{-1}\sum_{u=1}^{\infty}(ir+js)u^{-1}T_n(\zeta_n\pi_n^{u(ir+js)-1}) = -q_n^{-1}(ir+js)T_n(\zeta_n\pi_n^{-1}\log\eta_{ir+js})$$

$$= (ir+js)[\pi_n, \eta_{ir+js}]_n$$

$$= [\pi_n^{ir+js}, 1-\pi_n^{ir+js}]_n$$

$$\equiv 0 \qquad \bmod q_n\text{[6)]}.$$

Hence

$$[\eta_i, \eta_j]_n - \langle \eta_i, \eta_j\rangle_n$$

$$\equiv q_n^{-1}\sum_{r,s}{}' \sum_{u=1}^{\infty} s'(ir+js)(us)^{-1}T_n(\zeta_n\pi_n^{u(ir+js)-1}) \qquad \bmod q_n,$$

where the outer sum on the right is taken over all pairs of integers $r, s\geq 1$ with $(r, s)=1$ and with s divisible by p. We shall next show that each term on the right hand side of the above is divisible by q_n.

With r, s, and u fixed, let p^e be the exact power of p dividing us. Then $e\geq 1$ by the assumption on s. Since $j\geq((n+2)(p-1)+1)p^{n-1}$, we have

$$\nu_n(q_n^{-1}(us)^{-1}\zeta_n\pi_n^{u(ir+js)-1}) = u(ir+js)-1-(n+1+e)(p-1)p^n$$

$$\geq ((n+2)(p-1)+1)p^{n-1+e}-(n+1+e)(p-1)p^n.$$

For real x, let $f(x)=((n+2)(p-1)+1)p^{n-1+x}-(n+1+x)(p-1)p^n$. Then $f(1)=p^n$ and $f'(x)\geq 0$ for $x\geq 1$. Hence $f(x)\geq p^n$ for $x\geq 1$. Since $e\geq 1$, it follows that

$$\nu_n(q_n^{-1}(us)^{-1}\zeta_n\pi_n^{u(ir+js)-1})\geq p^n$$

so that

$$q_n^{-1}s'(ir+js)(us)^{-1}T_n(\zeta_n\pi_n^{u(ir+js)-1})\equiv 0 \qquad \bmod q_n.$$

Thus we obtain

$$[\eta_i, \eta_j]_n \equiv \langle \eta_i, \eta_j\rangle_n \qquad \bmod q_n,$$

and the lemma is proved. Note that $[\eta_i, \eta_j]_n$ is a p-adic integer, determined only mod q_n. Note also that if $j>((n+1)(p-1)+1)p^n$, then η_j is a q_n-th power of an element in $1+\mathfrak{p}_n^{p^n+1}$ and the equality of the lemma can be verified immedi-

6) In general, $(\alpha, 1-\alpha)_n=1$.

ately because both sides are simply zero.

LEMMA 9. *Let* $m \geqq 3n+1$. *Then*

$$[N_{n,m}(\alpha), \beta]_n = \langle \alpha, \beta \rangle_m, \qquad \alpha \in \Phi_m^*, \ \beta \in 1+\mathfrak{p}_n.$$

PROOF. We first note that $\langle \alpha, \beta \rangle_m$ is well determined mod q_n by Lemma 6, i). Let $k=((m-2n)(p-1)+1)p^m$. It follows from $m \geqq 3n+1$, $p \geqq 3$ that $k \geqq ((m+2)(p-1)+1)p^{m-1}$. Since β is in $1+\mathfrak{p}_n$, β^{p^n} is contained in $1+\mathfrak{p}_n^{p^n}$, and $\beta^{p^{m-n}}$ in $1+p^{m-2n}\mathfrak{p}_n^{p^n}=1+\mathfrak{p}_m^k$. Hence $\beta^{p^{m-n}}$ is the limit of a sequence of certain products of $\eta_j^{(n)}$, $j \geqq ((m+2)(p-1)+1)p^{m-1}$. Therefore it follows from Lemma 8 that

$$[\eta_i^{(m)}, \beta^{p^{m-n}}]_m = \langle \eta_i^{(m)}, \beta^{p^{m-n}} \rangle_m, \qquad i \geqq 1.$$

Using the well known properties of the norm residue symbol, we then have

$$\begin{aligned}(N_{n,m}(\eta_i^{(n)}), \beta)_n &= (\eta_i^{(m)}, \beta)_m^{p^{m-n}} = (\eta_i^{(m)}, \beta^{p^{m-n}})_m \\ &= \zeta_m^{[\eta_i^{(m)}, \beta^{p^{m-n}}]_m} = \zeta_m^{<\eta_i^{(m)}, \beta^{p^{m-n}}>_m} \\ &= \zeta_m^{p^{m-n}<\eta_i^{(m)}, \beta>_m}.\end{aligned}$$

Since the first term of the above is a q_n-th root of unity,

$$q_n p^{m-n} \langle \eta_i^{(m)}, \beta \rangle_m \equiv 0 \qquad \text{mod } q_m.$$

Therefore $\langle \eta_i^{(m)}, \beta \rangle_m$ is a p-adic integer, and the last term of the above equalities can be written as $\zeta_n^{<\eta_i^{(m)}, \beta>_m}$. Hence

$$[N_{n,m}(\eta_i^{(m)}), \beta]_n = \langle \eta_i^{(m)}, \beta \rangle_m, \qquad i \geqq 1.$$

Since $1+\mathfrak{p}_m$ is topologically generated by $\eta_i^{(m)}$, $i \geqq 1$, it follows that the lemma holds for any α in $1+\mathfrak{p}_m$.

If α is an element of the group V of $(p-1)$-*st* roots of unity in Φ_m, then $[N_{n,m}(\alpha^{p-1}), \beta]_n = \langle \alpha^{p-1}, \beta \rangle_m = 0$ so that $[N_{n,m}(\alpha), \beta]_n = \langle \alpha, \beta \rangle_m = 0$. Hence the lemma again holds for such α. Let $\alpha = \pi_m$. Then $N_{n,m}(\alpha) = \pi_n$ and $q_m^{-1} T_{n,m}(\delta_m(\alpha)) = q_m^{-1} T_{n,m}(\zeta_m \pi_m^{-1}) = q_n^{-1} \zeta_n \pi_n^{-1}$. Hence the equality of the lemma in this case is nothing but the formula of Artin-Hasse. Since the group Φ_m^* is generated by π_m, V, and $1+\mathfrak{p}_m$, the lemma is completely proved.

We are now able to prove the main result of the paper. For each $n \geqq 0$, let Φ_n' denote the intersection of all $N_{n,m}(\Phi_m^*)$, $m \geqq n$. Then

$$\Phi_n^* = U_0 \times \Phi_n', \qquad \Phi_n' = N_{n,m}(\Phi_m'), \qquad m \geqq n,$$

with $U_0 = 1+p\boldsymbol{Z}_p$[7]. It follows in particular that for each α in Φ_n' and for any $m \geqq n$, there exists an element α' in Φ_m' such that $\alpha = N_{n,m}(\alpha')$. Any such α'

7) See [6], § 2, Proposition 9.

will be denoted by $N_{n,m}^{-1}(\alpha)$.

THEOREM 1. *For α in Φ_n', β in $1+\mathfrak{p}_n$,*

$$[\alpha, \beta]_n = \langle N_{n,m}^{-1}(\alpha), \beta\rangle_m$$

for any $m \geqq 2n+1$. More explicitly,

$$(\alpha, \beta)_n = \zeta_n^{-q_m^{-1}T_m(\zeta_m \alpha'^{-1}(d\alpha'/d\pi_m)\log\beta)}$$

with $\alpha' \in \Phi_m'$, $N_{n,n}(\alpha')=\alpha$, $m \geqq 2n+1$.

PROOF. Let $l \geqq m$, $3n+1$ and $\alpha''=N_{m,l}^{-1}(\alpha')$ so that $\alpha'=N_{m,l}(\alpha'')$, $\alpha=N_{n,l}(\alpha'')$. By Lemma 9,

$$[\alpha, \beta]_n = \langle \alpha'', \beta\rangle_l,$$

and by Lemma 7,

$$\langle \alpha', \beta\rangle_m \equiv \langle \alpha'', \beta\rangle_l \mod q_n.$$

Hence

$$[\alpha, \beta]_n = \langle \alpha', \beta\rangle_m.$$

Let $\alpha=\zeta_n$. Then $N_{n,m}^{-1}(\alpha)=\zeta_m$. Since $\delta_m(\zeta_m)=\zeta_m\zeta_m^{-1}(d\xi_m/d\pi_m)=-1$, the theorem gives us the first formula of Artin-Hasse:

$$[\zeta_n, \beta]_n = q_n^{-1}T_n(\log\beta), \qquad \beta \in 1+\mathfrak{p}_n.$$

Next, let $\alpha=\pi_n$. Using the equalities for π_n and π_m given in the proof of Lemma 5, we see immediately that Theorem 1 in this case gives us the second formula of Artin-Hasse:

$$[\pi_n, \beta]_n = -q_n^{-1}T_n(\zeta_n\pi_n^{-1}\log\beta), \qquad \beta \in 1+\mathfrak{p}_n.$$

3. In Theorem 1, if $\beta-1$ is divisible by a higher power of $\mathfrak{p}_n$, then the formula $[\alpha, \beta]_n = \langle N_{n,m}^{-1}(\alpha), \beta\rangle_m$ holds with a smaller m. The proof of the theorem shows that

$$[\alpha, \beta]_n = \langle N_{n,n+1}^{-1}(\alpha), \beta\rangle_{n+1} \qquad \alpha \in \Phi_n',\ \beta \in 1+\mathfrak{p}_n^{p^n},$$

$$[\alpha, \beta]_n = \langle \alpha, \beta\rangle_n, \qquad \alpha \in \Phi_n',\ \beta \in 1+\mathfrak{p}_n^{2p^n},$$

We shall next prove that the second equality holds not only for α in Φ_n' but also for arbitrary α in Φ_n^*. Since $\Phi_n^* = U_0 \times \Phi_n'$, it is sufficient to prove it for α in $U_0 = 1+p\boldsymbol{Z}_p$.

For integers $a \geqq 1$, $n \geqq 0$, let

$$s_n(a) = p^{-n} \sum_{\substack{0 \leqq i \leqq a \\ p^n | i}} (-1)^i i \binom{a}{i} = \sum_{0 \leqq kp^n \leqq a} (-1)^k k \binom{a}{kp^n}.$$

Using

$$\binom{a}{kp^n} \equiv \binom{ap}{kp^{n+1}} \mod p^{8)},$$

one sees immediately that

$$s_n(a) \equiv s_{n+1}(ap) \quad \mod p.$$

Let $a>0$, $(a, p)=1$, $n>0$. Then $kp^n \leqq a$ implies $kp^n \leqq a-1$. Hence

$$s_n(a) = \sum_{0 \leqq kp^n \leqq a-1} (-1)^k \frac{a}{a-kp^n} k\binom{a-1}{kp^n}$$

$$= \sum_{0 \leqq kp^n \leqq a-1} (-1)^k k\binom{a-1}{kp^n} \quad \mod p,$$

namely,

$$s_n(a) \equiv s_n(a-1) \quad \mod p.$$

LEMMA 10. *Let $a \geqq 2p^n$, $n \geqq 0$. Then*

$$s_n(a) \equiv 0 \quad \mod p.$$

PROOF. We use induction on n. For $n=0$,

$$s_0(a) = \sum_{k=0}^{a} (-1)^k k\binom{a}{k} = \sum_{k=1}^{a} (-1)^k a\binom{a-1}{k-1}$$

$$= -a(1-1)^{a-1} = 0.$$

Let $n>0$ and $a=bp+c$, $b \geqq 0$, $p>c \geqq 0$. Then we have $bp \geqq 2p^n$, and we see from the above remarks and from the induction assumption on $s_{n-1}(b)$, $b \geqq 2p^{n-1}$ that

$$s_n(a) \equiv s_n(bp) \equiv s_{n-1}(b) \equiv 0 \quad \mod p.$$

LEMMA 11. *Let $a \geqq 2p^n$, $n \geqq 0$. Then*

$$T_n(a\zeta_n\pi_n^{a-1}) \equiv 0 \quad \mod p^{2n+1};$$

PROOF. Since

$$T_n(\zeta_n^i) = \begin{cases} 0, & p^n \nmid i, \\ -p^n, & p^n \mid i, \quad p^{n+1} \nmid i, \\ p^{n+1}-p^n, & p^{n+1} \mid i, \end{cases}$$

and since

$$a\zeta_n\pi_n^{a-1} = \sum_{i=0}^{a-1} (-1)^i a\binom{a-1}{i}\zeta_n^{i+1} = -\sum_{i=0}^{a} (-1)^i i\binom{a}{i}\zeta_n^i,$$

we obtain by Lemma 10

8) See [1], Hilfssatz 2.

$$T_n(a\zeta_n\pi_n^{a-1}) = p^n \sum_{\substack{0\leq i\leq a\\ p^n|i}} (-1)^i i\binom{a}{i} - p^{n+1} \sum_{\substack{0\leq i\leq a\\ p^{n+1}|i}} (-1)^i i\binom{a}{i}$$

$$\equiv p^{2n} s_n(a) \qquad \bmod p^{2n+1}$$

$$\equiv 0 \qquad \bmod p^{2n+1}.$$

LEMMA 12. *For any* ξ *in* $1+\mathfrak{p}_n^{2p^n}$,

$$T_n(\delta_n(\xi)) \equiv 0 \qquad \bmod p^{2n+1}.$$

PROOF. Note first that since $T_n(\mathfrak{d}_n) \equiv 0 \bmod p^{2n+1}$, $T_n(\delta_n(\xi))$ is well determind mod p^{2n+1} (for any ξ in Φ_n^*). For the proof of the lemma, it suffices to show that for each $a \geqq 2p^n$, there exists an element ξ_a such that $\nu_n(\xi_a - 1) = a$ and $T_n(\delta_n(\xi_a)) \equiv 0 \bmod p^{2n+1}$. Let $b = ap-(p-1)p^n$ and let

$$\alpha = 1-\pi_n^{ap}, \qquad \beta = \exp(p^{-1}\log\alpha).$$

We see from $ap > p^{n+1}$ that β is defined, $\alpha = \beta^p$, and $\nu_n(\beta-1) \geqq b$. Let $f(T)$ be a power series for $\beta-1$ and let

$$(1+f(T))^p = 1+g(T).$$

Then the coefficient of T^i vanishes for $0 \leqq i < b$ in both $f(T)$ and $g(T)$. Since $g(\pi_n) = -\pi_n^{ap}$, $ap \geqq b$, we obtain from the proof of Lemma 3 that

$$g'(\pi_n) = p(1+f(\pi_n))^{p-1} f'(\pi_n) \equiv -ap\pi_n^{ap-1} \qquad \bmod \mathfrak{p}_n^b \mathfrak{d}_n.$$

It then follows from $b \geqq 2p^{n+1}-(p-1)p^n > (p-1)p^n = \nu_n(p)$ that

$$\frac{f'(\pi_n)}{1+f(\pi_n)} \equiv -\frac{a\pi^{ap-1}}{1-\pi_n^{ap}} \qquad \bmod \mathfrak{d}_n$$

so that

$$\delta_n(\beta) \equiv -\sum_{i=1}^{\infty} a\zeta_n \pi_n^{api-1} \qquad \bmod \mathfrak{d}_n.$$

Let

$$\xi_a = (1-\pi_n^a)\beta^{-1}.$$

Then $\nu_n(\beta-1) \geqq b > a$ implies $\nu_n(\xi_a - 1) = a$. We also see from the above that

$$\delta_n(\xi_a) \equiv \delta_n(1-\pi_n^a) - \delta_n(\beta)$$

$$\equiv -\sum_{i=1}^{\infty} a\zeta_n\pi_n^{ai-1} + \sum_{i=1}^{\infty} a\zeta_n\pi_n^{api-1}$$

$$\equiv -\sum_{\substack{i=1\\ (i,p)=1}}^{\infty} a\zeta_n\pi_n^{ai-1} \qquad \bmod \mathfrak{d}_n$$

so that

$$T_n(\delta_n(\xi_a)) \equiv -\sum_{\substack{i=1\\ (i,p)=1}}^{\infty} T_n(a\zeta_n\pi_n^{ai-1}) \qquad \bmod p^{2n+1}.$$

However, it follows from Lemma 11 that if $(i, p)=1$, then

$$T_n(a\zeta_n\pi_n^{ai-1})=i^{-1}T_n(ai\zeta_n\pi_n^{ai-1})\equiv 0 \quad \mod p^{2n+1}.$$

Hence $T_n(\delta_n(\xi_a))\equiv 0 \mod p^{2n+1}$, and the lemma is proved.

THEOREM 2. *For any α in Φ_n^* and any β in $1+\mathfrak{p}_n^{2p^n}$, $n\geqq 0$,*

$$[\alpha, \beta]_n=\langle\alpha, \beta\rangle_n,$$

namely

$$(\alpha, \beta)_n=\zeta_n^{-q_n^{-1}T_n(\zeta_n\alpha^{-1}(d\alpha/d\pi_n))\log\beta}.$$

PROOF. As noted earlier, we need only to prove the theorem for α in U_0. Since $\delta_n(\alpha)=0$ in this case, we have to show

$$(\alpha, \beta)_n=1, \qquad \alpha\in U_0,\ \beta\in 1+\mathfrak{p}_n^{2p^n}.$$

Let

$$\beta=\xi\eta, \qquad \xi\in U_0,\ \eta\in\Phi_n'.$$

Then $(\alpha, \xi)_n=1$ because both α and ξ are powers of $1+p$ with exponents in Z_p and $(1+p, 1+p)_n=1$. Since η is in Φ_n' and α in $1+\mathfrak{p}_n^{2p^n}$, we know that

$$[\eta, \alpha]_n=\langle\eta, \alpha\rangle_n=-q_n^{-1}T_n(\delta_n(\eta)\log\alpha)$$
$$=-q_n^{-1}\log\alpha T_n(\delta_n(\eta))$$

with $\log\alpha$ in pZ_p. On the other hand, it follows from $\beta\equiv 1 \mod \mathfrak{p}_n^{2p^n}$, $\xi\equiv 1 \mod p$ that $\eta\equiv 1 \mod \mathfrak{p}_n^{2p^n}$. Hence $T_n(\delta_n(\eta))\equiv 0 \mod p^{2n+1}$ by Lemma 12. Therefore $[\eta, \alpha]_n\equiv 0 \mod q_n$, $(\eta, \alpha)_n=1$, and

$$(\alpha, \beta)_n=(\alpha, \xi)_n(\alpha, \eta)_n=(\alpha, \xi)_n(\eta, \alpha)_n^{-1}=1.$$

Hence the theorem is proved.

Obviously Theorem 2 is a generalization of the third formula stated in the introduction. We also note that as the proof of the theorem indicates, Lemma 12 is equivalent with the fact that the conductor of the abelian extension $\Phi_n((1+p)^{q_n^{-1}})/\Phi_n$ is a divisor of $\mathfrak{p}_n^{2p^n}$.

4. For $n\geqq 0$, let $\mathfrak{X}_n$ denote the set of all elements ξ in Φ_n such that

$$T_n(\xi\log(1+\mathfrak{p}_n))\equiv 0 \qquad \mod Z_p.$$

$\mathfrak{X}_n$ is a compact subgroup of the additive group of Φ_n. As we saw in [6], the compact modules $\mathfrak{X}_n$, $n\geqq 0$, play important roles in the theory of cyclotomic fields. In Proposition 14 of that paper, we proved that for each $n\geqq 0$, there exists a unique map

$$\psi_n: \Phi_n'\to\mathfrak{X}_n/q_n\mathfrak{X}_n$$

such that

$$(\alpha, \beta)_n=\zeta_n^{T_n(\psi_n(\alpha)\log\beta)},$$

for any α in Φ'_n and β in $1+\mathfrak{p}_n$. It was also proved that ψ_n is a surjective κ-homomorphism.

THEOREM 3. *For* $m \geqq 2n+1$,

$$\psi_n(\alpha) = -q_m^{-1}T_{n,m}(\delta_m(N_{n,m}^{-1}(\alpha))), \qquad \alpha \in \Phi'_n.$$

PROOF. Let $\psi'_n(\alpha)$ denote the right hand side of the above. By Theorem 1,

$$[\alpha, \beta]_n = T_n(\psi'_n(\alpha) \log \beta), \qquad \beta \in 1+\mathfrak{p}_n.$$

This shows that $\psi'_n(\alpha)$ is contained in $\mathfrak{X}_n$ and is well determined mod $q_n\mathfrak{X}_n$ for any choice of the value of $\delta_m(N_{n,m}^{-1}(\alpha))$. Hence we abtain a map

$$\psi' : \Phi'_n \to \mathfrak{X}_n / q_n\mathfrak{X}_n$$

such that

$$(\alpha, \beta)_n = \zeta_n^{T_n(\psi'_n(\alpha) \log \beta)}, \qquad \alpha \in \Phi'_n, \ \beta \in 1+\mathfrak{p}_n.$$

The uniqueness of ψ_n then implies $\psi'_n = \psi_n$.

For $m \geqq n \geqq 0$, let $\mathfrak{X}_{n,m}$ denote the set of all elements of the form

$$q_m^{-1}T_{n,m}(\delta_m(\xi))$$

with ξ ranging over Φ^*_m and with $\delta_m(\xi)$ taking all possible values of the multi-valued $\delta_m(\xi)$. $\mathfrak{X}_{n,m}$ is a subgroup of the additive group of Φ_n, containing the open subgroup $q_m^{-1}T_{n,m}(\mathfrak{d}_m) = q_n^{-1}\mathfrak{d}_n$. Since $\Phi^*_m = U_0 \times \Phi'_m$ and $\delta_m(U_0) \equiv 0 \bmod \mathfrak{d}_m$, we have

$$\mathfrak{X}_{n,m} = q_m^{-1}T_{n,m}(\delta_m(\Phi'_m)) = q_m^{-1}T_{n,m}(\delta_m(N_{n,m}^{-1}(\Phi'_n))).$$

Let $m \geqq 2n+1$. Since $\psi_n : \Phi'_n \to \mathfrak{X}_n/q_n\mathfrak{X}_n$ is surjective, it follows from Theorem 3 that

$$\mathfrak{X}_n = \mathfrak{X}_{n,m} + q_n\mathfrak{X}_n.$$

However, $\mathfrak{X}_n$ is a profinite p-group and $\mathfrak{X}_{n,m}$ is an open subgroup of $\mathfrak{X}_n$. Hence we obtain from the above that $\mathfrak{X}_n = \mathfrak{X}_{n,m}$ for $m \geqq 2n+1$.

THEOREM 4. *For any* $m \geqq n$,

$$\mathfrak{X}_n = \mathfrak{X}_{n,m}.$$

In particular, $\mathfrak{X}_n = \mathfrak{X}_{n,n}$, *and* $\mathfrak{X}_n$ *consists of all elements of the form*

$$q_n^{-1}\delta_n(\alpha) = q_n^{-1}\zeta_n\alpha^{-1}(d\alpha/d\pi_n), \qquad \alpha \in \Phi^*_n.$$

PROOF. It is sufficient to show that $\mathfrak{X}_{n,m} = \mathfrak{X}_{n,n}$ for $m \geqq n$. By Lemma 5,

$$q_m^{-1}T_{n,m}(\delta_m(\xi)) \equiv q_n^{-1}\delta_n(N_{n,m}(\xi)) \qquad \bmod q_n^{-1}\mathfrak{d}_n$$

for any ξ in Φ^*_m. Since $q_n^{-1}\mathfrak{d}_n$ is contained in $\mathfrak{X}_{n,n}$, we see that $\mathfrak{X}_{n,m}$ is a subgroup of $\mathfrak{X}_{n,n}$.

Let ξ be a unit of $\mathfrak{o}_n$ and let $\delta_n(\xi)$ be fixed. By Lemma 4, iii), one value of $\delta_m(\xi)$ is given by $p^{m-n}\delta_n(\xi)$. Hence $q_n^{-1}\delta_n(\xi) = q_m^{-1}T_{n,m}(\delta_m(\xi))$ is contained in

$\mathfrak{X}_{n,m}$. On the other hand, since

$$q_n^{-1}\delta_n(\pi_n) = q_n^{-1}\zeta_n\pi_n^{-1} = q_m^{-1}T_{n,m}(\delta_m(\pi_m)),$$

$q_n^{-1}\delta_n(\pi_n)$ is also contained in $\mathfrak{X}_{n,m}$. Therefore $\mathfrak{X}_{n,n}$ is a subgroup of $\mathfrak{X}_{n,m}$.

In [6], the compact module $\mathfrak{X}_n$ was explicitly described as follows[9]: Let G_n denote the Galois group of Φ_n/Q_p and let

$$\mu_n = q_n^{-1}\zeta_n\pi_n^{-1}$$

$$\theta_n = q_n^{-1}\sum_{i=0}^{n}\zeta_i^{-1}.$$

Then the elements θ_n^σ, $\sigma \in G_n$, form a normal basis of Φ_n over Q_p, and $\mathfrak{X}_n$ consists of all elements of the form

$$\sum_\sigma c_\sigma \mu_n^\sigma + \sum_\sigma d_\sigma \theta_n^\sigma$$

with arbitrary p-adic integers c_σ and d_σ satisfying $\sum_\sigma d_\sigma = 0$.

We like to note here that a similar description of the compact module $\log(1+\mathfrak{p}_n)$ can be given as follows: Let

$$\lambda_n = p^{-n}\sum_{i=0}^{n} p^i\pi_i.$$

Then the elements λ_n^σ, $\sigma \in G_n$, also form a normal basis of Φ_n over Q_p, and $\log(1+\mathfrak{p}_n)$ consists of all elements of the form

$$ap + \sum_\sigma a_\sigma \lambda_n^\sigma$$

with arbitrary p-adic integers a and a_σ satisfying

$$\sum_\sigma \kappa(\sigma)a_\sigma \equiv 0 \qquad \mathrm{mod}\ q_n\text{[10]}.$$

The proof is obtained easily from the result on $\mathfrak{X}_n$ stated above and from the fact that if we put

$$\theta_n' = p^{-n}(-1 + \sum_{i=0}^{n} p^i\zeta_i),$$

then we have

$$T_n(\theta_n^\sigma \theta_n'^\tau) = \begin{cases} 1, & \sigma = \tau, \\ 0, & \sigma \neq \tau,\ \sigma, \tau \in G_n. \end{cases}$$

Princeton University

---

9) See [6], §1, Theorem 1.

10) For σ in G_n, $\kappa(\sigma)$ is defined in the obvious manner. It is a p-adic integer well determined mod q_n.

Bibliography

[1] E. Artin and H. Hasse, Die beiden Ergänzungssätze zum Reziprozitätsgesetz der l^n-ten Potenzrests im Körper der l^n-ten Einheitswurzeln, Abh. Math. Sem. Univ. Hamburg, 6 (1928), 146–162.

[2] E. Artin and J. Tate, Class field theory, Lecture notes at Princeton University, 1951–1952.

[3] H. Hasse, Zum expliziten Reziprozitätsgesetz, Abh. Math. Sem. Univ. Hamburg, 7 (1930), 52–63.

[4] H. Hasse, Bericht über die neuere Untersuchungen und Probleme aus der Theorie der algebraischen Zahlkörper, I, Ia, II, Leipzig und Berlin, 1930.

[5] H. Hasse, Zum expliziten Reziprozitätsgesetz, Arch. Math., 13 (1962), 479–485.

[6] K. Iwasawa, On some modules in the theory of cyclotomic fields, J. Math. Soc. Japan, 16 (1964), 42–82.

[7] K. Iwasawa, Some results in the theory of cyclotomic fields, Proc. Symposia in Pure Math., Amer. Math. Soc., 8 (1965), 66–69.

[47] Analogies between number fields and function fields

"Some Recent Advances in the Basic Sciences, Proc. Annual Sci. Conf." (New York, 1965-1966), Vol. 2, Belfer Grad. School Sci., Yeshiva Univ., New York, 1969, pp. 203-208.

MASSACHUSETTS INSTITUTE OF TECHNOLOGY

As is well-known, there exist close analogies between the fields of algebraic numbers and the fields of algebraic functions (of one variable), and many important advancements have been made in the theories of these fields by pursuing such analogies. In recent years, the theory of function fields has been studied with great success by the method of algebraic geometry. In the present paper, we shall consider some such results on function fields and examine what their analogues might be for number fields.

Let $\mathscr{C}$ be a complete, nonsingular, algebraic curve of genus g, defined over an algebraically closed field k. Let J be the Jacobian variety of the curve $\mathscr{C}$; J is a g-dimensional abelian variety so that it has the structure of an abelian group as well as that of an algebraic variety. Let p be a prime number different from the characteristic of k and let $W_p = \mathbf{Q}_p/\mathbf{Z}_p$ where $\mathbf{Q}_p$ and $\mathbf{Z}_p$ denote the additive group of p-adic numbers and that of p-adic integers respectively. Let J_p be the subgroup of all elements in J whose orders are powers of p. Then J_p has the following simple group-theoretical structure:

$$J_p \cong W_p^{2g} = W_p \oplus \cdots \oplus W_p \tag{1}$$

with $2g$ copies of W_p. It follows that the ring of endomorphisms of J_p is isomorphic to the ring of $2g \times 2g$ matrices over the ring of p-adic integers, $\mathbf{Z}_p$. Let τ be an algebraic correspondence on the algebraic curve $\mathscr{C}$. Then τ defines an endomorphism of the abelian group J, and hence also an endomorphism of the characteristic subgroup J_p of J. Therefore if an isomorphism (1) is fixed, τ is associated with a $2g \times 2g$ matrix

203

$M(\tau)$ over $\mathbf{Z}_p$. The mapping $\tau \to M(\tau)$ then induces a faithful representation of the ring of complex multiplications on $\mathscr{C}$ in the ring of $2g \times 2g$ matrices over $\mathbf{Z}_p$.

Now suppose that k has a positive characteristic and that $\mathscr{C}$ is also defined over a finite subfield k_0 of k. The Frobenius automorphism of k/k_0 then defines an algebraic correspondence φ on $\mathscr{C}$. Let $f(X)$ denote the characteristic polynomial of the $2g \times 2g$ matrix $M(\varphi)^{-1}$. Then $f(X)$ is equal to the quotient of the zeta function of the algebraic curve $\mathscr{C}$ defined over k_0 and the zeta function of an algebraic curve of genus 0 defined over k_0.

These are some of the profound results obtained by A. Weil in his solution of the Riemann hypothesis and of Artin's conjecture on L-series for the function fields with finite fields of constants. It is also noted that if p is equal to the characteristic of k, we still have an isomorphism of the form (1) where the number of W_p on the right-hand side is at most equal to g.

In view of the fact as stated above, it seems quite interesting to find some analogue of J_p (or J) in the theory of number fields. Let F be a finite algebraic number field, i.e., a finite extension of the rational field $\mathbf{Q}$. Let K be the field of rational functions on $\mathscr{C}$ defined over k; it is a field of algebraic functions of one variable over the field of constants k. Algebraically speaking, J_p is the group of all divisor classes (of degree 0) in K with orders powers of p. Therefore under the well-known analogy between number fields and function fields, the analogue of J_p for F would be the Sylow p-subgroup S_p of the ideal-class group of F. However, the abelian group S_p appears to be quite different from the abelian group J_p; not only S_p is finite while J_p is in general infinite, but also the structure of S_p depends very much upon the given field F, and we do not know any simple group-theoretical result on S_p such as (1) for J_p. One of the reasons which account for the lack of similarity between J_p and S_p may be as follows. In the above, the field k of constants in the function field K is algebraically closed. Suppose, instead, that we take a function field K' over an arbitrary field of constants k'. Then we do not always have an isomorphism of the form (1) for the divisor-class group J'_p for K'. More precisely, an isomorphism such as (1) exists for J'_p only when the field k' is sufficiently large. For example, if k' is a finite field, then J'_p is a finite group and its structure depends upon K' as much as S_p upon F. Now, under the correspondence between number fields and function fields, the constants of a function field correspond to the roots of unity in a number field. Therefore the above fact on J'_p seems to suggest that the irregularity of the structure of S_p is caused by the lack of sufficiently many roots of unity in the finite algebraic number field F, which in fact contains only a finite number of them.

We are thus led to consider algebraic number fields whose degrees over the rational field $\mathbf{Q}$ are not necessarily finite. Let F now denote such a number field. We shall first generalize the definition of the group S_p for such F. Let $\{E_i\}$ be the family of all finite extensions of $\mathbf{Q}$ contained in F, and let $\bar{I}_i$ denote the ideal-class group of E_i. If E_i is a subfield of a field E_j in the same family, then there is a natural homomorphism $\bar{I}_i \to \bar{I}_j$. Let $\bar{I}$ be the direct limit of the family of groups $\bar{I}_i$ relative to the above homomorphisms $\bar{I}_i \to \bar{I}_j$. We may call $\bar{I}$ the ideal-class group of F. It is a periodic abelian group, and we denote the Sylow p-subgroup (the p-primary component) of $\bar{I}$ by S_p. We now ask: What is the structure of S_p? Is there any similarity between S_p and J_p? Of course, for general F, we do not know any simple result on the structure of S_p because we have no such result even for a finite algebraic number field. However, for a certain type of infinite algebraic number fields, we can obtain an isomorphism for S_p similar to the isomorphism (1) for J_p.[1]

Let E be a finite algebraic number field. With a prime number p fixed, let F be the number field obtained by adjoining to E all p^nth roots of unity, $n = 1, 2, 3, \ldots$. For the group S_p of such a field F, we can obtain the following result. For each integer $\mu \geqslant 0$, let $C(\mu)$ denote the direct sum of countably infinite copies of a cyclic group of order p^μ, and let

$$C(\mu_1, \ldots, \mu_s) = C(\mu_1) \oplus \cdots \oplus C(\mu_s)$$

for any integers $\mu_1, \ldots, \mu_s \geqslant 0$. Then there exist subgroups B, C of S_p and integers $\lambda, \mu_1, \ldots, \mu_s \geqslant 0$ $(s \geqslant 0)$ such that

(i) both $S_p/(B + C)$ and $B \cap C$ are finite groups,
(ii) $B \cong W_p^\lambda = W_p \oplus \cdots \oplus W_p$ with λ copies of W_p,
(iii) $C = C(\mu_1, \ldots, \mu_s)/D$ with a finite subgroup D.

In short, S_p is up to finite groups the direct sum of W_p^λ and $C(\mu_1, \ldots, \mu_s)$ for some $\lambda, \mu_1, \ldots, \mu_s \geqslant 0$. Since the orders of elements in C are bounded, it follows from the above that

$$p^n S_p = B \cong W_p^\lambda \qquad (2)$$

for every sufficiently large $n \geqslant 0$. Thus we see some similarities between the structure of the group J_p and that of S_p. In particular, it follows from (2) that the ring of endomorphisms of $B = p^n S_p$ is the ring of all $\lambda \times \lambda$ matrices over $\mathbf{Z}_p$. Let σ be an automorphism of the field F. Then σ defines an automorphism of S_p in the obvious manner, and hence it also induces an automorphism of B. Therefore, with an isomorphism (2)

[1] See K. Iwasawa, "On Γ-extensions of algebraic number fields," *Bull. Amer. Math. Soc.* 65, 183–226 (1959); J.-P. Serre, "Classes des corps cyclotomiques," *Seminaire Bourbaki*, Expose 174 (1958–1959).

fixed, each automorphism σ of F is associated with a $\lambda \times \lambda$ matrix $M(\sigma)$ over the ring $\mathbf{Z}_p$ just as each algebraic correspondence τ of the algebraic curve $\mathscr{C}$ is associated with a $2g \times 2g$ matrix $M(\tau)$ over $\mathbf{Z}_p$.

Let F_0 be the subfield of F generated over E either by all pth roots of unity or by all fourth roots of unity according as $p > 2$ or $p = 2$. Then F is a Galois extension of F_0 and its Galois group Γ is topologically isomorphic to the compact abelian group $\mathbf{Z}_p$. Obviously Γ acts on the group S_p so that the latter may be considered as a Γ-module. Let Λ denote the ring of formal power series in an indeterminate T with coefficients in $\mathbf{Z}_p$ and let γ be a fixed element of Γ which generates a cyclic subgroup everywhere dense in Γ. Let X be the dual of the discrete abelian group S_p. Then X is a compact Γ-module, and we can make it into a Λ-module in such a way that $\gamma x = (1 + T)x$ for any x in X. We then see that X is finitely generated over Λ and that it is, up to finite Λ-modules, the direct sum of a finite number of Λ-modules of the form Λ/N where N is an ideal of the ring Λ. The above-mentioned fact on the structure of S_p is deduced from this result on the Λ-module X. Hence the groups B, C, $C(\mu_1 \ldots \mu_s)$, and D are actually all Γ-modules, and the isomorphism in (iii) is of course a Γ-isomorphism. It also follows that the integers $\lambda, \mu_1, \ldots, \mu_s$ $(\mu_i > 0)$, and $\mu = \mu_1 + \cdots + \mu_s$ are invariants of the Γ-module S_p.

Now, for each integer $n \geqslant 0$, let F_n denote the unique extension of F_0 in F such that $[F_n : F_0] = p^n$. Let p^{e_n} be the order of the Sylow p-subgroup of the ideal-class group of F_n, namely, the highest power of p dividing the class number of F_n. Then we can prove that for every sufficiently large n,

$$e_n = \lambda n + \mu p^n + \nu \tag{3}$$

with λ and μ as defined above and with an integer ν also independent of n. This gives us an arithmetic interpretation of the numbers λ and μ. When a prime number p and a finite algebraic number field E are given, the integers λ and μ are uniquely determined either from the structure of S_p or from the above formula (3). Hence we put

$$\lambda = \lambda(E, p), \qquad \mu = \mu(E, p).$$

Seeking an analogue of J_p in the theory of number fields, we have come to consider the group S_p. Whether or not it is a good analogue of J_p, the above observation seems to indicate that the Γ-module S_p deserves further investigations in algebraic number theory. We may propose such problems as follows:

1) Study the arithmetic nature of the numerical invariants $\lambda(E, p)$ and $\mu(E, p)$. In particular, with fixed p, compare $\lambda(E, p)$ and $\mu(E, p)$ for

various E, and with fixed E, compare these invariants for various choice of p. Find when $\lambda(E, p) = 0$ or $\mu(E, p) = 0$. Note that $\mu(E, p) = 0$ means $C = 0$, namely, that S_p is isomorphic to W_p^{λ} up to a finite group, and this is the case where S_p gives us a very good analogue of J_p.

2) Study the action of Γ (or Λ) on B, namely, the representation of Γ by $\lambda \times \lambda$ matrices over $\mathbf{Z}_p$ provided by the Γ-module B. This may be considered as an analogue of the representation, defined by J_p, of the cyclic group generated by the Frobenius automorphism of k/k_0. More generally, if E is a Galois extension over a subfield E', then F is also a Galois extension of E'. Let G be the Galois group of F/E'. Study the representation of G defined by the G-module B.

Let $p > 2$ and let $E = \mathbf{Q}$ in the above. Then F_n $(n \geqslant 0)$ is the cyclotomic field of p^{n+1}th roots of unity over $\mathbf{Q}$, and F is the union of these fields F_n, $n \geqslant 0$. The study of the group S_p in this special case seems particularly interesting not only because the cyclotomic fields are important in algebraic number theory in general, but also because it may give us some information on S_p that would suggest what to expect in the general case. In this case, F is obviously a Galois extension over $\mathbf{Q}$ and its Galois group G is the direct product of Γ and a cyclic group Δ of order $p - 1$ which is canonically isomorphic to the Galois group of $F_0/\mathbf{Q}$. For each p-adic unit u, let σ_u denote the automorphism of F such that $\sigma_u(\zeta) = \zeta^u$ for every root of unity ζ in F with order a power of p. Then the mapping $u \to \sigma_u$ defines a topological isomorphism of the multiplicative group of p-adic unit onto the Galois group G, and it maps the group of 1-units in $\mathbf{Q}_p$ to the group Γ, and the group $\mathbf{V}$ of $(p - 1)$ roots of unity in $\mathbf{Q}_p$ to the group Δ. We may take as γ the automorphism σ_{1+p}. For each integer i, let A_i denote the subgroup of all a in $A = S_p$ such that $\sigma_v(a) = v^i a$ for every v in $\mathbf{V}$, and let X_i be the dual of A_i. Then A is the direct sum of the Γ-modules A_i, $0 \leqslant i < p - 1$, and X is the direct sum of the Λ-modules X_i, $0 \leqslant i < p - 1$. Thus our study of the structure of the Γ-module S_p is reduced to that of the Λ-modules X_i.

Now assume that $A_i = 0$ for every even index i; this is a well-known conjecture in the theory of cyclotomic fields and it is verified by Vandiver and others for all prime numbers $p \leqslant 4001$. Under that assumption, we can show that for each odd index i, X_i is a Λ-module of the form $\Lambda/(g_i)$ where (g_i) denotes the principal ideal of Λ generated by a power series $g_i(T)$.[2] By Weierstrass' preparation theorem, $g_i(T)$ can be written in the form

$$g_i(T) = p^{\mu_i} u_i(T) m_i(T),$$

[2] See K. Iwasawa, "On some modules in the theory of cyclotomic fields," *Jour. Math. Soc. Japan* **19**, 42–82 (1964).

where μ_i is an integer, $\mu_i \geqslant 0$, $u_i(T)$ is a unit of the ring Λ, and $m_i(T)$ is a polynomial in $\mathbf{Z}_p[T]$ with the highest coefficient 1. Let λ_i be the degree of $m_i(T)$. We see easily that $f_i(X) = m_i(X-1)$ is the characteristic polynomial of the matrix $M_i(\gamma)$ in the representation of Γ defined by a submodule $B_i \cong W_p^{\lambda_i}$ in A_i. Let

$$\lambda = \sum \lambda_i, \qquad \mu = \sum \mu_i,$$

and let $g(T)$, $u(T)$, and $m(T)$ denote the product of all $g_i(T)$, that of all $u_i(T)$, and that of all $m_i(T)$ respectively (i odd, $0 \leqslant i < p-1$). Then $\lambda = \lambda(\mathbf{Q}, p)$, $\mu = \mu(\mathbf{Q}, p)$, and

$$g(T) = p^{\mu} u(T) m(T).$$

Furthermore,

$$f(X) = m(X-1)$$

is the characteristic polynomial of the matrix $M(\gamma)$ in the representation of Γ defined by the submodule B of S_p.

The power series $g_i(T)$ in the above is determined by the Λ-module X_i only up to a factor which is a unit of Λ. However, for a suitable choice of $g_i(T)$, we have the following rather remarkable result:[3] Let i be odd, $i \neq 1$, and let χ_i be the character of integers mod p, with values in $\mathbf{Q}_p$, such that $\chi_i(a) \equiv a^{p-i} \bmod p$ for every integer a. Then

$$g_i((1+p)^{-s} - 1) = -L_p(s; \chi_i), \qquad s \in \mathbf{Z}_p,$$

where $L_p(s; \chi_i)$ denotes the p-adic L-function of Kubota–Leopoldt for the character χ_i. Let Φ be the local cyclotomic field of pth roots of unity over $\mathbf{Q}_p$, and let Ψ be the subfield of Φ with $[\Phi : \Psi] = 2$. Then it follows from the above that $g((1+p)^{-s} - 1)$ is, up to a factor ± 1, equal to the quotient of the p-adic zeta-function of Ψ and the p-adic zeta-function of $\mathbf{Q}_p$. Thus we find here a result which is parallel to that of Weil in the theory of function fields. At present we do not know how such a result obtained for this very special case can be further generalized in the case of the extension F over an arbitrary finite algebraic number field E. However, we feel sure that some deeper facts are still hidden from us in this connection and that they are not yet fully revealed even in the above special case of cyclotomic fields.

[3] The proof will be published elsewhere.

[48] On p-adic L-functions

Ann. of Math., (2) 89 (1969), 198-205.

Let p be a prime number and let $\mathbf{Z}_p$, $\mathbf{Q}_p$, and Ω_p denote the ring of p-adic integers, the field of p-adic numbers, and an algebraic closure of $\mathbf{Q}_p$, respectively. Let $L(s;\chi)$ be the classical L-function of Dirichlet for a character χ. It is known that the values of $L(s;\chi)$ for rational integers $s \leqq 0$ are algebraic numbers so that they may be considered as elements of the field Ω_p. Let $f(s)$ be a function which is defined and continuous for all $s \in \mathbf{Z}_p$, $s \neq 1$, and which takes values in Ω_p such that

$$f(s) = (1 - \chi(p)p^{-s})L(s;\chi)$$

for every integer $s \leqq 0$, $s \equiv 1 \bmod p-1$. Since these integers s are everywhere dense in $\mathbf{Z}_p$, such a function $f(s)$ is unique if it exists. The existence of such a function $f(s)$ was, in fact, proved by Kubota and Leopoldt [3], and they called it the p-adic L-function for the character χ; we shall denote it here by $L_p(s;\chi)$.

In the following, we construct the function $f(s)$ by a method different from that of Kubota-Leopoldt. We shall obtain $f(s)$ from a quotient of power series in an indeterminate T by substituting $(1+q_0)^s - 1$ for T, q_0 being a certain integer determined by the character χ. We shall also discuss some immediate consequences of our construction, including an application to the theory of cyclotomic fields.

1. Let k be a finite extension of $\mathbf{Q}_p$ in Ω_p and let $\mathfrak{o}$ be the maximal order in k. Let Λ denote the ring of power series in an indeterminate T with coefficients in $\mathfrak{o}$: $\Lambda = \mathfrak{o}[[T]]$. Λ is a regular local ring of dimension two and is also a compact topological ring with respect to the topology defined by the powers of the maximal ideal of Λ. Let $f(T)$ be a power series in Λ. Then $f(\alpha)$ is well defined for each α in Ω_p with $|\alpha| < 1$, $|\cdot|$ being a fixed absolute value in Ω_p. Using the preparation theorem of Weierstrass, we see that for $f(T) \not\equiv 0$, there exist only a finite number of $\alpha \in \Omega_p$, $|\alpha| < 1$, such that $f(\alpha) = 0$. Let L be the quotient field of Λ and let $g(T)$ be an element of L. It follows from the above that $g(\alpha)$ is well defined for all but a finite number of $\alpha \in \Omega_p$ with $|\alpha| < 1$. Furthermore, if $h(T)$ is another element of L, then $g(T) \equiv h(T)$ if and only if $g(\alpha) = h(\alpha)$ for infinitely many values of $\alpha \in \Omega_p$, $|\alpha| < 1$.

2. Let $\mathbf{Z}$ denote the ring of rational integers. Fix an integer $m_0 \geqq 1$,

$(m_0, p) = 1$. For $n \geqq 0$, let $q_n = m_0 p^{n+1}$ if $p > 2$ and $q_n = m_0 2^{n+2}$ if $p = 2$, and let $\sigma_n(a)$ denote the residue class of $a \in \mathbf{Z}$ mod q_n. Let G_n be the multiplicative group of the residue class ring $\mathbf{Z}/q_n\mathbf{Z}$; it consists of all $\sigma_n(a)$ with $a \in \mathbf{Z}$, $(a, q_0) = 1$. For $m \geqq n \geqq 0$, there exists a surjective homomorphism $G_m \to G_n$ mapping $\sigma_m(a)$ to $\sigma_n(a)$. Let Γ_n be the kernel of $G_n \to G_0$; it is a cyclic group of order p^n generated by $\sigma_n(1 + q_0)$. Let Δ_n be the subgroup of all $\sigma_n(a) \in G_n$ with a satisfying $a^{p-1} \equiv 1 \bmod p^{n+1} (p > 2)$ or $a \equiv \pm 1 \bmod 2^{n+2} (p = 2)$. Then $G_n = \Gamma_n \times \Delta_n$, and for $m \geqq n \geqq 0$, $G_m \to G_n$ maps Γ_m onto Γ_n and $\Delta_m \xrightarrow{\sim} \Delta_n$. Let G, Γ, and Δ denote the inverse limits of G_n, Γ_n, and Δ_n, $n \geqq 0$, respectively. Then $G = \Gamma \times \Delta$, $\Delta \cong \Delta_0$. For each $\sigma_n(a) \in G_n$, let $\gamma_n(a)$ be the projection of $\sigma_n(a)$ on Γ_n in the decomposition $G_n = \Gamma_n \times \Delta_n$, and let $\sigma(a)$ and $\gamma(a)$ denote the elements of G determined by $\sigma_n(a)$ and $\gamma_n(a)$, $n \geqq 0$, respectively. $\gamma(a)$ is obviously the projection of $\sigma(a)$ on Γ in the decomposition $G = \Gamma \times \Delta$. If $p > 2$, each p-adic unit x in $\mathbf{Z}_p$ can be uniquely written in the form $x = \omega(x)\langle x \rangle$ where $\omega(x)$ is a $(p-1)^{\text{st}}$ root of unity and $\langle x \rangle$ is in $1 + p\mathbf{Z}_p$. If $p = 2$, we have a similar decomposition for x with $\omega(x) = \pm 1$ and $\langle x \rangle$ in $1 + 4\mathbf{Z}_2$[1]. Let $a, b \in \mathbf{Z}$, $(a, q_0) = (b, q_0) = 1$. Then, for $p > 2$, $\gamma_n(a) = \gamma_n(b)$ holds if and only if $\langle a \rangle \equiv \langle b \rangle \bmod p^{n+1}$, and for $p = 2$, if and only if $\langle a \rangle \equiv \langle b \rangle \bmod 2^{n+2}$. Therefore $\gamma(a) = \gamma(b)$ if and only if $\langle a \rangle = \langle b \rangle$. It follows that there is a topological isomorphism of compact groups, $\Gamma \xrightarrow{\sim} 1 + p\mathbf{Z}_p$, $p > 2$, or $\Gamma \xrightarrow{\sim} 1 + 4\mathbf{Z}_2$, $p = 2$, mapping $\gamma(a)$ to $\langle a \rangle$ for every $a \in \mathbf{Z}$, $(a, q_0) = 1$.

For $n \geqq 0$, let $k[\Gamma_n]$ and $R_n = \mathfrak{o}[\Gamma_n]$ denote the group algebras of Γ_n over k and $\mathfrak{o}$ respectively. If $m \geqq n \geqq 0$, $\Gamma_m \to \Gamma_n$ defines homomorphisms of algebras $k[\Gamma_m] \to k[\Gamma_n]$ and $R_m \to R_n$ in the obvious manner. Let R denote the inverse limit of R_n, $n \geqq 0$. It is a compact commutative algebra over $\mathfrak{o}$. For each $n \geqq 0$, there is a unique isomorphism of $\mathfrak{o}$-algebras $R_n \xrightarrow{\sim} \Lambda/(1 - (1 + T)^{p^n})$ mapping $\gamma_n(1 + q_0)$ to the residue class of $1 + T$. Hence we also have a topological isomorphism of $\mathfrak{o}$-algebras $R \xrightarrow{\sim} \Lambda$ such that $\gamma(1 + q_0) \to 1 + T$. Let $\xi \in R$ and let $f(T)$ be the image of ξ under the above isomorphism $R \xrightarrow{\sim} \Lambda$. Let $\varphi: R \to \Omega_p$ be a continuous homomorphism of $\mathfrak{o}$-algebras, and let $\varphi(\gamma(1 + q_0)) = 1 + \alpha$, $\alpha \in \Omega_p$, $|\alpha| < 1$. Then we have $\varphi(\xi) = f(\alpha)$. For integers $s, n \geqq 0$, let $\varphi_{s,n}$ denote the unique homomorphism of $\mathfrak{o}$-algebras $R_n = \mathfrak{o}[\Gamma_n] \to \mathfrak{o}/q_n\mathfrak{o}$ such that $\gamma_n(a) \to \langle a \rangle^{-s} \bmod q_n\mathfrak{o}$. Then the limit of $\varphi_{s,n}$, for $n \to \infty$, defines a continuous homomorphism of $\mathfrak{o}$-algebras $\varphi_s: R \to \mathfrak{o}$ such that $\gamma(a) \to \langle a \rangle^{-s}$ for every $a \in \mathbf{Z}$, $(a, q_0) = 1$, and we have $\varphi_s(\xi) = f((1 + q_0)^{-s} - 1)$ for the ξ and $f(T)$ stated in the above.

---

[1] We follow the notation in [3].

3. In the following, we shall fix a Dirichlet character χ such that $\chi(-1)=-1$. Let N be the conductor of χ and let $N=m_0p^e$, $(m_0, p)=1$, $e \geqq 0$; by definition, χ is a primitive character of the multiplicative group of the residue class ring $\mathbf{Z}/N\mathbf{Z}$. We may suppose that the values of χ are contained in the algebraic closure Ω_p of $\mathbf{Q}_p$. We shall assume that the field k in the preceding section is large enough to contain all such values of χ.

For $n \geqq 0$, let q_n, G_n, Γ_n, etc. be defined as before with respect to the above integer m_0. We also define an element $\xi_n(\chi)$ of $k[\Gamma_n]$ by

$$\xi_n(\chi) = -\frac{1}{q_n}\sum_a a\chi(a)^{-1}\gamma_n(a)^{-1},$$

where a ranges over all integers such that $0 \leqq a < q_n$, $(a, q_0)=1$. Using the assumption $\chi(-1)=-1$, we see that $\xi_m(\chi)\to\xi_n(\chi)$ under the homomorphism $k[\Gamma_m]\to k[\Gamma_n]$ for $m \geqq n \geqq e$.[2] Now, fix an integer $c \neq \pm 1$ with $(c, q_0)=1$ and let

$$\eta_n = \eta_n(\chi, c) = (1 - c\chi(c)^{-1}\gamma_n(c)^{-1})\xi_n(\chi), \qquad n \geqq 0.$$

For each $a \in \mathbf{Z}$, let $a_n = a_n(c)$ denote the unique integer such that $a_n \equiv ac \bmod q_n$, $0 \leqq a_n < q_n$, and let

$$a_n = ac + r_n(a)q_n, \qquad r_n(a)\in\mathbf{Z}.$$

Then $\chi(a_n)=\chi(ac)$, $\gamma_n(a_n)=\gamma_n(ac)$ for $n \geqq e$, and if a ranges over all integers such that $0 \leqq a < q_n$, $(a, q_0)=1$, then so does a_n. Therefore we obtain

(1) $$\eta_n = -\chi(c)^{-1}\gamma_n(c)^{-1}\sum_a r_n(a)\chi(a)^{-1}\gamma_n(a)^{-1},$$

and we see that η_n is contained in $R_n = \mathfrak{o}[\Gamma_n]$ for every $n \geqq e$. Furthermore $\eta_m \to \eta_n$ under $R_m \to R_n$, $m \geqq n \geqq e$. Therefore those η_n, $n \geqq 0$, determine an element $\eta(\chi, c)$ in R. Let $f(T;\chi, c)$ denote the corresponding power series in Λ.

Let $s \in \mathbf{Z}$, $s \geqq 0$. It follows from (1) that

$$\begin{aligned}\varphi_{s,n}(\eta_n) &\equiv -\chi(c)^{-1}\langle c\rangle^s\sum_a r_n(a)\chi(a)^{-1}\langle a\rangle^s\\ &\equiv -\chi_1(c)c^s\sum_a r_n(a)\chi_1(a)a^s \qquad \bmod q_n\mathfrak{o},\end{aligned}$$

where χ_1 denotes the Dirichlet character defined by $\chi_1=\chi^{-1}\omega^{-s}$.[3] Multiplying $\chi_1(a)$ to both sides of the congruence

$$a_n^{s+1} \equiv a^{s+1}c^{s+1} + (s+1)a^sc^sr_n(a)q_n \qquad \bmod q_n^2,$$

---

[2] Here, and on similar occasions below, it is actually sufficient to assume that $q_n \geqq N$, namely, that $n \geqq e-1$ $(p>2)$ or $n \geqq e-2$ $(p=2)$.

[3] Note that ω may be regarded as a Dirichlet character with conductor p $(p>2)$ or 4 $(p=2)$.

and taking the sum over a, we obtain

$$\varphi_{s,n}(\eta_n) \equiv -\frac{1-\chi_1(c)c^{s+1}}{s+1}\frac{1}{q_n}\sum_a \chi_1(a)a^{s+1} \quad \bmod \frac{q_n}{s+1}\mathfrak{o}\,.$$

However, we know that for $n \to \infty$,

$$\frac{1}{q_n}\sum_a \chi_1(a)a^{s+1} \longrightarrow (1-\chi_1(p)p^s)B_{\chi_1}^{s+1}$$

with the generalized Bernoulli number $B_{\chi_1}^{s+1}$ introduced by Leopoldt[4]. Hence we obtain

$$(2) \qquad \varphi_s(\eta(\chi, c)) = -(1-\chi_1(c)c^{s+1})\frac{(1-\chi_1(p)p^s)}{s+1}B_{\chi_1}^{s+1}\,.$$

Let $u(T) = u(T;\chi, c)$ denote the power series in Λ which corresponds to $1 - c\chi(c)^{-1}\gamma(c)^{-1}$ in R. If we put $\langle c\rangle = (1+q_0)^d$ with $d \in \mathbf{Z}_p$, then

$$u(T;\chi, c) = 1 - c\chi(c)^{-1}(1+T)^{-d}\,.$$

Since $c \in \mathbf{Z}$, $c \neq \pm 1$, we see that $\langle c\rangle \neq 1$, $d \neq 0$, and $u(T) \not\equiv 0$. Hence we may define an element g of L by

$$g(T;\chi, c) = \frac{f(T;\chi, c)}{u(T, \chi, c)}\,.$$

In general, for each Dirichlet character χ, let χ^* denote the character such that $\chi\chi^* = \omega$. Then, for the above χ, $u((1+q_0)^{-s}-1;\chi, c) = 1-\chi^*(c)\langle c\rangle^{s+1}$ for every $s \in \mathbf{Z}_p$, $|s| \leqq 1$, and it vanishes only when $s=-1$, $\chi^*(c)=1$ because $\chi^*(c)^n = 1$ for some $n \geqq 1$. Since $\chi^*(c)\langle c\rangle^{s+1} = \chi_1(c)c^{s+1}$, we see from (2), by substituting $-s$ for s, that

$$\begin{aligned} g((1+q_0)^s - 1;\chi, c) &= -\frac{1-\chi_1(p)p^{-s}}{1-s}B_{\chi_1}^{1-s} \\ &= (1-\chi_1(p)p^{-s})L(s;\chi_1) \end{aligned}$$

with $\chi_1 = \chi^{-1}\omega^s$, for every $s \in \mathbf{Z}$, $s \leqq 0$.[5] In particular, if $s \equiv 1 \bmod p-1$, then $\chi_1 = \chi^{-1}\omega = \chi^*$ and

$$g((1+q_0)^s - 1;\chi, c) = (1-\chi^*(p)p^{-s})L(s;\chi^*)\,.$$

As the right hand side of the above is independent of the integer c, it follows from the remark in § 1 that $g(T;\chi, c)$ is also independent of c. Hence we shall denote it by $g(T;\chi)$. Since $u((1+q_0)^s - 1;\chi, c) \neq 0$ for $s \neq 1$, we also see from the above that $g((1+q_0)^s - 1;\chi)$ has all the properties which uniquely characterize the p-adic L-function $L_p(s;\chi^*)$. Thus we obtain the formula

---

[4] See [4] and [3, p. 336].

[5] See [4, p. 134].

$$(3) \qquad g((1+q_0)^s - 1; \chi) = L_p(s; \chi^*), \qquad s \in \mathbf{Z}_p,\ s \neq 1,$$

with $\chi\chi^* = \omega$.

Since $\chi^{**} = \chi$, $\chi(-1)\chi^*(-1) = -1$, we now see that each p-adic L-function $L_p(s; \chi)$ with $\chi(-1) = 1$ can be obtained from a quotient of power series $g(T; \chi^*)$, $\chi\chi^* = \omega$, by substituting $(1+q_0)^s - 1$ for T. However, if χ is a Dirichlet character with $\chi(-1) = -1$, then the p-adic L-function $L_p(s; \chi)$ vanishes identically in s, because $L(s; \chi) = 0$ for all odd integers $s \leqq 0$. Therefore the above result holds also in this case, if we only define

$$g(T; \chi) \equiv 0$$

for each Dirichlet character χ with $\chi(-1) = 1$.

Let, again, $\chi(-1) = -1$ and $\chi\chi^* = \omega$. If $\chi \neq \omega$, $\chi^* \neq 1$, we can find c such that $\chi^*(c) \neq 1$. Then $u((1+q_0)^{-s} - 1; \chi, c) = 1 - \chi^*(c)\langle c\rangle^{s+1} \neq 0$ even for $s = -1$. Hence $L_p(s, \chi^*) = g((1+q_0)^s - 1; \chi)$ is continuous also at $s = 1$, as was shown by Kubota-Leopoldt. Suppose furthermore that the order of χ^* is not a power of p. Then we can find c such that $1 - c\chi(c)^{-1} = 1 - \chi^*(c)\langle c\rangle$ is a unit of $\mathfrak{o}$. For such c, $u(T; \chi, c)$ is a unit of $\Lambda = \mathfrak{o}[[T]]$, and we see that $g(T; \chi)$ itself is a power series in Λ. Finally, we also note that $g(T; \chi)$ is uniquely characterized by (3) so that it is independent of the field k, so far as k contains all values of the character χ. In particular, $g(T; \chi)$ is a quotient of power series in T, of which all coefficients are integers of the local field which is generated over $\mathbf{Q}_p$ by the values of χ.

4. With the prime number p fixed, let us call a Dirichlet character χ a character of the first kind if the conductor of χ is not divisible by $p^2 (p > 2)$ or by $2^3 (p = 2)$, and let us call χ a character of the second kind if the conductor of χ is a power of p and if $\chi(a)$ depends only upon $\langle a\rangle$(for $a \in \mathbf{Z}$, $(a, p) = 1$). Then each Dirichlet character χ can be uniquely decomposed into a product

$$\chi = \lambda\mu,$$

where λ and μ are characters of the first and the second kind respectively.

As in the preceding section, let $\chi(-1) = -1$ and let $N = m_0p^e$ be the conductor of χ. For the above decomposition, we then have $\lambda(-1) = -1$, and the conductor of λ is either m_0, $m_0p (p > 2)$, or $m_0 2^2 (p = 2)$. Let

$$\zeta = \chi(1+q_0) = \mu(1+q_0).$$

Then ζ is a root of unity of order p^t, $0 \leqq t \leqq e$.

Let $s \in \mathbf{Z}$, $s \geqq 0$. For each $n \geqq e$, there is a unique homomorphism of $\mathfrak{o}$-algebras $\psi_{s,n}: R_n \to \mathfrak{o}/q_n\mathfrak{o}$ such that

$$\gamma_n(a) \longrightarrow \mu(a)\varphi_{s,n}(\gamma_n(a)) = \mu(a)\langle a\rangle^{-s} \qquad \bmod q_n\mathfrak{o},$$

and those $\psi_{s,n}$ define an $\mathfrak{o}$-homomorphism $\psi_s: R \to \mathfrak{o}$. Since $\psi_s(\gamma(1+q_0)) = \zeta(1+q_0)^{-s}$, we have

$$f(\zeta(1+q_0)^{-s}-1; \lambda, c) = \psi_s(\eta(\lambda, c)) = \lim \psi_{s,n}(\eta_n(\lambda, c)),$$

where

$$\begin{aligned}\psi_{s,n}(\eta_n(\lambda, c)) &= -\lambda(c)^{-1}\psi_{s,n}(\gamma_n(c))^{-1}\textstyle\sum_a r_n(a)\lambda(a)^{-1}\psi_{s,n}(\gamma_n(a))^{-1}\\ &= -\chi(c)^{-1}\varphi_{s,n}(\gamma_n(c))^{-1}\textstyle\sum_a r_n(a)\chi(a)^{-1}\varphi_{s,n}(\gamma_n(a))^{-1}\\ &= \varphi_{s,n}(\eta_n(\chi, c)).\end{aligned}$$

Hence

$$f(\zeta(1+q_0)^{-s}-1; \lambda, c) = \varphi_s(\eta(\chi, c)) = f((1+q_0)^{-s}-1; \chi, c)$$

for every $s \in \mathbf{Z}$, $s \geqq 0$, and it follows that

$$f(\zeta(1+T)-1; \lambda, c) = f(T; \chi, c).$$

On the other hand, we see immediately from the definition that

$$u(\zeta(1+T)-1; \lambda, c) = u(T; \chi, c).$$

Therefore we obtain the following formula:

$$g(\zeta(1+T)-1; \lambda) = g(T; \chi).$$

In particular,

$$g(\zeta-1; \lambda) = g(0; \chi).$$

The formula shows that the quotient $g(T; \chi)$ for a general Dirichlet character χ with $\chi(-1) = -1$ can be obtained from the $g(T; \lambda)$ for the first factor λ of χ by a simple change of variable $T \to \zeta(1+T)-1$. It is clear that we have a similar relation between $g(T; \chi)$ and $g(T; \chi')$ whenever χ and χ' have the same first factor λ.

These results, which are obtained for characters χ with $\chi(-1) = -1$, also hold trivially for χ with $\chi(-1) = 1$ because $g(T; \chi) \equiv g(T; \lambda) \equiv 0$ in such a case. We also note that we have the following formulas for p-adic L-functions:

$$L_p(s; \chi^*) = g(\zeta(1+q_0)^s - 1; \lambda),$$

and in particular

$$L_p(0; \chi^*) = g(\zeta - 1; \lambda).$$

It seems that no similar fact is known for the classical L-functions.

5. We shall next explain an implication of the above result in the theory of cyclotomic fields[6]. Let m_0 and q_n be defined as in § 2. For each $n \geqq 0$, let

[6] Cf. [1].

F_n denote the cyclotomic field of q_nth roots of unity over the rational field Q, and let F be the union of all F_n, $n \geqq 0$. We may identify the group G_n in § 2 with the Galois group of $F_n/\mathbf{Q}$ by identifying $\sigma_n(a)$ with the automorphism of F_n which maps each q_n^{th} root of unity in F_n to its ath power. The group G is then identified with the Galois group of $F/\mathbf{Q}$ in the obvious manner.

In the following, we consider the special case where $p > 2$, $m_0 = 1$, and $k = \mathbf{Q}$ so that $\mathfrak{o} = \mathbf{Z}_p$, $q_n = p^{n+1}$. For $a \in \mathbf{Z}$, $(a, p) = 1$, let $\delta_n(a)$ denote the projection of $\sigma_n(a)$ on Δ_n in the decomposition $G_n = \Gamma_n \times \Delta_n$. Then $\delta_n(a) = \delta_n(b)$ if and only if $\omega(a) = \omega(b)$, and $\delta_n(a) \rightarrow \omega(a)$ defines an isomorphism of Δ_n onto the group of $(p-1)^{\text{st}}$ roots of unity in $\mathbf{Q}_p$. For $1 \leqq i \leqq p-1$, let

$$\varepsilon_n^{(i)} = \frac{1}{p-1} \sum_{a=1}^{p-1} \omega(a)^{-i} \delta_n(a) .$$

Then these $\varepsilon_n^{(i)}$ form a complete set of indecomposable idempotents in the algebra $\mathbf{Z}_p[\Delta_n]$, and $\mathbf{Z}_p[G_n]$ is the direct sum of the ideals $R_n\varepsilon_n^{(i)}$, $1 \leqq i \leqq p-1$.

Let S_n be the Sylow p-subgroup of the ideal-class group of F_n. We may consider S_n as a $\mathbf{Z}_p[G_n]$-module in the obvious manner. Then it is the direct sum of the submodules $S_n^{(i)} = \varepsilon_n^{(i)} S_n$, $1 \leqq i \leqq p-1$. For $0 \leqq n \leqq m$, there is a natural homomorphism $S_n \rightarrow S_m$ induced by the imbedding of the ideal group of F_n in that of F_m. Let S denote the direct limit of S_n, $n \geqq 0$, with respect to the above homomorphisms, and let $S^{(i)}$ be the direct limit of $S_n^{(i)}$, $n \geqq 0$ $(1 \leqq i \leqq p-1)$. Since each S_n is an R_n-module, the limit S is an R-module, and it is the direct sum of the submodules $S^{(i)}$, $1 \leqq i \leqq p-1$. By means of the isomorphism $R \xrightarrow{\sim} \Lambda$, we can make S and $S^{(i)}$ into Λ-modules. Let X be the dual of the discrete abelian group S and let $\langle c, x \rangle$ be the pairing of S and X. Define the action of Λ on X so that

$$\langle fc, x \rangle = \langle c, fx \rangle$$

for all $c \in S$, $x \in X$, and $f = f(T) \in \Lambda$. Then X becomes a Λ-module and it is the direct sum of submodules $X^{(i)}$, $1 \leqq i \leqq p-1$, $X^{(i)}$ being the dual of $S^{(i)}$.

Let ξ_n be an element of $\mathbf{Q}[G_n]$ defined by

$$\xi_n = -\frac{1}{q_n} \sum_a a\sigma_n(a)^{-1} , \qquad 0 \leqq a < q_n,\ (a, p) = 1 .$$

It is easy to see that

$$\varepsilon_n^{(i)} \xi_n = \frac{1}{p-1} \varepsilon_n^{(i)} \xi_n(\omega^i)$$

with the $\xi_n(\omega^i)$ defined in § 3 and that $\xi_n(\omega^i)$ belongs to R_n for every $i \neq 1$ and $n \geqq 0$. For $i \neq 1$, let $\xi(\omega^i)$ denote the element of R defined by those $\xi_n(\omega^i)$,

$n \geqq 0$. It is known in the theory of cyclotomic fields[7] that $S^{(1)} = 0$, $X^{(1)} = 0$ and that $\varepsilon_n^{(i)}\xi_n S_n = \xi_n(\omega^i)S_n^{(i)} = 0$, $n \geqq 0$, so that $\xi(\omega^i)S^{(i)} = 0$, $\xi(\omega^i)X^{(i)} = 0$ for every $i \neq 1$. It then follows that if the so-called second factor of the class number of F_0 is prime to p, then

$$X^{(i)} = 0\,, \qquad \text{for } i \text{ even or } i = 1\,,$$
$$X^{(i)} \cong \Lambda/(g(T;\omega^i))\,, \qquad \text{for } i \text{ odd, } i \neq 1\,.$$

Here $(g(T;\omega^i))$ denotes the ideal of Λ generated by $g(T;\omega^i)$ and the isomorphism is of course an isomorphism of Λ-modules.

It is now clear that the formula (3) in § 3, combined with the above, proves what we announced in our previous paper [2], where we also explained in detail how such a result may be regarded as an analogue of a well-known theorem of A. Weil on zeta-functions of algebraic curves. We shall not repeat it here but make only the following remark. If we define the action of Λ on X so that $\langle \sigma c, \sigma x\rangle = \langle c, x\rangle$ for all $\sigma \in G$, $c \in S$, and $x \in X$, then we have to replace $g(T;\omega^i)$ by $g((1+T)^{-1} - 1;\omega^i)$ in the above isomorphism for $X^{(i)}$ (i odd and $i \neq 1$). This accounts for the slight difference between our formula (3) and the corresponding formula in [2] where we tacitly adopted the definition of the Λ-structure on X as stated immediately in the above.

PRINCETON UNIVERSITY

BIBLIOGRAPHY

[1] K. IWASAWA, *On some modules in the theory of cyclotomic fields.* J. Math. Soc. Japan **16** (1964), 42-82.

[2] ———, Analogies between number fields and function fields. in Proc. of the Science Conference, Yeshiva Univ., 1966.

[3] T. KUBOTA and H. W. LEOPOLDT, *Eine p-adische Theorie der Zetawerte*: I. J. reine u. angew. Math., 214/215 (1964), 328-339.

[4] H. W. LEOPOLDT, *Eine Verallgemeinering der Bernoullischen Zahlen.* Abh. Math. Seminar, Hamburg **22** (1958), 131-140.

(Received October 14, 1968)

[7] For the results on cyclotomic fields stated below, see [1, § 3].

[49] On some infinite Abelian extensions of algebraic number fields

"Actes du Congrés Int. Math." (Nice, 1970), Tome 1, Gauthier-Villars, Paris, 1971, pp. 391-394.

Let l be a prime number and let Z_l and Q_l denote the ring of l-adic integers and the field of l-adic numbers respectively; their additive groups will be also denoted by the same letters.

Now, an extension K of a field k is called a Z_l-extension if K/k is a Galois extension and its Galois group is topologically isomorphic to the additive group Z_l (1). For such an extension K/k, there exists a sequence of fields

$$k = k_0 \subset k_1 \subset \ldots \subset k_n \subset \ldots$$

such that each k_n/k is a cyclic extension of degree l^n and K is the union of all k_n, $n \geqq 0$. Conversely, if there is a sequence of cyclic extensions k_n/k such as mentioned above, then the union K of all k_n, $n \geqq 0$, is a Z_l-extension of k.

In the following, we shall consider Z_l-extensions of which the ground fields are finite algebraic number fields, i. e. finite extensions of the rational field Q. We first give some examples. For each $n \geqq 0$, let $P_{l,n}$ denote the cyclotomic field of l^{n+1}-th or 2^{n+2}-th roots of unity according as $l > 2$ or $l = 2$. Let $P_l = P_{l,0}$ and let $P_{l,\infty}$ be the union of all $P_{l,n}$, $n \geqq 0$. Then $P_{l,\infty}/P_l$ is a Z_l-extension with intermediate fields $P_{l,n}$, $n \geqq 0$. The field $P_{l,\infty}$ has a unique subfield $Q_{l,\infty}$ such that $P_l \cap Q_{l,\infty} = Q$, $P_l Q_{l,\infty} = P_{l,\infty}$, and this $Q_{l,\infty}$ gives us the unique Z_l-extension of the rational field Q. Furthermore, for any finite algebraic number field k, the composite $kQ_{l,\infty}$ is a Z_l-extension of k. Hence each k has at least one Z_l-extension over it.

Let K be a Z_l-extension of a finite algebraic number field k and let k_n, $n \geqq 0$, be the intermediate fields of k and K. Let C_n denote the ideal-class group of k_n, and A_n the Sylow l-subgroup of C_n. Denote by l^{e_n} *the* order of A_n, i. e. the highest power of l dividing the class number of k_n. Then, for all sufficiently large n, the exponent e_n is given by a formula

$$e_n = \lambda n + \mu l^n + \nu,$$

where λ, μ, and ν are integers (λ, $\mu \geqq 0$), independent of n. Since these integers are uniquely determined for given K/k by the above formula, we shall denote them by $\lambda(K/k)$, $\mu(K/k)$, and $\nu(K/k)$ respectively. For the special Z_l-extension $K = kQ_{l,\infty}$ over k, they will be denoted also by $\lambda_l(k)$, $\mu_l(k)$, and $\nu_l(k)$ respectively; furthermore we

---

(1) In earlier papers, the author called such extensions Γ-extensions.

simply put $\lambda_l = \lambda_l(P_l)$, $\mu_l = \mu_l(P_l)$, $\nu_l = \nu_l(P_l)$. Thus we obtain arithmetic invariants $\lambda_l(k)$, $\mu_l(k)$, $\nu_l(k)$ depending upon k and l, and λ_l, μ_l, ν_l for each prime number l.

Let O denote the ring of all algebraic integers in K, I the group of all invertible O-modules in K, and C the factor group of I modulo the principal O-modules ([2]); we may simply call I and C the ideal group and the ideal-class group of O in K, respectively. Let A be the Sylow l-subgroup (i. e. the l-primary component) of C. Then C is the direct limit of C_n, $n \geqq 0$, and A that of A_n, $n \geqq 0$, and the Galois group Gal (K/k) acts on C and A in the obvious manner. The above formula for e_n is obtained by analysing the structure of this Gal (K/k)-module A. Thus we see in particular that the Tate module $T_l(A)$ for the abelian l-group A is a free Z_l-module and its rank over Z_l is equal to the invariant λ.

At the present, little is known on the nature of the invariants $\lambda(K/k)$, $\mu(K/k)$, and $\nu(K/k)$ defined above. Yet it is clear that they play an essential role in the theory of Z_l-extensions. It seems particularly interesting to see when $\lambda = 0$ or $\mu = 0$ or $\lambda = \mu = 0$. It is easy to find a Z_l-extension K/k for which $\lambda(K/k)$ is arbitrary large. On the other hand, no example of K/k with $\mu(K/k) > 0$ is yet found. Although the number of such examples for which we have verified $\mu(K/k) = 0$ is quite limited, we are tempted to conjecture that $\mu(K/k) = 0$ for every Z_l-extension K/k, or at least that $\mu_l(k) = 0$ for every k and l or $\mu_l = 0$ for every l.

For the invariants λ_l and μ_l, we know that $\lambda_l = \mu_l = 0$ if and only if l is a regular prime, and if this is the case, then $\nu_l = 0$ also. Let P_l' denote the maximal real subfield of P_l and put

$$\lambda_l' = \lambda_l(P_l'), \quad \mu_l' = \mu_l(P_l'), \quad \nu_l' = \nu_l(P_l').$$

Then $\lambda_l' \leqq \lambda_l$, $\mu_l' \leqq \mu_l$. Now, a well-known conjecture of Vandiver states that the class number of P_l' is not divisible by l for every prime number l; it is checked by numerical computation for a large number of primes. For given l, the conjecture is equivalent with $\lambda_l' = \mu_l' = \nu_l' = 0$. Unlike what is said above for λ_l, μ_l, and ν_l, it is not known whether $\lambda_l' = \mu_l' = 0$ implies $\nu_l' = 0$ and, hence, Vandiver's conjecture. Nevertheless, it would be an interesting problem to find out if $\lambda_l' = \mu_l' = 0$ for every l.

Let $T_l(A)$ be the Tate module defined above and let V denote the tensor product of $T_l(A)$ and Q_l over Z_l: $V = V_l(A) = T_l(A) \otimes Q_l$. V is a vector space of dimension λ over Q_l and the Galois group Gal (K/k) acts on V continuously so that it defines an l-adic representation of Gal (K/k). It is clear that the definition of V above is completely parallel to the usual construction of such l-adic representations by means of, say, abelian varieties ([3]). Note also that if k' is a subfield of k such that K/k' is a Galois extension, then the same vector space V defines an l-adic representation of the larger group Gal (K/k').

Let $l > 2$ and $k = P_l$, $K = P_{l,\infty}$ in the above ([4]). Then V is decomposed into the direct sum of $l-1$ subspaces V_i, $0 \leqq i < l-1$, with respect to the action of

([2]) In other words, C = Pic (O). Note that the ring O is not noetherian.

([3]) See J.-P. SERRE, Abelian l-adic representations of elliptic curves, Benjamin (New York, Amsterdam), 1968.

([4]) For $l = 2$, slight modification is needed in what is said below about the decomposition of V and the definition of σ_0.

Gal (P_l/Q). Let W denote the group of all roots of unity in $P_{l,\infty}$ with order a power of l. Let σ_0 be the automorphism of K/k such that $\sigma_0(\zeta) = \zeta^{1+l}$ for every ζ in W and let $f_i(x)$ be the characteristic polynomial of σ_0 acting on V_i. Assuming Vandiver's conjecture for the prime l, we can describe the representation of Gal (K/k) on each V_i rather explicitly. It then follows ([5]) in particular that the characteristic polynomials $f_i(x)$, $0 \leqq i < l-1$, are closely related to the l-adic L-functions of Kubota-Leopoldt associated with the characters Gal $(P_l/Q) \to Z_l^\times$.

For a Z_l-extension K/k in general, we know very little on the structure of the l-adic representation Gal $(K/k) \to GL(V)$. However the following fact might be of some interest, in particular when viewed as an analogue of a similar result in algebraic geometry. Let k be any finite algebraic number field containing P_l and let

$$K = kQ_{l,\infty} = kP_{l,\infty}.$$

Let O' denote the ring of all l-integers in K, i. e. the union of all $l^{-n}O$, $n \geqq 0$. Let C' be the ideal class group of O' in K, and A' the Sylow l-subgroup of C'. Let $V' = T_l(A') \otimes Q_l$ over Z_l, where $T_l(A')$ denotes the Tate module for A'. As before, V' defines an l-adic representation of Gal (K/k), and the natural map $A \to A'$ induces an epimorphism $T_l(A) \to T_l(A')$ so that V' is a factor space of the representation space V. Now, an element α in K will be called l^n-hyperprimary ($n \geqq 0$) if α is an l^n-th power in the v-completion K_v for every place v of K lying above the place l of Q ([6]), and an O'-module $\mathfrak{a}$ of K will be called hyperprimary if, for some $n \geqq 0$, $\mathfrak{a}^{l^n} = (\alpha)$ with an l^n-hyperprimary element α in K. Let B' be the subgroup of all classes in A' which are represented by hyperprimary O'-modules and let $V'' = T_l(B') \otimes Q_l$ over Z_l. Then V'' again defines an l-adic representation of Gal (K/k), and $B' \to A'$ induces a monomorphism $T_l(B') \to T_l(A')$ so that V'' is a subspace of V'. Hence V'' is involved in the original representation space V. Let W be the group of roots of unity as defined above and let $V_0 = T_l(W) \otimes Q_l$ over Z_l. Then V_0 defines a one-dimensional l-adic representation of Gal (K/k) through its action on W.

Now, suppose that the ground field k is abelian either over the rational field or over an imaginary quadratic field. Then we can prove ([7]) that there exists a non-degenerate skew-symmetric Q_l-bilinear form

$$V'' \times V'' \to V_0$$

such that

$$\langle \sigma u, \sigma v \rangle = \sigma(\langle u, v \rangle)$$

for all u, v in V'' and σ in Gal (K/k). It follows that V'' is even dimensional and that the characteristic polynomial of each σ in Gal (K/k) acting on V'' satisfies a functional equation similar to the one for the zeta-function of an algebraic curve defined over a finite field ([8]).

([5]) See K. Iwasawa, On p-adic L-functions, *Ann. Math.*, 89 (1969), pp. 198-205.

([6]) There exist only a finite number of such places v in K.

([7]) The proof will be published elsewhere.

([8]) Actually we can prove the above result for a wider class of ground fields including those mentioned above. It seems likely that the same result holds for an arbitrary ground field k containing P_l, without any further assumption on k.

As already mentioned, the skew-symmetric form defined above is essentially an analogue of the classical Riemann forms on complex tori, of which a purely algebraic construction was given by Weil for abelian varieties of arbitrary characteristic ([9]). It would be interesting to pursue such analogy further in studying the structure of the representation space V''.

([9]) The original idea of Weil appears in his paper: Sur les fonctions algébriques à corps de constantes fini, *C. R. Paris*, 210 (1940), pp. 592-594.

Princeton University
Department of Mathematics,
Princeton, New Jersey 08540
(U. S. A.)

[50] Skew-symmetric forms for number fields

Proc. Sympos. Pure Math., Vol. XX, Amer. Math. Soc., 1971, p. 86.

Let H be the 1-dimensional real cohomology group of a compact Riemann surface X. For x_1, x_2 in H, put

$$\langle x_1, x_2 \rangle = \int_X \omega_1 \wedge \omega_2,$$

where ω_1 and ω_2 denote real closed 1-forms on X representing the cohomology classes x_1 and x_2 respectively. Then $\langle x_1, x_2 \rangle$ defines a nondegenerate skew-symmetric bilinear form on the real vector space H, which plays an important role in the classical theory of algebraic functions. Following the analogy between algebraic number fields and algebraic function fields, we construct a skew-symmetric bilinear form for a number field which may be regarded as an analogue of the above $\langle x_1, x_2 \rangle$.

Let l be a fixed prime number and let Z_l and Q_l denote the ring of l-adic integers and the field of l-adic numbers respectively. We consider a number field K which is generated over a certain type of finite algebraic number field k by all l^n-th roots of unity for all $n \geqq 0$. Let L be the maximal unramified abelian l-extension over K, and F the field generated over K by all l^n-th roots, $n \geqq 0$, of all units in K. Since the Galois group G of the extension $L/L \cap F$ is an abelian profinite l-group, it may be considered as a Z_l-module. Hence, let V denote the tensor product of V and Q_l over Z_l. Then we can prove that V is a finite (even) dimensional vector space over Q_l and that it admits a nondegenerate skew-symmetric bilinear form as mentioned above.

V is obviously an l-adic representation space for the Galois group Γ of K/k. Examining the action of Γ on the bilinear form, we see that the characteristic polynomial of the linear transformation representing an element of Γ satisfies a functional equation of well-known type.

PRINCETON UNIVERSITY
PRINCETON, NEW JERSEY

86

[51] On the μ-invariants of cyclotomic fields

Acta Arith., 21 (1972), 99-101.

Let p be an odd prime. For each $n \geqslant 0$, let k_n denote the cyclotomic field of p^{n+1}-th roots of unity and let p^{e_n}, $e_n \geqslant 0$, be the highest power of p which divides the class number of k_n. It is known (see [1]) that for all sufficiently large n, the exponent e_n is given by a formula

$$e_n = \lambda n + \mu p^n + \nu$$

where λ, μ, and ν are integers ($\lambda, \mu \geqslant 0$). independent of n. In the present paper, we shall prove that

$$\mu < p-1.$$

Let Z_p denote the ring of p-adic integers and let Λ be the ring of all formal power series in an indeterminate T with coefficients in Z_p: $\Lambda = Z_p[[T]]$. We shall first prove a lemma on Λ-modules (¹).

A Λ-module Y is called *elementary* if Y is the direct sum of a finite number of Λ-modules of the form Λ/P^m, $m \geqslant 0$, where P are prime ideals of height 1 in Λ. Let X be a noetherian torsion Λ-module. Then there exist an elementary Λ-module Y and a morphism

$$f\colon X \to Y$$

such that both the kernel and the cokernel of f are finite modules. Let

$$Y = \sum_i \Lambda/P_i^{m_i}$$

be the direct decomposition for Y and let

$$\mu = \sum{}' m_i,$$

where the sum is taken over all indices i such that $P_i = p\Lambda$. The integer μ is then uniquely determined for X by the above and hence is denoted by $\mu(X)$.

(¹) For the theory of Λ-modules, see [3].

LEMMA. *Let X be a noetherian torsion Λ-module with $\mu = \mu(X)$. Then the order of X/TX is at least equal to p^{μ}.*

Proof. Let $f: X \to Y$ be as above and let $Z = f(X)$, the image of f. It is clear that the order of X/TX is not less than the order of Z/TZ.

Now, if Y/TY is infinite, then so is X/TX and the lemma holds trivially. Hence we may assume that Y/TY is finite. In such a case, we see easily that $P_i \neq T\Lambda$ for every index i in the direct decomposition of Y so that the map

$$Y \to Y,$$
$$y \to Ty$$

is injective and that

$$Y/Z \simeq TY/TZ.$$

Since Y/Z and Y/TY are both finite, it follows that the order of Z/TZ is equal to that of Y/TY. Therefore it is sufficient to show that the order of Y/TY is at least equal to p^{μ}. However, this is an immediate consequence of the fact that if $U = \Lambda/p^m\Lambda$, $m \geqslant 0$, then the order of U/TU is equal to p^m.

Now, let $k = k_0$ and let K denote the union of all k_n, $n \geqslant 0$. K is a Galois extension of k and its Galois group is isomorphic to the additive group of the compact ring $\boldsymbol{Z}_p$. Let L be the maximal unramified abelian p-extension over K and let X be the Galois group of L/K. Since L/k is also a Galois extension, Γ acts on the abelian group X in the obvious manner. Fixing a topological generator γ of the compact group Γ, we can then make X into a Λ-module module so that $(1+T)x = \gamma x$ for every x in X. Furthermore, we can show (cf. [1] and [3]) that X is a noetherian torsion Λ-module and its invariant $\mu(X)$ is equal to the second coefficient μ in the formula for e_n mentioned above: $\mu = \mu(X)$.

Let J denote the automorphism of L which maps each a in L to its complex-conjugate $\bar{a}$. Clearly J also acts on X. Let X^+ (resp. X^-) be the set of all x in X such that $Jx = x$ (resp. $Jx = -x$). Then X^+ and X^- are Λ-submodules of X and

$$X = X^+ \oplus X^-.$$

Hence we have

$$\mu = \mu(X) = \mu^+ + \mu^-$$

where $\mu^+ = \mu(X^+)$ and $\mu^- = \mu(X^-)$. It is also known (cf. [2]) that

$$\mu^+ \leqslant \mu^-.$$

Therefore

$$\mu \leqslant 2\mu^-.$$

Let h^- denote the so-called first factor of the class number of k, the cyclotomic field of pth roots of unity. It is proved (see [1] and [2]) that the order of X^-/TX^- is just equal to the highest power of p which divides h^-. Hence, applying the above lemma for X^-, we see that

$$p^{\mu^-} \leqslant h^- .$$

On the other hand, the classical class number formula for k states that

$$h^- = 2p \prod_{\chi} \Big(-\frac{1}{2p} \sum_{a=1}^{p-1} a\chi(a) \Big),$$

where the product is taken over all Dirichlet characters χ defined mod p with $\chi(-1) = -1$. Since

$$\Big| \sum_{a=1}^{p-1} a\chi(a) \Big| < \sum_{a=1}^{p-1} a = \frac{(p-1)p}{2},$$

we have

$$h^- < 2^{2-p} p (p-1)^{(p-1)/2} \leqslant p^{(p-1)/2} .$$

It then follows that

$$\mu^- < (p-1)/2$$

so that

$$\mu < p-1,$$

q.e.d.

Instead of the above elementary argument, we may estimate h^- also by using

$$|L(1;\chi)| < 2 \log p, \quad \chi \neq 1 .$$

We then see that for any given real number $c > \frac{1}{2}$, there exists an integer $N(c)$ such that

$$\mu < c(p-1)$$

whenever $p \geqslant N(c)$. It is also clear that by the same method, we can find an upper bound for the μ-invariant of a so-called $\boldsymbol{Z}_p$-extension K/k in many special cases. In particular, if K has only one prime divisor which divides the rational prime p (as in the special case discussed above), then

$$\mu(K/k) \leqslant \log h / \log p,$$

where h is the class number of k.

References

[1] K. Iwasawa, *On Γ-extensions of algebraic number fields*, Bull. Amer. Math. Soc. 65 (1959), pp. 183–226.

[2] — *On the theory of cyclotomic fields*, Ann. of Math. 70 (1959), pp. 530–561.

[3] J.-P. Serre, *Classes des corps cyclotomiques*, Seminaire Bourbaki, Exposé 174 (1958/1959).

PRINCETON UNIVERSITY

Received on 21. 2. 1971

[52] On Z_ℓ-extensions of algebraic number fields

Ann. of Math., (2) 98 (1973), 246-326.

To ANDRÉ WEIL

Let l be a prime number which will be fixed throughout the following, and let $\mathbf{Z}_l$ denote the ring of all l-adic integers. A Galois extension K of a field k is called a $\mathbf{Z}_l$-extension over k if the Galois group Gal (K/k) is topologically isomorphic to the additive group of the compact ring $\mathbf{Z}_l$[1]. In the present paper, we shall study arithmetic properties of such $\mathbf{Z}_l$-extensions in the case where the ground fields are finite algebraic number fields.

The study of such extensions was originated in our earlier paper [4] where a class number formula was proved and certain invariants, λ and μ, were introduced for each $\mathbf{Z}_l$-extension K/k. In the succeeding papers [5] and [6], we investigated deeper arithmetic properties of the extension K/k in the special case where k is the cyclotomic field of l-th roots of unity, $l > 2$, and K is the field generated over the rational field by all l^n-th roots of unity for all $n \geqq 0$. In the present paper, we shall first show that most of the results obtained in that special case can be extended for a more general type of $\mathbf{Z}_l$-extensions called cyclotomic $\mathbf{Z}_l$-extensions, and we shall then study some further arithmetic properties of these extensions.

An outline of the paper is as follows. In the first five sections, §§1-5, we develop a general theory of $\mathbf{Z}_l$-extensions which may also be regarded as an introduction for what will be discussed in the rest of the paper. Since this part is more or less a direct generalization of what we have done in our earlier publications, the proof will be given briefly in most cases, in particular when we deal with purely algebraic problems. In the next part, §§6-8, we study cyclotomic $\mathbf{Z}_l$-extensions, namely, the extensions K/k where K contains all l^n-th roots of unity for all $n \geqq 0$. Because of the existence of such roots of unity in K, we can apply here both class field theory and Kummer theory to study the structure of the Galois groups of various abelian extensions over K. In classical algebraic number theory, the application of these two principles results in the theory of power residue symbols. For cyclotomic $\mathbf{Z}_l$-extensions, it provides us with certain skew-symmetric pairings of $\mathbf{Z}_l$-modules which may be regarded as an analogue of classical Riemann forms

[1] In some of our earlier publications, these extensions are called Γ-extensions (for the prime l).

on complex tori. A purely algebraic construction of such Riemann forms for algebraic function fields over finite fields of constants was first given by A. Weil in his well-known paper [11]. In §§9–11, we construct our pairings similarly, following the analogy between number fields and function fields[2]. In the last §12, we briefly discuss two typical examples of Z_l-extensions where the ground fields are not finite algebraic number fields; we shall show in particular how the analogy between number fields and function fields works in the case of Z_l-extensions.

As the reader will find, many important arithmetic problems on Z_l-extensions are left unsolved in this paper. However, we hope that the present paper will serve as a ground work for future investigations in this area.

Throughout the paper, **Z**, **Q**, and **C** denote the ring of rational integers, the field of rational numbers, and the field of complex numbers respectively. As already mentioned, Z_l denotes the ring of l-adic integers. The field of l-adic numbers will be denoted by Q_l. The additive groups of these rings and fields are denoted by the same letters; for example, Z_l will denote the additive group of the ring of l-adic integers as well as the ring itself. All algebraic extensions of **Q** are supposed to be contained in the complex field **C**. Thus, for example, when we say that K is a Z_l-extension of a finite algebraic number field k, it means that k is a finite algebraic extension of **Q** in **C** and K is a Z_l-extension of k, also contained in **C**. Most of the abelian groups we consider will be regarded as additive groups. However, in some cases, both additive and multiplicative notations are used for the same group, in particular when the group composition in that group is customarily written multiplicatively.

TABLE OF CONTENTS

[2] In this connection, we owe to Weil an important remark that the two forms, denoted by A and B in [11], are actually the same. This motivated our proof in §10.2. See §12.5 below.

1. Γ-modules and Λ-modules[3]

1.1. Let Γ denote a compact abelian (multiplicative) group isomorphic to the additive group of $\mathbf{Z}_l$; in later sections, Γ will be the Galois group of a $\mathbf{Z}_l$-extension K/k. Let Λ be the ring of all formal power series in an indeterminate T with coefficients in $\mathbf{Z}_l$: $\Lambda = \mathbf{Z}_l[[T]]$. Λ is a noetherian, regular, local domain of dimension 2 and is a compact topological ring in its $\mathfrak{m}$-adic topology, $\mathfrak{m}$ being the maximal ideal of Λ. In the present section, we consider profinite abelian l-groups on which Γ or Λ acts continuously. For simplicity, we shall call such groups compact Γ-modules or compact Λ-modules respectively.

Now, fix a topological generator of Γ, namely, an element γ_0 in Γ which generates a subgroup everywhere dense in Γ. Then each compact Γ-module X admits a unique structure of compact Λ-module such that

$$(1+T)x = \gamma_0 x$$

for every x in X. Conversely, each compact Λ-module uniquely determines a compact Γ-module satisfying the condition stated above. Thus the topological generator γ_0 determines an isomorphism of the category of compact Γ-modules onto that of compact Λ-modules, and any results on compact Γ-modules give us also information on the structure of compact Λ-modules, and vice versa.

1.2. Let X be a noetherian Λ-module, defined algebraically. Then there exists a unique way of topologizing X so that it becomes a compact Λ-module. In the following, we shall consider such noetherian compact Λ-modules. Note that a compact Λ-module X is noetherian if and only if $X/\mathfrak{m}X$ is finite.

Let $e_0, e_1, \cdots, e_s$ be any non-negative integers and let $\mathfrak{p}_1, \cdots, \mathfrak{p}_s$ be any prime ideals of height 1 in Λ, not necessarily distinct. Let

$$E(e_0; \mathfrak{p}_1^{e_1}, \cdots, \mathfrak{p}_s^{e_s}) = \Lambda^{e_0} \oplus \Lambda/\mathfrak{p}_1^{e_1} \oplus \cdots \oplus \Lambda/\mathfrak{p}_s^{e_s}$$

where Λ^{e_0} denotes the direct sum of e_0 copies of Λ while $\mathfrak{p}_i^{e_i}$ is the e_i-th power of the ideal $\mathfrak{p}_i$ in Λ. Clearly $E(e_0; \mathfrak{p}_1^{e_1}, \cdots, \mathfrak{p}_s^{e_s})$ is a noetherian Λ-module. We call such modules elementary Λ-modules. The integer e_0 and the ideals $\mathfrak{p}_1^{e_1}, \cdots, \mathfrak{p}_s^{e_s}$ constitute a complete set of invariants for $E(e_0; \mathfrak{p}_1^{e_1}, \cdots, \mathfrak{p}_s^{e_s})$; namely, two elementary Λ-modules are isomorphic if and only if they have the same set of such invariants. For simplicity, $E(0; \mathfrak{p}_1^{e_1}, \cdots, \mathfrak{p}_s^{e_s})$ will be denoted also by $E(\mathfrak{p}_1^{e_1}, \cdots, \mathfrak{p}_s^{e_s})$.

Let both X and X' be noetherian Λ-modules. A morphism $f: X \to X'$ is called a pseudo-isomorphism if the kernel and the cokernel of f are both finite modules. When there exists such a pseudo-isomorphism, we write $X \sim X'$.

---

[3] For this section, see [9].

If both X and X' are torsion Λ-modules, then $X \sim X'$ implies $X' \sim X$. However, this is not true in general. Now, a fundamental result on noetherian Λ-modules states that given such a Λ-module X, there exists a unique elementary Λ-module $E(e_0; \mathfrak{p}_1^{e_1}, \cdots, \mathfrak{p}_s^{e_s})$ such that

$$X \sim E(e_0; \mathfrak{p}_1^{e_1}, \cdots, \mathfrak{p}_s^{e_s}) .$$

We call the latter the elementary Λ-module associated with X. If $t(X)$ denotes the torsion Λ-submodule of X, then it follows from the above that

$$t(X) \sim E(\mathfrak{p}_1^{e_1}, \cdots, \mathfrak{p}_s^{e_s}) , \quad X/t(X) \sim E(e_0) = \Lambda^{e_0} .$$

The divisor of X in the divisor group of Λ is defined by

$$\operatorname{div} X = \sum_{i=1}^{s} e_i \mathfrak{p}_i ;$$

this is also the divisor of $t(X)$.

Let $\mathfrak{p}$ be a prime ideal of height 1 in Λ. Then either $\Lambda = (l)$, the principal ideal generated by l, or there exists a unique irreducible distinguished polynomial $p(T)$ in $Z_l[T]$ such that $\mathfrak{p} = (p(T))$. Define

$$\begin{aligned} &\lambda(\mathfrak{p}) = 0 , && \mu(\mathfrak{p}) = 1 , && \text{if } \mathfrak{p} = (l) , \\ &\lambda(\mathfrak{p}) = \deg p(T) > 0 , && \mu(\mathfrak{p}) = 0 , && \text{if } \mathfrak{p} = (p(T)) , \end{aligned}$$

and extend the definition linearly on the whole divisor group of Λ. For a noetherian Λ-module X, we then define the invariants $\lambda(X)$ and $\mu(X)$ by

$$\lambda(X) = \lambda(\operatorname{div} X) , \quad \mu(X) = \mu(\operatorname{div} X) .$$

Let

$$V = X \otimes_{Z_l} Q_l .$$

V is a vector space over Q_l, and $\dim V$ is finite if and only if X is a torsion Λ-module. In fact, in such a case,

$$\dim V = \lambda(X) .$$

1.3. Let X be a noetherian torsion Λ-module. For each prime ideal $\mathfrak{p}$ of height 1 in Λ, let

$$X_\mathfrak{p} = X \otimes_\Lambda \Lambda_\mathfrak{p} ,$$

where $\Lambda_\mathfrak{p}$ denotes the quotient ring of Λ with respect to $\mathfrak{p}$. Then $X_\mathfrak{p} \neq 0$ if and only if $\mathfrak{p}$ is contained in $\operatorname{div} X$. Hence $X_\mathfrak{p} \neq 0$ only for a finite number of $\mathfrak{p}$'s. Let X^0 and Y denote the kernel and the cokernel, respectively, of the morphism

$$X \longrightarrow \prod_\mathfrak{p} X_\mathfrak{p} ,$$

induced by the canonical maps $X \to X_\mathfrak{p}$, the product being taken over all $\mathfrak{p}$.

Then X^0 is the maximal finite Λ-submodule of X containing all finite Λ-submodules of X. Let

$$\alpha(X) = \mathrm{Hom}_{Z_l}(Y, Q_l/Z_l)$$

and make $\alpha(X)$ into a Λ-module by defining

$$(\xi \cdot \varphi)(y) = \varphi(\xi y)$$

for ξ in Λ, φ in $\alpha(X)$, and y in Y. We then see that $\alpha(X)$ is a noetherian torsion Λ-module and we call it the adjoint of X[4]. We shall next describe some of the properties of $\alpha(X)$ which we need later.

Clearly, $X \to \alpha(X)$ defines a contravariant functor on the category of noetherian torsion Λ-modules. If $X \sim X'$, then

$$\alpha(X) \sim \alpha(X') .$$

If E is an elementary torsion Λ-module, then

$$\alpha(E) \simeq E .$$

Hence, in general,

$$\alpha(X) \sim X .$$

Furthermore, $\alpha(X)^0 = 0$, i.e., $\alpha(X)$ has no non-trivial finite submodule.

To see the structure of $\alpha(X)$ explicitly, we have the following formula: given X, there always exists a sequence of non-zero elements $\{\pi_n\}$, $n \geqq 0$, in Λ such that

$$\pi_0 \in \mathfrak{m}\,, \qquad \pi_{n+1} \in \pi_n \mathfrak{m}\,, \qquad n \geqq 0\,,$$

and such that the principal divisors (π_n), $n \geqq 0$, are disjoint from div X. ($\mathfrak{m}$ denotes, as before, the maximal ideal of Λ). With such $\{\pi_n\}$,

$$\alpha(X) \simeq \varprojlim \mathrm{Hom}_{Z_l}(X/\pi_n X, Q_l/Z_l) .$$

Here the inverse limit is taken with respect to the morphisms induced by

$$X/\pi_n X \longrightarrow X/\pi_m X$$

$$x \bmod \pi_n X \longmapsto (\pi_m/\pi_n)x \bmod \pi_m X\,, \qquad m \geqq n \geqq 0\,,$$

and the Λ-structure of $\mathrm{Hom}_{Z_l}(X/\pi_n X, Q_l/Z_l)$ is defined similarly as for $\alpha(X)$ in the above[5].

1.4. Finally, we define a number of element in Λ which we shall frequently use in the following. Let

$$\omega_n = (1+T)^{l^n} - 1\,, \qquad n \geqq 0\,,$$

[4] This definition of adjoints is due to J. Tate; the original definition in [5] is slightly different.

[5] See [5]. The proof of these properties for $\alpha(X)$ may be given elsewhere.

$$\xi_0 = \omega_0 = T\,, \qquad \xi_n = \omega_n/\omega_{n-1} = \sum_{i=0}^{l-1} (1+T)^{il^{n-1}}\,, \qquad n \geqq 1\,,$$
$$\nu_{n,m} = \omega_m/\omega_n = \xi_{n+1} \cdots \xi_m\,, \qquad m \geqq n \geqq 0\,.$$

ξ_n is a distinguished polynomial in $\mathbf{Z}_l[T]$ and it is irreducible in $\mathbf{Z}_l[T]$ as well as in $\Lambda = \mathbf{Z}_l[[T]]$. Hence the principal ideal (ξ_n) is a prime ideal of height 1 in Λ, and $(\xi_n) \neq (\xi_m)$ if $n \neq m$.

Now, let X be a compact Γ-module and let X be made into a compact Λ-module by a fixed topological generator γ_0 of Γ. We shall next explain that some submodules and factor modules of X, which are defined by means of the elements in Λ introduced above, actually depend only upon the Γ-structure of X and are independent of the choice of γ_0.

For each $n \geqq 0$, let Γ_n denote the unique closed subgroup of Γ with index l^n; Γ/Γ_n is cyclic and Γ_n is topologically generated by $\gamma_0^{l^n}$. Since

$$\omega_n x = (\gamma_0^{l^n} - 1)x\,, \qquad x \in X\,,$$

we see that the submodule $\omega_n X$ is uniquely characterized as the minimal Γ-submodule of X such that Γ_n acts trivially on its factor module. Hence $\omega_n X$ does not depend upon the choice of the Λ-structure on X. Similarly, morphisms such as

$$\nu_{n,m}\colon X/\omega_m X \longrightarrow X/\omega_m X$$
$$x \bmod \omega_m X \longmapsto \nu_{n,m} x \bmod \omega_m X\,, \qquad m \geqq n \geqq 0\,,$$

also depend only upon the original Γ-structure of X, although they are defined by means of ω_n, $\nu_{n,m}$, etc. We also note that X is noetherian over Λ if and only if $X/\omega_0 X$ is noetherian over $\mathbf{Z}_l$ so that the fact whether or not X is noetherian over Λ again depends only upon the Γ-structure of X.

2. Existence of $\mathbf{Z}_l$-extensions

2.1. Let W denote the group consisting of all l^n-th roots of unity in $\mathbf{C}$ for all $n \geqq 0$. Let $\mathbf{P}$ be the cyclotomic field of l-th or 4-th roots of unity according as $l > 2$ or $l = 2$, and let

$$\mathbf{P}_\infty = \mathbf{Q}(W) = \mathbf{P}(W)\,.$$

Then $\mathbf{P}_\infty/\mathbf{P}$ is a $\mathbf{Z}_l$-extension. The field $\mathbf{P}_\infty$ has a unique subfield $\mathbf{Q}_\infty$ such that

$$\mathbf{P}\mathbf{Q}_\infty = \mathbf{P}_\infty\,, \qquad \mathbf{P} \cap \mathbf{Q}_\infty = \mathbf{Q}\,.$$

Clearly $\mathbf{Q}_\infty/\mathbf{Q}$ is also a $\mathbf{Z}_l$-extension. We shall see later that $\mathbf{Q}_\infty$ is the only $\mathbf{Z}_l$-extension over the rational field $\mathbf{Q}$.

In general, if k is a finite algebraic number field, $K = k\mathbf{Q}_\infty$ is a $\mathbf{Z}_l$-extension over k. Hence each k has at least one $\mathbf{Z}_l$-extension over it. In this section, we shall consider the variety of all $\mathbf{Z}_l$-extensions over a given field k.

2.2. In the following, when an algebraic extension of an algebraic number field (finite or infinite over Q) is unramified outside the set of all prime divisors of the ground field lying above the rational prime l, we shall simply say that the extension is unramified outside l.

Let k be a finite algebraic number field and let F be an abelian l-extension over k, not necessarily finite over k. For each prime divisor v of k, let T_v denote the inertia group of v for the extension F/k, and let k_v be the v-completions of k. If v does not lie above l, then the local abelian extension Fk_v/k_v is tamely ramified. Hence, by local class field theory, T_v is a finite group. This implies that if the Galois group $\mathrm{Gal}(F/k)$ is torsion free, then F/k is unramified outside l. In particular, we obtain the following theorem:

THEOREM 1. *Let F be an abelian extension of a finite algebraic number field k such that*

$$\mathrm{Gal}(F/k) \simeq \mathbf{Z}_l^a = \mathbf{Z}_l \oplus \cdots \oplus \mathbf{Z}_l$$

for some integer $a \geqq 0$. Then F/k is unramified outside l. In particular, a $\mathbf{Z}_l$-extension K/k is unramified outside l.

Now, let X be a profinite abelian l-group, i.e., a compact $\mathbf{Z}_l$-module. Clearly X/lX is a vector space over the prime field $\mathbf{F}_l = \mathbf{Z}_l/l\mathbf{Z}_l$. We call its dimension over $\mathbf{F}_l$ the rank of X. On the other hand, the dimension of the vector space $X \oplus_{\mathbf{Z}_l} \mathbf{Q}_l$ over $\mathbf{Q}_l$ will be called the essential rank of X. The rank X is finite if and only if X is noetherian over $\mathbf{Z}_l$. In such a case, the torsion $\mathbf{Z}_l$-submodule Y of X is finite and X/Y is a free $\mathbf{Z}_l$-module with rank equal to the essential rank of X.

THEOREM 2. *Let F be the maximal abelian l-extension over a finite algebraic number field k, unramified outside l. Then rank* $\mathrm{Gal}(F/k)$ *is finite and*

$$r_2 + 1 \leqq \text{ess. rank } \mathrm{Gal}(F/k) \leqq d$$

where $d = [k\colon \mathbf{Q}]$ and r_2 is the number of complex archimedean prime divisors of k.

Proof. For each prime divisor v of k lying above l, let U_v denote the group of local units α in the v-completion k_v such that

$$\alpha \equiv 1 \bmod \mathfrak{p}_v\,,$$

$\mathfrak{p}_v$ being the maximal ideal of k_v. Let U be the direct product of U_v for all such v, and let E be the group of units ε in k satisfying

$$\varepsilon \equiv 1 \bmod \mathfrak{p}_v$$

for every v as mentioned above. We imbed E in U in the obvious manner and denote by $\bar{E}$ the closure of E in U, the latter being a compact group in the topology defined by the $\mathfrak{p}_v$-adic topologies on U_v. Let F' denote the maximal unramified extension of k contained in F. By class field theory, Gal (F'/k) is finite and

$$\text{Gal}\,(F/F') \simeq U/\bar{E}\,.$$

On the other hand, it is also known that

$$\text{rank}\, U < \infty\,, \quad \text{ess. rank}\, U = d = r_1 + 2r_2\,,$$

r_1 being the number of real archimedean prime divisors of k. Furthermore, since E is the direct product of a finite group and $\mathbf{Z}^r$, $r = r_1 + r_2 - 1$, we have

$$\text{ess. rank}\, \bar{E} \leqq r_1 + r_2 - 1\,.$$

Hence rank Gal (F/k) is finite and

$$r_2 + 1 \leqq \text{ess. rank Gal}\,(F/k) \leqq d$$

because

$$\text{ess. rank Gal}\,(F/k) = \text{ess. rank Gal}\,(F/F') = \text{ess. rank}\, U - \text{ess. rank}\, \bar{E}.$$

Let S be a finite set of prime divisors on k including all those lying above the rational prime l, and let F_S denote the maximal abelian l-extension over k, unramified outside S. Using the remark on T_v mentioned above, we see easily that

$$\text{rank Gal}\,(F_S/k) < \infty\,, \quad \text{ess. rank Gal}\,(F_S/k) = \text{ess. rank Gal}\,(F/k)\,.$$

2.3. Let F/k be as stated in Theorem 2. We define an invariant $a_l(k)$ by

$$a_l(k) = \text{ess. rank Gal}\,(F/k)\,.$$

THEOREM 3. *Let $k_{\mathbf{Z}_l}$ denote the composite of all $\mathbf{Z}_l$-extensions over a fixed finite algebraic number field k. Then $k_{\mathbf{Z}_l}/k$ is abelian and*

$$\text{Gal}\,(k_{\mathbf{Z}_l}/k) \simeq \mathbf{Z}_l^a$$

with $a = a_l(k)$ defined above.

Proof. By Theorem 1, $k_{\mathbf{Z}_l}$ is contained in F. One sees immediately that Gal $(F/k_{\mathbf{Z}_l})$ is the torsion $\mathbf{Z}_l$-submodule of Gal (F/k).

The theorem states that $a_l(k)$ is the maximal number of independent $\mathbf{Z}_l$-extensions over k. By Theorem 2,

$$r_2 + 1 \leqq a_l(k) \leqq d\,.$$

Thus, for example, if $k = \mathbf{Q}$, then $d = 1$, $r_2 = 0$ so that

$$a_l(\mathbf{Q}) = 1\,.$$

This implies that $\mathbf{Q}_\infty$ is the only $\mathbf{Z}_l$-extension over Q.

Now, it is conjectured that

$$a_l(k) = r_2 + 1$$

for every k and l; as the proof of Theorem 2 indicates, this is equivalent to

$$\text{ess. rank } \bar{E} = r_1 + r_2 - 1 .$$

We call it Leopoldt's conjecture because when k is a totally real field ($r_2 = 0$), it is equivalent to the non-vanishing of the l-adic regulator of k, $R_l \neq 0$, conjectured by Leopoldt [8]. The conjecture was first verified by J. Ax for k with small degree. Brumer [1] then proved it in the case where k is abelian over Q or over an imaginary quadratic field. Note that for a totally real field k, it states that $k\mathbf{Q}_\infty$ is the only $\mathbf{Z}_l$-extension over k[6].

3. Associated modules

3.1. Let K be a $\mathbf{Z}_l$-extension over a finite algebraic number field k and let $\Gamma = \text{Gal}(K/k) \simeq \mathbf{Z}_l$. Corresponding to the subgroups Γ_n, $n \geqq 0$, of Γ defined in §1.3, there exists a sequence of fields

$$k = k_0 \subset k_1 \subset \cdots \subset k_n \subset \cdots \subset K$$

such that each k_n/k is a cyclic extension of degree l^n, $n \geqq 0$. K is the union of all these k_n, $n \geqq 0$, and there exists no other field between k and K.

Let F be an abelian l-extension over K such that F/k is also a Galois extension. For each $n \geqq 0$, let F_n denote the maximal abelian extension of k_n contained in F. Then

$$K \subseteq F_0 \subseteq F_1 \subseteq \cdots \subseteq F_n \subseteq \cdots \subseteq F$$

and F is the union of all F_n, $n \geqq 0$. Let

$$G = \text{Gal}(F/k) , \qquad K = \text{Gal}(F/K) .$$

Then X is a closed abelian normal subgroup of the profinite l-group G such that

$$G/X = \text{Gal}(K/k) = \Gamma .$$

Hence Γ acts on X in the obvious manner and it makes X into a compact Γ-module in the sense of §1.1. Fixing a topological generator γ_0 of Γ, we make X also into a compact Λ-module. The characterization of $\omega_n X$ given in §1.3 then shows that

$$\omega_n X = \text{Gal}(F/F_n) , \qquad X/\omega_n X = \text{Gal}(F_n/K) , \qquad n \geqq 0 .$$

[6] At the Seminar on Modern Methods in Number Theory in Tokyo, 1971, the solution of Leopoldt's conjecture was announced by Y. Akagawa. The details, however, are not yet published.

LEMMA 1. *X is noetherian over Λ if and only if rank* Gal (F_0/k) *is finite.*

Proof. We know that X is noetherian if and only if $X/\mathfrak{m}X$ is finite, $\mathfrak{m}$ being the maximal ideal of Λ. Let

$$Y = X/\omega_0 X = \mathrm{Gal}\,(F_0/K)\,, \qquad \omega_0 = T\,.$$

Then $X/\mathfrak{m}X = Y/lY$ so that $X/\mathfrak{m}X$ is finite if and only if rank Y is finite. However, $k \subseteq K \subseteq F_0$ and $\mathrm{Gal}\,(K/k) = \Gamma \simeq \mathbf{Z}_l$. Hence rank Y is finite if and only if rank $\mathrm{Gal}\,(F_0/k)$ is finite.

3.2. Let M denote the maximal abelian l-extension over K, unramified outside l. Clearly M/k is a Galois extension so that we may apply the result of § 3.1 to the extension M/k. Let M_n be the maximal abelian extension of k_n contained in M. We see easily that M_n is also the maximal abelian l-extension over k_n unramified outside l. We have

$$K \subseteq M_0 \subseteq M_1 \subseteq \cdots \subseteq M_n \subseteq \cdots \subseteq M$$

and M is the union of all M_n, $n \geqq 0$. Let

$$X = \mathrm{Gal}\,(M/K)\,.$$

Then X is a compact Γ-module as well as a compact Λ-module, and

$$\omega_n X = \mathrm{Gal}\,(M/M_n)\,, \qquad X/\omega_n X = \mathrm{Gal}\,(M_n/K)\,, \qquad n \geqq 0\,.$$

THEOREM 4. Gal (M/K) *is a noetherian Λ-module.*

Proof. Since M_0 is the maximal abelian l-extension over k, unramified outside l, it follows from Theorem 2 that rank $\mathrm{Gal}\,(M_0/k)$ is finite. Hence X is noetherian by Lemma 1.

Let S be a finite set of prime divisors on k containing all those which lie above the rational prime l and let M_S denote the maximal abelian l-extension over K such that M_S/k is unramified outside S. As in the above, we then see that the compact Λ-module $\mathrm{Gal}\,(M_S/K)$ is noetherian over Λ.

3.3. To prove the next theorem, we need some group-theoretical preliminaries. Let both A and B be abelian groups and let

$$\alpha\colon A \longrightarrow A\,, \quad \beta\colon B \longrightarrow B\,, \quad \varphi\colon A \longrightarrow B$$

be homomorphisms such that the following diagram is commutative:

$$\begin{array}{ccc} A & \xrightarrow{\alpha} & A \\ \downarrow{\scriptstyle\varphi} & & \downarrow{\scriptstyle\varphi} \\ B & \xrightarrow{\beta} & B\,. \end{array}$$

This defines a commutative diagram

$$\begin{array}{ccccc} \alpha(A) & \longrightarrow & A & \longrightarrow & A/\alpha(A) \\ \downarrow{\scriptstyle\varphi'} & & \downarrow{\scriptstyle\varphi} & & \downarrow{\scriptstyle\varphi''} \\ \beta(B) & \longrightarrow & B & \longrightarrow & B/\beta(B) . \end{array}$$

Denoting by $|G|$ the order of a group G, let,

$$a = \max(|\mathrm{Ker}\,\varphi|, \ |\mathrm{Coker}\,\varphi|) < \infty .$$

Then, by elementary argument, we can prove the following lemma:

LEMMA 2.

$$|\mathrm{Ker}\,\varphi'| , \ |\mathrm{Coker}\,\varphi'| , \ |\mathrm{Coker}\,\varphi''| \leqq a , \ |\mathrm{Ker}\,\varphi''| \leqq a^2 .$$

LEMMA 3. *Let X be a noetherian Λ-module. Then:*

(i) *rank $X/\omega_n X < \infty$ for every $n \geqq 0$,*

(ii) *X is torsion if and only if ess. rank $X < \infty$, and this is so if and only if ess. rank $X/\omega_n X$ is bounded for all $n \geqq 0$,*

(iii) *X is torsion with $\mu(X) = 0$ if and only if rank $X < \infty$, and this is so if and only if rank $X/\omega_n X$ is bounded for all $n \geqq 0$.*

Proof. Let $f\colon X \to E$ be a pseudo-isomorphism of X into an elementary Λ-module E (see §1.2) and let

$$f_n\colon X/\omega_n X \longrightarrow E/\omega_n E , \qquad n \geqq 0 ,$$

be the maps induced by f. By Lemma 2, $|\mathrm{Ker}\,f_n|$ and $|\mathrm{Coker}\,f_n|$ are bounded for all $n \geqq 0$. Hence it is sufficient to prove the lemma for E instead of X. Now, E is a direct sum of modules of the form Λ or $\Lambda/\mathfrak{p}^m$, $m \geqq 1$, $\mathfrak{p}$ being a prime ideal of height 1 in Λ. Therefore the proof is again reduced to the case where $X = \Lambda$ or $X = \Lambda/\mathfrak{p}^m$ with $\mathfrak{p} = (l)$ or $\mathfrak{p} = (p(T))$, and for such X, (i), (ii), and (iii) can be verified easily by simple computation.

3.4. As before, let K be a $\mathbf{Z}_l$-extension over a finite algebraic number field k and let k_n, $n \geqq 0$, be the intermediate fields of k and K. Let S_0 denote the set of all prime divisors of k which are ramified in K. Since K/k is an infinite abelian extension, S_0 is non-empty, and since K/k is unramified outside l, S_0 is a finite set. For each v in S_0, let T_v denote the inertia group of v for the extension K/k. Then T_v is a non-trivial closed subgroup of Γ so that $T_v = \Gamma_{n_v}$ for some $n_v \geqq 0$. Let

$$n_0 = n_0(K/k) = \max(n_v;\ v \in S_0) .$$

Then, if $n \geqq n_0$, each prime divisor of k_n is either unramified or fully ramified in K. Therefore the number of the prime divisors of k_n which are ramified in K, is the same for all $n \geqq n_0$, and it is equal to the number of prime divisors of K which are ramified for the extension K/k. We shall denote this number by

$$s = s(K/k) .$$

As S_0 is non-empty,

$$0 < s(K/k) < \infty .$$

Now, let L denote the maximal unramified abelian l-extension over K. L is contained in the field M of § 3.2 and L/k is again a Galois extension. Let L_n be the maximal abelian extension of k_n, contained in L. Then

$$K \subseteq L_0 \subseteq L_1 \subseteq \cdots \subseteq L_n \subseteq \cdots \subseteq L \subseteq M$$

and L is the union of all L_n, $n \geqq 0$. Let

$$X = \mathrm{Gal}\,(L/K) .$$

By § 3.1, X is a compact Γ-module and, hence, a compact Λ-module, and

$$\omega_n X = \mathrm{Gal}\,(L/L_n) , \qquad X/\omega_n X = \mathrm{Gal}\,(L_n/K) , \qquad n \geqq 0 .$$

For each $n \geqq 0$, let K_n denote the maximal unramified abelian l-extension over k_n. K_n is a finite extension of k_n and is the maximal unramified extension of k_n contained in L_n. It is easy to see that

$$k \subseteq K_0 \subseteq K_1 \subseteq \cdots \subseteq K_n \subseteq \cdots \subseteq L$$

and L is also the union of all K_n, $n \geqq 0$.

We now fix an integer $n \geqq n_0 = n_0(K/k)$ and denote by $w_1, \cdots, w_s$, $s = s(K/k)$, all the prime divisors of k_n which are ramified in K. For each i, $1 \leqq i \leqq s$, let T_i denote the inertia group of w_i for the abelian extension L_n/k_n. Since w_i is fully ramified in K and since L_n/K is an unramified extension, we see that

$$T_i \simeq \Gamma_n \simeq Z_l , \qquad i = 1, \cdots, s .$$

On the other hand, no prime divisor of k_n, different from $w_1, \cdots, w_s$, is ramified in L_n. Hence

$$\mathrm{Gal}\,(L_n/K_n) = T_1 T_2 \cdots T_s$$

for the maximal unramified extension K_n of k_n contained in L_n. Using the fact that $\mathrm{Gal}\,(K_n/k_n)$ is finite, we obtain

$$\text{ess. rank}\,\mathrm{Gal}\,(L_n/k) = \text{ess. rank}\,(T_1 T_2 \cdots T_s) \leqq s$$

so that

$$\text{ess. rank}\,\mathrm{Gal}\,(L_n/K) = \text{ess. rank}\,\mathrm{Gal}\,(L_n/k) - 1 \leqq s - 1 ,$$

i.e.,

$$\text{ess. rank}\, X/\omega_n X \leqq s - 1 .$$

This holds for all $n \geqq n_0$ and, hence, for all $n \geqq 0$. Therefore the following theorem is proved by Lemma 3:

THEOREM 5. *Let* $X = \mathrm{Gal}(L/K)$. *Then* X *is a noetherian torsion* Λ-*module and*

$$\text{ess. rank } X/\omega_n X \leqq s - 1 , \qquad \text{for all } n \geqq 0 ,$$

with $s = s(K/k)$ *defined above.*

In the above proof, we fixed an integer $n \geqq n_0$ and considered the inertia groups $T_1, \cdots, T_s$ in $\mathrm{Gal}(L_n/k_n)$. However, we can also see more precisely how these groups T_i vary with the index $n \geqq n_0$. In this manner we obtain the following result[7]:

THEOREM 6. *Let* $X = \mathrm{Gal}(L/K)$ *as above and let*

$$Y = \mathrm{Gal}(L/KK_{n_0}) , \qquad n_0 = n_0(K/k) .$$

Then, for $n \geqq n_0$,

$$K \cap K_n = k_n ,$$

$$\mathrm{Gal}(L/KK_n) = \nu_{n_0,n} Y ,$$

$$\mathrm{Gal}(K_n/k_n) \xrightarrow{\sim} \mathrm{Gal}(KK_n/K) = X/\nu_{n_0,n} Y .$$

For each x in $X = \mathrm{Gal}(L/K)$, let $x \mid K_n$ denote the restriction of x on the subfield K_n of L. The above isomorphism $\mathrm{Gal}(K_n/k_n) \xrightarrow{\sim} X/\nu_{n_0,n} Y$ states that the map

$$x \longmapsto x \mid K_n$$

defines a surjective morphism $X \longrightarrow \mathrm{Gal}(K_n/k_n)$ with kernel $\nu_{n_0,n} Y$. Note also that

$$X/Y = X/\nu_{n_0,n_0} Y \xrightarrow{\sim} \mathrm{Gal}(K_{n_0}/k_{n_0})$$

so that X/Y is a finite module. Hence Y, as well as X, is a noetherian torsion Λ-module and

$$X \sim Y$$

in the sense of § 1.2.

4. Fundamental diagrams

4.1. Let K/k, k_n, K_n, etc., be the same as in § 3. For each $n \geqq 0$, let I_n denote the ideal group of k_n, P_n the subgroup of principal ideals in I_n, C_n the ideal-class group of k_n, i.e., $C_n = I_n/P_n$, and A_n the Sylow l-subgroup

[7] See [4], § 7.4.

of C_n. By class field theory, we have a canonical isomorphism

$$A_n \xrightarrow{\sim} \mathrm{Gal}\,(K_n/k_n)\,,$$
$$c \longmapsto \left(\frac{K_n/k_n}{\mathfrak{a}}\right),$$

where $\left(\frac{K_n/k_n}{\mathfrak{a}}\right)$ denotes the Artin symbol for an ideal $\mathfrak{a}$ contained in the ideal-class c. For $n \geqq n_0$, the product of this and the morphism in Theorem 6 gives us a fundamental isomorphism

$$A_n \xrightarrow{\sim} X/\nu_{n_0,n}Y\,,$$
$$c \longmapsto x \bmod \nu_{n_0,n}Y$$

where c and x are related by

$$x \mid K_n = \left(\frac{K_n/k_n}{\mathfrak{a}}\right)$$

with $\mathfrak{a}$ in the ideal-class c.

Now, let $m \geqq n \geqq 0$. Then the natural injective map $I_n \to I_m$ induces morphisms

$$C_n \longrightarrow C_m\,, \qquad A_n \longrightarrow A_m\,.$$

On the other hand, the norm map $N_{n,m}$ from k_m to k_n defines

$$C_m \longrightarrow C_n\,, \qquad A_m \longrightarrow A_n\,.$$

If $m \geqq n \geqq n_0$, we also have two morphisms

$$\nu_{n,m}\colon X/\nu_{n_0,n}Y \longrightarrow X/\nu_{n_0,m}Y\,,$$
$$X/\nu_{n_0,m}Y \longrightarrow X/\nu_{n_0,n}Y\,,$$

defined respectively by

$$x \bmod \nu_{n_0,n}Y \longmapsto \nu_{n,m}x \bmod \nu_{n_0,m}Y\,,$$
$$x \bmod \nu_{n_0,m}Y \longmapsto x \bmod \nu_{n_0,n}Y\,.$$

THEOREM 7. *Let X and Y be the same as in Theorem 6. For $m \geqq n \geqq n_0 = n_0(K/k)$, the following diagrams are commutative:*

$$\begin{array}{ccc} A_n & \xrightarrow{\sim} & X/\nu_{n_0,n}Y \\ \downarrow & & \downarrow \\ A_m & \xrightarrow{\sim} & X/\nu_{n_0,m}Y\,, \end{array} \qquad \begin{array}{ccc} A_m & \xrightarrow{\sim} & X/\nu_{n_0,m}Y \\ \downarrow & & \downarrow \\ A_n & \xrightarrow{\sim} & X/\nu_{n_0,n}Y\,. \end{array}$$

Proof. Let x be an element in X. Find an ideal $\mathfrak{A}$ of k_m in an ideal-class of A_m such that

$$x \mid k_m = \left(\frac{K_m/k_m}{\mathfrak{A}}\right).$$

Since $K_n k_m \subseteq K_m$, we have

$$x \mid k_n = \left(\frac{K_n/k_n}{\mathfrak{a}}\right)$$

with $\mathfrak{a} = N_{n,m}\mathfrak{A}$. This proves the commutativity of the second diagram. On the other hand, if ρ ranges over all elements of Gal (k_m/k_n),

$$\left(\frac{K_m/k_m}{\mathfrak{a}}\right) = \prod_\rho \left(\frac{K_m/k_m}{\rho(\mathfrak{A})}\right) = \prod_\rho \rho\left(\frac{K_m/k_m}{\mathfrak{A}}\right)$$
$$= \prod_\rho \rho(x \mid K_m) = \nu_{n,m} x \mid K_m .$$

Hence the commutativity of the first diagram is also proved.

4.2. Let h_n and l^{e_n} denote the orders of the finite abelian groups C_n and A_n respectively ($n \geqq 0$); h_n is the class number of k_n and l^{e_n} is the highest power of l which divides h_n. By the fundamental isomorphism $A_n \cong X/\nu_{n_0,n} Y$, l^{e_n} is also the order of $X/\nu_{n_0,n}Y$. Let $\lambda(X)$ and $\mu(X)$ be the invariants of the noetherian torsion Λ-module X, defined in §1.2. Using the result on the structure of such Λ-modules stated in §1.2, we can prove the following class number formula for K/k[8]:

For every sufficiently large n, the exponent e_n is given by

$$e_n = \lambda n + \mu l^n + \nu ,$$

where $\lambda = \lambda(X)$, $\mu = \mu(X)$, and ν is another integer, also independent of n. Since these integers λ, μ, and ν are uniquely determined for the given Z_l-extension K/k, they are denoted by $\lambda(K/k)$, $\mu(K/k)$, and $\nu(K/k)$ respectively.

4.3. For $n \geqq 0$, let $\mathfrak{o}_n$ denote the ring of all algebraic integers in k_n, and $\mathfrak{o}'_n$ the ring of all l-integers in k_n, namely, the union of all $l^{-a}\mathfrak{o}_n$, $a \geqq 0$, in k_n. The invertible $\mathfrak{o}'_n$-submodules in k_n will be called l-ideals of k_n. Let I'_n denote the multiplicative group of such l-ideals of k_n, P'_n the subgroup of principal l-ideals in I'_n, C'_n the l-ideal-class group of k_n, i.e., $C'_n = I'_n/P'_n$, and A'_n the Sylow l-subgroup of C'_n. The morphism

$$I_n \longrightarrow I'_n ,$$
$$\mathfrak{a} \longmapsto \mathfrak{a}\mathfrak{o}'_n$$

is surjective and its kernel is the subgroup of I_n generated by the prime ideals of k_n dividing the rational prime l. Hence, if I^0_n denotes the subgroup of all ideals in I_n which are prime to l, then

[8] See [4], §7.5.

$$I_n^0 \xrightarrow{\sim} I'_n .$$

The morphism in the above also induces surjective maps

$$C_n \longrightarrow C'_n , \qquad A_n \longrightarrow A'_n .$$

It follows that there exists a field K'_n such that

$$k_n \subseteq K'_n \subseteq K_n$$

and such that

$$\begin{array}{ccc} A_n & \xrightarrow{\sim} & \mathrm{Gal}\,(K_n/k_n) \\ \downarrow & & \downarrow \\ A'_n & \xrightarrow{\sim} & \mathrm{Gal}\,(K'_n/k_n) \end{array}$$

is commutative. Here the vertical arrow on the right is the restriction map and the isomorphism below is given by

$$c' \longmapsto \left(\frac{K'_n/k_n}{\mathfrak{a}}\right) ,$$

$\mathfrak{a}$ being an ideal such that $\mathfrak{a}\mathfrak{o}'_n$ is contained in c'. K'_n is also characterized as the maximal extension of k_n in K_n, in which every prime divisor (prime ideal) of k_n lying above l (dividing l), is completely decomposed. For $m \geqq n \geqq 0$,

$$K'_n \subseteq K'_m ,$$

and the injections $I'_n \to I'_m$ and $N_{n,m}$: $I'_m \to I'_n$ again induce

$$C'_n \longrightarrow C'_m , \qquad A'_n \longrightarrow A'_m ,$$
$$C'_m \longrightarrow C'_n , \qquad A'_m \longrightarrow A'_n .$$

Now, let L' denote the maximal unramified abelian l-extension over K in which every prime divisor of K lying above l is completely decomposed. Then

$$k \subseteq K \subseteq L' \subseteq L ,$$
$$k \subseteq K'_0 \subseteq K'_1 \subseteq \cdots \subseteq K'_n \subseteq \cdots \subseteq L' ,$$

and L' is the union of all K'_n, $n \geqq 0$. Let

$$X' = \mathrm{Gal}\,(L'/K) .$$

Since L'/k is a Galois extension, X' is again a compact Γ-module, and hence a compact Λ-module. As usual, we denote by L'_n the maximal abelian extension of k_n contained in L'. Then

$$k \subseteq K \subseteq L'_0 \subseteq L'_1 \subseteq \cdots \subseteq L'_n \subseteq \cdots \subseteq L'$$

and L' is the union of all L'_n, $n \geqq 0$. Furthermore,

$$\omega_n X' = \mathrm{Gal}\,(L'/L'_n) , \qquad X'/\omega_n X' = \mathrm{Gal}\,(L'_n/K) , \qquad n \geqq 0 .$$

THEOREM 8. *X' is a noetherian torsion Λ-module. Let*

$$Y' = \mathrm{Gal}\,(L'/KK'_{n_0})\,, \qquad n_0 = n_0(K/k)\,.$$

Then

$$K \cap K'_n = k_n\,,$$

$$\mathrm{Gal}\,(L'/KK'_n) = \nu_{n_0,n} Y'\,,$$

$$\mathrm{Gal}\,(K'_n/k_n) \xrightarrow{\sim} \mathrm{Gal}\,(KK'_n/K) = X'/\nu_{n_0,n} Y'$$

for each $n \geqq n_0$. Furthermore, the diagrams

$$\begin{array}{ccc} A'_n & \xrightarrow{\sim} & X'/\nu_{n_0,n}Y' \\ \downarrow & & \downarrow \\ A'_m & \xrightarrow{\sim} & X'/\nu_{n_0,m}Y'\,, \end{array} \qquad \begin{array}{ccc} A'_m & \xrightarrow{\sim} & X'/\nu_{n_0,m}Y' \\ \downarrow & & \downarrow \\ A'_n & \xrightarrow{\sim} & X'/\nu_{n_0,n}Y' \end{array}$$

are commutative ($m \geqq n \geqq n_0$).

Proof. Since X' is a factor module of $X = \mathrm{Gal}\,(L/K)$, the first part is a consequence of Theorem 5. The proof of the remaining part is similar to that of Theorems 6 and 7.

4.4. We shall next consider the structure of $\mathrm{Gal}\,(L/L')$. Clearly

$$\mathrm{Gal}\,(L/L') = \varprojlim \mathrm{Gal}\,(K_n/K'_n)$$

where the inverse limit is taken with respect to the restriction maps

$$\mathrm{Gal}\,(K_m/K'_m) \longrightarrow \mathrm{Gal}\,(K_n/K'_n)\,, \qquad m \geqq n \geqq 0\,.$$

For $n \geqq 0$, let D_n denote the subgroup of C_n generated by the classes of prime ideals of k_n which divide l. Then $A_n \xrightarrow{\sim} \mathrm{Gal}\,(K_n/k_n)$ induces

$$A_n \cap D_n \xrightarrow{\sim} \mathrm{Gal}\,(K_n/K'_n)\,.$$

Furthermore the diagram

$$\begin{array}{ccc} A_m \cap D_m & \xrightarrow{\sim} & \mathrm{Gal}\,(K_m/K'_m) \\ \Big\downarrow N_{n,m} & & \Big\downarrow \\ A_n \cap D_n & \xrightarrow{\sim} & \mathrm{Gal}\,(K_n/K'_n)\,, \end{array} \qquad m \geqq n \geqq 0\,,$$

is commutative. Hence $\mathrm{Gal}\,(L/L')$ is isomorphic to the inverse limit of $A_n \cap D_n$, $n \geqq 0$.

Now, assume that every prime divisor of k, lying above the rational prime l, is ramified in K. Then, by the definition of the integer $n_0 = n_0(K/k)$ in §3.4, each k_n, $n \geqq n_0$, has the same number of prime ideals dividing l:

$$\mathfrak{l}_{n,1}, \cdots, \mathfrak{l}_{n,s}\,, \qquad s = s(K/k)\,,$$

and they satisfy, with a suitable choice of indices i,

$$N_{n,m}\mathfrak{l}_{m,i} = \mathfrak{l}_{m,i}^{l^{m-n}} = \mathfrak{l}_{n,i}$$

for $m \geqq n \geqq n_0$, $i = 1, \cdots, s$. The Galois group Γ permutes $\mathfrak{l}_{n,1}, \cdots, \mathfrak{l}_{n,s}$ among themselves and the subgroup Γ_{n_0} fixes each one of these. Hence we obtain the following result:

THEOREM 9. *Assume that every prime divisor of k, lying above the rational prime l, is ramified in K. Then the rank of the Z_l-module* Gal (L'/L) *is at most equal to $s = s(K/k)$ and the subgroup Γ_{n_0} of Γ, $n_0 = n_0(K/k)$, acts trivially on* Gal (L/L'). *The elementary Λ-module associated with* Gal (L/L') *is of the form $E(\mathfrak{p}_1, \cdots, \mathfrak{p}_a)$ where*

$$\mathfrak{p}_i = (\xi_{n_i}) , \qquad i = 1, \cdots, a ,$$
$$\sum_{i=1}^{a} \deg \xi_{n_i}(T) \leqq s = s(K/k) ,$$

with ξ_n as in §1.3.

We note that the assumption of the theorem is satisfied for every cyclotomic Z_l-extension defined in §6 below (see §6.2, Lemma 4).

Let $l^{e'_n}$ denote the order of the group A'_n, $n \geqq 0$. Similarly, as for e_n in §4.2, it can be proved that for all sufficiently large n,

$$e'_n = \lambda' n + \mu' l^n + \nu'$$

with $\lambda' = \lambda(X')$, $\mu' = \mu(X')$, and with an integer ν' also independent of n. λ', μ', and ν' give us another set of invariants for the Z_l-extension K/k, and we denote them by $\lambda'(K/k)$, $\mu'(K/k)$, and $\nu'(K/k)$ respectively. Under the same assumption as in Theorem 9, we then have

$$\lambda(K/k) - s \leqq \lambda'(K/k) \leqq \lambda(K/k) , \qquad \mu(K/k) = \mu'(K/k) .$$

5. Ideals and units

5.1. Let K/k and k_n, $n \geqq 0$, be the same as above. Let $\mathfrak{o}$ be the ring of all algebraic integers in K. As usual, invertible $\mathfrak{o}$-submodules of K are called ideals of K; these ideals form a multiplicative abelian group I, called the ideal group of K. Let P denote the subgroup of principal ideals in I and let $C = I/P$; C is called the ideal-class group of K. It is easy to see that

$$I = \varinjlim I_n$$

with respect to the maps $I_n \to I_m$, $m \geqq n \geqq 0$, in §4.1. Similarly

$$C = \varinjlim C_n .$$

This implies that each element in C has a finite order, although C may not be a finite group. Hence, let A denote the l-primary component of C, namely the subgroup of all elements in C with orders powers of l. Then

$$A = \varinjlim A_n$$

in the obvious sense.

Let $\mathfrak{o}'$ be the ring of all l-integers in K, i.e., the union of all $l^{-a}\mathfrak{o}$, $a \geqq 0$. Invertible $\mathfrak{o}'$-submodules in K are called l-ideals of K and the group of these l-ideals of K is denoted by I'. As in the above, we also define the subgroup P' of principal l-ideals in I', the l-ideal-class group $C' = I'/P'$ of K, and the l-primary component A' of C'. These I', C', and A' are the direct limits of I'_n, C'_n, and A'_n, $n \geqq 0$, respectively.

5.2. Let c be an element in the kernel of the natural morphism $C_n \rightarrow C_m$, $m \geqq n \geqq 0$, and let $\mathfrak{a}$ be an ideal in the class c. By the definition of the morphism, $\mathfrak{a}$ becomes a principal ideal in k_m: $\mathfrak{a} = (\alpha)$, $\alpha \in k_m$. Applying the norm map $N_{n,m}$ to both sides, we have

$$\mathfrak{a}^{l^{m-n}} = N_{n,m}\mathfrak{a} = (N_{n,m}(\alpha)) .$$

This shows that $c^{l^{m-n}} = 1$ so that c belongs to A_n. Hence

$$\operatorname{Ker}(C_n \longrightarrow C_m) = \operatorname{Ker}(A_n \longrightarrow A_m) , \qquad m \geqq n \geqq 0 .$$

Similarly,

$$\operatorname{Ker}(C'_n \longrightarrow C'_m) = \operatorname{Ker}(A'_n \longrightarrow A'_m) , \qquad m \geqq n \geqq 0 ,$$

and also

$$\operatorname{Ker}(C_n \longrightarrow C) = \operatorname{Ker}(A_n \longrightarrow A) ,$$
$$\operatorname{Ker}(C'_n \longrightarrow C') = \operatorname{Ker}(A'_n \longrightarrow A') , \qquad n \geqq 0 .$$

THEOREM 10. *The orders of the groups*

$$|\operatorname{Ker}(A_n \longrightarrow A_m)| , \qquad |\operatorname{Ker}(A_n \longrightarrow A)| ,$$
$$|\operatorname{Ker}(A'_n \longrightarrow A'_m)| , \qquad |\operatorname{Ker}(A'_n \longrightarrow A')|$$

are bounded for all $m \geqq n \geqq 0$.

Proof. We shall first prove that $|\operatorname{Ker}(A_n \rightarrow A_m)|$ is bounded for all $m \geqq n \geqq 0$. It is sufficient to prove this for $m \geqq n \geqq n_0 = n_0(K/k)$. By Theorem 7,

$$\begin{aligned}|\operatorname{Ker}(A_n \longrightarrow A_m)| &= |\operatorname{Ker}(X/\nu_{n_0,n} Y \longrightarrow X/\nu_{n_0,m} Y)| \\ &\leqq |X/Y|\,|\operatorname{Ker}(Y/\nu_{n_0,n} Y \longrightarrow Y/\nu_{n_0,m} Y)| .\end{aligned}$$

Let

$$f\colon Y \longrightarrow U$$

be a pseudo-isomorphism of Y into an elementary Λ-module U and let

$$a = \max(|\mathrm{Ker}\, f|,\ |\mathrm{Coker}\, f|)\,.$$

For each $n \geqq n_0$, f induces

$$f_n:\ Y/\nu_{n_0,n} Y \longrightarrow U/\nu_{n_0,n} U\,,$$

and by Lemma 2,

$$|\mathrm{Ker}\, f_n| \leqq a^2\,, \qquad |\mathrm{Coker}\, f_n| \leqq a\,.$$

Now, consider the commutative diagram

$$\begin{array}{ccc} Y/\nu_{n_0,n} Y & \xrightarrow{f_n} & U/\nu_{n_0,n} U \\ \downarrow \nu_{n,m} & & \downarrow \nu_{n,m} \\ Y/\nu_{n_0,m} Y & \xrightarrow{f_m} & U/\nu_{n_0,m} U\,. \end{array}$$

Since $Y/\nu_{n_0,n} Y$ is finite and $|\mathrm{Coker}\, f_n| \leqq a$, $U/\nu_{n_0,n} U$ is also a finite group. As we shall see below, this implies that

$$\nu_{n,m}:\ U/\nu_{n_0,n}\, U \longrightarrow U/\nu_{n_0,m} U$$

is injective. Hence it follows from the above diagram that

$$|\mathrm{Ker}\,(Y/\nu_{n_0,n} Y \longrightarrow Y/\nu_{n_0,m} Y)| \leqq |\mathrm{Ker}\, f_n| \leqq a^2$$

so that

$$|\mathrm{Ker}\,(A_n \longrightarrow A_m)| \leqq |X/Y|\, a^2$$

for $m \geqq n \geqq n_0$.

To see that the above morphism $\nu_{n,m}$ is injective, we may assume that the elementary module U is of the form $\Lambda/\mathfrak{p}^e$, $e \geqq 1$, with a prime ideal $\mathfrak{p}$ of height 1 in Λ. As in §1.2, let $\mathfrak{p} = (\pi)$ with $\pi = l$ or $p(T)$. The fact that $U/\nu_{n_0,n} U$ is finite shows that π does not divide $\nu_{n_0,n}$, $n \geqq n_0$, in the unique factorization domain $\Lambda = Z_l[[T]]$. We then see easily that

$$\nu_{n,m}:\ \Lambda/(\nu_{n_0,n}, \pi^e) \longrightarrow \Lambda/(\nu_{n_0,m}, \pi^e)$$

is injective for $m \geqq n \geqq n_0$. This completes the proof of the first part of the theorem.

Now, since $\mathrm{Ker}\,(A_n \to A)$ is the union of the increasing sequence of subgroups $\mathrm{Ker}\,(A_n \to A_m)$, $m \geqq n$, in the finite group A_n, we see that $\mathrm{Ker}\,(A_n \to A) = \mathrm{Ker}\,(A_n \to A_m)$ for all sufficiently large $m \geqq n$. Hence $|\mathrm{Ker}\,(A_n \to A)|$ is also bounded for all $n \geqq 0$. The proof for $|\mathrm{Ker}\,(A'_n \to A'_m)|$ and $|\mathrm{Ker}\,(A'_n \to A')|$ is similar.

5.3. Let

$$X = \mathrm{Gal}\,(L/K)\,, \qquad Y = \mathrm{Gal}\,(L/KK_{n_0})$$

as in §§3 and 4. By §4.1,

$$A = \varinjlim A_n \xrightarrow{\sim} \varinjlim X/\nu_{n_0,n} Y .$$

However, in the exact sequence

$$0 \longrightarrow \varinjlim Y/\nu_{n_0,n} Y \longrightarrow \varinjlim X/\nu_{n_0,n} Y \longrightarrow \varinjlim X/Y \longrightarrow 0 ,$$

we have

$$\varinjlim X/Y = 0$$

because the maps

$$\nu_{n,m}\colon X/Y \longrightarrow X/Y$$

are zero whenever m is sufficiently larger than n. Hence

$$A \xrightarrow{\sim} \varinjlim Y/\nu_{n_0,n} Y .$$

Now, let

$$\pi_n = \nu_{n_0,n_0+n} , \qquad n \geqq 0 .$$

The fact that $Y/\nu_{n_0,n} Y$ is finite shows that the principal divisors (π_n), $n \geqq 0$, are disjoint from $\operatorname{div} Y$. Therefore it follows from the above and from the formula for adjoints in §1.3 that

$$\begin{aligned} \operatorname{Hom}_{\mathbf{Z}_l}(A, \mathbf{Q}_l/\mathbf{Z}_l) &\xrightarrow{\sim} \varprojlim \operatorname{Hom}_{\mathbf{Z}_l}(Y/\pi_n Y, \mathbf{Q}_l/\mathbf{Z}_l) \\ &\simeq \alpha(Y) . \end{aligned}$$

This is a Λ-isomorphism if the Λ-structure on $\operatorname{Hom}_{\mathbf{Z}_l}(A, \mathbf{Q}_l/\mathbf{Z}_l)$ is defined as in §1.3. Thus:

THEOREM 11. *Let* $X = \operatorname{Gal}(L/K)$ *and* $Y = \operatorname{Gal}(L/KK_{n_0})$ *with* $n_0 = n_0(K/k)$. *Then*

$$\operatorname{Hom}_{\mathbf{Z}_l}(A, \mathbf{Q}_l/\mathbf{Z}_l) \simeq \alpha(Y) \sim X$$

as noetherian torsion Λ*-modules. Similarly,*

$$\operatorname{Hom}_{\mathbf{Z}_l}(A', \mathbf{Q}_l/\mathbf{Z}_l) \simeq \alpha(Y') \sim X'$$

for $X' = \operatorname{Gal}(L'/K)$, $Y' = \operatorname{Gal}(L'/KK'_{n_0})$.

Note that since A is l-primary, $\operatorname{Hom}_{\mathbf{Z}_l}(A, \mathbf{Q}_l/\mathbf{Z}_l)$ is isomorphic to the dual $\hat{A}$ of the discrete abelian group A in the sense of Pontrjagin. Similarly, $\operatorname{Hom}_{\mathbf{Z}_l}(A', \mathbf{Q}_l/\mathbf{Z}_l)$ is isomorphic to the dual $\hat{A}'$ of A'.

5.4. For each $n \geqq 0$, let E_n denote the group of all units in k_n, and E'_n the group of all l-units in k_n, i.e., the multiplicative group of the ring $\mathfrak{o}'_n$. Clearly, E_n is a subgroup of E'_n. We also have injective maps

$$E_n \longrightarrow E_m\,, \qquad E'_n \longrightarrow E'_m$$

for $m \geqq n \geqq 0$. Let

$$E = \varinjlim E_n\,, \qquad E' = \varinjlim E'_n\,.$$

E is nothing but the group of all units in K, and E' the group of all l-units in K, i.e., the multiplicative group of $\mathfrak{o}'$. Various Galois groups act on these groups in the obvious manner.

THEOREM 12.

$$\mathrm{Ker}\,(A'_n \longrightarrow A'_m) \xrightarrow{\simeq} H^1(\mathrm{Gal}\,(k_m/k_n), E'_m)\,,$$

$$\mathrm{Ker}\,(A'_n \longrightarrow A') \xrightarrow{\simeq} H^1(\mathrm{Gal}\,(K/k_n), E')\,, \qquad m \geqq n \geqq 0\,.$$

Proof. It is sufficient to prove the first isomorphism. Fix a generator σ of $\mathrm{Gal}\,(k_m/k_n)$. Let c be an element in $\mathrm{Ker}\,(A'_n \to A'_m)$ and let $\mathfrak{a}'$ be an l-ideal in the class c. Then $\mathfrak{a}'$ becomes principal in k_m so that $\mathfrak{a}' = \alpha\,\mathfrak{o}'_m$ with $\alpha \neq 0$ in k_m. Let $\varepsilon = \alpha^{\sigma-1}$. ε is an element of E'_m such that $N_{n,m}(\varepsilon) = 1$, and the cohomology class $\{\varepsilon\}$ determined by ε depends only upon c so that we have a morphism

$$\mathrm{Ker}\,(A'_n \longrightarrow A'_m) \longrightarrow H^1(\mathrm{Gal}\,(k_m/k_n), E'_m)$$
$$c \longmapsto \{\varepsilon\}\,.$$

It is easy to see that this is injective.

To see that it is also surjective, let $\{\varepsilon\}$ be a cohomology class represented by an element ε in E'_m such that $N_{n,m}(\varepsilon) = 1$. Let $\varepsilon = \alpha^{\sigma-1}$ with α in k_m and let $\mathfrak{a}$ be an ideal in I^0_m (prime to l) such that $\alpha\mathfrak{o}'_m = \mathfrak{a}\mathfrak{o}'_m$. Then $\mathfrak{a}^{\sigma-1}\mathfrak{o}'_m = \varepsilon\mathfrak{o}'_m = \mathfrak{o}'_m$ implies $\mathfrak{a}^\sigma = \mathfrak{a}$. Since no prime ideal of k_n, prime to l, is ramified in k_m, $\mathfrak{a}$ is an ideal in I_n. Let c be the class of the l-ideal $\mathfrak{a}' = \mathfrak{a}\mathfrak{o}'_n$ in I'_n. Then c belongs to $\mathrm{Ker}\,(A'_n \to A'_m)$ and c is mapped to $\{\varepsilon\}$ under the above morphism.

COROLLARY. *The orders*

$$|H^1(\mathrm{Gal}\,(k_m/k_n), E'_m)|\,, \qquad |H^1(\mathrm{Gal}\,(K/k_n), E')|$$

are bounded for all $m \geqq n \geqq 0$.

There exist similar morphisms

$$\mathrm{Ker}\,(A_n \longrightarrow A_m) \longrightarrow H^1(\mathrm{Gal}\,(k_m/k_n), E_m)\,,$$
$$\mathrm{Ker}\,(A_n \longrightarrow A) \longrightarrow H^1(\mathrm{Gal}\,(K/k_n), E)\,.$$

These are injective but not necessarily surjective.

6. Cyclotomic Z_l-extensions

6.1. As in §2.1, let $\mathbf{P}$ denote the cyclotomic field of l-th or 4-th roots of unity according as $l>2$ or $l=2$, and let $\mathbf{P}_\infty=\mathbf{Q}(W)=\mathbf{P}(W)$ where W is the group of all l^n-th roots of unity for all $n\geqq 0$. A $\mathbf{Z}_l$-extension K over a finite algebraic number field k is called a cyclotomic $\mathbf{Z}_l$-extension if

$$\mathbf{P}\subseteq k\,,\qquad K=k\mathbf{P}_\infty=k(W)\,.$$

A typical example is given of course by $\mathbf{P}_\infty/\mathbf{P}$. In the following, we shall study arithmetic properties of such cyclotomic $\mathbf{Z}_l$-extensions.

Let K/k be as stated above and let

$$\mathbf{P}=\mathbf{P}_0\subset\mathbf{P}_1\subset\cdots\subset\mathbf{P}_n\subset\cdots\subset\mathbf{P}_\infty\,,$$
$$k=k_0\subset k_1\subset\cdots\subset k_n\subset\cdots\subset K$$

be the sequences of intermediate fields for $\mathbf{P}_\infty/\mathbf{P}$ and K/k respectively (see §3.1.). By the assumption on k,

$$k\cap\mathbf{P}_\infty=\mathbf{P}_e$$

for some integer $e\geqq 0$. We then have

$$k_n\cap\mathbf{P}_\infty=\mathbf{P}_{n+e}\,,\qquad k_n=k\mathbf{P}_{n+e}\,,\qquad n\geqq 0\,.$$

Let $k_n^\times$ denote the multiplicative group of k_n and define

$$W_n=W\cap k_n^\times\,,\qquad q_n=|W_n|\,,\qquad n\geqq 0\,.$$

Then W is the union of the increasing sequence of subgroups W_n, $n\geqq 0$, and

$$q_n=l^nq\,,\qquad n\geqq 0$$

with $q=q_0=l^{e+1}$ or 2^{e+2} according as $l>2$ or $l=2$.

LEMMA 4. *A prime divisor v on k is ramified in K if and only if v lies above the rational prime l.*

Proof. By Theorem 1, it suffices to show that if v lies over l, then v is ramified in K. Let v_n denote a prime divisor of k_n, $n\geqq 0$, lying above v, and let w_n be the restriction (projection) of v_n on the subfield $\mathbf{P}_{n+e}$. For the ramification indices of these prime divisors, we then have

$$\begin{aligned}e(v_n/l)&=e(v_n/v)e(v/l)\\&=e(v_n/w_n)e(w_n/l)\,.\end{aligned}$$

Here $e(w_n/l)=[\mathbf{P}_{n+e}:\mathbf{Q}]=(l-1)l^{n+e}$ or 2^{n+e+1}. Hence if n is sufficiently large, then $e(v_n/v)>1$ so that v is ramified in K.

We note that the statement of the lemma is not necessarily true for a general $\mathbf{Z}_l$-extension.

6.2. Let n_0 be the integer defined in §3.4: $n_0=n_0(K/k)$. It follows from the above lemma that for $n\geqq n_0$, every prime divisor of k_n, lying above l, is fully ramified in K so that the number of prime divisors of K which lie above the rational prime l is $s = s(K/k)$. We fix such a prime divisor w of K and denote by K_w the completion of K with respect to w. For each $n \geqq 0$, let $k_{n,w}$ be the completion of k_n with respect to the restriction of w on k_n. Then we have

$$\mathbf{Q}_l \subseteq k_w = k_{0,w} \subseteq k_{1,w} \subseteq \cdots \subseteq k_{n,w} \subseteq \cdots \subseteq K_w .$$

LEMMA 5. *The union of all* $k_{n,w}$, $n \geqq 0$, *is the algebraic closure of* $\mathbf{Q}_l$ *in* K_w.

Proof. It is clear that the union is contained in the algebraic closure of $\mathbf{Q}_l$ in K_w. Let α be an element of K_w, algebraic over $\mathbf{Q}_l$, and let

$$\alpha = \lim_{i\to\infty} \alpha_i , \qquad \alpha_i \in K ,$$

in the w-topology on K_w. By a theorem of Krasner,

$$\mathbf{Q}_l(\alpha) \subseteq \mathbf{Q}_l(\alpha_i)$$

whenever i is sufficiently large. Fix such an i. Since K is the union of k_n, $n \geqq 0$, α_i is contained in some k_n. Then $\mathbf{Q}_l(\alpha_i) \subseteq k_{n,w}$ so that α belongs to $k_{n,w}$.

Let t be any non-negative integer. An element α in K will be called a local t-th power in K (at l) if α is a t-th power in K_w for every prime divisor w of K, lying above l. Similarly, an element β in k_n will be called a local t-th power in k_n (at l) if β is a t-th power in $k_{n,w'}$ for every prime divisor w' of k_n, lying above l.

LEMMA 6. *An element* α *in* K *is a local* t-th *power in* K *if and only if it is a local* t-th *power in* k_n *for some* $n \geqq 0$.

Proof. Let α be local t-th power in K and let $\alpha = \beta^t$ with β in K_w, w being a prime divisor of K as stated above. Then β is algebraic over $\mathbf{Q}_l$ so that it belongs to $k_{m,w}$ for some $m = m_w$. Let n be the maximum of m_w when w ranges over s prime divisors of K lying above l. Then α is a local t-th power in k_n. The converse is trivial.

6.3. Let E' and E'_n, $n \geqq 0$, be the groups of l-units as defined in §5.4. Let $\tilde{W}$ and $\tilde{W}_n$ denote the groups of all roots of unity in K and in k_n, respectively. It is easy to see that $\tilde{W}$ is the direct product of W and a finite group with order prime to l. Clearly $\tilde{W} \subseteq E'$ and $\tilde{W}_n = \tilde{W} \cap E'_n$. Hence, if we define

$$\mathcal{E}' = E'/\tilde{W} , \qquad \mathcal{E}'_n = E'_n/\tilde{W}_n , \qquad n \geqq 0 ,$$

$\mathcal{E}'$ is the union of the increasing sequence of subgroups $\mathcal{E}'_n$, $n \geqq 0$. The Galois group Γ acts on $\mathcal{E}'$ and $\mathcal{E}'_n$, $n \geqq 0$, in the obvious manner. Using the fact

$H^1(\Gamma, \tilde{W}) = 0$, one sees immediately that $\mathcal{E}'_n$ is the subgroup of all elements in $\mathcal{E}'$ fixed by $\Gamma_n = \mathrm{Gal}(K/k_n)$:

$$\mathcal{E}'_n = \mathcal{E}'^{\Gamma_n} .$$

For $n \geqq 0$, let

$$d = [k\colon \mathbf{Q}] , \quad d_n = [k_n\colon \mathbf{Q}] = dl^n ,$$

and let s_n be the number of prime divisors of k_n lying above l. We know that $s_n = s = s(K/k)$ if $n \geqq n_0$. By a classical result, $\mathcal{E}'_n$ is a free abelian group of rank $d_n/2 + s_n - 1$:

$$\mathcal{E}'_n \simeq \mathbf{Z}^{d_n/2+s_n-1} , \qquad n \geqq 0 .$$

For $m \geqq n \geqq 0$, we have

$$\mathcal{E}'_n = \mathcal{E}'^{\Gamma_n}_m , \quad \mathcal{E}'_m \simeq \mathbf{Z}^{d_m/2+s_m-1} .$$

Hence $\mathcal{E}'_n$ must be a direct summand in $\mathcal{E}'_m$, and it follows that $\mathcal{E}'$ is a free abelian group and that $\mathcal{E}'_n$ is also a direct summand of $\mathcal{E}'$.

We can obtain similar results for $\mathcal{E} = E/\tilde{W}$ and $\mathcal{E}_n = E_n/\tilde{W}_n$, $n \geqq 0$.

Now, let

$$D = E' \otimes_{\mathbf{Z}} (\mathbf{Q}_l/\mathbf{Z}_l) .$$

Since $\tilde{W} \otimes_{\mathbf{Z}} (\mathbf{Q}_l/\mathbf{Z}_l) = 0$, we may identify D with $\mathcal{E}' \otimes_{\mathbf{Z}} (\mathbf{Q}_l/\mathbf{Z}_l)$:

$$D = \mathcal{E}' \otimes_{\mathbf{Z}} (\mathbf{Q}_l/\mathbf{Z}_l) .$$

Γ acts on D through its action on the first factor E' or $\mathcal{E}'$. Let D^{Γ_n}, $n \geqq 0$, denote the subgroup of all elements in D which are fixed by the subgroup Γ_n. Since $\mathcal{E}'_n$ is a direct summand of $\mathcal{E}'$, we may consider $\mathcal{E}'_n \otimes_{\mathbf{Z}} (\mathbf{Q}_l/\mathbf{Z}_l)$ as a submodule of $D = \mathcal{E}' \otimes_{\mathbf{Z}} (\mathbf{Q}_l/\mathbf{Z}_l)$. As such,

$$\mathcal{E}'_n \otimes_{\mathbf{Z}} (\mathbf{Q}_l/Z_l) \subseteq D^{\Gamma_n} .$$

Let $\mathbf{Q}'$ denote the additive group of all rational numbers of which the denominators are powers of l. Since $\mathbf{Q}'/\mathbf{Z} \simeq \mathbf{Q}_l/\mathbf{Z}_l$ and since $\mathcal{E}'$ is free abelian, we have an exact sequence

$$0 \longrightarrow \mathcal{E}' \longrightarrow \mathcal{E}' \otimes_{\mathbf{Z}} \mathbf{Q}' \longrightarrow D \longrightarrow 0 .$$

As the group in the middle is divisible by l, we obtain from the above an exact sequence

$$\longrightarrow H^0(\Gamma_n, \mathcal{E}' \otimes_{\mathbf{Z}} \mathbf{Q}') \longrightarrow H^0(\Gamma_n, D) \longrightarrow H^1(\Gamma_n, \mathcal{E}') \longrightarrow 0$$

and hence an isomorphism

$$D^{\Gamma_n}/\mathcal{E}_n \otimes_{\mathbf{Z}} (\mathbf{Q}_l/\mathbf{Z}_l) \simeq H^1(\Gamma_n, \mathcal{E}') , \qquad n \geqq 0 .$$

However, it is also clear that

$$H^1(\Gamma_n, \mathcal{E}') \simeq H^1(\Gamma_n, E'), \qquad n \geqq 0.$$

Combining these facts with the corollary of Theorem 12, we obtain the following

LEMMA 7. *For each* $n \geqq 0$,

$$D^{\Gamma_n}/\mathcal{E}'_n \otimes_{\mathbf{Z}} (\mathbf{Q}_l/\mathbf{Z}_l) \simeq H^1(\Gamma_n, E').$$

Hence the index $[D^{\Gamma_n}: \mathcal{E}'_n \otimes_{\mathbf{Z}} (\mathbf{Q}_l/\mathbf{Z}_l)]$ *is bounded for all* $n \geqq 0$.

7. Kummer pairings

7.1. Let K/k be a cyclotomic $\mathbf{Z}_l$-extension and let K_{ab} denote the maximal abelian l-extension over K. Let

$$G = \mathrm{Gal}\,(K_{ab}/K).$$

Since K_{ab}/k is a Galois extension, G is a compact Γ-module as explained in §3.1. Let $K^\times$ denote the multiplicative group of K and let

$$\mathfrak{K} = K^\times \otimes_{\mathbf{Z}} (\mathbf{Q}_l/\mathbf{Z}_l).$$

We consider $\mathfrak{K}$ in discrete topology. The Galois group $\Gamma = \mathrm{Gal}\,(K/k)$ then acts continuously on the discrete abelian l-group $\mathfrak{K}$ through its action on $K^\times$. Each element x in $\mathfrak{K}$ can be written in the form

$$x = \alpha \otimes (l^{-a} \bmod \mathbf{Z}_l)$$

with some α in $K^\times$ and an integer $a \geqq 0$.

Now, since the field K contains all l^n-th roots of unity for all $n \geqq 0$, the well-known theory of Kummer extensions gives us a natural pairing

$$G \times \mathfrak{K} \longrightarrow W,$$

$$(\sigma, x) \longmapsto \langle \sigma, x \rangle,$$

with the following properties:

(i) This is an orthogonal pairing of the compact abelian group G and the discrete abelian group $\mathfrak{K}$ in the sense of Pontrjagin when the values $\langle \sigma, x \rangle$ are considered as complex numbers with modulus 1,

(ii) For any γ in Γ,

$$\langle \gamma\sigma, \gamma x \rangle = \gamma(\langle \sigma, x \rangle).$$

Note that $\gamma\sigma$ is defined by

$$\gamma\sigma = g\sigma g^{-1},$$

where g is an element of $\mathrm{Gal}\,(K_{ab}/k)$ with restriction γ on K,

(iii) If $x = \alpha \otimes (l^{-a} \bmod \mathbf{Z}_l)$, then

$$\langle \sigma, x \rangle = \sigma(\sqrt[l^a]{\alpha})/\sqrt[l^a]{\alpha},$$

$\sqrt[l^a]{\alpha}$ being any l^a-th root of α in K_{ab}.

Let H be a closed subgroup of G and let $H^\perp$ denote the annihilator of H in $\mathfrak{R}$ with respect to the above pairing. The orthogonality mentioned in (i) indicates that H is conversely the annihilator of $H^\perp$ in G and that the induced pairings

$$H \times \mathfrak{R}/H^\perp \longrightarrow W\,, \qquad G/H \times H^\perp \longrightarrow W$$

are again orthogonal. In the following, we shall consider various intermediate fields Φ of K and K_{ab}, and apply the above remark for

$$H = \mathrm{Gal}\,(K_{ab}/\Phi)\,, \qquad G/H = \mathrm{Gal}\,(\Phi/K)\,.$$

Note that if Φ/k is a Galois extension, then both H and $H^\perp$ are again Γ-modules and the induced pairings again have the property stated in (ii).

We need later the following simple lemma which is also a consequence of the above:

LEMMA 8. *Let x be an element of $\mathfrak{R}$ and let*

$$x = \alpha \otimes (l^{-a} \bmod \mathbf{Z}_l)\,, \qquad \alpha \in K^\times,\ a \geqq 0\,.$$

Then $x = 0$ if and only if $\alpha = \beta^{l^a}$ for some β in $K^\times$.

Proof. It is obvious that $\alpha = \beta^{l^a}$ implies $x = 0$. Assume $x = 0$. Then

$$\langle \sigma, x \rangle = \sigma(\sqrt[l^a]{\alpha}) \,/\, \sqrt[l^a]{\alpha} = 1$$

for every σ in G. Hence $\beta = \sqrt[l^a]{\alpha}$ is an element of K and $\alpha = \beta^{l^a}$, $\beta \in K^\times$.

7.2. Let M be the maximal abelian l-extension over K, unramified outside l. Let L denoted the maximal unramified abelian l-extension over K, and L' the maximal sub-extension of L/K in which every prime divisor of K lying above l is completely decomposed (see §§3.2, 3.4, and §4.3). We then have

$$k \subseteq K \subseteq L' \subseteq L \subseteq M \subseteq K_{ab}\,,$$

and M/k, L/k, and L'/k are all Galois extensions. Let

$$\mathfrak{M} = \mathrm{Gal}\,(K_{ab}/M)^\perp\,,$$

$$\mathfrak{L} = \mathrm{Gal}\,(K_{ab}/L)^\perp\,,$$

$$\mathfrak{L}' = \mathrm{Gal}\,(K_{ab}/L')^\perp\,.$$

These are Γ-submodules of $\mathfrak{R}$ and

$$\mathfrak{L}' \subseteq \mathfrak{L} \subseteq \mathfrak{M} \subseteq \mathfrak{R}\,.$$

We shall next describe $\mathfrak{M}$ and $\mathfrak{L}'$ explicitly.

Let Φ be a finite cyclic extension of K, contained in K_{ab}. Then

$$\Phi = K(\sqrt[l^a]{\alpha})$$

with some α in $K^\times$ and some integer $a \geqq 0$. Let $(\alpha)' = \alpha\mathfrak{o}'$ be the principal l-ideal in K generated by α (see §5.1).

LEMMA 9.

(i) $\Phi \subseteq M$ *if and only if* $(\alpha)' = \mathfrak{a}'^{l^a}$ *with an l-ideal* $\mathfrak{a}'$ *in* I',

(ii) $\Phi \subseteq L'$ *if and only if* $(\alpha)' = \mathfrak{a}'^{l^a}$, $\mathfrak{a}' \in I'$, *and* α *is a local* l^a-th *power in* K.

Proof. Find an integer $n \geqq a$ such that α is contained in k_n, and let $\Psi = k_n(\sqrt[l^a]{\alpha})$. Then Ψ/k_n is a cyclic extension with degree dividing l^a. Choosing n large enough, we may assume that Φ/k_n is the direct product of the extensions K/k_n and Ψ/k_n.

Suppose first that $\Phi \subseteq M$. Then Ψ/k_n is unramified outside l so that $\alpha\mathfrak{o}'_n = \mathfrak{a}'^{l^a}_n$ for some $\mathfrak{a}'_n$ in I'_n. Hence $(\alpha)' = \mathfrak{a}'^{l^a}$, $\mathfrak{a}' \in I'$. Suppose conversely that $(\alpha)' = \mathfrak{a}'^{l^a}$, $\mathfrak{a}' \in I'$. Since $I' = \varinjlim I'_m$, we may have chosen n large enough so that $\alpha\mathfrak{o}'_n = \mathfrak{a}'^{l^a}_n$, $\mathfrak{a}'_n \in I'_n$. It then follows that Ψ/k_n is unramified outside l so that $\Phi = K\cdot\Psi \subseteq M$. Suppose furthermore that α is a local l^a-th power in K. Because of Lemma 6, we may assume, by choosing n large enough that α is a local l^a-th power in k_n. Then every prime divisor of k_n, lying above l, is completely decomposed in Ψ so that $\Phi \subseteq L'$.

It now remains to show that if $\Phi \subseteq L'$, then α is a local l^a-th power in K. Choose the above integer n so that it also satisfies $n \geqq n_0 = n_0(K/k)$. Let v be a prime divisor of k_n, lying above l, and let Z_v denote the decomposition group of v for the abelian extension Φ/k_n. Since $\Phi \subseteq L'$ and since v is fully ramified for the extension K/k_n, we see that $Z_v \cap \mathrm{Gal}(\Phi/K) = 1$ and $Z_v \simeq \mathrm{Gal}(K/k_n) = \Gamma_n \simeq \mathbf{Z}_l$. Let $m \geqq n$ and let v' be the unique extension of v on k_m. Then it follows from the above that $l^{m-n}Z_v$ (or $Z_v^{l^{m-n}}$, as multiplicative group) is the decomposition group of v' for the extension Φ/k_m: $l^{m-n}Z_v = Z_v \cap \mathrm{Gal}(\Phi/k_m)$. Let $m = n + a$. Since $\mathrm{Gal}(\Phi/k_n) = \mathrm{Gal}(\Phi/K) \times \mathrm{Gal}(\Phi/\Psi)$, $l^a Z_v$ is contained in $\mathrm{Gal}(\Phi/\Psi)$ and, hence, also in $\mathrm{Gal}(\Phi/k_m\Psi)$. This means that v' is completely decomposed in $k_m\Psi = k_m(\sqrt[l^a]{\alpha})$. As this holds for every prime divisor v' of k_m lying above l, we see that α is a local l^a-th power in k_m and, hence, in K.

THEOREM 13.

(i) $\mathfrak{M}$ *is the submodule of* $\mathfrak{K} = K^\times \otimes_{\mathbf{Z}} (\mathbf{Q}_l/\mathbf{Z}_l)$ *consisting of all elements of the form*

$$\alpha \otimes (l^{-a} \bmod \mathbf{Z}_l), \qquad \alpha \in K^\times, \ a \geqq 0,$$

such that $(\alpha)' = \alpha\mathfrak{o}'$ *is an* l^a-th *power in* I',

(ii) *Similarly,* $\mathfrak{L}'$ *consists of all*

$$\alpha \otimes (l^{-a} \bmod \mathbf{Z}_l)\,, \qquad \alpha \in K^\times,\ a \geqq 0\,,$$

such that $(\alpha)' = \alpha\mathfrak{o}'$ *is an* l^a-th *power in* I' *and* α *is a local* l^a-th *power in* K.

Proof. Let $\mathfrak{M}_1$ denote the set of all elements $\alpha \otimes (l^{-a} \bmod \mathbf{Z}_l)$ as stated in (i). It is easy to see that $\mathfrak{M}_1$ is a submodule of $\mathfrak{K}$. Let σ be an element of $G = \mathrm{Gal}(K_{ab}/K)$. Using the formula (iii) for the fundamental pairing $G \times \mathfrak{K} \to W$, we see that $\langle \sigma, \mathfrak{M}_1 \rangle = 0$ if and only if

$$\sigma(\sqrt[l^a]{\alpha}) = \sqrt[l^a]{\alpha}$$

for all $\alpha \otimes (l^{-a} \bmod \mathbf{Z}_l)$ in $\mathfrak{M}_1$. It then follows from Lemma 9, (i) that $\langle \sigma, \mathfrak{M}_1 \rangle = 0$ if and only if σ belongs to $\mathrm{Gal}(K_{ab}/M)$. Hence $\mathfrak{M}_1^\perp = \mathrm{Gal}(K_{ab}/M)$ so that $\mathfrak{M}_1 = \mathrm{Gal}(K_{ab}/M) = \mathfrak{M}$. This proves the first part. The second part can be proved similarly by using Lemma 9, (ii).

7.3. As in §5.4, let E and E' denote the group of all units and the group of all l-units in K, respectively. Let N (resp. N') be the field generated over K by the l^a-th roots of all elements ε in E (resp. all elements ε' in E') for all integers $a \geqq 0$. Clearly

$$k \subseteq K \subseteq N \subseteq N' \subseteq K_{ab}\,,$$

and both N/k and N'/k are again Galois extensions. Let

$$\mathfrak{N} = \mathrm{Gal}(K_{ab}/N)^\perp\,,$$
$$\mathfrak{N}' = \mathrm{Gal}(K_{ab}/N')^\perp$$

so that

$$\mathfrak{N} \subseteq \mathfrak{N}' \subseteq \mathfrak{K}\,.$$

By an argument similar to that in the proof of Theorem 13, one sees immediately that $\mathfrak{N}$ consists of all elements of the form

$$\varepsilon \otimes (l^{-a} \bmod \mathbf{Z}_l)\,, \qquad \varepsilon \in E,\ a \geqq 0\,,$$

and $\mathfrak{N}'$ those of the form

$$\varepsilon' \otimes (l^{-a} \bmod \mathbf{Z}_l)\,, \qquad \varepsilon' \in E',\ a \geqq 0\,.$$

Since $\varepsilon'\mathfrak{o}' = \mathfrak{o}'$, it follows from Theorem 13 that

$$\mathfrak{N} \subseteq \mathfrak{N}' \subseteq \mathfrak{M}$$

so that

$$N \subseteq N' \subseteq M\,.$$

Since E is a subgroup of $K^\times$, the injection map $E \to K^\times$ induces

$$E \otimes_Z (Q_l/Z_l) \longrightarrow K^\times \otimes_Z (Q_l/Z_l) = \mathfrak{K} ,$$

and $\mathfrak{N}$ is nothing but the image of this morphism. Similarly, $\mathfrak{N}'$ is the image of the map

$$E' \otimes_Z (Q_l/Z_l) \longrightarrow K^\times \otimes_Z (Q_l/Z_l) = \mathfrak{K} .$$

Now, let x be an element of $\mathfrak{M}$ and let

$$x = \alpha \otimes (l^{-a} \bmod Z_l) , \qquad \alpha \in K^\times , \ a \geqq 0 ,$$

with $\alpha\mathfrak{o}' = \mathfrak{a}'^{l^a}$, $\mathfrak{a}' \in I'$. Let $c(\mathfrak{a}')$ denote the ideal-class of $\mathfrak{a}'$ in the l-ideal-class group $C' = I'/P'$. Since $c(\mathfrak{a}')^{l^a} = 1$, $c(\mathfrak{a}')$ belongs to the l-primary subgroup A' of C'. Suppose that x can be written also in the form

$$x = \beta \otimes (l^{-b} \bmod Z_l) , \qquad \beta \in K^\times, \ b \geqq 0 ,$$

with $\beta\mathfrak{o}' = \mathfrak{b}'^{l^b}$, $\mathfrak{b}' \in I'$. We shall next show that $c(\mathfrak{a}') = c(\mathfrak{b}')$. For this, suppose $b \geqq a$. Then

$$\alpha^{l^{b-a}}\beta^{-1} \otimes (l^{-b}Z_l) = 0 .$$

Hence, by Lemma 8,

$$\alpha^{l^{b-a}}\beta^{-1} = \theta^{l^b} , \qquad \theta \in K^\times ,$$

so that

$$\mathfrak{a}' = \theta\mathfrak{b}' , \qquad c(\mathfrak{a}') = c(\mathfrak{b}') .$$

Thus we can now define a morphism

$$\mathfrak{M} \longrightarrow A' ,$$

$$x = \alpha \otimes (l^{-a} \bmod Z_l) \longmapsto c(\mathfrak{a}') , \qquad \alpha\mathfrak{o}' = \mathfrak{a}'^{l^a} .$$

LEMMA 10. *The sequence*

$$0 \longrightarrow E' \otimes_Z (Q_l/Z_l) \longrightarrow \mathfrak{M} \longrightarrow A' \longrightarrow 0$$

is exact.

Proof. Let c be an element of A' represented by an l-ideal $\mathfrak{a}'$: $c = c(\mathfrak{a}')$. Let $c^{l^a} = 1$, $a \geqq 0$, so that $\mathfrak{a}'^{l^a} = \alpha\mathfrak{o}'$, $\alpha \in K^\times$. Then c is the image of $\alpha \otimes (l^{-a} \bmod Z_l)$ under $\mathfrak{M} \to A'$. Hence $\mathfrak{M} \to A'$ is surjective. It is clear that $E' \otimes_Z (Q_l/Z_l) \to \mathfrak{M} \to A'$ is exact. Let $u = \varepsilon' \otimes (l^{-a} \bmod Z_l)$ be an element of $E' \otimes_{Z_l} (Q_l/Z_l)$ such that the product $\varepsilon' \otimes (l^{-a} \bmod Z_l)$ is zero in $\mathfrak{M}$. By Lemma 8, $\varepsilon' = \beta^{l^a}$ for some β in $K^\times$. However, since ε' is an l-unit, β also must be such an l-unit in K: $\beta \in E'$. Hence $u = 0$ in $E' \otimes_Z (Q_l/Z_l)$, and the map $E' \otimes_Z (Q_l/Z_l) \to \mathfrak{M}$ is injective.

We note that by a similar argument, we can also obtain an exact sequence

$$0 \longrightarrow E \otimes_Z (Q_l/Z_l) \longrightarrow \mathfrak{M} \longrightarrow A \longrightarrow 0$$

so that we may identify $E \otimes_Z (\mathbf{Q}_l/\mathbf{Z}_l)$ and $E' \otimes_Z (\mathbf{Q}_l/\mathbf{Z}_l)$ with $\mathfrak{N}$ and $\mathfrak{N}'$ respectively:

$$\mathfrak{N} = E \otimes_Z (\mathbf{Q}_l/\mathbf{Z}_l), \qquad \mathfrak{N}' = E' \otimes_Z (\mathbf{Q}_l/\mathbf{Z}_l).$$

We then have a commutative diagram

$$\begin{array}{ccccccccc} 0 & \longrightarrow & \mathfrak{N} & \longrightarrow & \mathfrak{M} & \longrightarrow & A & \longrightarrow & 0 \\ & & \downarrow & & \downarrow & & \downarrow & & \\ 0 & \longrightarrow & \mathfrak{N}' & \longrightarrow & \mathfrak{M} & \longrightarrow & A' & \longrightarrow & 0 \end{array}$$

where the horizontal rows are exact and where $A \to A'$ is surjective. Obviously, this diagram induces another exact sequence

$$0 \longrightarrow \mathfrak{N}'/\mathfrak{N} \longrightarrow A \longrightarrow A' \longrightarrow 0.$$

Now, let B' denote the submodule of A' consisting of all classes which are represented by l-ideals $\mathfrak{a}'$ in I' such that

$$\mathfrak{a}'^{l^a} = \alpha\mathfrak{o}'$$

for some $a \geqq 0$ and for some local l^a-th power α in K. By Theorem 13, (ii), B' is the image of $\mathfrak{L}'$ under $\mathfrak{M} \to A'$ so that the above exact sequence induces another exact sequence

$$0 \longrightarrow \mathfrak{N}' \longrightarrow \mathfrak{L}' + \mathfrak{N}' \longrightarrow B' \longrightarrow 0.$$

Since

$$\mathfrak{M} = \mathrm{Gal}(K_{ab}/M)^{\perp}, \quad \mathfrak{L}' = \mathrm{Gal}(K_{ab}/L')^{\perp}, \quad \mathfrak{N}' = \mathrm{Gal}(K_{ab}/N')^{\perp},$$

the fundamental Kummer pairing

$$\mathrm{Gal}(K_{ab}/K) \times \mathfrak{K} \longrightarrow W$$

induces orthogonal pairings

$$\mathrm{Gal}(M/K) \times \mathfrak{M} \longrightarrow W,$$
$$\mathrm{Gal}(L'/K) \times \mathfrak{L}' \longrightarrow W,$$
$$\mathrm{Gal}(N'/K) \times \mathfrak{N}' \longrightarrow W.$$

Hence we obtain the following result:

THEOREM 14. *There exist canonical orthogonal pairings*

$$\mathrm{Gal}(M/N') \times A' \longrightarrow W,$$
$$\mathrm{Gal}(L'N'/N') \times B' \longrightarrow W,$$

satisfying

$$\langle\gamma\sigma, \gamma c\rangle = \gamma(\langle\sigma, c\rangle), \qquad \gamma \in \Gamma.$$

Note that the value of $\langle \sigma, c \rangle$ in the first pairing is given as follows. Let $\mathfrak{a}'$ be an l-ideal in the class c and let

$$\mathfrak{a}'^{l^a} = \alpha\mathfrak{o}' , \qquad a \geqq 0 , \ \alpha \in K^\times .$$

Then

$$\langle \sigma, c \rangle = \sigma(\sqrt[l^a]{\alpha}) / \sqrt[l^a]{\alpha} .$$

If α is a local l^a-th power in K, then $\sqrt[l^a]{\alpha}$ is contained in L', and the same formula gives us the value of $\langle \sigma, c \rangle$ in the second pairing.

8. The structure of Galois groups

8.1. We first discuss some further properties of Γ-modules.

For each γ in $\Gamma = \mathrm{Gal}(K/k)$, there exists a unique l-adic unit $\kappa(\gamma)$ such that

$$\gamma(\zeta) = \zeta^{\kappa(\gamma)}$$

for every ζ in W. We see easily that the map

$$\gamma \longmapsto \kappa(\gamma)$$

defines a topological isomorphism of Γ onto the subgroup $1 + q\mathbf{Z}_l$ of the multiplicative group of l-adic units in $\mathbf{Q}_l$;

$$\kappa \colon \Gamma \xrightarrow{\sim} 1 + q\mathbf{Z}_l , \qquad q = q_0 = |W_0| .$$

Let e^q be the l-adic unit defined by

$$e^q = \exp q = \sum_{n=0}^{\infty} \frac{q^n}{n!} .$$

Since $1 + q\mathbf{Z}_l$ is topologically generated by e^q, there exists a unique topological generator γ_0 of Γ such that

$$\kappa(\gamma_0) = e^q .$$

In the following, whenever we make a Γ-module into a Λ-module, we shall always do so by means of this fixed topological generator γ_0 in Γ.

Next, we define an element $\dot{T}$ in $\Lambda = \mathbf{Z}_l[[T]]$ by

$$(1 + T)(1 + \dot{T}) = e^q .$$

Since $e^q \equiv 1 \bmod q\mathbf{Z}_l$, $\dot{T}$ is in fact contained in the maximal ideal $\mathfrak{m}$ of Λ. Hence, for each power series $\xi(T)$ in Λ, $\xi(\dot{T})$ is well defined in Λ, and we denote it by $\dot{\xi}(T)$:

$$\dot{\xi}(T) = \xi(\dot{T}) .$$

Clearly

$$\Lambda \longrightarrow \Lambda ,$$
$$\xi \longmapsto \dot{\xi}$$

defines an involutive automorphism of the local ring Λ over $\mathbf{Z}_l$.

Now, let X be a compact Γ-module in the sense of §1.1. We define a new Γ-structure on the compact abelian l-group X by

$$\Gamma \times X \longrightarrow X\,,$$
$$(\gamma, x) \longmapsto \gamma \circ x = \kappa(\gamma)\gamma^{-1}x\,,$$

and we denote this new Γ-module by $\dot{X}$ to distinguish it from the original X. Obviously

$$\ddot{X} = X$$

and $X \longmapsto \dot{X}$ defines an automorphism of the category of compact Γ-modules. Suppose now that both X and $\dot{X}$ are made into Λ-modules by means of the fixed topological generator γ_0, and let

$$\Lambda \times X \longrightarrow X\,, \qquad \Lambda \times \dot{X} \longrightarrow \dot{X}\,,$$
$$(\xi, x) \longmapsto \xi x\,, \qquad (\xi, x) \longmapsto \xi \circ x$$

be the maps which define the Λ-structures on X and $\dot{X}$ respectively. Then, as one sees immediately,

$$\xi \circ x = \dot{\xi} x\,, \qquad \xi \in \Lambda\,.$$

Hence we obtain, in particular, that

$$(\Lambda/(\xi))^{\cdot} = \Lambda/(\dot{\xi})$$

for any ξ ($\xi = 0$ or $\xi \neq 0$) in Λ.

In the following, we shall also fix a $\mathbf{Z}_l$-isomorphism

$$W \xrightarrow{\sim} \mathbf{Q}_l/\mathbf{Z}_l\,.$$

Let

$$X \times D \longrightarrow W\,,$$
$$(x, d) \longmapsto \langle x, d\rangle$$

be an orthogonal pairing of a compact Γ-module X and a discrete Γ-module D such that

$$\langle \gamma x, \gamma d\rangle = \gamma\langle x, d\rangle\,, \qquad \gamma \in \Gamma\,.$$

Clearly the pairing induces a $\mathbf{Z}_l$-isomorphism

$$X \xrightarrow{\sim} \mathrm{Hom}_{\mathbf{Z}_l}(D, W)\,.$$

We define $\gamma\varphi$ for φ in $\mathrm{Hom}_{\mathbf{Z}_l}(D, W)$ by

$$(\gamma\varphi)(d) = \gamma\varphi(\gamma^{-1}d)\,, \qquad \gamma \in \Gamma\,,\ d \in D\,.$$

Then the above $\mathbf{Z}_l$-isomorphism becomes a Γ-isomorphism. Let

$$\dot{X} \times D \longrightarrow \mathbf{Q}_l/\mathbf{Z}_l ,$$
$$(x, d) \longmapsto \langle\!\langle x, d \rangle\!\rangle$$

be the product of $\dot{X} \times D = X \times D \to W$ and the isomorphism $W \cong \mathbf{Q}_l/\mathbf{Z}_l$ fixed in the above. Then we have

$$\langle\!\langle \gamma \circ x, d \rangle\!\rangle = \langle\!\langle x, \gamma d \rangle\!\rangle , \qquad \gamma \in \Gamma ,$$
$$\langle\!\langle \xi \circ x, d \rangle\!\rangle = \langle\!\langle x, \xi d \rangle\!\rangle , \qquad \xi \in \Lambda .$$

Here D is also made into a Λ-module by means of the fixed γ_0. It follows that the $\mathbf{Z}_l$-isomorphism

$$\dot{X} \xrightarrow{\sim} \mathrm{Hom}_{\mathbf{Z}_l} (D, \mathbf{Q}_l/\mathbf{Z}_l) ,$$

induced by the above pairing, is a Γ-isomorphism as well as a Λ-isomorphism when $\gamma\psi$ and $\xi\psi$, for ψ in $\mathrm{Hom}_{\mathbf{Z}_l} (D, \mathbf{Q}_l/\mathbf{Z}_l)$, are defined by

$$(\gamma\psi)(d) = \psi(\gamma d) , \qquad \gamma \in \Gamma ,$$
$$(\xi\psi)(d) = \psi(\xi d) , \qquad \xi \in \Lambda .$$

8.2. We need also that the following result on Λ-modules. For each noetherian Λ-module X, we define a function $i(n, a; X)$ of integers n and a, $n, a \geqq 0$, by

$$[X: \omega_n X + l^a X] = l^{i(n, a; X)} .$$

For example,

$$i(n, a; \Lambda) = a l^n ,$$
$$i(n, a; \Lambda/(l)^e) = \min (e, a) l^n , \qquad e \geqq 0 .$$

It is clear that $i(n, a; X)$ is additive in X. By Lemma 2, we also see that if $X \sim X'$, then the difference

$$i(n, a; X) - i(n, a; X')$$

is bounded, in absolute value, for all n, $a \geqq 0$.

Now, let $\lambda = \lambda(X)$ and $\mu = \mu(X)$ be the invariants of a noetherian Λ-module X and let

$$E(e_0; \mathfrak{p}_1^{e_1}, \cdots, \mathfrak{p}_t^{e_t}) , \qquad e_0 \geqq 0 , \ e_1, \cdots, e_t \geqq 1 ,$$

be the elementary Λ-module associated with X (see §1.2). Let

$$\alpha = \textstyle\sum_i \deg \xi_{n_i} ,$$

where the sum is taken over all indices i, $1 \leqq i \leqq t$, such that $\mathfrak{p}_i = (\xi_{n_i})$, for some integer $n_i \geqq 0$, ξ_n being the element of Λ defined in §1.4. Then:

LEMMA 11. *There exist an integer $n^* \geqq 0$ and an integer-valued function $a^*(n)$ defined for $n \geqq 0$ such that*

$$i(n, a; X) - \{(e_0 l^n + \alpha)a + \mu l^n + (\lambda - \alpha)n\}$$

is bounded, in absolute value, for all $n \geqq n^*$ and $a \geqq a^*(n)$.

Proof. By the remarks mentioned above, it is sufficient to prove the lemma for

$$X = \Lambda/\mathfrak{p}^e, \qquad e \geqq 1,$$

where $\mathfrak{p} = (\eta) \neq (l)$, η being a distinguished irreducible polynomial $p(T)$ in $\mathbf{Z}_l[T]$. Assume that $\eta \neq \xi_i$ for any $i \geqq 0$. Then η and $\omega_n = \xi_0\xi_1 \cdots \xi_n$ are relatively prime in Λ so that

$$[\Lambda: (\omega_n, \eta^e)] = l^{a^*}$$

with an integer $a^* = a^*(n) < \infty$. Let $a \geqq a^*(n)$. Then

$$X/(\omega_n X + l^a X) = \Lambda/(\omega_n, l^a, \eta^e) = \Lambda/(\omega_n, \eta^e) = X/\omega_n X.$$

However, we can see easily that for $n \gg 0$,

$$[X: \omega_n X] = l^{\lambda n + c}$$

with $\lambda = \lambda(X)$ and with a constant c independent of n. This proves the lemma for $X = \Lambda/\mathfrak{p}^e$, $\mathfrak{p} = (\eta) \neq (l), (\xi_i)$, $i \geqq 0$. In the case $\mathfrak{p} = (\xi_i)$ for some $i \geqq 0$, the proof is slightly more complicated. However, the principle of the proof is the same as the above so that we may omit the detail here.

8.3. We now consider the structure of the Galois group $\mathrm{Gal}(N'/K)$. We have seen in §7.3 that there exists a canonical pairing

$$\mathrm{Gal}(N'/K) \times \mathfrak{N}' \longrightarrow W.$$

Let

$$X = \mathrm{Gal}(N'/K)^{\cdot}, \qquad \dot{X} = \mathrm{Gal}(N'/K),$$

$$D = \mathfrak{N}' = E' \otimes_{\mathbf{Z}} (\mathbf{Q}_l/\mathbf{Z}_l) = \mathcal{E}' \otimes_{\mathbf{Z}} (\mathbf{Q}_l/\mathbf{Z}_l).$$

By the method explained in §8.1, the above canonical pairing induces an orthogonal pairing

$$X \times D \longrightarrow \mathbf{Q}_l/\mathbf{Z}_l$$

satisfying

$$\langle \gamma x, d\rangle = \langle x, \gamma d\rangle, \qquad \gamma \in \Gamma,$$

$$\langle \xi x, d\rangle = \langle x, \xi d\rangle, \qquad \xi \in \Lambda.$$

Let X_n, $n \geqq 0$, denote the annihilator of $\mathcal{E}'_n \otimes_{\mathbf{Z}} (\mathbf{Q}_l\mathbf{Z}_l)$ in X:

$$X_n = (\mathcal{E}'_n \otimes_{\mathbf{Z}} (\mathbf{Q}_l/\mathbf{Z}_l))^{\perp}.$$

Since

$$\omega_n X = (D^{\Gamma_n})^{\perp},$$

it follows from $\mathcal{E}'_n \otimes_Z (Q_l/Z_l) \subseteq D^{\Gamma_n}$ that

$$\omega_n X \subseteq X_n \subseteq X .$$

Furthermore, X/X_n and $X_n/\omega_n X$ are dual to $\mathcal{E}'_n \otimes_Z (Q_l/Z_l)$ and $D^{\Gamma_n}/\mathcal{E}'_n \otimes_Z (Q_l/Z_l)$, respectively in the sense of Pontrjagin. However, we see from §6.3 that

$$\mathcal{E}'_n \otimes_Z (Q_l/Z_l) \simeq (Q_l/Z_l)^{d_n/2+s_n-1}.$$

Hence, for $n \geqq n_0 = n_0(K/k)$,

$$X/X_n \simeq Z_l^{(d/2)l^n+s-1}$$

with $d = [k:Q]$ and $s = s(K/k)$. On the other hand, it also follows from Lemma 7 that $[X_n: \omega_n X]$ is bounded for all $n \geqq 0$. Therefore we see immediately that

$$i(n, a; X) - \left(\frac{d}{2} l^n + s - 1\right) a$$

is bounded for all $n \geqq n_0$ and $a \geqq 0$. Note that since $K \subseteq N' \subseteq M$ and $\mathrm{Gal}(M/K)$ is noetherian by Theorem 5, $X = \mathrm{Gal}(N'/K)^{\cdot}$ is also a noetherian Λ-module.

Now, let

$$f\colon X \longrightarrow E(e_0; \mathfrak{p}_1^{e_1}, \cdots, \mathfrak{p}_t^{e_t})$$

be a pseudo-isomorphism of X into the associated elementary Λ-module. Comparing the approximations of $i(n, a; X)$, for $n \gg 0$ and $a \gg n$, given in the above and in Lemma 11, we see that

$$e_0 = \frac{d}{2}, \qquad \lambda = \alpha = s - 1, \qquad \mu = 0$$

with λ, μ, and α as in Lemma 11. It then follows from the definition of these numbers that $e_i = 1$, $\mathfrak{p}_i = (\xi_{n_i})$ for every i, $1 \leqq i \leqq t$. On the other hand, since $\mathcal{E}'$ is a free abelian group, the dual X of $D = \mathcal{E}' \otimes_Z (Q_l/Z_l)$ is a free Z_l-module. Hence X has no non-trivial finite subgroup and the pseudo-isomorphism f is injective. As $\mathrm{Gal}(N'/K) = \dot{X}$, the following theorem is proved:

THEOREM 15. *Let N' be the field generated over K by l^n-th roots of all l-units in K for all $n \geqq 0$. Then $\mathrm{Gal}(N'/K)$ is a torsion-free Z_l-module and is contained as a Λ-submodule of finite index in an elementary Λ-module of the form*

$$E\left(\frac{d}{2};\ (\dot{\xi}_n), \cdots, (\dot{\xi}_{n_t})\right),$$

where $d = [k:\mathbf{Q}]$ *and*

$$\textstyle\sum_{i=1}^{t} \deg \xi_{n_i} = s(K/k) - 1\,.$$

8.4. We shall next consider the consequence of the canonical pairing

$$\mathrm{Gal}(M/N') \times A' \longrightarrow W\,,$$

obtained in Theorem 14. As explained in §8.1, this induces a pairing

$$\mathrm{Gal}(M/N')^{\cdot} \times A' \longrightarrow \mathbf{Q}_l/\mathbf{Z}_l$$

and an isomorphism

$$\mathrm{Gal}(M/N')^{\cdot} \xrightarrow{\ \sim\ } \mathrm{Hom}_{\mathbf{Z}_l}(A', \mathbf{Q}_l/\mathbf{Z}_l)\,.$$

However, we know by Theorem 11 that the last module is isomorphic to $\alpha(Y')$ and is pseudo-isomorphic to X'. Hence:

THEOREM 16. $\mathrm{Gal}(M/N')$ *is a noetherian torsion* Λ*-module and*

$$\begin{aligned}\mathrm{Gal}(M/N')^{\cdot} &\simeq \mathrm{Hom}_{\mathbf{Z}_l}(A', \mathbf{Q}_l/\mathbf{Z}_l)\\ &\simeq \alpha(\mathrm{Gal}(L'/KK'_{n_0})) \sim \mathrm{Gal}(L'/K)\,,\end{aligned}$$

with $n_0 = n_0(K/k)$.

The following result is an immediate consequence of Theorems 15 and 16:

THEOREM 17. *The factor torsion* Λ*-module for* $\mathrm{Gal}(M/K)$ *is the same as that for* $\mathrm{Gal}(N'/K)$ *and it is a submodule of finite index in* $\Lambda^{d/2}$, $d = [k:\mathbf{Q}]$.

Now, the isomorphism $\mathrm{Gal}(M/N') \simeq \alpha(Y')$ implies that $\mathrm{Gal}(M/N')^{\cdot}$ has no non-trivial finite Λ-submodule (see §1.3). Hence the same holds also for $\mathrm{Gal}(M/N')$. As $\mathrm{Gal}(N'/K)$ is free over $\mathbf{Z}_l$ by Theorem 15, the following theorem is proved:

THEOREM 18. $\mathrm{Gal}(M/K)$ *has no non-trivial finite* Γ*-submodule* (Λ*-submodule*).

8.5. Let $t(\mathrm{Gal}(N'/K))$ denote the torsion Λ-submodule of $\mathrm{Gal}(N'/K)$. We shall next see when the factor torsion Λ-module $\mathrm{Gal}(N'/K)/t(\mathrm{Gal}(N'/K))$ in Theorem 17 is actually equal to $\Lambda^{d/2}$. As in §8.3, let $X = \mathrm{Gal}(N'/K)^{\cdot}$, and suppose that X is contained as a submodule of finite index in

$$X' = E\left(\frac{d}{2};\ (\xi_{n_1}), \cdots, (\xi_{n_t})\right)$$

with

$$\textstyle\sum_{i=1}^{t} \deg \xi_{n_i} = s - 1\,, \qquad\qquad s = s(K/k)\,.$$

Let

$$a = \max(n_1, \cdots, n_t)\,.$$

LEMMA 12. $\mathrm{Gal}(N'/K)/t(\mathrm{Gal}(N'/K))$ *is isomorphic to* $\Lambda^{d/2}$ *if and only if*

$$H^1(\Gamma_n, E') = 0$$

for every $n \geqq a$.

Proof. Let X_n, $n \geqq 0$, be the submodule of X as defined in §8.1. The argument there shows that $X_n/\omega_n X$ is finite while X/X_n is free over Z_l so that $X_n/\omega_n X$ is the torsion Z_l-submodule of $X/\omega_n X$. Let $t(X)$ and $t(X')$ denote the torsion Λ-submodules of X and X' respectively. Then

$$X' = \Lambda^{d/2} \oplus t(X') ,$$

$$t(X') = E((\xi_{n_1}), \cdots, (\xi_{n_t})) ,$$

$$= \mathrm{Ker}\,(\omega_n\colon X' \longrightarrow X') ,$$

$$t(X) = X \cap t(X') = \mathrm{Ker}\,(\omega_n\colon X \longrightarrow X) ,$$

for every $n \geqq a$. It follows that

$$X'/(X + t(X')) \simeq \omega_n X'/\omega_n X , \qquad n \geqq a .$$

Hence, in particular, $X \cap \omega_n X'/\omega_n X$ is finite. On the other hand,

$$X/X \cap \omega_n X' \simeq (X + \omega_n X')/\omega_n X' , \qquad n \geqq a ,$$

and the right hand side is a submodule of $X'/\omega_n X' = (\Lambda/\omega_n\Lambda)^{d/2} \oplus t(X')$ which is free over Z_l. Therefore $X \cap \omega_n X'/\omega_n X$ is the torsion Z_l-submodule of $X/\omega_n X$ so that

$$X \cap \omega_n X' = X_n , \qquad n \geqq a .$$

Now, suppose that $\mathrm{Gal}(N'/K)/t(\mathrm{Gal}(N'/K)) \simeq \Lambda^{d/2}$. Then $X/t(X) \simeq \Lambda^{d/2}$ so that $X = \Lambda^{d/2} \oplus t(X)$ and

$$X/\omega_n X \simeq (\Lambda/\omega_n\Lambda)^{d/2} \oplus t(X) , \qquad n \geqq a .$$

This shows that $X/\omega_n X$ is free over Z_l and hence $X_n/\omega_n X = 0$ for $n \geqq a$. By Lemma 7, we then have

$$H^1(\Gamma_n, E') = 0 , \qquad n \geqq a .$$

Conversely, assume that the above equalities hold for $n \geqq a$, namely that

$$X \cap \omega_n X'/\omega_n X = X_n/\omega_n X = 0 , \qquad n \geqq a .$$

Since X'/X is finite, $\omega_n X' \subseteq X$ for sufficiently large n. Hence, for such n, the assumption implies that $\omega_n X' = \omega_n X$ so that

$$X'/(X + t(X)) = 0 .$$

Therefore

$$X/t(X) \simeq X'/t(X') \simeq \Lambda^{d/2}$$

and consequently

$$\mathrm{Gal}\,(N'/K)\,/\,t(\mathrm{Gal}\,(N'/K)) \simeq \Lambda^{d/2}, \qquad \text{Q.E.D.}$$

Let us consider the special case where $s = s(K/k) = 1$. In such a case, $t = a = 0$, $X' = \Lambda^{d/2}$ so that

$$X'/X \simeq \omega_0 X'/\omega_0 X.$$

It then follows from

$$(X + \omega_0 X')/X \simeq \omega_0 X'/(\omega_0 X' \cap X)$$

that

$$[X' : X + \omega_0 X'] = [\omega_0 X' \cap X : \omega_0 X].$$

Hence if $H^1(\Gamma_0, E') = 0$, then $X' = X + \omega_0 X'$, and this implies

$$X = X' = \Lambda^{d/2}.$$

Thus we see that $\mathrm{Gal}\,(N'/K)\,/\,t(\mathrm{Gal}\,(N'/K))$ is isomorphic to $\Lambda^{d/2}$ if and only if

$$H^1(\Gamma, E') = 0.$$

A typical example of K/k with $s(K/k) = 1$ is given by $\mathbf{P}_\infty/\mathbf{P}$. If l is a regular prime, i.e., if the class number of $\mathbf{P}$ is not divisible by l, then it can be seen easily that the above equality holds. More generally, if the class number of the maximal real subfield of $\mathbf{P}$ is not divisible by l, then $H^1(\Gamma, E') = 0$ again holds. The assumption just mentioned above is known as Vandiver's conjecture (for the prime l). It is still an open problem to prove $H^1(\Gamma, E') = 0$ in this special case without making any such assumption on l[9].

9. Some lemmas

9.1. In this section, we shall prove some preliminary lemmas which we need in the proof of our main result in the next section.

Let X be a noetherian torsion Λ-module. We call X an l-module if $\operatorname{div} X = n(l)$ for some $n \geqq 0$, and we call it an l'-module if $\operatorname{div} X$ is prime to the principal divisor (l). X is an l-module if and only if X is a torsion $\mathbf{Z}_l$-module, and when this is the case, then $l^a X = 0$ for some $a \geqq 0$. On the other hand, X is an l'-module if and only if X is noetherian over $\mathbf{Z}_l$. Thus the fact that X is an l-module or an l'-module can be seen from the structure of X as a $\mathbf{Z}_l$-module.

Let X be an arbitrary noetherian torsion Λ-module. Then X has a unique maximal l-submodule X' which contains all l-submodules of X; X' is nothing

[9] See [2].

but the torsion Z_l-submodule of X. Similiarly, X has a unique maximal l'-submodule X''. X/X' is an l'-module, X/X'' is an l-module, and both $X' \cap X''$ and $X/(X' + X'')$ are finite so that

$$X/X' \sim X'' , \qquad X/X'' \sim X' .$$

In fact, $X' \cap X''$ is the maximal finite Λ-submodule X^0 in X. Clearly div X' is the (l)-part of div X, and div X'' is the remaining part of div X.

Now, let

$$X \times D \longrightarrow \mathbf{Q}_l/\mathbf{Z}_l$$

be an orthogonal pairing of a compact Λ-module X and a discrete Λ-module D such that

$$\langle \xi x, d \rangle = \langle x, \xi d \rangle , \qquad \xi \in \Lambda .$$

Let $T(D)$ denote the Tate module for the l-primary abelian group D, namely,

$$T(D) = \varprojlim D^{(n)} ,$$

where $D^{(n)}$ denotes the submodule of all d in D such that $l^n d = 0$ and where the inverse limit is defined with respect to the maps

$$l^{m-n}\colon D^{(m)} \longrightarrow D^{(n)} ,$$
$$d \longmapsto l^{m-n} d , \qquad m \geqq n \geqq 0 .$$

Assume that X is noetherian and torsion over Λ and let $\alpha(X/X')$ be the adjoint of X/X' defined in §1.3.

LEMMA 13. *As Λ-modules,*

$$T(D) \simeq \alpha(X/X')$$

so that

$$T(D) \sim X/X' \sim X'' .$$

Proof. Let D' denote the annihilator of X' in the pairing $X \times D \to \mathbf{Q}_l/\mathbf{Z}_l$ so that the induced pairing

$$X/X' \times D' \longrightarrow \mathbf{Q}_l/\mathbf{Z}_l$$

is again orthogonal. Since div X/X' is prime to (l), we see that

$$\alpha(X/X') \simeq \varprojlim \mathrm{Hom}_{\mathbf{Z}_l}(X/(l^n X + X'), \mathbf{Q}_l/\mathbf{Z}_l) ,$$

by putting $\pi_n = l^n$, $n \geqq 0$, in the formula in §1.3. However, the above pairing induces

$$\mathrm{Hom}_{\mathbf{Z}_l}(X/(l^n X + X'), \mathbf{Q}_l/\mathbf{Z}_l) \simeq D'^{(n)} = D' \cap D^{(n)} .$$

Hence

$$\alpha(X/X') \simeq \varprojlim D'^{(n)} = T(D') .$$

However, since X' is the torsion $\mathbf{Z}_l$-submodule of X, D' is the maximal divisible subgroup of the l-primary abelian group D. Hence

$$T(D') \simeq T(D),$$

and the lemma is proved.

9.2. Let M, L', and N' be the extensions of K as defined in §§7.2 and 7.3. Let X denote the maximal l'-submodule of the noetherian torsion Λ-module $\mathrm{Gal}(M/N')$. By Theorem 18, $\mathrm{Gal}(M/N')$ has no non-trivial finite Λ-submodule. Hence X is a free $\mathbf{Z}_l$-module of rank $\lambda = \lambda(\mathrm{Gal}(M/N'))$.

Let x be an element of X and let a be any non-negative integer, $a \geqq 0$. Since $X/l^a X$ is finite, there exists an integer $n \geqq 0$ such that $\omega_n X \subseteq l^a X$. Hence

$$\omega_n x = l^a y$$

for some y in X. Let $y|K'_n$ denote the restriction of y in $\mathrm{Gal}(M/N')$ on the subfield K'_n of M defined in §4.3, and let

$$y \,|\, K'_n \longmapsto c_n \longmapsto c$$

under the maps

$$\mathrm{Gal}(K'_n/k_n) \xrightarrow{\ \simeq\ } A'_n \longrightarrow A'$$

defined in §§4.3 and 5.1. We shall next show that c depends only upon x and a, and is independent of the choice of y and n satisfying $\omega_n x = l^a y$. To see this, suppose that

$$\omega_m x = l^a y'$$

with y' in X and $m \geqq n$, and let

$$y' \,|\, K'_m \longmapsto c'_m \longmapsto c'$$

under

$$\mathrm{Gal}(K'_m/k_m) \xrightarrow{\ \simeq\ } A'_m \longrightarrow A' .$$

Since $\omega_m = \omega_n \nu_{n,m}$ and since X is torsion-free over $\mathbf{Z}_l$, we obtain from the above that $l^a(y' - \nu_{n,m} y) = 0$ so that

$$y' = \nu_{n,m} y .$$

Considering the restrictions $y \,|\, L'$ and $y' \,|\, L'$ in $\mathrm{Gal}(L'/K)$, we then see from Theorem 8 that $c_n \mapsto c'_m$ under $A'_n \to A'_m$. Hence $c = c'$, and our contention is proved. Thus we shall denote the above class c by $c(x, a)$:

$$c = c(x, a) .$$

Let t be a sufficiently large integer such that whenever a non-zero element α in k_n satisfies $\alpha \equiv 1 \bmod l^t$, α is a local l^a-th power in k_n. Let Φ be the ray class field mod l^t over k_n. Then $k_n \subseteq K'_n \subseteq \Phi \cap M_n \subseteq M_n$ and $[\Phi: \Phi \cap M_n]$ is prime to l. Let $\mathfrak{a}$ be an ideal in I_n, prime to l, such that

$$\left(\frac{\Phi \cap M_n/k_n}{\mathfrak{a}}\right) = y \mid \Phi \cap M_n$$

and such that the order of $\left(\frac{\Phi/k_n}{\mathfrak{a}}\right)$ is a power of l. Since $\mathrm{Gal}(M/M_n) = \omega_n \mathrm{Gal}(M/K)$ by §3.2, it follows from $\omega_n x = l^a y$ that

$$\left(\frac{\Phi \cap M_n/k_n}{\mathfrak{a}}\right)^{l^a} = l^a y \mid \Phi \cap M_n = 1$$

so that

$$\left(\frac{\Phi/k_n}{\mathfrak{a}}\right)^{l^a} = 1 .$$

Therefore $\mathfrak{a}^{l^a} = (\alpha)$ with a local l^a-th power α in k_n. As

$$\left(\frac{K'_n/k_n}{\mathfrak{a}}\right) = \left(\frac{\Phi \cap M_n/k_n}{\mathfrak{a}}\right)\Big| K'_n = y \mid K'_n ,$$

$c(x, a)$ is the class of $\mathfrak{a}\mathfrak{o}'$ in A'. But $(\mathfrak{a}\mathfrak{o}')^{l^a} = \alpha\mathfrak{o}'$ shows that $c(x, a)$ is contained in the submodule B' of A' defined in §7.3. Thus we obtain a map

$$X \longrightarrow B'^{(a)} ,$$
$$x \longmapsto c(x, a) ,$$

where $B'^{(a)}$ is the submodule of all c in B' satisfying $l^a c = 0$. It is clear that this is a Γ-homomorphism as well as a Λ-homomorphism. It also follows from $\omega_n x = l^b(l^{a-b}y)$, $a \geqq b \geqq 0$, that

$$c(x, b) = l^{a-b}c(x, a) , \qquad a \geqq b \geqq 0 ,$$

so that

$$c(x, a) \longmapsto c(x, b)$$

under

$$l^{a-b}\colon B'^{(a)} \longrightarrow B'^{(b)} .$$

Hence we obtain an important morphism

$$f\colon X \longrightarrow T(B')$$

where

$$T(B') = \varprojlim B'^{(a)}$$

is the Tate module for B'.

9.3. We shall next determine the kernel of f. Let

$$Y = X \cap \mathrm{Gal}(M/L'N')\,;$$

Y is the maximal l'-submodule of $\mathrm{Gal}(M/L'N')$. For $n \geqq 0$, let $\omega_n^{-1}(Y)$ denote the inverse image of Y under the endomorphism

$$\omega_n\colon X \longrightarrow X\,,$$

and let $\omega^{-1}(Y)$ be the union of the increasing sequence of Λ-submodules $\omega_n^{-1}(Y)$, $n \geqq 0$, in X. Since X is noetherian, $\omega^{-1}(Y) = \omega_n^{-1}(Y)$ whenever n is sufficiently large. We fix an integer $u \geqq 0$ such that

$$\omega^{-1}(Y) = \omega_u^{-1}(Y)\,.$$

Let Y' be the submodule of X containing $\omega^{-1}(Y)$ such that $Y'/\omega^{-1}(Y)$ is the torsion $\mathbf{Z}_l$-submodule of $X/\omega^{-1}(Y)$. Since X is noetherian over $\mathbf{Z}_l$, $Y'/\omega^{-1}(Y)$ is finite so that $\nu_{u,n}Y' \subseteq \omega^{-1}(Y)$ for some $n \geqq u$. It follows that $\omega_n Y' = \omega_u(\nu_{u,n}Y') \subseteq Y$ and $Y' \subseteq \omega_n^{-1}Y = \omega^{-1}(Y)$. Hence $Y' = \omega^{-1}(Y)$ and we see that $Y/\omega^{-1}(Y)$ is a free $\mathbf{Z}_l$-module.

LEMMA 14.

$$\mathrm{Ker}\,(f) = \omega^{-1}(Y)\,.$$

Proof. Let x be an element of $\omega^{-1}(Y) = \omega_u^{-1}(Y)$ and let a be any integer, $a \geqq 0$. Since Y/l^aY is finite, $\nu_{u,n}Y \subseteq l^aY$ for some $n \geqq u$. As $\omega_u x$ belongs to Y, we see that

$$\omega_n x = \nu_{u,n}(\omega_u x) = l^a y$$

for some y in Y. However, since $K_n' \subseteq L'$ and y is an element of $\mathrm{Gal}(M/L'N')$, $y\,|\,K_n'$ is the identity. Hence

$$c(x, a) = 0$$

by the definition of $c = c(x, a)$. As this holds for every $a \geqq 0$, we have $f(x) = 0$.

To prove the converse, we first fix an integer $v \geqq 0$ as follows. Let $\mathrm{Gal}(L'/K)^0$ denote the maximal finite Λ-submodule of $\mathrm{Gal}(L'/K)$. By Theorem 10, the orders of the groups $\mathrm{Ker}\,(A_n' \to A')$ are bounded for all $n \geqq 0$. Hence we choose $v \geqq 0$ so that

$$l^v\,\mathrm{Gal}(L'/K)^0 = 0\,, \qquad l^v\,\mathrm{Ker}\,(A_n' \longrightarrow A') = 0$$

for all $n \geqq 0$.

Now, suppose that $f(x) = 0$, i.e., $c(x, a) = 0$ for every $a \geqq 0$. Fix $a \geqq v$ and let

$$\omega_n x = l^a y,\ \ y\,|\,K_n' \mapsto c_n \mapsto c$$

as in §9.2. We may suppose that $n \geqq n_0$ where $n_0 = n_0(K/k)$ is the integer defined in §3.4. Since $c = c(x, a) = 0$, c_n belongs to $\mathrm{Ker}\,(A_n' \to A')$. Hence

$l^v c_n = 0$ and consequently

$$l^v y \mid K'_n = 0 .$$

Let

$$x' = x \mid L' , \qquad y' = y \mid L' .$$

Then $l^v y \mid K'_n = 0$ implies that $l^v y'$ is contained in $\mathrm{Gal}(L'/KK'_n)$. By Theorem 8, we then have

$$l^v y' = \nu_{n_0,n} z' , \qquad z' \in \mathrm{Gal}(L'/K) .$$

Since $\omega_n x' = l^a y'$, $a \geqq v$, it follows that $\omega_n x' = l^{a-v}\nu_{n_0,n} z'$, namely, that

$$\nu_{n_0,n}(\omega_n, x' - l^{a-v} z') = 0 .$$

However, since $\mathrm{Gal}(L'/K)/\nu_{n_0,n}\,\mathrm{Gal}(L'/K)$ is finite, the principal divisor $(\nu_{n_0,n})$ is disjoint from $\mathrm{div}\,\mathrm{Gal}(L'/K)$ so that

$$\mathrm{Ker}\,(\nu_{n_0,n}\colon \mathrm{Gal}(L'/K) \longrightarrow \mathrm{Gal}(L'/K))$$

is contained in $\mathrm{Gal}(L'/K)^0$. Hence it follows from the above that

$$l^v(\omega_{n_0} x - l^{a-v} z') = 0 ,$$

namely, that

$$l^v \omega_{n_0} x = l^a z' , \qquad z' \in \mathrm{Gal}(L'/K) .$$

Thus $l^v\omega_{n_0}x$ is contained in $l^a\,\mathrm{Gal}(L'/K)$ for every $a \geqq v$. Hence $l^v\omega_{n_0}x' = 0$; namely, $l^v\omega_{n_0}x$ is contained in $Y = X \cap \mathrm{Gal}(M/L'N')$. Therefore $l^v x$ is in $\omega^{-1}(Y)$, and by the remark mentioned earlier, x itself is contained in $\omega^{-1}(Y)$. This completes the proof of the lemma.

It is difficult to describe the image of $f\colon X \to T(B')$ exactly. However, we shall later determine $\mathrm{Im}(f)$ up to a finite factor module.

9.4. We shall next prove another lemma which will be used in the following section. In general, let m be an integer, $m \geqq 1$, and let Φ be a finite algebraic number field containing all m-th roots of unity. As usual, we denote by

$$\left(\frac{\alpha, \Phi}{\mathfrak{a}}\right)_m$$

the m-th power residue symbol for Φ[10]. We consider such symbols for the intermediate fields k_n, $n \geqq 0$, of the Z_l-extension K/k. Let W, $W_n = W \cap k_n^\times$, and $q_n = |W_n|$ be defined as in §6.1. For each $n \geqq 0$, let I_n^0 denote the subgroup of all ideals of I_n which are prime to l (see §4.3). Then

$$\left(\frac{\alpha, k_n}{\mathfrak{a}}\right)_{q_n}$$

10 For the theory of power residue symbols, see [3], II.

is defined for each α in $k_n^\times$ and for each $\mathfrak{a}$ in I_n^0, prime to α. Clearly, if $m \geqq n$, then

$$\left(\frac{\alpha, k_m}{\mathfrak{a}}\right)_{q_m}$$

is also defined for the same α and $\mathfrak{a}$. For $n \geqq 0$, let

$$\begin{cases} \delta_n = 1\,, & \text{if } l > 2\,, \\ \quad = 1 + \dfrac{q_n}{2}\,, & \text{if } l = 2\,. \end{cases}$$

Then we have the following lemma:

LEMMA 15.

$$\left(\frac{\alpha, k_n}{\mathfrak{a}}\right)_{q_n}^{\delta_n} = \left(\frac{\alpha, k_m}{\mathfrak{a}}\right)_{q_m}^{\delta_m}, \qquad m \geqq n \geqq 0\,.$$

Proof. It is sufficient to prove the lemma in the case where $m = n + 1$ and $\mathfrak{a}$ is a prime ideal of k_n, prime to α: $\mathfrak{a} = \mathfrak{p}$. Let $\mathfrak{P}$ be a prime ideal of k_{n+1}, dividing $\mathfrak{p}$, and let $N_n\mathfrak{p}$ and $N_{n+1}\mathfrak{P}$ denote the absolute norms of the ideals $\mathfrak{p}$ and $\mathfrak{P}$ respectively. Since k_{n+1}/k_n is unramified outside l, either $\mathfrak{p} = \mathfrak{P}$ or $\mathfrak{p} = N_{n,n+1}\mathfrak{P}$, the norm of $\mathfrak{P}$.

Consider first the case $\mathfrak{p} = \mathfrak{P}$. Since the residue class field of $\mathfrak{P}$ is an extension of degree l over that of $\mathfrak{p}$, generated by q_{n+1}-th roots of unity, we have

$$N_n\mathfrak{p} = 1 + aq_n$$

with an integer a, prime to l. Since

$$N_{n+1}\mathfrak{P} = (N_n\mathfrak{p})^l = (1 + aq_n)^l\,,$$

we see easily that

$$\frac{N_{n+1}\mathfrak{P} - 1}{q_{n+1}}\delta_{n+1} \equiv \frac{N_n\mathfrak{p} - 1}{q_n}\delta_n \bmod aq_n\,.$$

Hence

$$\left(\frac{\alpha, k_{n+1}}{\mathfrak{P}}\right)_{q_{n+1}}^{\delta_{n+1}} \equiv \alpha^{((N_{n+1}\mathfrak{P}-1)/q_{n+1})\delta_{n+1}} \equiv \alpha^{((N_n\mathfrak{p}-1)/q_n)\delta_n} \equiv \left(\frac{\alpha, k_n}{\mathfrak{p}}\right)_{q_n}^{\delta_n} \bmod \mathfrak{p}$$

because

$$\alpha^{aq_n} = \alpha^{N_n\mathfrak{p}-1} \equiv 1 \bmod \mathfrak{p}.$$

It then follows that

$$\left(\frac{\alpha, k_{n+1}}{\mathfrak{p}}\right)_{q_{n+1}}^{\delta_{n+1}} = \left(\frac{\alpha, k_{n+1}}{\mathfrak{P}}\right)_{q_{n+1}}^{\delta_{n+1}} = \left(\frac{\alpha, k_n}{\mathfrak{p}}\right)_{q_n}^{\delta_n}.$$

Suppose next that $\mathfrak{p} = N_{n,n+1}\mathfrak{P}$, namely, that

$$\mathfrak{p} = \prod_{i=0}^{l-1} \sigma_i(\mathfrak{P})$$

where σ_i denotes the automorphism in $\mathrm{Gal}(k_{n+1}/k_n)$ such that

$$\sigma_i(\zeta) = \zeta^{1+iq_n}, \qquad \zeta \in W_{n+1}.$$

It follows that

$$\left(\frac{\alpha, k_{n+1}}{\mathfrak{p}}\right)_{q_{n+1}}^{\delta_{n+1}} = \prod_{i=0}^{l-1}\left(\frac{\alpha, k_{n+1}}{\sigma_i(\mathfrak{P})}\right)_{q_{n+1}}^{\delta_{n+1}} = \prod_{i=0}^{l-1}\sigma_i\left(\frac{\alpha, k_{n+1}}{\mathfrak{P}}\right)_{q_{n+1}}^{\delta_{n+1}}$$
$$= \prod_{i=0}^{l-1}\left(\frac{\alpha, k_{n+1}}{\mathfrak{P}}\right)_{q_{n+1}}^{(1+iq_n)\delta_{n+1}}.$$

However,

$$\sum_{i=0}^{l-1}(1 + iq_n)\delta_{n+1} = \left(l + \frac{l(l-1)}{2}q_n\right)\delta_{n+1} \equiv l\delta_n \bmod q_{n+1}.$$

Hence

$$\left(\frac{\alpha, k_{n+1}}{\mathfrak{p}}\right)_{q_{n+1}}^{\delta_{n+1}} = \left(\frac{\alpha, k_{n+1}}{\mathfrak{P}}\right)_{q_{n+1}}^{l\delta_n} = \left(\frac{\alpha, k_{n+1}}{\mathfrak{P}}\right)_{q_n}^{\delta_n}$$
$$= \left(\frac{\alpha, k_n}{\mathfrak{p}}\right)_{q_n}^{\delta_n}, \qquad \text{Q.E.D.}$$

Now, let α be any non-zero element of K and let $\mathfrak{a}$ be an ideal of K, prime to l and α. Define

$$\left(\frac{\alpha, K/k}{\mathfrak{a}}\right) = \left(\frac{\alpha, k_n}{\mathfrak{a}}\right)_{q_n}^{\delta_n},$$

where n is any integer such that $\alpha \in k_n^\times$, $\mathfrak{a} \in I_n^c$; by the above lemma, $\left(\frac{\alpha, K/k}{\mathfrak{a}}\right)$ is indeed independent of the choice of such $n \geqq 0$. The symbol $\left(\frac{\alpha, K/k}{\mathfrak{a}}\right)$ takes its values in W and is multiplicative in α and $\mathfrak{a}$. Furthermore, it also satisfies formulas such as

$$\left(\frac{\gamma\alpha, K/k}{\gamma\mathfrak{a}}\right) = \gamma\left(\frac{\alpha, K/k}{\mathfrak{a}}\right), \qquad \gamma \in \Gamma.$$

10. Skew-symmetric pairings

10.1. We are now going to define for a given Z_l-extension K/k a skew-symmetric pairing stated in the introduction.

Let $T(W)$ denote the Tate module for the group W of l^n-th roots of unity for all $n \geqq 0$. By the definition,

$$T(W) = \varprojlim W^{(a)},$$

where $W^{(a)}$ is the subgroup of all l^a-th roots of unity in W. As Z_l-modules,

$$T(W) \simeq \mathbf{Z}_l\,,$$

and $\Gamma = \mathrm{Gal}(K/k)$ acts on $T(W)$ so that

$$\gamma w = \kappa(\gamma) w\,, \qquad \gamma \in \Gamma\,,\ w \in T(W)\,,$$

with $\kappa(\gamma)$ defined in §8.1. Hence

$$T(W) \simeq \Lambda/(1 - e^q + T)$$

as compact Λ-modules (see §8.1).

Now, let M, L', and N' be the same as in §9.2 and let

$$\mathrm{Gal}(M/N') \times A' \longrightarrow W\,,$$
$$(\sigma, c) \longmapsto \langle \sigma, c \rangle$$

be the natural orthogonal pairing given in Theorem 14. Let X again denote the maximal l'-submodule of $\mathrm{Gal}(M/N')$. For x_1, x_2 in X and for each integer $a \geqq 0$, let

$$[x_1, x_2]_a = \langle x_1, c(x_2, a) \rangle$$

where $c(x_1, a)$ is the element of A' defined in §9.2. Since $l^a c(x_2, a) = 0$, we obtain a pairing

$$X \times X \longrightarrow W^{(a)}\,,$$
$$(x_1, x_2) \longmapsto [x_1, x_2]_a\,.$$

Furthermore, $c(x, b) = l^{a-b} c(x, a)$, $a \geqq b \geqq 0$, implies

$$[x_1, x_2]_b = l^{a-b}[x_1, x_2]_a\,, \qquad a \geqq b \geqq 0\,.$$

Hence the above pairings, for all $a \geqq 0$, define a map

$$X \times X \longrightarrow T(W)\,,$$
$$(x_1, x_2) \longmapsto [x_1, x_2]$$

where

$$[x_1, x_2] = \varprojlim [x_1, x_2]_a\,.$$

This is a continuous $\mathbf{Z}_l$-bilinear pairing and it also satisfies

$$[\gamma x_1, \gamma x_2] = \gamma [x_1, x_2]\,, \qquad \gamma \in \Gamma\,,$$

because $\langle \sigma, c \rangle$ satisfies such a formula. We shall next prove that the pairing is skew-symmetric, i.e.,

$$[x_1, x_2] + [x_2, x_1] = 0\,, \qquad x_1, x_2 \in X\,.$$

10.2. We fix x_1, x_2 in X and an integer $a \geqq 0$, and show that

$$[x_1, x_2]_a + [x_2, x_1]_a = 0\,.$$

Since $X/l^a X$ is finite, we can find y_1, y_2 in X and an integer $n \geqq 0$ such that

$$\omega_n x_1 = l^a y_1 , \qquad \omega_n x_2 = l^a y_2 , \qquad l^a \,|\, q_n .$$

Similarly, we find z_1, z_2 in X and an integer $m \geqq n$ such that

$$\nu_{n,m} x_1 = q_n z_1 , \qquad \nu_{n,m} x_2 = q_n z_2 .$$

These elements y_1, y_2, z_1, z_2 in X and the integers n and m will also be fixed throughout the following proof. Since the proof is not short, we describe it in several steps.

(i) Let H denote the subgroup of all principal ideals (α) in I_n^0 such that α is a local q_m-th power (at l) in k_n. Since H contains the ray group mod l^u in k_n for any sufficiently large exponent u, I_n^0/H is a finite group. Let U and V be the subgroups of I_n^0 containing H such that U/H is the Sylow l-subgroup of I_n^0/H and V/H is the complement of U/H in I_n^0/H. By class field theory, there exists a subfield Φ of M_n containing k_n such that

$$\mathrm{Gal}\,(\Phi/k_n) \simeq I_n^0/V \simeq U/H$$

canonically. Since H is a subgroup of $I_n^0 \cap P_n$, we have

$$k_n \subseteq K_n' \subseteq K_n \subseteq \Phi \subseteq M_n .$$

Let t be an integer, $t \geqq 0$, and let Ψ be the intermediate field of k_m and M_m such that

$$\mathrm{Gal}\,(M_m/\Psi) = l^t\,\mathrm{Gal}\,(M_m/k_m) .$$

Choosing t large enough, we may suppose that

$$k_m \subseteq K_m \Phi \subseteq \Psi \subseteq M_m , \qquad q_m \,|\, l^t .$$

Note that the rank of $\mathrm{Gal}\,(M_m/k_m)$ is finite by Theorem 2 so that $\mathrm{Gal}\,(\Psi/k_m) = \mathrm{Gal}\,(M_m/k_m)/l^t\,\mathrm{Gal}\,(M_m/k_m)$ is a finite group.

(ii) Now, let $\mathfrak{A}_1$, $\mathfrak{A}_2$, $\mathfrak{B}_1$, $\mathfrak{B}_2$ be ideals of I_m^0 such that $\mathfrak{A}_1$, $\mathfrak{A}_2$, $\mathfrak{B}_1$, $\mathfrak{B}_2$ and all their conjugates over $\mathbf{Q}$ are prime to each other and such that

$$\left(\frac{\Psi/k_m}{\mathfrak{A}_i}\right) = x_i \,|\, \Psi , \quad \left(\frac{\Psi/k_m}{\mathfrak{B}_i}\right) = y_i \,|\, \Psi , \qquad i = 1, 2 .$$

Since $k_m \subseteq \Psi \subseteq M_m \subseteq M$ and since $x_i|\Psi$, $y_i\,|\,\Psi$, $i = 1, 2$, are elements of $\mathrm{Gal}\,(\Psi/k_m)$, such $\mathfrak{A}_i$, $\mathfrak{B}_i$, $i = 1, 2$, exist. Let

$$\mathfrak{a}_i = N_{n,m}\mathfrak{A}_i , \qquad \mathfrak{b}_i = N_{n,m}\mathfrak{B}_i , \qquad i = 1, 2 ,$$

where $N_{n,m}$ denotes the norm from k_m to k_n. Clearly $\mathfrak{a}_i$ and $\mathfrak{b}_i$ belong to I_n^0. However, replacing the original $\mathfrak{A}_i$, $\mathfrak{B}_i$ by their suitable powers $\mathfrak{A}_i^v$, $\mathfrak{B}_i^v$, we may assume that $\mathfrak{a}_i$ and $\mathfrak{b}_i$, $i = 1, 2$, are also contained in the subgroup U of I_n^0. It follows that the ideal-classes of $\mathfrak{a}_i$ and $\mathfrak{b}_i$ are in the Sylow l-subgroup

A_n of C_n and the l-ideal-classes of $\mathfrak{a}'_i = \mathfrak{a}_i\mathfrak{o}'_n$ and $\mathfrak{b}'_i = \mathfrak{b}_i\mathfrak{o}'_n$ are in the Sylow l-subgroup A'_n of C'_n.

Since $k_n \subseteq \Phi \subseteq k_m\Phi \subseteq \Psi$,

$$\left(\frac{\Phi/k_n}{\mathfrak{a}_i}\right) = \left(\frac{\Psi/k_m}{\mathfrak{A}_i}\right)\Big| \Phi = x_i \mid \Phi\,, \qquad i = 1, 2\,,$$

and similarly

$$\left(\frac{\Phi/k_n}{\mathfrak{b}_i}\right) = y_i \mid \Phi\,, \qquad i = 1, 2.$$

As $\Phi \subseteq M_n$ and $l^a y_i = \omega_n x_i$ belongs to $\omega_n \operatorname{Gal}(M/K) = \operatorname{Gal}(M/M_n)$, we have

$$\left(\frac{\Phi/k_n}{\mathfrak{b}_i^{l^a}}\right) = l^a y_i \mid \Phi = 1$$

so that $\mathfrak{b}_i^{l^a}$ is contained in $H = U \cap V$. Hence

$$\mathfrak{b}_i^{l^a} = (\beta_i)\,, \qquad i = 1, 2\,,$$

with a local q_m-th power β_i in k_n. Since $m \geqq n$ and $l^a \mid q_n$, β_i is also a local l^a-th power in k_n. Therefore

$$k_n \subseteq k_n(\sqrt[l^a]{\beta_i}) \subseteq K'_n \subseteq \Phi$$

and it implies that

$$\left(\frac{k_n(\sqrt[l^a]{\beta_i})\,/\,k_n}{\mathfrak{a}_j}\right) = \left(\frac{\Phi/k_n}{\mathfrak{a}_j}\right)\Big| k_n(\sqrt[l^a]{\beta_i}) = x_j \mid k_n(\sqrt[l^a]{\beta_i})$$

so that

$$\left(\frac{\beta_i,\, k_n}{\mathfrak{a}_j}\right)_{l^a} = x_j(\sqrt[l^a]{\beta_i})/\sqrt[l^a]{\beta_i}\,, \qquad i, j = 1, 2\,.$$

On the other hand,

$$\left(\frac{K'_n/k_n}{\mathfrak{b}_i}\right) = \left(\frac{\Phi/k_n}{\mathfrak{b}_i}\right)\Big| K'_n = y_i \mid K'_n\,.$$

This means that $y_i \mid K'_n$ is mapped to the class of $\mathfrak{b}_i\mathfrak{o}'_n$ under $\operatorname{Gal}(K'_n/k_n) \xrightarrow{\sim} A'_n$. Hence $c(x_i, a)$ is the class of $\mathfrak{b}_i\mathfrak{o}'$ in A'. Using the remark at the end of § 7.3, we then have

$$\begin{aligned}[x_1, x_2]_a = \langle x_1, c(x_2, a)\rangle &= x_1(\sqrt[l^a]{\beta_2})/\sqrt[l^a]{\beta_2}\\ &= \left(\frac{\beta_2,\, k_n}{\mathfrak{a}_1}\right)_{l^a}.\end{aligned}$$

Similarly

$$[x_2, x_1]_a = \left(\frac{\beta_1,\, k_n}{\mathfrak{a}_2}\right)_{l^a}.$$

(iii) Let ε be an l-unit in k_m: $\varepsilon \in E'_m$. Then $k_m(\sqrt[q_m]{\varepsilon})/k_m$ is a cyclic extension, unramified outside l, and its degree is a factor of q_m. Since $\mathrm{Gal}(M_m/\Psi) = l^t \mathrm{Gal}(M_m/k_m)$ and $q_m \mid l^t$, it follows that

$$k_m \subseteq k_m(\sqrt[q_m]{\varepsilon}) \subseteq \Psi \subseteq M_m$$

so that

$$\left(\frac{k_m(\sqrt[q_m]{\varepsilon})k_m}{\mathfrak{B}_i}\right) = \left(\frac{\Psi/k_m}{\mathfrak{B}_i}\right)\Big| k_m(\sqrt[q_m]{\varepsilon}) = y_i \mid k_m(\sqrt[q_m]{\varepsilon}), \qquad i = 1, 2 .$$

However, $k_m(\sqrt[q_m]{\varepsilon})$ is obviously a subfield of N' and y_i is an element of $\mathrm{Gal}(M/N')$. Hence

$$\left(\frac{k_m(\sqrt[q_m]{\varepsilon})/k_m}{\mathfrak{B}_i}\right) = y_i \mid k_m(\sqrt[q_m]{\varepsilon}) = 1$$

and consequently,

$$\left(\frac{\varepsilon, k_m}{\mathfrak{B}_i}\right)_{q_m} = 1 , \qquad i = 1, 2 ,$$

for every ε in E'_m. If σ is any automorphism in $\mathrm{Gal}(k_m/k_n)$, then

$$\left(\frac{\varepsilon, k_m}{\sigma(\mathfrak{B}_i)}\right)_{q_m} = \sigma\left(\frac{\sigma^{-1}(\varepsilon), k_m}{\mathfrak{B}_i}\right)_{q_m} = \sigma(1) = 1 .$$

Therefore we also have

$$\left(\frac{\varepsilon, k_m}{\mathfrak{b}_i}\right)_{q_m} = 1 . \qquad i = 1, 2 ,$$

for all ε in E'_m.

(iv) Let $\mathfrak{C}_1$ and $\mathfrak{C}_2$ be ideals of I_m^0 such that $\mathfrak{C}_1$, $\mathfrak{C}_2$ are prime to all conjugates of $\mathfrak{A}_1$, $\mathfrak{A}_2$, $\mathfrak{B}_1$, $\mathfrak{B}_2$ over $\mathbf{Q}$, such that

$$\left(\frac{K_m/k_m}{\mathfrak{C}_1}\right) = z_1 \mid K_m , \qquad \left(\frac{K_m/k_m}{\mathfrak{C}_2}\right) = z_2 \mid K_m ,$$

and such that the ideal-classes of $\mathfrak{C}_1$ and $\mathfrak{C}_2$ are in the subgroup A_m of C_m. Since $\nu_{n,m}x_i = q_n z_i$, $i = 1, 2$,

$$\left(\frac{K_m/k_m}{\mathfrak{C}_i^{q_n}}\right) = \nu_{n,m}x_i \mid K_m , \qquad i = 1, 2 .$$

However, $\mathfrak{a}_i = N_{n,m}\mathfrak{A}_i$ and

$$\left(\frac{K_m/k_m}{\mathfrak{A}_i}\right) = \left(\frac{\Psi/k_m}{\mathfrak{A}_i}\right)\Big| K_m = x_i \mid K_m$$

induce

$$\left(\frac{K_m/k_m}{\mathfrak{a}_i}\right) = \nu_{n,m}x_i \mid K_m .$$

Hence, it follows from $\mathrm{Gal}(K_m/k_m) \xrightarrow{\sim} A_m$ that

$$\mathfrak{a}_i = \mathfrak{C}_i^{q_n}(\theta_i)\,, \qquad \theta_i \in k_m^\times\,,\quad i=1,2\,.$$

On the other hand,

$$\omega_n x_i = l^a y_i\,, \qquad \nu_{n,m} x_i = q_n z_i\,, \qquad l^a \mid q_n$$

implies

$$l^a\left(\nu_{n,m} y_i - \frac{q_n}{l^a}\,\omega_n z_i\right) = 0\,.$$

Since X is a free $\mathbf{Z}_l$-module, we then have

$$\nu_{n,m} y_i = \frac{q_n}{l^a}\,\omega_n z_i\,.$$

Just as above, we obtain from this equality that

$$\left(\frac{K_m/k_m}{\mathfrak{b}_i}\right) = \left(\frac{K_m/k_m}{\mathfrak{C}_i^{(q_n/l^a)\omega_n}}\right)$$

so that

$$\mathfrak{b}_i = \mathfrak{C}_i^{(q_n/l^a)\omega_n}(\pi_i)\,, \qquad \pi_i \in k_m^\times\,,\quad i=1,2\,,$$

with the operator

$$\omega_n = \gamma_0^{l^n} - 1$$

for ideals[11]. Since $\gamma_0^{l^n}$ is a topological generator of $\Gamma_n = \mathrm{Gal}(K/k_n)$ and since $\mathfrak{a}_i$ is an ideal of k_n, we have

$$\mathfrak{a}_i^{\omega_n} = 1\,, \qquad \mathfrak{C}_i^{q_n\omega_n} = (\theta_i)^{-\omega_n}\,, \qquad i=1,2\,.$$

Hence it follows from the above that

$$(\beta_i) = \mathfrak{b}_i^{l^a} = \mathfrak{C}^{q_n\omega_n}(\pi_i)^{l^a} = (\theta_i)^{-\omega_n}(\pi_i)^{l^a}\,,$$

namely, that

$$\beta_i = \theta_i^{-\omega_n}\pi_i^{l^a}\varepsilon_i\,, \qquad i=1,2\,,$$

with a unit ε_i in E_m.

(v) Now, we know by (ii) that

$$[x_1, x_2]_a = \left(\frac{\beta_2,\, k_n}{\mathfrak{a}_1}\right)_{l^a} = \left(\frac{\beta_2,\, k_n}{\mathfrak{a}_1}\right)_{q_n}^{q_n/l^a}.$$

Hence, by Lemma 15,

$$[x_1, x_2]_a = \left(\frac{\beta_2,\, k_m}{\mathfrak{a}_1}\right)_{q_m}^{\delta q_n/l^a}$$

[11] Note that Γ acts on ideal groups but Λ does not.

where

$$\delta = \delta_m \delta_n^{-1}$$

with δ_n in § 9.4. Since $\mathfrak{a}_1 = \mathfrak{C}_1^{q_n}(\theta_1)$, we have

$$[x_1, x_2]_a = \left\{\left(\frac{\beta_2, k_m}{\mathfrak{C}_1^{q_n}}\right)_{q_m}^{q_n/l^a} \cdot \left(\frac{\beta_2, k_m}{\theta_1}\right)_{q_m}^{-q_n/l^a}\right\}^{\delta}.$$

Let

$$a_n = \kappa(\gamma_0^{l^n}) - 1 = e^{l^n q} - 1 = e^{q_n} - 1\,.$$

Then a_n/q_n is an l-adic unit and

$$\zeta^{\omega_n} = \zeta^{a_n}\,, \qquad \zeta \in W\,.$$

Since β_2 is an element of k_n,

$$\left(\frac{\beta_2, k_m}{\mathfrak{C}_1^{\omega_n}}\right)_{q_m} = \left(\frac{\beta_2, k_m}{\mathfrak{C}_1}\right)_{q_m}^{\omega_n} = \left(\frac{\beta_2, k_m}{\mathfrak{C}_1}\right)_{q_m}^{a_n} = \left(\frac{\beta_2, k_m}{\mathfrak{C}_1^{q_n}}\right)_{q_m}^{a_n/q_n}.$$

Hence

$$\begin{aligned}\left(\frac{\beta_2, k_m}{\mathfrak{C}_1^{q_n}}\right)_{q_m}^{q_n/l^a} &= \left(\frac{\beta_2, k_m}{\mathfrak{C}_1^{\omega_n}}\right)_{q_m}^{q_n/a_n \cdot q_n/l^a} = \left(\frac{\beta_2, k_m}{\mathfrak{C}_1^{(q_n/l^a)\omega_n}}\right)_{q_m}^{q_n/a_n} \\ &= \left\{\left(\frac{\beta_2, k_m}{\mathfrak{b}_1}\right)_{q_m}\left(\frac{\beta_2, k_m}{\pi_1}\right)_{q_m}^{-1}\right\}^{q_n/a_n}\end{aligned}$$

because $\mathfrak{C}_1^{(q_n/l^a)\omega_n} = \mathfrak{b}_1(\pi_1)^{-1}$.

On the other hand, β_2 is a local q_m-th power in k_n and, hence, in k_m, and it is prime to $(\theta_1) = \mathfrak{a}_1\mathfrak{C}_1^{-q_n}$. Therefore the reciprocity law for the power residue symbol[12] implies

$$\left(\frac{\beta_2, k_m}{\theta_1}\right)_{q_m} = \left(\frac{\theta_1, k_m}{\beta_2}\right)_{q_m} = \left(\frac{\theta_1, k_m}{\mathfrak{b}_2}\right)_{q_m}^{l^a}$$

so that

$$\left(\frac{\beta_2, k_m}{\theta_1}\right)_{q_m}^{q_n/l^a} = \left(\frac{\theta_1, k_m}{\mathfrak{b}_2}\right)_{q_m}^{q_n}.$$

However, as in the above,

$$\left(\frac{\theta_1^{\omega_n}, k_m}{\mathfrak{b}_2}\right)_{q_m} = \left(\frac{\theta_1, k_m}{\mathfrak{b}_2}\right)_{q_m}^{a_n}.$$

Hence we obtain

$$\begin{aligned}\left(\frac{\beta_2, k_m}{\theta_1}\right)_{q_m}^{q_n/l^a} &= \left(\frac{\theta_1^{\omega_n}, k_m}{\mathfrak{b}_2}\right)_{q_m}^{q_n/a_n} \\ &= \left\{\left(\frac{\beta_1, k_m}{\mathfrak{b}_2}\right)_{q_m}^{-1}\left(\frac{\pi_1^{l^a}, k_m}{\mathfrak{b}_2}\right)_{q_m}\right\}^{q_n/a_n} \\ &= \left\{\left(\frac{\beta_1, k_m}{\mathfrak{b}_2}\right)_{q_m}^{-1}\left(\frac{\pi_1, k_m}{\beta_2}\right)_{q_m}\right\}^{q_n/a_n},\end{aligned}$$

[12] See [3], II, § 12.

because $\theta_1^{\omega_n} = \beta_1^{-1}\pi_1^{l^a}\varepsilon_1$ and

$$\left(\frac{\varepsilon_1, k_m}{\mathfrak{b}_2}\right)_{q_m} = 1$$

as proved in (iii).

(vi) We now see from the above that

$$[x_1, x_2]_a = \left\{\left(\frac{\beta_2, k_m}{\mathfrak{b}_1}\right)_{q_m}\left(\frac{\beta_1, k_m}{\mathfrak{b}_2}\right)_{q_m}^{-1}\left(\frac{\beta_2, k_m}{\pi_1}\right)_{q_m}\left(\frac{\pi_1, k_m}{\beta_2}\right)_{q_m}^{-1}\right\}^{\delta q_n/a_n}.$$

However, as in (v), the reciprocity law induces that the product of the last two factors in the above bracket is 1. Hence

$$[x_1, x_2]_a = \left\{\left(\frac{\beta_2, k_m}{\mathfrak{b}_1}\right)^{\delta}\left(\frac{\beta_1, k_m}{\mathfrak{b}_2}\right)^{-\delta}\right\}^{q_n/a_n}$$
$$= \left\{\left(\frac{\beta_2, k_n}{\mathfrak{b}_1}\right)\left(\frac{\beta_1, k_n}{\mathfrak{b}_2}\right)^{-1}\right\}^{q_n/a_n}.$$

Since

$$q_n/a_n = q_n/(e^{q_n} - 1) \equiv \delta_n \bmod q_n\,,$$

we finally obtain the following formula:

$$[x_1, x_2]_a = \left(\frac{\beta_2, K/k}{\mathfrak{b}_1}\right)\left(\frac{\beta_1, K/k}{\mathfrak{b}_2}\right)^{-1}.$$

It is obvious that we have similarly

$$[x_2, x_1]_a = \left(\frac{\beta_1, K/k}{\mathfrak{b}_2}\right)\left(\frac{\beta_2, K/k}{\mathfrak{b}_1}\right)^{-1}.$$

Hence

$$[x_1, x_2]_a + [x_2, x_1]_a = 0\,,$$

if $W^{(a)}$ is regarded as an additive group. Thus the skew-symmetry of the pairing

$$X \times X \longrightarrow T(W)$$

is proved.

10.3. We shall next determine the kernel of the above pairing, namely, the set of all x in X such that $[x, X] = 0$ or, equivalently, such that $[X, x] = 0$.

Clearly, $[X, x] = 0$ is equivalent to the fact that

$$[X, x]_a = \langle X, c(x, a)\rangle = 0$$

for every $a \geqq 0$. Since X is the maximal l'-submodule of $\mathrm{Gal}\,(M/N')$, $\mathrm{Gal}\,(M/N')/X$ is an l-module so that

$$l^u\, \mathrm{Gal}\,(M/N') \subseteq X$$

for some integer $u \geqq 0$. Hence, if $\langle X, c(x, a+u)\rangle = 0$, then

$$\langle \mathrm{Gal}(M/N'),\ c(x, a)\rangle = \langle l^u \,\mathrm{Gal}(M/N'),\ c(x, a+u)\rangle = 0 .$$

As the pairing $\mathrm{Gal}(M/N') \times A' \to W$ is orthogonal, it follows that $[X, x] = 0$ if and only if $c(x, a) = 0$ for every $a \geqq 0$.

Now, let

$$f\colon X \longrightarrow T(B')$$

be the morphism defined in § 9.2. It is clear from the definition of f that $f(x) = 0$ if and only if $c(x, a) = 0$ for every $a \geqq 0$. Therefore we see that $[X, x] = 0$ is equivalent to $f(x) = 0$. It then follows from Lemma 14 that the kernel of the pairing

$$X \times X \longrightarrow T(W)$$

is the submodule $\omega^{-1}(Y)$ of X.

We summarize the results obtained in §§ 10.2 and 10.3 as follows:

THEOREM 19. *Let X be the maximal l'-submodule of the noetherian torsion Λ-module* $\mathrm{Gal}(M/N')$ *and let $T(W)$ denote the Tate module for W. The pairing*

$$X \times X \longrightarrow T(W)$$

defined in § 10.1 *is skew-symmetric and satisfies*

$$[\gamma x_1, \gamma x_2] = \gamma[x_1, x_2] = \kappa(\gamma)[x_1, x_2]$$

for x_1, x_2 in X and γ in $\Gamma = \mathrm{Gal}(K/k)$. *The kernel of the pairing consists of all elements x in X such that $\omega_n x$ is contained in* $\mathrm{Gal}(M/L'N')$ *for some $n \geqq 0$.*

10.4. Let

$$Z = X/\omega^{-1}(Y) .$$

By the above theorem, we have a non-degenerate skew-symmetric pairing

$$Z \times Z \longrightarrow T(W) .$$

This induces an injective morphism

$$Z \longrightarrow \mathrm{Hom}_{Z_l}(Z, T(W))$$

which maps each z in Z to the homomorphism φ on the right hand side defined by

$$\varphi(z') = [z, z'] , \qquad z' \in Z .$$

Note that the Γ-structure on $\mathrm{Hom}_{Z_l}(Z, T(W))$ is defined by

$$(\gamma\cdot\varphi)(z') = \gamma\varphi(\gamma^{-1}z') , \qquad z' \in Z .$$

Now,

$$T(W) = \varprojlim W^{(n)} = \varprojlim \mathrm{Hom}_{\mathbf{Z}_l}(l^{-n}\mathbf{Z}_l/\mathbf{Z}_l, W)$$
$$= \mathrm{Hom}_{\mathbf{Z}_l}(\mathbf{Q}_l/\mathbf{Z}_l, W) .$$

Hence

$$\begin{aligned}\mathrm{Hom}_{\mathbf{Z}_l}(Z, T(W)) &= \mathrm{Hom}_{\mathbf{Z}_l}(Z, \mathrm{Hom}_{\mathbf{Z}_l}(\mathbf{Q}_l/\mathbf{Z}_l), W) \\ &= \mathrm{Hom}_{\mathbf{Z}_l}(Z \otimes_{\mathbf{Z}_l} (\mathbf{Q}_l/Z_l), W) \\ &= (\mathrm{Hom}_{\mathbf{Z}_l}(Z \otimes_{\mathbf{Z}_l} (\mathbf{Q}_l/\mathbf{Z}_l), \mathbf{Q}_l/\mathbf{Z}_l))^{\cdot} ,\end{aligned}$$

as explained in §8.1. However, since Z is an l'-module, it is noetherian over $\mathbf{Z}_l$ and div Z is prime to the principal divisor (l). Therefore, applying the formula in §1.3 with $\pi_n = l^n$, $n \geqq 0$, we obtain

$$\begin{aligned}\mathrm{Hom}_{\mathbf{Z}_l}(Z \otimes_{\mathbf{Z}_l} (\mathbf{Q}_l/\mathbf{Z}_l), \mathbf{Q}_l/\mathbf{Z}_l) &= \mathrm{Hom}_{\mathbf{Z}_l}(\varinjlim Z \otimes (l^{-n}\mathbf{Z}_l/\mathbf{Z}_l), \mathbf{Q}_l/\mathbf{Z}_l) \\ &= \varprojlim \mathrm{Hom}_{\mathbf{Z}_l}(Z/l^n Z, \mathbf{Q}_l/\mathbf{Z}_l) \\ &\simeq \alpha(Z) .\end{aligned}$$

Thus there exists an injective morphism

$$Z \longrightarrow \alpha(Z)^{\cdot} .$$

Now, $\alpha(Z)^{\cdot}$, as well as Z and $\alpha(Z)$, is noetherian over $\mathbf{Z}_l$, and

$$\lambda(\alpha(Z)^{\cdot}) = \lambda(\alpha(Z)) = \lambda(Z)$$

for the λ-invariant defined in §1.2. Hence we see easily that the cokernel of the above injective map is finite so that

$$Z \sim \alpha(Z)^{\cdot} .$$

Since $\alpha(Z)^{\cdot} \sim \dot{Z}$, the following lemma is proved:

LEMMA 16. *Let X and $\omega^{-1}(Y)$ be the same as in* §9.3 *and let* $Z = X/\omega^{-1}(Y)$. *Then*

$$Z \sim \dot{Z} .$$

THEOREM 20. *Let* $\mathrm{Gal}(L'N'/N')_\omega$ *denote the submodule of all elements x in* $\mathrm{Gal}(L'N'/N')$ *such that $\omega_n x = 0$ for some $n \geqq 0$. Let*

$$U = \mathrm{Gal}(L'N'/N')/\mathrm{Gal}(L'N'/N')_\omega .$$

Then

$$U \sim \dot{U} .$$

Proof. Let U' and U'' denote the maximal l-submodule and the maximal l'-submodule of U, respectively. Consider X/Y as a submodule of $\mathrm{Gal}(L'N'/N')$ in the obvious manner $(Y = X \cap \mathrm{Gal}(M/L'N'))$. Then

$$\omega^{-1}(Y)/Y = (X/Y) \cap \mathrm{Gal}(L'N'/N')_\omega .$$

Hence $Z = X/\omega^{-1}(Y)$ may be regarded as a submodule of U. It is easy to see that $Z \sim U''$ so that

$$U'' \sim \dot{U}''$$

by the above lemma. On the other hand, for the l-module U', we have

$$U' \sim \dot{U}' .$$

It then follows that

$$U \sim \dot{U} , \qquad \text{Q.E.D.}$$

We shall next consider the image of

$$f\colon X \longrightarrow T(B')$$

defined in §9.2. By Theorem 14, we have a natural orthogonal pairing

$$\mathrm{Gal}\,(L'N'/N') \times B' \longrightarrow W .$$

As explained in §8.1, this induces a pairing

$$\mathrm{Gal}\,(L'N'/N')^{\cdot} \times B' \longrightarrow \mathbf{Q}_l/\mathbf{Z}_l .$$

We see easily that X/Y is pseudo-isomorphic to the maximal l'-submodule of $\mathrm{Gal}\,(L'N'/N')$ so that $(X/Y)^{\cdot}$ is pseudo-isomorphic to the maximal l'-submodule of $\mathrm{Gal}\,(L'N'/N')^{\cdot}$. Hence Lemma 13 implies the following

LEMMA 17.

$$T(B') \sim (X/Y)^{\cdot} = \dot{X}/\dot{Y} .$$

LEMMA 18. *Whenever n is sufficiently large, $\dot{\omega}_n T(B')$ is a submodule of finite index in* $\mathrm{Im}\,(f)$.

Proof. Let

$$E(\mathfrak{p}_1^{e_1}, \cdots, \mathfrak{p}_t^{e_t}) , \qquad e_i \geqq 1 ,$$

be the elementary Λ-module associated with X/Y. One sees easily that the elementary Λ-module associated with $Z = X/\omega^{-1}(Y)$ is

$$E(\mathfrak{p}_1^{e_1'}, \cdots, \mathfrak{p}_t^{e_t'})$$

where $e_i' = e_i - 1$ if $\mathfrak{p}_i = (\xi_{n_i})$ for some $n_i \geqq 0$, and $e_i' = e_i$ otherwise. By Lemmas 13 and 16, we have

$$\mathrm{Im}\,(f) \simeq \mathrm{Coim}\,(f) = Z \sim \dot{Z} \sim E(\dot{\mathfrak{p}}_1^{e_1'}, \cdots, \dot{\mathfrak{p}}_t^{e_t'}) .$$

On the other hand,

$$T(B') \sim (X/Y)^{\cdot} \sim E(\dot{\mathfrak{p}}_1^{e_1}, \cdots, \dot{\mathfrak{p}}_t^{e_t}) .$$

Since $\dot{\omega}_n E(\dot{\mathfrak{p}}_1^{e_1}, \cdots, \dot{\mathfrak{p}}_t^{e_t})$ is a submodule of finite index in $E(\dot{\mathfrak{p}}_1^{e_1'}, \cdots, \dot{\mathfrak{p}}_t^{e_t'})$ for every sufficiently large n, the lemma follows immediately from the above pseudo-isomorphisms.

10.5. Let

$$V = \mathrm{Gal}\,(L'N'/N') \otimes_{\mathbf{Z}_l} \mathbf{Q}_l\,.$$

As mentioned in §1.2, V is a finite dimensional vector space over $\mathbf{Q}_l$ with dimension

$$\lambda(\mathrm{Gal}\,(L'N'/N')) = \lambda(\mathrm{Gal}\,(L'/L' \cap N'))\,.$$

It is a Γ-module as well as a Λ-module. Since $\mathrm{Gal}\,(L'N'/N')/(X/Y)$ is an l-module, we may identify V with $(X/Y) \otimes_{\mathbf{Z}_l} \mathbf{Q}_l$:

$$V = (X/Y) \otimes_{\mathbf{Z}_l} \mathbf{Q}_l\,.$$

Similarly, we may define V_0 by

$$V_0 = \mathrm{Gal}\,(L'N'/N')_\omega \otimes_{\mathbf{Z}_l} \mathbf{Q}_l = (\omega^{-1}(Y)/Y) \otimes_{\mathbf{Z}_l} \mathbf{Q}_l\,.$$

V_0 is the subspace of V consisting of all vectors in V annihilated by some ω_n, $n \geqq 0$, namely, fixed by some Γ_n, $n \geqq 0$. It is also the kernel of the endomorphism

$$\omega_n \colon V \longrightarrow V$$

for any sufficiently large n. Let

$$V' = V/V_0\,.$$

Then

$$V' = (\mathrm{Gal}\,(L'N'/N')/\mathrm{Gal}\,(L'N'/N')_\omega) \otimes_{\mathbf{Z}_l} \mathbf{Q}_l$$
$$= (X/\omega^{-1}(Y)) \otimes_{\mathbf{Z}_l} \mathbf{Q}_l\,.$$

Let

$$V(W) = T(W) \otimes_{\mathbf{Z}_l} \mathbf{Q}_l\,.$$

$V(W)$ is a one-dimensional vector space over $\mathbf{Q}_l$ and Γ acts on $V(W)$ by

$$\gamma_v = \kappa(\gamma)v\,, \qquad \gamma \in \Gamma,\ v \in V(W)\,.$$

Since $X/\omega^{-1}(Y)$ is free over $\mathbf{Z}_l$, it is clear that we can obtain the following result from Theorem 19:

THEOREM 21. *There exists a non-degenerate skew-symmetric $\mathbf{Q}_l$-bilinear pairing*

$$V' \times V' \longrightarrow V(W)$$

satisfying

$$[\gamma v_1, \gamma v_2] = \gamma[v_1, v_2] = \kappa(\gamma)[v_1, v_2]$$

for v_1, v_2 in V' and γ in $\Gamma = \mathrm{Gal}\,(K/k)$.

Now, let

$$V(B') = T(B') \otimes_{Z_l} Q_l .$$

Since $T(B') \sim (X/Y)^{\cdot}$ by Lemma 17, $V(B')$ is a vector space over Q_l with dimension

$$\lambda(T(B')) = \lambda(X/Y) = \dim V .$$

Let $\dot{\omega} V(B')$ denote the intersection of the decreasing sequence of subspaces $\dot{\omega}_n V(B')$, $n \geqq 0$, in $V(B')$; since $V(B')$ is finite dimensional,

$$\dot{\omega} V(B') = \dot{\omega}_n V(B')$$

whenever n is sufficiently large. The following theorem is then an immediate consequence of Lemmas 14 and 18:

THEOREM 22. *The morphism* $f\colon X \rightarrow T(B')$ *defined in* § 9.3 *induces an isomorphism*

$$V' \xrightarrow{\sim} \dot{\omega} V(B') .$$

It follows that there exists a pairing

$$\dot{\omega} V(B') \times \dot{\omega} V(B') \longrightarrow V(W) ,$$

similar to the one stated in Theorem 21.

Let us now consider the finite dimensional vector space V' over Q_l. Since V' admits a non-degenerate skew-symmetric bilinear form, its dimension must be even:

$$\dim V' = 2g , \qquad g \geqq 0 .$$

Furthermore, the Galois group $\Gamma = \mathrm{Gal}(K/k)$ acts continuously on V' so that it defines a continuous homomorphism

$$\rho\colon \Gamma \longrightarrow GL(V') .$$

Using the fact that the determinant of a symplectic linear transformation is 1, we see from the equality in Theorem 21 that

$$\det \rho(\gamma) = \kappa(\gamma)^{2g} , \qquad \gamma \in \Gamma .$$

It also follows from the same equality by routine argument that the characteristic polynomial $f_\gamma(T)$ of $\rho(\gamma)$ satisfies a functional equation of well-known type:

$$T^{2g} f_\gamma(\kappa(\gamma) T^{-1}) = \kappa(\gamma)^g f_\gamma(T) .$$

ρ is of course an l-adic representation for k in the sense of Serre [10]. It is an analogue, for algebraic number fields, of such l-adic representations obtained from algebraic function fields or from abelian varieties; we shall explain this later in some detail. It would be an important problem to pursue such analogy further and study various properties of the bilinear form and the representation ρ on V' obtained in this section.

11. Some special cases

11.1. In this section, we shall discuss the results obtained in §10 in some special cases, assuming certain further conditions on the cyclotomic $\mathbf{Z}_l$-extension K/k.

We first consider the following property of $\mathrm{Gal}(L'N'/N')$:

(A) $\mathrm{Div}\,\mathrm{Gal}(L'N'/N')$ is disjoint from all divisors (ω_n), $n \geqq 0$.

Since $\omega_n = \xi_0\xi_1 \cdots \xi_n$, $n \geqq 0$, (A) is equivalent to saying that none of the prime divisors (ξ_n), $n \geqq 0$, is contained in $\mathrm{div}\,\mathrm{Gal}(L'N'/N')$. We also note that $\mathrm{Gal}(L'N'/N') \simeq \mathrm{Gal}(L'/L' \cap N')$ so that

$$\mathrm{div}\,\mathrm{Gal}(L'N'/N') = \mathrm{div}\,\mathrm{Gal}(L'/L' \cap N') .$$

LEMMA 19. (A) *is also equivalent to each one of the following statements*:

(i) $\mathrm{Gal}(L'N'/N') \sim \mathrm{Gal}(L'N'/N')^{\cdot}$

(ii) $\mathrm{div}(X/Y)$ *is disjoint from all* (ω_n), $n \geqq 0$,

(iii) $\mathrm{div}\,T(B')$ *is disjoint from all* $(\dot{\omega}_n)$, $n \geqq 0$,

(iv) $\mathrm{Gal}(L'N'/N')_\omega$ *is finite*,

(v) $\omega^{-1}(Y)/Y$ *is finite*,

(vi) $T(B')/\dot{\omega}_n T(B')$ *is finite for every* $n \geqq 0$.

Here X/Y, $T(B')$, *etc., are the same as in the preceding sections.*

Proof. Since $\mathrm{Gal}(L'N'/N')/(X/Y)$ is an l-module, the difference of $\mathrm{div}\,\mathrm{Gal}(L'N'/N')$ and $\mathrm{div}(X/Y)$ is a multiple of (l). Hence (ii) is equivalent to (A). The equivalence of (ii) and (iii) follows from Lemma 17. That (iv) is equivalent to (A) follows easily from the definition of the module $\mathrm{Gal}(L'N'/N')_\omega$. Similarly, (v) is equivalent to (ii), and (vi) is equivalent to (iii). By Theorem 20, (iv) induces (i). By the same theorem, (i) also induces

$$\mathrm{div}\,\mathrm{Gal}(L'N'/N')_\omega = \mathrm{div}\,\mathrm{Gal}(L'N'/N')^{\cdot}_\omega .$$

Since $\omega_n\,\mathrm{Gal}(L'N'/N')_\omega = \dot{\omega}_n\,\mathrm{Gal}(L'N'/N')^{\cdot}_\omega = 0$ for some large n, the above equality holds only when $\mathrm{Gal}(L'N'/N')_\omega$ is finite. This completes the proof of the lemma.

LEMMA 20. *If* $\mathrm{Gal}(L'N'/N')$ *has the property* (A), *then*

$$V_0 = 0 , \qquad V' = V , \qquad \dot{\omega} V(B') = V(B')$$

for the vector spaces V, V_0, etc., defined in §10.5, *and the morphism* $f: X \to T(B')$ *induces a pseudo-isomorphism*

$$X/Y \longrightarrow T(B') .$$

Proof. By Lemma 19, $\mathrm{Gal}(L'N'/N')_\omega$ is finite. Hence

$$V_0 = \mathrm{Gal}(L'F'/F')_\omega \otimes_{Z_l} Q_l = 0 , \qquad V' = V/V_0 = V .$$

Similarly, since $T(B')/\dot\omega_n T(B')$ is finite,

$$\dot\omega V(B') = \dot\omega_n V(B') = \dot\omega_n T(B') \otimes_{Z_l} Q_l = T(B') \otimes_{Z_l} Q_l = V(B')$$

for large n. By Lemma 14, the kernel of $X/Y \to T(B')$ is $\omega^{-1}(Y)/Y$, and by Lemma 18, the cokernel is a factor module of $T(B')/\dot\omega_n T(B')$ if n is sufficiently large. By Lemma 19, both of these are finite modules under the assumption (A).

It is now clear that we have the following theorem from Theorems 21 and 22.

THEOREM 23. *Suppose that* $\mathrm{div\,Gal}(L'N'/N')$ $(= \mathrm{div\,Gal}(L'/L' \cap N'))$ *is disjoint from all principal divisors* (ω_n), $n \geqq 0$. *Then there exists a natural isomorphism*

$$V = \mathrm{Gal}(L'N'/N') \otimes_{Z_l} Q_l \xrightarrow{\sim} V(B') = T(B') \otimes_{Z_l} Q_l$$

and a non-degenerate skew-symmetric Q_l*-bilinear pairing*

$$V \times V \longrightarrow V(W) = T(W) \otimes Q_l ,$$

satisfying

$$[\gamma v_1, \gamma v_2] = \gamma[v_1, v_2] = \kappa(\gamma)[v_1, v_2]$$

for v_1, v_2 in V and γ in $\Gamma = \mathrm{Gal}(K/k)$.

11.2. We shall next see what properties of a cyclotomic Z_l-extension K/k imply the property (A) for the Galois group $\mathrm{Gal}(L'N'/N')$.

LEMMA 21. *Suppose that Leopoldt's conjecture stated in* §2.3 *holds for all intermediate fields k_n, $n \geqq 0$, of k and K. Then* (A) *is satisfied for* $\mathrm{Gal}(L'N'/N')$.

Proof. Let M be, as before, the maximal abelian l-extension over K, unramified outside l, and let

$$Q = \mathrm{Gal}(M/K) .$$

Let

$$Q \longrightarrow Q'$$

be a pseudo-isomorphism of Q into the associated elementary Λ-module

$$Q' = E(e_0; \mathfrak{p}_1^{e_1}, \cdots, \mathfrak{p}_t^{e_t}), \qquad e_1, \cdots, e_t \geqq 1.$$

For each $n \geqq 0$, it induces a pseudo-isomorphism

$$\mathrm{Gal}(M_n/K) = Q/\omega_n Q \longrightarrow Q'/\omega_n Q',$$

where M_n denotes the maximal abelian extension of k_n contained in M (see § 3.2). It follows that

$$\begin{aligned}\text{ess. rank}\,\mathrm{Gal}(M_n/k_n) &= 1 + \text{ess. rank}\,\mathrm{Gal}(M_n/K)\\ &= 1 + \text{ess. rank}\,Q'/\omega_n Q'\end{aligned}$$

so that Leopoldt's conjecture holds for k_n if and only if

$$\text{ess. rank}\,Q'/\omega_n Q' = \frac{d_n}{2}$$

where $d_n = [k_n: \mathbf{Q}] = l^n d$ with $d = [k: \mathbf{Q}]$.

Now, let

$$u_n = \sum{}' \lambda(\mathfrak{p}_i),$$

where the sum is taken over all $\mathfrak{p}_i$, $1 \leqq i \leqq t$, such that $\mathfrak{p}_i = (\xi_{n_i})$ for some integer $n_i \leqq n$. Then one sees by a simple computation that

$$\text{ess. rank}\,Q'/\omega_n Q' = e_0 l^n + u_n.$$

However, it follows from Theorem 17 that $e_0 = d/2$. Hence we see that Leopoldt's conjecture for k_n is equivalent to $u_n = 0$. It is now clear that Leopoldt's conjecture holds for all k_n, $n \geqq 0$, if and only if $\mathrm{div}\,\mathrm{Gal}(M/K)$ is disjoint from all (ω_n), $n \geqq 0$. Since $\mathrm{div}\,\mathrm{Gal}(L'N'/N')$ is a part of $\mathrm{div}\,\mathrm{Gal}(M/F)$, the lemma is proved.

LEMMA 22. *Suppose that each* k_n, $n \geqq 0$, *has only one prime divisor (prime ideal) lying above* l *(dividing* l*). Then* (A) *is again satisfied for* $\mathrm{Gal}(L'N'/N')$.

Proof. Let K_n be the maximal unramified abelian l-extension over k_n. Let L be the maximal unramified abelian l-extension over K, and L_n the maximal abelian extension of k_n contained in L (see § 3.4). Under the assumption of the lemma, one sees easily that for sufficiently large n,

$$KK_n = L_n, \qquad K \cap K_n = k_n$$

so that

$$\mathrm{Gal}(L_n/K) \simeq \mathrm{Gal}(K_n/k_n).$$

However, by § 3.4,

$$\mathrm{Gal}(L_n/K) = \mathrm{Gal}(L/K)/\omega_n\,\mathrm{Gal}(L/K).$$

Since K_n/k_n is a finite extension, it follows that $\mathrm{Gal}(L/K)/\omega_n\,\mathrm{Gal}(L/K)$ is finite for every $n \geqq 0$, and this implies that div Gal (L/K) is disjoint from all (ω_n), $n \geqq 0$. As

$$K \subseteq L' \cap N' \subseteq L' \subseteq L \,,$$

div Gal $(L'/L' \cap N')$ is also disjoint from all (ω_n), $n \geqq 0$, and the lemma is proved.

Suppose now that the ground field k is an abelian extension of the rational field $\mathbf{Q}$. In such a case, each k_n is also abelian over $\mathbf{Q}$, and by Brumer's theorem, Leopoldt's conjecture holds for k_n. Hence we see from Lemma 21 that Gal $(L'N'/N')$ has the property (A) so that conclusions in § 11.1 hold for such a $\mathbf{Z}_l$-extension K/k. Of course, we have a similar result in the case where k is an abelian extension of an imaginary quadratic field. There are also other cases in which the assumption of Lemma 22 is satisfied. It seems quite likely that Gal $(L'N'/N')$ has the property (A) for every cyclotomic $\mathbf{Z}_l$-extension K/k so that we always have the results mentioned at the end of § 11.1.

11.3. Let J denote the automorphism of the complex field $\mathbf{C}$ which maps each complex number to its complex conjugate. An algebraic number field Φ, finite or infinite over $\mathbf{Q}$, will be called a J-field if $J(\Phi) = \Phi$ and if

$$\iota J = J\iota$$

for any isomorphism ι of Φ into $\mathbf{C}$. It is easy to see that Φ is a J-field if and only if Φ is either totally real or a totally imaginary quadratic extension of a totally real subfield. If Φ is abelian over $\mathbf{Q}$, then Φ is a J-field, and if both Φ and Ψ are J-fields, then so is the composite $\Phi\Psi$.

We shall next consider a cyclotomic $\mathbf{Z}_l$-extension K/k where the ground field k is a J-field. Since $K = k\mathbf{P}_\infty$, K is then also a J-field, and so are the intermediate fields k_n, $n \geqq 0$. Let k_+, $k_{n,+}$ and K_+ denote the maximal real subfields of k, k_n, and K, respectively. These are totally real fields and

$$[k\colon k_+] = [k_n\colon k_{n,+}] = [K\colon K_+] = 2 \,.$$

$K = k\mathbf{P}_\infty$ also shows that K/k_+ is an abelian extension. Its Galois group Gal (K/k_+) is the direct product of $\Gamma =$ Gal (K/k) and Gal (K/K_+) which is generated by $J|K$. Clearly K_+/k_+ is a $\mathbf{Z}_l$-extension with intermediate fields $k_{n,+}$, $n \geqq 0$; in fact $K_+ = k_+\mathbf{Q}_\infty$ with $\mathbf{Q}_\infty$ as in § 2.1.

Most of the modules considered in the preceding sections are not only Γ-modules but also Gal (K/k_+)-modules so that we may consider the action of J on these modules. In order to study the structure of such modules, the following remark is often quite useful. Let X be a module on which J acts as an involutive automorphism. Let X^- and X^+ denote the kernels of the endomorphisms $1+J: X \to X$ and $1-J: X \to X$ respectively so that

$$X_+ = X/X^- \xrightarrow{\sim} (1+J)X\,, \qquad X_- = X/X^+ \xrightarrow{\sim} (1-J)X\,.$$

Then

$$2X^+ \subseteq (1+J)X \subseteq X^+\,, \qquad 2X^- \subseteq (1-J)X \subseteq X^-\,,$$

$$2(X^+ \cap X^-) = 0\,, \qquad 2x \subseteq (1+J)X + (1-J)X \subseteq X^+ + X^-\,.$$

Hence, if X is a $\mathbf{Z}_l$-module with $l > 2$, then

$$X^+ = (1+J)X\,, \qquad X^- = (1-J)X\,, \qquad X = X^+ \oplus X^-$$

and

$$X_+ \xrightarrow{\sim} X^+\,, \qquad X_- \xrightarrow{\sim} X^-\,.$$

Now, it is clear from the definition that L' is a Galois extension over k_+ so that $\mathrm{Gal}\,(L'/K)$ is a $\mathrm{Gal}\,(K/k_+)$-module. Let L'^+ denote the maximal abelian extension of K_+ contained in L'. Then $K \subseteq L'^+ \subseteq L'$, and one sees immediately that

$$\mathrm{Gal}\,(L'/L'^+) = (1-J)\,\mathrm{Gal}\,(L'/K)\,.$$

On the other hand, let L'_+ denote the maximal unramified abelian l-extension over K_+ in which every prime divisor of K_+, lying above the rational prime l, is completely decomposed; L'_+ is the extension of K_+ defined similarly as L' over K. Clearly

$$K_+ \subseteq L'_+ \subseteq L'^+\,.$$

LEMMA 23.

$$2\,\mathrm{Gal}\,(L'^+/L'_+) = 0\,.$$

Proof. For each prime divisor v of K_+, let Z_v and T_v denote the decomposition group and the inertia group, respectively, of v for the abelian extension L'^+/K_+. It follows from the definition of L'_+ that $\mathrm{Gal}\,(L'^+/L'_+)$ is the subgroup of $\mathrm{Gal}\,(L'^+/K_+)$ generated by T_v for all prime divisors v of K_+ and also by Z_v for all those v which lie above the rational prime l. However, since L'^+/K is unramified, $\mathrm{Gal}\,(L'^+/K) \cap T_v = 0$ so that the order of T_v is 1 or 2. Similarly, Z_v, for v above l, has order 1 or 2. Hence $2\,\mathrm{Gal}\,(L'^+/L'_+) = 0$.

Let

$$\varphi: \mathrm{Gal}\,(L'^+/K) \longrightarrow \mathrm{Gal}\,(L'_+/K_+)\,,$$
$$\sigma \longmapsto \sigma\,|\,L'_+$$

so that

$$\operatorname{Ker}\varphi = \operatorname{Gal}(L'^{+}/KL'_{+}), \qquad \operatorname{Coker}\varphi = \operatorname{Gal}(K \cap L'_{+}/K_{+}).$$

By Lemma 23 and by $[K: K_{+}] = 2$, we have

$$2\operatorname{Ker}\varphi = 0, \qquad 2\operatorname{Coker}\varphi = 0.$$

Since $\operatorname{Ker}\varphi$ and $\operatorname{Coker}\varphi$ are both $\mathbf{Z}_l$-modules, it follows that if $l > 2$, then

$$\varphi: \operatorname{Gal}(L'^{+}/K) \xrightarrow{\sim} \operatorname{Gal}(L'_{+}/K_{+}).$$

We have similar results for the Galois groups of similarly defined extensions L^{+}, L_{+}, M^{+}, and M_{+} over K_{+}.

11.4. Let N'^{+} denote the maximal abelian extension of K_{+} contained in N'. As above,

$$K \subseteq N'^{+} \subseteq N', \qquad \operatorname{Gal}(N'/N'^{+}) = (1 - J)\operatorname{Gal}(N'/K).$$

As before, let

$$D = E' \otimes_{\mathbf{Z}} (\mathbf{Q}_l/\mathbf{Z}_l)\ (= \mathfrak{N}'),$$

and let

$$\operatorname{Gal}(N'/K) \times D \longrightarrow W$$

be the orthogonal pairing defined in §7.3. Clearly D is also a $\operatorname{Gal}(K/k_{+})$-module and the pairing satisfies

$$\langle Jx, Jd\rangle = J(\langle x, d\rangle) = -\langle x, d\rangle,$$

namely,

$$\langle Jx, d\rangle = -\langle x, Jd\rangle.$$

Let D^{-} denote the kernel of $1 + J: D \to D$. It follows from the above that D^{-} is the annihilator of $\operatorname{Gal}(N'/N'^{+})$ in D:

$$D^{-} = \operatorname{Gal}(N'/N'^{+})^{\perp}.$$

Let E'^{-} be the kernel of $1 + J: E' \to E'$ and let

$$D' = E'^{-} \otimes_{\mathbf{Z}} (\mathbf{Q}_l/\mathbf{Z}_l).$$

The injection $E'^{-} \to E'$ induces a morphism

$$D' \longrightarrow D.$$

Let

$$x = \varepsilon \otimes (l^{-a} \bmod \mathbf{Z}_l), \qquad \varepsilon \in E'^{-},\ a \geqq 0,$$

be an element of D', mapped to zero in D. By Lemma 8[13], we then have

$$\varepsilon = \eta^{l^a}$$

[13] Note that D is imbedded in $\mathfrak{K}$ in §7.3.

with η in $K^\times$. Clearly η belongs to E'. Furthermore

$$(\eta^{1+J})^{l^a} = \varepsilon^{1+J} = 1$$

induces

$$\eta^{1+J} = 1$$

because $\eta^{1+J} > 0$. Hence η belongs to E'^- so that $x = 0$ in D'. This shows that $D' \to D$ is injective so that D' may be considered as a submodule of D. Clearly $D' \subseteq D^-$. Let

$$x = \varepsilon \otimes (l^{-a} \bmod \mathbf{Z}_l) , \qquad \varepsilon \in E', \ a \geqq 0 ,$$

be an element of D^-. As above, $(1 + J)x = 0$ implies

$$\varepsilon^{1+J} = \eta^{l^a}$$

with some η in E'. Since

$$\varepsilon^{2(1+J)} = \varepsilon^{(1+J)^2} = \eta^{l^a(1+J)} ,$$

we see that $\varepsilon^2\eta^{-l^a}$ is an element of E'^- so that $2x$ is contained in D'. Hence we have

$$2D^- \subseteq D' \subseteq D^- .$$

Let N'' denote the field generated over K by l^n-th roots, for all $n \geqq 0$, of all elements in E'^-. Then

$$K \subseteq N'' \subseteq N' , \qquad D' = \mathrm{Gal}\,(N'/N'')^\perp$$

and it follows from the above that

$$N'' \subseteq N'^+ , \qquad 2\,\mathrm{Gal}\,(N'^+/N'') = 0 .$$

Let $n_0 = n_0(K/k)$ be the integer defined in § 3.4 and let

$$E'^-_{n_0} = E'^- \cap E'_{n_0} .$$

Take any element ε in E'^-; ε belongs to E'_n for some $n \geqq n_0$. Let σ be a generator of $\mathrm{Gal}\,(k_n/k_{n_0})$. Since every prime ideal of k_n, dividing l, is fixed by σ, $\varepsilon^{1-\sigma}$ is a unit of k_n:

$$\eta = \varepsilon^{1-\sigma} \in E_n .$$

It then follows from $\varepsilon^{1+J} = 1$ and $\sigma J = J\sigma$ that

$$\eta^{1+J} = 1 .$$

As k_n is a J-field, this implies that the absolute value of every conjugate of η in E_n is 1. Hence η is a root of unity: $\eta \in \tilde{W}_n$. Since the norm of η from k_n to k_{n_0} is 1, we see that

$$\eta = \zeta^{1-\sigma} , \qquad \zeta \in \tilde{W}_n$$

so that

$$\varepsilon = \zeta\varepsilon_0$$

with some ε_0 in $E'^{-}_{n_0} = E'^{-} \cap E'_{n_0}$. This proves the following

LEMMA 24.

$$E'^{-} = \tilde{W} E'^{-}_{n_0}, \qquad n_0 = n_0(K/k).$$

Now, let $s = s(K/k)$ denote as before the number of prime divisors on K which are ramified for the $\mathbf{Z}_l$-extension K/k, namely, the number of prime divisors v on K lying above the rational prime l. Let s_1 be the number of v's such that $Jv = v$ and let s_2 be the number of pairs (v, Jv) such that $Jv \neq v$. Clearly

$$s = s_1 + 2s_2 .$$

Furthermore, $s_+ = s_1 + s_2$ is the number of prime divisors on K_+ which are ramified for the extension K_+/k_+:

$$s_+ = s_1 + s_2 = s(K_+/k_+) .$$

Let $E'^{+}_{n_0} = E'_{n_0} \cap k^{\times}_{n_0,+}$; this is the group of all l-adic units in the real field $k_{n_0,+}$. We know that $E'_{n_0}/\tilde{W}_{n_0}$ is a free abelian group of rank $d_{n_0}/2 + s - 1$, and $E'^{+}_{n_0}/\{\pm 1\}$ that of rank $d_{n_0}/2 + s_+ - 1$. On the other hand, the remark at the beginning of §11.3 implies that both $E'_{n_0}/E'^{+}_{n_0} \cdot E'^{-}_{n_0}$ and $E'^{+}_{n_0} \cap E'^{-}_{n_0}$ are finite groups. It follows that

$$E'^{-}_{n_0}/\tilde{W}_{n_0} \simeq \mathbf{Z}^{s_2}$$

so that

$$D' = E'^{-} \otimes_{\mathbf{Z}} (\mathbf{Q}_l/\mathbf{Z}_l) = \tilde{W} E'^{-}_{n_0} \otimes_{\mathbf{Z}} (\mathbf{Q}_l/\mathbf{Z}_l)$$
$$\simeq (E'^{-}_{n_0}/\tilde{W}_{n_0}) \otimes_{\mathbf{Z}} (\mathbf{Q}_l/\mathbf{Z}_l) \simeq (\mathbf{Q}_l/\mathbf{Z}_l)^{s_2} .$$

Since there is an orthogonal pairing

$$\mathrm{Gal}\,(N''/K) \times D' \longrightarrow W ,$$

we see from the above that

$$\mathrm{Gal}\,(N''/K) \simeq \mathbf{Z}_l^{s_2}$$

as $\mathbf{Z}_l$-modules. Furthermore, as $\omega_{n_0}(E''/\tilde{W}_{n_0}) = 0$ induces $\omega_{n_0} D' = 0$, the same pairing also gives us

$$\dot{\omega}_{n_0}\,\mathrm{Gal}\,(N''/K) = 0 .$$

We summarize these results as follows:

LEMMA 25. *Let N'^{+} denote the maximal abelian extension of K_+ contained in N' and let N'' be the field generated over K by l^n-th roots, for all $n \geqq 0$, for all l-units ε in K such that $\varepsilon^{1+J} = 1$. Then*

$$K \subseteq N'' \subseteq N'^{+} \subseteq N' ,$$

$$2\,\mathrm{Gal}(N'^{+}/N'') = 0 , \quad \mathrm{Gal}(N''/K) \simeq \mathbf{Z}_l^{s_2} , \quad \dot{\omega}_{n_0}\,\mathrm{Gal}(N''/K) = 0 .$$

Here $n_0 = n_0(K/k)$ *and* s_2 *is the number of pairs* (v, Jv) *of prime divisors of* K *lying above* l *such that* $v \neq Jv$.

Note that N'' is uniquely characterized as the intermediate field of K and N'^{+} such that $\mathrm{Gal}(N'^{+}/N'')$ is the torsion subgroup of $\mathrm{Gal}(N'^{+}/K)$. Note also that if $l > 2$, then we have simply

$$N'' = N'^{+} .$$

11.5. We shall next prove the following

LEMMA 26. *$L' \cap N''$ is a finite extension over K.*

Proof. Let A_{n_0} denote as before the Sylow l-subgroup of the ideal-class group C_{n_0} of k_{n_0} and let l^u, $u \geqq 0$, be the exponent of A_{n_0}:

$$l^u A_{n_0} = 0 .$$

Suppose that $L' \cap N''$ contains a cyclic extension Φ of degree l^a, $a > u + 1$, over K. Let

$$\Phi = K(\sqrt[l^a]{\varepsilon}) , \qquad \varepsilon \in E'' = E'^{-}_{n_0} .$$

Since Φ is contained in L', we can find an integer $n \geqq n_0$ such that ε is a local l^a-th power in k_n. Let v be a prime divisor of K lying above l, and let $k_{n_0,v}$ and $k_{n,v}$ denote the v-completions of k_{n_0} and k_n respectively. By the assumption, ε is an l^a-th power in $k_{n,v}$. However, since $k_{n,v} = k_{n_0,v}(W_n)$, this implies that

$$\varepsilon = \zeta \alpha^{l^a} , \qquad \zeta \in W_{n_0} , \ \alpha \in k^{\times}_{n_0,v} .$$

Let $\mathfrak{l}$ be the prime ideal of k_{n_0} corresponding to the prime divisor v. The above equality shows that the order of ε with respect to $\mathfrak{l}$ is divisible by l^a. Since ε is an l-unit in k_{n_0}, we then have

$$(\varepsilon) = \mathfrak{a}^{l^a}$$

with an ideal $\mathfrak{a}$ in k_{n_0}. Clearly the ideal-class of $\mathfrak{a}$ belongs to A_{n_0} Hence, by the choice of u,

$$\mathfrak{a}^{l^u} = (\alpha) , \qquad \alpha \in k^{\times}_{n_0} ,$$

so that

$$\varepsilon = \eta \alpha^{l^{a-u}} , \qquad \eta \in E_{n_0} .$$

As $\varepsilon^{1+J} = 1$, it follows that

$$\varepsilon^2 = \eta^{1-J} \alpha^{(1-J) l^{a-u}} .$$

However, η^{1-J} is a root of unity because it is a unit in the J-field k_{n_0} with $\eta^{(1-J)(1+J)} = 1$. Therefore the field

$$\Phi' = K(\sqrt[l^a]{\varepsilon^2}) = K(\sqrt[l^u]{\alpha^{1-J}})$$

has the degree at most l^u over K. Since the degree of the extension Φ/Φ' is either 1 or 2, this contradicts our assumption that

$$[\Phi: K] = l^a, \qquad a > u+1.$$

It is now proved that

$$l^{u+1}\,\mathrm{Gal}(L' \cap N''/K) = 0.$$

Since $\mathrm{Gal}(L' \cap N''/K)$ is a factor group of $\mathrm{Gal}(N''/K) \simeq \mathbf{Z}_l^{s_2}$, we see that $L' \cap N''/K$ is a finite extension with degree dividing $l^{(u+1)s_2}$.

Let $(L' \cap N')^+$ denote the maximal abelian extension of K_+ contained in $L' \cap N'$. Then

$$(L' \cap N')^+ = L'^+ \cap N' = L' \cap N'^+.$$

By Lemmas 25 and 26,

$$2\,\mathrm{Gal}(N'^+/N'') = 0, \qquad l^{u+1}\,\mathrm{Gal}(L' \cap N''/K) = 0.$$

Hence it follows from $K \subseteq L' \cap N'' \subseteq L' \cap N'^+$ that

$$2l^{u+1}\,\mathrm{Gal}\big((L' \cap N')^+/K\big) = 0$$

with the integer u introduced in the proof of Lemma 25.

11.6. Now, as in §10.5, let

$$V = \mathrm{Gal}(L'N'/N') \otimes \mathbf{Q}_l;$$

here and in the following, the tensor product with $\mathbf{Q}_l$ is always taken over $\mathbf{Z}_l$ so that we omit the suffix $\mathbf{Z}_l$ in such a product. By means of the canonical isomorphism

$$\mathrm{Gal}(L'N'/N') \xrightarrow{\sim} \mathrm{Gal}(L'/L' \cap N'),$$

we may identify V with $\mathrm{Gal}(L'/L' \cap N') \otimes \mathbf{Q}_l$:

$$V = \mathrm{Gal}(L'/L' \cap N') \otimes \mathbf{Q}_l.$$

Since V is a $\mathrm{Gal}(K/k_+)$-module, V^+ and V^- are defined as usual, and it follows from the remark in §11.3 that

$$V^+ = (1+J)V, \qquad V^- = (1-J)V, \qquad V = V^+ \oplus V^-.$$

THEOREM 24. *There exists a canonical isomorphism*

$$V^+ \simeq \mathrm{Gal}(L'_+/K_+) \otimes \mathbf{Q}_l.$$

Proof. Let

$$X = \mathrm{Gal}(L'/K), \qquad Y = \mathrm{Gal}(L'/L' \cap N')$$

so that $Y \subseteq X$ and

$$X/Y = \mathrm{Gal}\,(L' \cap N'/K)\,.$$

Then

$$X/((1-J)X + Y) = \mathrm{Gal}\,((L' \cap N')^{+}/K)\,,$$

and we know that this is a torsion $\mathbf{Z}_l$-module. Applying $1+J$ to the left hand side, we see that $(1+J)X/(1+J)Y$ is also a torsion $\mathbf{Z}_l$-module. Hence

$$V^{+} = (1+J)Y \otimes \mathbf{Q}_l \xrightarrow{\sim} (1+J)X \otimes \mathbf{Q}_l\,.$$

On the other hand, $2X^{-} \subseteq (1-J)X \subseteq X^{-}$ implies

$$(X/(1-J)X) \otimes \mathbf{Q}_l \xrightarrow{\sim} (X/X^{-}) \otimes \mathbf{Q}_l \xrightarrow{\sim} (1+J)X \otimes \mathbf{Q}_l\,.$$

Here, $(X/(1-J)X) \otimes \mathbf{Q}_l$ is equal to $\mathrm{Gal}\,(L'^{+}/K) \otimes Q_l$. By the remark after Lemma 23, the morphism

$$\mathrm{Gal}\,(L'^{+}/K) \longrightarrow \mathrm{Gal}\,(L'_{+}/K_{+})$$

induces an isomorphism

$$\mathrm{Gal}\,(L'^{+}/K) \otimes \mathbf{Q}_l \xrightarrow{\sim} \mathrm{Gal}\,(L'_{+}/K_{+}) \otimes \mathbf{Q}_l\,.$$

Hence the theorem is proved.

Now, assume, for simplicity, that $\mathrm{Gal}\,(L'N'/N')$ has the property (A) mentioned in §11.1. Then we have a non-degenerate skew-symmetric pairing

$$V \times V \longrightarrow V(W)\,,$$

as stated in Theorem 23. In the present case, it also satisfies

$$[Jv_1, Jv_2] = J([v_1, v_2]) = -[v_1, v_2]$$

for v_1, v_2 in V. Hence it follows that

$$[V^{+}, V^{+}] = [V^{-}, V^{-}] = 0\,.$$

Since $V = V^{+} \otimes V^{-}$, we see that

$$\dim V^{+} = \dim V^{-} = \frac{1}{2}\dim V$$

and that both V^{+} and V^{-} are maximal isotropic subspaces of V. By Theorem 24,

$$\dim V^{+} = \lambda(\mathrm{Gal}\,(L'_{+}/K_{+})) = \lambda'(K_{+}/k_{+})$$

(see §4.4) so that

$$\dim V = 2\lambda'(K_{+}/k_{+})\,.$$

These facts give us some insight into the nature of the fundamental pairing

$$V \times V \longrightarrow V(W) .$$

Without assuming that Gal $(L'F'/F')$ has the property (A), we still have similar results, by Theorem 21, for the factor space

$$V' = V/V_0 .$$

Note that

$$V' = V'^{+} \oplus V'^{-} , \quad V'^{+} = V^{+}/V_0^{+} , \quad V'^{-} = V^{-}/V_0^{-}$$

and that

$$V_0^{+} \xrightarrow{\sim} \mathrm{Gal}\,(L'_{+}/K_{+})_{\omega} \otimes \mathbf{Q}_l ,$$

where Gal $(L'_{+}/K_{+})_{\omega}$ is defined similarly as Gal $(L'F'/F')_{\omega}$ in Theorem 20.

11.7. We would like to include here a simple consequence of Lemma 24, namely $E'^{-} = \tilde{W}E'^{-}_{n_0}$. Let

$$v_1, \cdots, v_{s_2} , \qquad Jv_1, \cdots, Jv_{s_2}$$

denote all the prime divisors v on K, lying above l, such that $v \neq Jv$. Let $n \geqq n_0 = n_0(K/k)$. For each i, $l \leqq i \leqq s_2$, let $\mathfrak{l}_{i,n}$ denote the prime ideal of k_n determined by v_i, and let H_n denote the group of ideals generated by

$$\mathfrak{a}_{i,n} = \mathfrak{l}_{i,n}^{1-J} , \qquad i = 1, \cdots, s_2 .$$

H_n is a free abelian group of rank s_2, and $\mathfrak{l}_{i,n+1}^{l} = \mathfrak{l}_{i,n}$ implies

$$H_{n+1}^{l} = H_n , \qquad n \geqq n_0 .$$

Hence, if H denotes the union of the increasing sequence of subgroups H_n, $n \geqq n_0$, in the ideal group I of K, then

$$H/H_{n_0} \simeq (\mathbf{Q}_l/\mathbf{Z}_l)^{s_2} .$$

Now, let $\mathfrak{a}$ be a principal ideal in H:

$$\mathfrak{a} = (\alpha) , \qquad \alpha \in K^{\times} .$$

Since $\mathfrak{a}^{1+J} = 1$, we have

$$\mathfrak{a}^2 = \mathfrak{a}^{1-J} = (\alpha^{1-J}) .$$

Here α^{1-J} is obviously contained in $E'^{-} = \tilde{W}E'^{-}_{n_0}$. Hence

$$\mathfrak{a}^2 = (\varepsilon) , \qquad \varepsilon \in E'^{-}_{n_0} ,$$

which shows that $\mathfrak{a}^2$ is an ideal in H_{n_0}, so that $\mathfrak{a}$ itself is contained in H_{n_0+1}. (If $l > 2$, $\mathfrak{a}$ is even contained in H_{n_0}.) Thus

$$H \cap P \subseteq H_{n_0+1} ,$$

where P denotes the subgroup of principal ideals in I, and it follows that $C = I/P$ has a subgroup isomorphic to $(\mathbf{Q}_l/\mathbf{Z}_l)^{s_2}$. Clearly this subgroup is contained in the l-primary component A of C. Hence it follows from Theorem 11 that $\mathrm{Gal}(L/K)$ has a factor group isomorphic to $\mathbf{Z}_l^{s_2}$. Therefore

$$\lambda(K/k) = \dim \mathrm{Gal}(L/K) \otimes_{\mathbf{Z}_l} \mathbf{Q}_l \geqq s_2 .$$

Now, given any integer $a \geqq 1$, we can easily find a finite algebraic number field k containing $\mathbf{P}$ such that k is a J-field and such that the number of prime ideals $\mathfrak{l}$ of k, which divide l and satisfy $\mathfrak{l}^J \neq \mathfrak{l}$, is at least equal to 2a. Then the integer s_2 for the cyclotomic $\mathbf{Z}_l$-extension K/k, $K = k\mathbf{P}_\infty$, is obviously at least equal to a. Hence, by the above, we have

$$\lambda(K/k) \geqq a .$$

Thus we see that there exist cyclotomic $\mathbf{Z}_l$-extensions for which the λ-invariants are arbitrarily large. On the other hand, we have found so far no example of a cyclotomic $\mathbf{Z}_l$-extension K/k where k and K are J-fields and where

$$\dim V^+ = \lambda'(K_+/k_+) > 0 .$$

It would be an important problem to find out when $\dim V^+ > 0$.

12. Local fields and function fields

12.1. In this last section, we shall briefly discuss $\mathbf{Z}_l$-extensions for which the ground fields are different from finite algebraic number fields. As we shall see, in these cases, more definite results are obtained by arguments simpler than those in the preceding. We also note that the notations of this section are independent of those introduced in the preceding sections.

We first consider the case where the ground fields are local fields of characteristic zero; the results obtained here may also be useful in future investigations of global $\mathbf{Z}_l$-extensions.

Let p be a prime number which may or may not be equal to l. We fix an algebraic closure Ω_p of the p-adic number field $\mathbf{Q}_p$ and always consider algebraic extensions of $\mathbf{Q}_p$ contained in Ω_p.

Let k be a finite algebraic extension of $\mathbf{Q}_p$ (in Ω_p) with

$$d = [k : \mathbf{Q}_p] .$$

Let k_{ab} denote the maximal abelian l-extension over k (in Ω_p). By local class field theory, there is a canonical isomorphism

$$\mathrm{Gal}(k_{ab}/k) \xrightarrow{\sim} A_k ,$$

where A_k denotes the Sylow l-subgroup of a certain compactification of the multiplicative group $k^\times$. Let W_k be the group of all roots of unity in k with orders powers of l. Then W_k is a subgroup of A_k and

$$\begin{cases} A_k/W_k \simeq Z_l\,, & \text{if } l \neq p\,, \\ \quad\ \ \simeq Z_l^{d+1}\,, & \text{if } l = p\,. \end{cases}$$

Hence

$$\begin{cases} \mathrm{Gal}\,(k_{ab}/k) \simeq W_k \oplus Z_l\,, & \text{if } l \neq p\,, \\ \qquad\qquad\ \simeq W_k \oplus Z_l^{d+1}\,, & \text{if } l = p\,. \end{cases}$$

Let k_{ur} denote the maximal unramified abelian l-extension over k. It is well known that k_{ur}/k is a Z_l-extension:

$$\mathrm{Gal}\,(k_{ur}/k) \simeq Z_l\,.$$

It follows from the above that if $l \neq p$, then k_{ur} is the only Z_l-extension over k. On the other hand, if $l = p$, then

$$\mathrm{Gal}\,(F/k) \simeq Z_l^{d+1}$$

for the composite F of all Z_l-extensions over k. In particular, there exist Z_l-extensions over k, different from k_{ur}/k.

Now, let K be a fixed Z_l-extension over k, let $\Gamma = \mathrm{Gal}\,(K/k)$, and let k_n, $n \geqq 0$, denote the intermediate fields of k and K. Let $L(= K_{ur})$ and K_n $(= k_{n,ur})$ denote the maximal unramified abelian l-extensions of K and k_n, respectively. Then $K_n = k_n k_{ur}$ for $n \geqq 0$. Hence we have

$$L = K k_{ur}\,.$$

Therefore, if $K = k_{ur}$, then

$$L = K\,.$$

This is always the case when $l \neq p$. On the other hand, if $K \neq k_{ur}$, then

$$\mathrm{Gal}\,(L/K) \simeq Z_l\,.$$

Since L/k is an abelian extension, Γ acts trivially on $\mathrm{Gal}\,(L/K)$.

12.2. Let M $(= K_{ab})$ and M_n $(= k_{n,ab})$ denote the maximal abelian l-extensions of K and k_n respectively. Then

$$k \subseteq K \subseteq M_0 \subseteq \cdots \subseteq M_n \subseteq \cdots \subseteq M$$

and M is the union of all M_n, $n \geqq 0$. Let

$$X = \mathrm{Gal}\,(M/K)\,.$$

As in the case of global fields, X is a compact Γ-module and, hence, also a compact Λ-module. We shall next study the structure of the module X.

As in § 3.1, we have

$$\omega_n X = \mathrm{Gal}\,(M/M_n)\,, \qquad X/\omega_n X = \mathrm{Gal}\,(M_n K)\,, \qquad n \geqq 0\,.$$

We have seen in the above that $\mathrm{Gal}\,(M_0/k) = \mathrm{Gal}\,(k_{ab}/k)$ is noetherian over

$\mathbf{Z}_l$. Hence it follows from Lemma 1 that X is noetherian over Λ.

Let W denote the group of all roots of unity in Ω_p with orders powers of l, and let

$$W_K = W \cap K^\times, \qquad W_n = W_{k_n} = W \cap k_n^\times, \qquad n \geqq 0.$$

Clearly W_K is the union of the increasing sequence of finite subgroups W_n, $n \geqq 0$. Hence either W_K is finite and $W_K = W_n$ for large n, or $W_K = W$ and $K = k(W)$, namely, K is the (local) cyclotomic $\mathbf{Z}_l$-extension over k.

Since $M_n = k_{n,ab}$, $\mathrm{Gal}(M_n/k_n)$ and, hence, $\mathrm{Gal}(M_n/K)$ are both noetherian $\mathbf{Z}_l$-modules. Let X_n be the submodule of X containing $\omega_n X$ such that $X_n/\omega_n X$ is the torsion $\mathbf{Z}_l$-submodule of $X/\omega_n X = \mathrm{Gal}(M_n/K)$. The result in §12.1 then shows that there exists a canonical isomorphism

$$X_n/\omega_n X \xrightarrow{\simeq} W_n$$

and that

$$\begin{cases} X/X_n = 0, & \text{if } l \neq p, \\ \qquad\; \simeq \mathbf{Z}_l^{d_n}, & \text{if } l = p, \end{cases}$$

where

$$d_n = dl^n = [k_n : \mathbf{Q}_p].$$

For $m \geqq n \geqq 0$, X_m is contained in X_n and we have a commutative diagram

$$\begin{array}{ccc} X_m/\omega_m X & \xrightarrow{\simeq} & W_m \\ \downarrow & & \downarrow \\ X_n/\omega_n X & \xrightarrow{\simeq} & W_n \end{array}$$

where

$$N_{n,m} : W_m \longrightarrow W_n$$

is the norm map from k_m to k_n.

THEOREM 25. *The structure of the* Γ*-module* (Λ*-module*) $X = \mathrm{Gal}(M/K)$ *is given as follows:*

(i) *Suppose that* $K \neq k(W)$, *i.e.,* W_K *is finite. Then*

$$W_K = 1, \quad X = 0, \qquad \text{if } l \neq p,$$

$$X \subseteq \Lambda^d, \quad \Lambda^d/X \simeq W_K, \quad \text{if } l = p.$$

(ii) *Suppose that* $K = k(W)$, $W_K = W$. *Then*

$$\begin{cases} X \simeq T(W), & \text{if } l \neq p, \\ \;\; \simeq T(W) \oplus \Lambda^d, & \text{if } l = p. \end{cases}$$

Here $T(W)$ *denotes the Tate module for* W *(see* §10.1*).*

Proof. Suppose first that $l \neq p$ so that $K = k_{ur}$. By the above,

$$X = \varprojlim X/\omega_n X \xrightarrow{\sim} \varprojlim W_n .$$

Let r be the number of elements in the residue field for the local field k. Since k_n/k is unramified, the number of elements in the residue field for k_n is r^{l^n}. Suppose that $r \not\equiv 1 \bmod l$. Then $r^{l^n} \not\equiv 1 \bmod l$ so that $W_K = W_n = 1$ for all $n \geqq 0$. Hence $X = 0$. On the other hand, if $r \equiv 1 \bmod l$, then $r^{l^n} \equiv 1 \bmod l^{n+1}$. Hence $W_K = W$, $K = k(W)$. In this case, $N_{n,m}\colon W_m \to W_n$ is surjective, and one sees easily that $X \simeq T(W)$.

Suppose next that $l = p$. Assume that $K \neq k(W)$ and $W_K = W_n$ for all sufficiently large $n \geqq 0$. Since

$$X/\omega_n X \simeq W_n \oplus Z_l^{d_n} ,$$

we see immediately that if we define $i(n, a; X)$ as in §8.2, then

$$i(n, a; X) - dl^n a$$

is bounded for all n, $a \geqq 0$. It then follows from Lemma 11 that

$$X \sim \Lambda^d .$$

However, since $W_n = W_K$ for all large n,

$$X = \varprojlim X/\omega_n X = \varprojlim X/X_n$$

so that X is a torsion-free Z_l-module. Therefore the pseudo-isomorphism $X \to \Lambda^d$ is injective, and we may consider X as a submodule of finite index in Λ^d. For sufficiently large n, we then have $\omega_n \Lambda^d \subseteq X$ so that

$$X/\omega_n \Lambda^d \subseteq \Lambda^d/\omega_n \Lambda^d \simeq Z_l^{d_n} , \qquad \omega_n \Lambda^d/\omega_n X \simeq \Lambda^d/X .$$

This implies $X_n = \omega_n \Lambda^d$ and, consequently,

$$\Lambda^d/X \simeq X_n/\omega_n X \simeq W_n \simeq W_K .$$

Assume finally that $K = k(W)$, $W_K = W$ (and $l = p$). In this case, $N_{n,m}\colon W_m \to W_n$ is surjective so that

$$X_n = X_m + \omega_n X , \qquad m \geqq n \geqq 0 .$$

Let Y denote the intersection of all X_n, $n \geqq 0$, and let $X' = X/Y$. Since X is a compact group, the above equalities imply

$$X_n = Y + \omega_n X , \qquad n \geqq 0 .$$

Hence

$$Y = \varprojlim Y/(\omega_n X \cap Y) \simeq \varprojlim X_n/\omega_n X \simeq \varprojlim W_n = T(W) .$$

It also follows that

$$X'/\omega_n X' \simeq X/\omega_n X \simeq \mathbf{Z}_l^{d_n}, \qquad n \geqq 0 .$$

Similarly as in the case $K \neq k(W)$, we see from this that $X' \sim \Lambda^d$ and then also that

$$X' \simeq \Lambda^d .$$

It is now clear that

$$X \simeq T(W) \oplus \Lambda^d ,$$

and the theorem is completely proved.

Let $l \neq p$ and $K = k_{ur}$. Let γ_0 denote the Frobenius automorphism for the unramified extension K/k. Since γ_0 is a topological generator of $\Gamma = \mathrm{Gal}(K/k)$, we may define the Λ-structure on X by means of this γ_0. We then see from the above theorem that

$$X \simeq \Lambda/(1 - r + T)$$

in both cases $K \neq k(W)$ and $K = k(W)$. Here r denotes, as in the above, the number of elements in the residue field for k. Note that if $r \not\equiv 1 \bmod l$, then $1 - r + T$ is invertible in $\Lambda = \mathbf{Z}_l[[T]]$ so that $\Lambda = (1 - r + T)$.

We also note that the isomorphism $\Lambda^d/X \simeq W_K$ uniquely determines the structure of X.

Now, for each $n \geqq 0$, let $\mathfrak{p}_n$ denote the maximal ideal of the local field k_n and let U_n be the multiplicative group of all units α in k_n such that $\alpha \equiv 1 \bmod \mathfrak{p}_n$. U_n is a $\mathbf{Z}_p$-module in the obvious manner. Hence it is also an R_n-module where R_n denotes the group algebra of $\mathrm{Gal}(k_n/k)$ over $\mathbf{Z}_p$. Using the canonical isomorphism in local class field theory, we may obtain from Theorem 25 also information on the structure of the R_n-module U_n.

As an example, let us consider the case where $l = p$, $K = k(W)$, and $K \cap k_{ur} = k$. Then we can show that

$$U_n \sim W_n \oplus (R_n/(\sigma_n - 1)R_n) \oplus (\sigma_n - 1)R_n \oplus R_n^{d-1},$$

where $d = [k: \mathbf{Q}_p]$ as before and where σ_n denotes a generator of the cyclic group $\mathrm{Gal}(k_n/k)$. We can deduce similar results also for other cases[14].

12.3. We shall next consider a special type of $\mathbf{Z}_l$-extensions for which the ground fields are algebraic function fields of one variable over finite fields of constants. Many important facts are known for such function fields by algebro-geometric method. Our main purpose here is not to deduce from our theory any further results on such fields but to show how the analogy between number fields and function fields works in this case.

Let $\mathbf{F}$ be a finite field containing l-th roots of unity different from 1 (so

[14] For the proof, see [7] where a special case is treated.

that the characteristic of $\mathbf{F}$ is different from l). Let k be an algebraic function field of one variable over the constant field $\mathbf{F}$. Let J denote the Jacobian variety associated with k, A the subgroup of all points in J with orders powers of l, and $\mathbf{F}_\infty$ the field generated over $\mathbf{F}$ by all such points in A. As usual, we fix a universal domain Ω over $\mathbf{F}$ and consider extensions of $\mathbf{F}$ such as k and $\mathbf{F}_\infty$ as subfields of Ω. Replacing $\mathbf{F}$, if necessary, by a suitable finite extension of $\mathbf{F}$, we may assume that every point c in A with $lc = 0$ is rational over $\mathbf{F}$. We then have

$$\mathbf{F}_\infty = \mathbf{F}(W) , \qquad \mathrm{Gal}\,(\mathbf{F}_\infty/\mathbf{F}) \simeq \mathbf{Z}_l ,$$

where W denotes the group of all l^n-th roots of unity in Ω for all $n \geqq 0$. Let

$$K = k\mathbf{F}_\infty = k(W) .$$

K is an algebraic function field over the constant field $\mathbf{F}_\infty$ and A is the l-primary component of the divisor-class group of divisors of degree zero in K. Since $k \cap \mathbf{F}_\infty = \mathbf{F}$,

$$\Gamma = \mathrm{Gal}\,(K/k) \simeq \mathrm{Gal}\,(\mathbf{F}_\infty/\mathbf{F})$$

so that both K/k and $\mathbf{F}_\infty/\mathbf{F}$ are $\mathbf{Z}_l$-extensions. For the intermediate fields k_n and $\mathbf{F}_n$, $n \geqq 0$, we then have

$$k_n = k\mathbf{F}_n , \qquad k_n \cap \mathbf{F}_\infty = F_n , \qquad n \geqq 0 .$$

The Galois group $\Gamma = \mathrm{Gal}\,(K/k)$ (or $\mathrm{Gal}\,(\mathbf{F}_\infty/\mathbf{F})$) acts on A in the obvious manner so that A is a discrete Γ-module. Let A_n be the submodule of all points in A which are fixed by $\Gamma_n = \mathrm{Gal}\,(K/k_n)$. A_n is a finite group and is naturally identified with the Sylow l-subgroup of the divisor-class group of divisors of degree zero in k_n. Let γ_0 denote the Frobenius automorphism for the unramified extension K/k. Since γ_0 is a topological generator of $\Gamma = \mathrm{Gal}\,(K/k)$, we can make Γ-modules, such as A, into Λ-modules by means of γ_0.

Now, let L $(= M)$ denote the maximal unramified abelian l-extension over K (in Ω) and let L_n be the maximal abelian extension of k_n contained in L. Since K/k is unramified, L is an unramified extension of k and L_n coincides with the maximal unramified abelian l-extension K_n over k_n (in Ω). Clearly

$$k \subseteq K \subseteq L_0 \subseteq L_1 \subseteq \cdots \subseteq L_n \subseteq \cdots \subseteq L$$

and L is the union of all L_n, $n \geqq 0$. Let

$$X = \mathrm{Gal}\,(L/K) .$$

As for number fields, X is a compact Γ-module, and hence a compact Λ-module, and

$$\omega_n X = \mathrm{Gal}\,(L/L_n) , \qquad X/\omega_n X = \mathrm{Gal}\,(L_n/K) , \qquad n \geqq 0 .$$

Since $L_n = K_n$, there exists, by class field theory for function fields, a canonical isomorphism

$$X/\omega_n X = \mathrm{Gal}\,(L_n/K) \xrightarrow{\sim} A_n$$

for each $n \geqq 0$ such that

$$\begin{array}{ccc} X/\omega_n X & \xrightarrow{\sim} & A_n \\ \downarrow{\scriptstyle \nu_{n,m}} & & \downarrow \\ X/\omega_m X & \xrightarrow{\sim} & A_m\,, \end{array} \qquad \begin{array}{ccc} X/\omega_m X & \xrightarrow{\sim} & A_m \\ \downarrow & & \downarrow{\scriptstyle \nu_{n,m}} \\ X/\omega_n X & \xrightarrow{\sim} & A_n\,, \end{array} \qquad m \geqq n \geqq 0\,,$$

are commutative. In particular, $X/\omega_n X$ is finite for every $n \geqq 0$, and this implies that X is a noetherian torsion Λ-module with $\operatorname{div} X$ disjoint from all (ω_n), $n \geqq 0$. It then follows from the above isomorphisms that

$$\alpha(X) \simeq \varprojlim \mathrm{Hom}_{\mathbf{Z}_l}(X/\omega_n X, \mathbf{Q}_l/\mathbf{Z}_l) \simeq \mathrm{Hom}_{\mathbf{Z}_l}(A, \mathbf{Q}_l/\mathbf{Z}_l)\,.$$

12.4. Since $\mathbf{F}_\infty$ has no extension of degree l, each element in $\mathbf{F}_\infty$ is an l-th power in $\mathbf{F}_\infty$. Hence the extensions of K, which correspond to N/K and N'/K defined in § 7.3, are trivial extensions. Therefore, instead of Theorem 14, we obtain a similar orthogonal pairing

$$X \times A \longrightarrow W\,,$$

and it induces

$$X \xrightarrow{\sim} \mathrm{Hom}_{\mathbf{Z}_l}(A, W) \xrightarrow{\sim} (\mathrm{Hom}_{\mathbf{Z}_l}(A, \mathbf{Q}_l/\mathbf{Z}_l))^{\boldsymbol{\cdot}}\,.$$

Now, a fundamental theorem of Weil states that as $\mathbf{Z}_l$-modules,

$$A \simeq (\mathbf{Q}_l/\mathbf{Z}_l)^{2g}$$

where g denotes the genus of the function field k. Hence it follows from the above that as compact $\mathbf{Z}_l$-modules,

$$X \simeq \mathbf{Z}_l^{2g}\,.$$

This shows that the Λ-module X is an l'-module, i.e.,

$$\mu(X) = 0\,.$$

For each $a \geqq 0$, let $A^{(a)}$ denote the submodule of all c in A satisfying $l^a c = 0$. Define a morphism

$$X \longrightarrow A^{(a)}\,,$$
$$x \longmapsto c(x, a)\,,$$

in much the same way as we defined $X \to B'^{(a)}$ in § 9.2. We then again have

$$f\colon X \longrightarrow T(A)$$

where $T(A)$ denotes the Tate module for A:

$$T(A) = \varprojlim A^{(a)} .$$

However, since $X \simeq \mathbf{Z}_l^{2g}$ and since div X is disjoint from (ω_n), we see that the morphism

$$\omega_n \colon X \longrightarrow X , \qquad n \geqq 0 ,$$

is injective. Hence it follows immediately from the definition that $X \to A^{(a)}$ induces an isomorphism

$$X/l^a X \xrightarrow{\sim} A^{(a)}$$

for every $a \geqq 0$ so that f is also an isomorphism:

$$f\colon X \xrightarrow{\sim} T(A) .$$

By means of $X \times A \to W$, we can now define the pairings

$$X \times X \longrightarrow W^{(a)} , \qquad a \geqq 0$$

and

$$X \times X \longrightarrow T(W)$$

in a way similar to §10.1. In this case, it is almost trivial that the kernels of $X \times X \to T(A)$ are zero. By the isomorphism f, we then also obtain a pairing

$$(1) \qquad T(A) \times T(A) \longrightarrow T(W)$$

with kernels zero.

Let $a \geqq 0$ and let c and d be elements of $A^{(a)}$. Choose divisors of degree zero in K, $\mathfrak{c}$ and $\mathfrak{d}$, which are disjoint from each other and represent the divisor-classes c and d, respectively. Let

$$\mathfrak{c}^{l^a} = (\varphi) , \qquad \mathfrak{d}^{l^a} = (\psi)$$

with φ and ψ in $K^\times$, and define

$$[c, d]_a = \psi(\mathfrak{c})\varphi(\mathfrak{d})^{-1}$$

where $\psi(\mathfrak{c})$ and $\varphi(\mathfrak{d})$ are roots of unity, i.e., elements in the algebraic closure of F in Ω, defined in the standard manner (see [11]); we see immediately that the right hand side of the above depends, in fact, only upon c and d, and is an l^a-th root of unity. Hence we obtain a pairing

$$A^{(a)} \times A^{(a)} \longrightarrow W^{(a)} ,$$
$$(c, d) \longmapsto [c, d]_a .$$

Since

$$[lc, ld]_{a-1} = l[c, d]_a , \qquad a \geqq 1 ,$$

these pairings, for all $a \geqq 0$, define

$$T(A) \times T(A) \longrightarrow T(W) . \tag{2}$$

It is obvious that this pairing is skew-symmetric.

12.5. We shall next show that the two pairings (1) and (2) are the same[15]. The argument is similar to that in §10.2, but is simpler in this case. Let $a \geqq 0$ and let x_1 and x_2 be elements of X. We find integers m and n, $m \geqq n \geqq 0$, and elements y and z in X such that $W^{(a)}$ is contained in $k_n^\times$ and

$$\nu_{n,m} x_1 = l^a y , \qquad \omega_n x_2 = l^a z .$$

Let

$$x_1 \mid L_n \longmapsto c , \qquad z \mid L_n \longmapsto d$$

under

$$X/\omega_n X = \mathrm{Gal}\,(L_n/K) \xrightarrow{\sim} A_n ,$$

and let $\mathfrak{c}$ and $\mathfrak{d}$ be divisors of k_n, disjoint from each other and representing the divisor-classes c and d, respectively. Since $l^a z$ belongs to $\omega_n X$, $l^a d = 0$ in A_n so that

$$\mathfrak{d}^{l^a} = (\psi)$$

for some ψ in $k_n^\times$. As

$$d = c(x_2, a) ,$$

by the definition of $X \to A^{(a)}$, we have

$$\begin{aligned} [x_1, x_2]_a &= \langle x_1, c(x_2, a) \rangle = (\sqrt[l^a]{\psi})^{x_1 - 1} \\ &= \left(\frac{\psi, k_n}{\mathfrak{c}}\right)_{l^a} = \psi(\mathfrak{c})^{(r_n - 1)/l^a} \end{aligned}$$

for the value of $[x_1, x_2]_a$ in $X \times X \to W^{(a)}$. Here $(\psi, k_n/\mathfrak{c})_{l^a}$ denotes the l^a-th power residue symbol in k_n and r_n is the number of elements in the constant field F_n of k_n. Let

$$y \mid L_m \longmapsto c'$$

under

$$\mathrm{Gal}\,(L_m/K) \xrightarrow{\sim} A_m$$

and let $\mathfrak{a}$ be a divisor of k_m, disjoint from $\mathfrak{d}$, representing c' in A_m. Since $\nu_{n,m} x_1 \mid L_m$ is mapped to the divisor-class of $\mathfrak{c}$ under the same isomorphism, it follows from $\nu_{n,m} x_1 = l^a y$ that

$$\mathfrak{c} = \mathfrak{a}^{l^a}(\theta)$$

for some θ in $k_m^\times$. Using the well-known reciprocity:

---

[15] This is the result of Weil, $A = B$, mentioned in the footnote 2.

$$\psi((\theta)) = \theta((\psi)),$$

we then see that

$$\psi(\mathfrak{c})^{(r_n-1)/l^a} = \psi(\mathfrak{a})^{r_n-1}\psi((\theta))^{(r_n-1)/l^a} = \psi(\mathfrak{a})^{r_n-1}\theta((\psi))^{(r_n-1)/l^a}$$
$$= \psi(\mathfrak{a})^{r_n-1}\theta(\mathfrak{d})^{r_n-1} = \psi(\mathfrak{a}^{\omega_n})\theta^{\omega_n}((\psi))$$

with

$$\omega_n = \gamma_0^{l^n} - 1 .$$

However, $\nu_{n,m}x_1 = l^a y$ implies

$$\omega_m x_1 = l^a \omega_n y$$

so that

$$\omega_n c' = c(x_1, a) .$$

As $\mathfrak{a}^{\omega_n}$ represents the divisor-class $\omega_n c'$ and as

$$(\mathfrak{a}^{\omega_n})^{l^a} = (\theta^{-\omega_n}),$$

it follows from the above that

$$[x_1, x_2]_a = \psi(\mathfrak{a}^{\omega_n})\theta^{\omega_n}(\mathfrak{d}) = [c(x_1, a), c(x_2, a)]_a ,$$

the last term being the value of the pairing in $A^{(a)} \times A^{(a)} \to W^{(a)}$. This proves our contention that the two pairings

$$T(A) \times T(A) \longrightarrow T(W)$$

are the same.

12.6. Finally, let

$$V = X \otimes_{Z_l} \mathbf{Q}_l .$$

V is a $2g$-dimensional vector space over $\mathbf{Q}_l$ and it defines an l-adic representation

$$\rho\colon \Gamma \longrightarrow GL(V) .$$

The isomorphism $f\colon X \xrightarrow{\sim} T(A)$ induces an isomorphism of representation spaces for Γ:

$$V = X \otimes_{Z_l} \mathbf{Q}_l \xrightarrow{\sim} V(A) = T(A) \otimes_{Z_l} \mathbf{Q}_l .$$

Hence the characteristic polynomial of $\rho(\gamma_0)$ gives us the essential part of the zeta-function for k. Thus we see that the l-adic representation constructed in §10.5 is an analogue, for algebraic number fields, of such a representation obtained from the abelian variety J in the standard manner and that the functional equation for $f_\gamma(T)$ mentioned in §10.5 is the corresponding analogue of the functional equation for the zeta-function of k.

PRINCETON UNIVERSITY

BIBLIOGRAPHY

[1] A. BRUMER, On the units of algebraic number fields, Mathematika, **14** (1967), 121-124.
[2] J. COATES, On K_2 and some classical conjectures in algebraic number theory, Ann. Math., **95** (1972), 99-116.
[3] H. HASSE, Bericht über neuere Untersuchungen und Probleme aus der Theorie der algebraischen Zahlkörper, I, Ia, II, B. G. Teubner, Berlin-Leipzig, 1930.
[4] K. IWASAWA, On Γ-extensions of algebraic number fields, Bull. Amer. Math. Soc., **65** (1959), 183-226.
[5] ———, On some properties of Γ-finite modules, Ann. Math., **70** (1959), 291-312.
[6] ———, On the theory of cyclotomic fields, Ann. Math., **70** (1959), 530-561.
[7] ———, On local cyclotomic fields, J. Math. Soc. Japan, **12**, (1960), 16-21.
[8] H. W. LEOPOLDT, Zur Arithmetik in abelschen Zahlkörpern, J. reine u. ang. Math., **209** (1962), 54-71.
[9] J.-P. SERRE, Classes des corps cyclotomiques, Seminaire Bourbaki, Expose **174** (1958-1959).
[10] ———, *Abelian l-adic representations of elliptic curves*, W. A. Benjamin, Inc., New York-Amsterdam, 1968.
[11] A. WEIL, Sur les fonctions algébriques à corps de constantes fini, C. R., Paris, **210** (1940), 592-594.

(Received October 1, 1972)

[53] On the μ-invariants of Z_ℓ-extensions

"Number Theory, Algebraic Geometry and Commutative Algebra, in honor of Yasuo Akizuki", Kinokuniya, Tokyo, 1973, pp. 1-11.

(Received March 1, 1972)

Let l be a prime number and let Z_l denote the ring of l-adic integers. A Galois extension K over a field k is called a Z_l-extension if the Galois group of K/k is topologically isomorphic to the additive group of the compact ring Z_l:

$$\mathrm{Gal}\,(K/k) \simeq Z_l .$$

For such an extension K/k, there exists a unique sequence of fields

$$k = k_0 \subset k_1 \subset \cdots \subset k_n \subset \cdots \subset K$$

such that each k_n/k is a cyclic extension of degree l^n; K is then the union of the increasing sequence of fields k_n, $n \geqq 0$.

Suppose now that the ground field k is a finite algebraic number field so that each k_n is also a finite extension of the rational field Q. Let l^{e_n} denote the highest power of l which divides the class number h_n of k_n. Then it is known[1] that for all sufficiently large n (i.e., for $n \geqq n_0$), the exponent e_n is given by a formula

$$e_n = \lambda n + \mu l^n + \nu$$

where λ, μ, and ν are integers independent of n (λ, $\mu \geqq 0$). Since these integers are uniquely determined by K/k in the manner as described above, they are denoted by $\lambda(K/k)$, $\mu(K/k)$, and $\nu(K/k)$ respectively.

The numerical values of the invariants λ and μ are computed for various Z_l-extensions. Viewing those examples, we were led to conjecture that

$$\mu(K/k) = 0$$

[1] See K. Iwasawa, On Γ-extensions of algebraic number fields, Bull. Amer. Math. Soc., **65** (1959), 183–226. The formula is proved in Theorem 11, §7.5. (Z_l-extensions are called Γ-extensions in this paper.)

for every Z_l-extension K/k[2]. In the present paper, however, we shall show that there exists a Z_l-extension K/k for which $\mu(K/k)$ is arbitrarily large. Hence the above conjecture is not true in general. On the other hand, there is an important family of Z_l-extensions, called basic Z_l-extensions, for which the conjecture $\mu = 0$ still appears to be true. In the second part of the paper, we shall prove some results which seem to support the conjecture restricted to such basic Z_l-extensions.

§1. Let K be a Z_l-extension over a finite algebraic number field k. Let $\mathfrak{p}$ be a prime ideal of k and let $Z_{\mathfrak{p}}$ denote the decomposition group of $\mathfrak{p}$ for the extension K/k; $Z_{\mathfrak{p}}$ is the inverse limit of the decomposition groups of $\mathfrak{p}$ for k_n/k, $n \geqq 0$. As a closed subgroup of Gal $(K/k) \simeq Z_l$, $Z_{\mathfrak{p}}$ is either the identity or a subgroup of finite index in Gal (K/k). It follows that if g_n denotes the number of the prime ideals of k_n dividing $\mathfrak{p}$, then either $g_n = l^n$ for every $n \geqq 0$ or g_n is bounded for all $n \geqq 0$. We shall say that $\mathfrak{p}$ is fully decomposed or finitely decomposed in K according as the first or the second case occurs.

Assume now that the ground field k contains primitive l-th roots of unity and that there exist prime ideals $\mathfrak{p}_1, \cdots, \mathfrak{p}_t$, $t \geqq 1$, in k which are prime to l and are fully decomposed in K. Let α be a non-zero element of k which is divisible exactly by the first power of $\mathfrak{p}_i$ for $i = 1, \cdots, t$, and let

$$k' = k(\sqrt[l]{\alpha}), \qquad K' = Kk'.$$

One sees immediately that $K \cap k' = k$ and that K'/k' is a Z_l-extension. Let

$$k' = k'_0 \subset k'_1 \subset \cdots \subset k'_n \subset \cdots \subset K'$$

be the sequence of intermediate fields for K'/k'. Then

$$k'_n = k_n k' = k'_n(\sqrt[l]{\alpha}), \qquad n \geqq 0$$

so that k'_n/k_n is a cyclic extension of degree l. Let $l^{e'_n}$ denote the highest power of l dividing the class number h'_n of k'_n. Then, for all sufficiently large n,

$$e'_n = \lambda' n + \mu' l^n + \nu'$$

with $\lambda' = \lambda(K'/k')$, $\mu' = \mu(K'/k')$, and $\nu' = \nu(K'/k')$.

Let s_n denote the number of the prime *divisors* of k_n which are ramified in k'_n. By the assumption, each $\mathfrak{p}_i$ is fully decomposed in K so that it has l^n

2) See K. Iwasawa, On some infinite abelian extensions of algebraic number fields, Proc. of the International Congress of Math., Nice, 1970, vol. 1, 391–394.

prime ideal factors in k_n. Since these prime ideals are ramified in k'_n, we have

$$tl^n \leqq s_n\,, \qquad n \geqq 0\,.$$

Let C'_n denote the ideal-class group of k'_n and let σ be a generator of the cyclic group Gal (k'_n/k_n). A well-known classical formula in class field theory states that

$$(1) \qquad [C'_n : C'^{1-\sigma}_n] = h_n l^{s_n-1}[E_n : E^*_n]^{-1}$$

where E_n denotes the group of all units in k_n, and E^*_n the subgroup of those units in E_n which are norms of elements in k'_n. Let

$$d = [k : Q]\,, \qquad d_n = l^n d = [k_n : Q]\,.$$

Since E_n is an abelian group generated by at most d_n elements and since E^l_n is contained in E^*_n, the index $[E_n : E^*_n]$ is a factor of l^{d_n}. Also, the left hand side of (1) is a factor of $h'_n = [C'_n : 1]$. Hence, comparing the l-parts of the integers on the both sides of (1), we obtain

$$\begin{aligned} e'_n &\geqq e_n + s_n - 1 - d_n \qquad (e_n \geqq 0) \\ &\geqq (t-d)l^n\,, \qquad\qquad n \geqq 0\,. \end{aligned}$$

It then follows from the formula for e'_n that

$$\mu(K'/k') \geqq t - d\,.$$

§2. In this section, we let k be the cyclotomic field of l-th or 4-th roots of unity according as $l > 2$ or $l = 2$. Hence

$$d = [k : Q] = l - 1 \quad \text{or} \quad 2\,.$$

The rational prime l is fully ramified in k so that $(l) = \mathfrak{l}^d$ with a prime ideal $\mathfrak{l}$ of k. Let $k_\mathfrak{l}$ denote the $\mathfrak{l}$-completion of k. $k_\mathfrak{l}$ is a Galois extension of the l-adic number field Q_l such that

$$k_\mathfrak{l} = kQ_l\,, \qquad k \cap Q_l = Q\,.$$

Hence the restriction from $k_\mathfrak{l}$ to k defines a canonical isomorphism

$$\mathrm{Gal}\,(k_\mathfrak{l}/Q_l) \cong \mathrm{Gal}\,(k/Q)\,.$$

In general, if a subfield F of the complex field is invariant under the complex-conjugation, we shall denote by J_F, or often simply by J, the automorphism of F which maps each complex number in F to its complex-conjugate. The automorphism of $k_{\mathfrak{l}}/Q_l$, which is mapped to $J = J_k$ under the above isomorphism, will also be denoted by J.

Let U be the multiplicative group of all local units in the local field $k_{\mathfrak{l}}$; U is a compact abelian group on which Gal $(k_{\mathfrak{l}}/Q_l)$ acts in the obvious manner. The action of Gal $(k_{\mathfrak{l}}/Q_l)$ can be studied by means of the l-adic log function which maps a subgroup of finite index in U into the additive group of $k_{\mathfrak{l}}$. Thus one sees that there exists a closed subgroup V of U such that

$$(2) \qquad U^{1+J} \subseteq V \subseteq U\,, \qquad U/V \simeq Z_l\,.$$

Let E denote the group of all (global) units in k, and E_+ the subgroup of all real units in E. Then

$$E_+^{1+J} = E_+^2 \subseteq E_+ \subseteq E \subseteq U$$

and both E/E_+ and E_+/E_+^2 are finite groups. Hence EU^{1+J}/U^{1+J} is also finite, and it follows from (2) that

$$\bar{E} \subseteq EU^{1+J} \subseteq V \subseteq U\,, \qquad (U/V)^{1+J} = 1$$

where $\bar{E}$ denotes the closure of E in U.

Now, let L be Hilbert's class field over k, i.e., the maximal unramified abelian extension over k, and let M be the maximal abelian extension over k in which every prime ideal of k, different from $\mathfrak{l}$, is unramified. Clearly

$$Q \subseteq k \subseteq L \subseteq M$$

and both L/Q and M/Q are Galois extensions. By class field theory, Gal (L/k) is isomorphic to the ideal-class group C_k of k so that L/k is a finite extension. There also exists a canonical isomorphism

$$\mathrm{Gal}\,(M/L) \rightrightarrows U/\bar{E}\,.$$

Since M is invariant under the complex-conjugation, $J = J_M$ is an element of Gal (M/Q). As Gal (M/L) is a normal subgroup of Gal (M/Q), we can make J act on Gal (M/L) by

$$\sigma^J = J\sigma J^{-1}\,, \qquad \sigma \in \mathrm{Gal}\,(M/L)\,.$$

With the action of J on $U/\bar{E}$ already mentioned, the above isomorphism from Gal (M/L) to $U/\bar{E}$ is a J-isomorphism.

Let F be the field such that

$$L \subseteq F \subseteq M\,, \qquad \mathrm{Gal}\,(M/F) \rightrightarrows V/\bar{E}$$

under the above isomorphism. Since $U^{1+J} \subseteq V$, we have $V^J = V$ so that $\mathrm{Gal}(M/F)^J = \mathrm{Gal}(M/F)$, $F^J = F$. Hence $J = J_F$ acts on $\mathrm{Gal}(F/L)$ in the same manner as J_M does on $\mathrm{Gal}(M/L)$, and

$$(3) \qquad \mathrm{Gal}(F/L) \simeq Z_l\,, \qquad \mathrm{Gal}(F/L)^{1+J} = 1\,.$$

Let K be the field such that

$$k \subseteq K \subseteq F$$

and such that $\mathrm{Gal}(F/K)$ is the torsion subgroup of the abelian group $\mathrm{Gal}(F/k)$. Clearly $K^J = K$. Since $\mathrm{Gal}(L/k)$ is finite, it follows immediately from (3) that

$$\mathrm{Gal}(K/k) \simeq Z_l\,, \qquad \mathrm{Gal}(K/k)^{1+J} = 1$$

with $J = J_K$. Thus K is a Z_l-extension over k. Hence, as before, let

$$k = k_0 \subset k_1 \subset \cdots \subset k_n \subset \cdots \subset K$$

be the sequence of intermediate fields for K/k. Let k_+ denote the maximal real subfield of k, i.e., the fixed field of J in k. The fact $K^J = K$ induces that K is a Galois extension over k_+. Therefore each k_n is also a Galois extension over k_+. Let

$$G_n = \mathrm{Gal}(k_n/k_+)\,, \quad H_n = \mathrm{Gal}(k_n/k)\,, \quad n \geqq 0\,.$$

Then G_n is the semi-direct product of the normal subgroup H_n and the cyclic subgroup of order two generated by $J = J_{k_n}$. Furthermore, $\mathrm{Gal}(K/k)^{1+J} = 1$ implies $H_n^{1+J} = 1$, namely, that

$$J\sigma J^{-1} = \sigma^{-1}\,, \qquad \sigma \in H_n\,.$$

Thus G_n is a dihedral group of order $2l^n$.

Now, let $\mathfrak{p}_+$ be a prime ideal of k_+, unramified and undecomposed in k, and let $\mathfrak{p}$ be the unique prime ideal of k dividing $\mathfrak{p}_+$. Since $\mathfrak{p}_+$ is prime to l and $k_+ \subseteq k \subseteq k_n \subseteq K \subseteq M$, $\mathfrak{p}_+$ is also unramified in k_n. Let $\mathfrak{p}_n$ be a prime ideal of k_n dividing $\mathfrak{p}$ and let Z_n denote the decomposition group of $\mathfrak{p}_n$ for the Galois extension k_n/k_+. Since $\mathfrak{p}_+$ is unramified in k_n and is undecomposed in k, Z_n is a cyclic subgroup of G_n such that

$$G_n = Z_n H_n\,.$$

As G_n is a dihedral group, it follows that Z_n is a cyclic group of order two satisfying

$$Z_n \cap H_n = 1 .$$

However, $Z_n \cap H_n$ is nothing but the decomposition group of $\mathfrak{p}_n$ for the extension k_n/k. Hence $Z_n \cap H_n = 1$ means that $\mathfrak{p}$ is fully decomposed in k_n. Since this holds for every $n \geqq 0$, we see that the prime ideal $\mathfrak{p}$ of k is fully decomposed in K.

It is well-known that there exist infinitely many prime ideals $\mathfrak{p}_+$ in k_+ which are unramified and undecomposed in k. Therefore there also exist infinitely many prime ideals $\mathfrak{p}$ in k which are fully decomposed in K. From what we have observed in the preceding section, we now obtain the following result:

Theorem 1. *Let k be the cyclotomic field of l-th or 4-th roots of unity according as $l > 2$ or $l = 2$. For any given integer $N \geqq 1$, there exist a cyclic extension k' of degree l over k and a Z_l-extension K' over k' such that*

$$\mu(K'/k') \geqq N .$$

We note that if l is a regular prime, i.e., if the class number of k is not divisible by l, then

$$\mu(K/k) = 0$$

for the Z_l-extension K/k constructed above (and, in fact, for any Z_l-extension over k).

§3. We shall next show that there exist many Z_l-extensions for which the μ-invariants are zero. We first make some remarks on finite abelian groups. Let A be a finite abelian group with order a power of l. Let F_l denote the prime field of characteristic l. As usual, the dimension of the vector space $A \otimes F_l$ over F_l is called the rank of A, and it will be denoted by rank A. If B is a subgroup of A, then

$$\text{rank } B,\ \text{rank } A/B \leqq \text{rank } A \leqq \text{rank } B + \text{rank } A/B .$$

Let G be a cyclic group of order l^e, $e \geqq 0$, generated by an element σ. Suppose that G acts on A. Then we have

$$\text{rank } A \leqq l^e \text{ rank } A/A^{\sigma-1} .$$

This can be proved by observing that the l^e-th power of the linear transformation defined by $\sigma - 1$ on the vector space $A \otimes F_l$ is zero.

Now, let k be a finite algebraic number field and let k'/k be a cyclic extension of degree l. Let A and A' denote the Sylow l-subgroups of the ideal-class groups of k and k' respectively. Let

$$r = \text{rank } A\,, \qquad r' = \text{rank } A'$$

and let s denote the number of the prime divisors of k which are ramified in k'. We shall prove the following formula:

$$(4) \qquad r - 1 \leqq r' \leqq l(r + s)\,.$$

For the proof, let L and L' denote the maximal unramified abelian l-extensions (i.e., Hilbert's l-class fields) over k and k' respectively. By class field theory,

$$A \simeq \text{Gal}\,(L/k)\,, \qquad A' \simeq \text{Gal}\,(L'/k')$$

so that

$$r = \text{rank Gal}\,(L/k)\,, \qquad r' = \text{rank Gal}\,(L'/k')\,.$$

Clearly L is contained in L'. Let M denote the maximal abelian extension of k contained in L'. Then

$$k \subseteq L\,, \qquad k' \subseteq M \subseteq L'\,,$$

and it is well-known that

$$\text{Gal}\,(L'/M) = \text{Gal}\,(L'/k')^{\sigma-1}$$

where σ denotes a generator of Gal (k'/k) which acts on Gal (L'/k') in the obvious manner. Since Gal $(M/k') =$ Gal $(L'/k')/$Gal (L'/M), it follows from the remark mentioned above that

$$(5) \qquad r' = \text{rank Gal}\,(L'/k') \leqq l \text{ rank Gal}\,(M/k')\,.$$

Let $v_1, \cdots, v_s$ denote the prime divisors of k which are ramified in k' and let T_i be the inertia group of v_i for the abelian extension M/k $(i = 1, \cdots, s)$. Since M/L is unramified, each T_i is a cyclic subgroup of order l in Gal (M/k), and no other prime divisor of k is ramified in M. As L is obviously the maximal unramified extension of k contained in M, we see immediately that

$$\text{Gal}\,(M/L) = T_1 \cdots T_s$$

so that

$$\text{rank Gal}\,(M/L) \leqq s\,.$$

Hence

$$\begin{aligned}\text{rank Gal}(M/k') &\leqq \text{rank Gal}(M/k)\\ &\leqq \text{rank Gal}(L/k) + \text{rank Gal}(M/L)\\ &\leqq r+s\end{aligned}$$

and consequently, by (5),

$$r' \leqq l(r+s)\,.$$

On the other hand,

$$\begin{aligned}\text{rank Gal}(L/k) &\leqq \text{rank Gal}(M/k)\\ &\leqq \text{rank Gal}(k'/k) + \text{rank Gal}(M/k')\\ &\leqq 1 + \text{rank Gal}(L'/k')\,,\end{aligned}$$

namely,

$$r \leqq 1 + r'\,.$$

Hence our formula (4) is proved.

We now prove the following result:

Theorem 2. *Let K be a Z_l-extension over a finite algebraic number field k and let k be a totally imaginary field if $l=2$. Let k' be a finite Galois extension over k with degree a power of l and let $K' = Kk'$. Suppose that every prime ideal of k, which is ramified in k', is finitely decomposed in K. Then*

$$\mu(K/k) = 0$$

if and only if

$$\mu(K'/k') = 0\,.$$

For the proof, let k'' be any field between k and k': $k \subseteq k'' \subseteq k'$, and let $K'' = K\cdot k''$ so that $K' = K''\cdot k'$. It follows from the assumption that if a prime ideal of k'' is ramified in k', then it is finitely decomposed in K''. Since $\text{Gal}(k'/k)$ is a finite l-group, we see immediately that it suffices to prove the theorem in the case where k'/k is a cyclic extension of degree l. Assume, therefore, that k'/k is such an extension and let

$$\begin{aligned}k = k_0 \subset k_1 \subset \cdots \subset k_n \subset \cdots \subset K\\ k' = k'_0 \subset k'_1 \subset \cdots \subset k'_n \subset \cdots \subset K'\end{aligned}$$

be the sequences of intermediate fields for K/k and K'/k' respectively. Then either $K \cap k' = k$ or $K \cap k' = k_1$. In the latter case, we have

$$K' = K\,, \quad k'_n = k_{n+1}\,, \quad n \geqq 0$$

so that

$$\mu(K'/k') = l\mu(K/k)$$

by the definition of the μ-invariants. Hence the theorem is clear in this case. Therefore, we may assume that

$$K \cap k' = k\,.$$

It then follows that $k'_n = k_n k'$ and that k'_n/k_n is a cyclic extension of degree l. Let A_n and A'_n denote the Sylow l-subgroups of the ideal-class groups of k_n and k'_n respectively. Let

$$r_n = \text{rank } A_n\,, \qquad r'_n = \text{rank } A'_n$$

and let s_n denote the number of the prime divisors of k_n which are ramified in k'_n. By the formula (4), we then have

$$r_n - 1 \leqq r'_n \leqq l(r_n + s_n)\,, \qquad n \geqq 0\,.$$

If $l > 2$, then no archimedean prime divisor of k_n is ramified in k'_n because k'_n/k_n is a cyclic extension of degree l. On the other hand, if $l = 2$, then k is totally imaginary by the assumption. Hence k_n is also a totally imaginary field, and again no archimedean prime divisor of k_n is ramified in k'_n. Therefore s_n is also the number of the prime ideals of k_n which are ramified in k'_n. Let $\mathfrak{q}$ be such an ideal of k_n. Since $k'_n = k_n k'$, $\mathfrak{q}$ must be a factor of a prime ideal $\mathfrak{p}$ of k which is ramified in k'. However, by the assumption, such a prime ideal $\mathfrak{p}$ is finitely decomposed in K so that the number of the prime ideal factors of $\mathfrak{p}$ in k_n is bounded for all $n \geqq 0$. Therefore s_n is also bounded for all $n \geqq 0$, and it follows from the above inequalities that r_n is bounded for all $n \geqq 0$ if and only if r'_n is bounded for all $n \geqq 0$. Now, it is known in the theory of Z_l-extensions[3] that $\mu(K/k) = 0$ (resp. $\mu(K'/k') = 0$) if and only if r_n (resp. r'_n) is bounded for all $n \geqq 0$. Hence the theorem is proved.

Let K/k be a Z_l-extension, k'/k a finite extension, and $K' = Kk'$. Then it can be shown[4] that

---

3) See the proof of Theorem 11 in the paper referred to in 1).
4) Loc. cit. 1), §7.3.

$$\lambda(K/k) \leqq \lambda(K'/k'), \quad \mu(K/k) \leqq \mu(K'/k').$$

Therefore the part of the theorem which states that $\mu(K'/k') = 0$ induces $\mu(K/k) = 0$, can actually be proved without any assumption on the extension k'/k.

§4. It is known[5)] that there exists a unique Z_l-extension K^0 over the rational field Q and that

$$\lambda(K^0/Q) = \mu(K^0/Q) = 0.$$

For each finite algebraic number field k, the composite K^0k then gives us a Z_l-extension over k. We call K^0k/k the basic Z_l-extension over k and denote its invariants by $\lambda_l(k)$, $\mu_l(k)$, and $\nu_l(k)$:

$$\lambda_l(k) = \lambda(K^0k/k), \quad \mu_l(k) = \mu(K^0k/k), \quad \nu_l(k) = \nu(K^0k/k).$$

A significant property of such a basic Z_l-extension is that every prime ideal of the ground field k is finitely decomposed in $K = K^0k$; this can be seen easily by first proving it for K^0/Q. Hence we obtain the following result from Theorem 2:

Theorem 3. *Let k be a finite algebraic number field and let k be totally imaginary if $l = 2$. Let k' be a finite Galois extension over k with degree a power of l. Then*

$$\mu_l(k) = 0$$

if and only if

$$\mu_l(k') = 0.$$

Let k be a finite Galois extension over the rational field Q with degree a power of l. Assume first that $l > 2$. Since $\mu_l(Q) = \mu(K^0/Q) = 0$, it follows immediately from the above theorem that

$$\mu_l(k) = 0.$$

Suppose next that $l = 2$. In this case, we may not apply Theorem 3 for the extension k/Q. However, we can still have the same conclusion by the following argument. Let

$$k' = k(\sqrt{-1}).$$

[5)] Loc. cit. 1), §7.7.

Then k' is a finite Galois extension of $Q(\sqrt{-1})$ with degree a power of 2. Since $Q(\sqrt{-1})$ is totally imaginary and since $\mu_2(Q(\sqrt{-1}))=0$[6], we see from Theorem 3 that $\mu_2(k')=0$. It then follows from the remark at the end of § 3 that

$$\mu_2(k)=0\ .$$

We now know in particular that if k is a cyclic extension of degree l over the rational field Q, then

$$\mu_l(k)=0\ .$$

For example,

$$\mu_2(k)=0$$

for every quadratic field k. These results, however limited, seem to support the conjecture that

$$\mu_l(k)=0$$

for every l and k. In any case, the study in the present paper suggests that for a Z_l-extension K/k, the vanishing of the μ-invariant, $\mu(K/k)=0$, is closely related to the non-existence of prime ideals in the ground field k which are infinitely decomposed in K.

PRINCETON UNIVERSITY

6) Loc. cit. 5).

[54] A note on Jacobi sums

"Convegno di Strutture in Corpi Algebrici" (Rome, 1973), INDAM, Symposia Mathematica, Vol. XV, Academic Press, London, 1975, pp. 447-459.

In the present paper, we shall make some simple remarks on Jacobi sums for a fixed odd prime number l. As an application, we shall also discuss relations between such Jacobi sums and the class number of the cyclotomic field of l-th roots of unity. Throughout the following, $\mathbb{Z}$, $\mathbb{Z}_l$, $\mathbb{Q}$, and $\mathbb{Q}_l$ will denote the ring of rational integers, the ring of l-adic integers, the field of rational numbers, and the field of l-adic numbers respectively, and for each positive integer m, C_m will denote the cyclotomic field of m-th roots of unity.

§ 1. Let m be an integer, $m > 1$. Let $\mathfrak{o}$ denote the ring of all algebraic integers in the cyclotomic field C_m and let I be the multiplicative group of all ideals of C_m which are prime to m. For each r-tuple of integers $a = (a_1, \ldots, a_r)$, $a_i \in \mathbb{Z}$, $1 \leqslant i \leqslant r$, we define a homomorphism J_a of I into the multiplicative group $C_m^\times$ of the field C_m as follows. For each prime ideal $\mathfrak{p}$ in I, let $\chi_\mathfrak{p}(x) (= (x/\mathfrak{p})_m)$ denote the m-th power residue symbol mod $\mathfrak{p}$ in C_m and let

$$J_a(\mathfrak{p}) = (-1)^{r+1} \sum_{x_i} \chi_\mathfrak{p}(x_1)^{a_1} \ldots \chi_\mathfrak{p}(x_r)^{a_r}$$

where $x_1, \ldots, x_r$ run through all representatives of $\mathfrak{o}$ mod $\mathfrak{p}$ subject to the condition $x_1 + \ldots + x_r + 1 \equiv 0 \bmod \mathfrak{p}$. Then $J_a(\mathfrak{p})$ is a non-zero algebraic integer in $\mathfrak{o}$ so that the map $\mathfrak{p} \mapsto J_a(\mathfrak{p})$ for the prime ideals in I can be uniquely extended to a homomorphism

$$J_a \colon I \to C_m^\times .$$

It is clear from the definition that J_a, $a = (a_1, \ldots, a_r)$, depends only upon the residue classes of a_i mod m. Let G denote the Galois group

(*) I risultati conseguiti in questo lavoro sono stati esposti nella conferenza tenuta il 9 aprile 1973.

of $C_m/\mathbf{Q}$. Then G acts on both I and $C_m^\times$ in the obvious manner and one sees immediately that J_a is a G-homomorphism. The numbers $J_a(\mathfrak{a})$, $\mathfrak{a} \in I$, are called Jacobi sums for the integer m [1].

Let J denote the subgroup of $\mathrm{Hom}_G(I, C_m^\times)$ generated by J_a for all $a = (a_1, \ldots, a_r) \not\equiv (0, \ldots, 0) \bmod m$, $a_i \in \mathbb{Z}$, $r = 1, 2, \ldots$ Then we have a pairing

$$J \times I \to C_m^\times .$$

For each $\mathfrak{a}$ in I, we denote by $J(\mathfrak{a})$ the image of $J \times \mathfrak{a}$ under the above pairing, and by $J(I)$ the subgroup of $C_m^\times$ generated by $J(\mathfrak{a})$ for all $\mathfrak{a}$ in I. $J(I)$ is nothing but the subgroup of $C_m^\times$ generated by all $J_a(\mathfrak{a})$, $\mathfrak{a} \in I$, $a = (a_1, \ldots, a_r) \not\equiv (0, \ldots, 0) \bmod m$.

LEMMA 1: Let $m > 2$. Then J is generated by $J_{(u,v)}$ for all $(u, v) \not\equiv (0, 0) \bmod m$, $u, v \in \mathbb{Z}$. For $m = 2$, J is generated by all such $J_{(u,v)}$ together with the norm map

$$N: I \to C_m^\times ,$$

$$\mathfrak{a} \mapsto N(\mathfrak{a}) = \text{the norm of } \mathfrak{a} .$$

PROOF: By [W], $J_{(1,-1,1)}(\mathfrak{a}) = J_{-1}(\mathfrak{a}) J_1(\mathfrak{a}) N(\mathfrak{a})$. Hence N belongs to J. Let J' denote the subgroup of J generated by $J_{(u,v)}$ for all $(u, v) \not\equiv (0, 0) \bmod m$. Then, by [W], $J_u = J_{(u,0)} \in J'$ for all $u \not\equiv 0 \bmod m$. Using the formulas in [W], p. 492, one sees easily by induction on r that every J_a, $a = (a_1, \ldots, a_r) \not\equiv (0, \ldots, 0) \bmod m$, $r \geqslant 3$, belongs to the subgroup of $\mathrm{Hom}_G(I_m, C_m^\times)$ generated by J' and N. Hence J is generated by J' and N. Let $m > 2$. By [W], $J_{(1,1,-1)} = J_{(1,-1,1)}$ so that

$$J_2 J_{(1,1)} J_{(2,-1)} = J_{-1} J_1 N^{-1} .$$

Hence $N \in J'$ and $J = J'$.

REMARK: Let $m = 2$. Then

$$J_{(1,1)} = J_{(1,0)} = J_{(0,1)} = J_1 = \pm 1 .$$

Hence $N \notin J'$.

(1) Cf. A. WEIL, *Jacobi sums as « Grössencharaktere »*, Trans. Amer. Math. Soc., 73 (1952), 487-495. In the following, this paper will be simply referred to as [W]. The reader will find there all fundamental properties of Jacobi sums which we need in our proofs.

For each prime ideal $\mathfrak{p}$ in I, we fix a non-trivial character $\psi_{\mathfrak{p}}$ of the additive group of the finite field $\mathfrak{o}/\mathfrak{p}$ and define, for each integer u, the Gaussian sum

$$\tau_u(\mathfrak{p}) = -\sum_x \chi_{\mathfrak{p}}(x)^u \psi_{\mathfrak{p}}(x)$$

where x runs through all representatives of $\mathfrak{o}$ mod $\mathfrak{p}$. If p is the prime number contained in $\mathfrak{p}$, then $\tau_u(\mathfrak{p})$ is a number in the cyclotomic field C_{mp}, $(m, p) = 1$. Obviously $\tau_u(\mathfrak{p})$ depends only upon the residue class of u mod m and it is also known that $\tau_u(\mathfrak{p}) \neq 0$. Let C'_m denote the composite of the cyclotomic fields C_n for all $n > 1$ with $(m, n) = 1$ and let $C''_m = C_m C'_m$. Since $C_m \cap C'_m = \mathbf{Q}$, the Galois group $G = \mathrm{Gal}(C_m/\mathbf{Q})$ can be identified with Gal (C''_m/C'_m). Choosing $\psi_{\mathfrak{p}}(x)$ so that $\psi_{\sigma(\mathfrak{p})}(\sigma(x)) = \psi_{\mathfrak{p}}(x)$, $\sigma \in G$, we may extend the map $\mathfrak{p} \mapsto \tau_u(\mathfrak{p})$ to a G-homomorphism

$$\tau_u : I \to (C''_m)^\times .$$

Thus τ_u is an element of $\mathrm{Hom}_G(I, (C''_m)^\times)$ which contains $\mathrm{Hom}_G(I, C_m^\times)$.

LEMMA 2: Let $m > 2$. Then J is generated by $\tau_{-r}\tau_1^r$, $1 \leqslant r \leqslant m$. For $m = 2$. J is generated by N and $\tau_{-1}\tau_1 = \tau_{-2}\tau_1^2 = \tau_1^2$.

PROOF: For each integer $r \geqslant 1$, let $r \times 1$ denote the r-tuple $(1, 1, \ldots, 1)$. Then, by [W],

$$J_{r\times 1} = N^{-1}\tau_{-r}\tau_1^r$$

so that $\tau_{-r}\tau_1^r$ belongs to J. Let J'' denote the subgroup of J generated by $\tau_{-r}\tau_1^r$, $1 \leqslant r \leqslant m$. Since $\tau_{-m} = 1$, $\tau_1^m \in J''$, J'' contains $\tau_{-r}\tau_1^r$ for any integer r. Again by [W],

$$\begin{aligned} J_{(u,v)} &= N^{-1}\tau_{-u-v}\tau_u\tau_v \\ &= N^{-1}(\tau_{-u-v}\tau_1^{u+v})(\tau_{-u}\tau_1^u)(\tau_{-v}\tau_1^v) \end{aligned}$$

for any $(u, v) \not\equiv (0, 0)$ mod m. Hence it follows from Lemma 1 that J is generated by N and J''. This proved the lemma for $m = 2$. Let $m > 2$ so that $(2, 0) \not\equiv (0, 0)$ mod m. The above argument shows that $J_{(2,0)} \equiv N^{-1}$ mod J''. However, for every prime ideal $\mathfrak{p}$ in I, $J_{(2,0)}(\mathfrak{p}) = J_2(\mathfrak{p}) = \chi_{\mathfrak{p}}(-1)^2 = 1$ so that $J_{(2,0)} = 1$. Hence N belongs to J'' and we have $J = J''$.

REMARK: Let $m=2$, $C_m=C_2=\mathbf{Q}$, and let $\mathfrak{p}=(p)$ with a prime number p such that $p\equiv 3 \mod 4$. Then

$$J_{(1,-1)}(\mathfrak{p})=N\mathfrak{p}^{-1}\tau_0(\mathfrak{p})\tau_1(\mathfrak{p})\tau_{-1}(\mathfrak{p})=N\mathfrak{p}^{-1}\tau_1(\mathfrak{p})^2$$
$$=J_{(1,0)}(\mathfrak{p})=J_1(\mathfrak{p})=\chi_{\mathfrak{p}}(-1)=(-1/p)=-1\,,$$

and we see that $N\notin J''$.

§2. We now fix an odd prime number l and let $m=l$ in the above. Let k be the cyclotomic field of l-th roots of unity, i.e., $k=C_l$, and let ζ be a fixed primitive l-th root of unity in k so that $k=\mathbf{Q}(\zeta)$. There is a unique prime ideal $\mathfrak{l}$ in k which divides the rational prime l and one has $\mathfrak{l}=(1-\zeta)$, $\mathfrak{l}^{l-1}=(l)$. As in the above, let $G=\mathrm{Gal}(k/\mathbf{Q})=$ $=\mathrm{Gal}(C''_l/C'_l)$. For each integer t, prime to l, there is a unique automorphism σ_t in G such that $\zeta^{\sigma_t}=\zeta^t$. σ_t depends only upon the residue class of $t \mod l$ and G consists of $l-1$ such σ_t's.

For simplicity, we shall denote τ_1 in §1 by τ. Thus if $\mathfrak{p}$ is a prime ideal in I (i.e., $\mathfrak{p}\neq\mathfrak{l}$), then

$$\tau(\mathfrak{p})=\tau_1(\mathfrak{p})=-\sum\chi_{\mathfrak{p}}(x)\psi_{\mathfrak{p}}(x)\,.$$

Since

$$\chi_{\mathfrak{p}}(x)^{\sigma_t}=\chi_{\mathfrak{p}}(x)^t\,,\qquad \psi_{\mathfrak{p}}(x)^{\sigma_t}=\psi_{\mathfrak{p}}(x)\,,$$

we see that

$$\tau(\mathfrak{p})^{\sigma_t}=\tau_t(\mathfrak{p})\,,\qquad \tau_{-t}(\mathfrak{p})\tau_1(\mathfrak{p})^t=\tau(\mathfrak{p})^{\sigma_{-t}+t}$$

for $(t,l)=1$. Hence we may write $\tau^{\sigma_{-t}+t}$ for the homomorphism $\tau_{-t}\tau^t$.

LEMMA 3: J is generated by $\tau^{t-\sigma_t}$ for all integers t, prime to l.

PROOF: This follows from Lemma 2 and from

$$\tau_{-l}(\mathfrak{p})\tau_1(\mathfrak{p})^l=\tau(\mathfrak{p})^l=\tau(\mathfrak{p})^{1+l-\sigma_{1+l}}\,.$$

THEOREM 1: $J(I)\equiv 1 \mod \mathfrak{l}^2$, namely, $\alpha\equiv 1 \mod \mathfrak{l}^2$ for every α in $J(I)$.

PROOF (2): It is sufficient to show that $J(\mathfrak{p})\equiv 1 \mod \mathfrak{l}^2$ for every prime ideal $\mathfrak{p}\neq\mathfrak{l}$. For $x\in\mathfrak{o}$, $x\notin\mathfrak{p}$, $\chi_{\mathfrak{p}}(x)$ is an l-th root of unity in k

(2) Cf. H. HASSE, *Zetafunktion und L-Funktionen zu einem arithmetischen Funktionenkörper von Fermatschen Typus*, Abh. Akad. Berlin, III, Heft 4, 1954.

so that $\chi_{\mathfrak{p}}(x) \equiv 1 \mod \mathfrak{l}$. Hence

$$\tau(\mathfrak{p}) \equiv -\sum_{x \notin \mathfrak{p}} \psi_{\mathfrak{p}}(x) \equiv 1 \mod \mathfrak{l}.$$

Let p be the prime number contained in $\mathfrak{p}$ and let $\mathfrak{L}$ be a prime ideal of C_{lp} dividing the prime ideal $\mathfrak{l}$ in $k = C_l$. Then $\mathfrak{L}$ is unramified for the extension C_{lp}/C_l and is fully ramified for the extension C_{lp}/C_p. Hence

$$\tau(\mathfrak{p}) \equiv 1 + y(\zeta - 1) \mod \mathfrak{L}^2$$

with an element y in C_p. Since

$$\zeta^{\sigma_t} = \zeta^t \equiv 1 + t(\zeta - 1) \mod \mathfrak{l}^2,$$

it follows that

$$\tau(\mathfrak{p})^{\sigma_t} \equiv 1 + ty(\zeta - 1) \equiv \tau(\mathfrak{p})^t \mod \mathfrak{L}^2$$

for every integer t, prime to l, and for every $\mathfrak{L}$ as stated above. Hence

$$\tau(\mathfrak{p})^{t-\sigma_t} \equiv 1 \mod \mathfrak{l}^2, \qquad (t, l) = 1$$

and by Lemma 3,

$$J(\mathfrak{p}) \equiv 1 \mod \mathfrak{l}^2 .$$

REMARK: As we shall see below, if $l > 3$, then $J(I) = 1 \mod \mathfrak{l}^3$.

Now, for each real number α, let $\langle\alpha\rangle = \alpha - [\alpha]$ where $[\alpha]$ denotes the largest integer not greater than α, and for each integer u, let

$$\theta(u) = \sum_{t=1}^{l-1} \left\langle \frac{ut}{l} \right\rangle \sigma_{-t}^{-1} ;$$

$\theta(u)$ is an element of the group ring $\mathbf{Q}[G]$. For any r-tuple $a = (a_1, \ldots, a_r)$, $a_i \in \mathbf{Z}$, let $a_0 = -\sum_{i=1}^{r} a_i$ and let

$$\omega(a) = \omega(a_1, \ldots, a_r) = \sum_{i=0}^{r} \theta(a_i) - \sum_{t=1}^{l-1} \sigma_t .$$

Clearly $\theta(u)$ and $\omega(a_1, \ldots, a_r)$ depend only upon the residue classes of u and $a_i \mod l$. One also sees easily that $\omega(a)$ is an element of the group ring $\mathbf{Z}[G]$.

THEOREM 2: Let $a=(a_1, \dots, a_r) \not\equiv (0, \dots, 0) \bmod l$. Let $\mathfrak{a}$ be a principal ideal in I and let

$$\mathfrak{a}=(\alpha), \quad \alpha \equiv 1 \bmod \mathfrak{l}^2, \quad \alpha \in k^\times.$$

Then

$$J_a(\mathfrak{a})=\alpha^{\omega(a)}.$$

PROOF: By [W], $(J_a(\mathfrak{a}))=\mathfrak{a}^{\omega(a)}=(\alpha^{\omega(a)})$ so that

$$J_a(\mathfrak{a})=\varepsilon\alpha^{\omega(a)}$$

with a unit ε in k. Let s be the number of integers in $a_0, a_1, \dots, a_r$ which are not divisible by l. Then it is also known that

$$|J_a(\mathfrak{a})|^2 = N\mathfrak{a}^{s-2}=N\alpha^{s-2}.$$

On the other hand, since

$$(\sigma_1+\sigma_{-1})\theta(u)=\sum_{t=1}^{l-1}\sigma_t, \quad \text{if } u \not\equiv 0 \bmod l,$$

$$=0, \quad \text{if } u \equiv 0 \bmod l,$$

we see that

$$(\sigma_1+\sigma_{-1})\,\omega(a_1, \dots, a_r)=(s-2)\sum_t \sigma_t$$

so that

$$|\alpha^{\omega(a)}|^2=\alpha^{(\sigma_1+\sigma_{-1})\omega(a)}=N\alpha^{s-2}.$$

Therefore we obtain $|\varepsilon|=1$. Since ε is a unit of k, this induces that $\varepsilon=\pm\zeta^c$ with an integer c. However, by Theorem 1, $J_a(\mathfrak{a})\equiv 1 \bmod \mathfrak{l}^2$, and clearly $\alpha^{\omega(a)}\equiv 1 \bmod \mathfrak{l}^2$. Therefore $\varepsilon=\pm\zeta^c\equiv 1 \bmod \mathfrak{l}^2$ and it follows that $\varepsilon=1$, $J_a(\mathfrak{a})=\alpha^{\omega(a)}$.

Let $k_\mathfrak{l}$ denote the $\mathfrak{l}$-completion of $k=\mathbf{Q}(\zeta)$. Then $k_\mathfrak{l}=k\mathbf{Q}_l=\mathbf{Q}_l(\zeta)$ and it is the local cyclotomic field of l-th roots of unity over the l-adic number field $\mathbf{Q}_l$. The maximal ideal of the local field $k_\mathfrak{l}$ is the closure of the prime ideal $\mathfrak{l}$ of k. For simplicity, we shall denote this maximal ideal of $k_\mathfrak{l}$ again by $\mathfrak{l}$, if there is no risk of misunderstanding. Let U be the multiplicative group of all local units in $k_\mathfrak{l}$ and let U_n, for $n\geqslant 0$, denote the subgroup of all α in U such that $\alpha\equiv 1 \bmod \mathfrak{l}^n$. Then

$$\dots \subseteq U_n \subseteq U_{n-1} \subseteq \dots \subseteq U_1 \subseteq U_0 = U.$$

Each U_n, $n \geqslant 1$, is a profinite abelian l-group and hence is a compact $\mathbb{Z}_l$-module. Let $G = \mathrm{Gal}(k/\mathbb{Q})$ be identified with $\mathrm{Gal}(k_{\mathfrak{l}}/\mathbb{Q}_l)$ in the obvious manner and let R denote the group ring of G over $\mathbb{Z}_l$:

$$R = \mathbb{Z}_l[G] .$$

The G acts on each U_n, $n \geqslant 0$, so that U_1 is a compact R-module and U_n, $n \geqslant 1$, are closed R-submodules of U_1.

Now, let P denote the subgroup of all principal ideals in I, namely, the group of all principal ideals of k, prime to l. Let $J(P)$ be the subgroup of $J(I)$ generated by $J(\mathfrak{a})$ for all $\mathfrak{a}$ in P. By Theorem 1,

$$J(P) \subseteq J(I) \subseteq U_2 .$$

Hence

$$\bar{J}(P) \subseteq \bar{J}(I) \subseteq U_2$$

where $\bar{J}(I)$ and $\bar{J}(P)$ denote the closures of $J(I)$ and $J(P)$ in U_2, respectively. It is clear that both $\bar{J}(I)$ and $\bar{J}(P)$ are closed R-submodules of U_2. Let S denote the $\mathbb{Z}_l$–submodule of $R = \mathbb{Z}_l[G]$ generated by the elements $\omega(a)$ for all $a = (a_1, \ldots, a_r)$, $a_i \in \mathbb{Z}$, $r = 1, 2, \ldots$. If $a = (a_1, \ldots, a_r) \equiv (0, \ldots, 0) \bmod l$, then

$$\omega(a) = -\sum_t \sigma_t = -\big(\omega(1, 1) + \omega(-1, -1)\big) .$$

Hence S is also generated over $\mathbb{Z}_l$ by all $\omega(a)$ such that $a = (a_1, \ldots, a_r) \not\equiv (0, \ldots, 0) \bmod l$. In the next section, we shall prove a lemma which includes the following

THEOREM 3:

$$\bar{J}(P) = U_2^S (= S \cdot U_2) .$$

§ 3. Let

$$\theta = \theta(1) = \frac{1}{l} \sum_{t=1}^{l-1} t\, \sigma_{-t}^{-1} .$$

We consider θ as an element in the group ring $\mathbb{Q}_l[G]$ which contains R. We also fix a primitive root g mod l^2, $g \geqslant 1$, and define

$$\theta_0 = \omega(g \times 1) = (g - \sigma_g)\theta$$

where $g \times 1 = (1, 1, \ldots, 1)$.

LEMMA 4: S is generated over $\mathbf{Z}_l$ by $(u-\sigma_u)\theta$ for all integers u, prime to l, and

$$S = R \cap R\theta = R\theta_0 .$$

PROOF: Let S' denote the $\mathbf{Z}_l$-submodule of $\mathbf{Q}_l[G]$ generated by all $(u-\sigma_u)\theta$, $(u, l)=1$. If $u \geqslant 1$, $(u, l)=1$, then $(u-\sigma_u)\theta = \omega)u \times 1) \in$ $\in S$. Using $l\theta = ((l+1)-\sigma_{l+1})\theta = \omega((l+1)\times 1)$, one sees easily that $(u-\sigma_u)\theta$, for $u<0$, $(u, l)=1$, also belongs to S. Hence $S' \subseteq S$. Since

$$\sum_{t=1}^{l-1} \sigma_t = \theta(1) + \theta(-1) = (\sigma_1 - 1)\theta + (\sigma_{-1}+1)\theta ,$$

$\sum_t \sigma_t$ belongs to S'. Let $a=(a_1, \dots, a_r)$, $a_0 = -\sum_{i=1}^{r} a_i$. Then

$$\omega(a) = \sum_{i=0}^{r} (\theta(a_i) - a_i\theta(1)) - \sum_t \sigma_t$$

where

$$\begin{aligned} \theta(a_i) - a_i\theta(1) &= (\sigma_{a_i} - a_i)\theta , & \text{if } a_i \not\equiv 0 \bmod l, \\ &= a_i l^{-1}(\sigma_{l+1} - (l+1))\theta , & \text{if } a_i \equiv 0 \bmod l. \end{aligned}$$

Hence $S \subseteq S'$ so that $S = S'$.

Since $S = S'$, it is now clear that $S \subseteq R \cap R\theta$. Let ξ be an element of $R \cap R\theta$ and let

$$\xi = \Big(\sum_{t=1}^{l-1} c_t \sigma_t\Big)\theta, \quad c_t \in \mathbf{Z}_l .$$

Then

$$\xi = \sum_{t=1}^{l-1} c_t(\sigma_t - t)\theta + c\theta$$

where $c = \sum_t tc_t$. As ξ belongs to R, c must be divisible by l so that

$$c\theta = cl^{-1}((l+1) - \sigma_{l+1})\theta .$$

Hence ξ is contained in $S'(=S)$. This proves that $S = R \cap R\theta$. In particular, we see that S is an ideal of the ring R so that $R\theta_0 \subseteq S$ because θ_0 is obviously an element of $S'(=S)$. Now, since g is a primitive root mod l^2, it is also a primitive root mod l^n for every $n \geqslant 1$.

Hence, given any integer u, prime to l, there exists an integer $\nu \geqslant 1$ such that

$$u \equiv g^{\nu} \mod l^n .$$

Then we have

$$\begin{aligned} u - \sigma_u &= (g^{\nu} - \sigma_g^{\nu}) + cl^n \\ &= \xi(g - \sigma_g) + cl^n , \quad \xi \in R,\ c \in \mathbb{Z}. \end{aligned}$$

Since this holds every $n \geqslant 1$, we see that $u - \sigma_u$ belongs to the closure of $R(g - \sigma_g)$ in R, namely, $R(g - \sigma_g)$ itself. Hence we have $(u - \sigma_u)\theta \in R\theta_0$ so that $S = R\theta_0$.

Now, for each $n \geqslant 0$, let P_n be the subgroup of all principal ideals $\mathfrak{a} = (\alpha)$ in P such that $\alpha \equiv 1 \mod \mathfrak{l}^n$. Since each coset of U/U_2 is represented by a unit in k, we have

$$\ldots \subseteq P_n \subseteq P_{n-1} \subseteq \ldots \subseteq P_2 = P_1 = P \subseteq I .$$

Let I' denote the subgroup of I containing P such that I'/P is the Sylow l-subgroup of the ideal-class group I/P of k. Let

$$X = \varprojlim I/P_n , \quad Y = \varprojlim I'/P_n , \quad Z = \varprojlim P/P_n$$

so that $Z \subseteq Y \subseteq X$. These are profinite abelian groups and Y is the Sylow l-subgroup of X. I is contained in X as an everywhere dense subgroup of X.

We consider the homomorphism

$$J_{g\times 1}\colon I \to U_2 ,$$
$$\mathfrak{a} \mapsto J_{g\times 1}(\mathfrak{a}) ,$$

where g is the integer fixed above. Since $(g, l) = 1$, it follows from[W] that

$$J_{g\times 1} = N^{-1}\tau_{-g}\tau_1^g = \tau_g^{-1}\tau_1^g = \tau^{g-\sigma_g} .$$

By Theorem 2, if $\mathfrak{a} = (\alpha)$, $\alpha \equiv 1 \mod \mathfrak{l}^n$, $n \geqslant 2$, then $J_{g\times 1}(\mathfrak{a}) \equiv 1 \mod \mathfrak{l}^n$. Hence $J_{g\times 1}$ can be uniquely extended to a continuous G-homomorphism

$$f\colon X \to U_2 .$$

The restriction of f on the submodule Y:

$$f\colon Y \to U_2$$

is a morphism of R-modules.

LEMMA 5:

$$f(X) = f(Y) = \bar{J}(I), \quad f(Z) = \bar{J}(P) = U_2^S.$$

PROOF: Since X/Y is a finite group of order prime to l and since U_2 is a profinite l-group, we have $f(X) = f(Y)$. Since X is compact, $f(X)$ is the closure of $J_{g\times 1}(I)$ in U_2 so that $f(X) \subseteq \bar{J}(I)$. Let $\mathfrak{a}$ be an ideal of I and let u be an integer, prime to l. As shown in the proof of Lemma 4, $u - \sigma_u$ is contained in $R(g - \sigma_g)$. Since $J_{g\times 1}(\mathfrak{a}) = \tau(\mathfrak{a})^{g-\sigma_g}$ and since $f(X)$ is a closed R-submodule of U_2 containing $J_{g\times 1}(I)$, it follows that $\tau(\mathfrak{a})^{u-\sigma_u}$ belongs to $f(X)$. By Lemma 3, we then have $J(\mathfrak{a}) \subseteq f(X)$. Therefore $J(I) \subseteq f(X)$ so that $f(X) = \bar{J}(I)$.

In a similar way, we see that $f(Z)$ is the closure of $J_{g\times 1}(P)$ and $f(Z) \subseteq \bar{J}(P)$. Let $\mathfrak{a}$ be an ideal of $P = P_2$ and let $\mathfrak{a} = (\alpha)$, $\alpha \equiv 1 \bmod \mathfrak{l}^2$. By Theorem 2,

$$J_{g\times 1}(\mathfrak{a}) = \mathfrak{a}^{\omega(g\times 1)} = \alpha^{\theta_0}.$$

It follows that

$$f(Z) = U_2^{\theta_0} = U_2^{R\theta_0} = U_2^S.$$

Let $a = (a_1, \ldots, a_r) \not\equiv (0, \ldots, 0) \bmod l$. For the above $\mathfrak{a} = (\alpha)$, we obtain from Theorem 2 that

$$J_a(\mathfrak{a}) = \alpha^{\omega(a)} \in U_2^S = f(Z).$$

Therefore $J(P) \subseteq f(Z)$ so that $\bar{J}(P) = f(Z) = U_2^S$. Theorem 3 in § 2 is now also proved.

§ 4. Define the elements $\varepsilon^{\pm}$ in $R = \mathbf{Z}_l[G]$ by

$$\varepsilon^{\pm} = \tfrac{1}{2}(\sigma_1 \pm \sigma_{-1}).$$

Let M be an R-module and let

$$M^+ = \varepsilon^+ M, \quad M^- = \varepsilon^- M.$$

M^+ (resp. M^-) is the submodule of M consisting of all x in M such that $\sigma_{-1}x = x$ (resp. $\sigma_{-1}x = -x$), and

$$M = M^+ \oplus M^-.$$

We shall next consider such decompositions for the R-modules $\bar{J}(I)$ and $\bar{J}(P)$.

THEOREM 4:

$$\bar{J}(I)^+ = \bar{J}(P)^+ = 1 + l\mathbf{Z}_l\,.$$

PROOF: Since

$$\varepsilon^+\theta = \sum_t \sigma_t\,, \quad \varepsilon^+\theta_0 = (g-1)\sum_t \sigma_t$$

and since $(g-1, l) = 1$, we obtain from Theorem 3 that

$$\bar{J}(P)^+ = U_2^{\varepsilon^+\theta_0} = N(U_2) = 1 + l\mathbf{Z}_l$$

where N now denotes the norm from $k_\mathfrak{l}$ to $\mathbf{Q}_l$. By [W], $J_a(\mathfrak{a})^{2\varepsilon^+} = |J_a(\mathfrak{a})|^2$ is a power of $N(\mathfrak{a})$. Since $N(\mathfrak{a}) \equiv 1 \mod l$ for $\mathfrak{a}$ in I, we see that $J(I)^+ \subseteq 1 + l\mathbf{Z}_l$. Therefore $\bar{J}(I)^+ = \bar{J}(P)^+ = 1 + l\mathbf{Z}_l$.

REMARK: Let $l > 3$. Then $\bar{J}(I)^+ = 1 + l\mathbf{Z}_l \subseteq U_{l-1} \subseteq U_3$. On the other hand, $\bar{J}(I)^- \subseteq U_2^- \subseteq U_3$. Hence $\bar{J}(I) \subseteq U_3$, i.e., $J(I) \equiv 1 \mod \mathfrak{l}^3$.

THEOREM 5: As R-modules,

$$U_2^-/\bar{J}(P)^- \simeq R^-/S^-\,.$$

PROOF: One sees easily that $U_2 \simeq R$ as R-modules. Hence Theorem 5 is an immediate consequence of Theorem 3.

§ 5. Let h denote the class number of $k = C_l$ and let h^+ be the class number of the maximal real subfield of k. It is well known that h is divisible by h^+ so that

$$h = h^- h^+$$

with an integer h^-. h^- is called the first factor of the class number h, and h^+ the second factor of h. On the other hand, it is also known (3) that the order of R^-/S^- is equal to the highest power of l dividing the first factor h^-, namely, the order of $Y^-/Z^- = (I'/P)^-$. Hence we obtain the following result from Theorem 5:

THEOREM 6:

$$[U_2^- : \bar{J}(P)^-] = [Y^- : Z^-]\,,$$

(3) Cf. K. IWASAWA, *A class number formula for cyclotomic fields*, Ann. Math., 76 (1962), 171-179.

and this is also equal to the highest power of l dividing the first factor h^- of the class number h of $k = C_l$.

It follows in particular that both $U_2^-/\bar{J}(I)^-$ and $U_2^-/\bar{J}(P)^-$ are finite groups. Since U_2^- is a free $\mathbb{Z}_l$-module of rank $(l-1)/2$, we see that both $\bar{J}(I)^-$ and $\bar{J}(P)^-$ are free $\mathbb{Z}_l$-modules of the same rank $(l-1)/2$. Therefore, by Theorem 4, both $\bar{J}(I)$ and $\bar{J}(P)$ are free $\mathbb{Z}_l$-modules of rank $(l+1)/2$.

LEMMA 6: The kernel of the R-homomorphism

$$f\colon Y \to U_2$$

defined in § 3 is the torsion $\mathbb{Z}_l$-submodule T of Y.

PROOF: Let T' denote the kernel of f. Then $Y/T' \simeq f(Y) = \bar{J}(I)$ so that Y/T' is a free $\mathbb{Z}_l$-module of rank $(l+1)/2$. Since T is a finite module, we see that $T \subseteq T'$. On the other hand, by a theorem of Brumer ([4]), Y/T is also a free $\mathbb{Z}_l$-module of rank $(l+1)/2$. Since $T \subseteq T'$, we obtain $T' = T$.

It follows from the lemma that f induces R-isomorphisms

$$Y/T \xrightarrow{\sim} \bar{J}(I),$$

$$Y^+/T^+ \xrightarrow{\sim} \bar{J}(I)^+, \qquad Y^-/T^- \xrightarrow{\sim} \bar{J}(I)^-.$$

THEOREM 7:

$$Y^-/Z^-T^- \simeq \bar{J}(I)^-/\bar{J}(P)^-,$$

$$[T^- : 1] = [U_2^- : \bar{J}(I)^-].$$

PROOF: Since $f(Z^-) = \bar{J}(P)^-$ by Lemma 5, the isomorphism in the theorem is a consequence of $Y^-/T^- \xrightarrow{\sim} \bar{J}(I)^-$. On the other hand, there is an isomorphism $U_2^- \xrightarrow{\sim} Z^-$ which extends the map $U_2^- \cap k \to P$ defined by $\alpha \mapsto (\alpha)$. Hence Z^- is a free $\mathbb{Z}_l$-module and $Z^- \cap T^- = 1$. Therefore, by Theorem 6 and by the isomorphism just proved,

$$[T^- : 1] = [Z^-T^- : Z^-] = [U_2^- : \bar{J}(I)^-].$$

In his study on Fermat's problem, S. Vandiver conjectured that for any odd prime l, the second factor h^+ of the class number h

([4]) Cf. A. BRUMER, *On the units of algebraic number fields*, Mathematika, 14 (1967), 121-124.

of k $(=C_l)$ is prime to l. The conjecture is highly interesting not only because of its applications on Fermat's problem but also because of its many important consequences in the theory of cyclotomic fields, and it has been verified by numerical computations for a large number of primes. Now, by a classical argument in class field theory, one can show without much difficulty that h^+ is prime to l if and only if $T^- = 1$. Therefore we obtain the following result from Theorem 7:

THEOREM 8: The conjecture of Vandiver is true for the prime number l if and only if

$$\bar{J}(I)^- = U_2^- .$$

Since U_2 is a profinite abelian l-group, the equality in the theorem is equivalent with

$$U_2^- \subseteq J(I)\, U_2^l (= J(I)\, U_{l+1}) .$$

One sees easily that it is also equivalent with the fact that the rank of the elementary abelian group $J(I)\, U_2^l / U_2^l$ is $(l+1)/2$.

Testo pervenuto il 5 aprile 1973.
Bozze licenziate il 5 febbraio 1975.

[55] A note on cyclotomic fields

Invent. math., 36 (1976), 115-123.

To Jean-Pierre Serre

Let p be an odd prime, $p>2$. Let K be the cyclotomic field of p-th roots of unity, and K^+ the maximal real subfield of K. It is well known that the class number h of K is divisible by the class number h^+ of K^+:

$$h=h^+h^-$$

and that the factors h^+ and h^- are given by the following formulas:

$$h^+=[E:E_0], \tag{I}$$

$$h^-=2p\prod_{\chi}(-\tfrac{1}{2}B_\chi). \tag{II}$$

Here in the first formula, E denotes the group of all units in K^+, and E_0 the subgroup of E generated by so-called circular units; and in the second formula, the product is taken over all Dirichlet characters χ to the modulus p with $\chi(-1)=-1$ and B_χ is given by

$$B_\chi=\frac{1}{p}\sum_{a=1}^{p-1}a\chi(a).$$

These formulas are obtained by means of analytic method, computing the values at $s=1$ of Dirichlet's L-functions $L(s:\chi)$, $\chi\neq 1$, associated with K, and no purely algebraic proof is known. Let (I_p) denote the p-part of (I), namely, the formula which states that the exact power of p dividing h^+ is equal to that dividing $[E:E_0]$, and let (II_p) be the p-part of (II), derived similarly from (II). Then it follows from (I_p) that h^+ is prime to p if and only if $E=E_0E^p$. This statement will be denoted by (I'_p). In the present note, we shall show by essentially algebraic method *that* (I'_p) *can be deduced from* (II_p) *without resorting to the original formula* (I) *and that if the equivalent conditions* $(h^+,p)=1$ *and* $E=E_0E^p$ *in* (I'_p) *are both satisfied, then* (II_p) *can be proved similarly, without using* (II)[1]. Although nothing new can be obtained by such an argument, it will give us some further insight into the mysterious classical formulas (I) and (II).

1. Let $\mathbb{Z}$, $\mathbb{Z}_p$, $\mathbb{Q}$, and $\mathbb{Q}_p$ denote the ring of rational integers, the ring of p-adic integers, the field of rational numbers, and the field of p-adic numbers respectively. Let ζ be a fixed primitive p-th root of unity in K so that $K=\mathbb{Q}(\zeta)$. For each a in $\mathbb{Z}$,

[1] For the second statement, see [3], § 3.5.

prime to p, there is a unique automorphism σ_a in the Galois group $G=\mathrm{Gal}(K/\mathbb{Q})$ such that $\sigma_a(\zeta)=\zeta^a$, and the map $a\mapsto\sigma_a$ induces an isomorphism $(\mathbb{Z}/p\mathbb{Z})^\times \xrightarrow{\sim} G$. By means of this isomorphism, we identify the corresponding characters of $(\mathbb{Z}/p\mathbb{Z})^\times$ and G so that if χ is such a character, we have $\chi(a)=\chi(\sigma_a)$ for every a in $\mathbb{Z}$, prime to p. Note that the characters of the multiplicative group $(\mathbb{Z}/p\mathbb{Z})^\times$ are nothing but Dirichlet characters to the modulus p in the classical sense. Let $J=\sigma_{-1}$; J is the automorphism of K which maps each α in K to its complex conjugate $\bar{\alpha}$, and $\chi(J)=\chi(-1)=\pm 1$ for every character χ. The group $\mathbb{V}$ of all roots of unity in $\mathbb{Q}_p$ is a cyclic group of order $p-1$ so that we may identify $\mathbb{V}$, by means of a fixed isomorphism, with the group of all (ordinary) $(p-1)$-st roots of unity in the complex field. The characters of $(\mathbb{Z}/p\mathbb{Z})^\times$ and G are then simply the homomorphisms from these groups into $\mathbb{V}$. For each integer a, prime to p, there exists a unique element $\omega(a)$ in $\mathbb{V}$ such that $\omega(a)\equiv a \bmod p\mathbb{Z}_p$, and the map $a\mapsto\omega(a)$ defines a homomorphism of $(\mathbb{Z}/p\mathbb{Z})^\times$ into $\mathbb{V}$, namely, a character ω of $(\mathbb{Z}/p\mathbb{Z})^\times$ (and of G). Since the order of ω is $p-1$, each character χ of $(\mathbb{Z}/p\mathbb{Z})^\times$ is a power of ω: $\chi=\omega^i$.

Let R denote the group ring of $G=\mathrm{Gal}(K/\mathbb{Q})$ over $\mathbb{Z}_p$: $R=\mathbb{Z}_p[G]$. For each character χ of G, we define an element ε_χ in R by

$$\varepsilon_\chi=\frac{1}{p-1}\sum_{\sigma\in G}\chi(\sigma)^{-1}\sigma=\frac{1}{p-1}\sum_{a=1}^{p-1}\chi(a)^{-1}\sigma_a.$$

Then $\sigma\varepsilon_\chi=\chi(\sigma)\varepsilon_\chi$ for every σ in G, and these ε_χ, for all characters χ of G, give us a complete set of mutually orthogonal idempotents in the ring R with the sum $\sum_\chi \varepsilon_\chi=1$. Hence if M is any R-module, then $M_\chi=\varepsilon_\chi M$ consists of all elements x in M such that $\sigma(x)=\chi(\sigma)x$ for every σ in G, and M is the direct sum of all such submodules M_χ. Let $M^\pm=\frac{1}{2}(1\pm J)M$. Then M^+ (resp. M^-) is the direct sum of M_χ with $\chi(J)=\chi(-1)=1$ (resp. $\chi(J)=\chi(-1)=-1$), and $M=M^+\oplus M^-$.

2. Let $\pi=1-\zeta$. The principal ideal (π) is a prime ideal of K with $(\pi)^{p-1}=(p)$ and the (π)-completion of K is the local cyclotomic field $F=\mathbb{Q}_p(\zeta)$. Since $\mathrm{Gal}(F/\mathbb{Q}_p)$ can be canonically identified with $G=\mathrm{Gal}(K/\mathbb{Q})$, F is an R-module in the obvious manner, and for each character χ of G, we have $F_\chi=\varepsilon_\chi F=\mathbb{Q}_p\gamma_\chi$ with

$$\gamma_\chi=\varepsilon_\chi(\zeta)=\frac{1}{p-1}\sum_{a=1}^{p-1}\chi(a)^{-1}\zeta^a.$$

Up to the factor $(p-1)^{-1}$, γ_χ is a Gauss sum, and if $\chi\neq 1$, then

$$\gamma_\chi\gamma_{\chi^{-1}}=\chi(-1)(p-1)^{-2}p.$$

We also have for $\chi\neq 1$ that

$$\varepsilon_\chi\left(-\frac{\zeta}{\pi}\right)=\varepsilon_\chi\left(1-\frac{1}{1-\zeta}\right)=\frac{1}{p}\sum_{a=1}^{p-1}a\varepsilon_\chi(\zeta^a)=\frac{1}{p}\sum_{a=1}^{p-1}a\chi(a)\gamma_\chi,$$

namely,

$$\varepsilon_\chi\left(-\frac{\zeta}{\pi}\right)=B_\chi\gamma_\chi$$

by using $(1-\zeta)(\zeta+2\zeta^2+\cdots+(p-1)\zeta^{p-1})=-p$.

Let ν denote the valuation on F which extends the standard p-adic valuation on $\mathbb{Q}_p$. Then $\nu(p)=1$, $\nu(\pi)=(p-1)^{-1}$ and it is also well known that $\nu(\gamma_\chi)=i(p-1)^{-1}$ for $\chi=\omega^i$, $0\leqq i<p-1$. Let

$$e=\nu(h^-), \qquad e_\chi=\nu(B_{\chi^{-1}}).$$

Since $\omega(a)\equiv a \bmod p\mathbb{Z}_p$, we find immediately that $e_\chi\geqq 0$ for $\chi\neq\omega$, while $e_\omega=-1$. Hence the formula (II_p) stated in the introduction can be written in the form

$$e=\sum_\chi e_\chi \tag{II_p}$$

where χ ranges over all characters such that $\chi(-1)=-1$, $\chi\neq\omega$. Note that (II_p) induces that $e_\chi<+\infty$, i.e., $B_{1,\chi^{-1}}\neq 0$, for every χ with $\chi(-1)=-1$. However, in the most part of the following, we shall not assume (II_p) so that the possibility $e_\chi=+\infty$ will not be excluded.

3. Let I be the multiplicative group of all ideals of K which are prime to (π) and let P denote the subgroup of all principal ideals in I so that I/P is (isomorphic to) the ideal class group of K. Let I' be the subgroup of I containing P such that

$$X=I'/P$$

is the Sylow p-subgroup of I/P. Since X is a finite abelian p-group and since $G=\mathrm{Gal}(K/\mathbb{Q})$ acts on it in the obvious manner, X is an R-module, and the submodules X_χ are defined for all characters χ of G. For each χ, let p^{f_χ}, $0\leqq f_\chi<+\infty$, be the order of X_χ. The imbedding $K^+\to K$ of the maximal real subfield K^+ in K induces a homomorphism of the ideal class group of K^+ into that of K, and we see easily that it maps the Sylow p-subgroup of the former group isomorphically onto the submodule $X^+=\frac{1}{2}(1+J)X$ in X. Hence it follows from $h=h^+h^-$, $X=X^+\oplus X^-$, and $e=\nu(h^-)$ that the order of X^- is p^e. Since X^- is the direct sum of X_χ with $\chi(-1)=-1$ and since $X_\omega=0$ as we shall see below, we obtain

$$e=\sum_\chi f_\chi \tag{1}$$

where χ again ranges over all characters such that $\chi(-1)=-1$, $\chi\neq\omega$.

Let

$$\theta=\frac{1}{p}\sum_{a=1}^{p-1} a\sigma_{-a}^{-1}.$$

θ is an element of the group ring $\mathbb{Q}[G]$ and for any integer t, prime to p, $(\sigma_t-t)\theta$ is contained in $\mathbb{Z}[G]$. A classical theorem of Stickelberger states that such an element in $\mathbb{Z}[G]$ annihilates the ideal class group I/P and, hence, $X=I'/P$:

$$((\sigma_t-t)\theta)\,X=0. \tag{2}$$

Taking $1+p$ for t, we find in particular that

$$(p\theta)X=\left(\sum_{a=1}^{p-1} a\sigma_{-a}^{-1}\right)X=0. \tag{3}$$

Let x be an element of X_ω so that $\sigma_a(x)=\omega(a)x$. Then $(p\theta)=\left(\sum_{a=1}^{p-1} a\omega(-a)^{-1}\right)x$ with $\sum_{a=1}^{p-1} a\omega(-a)^{-1}\equiv-(p-1)\equiv 1 \bmod p\mathbb{Z}_p$. Hence it follows from (3) that $x=0$, and this proves $X_\omega=0$, as mentioned above. On the other hand, for $\chi\neq\omega$, $\chi(-1)=-1$,

$$\varepsilon_\chi=\frac{1}{p-1}\sum_{a=1}^{p-1}(\chi(a)^{-1}(\sigma_a-a)+\chi(a)^{-1}a)$$

where $\sum_{a=1}^{p-1}\chi^{-1}(a)a\equiv 0 \bmod p\mathbb{Z}_p$. Hence it follows from (2) and (3) that

$$(\theta\varepsilon_\chi)X_\chi=(\theta\varepsilon_\chi)X=0$$

with

$$\theta\varepsilon_\chi=\frac{1}{p}\sum_{a=1}^{p-1}a\chi(-a)^{-1}\varepsilon_\chi=-B_{\chi^{-1}}\varepsilon_\chi.$$

Since $e_\chi=\nu(B_{\chi^{-1}})$, we obtain

$$p^{e_\chi}X_\chi=0 \tag{4}$$

for every $\chi\neq\omega$, $\chi(-1)=-1$. Here and in the following, it is to be understood that equalities $p^\infty M=p^\infty x=0$ hold for any finite abelian p-group M and for any element x in such a group. With such an understanding, the above formula holds even when we do not yet know that $e_\chi<+\infty$.

Let $\mathfrak{o}_p$ be the ring of all π-adic integers in the local field F, and for $n\geqq 1$, let $U_n=1+\pi^n\mathfrak{o}_p$. Then these U_n, $n\geqq 1$, are subgroups of the group of local units in F and we have $U_1^p=U_{p+1}$. Obviously $G=\mathrm{Gal}(F/\mathbb{Q}_p)$ acts on U_1 so that U_1/U_{p+1} $(=U_1/U_1^p)$ is an R-module, and we see easily that $(U_1/U_{p+1})_\omega$ is an abelian group of type (p,p) and that $(U_1/U_{p+1})_\chi$ is a cyclic group of order p for every other character $\chi\neq\omega$. For each integer a, prime to p, $(1-\zeta^a)(1-\zeta)^{-1}$ is a (global) unit of K such that $(1-\zeta^a)(1-\zeta)^{-1}\equiv a \bmod (\pi)$. Hence each principal ideal in P is generated by an element α in K such that $\alpha\equiv 1 \bmod (\pi)$. Let P' denote the subgroup of all principal ideals (α) in P such that $\alpha\equiv 1 \bmod (\pi)^{p+1}$, and let

$$Y=I'/P',\qquad Z=P/P'$$

so that $X=Y/Z$. Since there is an obvious surjective homomorphism $U_1/U_{p+1}\to Z$, both Z and Y are finite abelian p-groups. Hence Y is an R-module, Z is a submodule of Y, and $X_\chi=Y_\chi/Z_\chi$ for every character χ. Using the fact that each unit in K is a product of a power of ζ and a real unit, we see that $U_1/U_{p+1}\to Z$ induces an isomorphism $(U_1/U_{p+1})_\chi\xrightarrow{\sim}Z_\chi$ for every $\chi\neq\omega$, $\chi(-1)=-1$, and that the kernel of $(U_1/U_{p+1})_\omega\to Z_\omega$ is a group of order p. Hence for each χ with $\chi(-1)=-1$, Z_χ is a cyclic group of order p so that Y_χ is a group of order $p^{f_\chi+1}$. Note that since $X_\omega=0$, we have $Y_\omega=Z_\omega$, and Y_ω is a cyclic group of order p.

4. The group E of all units in the maximal real subfield K^+ of K is the direct product of ± 1 and a free abelian group of rank $\dfrac{p-3}{2}$, and by a theorem of

Minkowski, it contains a unit η such that its conjugates η^τ, $\tau \in \mathrm{Gal}(K^+/\mathbb{Q})$, generate a subgroup E' of finite index in E[2]. Replacing η by η^2 if necessary, we may assume that the norm of η is 1. Let $H = \mathrm{Gal}(K^+/\mathbb{Q})$. Then the map $\tau \mapsto \eta^\tau$, $\tau \in H$, induces a surjective homomorphism from the group ring $\mathbb{Z}[H]$ onto E' with kernel generated over $\mathbb{Z}$ by the element $\sum \tau$, $\tau \in H$. The group G acts on E and E' through the canonical homomorphism $G = \mathrm{Gal}(K/\mathbb{Q}) \to H = \mathrm{Gal}(K^+/\mathbb{Q})$ so that both E/E^p and E'/E'^p are R-modules, and it follows from the above that $(E'/E'^p)_\chi \neq 1$ for $\chi \neq 1$, $\chi(-1) = 1$. Let E'' be the subgroup of E containing E' such that E''/E' is the Sylow p-subgroup of E/E'. Then E''/E'^p is an R-module such that $(E''/E'^p)_\chi \neq 1$ for $\chi \neq 1$, $\chi(-1) = 1$. Since $E/E^p \simeq E''/E''^p$ as R-modules and since E/E^p is an abelian group of type $(p, \dots, p)$ with rank $\frac{p-3}{2}$, it follows that $(E/E^p)_\chi$ is a cyclic group of order p for $\chi \neq 1$ with $\chi(-1) = 1$ and that $(E/E^p)_\chi = 1$ for all other characters χ of G.

By definition, the subgroup E_0 of E in the formula (I) is the intersection of E with the multiplicative group generated by $\pm\pi^\sigma$ for all σ in G. Let ψ be a character of G such that $\psi(-1) = 1$, $\psi \neq 1$. Since $\sum_{a=1}^{p-1} \psi(a)^{-1} = 0$, there exists an element $\mu = \sum_{a=1}^{p-1} m_a \sigma_a$, $m_a \in \mathbb{Z}$, in $\mathbb{Z}[G]$ such that

$$\sum_{a=1}^{p-1} m_a = 0, \qquad \mu \equiv \varepsilon_\psi \bmod pR.$$

Let

$$\eta_\psi = \pi^\mu.$$

Then η_ψ is a unit in E_0 and the coset of η_ψ mod E^p, which we shall denote by r_ψ, depends only upon ψ and is independent of the choice of μ as mentioned above. It is clear that $(E_0 E^p/E^p)_\psi$ is generated by r_ψ. Since $(E/E^p)_\psi$ is a group of order p, it follows that $(E/E_0 E^p)_\psi \neq 1$ if and only if $r_\psi = 1$, namely, if and only if η_ψ is a p-th power in K^+ or, equivalently, in K.

5. For any unit η in K and for any ideal $\mathfrak{a}$ in the ideal group I, the p-th power residue symbol $\left(\frac{\eta}{\mathfrak{a}}\right)$ is defined[3]. Since it is a p-th root of unity in K, we may write it in the form

$$\left(\frac{\eta}{\mathfrak{a}}\right) = \zeta^{[\eta, \mathfrak{a}]}$$

with an integer (or a p-adic integer) $[\eta, \mathfrak{a}]$, well determined mod p (or mod $p\mathbb{Z}_p$). For η in E and $\mathfrak{a}$ in I', $[\eta, \mathfrak{a}]$ then defines a pairing of E and I' into $\mathbb{Z}/p\mathbb{Z} (= \mathbb{Z}_p/p\mathbb{Z}_p)$ such that $[E^p, \mathfrak{a}] = [\eta, P'] = 0$. Hence it also induces a pairing of the R-modules E/E^p and $Y = I'/P'$ into $\mathbb{Z}/p\mathbb{Z}$. Since $\left(\frac{\eta^\sigma}{\mathfrak{a}^\sigma}\right) = \left(\frac{\eta}{\mathfrak{a}}\right)^{\omega(a)}$ for $\sigma = \sigma_a$ in G, we see easily that unless $\psi\chi = \omega$, $(E/E^p)_\psi$ and Y_χ are orthogonal in the above pairing.

---

[2] It seems that we need here some real analysis in the form of Minkowski's lattice theorem.

[3] For the properties of power residue symbol needed below, see [2], Abschnitt III.

Let ψ be a character of G such that $\psi(-1)=1$, $\psi \neq 1$, and let $\chi=\omega\psi^{-1}$ so that $\chi(-1)=-1$, $\chi \neq \omega$. We shall next calculate $[r_\psi, y]$ for the r_ψ defined above and for arbitrary y in Y. Let $\mathfrak{a}$ be an ideal of I', contained in the ideal class y. Then

$$\mathfrak{a}^{p^n}=(\alpha)$$

for some integer $n \geqq 0$ and for some α in K satisfying $\alpha \equiv 1 \bmod (\pi)$. As elements of the local field F, both α and $\alpha^{\varepsilon_\chi}$ belong to $U_1=1+\pi\mathfrak{o}_p$ so that the p-adic logarithm is defined for these elements. Since $\varepsilon_\chi F=\mathbb{Q}_p\gamma_\chi$, we have

$$\log \alpha^{\varepsilon_\chi}=\varepsilon_\chi(\log\alpha)=u\gamma_\chi$$

with some u in $\mathbb{Q}_p$.

Lemma 1. *With the notations introduced above,*

$$[r_\psi, y] \equiv \frac{1}{(p-1)p^n}\chi(-1)B_{\chi^{-1}}u \qquad \bmod p\mathbb{Z}_p.$$

Proof. Let ζ_n be a primitive p^{n+1}-th root of unity such that $\zeta_n^{p^n}=\zeta$ and let $K_n=\mathbb{Q}(\zeta_n)$ so that $\pi=1-\zeta$ is the norm of $\pi_n=1-\zeta_n$ from K_n to K. We identify $G=\mathrm{Gal}(K/\mathbb{Q})$ with the subgroup of order $p-1$ in $\mathrm{Gal}(K_n/\mathbb{Q})$ in the obvious manner. Choose an element $\mu=\sum_{a=1}^{p-1} m_a\sigma_a$ in $\mathbb{Z}[G]$ as mentioned in § 4. Then π_n^μ is a unit of K_n and its norm from K_n to K is $\eta_\psi=\pi^\mu$. Hence

$$\left(\frac{\eta_\psi}{\mathfrak{a}}\right)=\left(\frac{\pi_n^\mu}{\mathfrak{a}}\right)=\left(\frac{\pi_n^\mu}{\mathfrak{a}}\right)_n^{p^n}=\left(\frac{\pi_n^\mu}{\alpha}\right)_n$$

where the second term denotes the p-th power residue symbol in K_n and the last two terms are the p^{n+1}-th power residue symbol in the same field. By a theorem of Artin-Hasse[4],

$$\left(\frac{\pi_n}{\alpha}\right)_n=\zeta_n^A, \quad A=\frac{1}{p^{n+1}}T_n\left(-\frac{\zeta_n}{\pi_n}\log\alpha\right)=\frac{1}{p}T\left(-\frac{\zeta}{\pi}\log\alpha\right)$$

with T (resp. T_n) denoting the trace from $F=\mathbb{Q}_p(\zeta)$ to $\mathbb{Q}_p$ (resp. from $\mathbb{Q}_p(\zeta_n)$ to $\mathbb{Q}_p$). Let σ_a, $(a,p)=1$, be any element of G. As an automorphism of K_n, σ_a is determined by $\sigma_a(\zeta_n)=\zeta_n^{\omega(a)}$. Hence we obtain from the above that

$$\left(\frac{\pi_n^\mu}{\alpha}\right)_n=\zeta_n^B, \quad B=\frac{1}{p}T\left(\mu'\left(-\frac{\zeta}{\pi}\right)\log\alpha\right)$$

with

$$\mu'=\sum_{a=1}^{p-1} m_a\omega(a)\sigma_a.$$

Since μ is any element in $\mathbb{Z}[G]$ such that $\sum_{a=1}^{p-1} m_a=0$, $\mu\equiv\varepsilon_\psi \bmod pR$, we can choose it in such a way that μ' is arbitrarily close to $\varepsilon_{\chi^{-1}}=(p-1)^{-1}\sum_{a=1}^{p-1}\psi(a)^{-1}\omega(a)\sigma_a$ in

[4] See [1].

the p-adic topology of $R=\mathbb{Z}_p[G]$. Hence in the above formula for $\left(\frac{\pi_n^{\mu}}{\alpha}\right)_n$, μ' in B can be replaced by $\varepsilon_{\chi^{-1}}$. It then follows from $\zeta_n^{p^n}=\zeta$ that

$$[r_\psi, y] \equiv \frac{1}{p^{n+1}} T\left(\varepsilon_{\chi^{-1}}\left(-\frac{\zeta}{\pi}\right) \log \alpha\right) \mod p\mathbb{Z}_p.$$

However, since $T=(p-1)\varepsilon_1$ as operators on F, we see that

$$T\left(\varepsilon_{\chi^{-1}}\left(-\frac{\zeta}{\pi}\right) \log \alpha\right)=(p-1)\,\varepsilon_{\chi^{-1}}\left(-\frac{\zeta}{\pi}\right) \varepsilon_\chi(\log \alpha)$$

where

$$\varepsilon_{\chi^{-1}}\left(-\frac{\zeta}{\pi}\right) \varepsilon_\chi(\log \alpha)=B_{\chi^{-1}} \gamma_{\chi^{-1}} u \gamma_\chi=\chi(-1)(p-1)^{-2} p B_{\chi^{-1}} u$$

by the remarks in § 2. Hence the formula of the lemma is proved.

Lemma 2. *Let ψ and $\chi=\omega\psi^{-1}$ be as in Lemma 1 and let y be an element of Y_χ. Then $[r_\psi, y]=0$ if and only if $p^{e_\chi} y=0$. In fact, if $[r_\psi, y]\neq 0$, then $e_\chi<+\infty$ and $p^{e_\chi} y$ is a non-zero element of Z_χ.*

Proof. If $[r_\psi, y]\neq 0$, then $B_{\chi^{-1}}\neq 0$, namely, $e_\chi<+\infty$, by Lemma 1. On the other hand, if $[r_\psi, y]=0$ and $e_\chi=+\infty$, then $p^{e_\chi} y=0$ by the definition. Hence in the following proof, we may assume that $e_\chi<+\infty$. As $p^{e_\chi} X_\chi=0$ by (4), $p^{e_\chi} y$ is an element of Z_χ, and if $\mathfrak{a}$ is an ideal in the ideal class y, the p^{e_χ}-th power of $\mathfrak{a}$ is a principal ideal (α) with $\alpha\equiv 1 \bmod (\pi)$. Hence we may apply Lemma 1 for such $\mathfrak{a}$ and α with $n=e_\chi$. Since $p^{-n} B_{\chi^{-1}}$ is a p-adic unit in this case, we then see that $[r_\psi, y]=0$ if and only if $u\equiv 0 \bmod p\mathbb{Z}_p$ for u defined by $\log \alpha^{\varepsilon_\chi}=u\gamma_\chi$. Since $\chi\neq\omega$, let $\chi=\omega^i$, $1<i<p-1$, so that $v(\gamma_\chi)=i(p-1)^{-1}$. Then $u\equiv 0 \bmod p\mathbb{Z}_p$ if and only if $\log \alpha^{\varepsilon_\chi}\equiv 0 \bmod (\pi)^{p+1}$, namely, if and only if $\alpha^{\varepsilon_\chi}$ belongs to U_{p+1}. Clearly for any ε in $\mathbb{Z}[G]$, $(\mathfrak{a}^\varepsilon)^{p^n}=(\alpha^\varepsilon)$ with $n=e_\chi$. If ε is sufficiently close to ε_χ in the p-adic topology of R, then $\mathfrak{a}^\varepsilon$ belongs to the ideal class $\varepsilon_\chi(y)$ $(=y^{\varepsilon_\chi})=y$ and $\alpha^\varepsilon\equiv\alpha^{\varepsilon_\chi} \bmod U_{p+1}$. The isomorphism $(U_1/U_{p+1})_\chi \xrightarrow{\sim} Z_\chi$ then shows that $\alpha^{\varepsilon_\chi}$ belongs to U_{p+1} if and only if (α^ε) is contained in P', namely, if and only if $p^{e_\chi} y=0$. Thus $[r_\psi, y]=0$ if and only if $p^{e_\chi} y=0$.

6. The next lemma is the core of the whole argument in the present paper.

Lemma 3. *Let ψ be a character of G such that $\psi(-1)=1$, $\psi\neq 1$, and let $\chi=\omega\psi^{-1}$ so that $\chi(-1)=-1$, $\chi\neq\omega$. Then $(E/E_0E^p)_\psi\neq 1$ if and only if $p^{e_\chi} Y_\chi=0$. In fact, if $(E/E_0E^p)_\psi=1$, then $e_\chi<+\infty$ and Y_χ has a cyclic subgroup of order $p^{e_\chi+1}$ containing Z_χ.*

Proof. As mentioned in § 4, if $(E/E_0E^p)_\psi=1$, then $r_\psi\neq 1$ and η_ψ is not a p-th power in K. Hence, in this case, there exists an ideal $\mathfrak{a}$ in I such that $\left(\frac{\eta_\psi}{\mathfrak{a}}\right)\neq 1$. Since $[I:I']$ is prime to p, we may assume that $\mathfrak{a}$ is contained in I'. Then $[r_\psi, y]\neq 0$ for the coset y of $\mathfrak{a}$ mod P'. Since r_ψ is an element of $(E/E^p)_\psi$, $[r_\psi, y]=[r_\psi, \varepsilon_\chi(y)]$

by a remark in § 5. Hence, replacing y by $\varepsilon_\chi(y)$, we may suppose that $[r_\psi, y] \neq 0$ for an element y in Y_χ. It then follows from Lemma 2 that $e_\chi < +\infty$ and that y generates a cyclic subgroup of order $p^{e_\chi+1}$ containing Z_χ. Assume next that $(E/E_0 E^p)_\psi \neq 1$. Then $r_\psi = 1$ by § 4 so that $[r_\psi, y] = 0$ for every y in Y_χ. Hence $p^{e_\chi} Y_\chi = 0$ by Lemma 2.

Corollary. *If* $(E/E_0 E^p)_\psi = 1$, *then* $e_\chi \leqq f_\chi$. *Hence if* $E = E_0 E^p$, *then*

$$\sum_\chi e_\chi \leqq e$$

where χ *ranges over all characters of* G *such that* $\chi(-1) = -1$, $\chi \neq \omega$.

Proof. This follows immediately from (1) and from the above lemma.

Lemma 4. $(h^+, p) = 1$ *if and only if* Y_χ *is a cyclic group for every character* χ *of* G *such that* $\chi(-1) = -1$.

Proof. Since this is more or less a well-known fact in the theory of cyclotomic fields, we shall briefly describe the proof. By class field theory, h^+ is divisible by p if and only if there exists an unramified cyclic extension L/K^+ with degree p. If such L exists, then $KL = K(\sqrt[p]{\alpha})$ with an element α in K such that $\alpha^{1+J} = 1$, $\alpha \equiv 1 \bmod (\pi)^{p+1}$, and such that $(\alpha) = \mathfrak{a}^p$ with an ideal $\mathfrak{a}$ in I'. Furthermore, using the first two properties of α, we can see that such an ideal $\mathfrak{a}$ is nonprincipal. Hence if y denotes the coset of $\mathfrak{a}$ mod P', then $y \in Y^-$, $py = 0$, and $y \notin Z$, and it follows that Y_χ is non-cyclic for some χ with $\chi(-1) = -1$. Conversely, if Y_χ is non-cyclic for some χ with $\chi(-1) = -1$, then reversing the above argument step by step, we see that there is an unramified cyclic extension L/K^+ with degree p.

Corollary. *Let* $(h^+, p) = 1$. *Then* $f_\chi \leqq e_\chi$ *for every character* χ *of* G *such that* $\chi(-1) = -1$, $\chi \neq \omega$. *Hence*

$$e \leqq \sum_\chi e_\chi$$

where the sum is taken over all characters χ *as mentioned above.*

Proof. By the above lemma, Y_χ is cyclic so that X_χ is a cyclic group of order p^{f_χ}. Hence by (4), $f_\chi \leqq e_\chi$. The second part then follows from (1).

We are now able to prove the results mentioned in the introduction. Assume first that $(h^+, p) = 1$ and $E = E_0 E^p$. By the corollaries of Lemmas 3, 4, we then have $e = \sum e_\chi$, namely, the formula (II_p). Assume next that (II_p) holds and $(h^+, p) = 1$. By Lemma 4 and its corollary, we then see that for every $\chi \neq \omega$ with $\chi(-1) = -1$, $e_\chi = f_\chi$ and Y_χ is a cyclic group of order $p^{e_\chi+1}$. Hence by Lemma 3, $(E/E_0 E^p)_\psi = 1$ for every $\psi \neq 1$ with $\psi(-1) = 1$, and consequently $E = E_0 E^p$. Finally, assume that (II_p) holds and $E = E_0 E^p$. By Lemma 3 and its corollary, we then see that for every $\chi \neq \omega$ with $\chi(-1) = -1$, $e_\chi = f_\chi$ and Y_χ is a cyclic group. Since $Y_\omega (= Z_\omega)$ is also cyclic, it follows from Lemma 4 that $(h^+, p) = 1$. Thus the assertions state in the introduction are completely proved.

References

1. Artin, E., Hasse, H.: Die beiden Ergänzungssätze zum Reziprozitätsgesetz der l^n-ten Potenzreste im Körper der l^n-ten Einheitswurzeln. Abh. Math. Sem. (Hamburg) **6**, 146–162 (1928)
2. Hasse, H.: Bericht über die neuere Untersuchungen und Probleme aus der Theorie der algebraischen Zahlkörper, Teil II. Jahresbericht D.M.-V., Erg.-Bd. **6** (1930)
3. Iwasawa, K.: On some modules in the theory of cyclotomic fields. J. Math. Soc. (Japan) **16**, 42–82 (1964)

Received April 5, 1975

Kenkichi Iwasawa
Princeton University
Department of Mathematics
Princeton, N.Y. 08540
USA

[56] Some remarks on Hecke characters

"Algebraic Number Theory" (Kyoto, 1976), Japan Soc. Promotion Sci., Tokyo, 1977, pp. 99-108.

In the present paper, we shall make some simple remarks on Hecke characters of type (A_0) for a special type of finite algebraic number fields. For imaginary abelian extensions over the rational field, examples of such characters are provided by Jacobi sums and these will also be discussed briefly.[1]

§1. Let j denote the automorphism of the complex field C mapping each α in C to its complex-conjugate $\bar{\alpha}$. An algebraic number field k, i.e., an algebraic extension of the rational field Q contained in C, will be called a j-field if k is invariant under j and $\sigma j = j\sigma$ for every isomorphism σ of k into C. One sees immediately that k is a j-field if and only if k is either a totally real field or a totally imaginary quadratic extension of a totally real subfield.

In the following, we shall consider Hecke characters of type (A_0) for a field k which is an imaginary j-field and is also a finite Galois extension over Q. Let $G = \mathrm{Gal}(k/Q)$. The restriction of j on k, which will simply be denoted again by j, is an element of order 2 in the center of G. Let I denote the idele group of k. Then $I = I_0 \times I_\infty$ where I_0 and I_∞ are the finite part and the infinite part of I respectively. The multiplicative group $k^\times$ of the field k is naturally imbedded in I as a discrete subgroup of the locally compact abelian group I. Hence each α in $k^\times$ can be uniquely written in the form $\alpha = \alpha_0\alpha_\infty$ with $\alpha_0 \in I_0$, $\alpha_\infty \in I_\infty$. Now, a Hecke character of k (for ideles) is, by definition, a continuous homomorphism $\chi: I \to C^\times$ such that $\chi(k^\times) = 1$; it is called a Hecke character of type (A_0) if there exists an element ω in the group ring $R = Z[G]$ of G over the ring of rational integers Z with the property that

$$\chi(\alpha_\infty) = \alpha^{-\omega}$$

1) For Hecke characters of type (A_0) in general and for Jacobi sums in particular, see Weil [2a], [2b], [2c].

for every α in $k^\times$. Such ω is uniquely determined for χ by the above equality and is denoted by ω_χ. The set H of all Hecke characters of type (A_0) on k forms a multiplicative abelian group in the obvious manner and the map $\chi \mapsto \omega_\chi$ defines a homomorphism

$$\varphi: H \longrightarrow R$$

from the multiplicative group H into the additive group of the group ring $R = Z[G]$. One sees easily that the kernel of φ is the torsion subgroup T of H which is, by class field theory, dual to the Galois group of the maximal abelian extension over k. Let A denote the image of φ in R so that

$$H/T \xrightarrow{\sim} A .$$

Lemma 1. *Let*

$$\theta = \sum_{\sigma \in G} \sigma .$$

Then A consists of all elements ω in R such that

$$(1 + j)\omega = a\theta$$

for some integer a. In particular, A is a two-sided ideal of R containing $(1 - j)R$.

Proof. Let $\|\xi\|$ denote the norm of an idele ξ in I defined in the usual manner. It is well known that $\xi \mapsto \|\xi\|$ defines a surjective homomorphism of I onto the multiplicative group of real numbers, that $k^\times$ is contained in the kernel I_1 of the homomorphism, and that $I_1/k^\times$ is a compact group. It then follows that for each Hecke character χ, there exists a real number r such that $|\chi(\xi)| = \|\xi\|^r$ for every idele ξ. The lemma follows from this and from the definition of Hecke characters of type (A_0).

Let $[G: 1] = [k: Q] = 2n$ and let

$$G = \{\sigma_1, \cdots, \sigma_n, j\sigma_1, \cdots, j\sigma_n\} , \qquad \theta' = \sum_{i=1}^{n} \sigma_i .$$

Then $(1 + j)\theta' = \theta$ and it follows immediately from the above lemma that

$$A = (1 - j)R \oplus Z\theta' .$$

Hence A is a free abelian group of rank $n + 1$. Note also that A is generated over Z by the sums $\theta' = \sum_{i=1}^{n} \sigma_i$ when $\{\sigma_1, \cdots, \sigma_n\}$ ranges over all subsets of G such that $G = \{\sigma_1, \cdots, \sigma_n, j\sigma_1, \cdots, j\sigma_n\}$.

Now, for each integral ideal $\mathfrak{m}$ of k, let $\mathfrak{J}_{\mathfrak{m}}$ denote the multiplicative group of all ideals of k which are prime to $\mathfrak{m}$. Let χ be a Hecke character of k. Then it follows from the continuity of χ that for a suitable integral ideal $\mathfrak{m}$, χ induces a homomorphism

$$\tilde{\chi}: \mathfrak{J}_{\mathfrak{m}} \longrightarrow C^{\times}$$

with the property that

$$\tilde{\chi}((\alpha)) = \chi(\alpha_0) = \chi(\alpha_\infty)^{-1}$$

for every α in $k^{\times}$ satisfying $\alpha \equiv 1 \bmod^{\times} \mathfrak{m}$. Furthermore, if χ is of type (A_0), the sets $\chi(I_0)$ and $\tilde{\chi}(\mathfrak{J}_{\mathfrak{m}})$ generate over k the same field k_χ:

$$k_\chi = k(\chi(I_0)) = k(\tilde{\chi}(\mathfrak{J}_{\mathfrak{m}})) .$$

Let $\omega = \omega_\chi$ and $(1 + j)\omega = a\theta$, $a \in Z$, as stated in Lemma 1.

Lemma 2. *k_χ is a j-field, finite over k, and for each ideal $\mathfrak{a}$ in $\mathfrak{J}_{\mathfrak{m}}$,*

$$\tilde{\chi}(\mathfrak{a})^{1+j} = N(\mathfrak{a})^a , \qquad \mathfrak{a}^\omega = (\tilde{\chi}(\mathfrak{a}))$$

where $N(\mathfrak{a})$ denotes the norm of $\mathfrak{a}$ over Q and $(\tilde{\chi}(\mathfrak{a}))$ is the principal ideal of k_χ generated by $\tilde{\chi}(\mathfrak{a})$.

Proof. If $\alpha \equiv 1 \bmod^{\times} \mathfrak{m}$, then $\tilde{\chi}((\alpha)) = \chi(\alpha_\infty)^{-1} = \alpha^\omega$. The two equalities of the lemma follow from this and from the fact that the ray class group mod $\mathfrak{m}$ is a finite group. Let $\xi = \tilde{\chi}(\mathfrak{a})$ for an ideal $\mathfrak{a}$ in $\mathfrak{J}_{\mathfrak{m}}$ and let $\mathfrak{a}^h = (\alpha)$, $h \geqslant 1$, $\alpha \in k^{\times}$, $\alpha \equiv 1 \bmod^{\times} \mathfrak{m}$. Then $\xi^h = \tilde{\chi}((\alpha)) = \alpha^\omega$ belongs to k while $\xi^{1+j} = N(\mathfrak{a})^a$ is a rational number. Hence $k(\xi)$ is a finite extension of k, invariant under the complex-conjugation. Let σ be any isomorphism of $k(\xi)$ into C. Since k is a j-field, $\sigma j = j\sigma$ on k. In particular, $(\xi^h)^{\sigma(1+j)} = (\xi^h)^{(1+j)\sigma}$, namely, $(\xi^{1+j})^{h\sigma} = (\xi^\sigma)^{h(1+j)}$. Since $(\xi^{1+j})^\sigma = N(\mathfrak{a})^{a\sigma} > 0$ and $(\xi^\sigma)^{1+j} > 0$, it follows that $(\xi^{1+j})^\sigma = (\xi^\sigma)^{1+j}$ so that $\sigma j = j\sigma$ on $k(\xi)$. Therefore $k(\tilde{\chi}(\mathfrak{a})) = k(\xi)$ is a j-field, finite over k. As the ray class group mod $\mathfrak{m}$ is finite, k_χ is the composite of a finite number of j-fields such as $k(\xi)$. Hence k_χ is again a j-field, finite over k.

By a similar argument, one can prove the following result which may be of some interest for itself: For each j-field k, finite over Q, there exists a j-field F, finite over k, such that every ideal of k becomes a principal ideal in F.

§ 2. Let F be an algebraic number field containing k and let A_F denote the set of all ω in A such that for every ideal $\mathfrak{a}$ of k, $\mathfrak{a}^{\omega}$ becomes a principal ideal in F. On the other hand, let B_F be the set of all $\omega_\chi = \varphi(\chi)$ for Hecke characters χ of type (A_0) with the property that $k_\chi \subseteq F$. Then A_F and B_F are additive subgroups of A and by Lemma 2

$$B_F \subseteq A_F \subseteq A .$$

From now on, we shall assume that F is a j-field containing k; since k is imaginary, F also is an imaginary j-field. Let E, E_r, and E_+ denote the group of all units in F, the subgroup of all real units in E, and the subgroup of all totally positive real units in E_r, respectively. Clearly $E_r^2 \subseteq E^{1+j} \subseteq E_+ \subseteq E_r \subseteq E$. Let W be the group of all roots of unity contained in F. Since F is a j-field, the index $[E: WE_r]$ is either 1 or 2, and so is the index $[E^{1+j}: E_r^2]$. Let

$$\bar{E}_F = \bar{E} = E_+/E^{1+j} .$$

Obviously $\bar{E}$ is an abelian group with exponent (at most) 2.

Now, let ω be an element of A_F and let $\mathfrak{a}$ be any ideal of k. By the definition of A_F, $\mathfrak{a}^{\omega}$ is a principal ideal in F so that

$$\mathfrak{a}^{\omega} = (\mu)$$

with an element μ in $F^{\times}$. Since $A_F \subseteq A$, it follows from Lemma 1 that

$$(1 + j)\omega = a\theta$$

with an integer a. Hence

$$N(\mathfrak{a})^a = \mathfrak{a}^{a\theta} = (\mu^{1+j})$$

and we see that

$$\varepsilon = N(\mathfrak{a})^{-a}\mu^{1+j}$$

is a unit of F and, indeed, a totally positive real unit in E_+. If μ is replaced by $\mu\eta$ with η in E, then ε is replaced by $\varepsilon\eta^{1+j}$. Therefore the coset of ε mod E^{1+j} is uniquely determined by ω and $\mathfrak{a}$ so that it may be denoted by $[\omega, \mathfrak{a}]$:

$$[\omega, \mathfrak{a}] = N(\mathfrak{a})^{-a}\mu^{1+j} \bmod E^{1+j} .$$

It is clear that $[\omega, \mathfrak{a}]$ defines a pairing of A_F and the ideal group of k into $\bar{E}$. Furthermore, if $\mathfrak{a} = (\alpha)$, $\alpha \in k^{\times}$, then $\mathfrak{a}^{\omega} = (\mu)$ with $\mu = \alpha^{\omega}$ so that $\varepsilon = N(\mathfrak{a})^{-a}\mu^{1+j}$

$= 1$. Hence $[\omega, \mathfrak{a}]$ depends only upon the ideal class of $\mathfrak{a}$ and it also defines a pairing of A_F and the ideal class group C_k of k into $\bar{E} = \bar{E}_F$:

$$A_F \times C_k \longrightarrow \bar{E}_F \ .$$

Theorem. *B_F is the annihilator of C_k in A_F in the above pairing so that there is a monomorphism*

$$A_F/B_F \longrightarrow \mathrm{Hom}\,(C_k, \bar{E}_F) \ .$$

Proof. In the above, F is not necessarily a finite extension over k. However, the proof for the general case can be easily reduced to that of the special case where F is a finite extension of k. Therefore we shall assume in the following that F is finite over k and, hence, also finite over Q.

It is clear from Lemma 2 that B_F is contained in the annihilator of C_k in A_F. To prove the converse, let ω be any element of A_F which annihilates C_k in the above pairing. Let w denote the order of the finite group W consisting of all roots of unity in F. We fix a prime ideal $\mathfrak{p}$ of k, prime to w, and denote the norm of $\mathfrak{p}$ over Q by q: $q = N(\mathfrak{p})$. The residue class field of $\mathfrak{p}$ then contains a subgroup canonically isomorphic to W so that $q - 1$ is divisible by w. Let $\mathfrak{a}$ be any ideal of the ideal group $\mathfrak{J}_q$ in k and let $\mathfrak{a}^{\omega} = (\mu)$, $\mu \in F^\times$, as stated above. Since $[\omega, \mathfrak{a}] = 1$ by the assumption on ω, $\varepsilon = N(\mathfrak{a})^{-a}\mu^{1+j}$ is contained in E^{1+j}: $\varepsilon = \eta^{1+j}$, $\eta \in E$. Hence, replacing μ by $\mu\eta$, we may assume that $N(\mathfrak{a})^a = \mu^{1+j}$. As $\mathfrak{a}$ belongs to $\mathfrak{J}_q$, μ is prime to $\mathfrak{p}$, and there exists a root of unity ζ in W satisfying $\mu^{(q-1)/w} \equiv \zeta \bmod^\times \mathfrak{p}$. Let

$$\omega' = \frac{q-1}{w}\omega \ , \quad \nu = \mu^{(q-1)/w}\zeta^{-1} \ .$$

Then

$$\mathfrak{a}^{\omega'} = (\nu) \ , \quad N(\mathfrak{a})^{((q-1)/w)a} = \nu^{1+j} \ , \quad \nu \equiv 1 \bmod^\times \mathfrak{p} \ .$$

We shall next show that for each $\mathfrak{a}$ in $\mathfrak{J}_q$, there exists only one ν in $F^\times$ satisfying the above conditions. Indeed, let ν' be another element in $F^\times$ satisfying the same conditions and let $\nu' = \nu\eta$. Then η is a unit of F such that $\eta^{1+j} = 1$, $\eta \equiv 1 \bmod^\times \mathfrak{p}$. Since $[E : WE_r] = 1$ or 2, let $\eta^2 = \zeta'\eta_0$ with $\zeta' \in W$, $\eta_0 \in E_r$. From $\eta^{2(1+j)} = 1$, we then see that $\eta_0^2 = 1$, $\eta_0 = \pm 1$ so that $\eta = \pm\zeta'$ belongs to W. It then follows from $\eta \equiv 1 \bmod^\times \mathfrak{p}$ that $\eta = 1$, $\nu' = \nu$.

Now, since ν is unique for $\mathfrak{a}$, the map $\mathfrak{a} \mapsto \nu$ defines a homomorphism

$$\rho : \mathfrak{J}_q \longrightarrow C^\times .$$

If α is an element of $k^\times$ satisfying $\alpha \equiv 1 \bmod^\times \mathfrak{q}$, then $\nu = \alpha^{\omega'}$ obviously satisfies the above conditions for the principal ideal $\mathfrak{a} = (\alpha)$. Hence

$$\rho((\alpha)) = \alpha^{\omega'}$$

for such an ideal (α). By [2a], we then know that there exists a Hecke character χ on k which induces the homomorphism ρ on $\mathfrak{J}_q$: $\tilde{\chi} = \rho$. We also see immediately that χ is of type (A_0) and $\omega' = \omega_\chi = \varphi(\chi)$. Since $\tilde{\chi}(\mathfrak{J}_q) = \rho(\mathfrak{J}_q) \subseteq F^\times$, $\omega' = ((q-1)/w)\omega$ is contained in B_F.

In the above, we have fixed a prime ideal $\mathfrak{p}$ of F, prime to w. However, it is easy to see by class field theory that when $\mathfrak{p}$ ranges over all prime ideals of k, prime to w, then the g.c.d. of $N(\mathfrak{p}) - 1$ is w. Therefore it follows from the above that ω itself is contained in B_F, and this completes the proof of the theorem.

§3. We shall next make some remarks on the results mentioned above.

Let F be an arbitrary j-field containing k. Since A_F is a subgroup of A which is a free abelian group of rank $n+1$ and since $\bar{E}^2 = 1$, it follows from the theorem that A_F/B_F is a finite abelian group of type $(2, \cdots, 2)$ with rank at most equal to $n+1$. In particular,

$$2A_F \subseteq B_F \subseteq A_F .$$

Note also that if $\bar{E} = 1$, i.e., if $E_+ = E^{1+j}$, then $A_F = B_F$.

Let $F = k$. In this case, A_k is the subgroup of all ω in A such that $C_k^\omega = 1$, namely, the subgroup of all "relations" on the R-module C_k contained in A. The theorem then states that up to a finite factor group of exponent 2, all such relations are provided by Hecke characters χ of type (A_0) on k with the property $k_\chi = k$, i.e., $\chi(I_0) \subseteq k$.

Let L denote the field generated over k by $\chi(I_0)$ for all Hecke characters χ of type (A_0) on k and let K be the subfield of L generated by $\chi(I_0)$ for all χ in the torsion subgroup T of H:

$$k \subseteq K \subseteq L .$$

As one sees immediately, K is the field generated over k by all roots of unity in C. By the definition, L is the composite of the fields k_χ for all χ in H. Hence, by Lemma 2, L is a j-field. It is also easy to show that L is a Galois extension of the rational field Q. Furthermore, since $H/T \simeq A$ and A is a finitely generated abelian group, it follows from the proof of Lemma 2 that L/K is a finite abelian extension. It seems that the extension L/K, which is thus canonically associated with the field k, has some significance for the arithmetic of the ground field k. Here we note only the following simple fact. Let ω be any element of $A = \varphi(H)$ and let $\omega = \omega_\chi$ with χ in H. For any σ in Gal (L/K), define

$$\chi^{\sigma-1}: I \longrightarrow C^\times$$

by $\chi^{\sigma-1}(\xi) = \chi(\xi)^{\sigma-1}$ for $\xi \in I$. Then $\chi^{\sigma-1}$ is a Hecke character in the torsion subgroup T of H and it depends only upon ω and σ. The map $(\omega, \sigma) \mapsto \chi^{\sigma-1}$ then defines a pairing of A and Gal (L/K) into T and this induces a non-degenerate pairing

$$A/B_K \times \mathrm{Gal}\,(L/K) \longrightarrow T\ .$$

Therefore there exist monomorphisms

$$A/B_K \longrightarrow \mathrm{Hom}\,(\mathrm{Gal}\,(L/K), T)\ , \qquad \mathrm{Gal}\,(L/K) \longrightarrow \mathrm{Hom}\,(A/B_K, T)\ .$$

§4. Important examples of Hecke characters χ of type (A_0) with $k_\chi = k$ are provided by Jacobi sums when k is an abelian extension over the rational field.[2] We add here some further remarks on such Hecke characters in connection with what has been discussed above.

For each integer $m \geqslant 1$, let K_m denote the cyclotomic field of m-th roots of unity and let $G_m = \mathrm{Gal}\,(K_m/Q)$ and $R_m = Z[G_m]$. For any real number α, let $\langle\alpha\rangle = \alpha - [\alpha]$ where $[\alpha]$ denotes the largest integer $\leqslant \alpha$, and for any integer t, prime to m, let σ_t denote the automorphism of K_m which maps every m-th root of unity in K_m to its t-th power. We then define

$$\theta_m(a) = \sum_{\substack{t \bmod m \\ (t,m)=1}} \left\langle -\frac{at}{m} \right\rangle \sigma_t^{-1}$$

for any integer a. $\theta_m(a)$ is an element of $Q[G_m]$ such that $m\theta_m(a) \in R_m$, and it depends only upon the residue class of $a \bmod m$. Let $r = (r_1, \cdots, r_s)$ be any finite sequence of elements r_i in Q/Z such that $mr = 0$ and let $r_i = a_i/m$ (mod Z) with $a_i \in Z$, $1 \leqslant i \leqslant s$. For such a sequence r, we put

$$\gamma(m, r) = \sum_{i=1}^{s} \theta_m(a_i)\ .$$

Again $\gamma(m, r)$ is an element of $Q[G_m]$ such that $m\gamma(m, r) \in R_m$. In [2c], a

2) See [2b], [2c].

"modified" Gauss sum $J_m(r, \mathfrak{A})$ is defined for each sequence $r=(r_1, \cdots, r_s)$ with $mr=0$ and for each ideal $\mathfrak{A}$ of K_m, prime to m. It is an algebraic number of which the m-th power is contained in K_m, and

$$(J_m(r, \mathfrak{A})^m) = \mathfrak{A}^{m\gamma(m,r)}$$

for the principal ideal $(J_m(r, \mathfrak{A})^m)$ in K_m.

In general, let $Q \subseteq k' \subseteq k$ and let both k/Q and k'/Q be finite Galois extensions. Let $G = \mathrm{Gal}(k/Q)$, $G' = \mathrm{Gal}(k'/Q)$ and $R = Z[G]$, $R' = Z[G']$. Then the canonical homomorphism $f_{k/k'}: G \to G'$ induces a ring homomorphism $R \to R'$ and this will be denoted again by $f_{k/k'}$. There also exists an additive homomorphism $f_{k'/k}: R' \to R$ which maps each σ' in G' to the sum of all σ in G such that $f_{k/k'}(\sigma) = \sigma'$. $f_{k/k'}$ and $f_{k'/k}$ can be extended to homomorphisms $Q[G] \to Q[G']$ and $Q[G'] \to Q[G]$ respectively in the obvious manner. When one or both of k and k' are cyclotomic fields, e.g., $k = K_m$, we shall write simply $f_{m/k'}$ and $f_{k'/m}$ for $f_{k/k'}$ and $f_{k'/k}$ respectively.

Now, let k be an arbitrary finite abelian extension over the rational field Q and let m be any positive integer. Let k' be any subfield of $k \cap K_m$ and let $r=(r_1, \cdots, r_s)$ be a sequence such that $mr=0$ and such that $f_{m/k'}(\gamma(m,r))$ is contained in the group ring $R' = Z[G']$ of $G' = \mathrm{Gal}(k'/Q)$[3]. Then $J_m(r, N_{k/k'}(\mathfrak{a}))$ is contained in k (in fact, in k') for any ideal $\mathfrak{a}$ of k, prime to m. Such an element of k is called a (generalized) Jacobi sum for the field k because in the special case where $k = K_m = k'$, it coincides with a classical Jacobi sum for K_m studied in [2b]. Let a be any integer. The main theorem in [2c] states that there is a Hecke character χ of type (A_0) on k which induces the homomorphism $\tilde{\chi}: \mathfrak{I}_{2m} \longrightarrow C^\times$ defined by

$$\tilde{\chi}(\mathfrak{a}) = J_m(r, N_{k/k'}(\mathfrak{a}))N_{k/Q}(\mathfrak{a})^a .$$

From now on, let us assume that k is an imaginary abelian extension over Q. We then see that $k_\chi = k$ for the Hecke character χ mentioned above and that $\omega_\chi = \varphi(\chi)$ in $R = Z[G]$ is given by

$$\delta_k(m, k', r, a) = f_{k'/k}(f_{m/k'}(\gamma(m,r)) + a\theta$$

where θ denotes as before the sum of all elements in $G = \mathrm{Gal}(k/Q)$. With k fixed, let S denote the submodule of R generated over Z by such $\delta_k(m, k', r, a)$ for all possible choices of m, k', r, and a (i.e., $m, a \in Z$, $m \geqslant 1$, $k' \subseteq k \cap K_m$,

3) One checks easily that this condition on r is equivalent to the condition $d|r|=0$ in the lemma on p. 6 of [2c].

$mr = 0$, and $f_{m/k'}(\gamma(m, r)) \in R' = Z[G']$ where $G' = \mathrm{Gal}(k'/Q)$. Then S is an ideal of R contained in B_k:

$$S \subseteq B_k \subseteq A_k \subseteq A \subseteq R .$$

The elements in S may be called Stickelberger operators for k because in the special case where $k = K_m$, they appear in the classical theorem of Stickelberger. Using the fact that Dirichlet's L-functions $L(s; \psi)$ do not vanish at $s = 1$, we can show that

$$[A : S] < +\infty .$$

Now the question arises: What are the indices

$$[A : S], \ [A_k : S] = [A_k : B_k][B_k : S] .$$

In the simplest case, namely, in the case where k is an imaginary quadratic field, one can compute $[A : A_k]$ and $[A : S]$ without much difficulty and find that $[A : A_k]$ is the exponent of the ideal class group C_k of k while $[A : S]$ is the order of C_k, namely, the class number of k. Hence, in general, A_k/S is not a 2-group like A_k/B_k. On the other hand, one can also prove that if k is the cyclotomic field of m-th roots of unity, $k = K_m$, and if m is divisible by at most two distinct prime numbers, then $[A : S]$ is equal to the first factor of the class number of k[4]. Although the number of known examples is limited, this seems to suggest that the same equality might hold for any imaginary abelian extension k over Q or at least for all cyclotomic fields K_m[5]. The proof of such a conjecture, if true, may require some intrinsic characterization of the elements of S among the relations on C_k given by the elements of A_k or B_k, and such a characterization in turn may enable us to define Stickelberger operators for an arbitrary Galois j-field k which is not necessarily abelian over Q.

Bibliography

[1] Iwasawa, K., A class number formula for cyclotomic fields, Ann. Math. 76 (1962), 171–179.

[2] Weil, A., (a) On a certain type of characters of the idele-class group, Proc. Int. Symp. Alg. Number Theory, Tokyo-Nikko, 1955, 1–7; (b) Jacobi sums as "Grössencharak-

4) See [1] for the case where m is a power of an odd prime.

5) *Added in proof*. Indeed, such an equality for K_m has since been proved by W. M. Sinnott with an additional factor which is a power of 2.

tere", Trans. Amer. Math. Soc. 73 (1952), 487–495; (c) Sommes de Jacobi et caracteres de Hecke, Nachrich. der Akad. Göttingen 1974, 1–14.

Department of Mathematics
Princeton University
Princeton, N.J. 08540
U.S.A.

[57] On p-adic representations associated with $\mathbb{Z}_p$-extensions

"Automorphic Forms, Representation Theory and Arithmetic" (Bombay, 1979), Tata Inst. Fund. Res., Bombay, 1981, pp. 141-153.

IN THE PRESENT paper, we shall discuss some results on the p-adic representations of Galois groups, associated with so-called cyclotomic $\mathbb{Z}_p$-extensions of finite algebraic number fields.

1. Let p be a prime number which will be fixed throughout the following, and let $\mathbb{Z}_p$ and $\mathbb{Q}_p$ denote the ring of p-adic integers and the field of p-adic numbers respectively. A Galois extension K/k is called a $\mathbb{Z}_p$-extension if its Galois group is isomorphic to the additive group of the compact ring $\mathbb{Z}_p$ [1]. Let Ω denote the field of all algebraic numbers, i.e., the algebraic closure of the rational field $\mathbb{Q}$ in the field $\mathbb{C}$ of all complex numbers, and let W_∞ be the group of all p^n-th roots of unity in Ω for all $n \geqslant 0$. Then the field $\mathbb{Q}(W_\infty)$ contains a unique subfield $\mathbb{Q}_\infty$ which is a $\mathbb{Z}_p$-extension over $\mathbb{Q}$. In fact, $\mathbb{Q}_\infty$ is the unique $\mathbb{Z}_p$-extension over $\mathbb{Q}$ contained in Ω, and the degree of the extension $\mathbb{Q}(W_\infty)/\mathbb{Q}_\infty$ is either $p-1$ or 2 according as $p > 2$ or $p = 2$. For any finite extension k of $\mathbb{Q}$, the composite $k_\infty = k\mathbb{Q}_\infty$ is then a $\mathbb{Z}_p$-extension over k and it is called the cyclotomic $\mathbb{Z}_p$-extension over k. For each integer $n \geqslant 0$, there then exists a unique intermediate field k_n with $[k_n : k] = p^n$, and

$$k = k_0 \subset k_1 \subset \ldots \subset k_n \subset \ldots \subset k_\infty = \bigcup_{n \geq 0} k_n .$$

Let C_n denote the Sylow p-subgroup of the ideal class group of k_n. For $n \leqslant m$, $k_n \subseteq k_m$, there exists a natural homomorphism $C_n \to C_m$, and these homomorphisms define the direct limit

$$C_\infty = \varinjlim C_n .$$

Clearly C_∞ is a p-primary abelian group and its Tate module $T(C_\infty)$ is a $\mathbb{Z}_p$-module. It is known that

$$T(C_\infty) \simeq \mathbb{Z}_p^\lambda$$

---

1) For various definitions and results on $\mathbb{Z}_p$-extensions referred to throughout the following, see Iwasawa [5] or Lang [6].

where $\lambda = \lambda_p(k)$ is a non-negative integer, called the λ-invariant of k for the prime number p. Hence

$$V = T(C_\infty) \otimes_{Z_p} Q_p$$

is a λ-dimensional vector space over Q_p. Let

$$\Gamma = \mathrm{Gal}(k_\infty/k) = \varprojlim \mathrm{Gal}(k_n/k)$$

so that $\Gamma \simeq Z_p$. Clearly $\mathrm{Gal}(k_n/k)$ acts on C_n for each $n \geqslant 0$ and hence Γ acts on $C_\infty = \varinjlim C_n$ in the natural manner. Therefore Γ acts also on $T(C_\infty)$ and V. Thus we have a natural continuous finite dimensional p-adic representation of the Galois group $\Gamma = \mathrm{Gal}(k_\infty/k)$ on the λ-dimensional vector space V over Q_p. We shall next investigate the properties of the p-adic representation space V for Γ.

2. Let us first consider the special case where $p > 2$ and where $k = Q(\sqrt[p]{1})$ = the cyclotomic field of p-th roots of unity.

Let

$$K = k_\infty = kQ_\infty = Q(W_\infty).$$

In this case, K/Q is an abelian extension and

$$G = \mathrm{Gal}(K/Q) = \Gamma \times \Delta$$

where $\Gamma = \mathrm{Gal}(K/k) \simeq Z_p$ and where $\Delta = \mathrm{Gal}(K/Q_\infty) = \mathrm{Gal}(k/Q)$ is a cyclic group of order $p - 1$. Let $\hat{\Delta}$ denote the character group of Δ; we may identify $\hat{\Delta}$ with $\mathrm{Hom}(\Delta, Z_p^\times)$ where $Z_p^\times$ denotes the multiplicative group of all p-adic units in Q_p. It is well known that $\hat{\Delta}$ may be identified also with the group of all Dirichlet characters to the modulus p and that it is generated by a special character ω called the Teichmuller character for p. A character χ in $\hat{\Delta}$ is called even or odd according as $\chi(-1) = 1$ or $\chi(-1) = -1$ respectively.

As one sees immediately, in this special case, not only $\Gamma = \mathrm{Gal}(K/k)$ but $G = \mathrm{Gal}(K/Q)$ also acts on C_∞, $T(C_\infty)$, and $V = T(C_\infty) \otimes_{Z_p} Q_p$ naturally. Hence V is again a p-adic representation space for G. For each χ in $\hat{\Delta}$, let

$$V_\chi = \{\, v \mid v \in V,\ \delta . v = \chi(\delta) v \text{ for all } \delta \text{ in } \Delta \,\}.$$

Since $G = \Gamma \times \Delta$, V_χ is then a Γ-subspace of V and

$$V = \bigotimes_\chi V_\chi, \quad \chi \in \hat{\Delta}.$$

Let γ_0 denote the element of Γ such that $\gamma_0(\zeta) = \zeta^{1+p}$ for all ζ in W_∞. γ_0 is a topological generator of Γ; namely, the cyclic subgroup generated by γ_0 is dense in Γ. For each χ in $\hat{\Delta}$, let

$$g_\chi(X) = \text{the characteristic polynomial of } \gamma_0 - 1 \text{ acting on } V_\chi$$

and let

$$g(X) = \text{the characteristic polynomial of } \gamma_0 - 1 \text{ acting on V} = \prod_\chi g_\chi(X).$$

On the other hand, let $L_p(s;\chi)$ denote the p-adic L-function for the Dirichlet character χ in $\hat{\Delta}$. It is known in the theory of p-adic L-functions[2] that for each such χ, there exists a power series $\xi_\chi(T)$ in the ring $\mathbf{Z}_p[[T]]$ of all formal power series in an indeterminate T with coefficients in $\mathbf{Z}_p$ such that

$$L_p(s;\chi) \begin{cases} = \xi_{\omega\chi^{-1}}((1+p)^s - 1) & , \text{for } \chi \neq 1, s \in \mathbf{Z}_p, \\ = \xi_\omega((1+p)^s - 1)/((1+p)^{1-s} - 1) & , \text{for } \chi = 1, s \in \mathbf{Z}_p, s \neq 1. \end{cases}$$

Since $L_p(s;\chi) \not\equiv 0$ if χ is even but $L_p(s;\chi) \equiv 0$ if χ is odd, $\xi_\chi(T) \equiv 0$ if χ is even and $\xi_\chi(T) \not\equiv 0$ if χ is odd. By Weierstrass' preparation theorem, $\xi_\chi(T)$ for odd χ can be uniquely written in the form

$$\xi_\chi(T) = \eta_\chi(T) p^{e_\chi} f_\chi(T)$$

where $\eta_\chi(T)$ is an invertible power series in the ring $\mathbf{Z}_p[[T]]$, e_χ is a non-negative integer[3], and $f_\chi(T)$ is a so-called distinguished polynomial in $\mathbf{Z}_p[T]$. The next theorem tells us that there exists a relation between the p-adic representation of $\Gamma = \mathrm{Gal}(K/k)$ on V and the p-adic L-functions $L_p(s;\chi)$ for the characters χ in $\hat{\Delta}$, or, more precisely, between the polynomials $g_\chi(X)$ and $f_\chi(T)$ defined above. Namely, we have the following result[4]:

2) See Iwasawa [4] or Lang [6].

3) A recent theo:em of B. Ferrero and L. Washington implies that $e_\chi = 0$ for all odd χ.

4) See Iwasawa [3].

THEOREM *1. Let k^+ denote the maximal real subfield of the cyclotomic field $k = Q(\sqrt[p]{1})$ and let h^+ be the class number of k^+. Assume that h^+ is not divisible by p. Then*

$$g_\chi(X) = 1, V_\chi = 0, \quad \text{for all even } \chi \text{ in } \hat{\Delta},$$
$$g_\chi(X) = f_\chi(X), \quad \text{for all odd } \chi \text{ in } \hat{\Delta}.$$

The assumption $p \nmid h^+$ in the theorem is known as Vandiver's conjecture. It has been verified by numerical computation for all primes $p < 125{,}000$, and no counter example is yet found. On the other hand, if we define, following Leopoldt, the p-adic zeta function $\zeta_p(s; k^+)$ of the totally real field k^+ by

$$\zeta_p(s;k^+) = \prod_\chi{}^+ L_p(s;\chi), \quad \chi \in \hat{\Delta}, \chi(-1) = 1,$$

then the theorem implies that under the assumption $p \nmid h^+$, $\zeta_p(s;k^+)$ is essentially equal to the characteristic polynomial g(X) of $\gamma_0 - 1$ acting on the representation space V over Q_p, up to the change of variables $s \to (1+p)^s$. The result is mysteriously analogous to a well known theorem of A. Weil which states that a similar relation exists between the zeta function of an algebraic curve defined over a finite field and the characteristic polynomial of the Frobenius endomorphism acting on the p-adic representation space defined by the Jacobian variety of that curve.

Now, although the above theorem is proved only for a very special case (and even that under the assumption $p \nmid h^+$), we feel that it is not just an isolated fact for $k = Q(\sqrt[p]{1})$, but is rather a part of a much more general result on the cyclotomic Z_p-extensions over finite algebraic number fields. In fact, Greenberg [2] generalizes Theorem 1 to the case where the ground field k is a certain type of finite abelian extension over the rational field Q, and Coates [1] also discusses such a generalization for an abelian extension k of an arbitrary totally real field. In the following, we shall report some results on cyclotomic Z_p-extensions, related to some further generalization of the above Theorem 1.

3. We now assume that p is an odd prime, $p > 2$[5], and consider as our

5) The case $p = 2$ can be treated similarly but with some modifications.

ground field a finite algebraic number field k with the following properties:

i) k is a Galois extension of the rational field,

ii) k contains primitive p-th roots of unity so that it is a totally imaginary field,

iii) k also contains a totally real subfield k^+ with $[k:k^+] = 2$; namely, k is a number field of C-M type.

In general, let J denote the automorphism of the complex field $\mathbb{C}$ which maps each complex number α to its complex conjugate $\bar{\alpha}$. For simplicity, the restriction of J on any subfield of $\mathbb{C}$, invariant under J, will be denoted again by J. Let

$$\Delta = \mathrm{Gal}(k/\mathbb{Q})$$

for the field k mentioned above. Then by ii) and iii), J is an element in the center of Δ and $J \neq 1$, $J^2 = 1$. As in §1, let $K = k_\infty = k\mathbb{Q}_\infty$ denote the cyclotomic $\mathbb{Z}_p$-extension over k. Since k contains p-th roots of unity, $K = k(W_\infty)$. Similarly, let $K^+ = k_\infty^+ = k^+\mathbb{Q}_\infty$ be the cyclotomic $\mathbb{Z}_p$-extension over k^+. Then K^+ is a totally real subfield of the totally imaginary field K with $[K:K^+] = 2$. Clearly $K/\mathbb{Q}$ is a Galois extension because both $k/\mathbb{Q}$ and $\mathbb{Q}_\infty/\mathbb{Q}$ are Galois extensions. Hence, let

$$G = \mathrm{Gal}(K/\mathbb{Q}), \quad \Gamma = \mathrm{Gal}(K/k) \simeq \mathbb{Z}_p .$$

Then we see immediately that Γ is a central subgroup of G and

$$\Delta = \mathrm{Gal}(k/\mathbb{Q}) = G/\Gamma.$$

As in the special case of §2, the Galois group G acts on C_∞, $T(C_\infty)$, and $V = T(C_\infty) \otimes_{\mathbb{Z}_p} \mathbb{Q}_p$ so that V provides us with a finite dimensional p-adic representation space for $G = \mathrm{Gal}(K/\mathbb{Q})$.

THEOREM 2. *Assume that $\lambda_p(k^+) = 0$ and that the so-called Leopoldt's conjecture holds for all intermediate fields k_n^+, $n \geqslant 0$, of the extension K^+/k^+. Then $V = T(C_\infty) \otimes_{\mathbb{Z}_p} \mathbb{Q}_p$ is cyclic over $G = Gal(K/\mathbb{Q})$; namely, there exists a vector v_0 in V such that the whole space V is spanned over $\mathbb{Q}_p$ by the vectors $\sigma \cdot v_0$, $\sigma \in G$.*

Recall that $\lambda_p(k^+)$ denotes the λ-invariant of the totally real field k^+ for the prime p and that Leopoldt's conjecture for k_n^+ states that any set of units in k_n^+, multiplicatively linearly independent over the ring of rational integers $\mathbb{Z}$, remains multiplicatively linearly independent over $\mathbb{Z}_p$ when these units are imbedded in the multiplicative group of the algebra $k^+ \otimes_{\mathbb{Q}} \mathbb{Q}_p$. We note that both these assumptions are conjectured to be true for any totally real number field k^+. Note also that since $T(C_\infty) \simeq \mathbb{Z}_p^\lambda$, the conclusion of the theorem is equivalent to say that there exists an element v_0 in $T(C_\infty)$ such that the elements of the form $\sigma \cdot v_0$, $\sigma \in G$, generate over $\mathbb{Z}_p$ a submodule of finite index in $T(C_\infty)$. The proof of the theorem will be briefly indicated in the next section.

In general, let G be any profinite group and let $G = \varprojlim G_i$ with a family of finite groups $\{G_i\}$. The homomorphisms $G_j \to G_i$, $i \leqslant j$, which define the inverse limit, induce the homomorphisms $\mathbb{Z}_p[G_j] \to \mathbb{Z}_p[G_i]$ of the group rings of finite groups over $\mathbb{Z}_p$, and they in turn define

$$\mathbb{Z}_p[[G]] = \varprojlim \mathbb{Z}_p[G_i].$$

$\mathbb{Z}_p[[G]]$ is a compact topological algebra over $\mathbb{Z}_p$ and it depends only upon G and is independent of the family $\{G_i\}$ such that $G = \varprojlim G_i$.

We apply the above general remark for $G = \mathrm{Gal}(K/\mathbb{Q})$ in Theorem 2 and define

$$R = \mathbb{Z}_p[[G]], \quad R' = R \otimes_{\mathbb{Z}_p} \mathbb{Q}_p.$$

Let $G_n = \mathrm{Gal}(k_n/\mathbb{Q})$, $R_n = \mathbb{Z}_p[G_n]$, $n \geqslant 0$. Since $G = \varinjlim G_n$, we then have

$$R = \varprojlim R_n.$$

Since C_n is an R_n-module in the obvious manner, $C_\infty = \varprojlim C_n$ is an R-module. Hence $T(C_\infty)$ also is an R-module and $V = T(C_\infty) \otimes_{\mathbb{Z}_p} \mathbb{Q}_p$ is an R'-module. We next define a subset A_n of R_n by

$$A_n = \{\alpha \mid \alpha \in (1-J)R_n, \alpha \cdot C_n = 0\}.$$

Note that $J = J|k_n$ is contained in the center of G_n so that A_n is a two-sided ideal of R_n, contained in $(1-J)R_n$. Furthermore, if n is large enough and $m \geqslant n$, then the homomorphism $R_m \to R_n$ maps A_m into A_n. Therefore

$$A = \varprojlim A_n$$

is defined and it is a two-sided ideal of R, contained in $(1-J)R$. Let

$$A' = A \otimes_{Z_p} \mathbb{Q}_p.$$

Clearly A' is a two-sided ideal of $R' = R \otimes_{Z_p} \mathbb{Q}_p$, contained in $(1-J)R'$. Moreover, it can also be proved that

$$A' = \{\alpha' \mid \alpha' \in (1-J)R', \alpha' \cdot V = 0\},$$

namely, that A' is the annihilator of the R'-module V in $(1-J)R'$. Let

$$d = [k:\mathbb{Q}].$$

Using Theorem 2, we can then easily prove the following

THEOREM 3. *Let*

$$V'' = (1-J)R'/A'.$$

Under the same assumptions as in Theorem 2, there exist exact sequences of R'-modules

$$V'' \to V \to 0, \quad 0 \to V'' \to V^d.$$

In particular, V'' is a finite dimensional vector space over $\mathbb{Q}_p$, and as p-adic representation spaces for $G = Gal(K/\mathbb{Q})$, V and V'' have the same com position factors.

At this point, let us consider again the special case where $k = \mathbb{Q}(\sqrt[p]{1})$, $p > 2$; the field $\mathbb{Q}(\sqrt[p]{1})$ certainly satisfies the conditions i), ii), and iii) stated at the beginning of this section. In this case, $K = k_\infty$, $K^+ = k_\infty^+$, and k_n^+, $n \geqslant 0$, are all abelian extensions over $\mathbb{Q}$, and Leopoldt's conjecture for k_n^+ is known to be true by a theorem of Brumer. On the other hand, it is easy to deduce $\lambda_p(k^+) = 0$ from Vandiver's conjecture $p \nmid h^+$ for the class number h^+ of k^+. Therefore we know by Theorem 2 that under the assumption $p \nmid h^+$, V is cyclic over $G = Gal(K/\mathbb{Q})$, namely,

$$V = R'v_0$$

with some vector v_0 in V. Now, $\lambda_p(k^+) = 0$ also implies $V = (1-J)V$ so that $V = (1-J)R'v_0$. Since $G = Gal(K/\mathbb{Q})$ is an abelian group in this case, both R and R' are commutative rings. Hence it follows from the above that the map $\alpha' \to \alpha' v_0, \alpha' \in (1-J)R'$, induces an

R′-isomorphism

$$V' = (1-J)R'/A' \xrightarrow{\sim} V.$$

Furthermore, we know in this special case that there are many explicitly described elements in the ideal A_n of R_n, $n \geqslant 0$, called Stickelberger operators for k_n, and that the p-adic L-functions $L_p(s; \chi)$ for χ in $\hat{\Delta} = \mathrm{Hom}(\Delta, \Omega_p^\times)$ can be constructed by means of such Stickelberger operators[6]. Thus we obtain a relation between the p-adic representation space V′ and the p-adic L-functions $L_p(s; \chi)$, and hence between V and $L_p(s; \chi)$ through the above isomorphism. This is the way how Theorem 1 is proved, and the proof is similar for Greenberg's generalization.

We now consider again the general case where k is any finite algebraic number field satisfying the conditions i), ii), and iii). For each C-M subfield k′ of k such that k/k′ is abelian, Stickelberger operators for k_n/k' are still defined, and it is proved by Deligne and Ribet that such Stickelberger operators are related to abelian p-adic L-functions for $k' \cap k^+$ in much the same way as in the special case mentioned above. However, it is not known whether such general Stickelberger operators belong to the ideal A_n and provide us with any essential part of A_n defined above[7]. This prevents us from obtaining any nice relation between the p-adic representation space V′ and p-adic L-functions. On the other hand, we can find examples of $k/\mathbb{Q}$, satisfying i), ii), iii) and also the assumptions in Theorem 2, such that the representation spaces V and V′ for $G = \mathrm{Gal}(K/\mathbb{Q})$ in Theorem 3 are not isomorphic to each other. Thus we see that the results of Theorems 2, 3 tell us much less on the nature of the p-adic representation space V for $G = \mathrm{Gal}(K/\mathbb{Q})$ than Theorem 1 for the special case $k = \mathbb{Q}(\sqrt[p]{1})$. Nevertheless, we still feel and hope that those theorems would be of some use in the future investigations to obtain a full generalization of Theorem 1 in §2.

We also note in this connection that in such a generalization of Theorem 1, one has certainly to consider p-adic (non-abelian) Artin L-functions. Given any Galois extension L/K of totally real finite algebraic number fields, it is not difficult to define p-adic Artin L-function $L_p(s; \chi)$ for

6) See Iwasawa [4] or Lang [6].
7) See the discussions in Coates [1].

each character χ of the Galois group Gal (L/K) so that $L_p(s;\chi)$ is related to the classical Artin L-function $L(s;\chi)$ in the usual manner and that those $L_p(s;\chi)$ share with the classical functions $L(s;\chi)$ all essential formal properties such as the formula concerning induced characters. One can even formulate the p-adic Artin conjecture for such L-functions; the conjecture is not yet verified and, in fact, it is closely related to the above mentioned problem of generalizing Theorem 1. For all these, we refer the reader to forth-coming papers by R. Greenberg and B. Gross, noting here only that Weil's solution of Artin's conjecture for L-functions of algebraic curves defined over finite fields is based upon the study of the representations of Galois groups on the spaces similar to V mentioned above.

4. We shall next briefly indicate an outline of the proof of Theorem 2[8]. Following the general definition in § 3, let

$$\Lambda = \mathbf{Z}_p[[\Gamma]]$$

for the profinite group $\Gamma = \mathrm{Gal}(K/k)$, and let γ_0 be any topological generator of $\Gamma \simeq \mathbf{Z}_p$. Let $\mathbf{Z}_p[[T]]$ denote as in § 2 the ring of all formal power series in T with coefficients in $\mathbf{Z}_p$. Then it is known that there is a unique isomorphism of compact algebras over $\mathbf{Z}_p$:

$$\Lambda = \mathbf{Z}_p[[\Gamma]] \xrightarrow{\sim} \mathbf{Z}_p[[T]]$$

such that $\gamma_0 \to 1 + T$. Hence fixing a topological generator γ_0, we may identify $\Lambda = \mathbf{Z}_p[[\Gamma]]$ with $\mathbf{Z}_p[[T]]$ so that $\gamma_0 = 1 + T$. Then

$$\Lambda' = \Lambda \otimes_{\mathbf{Z}_p} \mathbf{Q}_p = \mathbf{Z}_p[[T]] \otimes_{\mathbf{Z}_p} \mathbf{Q}_p$$

and it is easy to see that Λ' is a principal ideal domain. One also proves immediately that $\Lambda = \mathbf{Z}_p[[\Gamma]]$ is a central subalgebra of $R = \mathbf{Z}_p[[G]]$ and that the latter is a free Λ-module of rank $d = [k:\mathbf{Q}] = [G:\Gamma]$. Hence $R' = R \otimes_{\mathbf{Z}_p} \mathbf{Q}_p$ is an algebra over $\Lambda' = \Lambda \otimes_{\mathbf{Z}_p} \mathbf{Q}_p$ and it is a free module of rank d over the principal ideal domain Λ'.

Now, let L denote the maximal unramified abelian p-extension (i.e., Hilbert's p-class field) over K, and M the maximal p-ramified abelian

8) Cf. the proof of Theorem 5 in Greenberg [2].

p-extension over K. Then

$$\mathbb{Q} \subseteq k \subseteq K \subseteq L \subseteq M$$

and both $L/\mathbb{Q}$ and $M/\mathbb{Q}$ are Galois extensions. Let

$$X = \mathrm{Gal}(L/K), \qquad Y = \mathrm{Gal}(M/K).$$

These are abelian pro-p-groups and, hence, are $\mathbb{Z}_p$-modules in the natural manner. Since $G = \mathrm{Gal}(K/\mathbb{Q})$ acts on X and Y in the obvious way, we see that both X and Y are R-modules and, consequently, also Λ-modules. It is known that X is a torsion Λ-module so that there is an element $\xi \neq 0$ in Λ such that

$$\xi \cdot X = 0.$$

Let

$$X' = X \otimes_{\mathbb{Z}_p} \mathbb{Q}_p.$$

It is clear that X' is an R'-module. However it is also known in the theory of $\mathbb{Z}_p$-extensions that

$$V = T(C_\infty) \otimes_{\mathbb{Z}_p} \mathbb{Q}_p \stackrel{\sim}{\rightarrow} X'$$

as modules over R'. Hence, in order to prove Theorem 2, we have only to show that X' is cyclic over R' under the assumptions of that theorem.

Let Y^- denote the submodule of all y in Y satisfying $(1 + J)y = 0$. Since $p > 2$, $Y^- = (1-J)Y$ and since $J = J|K$ is contained in the center of $G = \mathrm{Gal}(K/\mathbb{Q})$, Y^- is an R-submodule of Y. Let $t(Y^-)$ denote the torsion Λ-submodule of the Λ-module Y^- and let

$$Z = Y^-/t(Y^-), \qquad Z' = Z \otimes_{\mathbb{Z}_p} \mathbb{Q}_p.$$

Then Z is an R-module, and Z' an R'-module. Furthermore, we can prove by using the assumptions of Theorem 2 that there is an exact sequence of R'-modules

$$Z'/\xi Z' \rightarrow X' \rightarrow 0.$$

Therefore the proof is now reduced to show that $Z'/\xi Z'$ is cyclic over R'.

Let

$$R'^{-} = (1-J)R' = R'(1-J).$$

Then we have the following two lemmas:

LEMMA *1. Both Z' and R'^{-} are free Λ'-modules with the same rank $\frac{d}{2}$ and*

$$Z'/TZ' \simeq R'^{-}/TR'^{-}$$

as modules over R'.

LEMMA *2. Let A and B be R'-modules which are free and of the same finite rank over Λ', and let*

$$A/TA \simeq B/TB$$

as R'-modules. Then, as modules over R',

$$A/\mathfrak{p}A \simeq B/\mathfrak{p}B$$

for any non-zero prime ideal $\mathfrak{p}$ of the principal ideal domain Λ'.

That Z' is a free Λ'-module of rank $\frac{d}{2}$, where d = [k : Q], is a known fact in the theory of Z_p-extensions. The rest of Lemma 1 can be proved by considering the Galois group of the maximal p-ramified abelian p-extension over k. To see the proof of Lemma 2, let us first assume for simplicity that

$$k \cap Q_\infty = Q.$$

In this case, $G = \Gamma \times \Delta$ where $\Delta = \mathrm{Gal}(k/Q) = \mathrm{Gal}(K/Q_\infty)$, and $R = Z_p[[G]]$ is nothing but the group ring of the finite group Δ over $\Lambda = Z_p[[\Gamma]]$:

$$R = \Lambda[\Delta].$$

Hence

$$R' = \Lambda'[\Delta]$$

where $\Lambda' = \Lambda \otimes_{Z_p} Q_p$ is a principal ideal domain. The lemma then follows easily from the results of Swan on the group ring of finite groups over Dedekind domains[9]. The case $k \cap \mathbf{Q}_\infty \neq \mathbf{Q}$ can be proved similarly by reducing it to the above mentioned special case.

Now, since $R'^- = R'(1-J)$ is cyclic over R', we see from the above two lemmas that Z'/pZ' is cyclic over R' for all p as stated in Lemma 2. As ξ is a non-zero element of the principal ideal domain Λ', it follows that $Z'/\xi Z'$ also is cyclic over R'. This completes the proof of Theorem 2.

Finally, we would like to mention here also the following result which can be proved by similar arguments as described above. Namely, changing the notations from the above, let

k = an arbitrary (totally) real finite Galois extension over $\mathbf{Q}$,
$K = k_\infty = k\mathbf{Q}_\infty$ = the cyclotomic $\mathbf{Z}_p$-extension over k,
L' = the maximal unramified abelian p-extension over K in which every p-spot of K is completely decomposed,
M = the maximal p-ramified abelian p-extension over K.

Then, again,

$$\mathbf{Q} \subseteq k \subseteq K \subseteq L' \subseteq M$$

and $K/\mathbf{Q}$, $L'/\mathbf{Q}$, and $M/\mathbf{Q}$ are Galois extensions. Let

$$G = \mathrm{Gal}(K/\mathbf{Q}), \quad R = \mathbf{Z}_p[[G]], \quad R' = R \otimes_{\mathbf{Z}_p} \mathbf{Q}_p.$$

As in the case discussed above, the Galois groups Gal(M/K) and Gal(M/L′) are modules over R so that $\mathrm{Gal}(M/K) \otimes_{\mathbf{Z}_p} \mathbf{Q}_p$ and $\mathrm{Gal}(M/L') \otimes_{\mathbf{Z}_p} \mathbf{Q}_p$ are R'-modules.

THEOREM 4. *$\mathrm{Gal}(M/L') \otimes_{\mathbf{Z}_p} \mathbf{Q}_p$ is cyclic over R'. If in particular $\lambda_p(k) = 0$, then $\mathrm{Gal}(M/K) \otimes_{\mathbf{Z}_p} \mathbf{Q}_p$ also is cyclic over R'.*

The first part of the theorem is proved without any assumption, and the second part without assuming Leopoldt's conjecture. Hence the theorem might be more useful in some applications than Theorem 2.

9) See Swan [7].

P-ADIC REPRESENTATIONS

BIBLIOGRAPHY

[1] COATES, J. p-adic L-functions and Iwasawa's theory, *Algebraic Number Fields*, Acad. Press, London, 1977, 269–353.

[2] GREENBERG, R. On the Iwasawa invariants of totally real number fields, *Amer. Jour. Math.* 93(1976), 263–284.

[3] IWASAWA, K. On p-adic L-functions, *Ann. Math.* 89(1969), 198–205.

[4] IWASAWA, K. *Lectures on* p-*adic* L-*functions*, Princeton University Press, Princeton, 1972.

[5] IWASAWA, K. On $\mathbb{Z}_l$-extensions of algebraic number fields, *Ann. Math.* 98(1973), 246–326.

[6] LANG, S. *Cyclotomic fields*, Springer-Verlag, New York-Heidelberg-Berlin, 1978.

[7] SWAN, R. Induced representations and projective modules, *Ann. Math.* 71 (1960), 552–578.

[58] Riemann-Hurwitz formula and p-adic Galois representations for number fields

Tôhoku Math. J., (2) 33 (1981), 263-288.

(Received July 29, 1980)

Let $f: R' \to R$ be an n-fold covering of compact, connected Riemann surfaces and let g and g' denote the genara of R and R' respectively. The classical formula of Riemann-Hurwitz then states that

$$2g' - 2 = (2g - 2)n + \sum (e(P') - 1)$$

where the sum is taken over all points P' on R' and $e(P')$ denotes the ramification index of P' for the covering f. In a recent paper [6], Kida proved a highly interesting analogue of the above forumla for algebraic number fields.[1] In the present paper, we shall give an alternate proof for the theorem of Kida from a different point of view. Namely, assuming that the covering f is regular, let G denote the group of all covering transformations for f. Then the finite group G acts naturally on the space of all differentials of the first kind on R', and the representation of G thus defined was completely determined by Chevalley-Weil [2]. In the following, we shall study certain p-adic representations of Galois groups which may be regarded as analogues, for algebraic number fields, of the representation of G mentioned above, and we shall prove a result for such p-adic representations, quite similar to the theorem of Chevalley-Weil for the representation of G. The formula of Kida will then follow from this by comparing the degrees of the representations. Our proof is based essentially upon Galois cohomology theory for algebraic number fields which are not necessarily finite over the rational field. Hence some preliminary results in that theory will be discussed in the earlier part of the paper.[2] In the last section of the paper, we shall also indicate briefly another approach to Kida's formula which is slightly different from what is described above; this might be of some interest because it applies also for, e.g., totally real algebraic number fields.

1. Throughout the following, let $\boldsymbol{Z}$, $\boldsymbol{Q}$, $\boldsymbol{R}$, and $\boldsymbol{C}$ denote the ring of rational integers, the field of rational numbers, the field of real numbers, and that of complex numbers, respectively. By a number field K, we

[1] See the formula (10) in § 9 below.

[2] See also [1], [9] for Galois cohomology theory.

shall mean any algebraic extension of $\boldsymbol{Q}$ in $\boldsymbol{C}$, not necessarily finite over $\boldsymbol{Q}$. For such a field K, let $\mathfrak{o}_K$ denote the ring of all algebraic integers in K. The invertible $\mathfrak{o}_K$-submodules of K are called the ideals of K, and they form a multiplicative group I_K, the ideal group of K. Let P_K denote the subgroup of principal ideals (α), $\alpha \in K^\times$. The ideal-class group $C_K = I_K/P_K$ is then a torsion abelian group so that

$$C_K = \bigoplus_q C_K(q)$$

where q ranges over all prime numbers and $C_K(q)$ denotes the q-primary component of C_K. If K is a subfield of a number field L, there is a natural imbedding $I_K \to I_L$, and it induces homomorphisms $C_K \to C_L$, $C_K(q) \to C_L(q)$. Let v be a finite, i.e., non-archimedean, place on K and let $\mathfrak{p}_v$ denote the associated maximal ideal of $\mathfrak{o}_K$. If $K/\boldsymbol{Q}$ is finite, then $\mathfrak{p}_v$ belongs to I_K and generates a cyclic subgroup $\langle \mathfrak{p}_v \rangle$ of I_K. For infinite $K/\boldsymbol{Q}$, this is not true in general. However, in some special cases, we can still define a subgroup I_v of I_K, similar to $\langle \mathfrak{p}_v \rangle$ mentioned above. Namely, assume that K has a subfield k, finite over $\boldsymbol{Q}$, such that v is the unique extension of $v|k$ on K, $v|k$ being the restriction (projection) of v on the subfield k. Let $K = \varinjlim k_i$ where $k \subseteq k_i \subseteq K$ and $k_i/\boldsymbol{Q}$ is finite, and let $v_i = v|k_i$. Then $\langle \mathfrak{p}_{v_i} \rangle$ is contained in $\langle \mathfrak{p}_{v_j} \rangle$ for $k_i \subseteq k_j$ so that a subgroup I_v of $I_K = \varinjlim I_{k_i}$ is defined by

$$I_v = \varinjlim \langle \mathfrak{p}_{v_i} \rangle .$$

It is isomorphic to a subgroup of the additive group of $\boldsymbol{Q}$, containing $\boldsymbol{Z}$. In particular, if there exists a subfield k such that v is unramified for the extension K/k, then $\mathfrak{p}_v$ belongs to I_K and $I_v = \langle \mathfrak{p}_v \rangle \cong \boldsymbol{Z}$. We note here in passing that the ramification theory can be reasonably extended to places on extensions of number fields which are not necessarily finite over $\boldsymbol{Q}$; for example, the ramification indices are defined for such places so that they agree with the classical definition for finite algebraic number fields and satisfy the chain rule for $K \subseteq L \subseteq M$.

Let p be a fixed prime number and let $\boldsymbol{Z}_p$ and $\boldsymbol{Q}_p$ denote the ring of p-adic integers and the field of p-adic numbers respectively. Let $\boldsymbol{Q}_\infty$ be the unique $\boldsymbol{Z}_p$-extension over $\boldsymbol{Q}$ in $\boldsymbol{C}$.[3)] For each number field k, finite over $\boldsymbol{Q}$, the composite $k_\infty = k\boldsymbol{Q}_\infty$ is then a $\boldsymbol{Z}_p$-extension over k, and it is clearly a finite extension of $\boldsymbol{Q}_\infty$. Conversely, if K is a finite extension over $\boldsymbol{Q}_\infty$ in $\boldsymbol{C}$, then there exists k, finite over $\boldsymbol{Q}$, such that $K = k_\infty$. In the following, we shall call such a number field K simply a $\boldsymbol{Z}_p$-field.

---

3) For $\boldsymbol{Z}_p$-extensions of algebraic number fields, see [5] and the papers in the bibliography of [5].

Let q be any prime number and let v be a q-place on a Z_p-field K, i.e., a finite place on K with $v|Q = q$. From $K = k_\infty = kQ_\infty$, we then see easily that the subgroup I_v of I_K is defined as explained above and that if $q \neq p$, then $I_v = \langle \mathfrak{p}_v \rangle \cong Z$, and if $q = p$, then I_v is isomorphic to the union $\bigcup_{n \geq 0} p^{-n}Z$. Furthermore, for each prime number q, there exist only a finite number of q-places on K, and

$$I_K = \bigoplus_v I_v$$

where v ranges over all finite places on K.

Now, let L be a finite Galois extension of a Z_p-field K with $G = \mathrm{Gal}(L/K)$; L itself is then a Z_p-field. For each finite place v on K, let

$$I_{L,v} = \bigoplus_w I_w$$

where w runs over all extensions of v on L and where I_w denotes the subgroup of I_L defined similarly as I_v for K. Clearly $I_{L,v}$ is a subgroup of I_L, invariant under G, and

$$I_L = \bigoplus_v I_{L,v}$$

with v ranging over all finite places of K. Writing $H^n(L/K, \ )$ for the cohomology group $H^n(G, \ )$, we then have

$$H^n(L/K, I_L) = \bigoplus_v H^n(L/K, I_{L,v}) , \quad \text{for} \quad n \geqq 0 .$$

We shall next consider $H^n(L/K, I_{L,v})$ for each finite place v on K.

First, let v be a non-p-place, i.e., $v|Q = q \neq p$. For an extension w of v on L, we can define the decomposition group $Z = Z(w/v)$ and the inertia group $T = T(w/v)$ as usual so that $T \subseteq Z \subseteq G$. Let $e = e(w/v) = [T: 1]$, $f = f(w/v) = [Z: T]$, $g = g(w/v) = [G: Z]$ so that $efg = [G: 1] = [L: K]$. Then, as in the classical theory for finite algebraic number fields, one proves the following:

(a) All extensions of v on L are given by $\sigma(w)$, $\sigma \in G$, so that g is the number of distinct extensions of v on L. Hence, as G-modules,

$$I_{L,v} \xrightarrow{\sim} Z[G/Z]$$

where $Z[G/Z]$ denotes the module of all linear combinations of the left cosets of G modulo Z with coefficients in Z and where the isomorphism is defined by $\sigma(\mathfrak{p}_w) \mapsto \sigma Z$, for $\sigma \in G$. Consequently

$$H^n(L/K, I_{L,v}) \xrightarrow{\sim} H^n(G, Z[G/Z]) \xrightarrow{\sim} H^n(Z, Z) , \quad n \geqq 0$$

with trivial Z-action on Z.[4)]

[4)] See [7], Chap. II, Théorème 6.

(b) Let $w_1, \cdots, w_g$ be the distinct extensions of v on L. Then, in the ideal group I_L,

$$\mathfrak{p}_v = \prod_\sigma \sigma(\mathfrak{p}_w) = \left(\prod_{i=1}^{g} \mathfrak{p}_{w_i}\right)^e, \quad \sigma \in G.$$

Hence

$$I_{L\,v}^G = H^0(L/K, I_{L,v}) = \langle \mathfrak{p}_v^{1/e} \rangle \cong \boldsymbol{Z}.$$

(c) T is a normal subgroup of Z and

$$Z/T \simeq \mathrm{Gal}\,(\mathfrak{k}_w/\mathfrak{k}_v)$$

where $\mathfrak{k}_v = \mathfrak{o}_K/\mathfrak{p}_v$ and $\mathfrak{k}_w = \mathfrak{o}_L/\mathfrak{p}_w$ are the residue fields of v and w respectively. In particular, $f = [\mathfrak{k}_w : \mathfrak{k}_v]$. Note that both $\mathfrak{k}_v$ and $\mathfrak{k}_w$ are algebraic extensions of the finite field F_q with q elements and that $[\mathfrak{k}_v : F_q]$ is divisible by p^∞ because $K = k_\infty$.

Assume now that L/K is a p-extension. Then Z/T is a p-group, and the remark just mensioned above implies that $f = [\mathfrak{k}_w : \mathfrak{k}_v] = 1$, $Z = T$. However, since $v | Q \neq p$, v is tamely ramified in L so that T is a cyclic group of order e. Hence, in this case, we obtain from (a) above that

$$\begin{aligned} H^n(L/K, I_{L,v}) &\cong \boldsymbol{Z}/e\boldsymbol{Z}, \quad \text{for all even} \quad n \geqq 2, \\ &= 0, \quad \text{for all odd} \quad n \geqq 1, \end{aligned}$$

while $H^0(L/K, I_{L,v})$ is given in general by (b).

Next, let v be a p-place on K. Then $I_{L,v}$ is uniquely divisible by p as mentioned earlier. Hence, for a p-extension L/K, we see easily that

$$\begin{aligned} I_{L,v}^G &= H^0(L/K, I_{L,v}) = I_v, \\ &= H^n(L/K, I_{L,v}) = 0, \quad \text{for} \quad n \geqq 1. \end{aligned}$$

2. Let K be again a $\boldsymbol{Z}_p$-field and let L/K be a p-extension with $G = \mathrm{Gal}\,(L/K)$. We do not assume here that L/K is a finite extension. Hence L is not necessarily a $\boldsymbol{Z}_p$-field. However, if $\{L_i\}$ denotes the family of all finite Galois extensions of K contained in L, then each L_i is a $\boldsymbol{Z}_p$-field and

$$L = \varinjlim L_i.$$

For a finite place v on K, let

$$I_{L,v} = \varinjlim I_{L_i,v}$$

where $I_{L_i,v}$ is defined as in § 1 and where the limit is taken with respect to $I_{L_i,v} \to I_{L_j,v}$ for $L_i \subseteqq L_j$. Then $I_{L,v}$ is a subgroup of I_L, invariant under G, and

$$I_L = \bigoplus_v I_{L,v}.$$

Hence

$$H^n(L/K, I_L) = \bigoplus_v H^n(L/K, I_{L,v}),$$

$$H^n(L/K, I_{L,v}) = \varinjlim H^n(L_i/K, I_{L_i,v}), \quad n \geqq 0.$$

LEMMA 1. *Let $v|Q = p$. Then*

$$I_{L,v}^G = H^0(L/K, I_{L,v}) = I_v,$$
$$H^n(L/K, I_{L,v}) = 0, \quad \textit{for} \quad n \geqq 1.$$

Let $v|Q \neq p$. Let w be an extension of v on L and let $e = e(w/v)$ denote the ramification index of w/v, i.e., the order of the inertia group $T(w/v)$. If e is finite, then $T(w/v) \cong Z/eZ$ and

$$I_{L,v}^G = H^0(L/K, I_{L,v}) = \langle \mathfrak{p}_v^{1/e} \rangle \cong Z,$$
$$H^n(L/K, I_{L,v}) \cong Z/eZ, \quad \textit{for even} \quad n \geqq 2,$$
$$= 0, \quad \textit{for odd} \quad n \geqq 1.$$

On the other hand, if e is infinite, then $T(w/v) \cong Z_p$ and

$$I_{L,v}^G = H^0(L/K, I_{L,v}) = \{\mathfrak{p}_v^r | r \in R\} \cong R = \bigcup_{n \geqq 0} p^{-n}Z,$$
$$H^n(L/K, I_{L,v}) = 0, \quad \textit{for} \quad n \geqq 1.$$

PROOF. The first part is an immediate consequence of the results for $H^n(L_i/K, I_{L_i,v})$ mentioned in §1. Let $v|Q \neq p$ and let $w_i = w|L_i$, $e_i = e(w_i/v)$, $T_i = T(w_i/v) = Z(w_i/v)$ for each i. Then $T = T(w/v) = \varprojlim T_i$, and the argument in §1 shows that the following diagram is commutative for $L_i \subseteq L_j$:

$$\begin{array}{ccc} H^n(L_i/K, I_{L_i,v}) & \xrightarrow{\sim} & H^n(T_i, Z) \\ \downarrow & & \downarrow \\ H^n(L_j/K, I_{L_j,v}) & \xrightarrow{\sim} & H^n(T_j, Z). \end{array}$$

Here the vertical map on the right is defined by the natural homomorphism $T_j \to T_i$ and by the endomorphism $a \mapsto (e_j/e_i)a$ of Z. Hence it follows that

$$H^n(L/K, I_{L,v}) \xrightarrow{\sim} H^n(T, \varinjlim Z), \quad n \geqq 0.$$

As each T_i is a cyclic group of order e_i, the statements in the second part are consequences of the above remarks and of the results in §1.

When $e(w/v)$ is infinite and $T(w/v) \cong Z_p$, we shall say that the finite non-p-place v is infinitely ramified in L; note that if w' is another extension of v on L, then $e(w'/v) = e(w/v)$, $T(w'/v) \cong T(w/v)$ so that the definition is in fact independent of the choice of w. The following result is then an immediate consequence of Lemma 1:

LEMMA 2. *Let L be a p-extension over a Z_p-field K with $G = \mathrm{Gal}(L/K)$. Suppose that each finite non-p-place on K is either unramified in L or infinitely ramified in L. Then*

$$H^n(L/K, I_L) = 0, \quad \text{for all} \quad n \geqq 1,$$

and

$$I_L^G/I_K = \bigoplus_v A_v$$

where v ranges over all finite non-p-places on K, ramified in L and where, for each v,

$$A_v = \mathfrak{p}_v^R/\mathfrak{p}_v \cong R/Z \cong Q_p/Z_p$$

with $R = \bigcup_{n \geq 0} p^{-n}Z$ in Q.

3. Let K be a Z_p-field, v a finite non-p-place on K, and $\mathfrak{k}_v$ the residue field of v: $\mathfrak{k}_v = \mathfrak{o}_K/\mathfrak{p}_v$. Since $\mathfrak{k}_v$ is an algebraic extension of a finite field and $K = k_\infty = kQ_\infty$ with k finite over Q, we see easily that the multiplicative group $\mathfrak{k}_v^\times$ is a torsion abelian group and that the order of $\mathfrak{k}_v^\times$ is either divisible by p^∞ or prime to p, i.e., the p-primary component of $\mathfrak{k}_v^\times$ is either infinite or the identity group.

LEMMA 3. *Let the order of $\mathfrak{k}_v^\times$ be prime to p and let L/K be a p-extension. Then v is unramified in L.*

PROOF. We may assume that L/K is finite. Then we can find a p-extension of finite algebraic number fields, k'/k, such that $K = k_\infty$, $L = k'_\infty$. Let $v_0 = v|k$. Then v_0 is a non-p-place on k and the order of $\mathfrak{k}_{v_0}^\times$ is prime to p. Since k'/k is a p-extension, it follows that v_0 is unramified in k'. Hence v also is unramified in L.

LEMMA 4. *Let the order of $\mathfrak{k}_v^\times$ be divisible by p^∞ and let S be a set of places on K, including v and all p-places of K. Let L denote the maximal abelian p-extension over K, unramified outside S. Then v is infinitely ramified in L.*

PROOF. Let w be an extension of v on L and assume that

$$e(w/v) = p^a < \infty, \quad a \geqq 0.$$

Let k be a finite algebraic number field such that $K = k_\infty$ and that v is the unique extension of $v|k$ on K, and let k_n, $n \geqq 0$, be the intermediate fields of the Z_p-extension K/k:

$$k = k_0 \subset \cdots \subset k_n \subset \cdots \subset k_\infty = K, \quad K = \bigcup_{n \geq 0} k_n.$$

Let S_n denote the union of $v_n = v|k_n$ and all p-places on k_n, and F_n the maximal abelian p-extension over k_n, unramified outside S_n. Then $k_n \subsetneqq K \subsetneqq F_n \subsetneqq L$ so that we put $w_n = w|F_n$. As v_n is unramified in K and $e(v/v_n) = 1$, we have

$$e(w_n/v_n) = e(w_n/v) \leqq e(w/v) = p^a .$$

For each place z in S_n, let $k_{n,z}$ denote the z-completion of k_n, and $U_{n,z}$ the multiplicative group of the local units in $k_{n,z}$. Let U_n be the direct product of all such $U_{n,z}$, $z \in S_n$, and let E_n be the group of global units in k_n. Then there is a natural imbedding $E_n \to U_n$, and we shall denote by $\bar{E}_n$ the closure of E_n in the compact group U_n. On the other hand, let V_n denote the Sylow p-subgroup of the profinite abelian group U_{n,v_n}. Since $v|Q \neq p$, V_n is isomorphic to the Sylow p-subgroup of $\mathfrak{k}_{v_n}^\times$, and since the order of $\mathfrak{k}_v^\times$ is divisible by p^∞, V_n is a finite cyclic group with order divisible by p^{n+1}.

Now, applying class field theory for F_n/k_n, one sees that $e(w_n/v_n)$ is the order of the image of the product of maps $V_n \to U_n \to U_n/\bar{E}_n$. Hence the order of the image of $V_n \to U_n/E_n U_n^{p^{n+1}}$ is at most equal to $e(w/v) = p^a$, and it follows that

$$V_n^{p^a} \times 1 \times \cdots \times 1 \subseteqq E_n U_n^{p^{n+1}} \subsetneqq U_n = \prod_z U_{n,z} .$$

Consequently, if $n \geqq a$, then there exists a unit ε in E_n such that ε is a p^{n+1}-th power in $k_{n,z}$ for every p-place z on k_n, but is at most a p^a-th power in k_{n,v_n}. Let ζ be a primitive $2p$-th root of unity in C and let

$$k' = k(\zeta) , \qquad K' = k'_\infty = K(\zeta) .$$

Replacing k by $K \cap k'$ if necessary, we may assume that $K \cap k' = k$ so that $k_n(\zeta)$ is the n-th intermediate field of the Z_p-extension K'/k' for all $n \geqq 0$: $k'_n = k_n(\zeta)$. Let v' be an extension of v on K', let $v'_n = v'|k'_n$, and let V'_n be the group defined for the local field $k'_{n,v'}$ similarly as V_n for k_n. Since the degree $[\mathfrak{k}_{v'_n} : \mathfrak{k}_{v_n}]$ is a factor of $b = [k' : k] = [k'_n : k_n]$ and since $b | p-1$ if $p > 2$ and $b = 1$ or 2 if $p = 2$, it follows that $V'_n = V_n$ for $p > 2$ and $[V'_n : V_n] = 1$ or 2 for $p = 2$, $n \geqq 1$. Consequently, we see that ε is at most a p^{a+1}-th power in the local field k'_{n,v'_n}, for $n > a$. Let ε_n denote a p^{n+1}-th root of ε in C. As k'_n contains primitive p^{n+1}-th roots of unity, it follows from the above that $k'_n(\varepsilon_n)/k'_n$ is an unramified cyclic p-extension, of which the local degree at v'_n is divisible by $p^{n+1}p^{-a-1} = p^{n-a}$. Therefore, by class field theory, the order of the ideal-class of $\mathfrak{p}_{v'_n}$ in k'_n is divisible by p^{n-a} for all $n > a$. However, this is impossible because, for a non-p-place v', the order of the ideal-class of $p_{v'_n}$ in k'_n is equal to the order of the ideal-class of $\mathfrak{p}_{v'}$ in K' whenever n is large enough.

The contradiction proves that $e(w/v)$ is infinite, namely, that v is infinitely ramified in L.

Now, let S be any set of places on a Z_p-field K, containing all p-places of K, and let S_0 denote the subset of all finite non-p-places v in S such that the order of $\mathfrak{k}_v^\times$ is divisible by p^∞. Let L_0 denote the maximal abelian p-extension over K, unramified outside S, and let L be any p-extension over K, containing L_0 and unramified outside S: $K \subseteqq L_0 \subseteqq L$. Let v be a finite non-p-place on K. Then it follows from Lemmas 3, 4 that v is unramified in L or infinitely ramified in L according as $v \notin S_0$ or $v \in S_0$. Hence we may apply Lemma 2 for such an extension L/K.

For each integer $n \geqq 1$, let W_n denote the group of all n-th roots of unity in C, and let

$$W(p) = \bigcup_{n \geqq 0} W_{p^n} .$$

Then a Z_p-field K contains $W(p)$ if and only if it contains W_{2p}. When this happens, we shall call K a cyclotomic Z_p-field. Let v be any finite non-p-place on such a cyclotomic Z_p-field K. It is clear that the order of $\mathfrak{k}_v^\times$ is then divisible by p^∞. Let $\mathfrak{p}_v^a = (\alpha)$ with $\alpha \in K^\times$, $a \geqq 1$, and let α_n be a p^n-th root of α in C for $n \geqq 1$. Then one sees easily that the field L in Lemma 4 contains all α_n, $n \geqq 1$, so that v is infinitely ramified in L. Thus the proof of Lemma 4 is much simpler in this case.

4. We need another lemma as follows:

LEMMA 5. *Let K be a number field containing Q_∞ (e.g., a Z_p-field) and let L/K be a p-extension, unramified at every infinite, i.e., archimedean, place on K. Then*

$$H^n(L/K, L^\times) = 0\,, \quad \textit{for all} \quad n \geqq 1\,.$$

PROOF. We may assume that L/K is finite. Since $H^1(L/K, L^\times) = 0$, it is sufficient to prove $H^2(L/K, L^\times) = 0$.[5] Let $K = \varinjlim k_n$, $L = \varinjlim k'_n$ where for each $n \geqq 0$, k'_n/k_n is a Galois extension of finite algebraic number fields such that $k'_n K = L$, $k'_n \cap K = k_n$. Then it follows from the assumption that whenever n is large enough, k'_n/k_n is unramified at every infinite place of k_n. On the other hand, since K contains Q_∞, the local degree of the extension K/Q at each finite place of K is divisible by p^∞. Hence we obtain $H^2(L/K, L^\times) = 0$ by the same argument as in the proof of [9], Chap. II, Proposition 9.

Now, in general, for each number field K, let E_K and W_K denote the group of all units of K and the group of all roots of unity in K,

[5] See [8], Chap. IX, Théorème 8.

respectively. Let M/K be an arbitrary Galois extension of number fields and let $G = \mathrm{Gal}(M/K)$. Then we have short exact sequences

$$(1) \quad \begin{gathered} 0 \to P_M \to I_M \to C_M \to 0\,, \\ 0 \to E_M \to M^\times \to P_M \to 0\,, \end{gathered}$$

and they induce long exact sequences for the cohomology groups $H^n(M/K, \ ) = H^n(G, \ )$, $n \geqq 0$.

THEOREM 1. *Let K be a Z_p-field and let M be a p-extension over K with the following properties*: (i) $C_M(p) = 0$ *and M/K is unramified at every infinite place of K*, (ii) *each finite non-p-place on K is either unramified in M or infinitely ramified in M. Then*

$$H^n(M/K, E_M) = 0\,, \quad \textit{for all} \quad n \geqq 2\,,$$

and there exists an exact sequence

$$0 \to C_K(p) \to H^1(M/K, E_M) \to \bigoplus_v A_v \to 0$$

where v ranges over all non-p-places v on K, ramified in M, and where, for each v,

$$A_v = \mathfrak{p}_v^R/\mathfrak{p}_v^Z \cong \boldsymbol{Q}_p/\boldsymbol{Z}_p\,, \quad R = \bigcup_{n \geqq 0} p^{-n}\boldsymbol{Z}\,.$$

PROOF. Since $G = \mathrm{Gal}(M/K)$ is a pro-p-group,

$$H^n(M/K, C_M) = H^n(M/K, C_M(p)) = 0\,, \quad \text{for} \quad n \geqq 1\,.$$

By Lemmas 2, 5, we also have

$$H^n(M/K, I_M) = 0\,, \qquad H^n(M/K, M^\times) = 0\,, \quad \text{for} \quad n \geqq 1\,.$$

Therefore (1) induces exact sequences and isomorphisms as follows:

$$\begin{gathered} I_M^G \to C_M^G \to H^1(M/K, P_M) \to 0\,, \qquad H^n(M/K, P_M) = 0\,, \quad n \geqq 2\,, \\ K^\times \to P_M^G \to H^1(M/K, E_M) \to 0\,, \qquad H^n(M/K, P_M) \xrightarrow{\sim} H^{n+1}(M/K, E_M)\,, \quad n \geqq 1\,. \end{gathered}$$

However, since $H^1(M/K, P_M)$ is a p-primary abelian group, the first exact sequence and $C_M(p) = 0$ induce $H^1(M/K, P_M) = 0$. Hence

$$H^n(M/K, E_M) = 0\,, \quad \text{for all} \quad n \geqq 2\,.$$

We also obtain from the second exact sequence that

$$(2) \quad H^1(M/K, E_M) \cong P_M^G/P_K = (I_M^G \cap P_M)/P_K\,.$$

From $P_K \subseteqq I_K \cap P_M \subseteqq I_M^G \cap P_M$, we then see that both $I_K \cap P_M/P_K$ and $I_M^G \cap P_M/I_K \cap P_M$ are p-primary abelian groups. Now, in the exact sequence

$$0 \to I_K \cap P_M/P_K \to I_K/P_K \to I_K/I_K \cap P_M \to 0\,,$$

we have $I_K/I_K \cap P_M = I_K P_M/P_M \subseteqq I_M/P_M$. Hence we obtain an exact sequence

$$0 \to I_K \cap P_M/P_K \to C_K(p) \to C_M(p) .$$

On the other hand, we also have the exact sequence

$$0 \to I_M^G \cap P_M/I_K \cap P_M \to I_M^G/I_K \to I_M^G/(I_M^G \cap P_M)I_K \to 0 .$$

Since I_M^G/I_K is easily seen to be p-primary and since

$$I_M^G/I_M^G \cap P_M = I_M^G P_M/P_M \subseteqq I_M/P_M ,$$

we know that $I_M^G/(I_M^G \cap P_M)I_K$ is a factor group of a subgroup of $C_M(p)$. The assumption $C_M(p) = 0$ then implies

$$I_K \cap P_M/P_K \cong C_K(p) , \qquad I_M^G \cap P_M/I_K \cap P_M \cong I_M^G/I_K .$$

Therefore, by Lemma 2 and the isomorphism (2) above, one obtains the exact sequence for $H^1(M/K, E_M)$ stated in the theorem.

REMARK. For any Galois extension L/K of number fields, we see easily that $H^1(L/K, I_L) = 0$. Since $H^1(L/K, L^\times) = 0$, it follows from (1) that

$$H^1(L/K, E_L) \cong P_L^G/P_K ,$$
$$\mathrm{Ker}\,(H^2(L/K, E_L) \to H^2(L/K, L^\times)) \cong \mathrm{Coker}\,(I_L^G \to C_L^G) .$$

When both K and L are finite over $\boldsymbol{Q}$, these isomorphisms are well known in the classical literature.

5. Let K be a number field. In the following, we shall denote the p-primary part $C_K(p)$ of the ideal-class group C_K simply by A_K:

$$A_K = C_K(p) .$$

As a p-primary abelian group, A_K is a torsion $\boldsymbol{Z}_p$-module. For a $\boldsymbol{Z}_p$-field K, the following facts are known on the structure of the $\boldsymbol{Z}_p$-module A_K.[6] Namely, let $K = k_\infty = k\boldsymbol{Q}_\infty$ with k finite over $\boldsymbol{Q}$ and let $\lambda = \lambda(K/k)$ and $\mu = \mu(K/k)$ denote respectively the so-called λ- and μ-invariant of the $\boldsymbol{Z}_p$-extension K/k. Then

$$A_K \cong (\boldsymbol{Q}_p/\boldsymbol{Z}_p)^\lambda \oplus A'$$

with A' a $\boldsymbol{Z}_p$-module of bounded exponent: $p^a A' = 0$, $a \geqq 0$. Furthermore, if $\mu = 0$, then $A' = 0$ so that

$$A_K \cong (\boldsymbol{Q}_p/\boldsymbol{Z}_p)^\lambda , \qquad A_K/pA_K = 0 ;$$

and if $\mu > 0$, then A' is infinite so that A_K/pA_K is an infinite $\boldsymbol{Z}_p$-module.

[6] For the results stated below, see [4], [5].

Now, it follows from the above that $\lambda = \lambda(K/k)$ depends only upon the structure of the Z_p-module A_K and is independent of the choice of k such that $K = k_\infty$. Hence it may be denoted by λ_K: $\lambda_K = \lambda(K/k)$. Similarly, since A_K/pA_K is zero or infinite according as $\mu = 0$ or $\mu > 0$, the fact that $\mu(K/k) = 0$ is actually a property of K so that it may be simply written as

$$\mu_K = 0 .$$

The isomorphism for A_K shows that in the case $\mu_K = 0$, A_K gives us a good analogue of the p-power division points of the Jacobian variety of an algebraic curve and, hence, λ_K an analogue of twice the genus of that curve.

In general, for each torsion Z_p-module A, let

$$V(A) = \mathrm{Hom}_{Z_p}(A, Q_p/Z_p) \otimes_{Z_p} Q_p .$$

Then $A \mapsto V(A)$ defines an exact contravariant functor from torsion Z_p-modules into vector spaces over Q_p. For $A = A_K$, let

$$V_K = V(A_K) .$$

Since $V(A') = 0$ for the above Z_p-module A', we then have

$$V_K \cong Q_p^\lambda , \qquad \lambda_K = \dim_{Q_p} V_K .$$

Now, it has been conjectured for some time that $\mu_K = 0$ for all Z_p-fields K. The conjecture was proved by Ferrero-Washington [3] in the case where K is an abelian extension over the rational field Q, but the problem is yet unsolved in the general case. As an application of Theorem 1, we shall next give a necessary and sufficient condition for $\mu_k = 0$ when K is a cyclotomic Z_p-field.

THEOREM 2. *Let K be a cyclotomic Z_p-field, S a finite set of places on K, including all p-places of K, and M the maximal p-extension over K, unramified outside S. Then $\mu_K = 0$ if and only if* $\mathrm{Gal}(M/K)$ *is a free pro-p-group.*

PROOF.[7] Since there is no non-trivial unramified abelian p-extension over M, we see easily that $A_M = C_M(p) = 0$. As K contains $W(p)$, it is a totally imaginary field so that M/K is unramified at every infinite place of K. By the remark after Lemma 4, each finite non-p-place on K is either unramified or infinitely ramified in M. Therefore both conditions (i) and (ii) in Theorem 1 are satisfied for the extension M/K. On the

[7] This is essentially a known result. Cf. [1], Corollary 2.5. A proof is included here to clarify, in this simpler case, the main idea of arguments which will again be applied later to prove more elaborate results.

other hand, we also see from the definition of M that the map $\varepsilon \mapsto \varepsilon^p$ for ε in E_M defines an exact sequence

$$0 \to W_p \to E_M \xrightarrow{p} E_M \to 0 ,$$

and this induces the exact sequence

$$H^1(M/K, E_M) \xrightarrow{p} H^1(M/K, E_M) \to H^2(M/K, W_p) \to H^2(M/K, E_M)$$

where $H^2(M/K, E_M) = 0$ by Theorem 1. Hence

$$H^2(M/K, W_p) \cong H^1(M/K, E_M)/pH^1(M/K, E_M) .$$

However, by the same theorem, there exists an exact sequence

$$0 \to A_K \to H^1(M/K, E_M) \to \bigoplus_v A_v \to 0$$

where v ranges over a finite set and $A_v \cong \boldsymbol{Q}_p/\boldsymbol{Z}_p$ for every such v. Since A_K/pA_K is zero or infinite according as $\mu_K = 0$ or $\mu_K > 0$, it follows from the above exact sequence that $\mu_K = 0$ if and only if $H^2(M/K, W_p) = 0$, i.e., $H^2(\mathrm{Gal}(M/K), \boldsymbol{Z}/p\boldsymbol{Z}) = 0$. As $\mathrm{Gal}(M/K)$ is a pro-p-group, the last equality means that $\mathrm{Gal}(M/K)$ is a free pro-p-group.[8] Hence the theorem is proved. Note that the finiteness of the set S is used only to deduce $\mu_K = 0$ from $H^2(M/K, W_p) = 0$, and not in the opposite deduction.

6. Let J denote the complex-conjugation of the complex field $\boldsymbol{C}$. When a number field K is invariant under the automorphism J of $\boldsymbol{C}$, the restriction of J on the subfield K will be simply denoted again by J. In such a case, J acts on various abelian groups canonically associated with K, e.g., $K^\times$, I_K, C_K, etc. In general, when J acts on an abelian group A, we define the subgroups A^+ and A^- by

$$A^\pm = \{a \mid a \in A, J(a) = \pm a\} .$$

One sees immediately that if A is uniquely divisible by 2, e.g., a $\boldsymbol{Z}_p$-module with $p > 2$, then

$$A = A^+ \oplus A^- .$$

For each number field K, let K^+ denote the maximal real subfield of K: $K^+ = K \cap \boldsymbol{R}$. K is called a number field of C-M type if K^+ is totally real, K is totally imaginary, and $[K: K^+] = 2$. K is then invariant under J, and

$$\mathrm{Gal}(K/K^+) = \{1, J\} .$$

It is also known that for such K,

(3) $\quad E_K^+ = E_{K^+} , \quad E_K^- = W_K , \quad [E_K : W_K E_{K^+}] = 1$ or 2 .

[8] See [9], Chap. I, § 4.

We now assume that $p > 2$ and K is a $\boldsymbol{Z}_p$-field of C-M type. We can then fined a number field k of C-M type, finite over $\boldsymbol{Q}$, such that $K = k_\infty$, $K^+ = k_\infty^+$. For such K/k, we can define four invariants $\lambda(K/k)^\pm$, $\mu(K/k)^\pm$ which have similar properties as explained in § 5 for $\lambda(K/k)$ and $\mu(K/k)$.[9] For example, $\lambda(K/k)^\pm$ depend only upon K so that they may be denoted by $\lambda_K^\pm$. In fact, we have

$$A_K = A_K^+ \oplus A_K^- , \qquad A_K^+ = A_{K^+} ,$$
$$V_K = V_K^+ \oplus V_K^- , \qquad V_K^\pm = V(A_K^\pm) , \quad \dim_{\boldsymbol{Q}_p} V_K^\pm = \lambda_K^\pm .$$

Furthermore, if $\mu_K^- = 0$, then

$$A_K^- \cong (\boldsymbol{Q}_p/\boldsymbol{Z}_p)^{\lambda_K^-} ;$$

and A_K^-/pA_K^- is zero or infinite according as $\mu_K^- = 0$ or $\mu_K^- > 0$.

In addition to the above, let us now also assume that K is a cyclotomic $\boldsymbol{Z}_p$-field, i.e., $W(p) \subseteq K$. Let S^+ be a set of finite places on $K^+ = K \cap \boldsymbol{R}$, containing all p-places of K^+, and let M^+ denote the maximal p-extension over K^+, unramified outside S^+. Let

$$M = KM^+ .$$

Since M^+/K^+ is unramified at every infinite place on K^+, M^+ is totally real, and it follows that M is a number field of C-M type and that $M^+ = M \cap \boldsymbol{R}$, justifying the notation M^+. It is easy to see that M/K^+ is a Galois extension, that $K \cap M^+ = K^+$, and that

$$\text{(4)} \qquad \begin{aligned} &\mathrm{Gal}\,(M/K^+) = \mathrm{Gal}\,(M/K) \times \mathrm{Gal}\,(M/M^+) , \\ &\mathrm{Gal}\,(M/K) = \mathrm{Gal}\,(M^+/K^+) . \end{aligned}$$

It is also known that $\mu_K = 0$ if and only if $\mu_K^- = 0$.

LEMMA 6. *The extension M/K has the properties* (i), (ii) *stated in Theorem* 1 *so that the cohomology groups $H^n(M/K, E_M)$, $n \geqq 1$, are given by that theorem.*

PROOF. As in the proof of Theorem 2, $A_M^+ = A_{M^+} = 0$ for the maximal p-extension over K^+, unramified outside S^+. Since K is totally imaginary, M/K is unramified at every infinite place of K. (This follows also from the fact that M/K is a p-extension with $p > 2$.) Let $c \in A_M^-$, $pc = 0$. Let $\mathfrak{a}$ be an ideal of I_M in the ideal-class c and let $\mathfrak{a}^p = (\alpha)$, $\alpha \in M^\times$. Then $\mathfrak{b}^p = (\beta)$ with $\mathfrak{b} = \mathfrak{a}^{1-J}$, $\beta = \alpha^{1-J}$. Since M contains W_p, $M(\sqrt[p]{\beta})/M$ is a cyclic extension of degree 1 or p. However, $\beta^{1+J} = 1$ implies that $M(\sqrt[p]{\beta})/M^+$ is abelian. Hence there exists a cyclic extension M'/M^+ which is of degree 1 or p, unramified outside p-places of M^+,

9) See [5].

and satisfies $M(\sqrt[p]{\beta}) = MM'$. It then follows from the definition of M^+ that $M' = M^+$, $M(\sqrt[p]{\beta}) = M$ so that $\mathfrak{b} = (\sqrt[p]{\beta})$ with $\sqrt[p]{\beta} \in M^\times$. Thus $(1-J)c = 0$, $c \in A_M^- = 0$, and consequently $c = 0$. Therefere $A_M^- = 0$, $A_M = A_M^+ \oplus A_M^- = 0$, and the property (i) of Theorem 1 is verified for M/K. Next, let v be a non-p-place on K and let $v^+ = v|K^+$. Since $K^+ = k_\infty^+$ is a Z_p-field, it follows from the remark after Lemma 4 that v^+ is either unramified in M^+ or infinitely ramified in M^+. Using the fact that $\mathrm{Gal}(M^+/K^+)$ is a pro-p-group with $p > 2$, we then see from (4) that v is unramified or infinitely ramified in M according as v^+ is unramified or infinitely ramified in M^+. Hence M/K has also the property (ii) of Theorem 1.

Let v^+ be any place on K^+ and let v be an extension of v^+ on K. Then v and $\bar{v} = J(v)$ are the only extensions of v^+ on K. As usual, we shall say that v^+ splits in K if $v \neq \bar{v}$. Since each place on K^+ has at most two extensions on K, the above proof shows that if S^+ is a finite set, then there exist only a finite number of finite non-p-places v on K which are ramified in M; namely, the direct sum $\bigoplus_v A_v$ in the exact sequence for $H^1(M/K, E_M)$ in Theorem 1 is a finite sum.

THEOREM 3. *Let K be a cyclotomic Z_p-field of* C-M *type for $p > 2$ and let S^+ be a finite set of finite places on $K^+ = K \cap R$, including all p-places of K^+. Let M^+ be the maximal p-extension over K^+, unramified outside S^+, and let s denote the number of finite non-p-places on K^+ which split in K and are ramified in M^+. Then $\mu_K = 0$ if and only if* $\mathrm{Gal}(M^+/K^+)$ *is a free pro-p-group, and if this is the case, then the minimal number of generators for* $\mathrm{Gal}(M^+/K^+)$ *is $\lambda_K^- + s$.*

PROOF. As stated above, let $M = KM^+$. Then $\mathrm{Gal}(M/K^+)$ acts on E_M so that J also acts on $H^n(M/K, E_M)$. Hence $H^n(M/K, E_M)^\pm$ are defined. Since $p > 2$, it follows from (3), (4) and $W(p) \subseteq W_M$ that

$$H^n(M/K, E_M)^- = H^n(M/K, W_M) = H^n(M/K, W(p)) , \quad \text{for} \quad n \geqq 1 .$$

Therefore, by Theorem 1 and Lemma 6, we see that

$$H^n(M/K, W(p)) = 0 , \quad \text{for} \quad n \geqq 2 .$$

Furthermore, the exact sequence in Theorem 1 also implies an exact sequence

$$0 \to A_K^- \to H^1(M/K, W(p)) \to (\bigoplus_v A_v)^- \to 0$$

where, as easily seen from the definition of A_v,

$$(\bigoplus_v A_v)^- \cong (Q_p/Z_p)^s .$$

Now, from $H^2(M/K, W(p)) = 0$ and from the exact sequence

$$0 \to W_p \to W(p) \xrightarrow{p} W(p) \to 0 ,$$

one obtains

$$H^2(M/K, W_p) \cong H^1(M/K, W(p))/pH^1(M/K, W(p)) .$$

On the other hand, A_K^-/pA_K^- is zero or infinite according as $\mu_K^- = 0$ or $\mu_K^- > 0$. Therefore, similarly as in the proof of Theorem 2, we conclude from the above exact sequence for $H^1(M/K, W(p))$ that $\mu_K^- = 0$ if and only if $H^2(M/K, W_p) = 0$, namely, that $\mu_K = 0$ if and only if $\mathrm{Gal}(M^+/K^+) = \mathrm{Gal}(M/K)$ is a free pro-p-group. Furthermore, when this is the case, then $A_K^- \cong (\boldsymbol{Q}_p/\boldsymbol{Z}_p)^{\lambda^-}$ so that

$$H^1(M/K, W(p)) \cong (\boldsymbol{Q}_p/\boldsymbol{Z}_p)^{\lambda^- + s} .$$

However, since $W(p) \subseteq K$, $H^1(M/K, W(p))$ is the Pontrjagin dual of the factor commutator group of $\mathrm{Gal}(M/K)$. Therefore the above isomorphism indicates that $\lambda_K^- + s$ is the minimal number of generators for the free pro-p-group $\mathrm{Gal}(M^+/K^+)$.

We note here that an essential difference between Theorem 2 and Theorem 3 lies in that the free pro-p-group $\mathrm{Gal}(M^+/K^+)$ in Theorem 3 is finitely generated while this is not so for the free pro-p-group $\mathrm{Gal}(M/K)$ in Theorem 2.

We shall next also briefly explain a generalization of Theorem 3. Let K' be a totally real $\boldsymbol{Z}_p$-field with $p > 2$. Let S' be a finite set of finite places on K', including all p-places of K', and let M' denote the maximal p-extension over K', unramified outside S'. Let

$$K = K'(W_p) .$$

Then K is a cyclotomic $\boldsymbol{Z}_p$-field of C-M type and K/K' is a cyclic extension with degree a factor of $p - 1$. In fact, there exists a monomorphism

$$\chi : \mathrm{Gal}(K/K') \to \boldsymbol{Z}_p^\times$$

such that $\sigma(\zeta) = \zeta^{\chi(\sigma)}$ for all $\sigma \in \mathrm{Gal}(K/K')$ and $\zeta \in W(p) \subseteq K$. Let $A_K^{(1)}$ denote the submodule of $A_K = C_K(p)$, consisting of all a in A_K such that $\sigma(a) = \chi(\sigma)a$ for every σ in $\mathrm{Gal}(K/K')$. The $\boldsymbol{Z}_p$-module $A_K^{(1)}$ is then a direct summand of A_K and it has similar properties as mentioned earlier for A_K and $A_K^\pm$; for example, if $A_K^{(1)} = pA_K^{(1)}$, then

$$A_K^{(1)} \cong (\boldsymbol{Q}_p/\boldsymbol{Z}_p)^\lambda$$

with an integer $\lambda = \lambda_K^{(1)} \geqq 0$. Let s be the number of finite non-p-places on K' which are ramified in M' and which split completely in K.

Now, it can be proved that the Galois group $\mathrm{Gal}(M'/K')$ is a free pro-p-group if and only if $A_K^{(1)} = pA_K^{(1)}$, and if this is the case, then the minimal number of generators for $\mathrm{Gal}(M'/K')$ is $\lambda_K^{(1)} + s$. The proof is similar to that of Theorem 3, the key fact in this case being $C_M(p)^{(1)} = 0$ for $M = KM'$. However, we omit the detail. It is clear that $A_K^{(1)} = pA_K^{(1)}$ follows from $A_K = pA_K$, namely, from $\mu_K = 0$. Therefore $\mu_K = 0$ for $K = K'(W_p)$ implies that $\mathrm{Gal}(M'/K')$ is a free pro-p-group. Hence it follows from the theorem of Ferrero-Washington [3] that for any Z_p-field K', $p > 2$, which is real and abelian over Q, the Galois group $\mathrm{Gal}(M'/K')$ is always a free pro-p-group with a finite number of generators.

7. In general, let K be a Z_p-field, and L/K a finite Galois extension with $G = \mathrm{Gal}(L/K)$. Then L is again a Z_p-field and G acts on $A_L = C_L(p)$ in the natural manner. Therefore G acts also on the vector space $V_L = V(A_L)$ defined in § 5 and we obtain a λ_L-dimensional p-adic representation

$$\pi_{L/K}: G = \mathrm{Gal}(L/K) \to GL(V_L) .$$

Let $p > 2$ and assume that both K and L are number fields of C-M type. Then L/K^+ is a Galois extension and

$$\mathrm{Gal}(L/K^+) = \mathrm{Gal}(L/K) \times \mathrm{Gal}(L/L^+) , \qquad \mathrm{Gal}(L/K) = \mathrm{Gal}(L^+/K^+) .$$

Consequently, we have the decomposition for $\pi_{L/K}$:

$$\pi_{L/K} = \pi_{L/K}^+ \oplus \pi_{L/K}^-$$

where

$$\pi_{L/K}^{\pm}: G \to GL(V_L^{\pm}) .$$

Let us next investigate the above representation $\pi_{L/K}^-$ in the case where L/K is a finite p-extension and K is a cyclotomic Z_p-field, i.e., $W(p) \subseteq K$. We first note that there exist only a finite number of finite places on K^+ which are ramified in L^+; in fact, this is true in general for any finite extension of Z_p-fields. Since L^+/K^+ is unramified at every infinite place of K^+, there exists a finite set S^+ of finite places on K^+, including all p-places of K^+, such that L^+ is contained in the maximal p-extension M^+ over K^+, unramified outside S^+:

$$K^+ \subseteq L^+ \subseteq M^+ .$$

For $M = KM^+$, we then have

$$K \subseteq L \subseteq M .$$

It is clear that M^+ is the maximal p-extension over L^+, unramified outside the set of extensions, on L^+, of all v^+ in S^+. Thus M^+/L^+ is the same kind of extension as M^+/K^+ in Lemma 6 so that there exists an exact sequence

$$0 \to A_L \to H^1(M/L, E_M) \to \bigoplus_w A_w \to 0$$

where w ranges over all finite non-p-places on L, ramified in M. Furthermore, in this case, the maps in the exact sequence are homomorphisms over Gal (L/K^+). Hence it induces an exact sequence of G-modules:

$$0 \to A_L^- \to H^1(M/L, E_M)^- \to (\bigoplus_w A_w)^- \to 0\,,$$

where $H^1(M/L, E_M)^- = H^1(M/L, W(p))$ as explained in the proof of Theorem 3. Let $v^+ = w|K^+$. By the remark after Lemma 4, v^+ is either unramified in M^+ or infinitely ramified in M^+. Therefore, if w is ramified in M, then v^+ is ramified in M^+, and the converse is alse true. In general, for each finite non-p-place v^+ on K^+, define a module B_{v^+} over $G =$ Gal (L^+/K^+) by

$$B_{v^+} = \bigoplus_{w^+} A_{w^+}\,,$$

with w^+ ranging over all extensions of v^+ on L^+. Then we obtain from the above an exact sequence of G-modules:

$$0 \to A_L^- \to H^1(M/L, W(p)) \to \bigoplus_{v^+} B_{v^+} \to 0\,,$$

where v^+ now runs over all finite non-p-places on K^+, ramified in M^+ and split in K. Let

$$V_0 = V(H^1(M/L, W(p))$$

with the functor V defined in § 5. Since V is exact, it follows from the above that

$$V_0 \cong V_L^- \oplus (\bigoplus_{v^+} V(B_{v^+}))$$

as representation spaces for the finite group G. Thus

$$\pi_0 = \pi_{L/K}^- \oplus (\bigoplus_{v^+} \pi_{v^+}) \tag{5}$$

with π_0 and π_{v^+} denoting the representations of G on the spaces V_0 and $V(B_{v^+})$ respectively. Now, for each v^+, fix an extension w^+ of v^+ on L^+, and let $T = T(w^+/v^+)$. Then $T = Z(w^+/v^+)$ by § 1 so that as G-modules,

$$B_{v^+} \cong (\boldsymbol{Q}_p/\boldsymbol{Z}_p)[G/T]$$

where the right hand side denotes the linear combinations of the left cosets of G modulo T with coefficients in $\boldsymbol{Q}_p/\boldsymbol{Z}_p$. Consequently

$$V(B_{v^+}) \cong \boldsymbol{Q}_p[G/T]\,,$$

and we see that π_{v^+} is isomorphic to the representation of G over the field Q_p, induced by the trivial one-dimensional representation of the subgroup $T = T(w^+/v^+)$. It is then clear from (5) that the representation $\pi^-_{L/K}$ is completely determined if we know the representation π_0 of G on the space $V_0 = V(H^1(M/L, W(p))$.

Let

$$X = \mathrm{Gal}\,(M^+/K^+) = \mathrm{Gal}\,(M/K)\,, \qquad Y = \mathrm{Gal}\,(M^+/L^+) = \mathrm{Gal}\,(M/L)\,.$$

Then Y is a closed normal subgroup of the pro-p-group X and

$$X/Y = \mathrm{Gal}\,(L^+/K^+) = \mathrm{Gal}\,(L/K) = G\,.$$

As explained in the proof of Theorem 3, $H^1(M/L, W(p))$ is the Pontrjagin dual of Y^{ab}, the factor commutator group of Y. Since $W(p) \subseteqq K$, we see that as G-modules,

$$V_0 \cong Y^{ab} \otimes_{Z_p} Q_p\,,$$

where the action of $G = X/Y$ on the right is given by conjugations in X. Let us now assume that $\mu_K = 0$. Then it follows from Theorem 3 that X is a free pro-p-group with $\lambda_K^- + s$ generators, where s denotes the number of places v^+ in the direct sum in (5). Therefore the representation π_0 can be described by the following purely group-theoretical lemma:

LEMMA 7. *Let X be a free pro-p-group with m generators, $m \geqq 1$, and let Y be a closed normal subgroup with finite index in X. Then the representation π_0 of the finite group $G = X/Y$ on the vector space $Y^{ab} \otimes_{Z_p} Q_p$ is given by*

$$\pi_0 = \pi_1 \oplus (m-1)\pi_G\,,$$

where π_1 denotes the one-dimensional trivial representation of G over Q_p, and π_G the regular representation of G over Q_p.

PROOF. Let I_G denote the augmentation ideal of the group ring $Z_p[G]$ so that

$$0 \to I_G \to Z_p[G] \to Z_p \to 0$$

is exact. Then there exists[10] an exact sequence of $Z_p[G]$-modules:

$$0 \to Y^{ab} \to Z_p[G]^m \to I_G \to 0\,.$$

The lemma follows from this by replacing each term of the above by its tensor product with Q_p over Z_p.

We now consider again the case where $X = \mathrm{Gal}\,(M/K)$, $Y = \mathrm{Gal}\,(M/L)$,

---

10) See [10]. Note that m need not be the minimal number of generators for G.

and $G = X/Y = \mathrm{Gal}(L/K)$. We know that X is a free pro-p-group with $m = \lambda_K^- + s$ generators. Assume that $G \neq 1$, i.e., $K \neq L$. Then $X \neq 1$, $m \geqq 1$ so that we obtain from (5) and Lemma 7 that

$$\pi_1 \oplus (\lambda_K^- + s - 1)\pi_G = \pi_{L/K}^- \oplus (\bigoplus_{v^+} \pi_{v^+}) ,$$

where s is exactly the number of places v^+ which appear on the right. For such a place v^+, let w^+ denote an extension of v^+ on L^+ and let $T = T(w^+/v^+)$. Then π_{v^+} is the representation of G over $\boldsymbol{Q}_p$, induced by the trivial representation $\pi_1 | T$ of the subgroup T. Therefore

$$\pi_G = \pi_{v^+} \oplus \pi'_{v^+}$$

where π'_{v^+} is the representation of G over $\boldsymbol{Q}_p$, induced by the representation $\pi_T - \pi_1 | T$ for T. Note that when we change the extension w^+ of v^+ on L^+, the inertia group $T(w^+/v^+)$ is relaced by its conjugate in G so that π_{v^+} and π'_{v^+} are unchanged (up to isomorphisms). We now obtain from the above that

$$\pi_{L/K}^- = \pi_1 \oplus (\lambda_K^- - 1)\pi_G \oplus (\bigoplus_{v^+} \pi'_{v^+}) \tag{6}$$

where v^+ ranges over all finite non-p-places on K^+ which are ramified in M^+ and split in K. However, if v^+ is unramified in M^+, then $T(w^+/v^+) = 1$ so that $\pi_{v^+} = \pi_G$, $\pi'_{v^+} = 0$. Hence the sum in (6) may be taken over all v^+ which split in K. Note also that if $\lambda_K^- = 0$, then the right hand side of (6) should be interpreted as the difference of the sum over v^+ and the representation $\pi_G - \pi_1$.

Finally, if $G = 1$, then $K = L$, $V_K^- = V_L^-$, and $\pi_{L/K}^- = \lambda_K^- \pi_1$. Since $\pi'_{v^+} = 0$ for all v^+, the equality (6) still holds in this case. Thus we have proved the following

THEOREM 4. *Let $p > 2$ and let L/K be a finite p-extension of cyclotonic $\boldsymbol{Z}_p$-fields of* C-M *type. Let $\pi_{L/K}^- : G \to GL(V_L^-)$ be the representation of $G =$* Gal(L/K) *on the vector space $V_L^- = V(A_L^-)$ over $\boldsymbol{Q}_p$. Assume that $\mu_K = 0$. Then*

$$\pi_{L/K}^- = \pi_1 \oplus (\lambda_K^- - 1)\pi_G \oplus (\bigoplus_{v^+} \pi'_{v^+}) .$$

Here π_1 is the one-dimensional trivial representation of G over $\boldsymbol{Q}_p$, π_G is the regular representation of G over $\boldsymbol{Q}_p$, v^+ ranges over all finite non-p-places on $K^+ = K \cap \boldsymbol{R}$ which split in K, and π'_{v^+} denotes the complement in π_G of the representation π_{v^+} of $G =$ Gal(L^+/K^+), *induced by the one-dimensional trivial representation, over $\boldsymbol{Q}_p$, of the inertia group $T(w^+/v^+)$ of v^+ for the Galois extension L^+/K^+.*

As mentioned above, when $\lambda_K^- = 0$, the right hand side of the formula should be interpreted as the difference of the sum over v^+ and $\pi_G - \pi_1$.

8. Again, let $p > 2$ and let L/K be a finite p-extension of Z_p-fields of C-M type. We shall next study the representation $\pi_{L/K}^-$ in the case where K is not cyclotomic. Since $p > 2$, $[Q(W_p): Q] = p - 1$, L also is then non-cyclotomic and W_p is not contained in W_L. We can find a finite set S of finite places on K, including all p-places of K, such that L is contained in the maximal p-extension M over K, unramified outside $S: K \subseteqq L \subseteqq M$. Furthermore, we may choose S so that it is invariant under the automorphism J of K/K^+. Then M/K^+ is a Galois extension. Clearly, if S' denotes the set of all extensions, on L, of the places in S, then M is the maximal p-extension over L, unramified outside S'. Therefore, as in the proof of Theorem 2, we can see that both extensions M/K and M/L satisfy the conditions (i), (ii) of Theorem 1; here one has only to note that both K and L are totally imaginary because they are number fields of C-M type. It then follows from Theorem 1 that

$$H^n(M/L, E_M) = 0\,, \quad \text{for all} \quad n \geqq 2\,.$$

Therefore the Hochschild-Serre spectral sequence for $K \subseteqq L \subseteqq M$ induces[11] an exact sequence

$$\begin{aligned} 0 &\to H^1(L/K, E_L) \to H^1(M/K, E_M) \to H^0(L/K, H^1(M/L, E_M)) \\ &\to H^2(L/K, E_L) \to H^2(M/K, E_M) \to H^1(L/K, H^1(M/L, E_M)) \\ &\to \cdots\,. \end{aligned}$$

Since M/K^+ and M/L^+ are Galois extensions, the automorphism J acts on each term of the above exact sequence, and we obtain similar exact sequences where $H^n(\ ,\ )$ are relaced by $H^n(\ ,\ )^{\pm}$. By Theorem 1,

$$H^n(M/K, E_M) = H^n(M/K, E_M)^{\pm} = 0\,, \quad \text{for} \quad n \geqq 2\,.$$

Since W_p is not contained in W_L,

$$H^n(L/K, E_L)^- = H^n(L/K, W_L) = 0\,, \quad \text{for} \quad n \geqq 1\,.$$

Therefore it follows from the exact sequence for $H^n(\ ,\ )^-$ that

$$A^G \cong H^1(M/K, E_M)^-\,, \quad H^n(G, A) = 0\,, \quad \text{for} \quad n \geqq 1$$

where we put

$$G = \mathrm{Gal}\,(L/K)\,, \qquad A = H^1(M/L, E_M)^-\,.$$

---

11) See [7], Chap, VI, Théorème 4. Of course, one has to consider here Galois cohomology groups instead of cohomology groups of discrete groups.

Let s now denote the number of finite non-p-places v^+ on K^+ which split in K and are ramified in M. Similarly as in § 7, Theorem 1 for M/K and M/L then give us the exact sequences

$$0 \to A_K^- \to H^1(M/K, E_M)^- \to \bigoplus_{v^+} A_{v^+} \to 0\,,$$

$$(7) \qquad 0 \to A_L^- \longrightarrow A \longrightarrow \bigoplus_{v^+} B_{v^+} \to 0\,,$$

where v^+ ranges over the places on K^+ as mentioned above and where A_{v^+} and B_{v^+} are defined in the same manner as in § 7.

Assume now that $\mu_K = 0$. It is known in general that this induces $\mu_L = 0$ for a finite p-extension L/K. Therefore $\mu_K^- = \mu_L^- = 0$ so that

$$A_K^- \cong (Q_p/Z_p)^{\lambda_K^-}\,, \qquad A_L^- \cong (Q_p/Z_p)^{\lambda_L^-}\,,$$

as explained in § 6. Hence we obtain from the above that

$$(8) \qquad A^G \cong (Q_p/Z_p)^{\lambda_K^- + s} \qquad A \cong (Q_p/Z_p)^t$$

with s defined above and with a certain integer $t \geqq \lambda_L^-$. Let

$$A_p = \mathrm{Ker}\,(p\colon A \to A)$$

so that $0 \to A_p \to A \xrightarrow{p} A \to 0$ is exact. Then $H^n(G, A) = 0$, for $n \geqq 1$, imply

$$H^n(G, A_p) = 0\,, \quad \text{for all} \quad n \geqq 2\,.$$

Since G is a finite p-group, it follows that A_p is free over $(Z/pZ)[G]$ [12]. From (8), we then see that

$$A \cong (Q_p/Z_p)[G]^{\lambda_K^- + s}\,.$$

Let π_0 denote as before the representation of G on the vector space $V_0 = V(A)$ over Q_p. Then the above isomorphism implies

$$V_0 \cong Q_p[G]^{\lambda_K^- + s}\,, \qquad \pi_0 = (\lambda_K^- + s)\pi_G$$

with π_G denoting again the regular representation of G over Q_p. From the exact sequence (7), we again obtain the equality (5) in § 7, but with the representation π_0 as mentioned above. Therefore, just as in § 7, we can prove the following result:

THEOREM 5. *Let $p > 2$ and let L/K be a finite p-extension of non-cyclotomic Z_p-fields of* C-M *type. Let $\pi_{L/K}^-\colon G \to GL(V_L^-)$ be the representation of $G = \mathrm{Gal}\,(L/K)$ on the vector space $V_L^- = V(A_L^-)$ over Q_p. Assume that $\mu_K = 0$. Then*

$$\pi_{L/K}^- = \lambda_K^- \pi_G \oplus \Big(\bigoplus_{v^+} \pi'_{v^+}\Big)$$

[12] See [8], Chap. IX, Théorème 5.

where π_G *and the sum over* v^+ *are defined in the same manner as in Theorem* 4.

Note that in both Theorem 4 and Theorem 5, $\pi'_{v^+}=0$ if v^+ is unramified in L^+. Hence the sum over v^+ is actually a finite sum and it is zero if and only if L^+/K^+ is unramified outside the p-places of K^+.

9. In general, for each Z_p-field K, let

$$\delta_K = 1 \quad \text{or} \quad 0$$

according as K is cyclotomic or not. As in Theorems 4, 5, let L/K be a finite p-extension of Z_p-fields of C-M type. Then the formulae in those theorems can be uniformly written as

$$\pi^-_{L/K} = \delta_K\pi_1 \oplus (\lambda^-_K - \delta_K)\pi_G \oplus (\bigoplus_{v^+} \pi'_{v^+}) . \tag{9}$$

We shall next compare the degrees of the representations on the both sides. It is clear that

$$\deg(\pi^-_{L/K}) = \lambda^-_L , \qquad \deg(\pi_1) = 1 , \qquad \deg(\pi_G) = n = [L: K] .$$

Since π_{v^+} is induced by the trivial representation of the inertia group $T(w^+/v^+) = Z(w^+/v^+)$, we have

$$\deg(\pi_{v^+}) = n/e = g , \qquad \deg(\pi'_{v^+}) = n - g = g(e-1)$$

where $e = e(w^+/v^+) = [T(w^+/v^+): 1]$ is the ramification index of w^+ for the extension L^+/K^+ and where $g = g(w^+/v^+)$ is the number of the extensions w^+ of v^+ on L^+. Hence it follows from (9) that

$$\lambda^-_L = \delta_K + (\lambda^-_K - \delta_K)[L : K] + \sum_{w^+} (e(w^+/v^+) - 1) ,$$

where the sum on the right is now taken over all finite non-p-places w^+ on L^+, split in L. The same formula may be also written as

$$2(\lambda^-_L - \delta_L) = 2(\lambda^-_K - \delta_K)[L : K] + \sum_{w} (e(w/v) - 1) \tag{10}$$

with w ranging over all finite non-p-places on L, split for the extension L/L^+. This is the formula of Kida [6], mentioned in the introduction. We note that it is easy to deduce from (9) a similar formula for L'/K where L' is an arbitrary intermediate field of K and L, not necessarily a Galois extension over K.

The proof of the above formulae for $\pi^-_{L/K}$ and λ^-_L does not work to obtain similar results for $\pi^+_{L/K}$ and λ^+_L. The main reason is that while Galois cohomology groups with values in $E^-_L = W_L$ can be quite simply calculated, it is not so for $E^+_L = E_{L^+}$. However, less explicit formulae involving some cohomological invariants still can be proved by similar arguments in certain cases. We shall next explain it briefly.

In what follows, let p denote an arbitrary prime number, possibly 2, and let L be a cyclic extension of degree p over a Z_p-field K, unramified at every infinite place on K. Let S be the set of all finite non-p-places on K, ramified in L, and let $I_{L,S}$ and $P_{L,S}$ denote the group of all S-ideals of L and the subgroup of principal ideals in $I_{L,S}$, respectively.[13] Then we have exact sequences of modules over $G = \mathrm{Gal}(L/K)$:

$$0 \to P_{L,S} \to I_{L,S} \to C_{L,S} \to 0 ,$$
$$0 \to E_{L,S} \to L^\times \to P_{L,S} \to 0 ,$$

where $C_{L,S}$ is the ideal-class group of S-ideals of L, and $E_{L,S}$ the group of S-units in L. It follows from Lemmas 1, 5 that

$$H^n(L/K, I_{L,S}) = 0 , \quad H^n(L/K, L^\times) = 0 , \quad \text{for} \quad n \geqq 1 .$$

Hence the above exact sequences imply that for $n \geqq 1$,

$$H^n(L/K, C_{L,S}) \cong H^{n+1}(L/K, P_{L,S}) , \qquad H^n(L/K, P_{L,S}) \cong H^{n+1}(L/K, E_{L,S}) .$$

As G is cyclic, we then obtain

$$H^n(L/K, A_{L,S}) \cong H^n(L/K, E_{L,S}) , \quad \text{for} \quad n \geqq 1 ,$$

where we put

$$A_{L,S} = C_{L,S}(p) .$$

We now assume that $\mu_K = 0$ so that $\mu_L = 0$. Since $A_{L,S}$ is a factor group of $A_L = C_L(p)$ modulo a finite subgroup, it follows from the remark in § 5 that

$$A_{L,S} \cong (\boldsymbol{Q}_p/\boldsymbol{Z}_p)^{\lambda_L} .$$

The structure of such a G-module $A_{L,S}$ can be described as follows. Namely, let σ be a generator of G and let

$$X_p = \boldsymbol{Z}_p[G] , \quad X_{p-1} = (1-\sigma)X_p , \quad X_1 = X_p/X_{p-1} = \boldsymbol{Z}_p ,$$
$$A_i = \mathrm{Hom}_{\boldsymbol{Z}_p}(X_i, \boldsymbol{Q}_p/\boldsymbol{Z}_p) , \quad \text{for} \quad i = 1,\ p-1,\ p .$$

Then

$$A_{L,S} = A_1^{a_1} \oplus A_{p-1}^{a_{p-1}} \oplus A_p^{a_p} \tag{11}$$

with some integers $a_1, a_{p-1}, a_p \geqq 0$[14]. With the functor V defined in § 5, let

$$V_i = V(A_i) , \quad \pi_i \colon G \to GL(V_i) , \quad \text{for} \quad i = 1,\ p-1,\ p .$$

---

[13] Cf. [1].

[14] This is a well-known result on the integral representations of G over the ring $\boldsymbol{Z}_p$.

Then π_1 is the trivial representation of G over Q_p, π_{p-1} is the unique faithful irreducible representation of G over Q_p, and

$$\pi_p = \pi_1 \oplus \pi_{p-1} = \pi_G .$$

Since

$$V_L = V(A_L) = V(A_{L,S}) ,$$

it follows from the above that

$$\pi_{L/K} = a_1\pi_1 \oplus a_{p-1}\pi_{p-1} \oplus a_p\pi_p$$

for the representation $\pi_{L/K}$ of $G = \mathrm{Gal}(L/K)$ on the space V_L. We shall next compute the integers a_1, a_{p-1}, and a_p.

Since G is a cyclic group of order p, the cohomology groups of G are abelian groups of exponent p. One checks easily that

$$r(H^1(G, A_1)) = 1 , \quad r(H^1(G, A_{p-1})) = 0 , \quad r(H^1(G, A_p)) = 0 ,$$
$$r(H^2(G, A_1)) = 0 , \quad r(H^2(G, A_{p-1})) = 1 , \quad r(H^2(G, A_p)) = 0$$

for the ranks of the abelian groups $H^n(G, A_i)$. Let

$$r_n = r(H^n(L/K, A_{L,S})) = r(H^n(L/K, E_{L,S})) , \quad n \geqq 1 .$$

Then it follows from (11) that

$$r_1 = a_1 , \qquad r_2 = a_{p-1} .$$

In particular, we see that both $H^1(L/K, E_{L,S})$ and $H^2(L/K, E_{L,S})$ are finite groups so that the Herbrand quotient $q(E_{L,S})$ of the G-module $E_{L,S}$ is defined. Let d denote the number of places on K in the finite set S and let

$$h_n = r(H^n(L/K, E_L)) , \quad n \geqq 1$$

for the G-submodule E_L of $E_{L,S}$. Then it can be seen easily that

$$q(E_{L,S}/E_L) = p^d .$$

Hence, by Herbrand's lemma, $q(E_L)$ is defind and

$$q(E_{L,S}) = q(E_L)p^d .$$

Thus both h_1 and h_2 are finite and they satisfy

$$a_{p-1} - a_1 = h_2 - h_1 + d .$$

Now, the assumption $\mu_K = 0$ implies

$$A_{K,S} = C_{K,S}(p) \cong (Q_p/Z_p)^{\lambda_K} .$$

On the other hand, since $H^2(L/K, L^\times) = 0$ by Lemma 5, the remark at the end of § 4, applied for $E_{L,S}$, gives us

$$H^1(L/K, E_{L,S}) \cong P_{L,S}^G/P_{K,S}\,,$$
$$H^2(L/K, E_{L,S}) \cong \operatorname{Coker}(I_{L,S}^G \to C_{L,S}^G)\,.$$

As the cohomology groups on the left are finite, it follows that both the kernel and the cokernel of $A_{K,S} \to A_{L,S}^G$ are finite groups. Hence we see from (11) that

$$\lambda_K = a_1 + a_p\,.$$

Therefore

$$\begin{aligned}\pi_{L/K} &= a_1\pi_1 \oplus a_{p-1}\pi_{p-1} \oplus a_p\pi_p \\ &= (a_1 + a_p)\pi_p \oplus (a_{p-1} - a_1)\pi_{p-1} \\ &= \lambda_K\pi_G \oplus d\pi_{p-1} \oplus (h_2 - h_1)\pi_{p-1}\,.\end{aligned}$$

For each non-p-place v on K, let the representations π_v and π'_v be defined similarly as π_{v^+} and π'_{v^+} in § 7. Then for the cyclic extension L/K of degree p,

$$\pi'_v = \pi_{p-1} \quad \text{or} \quad 0$$

according as v is ramified or unramified in L. Hence the following theorem is proved:

THEOREM 6. *Let $p \geqq 2$ and let L be a cyclic extension of degree p over a Z_p-field K, unramified at every infinite place of K. Assume that $\mu_K = 0$. Then*

$$\pi_{L/K} = \lambda_K\pi_G \oplus \Big(\bigoplus_v \pi'_v\Big) \oplus (h_2 - h_1)\pi_{p-1}$$

for the representation $\pi_{L/K}$ of $G = \operatorname{Gal}(L/K)$ on the vector space V_L over Q_p. Here, h_i denotes the rank of the abelian group $H^i(L/K, E_L)$ for $i = 1, 2$, π_G is the regular representation of G over Q_p, π_{p-1} the unique faithful irreducible representation of G over Q_p with degree $p-1$, and v ranges over all non-p-places of K, with π'_v as mentioned above.

Of course, if $h_2 - h_1$ is negative, the right hand side of the formula should be interpreted as a difference of two representations.

Now, comparing the degrees of the representations in Theorem 6, we immediately obtain the following formula for λ_L and λ_K:

$$\lambda_L = p\lambda_K + \sum_w (e(w/v) - 1) + (p-1)(h_2 - h_1)\,,$$

where w ranges over all non-p-places on L. Checking the proof of Theorem 6, one also finds that in the case where $p > 2$ and L/K is an extension of Z_p-fields of C-M type, similar results can be proved for $\pi_{L/K}^-$ and λ_L^- by the same method. These are of course the special cases of the formulae (9) and (10) for a cyclic extension L/K of degree p. However, as Kida [6] pointed out, his formula in the general case is actually an easy consequence of the special case mentioned above. Thus the above method provides another proof of Kida's formula (10).

BIBLIOGRAPHY

[1] A. Brumer, Galois groups of extensions of algebraic number fields with given ramification, Mich. Jour. Math. 13 (1966), 33-40.

[2] C. Chevalley and A. Weil, Über das Verhalten der Integrale erster Gattung bei Automorphismen des Funktionenkörpers, Hamb. Abh. 10 (1934), 358-361.

[3] B. Ferrero and L. Washington, The Iwasawa invariant μ_p vanishes for abelian number fields, Ann. Math. 109 (1979), 377-395.

[4] K. Iwasawa, On Γ-extensions of algebraic number fields, Bull. Amer. Math. Soc. 65 (1959), 183-226.

[5] K. Iwasawa, On Z_l-extensions of algebraic number fields, Ann. Math. 98 (1973), 246-326.

[6] Y. Kida, l-extensions of CM-fields and cyclotomic invariants, J. Number Theory 12 (1980), 519-528.

[7] S. Lang, Rapport sur la Cohomologie des Groupes, W. A. Benjamin, Inc., New York-Amsterdam, 1966.

[8] J.-P. Serre, Corps Locaux, Hermann, Paris, 1962.

[9] J.-P. Serre, Cohomologie Galoisienne, Lecture Notes in Math. 5, Springer-Verlag, Berlin-Heidelberg-New York, 1964.

[10] K. Wingberg, Die Einheitengruppe von p-Erweiterungen regulär p-adischer Zahlkörper als Galoismodul, Jour. für die reine u. angew. Math. 305 (1979), 206-214.

Department of Mathematics
Princeton University
Princeton, New Jersey 08544
U.S.A.

[59] On cohomology groups of $\mathbb{Z}_p$-extensions

"$\mathbb{Z}_p$-extensions and Related Topics" (Kyoto, 6/1981; G. Fujisaki, ed.), Kokyuroku, No. 440, RIMS, Kyoto Univ., 1981, pp. 76-84

Abstract

These are notes by G. Fujisaki based on the lecture of Iwasawa in the workshop "$\mathbb{Z}_p$-extensions and Related Topics" at RIMS (the Research Institute of Mathematical Sciences), Kyoto University, June 1981. Their content is basically contained in the paper [60] (1983) except for the following statement (Theorem 3) on Greenberg's conjecture. For a totally real number field k, let K/k be the cyclotomic $\mathbb{Z}_p$-extension. Then, Greenberg's conjecture on the Iwasawa invariants of K/k claims that one has $\lambda = \mu = 0$; in other words, the p-primary component of the ideal class group of K is trivial (or, in the notation of [60], $C(p) = 0$). Theorem 3 of this paper says, in the notation of [60], that Greenberg's conjecture for K/k is true if and only if

1) $\mathrm{Ker}(\varphi_S : H^2(K/k, E_S) \to \mathrm{Br}(k)) = 0$ for any finite set S of finite primes in k, and
2) $H^2(K/k_n, E) = (\mathbb{Q}_p/\mathbb{Z}_p)^{s_n - 1}$ for any $n \geq 0$, where k_n is the intermediate field of K/k with $[k_n : k] = p^n$ and s_n is the number of the places of k_n lying over p.
(For the proof of this theorem the reader may consult the remark in [60], §4.)

$\mathbb{Z}_p$ 拡大のコホモロジー群について

1° 以下, 考える体はすべて代数体 (複素数体 $\mathbb{C}$ の部分体であって, 有理数体 $\mathbb{Q}$ の必ずしも有限次とは限らない代数的拡大体) である.

L/K を代数体 K の Galois 拡大, $G = \mathrm{Gal}(L/K)$ をその Galois 群, A を G 加群とすれば, Galois コホモロジー群 $H^n(G, A)$ が定義される. ただし, A は discrete な加群で, G の A への作用 $G \times A \to A$ は連続である. $H^n(G, A)$ をまた $H^n(L/K, A)$ あるいは 単に $H^n(A)$ と表すことが多い.

一例として, $G(= \mathrm{Gal}(L/K))$ 加群の完全系列

$$0 \to E_L \to L^\times \to P_L \to 0$$
$$0 \to P_L \to I_L \to C_L \to 0$$

(ただし, $E_L = L$ の単数群, $P_L = L$ の単項 ideal 群, $I_L = L$ の ideal 群, $C_L = I_L/P_L$) を考えれば, これらの完全系列から次の 2 つの完全系列が導かれる.

$$0 \to E_K \to K^\times \to P_L{}^G \to H^1(E_L) \to H^1(L^\times) \to H^1(P_L) \to H^2(E_L) \to H^2(L^\times) \to \cdots$$

および

$$\cdots \to {I_L}^G \to {C_L}^G \to H^1(P_L) \to H^1(I_L) \to \cdots \quad .$$

上の 2 つの完全系列において $H^1(L^\times)=0$, $H^1(I_L)=0$ であることに注意すれば

$$H^1(L/K, E_L) \simeq {P_L}^G/P_K,$$

$$\mathrm{Ker}(H^2(L/K, E_L) \to H^2(L/K, L^\times)) \simeq \mathrm{Coker}({I_L}^G \to {C_L}^G)$$

が導かれる.

2° k を有限次代数体, K/k を $\mathbb{Z}_p$ 拡大, $\Gamma = \mathrm{Gal}(K/k)(\simeq \mathbb{Z}_p)$ とする. $\Gamma \simeq \mathbb{Z}_p$ のコホモロジー次元から, 任意の Γ 加群 A に対して

$$H^n(K/k, A) = H^n(\Gamma, A) = 0 \quad (n \geq 3),$$

また, 任意の torsion Γ 加群 A に対して

$$H^n(K/k, A) = H^n(\Gamma, A) = 0 \quad (n \geq 2)$$

となることは知られている. したがって, $H^n(K/k, A)$ $(n \geq 0)$ において実際考察の対象となるのは $n = 0, 1, 2$ の場合である.

特に, 整数論において大事な $A = E_K$ の場合に $H^2(E_K)$ を考えれば, 予想される形は

$$H^2(K/k, E_K) \simeq (\mathbb{Q}_p/\mathbb{Z}_p)^a, \quad a \geq 0$$

であり, $a = a(K/k)$ は K/k で一意的に定まる $\mathbb{Z}_p$ 拡大 K/k の 1 つの不変量である.

この不変量 $a = a(K/k)$ と K/k の他の不変量との関係, または a と k の不変量との関係を調べることは大事な問題であろうと考えられる.

3° K, k, Γ は 2° におけるとおりとして

$$S = k \text{ の有限素点の任意の有限集合}$$

$$S_0 = K/k \text{ で分岐する } k \text{ の素点全体}$$

とする. $v \in S_0 \Longrightarrow v|p$ である.

いま, $E_S = E_{K,S}$ を K の S 単数群（すなわち, S の素点の上にない K のすべての非 Archimedes 素点で単数となる K の元全体のつくる群）として

$$\varphi_S : H^2(K/k, E_S) \to H^2(K/k, K^\times) \hookrightarrow \mathrm{Br}(k)$$

($\mathrm{Br}(k) = k$ の Brauer 群) を考えれば

$$\mathrm{Ker}(\varphi_S) \simeq \mathrm{Coker}({I_{K,S}}^\Gamma \to {C_{K,S}}^\Gamma) .$$

ここで, $I_{K,S} = K$ の S ideal 群, $C_{K,S} = I_{K,S}/P_{K,S}$ である.

$H^2(K/k, E_S)$ については次の定理が成り立つ.

定理 1. $S_0 \subseteq S$ ならば

$$H^2(K/k, E_S) \simeq (\mathbb{Q}_p/\mathbb{Z}_p)^{t-1}.$$

ここで, t は S に含まれていて K において "有限分解" する k の素点の個数である.

v を k の素点, Z_v を v の K/k に関する分解群とすれば Z_v は $\Gamma = \mathrm{Gal}(K/k)(\simeq \mathbb{Z}_p)$ の閉部分群であるから

$$Z_v = \{1\} \Longleftrightarrow v \text{ は } K/k \text{ で完全分解}$$

または

$$\Gamma/Z_v = \text{有限巡回群} \Longleftrightarrow v \text{ は } K/k \text{ で有限分解}$$

となる. $S_0 \subseteq S$ のとき

$$\begin{array}{rcl} \varphi_S : H^2(K/k, E_S) & \to & \mathrm{Br}(k) \\ c & \mapsto & \varphi_S(c) = c' = (\cdots, \mathrm{inv}_v(c'), \cdots) \end{array}$$

とすれば, $\Gamma \simeq \mathbb{Z}_p$ pro-p-群 $\Longrightarrow \mathrm{inv}_v(c') \in \mathbb{Q}_p/\mathbb{Z}_p$ であり, また $v \notin S$ であるか, あるいは $v \in S$ かつ v は完全分解ならば $\mathrm{inv}_v(c') = 0$ である. さらに, 和公式 $\sum_v \mathrm{inv}_v(c') = 0$ が成り立つから, t が有限分解する $v \in S$ の個数ならば

$$\mathrm{Im}(\varphi_S) \subseteq (\mathbb{Q}_p/\mathbb{Z}_p)^{t-1} \subseteq \mathrm{Br}(k).$$

それゆえ, 定理 1 により, $S_0 \subseteq S$ であるとき

$$\mathrm{Ker}(\varphi_S) = \text{有限} \Longleftrightarrow \mathrm{Im}(\varphi_S) \simeq (\mathbb{Q}_p/\mathbb{Z}_p)^{t-1}$$

である.

Lemma 1. 次の i), ii), iii) は同値である.

i) $\mathrm{Ker}(\varphi_S) =$ 有限 ($\forall S$ に対して),
ii) $\mathrm{Ker}(\varphi_S) =$ 有限 ($\forall S \supseteq S_0$ に対して),
iii) $\mathrm{Ker}(\varphi_{S_0}) =$ 有限.

問題 1. $\quad Ker(\varphi_S) =$ 有限 $(\forall S)$?

$\mathrm{Ker}(\varphi_S) = 0$ かどうかについては次の定理が成り立つ.

定理 2. S_0 を含むすべての S に対して $\mathrm{Ker}\ (\varphi_S) = 0 \Longleftrightarrow C_{K,S_0}(p) = 0$.
ここで, $C_{K,S_0}(p)$ は $C_{K,S_0} = I_{K,S_0}/P_{K,S_0}$ の p 成分である.

$\mathrm{Ker}(\varphi_S) \neq 0$ となる例: $k = \mathbb{Q}(\sqrt[p]{1})$, $K = \mathbb{Q}(\sqrt[p^\infty]{1})$ $(p > 2)$ とすれば K/k は $\mathbb{Z}_p$ 拡大であり, $S_0 = \{v_p\}$ (v_p は p を割る k の唯 1 つの素点), $C_{K,S_0} = C_K$ である.
とくに, p を irregular な素数とすれば, $C_{K,S_0}(p) = C_K(p) \neq 0$ (素数 p について, p irregular $\Longleftrightarrow C_K(p) \neq 0$) であるから, 定理 2 により, ある $S \supseteq S_0$ で $\mathrm{Ker}(\varphi_S) \neq 0$ となるものが存在する.

4° 以下, k は総実な有限次代数体, $K = k_\infty$ は k の basic $\mathbb{Z}_p$ 拡大体であるとする. このとき, 問題 1 と関連して次の補題が成り立つ.

Lemma 2.

$$C_K(p)^\Gamma = \text{有限} \implies C_{K,S}(p)^\Gamma = \text{有限} \quad (\forall S)$$

$$\implies \text{問題 1（が肯定的に解ける）}.$$

また, 次のことも証明される.

Lemma 3.

$$C_K(p)^\Gamma = \text{有限} \implies H^2(K/k, E_K) = (\mathbb{Q}_p/\mathbb{Z}_p)^{s-1} \quad (s = \sharp(S_0))$$

$\iff$ 有理素数 p を割る k の素 ideal $\mathfrak{p}$ は K においてすべて単項 ideal となる.

問題 2. $C_K(p)^\Gamma =$ 有限 ?

問題 3. $a(K/k) = s - 1$? $(s = \sharp(S_0))$,
すなわち, $H^2(K/k, E_K) \simeq (\mathbb{Q}_p/\mathbb{Z}_p)^{s-1}$?

問題 2, 3 においては, $k =$ 総実な有限次代数体, $K/k =$ basic $\mathbb{Z}_p$ 拡大である. 問題 2 は総実でない k に対しては一般に成り立たない.

5° $k =$ 総実な有限次代数体の場合には, Leopoldt 予想は次のように述べることができる.

Leopoldt 予想 (L.C.) : $K = k_\infty$ （k の basic $\mathbb{Z}_p$ 拡大体）が k 上の唯 1 つの $\mathbb{Z}_p$ 拡大である (?)

また, 総実な k に対しては次の予想がある.

Greenberg 予想 (G.C.) : K/k が basic $\mathbb{Z}_p$ 拡大ならば, $\lambda(K/k) = \mu(K/k) = 0$? ($\iff C_K(p) = 0$?)

上に述べた 2 つの予想と問題との間には次のような関連がある ($k =$ 総実有限次代数体).

L.C. ⇘ 問題 2 ⇗ 問題 1
G.C. ⇗ 問題 2 ⇘ 問題 3

さらに, 次のことが証明される.

$$\text{G.C.} \implies \operatorname{Ker}(\varphi_S) = 0 \quad (\forall S) \implies C_{K,S_0}(p) = 0,$$

$C_K(p) = C_{K,S_0}(p) \iff$ 問題 3 がすべての K/k_n に対して肯定的に解ける.

（ここで, k_n は $\mathbb{Z}_p$ 拡大 K/k のすべての中間体を表す.）

それゆえ, 次の定理が成り立つ.

定理 3. $k =$ 総実な有限次代数体, $K/k =$ basic $\mathbb{Z}_p$ 拡大とするとき,

G.C. 正しい $\Longleftrightarrow$ 1) $\mathrm{Ker}(\varphi_S) = 0 \quad (\forall S)$.
2) $k = k_0 \subseteq k_1 \subseteq \cdots \subseteq k_n \subseteq \cdots \subseteq K$ とするとき, 問題 3 がすべての K/k_n に対して肯定的に解ける.

以上に述べたことからわかるとおり, 問題 2 および問題 1, 3 は Leopoldt 予想あるいは Greenberg 予想の必要条件である. したがって, L.C. あるいは G.C. が正しいかどうかを確かめるために（も）問題 1, 2, 3 を考えることは意味があると思われる.（なお, Leopoldt 予想は総実な k に対して essential である.）

附記. この講演記録は, 藤﨑が, 岩澤教授の講演の際とったノートにもとづいて作製したものです. したがって, 思いちがいや誤りは一切藤﨑に責任があります.

[60] On cohomology groups of units for $\mathbf{Z}_p$-extensions

Amer. J. Math., 105 (1983), 189-200; "Geometry and Number Theory, A Volume in Honor of André Weil" (J.-P. Serre and G. Shimura eds.), The Johns Hopkins University Press, 1983, pp. 189-200.

To André Weil on his 77th birthday

Let K/k be a Galois extension of number fields, S a set of finite places on the ground field k, and E_S the group of S-units in K. The cohomology groups $H^n(K/k, E_S)$, i.e., $H^n(\mathrm{Gal}(K/k), E_S)$, have been studied for various K/k and S, and found to be an important device to study the arithmetic properties of the extension K/k. In the present paper, we shall make some simple remarks on $H^n(K/k, E_S)$ in the special case where K is a so-called $\mathbf{Z}_p$-extension over a finite algebraic number field k.

1. We first consider the general case where K/k is an arbitrary Galois extension of number fields[1] and S is any set of finite, i.e., non-archimedean, places on k. As stated above, let E_S denote the group of S-units in K. Further, let I_S, P_S, and C_S denote the group of S-ideals of K, the subgroup of principal S-ideals in I_S, and the S-ideal-class group of K, respectively.[2] For simplicity, the cohomology group of Gal(K/k) acting on an abelian group A such as mentioned above, will be denoted by $H^n(A)$; thus, e.g., $H^n(E_S) = H^n(K/k, E_S)$. For such cohomology groups, we shall next prove an exact sequence which is essentially the same as the special case for number fields, of the seven term exact sequence found independently by Auslander-Brumer [1] and Chase-Harrison-Rosenberg [3][3]. A proof will be given here because it is quite simple in this special case, and also because the exact sequence plays an important role in the subsequent arguments.

Now, it can be shown easily that $H^1(I_S) = 0$. Since $H^1(K^\times) = 0$, we obtain from $P_S = K^\times/E_S$ and $C_S = I_S/P_S$ the exact sequences as follows:

$$0 \to E_S^\Gamma \to k^\times \to P_S^\Gamma \to H^1(E_S) \to 0,$$

$$0 \to H^1(P_S) \to H^2(E_S) \to H^2(K^\times) \to H^2(P_S) \to H^3(E_S) \to \ldots,$$

Manuscript received August 17, 1982

$$0 \to P_S^\Gamma \to I_S^\Gamma \to C_S^\Gamma \to H^1(P_S) \to 0,$$
$$0 \to H^1(C_S) \to H^2(P_S) \to H^2(I_S) \to \cdots\cdots$$

where $\Gamma = \mathrm{Gal}(K/k)$. Let $P_{k,S}$ denote the group of principal S-ideals of the ground field k. Then the first exact sequence in the above yields

$$H^1(E_S) \simeq P_S^\Gamma / P_{k,S}.$$

Hence the third one can be written as

(1) $$0 \to H^1(E_S) \to I_S^\Gamma / P_{k,S} \to C_S^\Gamma \to H^1(P_S) \to 0.$$

In the second exact sequence above, let

$$X = \mathrm{Im}(H^2(E_S) \to H^2(K^\times)),$$

$$Y = \mathrm{Im}(H^2(K^\times) \to H^2(P_S)) = \mathrm{Ker}(H^2(P_S) \to H^3(E_S))$$

so that

(2) $$0 \to H^1(P_S) \to H^2(E_S) \to X \to 0,$$

(3) $$0 \to X \to H^2(K^\times) \to Y \to 0$$

are exact. By the fourth exact sequence, we may regard $H^1(C_S)$ as a subgroup of $H^2(P_S)$ so that

(4) $$H^1(C_S) = \mathrm{Ker}(H^2(P_S) \to H^2(I_S)).$$

Then

(5) $$0 \to Y \cap H^1(C_S) \to H^1(C_S) \to H^3(E_S)$$

is exact. Let

$$B_S = \mathrm{Ker}(H^2(K^\times) \to H^2(I_S))$$

where the map is induced by $K^\times \to I_S$, $\alpha \mapsto (\alpha)$. Since that is the product of maps

$$H^2(K^\times) \to H^2(P_S) \to H^2(I_S),$$

(3) and (4) yield an exact sequence

$$0 \to X \to B_S \to Y \cap H^1(C_S) \to 0.$$

Hence, by (2) and (5), we have another exact sequence

$$0 \to H^1(P_S) \to H^2(E_S) \to B_S \to H^1(C_S) \to H^3(E_S).$$

Finally, combining this with (1), we obtain the following

PROPOSITION 1. *There exists a seven term exact sequence*

$$0 \to H^1(E_S) \to I_S^\Gamma/P_{k,S} \to C_S^\Gamma \to H^2(E_S)$$

$$\to B_S \to H^1(C_S) \to H^3(E_S),$$

where $B_S = \mathrm{Ker}(H^2(K^\times) \to H^2(I_S))$, $\Gamma = \mathrm{Gal}(K/k)$.

2. Let **Z** and **Q** denote as usual the ring of rational integers and the field of rational numbers respectively. Throughout the following, we fix a prime number p and denote by $\mathbf{Z}_p$ and $\mathbf{Q}_p$ the ring of p-adic integers and the field of p-adic numbers respectively. For simplicity, the additive groups of those rings and fields will be denoted again by the same letters. In general, if A is an abelian group, let $A(p)$ denote the subgroup of all elements in A with p-power orders. In particular, we put

$$R = (\mathbf{Q}/\mathbf{Z})(p) = \mathbf{Q}_p/\mathbf{Z}_p.$$

If A is the direct sum of a finite group and R^n, the direct sum of n copies of A, then we write

$$A \sim R^n, \qquad n = \rho(A).$$

From now on, we always suppose that K is a $\mathbf{Z}_p$-extension over a finite algebraic number field k and that S is a finite set of finite places on the ground field k. By the definition of a $\mathbf{Z}_p$-extension,

$$\Gamma = \mathrm{Gal}(K/k) \simeq \mathbf{Z}_p$$

so that Γ is a free pro-p-group. Hence if $n \geq 3$, then $H^n(\Gamma, A) = 0$ for every discrete Γ-module A, and $H^2(\Gamma, A) = 0$ if A is in particular a torsion abelian group. This implies that for any discrete Γ-module A, the map $x \mapsto px$ induces a surjective homomorphism

$$p : H^2(\Gamma, A) \to H^2(\Gamma, A).$$

Now, since $H^3(E_S) = H^3(\Gamma, E_S) = 0$, Proposition 1 gives us the following exact sequence:

$$(6) \qquad 0 \to H^1(E_S) \to I_S^\Gamma/P_{k,S} \to C_S^\Gamma$$
$$\to H^2(E_S) \to B_S \to H^1(C_S) \to 0.$$

Here $I_S^\Gamma/P_{k,S}$ and C_S^Γ may be replaced by $(I_S^\Gamma/P_{k,S})(p)$ and $C_S^\Gamma(p) = C_S(p)^\Gamma$ respectively, because all other terms in (6) are p-primary abelian groups. We shall next show that $\rho(\quad)$ is defined for every group in the modified exact sequence.

Let S_0 denote the set of all places on k which are ramified in K. Since only p-places on k can be ramified in the $\mathbf{Z}_p$-extension K over k, S_0 is a finite set of finite places on k. Let S' be the union of S and S_0, and let S_0' denote the difference of S' and S:

$$S' = S \cup S_0 = S \cup S_0', \qquad S \cap S_0' = \phi, \qquad S_0' \subseteq S_0.$$

Let u be the number of places contained in S_0'. Then one sees easily from the definition of I_S^Γ that

$$(I_S^\Gamma/P_{k,S})(p) \sim R^u, \qquad \rho((I_S^\Gamma/P_{k,S})(p)) = u.$$

On the other hand, it is known in the theory of $\mathbf{Z}_p$-extensions that $C_S(p)$ is the dual of a compact, neotherian, torsion module over the group ring $\Lambda = \mathbf{Z}_p[[\Gamma]]$ so that both $\rho(C_S(p)^\Gamma)$ and $\rho(H^1(C_S))$ are well-defined and they are equal[4]:

$$\rho(C_S(p)^\Gamma) = \rho(H^1(C_S)).$$

In general, for each place v on k, let Z_v denote the decomposition group of v for the extension K/k. Since $\Gamma = \mathrm{Gal}(K/k) \simeq \mathbf{Z}_p$, either $Z_v = 1$ or Z_v is a closed subgroup of finite index in Γ. According to the cases, we say that v is completely, or, finitely decomposed in K. For example, an infinite place is completely decomposed.

LEMMA 1. *Let t denote the number of places in S which are finitely decomposed in K. Then*

$$B_S \simeq R^{t+u-1}, \qquad \rho(B_S) = t + u - 1.$$

Proof. Let k_v and K_v denote the v-completions of k and K respectively so that $Z_v = \mathrm{Gal}(K_v/k_v)$. Then

$$H^2(K_v/k_v, K_v^\times) = 0 \text{ or } R$$

according as v is completely or finitely decomposed in K. Let v be a finite place on k, *not* contained in S, and let $\mathfrak{p}_v$ be the prime ideal in $I_{k,S}$, associated with v, $I_{k,S}$ being the group of S-ideals in k. Let I_v denote the subgroup of I_S, generated by the ideals of I_S which divide $\mathfrak{p}_v$ (in the obvious sense). Then

$$I_S = \bigoplus_v I_v, \qquad H^2(I_S) = \bigoplus_v H^2(K/k, I_v)$$

where the direct sums are taken over all v as stated above. In the last sum, if v is unramified in K, then

$$H^2(K/k, I_v) = H^2(Z_v, \mathbf{Z}) \simeq H^2(K_v/k_v, K_v^\times),$$

but if v is ramified in K, i.e., $v \in S_0'$, then

$$H^2(K/k, I_v) = 0$$

because I_v is uniquely divisible by p in this case. Now, let c be an element of $H^2(K^\times)$ and let

$$c = (\ldots\ldots, \mathrm{inv}(c_v), \ldots\ldots)$$

with local invariants $\mathrm{inv}(c_v)$ in $H^2(K_v/k_v, K_v^\times)$. From what is mentioned in the above, we then see that c belongs to $B_S = \mathrm{Ker}(H^2(K^\times) \to H^2(I_S))$ if and only if $\mathrm{inv}(c_v) = 0$ for all v, not contained in $S' = S \cup S_0$. Since $\mathrm{inv}(c_v)$

$= 0$ any way if v is completely decomposed in K and since the sum of $\mathrm{inv}(c_v)$ for all v is zero, we have $B_S \simeq R^{t+u-1}$ as stated in the lemma.

We now see from the exact sequence (6) that

$$r = \rho(H^1(E_S)), \qquad s = \rho(H^2(E_S))$$

are well defined and that

$$0 \leq r \leq u, \qquad r - u + a - s + (t + u - 1) - a = 0$$

with $a = \rho(C_S(p)^\Gamma) = \rho(H^1(C_S))$. Thus

$$s = r + t - 1.$$

Since $p : H^2(E_S) \to H^2(E_S)$ is surjective, we obtain the following

PROPOSITION 2. *Let t denote the number of places in S, finitely decomposed in K, and let u be the number of places on k, ramified in K, but not contained in S. Then*

$$H^1(K/k, E_S) \sim R^r, \qquad H^2(K/k, E_S) \simeq R^s$$

with integers $r = r(K/k, S)$ and $s = s(K/k, S)$ satisfying

$$0 \leq r \leq u, \qquad s = r + t - 1.$$

3. In most cases, the integers t and u in Proposition 2 are easy to determine for the given K/k and S. But, because of the inequalities $0 \leq r \leq u$, the proposition does not exactly describe the structure of $H^1(K/k, E_S)$ and $H^2(K/k, E_S)$ by means of those integers t and u. However, there are cases where $r = r(K/k, S)$ and $s = s(K/k, S)$ are in fact well determined by t and u, and we shall next discuss such cases.

PROPOSITION 3. *Suppose that S contains all places on k which are ramified in K, i.e., $S_0 \subseteq S$. Then*

$$H^1(K/k, E_S) \sim 0, \qquad H^2(K/k, E_S) \simeq R^{t-1}.$$

Proof. This is clear from Proposition 2 because, in this case, $u = 0$ so that $r = 0$.

In general, let φ_S denote the map from $H^2(E_S)$ to B_S which appears in the exact sequence (6):

$$\varphi_S: H^2(K/k, E_S) \to B_S.$$

LEMMA 2. *The following statements* i), ii), iii) *are equivalent:*

i) *φ_S is surjective,*
ii) *$C_S(p)^\Gamma$ is finite,*
iii) $H^1(K/k, C_S) = 0$.

These i), ii), iii) *imply that* $\mathrm{Ker}(\varphi_S)$ *is finite; and in the case $S_0 \subseteq S$, the converse is also true.*

Proof. As mentioned earlier, $\rho(C_S(p)^\Gamma) = \rho(H^1(K/k, C_S))$. Hence, if $C_S(p)^\Gamma$ is finite, then so is $H^1(K/k, C_S)$. However, by (6) and Lemma 1, $H^1(K/k, C_S)$ is a homomorphic image of $B_S \simeq R^{t+u-1}$. Hence $H^1(K/k, C_S)$ is finite only when $H^1(K/k, C_S) = 0$. Thus ii) and iii) are equivalent. The equivalence of i) and iii) is clear from (6). It also follows from (6) that if $C_S(p)^\Gamma$ is finite, then $\mathrm{Ker}(\varphi_S)$ is finite too. In the case $S_0 \subseteq S$, the integer u in Lemma 1 and Proposition 2 is zero so that

$$H^2(K/k, E_S) \simeq B_S \simeq R^{t-1}.$$

Hence φ_S is surjective if and only if $\mathrm{Ker}(\varphi_S)$ is finite.

LEMMA 3. *If φ_T is surjective for a subset T of S, then φ_S is also surjective.*

Proof. Since $T \subseteq S$, there exist natural homomorphisms $C_T \to C_S$, $C_T(p) \to C_S(p)$. As $C_T(p)$ and $C_S(p)$ are both duals of compact, noetherian, torsion Λ-modules, it follows from the theory of such modules that $C_T(p)^\Gamma \sim 0$ implies $C_S(p)^\Gamma \sim 0$. The lemma therefore follows from the preceding Lemma 2.

PROPOSITION 4. *Suppose that φ_S is surjective. Then* $\mathrm{Ker}(\varphi_S)$ *is finite and*

$$H^1(K/k, E_S) \sim R^u, \qquad H^2(K/k, E_S) \simeq R^{t+u-1}$$

with integers t and u as stated in Proposition 1.

Proof. The assumption is equivalent to that $a = \rho(C_S(p)^\Gamma) = \rho(H^1(K/k, C_S)) = 0$. The proof of Proposition 2 then shows that $r = u$, $s = t + u - 1$.

Let us next consider the important special case where S is the empty set: $S = \emptyset$. In this case, we shall denote $I_\emptyset$, $C_\emptyset$, $E_\emptyset$, etc. simply by I, C, E, etc. Thus C is the ideal class group of K and E is the group of all units in K in the usual sense. We also write φ for $\varphi_\emptyset$:

$$\varphi: H^2(K/k, E) \to B = B_\emptyset.$$

PROPOSITION 5. *Let $u(K/k)$ denote the number of places on k which are ramified in K. Then*

$$H^1(K/k, E) \sim R^r, \qquad H^2(K/k, E) \simeq R^{r-1}$$

with an integer r satisfying $1 \leq r \leq u(K/k)$. If φ is surjective, then $\mathrm{Ker}(\varphi)$ *is finite and the equality*

$$r = u(K/k)$$

holds. Furthermore, in this case, φ_S is surjective for every S so that $H^1(K/k, E_S)$ and $H^2(K/k, E_S)$ are given by Proposition 4.

Proof. In the case $S = \emptyset$, we have $u = u(K/k)$, $t = 0$ in Proposition 2. Hence $r - 1 = s \geq 0$, and the first part is clear. The other part is also an immediate consequence of Proposition 4 and Lemma 2.

4. Now the question arises: when is the map φ_S surjective? Recall first that a $\mathbf{Z}_p$-extension K/k is called basic (or cyclotomic) if $K = k\mathbf{Q}_\infty$ where $\mathbf{Q}_\infty$ is the unique $\mathbf{Z}_p$-extension over the rational field $\mathbf{Q}$, and that it is called a $\mathbf{Z}_p$-extension of *C-M* type if both k and K are number fields of *C-M* type. We know by Lemma 2 that φ is surjective if and only if $C(p)^\Gamma$ is finite. But the latter is a rather strong condition on the $\mathbf{Z}_p$-extension K/k, and one can easily find $\mathbf{Z}_p$-extensions, in fact, basic $\mathbf{Z}_p$-extensions of *C-M* type, for which $C(p)^\Gamma$ are infinite. Thus φ is in general not surjective, even for the nice family of basic $\mathbf{Z}_p$-extensions of *C-M* type, and the exact evaluation of the integer r in Proposition 5, instead of $1 \leq r \leq u(K/k)$, provides us an interesting open problem.

Next, consider φ_S: $H^2(K/k, E_S) \to B_S$ in the case $S_0 \subseteq S$. In such a case, both $H^2(K/k, E_S)$ and B_S are isomorphic to R^{t-1} so that there seems to be a better chance for φ_S to be surjective (or, even an isomorphism). However, it has recently been found that there are non-basic $\mathbf{Z}_p$-extensions

K/k for which φ_{S_0} is not surjective, i.e., for which $C_{S_0}(p)^\Gamma$ is infinite.[5] Thus again φ_S is in general not surjective even under the assumption $S_0 \subseteq S$. On the other hand, there are also many cases in which φ_{S_0} are surjective. We shall discuss some of these below.

As before, let $u(K/k)$ denote the number of places on k, ramified in K. Now, fix the ground field k and find a $\mathbf{Z}_p$-extension K over k for which $u(K/k)$ attains the minimum in the family of all $\mathbf{Z}_p$-extensions over k. By the theory of $\mathbf{Z}_p$-extensions, we then see easily that $C(p)^\Gamma$ is finite for this K/k so that φ is surjective. Thus, for each finite algebraic number field k, there exists at least one $\mathbf{Z}_p$-extension K over k for which we may apply Proposition 5; in particular, φ_S is surjective for every S for such K/k.

Let us now suppose that k is a totally real field and let k_∞ denote the basic $\mathbf{Z}_p$-extension over k: $k_\infty = k\mathbf{Q}_\infty$. A well-known conjecture of Leopoldt then states that k_∞ is the only $\mathbf{Z}_p$-extension over k. Therefore, if the conjecture is true for k (and for the fixed prime p), then it follows from the above remark that φ_S is surjective for every S so that $H^1(k_\infty/k, E_S)$ and $H^2(k_\infty/k, E_S)$ are given by Proposition 4. In particular, we have

$$H^1(k_\infty/k, E) \sim R^u, \qquad H^2(k_\infty/k, E) \simeq R^{u-1}$$

with $u = u(k_\infty/k)$ as stated above. It should be quite interesting if one could prove the last result without assuming Leopoldt's conjecture.

Next, let K/k be a basic $\mathbf{Z}_p$-extension of C-M type and let K^+/k^+ be the corresponding $\mathbf{Z}_p$-extension of maximal real subfields. If the set S of places on k is invariant under the action of the complex conjugation J, then φ_S induces two maps

$$\varphi_S^\pm : H^2(K/k, E_S)^\pm \to B_S^\pm$$

where $\pm$ for groups indicate the subgroups on which J acts as multiplication by ± 1. For odd p, φ_S is the direct sum of φ_S^+ and φ_S^-, and φ_S^+ is nothing but the map for totally real K^+/k^+ just discussed in the above. On the other hand, that φ_S^- is surjective for $S = S_0$ (and hence for all S with $S_0 \subseteq S$) is the conjecture of Gross in [6], and this has been verified in the case k is abelian over $\mathbf{Q}$[6]. Since Leopoldt's conjecture is also proved in that case, it follows that for a basic $\mathbf{Z}_p$-extension k_∞/k of abelian fields and for $p > 2$, φ_S is always surjective for all S containing S_0. Of course, the same result can be obtained in general for a basic $\mathbf{Z}_p$-extension k_∞/k, $p > 2$, if the conjectures of Leopoldt and Gross are both true for the ground field k[7].

In the above, we have not discussed whether or not the map $\varphi_S : H^2(K/k, E_S) \to B_S$ is injective. We now briefly take up this question. The basic fact in this situation is the following isomorphism obtained from (6):

$$\mathrm{Ker}(\varphi_S) \xrightarrow{\sim} \mathrm{Coker}(I_S^\Gamma \to C_S^\Gamma)(p).$$

Assume first that $C_{S_0}(p) = 0$. Then $C_S(p) = 0$ for every S with $S_0 \subseteq S$ and it follows from the above that $\mathrm{Ker}(\varphi_S) = 0$ for all such S. Suppose conversely that $\mathrm{Ker}(\varphi_S) = 0$ for all S containing S_0. One sees easily that there exists an S containing S_0 such that $\mathrm{Im}(I_S^\Gamma \to C_S^\Gamma) = 0$ and that every place v in S is finitely decomposed in K. As $\mathrm{Ker}(\varphi_S) = 0$ by the assumption, it then follows again from the above that $C_S(p)^\Gamma = 0$ for this set S. Since Γ is a pro-p-group, this implies $C_S(p) = 0$. However, by the choice of S, $\mathrm{Ker}(C_{S_0}(p) \to C_S(p))$ is finite. Hence $C_{S_0}(p)$ must be finite, and it follows from the theory of $\mathbf{Z}_p$-extensions that $C_{S_0}(p) = 0^8$. Thus it is proved that $\mathrm{Ker}(\varphi_S) = 0$ for every S containing S_0 if and only if $C_{S_0}(p) = 0$. As an example, let K be the basic $\mathbf{Z}_p$-extension over the cyclotomic field of $2p$-th roots of unity. In this case, it is known that $C(p) = C_{S_0}(p)$ and that $C(p) = 0$ if and only if p is a regular prime. Hence, if p is regular, then φ_S is injective for all S, $S_0 \subseteq S$, but if p is irregular, then φ_S is not injective for some S containing S_0.

5. Finally, we shall briefly sketch two additional proofs of Proposition 2 which might be of some interest in other connections.

Keeping the notation introduced in Section 2, let $S' = S \cup S_0$ and let M denote the maximal p-extension over k, unramified outside S'. Let E_S' denote the group of S-units in M. Then, since $k \subseteq K \subseteq M$, the Hochschild-Serre spectral sequence

$$E_2^{m,n} = H^m(K/k, H^n(M/K, E_S')), \qquad m, n \geq 0$$

is defined. As $\Gamma = \mathrm{Gal}(K/k)$ is a free pro-p-group, most of the terms $E_2^{m,n}$ are zero. Furthermore, using the results of Brumer [2], one can compute $\rho(H^2(M/k, E_S'))$ and $\rho(E_2^{0,2}) = \rho(H^2(M/K, E_S')^\Gamma)$, and one also sees easily that $\rho(H^1(M/k, E_S'))$ is equal to the integer u in Proposition 2. The convergence of the spectral sequence then gives us the formula $s = r + t - 1$ where $r = \rho(E_2^{1,0})$, $s = \rho(E_2^{2,0})$.

For another proof of Proposition 2, let k_n denote, for each $n \geq 0$, the unique intermediate field of k and K with degree p^n over k, and let E_n be the group of S-units in k_n. Since Gal(k_n/k) is a cyclic group of order p^n, the inflation maps

$$H^1(k_n/k, E_n) \to H^1(K/k, E_S), \qquad H^2(k_n/k, E_n) \to H^2(K/k, E_S)$$

induce homomorphisms

$$\lambda_n: H^1(k_n/k, E_n) \to H^1(K/k, E_S)_{p^n},$$

$$\mu_n: H^2(k_n/k, E_n) \to H^2(K/k, E_S)_{p^n}$$

where the right hand sides are the subgroups of $H^1(K/k, E_S)$ and $H^2(K/k, E_S)$, respectively, consisting of all those elements with orders dividing p^n. It is well known that λ_n is injective. But, we can also prove that the orders of the groups

$$\mathrm{Ker}(\lambda_n), \qquad \mathrm{Coker}(\lambda_n), \qquad \mathrm{Ker}(\mu_n), \qquad \mathrm{Coker}(\mu_n)$$

are bounded for all $n \geq 0$; one essential arithmetic fact needed for the proof is a result on $\mathbf{Z}_p$-extensions that if C_n denotes the S-ideal class group of k_n, then the order of $\mathrm{Ker}(C_n \to C_S)$ is bounded for all $n \geq 0$[9]. Let $r = \rho(H^1(K/k, E_S))$, $s = \rho(H^2(K/k, E_S))$ as before and let p^{e_n} and p^{f_n} denote the orders of $H^1(k_n/k, E_n)$ and $H^2(k_n/k, E_n)$ respectively. Then it follows from the above that

$$| e_n - rn |, \qquad | f_n - sn |$$

are bounded for all $n \geq 0$. However, the well-known formula on the Herbrand quotient of E_n for the cyclic extension k_n/k tells us that $| f_n - e_n - (t-1)n |$ is also bounded for all $n \geq 0$. Hence we obtain immediately the equality $s = r + t - 1$. Although this proof is more involved than the others when details are carried out, it gives us additional information on the relation between $H^1(K/k, E_S)$, $H^2(K/k, E_S)$ and $H^1(k_n/k, E_n)$, $H^2(k_n/k, E_n)$, $n \geq 0$. In particular, it shows that the formula $s = r + t - 1$ in Proposition 2 is, in a sense, an analogue, for the $\mathbf{Z}_p$-extension K/k, of the classical formula on the Herbrand quotient of units for a cyclic extension of finite algebraic number fields.

PRINCETON UNIVERSITY

REFERENCES

[1] M. Auslander and A. Brumer, Galois cohomology and the Brauer group of commutative rings (unpublished).
[2] A. Brumer, Galois groups of extensions of algebraic number fields with given ramification. *Michigan Math. Jour.* **13** (1966), 33–40.
[3] S. U. Chase, D. K. Harrison, and A. Rosenberg, Galois theory and cohomology of commutative rings. *Mem. Amer. Math.* Soc. No. **52** (1965).
[4] L. Federer and B. Gross, Regulators and Iwasawa modules. *Inv. Math.* **62** (1981), 443–457.
[5] R. Greenberg, On a certain l-adic representation. *Inv. Math.* **21** (1973), 117–124.
[6] B. Gross, p-adic L-series at $s = 0$. *Jour. Fac. Science, Tokyo University*, Sec. IA, Math. **28** (1981), 979–994.
[7] K. Iwasawa, On $\mathbf{Z}_l$-extensions of algebraic number fields. *Ann. Math.* **98** (1973), 246–326.
[8] J. F. Jaulent, Sur la théorie des genres dans une extension procyclique métabélienne sur un sous-corps. To appear.
[9] H. Kisilevsky, Some non-semi-simple Iwasawa modules. To appear in *Compositio Mathematica*.

[1] By a number field, we mean an algebraic, but not necessarily finite, extension of the rational field.

[2] For the definition of the groups E_S, I_S, etc., see [2].

[3] Note, however, that the extension K/k is not necessarily unramified outside S.

[4] Cf. [7], Theorem 11.

[5] See Jaulent [8], Kisilevsky [9].

[6] See [5], [6]. See also [4] for the fact that the conjecture stated in [6] is in fact equivalent to the surjectivity of φ_S^-, i.e., the finiteness of $C_S(p)^\Gamma$, for $S = S_0$.

[7] Note that the examples in [8], [9] are non-basic $\mathbf{Z}_p$-extensions.

[8] Loc. cit.[4].

[9] Cf. [7], Theorem 10.

[61] A simple remark on Leopoldt's conjecture

"Report of the Conference on Algebraic Number theory"
(RIMS, Kyoto, 7/1984; K. Shiratani, ed.), 1984, pp. 45-54

Abstract

These are notes by G. Fujisaki and K. Tateyama based on the lecture of Iwasawa in the "Conference on Algebraic Number Theory" at RIMS, Kyoto University, July 1984. A more detailed account can be found in his unpublished article [U4]. The main theorem of this article is the same as Theorem 1 in [U4], but stated here without using the terminology "q-field". The proof given here is rather sketchy; but in the closing comments he says that there are several possible proofs. Actually he first obtained a proof by using cohomology theory and later found an elementary one as given here (and also in [U4]) depending on a theorem of Chevalley (J. Math. Soc. Japan, **3** (1951)).

Leopoldt 予想についての簡単な注意

k を有限次代数体, p を素数として以下 fix する.

E を k の単数群, n, m を自然数とするとき E の部分群 E_m, E^n を次の様に定義する.

$$\begin{aligned} E_m &= \{\varepsilon \in E \mid \varepsilon \equiv 1 \mod p^m\}, \\ E^n &= \{\varepsilon^n \mid \varepsilon \in E\}. \end{aligned}$$

このとき, Leopoldt conjecture (以下 L.C. と略記する) は, 次の様にいうことができる.

「n が p 巾のとき, m を十分大とすると $E_m \subseteq E^n$.」

その他, $\mathbb{Z}_p$ 拡大, cohomology 等を用いた formulation がある. 例えば, cohomology を用いると, 次の様になる.

M : maximal pro-p-extension over k unramified outside p

とする. このとき ($p>2$ または $p=2,\ \sqrt{-1}\in k$ ならば)

「L.C. $\Longleftrightarrow$ $H^2(M/k, \mathbb{Q}_p/\mathbb{Z}_p) = 0$.」

ここでの目標は, 次に述べる単数群に関する Chevalley の定理を用いて L.C. の言い換えをすることである.

定理 (Chevalley, J. Math. Soc. Japan, 1951 高木記念号).
$\forall N,\ n \geq 1$ に対して, 次の様な自然数 a が存在する.
$(a, N) = 1$ であり, $\varepsilon \in E$ が $\varepsilon \equiv 1 \bmod a$ ならば, $\varepsilon \in E^n$ である.

a は, もちろん n に依存するわけであるが Chevalley の証明からは, 関係がよくわからない. しかし L.C. の観点から考えると

「L.C. $\iff$ Chevalley の定理において $N = 1$,
$n = p$ 巾のとき, a として p 巾がとれる.」

以下, Chevalley の定理を用いて L.C. の言い換えを述べる.

まず, 次の記号を導入する.
$\mathfrak{q} : p$ と素な k の素イデアル,
$k_{\mathfrak{q}}$: 剰余体,
$N(\mathfrak{q}) = \sharp(k_{\mathfrak{q}})$.
このとき
$N(\mathfrak{q}) - 1 = \sharp(k_{\mathfrak{q}}^{\times}) = d(\mathfrak{q})e(\mathfrak{q}),\quad p \nmid d(\mathfrak{q}),\ e(\mathfrak{q}) : p$ 巾
とする.
また K/k を有限次アーベル拡大とするとき
$e(\mathfrak{q} : K/k) = \mathfrak{q}$ の K/k における分岐指数,
$e(\mathfrak{q} : K/k)_p = e(\mathfrak{q} : K/k)$ の p 成分
とする. $\mathfrak{q} \nmid p$ より $e(\mathfrak{q} : K/k)_p \mid e(\mathfrak{q})$ である.

定義.

$$e^*(\mathfrak{q}) = \max_K \{\ e(\mathfrak{q} : K/k)_p \mid K/k : p\mathfrak{q} \text{ の外で不分岐な有限次アーベル拡大 } \}.$$

上に注意したことから, $e^*(\mathfrak{q}) \mid e(\mathfrak{q})$ である.

定理. L.C. $\iff$ $e^*(\mathfrak{q}) = e(\mathfrak{q})\quad \forall \mathfrak{q} \nmid p$.

証明. $m, n \in \mathbb{N} \cup \{0\}$ に対して,
$K_{p^m\mathfrak{q}^n} : k$ 上の ray class field mod $p^m\mathfrak{q}^n$
とする. このとき $K_{p^m\mathfrak{q}^n}$ は k 上有限次アーベル拡大で $p\mathfrak{q}$ の外では不分岐である. また K/k が有限次アーベル拡大で, $p\mathfrak{q}$ の外で不分岐ならば

$$k \subseteq K \subseteq K_{p^m\mathfrak{q}^n} \qquad m, n \gg 0.$$

よって

$$\begin{aligned} e^*(\mathfrak{q}) &= \max\{\ e(\mathfrak{q} : K_{p^m\mathfrak{q}^n}/k)_p \mid m, n \geq 1\ \} \\ &= \max\{\ e(\mathfrak{q} : K_{p^m\mathfrak{q}}/k)_p \mid m \geq 1\ \} \\ &= e(\mathfrak{q} : K_{p^m\mathfrak{q}}/k)_p \quad m \gg 0. \end{aligned}$$

まず次の補題を証明する.

補題. $e^*(\mathfrak{q}) = e(\mathfrak{q}) \iff {E_m}^{d(\mathfrak{q})} \equiv 1 \bmod \mathfrak{q}\qquad$ for $m \gg 0$.

証明. 上の考察より, $m \gg 0$ のとき $e^*(\mathfrak{q}) = e(\mathfrak{q} : K_{p^m\mathfrak{q}}/k)_p$.

また K_{p^m}/k で $\mathfrak{q}$ は不分岐であるから $e(\mathfrak{q}:K_{p^m\mathfrak{q}}/k)=[K_{p^m\mathfrak{q}}:K_{p^m}]$.
ここで

$$A=\{\ \alpha\in k^\times|\ (\alpha,p\mathfrak{q})=1 \text{ and } \alpha\equiv 1 \mod p^m\ \},$$
$$B=\{\ \alpha\in k^\times|\ (\alpha,p\mathfrak{q})=1 \text{ and } \alpha\equiv 1 \mod p^m\mathfrak{q}\ \},$$
$$(A)=AE/E,\ \ (B)=BE/E$$

とすれば類体論より

$$[K_{p^m\mathfrak{q}}:K_{p^m}]=[(A):(B)]=[AE:BE],$$
$$AE/BE\cong A/A\cap BE=A/B(A\cap E)=A/BE_m.$$

よって, $e^*(\mathfrak{q})=e(\mathfrak{q}:K_{p^m\mathfrak{q}}/k)_p=[A:BE_m]_p$.
また, $B\subset BE_m\subset A$, $A/B\cong(\mathfrak{O}/\mathfrak{q})^\times$ より

$$e^*(\mathfrak{q})=e(\mathfrak{q}) \iff p\nmid[BE_m:B].$$

$A/B\cong(\mathfrak{O}/\mathfrak{q})^\times$ は order $d(\mathfrak{q})e(\mathfrak{q})$ の cyclic group であるから

$$p\nmid[BE_m:B]\iff(BE_m)^{d(\mathfrak{q})}\subset B \iff {E_m}^{d(\mathfrak{q})}\subset B$$
$$\iff {E_m}^{d(\mathfrak{q})}\equiv 1 \mod \mathfrak{q}.$$

よって補題は証明された.

定理の証明にもどると,
(L.C. $\Longrightarrow$)

$$\begin{aligned}\text{L.C.} &\Longrightarrow\ m\gg 0 \text{ ならば } E_m\subseteq E^{e(\mathfrak{q})}\\ &\Longrightarrow\ {E_m}^{d(\mathfrak{q})}\subseteq E^{d(\mathfrak{q})e(\mathfrak{q})}=E^{N\mathfrak{q}-1}\equiv 1 \mod \mathfrak{q}\\ &\Longrightarrow\ {E_m}^{d(\mathfrak{q})}\equiv 1 \mod \mathfrak{q}.\end{aligned}$$

したがって, 補題により $e^*(\mathfrak{q})=e(\mathfrak{q})$.

(L.C. $\Longleftarrow$)
Chevalley の定理において, $N=p$, $n=p$ 巾 とすると,

$$\exists a\in\mathbb{N} \quad \text{s.t.}\begin{cases}(a,p)=1\\ \varepsilon\in E,\ \varepsilon\equiv 1 \bmod a \Longrightarrow \varepsilon\in E^n.\end{cases}$$

(a) の k における素イデアル分解を, $(a)=\prod_{i=1}^r {\mathfrak{q}_i}^{\nu_i}$, $(\mathfrak{q}_i,p)=1$, とする.

仮定 $\Longrightarrow e^*(\mathfrak{q}_i)=e(\mathfrak{q}_i)\Longrightarrow$ 補題により ${E_m}^{d_i}\equiv 1(\mathfrak{q}_i)\quad m\gg 0, d_i=d(\mathfrak{q}_i)$

m は, 十分大きくとることによって, すべての i に共通にとることができる.
ここで, ${d_i}'=N(\mathfrak{q}_i)^{\nu_i-1}$, $d=\prod d_i{d_i}'$ とすると

$$\begin{aligned}&{E_m}^{d_i{d_i}'}\equiv 1 \mod {\mathfrak{q}_i}^{\nu_i}\\ \Longrightarrow\ &{E_m}^{d}\equiv 1 \mod {\mathfrak{q}_i}^{\nu_i} \text{ for } \forall\, i\\ \Longrightarrow\ &{E_m}^{d}\equiv 1 \mod a.\end{aligned}$$

故に, Chevalley の定理より $E_m{}^d \subset E^n$.
$(d,p)=1,\ \ n=p$ 巾より $E_m \subset E^n$.
よって定理は証明された.

上で証明したことより

$$\text{L.C.} \iff e^*(\mathfrak{q}) = e(\mathfrak{q}),\ \ \forall \mathfrak{q} \nmid p$$

$\iff$ 任意の $\mathfrak{q}$ $(\mathfrak{q} \nmid p)$ に対して, 次のような K/k が存在する.

$$\begin{cases} K/k: \text{有限次アーベル拡大で},\ p\mathfrak{q}\ \text{の外で不分岐}, \\ e(\mathfrak{q}: K/k)_p = e(\mathfrak{q}). \end{cases}$$

系.

L.C. $\iff$ 任意の $\mathfrak{q}_1, \ldots, \mathfrak{q}_s$ $\mathfrak{q}_i \nmid p$ $(i=1,\ldots,s)$ に対して, 次の様な K/k が存在する.

$$\begin{cases} K/k: \text{有限次アーベル拡大で},\ p\mathfrak{q}_1\cdots\mathfrak{q}_s \text{の外で不分岐}, \\ e(\mathfrak{q}_i: K/k)_p = e(\mathfrak{q}_i)\ \ (i=1,\ldots,s). \end{cases}$$

Comments.

1. 上の定理には, 何通りもの証明がある. 例えば, $\mathbb{Z}_p$ 拡大, cohomology 等を用いて証明できる. 最初は, cohomology の理論を用いて証明したが, Chevalley の定理をつかうと, 上記のような初等的な証明ができることがわかった.
2. $p>2$, $\zeta^p=1$, $\zeta \neq 1$, $\zeta \in k$ のとき定理の証明は簡単になり, 又結果もいくらか良くなる. (Chevalley の論文参照)
3. 上のような体 K の存在を直接示すことが, 本来の L.C. を証明することより簡単であるかどうかはわからない. しかし, Mazur-Wiles の仕事などから見て、K の存在を示すことが何等かの方法で出来るかもしれない. ただし, Mazur-Wiles の構成した体とは, 少しちがうようである.
4. k が $\mathbb{Q}$ 上アーベル拡大のときは, L.C. が証明されている. したがって, 上の体 K の存在も示されているわけであるが, このことは, 内容が代数的であるにもかかわらず, 古典的な類体論の範囲外の結果と思われる.
5. $k=\mathbb{Q}$ のとき, 体 K の存在は自明である. 実際 K は $\mathbb{Q}$ に適当な Gauss の和を添加することによって得られる. よって一般の場合に K の存在を示すことは, Gauss (及び, Jacobi) の和の一般化につながる可能性があるのではないか? 特に 4 の場合に K の性質を更に研究することを薦めます.

附記. 以上の岩澤教授の講演記録は, 藤崎, 館山, 堀江, 3 名のノートに基づき, 藤崎, 館山両名が草稿を作成し, それを岩澤教授に補筆, 訂正して頂いたものを館山が清書したものであります. ただし, 考え違い, 誤記などの責任はすべて藤崎, 館山の両名にあります.

[62] Some problems on cyclotomic fields

"Algebraic Number Theory" (Kyoto, 12/1987; I. Satake and Y. Morita, eds.), Kokyuroku, No. 658, RIMS, Kyoto Univ., 1988, pp. 43-55

Abstract

This is a summary (by Iwasawa himself) of his lecture in the workshop "Algebraic Number Theory" at RIMS, Kyoto University, December 1987. Here he poses several problems (conjectures) in the theory of cyclotomic fields and discusses the relations between them. The content of this article is basically contained in the (unpublished) English version [U3] except for some differences in the presentation of the problems. (The first page of his hand-written manuscript is reproduced in the front page of this volume.)

円分体に関するいくつかの問題

p を任意の奇素数, k を円の p 分体とし, k に関するいくつかの問題を簡単に説明します. 殆ど皆よく知られたことばかりですが, この研究集会のような機会にそれらの問題を一応整理しておくのもいくらか意義があるかと考える次第です.

§1. 上記 k の有理数体上のガロア群を Δ, Δ の指標群を $\widehat{\Delta}$, k のイデヤル類群を C とする. 周知のように $\widehat{\Delta}$ は mod p の Dirichlet 指標の成す群と同一視される. Δ は C の上に自然に作用するから, k の complex conjugation を J とする時 $(J \in \Delta)$ C の部分群 $C^{\pm}$ を

$$C^{+} = \mathrm{Ker}(1 - J : C \to C), \quad C^{-} = \mathrm{Ker}(1 + J : C \to C) \tag{1}$$

により定義する. k の最大実部分体を k^{+} とし, k^{+} のイデヤル類群を C_{+} とすれば, 自然な同型 : $C/C^{-} \simeq C_{+}$ が存在する.[1)] よって C, C^{-}, C_{+} の位数を h, h^{-}, h^{+} とすると

$$h = h^{-}h^{+}.$$

定義により h, h^{+} はそれぞれ代数体 k, k^{+} の類数である. この $h^{\pm}$ について次の古典的な類数公式が知られている :

I. $$h^{-} = 2p \prod_{\chi} (\frac{1}{2} h_{\chi}).$$

ここに右辺の $\prod$ は $\chi(-1) = \chi(J) = -1$ を満足する凡ての $\chi \in \widehat{\Delta}$ の上に亘る積で，又

$$(2) \qquad h_\chi = -\frac{1}{p}\sum_{a=1}^{p-1}\chi(a)^{-1}a.$$

$$\text{II.} \qquad h^+ = [E^+ : E_c].$$

但し E^+ は k^+ の単数群，E_c は E^+ に含まれる円単数から成る E^+ の部分群である．

上の I, II は解析的方法，詳しく言えば k の zeta 函数や Dirichlet の L 函数を用いて証明される公式であって，それの純数論的証明は今の所知られていない．そこで公式 I,II の持つ数論的な内容をもう少し深く調べて見たいと言うのが以下に述べるいくつかの問題の出発点である．

§2. 有理整数環 $\mathbb{Z}$ 上の Δ の群環を $\mathfrak{R}$ とする：$\mathfrak{R} = \mathbb{Z}[\Delta]$. $\mathfrak{R}$ は明らかに C の上に作用し，C は $\mathfrak{R}$-加群と考えられる．$\mathfrak{R}$ の Stickelberger イデヤルを $\mathfrak{S}$ とすれば $\mathfrak{S}\cdot C = 0$.[2] $\mathfrak{R}$, $\mathfrak{S}$ の部分群 $\mathfrak{R}^\pm$, $\mathfrak{S}^\pm$ を (1) の $C^\pm$ と同様に定義すると次の Lemma が成り立つ：

Lemma 1. I の右辺 $= [\mathfrak{R}^- : \mathfrak{S}^-]$.[3]

よって公式 I を次のように二つの有限アーベル群の位数の間の等式として書き直すことが出来る：

$$\text{I.} \qquad |C^-| = |\mathfrak{R}^-/\mathfrak{S}^-|.$$

同様に II は次の如く書ける：

$$\text{II.} \qquad |C_+| = |E^+/E_c|.$$

I を上のように書いて見れば，誰でも思いつくのは，C^- と $\mathfrak{R}^-/\mathfrak{S}^-$ との間には単に位数が等しいと言うばかりでなくもっと深い群論的な関係，例えば同型関係が存在するのではなかろうか，と言うことである．ガロア群 Δ は C^- にも $\mathfrak{R}^-/\mathfrak{S}^-$ にも自然に作用するから

C^- と $\mathfrak{R}^-/\mathfrak{S}^-$ とは Δ-同型ではなかろうか？

と言う推測も生まれる．同様に

C_+ と E^+/E_c とは Δ-同型か？

然しこれらの推測は実はいずれも成立しない：はじめの方は例えば $p = 3299$ に対し，又後の方は $p = 32009$ に対して成立しないことが容易に示される．(このような p は沢山ある.)

§3. そこで今度は公式 I, II に出てくる有限アーベル群の p-Sylow 群を考える.[4] その為 p 進整数環 $\mathbb{Z}_p$ 上の Δ の群環を R とする：$R = \mathbb{Z}_p[\Delta]$. $R = \mathfrak{R} \otimes \mathbb{Z}_p$ であるから $S = \mathfrak{S} \otimes \mathbb{Z}_p$ は R のイデヤルとなる．又 Δ の位数は $p-1$ であるから，任意の $\sigma \in \Delta$, $\chi \in \widehat{\Delta}$ に対し $\chi(\sigma)$ は 1 の $p-1$ 乗根であるが，$\mathbb{Z}_p$ は 1 の $p-1$ 乗根を $p-1$ 個含むから，複素数の $p-1$ 乗根の全体を予め $\mathbb{Z}_p$ に埋め込んでおけば $\chi(\sigma)$ は

$\mathbb{Z}_p$ の元と考えることが出来る. 以下 $\chi(\sigma)$ はいつもこのように解釈する. さて任意の R-加群 M が与えられた時, $M^{\pm}$ を (1) の $C^{\pm}$ と同様に定義し, 又 $\chi \in \widehat{\Delta}$ に対して

$$M_\chi = \{x \in M \mid \sigma \cdot x = \chi(\sigma)x, \ \forall \sigma \in \Delta\}$$

とおけば, M は次のように直和分解される：

$$M = M^+ \oplus M^- = \bigoplus_{\chi \in \hat{\Delta}} M_\chi.$$

この分解を R-加群である R, S に適用すれば

$$R = R^+ \oplus R^- = \bigoplus_\chi R_\chi, \quad S = S^+ \oplus S^- = \bigoplus_\chi S_\chi$$

を得る. 更にイデヤル類群 C の p-Sylow 群 A も自然に R-加群となるから

$$A = A^+ \oplus A^- = \bigoplus_\chi A_\chi.$$

以上の準備をして, 又 C_+ と C^+ の p-Sylow 群が一致することに注意すれば, I, II から直ちに次の公式が得られる：

I_p. $$|A^-| = |R^-/S^-|.$$

II_p. $$|A^+| = |(E^+/E_c)(p)|.$$

但し $(E^+/E_c)(p)$ は E^+/E_c の p-Sylow 群をあらわす. そこでこの I_p, II_p から §2 の終りに述べたと同様な考えで次の問題が導かれる：

P1. (問題 1) A^- と R^-/S^- とは R-同型か？

P2. (問題 2) A^+ と $(E^+/E_c)(p)$ とは R-同型か？

以下この二つの問題についていくつかの comments を述べよう.

注意. 同様にして, 任意の素数 q に対し I, II の有限アーベル群の q-Sylow 群の間の Δ-同型を問題にすることも出来るが, $q \neq p$ の場合これは不成立ではなかろうか. (q を与えた時, 不成立であるような $p(\neq q)$ が存在するとゆう意味.) これも一つの問題である.

§4. 先ず次の Lemma は容易にわかる：

Lemma 2. P1 は次の i), ii) のどちらとも同値である：
i) $\mathbb{Z}_p$-同型：$A_\chi \simeq R_\chi/S_\chi$ が凡ての $\chi \in \widehat{\Delta}$, 但し $\chi(-1) = -1$, に対して成り立つ.
ii) $A^- = R \cdot x$ を満足する A^- の元 x が存在する.

次に ω を $\bmod p$ の Teichmüller 指標とする：$\omega \in \widehat{\Delta}$, $\omega(-1) = -1$, $\widehat{\Delta} = \langle \omega \rangle$. この ω に対しては $A_\omega = 0$, $R_\omega/S_\omega = 0$ が直ちに言えるから, i) において $\chi \neq \omega$ としてもよい.

Lemma 3. $\chi \in \widehat{\Delta}$, $\chi(-1) = -1$, $\chi \neq \omega$ とすれば, (2) の h_χ は 0 でない p 進整数であって,

$$h_\chi \cdot A_\chi = 0, \quad R_\chi/S_\chi \simeq \mathbb{Z}_p/h_\chi\mathbb{Z}_p.$$

この Lemma により

$$P1 \Longleftrightarrow A_\chi \simeq \mathbb{Z}_p/h_\chi\mathbb{Z}_p, \quad \forall\chi \in \widehat{\Delta},\ \chi(-1) = -1,\ \chi \neq \omega.$$

右辺の同型が成立すれば明らかに

$$|A_\chi| = h_\chi \text{ を割る最高の } p \text{ 巾}, \quad \forall\chi \in \widehat{\Delta},\ \chi(-1) = -1,\ \chi \neq \omega$$

となる. 即ち P1 から上の等式が得られるわけであるが, 実はこの等式は Mazur-Wiles の基本定理の特別な場合として P1 とは independent に既に証明されている. これは P1 を支持する一つの証と見られるであろう.

上のように $\chi \in \widehat{\Delta}$, $\chi(-1) = -1$, $\chi \neq \omega$ とし, $\omega\chi^{-1}$ に属する Leopoldt の p 進 L 函数を $L_p(s;\omega\chi^{-1})$ とする. ($\omega\chi^{-1} \in \widehat{\Delta}$, $\omega\chi^{-1}(-1) = 1$ に注意.)

Lemma 4. 与えられた χ に対し, 次の條件を満足する $\mathbb{Z}_p[[T]]$ の巾級数 $\xi_\chi(T)$ が唯一つ存在する：

$$\xi_\chi((1+p)^s - 1) = L_p(s;\omega\chi^{-1}), \quad \forall s \in \mathbb{Z}_p.$$

これは p 進 L 函数の基本的な性質の一つである. 勿論 $\xi_\chi(T) \neq 0$ であるから Weierstrass の preparation theorem により $\xi_\chi(T)$ は次のように一意的に書かれる：

$$\xi_\chi(T) = \eta_\chi(T)p^{\mu_\chi}f_\chi(T).$$

ここに $\eta_\chi(T)$ は $\mathbb{Z}_p[[T]]$ の巾級数でその定数項 $\eta_\chi(0)$ が p で割れぬもの, μ_χ は一般には 0 又は正の整数であるが今の場合 $\xi_\chi(T)$ に対しては Ferrero-Washington の定理により $\mu_\chi = 0$ であることが知られている. 又 $f_\chi(T)$ は次のような $\mathbb{Z}_p[T]$ の多項式 (所謂 distinguished polynomial) である：

$$f_\chi(T) = T^n + a_1T^{n-1} + \cdots + a_n, \quad n \geq 0,\ a_i \in p\mathbb{Z}_p,\ i = 1, \cdots, n.$$

この $f_\chi(T)$ に関して次の問題がある：

$(\mathbf{P3})_\chi$. $f_\chi(T)$ は $\mathbb{Q}_p[T]$ で既約か？

何故このようなことが問題になるかと言えば次の Lemma が成立するからである：

Lemma 5. $(\mathrm{P3})_\chi$, $\forall\chi \in \widehat{\Delta}$, $\chi(-1) = -1$, $\chi \neq \omega \implies$ P1.

注意. Mazur-Wiles の定理を使えばもっと精密に次のことも言われる. 即ち χ を一つ定めた時

$$(\mathrm{P3})_\chi \Longrightarrow A_\chi \simeq R_\chi/S_\chi.$$

又 $f_\chi(T)$ が必ずしも既約でなくても, 重複根を持たなければ同じ結論が得られる.

次に $(\mathrm{P3})_\chi$ が成立する為の十分條件として次の二つの問題も考えられる：

(P4)$_\chi$. $f_\chi(T)$ は $\mathbb{Z}_p[T]$ の Eisenstein 多項式か？

(P5)$_\chi$. $\deg f_\chi(T) \leq 1$？

この (P4)$_\chi$, (P5)$_\chi$ を特に掲げたのはこれらが Bernoulli 数 B_n に関する條件として言うことが出来るからである. χ を上の通りとする時

$$\chi = \omega^{1-i}, \ \ 0 < i < p-1$$

を満足する偶数 i が一意的に定まる. この i を用いて：

Lemma 6. (P4)$_\chi$ $\iff$ $B_{1+(i-1)p} \not\equiv 0 \bmod p^2$.

Lemma 7. (P5)$_\chi$ $\iff$ $\frac{B_i}{i} \not\equiv \frac{B_{i+p-1}}{i+p-1} \bmod p^2$.

上述により, もし P1 が成り立たなければ (P3)$_\chi$ が不成立であるような χ が存在する. 従ってその χ に対しては (P4)$_\chi$ も (P5)$_\chi$ も不成立となるから上の Lemma 6, 7 を用いれば

$$\frac{B_{i+\nu(p-1)}}{i+\nu(p-1)} \equiv 0 \bmod p^2, \ \ \forall \nu = 0, 1, 2, \cdots$$

が得られる. 特に $B_i \equiv 0 \bmod p^2$ でなければならない. 然しこのような p と i は今の所発見されていない. 実際 Wagstaff は Bernoulli 数を計算して, 凡ての $p < 125000$ と凡ての $\chi \in \widehat{\Delta}$, $\chi(-1) = -1$, $\chi \neq \omega$, に対し (P4)$_\chi$, (P5)$_\chi$ が共に成立することを確かめた.

さて Lemma 7 の右辺の合同式は mod p では常に成立する. これは Kummer の証明した Bernoulli 数に関する一連の合同式の一つであるが, 他の合同式, 例えば三個の Bernoulli 数の間の mod p^2 の合同式が $\deg f_\chi(T) \leq 2$ と関係があるかどうか, と言うようなことも一応考えて見る価値があろう. 更に, 多項式 $f_\chi(T)$ の根 α は p 進 L 函数 $L_p(s; \omega\chi^{-1})$ の零点を与えるものであるから, 函数体との類似を考えれば大いに興味のある対象である. 例えば (P3)$_\chi$ が成立する時, p 進数体 $\mathbb{Q}_p(\alpha)$ はどんな意味を持っているか, 等々.

§5. 以上専ら P1 について考えてきたが, 次に P2 に関係のあることを少し述べる.

P6. (Kummer-Vandiver の予想) $p \nmid h^+$？

この P6 は Fermat の問題に関聯してよく知られた予想であるが, 我々にとっても興味があるのは

$$\text{P6} \implies \text{P1, P2}$$

が容易に証明されるからである.

次に k の最大実部分体 k^+ の上の円分 $\mathbb{Z}_p$ 拡大体を K^+ とし, K^+/k^+ の不変量を λ, μ とする. 一般に $\mathbb{Z}_p$ 拡大の不変量は 0 又は正の整数であるが, 今の場合には Ferrero-Washington の定理により $\mu = 0$. ところで

P7. λ も 0 ではないか： $\lambda = 0$？

このP7はGreenbergの予想と呼ばれる予想のspecial caseであるが, P6 $\Longrightarrow$ P7は直ちに証明される. P6とP7との関係を更によく見る為に, 任意の $n=0,1,2,\cdots$ に対し円の p^{n+1} 分体を k_n とし, k_n のイデヤル類群, 単数群をそれぞれ C_n, E_n とする. $m \geq n \geq 0$ であれば $k_n \subseteq k_m$ であるから自然な準同型 $C_n \to C_m$, $E_m \to E_n$ が定義される.(後者はノルム写像.) ここに

Lemma 8. $C_n \to C_m$ が単射 $\Longleftrightarrow$ $E_m \to E_n$ が全射.

これは容易に言えるが, そこで次の問題：

P8. 凡ての $m \geq n \geq 0$ に対し $C_n \to C_m$ は単射か？

が生れる. このP8と先のP6, P7との関係は次のLemmaによる：

Lemma 9. P6 $\Longleftrightarrow$ P7 & P8.

更にP2との関聯について言えば

Lemma 10. P1 & P8 $\Longrightarrow$ P2

も証明される. 即ちP8が成立すればP2はP1に含まれる. このようにP8は中々面白い問題であるが, 一番簡単な場合, 即ち $m=1$, $n=0$ の場合に $C_0 \to C_1$ が単射であることすら未だ解決されていない. もっともこの場合に証明出来れば一般の場合にも同様にして証明が得られそうな気がするが.

§6. 元来円の p 分体 k はKummerがFermatの問題を解こうとして研究した代数体であって, それが今日の代数的整数論の端緒となったことはよく知られている. よって終りに, 先のP1, P2とは無関係だが, Fermatの問題といくらか関わりのある k についての問題を一つ付け加えておく. 即ち：

P9. k 上の p-class field towerは有限か？

一般に任意の有限次代数体上のclass field towerが有限で切れるであろうと言うのは古典的類体論が完成した頃からの有名な予想であったが, 1960年代になってそれはGolod-Šafarevičにより否定的に解決された. その時の方法を用いれば直ちに次のことが言われる. 即ち, 前のように k のイデヤル類群 C の p-Sylow群を A とし, A のrankを r とする時

$$\text{(3)} \qquad \text{P9} \quad \Longrightarrow \quad r < 2 + \sqrt{2(p+1)}.$$

この r は歴史的にFermatの問題と関係の深い数で, 例えば $r=0$, 即ち $p \nmid h$ ならばその p についてのFermatの問題が解決されると言うKummerの定理はよく知られている. r とFermatの問題に関する多くの結果のうちで, 次のEichlerの定理は特に重要である：

$$\text{(4)} \qquad r < [\sqrt{p}\,] - 1 \quad \Longrightarrow \quad \text{the first case of Fermat's problem.}$$

明らかに $[\sqrt{p}]-1<2+\sqrt{2(p+1)}$ であるが, これらの二数はどちらも大体 $\sqrt{p}$ の大きさの数である. それ故例えば Golod-Šafarevič の結果を特に円分体 k の場合に精密化して (3) の限界 $2+\sqrt{2(p+1)}$ を $[\sqrt{p}]-1$ 迄引下げることは出来ないだろうか. 乃至は Eichler の定理 (4) の限界 $[\sqrt{p}]-1$ を何とかして $2+\sqrt{2(p+1)}$ 迄引上げることは出来ないだろうか. もしそれが出来れば

$$\text{P9} \quad \Longrightarrow \quad \text{the first case of Fermat's problem}$$

となるわけで, これが上に述べた P9 と Fermat の問題との関聯である. 然しこのような (希望的) 関係を度外視しても, P9 はそれ自身十分興味ある問題であって, 又たとえそれが否定的に解決されても面白い結果であると思う. もっと一般に, F を任意の有限次代数体, L を F 上の最大不分岐 (ガロア) p 拡大とする時, L/F が無限次拡大である場合 (即ち F 上の p-class field tower が無限になる場合) に拡大 L/F の数論的性質を調べるという問題もあるが, このような一般的な問題に対しても $F=k$ の special case は critical であると思われる.

以上説明してきた円の p 分体 k に関する問題のなかには普通 "予想" と呼ばれているものも含まれていますが, その大部分は私の極く大雑把な感じだけから, "こうなるのではなかろうか" とか "こうなって欲しい" と言うようなことを述べたものであって, 私自身その成否についてそれ程確信があるわけではありません. 特に $(\text{P4})_\chi$, $(\text{P5})_\chi$ などはたとえそれが計算により $p<125000$ 迄確かめられたとしても, 不成立の可能性も十分考えるべきかと思います. そのような問題を持ち出したことについては, 「数論の面白さは予測の当らない所にある」と言うことにして言いわけとします.

註

1) C^+ と C_+ とは 2-Sylow 群を除いて一致するが, 必ずしも常に同型ではない. 例 : $p=29$.
2) $\mathfrak{S}$ の定義その他については Washington : Introduction to Cyclotomic Fields を参照. 以下に用いた円分体に関する色々な結果に関しても同様.
3) この Lemma は W. Sinott により一般の円分体に対して拡張されている.
4) 何故 p-Sylow 群を考えるかと言うことについてはこの § の終りの注意参照.

[63] A note on capitulation problem for number fields

Proc. Japan Acad., 65-2, Ser. A (1989), 59-61.

Princeton University

(Communicated by Shokichi IYANAGA, M. J. A., Feb. 13, 1989)

Let F be a finite extension of a finite algebraic number field k and let C_k and C_F denote the ideal class groups of k and F respectively. A subgroup A of C_k is said to *capitulates* in F if $A \to 1$ under the natural homomorphism $C_k \to C_F$. The principal ideal theorem of class field theory states that C_k always capitulates in Hilbert's class field K over k. However, as shown in Heider-Schmithals [1], for some k, C_k capitulates already in a proper subfield M of K: $k \subseteq M \subseteq K$, $M \neq K$. In the present note, we shall give further simple examples of such number fields k for which the capitulation of C_k occurs in a proper subfield M of Hilbert's class field K over k*).

1. Let L be a finite abelian (or nilpotent) extension over k. For each prime number p, let L_p denote the maximal p-extension over k contained in L, and let $C_{k,p}$ be the p-class group of k, i.e., the Sylow p-subgroup of C_k. It is then easy to see that C_k capitulates in L if and only if $C_{k,p}$ capitulates in L_p for every prime number p. Applying this for Hilbert's class field K over k, we see that a number field M such as stated in the introduction exists if and only if there is a prime number p such that $C_{k,p}$ capitulates in a proper subfield F of Hilbert's p-class field K_p over k: $k \subseteq F \subseteq K_p$, $F \neq K_p$. In what follows, we shall find k such that the 2-class group $C_{k,2}$ capitulates in a proper subfield of Hilbert's 2-class field K_2 over k.

2. Let p, p_1, p_2 be three distinct prime numbers such that

i) $p \equiv p_1 \equiv p_2 \equiv 1 \bmod 4$, $(p/p_1) = (p/p_2) = -1$, the brackets being Legendre's symbol, and that

ii) the norm of the fundamental unit of the real quadratic field $k' = Q(\sqrt{p_1 p_2})$ is 1.

Let

$$k = Q(\sqrt{p p_1 p_2}).$$

By Iyanaga [3], p. 12, we know for the real quadratic field k that

iii) the 2-class group $C_{k,2}$ is an abelian group of type (2, 2) and that

iv) the norm of the fundamental unit of k is -1.

Since $[K_2 : k] = |C_{k,2}| = 4$ for Hilbert's 2-class field K_2 over k, we see immediately that

$$K_2 = Q(\sqrt{p}, \sqrt{p_1}, \sqrt{p_2}).$$

---

*) The author was informed by Prof. S. Iyanaga, that he had been reminded of the problem of finding such number fields k by Dr. Li Delang at Sichuan University, China. For various aspects of capitulation problem in general, see Miyake [4].

Let

$$F=kk'=Q(\sqrt{p}, \sqrt{p_1p_2}).$$

Then

$$k\subseteq F\subseteq K_2, \qquad [K_2:F]=[F:k]=2.$$

Proposition. *For the above k, $C_{k,2}$ capitulates in the proper subfield F of K_2. Consequently, the ideal class group C_k of k capitulates in a proper subfield M of Hilbert's class field K over k: $k\subseteq M\subseteq K$, $M\neq K$.*

Proof. Let E_k, $E_{k'}$, and E_F denote the groups of units in k, k', and F respectively. By ii), iv),

$$N_{k'/Q}(E_{k'})=\{1\}, \qquad N_{k/Q}(E_k)=\{\pm 1\}.$$

Hence it follows from

$$N_{k/Q}(N_{F/k}(E_F))=N_{k'/Q}(N_{F/k'}(E_F)) \qquad (=N_{F/Q}(E_F))$$

that

$$N_{F/k}(E_F)\neq E_F, \quad \text{i.e.,} \quad H^2(F/k, E_F)\neq 1$$

for the Galois cohomology group of F/k for E_F. Since F/k is a cyclic extension of degree two, we have

$$|H^1(F/k, E_F)|=2|H^2(F/k, E_F|$$

for the orders of the cohomology groups. Hence we obtain from the above that

$$|H^1(F/k, E_F)|>2.$$

On the other hand, since F/k is an unramified 2-extension (see [2]),

$$H^1(F/k, E_F)\simeq \mathrm{Ker}\,(C_{k,2}\longrightarrow C_F).$$

Therefore it follows from $|C_{k,2}|=4$ that

$$\mathrm{Ker}\,(C_{k,2}\longrightarrow C_F)=C_{k,2},$$

namely, that $C_{k,2}$ capitulates in F. Q.E.D.

3. There are many pairs of prime numbers (p_1, p_2), $p_1\equiv p_2\equiv 1 \bmod 4$, satisfying the condition ii) in §2:

$$(p_1, p_2)=(5, 41), \quad (5, 61), \quad (13, 17), \quad \text{etc.}^{**)}$$

Fix one such pair (p_1, p_2). Then, by Dirichlet's theorem on primes in an arithmetic progression, there exist infinitely many p's satisfying the condition i) in §2. Hence there are infinitely many real quadratic fields k of the form $Q(\sqrt{pp_1p_2})$ such that C_k capitulates in a proper subfield of Hilbert's class field K over k.

Example. Let

$$(p_1, p_2)=(13, 17), \qquad p_1p_2=221.$$

Since

$$\left(\frac{5}{13}\right)=\left(\frac{5}{17}\right)=-1,$$

any prime number p satisfying

$$p\equiv 5 \bmod 884, \qquad 884=4\cdot 13\cdot 17,$$

provides us a real quadratic field

$$k=Q(\sqrt{221\cdot p})$$

**) In fact, it seems likely that there exist infinitely many such pairs (p_1, p_2).

such that C_k capitulates in a proper subfield of K. In particular, for $p=5$,

$$k=Q(\sqrt{1105})$$

has class number 4. Hence, in this case,

$$C_k=C_{k,2}, \qquad K=K_2=Q(\sqrt{5}, \sqrt{13}, \sqrt{17})$$

and C_k capitulates in the proper subfield $F=Q(\sqrt{5}, \sqrt{1105})$ of K. H. Wada kindly informed the author that by computing unit groups explicitly, he found that in the above case, C_k capitulates also in $Q(\sqrt{13}, \sqrt{1105})$ as well as in $Q(\sqrt{17}, \sqrt{1105})$. G. Fujisaki then proved in general that the 2-class group $C_{k,2}$ of $k=Q(\sqrt{221\cdot p})$, $p\equiv 5 \bmod 884$, capitulates in all three quadratic subextensions of Hilbert's 2-class field K_2 over k.

Remark. In general, $C_k \neq C_{k,2}$ for $k=Q(\sqrt{pp_1p_2})$ with (p, p_1, p_2) satisfying i), ii) in § 2. For example, the class number of $Q(\sqrt{53\cdot 5\cdot 41})$ is 12.

References

[1] F.-P. Heider und B. Schmithals: Zur Kapitulation der Idealklassen in unverzweigten primzyklischen Erweiterungen. J. reine angew. Math., **336**, 1–25 (1982).

[2] K. Iwasawa: A note on the group of units of an algebraic number field. J. Math. pures appl., **35**, 189–192 (1956).

[3] S. Iyanaga: Sur les classes d'idéaux dans les corps quadratiques. Actualités Sci. Indust., no. 197, Hermann, Paris (1935).

[4] K. Miyake: On capitulation problem. Sugaku, **37**, 128–143 (1985) (in Japanese).

[64] A note on capitulation problem for number fields. II

Proc. Japan Acad., 65-6, Ser. A (1989), 183-186.

Princeton University

(Communicated by Shokichi IYANAGA, M. J. A., June 13, 1989)

In the present note, we shall again consider a capitulation problem for number fields which we discussed in our earlier paper [2]. Using some properties of Z_p-extensions of number fields, we shall prove the following:

Proposition. *For each prime number $p \geqslant 2$, there exist infinitely many finite algebraic number fields k such that the p-class group of k capitulates in a proper subfield of Hilbert's p-class field over k.*

We note that in the special case $p=2$, the proposition was proved in [2] by elementary argument.

1. Let M be any number field, finite or infinite over the rational field Q. Throughout the following, we fix a prime number $p \geqslant 2$ and denote by $A(M)$ the p-primary component of the ideal class group of M; if M is finite over Q, this is the p-class group of M, denoted by $C_{M,p}$ in [2].

Lemma 1. *Let k' be an unramified cyclic extension of degree p over a finite algebraic number field k. Then $A(k)$ capitulates in k' if and only if the following* a), b) *hold:*

a) *there exists a prime ideal of k which is undecomposed and principal in k',*

b) *if the class of an ideal $\mathfrak{a}'$ of k' belongs to $A(k')$, the norm $N_{k'/k}(\mathfrak{a}')$ is a principal ideal in k'.*

Proof. Let K and K' denote Hilbert's p-class fields over k and k' respectively: $k \subseteq k' \subseteq K \subseteq K'$. Let t: $\mathrm{Gal}(K/k) \to \mathrm{Gal}(K'/k')$ be the transfer map. Fix an element σ of $\mathrm{Gal}(K'/k)$ such that the restriction $\sigma|k'$ is a generator of $\mathrm{Gal}(k'/k)$. Then, for any τ in $\mathrm{Gal}(K'/k')$, we have

$$t(\tau|K) = \prod_{i=0}^{p-1} \sigma^i \tau \sigma^{-i}.$$

By Artin [1], $A(k)$ capitulates in k' if and only if $\mathrm{Im}(t)=1$. Hence the lemma follows from the fact that a) is equivalent with $t(\sigma|K)=1$ and b) with $t(\tau|K)=1$ for all τ in $\mathrm{Gal}(K'/k')$.

2. Let Q_∞ denote the unique Z_p-extension over Q: $\mathrm{Gal}(Q_\infty/Q) \simeq Z_p$, and let

$$Q = Q_0 \subset Q_1 \subset \cdots \subset Q_n \subset \cdots \subset Q_\infty$$

be the sequence of intermediate fields for Q_∞/Q. For each $n \geqslant 0$, let $\mathfrak{p}_n$ be the unique prime ideal of Q_n, dividing the rational prime p; $\mathfrak{p}_n$ is a principal ideal in Q_n and $\mathfrak{p}_{n+1}^p = \mathfrak{p}_n$ for $n \geqslant 0$.

Let F be a real cyclic extension of degree p over Q such that

i) (p) is a prime ideal in F,

ii) the class number of F is divisible by p,

iii) $A(F_\infty)=0$ for $F_\infty=FQ_\infty$.

Let $F_n=FQ_n$ for $n\geqslant 0$. Then it follows from i) that $\mathfrak{p}_n$ is a prime ideal in F_n, $n\geqslant 0$, that $F\cap Q_\infty=Q$, and that

$$F=F_0\subset F_1\subset\cdots\subset F_n\subset\cdots\subset F_\infty$$

is the sequence of intermediate fields for the Z_p-extension F_∞/F. For each $n\geqslant 0$, let K_n denote Hilbert's p-class field over F_n. Since the prime ideal (p) of F is fully ramified in F_n, we see that for $0\leqslant m\leqslant n$, $F_n\cap K_m=F_m$ and $[K_m:F_m]$ divides $[K_n:F_n]$. Hence it follows from ii) and iii) that

iv) for any $n\geqslant 0$, the class number of F_n is divisible by p,

v) for sufficiently large $n\geqslant 0$, $F_{n+1}K_n=K_{n+1}$ so that $\mathrm{Gal}(F_{n+1}/F_n)$ acts trivially on $\mathrm{Gal}(K_{n+1}/F_{n+1})$ and, hence, on $A(F_{n+1})$.

Lemma 2. *Let n be an integer such that $\mathrm{Gal}(F_{n+1}/F_n)$ acts trivially on $A(F_{n+1})$. Let $k'=F_{n+1}$ and let k be a field such that*

$$Q_n\subseteq k\subseteq k',\quad [k':k]=p,\quad k'\neq Q_{n+1}, F_n.$$

Then k' is a proper subfield of Hilbert's p-class field over k and $A(k)$ capitulates in k'.

Proof. We first note that since F_{n+1}/Q_n is an abelian extension of type (p,p), there exist $p-1$ fields k as mentioned above. It is also easy to see that k'/k is an unramified cyclic extension of degree p and that there exists a prime ideal $\mathfrak{p}$ of k such that $\mathfrak{p}_n=\mathfrak{p}^p$ in k and $\mathfrak{p}=\mathfrak{p}_{n+1}$ in k', the latter equality being a consequence of i). Thus the condition a) of Lemma 1 is satisfied for k'/k. Let $\mathfrak{a}'$ be as stated in the condition b) of the same lemma. Since $\mathrm{Gal}(F_{n+1}/F_n)$ acts trivially on $A(k')$ $(=A(F_{n+1}))$, $N_{k'/k}(\mathfrak{a}')$ and $N_{k'/Q_{n+1}}(\mathfrak{a}')$ lie in the same ideal class as ideals of k'. However, as is well known, the class number of Q_{n+1} is prime to p. Therefore $N_{k'/Q_{n+1}}(\mathfrak{a}')$ is principal in Q_{n+1} and, consequently, $N_{k'/k}(\mathfrak{a}')$ is a principal ideal in k'. Thus the condition b) in Lemma 1 is also satisfied, and $A(k)$ capitulates in k' by that lemma. That k' is a proper subfield of Hilbert's p-class field over k follows from iv) above.

3. To find number fields F with properties i), ii), iii) in § 2, we need two more lemmas.

Lemma 3. *Let L be a number field, finite over Q, and let L'/L be a cyclic extension of degree p, unramified at infinity. Let $L_\infty=LQ_\infty$, $L'_\infty=L'Q_\infty$. Suppose that*

1) *L_∞ has a unique p-place,*

2) *every prime ideal of L, prime to p and ramified in L', is undecomposed in L_∞,*

3) *$A(L)=0$, $A(L')=0$.*

Then $A(L'_\infty)=0$.

We omit the proof here, noting only that the essential step is to show that $H^2(L'_\infty/L_\infty, E)=0$ for the group E of units in L'_∞.

From now on, we assume that p is an odd prime: $p>2$. For each prime number q with $q\equiv 1 \bmod p$, there is a unique subfield C_q of the cyclotomic field of q-th roots of unity such that C_q/Q is a cyclic extension of degree p; q is then the unique rational prime ramified in C_q and, consequently, $A(C_q)=0$.

Lemma 4. *There exist infinitely many pairs of prime numbers* (q_1, q_2) *with the following properties*:

1) $q_1 \equiv q_2 \equiv 1 \bmod p$ *and both* q_1 *and* q_2 *are undecomposed in* Q_∞,

2) *for* $M_1=C_{q_1}$, $M_2=C_{q_2}$,

i) p *is undecomposed in* M_1, *but* q_2 *is decomposed in* M_1,

ii) q_1 *is undecomposed in* M_2.

Proof. Let P and P' denote the cyclotomic fields of p-th and p^2-th roots of unity, respectively. Then P' and $P(\sqrt[p]{p})$ are cyclic extensions of degree p over P and there exists a prime ideal $\mathfrak{q}_1$ of P with absolute degree 1 such that $\mathfrak{q}_1$ is undecomposed in both P' and $P(\sqrt[p]{p})$. Let $q_1=N_{P/Q}(\mathfrak{q}_1)$. Then $q_1 \equiv 1 \bmod p$, p is undecomposed in $M_1=C_{q_1}$, and q_1 is undecomposed in Q_1 and, hence, in Q_∞. Now, P', PM_1, and $P(\sqrt[p]{q_1})$ are independent cyclic extensions of degree p over P. Hence there is a prime ideal $\mathfrak{q}_2$ of P with absolute degree 1 such that $\mathfrak{q}_2$ is undecomposed in P' and $P(\sqrt[p]{q_1})$, but is decomposed in PM_1. Let $q_2=N_{P/Q}(\mathfrak{q}_2)$. Then $q_2 \equiv 1 \bmod p$, q_2 is undecomposed in Q_∞, but is decomposed in M_1, and q_1 is undecomposed in $M_2=C_{q_2}$. Since there are infinitely many choices for $\mathfrak{q}_1$, $\mathfrak{q}_2$ above, there exist infinitely many pairs (q_1, q_2).

4. Let p still be an odd prime: $p>2$, and let (q_1, q_2) and M_1, M_2 be as stated in Lemma 4. Let $L=M_1$, $L'=M_1M_2$. Clearly L' is a totally real cyclic extension of degree p over L. For the extension L'/L, the conditions 1), 2) of Lemma 3 follow easily from Lemma 4, 1) and 2)-i). Since $L=M_1=C_{q_1}$, $A(L)=0$. By Lemma 4, 2)-ii), (q_1) is a prime ideal of M_2 and it is the unique prime ideal of M_2, ramified in $L'=M_1M_2$. Hence $A(M_2)=0$ implies $A(L')=0$. Thus 1), 2), 3) of Lemma 3 are satisfied for L'/L and it follows from that lemma that $A(L'_\infty)=0$.

Now, L'/Q is an abelian extension of type (p, p), unramified outside (q_1, q_2), and by Lemma 4, 2)-i), the decomposition field of p for the extension L'/Q is a cyclic extension of degree p over Q. Since $p>2$, there exists a cyclic extension F/Q of degree p such that $Q \subseteq F \subseteq L'$ and that F is different from M_1, M_2, and the decomposition field of p for L'/Q. Clearly (p) is then undecomposed in F. Since $F \neq M_1, M_2$, L'/F is an unramified extension of degree p and the class number of F is divisible by p. Furthermore $A(L'_\infty)=0$ implies $A(F_\infty)=0$ for $F_\infty=FQ_\infty \subseteq L'Q_\infty=L'_\infty$. Thus F is a number field satisfying i), ii), iii) of § 2. Since there exist infinitely many pairs (q_1, q_2) by Lemma 4, there also exist infinitely many number fields F such as defined above, and it follows from v) and Lemma 2 in § 2 that for each F, there exist infinitely many finite algebraic number fields k such that the p-class group $A(k)$ of k capitulates in a proper subfield of Hilbert's p-class field over k. This completes the proof of the proposition in the introduction for $p>2$. Since the case $p=2$ was already treated in [2], the proposition is now proved for any prime number $p \geqslant 2$.

Remark. Greenberg's conjecture for Z_p-extensions ($p \geqslant 2$) states that if F is a totally real finite algebraic number field, then $A(F_\infty)=0$ for $F_\infty = FQ_\infty$. It is clear from the above argument that if this conjecture is assumed, we can easily find many examples of number fields F satisfying i), ii), iii) in § 2, and hence also many examples of finite algebraic number fields k, having the property mentioned in the proposition. In fact, we can find a number field k such that $A(k)$ is an abelian group of type $(p, \cdots, p)$ with arbitrarily large rank and that $A(k)$ capitulates in an unramified cyclic extension of degree p over k.

References

[1] E. Artin: Idealklassen in Oberkörpern und allgemeines Reziprozitätsgesetz. Hamb. Abh., 8, 46–51 (1930).

[2] K. Iwasawa: A note on capitulation problem for number fields. Proc. Japan Acad., 65A, 59–61 (1989).

[65] On papers of Takagi in number theory

"Teiji Takagi Collected Papers", Springer-Verlag Tokyo, 1990, Appendices, I, pp. 342-351.

1. Teiji Takagi graduated from the Imperial University of Tokyo in 1897, the year Hilbert's Zahlbericht [15] was published. The following year he went to Germany, first to Berlin and then to Göttingen, where he studied under Hilbert for two years. The library cards kept at the University of Tokyo show that Takagi borrowed the Zahlbericht and other classical texts on number theory and elliptic functions from the library before going abroad. So he must have been well prepared for his study in Germany. The work of Takagi in Göttingen resulted in his doctoral thesis **6*** published in Tokyo in 1903, his first important contribution to number theory.

In a paper [22] of 1853, Kronecker announced that every abelian extension of the rational field $\mathbb{Q}$ is a subfield of a cyclotomic field. He also stated there that all abelian extensions of the quadratic field $\mathbb{Q}(\sqrt{-1})$ can be obtained similarly by dividing the lemniscate instead of the circle. This is the origin of what is now called Kronecker's Jugendtraum, namely, his conjecture that all abelian extensions of an imaginary quadratic field k can be generated by the singular values of the elliptic modular function $j(u)$ and the division values of elliptic functions which have complex multiplication in k. In his thesis **6**, Takagi proved Kronecker's statement on $\mathbb{Q}(\sqrt{-1})$.

Let $k = \mathbb{Q}(\sqrt{-1})$ and let p^h be any power of a prime number p ($h \geqq 1$). By evaluating Jacobi's elliptic function $sn(u)$ and Weierstrass' function $\wp(u)$ ($= sn(u)^{-2}$) associated with k at fractional multiples of their periods, Takagi obtained abelian extensions $L(p^h, \mu)$ and $M(p^h)$ over k satisfying the following conditions[1]:

i) $L(p^h, \mu)$ is defined for each prime element μ in $\mathbb{Z}[\sqrt{-1}]$ with norm m satisfing $m \equiv 1 \bmod p^h$ (resp. $m \equiv 1 \bmod 2^{h+2}$ if $p = 2$), and is a cyclic extension of degree p^h over k such that (μ) is the only prime ideal of k ramified in $L(p^h, \mu)$.
ii) $M(p^h)$ is an abelian extension of type (p^h, p^h) over k such that no prime ideal of k, prime to p, is ramified in $M(p^h)$.

Now, let K be any finite extension of k. Fixing a prime number p, let n_K denote the number of prime ideals of k which are prime to p and are ramified in K. For a cyclic extension K/k with p-power degree, Takagi proved that a) if $n_K \geqq 1$, then K is contained in a composite $K'L$ where K'/k is a cyclic extension of p-power degree with $n_{K'} < n_K$ and

* The figure in bold face refers to the article in the text.
[1] *cf.*, pp. 25–26 and 36–37.

$L = L(p^h, \mu)$ for some $h \geqq 1$ and μ, and that b) if $n_K = 0$, then K is contained in $M(p^h)$ for some $h \geqq 1$. It is clear that Kronecker's statement on $k = \mathbb{Q}(\sqrt{-1})$ follows immediately from a) and b)[2].

Earlier in 1896, Hilbert [16] gave a simple proof of the theorem of Kronecker on the abelian extensions of $\mathbb{Q}$. Takagi's proof sketched above followed the idea of this paper of Hilbert. However, in Hilbert's case, abelian extensions over $\mathbb{Q}$ similar to the extensions $L(p^h, \mu)$ and $M(p^h)$ over k mentioned above are immediately obtained as subfields of cyclotomic fields, whereas in Takagi's case, construction of the fields L and M required careful arithmetic study of the division fields of the elliptic functions $\mathrm{sn}(u)$ and $\wp(u)$, based on the classical work of Abel and Eisenstein. In retrospect, this beautiful thesis seems to have been a harbinger of the outstanding future of the young number-theorist.

2. Takagi returned to Tokyo from Göttingen in 1901. At the University of Tokyo, he continued his research in algebraic number theory and it culminated in his two major papers on class field theory **13** in 1920 and **17** in 1922. As mentioned earlier, Takagi's thesis **6** was published in 1903. Between 1903 and 1920, Takagi published six papers **7–12**, on number theory. Of these, the first **7**, is acomplement to his thesis **6** and it proves a generalization of a theorem of Eisenstein, namely, that if k is an imaginary quadratic field and $\mathfrak{p}^n$ is a power of an odd prime ideal $\mathfrak{p}$ of k, then the polynomial for the $\mathfrak{p}^n$-division values of the elliptic function $S(u)$[3], associated with k, is an Eisenstein polynomial for $\mathfrak{p}$. The other five papers **8–12**, are preliminary reports on Takagi's monumental accomplishments on class field theory which were later presented in detail in **13** and **17**. Before reviewing the contents of these two papers, we shall first briefly describe some historical background for Takagi's work[4].

The concept of class fields had been gradually evolved by Kronecker, Weber, and Hilbert by the study of complex multiplication of elliptic functions and by following the analogy with classical algebraic function fields. It seems that the terminology "class field" was proposed for the first time by Weber [27, 28]. Let k be a number field, and $\mathfrak{m}$ an integral divisor of k. Let $I_{\mathfrak{m}}$ denote the group of all ideals of k, prime to (the ideal part of) $\mathfrak{m}$, and $S_{\mathfrak{m}}$, the subgroup of all principal ideals (α) in $I_{\mathfrak{m}}$ with $\alpha \equiv 1 \bmod \mathfrak{m}$. A factor group of the form $I_{\mathfrak{m}}/H_{\mathfrak{m}}$, where $S_{\mathfrak{m}} \subseteq H_{\mathfrak{m}} \subseteq I_{\mathfrak{m}}$, is called a congruence ideal class group mod $\mathfrak{m}$ in k. Suppose that such a group $I_{\mathfrak{m}}/H_{\mathfrak{m}}$ is given. Weber called a finite extension K of k a class field over k for $I_{\mathfrak{m}}/H_{\mathfrak{m}}$ if a prime ideal $\mathfrak{p}$ of absolute degree 1 in $I_{\mathfrak{m}}$ is completely decomposed in K exactly when $\mathfrak{p}$ belongs to the subgroup $H_{\mathfrak{m}}$. Following the idea of Dirichlet, he then proved by using analytic properties of L-series that

$$[I_{\mathfrak{m}} : H_{\mathfrak{m}}] \leqq [K : k]$$

for such a class field K. Although Weber did not prove the existence of a class field K over k for a given congruence ideal class group $I_{\mathfrak{m}}/H_{\mathfrak{m}}$, he showed that if such K exists, it is unique for $I_{\mathfrak{m}}/H_{\mathfrak{m}}$ and that the existence of K implies the existence of infinitely many

[2] Hülfssatz 1) on p. 29 is false. This is used in §14 for the proof of ii) above in the case $p \equiv 3 \bmod 4$. However, it can be proved without using that Hülfssatz, by slightly modifying the argument in §14.
[3] *cf.*, Weber [28], §157.
[4] For the history of class field theory, see Hasse [14].

prime ideals of absolute degree 1 in each coset of the factor group $I_{\mathfrak{m}}/H_{\mathfrak{m}}$, a generalization of the classical theorem of Dirichlet on the prime numbers in an arithmetic progression.

In the introduction of his Zahlbericht [15], Hilbert stated that to him the theory of abelian extensions of number fields, initiated by Kummer and Kronecker, was the most richly endowed part of the edifice of algebraic number theory, with many precious treasures waiting to be discovered. As the simplest case of such abelian extensions, Hilbert [17–19] studied quadratic extensions of number fields. Using those results and imposing rather strong conditions on the ground field k, he then proved in the same papers that there is an unramified abelian extension K over k (where the ramification of archimedean prime divisors is allowed) such that the Galois group of K/k is isomorphic to the ideal class group of k in the narrow sense, and that K is the class field over k for that ideal class group in the sense of Weber. Hence Hilbert simply named K the class field of k. Although he treated only some very special cases, Hilbert was quite convinced that the class field K of k, such as described above, should exist for every number field k, and he conjectured important properties of the extension K/k, e.g., the law of decomposition for a prime ideal of k in the extension K.

As a part of his work on quadratic extensions mentioned above, Hilbert studied the reciprocity law for the power residue symbol and the norm residue symbol with exponent 2. These results of Hilbert were generalized by Furtwängler [6] for symbols with arbitrary prime exponent $l \neq 2$ (under the assumption that the class number of the ground field is prime to l). Following the idea of Hilbert, Furtwängler then went on to prove in [7] in 1907 the existence of Hilbert's class field K over an arbitrary ground field k, as predicted by Hilbert.

3. We now discuss the paper **13** of Takagi, published in 1920. In this paper, Takagi started with a new definition of class fields as follows. Let K be a finite Galois extension of a number field k with degree $[K:k] = n$. For each integral divisor $\mathfrak{m}$ of k, the extension K/k defines a congruence ideal class group $I_{\mathfrak{m}}/H_{\mathfrak{m}}$ in k, where $H_{\mathfrak{m}}$ is the subgroup of $I_{\mathfrak{m}}$ generated by $S_{\mathfrak{m}}$ and by all norms $N_{K/k}(\mathfrak{A})$ of ideals $\mathfrak{A}$ of K, prime to $\mathfrak{m}$. If an integral divisor $\mathfrak{n}$ is a factor of $\mathfrak{m}$, then the injection $I_{\mathfrak{m}} \to I_{\mathfrak{n}}$ induces a surjective homomorphism $I_{\mathfrak{m}}/H_{\mathfrak{m}} \to I_{\mathfrak{n}}/H_{\mathfrak{n}}$. Let $C_{K/k}$ ($= C$) denote the inverse limit of $I_{\mathfrak{m}}/H_{\mathfrak{m}}$ for all $\mathfrak{m}$, with respect to the maps $I_{\mathfrak{m}}/H_{\mathfrak{m}} \to I_{\mathfrak{n}}/H_{\mathfrak{n}}$. By the method of Weber, one sees that $[I_{\mathfrak{m}} : H_{\mathfrak{m}}] \leqq [K:k]$ for all $\mathfrak{m}$. Hence there is a natural isomorphism $C \cong I_{\mathfrak{m}}/H_{\mathfrak{m}}$ whenever $\mathfrak{m}$ is "large" enough. In fact, there exists an integral divisor $\mathfrak{f}$ such that $C \cong I_{\mathfrak{m}}/H_{\mathfrak{m}}$ holds if and only if $\mathfrak{m}$ is a multiple of $\mathfrak{f}$ and for any such $\mathfrak{m}$, C can be canonically identified with $I_{\mathfrak{m}}/H_{\mathfrak{m}} : C = I_{\mathfrak{m}}/H_{\mathfrak{m}}$. The group $C_{K/k}$ ($= C$) is called the ideal class group of k, associated with the Galois extension K/k, and the integral divisor $\mathfrak{f}$ is called the conductor of K/k. Let h be the order of $C_{K/k} : h = [C:1]$. Then we obtain the so-called second fundamental inequality of class field theory:

$$h \leqq n = [K:k].$$

Now, Takagi called the Galois extension K of k a class field over k when the equality

$$h = n$$

holds for the h and n above. If $\mathfrak{m}$ is an integral divisor of k such that $C_{K/k} = I_{\mathfrak{m}}/H_{\mathfrak{m}}$, the above equality means $[I_{\mathfrak{m}} : H_{\mathfrak{m}}] = [K : k]$, and K is then called a class field over k for the ideal class group $I_{\mathfrak{m}}/H_{\mathfrak{m}}$. For his class fields, Takagi **13** proved the following fundamental results:

1) A finite Galois extension K of a number field k is a class field over k if and only if K/k is an abelian extension.

2) (Existence Theorem) Given any congruence ideal class group $I_{\mathfrak{m}}/H_{\mathfrak{m}}$ in k, there exists a class field K over k for $I_{\mathfrak{m}}/H_{\mathfrak{m}}$. As a consequence, each coset of $I_{\mathfrak{m}}/H_{\mathfrak{m}}$ contains infinitely many prime ideals of k with absolute degree 1.

3) (Uniqueness Theorem) Let K and K' be class fields over k and let $C_{K/k} = I_{\mathfrak{m}}/H_{\mathfrak{m}}$, $C_{K'/k} = I_{\mathfrak{m}}/H'_{\mathfrak{m}}$ for $\mathfrak{m}$ divisible by the conductors of K/k and K'/k. Then $k \subseteq K' \subseteq K$ if and only if $H_{\mathfrak{m}} \subseteq H'_{\mathfrak{m}} \subseteq I_{\mathfrak{m}}$. In particular, the class field K over k in 2) is unique for the given ideal class group $I_{\mathfrak{m}}/H_{\mathfrak{m}}$.

4) Let K be a class field over k. Then:

i) (Isomorphism Theorem) The Galois group of K/k is isomorphic to the ideal class group $C_{K/k}$ associated with K/k.
ii) (Conductor Theorem) A prime divisor of k is ramified in K if and only if it divides the conductor $\mathfrak{f}$ of K/k.
iii) (Decomposition Theorem) The relative degree of a prime ideal $\mathfrak{p}$ of k, unramified in K, is equal to the order of the class of $\mathfrak{p}$ in the ideal class group $C_{K/k} = I_{\mathfrak{f}}/H_{\mathfrak{f}}$.

Now, let $I_{\mathfrak{m}}/H_{\mathfrak{m}}$ be a congruence ideal class group in k. It follows from the above and from the uniqueness theoem for Weber's class fields that an extension K of a number field k is a class field over k for $I_{\mathfrak{m}}/H_{\mathfrak{m}}$ in Weber's sense if and only if it is a class field over k for $I_{\mathfrak{m}}/H_{\mathfrak{m}}$ in the sense of Takagi. Thus both definitions of class fields turn out to be equivalent. However, as Takagi remarked, his definition of class field by postulation of the equality $h = n$ proved better suited for the proof of the Existence Theorem and other results. Needless to say, Hilbert's class field K of k is a special case of Takagi's class field, and the existence of K, as well as all but one of the properties of K conjectured by Hilbert, follows immediately from the above theorems[5].

In Takagi's proof of those theorems in **13** the key steps were the proofs of the following two statements:

a) Let l be an odd prime and let K be a cyclic extension of degree l over k with discriminant $\mathfrak{d} = \mathfrak{f}^{l-1}$. Then K is a class field over k and its conductor is a factor of the ideal $\mathfrak{f}$ of k.

b) Suppose that the ground field k contains a primitive l-th root of unity, l being an odd prime as above. Then, for each congruence ideal class group $I_{\mathfrak{m}}/H_{\mathfrak{m}}$ in k with order l, there exists a cyclic extension K of degree l over k such that K is a class field over k for the given $I_{\mathfrak{m}}/H_{\mathfrak{m}}$.

The proof of a) was carried out by computing (in modern terminology) the orders

[5] *cf.*, §8 below for the exception.

of the cohomology groups of the cyclic Galois group of K/k, acting on various abelian groups such as the unit group and the ideal class group of K. The computation gave the first fundamental inequality of class field theory,

$$h \geqq n,$$

for the extension K/k, and hence the equality $h = n$. In proving $h \geqq n$, Takagi also obtained the Norm Theorem for K/k which states that an element α of k is the norm of an element in K if and only if α is a norm for every local extension associated with K/k. For the proof of b), Takagi fixed an integral divisor $\mathfrak{m}$ of k and counted the number N of congruence ideal class groups $I_{\mathfrak{m}}/H_{\mathfrak{m}}$ in k with order $[I_{\mathfrak{m}} : H_{\mathfrak{m}}] = l$. On the other hand, using a) and the theory of Kummer extensions of Hilbert [15], he showed that there exist at least N class fields K of degree l over k with conductor dividing $\mathfrak{m}$. This of course proved *b*). At the same time, the argument also yielded that the ideal $\mathfrak{f}$ is actually the conductor of the extension K/k in *a*). The relation $\mathfrak{d} = \mathfrak{f}^{l-1}$ between the discriminant $\mathfrak{d}$ and the conductor $\mathfrak{f}$ of K/k was later generalized by Hasse [13] for an arbitrary class field K over k (the Conductor-Discriminant Theorem).

4. In the last chapter of **13**, Takagi discussed applications of his theory to the proof of Kronecker's statements mentioned in §1. First, he described briefly how the theorem of Kronecker on abelian extensions of the rational field $\mathbb{Q}$ follows immediately from his theory when applied to the ground field $k = \mathbb{Q}$. Then he discussed in more detail the application to Kronecker's Jugendtraum on abelian extensions of imaginary quadratic fields. Let k be such a field with discriminant $d < 0$. For each integer $m \geqq 1$, let $L(m) = k(\zeta_m, j(\alpha))$ where ζ_m is a primitive m-th root of unity and $j(\alpha)$ is the value of the elliptic modular function for a number α in k with discriminant $m^2 d$. As Weber [28] already knew, $L(m)$ is the class field over k for the ideal class group I_m/H_m where H_m consists of all principal ideals (α) in k such that $\alpha \equiv r \bmod m$ for some rational number r satisfying $r^2 \equiv \pm 1 \bmod m$. This implies, in particular, that every abelian extension of odd degree over k is contained in $L(m)$ for some $m \geqq 1$, a fact proved by Fueter [5] in 1914. As Fueter also pointed out there, not all abelian extensions of k are subfields of the fields $L(m)$, $m \geqq 1$. So, for each odd integer $n \geqq 1$, Takagi introduced the n-division field $T(n)$ of Jacobi's elliptic function $sn(u)$ associated with k, and using his theory, he then showed that $T(n)$ is a class field over k and that every abelian extension of k is contained in the composite $L(m)T(n)$ for some m and n. Thus Kronecker's Jugendtraum was verified. This achievement must have been particularly satisfying for Takagi who started his number-theoretical research by proving a special case of the above result in his thesis **6**.

Following an idea of Weber[6], Hasse [11] later gave a simpler proof for Kronecker's Jugendtraum by using Weber's elliptic function $\tau(u)$. To solve the Riemann hypothesis for elliptic curves over finite fields, Hasse also studied complex multiplication of such curves from a purely algebraic point of view. Weil [29] then developed an algebraic theory of abelian varieties over arbitrary ground fields and proved the Riemann hypothesis for any algebraic curve defined over a finite field. Based on Weil's theory, the work [25] of

[6] *cf.*, Weber [28], §155.

Shimura-Taniyama generalized the classical theory of complex multiplication to abelian varieties with complex multiplication in number fields.

5. Takagi's other major paper **17** was actually written in 1920 as a sequel to **13**. Following the work of Hilbert and Furtwängler, he presented here his theory of the power residue symbol and the norm residue symbol for a prime exponent l. Since full results on class fields were available to Takagi, his proofs are simpler and the results are more general than those of earlier work. Let k be a number field containing a primitive l-th root of unity. Let μ be a so-called primary number in k and ν a number of k, prime to μ and l. For the power residue symbol with exponent l, Takagi proved that if l is odd, then

$$\left(\frac{\mu}{\nu}\right) = \left(\frac{\nu}{\mu}\right),$$

and if $l = 2$, a similar equality still holds with an additional factor on the right which depends on the real archimedean prime divisors of k. Anticipating Artin, he also proved that the symbol $\left(\frac{\mu}{\mathfrak{a}}\right)$ depends only on the class of $\mathfrak{a}$ in the ideal class group $C = I_{\mathfrak{f}}/H_{\mathfrak{f}}$, associated with the class field $K = k(\sqrt[l]{\mu})$ over k, and made a remark that this fact is the most essential point of the reciprocity law. As for Hilbert's norm residue symbol, Takagi defined it by means of the power residue symbol, as was done by Hilbert in [15], stated the product formula, and proved in an important case that $\left(\frac{\nu, \mu}{\mathfrak{p}}\right) = 1$ if and only if ν is a local norm at $\mathfrak{p}$ for the extension $k(\sqrt[l]{\mu})/k$.

After Takagi's work **17**, a decisive advance in the theory of the reciprocity law was made by Artin [1] in 1927. Let K be a class field over a number field k with conductor $\mathfrak{f}$. For each prime ideal $\mathfrak{p}$ of k unramified in K, let $\sigma_{\mathfrak{p}}$ denote the Frobenius automorphism of $\mathfrak{p}$ for the abelian extension K/k. Artin proved that the homomorphism of the ideal group $I_{\mathfrak{f}}$ into the Galois group G of K/k, defined by $\mathfrak{p} \mapsto \sigma_{\mathfrak{p}}$, induces a canonical isomorphism:

$$C_{K/k} = I_{\mathfrak{f}}/H_{\mathfrak{f}} \simeq G.$$

This is called Artin's general reciprocity law because, as Artin showed, it immediately induces the reciprocity formula for the power residue symbol with arbitrary exponent m. It was proved by Hasse [12] soon afterward that the product formula for Hilbert's norm residue symbol with exponent m, too, can be deduced from Artin's reciprocity law. Note also that Artin's result includes the Isomorphism Theorem and the Decomposition Theorem of Takagi and it rendered those theorems more precise. It may be said that the classical class field theory, initiated by Takagi, was completed by Artin.

6. In 1920, two other papers of Takagi on number theory, **14**, **15**, were also published. **14** is the report on his talk at the International Congress of Mathematicians in 1920 at Strassburg. In three pages, Takagi gave a beautiful summary of his work on class field theory reviewed above. At the end of the paper, he called particular attention to the importance of the problem of generalizing his theory to non-abelian Galois extensions of number fields. In the other paper, **15**, a class of such Galois extensions was studied. In [4], Dedekind found an interesting relation between non-cyclic cubic fields and binary quadratic forms with rational integral coefficients. Namely, let k be such a cubic field with discriminant D. Then the number of classes of quadratic forms with discriminant D is divisible by 3 and one third of the classes are characterized by the fact that a prime number p, prime to D and a quadratic residue mod D, is completely decomposed in k if and only if p can be represented by a quadratic form in one of those classes. Actually, Dedekind proved the above result only in the case k is a pure cubic field. But he checked it for many examples and conjectured its truth in the general case. In **15**, Takagi gave a simple proof of the conjecture by applying his class field theory. In general, let k be a non-cyclic extension of $\mathbb{Q}$ of degree l, an odd prime number, and let K be the Galois extension over $\mathbb{Q}$ generated by all conjugates of k. Suppose that $K = kK_0$ with a cyclic extension K_0 over $\mathbb{Q}$, so that K/K_0 is a cyclic extension of degree l. Using the fact that the Galois group of $K/\mathbb{Q}$ is a special type of meta-cyclic group, Takagi studied the decomposition of a rational prime p in K and determined the conductor of the class field K over K_0. Applying these results for the special case $l = 3$, he then proved the conjecture of Dedekind.

7. The paper **23** of 1927 on cyclotomic fields, a part of which had been reported earlier in **18** in 1922, was Takagi s last important contribution to number theory, although he later published another paper **24** in 1928, an expository paper on binary quadratic forms. Let k denote the cyclotomic field of l-th roots of unity for an odd prime l, and let

$$\{\mu, \nu\} = \left(\frac{\mu}{\nu}\right)\left(\frac{\nu}{\mu}\right)^{-1},$$

where $(-)$ is the power residue symbol of exponent l in k. In the first half of **23**, Takagi gave a beautiful explicit formula for $\{\mu, \nu\}$ as follows. Let ζ be a fixed primitive l-th root of unity in k and let $\lambda = 1 - \zeta$ so that $(l) = (\lambda)^{l-1}$. Let A denote the multiplicative group of all numbers α in k such that $\alpha \equiv 1 \bmod \lambda$, and let B be the subgroup of all β in A satisfying $\beta \equiv 1 \bmod \lambda^l$. Then the Galois group Δ of $k/\mathbb{Q}$ acts naturally on the factor group A/B which is an abelian group of type $(l, \ldots, l)$ of rank $l-1$, namely, a vector space of dimension $l-1$ over the finite field with l elements. Since Δ is a cyclic group of order $l-1$, A/B decomposes into the direct sum of $l-1$ one-dimensional subspaces, each invariant under Δ. Using this fact, Takagi found a basis $\{\kappa_1, \ldots, \kappa_{l-1}\}$ (mod B) of A/B such that

$$\kappa_i \equiv 1 - \lambda^i \mod \lambda^{i+1}, \qquad \kappa_i^\sigma \equiv \kappa_i^{a^i} \mod \lambda^l,$$

σ being any element of Δ and a, an integer satisfying $\zeta^\sigma = \zeta^a$. Now for each μ in k, prime to l, one has a congruence

$$\mu \equiv \mu^l \prod_{i=1}^{l-1} \kappa^{t_i} \mod \lambda^l$$

with integers $t_i = t_i(\mu)$, $1 \leqq i \leqq l-1$, uniquely determined mod l by μ. With such t_i's, Takagi proved that

$$\{\mu, \nu\} = \zeta^u, \quad \text{where} \quad u = -\sum_{i=1}^{l-1} i t_i(\mu) t_{l-i}(\nu).$$

He also examined the relation between $t_i(\mu)$, $1 \leqq i \leqq l-1$, and Kummer's logarithmic differential quotients, $l_i(\mu)$, $1 \leqq i \leqq l-1$, and deduced from his formula Kummer's criterion for Fermat's problem to be found in [23].

In the second half of **23**, Takagi studied the l-rank of the ideal class group of the cyclotomic field k. Let k_0 be the maximal real subfield of k and let t and t_0 denote the l-ranks of the ideal class groups of k and k_0 respectively. Takagi proved that $2t_0 \leqq t$ and also that if $t_0 = 0$, then $t \leqq \frac{l-3}{2}$. Furthermore, he investigated the interesting relations between the units of k and the ideal classes of exponent l in k, again by decomposing abelian groups like A/B under the action of the Galois group Δ[7]. The same kind of problem as treated in this paper has been deeply studied in recent years in the theory of cyclotomic fields, in particular, by Mazur-Wiles [24]. The paper **23** is one of the original sources for the modern development of that theory.

8. The work of Takagi and Artin confirmed what Hilbert foresaw in 1897 in the introduction of [15], and class field theory has since become a major discipline in algebraic number theory. In concluding this review of Takagi's papers in number theory, we would like to add here a brief account of some of the further developments in that theory after Takagi-Artin.

With the appearance of Takagi's paper **13**, only one of the properties of class fields, predicted by Hilbert, remained unproved, namely, the fact that every ideal of a number field k becomes a principal ideal in Hilbert's class field K over k (the Principal Ideal Theorem). Applying his reciprocity law, Artin [2] reduced the above theorem on K/k to a purely group-theoretical statement on meta-abelian groups, and this was then proved by Furtwängler [8] by computation in group rings. A shorter, more conceptual proof of the same result was given later by Iyanaga [21] in 1934. In 1926–1930, Hasse's Zahlbericht [10] was published. This is an exposition of the theory of Takagi-Artin, including Hasse's own work on power and norm residue symbols, and it served to introduce class field theory to a wider audience. In the 1930's, simplification and reorganization of the original proofs in Takagi **13** were sought in order to obtain better and deeper insight into class field theory. At the same time, extensions of class field theory to local fields and function fields of one variable over finite constant fields were attempted, and both programs were successfully carried out by Hasse, Chevalley, Herbrand, F.K. Shmidt, and others. Another important problem around that time was the construction of class field theory by purely algebraic methods. In the theory of Weber-Takagi, the second fundamental inequality, $h \leqq n$, was obtained by using analytic properties of L-series. The problem was to prove this inequality by arithmetic arguments alone. This was done by Chevalley [3] in 1940, where he also introduced the notion of ideles which replaced the ideals in number fields. Chevalley's idea of using topological groups and infinite Galois extensions of number fields has had great impact in the later development of algebraic number theory in general. For class field theory, the era after the Second World War is marked by the introduction of another new method, namely, the application of the cohomology theory of groups. Following the pioneering work of Weil [30] and Hochschild-Nakayama [20],

[7] The integer t_0' on p. 253 is actually zero.

the cohomological approach was fully developed by Tate [26], bringing more clarity and generality to class field theory. With this method, for example, Golod-Šafarevič [9] succeeded in proving the classical conjecture on class field towers, namely, that for each prime number p there exist number fields over which the p-class field towers are infinite.

One of the outstanding problems in number theory today is the generalization of class field theory to (non-abelian) Galois extensions of number fields. As mentioned earlier, this problem was already proposed by Takagi in his talk **14** at Strassburg. Significant progress has been made in recent years by Langlands and others, but the final goal is as yet unattained. We hope that a class field theory for Galois extensions, in whatever formulation, will be helpful in solving difficult problems in number theory which are beyond our reach at present.

References

[1] Artin E (1927) Beweis des allgemeinen Reziprozitätsgesetzes. Abh Math Semin Univ Hamburg 5: 353–363
[2] Artin E (1930) Idealklassen in Oberkörpern und allgemeines Reziprozitätsgesetz. Abh Math Semin Univ Hamburg 7: 46–51
[3] Chevalley C (1940) La théorie du corps de classes. Ann Math 41: 394–417
[4] Dedekind R (1900) Über die Anzahl der Idealklassen in reinen kubischen Zahlkörpern. J reine angew Math 121: 40–123
[5] Fueter R (1914) Abel'sche Gleichungen in quadratisch-imaginären Zahlkörpern. Math Ann 75: 177–255
[6] Furtwängler Ph (1902) Über die Reziprozitätsgesetze zwischen l^{ten} Potenzresten in algebraischen Zahlkörpern, wenn l eine ungerade Primzahl bedeutet. Abh K Ges Wiss Göttingen 2: 1–82
[7] Furtwängler Ph (1907) Allgemeiner Existenzbeweis für den Klassenkörper eines beliebigen algebraischen Zahlkörpers. Math Ann 63: 1–37
[8] Furtwängler Ph (1930) Beweis des Hauptidealsatzes für Klassenkörper algebraischer Zahlkörper. Abh Math Semin Univ Hamburg 7: 14–36
[9] Golod ES, Šafarevič IR (1964) On class field towers (in Russian) Izv Akad Nauk SSSR 28: 261–272
[10] Hasse H (1926) Bericht über neuere Untersuchungen und Probleme aus der Theorie der algebraischen Zahlkörper I, Ia, II, Jber. dt Math Verein 35: 1–55; 36 (1927): 233–311; Exg Bd 6(1930a): 1–204
[11] Hasse H (1927) Neue Begründung der komplexen Multiplikation I, II. J reine angew Math 157: 115–139; 165(1931): 64–88
[12] Hasse H (1927) Über das Reziprozitätzgesetz der m-ten Potenzreste. J reine angew Math 158: 228–259
[13] Hasse H (1934) Normenresttheorie, Führer und Diskriminante Abelscher Zahlkörper. J Fac Sci Univ Tokyo Sec I 2: 477–498
[14] Hasse H (1967) History of class field theory. In: Algebraic Number Theory. Thompson Book, Washington, pp 266–279
[15] Hilbert D (1897) Die Theorie der algebraischen Zahlkörper. Jber dt Math Verein 4: 175–546
[16] Hilbert D (1896) Neuer Beweis des Kronecker'schen Fundamentalsatzes über Abel'sche Zahlkörper. Nachr Ges Wiss Göttingen 29–39
[17] Hilbert D (1899) Über die Theorie der relativquadratischen Zahlkörper. Jber dt Math Verein 6: 88–94
[18] Hilbert D (1899) Über die Theorie des relativquadratischen Zahlkörpers. Math Ann 51: 1–127
[19] Hilbert D (1902) Über die Theorie der relativ-Abel'schen Zahlkörper. Acta Math 26: 99–132

[20] Hochschild G, Nakayama T (1952) Cohomology in class field theory. Ann Math 55: 348–366
[21] Iyanaga S (1934) Zum Beweis des Hauptidealsatzes. Abh Math Univ Hamburg 10: 349–357
[22] Kronecker L (1853) Über die algebraisch auflösbaren Gleichungen I. Sber preuss Acad Wiss 365–374
[23] Kummer E (1857) Einige Sätze über die aus den Wurzeln der Gleichung $\alpha^{\lambda} = 1$ gebildeten komplexen Zahlen für den Fall, dass die Klassenzahl durch λ teilbar ist, nebst Anwendungen derselben auf einen weiteren Beweis des Fermatschen Lehrsatzes. Abh Akad Wiss Berlin 41–74
[24] Mazur B, Wiles A (1984) Class fields of abelian extensions of $\mathbb{Q}$. Inv Math 76: 179–330
[25] Shimura G, Taniyama Y (1961) Complex multiplication of abelian varieties and its applications to number theory. Publ Math Soc Japan, no 6
[26] Tate J (1952) The higher dimensional cohomology groups of class field theory. Ann Math 56: 294–297
[27] Weber H (1897) Über Zahlgruppen in algebraischen Körpern I, II, III. Math Ann 48: 433–473; 49(1897): 83–100; 50(1898): 1–26
[28] Weber H (1908) Lehrbuch der Algebra III. Braunschweig
[29] Weil A (1948) Variétés abéliennes et courbes algébriques. Hermann, Paris
[30] Weil A (1951) Sur la théorie du corps de classes. J Math Soc Jpn 3: 1–35

[66] A letter to J. Dieudonné

"Zeta Functions in Geometry" (Tokyo, 1990), Adv. St. in Pure Math., 21, 1992, pp. 445-450.

April 8, 1952

Professor J. Dieudonne
26 Rue Saint-Michel, Nancy
France

Dear Professor Dieudonne:

A few days ago, I received a letter from Professor A. Weil, asking me to send you a copy of a letter I wrote him the other day and to give you a brief account of my result on L-functions. I, therefore, enclose here a copy of that letter and write an outline of my idea on L-functions.

Let k be a finite algebraic number field, J the idele group of k, topologized as in a recent paper of Weil. J is a locally compact abelian group containing the principal idele group P as a discrete subgroup. We denote by J_0 the subgroup of J consisting of ideles $\mathfrak{a} = (a_p)$ such that $a_p = 1$ for all infinite (i.e. archimedean) primes P. We call J_0 the finite part of J and define the infinite part J_∞ similarly, so that we have

$$J = J_0 \times J_\infty, \quad \mathfrak{a} = \mathfrak{a}_0\mathfrak{a}_\infty, \quad \mathfrak{a}_0 \in J_0, \quad \mathfrak{a}_\infty \in J_\infty.$$

We also denote by U the compact subgroup of J consisting of ideles $\mathfrak{a} = (a_p)$ such that the absolute value $|a_p|_p = 1$ for every prime P. $U_0 = U \cap J_0$ is then an open, compact subgroup of J_0 and J_0/U_0 is canonically isomorphic to the ideal group I of k. According to Artin-Whaples, we can choose the absolute values $|a_p|_p$ so that the volume function $V(\mathfrak{a}) = \Pi_p |a_p|_p$ $(\mathfrak{a} = (a_p))$ has the value 1 at every principal idele $\alpha \in P$ (the product formula) and that $V(\mathfrak{a}_0)^{-1}$ is equal to the absolute norm $N(\tilde{\mathfrak{a}}_0)$ of the ideal $\tilde{\mathfrak{a}}_0$, which corresponds to $\mathfrak{a}_0$ by the above isomorphism between J_0/U_0 and I.

We now define a function $\varphi(\mathfrak{a})$ by

$$\varphi(\mathfrak{a}) = \varphi(\mathfrak{a}_0)\varphi(\mathfrak{a}_\infty), \quad \mathfrak{a} = \mathfrak{a}_0\mathfrak{a}_\infty,$$

$$\varphi(\mathfrak{a}_0) = \begin{cases} 1, & \text{if } \tilde{\mathfrak{a}}_0 \text{ is an integral ideal,} \\ 0, & \text{otherwise} \end{cases}$$

$$\varphi(\mathfrak{a}_\infty) = \exp\,(-\frac{\pi}{\sqrt[n]{\Delta}}\sum_{i=1}^{r} e_i|a_{p_{\infty,i}}|^2),$$

where n is the absolute degree of k, Δ is the discriminant of k, $a_{p_{\infty,i}}$ are the components of $\mathfrak{a}$ at the infinite primes $P_{\infty,i}$ and $e_i = 1$ or 2 according as $P_{\infty,i}$ is real or complex. Since U_0 is open in J_0, $\varphi(\mathfrak{a})$ is a continuous function on J and we define a function $\xi(s)$ by

$$\xi(s) = \int_J \varphi(\mathfrak{a})V(\mathfrak{a})^s\,d\mu(\mathfrak{a}), \qquad \text{for } s > 1. \tag{1}$$

Here $\mu(\mathfrak{a})$ denotes a Haar measure of the locally compact group J. We shall calculate this integral in two different ways.

First, using $J = J_0 \times J_\infty$, $\varphi(\mathfrak{a}) = \varphi(\mathfrak{a}_0)\varphi(\mathfrak{a}_\infty)$ and $V(\mathfrak{a}) = V(\mathfrak{a}_0)V(\mathfrak{a}_\infty)$, we have

$$\xi(s) = \int_{J_0} \varphi(\mathfrak{a}_0)V(\mathfrak{a}_0)^s\,d\mu(\mathfrak{a}_0)\int_{J_\infty} \varphi(\mathfrak{a}_\infty)V(\mathfrak{a}_\infty)^s\,d\mu(\mathfrak{a}_\infty).$$

If we note that U_0 is an open, compact subgroup of J_0 and $J_0/U_0 = I$, we see immediately that the first integral on the right-hand side is equal to (up to a positive constant) the zeta-function $\zeta(s) = \sum N(\tilde{\mathfrak{a}})^{-s}$ ($\tilde{\mathfrak{a}} =$ integral ideal) of k. On the other hand, J_∞ being the direct product of r copies of the multiplicative group K^* of the real or complex number-field K, the second integral is the product of integrals of the form

$$\int_{K*} \exp(-\frac{\pi}{\sqrt[n]{\Delta}}e|t|^2)|t|^s\,d\mu_k(t), \quad e = 1 \text{ or } 2,$$

which can be easily calculated to be equal to

$$\Delta^{\frac{s}{2n}}\pi^{\frac{-s}{2}}\Gamma(\frac{s}{2}) \quad \text{or} \quad \Delta^{\frac{s}{n}}2^{-s}\pi^{-s}\Gamma(s),$$

according as K is real or complex. We have therefore

$$\xi(s) = \text{const.}2^{-r_2 s}\Delta^{\frac{s}{2}}\pi^{-\frac{ns}{2}}\Gamma(\frac{s}{2})^{r_1}\Gamma(s)^{r_2}\zeta(s). \tag{2}$$

The above calculation also shows that the integral (1) actually converges for $s > 1$.

We now transform the same integral (1) in another way. Namely, we first integrate the function $f(\mathfrak{a}) = \varphi(\mathfrak{a})V(\mathfrak{a})^s$ on the subgroup P and then on the factor group $\bar{J} = J/P = \{\bar{\mathfrak{a}}\}$;

$$\int_J f(\mathfrak{a})d\mu(\mathfrak{a}) = \int_{\bar{J}} \{ \int_P f(\mathfrak{a}\alpha)d\mu(\alpha)\} d\mu(\bar{\mathfrak{a}}).$$

However, since P is discrete and $V(\mathfrak{a}\alpha) = V(\mathfrak{a})V(\alpha) = V(\mathfrak{a}) = V(\bar{\mathfrak{a}})$, we have

$$\int_P f(\mathfrak{a}\alpha)d\mu(\alpha) = (\sum_{\alpha\in P} \varphi(\mathfrak{a}\alpha))V(\bar{\mathfrak{a}})^s,$$

and if we put

$$\bar{\varphi}(\bar{\mathfrak{a}}) = \sum_{\alpha\in P} \varphi(\mathfrak{a}\alpha),$$

$$\Theta(\bar{\mathfrak{a}}) = 1 + \bar{\varphi}(\bar{\mathfrak{a}}) = \sum_{\alpha\in k} \varphi(\mathfrak{a}\alpha),$$

the theta-formula

$$\Theta(\bar{\mathfrak{a}}) = V(\bar{\mathfrak{a}})^{-1}\Theta(\bar{\vartheta}\bar{\mathfrak{a}}^{-1}) \text{ or } \bar{\varphi}(\bar{\mathfrak{a}}) = V(\bar{\mathfrak{a}})^{-1}\bar{\varphi}(\bar{\vartheta}\bar{\mathfrak{a}}^{-1}) + V(\bar{\mathfrak{a}})^{-1} - 1$$

holds, where ϑ is an idele of volume 1 such that $\tilde{\vartheta}_0$ is the different of k and its infinite components are all equal to $\sqrt[n]{\Delta}$. We have now

$$\xi(s) = \int_{\bar{J}} \bar{\varphi}(\bar{\mathfrak{a}})V(\bar{\mathfrak{a}})^s \, d\mu(\bar{\mathfrak{a}}) = \int_{V(\bar{\mathfrak{a}})\geqq 1} + \int_{V(\bar{\mathfrak{a}})\leqq 1},$$

and here the first integral on the right-hand side

$$\psi(s) = \int_{V(\bar{\mathfrak{a}})\geqq 1} \bar{\varphi}(\bar{\mathfrak{a}})V(\bar{\mathfrak{a}})^s \, d\mu(\bar{\mathfrak{a}})$$

gives an integral function of s, for this integral converges absolutely for every complex value s, because of the convergence of (1) for $s > 1$ and because of $V(\bar{\mathfrak{a}}) \geqq 1$. Using the theta-formula and the invariance of

Haar measures, we can transform the second integral as follows:

$$\begin{aligned}
&\int_{V(\bar{\mathfrak{a}})\leqq 1} \\
&= \int_{V(\bar{\mathfrak{a}})\leqq 1} (V(\bar{\mathfrak{a}})^{-1}\bar{\varphi}(\bar{\vartheta}\bar{\mathfrak{a}}^{-1}) + V(\bar{\mathfrak{a}})^{-1} - 1)V(\bar{\mathfrak{a}})^{s}\, d\mu(\bar{\mathfrak{a}}) \\
&= \int_{V(\bar{\mathfrak{a}})\geqq 1} (\bar{\varphi}(\bar{\vartheta}\bar{\mathfrak{a}})V(\bar{\mathfrak{a}})^{1-s} + V(\bar{\mathfrak{a}})^{1-s} - V(\bar{\mathfrak{a}})^{-s})d\mu(\bar{\mathfrak{a}}) \\
&\qquad\qquad\qquad (\text{by } \bar{\mathfrak{a}} \to \bar{\mathfrak{a}}^{-1}) \\
&= \int_{V(\bar{\mathfrak{a}})\geqq 1} \bar{\varphi}(\bar{\vartheta}\bar{\mathfrak{a}})V(\bar{\mathfrak{a}})^{1-s}\, d\mu(\bar{\mathfrak{a}}) + \int_{V(\bar{\mathfrak{a}})\geqq 1} (V(\bar{\mathfrak{a}})^{1-s} - V(\bar{\mathfrak{a}})^{-s})d\mu(\bar{\mathfrak{a}}) \\
&= \int_{V(\bar{\mathfrak{a}})\geqq 1} \bar{\varphi}(\bar{\mathfrak{a}})V(\bar{\mathfrak{a}})^{1-s}\, d\mu(\bar{\mathfrak{a}}) + \int_{V(\bar{\mathfrak{a}})\geqq 1} (V(\bar{\mathfrak{a}})^{1-s} - V(\bar{\mathfrak{a}})^{-s})d\mu(\bar{\mathfrak{a}}) \\
&\qquad\qquad\qquad (\text{by } \bar{\mathfrak{a}} \to \bar{\mathfrak{a}}^{-1}\bar{\mathfrak{a}} \text{ and } V(\bar{\mathfrak{a}}) = 1) \\
&= \psi(1-s) + \int_{V(\bar{\mathfrak{a}})\geqq 1} (V(\bar{\mathfrak{a}})^{1-s} - V(\bar{\mathfrak{a}})^{-s})d\mu(\bar{\mathfrak{a}}).
\end{aligned}$$

Now, the set of all ideles $\mathfrak{a}$ such that $V(\mathfrak{a}) = 1$ forms a closed subgroup J_1 of J and it can be seen easily that J is the direct product of $\bar{J}_1 = J_1/P$ and a subgroup S which is canonically isomorphic to the multiplicative group $T = \{t = V(\bar{\mathfrak{a}})\}$ of positive real numbers. Hence we have

$$\begin{aligned}
\int_{V(\bar{\mathfrak{a}})\geqq 1} (V(\bar{\mathfrak{a}})^{1-s} - V(\bar{\mathfrak{a}})^{-s})d\mu(\bar{\mathfrak{a}}) &= \int_{\bar{J}_1} \times \int_{S, V(\bar{\mathfrak{a}})\geqq 1} \\
&= \mu(\bar{J}_1) \int_{t\geqq 1} (t^{1-s} - t^{-s})\frac{dt}{t} \\
&= \mu(\bar{J}_1)(\frac{1}{s-1} - \frac{1}{s}).
\end{aligned}$$

We have, therefore, the formula

$$(3) \qquad \xi(s) = \psi(s) + \psi(1-s) + \mu(\bar{J}_1)(\frac{1}{s-1} - \frac{1}{s}), \quad (s > 1).$$

It then follows immediately that $\xi(s)$ is a regular analytic function of s on the whole s-plane except for simple poles at $s = 0, 1$ and it satisfies the equation

$$\xi(s) = \xi(1-s),$$

which is nothing but the functional equation of the zeta-function $\zeta(s)$ (cf. (2)).

The formula (3) also shows that the measure $\mu(\bar{J}_1)$ of $\bar{J}_1$ is finite. Since $\bar{J}_1$ is a locally compact group, this means that $\bar{J}_1$ is compact. Now, we put $H = (U_0 \times J_\infty) \cap J_1$ and consider the sequence of groups

$$J_1 \supset HP \supset UP \supset P.$$

Since U is compact UP is closed in J_1, and, since $U_0 \times J_\infty$ is open in J, H and HP are open subgroups of J_1. It then follows from the compactness of $\bar{J}_1 = J_1/P$ that J_1/HP and HP/UP are both compact groups. But, as HP is open and J_1/HP is discrete, J_1/HP must be finite. Consequently, the group $J/(U_0 \times J_\infty)P$, which is easily seen to be isomorphic to J_1/HP, is a finite group and this proves the finiteness of the ideal classes of k. Now, H/U is isomorphic to $(J_1 \cap J_\infty)/(U \cap J_\infty)$ and hence is an $(r-1)$-dimensional vector group. On the other hand, we see from the isomorphisms

$$HP/UP = H/U(H \cap P), \quad U(H \cap P)/U = H \cap P/U \cap P,$$

that $H/U(H \cap P)$ is compact and $U(H \cap P)/U$ is discrete. Since H/U is a vector group, this implies that $U(H \cap P)/U$ is an $(r-1)$-dimensional lattice in H/U and, consequently, that $H \cap P/U \cap P$ is a free abelian group with $r-1$ generators. However, as is readily seen, $H \cap P$ and $U \cap P$ are the unit group and the group of roots of unity in k. Hence the classical Dirichlet's unit theorem has been proved.

The above method of proving the functional equation can be also applied to Hecke's L-functions with "Grössencharakteren", for such a character X is a conitnuous character of $\bar{J}$ which is trivial on S. The integrand of (1) must be then replaced by

$$X(\mathfrak{a})\varphi(\mathfrak{a}, X)V(\mathfrak{a})^s,$$

where $\varphi(\mathfrak{a}, X)$ is a similar function to $\varphi(\mathfrak{a})$, depending on X. The zeta-function (or L-functions) of a division algebra over a finite algebraic number-field can also be treated in a similar way, though here integrations over linear groups appear and calculations are more complicated.

For the above proof of the functional equation of $\zeta(s)$, two group-theoretical facts seem to be essential. One is the topological structure of the group J, that of its subgroups and factor groups, together with the invariance of Haar measures on them, and the other is the theta-formula, which is an analytical expression for the self-duality of the additive group of the ring R of valuation vectors (= additive ideles) of k. J being exactly the multiplicative group of R, here the additive and multiplicative properties of R are subtly mixed up and it seems to me likely that something essential to the arithmetic of k is still hidden in this connection, though I only know that the usual topology of J coincides with the one which is obtained by considering J as a group of automorphisms of the additive group of R in the sense of Braconnier.

I am leaving the United States at the beginning of May and going back to Japan by way of Europe. I shall be in Paris about one week around the 12th of May. I hope I shall have enough time to go to Nancy to see you and others, though I am not sure of it yet.

Very sincerely yours,

Kenkichi Iwasawa

[U1] A note on unitary representations

A fundamental theorem of Yamabe on the structure of locally compact groups states that every connected locally compact group is a projective limit of Lie groups; this is equivalent to saying that each small neighborhood of the identity in such a group G contains a compact normal subgroup N such that G/N is a Lie group.[1)] In the following, we shall discuss a simple application of this theorem to the unitary representations of localy compact groups.

A unitary representation of a topological group is, by definition, a weakly continuous homomorphism of that group into the group of unitary operators on a Hilbert space. Let K be a compact group. It is well-known[2)] that every irreducible unitary representation of K is finite dimensional and that an arbitrary unitary representation κ of K is decomposed into the direct sum of a family of such finite dimensional irreducible representations κ_i, $i \in I$:

$$\kappa = \bigoplus_i \kappa_i.$$

Let H_i denote the representation space for κ_i so that the representation space H for κ is the direct sum of the Hilbert spaces H_i, $i \in I$:

$$H = \bigoplus_i H_i.$$[2a)]

Let C denote the family of all classes of mutually equivalent irreducible unitary representations of K and let γ be a class in C. A representation space for K will be called of type γ if the representation defined on that space belongs to the class γ. Let H_γ denote the direct sum of all H_i in the above which are of type γ. Then H is the direct sum of such subspaces H_γ where γ ranges over all classes in C :

$$H = \bigoplus H_\gamma, \qquad \gamma \in C.$$

Furthermore, one sees easily that H_γ is the closed subspace of H generated by all subspaces of type γ in H. Let κ_γ denote the representation of K defined by H_γ so that

$$\kappa = \bigoplus \kappa_\gamma, \qquad \gamma \in C.$$

For each H_i of type γ, we then have

$$\text{Ker } \kappa_\gamma = \text{Ker } \kappa_i.$$

Since κ_i is of finite dimensional representation, it follows that $K/\text{Ker}(\kappa_\gamma)$ is a compact Lie group.

We now prove the following result:

Theorem. Let ρ be a unitary representation of a connected locally compact group G. Then ρ is the direct sum of a family of unitary representations ρ_γ :

$$\rho = \bigoplus \rho_\gamma$$

such that each factor group $G/\mathrm{Ker}(\rho_\gamma)$ is a Lie group.

Collorary. Let ρ be an irreducible unitary representation of a connected locally compact group G. Then $G/\mathrm{Ker}(\rho)$ is a Lie group.

In the proof of the theorem, let H denote the representation space of ρ. By the theorem of Yamabe, there exists a compact normal subgroup N in G such that G/N is a Lie group. We apply the above remark for $K = N$ and for the restriction κ of ρ on the subgroup N, and obtain the decompositions

$$\kappa = \bigoplus \kappa_\gamma, \quad H = \bigoplus H_\gamma, \qquad \gamma \in C,$$

as described in the above. We shall next show that each H_γ is invariant under ρ. Let γ be a class in C and let ω be any irreducible unitary representation of N belonging to the class γ. Let σ be a fixed element in G. Since N is normal in G, the map ω^σ given by

$$\omega^\sigma(\tau) = \omega(\sigma\tau\sigma^{-1}), \qquad \text{for} \quad \tau \in N,$$

defines again an irreducible unitary representation of N, and it belongs to a class γ^σ which depends only upon γ and σ (and is independent of the choice of ω in γ). Let U be a subspace of type γ in H, i.e., a subspace of H which is invariant under κ and is of type γ as a representation space of N. Then we see immediately the $\rho(\sigma)U$ is a subspace of type γ^σ in a similar sense. Therefore it follows from the second definition of H_γ stated above that

$$\rho(\sigma)H_\gamma = H_{\gamma^\sigma}$$

so that, either $\rho(\sigma)H_\gamma = H_\gamma$ or $\rho(\sigma)H_\gamma$ is orthogonal to H_γ in H. Now, assume that $H_\gamma \neq 0$ and choose a vector $v \neq 0$ in H_γ. Since ρ is weakly continuous, the set W of all σ in G satisfying

$$(\rho(\sigma)v, v) \neq 0$$

is a neighborhood of the identity in G. For σ in W, $\rho(\sigma)H_\gamma$ is certainly not orthogonal to H_γ. Hence it follows from the above that

$$\rho(\sigma)H_\gamma = H_\gamma, \quad \gamma^\sigma = \gamma$$

for every σ in W. As G is connected and is generated by W, the same equalities hold also for every σ in G. Hence H_γ is invariant under ρ.

Now, let ρ_γ denote the unitary representation of G defined by the subspace H_γ so that

$$\rho = \bigoplus \rho_\gamma, \qquad \gamma \in C.$$

Clearly κ_γ is the restriction of ρ_γ on the subgroup N. By a remark mentioned ealier, $N/\mathrm{Ker}(\kappa_\gamma)$ is a compact Lie group. The equalities $\rho(\sigma)\ H_\gamma = H_\gamma$, for all σ in G, also shows that $\mathrm{Ker}(\kappa_\gamma)$ is a normal subgroup of G. Since both G/N

and $N/\mathrm{Ker}(\kappa_\gamma)$ are Lie groups, $G/\mathrm{Ker}(\kappa_\gamma)$ is also a Lie group.[3)] As $\mathrm{Ker}(\kappa_\gamma)$ is contained in $\mathrm{Ker}(\rho_\gamma)$ and $G/\mathrm{Ker}(\rho_\gamma)$ is a factor group of $G/\mathrm{Ker}(\kappa_\gamma)$, $G/\mathrm{Ker}(\rho_\gamma)$ is again a Lie group; and this completes the proof of our theorem.

We shall next show that the above corollary induces conversely the theorem of Yamabe so that these two are actually equivalent to each other. Let G be a connected locally compact group and let V_0 be any small open neighborhood of the identity in G such that $V_0^{-1} = V_0$ and such that the closure V of V_0 in G is compact. Let X denote the set-theoretical difference of $V^3 = VVV$ and V_0. By a well-known theorem of Gelfand-Raikov,[4)] there exists for each σ in X an irreducible unitary representation ρ of G such that $\rho(\sigma) \neq \rho(1)$. Since X is a compact subset of G, there exist a finite number of irreducible unitary representations of G, say, ρ_1, ..., ρ_s, such that the intersection M of the kernels $\mathrm{Ker}(\rho_i)$, $i = 1, ..., s$, is disjoint from the set X. Let

$$N = M \cap V_0 = M \cap V^3.$$

Since

$$NN^{-1} \subset M \cap V^2 \subset N, \quad \sigma N \sigma^{-1} \subset M \cap V^3 = N$$

for every σ in $V_0 = V_0^{-1}$, and since G is generated by V_0, we see that N is a compact normal subgroup of G. Furthermore

$$M \cap V_0 N \subset M \cap V^2 \subset N$$

indicates that G/M is locally isomorphic to G/N.

Assume now that the corollary is true for our group G so that each $G/\mathrm{Ker}(\rho_i)$, $i = 1, ..., s$, is a Lie group. Since M is the intersection of $\mathrm{Ker}(\rho_i)$, $i = 1, ..., s$, there exists a continuous monomorphism of G/M into the direct product of the Lie groups $G/\mathrm{Ker}(\rho_i)$. Hence, by a classical theorem of E. Cartan, G/M is a Lie group; and so is G/N because it is locally isomorphic to G/M. Thus we see that each small neighborhood V_0 of the identity in G contains a compact normal subgroup N such that G/N is a Lie group, and the theorem of Yamabe is proved.

It would be interesting to examine further the proof of Yamabe's theorem from the view-point of the theory of unitary representations.

Footnotes

1) See H. Yamabe, On the conjecture of Iwasawa and Gleason, Ann. of Math., **58** (1953), 48-54.
2) [*No reference is given in the text, but the results stated here follow easily from the Peter-Weyl theorem. See, e.g., A. W. Knapp, Representation Theory of Semisimple Groups, Princeton Univ. Press, 1986, p.17, or R. Carter, G. Segal, and I. Macdonald, Lectures on Lie Groups and Lie Algebras, London Math. Soc. Student Text, **32**, Cambridge Univ. Press, 1995, pp.91-92.]

*2a) [*By these notations it is meant that κ and H are the *topological* direct sum of the κ_i and H_i (i.e., H is the closure of the algebraic direct sum of the H_i).]

3) See M. Kuranishi, On local euclidean groups satisfying certain conditions, Proc. Amer. Math. Soc., **1** (1950), 372-380.
4) [*See I. Gelfand and D. Raikov, Irreducible unitary representations of locally bicompact groups, Mat. Sbornik, N.S. **13** (1943), 301-316; C.R. Doklady Acad. Sci. URSS, **42**-5 (1944), 199-201.]

Editors' Notes

This (undated) note was possibly written in the early 1950s. (It would be worth noting that Iwasawa gave a series of lectures on unitary representations, including the theorem of Gelfand-Raikov, at the University of Tokyo in the spring of 1950 just before his departure to the United States.) Later in 1987, the year he returned to Japan, the result of this note was given in a slightly extended form in his Colloquium lecture, entitled

"Unitary representations of locally compact groups and Yamabe's theorem", at Kyoto University on December 10, 1987. While in the text he proves the main result (i.e., the theorem and the equivalence of its corollary to Yamabe's theorem), assuming G to be a connected locally compact group, the same result is proved in his talk in Kyoto under a weaker assumption that G is a locally compact group such that G/G_0 is compact, G_0 denoting the connected component of the identity of G.

[U2] Automorphisms of Galois groups over number fields

Let $\overline{\mathbf{Q}}$ be a fixed algebraic closure of the rational field $\mathbf{Q}$, e.g., the field of all algebraic numbers, and let $\tilde{\mathbf{Q}}$ denote the maximal solvable extension of $\mathbf{Q}$ contained in $\overline{\mathbf{Q}}$. In his study on the Galois groups $\mathrm{Gal}(\overline{\mathbf{Q}}/\mathbf{Q})$ and $\mathrm{Gal}(\tilde{\mathbf{Q}}/\mathbf{Q})$, Neukirch [4], [5] proved, among other interesting results, the following theorem: Let G denote one of the above Galois groups, let G_1 and G_2 be open subgroups of G, and let (at least) one of G_1 and G_2 be a normal subgroup of G. Then G_1 is isomorphic to G_2 only when $G_1 = G_2$. An immediate consequence of the theorem is that an automorphism of G maps each closed normal subgroup of G onto itself. Based on these facts, Neukirch then proposed the problem to see whether a) every automorphism of G is an inner automorphism and whether b) two isomorphic open subgroups of G are always conjugate in G.

In a recent paper [2], Ikeda succeeded in proving a) by using a result, obtained earlier by Ikeda [1] and Komatsu [3] independently, that an automorphism of G leaves each conjugate class of G invariant. In what follows, we shall prove the theorem of Ikeda, following mainly his original ideas in [2] but modifying some of the delicate arithmetic proofs with simpler group theoretical arguments; it will enable us to obtain slightly more results as follows.[1)] In general, for each subfield K of $\overline{\mathbf{Q}}$, let K_{ab} denote the maximal abelian extension of K in $\overline{\mathbf{Q}}$. It will be proved that if F is any Galois extension of $\mathbf{Q}$ in $\overline{\mathbf{Q}}$ such that $F_{ab} = F$, then both statements a) and b) in the above are true for the Galois group $G = \mathrm{Gal}(F/\mathbf{Q})$; of course, $\overline{\mathbf{Q}}$ and $\tilde{\mathbf{Q}}$ are typical examples of such a field F.

In the following, all Galois groups are considered as compact groups in Krull topology and morphisms of such Galois groups arc always supposed to be continuous.

§ 1. We first prove two lemmas on topological groups.

Let p be a prime number. Let G be a compact group, N a closed normal subgroup of G, and $\Gamma = G/N$. Assume that there exist isomorphisms

$$\mathbf{Z}_p \xrightarrow{\sim} N, \quad \mathbf{U}_p \xrightarrow{\sim} \Gamma$$
$$a \mapsto \nu(a) \quad u \mapsto \sigma(u)$$

where $\mathbf{Z}_p$ and $\mathbf{U}_p$ denote the additive group of all p-adic integers and the multiplicative group of all p-adic units respectively. Since N is abelian, $\Gamma = G/N$ acts on N in the usual manner, and we further assume that the action of Γ on N is given by

$$\nu(a)^{\sigma(u)} = \nu(ua), \qquad a \in \mathbf{Z}_p,\ u \in \mathbf{U}_p.$$

Let H be a closed subgroup of G such that $H \cap N$ is open in N, and let α be an automorphism of H which maps $H \cap N$ onto itself.

Lemma 1. *α induces the identity automorphism on $H/H\cap N$.*

Proof. Since $H\cap N$ is open in N and $N\simeq \mathbf{Z}_p$, there exists an element u_0 in $\mathbf{U}_p$ such that $\nu(a)^\alpha=\nu(u_0a)$ for all $\nu(a)$ in $H\cap N$. Let $\sigma(u)$ be any element of $HN/N=H/H\cap N$ and let $\sigma(u)^\alpha=\sigma(u')$. Then

$$\begin{aligned}(\nu(a)^{\sigma(u)})^\alpha &= \nu(u_0a)^{\sigma(u')}=\nu(u'u_0a)\\ &=(\nu(ua))^\alpha=\nu(u_0ua)\end{aligned}$$

so that $u'u_0a=u_0ua$ for all $\nu(a)$ in $H\cap N$. Hence $u'=u$, $\sigma(u)^\alpha=\sigma(u)$ for $\sigma(u)$ in $H/H\cap N$.

An arithmetic example of the compact group G considered above is given as follows. Let W_p denote the set of all roots of unity in $\overline{\mathbf{Q}}$ with orders powers of p. For any a in $\mathbf{Z}_p$ and ω in W_p, ω^a is well-defined and is again an element of W_p. Let $C_p=\mathbf{Q}(W_p)$. C_p is an abelian extension of $\mathbf{Q}$ and there exists an isomorphism

$$\begin{aligned}\mathbf{U}_p \xrightarrow{\sim} \Gamma &= \mathrm{Gal}(C_p/\mathbf{Q}),\\ u &\mapsto \sigma(u)\end{aligned}$$

such that $\omega^{\sigma(u)}=\omega^u$ for every ω in W_p. Let q be another prime number, $q\neq p$, and let $C_{p,q}$ denote the field generated over C_p by all p^n-th roots of q for all $n\geq 0$. Then $C_{p,q}/\mathbf{Q}$ is a Galois extension and $N=\mathrm{Gal}(C_{p,q}/C_p)$ is a closed normal subgroup of $G=\mathrm{Gal}(C_{p,q}/\mathbf{Q})$. For $n\geq 0$, we choose a primitive p^n-th root of 1, ω_n, and a p^n-th root of q, π_n, in $C_{p,q}$ so that $\omega_{n+1}^p=\omega_n$, $\pi_{n+1}^p=\pi_n$. Then we see easily that for each a in $\mathbf{Z}_p$, there exists a unique element ν in N satisfying $\pi_n^{\nu-1}=\omega_n^a$ for all $n\geq 0$ and that $a\mapsto\nu$ defines an isomorphism

$$\begin{aligned}\mathbf{Z}_p &\to N,\\ a &\mapsto \nu(a).\end{aligned}$$

It then follows immediately from the definition that $\nu(a)^{\sigma(u)}=\nu(ua)$ for any a in $\mathbf{Z}_p$ and u in $\mathbf{U}_p$.

Lemma 2. *Suppose that a topological group G is the set-theoretical union of a finite number of closed subgroups U_i. Then either $G=U_i$ for some U_i or there exist two subgroups U_i and U_j, $U_i\neq U_j$, which are open in G.*

Proof. This is an immediate consequence of the following observation: if A is any non-empty open subset of G, then for each i, either $A\subseteq U_i$ so that U_i is open in G, or the difference $A-U_i$ is again a non-empty open subset of G.

We shall later apply the above lemma to a compact group G. In such a case, it also follows from the following similar fact on abstract groups: Suppose that a group G is the union of a finite number of subgroups U_i. Then either $G=U_i$ for some U_i or there exist U_i and U_j, $U_i\neq U_j$, which have finite indices in G.[2)]

§ 2. In the following, we fix a Galois extension F of $\mathbf{Q}$ in $\overline{\mathbf{Q}}$ such that $F=F_{ab}$, and denote its Galois group by G: $G = \mathrm{Gal}(F/\mathbf{Q})$. An essential property of such a field F is that if K is any subfield of F, then $K_{ab} \subseteq F_{ab} = F$. In particular, F contains $\mathbf{Q}_{ab}$, the field generated over $\mathbf{Q}$ by all roots of unity.

Let α be an automorphism of an open normal subgroup H of G. Let N be an arbitrary open normal subgroup of G and let $U = H \cap N$. Then U^α is an open subgroup of G, isomorphic to U. Now, the theorem of Neukirch mentioned in the introduction holds also for open subgroups G_1, G_2 of the Galois group $G = \mathrm{Gal}(F/\mathbf{Q})$; one checks easily that all arguments in [5] are valid for $F = F_{ab}$ as well as for the special case where $F = \tilde{\mathbf{Q}}$. Since U is normal in G, it follows that $U^\alpha = U$. It then also follows that for any closed normal subgroup M of G, α maps $H \cap M$ onto itself and induces an automorphism of $H/H \cap M$.

Lemma 3. *Let h be any element of H and let ζ be a root of unity in $\mathbf{Q}_{ab}$. Then $h(\zeta) = h^\alpha(\zeta)$.*

Proof. We may assume that the order of ζ is a power of a prime number p so that ζ is contained in the field C_p defined in § 1. Since $F = F_{ab}$, we have $\mathbf{Q} \subseteq C_p \subseteq C_{p,q} \subseteq F$ for the field $C_{p,q}$ in § 1. Let $N = \mathrm{Gal}(F/C_p), M = \mathrm{Gal}(F/C_{p,q})$ and let $G_1 = G/M = \mathrm{Gal}(C_{p,q}/\mathbf{Q}), N_1 = N/M = \mathrm{Gal}(C_{p,q}/C_p)$. By the above remark, α induces an automorphism on $H_1 = HM/M = H/H \cap M$ and it maps $H_1 \cap N_1 = H \cap N/H \cap M$ onto itself. Since H is open in G and $H_1 \cap N_1$ is open in N_1, it follows from Lemma 1 that α induces the identity automorphism on $H_1/H_1 \cap N_1 = H/H \cap N = \mathrm{Gal}(KC_p/K)$. (* K is the subfield of F corresponding to H, i.e., $H = \mathrm{Gal}(F/K)$.) This means that h and h^α define the same Galois automorphism on C_p. Therefore $h(\zeta) = h^\alpha(\zeta)$ for ζ in C_p.

For the proof of the next lemma, we first make the following remark. Let w be a prime spot on F lying above a rational prime p and let Z, T, and φ denote respectively the decomposition group, the inertia group, and a Frobenius substitution of w/p for the Galois extension $F/\mathbf{Q}$. Since $\mathbf{Q}_{ab} \subseteq F$, φ is uniquely characterized modulo T by the property that φ is an element of Z and $\varphi(\zeta) = \zeta^p$ for every root of unity ζ in $\mathbf{Q}_{ab}$ with order prime to p.

Lemma 4. *For each h in H, h and h^α are conjugate in G.*

Proof. Let U be an arbitrary open normal subgroup of G contained in H and let K and L be the fixed fields of H and U, respectively, in F: $H = \mathrm{Gal}(F/K)$, $U = \mathrm{Gal}(F/L)$, $\mathbf{Q} \subseteq K \subseteq L \subseteq F$, $U \subseteq H \subseteq G$. Let v be a prime spot on L lying above a rational prime p such that v/p is unramified and its Frobenius substitution for the Galois extension $L/\mathbf{Q}$ is $h|_L$, the restriction of h on L ; it is well-known that such v/p exists. Let w be an extension of v on F and let Z, T, and φ be as stated in the above. Then $Z \subseteq H$; $T \subseteq U$, and $\varphi|_L = h|_L$, namely, $\varphi \equiv h \bmod U$. Now, since $Z^\alpha \simeq Z$ in H, it follows from Theorem 1 in [5] that Z^α is the decomposition group of a prime spot w' on F lying above p, for the extension $F/\mathbf{Q}$; this theorem is stated in [5] only in the special case where $F = \tilde{\mathbf{Q}}$, but the proof again applies as well for any F with $F_{ab} = F$. Since w and w' lie above the same p, it follows that $w' = w^s$ for some s in $G = \mathrm{Gal}(F/\mathbf{Q})$. Consequently, $Z^\alpha = s^{-1}Zs$, $s^{-1}Ts$ is the inertia group of w'/p, and $s^{-1}\varphi s$ is a Frobenius substitution of w'/p for $F/\mathbf{Q}$.

Clearly φ^α is an element of Z^α and by Lemma 3, $\varphi^\alpha(\zeta) = \varphi(\zeta) = \zeta^p$ for every root of unity ζ in $\mathbf{Q}_{ab}$ with order prime to p. Hence φ^α is also a Frobenius substitution of w'/p and it follows that $\varphi^\alpha \equiv s^{-1}\varphi s \mod s^{-1}Ts$ so that $\varphi^\alpha \equiv s^{-1}\varphi s \mod U$ because $T \subseteq U$ and U is normal in G. On the other hand, since α maps the open normal subgroup U of G onto itself, $\varphi \equiv h \mod U$ induces $\varphi^\alpha \equiv h^\alpha \mod U$. Hence we obtain

$$h^\alpha \equiv \varphi^\alpha \equiv s^{-1}\varphi s \equiv s^{-1}hs \mod U$$

with an element s in G. As U is an arbitrary open normal subgroup of G contained in H, the above congruence shows that h^α is contained in the closure of the conjugate class of h in G. However, each conjugate class of the compact group G is closed. Therefore h and h^α are conjugate in G.

The above proof is essentially the same as that in Komatsu [3] for the special case where $H = G = \mathrm{Gal}(\overline{\mathbf{Q}}/\mathbf{Q})$.

§ 3. Let $F/\mathbf{Q}$ and $G = \mathrm{Gal}(F/\mathbf{Q})$ be as in § 2: $\mathbf{Q} \subseteq F = F_{ab} \subseteq \overline{\mathbf{Q}}$. Let K be a finite Galois extension of $\mathbf{Q}$ contained in F. Then $K_{ab}/\mathbf{Q}$ is also a Galois extension and $\mathbf{Q} \subseteq K \subseteq K_{ab} \subseteq F$. Let $U = \mathrm{Gal}(K_{ab}/\mathbf{Q})$, $N = \mathrm{Gal}(K_{ab}/K)$, and $\Gamma = U/N = \mathrm{Gal}(K/\mathbf{Q})$. Since N is an abelian normal subgroup of U, Γ acts on N by $x^\sigma = s^{-1}xs$ for $x \in N$, $s \in U$, and $\sigma = sN \in \Gamma$. For each σ in Γ, let N_σ denote the set of all x in N such that $x^\sigma = x$; N_σ is a closed subgroup of the compact group N.

Lemma 5. *Suppose that N_σ is open in N, i.e., $[N : N_\sigma] < \infty$. Then $\sigma = 1$.*

Proof. Let J and C be the idele group and the idele class group of K respectively and let C_0 denote the connected component of 1 in C. $\Gamma = \mathrm{Gal}(K/\mathbf{Q})$ acts on J, C, and $C' = C/C_0$ in a natural manner and by class field theory, there is a canonical Γ-isomorphism $N = \mathrm{Gal}(K_{ab}/K) \xrightarrow{\sim} C'$. Let C'_σ denote the subgroup of all ξ in C' such that $\xi^\sigma = \xi$. Then it follows from the assumption that $[C' : C'_\sigma] = [N : N_\sigma] < \infty$. Let v be a prime spot on K lying above a rational prime p such that v/p is unramified and σ is the Frobenius substitution of v/p. Then the v-completion K_v of K is a Galois extension of $\mathbf{Q}_p$ and $\mathrm{Gal}(K_v/\mathbf{Q}_p)$ is generated by σ. Furthermore, the multiplicative group $K_v^\times$ of the local field K_v is imbedded in the idele group J in the obvious manner and it is known that the product of the natural maps $K_v^\times \to J \to C \to C'$ is injective.[3)] Therefore we may consider $K_v^\times$ as a subgroup of C' by means of the map $K_v^\times \to C'$. It then follows that $K_v^\times \cap C'_\sigma = \mathbf{Q}_p^\times$ so that $[K_v^\times : \mathbf{Q}_p^\times] \leq [C' : C'_\sigma] < \infty$. We then see immediately that $K_v = \mathbf{Q}_p$, $\mathrm{Gal}(K_v/\mathbf{Q}_p) = 1$ and consequently that $\sigma = 1$.

Corollary. *The centralizer of N in U is N itself.*

Now, let U_1 and U_2 be open subgroups of U containing N and let $\Gamma_1 = U_1/N$, $\Gamma_2 = U_2/N$. Let α be an isomorphism of U_1 onto U_2 such that $N^\alpha = N$. Then α induces an isomorphism of Γ_1 onto Γ_2.

Lemma 6. *Suppose that there exists an element σ in $\Gamma = U/N$ such that*

$x^\alpha = x^\sigma$ for all x in N. Then

$$\gamma^\alpha = \sigma^{-1}\gamma\sigma$$

for all γ in Γ_1.

Proof. Let $g \in U_1$ and $x \in N$. Since $g^{-1}xg$ is again an element of N,

$$\begin{aligned}(g^{-1}xg)^\alpha &= (g^{-1}xg)^\sigma = g^{-\sigma}x^\sigma g^\sigma \\ &= g^{-\alpha}x^\alpha g^\alpha = g^{-\alpha}x^\sigma g^\alpha.\end{aligned}$$

This holds for all x^σ in $N^\sigma = N$. Hence, by the corollary of Lemma 5, we obtain $g^\alpha \equiv g^\sigma \bmod N$, namely, $\gamma^\alpha = \sigma^{-1}\gamma\sigma$ for $\gamma = gN$ in $\Gamma_1 = U_1/N$.

The lemma tells us how the effect of the isomorphism α on N determines the induced isomorphism $\Gamma_1 \xrightarrow{\sim} \Gamma_2$, and this is one of the key steps in the proof of Ikeda [2].

§ 4. We are now going to prove the main result.

Theorem. *Let F be a Galois extension of $\mathbf{Q}$ in $\overline{\mathbf{Q}}$ such that $F_{ab} = F$ and let $G = \mathrm{Gal}(F/\mathbf{Q})$. Let G_1 and G_2 be open subgroups of G and let α be an isomorphism of G_1 onto G_2. Then α is induced by an inner automorphism of G.*

Proof. Take an open normal subgroup H of G contained in both G_1 and G_2: $H \subseteq G_1, G_2 \subseteq G$, and let $H = \mathrm{Gal}(F/K)$ with a finite Galois extension K of $\mathbf{Q}$ in F. Then $K_{ab} \subseteq F$ and $H' = \mathrm{Gal}(F/K_{ab})$ is the topological commutator subgroup of H. Let $U = \mathrm{Gal}(K_{ab}/\mathbf{Q})$, $N = \mathrm{Gal}(K_{ab}/K)$, $\Gamma = U/N = \mathrm{Gal}(K/\mathbf{Q})$ as in § 3 and let $U_i = G_i/H'$, $i = 1, 2$. Then $U = G/H'$, $N = H/H'$, $\Gamma = G/H$, and $N \subseteq U_1, U_2 \subseteq U$. Since $[G_2 : H^\alpha] = [G_1 : H] < \infty$, H^α is open in G. Hence $H^\alpha = H$ by the theorem of Neukirch mentioned earlier. It follows that $H'^\alpha = H'$ so that α induces an isomorphism of U_1 onto U_2 such that $N^\alpha = N$. Consequently we also have an isomorphism $\Gamma_1 \xrightarrow{\sim} \Gamma_2$ where $\Gamma_i = U_i/N = G_i/H'$, $i = 1, 2$. Now, for each σ in Γ, let M_σ denote the set of all x in N such that $x^\alpha = x^\sigma$, x^σ being defined as in § 3 ; M_σ is a closed subgroup of N. By Lemma 4, N is the union of M_σ when σ ranges over the finite group Γ. Hence by Lemma 2, either $N = M_\sigma$ for some σ or there exist σ, τ in Γ, $\sigma \neq \tau$, such that both M_σ and M_τ are open in N. Assuming the second alternative, let $M = M_\sigma \cap M_\tau$. Then M is an open subgroup of N and $x^\alpha = x^\sigma = x^\tau$ for all x in M. It then follows that $x^\rho = x$ for $\rho = \sigma \circ \tau^{-1}$ and for all x in M. As M is open in N and $\rho \neq 1$, this contradicts Lemma 5. Thus we see that $N = M_\sigma$ for some σ in Γ, namely, that $x^\alpha = x^\sigma$ for all x in $N = H/H'$ and for a fixed σ in Γ. Furthermore, by the corollary of Lemma 5, such σ is unique in $\Gamma = G/H$, and by Lemma 6, $\gamma^\alpha = \sigma^{-1}\gamma\sigma$ for all γ in $\Gamma_1 = G_1/H$. Let H_0 be an open normal subgroup of G contained in H. Then, as for H, there exists a unique element σ_0 in G/H_0 such that $y^\alpha = y^{\sigma_0}$ for all y in H_0/H_0' and it satisfies $\delta^\alpha = \sigma_0^{-1}\delta\sigma_0$ for all δ in G_1/H_0. From the uniqueness of σ, one sees easily that $\sigma_0 \mapsto \sigma$ under the canonical morphism $G/H_0 \to G/H$. Since this holds for arbitrary H and H_0 as mentioned above, there exists an element s

in G such that $s \mapsto \sigma$ under $G \to G/H$ for every open normal subgroup H of G contained in G_1 and G_2, and it satisfies $g^\alpha = g^s = s^{-1}gs$ for all g in G_1. Thus the theorem is proved.

It is now clear that both a) and b) mentioned in the introduction are true for $G = \mathrm{Gal}(F/\mathbf{Q})$. As Ikeda points out in [2], it also follows easily from the corollary of Lemma 5 that the center of the group G is 1 so that G is a so-called complete group. Finally, note that for $F = \overline{\mathbf{Q}}$, b) is equivalent with the following statement: Let k_1 and k_2 be finite algebraic number fields, i.e., finite extensions of $\mathbf{Q}$ in $\overline{\mathbf{Q}}$. Then $k_1 \simeq k_2$ (over $\mathbf{Q}$) if and only if $\mathrm{Gal}(\overline{\mathbf{Q}}/k_1) \simeq \mathrm{Gal}(\overline{\mathbf{Q}}/k_2)$.

Bibliography

[1] M. Ikeda, On the group automorphisms of the absolute Galois group of the rational number field. Arch. d. Math., **26** (1975), 250–252.
[2] M. Ikeda, Completeness of the absolute Galois group of the rational number field. To appear in Jour. reine angew. Math.
[3] K. Komatsu, A remark on a Neukirch's conjecture. Proc. Acad. Japan, 50(1974), 253–255.
[4] J. Neukirch, Kennzeichnung der p-adischen und der endlichen algebraischen Zahlkörper. Inv. Math., **6** (1969), 296–314.
[5] J. Neukirch, Kennzeichnung der endlich-algebraischen Zahlkörper durch die Galoisgruppe der maximal auflösbaren Erweiterungen. Jour. reine angew. Math., **238** (1970), 135–147.

Footnotes

1) We were informed at the Symposium that the same result was obtained by K. Uchida.
2) See J. Sonn, *Groups that are the union of finitely many subgroups*, to appear in Amer. Math. Monthly. See also N. Bourbaki, *Algèbre commutative*, Chap. 2, § 1, Exercise 1.
3) See, e.g., A. Weil, *Basic number theory*, Springer-Verlag, Berlin-Heidelberg-New York, 3rd ed. (1974), Chap. XIII, § 8, Proposition 13.

Editors' Notes

Apparently Iwasawa prepared this manuscript for his lecture in the "Symposium on L-functions and Galois properties of algebraic number fields" at the University of Durham, September 1975. (However, his manuscript was not submitted to the proceedings of the symposium: A. Fröhlich, ed., "Algebraic Number Theory", Acad. Press, 1977.) The result of K. Uchida quoted in the footnote 1) was published in

K. Uchida, Isomorphisms of Galois groups, J. Math. Soc. Japan, **28** (1976), 617-620.

Note also that M. Ikeda's paper [2] in the Bibliography has appeared in J. reine angew. Math., **291** (1977), 1-22.

[U3] Some problems on cyclotomic fields and $\mathbb{Z}_p$-extensions

Let p be a prime number and let $\mathbb{Z}_p$ and $\mathbb{Q}_p$ denote the ring of p-adic integers and the field of p-adic numbers respectively. As usual, $\mathbb{Q}$, $\mathbb{R}$, and $\mathbb{C}$ will denote the rational field, the real field, and the complex field respectively, and all algebraic number fields will be regarded as subfields of $\mathbb{C}$. In the present paper, we shall discuss some open problems, or, conjectures, on the cyclotomic field of p-th roots of unity and also on the $\mathbb{Z}_p$-extensions of finite algebraic number fields in general. Most of these problems are well known for those who are working on the subjects. However, we hope that such a review would renew their interest and attract the attention of others, if not serve for the solution of the problems.

1. Let p be an odd prime and let k_p denote the cyclotomic field of p-th roots of unity:

$$k_p = \mathbb{Q}(\sqrt[p]{1}), \qquad p > 2.$$

The class number h of k_p is then the product of two integers:

$$h = h^- h^+$$

where the so-called second factor h^+ is the class number of the maximal real subfield $k_p^+ = k_p \cap \mathbb{R}$ in k_p.[1)] The classical class number formula for k_p states that

$$h^- = 2p \prod_{\chi} (-\tfrac{1}{2} B_{1,\chi}), \tag{1}$$

$$h^+ = [E : C] \tag{2}$$

where the factors in (1):

$$B_{1,\chi} = \frac{1}{p} \sum_{a=1}^{p-1} \chi(a) a$$

are the generalized Bernoulli numbers for odd Dirichlet characters χ defined mod p, E in (2) is the group of all units in k_p^+, and C the subgroup of all circular units in E. It is an important general problem to examine the arithmetic contents of those classical formulas which are proved only by means of analytic method.[2)] However, we shall discuss here the following more specific problem.

Namely, let $\Delta = \mathrm{Gal}(k_p/\mathbb{Q})$ be the Galois group of the cyclic extension $k_p/\mathbb{Q}$, and $R = \mathbb{Z}_p[\Delta]$ the group ring of Δ over $\mathbb{Z}_p$. The character group $\widehat{\Delta}$ of Δ can be naturally identified with the group of all Dirichlet characters defined mod p. It is known[3)] that any R-module X can be decomposed into direct sums:

$$X = X^+ \oplus X^- = \bigoplus_{\chi} X_\chi, \qquad \chi \in \widehat{\Delta},$$

with respect to the action of Δ. Let S denote the so-called Stickelberger ideal of the group ring $R = \mathbb{Z}_p[\Delta]$ and let

$$R = R^+ \oplus R^- = \bigoplus_\chi R_\chi,$$
$$S = S^+ \oplus S^- = \bigoplus_\chi S_\chi$$

for the R-modules R and S. On the other hand, let A be the Sylow p-subgroup of the ideal class group of k_p. Then A is also an R-module so that

$$A = A^+ \oplus A^- = \bigoplus_\chi A_\chi.$$

Let ω denote the Teichmüller character for p in $\widehat{\Delta}$. Then $pB_{1,\omega^{-1}}$ is a p-adic unit, while for $\chi \in \widehat{\Delta}$, $\chi(-1) = -1$, $\chi \neq \omega$, $B_{1,\chi^{-1}}$ is a non-zero p-adic integer and $[R_\chi : S_\chi]$ is equal to $[B_{1,\chi^{-1}}]_p$, the highest power of p dividing $B_{1,\chi^{-1}}$. Thus it follows easily from (1) that

(3) $$|A^-| = [R^- : S^-]$$

where $|A^-|$ denotes the order of the finite group A^-. This equality then leads us to the following conjecture:

C.1. $A^- \simeq R^-/S^-$ as modules over R.

Obviously C.1 implies (3) which is, as can be seen from the above remark, nothing but the p-part of the class number formula (1). Let $(E/C)(p)$ denote the Sylow p-subgroup of the finite abelian group E/C. Then it can be proved that C.1 also induces that

$$A^+ \simeq (E/C)(p) \quad \text{as modules over } R.$$

Clearly this isomorphism implies the p-part of the class number formula (2). On the other hand, it is easy to see that C.1 is equivalent to each one of the following statements:

A^- is cyclic as module over R,
A_χ is cyclic over $\mathbb{Z}_p$ for all $\chi \in \widehat{\Delta}$, $\chi(-1) = -1$, $\chi \neq \omega$,
$A_\chi \simeq R_\chi/S_\chi$ for all $\chi \in \widehat{\Delta}$, $\chi(-1) = -1$, $\chi \neq \omega$.

In particular, it follows from C.1 that

C.2. $|A_\chi| = [B_{1,\chi^{-1}}]_p$ for all $\chi \in \widehat{\Delta}$, $\chi(-1) = -1$, $\chi \neq \omega$.

This conjecture includes the theorem of Herbrand-Ribet:

$$A_\chi = 0 \iff p \nmid B_{1,\chi^{-1}}.$$

Since (3) may be written in the form

$$\prod_\chi |A_\chi| = \prod_\chi [B_{1,\chi^{-1}}]_p, \qquad \chi \in \widehat{\Delta},\ \chi(-1) = -1,\ \chi \neq \omega,$$

C.2 will be proved if one can show either that $|A_\chi|$ divides $[B_{1,\chi^{-1}}]_p$ for all χ, or that $[B_{1,\chi^{-1}}]_p$ divides $|A_\chi|$ for all χ as stated.

Now, the following conjecture of S. Vandiver is well known:

C.3. $p \nmid h^+$ for all $p > 2$.

By numerical computation, the conjecture has been verified by Wagstaff [9] for all k_p with $p < 125,000$. Vandiver conjectured C.3 in his research on Fermat's problem. For us, C.3 is important because

$$\text{C.3} \implies \text{C.1}.$$

For each integer $n \geq 0$, let C_n and E_n denote respectively the ideal class group and the group of units of the cyclotomic field $\mathbb{Q}(\sqrt[p^{n+1}]{1})$. If $m \geq n \geq 0$, the injection of the ideal group of $\mathbb{Q}(\sqrt[p^{n+1}]{1})$ into that of $\mathbb{Q}(\sqrt[p^{m+1}]{1})$ defines $C_n \to C_m$, and the norm map from $\mathbb{Q}(\sqrt[p^{m+1}]{1})$ to $\mathbb{Q}(\sqrt[p^{n+1}]{1})$ defines $E_m \to E_n$. Now the conjecture C.3 implies the following mutually equivalent statements on $C_n \to C_m$ and $E_m \to E_n$:

C.4. For any $m \geq n \geq 0$, $C_n \to C_m$ is injective and $E_m \to E_n$ is surjective.

Let $\lambda_p(k_p^+)$ and $\mu_p(k_p^+)$ denote respectively the λ-invariant and the μ-invariant of $k_p^+ = k_p \cap \mathbb{R}$ for the prime p. Then we see easily that C.3 implies

$$\lambda_p(k_p^+) = \mu_p(k_p^+) = 0.^{4)}$$

In fact, it can be proved that

$$\text{C.3} \iff \text{C.4 and } \lambda_p(k_p^+) = \mu_p(k_p^+) = 0.$$

Note that $\mu_p(k_p^+) = 0$ is proved to be true by Ferrero-Washington [3]. On the other hand, we also note that C.4 is not yet proved even for $m = 1, n = 0$.

2. The following C.5 and C.6 are the facts numerically verified by Wagstaff [9] for all k_p with $p < 125,000$, but there is no other reasonable ground for that they are true in general for all p. However, we mention them here because they have interesting consequences and seem to be true at least for a large number of primes.

C.5. $p^2 \nmid B_{1,\chi^{-1}}$ for all $\chi \in \widehat{\Delta}$, $\chi(-1) = -1$, $\chi \neq \omega$.

Note that if we put $\chi = \omega^{-j}$, $0 < j < p-2$, $j =$ odd, then $p^2 \nmid B_{1,\chi^{-1}}$ is equivalent to $p^2 \nmid B_{1+jp}$ for the ordinary Bernoulli number B_{1+jp}. Using the above theorem of Herbrand-Ribet and the remark mentioned thereafter, we can also show that

$$\text{C.5} \implies \text{C.1}.$$

In particular, C.5 induces

C.5′. $pA^- = 0.$

However, we do not know whether C.5′ is true, nor that C.5′ implies C.5.

Let $L_p(s;\chi)$ denote the p-adic L-function for a Dirichlet character χ in $\widehat{\Delta}$ and let $\chi(-1) = -1$, $\chi \neq \omega$. Then there exists a non-zero power series $\xi_\chi(T)$ in $\mathbb{Z}_p[[T]]$, the ring of all formal power series in T with coefficients in $\mathbb{Z}_p$, such that

$$L_p(s;\omega\chi^{-1}) = \xi_\chi((1+p)^s - 1), \quad \text{for all } s \in \mathbb{Z}_p.$$

By Weierstrass' preparation theorem, we have

$$\xi_\chi(T) = \eta_\chi(T) p^{e_\chi} f_\chi(T)$$

where $\eta_\chi(T)$ is an invertible power series in $\mathbb{Z}_p[[T]]$ (i.e., $p \nmid \eta_\chi(0)$), e_χ is a non-negative integer, and $f_\chi(T)$ is a distinguished polynomial in $\mathbb{Z}_p[T]$. Actually $e_\chi = 0$ by the theorem of Ferrero-Washington [3]. As mentioned earlier, it was verified for all k_p with $p < 125,000$ that

C.6. $\deg(f_\chi(T)) \leq 1 \quad \text{for all } \chi \in \widehat{\Delta},\ \chi(-1) = -1,\ \chi \neq \omega.$

Let $\chi = \omega^{1-i}$, $0 < i < p-1$, $i =$ even. Using $e_\chi = 0$ and

(4) $$\xi_\chi(0) = L_p(0;\omega\chi^{-1}) = -B_{1,\chi^{-1}},$$

we see that $\deg(f_\chi(T)) = 0$ if and only if $p \nmid B_i$ for the Bernoulli number B_i. Let $p|B_i$. Then $\deg(f_\chi(T)) = 1$ means that the coefficient of T in $\xi_\chi(T)$ is not divisible by p, and this is equivalent to

$$\frac{B_i}{i} \not\equiv \frac{B_{i+p-1}}{i+p-1} \mod p^2.$$

Note that the congruence holds mod p by Kummer's theorem.

Now, it follows from (4) and C.5 that the polynomials $f_\chi(T)$ as stated above are Eisenstein polynomials. Hence either one of C.5 and C.6 implies that $f_\chi(T)$ are irreducible polynomials in $\mathbb{Q}_p[T]$. Therefore we also see easily that if $f_\chi(T)$ is reducible for $\chi = \omega^{1-i}$, $0 < i < p-1$, $i =$ even, then $B_i \equiv 0 \mod p^2$. It would be interesting to study $k_p = \mathbb{Q}(\sqrt[p]{1})$ under the assumption that all those $f_\chi(T)$ are irreducible; e.g., to see whether C.4 is true in such a case.

3. We now consider $\mathbb{Z}_p$-extensions of number fields.[5)] Let p now denote any prime number, $p \geq 2$, k an arbitrary finite algebraic number field, and let

$$K = k_\infty = k\mathbb{Q}_\infty$$

be the basic $\mathbb{Z}_p$-extension over k, $\mathbb{Q}_\infty$ being the unique $\mathbb{Z}_p$-extension over the rational field $\mathbb{Q}$. Let $\lambda_p(k)$ and $\mu_p(k)$ denote respectively the λ-invariant and the μ-invariant of k for the prime p, i.e., the invariants of the $\mathbb{Z}_p$-extension K/k. It has been conjectured for sometime that

C.7. $\mu_p(k) = 0$ for all p and k.

A great progress on this problem was made last year by B. Ferrero and L. Washington [3] who proved C.7 in the case where $k/\mathbb{Q}$ is an abelian extension. Since their work and related results are to be discussed in this conference, we would make here only the following remark. Namely, suppose that k contains primitive $2p$-th roots of unity and let F denote the maximal p-ramified p-extension over k so that $k \subseteq K \subseteq F$. Then, as A. Brumer once pointed out, $\mu_p(k) = 0$ if and only if $\mathrm{Gal}(F/K)$ is a free pro-p-group.[6] Note that in general, if k' is a subfield of k, then $\mu_p(k') \leq \mu_p(k)$. Hence, to prove C.7 for all k, we may actually consider only those k which contain primitive $2p$-th roots of unity.

The next problem is to see whether

C.8. $\lambda_p(k) = \mu_p(k) = 0$ for all p and all totally real k.

This was conjectured by Greenberg [4] who verified it in many numerically given special cases. Also, as already mentioned, the Vandiver conjecture C.3 implies C.8 in the case where k is the maximal real subfield of $k_p = \mathbb{Q}(\sqrt[p]{1})$, $p > 2$. On the other hand, however, there is no theorem yet which proves C.8 for a sufficiently large class of totally real fields k. For example, we do not know yet whether C.8 is true for all real quadratic fields (even in the case $p = 2$). Let $\lambda'_p(k)$ and $\mu'_p(k)$ denote respectively the modified λ-invariant and the modified μ-invariant of k for the prime p, defined by means of the ideal class groups of p-ideals.[7] Since $\lambda'_p(k) \leq \lambda_p(k)$, $\mu'_p(k) \leq \mu_p(k)$ in general, C.8 of course induces the following conjecture:

C.8′. $\lambda'_p(k) = \mu'_p(k) = 0$ for all p and all totally real k.

Let k be again an arbitrary finite algebraic number field and let

$$\Gamma = \mathrm{Gal}(K/k)$$

for $K = k_\infty = k\mathbb{Q}_\infty$. Let I_K denote the ideal group of K, P_K the subgroup of all principal ideals in I_K, and $C_K = I_K/P_K$ the ideal class group of K. We denote by $C_K(p)$ the p-primary component of the (possibly infinite) abelian group C_K, and by $C_K(p)^\Gamma$ the subgroup of all elements in $C_K(p)$ which are fixed by the Galois group Γ. Using Theorem 11 of [5], we can then prove easily the following propositions:

Proposition 1. $\mu_p(k) = 0$ if and only if $pC_K(p) = C_K(p)$ for the additive group $C_K(p)$; and if this is the case, then

$$C_K(p) \simeq (\mathbb{Q}_p/\mathbb{Z}_p)^\lambda$$

with $\lambda = \lambda_p(k)$.

Proposition 2. $\lambda_p(k) = \mu_p(k) = 0$ if and only if $C_K(p) = 0$. Furthermore, for $C_K(p) = 0$, it is sufficient that $C_K(p)^\Gamma = 0$ or that $C_K(p)$ is a finite group.

Therefore C.7 and C.8 may be stated as properties of group $C_K(p)$. Let I'_K denote the group of all p-ideals of K, P'_K the subgroup of all principal ideals in I'_K, and let $C'_K = I'_K/P'_K$. Then the results similar to the above propositions can be proved for $\lambda'_p(k)$, $\mu'_p(k)$, and $C'_K(p)$. Hence C.8′ also may be formulated as properties of $C'_K(p)$, e.g., as $C'_K(p) = 0$. Some further discussions on C.8 and C.8′ will be given in the next section.

For any ground field k, the Tate module of $C_K(p)$ is always isomorphic to $\mathbb{Z}_p^\lambda$ with $\lambda = \lambda_p(k)$, and it provides us with a natural p-adic representation of the Galois group $\Gamma = \mathrm{Gal}(K/k)$. Let $g(T)$ denote the characteristic polynomial of a suitably chosen topological generator of Γ in the above mentioned p-adic representation. In the special case where $k = k_p = \mathbb{Q}(\sqrt[p]{1})$, $p > 2$, the author found that if C.3 is true for k_p, then the characteristic polynomial $g(T)$ is the product of the polynomials $f_\chi(T)$ defined in §2; and the result was later generalized by Greenberg [4] in the case where k is an abelian extension over the rational field $\mathbb{Q}$. Thus it is now conjectured that

C.9. A similar relation exists between $g(T)$ and certain p-adic L-functions for a wider class of ground fields k.

this is called the main conjecture in Coates [2] where the precise formulation of the conjecture is given and where relevant properties of general p-adic L-functions are also discussed. Therefore we note here only that in addition to the solution of C.8, the remaining key problem in this connection seems to be to find Stickelberger-like operators in the general case where k is not necessarily abelian over $\mathbb{Q}$.

Now, the well-known conjecture of Leopoldt states

C.10. The p-adic regulator of a number field k never vanishes.

Although the above statement does not involve $\mathbb{Z}_p$-extensions, it is actually equivalent to say that if the number of imaginary archimedean prime spots on k is r_2, then there exist at most $r_2 + 1$ independent $\mathbb{Z}_p$-extensions over k. The conjecture is quite important and has many consequences in the theory of $\mathbb{Z}_p$-extensions. It was verified by Brumer [1] in the case where k is abelian over the rational field $\mathbb{Q}$ or over an imaginary quadratic field. In a recent paper of Serre [8], relations between C.10 and p-adic zeta functions are discussed for totally real k. The result seems to support the conjecture C.9.

4. In this section, we shall always assume that k is a totally real field and shall further discuss some of the conjectures mentioned in §3 in this special case. We first note that for totally real k, that conjecture C.10 is equivalent to the following

C.10′. The basic $\mathbb{Z}_p$-extension $K = k_\infty = k\mathbb{Q}_\infty$ is the unique $\mathbb{Z}_p$-extension over k.

The conjecture is equivalent to say that there is no Galois extension F/k such that $k \subseteq K \subseteq F$, $\mathrm{Gal}(F/k) \simeq \mathbb{Z}_p{}^2$. Hence, using Theorem 11 of [5] again, one sees that C.10′ implies the following

C.11. $\quad C_K(p)^\Gamma$ is a finite group.

Therefore, by Proposition 2, we know that either one of C.8 and C.10 induces the conjecture C.11. Similarly, either one of C.8′ and C.10 induces

C.11′. $\quad C'_K(p)^\Gamma$ is a finite group.

Of course, C.11 ⇒ C.11′. Let s denote the number of prime ideals of k, dividing the rational prime p. If $s = 1$, then it is easy to show that both C.11 and C.11′ are true. However, in the general case, one cannot yet prove even the following statement which is an immediate consequence of C.11:

C.12. Every prime ideal of k, dividing the rational prime p, becomes a principal ideal in K.[8)]

Assume that C.12 is true for all totally real k, in particular, for all intermediate fields of K/k. Then we see that

$$C_K = C'_K$$

so that C.8 ⇔ C.8′, C.11 ⇔ C.11′. One can also show that C.12 is equivalent to either one of the following statements on the cohomology groups:

$$H^1(\Gamma, E_K) \sim (\mathbb{Q}_p/\mathbb{Z}_p)^s, \quad H^2(\Gamma, E_K) \sim (\mathbb{Q}_p/\mathbb{Z}_p)^{s-1}.$$

Here E_K denotes the group of all units in K and $\sim$ indicates that the groups are isomorphic modulo finite groups. Since $pH^2(\Gamma, E_K) = H^2(\Gamma, E_K)$, the second statement may be replaced by the isomorphism $H^2(\Gamma, E_K) \simeq (\mathbb{Q}_p/\mathbb{Z}_p)^{s-1}$. On the other hand, for the group E'_K of all p-units in K,[9)] one can prove the following

Proposition 3. $H^1(\Gamma, E'_K) \sim 0, \quad H^2(\Gamma, E'_K) \simeq (\mathbb{Q}_p/\mathbb{Z}_p)^{s-1}$.

Now, it is easy to show that

(5) $\quad \mathrm{Ker}(H^2(\Gamma, E_K) \to H^2(\Gamma, K^\times)) \simeq \mathrm{Coker}({I_K}^\Gamma \to C_K(p)^\Gamma)$,

(6) $\quad \mathrm{Ker}(H^2(\Gamma, E'_K) \to H^2(\Gamma, K^\times)) \simeq \mathrm{Coker}({I'_K}^\Gamma \to C'_K(p)^\Gamma)$,

where ${I_K}^\Gamma \to C_K(p)^\Gamma$ is the map induced by $I_K \to C_K = I_K/P_K$ and the projection $C_K \to C_K(p)$, and ${I'_K}^\Gamma \to C'_K(p)^\Gamma$ is defined similarly.[10)] Hence C.11 implies that the kernel in (5) is a finite group. Since $\mathrm{Im}({I'_K}^\Gamma \to C'_K(p)^\Gamma)$ is always finite, C.11′ is even equivalent to the finiteness of the kernel in (6). By Proposition 3, the latter statement is also equivalent to that

$$\mathrm{Im}(H^2(\Gamma, E'_K) \to H^2(\Gamma, K^\times)) \simeq (\mathbb{Q}_p/\mathbb{Z}_p)^{s-1},$$

a fact which can be formulated also as a property on the values of the local invariants of the cohomology classes in the Brauer group of k, represented by the elements of E'_K.

It is clear that the conjecture C.8, which is stronger than C.11, similarly induces the following stronger conjecture:

C.13. The map $H^2(\Gamma, E_K) \to H^2(\Gamma, K^\times)$ is injective.

Similarly C.8′ induces that $H^2(\Gamma, E'_K) \to H^2(\Gamma, K^\times)$ is injective. More generally, let S be any finite set of prime spots on k, including all those lying above the rational prime p, and let $E_{K,S}$ denote the group of all S-units in K defined as usual. Then C.8′ for k implies that

$$H^2(\Gamma, E_{K,S}) \to H^2(\Gamma, K^\times) \tag{7}$$

is injective. However, we can also prove the following

Proposition 4. Let S be chosen so that the ideal class group of k is generated by the classes of the prime ideals in S. Then the map (7) is injective if and only if $\lambda'_p(k) = \mu'_p(k) = 0$.

Actually the proposition can be proved for an arbitrary number field k, not necessarily totally real. Hence, applying it for $k_p = \mathbb{Q}(\sqrt[p]{1})$ with an irregular prime p, one sees that in such a case, (7) is not injective for a suitable choice of S because $\lambda_p(k) = \lambda'_p(k) \geq 1$.

5. Finally, to conclude our discussions, let us make a simple remark on Fermat's problem in connection with the cyclotomic field $k_p = \mathbb{Q}(\sqrt[p]{1})$, $p > 2$, in §1. As in that section, let A denote the Sylow p-subgroup of the ideal class group of k_p, and let r be the rank of the finite abelian group A. We know various results on Fermat's problem to the effect that if the rank r is "small", then the Fermat equation $x^p + y^p + z^p = 0$ has no integer solution with $xyz \neq 0$. The classical theorem of Kummer on regular primes is of course a special case of such results, namely, the case where $r = 0$. Now, an important theorem of Eichler states that if

$$r < [\sqrt{p}] - 1,$$

then the Fermat equation has no integral solution with $p \nmid xyz$. On the other hand, a well-known theorem of Golod-Safarevic says that if the p-class field tower over k_p is finite, then

$$r < 2 + \sqrt{2(p+1)}.$$

Although $[\sqrt{p}] - 1 < 2 + \sqrt{2(p+1)}$, we wonder if it be possible to improve either the theorem of Eichler or the theorem of Golod-Safarevic in this special case so that the finiteness of the p-class field tower over k_p induces the so-called first case of Fermat's problem.[11] In any case, it seems to be an interesting problem to find whether or not the p-class field tower over the cyclotomic field $k_p = \mathbb{Q}(\sqrt[p]{1})$, $p > 2$, is indeed finite.

References

[1] A. Brumer, On the units of algebraic number fields, Mathematika, **14** (1967), 121-124.
[2] J. Coates, p-adic L-functions and Iwasawa's theory, Algebraic number fields, Acad. Press, London, 1977, 269-353.
[3] B. Ferrero and L. Washington, The Iwasawa invariant μ_p vanishes for abelian number fields, Ann. Math., **109** (1979), 377-395.
[4] R. Greenberg, On the Iwasawa invariants of totally real number fields, Amer. Jour. Math., **98** (1976), 263-284.
[5] K. Iwasawa, On $\mathbb{Z}_l$-extensions of algebraic number fields, Ann. Math., **98** (1973), 246-326.
[6] K. Iwasawa, A note on cyclotomic fields, Inv. math., **36** (1976), 115-123.
[7] S. Lang, Cyclotomic fields, Springer-Verlag, New York-Heidelberg-Berlin, 1978.
[8] J.-P. Serre, Sur le résidu de la fonction zêta p-adique d'un corps de nombres, C. R. Paris, **287** (1978), 183-188.
[9] S. Wagstaff, Jr., The irregular primes to 125000, Math. Comp., **32** (1978), 583-591.

Footnotes

1) Throughout the following, we assume that the reader is familiar with the theory of cyclotomic fields and p-adic L-functions such as described in Lang [7].
2) Cf. [6].
3) Cf. [6].
4) See the conjecture C.8 below.
5) For the theory of $\mathbb{Z}_p$-extensions of number fields, see [5], [7].
6) Thus, for example, $\mathrm{Gal}(F/K)$ is a free pro-p-group for $k = k_p$ in §1. We can also show that the same is true for the Galois group $\mathrm{Gal}(F/K)$ defined similarly over $k_p^+ = k_p \cap \mathbb{R}$.
7) See [5], §5 for the definition.
8) This statement is false in general if k is not totally real.
9) Cf. loc. cit. 7) for the definition of p-units.
10) To prove (5), (6), the fact that k is totally real is not needed.
11) It looks that r is actually much smaller than $\sqrt{p}$. Cf. [9].

Editors' Notes

This manuscript was probably prepared for the workshop "Arithmetik der ABELschen Zahlkörper und Klassenkörper der komplexen Multiplikation" at Mathematisches Forschungsinstitut Oberwolfach, March 1979. Later in 1987, a part of its content (mainly the first half) was given in his lecture in the workshop "Algebraic Number theory" at RIMS, Kyoto University, which was published in Japanese as [62]. Note that conjecture (C.2) has been solved affirmatively as a corollary of "Iwasawa's Main Conjecture" (cf. Mazur and Wiles, "Class fields of abelian extensions of $\mathbb{Q}$", Invent. math., **76** (1984), Chap. 1, § 10, Th.2).

[U4] A simple remark on Leopoldt's conjecture

1. Let k be a number field, finite over the rational field, and let E denote the group of all units in k. For any integers $m, n \geq 1$, let

$$\begin{aligned} E^m &= \{ \varepsilon^m \mid \varepsilon \in E \}, \\ E_n &= \{ \varepsilon \in E \mid \varepsilon \equiv 1 \bmod n \}. \end{aligned}$$

Let p be a prime number. By Leopoldt's conjecture for k and p we mean the conjecture that the following statement is true:

LC. Given any p-power $m = p^a$, $a \geq 1$, there exists another p-power $n = p^b$, $b \geq 1$ such that

$$E_n \subseteq E^m.$$

If k is a totally real field, LC is equivalent to the usual statement of Leopoldt's conjecture, namely, the non-vanishing of the p-adic regulator of k. The statement LC reminds us the following theorem of Chevalley [1].

Theorem. Given any integer m, $N \geq 1$, there exists an integer $n \geq 1$ such that

$$(n, N) = 1, \quad E_n \subseteq E^m.$$

Thus LC states that in Chevalley's theorem, one can choose a p-power as n if m is a p-power. In the following, we shall make some simple remarks on LC, using Chevalley's theorem.

2. Let k and p be fixed. Let $\mathfrak{q}$ be a prime ideal of k, prime to p, and let $N(\mathfrak{q})$ denote its absolute norm. In general, if n is any natural number, we shall denote by $(n)_p$ the highest power of p dividing n. Let

$$e(\mathfrak{q}) = (N(\mathfrak{q}) - 1)_p, \quad N(\mathfrak{q}) - 1 = d(\mathfrak{q})e(\mathfrak{q}), \quad (d(\mathfrak{q}), p) = 1.$$

For a finite abelian extension K over k, we also denote by $e(\mathfrak{q}; K/k)$ the ramification index of the prime ideal $\mathfrak{q}$ for the extension K/k. Then

$$e(\mathfrak{q}; K/k)_p \leq e(\mathfrak{q}) = (N(\mathfrak{q}) - 1)_p.$$

We say that $\mathfrak{q}$ is fully ramified in K for p when the equality holds in the above.

Definition. A finite abelian extension K over k is called a $\mathfrak{q}$-field if

i) K/k is unramified outside $p\mathfrak{q}$ (i.e., every place v on k, $v \nmid p\mathfrak{q}$ is unramified in K) and

ii) $\mathfrak{q}$ is fully ramified in K for p.

Theorem 1. LC for k and p holds if and only if a $\mathfrak{q}$-field exists for every prime ideal $\mathfrak{q}$ of k, $\mathfrak{q} \nmid p$.

3. For the proof of the theorem, let

$$K_{a,b} = \text{the ray class field mod } \mathfrak{q}^a p^b, \quad \text{for } a, b \geq 0.$$

$K_{a,b}$ is a finite extension over k, unramified outside $p\mathfrak{q}$, and if K/k is any finite abelian extension unramified outside $p\mathfrak{q}$, then

$$k \subseteq K \subseteq K_{a,b}$$

for some sufficiently large a and b. Since $[K_{a,b} : K_{1,b}]$ is prime to p for $a \geq 1$,

$$\text{a } \mathfrak{q}\text{-field exists} \Longleftrightarrow K_{1,b} \text{ is a } \mathfrak{q}\text{-field for sufficiently large } b.$$

However, $[K_{1,b} : K_{0,b}] = e(\mathfrak{q}; K_{1,b}/k)$. Hence

$$\text{a } \mathfrak{q}\text{-field exists} \Longleftrightarrow [K_{1,b} : K_{0,b}]_p = e(\mathfrak{q}) \quad \text{for } b \gg 0.$$

Fixing $b \geq 0$, let

$$\begin{aligned} A &= \{\alpha \in k^\times \mid \alpha \text{ is prime to } p\mathfrak{q},\ \alpha \equiv 1 \bmod p^b\}, \\ B &= \{\beta \in k^\times \mid \beta \text{ is prime to } p\mathfrak{q},\ \beta \equiv 1 \bmod \mathfrak{q}p^b\}. \end{aligned}$$

For the groups of principal ideals $(A) = \{(\alpha) \mid \alpha \in A\}$ and $(B) = \{(\beta) \mid \beta \in B\}$, we then have

$$\begin{aligned} [K_{1,b} : K_{0,b}] &= [(A) : (B)] = [AE : BE] = [A : A \cap BE] \\ &= [A : B(A \cap E)] = [A : BE_n],\ n = p^b, \end{aligned}$$

where E and E_n are defined as in §1. Let $\mathfrak{O}$ be the ring of all algebraic integers in k. Then

$$B \subseteq BE_n \subseteq A, \quad A/B \cong (\mathfrak{O}/\mathfrak{q})^\times$$

where $(\mathfrak{O}/\mathfrak{q})^\times$ is a cyclic group of order $N(\mathfrak{q}) - 1 = d(\mathfrak{q})e(\mathfrak{q})$. Therefore it follows from the above that

$$\begin{aligned} [K_{1,b} : K_{0,b}]_p = e(\mathfrak{q}) &\Longleftrightarrow [A : BE_n]_p = (N(\mathfrak{q}) - 1)_p \\ &\Longleftrightarrow p \nmid [BE_n : B] \\ &\Longleftrightarrow (BE_n)^{d(\mathfrak{q})} \subseteq B \\ &\Longleftrightarrow {E_n}^{d(\mathfrak{q})} \subseteq B. \end{aligned}$$

Hence the following lemma is proved :

Lemma 1. A $\mathfrak{q}$-field exists if and only if ${E_n}^{d(\mathfrak{q})} \equiv 1 \bmod \mathfrak{q}$ for $n = p^b$ with sufficiently large b.

Assume now LC for k and p. Since $e(\mathfrak{q})$ is a power of p,

$$E_n \subseteq E^{e(\mathfrak{q})}, \quad \text{for } n = p^b,\ b \gg 0.$$

It follows that

$$E_n{}^{d(\mathfrak{q})} \subseteq E^{N(\mathfrak{q})-1}, \quad \text{where } E^{N(\mathfrak{q})-1} \equiv 1 \bmod \mathfrak{q}.$$

Hence $E_n{}^{d(\mathfrak{q})} \equiv 1 \bmod \mathfrak{q}$, and a $\mathfrak{q}$-field exists by the lemma.

Assume, conversely, that a $\mathfrak{q}$-field exists for every $\mathfrak{q}$, $\mathfrak{q} \nmid p$. Let m be any p-power : $m = p^a$, $a \geq 1$. By the theorem of Chevalley, there exists an integer $t \geq 1$ such that

$$(t, p) = 1, \ E_t \subseteq E^m.$$

Let

$$(t) = \prod_{i=1}^{r} \mathfrak{q}_i{}^{e_i}$$

where $\mathfrak{q}_i$'s are prime ideals of k, $\mathfrak{q}_i \nmid p$, and $e_i \geq 1$. Let $d_i = d(\mathfrak{q}_i)$, $1 \leq i \leq r$. Applying the lemma for $\mathfrak{q}_1, \cdots, \mathfrak{q}_r$, we see that there is a power $n = p^b$, $b \geq 1$, such that

$$E_n{}^{d_i} \equiv 1 \bmod \mathfrak{q}_i, \quad \text{for } 1 \leq i \leq r.$$

Let

$$d_i' = N(\mathfrak{q}_i)^{e_i - 1}, \quad d = \prod_{i=1}^{r} d_i d_i'.$$

Then the above congruences induce, successively,

$$E_n{}^{d_i d_i'} \equiv 1 \bmod \mathfrak{q}_i{}^{e_i}, \ 1 \leq i \leq r,$$

$$E_n{}^{d} \equiv 1 \bmod \mathfrak{q}_i{}^{e_i}, \ 1 \leq i \leq r,$$

$$E_n{}^{d} \equiv 1 \bmod t.$$

Hence

$$E_n{}^{d} \subseteq E_t \subseteq E^m.$$

Since m is a p-power and $(d, p) = 1$, it follows that

$$E_n \subseteq E^m, \quad \text{where } n = p^b, \ b \geq 1.$$

Therefore LC holds. This completes the proof of the theorem.

4. We now refine theorem 1 as follows. Fix an integer $a \geq 0$ and let

$$m = p^a.$$

By LC(a) we mean the following statement :

LC(a). There exists a p-power $n = p^b$, $b \geq 0$, such that

$$E_n \subseteq E^m.$$

On the other hand, for each prime ideal $\mathfrak{q}$ of k, $\mathfrak{q} \nmid p$, let

$$e(\mathfrak{q}, a) = \text{g.c.d.}(p^a, e(\mathfrak{q})).$$

A finite abelian extension K over k will be called a $(\mathfrak{q}, a)$-field if K/k is unramified outside $p\mathfrak{q}$ and if

$$e(\mathfrak{q}, a) \leq e(\mathfrak{q}; K/k)_p, \text{ i.e. } e(\mathfrak{q}, a)|e(\mathfrak{q}; K/k).$$

Theorem 2. Let $p > 2$. Then LC(a) holds if and only if a $(\mathfrak{q}, a)$-field exists for every prime ideal $\mathfrak{q}$ of k, $\mathfrak{q} \nmid p$.

We first prove the following lemma which holds also for $p = 2$:

Lemma 2. Let $k_\mathfrak{q}$ denote the $\mathfrak{q}$-completion of k. A $(\mathfrak{q}, a)$-field exists if and only if

$$E_n \subseteq (k_\mathfrak{q}{}^\times)^m$$

for some $n = p^b$, $b \geq 0$.

Proof. The proof of Lemma 1 shows that a $(\mathfrak{q}, a)$-fields exists if and only if

$$e(\mathfrak{q}, a) \mid [A : BE_n]$$

for some $n = p^b$, $b \geq 0$. Let $e(\mathfrak{q}, a) = e(\mathfrak{q})$. Then

$$\begin{aligned} e(\mathfrak{q}, a) \mid [A : BE_n] &\iff E_n{}^{d(\mathfrak{q})} \equiv 1 \mod \mathfrak{q} \\ &\iff E_n \subseteq (k_\mathfrak{q}{}^\times)^{e(\mathfrak{q})} \\ &\iff E_n \subseteq (k_\mathfrak{q}{}^\times)^m. \end{aligned}$$

Let $e(\mathfrak{q}, a) = p^a = m < e(\mathfrak{q})$. Then

$$\begin{aligned} e(\mathfrak{q}, a) \mid [A : BE_n] &\iff E_n \subseteq A^m B \\ &\iff E_n \subseteq (k_\mathfrak{q}{}^\times)^m. \end{aligned}$$

It is clear from Lemma 2 that if LC(a) holds then a $(\mathfrak{q}, a)$-field exists for every $\mathfrak{q}$, $\mathfrak{q} \nmid p$. Assume, conversely, that a $(\mathfrak{q}, a)$-field exists for every $\mathfrak{q}$, $\mathfrak{q} \nmid p$. Let

$$k' = k(\zeta)$$

where ζ is a primitive m-th root of unity, and let $\{K_i\}$, $1 \leq i \leq s$, denote the family of all fields of the form $k'(\sqrt[m]{\varepsilon}) \neq k'$ with $\varepsilon \in E$. Since each K_i/k' is a cyclic extension, there exists a prime ideal $\mathfrak{q}'_i$ of k' such that $\mathfrak{q}'_i$ is prime to p and is unramified and undecomposed in K_i. Let $\mathfrak{q}_i = \mathfrak{q}'_i \cap k$. Then $\mathfrak{q}_i$ is a prime ideal of k, $\mathfrak{q}_i \nmid p$. By Lemma 2, there is a p-power n_i such that

$$E_{n_i} \subseteq (k_{\mathfrak{q}_i})^m.$$

Let

$$n = \text{l.c.m.}(n_1, \cdots, n_s).$$

Let $\varepsilon \in E_n$ and $K' = k'(\sqrt[m]{\varepsilon})$. Since $E_n \subseteq (k_{\mathfrak{q}_i})^m$, $\mathfrak{q}_i$ is completely decomposed in K'. Hence $K' \neq K_i$ for $1 \leq i \leq s$, so that $K' = k'$, i.e, that ε is an m-th power of an element of $k' = k(\zeta)$. It then follows from a remark of Chevalley [1], I.5 that ε is an m-th power in $k : \varepsilon \in E^m$. Hence

$$E_n \subseteq E^m.$$

This completes the proof of Theorem 2.

Now, let $p = 2$. The above proof shows that LC(a) still implies the existence of $(\mathfrak{q}, a)$-fields. It also proves that if $(\mathfrak{q}, a)$-fields exist, there is a p-power n such that every element ε in E_n is an m-th power in $k' = k(\zeta)$. By the remark of Chevalley, if $\sqrt{-1} \in k$, then such an ε is an m-th power in k. Hence $E_n \subseteq E^m$ and Theorem 2 holds in this case without any modification. Let $\sqrt{-1} \notin k$ and let $c \geq 2$ be the minimal integer such that $k(\sqrt{-1})$ contains a primitive 2^c-th root of unity. Assume that a $(\mathfrak{q}, a + c)$-field exists for every $\mathfrak{q}$, $\mathfrak{q} \nmid p$. Then there is a p-power n such that every ε in E_n is a p^{a+c}-th power in $k' = k(\zeta)$ and, hence, in $k(\sqrt{-1})$. By Chevalley [1], I.3, ε is then a p^a-th power in k. Thus, in this case, the existence of $(\mathfrak{q}, a + c)$-fields, instead of $(\mathfrak{q}, a)$-fields, implies LC(a).

5. It is clear that LC holds if and only if LC(a) holds for all $a \geq 0$ and that a $\mathfrak{q}$-field exists if and only if a $(\mathfrak{q}, a)$-field exists for all $a \geq 0$. Hence Theorem 2, together with the above remark for $p = 2$, gives us another proof of Theorem 1. We also note that LC(a), for sufficiently large a, implies LC. In fact, for each prime ideal $\mathfrak{p}$ of k dividing p, let $c_{\mathfrak{p}}$ denote the highest power of p dividing the order of the group of roots of unity in the $\mathfrak{p}$-completion $k_{\mathfrak{p}}$. If

$$m = p^a > c_{\mathfrak{p}}, \quad \text{for all } \mathfrak{p},\ \mathfrak{p}|p,$$

then LC follows from LC(a). Thus, in order to prove LC, it is sufficient to show the existence of $(\mathfrak{q}, a)$-fields for sufficiently large a. For some special kind of ground fields k, even the existence of $(\mathfrak{q}, 1)$-fields for all $\mathfrak{q}$, $\mathfrak{q} \nmid p$, implies LC. For example, suppose that k contains a primitive p-th root of unity (resp. $\sqrt{-1} \in k$ if $p = 2$) and that there is a unique prime ideal of k dividing p.

Assume that a $(\mathfrak{q}, 1)$-fields exists for every prime ideal of k, $\mathfrak{q} \nmid p$. Then, by the above, LC(1) holds, namely, there exists a p-power $n = p^b$, $b \geq 0$, such that

$$E_n \subseteq E^p.$$

Let ν be the unique valuation on k such that $\nu(p) = 1$. Let c be any integer, $c \geq 0$ and let $\varepsilon \in E_{np^c}$. Then $\varepsilon = \eta^p$ for some $\eta \in E$ and

$$\nu(1 - \varepsilon) \geq np^c, \ 1 - \varepsilon = \prod_{i=0}^{p-1} (1 - \zeta^i \eta),$$

where ζ is a primitive p-th root of unity in k. Hence, if $c \geq 1$, then

$$\nu(1 - \zeta^i \eta) \geq np^{c-1}, \quad \text{for some } i.$$

Replacing η by $\zeta^i \eta$, we have

$$\varepsilon = \eta^p, \quad \eta \in E_{np^{c-1}},$$

namely,

$$E_{np^c} \subseteq E^p_{np^{c-1}}.$$

Therefore

$$E_{np^c} \subseteq E_n{}^{p^c}, \quad c \geq 0,$$

which shows that LC holds for k.

As another example, let $p > 2$ and let k be a real field. Let $k' = k(\sqrt[p]{1})$ and assume that every prime ideal $\mathfrak{p}$ of k' dividing p, equals its complex conjugate $\bar{\mathfrak{p}}$: $\mathfrak{p} = \bar{\mathfrak{p}}$. By a similar argument as above, we can see that the existence of $(\mathfrak{q}, 1)$-fields for all $\mathfrak{q}$, $\mathfrak{q} \nmid p$ implies LC.

6. It now seems that $\mathfrak{q}$-fields and $(\mathfrak{q}, a)$-fields are interesting families of abelian extensions of a number field, deserving further investigations. Here, however, we can make only a simple remark as follows.

Let k and $m = p^a$, $a \geq 0$, be as above and let T denote the set of all p-places and ∞-places on k. Let K/k be any finite abelian extension unramified outside $p\infty$ (i.e. outside T), G the Galois group of K/k, and e_K the number of roots of unity in K. Define (cf. Tate [4] , p.86)

$$\theta_{K,p}(s) = \theta_{K/k,T}(s) = \sum_{(\mathfrak{a},p)=1} N(\mathfrak{a})^{-1}\sigma_{\mathfrak{a}}{}^{-1} = \sum_{\sigma \in G} \zeta(s,\sigma)\sigma^{-1}$$

where $\sigma_{\mathfrak{a}} = (\mathfrak{a}, K/k)$ and the $\zeta(s,\sigma)$'s are the partial zeta functions of k associated with the element of the group ring $\mathbb{C}[G]$. It is known (cf. Tate [4], p.107) that

$$e_K\theta_{K,p}(0) \in \mathbb{Z}[G].$$

Assume now that k is a totally real field and that

a) the p-adic zeta function $\zeta_p(k,s)$ has a pole at $s = 1$.

By Serre [3], 3.13-3.17, there exists a finite abelian extension F/k, unramified outside p, such that

(1) $$p^a\theta_{F,p}(0) \notin \mathbb{Z}[\mathrm{Gal}(F/k)].$$

Let

$$L = F(\zeta)$$

where ζ is a primitive p^a-th root of unity. L/k is an abelian extension, unramified outside $p\infty$ and e_L is divisible by p^a. Since $k \subseteq F \subseteq L$, it follows from (1) that

(2) $$p^a\theta_{L,p}(0) \notin \mathbb{Z}[\mathrm{Gal}(L/k)].$$

Assume that

b) the Brumer-Stark conjecture BS(L,T) holds (cf. Tate [4], p.107).

Let $\mathfrak{q}$ be a prime ideal of k, $\mathfrak{q} \nmid p$, completely decomposed in the abelian extension L, and let $\mathfrak{Q}$ be a prime ideal of L, dividing $\mathfrak{q}$. By BS(L,T), there exists an element α in L such that

$$(\alpha) = \mathfrak{Q}^{e\theta(0)} \text{ where } e = e_L, \ \theta = \theta_{L,p}, \text{ and}$$

$$M = L(\sqrt[e]{\alpha}) \text{ is an abelian extension over } k.$$

Then M/k is unramified outside $p\mathfrak{q}\infty$ and it follows from (2) that

$$e(\mathfrak{q}; M/k)_p \geq p^{a+1}.$$

Hence there is an intermediate field K of k and M, $k \subseteq K \subseteq M$ such that K/k is a cyclic p-extension with

$$e(\mathfrak{q}; K/k) \geq p^{a+1}.$$

If $p > 2$, then K/k is unramified outside $p\mathfrak{q}$ so that K is a $(\mathfrak{q}, a)$-field (in fact, a $(\mathfrak{q}, a+1)$-field) over k. If $p = 2$, $k \subseteq K' \subseteq K$, $[K : K'] = 2$, then K'/k is unramified outside $p\mathfrak{q}$ and $e(\mathfrak{q}; K'/k) \geq 2^a$. Hence K' is a $(\mathfrak{q}, a)$-field over k.

It is known that the statement a) actually implies the Leopoldt conjecture LC. Hence the existence of a $(\mathfrak{q}, a)$-field, for all $\mathfrak{q}$, $\mathfrak{q} \nmid p$, and all $a \geq 0$, is guaranteed, by Theorems 1 and 2. So the main interest in the above remark lies in that under the assumptions a), b), a $(\mathfrak{q}, a)$-field, for some $\mathfrak{q}$, is given by a subfield of an abelian extension over k, predicted by the Brumer-Stark conjecture, thus providing an information on the nature of $(\mathfrak{q}, a)$-fields. It also raises a question such as whether there is any relation between the Leopoldt conjecture and the Brumer-Stark conjecture via $(\mathfrak{q}, a)$-fields. But, at the moment, we do not know any substantial result in this connection.

Remark. Let k be totally real and let $\{K_i\}$, $1 \leq i \leq s$, be the family of abelian extensions over $k' = k(\zeta)$ considered in the proof of Theorem 2. Assume $K_i \cap L \neq k'$ for some i and for the field $L = F(\zeta)$ above. Then there is an element $\eta \in E$, $\eta \notin E^p$, such that $k'(\sqrt[p]{\eta}) \subseteq L$. Since L/k is abelian, $k(\sqrt[p]{\eta})/k$ must also be abelian. As k is totally real and $p > 2$, this is a contradiction. Hence $K_i \cap L = k'$ for all i. It follows that there is a prime ideal $\mathfrak{q}_i'$ of K_i, $\mathfrak{q}_i' \nmid p$, with absolute degree 1 such that $\mathfrak{q}_i'$ is unramified and undecomposed in K_i but is completely decomposed in L. The prime ideal $\mathfrak{q}_i = \mathfrak{q}_i' \cap k$ of k is then completely decomposed in L, so that a $(\mathfrak{q}_i, a)$-field over k is provided in the manner as mentioned above (under the assumptions a), b)).

Appendix

In this appendix, we shall show that LC in §1 is equivalent to a statement usually known as Leopoldt's conjecture.

For each prime ideal $\mathfrak{p}$ of k, let $\mathfrak{O}_\mathfrak{p}$ denote the ring of integers in the local field $k_\mathfrak{p}$, the $\mathfrak{p}$-completion of k, and let $U_\mathfrak{p}$ be the group of units in $k_\mathfrak{p}$: $U_\mathfrak{p} = \mathfrak{O}_\mathfrak{p}^\times$. Let

$$U = \prod_\mathfrak{p} U_\mathfrak{p}, \quad U' = \prod_\mathfrak{p} (1 + \mathfrak{p}\mathfrak{O}_\mathfrak{p})$$

where the product is taken over all prime ideals $\mathfrak{p}$ dividing the rational prime p, U and U' are compact abelian groups and U' is also a finitely generated $\mathbb{Z}_p$-module. As before, let E be the group of all units in k. By Dirichlet's theorem,

$$E/W_k \cong \mathbb{Z}^{r-1}$$

where W_k denotes the group of all roots of unity in k and r the number of archimedean places on k. E can be imbedded in each $U_\mathfrak{p}$ and, hence, in U : $E \subseteq U$. Let

$$E' = E \cap U'$$

and let $\overline{E'}$ denote the closure of E' in the compact group U'. Let T denote the torsion submodule of the finitely generated $\mathbb{Z}_p$-submodule $\overline{E'}$ in U'. The usual statement of Leopoldt's conjecture for k and p says that

$$\overline{E'}/T \cong \mathbb{Z}_p{}^{r-1}. \tag{2}$$

This is equivalent to say that there exist $r-1$ units in E' which are linearly independent over $\mathbb{Z}_p$ in U'. For a totally real field k, it is also equivalent to the non-vanishing of the p-adic regulator of k.

Now, let

$$\widehat{E'} = \varprojlim E'/E'^{p^n}, \quad n \geq 0$$

where the inverse limit is taken with respect to the maps $E'/E'^{p^m} \to E'/E'^{p^n}$ for $m \geq n \geq 0$. $\widehat{E'}$ is an abelian pro-p-group and, hence, is a compact $\mathbb{Z}_p$-module. We see from (1) that

$$E'/W'_k \cong \mathbb{Z}^{r-1} \tag{3}$$

where W'_k ($= E' \cap W_k$) is the group of all p-power roots of unity in k. It then follows that E' is a dense subgroup of $\widehat{E'}$ and

$$\widehat{E'}/W'_k \cong \mathbb{Z}_p{}^{r-1}. \tag{4}$$

Now, given any integer $a \geq 0$, there exists $n \geq 0$ such that

$$E'^{p^n} \subseteq \prod_{\mathfrak{p}} (1 + \mathfrak{p}^a \mathfrak{O}_\mathfrak{p}), \quad \mathfrak{p}|p.$$

Hence the injection $E' \to U'$ can be uniquely extended to a continuous homomorphism

$$f : \widehat{E'} \to U'.$$

Clearly $Im(f) = \overline{E'}$, and by (4), $\mathrm{Ker}(f) \cong \mathbb{Z}_p{}^t$ for some t, $0 \leq t \leq r-1$. Hence

$$\overline{E'}/T \cong \mathbb{Z}_p{}^s, \quad s = r-1-t, \ 0 \leq s \leq r-1.$$

Therefore

$$\begin{aligned} (2) &\iff \mathrm{Ker}(f) = 0 \\ &\iff f \text{ induces a topological isomorphism } : \widehat{E'} \xrightarrow{\sim} \overline{E'}. \end{aligned}$$

Since E' is contained in both $\widehat{E'}$ and $\overline{E'}$ as dense subgroups, $\widehat{E'} \xrightarrow{\sim} \overline{E'}$ means that the topologies on E' induced by those of $\widehat{E'}$ and $\overline{E'}$ are the same. This is then equivalent to the statement that given any p-power $m = p^a$, $a \geq 0$, there exists a p-power $n = p^b$, $b \geq 1$, such that

$$E_n (= E'_n) \subseteq E'^m.$$

However, as $[E : E']$ is prime to p, the above statement is equivalent to LC. Thus the equivalence of (2) and LC is proved.

Bibliography

[1] C. Chevalley, Deux théorèmes d'Arithmétique, Jour. Math. Soc. Japan, **3** (1951), 36-44.

[2] J. Coates, p-adic L-functions and Iwasawa's theory, Alg. Number Fields, Acad. Press, London-New York - San Francisco, 1977, 269-354.

[3] J. P. Serre, Sur le résidu de la fonction zêta p-adique d'un corps de nombres, C. R. Acad. Sci. Paris, t. **287** (1978), 183-188.

[4] J. Tate, Les conjectures de Stark sur les Fonctions L d'Artin en $s = 0$, Birkhäuser, Boston-Basel-Stuttgart, 1984.

Editors' Notes

The first part of this paper (i.e., §§1-3) was the subject of a lecture by Iwasawa (with the same title) in the workshop "Conference on Algebraic Number Theory" at RIMS, Kyoto University, July 1984, which was published in Japanese as [61]. In this (unpublished) English version a more complete account is given. In particular, the notions of $\mathfrak{q}$-*fields* and $(\mathfrak{q}, a)$-*fields*, which the author says are interesting families of abelian extensions of a number field, are defined and studied. Theorem 1, which is also stated in [61], is then given a full proof.

[U5] On integral representations of some finite groups

1. Let G be a group, n a positive integer, $\mathfrak{o}$ an integral domain, and M an $\mathfrak{o}[G]$-module. M will be called a $(G, n, \mathfrak{o})$-module if it is free and of rank n over $\mathfrak{o} : M \simeq \mathfrak{o}^n$. Each basis of M over $\mathfrak{o}$ then defines in the obvious manner a homomorphism $f : G \to GL(n, \mathfrak{o})$, namely, an integral representation of G of degree n over $\mathfrak{o}$. In the following, such an f will be simply called a $(G, n, \mathfrak{o})$-representation. Changing the basis of M over $\mathfrak{o}$, we obtain a whole class of equivalent $(G, n, \mathfrak{o})$-representations, and the latter conversely determines the $(G, n, \mathfrak{o})$-module M up to an $\mathfrak{o}[G]$-isomorphism.

Let k be the quotient field of $\mathfrak{o}$ and let V be a (G, n, k)-module, i.e., an n-dimensional representation space over k for G. We denote by $\mathfrak{M}(V)$ the family of all $(G, n, \mathfrak{o})$-modules M such that $M \otimes_{\mathfrak{o}} k \simeq V$ over $k[G]$. The number of classes of mutually isomorphic $(G, n, \mathfrak{o})$-modules in the family $\mathfrak{M}(V)$ is denoted by $c(V)$ and is called the class number of the $(G, n, \mathfrak{o})$-modules in $\mathfrak{M}(V)$; it is also called the class number of the family of $(G, n, \mathfrak{o})$-representations, associated with the $(G, n, \mathfrak{o})$-modules in $\mathfrak{M}(V)$. For example, let G be a cyclic group of order p, a prime number, and let $\mathfrak{o} = \mathbf{Z}, k = \mathbf{Q}$. Then $\mathbf{Q}[G] = V_0 \oplus V_1$ where V_0 and V_1 are irreducible G-invariant subspaces of $\mathbf{Q}[G]$ with dimensions 1 and $p-1$ respectively, and it is known*) that $c(V_0) = 1$ and $c(V_1)$ equals the class number of the cyclotomic field of p-th roots of unity. We shall next discuss a generalization of the above fact on $c(V_1)$, showing that there exist similar relations between the class numbers $c(V)$ for certain $(G, n, \mathfrak{o})$-modules and the class numbers of some algebraic number fields.

2. Let k be any number field, finite over $\mathbf{Q}$. Throughout the following, we shall denote by $\mathfrak{o}_k$, I_k, P_k, C_k, and h_k, the maximal order, the ideal group, the group of principal ideals, the ideal class group, and the class number, respectively, of k.

Let k be as above and let g be any positive integer. Let $K = k(\mu_g)$ where μ_g denotes the group of all complex g-th roots of unity; K/k is of course a finite abelian extension. Let F be any intermediate field of k and $K : k \subseteq F \subseteq K$, and let

$$n = [K : k], \quad m = [K : F], \quad \Gamma = \mathrm{Gal}(K/F).$$

Let G denote the semi-direct product of Γ and μ_g with respect to the natural action of the Galois group Γ on μ_g ; G is the split extension of μ_g by Γ such that for any σ in Γ and ζ in μ_g, we have $\sigma\zeta\sigma^{-1} = \sigma(\zeta)$ ($=$ the image of ζ under the automorphism σ of K). G is a meta-abelian group of order mg. Putting

$$\mathfrak{o} = \mathfrak{o}_k,$$

we shall consider the $(G, n, \mathfrak{o})$-modules and $(G, n, \mathfrak{o})$-representations defined in §1.

Let the field K be denoted by V when K is regarded as an n-dimensional vector space over $k: V = K$. We define the action of Γ and μ_g on V by

$$\begin{aligned} \sigma \cdot \xi &= \sigma(\xi), && \text{for } \sigma \in \Gamma, \xi \in V, \\ \zeta \cdot \xi &= \text{the product of } \zeta \text{ and } \xi \text{ in } K, && \text{for } \zeta \in \mu_g, \xi \in V. \end{aligned}$$

Then one checks immediately that $\sigma \cdot \zeta \cdot \sigma^{-1} \cdot \xi = \sigma(\zeta) \cdot \xi$ so that the above action of Γ and μ_g on V can be extended to the action of $G = \Gamma \cdot \mu_g$ on V; and V becomes an n-dimensional representation space over k for G. Let I_K^Γ denote the subgroup of all ideals $\mathfrak{A}$ in I_K such that $\sigma(\mathfrak{A}) = \mathfrak{A}$ for every σ in Γ, and let

$$A = \{\mathfrak{A} \in I_K^\Gamma \mid \mathfrak{A} \simeq \mathfrak{o}^n \text{ as } \mathfrak{o}\text{-modules}\}.$$

It is clear that each ideal $\mathfrak{A}$ in the set A provides a $(G, n, \mathfrak{o})$-module, belonging to the family $\mathfrak{M}(V)$ for the above V.

We now assume that

$$\mathfrak{o}_K = \mathfrak{o}_k[\mu_g]$$

for the maximal orders $\mathfrak{o}_K$ and $\mathfrak{o}_k (= \mathfrak{o})$. This condition is satisfied, for example, if *i*) the discriminant of k is prime to g, or if *ii*) k and, hence, K are cyclotomic fields. Using the assumption, we can easily prove the following :

Lemma. 1) *For each $(G, n, \mathfrak{o})$-module M in the family $\mathfrak{M}(V)$, there exists an ideal $\mathfrak{A}$ in the set A such that $M \simeq \mathfrak{A}$ over $\mathfrak{o}[G]$.*

2) *Two ideals $\mathfrak{A}$ and $\mathfrak{B}$ in A are isomorphic over $\mathfrak{o}[G]$ if and only if there exists an element $\lambda \neq 0$ in F such that $\mathfrak{B} = \mathfrak{A}\lambda$.*

3. The condition $\mathfrak{A} \simeq \mathfrak{o}^n$ in the definition of the set A says that the ideal $\mathfrak{A}$ of K has a basis over $\mathfrak{o} = \mathfrak{o}_k$. For such a condition, we know a theorem of E. Artin as follows**). In general, let k be any number field, finite over $\mathbf{Q}$, and let K be any finite extension over k. For each ω in K with $K = k(\omega)$, let $D(\omega)$ denote the discriminant of ω for the extension K/k, and let $D(K/k)$ be the discriminant of K/k. It is well known that there is an ideal $\mathfrak{a}_\omega$ of k such that

$$(D(\omega)) = \mathfrak{a}_\omega^2 D(K/k)$$

and that the ideal class $c(K/k)$ of $\mathfrak{a}_\omega$ in k depends only on the extension K/k and is independent of the choice of ω as described above. Now, the theorem of Artin states:

An ideal $\mathfrak{A}$ of K has a basis over $\mathfrak{o}_k$ if and only if the norm $N_{K/k}(\mathfrak{A})$ belongs to the ideal class $c(K/k)$. In particular, the maximal order $\mathfrak{o}_K$ of K has a basis over $\mathfrak{o}_k$ if and only if $c(K/k) = 1$ in the ideal class group C_k.

4. We apply the above theorem for the special case $K = k(\mu_g)$ in §2. It follows from our assumption $\mathfrak{o}_K = \mathfrak{o}_k[\mu_g]$ that $\mathfrak{o}_K$ has a basis over $\mathfrak{o}_k$. Hence $c(K/k) = 1$ in this case. Therefore an ideal $\mathfrak{A}$ of K has a basis over $\mathfrak{o}_k (= \mathfrak{o})$ if and only if $N_{K/k}(\mathfrak{A})$ is a principal ideal of k and, consequently, the set A in §2 can be described as follows:

$$A = \{\mathfrak{A} \in I_K^\Gamma \mid N_{K/k}(\mathfrak{A}) \in P_k\}.$$

It is now clear that A is a subgroup of I_K^Γ, containing P_F :

$$P_F \subseteq A \subseteq I_K^\Gamma \subseteq I_K,$$

and we obtain immediately the following result from the lemma in §2:

Proposition. *Let V be the representation space for G as defined in §2. Then the class number $c(V)$ of the $(G, n, \mathfrak{o})$-modules belonging to the family $\mathfrak{M}(V)$ is given by*

$$c(V) = [A : P_F].$$

Let s be the number of prime ideals of F, ramified in K. Then $[I_K^\Gamma : I_F] \leq m^s$. Hence

$$c(V) = [A : P_F] \leq [I_K^\Gamma : P_F] = [I_K^\Gamma : I_F][I_F : P_F] \leq m^s h_F < +\infty.$$

5. We now apply the above proposition for some special cases and obtain more precise information on $c(V)$, relating it to the class numbers of number fields.

a) Let $F = K, \Gamma = 1, G = \mu_g$. In this case,

$$A = \{\mathfrak{A} \in I_K \mid N_{K/k}(\mathfrak{A}) \in P_k\}. \quad P_K \subseteq A \subseteq I_K.$$

Therefore

$$A/P_K = \mathrm{Ker}(N_{K/k} : C_K \to C_k).$$

By class field theory,

$$C_K/\mathrm{Ker}(N_{K/k} : C_K \to C_k) \simeq \mathrm{Gal}(KL/K),$$

where L denotes Hilbert's class field over k. Hence it follows from the proposition that

$$c(V) = [A : P_K] = [K \cap L : k] h_K h_k^{-1},$$

where $K \cap L$ is the maximal unramified abelian extension over k, contained in K. In particular, if $K \cap L = k$, then

$$c(V) = h_K h_k^{-1}.$$

Example. Let $k = \mathbf{Q}$. Then $K \cap L = \mathbf{Q}$ by the theorem of Minkowski. Hence

$$c(V) = h_K h_{\mathbf{Q}}^{-1} = h_K.$$

Thus the class number h_K of any cyclotomic field $K = \mathbf{Q}(\mu_g), g \geq 1$, is always equal to the class number of a certain family of integral representations of $G(= \mu_g)$ over $\mathbf{Z}$. For $g = p$, a prime number, this is the result mentioned in §1.

b) Let $F = k, \Gamma = \mathrm{Gal}(K/k), G = \mathrm{Gal}(K/k) \cdot \mu_g, |G| = ng$ where $n = [K : k]$. Then

$$A = \{\mathfrak{A} \in I_K^\Gamma \mid N_{K/k}(\mathfrak{A}) \in P_k\} = \{\mathfrak{A} \in I_K \mid \mathfrak{A}^n \in P_k\}.$$

Example. Let $g = p^2$ for a prime p and let $k = \mathbf{Q}(\mu_p), K = k(\mu_{p^2}) = \mathbf{Q}(\mu_{p^2})$. In this case, $n = [K : k] = p, |G| = p^3$, and each ideal $\mathfrak{A}$ in I_K^Γ can be uniquely written in the form $\mathfrak{A} = \mathfrak{a}\mathfrak{P}^i$ where $\mathfrak{a} \in I_k, 0 \leq i < p$, and $\mathfrak{P}$ is the unique prime ideal of K, dividing p. Since $\mathfrak{P}^p$ is an ideal in P_k, it follows from the above that

$$c(V) = [A : P_k] = ph_k^{(p)}$$

where $h_k^{(p)}$ is the order of the group $C_k^{(p)}$, consisting of all classes c in C_k such that $c^p = 1$. One checks easily that $c(V)$ in this case is the class number of the family of integral representations

$$\{f : G \to GL(p, \mathfrak{o}_k) \mid f(\zeta) = \zeta \cdot 1_p \text{ for all } \zeta \in \mu_p\},$$

where 1_p denotes the $p \times p$ identity matrix.

c) Assume that $h_k = 1$. Then

$$A = I_K^\Gamma, \quad c(V) = [A : P_F] = [I_K^\Gamma : I_F]h_F.$$

Example. Let $k = \mathbf{Q}$, $g = p$, an odd prime $(p > 2)$, and let $K = k(\mu_g) = \mathbf{Q}(\mu_p)$. Further, let $\mathbf{Q} \subseteq F \subseteq K, m = [K : F], G = \mathrm{Gal}(K/F) \cdot \mu_p, |G| = mp$. In this case, $[I_K^\Gamma : I_F] = m$ so that

$$c(V) = m\,h_F.$$

Again one sees easily that $c(V)$ is the class number of the family of integral representations

$$\{f : G \to GL(p-1, \mathbf{Z}) \mid f : \text{injective}\}.$$

In particular, if F is the maximal real subfield of $K = \mathbf{Q}(\mu_p)$, then $m = 2$ and G is the dihedral group of order $2p$; and $c(V) = 2h_F$ where h_F is the so-called second factor of the class number h_K of K.

Footnotes

*) See F. K. Diederichsen, Über die Ausreduktion ganzzahliger Gruppendarstellungen bei arithmetischer Äquivalenz. Hamb. Abh. **14** (1938), 357–412.

**) See E. Artin, Questions de base minimale dans la théorie des nombres algébriques. Colloque international du CNRS, Paris (1950), 19–20, also in Collected papers, 229–231.

Editors' Notes

The results of this manuscript were given by Iwasawa in his lecture on June 3, 1989 in Komaba Seminar at the University of Tokyo.

Bibliography

Papers

[1] Ueber die Struktur der endlichen Gruppen, deren echte Untergruppen sämtlich nilpotent sind, Proc. Phys.-Math. Soc. Japan, **23**-1 (1941), 1-4.

[2] Über die Einfachheit der speziellen projektiven Gruppen, Proc. Imp. Acad. Japan, **17**-3 (1941), 57-59.

[3] Über die endlichen Gruppen und die Verbände ihrer Untergruppen, J. Fac. Sci. Imp. Univ. Tokyo, Sec. I. **4**-3 (1941), 171-199.

[4] On almost periodic functions (概周期函数に就いて), Iso Sugaku (位相数学), **4**-2 (10/1942), 56-60.

[5] On normed rings and a theorem of Segal, I (1088. ノルム環ト Segal ノ定理ニツイテ I), Zenkoku Shijo Sugaku Danwakai (全国紙上数学談話会), **246** (12/14/1942), 1522-1555.

[6] On the structure of infinite *M*-groups, Japanese J. Math., **18** (1943), 709-728.

[7] On the structure of conditionally complete lattice-groups, Japanese J. Math., **18** (1943), 777-789.

[8] Einige Sätze über freie Gruppen, Proc. Imp. Acad. Japan, **19**-6 (1943), 272-274.

[9] On one-parameter families of probability laws (確率法則の一径数集合について), J. Phys.-Math. Soc. Japan (日本数学物理学会誌), **17**-6 (1943), 217-220.

[10] On normed rings and a theorem of Segal, II (1109. ノルム環ト Segal ノ定理ニツイテ II), Zenkoku Shijo Sugaku Danwakai (全国紙上数学談話会), **251** (3/19/1943), 167-186.

[11] On group rings of topological groups, Proc. Imp. Acad. Japan, **20**-2 (1944), 67-70.

[12] Über nilpotente topologische Gruppen I, Proc. Japan Acad., **21**-3 (1945), 124-137.

[13] Der Bezoutsche Satz in zweifach projektiven Räumen, Proc. Japan Acad., **21**-4 (1945), 213-222.

[14] Zur Theorie der algebraischen Korrespondenzen I, Schnittpunktgruppen von Korrespondenzen, Proc. Japan Acad., **21**-4 (1945), 204-212.

[15] Zur Theorie der algebraischen Korrespondenzen II, Multiplikation der Korrespondenzen, Proc. Japan Acad., **21**-9 (1945), 411-418.

[16] On the representation of Lie algebras, Japanese J. Math., **19** (1948), 405-426.

[17] On linearly ordered groups, J. Math. Soc. Japan, **1**-1 (9/1948), 1-9.

[18] Finite groups and compact groups (有限群と compact 群), Sugaku (数学), **1**-2 (3/1948), 94-95.

[19] Hilbert's fifth problem (Hilbert の第五の問題、可解位相群の構造について), Sugaku (数学), **1**-3 (11/1948), 161-171.

[20] (with T. Tamagawa) Automorphisms of a function field (代数函数体の自己同型置換), Sugaku (数学), **1**-4 (1/1949), 315-316.

[21] On some types of topological groups, Ann. of Math., (2) **50** (1949), 507-558.

[22] Topological groups with invariant compact neighborhoods of the identity, Ann. of Math., (2) **54** (1951), 345-348.

[23] A note on *L*-functions, "Proc. Int. Congress of Math." (Cambridge, Mass., 1950), Vol. 1, Amer. Math. Soc., 1952, p. 322.

[24] Some properties of (*L*)-groups, "Proc. Int. Congress of Math." (Cambridge, Mass., 1950), Vol. 2, Amer. Math. Soc., 1952, pp. 447-450.

[25] (with T. Tamagawa) On the group of automorphisms of a function field, J. Math. Soc. Japan, **3** (1951), 137-147. Corrections, ibid., **4** (1952), 100-101, 203-204.

[26] On the rings of valuation vectors, Ann. of Math., (2) **57** (1953), 331-356.

[27] On solvable extensions of algebraic number fields, Ann. of Math., (2) **58** (1953), 548-572.

[28] A note on Kummer extensions, J. Math. Soc. Japan, **5** (1953), 253-262.

[29] On Galois groups of local fields, Trans. Amer. Math. Soc., **80** (1955), 448-469.

[30] Galois groups acting on the multiplicative groups of local fields, "Proc. Int. Sympos. on Algebraic Number Theory" (Tokyo and Nikko, 1955), Science Council of Japan, Tokyo, 1956, pp. 63-64.

[31] A note on the group of units of an algebraic number field, J. Math. Pure and Appl., (9) **35** (1956), 189-192.

[32] A note on class numbers of algebraic number fields, Abh. Math. Sem. Univ. Hamburg, **20** (1956), 257-258.

[33] On some invariants of cyclotomic fields, Amer. J. Math., **80** (1958), 773-783. Erratum, ibid., **81** (1959), 280.

[34] Sheaves for algebraic number fields, Ann. of Math., (2) **69** (1959), 408-413.

[35] On Γ-extensions of algebraic number fields, Bull. Amer. Math. Soc., **65** (1959), 183-226.

[36] On some properties of Γ-finite modules, Ann. of Math., (2) **70** (1959), 291-312.

[37] On the theory of cyclotomic fields, Ann. of Math., (2) **70** (1959), 530-561.

[38] On local cyclotomic fields, J. Math. Soc. Japan, **12** (1960), 16-21.

[39] A class number formula for cyclotomic fields, Ann. of Math., (2) **76** (1962), 171-179.

[40] On a certain analogy between algebraic number fields and function fields (代数体と函数体とのある類似について), Sugaku (数学), **15**-2 (10/1963), 65-67.

[41] On some modules in the theory of cyclotomic fields, J. Math. Soc. Japan, **16** (1964), 42-82.

[42] Some results in the theory of cyclotomic fields, Proc. Sympos. Pure Math., Vol. VIII, Amer. Math. Soc., 1965, pp. 66-69.

[43] (with C. C. Sims) Computation of invariants in the theory of cyclotomic fields, J. Math. Soc. Japan, **18** (1966), 86-96.

[44] A note on ideal class groups, Nagoya Math J., **27** (1966), 239-247.

[45] Some modules in local cyclotomic fields, "Les Tendances Géom. en Algèbre et Théorie des Nombres", Éditions du Centre National de la Recherche Scientifique, Paris, 1966, pp. 87-96.

[46] On explicit formulas for the norm residue symbol, J. Math. Soc. Japan, **20** (1968), 151-165.

[47] Analogies between number fields and function fields, "Some Recent Advances in the Basic Sciences, Proc. Annual Sci. Conf." (New York, 1965-1966), Vol. 2, Belfer Grad. School Sci., Yeshiva Univ., New York, 1969, pp. 203-208.

[48] On p-adic L-functions, Ann. of Math., (2) **89** (1969), 198-205.

[49] On some infinite Abelian extensions of algebraic number fields, "Actes du Congrés Int. Math." (Nice, 1970), Tome 1, Gauthier-Villars, Paris, 1971, pp. 391-394.

[50] Skew-symmetric forms for number fields, Proc. Sympos. Pure Math., Vol. XX, Amer. Math. Soc., 1971, p. 86.

[51] On the μ-invariants of cyclotomic fields, Acta Arith., **21** (1972), pp. 99-101.

[52] On $\mathbf{Z}_\ell$-extensions of algebraic number fields, Ann. of Math., (2) **98** (1973), 246-326.

[53] On the μ-invariants of $\mathbf{Z}_\ell$-extensions, "Number Theory, Algebraic Geometry and Commutative Algebra, in honor of Yasuo Akizuki", Kinokuniya, Tokyo, 1973, pp. 1-11.

[54] A note on Jacobi sums, "Convegno di Strutture in Corpi Algebrici" (Rome, 1973), INDAM, Symposia Mathematica, Vol. XV, Academic Press, London, 1975, pp. 447-459.

[55] A note on cyclotomic fields, Invent. math., **36** (1976), 115-123.

[56] Some remarks on Hecke characters, "Algebraic Number Theory" (Kyoto, 1976), Japan Soc. Promotion Sci., Tokyo, 1977, pp. 99-108.

[57] On p-adic representations associated with $\mathbb{Z}_p$-extensions, "Automorphic Forms, Representation Theory and Arithmetic" (Bombay, 1979), Tata Inst. Fund. Res., Bombay, 1981, pp. 141-153.

[58] Riemann-Hurwitz formula and p-adic Galois representations for number fields, Tôhoku Math. J., (2) **33** (1981), 263-288.

[59] On cohomology groups of $\mathbb{Z}_p$-extensions ($\mathbb{Z}_p$-拡大のコホモロジー群について) (Notes by G. Fujisaki), "$\mathbb{Z}_p$-extensions and Related topics" ($\mathbb{Z}_p$-拡大およびその関連理論の研究) (Kyoto, 6/1981; G. Fujisaki, ed.), Kokyuroku (数理解析研究所講究録), No. **440**, Res. Inst. Math. Sci., Kyoto Univ., 1981, pp. 76-84.

[60] On cohomology groups of units for $\mathbf{Z}_p$-extensions, Amer. J. Math., **105** (1983), 189-200; "Geometry and Number Theory, A Volume in Honor of André Weil" (J.-P. Serre and G. Shimura, eds.), The Johns Hopkins Univ. Press, 1983, pp. 189-200.

[61] A simple remark on Leopoldt's conjecture (Leopoldt 予想についての簡単な注意), (Notes by G. Fujisaki and K. Tateyama), "Report of the Conference on Algebraic Number Theory" (代数的整数論研究集会報告集) (Res. Inst. Math. Sci., Kyoto Univ., 7/1984; K. Shiratani, ed.), 1984, pp. 45-54.

[62] Some problems on cyclotomic fields (円分体に関するいくつかの問題), "Algebraic Number Theory" (代数的整数論, 最近の種々の話題について) (Kyoto, 12/1987; I. Satake and Y. Morita, eds.), Kokyuroku (数理解析研究所講究録), No. **658**, Res. Inst. Math. Sci., Kyoto Univ., 1988, pp. 43-55.

[63] A note on capitulation problem for number fields, Proc. Japan Acad., **65**-2, Ser. A (1989), 59-61.

[64] A note on capitulation problem for number fields. II, Proc. Japan Acad., **65**-6, Ser. A (1989), 183-186.

[65] On papers of Takagi in number theory, "Teiji Takagi Collected Papers", Springer-Verlag Tokyo, 1990, Appendices, I, pp. 342-351.

[66] A letter to J. Dieudonné, "Zeta Functions in Geometry" (Tokyo, 1990), Adv. St. in Pure Math., **21**, 1992, pp. 445-450.

Unpublished papers

[U1] A note on unitary representations, (lectured in a Colloquium at Kyoto Univ., Dec. 1987).

[U2] Automorphisms of Galois groups over number fields, (lectured in the "Symposium on L-functions and Galois properties of algebraic number fields", the Univ. of Durham, Sept. 1975).

[U3] Some problems on cyclotomic fields and $\mathbb{Z}_p$-extensions, (lectured in "Arithmetik der ABELschen Zahlkörper und Klassenkörper der komplexen Multiplikation", Math. Forschungsinstitut Oberwolfach, March 1979).

[U4] A simple remark on Leopoldt's conjecture, (lectured in the "Conference on Algebraic Number Theory", Res. Inst. Math. Sci., Kyoto Univ., July 1984).

[U5] On integral representations of some finite groups, (lectured in Komaba seminar at the Univ. of Tokyo, June 1989).

Books

[B1] Theory of Algebraic Functions (代数函数論), Iwanami Shoten(岩波書店), Tokyo, 1952, revised ed. 1973, 380 pp; English transl. by G. Kato "Algebraic Functions", Transl. of Math. Monographs, Amer. Math. Soc. Vol. **118**, 1993, 287 pp.

[B2] (with A. Borel, S. D. Chowla, C. S. Hertz and J.-P. Serre) Seminar on Complex Multiplication, Seminar held at the Institute for Advanced Study, Princeton, 1957-58. Lecture Notes in Math. No. **21**, Springer-Verlag, Berlin-Heidelberg-New York, 1966, 102 pp.

[B3] Lectures on p-adic L-functions, Annals of Math. Studies, No. **74**, Princeton Univ. Press, Princeton, N. J. and Univ. of Tokyo Press, Tokyo, 1972, 106 pp.

[B4] Local Class Field Theory (局所類体論), Iwanami Shoten(岩波書店), Tokyo, 1980, 184 pp; Russian transl. by A. A. Bel'skii, "Mir", Moscow, 1983, 184 pp.

[B5] Local Class Field Theory*, Oxford Math. Monographs, Oxford Univ. Press, New York, Clarendon Press, Oxford, 1986, 155 pp.

Articles not included in these volumes

[A1] On groups and the associated lattices (909. 群トソノ lattice ニツイテ), ZSSD**, **211** (3/15/1941), 76-81.

[A2] On groups and the associated lattices II (群とその lattice について II), Iso Sugaku (位相数学), **4-1** (2/1942), 47-50.

[A3] On groups and the associated lattices III (975. 群トソノ lattice ニツイテ III), ZSSD, **225** (10/27/1941), 519-542.

[A4] A remark on lattice groups (1029. 束群ニ関する一注意), ZSSD, **234** (3/23/1942), 912-914.

[A5] On lattice groups (1044. 束群ニツイテ), ZSSD, **235** (4/16/1942), 1030-1048.

---

* [B5] is not an English translation of [B4]; it is an entirely new original book in English.

**ZSSD is an abbreviation of Zenkoku Shijo Sugaku Danwakai (全国紙上数学談話会).

[A6] On a generalization of central series (1089. 核心群列ノ一拡張ニツイテ), ZSSD, **246** (12/14/1942), 1555-1563.

[A7] On the correspondence between groups and ideals in the group rings (1090. 群ト群環ニ於ケル Ideal トノ対応ニ就イテ), ZSSD, **246** (12/14/1942), 1563-1591.

[A8] A remark on solvable groups (1149. 可解群ニ関スル一注意), ZSSD, **258** (10/25/1943), 539-544.

[A9] An example of a group (1168. 群ノ一例), ZSSD, **262** (3/20/1944), 67-69.

[A10] On faithful representations of a Lie algebra (23. Lie 環ノ同型表現ニツイテ), ZSSD (Ser.2), **3** (2/20/1947), 1-6.

[A11] Review of "Theory of differential forms - Grassmann algebras and Lie groups (微分式論 – グラスマン代数とリー群)" by Yukiyoshi Kawada (Kawade Shobo (河出書房), Tokyo, 1951), Math. Reviews **14** (1953), 410.

[A12] Review of "Allgemeine Theorie der algebraischen Zahlen" by Philipp Furtwängler (B. G. Teubner Verlagsgesellschaft, Leipzig, 1953), Math. Reviews **15** (1954), 404.

[A13] Review of "Introduction to algebraic number theory" by Henry B. Mann (The Ohio State Univ. Press, Columbus, Ohio, 1955), Math. Reviews **17** (1956), 240-241.

[A14] Review of "Topological transformation groups" by Deane Montgomery and Leo Zippin (Interscience Publishers, New York - London, 1955), Math. Reviews **17** (1956), 383-384.

[A15] Review of "Die Berechnung der Klasssenzahl Abelscher Körper über quadratischen Zahlkörpern" by Curt Meyer (Akademie-Verlag, Berlin, 1957), Bull. Amer. Math. Soc., **64** (1958), 211-212.

[A16] Impressions of the International Cogress of Mathematicians, I (Congress 印象記 I), Sugaku (数学), **14**-3 (2/1963), 174-175.

[A17] Review of "Introduction to quadratic forms" by O. T. O'Meara (Academic Press, Inc., Publishers, New York; Springer-Verlag, Berlin-Göttingen-Heidelberg, 1963), Math. Reviews **27** (1964), 2485-2486.

[A18] On the mathematical works of Prof. Hiraku Toyama – around the thesis "Non-abelian theory of algebraic functions" (遠山啓教授の数学的業績 – 学位論文「代数関数の非アーベル的理論」を中心に), Sugaku Seminar (数学セミナー) (3/1980), 24-28.

[A19] 120 minutes with Prof. Kenkichi Iwasawa (Interview by the editors of Sugaku) (岩澤健吉先生のお話を伺った 120 分 (編集部)), Sugaku (数学), **45**-4 (10/1990), 366-372.